AF864415

SPECTRO-CHROME MAGAZINES

(Abridged)

Dinshah P. Ghadiali

Dinshah Health Society
P O Box 707
Malaga NJ 08328

ISBN 0-933917-14-7
Typeset by D.H.S.

Edited by
Darius Dinshah, Pres.
Printed and bound in the U.S.A.

INTRODUCTION

1. Dinshah P. Ghadiali delivered his first public lecture on the Spectro-Chrome version of color therapy on April 26, 1920. He trained hundreds of professionals and laypersons but as might be expected, soon met with severe and increasing opposition. The first salvo was a long, derogatory article in the American Medical Association Journal (1924). The next assault was a criminal lawsuit (Buffalo NY, 1931), won with testimony of loyal Spectro-Chrome graduates. A seemingly endless series of legal entanglements followed, most of which were lost.

2. In 1947, loss of a case motivated by the Food and Drug Administration resulted in Dinshah being placed on probation for five years; one stipulation was that all material in his possession pertaining to Spectro-Chrome be surrendered for destruction, except for a set of his writings in his personal library. Much information was rendered out-of-print.

3. From 1922 through 1947 Dinshah edited the **Spectro-Chrome** monthly magazine. Complete seven-volume sets (240 issues, **History of Spectro-Chrome**) were made by photocopying in 1990 by Dinshah Health Society. Those seven volumes spanned considerably more than 6000 pages which made the set very informative but expensive, and more than many readers could digest. In this abridged version, the earlier years are considerably emphasized because there was a greater diversity of views and reports from professionals compared to later years.

4. The 496 pages in this book are numbered with those of the 41 original issues. They have been edited: most blank pages/cartoons/political pages/copyrights/etc. were deleted so page numbers are either on the outer or inner margin; page size was reduced about 2%. Many pages are out-of-square which is not unusual with Linotyped productions, and compounded by steps taken in copying; it is probably more evident here due to narrower margins.

5. Dinshah used the word "occultism" in a few places; its true (and his intended) meaning is "hidden knowledge", and nothing more. Current usage for devil-worship, etc. is entirely incorrect. An eclipse is an "occultation", resulting in a partially or completely <u>hidden</u> celestial object.

6. Spectro-Chrome technical aspects have evolved through the years. Though the results of Spectro-Chrome usage related here are for the most part very favorable, not everything in this book agrees with our present-day recommendations. As with many therapies, it is solely your responsibility how you proceed or utilize the information presented in this book.

DINSHAH HEALTH SOCIETY
Darius Dinshah, Pres.

SPECTRO-CHROME

JUNE 1922
TURQUOISE 1922

VOLUME 1
NUMBER 1

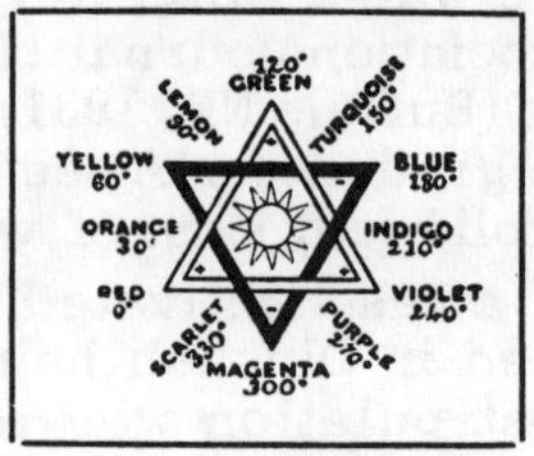

Devoted to: Spectro-Chrome Therapy **Dedicated to: Service of Humanity**

CONTENTS

Page

The Birth of Spectro-Chrome By Irene Hoger, S.C.T. . . . . Inside front cover
Dinshah's Prayer . . . . Inside front cover
The World Through the Eyes of Dinshah---
Brain Tumor---Why Prescribe Poison?---Washington Shows Sense---Middle Ages or Civilization?---Sanitation---Where is it?---America is Narrowing---Surgery in Agony---America the Free Gives Opium to China the Fettered as Gift---Sugar or Dross?---Spectro-Chrome in Measles---Areas fatal in X-Ray Work---Mathematics in Love---Duty ---Dinshah's Platform. 1-6
Hot Dogs!. A Hot Sketch on the Knickerbocker Girl vs. the Short Skirt Flapper by Anna C. Tilley, S. C. T. 7
An Attack of the Grip by Herbert M. Horn, M. D., D. C., S. C. T. 8
Foods and Sex by Anna C. De LaMotte, M. D., S. C. T. 10
Spectro-Chrome, Art and Drama by Madlaine Traverse, S. C. T. 11
The Blood-Boat by LaVerne A. Barber, M. D. 12
No Proof of Anything but Results by Kate W. Baldwin, M. D., F. A. C. S. 14
Spectro-Chrome in the Dental Office by Alice M. Norton, D. D. S. 16
Spectro-Chrome, The Harp of Health by Kashmira Dinshah, S. C. T. 17
Magenta by Henry C. Lanneau, S. C. T. 20
Area Chart---Therapeutic System 24
Miscellaneous Announcements . . . . Inside back cover
Spectro-Chrome Institute . . . . Outside back cover

Editor
Dinshah P. Ghadiali

In course of life you reach a time,
When Doctor, Lawyer, Priest alike,
Does fail to give the sought relief,
To troubled mind or burdened heart—
Then come to me;
I help you to help yourself.

SPECTRO-CHROME.

Subscription: June, 1922, to December, 1922 $1.75
Thereafter, January to December $3.00
Single Copy, 25 Cents

IRENE HOGER, S. C. T.

The Artist who Sketched

THE BIRTH OF SPECTRO-CHROME

Dinshah's Prayer to the Grand Architect of the Universe and the Benign Masters of Wisdom

Offered at Bombay, India,
Tuesday, February 10th, 1891

O Ahuramazda! Grant me Thy powers!! Grant me that which was in me but was suppressed!!! Grant me the tongue that resounds the world, the eye that enfeebles the wicked, the nose that scents the truth, the ear that hears the ant, the touch that feels the ether the intuition that knows the Norns, the health that helps the needy, the wealth that wields the worldly; all so that during life I exist selfless and at passage to the Higher Planes leave behind a lighted trail to be followed by the Searcher for the Beyond.

A fructification of this prayer is The Birth of Spectro-Chrome.

SPECTRO-CHROME

JUNE 1922 · VOLUME 1

TURQUOISE 1922 · NUMBER 1

Published Monthly by Spectro-Chrome Institute

2401 N. BROAD STREET, PHILADELPHIA, PENNA.

United States of America

Application for Entry as Second Class Matter at the Post Office at Philadelphia, Pa., pending

Devoted to the Latest Revelation in the Healing Art

SPECTRO-CHROME THERAPY

Restoration of the Human Radio-Active and Radio-Emanative Equilibrium

(Treatment of Dis-eases) by

ATTUNED COLOR WAVES

No Drugs - No Manipulation - No Surgery

Conceived, Originated, Developed, Applied and Copyrighted, 1920, by

Colonel Dinshah P. Ghadiali, M. D., D. C., Ph. D., LL. D., &c.
Metaphysician and Psychologist

Attuned Red Wave

(The World Through the Eyes of Dinshah)

The Birth of Spectro-Chrome

After long years of laborious investigation and original research, having succeeded in developing in 1920 both the theoretical scientific and practical applicative phases of the entirely new Healing Art now known and recognized as Spectro-Chrome Therapy, I began to organize the Spectro-Chrome Institute to train by personally conducted resident courses, those who were willing to study the system and use it for the alleviation of human dis-eases.

The Pioneer Class was formed at New York, N. Y., followed by Philadelphia, Pa., Atlantic City, N. J., Baltimore, Md., again Philadelphia, Pa., Washington, D. C., and a second class in New York, N. Y. All this was accomplished in less than 18 months and the graduates of the Spectro-Chrome Institute spread out so widely and produced such beneficial clinical results that they demanded a medium of communication not only among themselves, but between them and the outside world. I bowed to this demand and with this issue, SPECTRO-CHROME sees the light of day to enlighten the world for therapeutical emancipation.

Although essentially devoted to Spectro-Chrome Therapy, the columns of SPECTRO-CHROME are open to all for exchange of views on everything pertaining to human progress. Its platform is universal and broad in the broadest sense. SPECTRO-CHROME will be conducted on the principles of an open newspaper owned by the whole world as it were and what none may dare print because of partisan and pusillanimous views, it will publish boldly without prejudice. Its aim will be found to be the most impartial and independent public servant—ungraftable, incorruptible and fearless in the performance of its duty by the exponence of the truth.

BRAIN TUMOR

Henry P. Davison, a partner of J. P. Morgan, died under second surgical operation for brain tumor. It is reported that the case was considered utterly hopeless from the beginning. The tumor was said to be pressing on the auditory (eighth) nerve and was causing him pain and sleeplessness. Such are cases as are especially suited for Spectro-Chrome treatment, because where everything else fails Spectro-Chrome starts. A systemic Attuned Green followed by Attuned Indigo would bring about better results for such symptoms than all surgery could do, with a better opportunity for living through the danger. Even during surgical operation an Attuned Indigo Wave is a great standby for hemorrhage.

WHY PRESCRIBE POISON?

Dr. Blair Spencer, Assistant Director of Public Welfare and a member of the medical staff of the House of Correction of Philadelphia, Pa., is reported as stating that no doctor now knows when he makes a legal purchase of the article whether he is getting whiskey or wood alcohol. He blames it all on the Prohibition Laws and wishes the act repealed. No, doctor dear, let the law stay; you stop prescribing alcohol and the House of Correction will stay empty.

WASHINGTON SHOWS SENSE

All Government departments inaugurated the plan of working an earlier hour from May 15th, thus making a saving of day light by 3,600 seconds each day. Very creditable plan this is, compared to the silly idea of bungling astronomy by pushing the clock ahead 60 minutes. The cock will crow at the same time no matter how much you monkey the clock.

MIDDLE AGES OR CIVILIZATION?

On May 5 three negroes were lynched by a mob of 500 and burned to death in Kirvin, Texas, on their confession of having raped and murdered a seven-year-old white girl, Eula Awsley. No firearms were discharged and the lynching was deliberately carried out. There certainly is no defense in favor of the conduct of the negroes, but what can defend the inhuman atrocity of the more advanced white people who would burn human beings at stake, and what can justify the absence of legal processes in such cases? Retaliating murder with a worse murder will never bring about reformation.

SANITATION—WHERE IS IT?

Health Commissioner Dr. Royal Copeland, of New York City, shut down the milk plant of Dairymen's League Cooperative Association for conditions violating the Sanitary Code. It would be interesting to learn what sanitary conditions obtain in butcher shops and fish markets where the carcasses appeal with gruesome openness to the optic and olfactory senses. Milk! Where **IS** it?

AMERICA IS NARROWING

The conferees of the State and House reached an agreement for extension of the 3 per cent Immigration Restriction Law to two years more. Page the Pilgrim Fathers and the psyche of George Washington! America, they say, is overcrowded, and yet there are no human residences nearer than 20 miles of one another in some Southern States!

SURGERY IN AGONY

Prize for Tonsils.

London England specialists in major surgery are in despair over the attitude of the general medical practitioners and the public toward them. It appears that before the war a certain class of people was not satisfied unless a knife was stuck into their vitals for one reason or another periodically. This applied more to removal of adenoids and tonsils which to the specialists was a veritable mint. After the war the tide turned and the patients took to other things, forgetting many imaginary ills for which they previously ran to the specialists. Commenting on this, Dr. John Kynaston, an English surgeon, protests that 90 per cent of such operations are needless and insists on their danger. He shows that a large number has fatal issues, and maintains that the extirpation of tonsils is followed by many throat troubles. This is certainly worth looking into and with a view to encourage clear knowledge of the functions and purposes of the tonsils which surgeons generally believe to be useless, SPECTRO-CHROME offers a prize of Ten Dollars Cash for the best 1000-word essay on the subject. Competitors should send their articles by the 1st of August, 1922. Essays must be signed and certified as original. Rejected manuscripts will be returned if accompanied by stamped envelope. The prize-winning article will become the property of the Spectro-Chrome Institute.

This sign indicates editorial remarks. Contributors must give their full names and addresses for information, even if not for publication.

AMERICA THE FREE GIVES OPIUM TO CHINA THE FETTERED AS GIFT

Interchange of questions and answers between Miller and Representative Dyer, Republican, of Missouri, revealed that the principal source of the opium supply in China was in the United States. The drug first was imported into the United States and then shipped to Japan, whence it was continuously poured into China in flagrant violation of Chinese custom laws.

Dyer declared, "There are two or three concerns in this country that do or have been doing this very thing. One of them is located in Philadelphia." Miller replied that he thought "that condition will be effectively blocked by the enactment of legislation."

Legislation nothing! Expose the name of this Philadelphia firm and enforce laws already in existence. There are, according to the latest information, over 39,000 laws on the statute books of Pennsylvania and if a few were enforced all scandal would cease. The trouble lies not in want of legislation, but in political graft. Clean the gangs.

SUGAR OR DROSS?

A few hundred years ago sugar was considered a luxury in Europe and only medical practitioners and those with bulging pockets could afford to use it. Now it is recognized as an article of diet, but meddlesome chemists have converted it from a pure natural product into worthless concoction treated with bone ash, lime, sulphuric acid and the like to whiten it. White sugar is no food; let the sensible housewife use the variety which may perhaps be displeasing to the eye, but is the real[3] nutrient. It is the stomach that has to be the judge.

SPECTRO-CHROME IN MEASLES

The following interesting article was printed in The World, of New York:

MEASLES SERUM SEEMS TO PROMISE REAL IMMUNITY.

Dr. Rudolf Degkwitz Reports Successful Experiments with New Preventive.

Manchester, England, April 5.— Dr. Rudolf Degkwitz announces in the Deutsche Medizinische Wochenschrift (an authoritative publication) that he has succeeded in conferring immunity to measles by injecting a serum prepared from convalescent patients.

"This may prove to be one of the greatest epochs in preventive medicine," writes a medical correspondent of the Manchester Guardian. "In his last series of 1,700 cases Degkwitz claims that all the contacts inoculated were completely immunized or so protected that the disease was developed only in its mildest form and there was no untoward result.

"If these results are confirmed and the method can be effectively and generally applied, the result should be an even greater saving of life than has accrued from Jenner's discovery of vaccination.

"Measles, the most deadly of all the diseases of childhood, continues to be the least controllable of the scourges which regularly affect our people," the Guardian's correspondent says warningly. "With the possible exception of influenza there is no form of epidemic disease in face of which preventive medicine is so helpless. The causes for this are mainly two. In the first place the disease is not only highly infective but its infectivity is highest before the symptoms can be recognized. The child which contracts measles has three or four days in which to infect his family, his friends and his school-fellows before it is known definitely that he is suffering from measles at all. The second cause is public apathy and ignorance of the dangerous nature of the complaint.

"Half a century ago there appears to have been some justification for regarding measles as a trifling disorder, something that every child must 'get over,' and the sooner the better. Owing to the operation of the still mysterious laws which govern changes in type of disease, scarlet fever, which was justly dreaded has declined in virulence and ceased to be one of the major dangers of childhood; diphtheria is also much milder in form and its dangers can almost be eliminated by the early use of antitoxin. But measles has become more deadly year by year.

"The present writer can recall one recent epidemic of measles in a Lancashire town which slew in six months precisely the number of children that had succumbed to scarlet fever in ten years. Yet it is still not an unknown thing for a mother to put the other children in bed with a case of measles so that they may all 'get over it together,' and it is the constant complaint of doctors that they are not called in until the child is blue from pneumonia and beyond all aid.

"Compare this apathy with the horror of smallpox. Yet, before the introduction of vaccination, when smallpox was regarded as the national plague, the death rate from this cause averaged 420 per million. In 1915 the death rate from measles was 1,398 per million. Neither the Government nor the local authorities seem to have waked up to the facts represented by these figures. We continue to lavish enormous sums of money on the hospital isolation of scarlet fever with very little effect on the prevalence of the disease, but hospital accommodation for measles is very rarely provided.

"Scarlet fever is compulsorily notifiable; the notification of measles is optional and, in these days of anti-spending as opposed to true economy, practically in abeyance."

Having given the views of "the other side" as above, without any alteration, let us say in a few words that all the misery and deadliness of the scourge called "measles" may be saved by the use of the Spectro-Chrome alone without any serum or vaccine. Treat the patient with a full systemic Attuned Green Color Wave and guard the circulation by Attuned Magenta on Area 4. Should there be marked fever, treat with systemic Turquoise instead of Green. The results in each case will surpass your most sanguine expectations. Serums and vaccines are filthy—for treating dis-eases of filth, use Spectro-Chrome exclusively; it will cleanse the entire system and restore equilibrium as no other rational system can.

AREAS FATAL IN X RAY WORK

Many experiments conducted on dogs (O poor dogs), at the University of California Medical School by Drs. L. S. Warren and G. H. Whipple aim at proving that Roentgen Rays administered over the abdominal region are fatal in a majority of cases. The same rays thrown over the thorax produce no untoward effects. While they are "searching" for the causes and effects, innocent dogs will be uselessly sacrificed. Any Spectro-Chrome Therapist will be able to say without such torturing experiments that the fatal influence is owing to the X Ray being on the Ultraviolet section of the Spectrum. Thrown on the abdomen, the high potency acts through the Suprarenal Capsules and interferes with the Magenta which works on the heart therefrom, while through the thorax no such detrimental absorption takes place, the utmost that could happen being interference with metabolism. The physiology of animals does not coincide with that of man, their psychical and higher construction being different. Vivisection is a barbarous practice in modern civilization and at its best is worthless and a disgrace to the healing art. Spectro-Chrome study avoids it and is the cleanest and the most accurate healing art, in its positive conformity to the laws of God.

MATHEMATICS IN LOVE

An English psycho-analyst, Dr. David Forsyth, says that a man loving two women at the same time is one-half for each, and if loving five synchronously is only 20 per cent for each. He thus measures love by rule of proportion, forgetting that though the sun shines on the whole world, he shines no less or more according to the population shone upon. Love's irradiation is so brilliant that it is utterly impossible for the casual observer to fathom the fundamentals of its potency. The average thinker mixes the feelings of love and lust, and not being aware of the true difference, faulty deductions are likely to occur. Were love in its prime aspects an arithmetical calculation only, God and the Benign Masters would lose their glory. Love is the Light of the Lord; its expression in the Cosmos is universal; it is the sacrifice of personal interests for the good of others in whom we be interested; it asks for nothing; it gives and does not expect even to receive; it has no condition or obligation, no end to serve; it is all devotional. But, as comprehended today, so-called Love is nothing but the chasing of a chimera of conceived ideas of happiness, a material shell clothed in the aura of the semi-high and at its best is only a name without the soul of impersonality behind it to activate and energize it into the effulgence of the eternal self. What appears in the law courts as love shattered, never was love, or there

would be no mud-slinging, no blackwashing of character, no alimony demands, no aggrandizement, no ill feelings, no childish backtalks of lawyers, no desertions, no exactions of a mundane type, the root of all of which is nothing but downright jealousy of some third soul entering a human life as the answer to the unchangeable laws of Retributive Justice from own misdeeds or otherwise.

DUTY

Duty is its own reward. Its performance as dictated by one's conscience is its greatest fruit. Duty exacts no condition, yet in its fulfillment without regard to the imposition of any condition lies the evolution of man. Man, the glorious emblem of the Deity, is the essence of creation, protection and destruction; he is the highest image of his Creator, being represented without from within.

Humble Servant of Suffering Humanity
DINSHAH P. GHADIALI, Originator of Spectro-Chrome Therapy

DINSHAH'S PLATFORM AS ADOPTED BY HIM SINCE 1891

The Boundless Vibratory Ocean of Thought is Essentially Universal and All-pervading; it is the Individualized Monopoly of NO Person, and is the Common Heritage of Humanity's Evolution: thus, what a Development of Unrevealed Ages has given unto Me in the Form of Knowledge in My Present Incarnation is No Distinctive Acquisition of Mine for My Sole Use, Benefit or Elevation, but is All for Thee and Is Thine Without Condition, Without Obligation, Without Expectation. I Fear No One; Only God Above and Conscience Below, and From Them I have Nothing to Fear.

Attuned Orange Wave

(Humor, Wit, Fun, Satire, Tickle, Irony)

HOT DOGS!!!

A Hot Sketch on the Knickerbocker Girl vs. the Short Skirt Flapper, by Mrs. Anna Tilley, S. C. T.

"Hot Dogs" at last! She arrived, with her knicke rs, a charmingly picturesque treat, and to be sure, a fascinating person she is, from the crown of her bobbed head to the sole of her trim two-colored shoes, everything about her intrigues the fancy. And not the least of her charms is the fact that she is so wise in the way of costumes.

However indignant we grow at the modern generation's latest prank we know it is just a sensible style.

Why is it? Will some doctor, psychologist, or perhaps some wise layman tell me, is there a screw loose or is it just a kink in the brain that makes a certain class of people yell "Hot Dogs," snigger, or convulse with laughter when they see what they consider an oddity in dress? Do these silly flappers think because they wear short skirts and sheer hose, that that is the only style, or the most advanced style?

Please, gentle reader, do not misunderstand me. I am not just deriding short skirts or silk stockings. I am merely putting in a word for the girl who chooses to wear knickers instead of short skirts. Not, perhaps, because it is ultra-modern, but because it is more sensible to my way of thinking, more comfortable and surely more modest. I do not think that anybody will deny that knickers worn below the knee with wool stockings are more modest than short skirts worn above the knee with silk stockings, or as in some cases with socks.

It is well to notice that in those countries where modesty is still considered a virtue women have always worn trousers and are still wearing them.

I wonder, dear friends of the fair sex, and you, their husbands, brothers or sweethearts, if in looking up the history of skirts you would still laugh at the knicker girl, or would there be a rush to the clothiers for knicker suits? I think the latter would happen.

I know there are thousands of women who would like very much to wear knickers for the comfort they give, but these dear sisters are afraid of being laughed at. What difference does it make if people laugh, if you know and feel that you are comfortable, and just as well

dressed as the next one, if not better?

After all, propriety or impropriety is only comparative. It is the conception of the individual, the nation or race he or she belongs to; for instance, in the Morning World dated May 3rd a small article appeared of which the heading was "Modern Both Ways"—Albanian women are discarding trousers to become modern, and American women for the same reason are putting them on. So I suppose these women of Albania think they have been wrong and are right now. And we who wear knickers also think we were wrong before, and are right now. It is all comparative; what is right in one country will appear wrong in another, and what is right in one individual will seem wrong in another.

Last, but not least, YOU FLAPPERS who wear petticoats and envelope chemise, and presume to laugh at the knicker girl, if you could see yourself with her eyes when the flap of your chemise is flapping or your petticoat shows two inches below your very much abused short skirt, you would think twice before you would dare to laugh.

AN ATTACK OF THE GRIP

By Herbert M. Horn, M.D., D.C., S.C.T.

Say! Did you ever go joy riding with an attack of the grip? You may take it from me that it has all other forms of amusement beaten to a standstill. The grip germ is the brainiest little bug that was ever chased by a doctor.

I was sitting quietly at home reading Einstein's masterly treatment on the "Bending of Rays," when I suddenly began sneezing like a Bolshevik regiment answering roll call. A friend happened to be visiting at the time in my house. He was deeply absorbed in the mysteries of Darwin's latest achievements, "Monkey Glands and Monkey Shines." He acknowledged my first sneeze with the customary "Gesundheit," then he trailed bravely along with his responses for several minutes, but it was useless; I had more sneezes in my system than there are "Gesundheits" in the entire German army. "Mr. Johnson," he ventured after a time, "you are getting a cold." "I'm not getting it," I sniffed, "I have it now."

What a mean contemptible little creature a grip germ must be! Absolutely without any of the finer feelings and instincts of the more aristocratic germs, it sneaks into people's systems disguised as an ordinary cold. It isn't straightforward like the rheumatism germs or the toothache bacilli, both of which are brave and fearless, and will walk right up to you and kick you on the shins, big as you are.

I discovered shortly after I had sneezed myself into a condition of fearful profanity that a newly married couple of grip germs had taken a notion to build a nest in my solar plexus, and two hours later they had about 1,857 children attending school in my pituitary gland. Every time they came out for recess they would create such a rumpus in my coco that I would go up in the air and hit the ceiling with my nose. Before daylight came all these children had graduated from school, and after they tore down the schoolhouse, the whole bunch was married and had large families of their own; then all of them went out paddling their canoes on my alimentary canal. That was the fiercest night I ever spent.

Then the doctor arrived on the scene. He threw his saws behind the mantelpiece and put a bunch of fierce-looking instruments on the table; then he felt my pulse. He took my temperature and remarked that my tongue looked like a currycomb. Then he said "Ah!" and look at my leg very hard. "Say, Doc," I whispered, "It is no use cutting my leg off, because the germs will hide in my elbow." But the doctor disdained even to take notice of this ghastly attempt at humor; he began shooting questions at me faster than I could answer them. "Do you feel shooting pains in the back of the head? Have you a buzzing in the ears, and a confused sound like distant laughter in your abdomen? Does your tongue feel rare and high priced like a porterhouse steak at a summer resort? When you look at the wall paper does your brain sort of turn topsy turvy, and cause you to meld a 100 aces, and double pinochle, and eighty kings? When you lie on your left side, do you have an impulse to be on your right side, and then do you feel like jumping out of bed to throw stones at a policeman?"

I answered all these questions in the affirmative, because he had really mentioned all my symptoms. "Tell me the truth, doctor," I asked with palpitating heart, "what is it, the bubonic plague?" "You have something worse, you have the grip," the doctor whispered gently. Then he got out his pencil and paper and became busy with it, and left me enough prescriptions to keep the druggist in pocket money throughout the winter.

Then my friends and relatives began to drop in and annoy me with suggestions. Uncle Louis sat by my bedside and after I had barked for him two or three times he decided on pneumonia and was insistent that I tie a rubber band around my chest and take a bath in gasoline. I told my uncle I had no desire to become a human automobile, so he became mad and went home. But before he became mad he drank three bottles of home brew and before he went home he borrowed my Sunday suit of clothes. An hour later my friend John dropped in. John is a medical student. So he looked me over and decided I had galloping tuberculosis, incipient asthma, tonsilitis, croup and the measles. Outside of that, he solemnly assured me, I was all right. He insisted that I take 50 grains of quinine, ten grains of aspirin, rub the back of my neck with benzine, soak my ankles in kerosene, then swallow a little phenacetin, and a hot whiskey toddy every half hour. If I found it hard to take the whiskey toddy, he volunteered to run in every half hour and help me.

I obeyed the doctor's orders religiously for three consecutive days. But the only result that I got was that I was getting weaker every hour, and there was absolutely no improvement in my condition. So I decided to call in another doctor.

Now, in our neighborhood there lived a doctor who was considered a faker by everybody. He had certain queer ideas in regard to the treatment of dis-ease. He used to say that drugs were unnecessary for the treatment of dis-ease. He treated his patients with some sort of a crazy arrangement of colored lights. But the funniest part of the thing was that the blessed thing worked. This doctor had set right quite a number of people, who had been given up as hopeless by the other doctors.

So I sent for this "doctor." He came in and felt my pulse and took my temperature. "H'm," says he, "you need Spectro-Chrome Turquoise." "Doctor, is this fatal?" "Is what fatal?" "The dis-ease you just mentioned." "Ha, ha, ha! This is not a dis-ease, this is a treatment." "But, doctor, is this medicine going to kill the grip germ?" "So the doctors have got you convinced that an army of germs have invaded your body and are very gradually chewing you up. According to the blindness of this theory, the whole organic domain is fearfully wormy, and man, the highest product of physical life, is a seething army of vicious parasites which are consuming his vitals, some of which many physicians insist should be cut out. According to this theory God made man to be consumed by parasites and physicians, and to work like the devil all his life to pay physicians for catching and killing these little worms. Now I have something far better to right you. I am not going to pour any poisonous substance into your insides in order to kill the germs, for the simple reason that whatever is strong enough to kill the germs is also strong enough to destroy the live cells of your body. I am going to shed an Attuned Color upon you that will reduce your fever and make you feel perfectly all right in a couple of days. What have you been eating?" "Nothing at all, doctor; just a cup of hot broth every hour, and a soft-boiled egg and toast every once in a while, and a drink of whiskey occasionally to keep my strength up." "Now, all you are permitted to have is a little milk every once in a while. You must not take any such food or drugs, and take these treatments as I shall direct you."

Well, I did as this freak doctor directed, and would you believe it, in two days I was perfectly well? I swore off regular doctors for life, and I refuse to take the advice of relatives and friends. Give me my "freak doctor" every time.

Attuned Yellow Wave

(Correspondence, Query, Catechism, Inquiry, Interrogatory)

FOODS AND SEX

To the Editor of Spectro-Chrome:

Reviewing events of the last decade, it would almost appear as if civilization is retrograding instead of progressing.

The world is a school to teach us how to think and feel aright, so that we may be qualified to use those two subtler forces of thought and the power of emotion. Poverty, sickness, privations and tribulations are all lessons for not complying with law. The greater part of humanity drifts with the tide. The more man and woman takes cognizance of life and its complexity, the more they realize that the perverted use of the sex function (with all its concomittant sins) and perverted appetites (alcoholism, narcotics and flesh foods) are the main causes of slow evolution.

Inertia and indifference go hand in hand. In the individual, lack of faith (fear and worry) and self pity are the greatest hindrance for development. In each life we learn different lessons and self-mastery is the goal, for no amount of work can so tire the physical body and wreck it as when the enormous energy of the desire body is let loose in a fit of temper.

As there never can be universal happiness till sex life is properly regulated, explain in your columns the double function of the sex faculty, Procreation and Regeneration.

As service is our keynote, may the "Spectro-Chrome" be a torch and a beacon to humanity! "God's Kingdom come!"

Yours for the hastening of Brotherhood,

ANNA C. DE LA MOTTE,
M.D., S.C.T.

SPECTRO-CHROME cordially invites articles on the stated aspects of Sex, and also on Rational Human Diet.

SPECTRO-CHROME, ART AND DRAMA

My dear Col. Dinshah:

As a pupil of your classes in Spectro-Chrome Therapy and Higher and Occult Aspects of Human Life, for the benefit of the readers of the "Spectro-Chrome" (your new magazine), many of whom are among those who wrote me letters of appreciation regarding my art upon the screen and asked me questions about my habits and activities in daily life, I desire to write a few lines expressing my sincere views.

The acme of any attainment cannot be reached without a well-balanced body and mind; it is particularly so in the field of the regular as well as the silent drama. In fact, the silent drama exacts more from the artist because the entire expression has to come direct to the eyes of the spectators without appeal to other senses; herein lies one very great difficulty of the Motion Picture artist. She has to have a perfect body and a perfectly balanced mind working in it to score the success desired.

The Silent Drama draws forth the inner consciousness of the true professional: it illumines the features through the innermost Soul Spark; that is why I became a student of Applied and Practical Psychology and why I studied in your classes, as you can so thoroughly express and impart the truth about the inner self of human nature. And after going through your courses of instruction I believe that only now am I fitted for greater and higher expressions in dramatic work; with the right opportunity I shall be able to unfold the truth of it, the real art of it—and I shall be able proudly to assert Dinshah did it—he showed me the way!

With the greatest appreciation to you, my guide and teacher, I remain heartily and most sincerely yours,

MADLAINE TRAVERSE.

MISS MADLAINE TRAVERSE

The Well-Known Motion Picture Star is the first of her Profession to Graduate as a SPECTRO-CHROME THERAPIST

Dear Friend: Your kind words do me good and inspire me to greater efforts for the benefit and in the service of humanity. I am no less proud to have such a gifted person as you are in my classes and your beautiful presence there does certainly impart a further charm to the atmosphere. It is the grateful appreciation of his students that makes for the success of any teacher and your noble words do credit to you and your refinement. Thank you cordially.

Attuned Lemon Wave

(Medicine, Surgery, Science, Osteopathy, Chiropractic, Philosophy)

THE BLOOD-BOAT

Polarity of the Tissue Cell

By LaVerne A. Barber, M.D.,

The electron is the base of atomic structure, bodily cells are molecular masses, man is a human dynamo. In health, man functions as a Positive; dis-eased, he becomes a Negative.

The brain is divided into two equal hemispheres because of polarity. Polarically the left side, it is said, functions Positively; the right, Negatively. This is reversed in the body, where the left side is Negative; the right, Positive. There are exceptions, but this is the rule.

Why? The answer is Polarity.

The Medulla Oblongata is a commisure wherein the wires are switched across and the nerve-currents shunted over to a changed polarity and a quickened vitality. Thus it is that the slightest lesion in the medulla is fatal, because polarity is disturbed at its vital center and the human dynamo is short-circuited, "shocked" beyond repair.

Imperfect polaric exchange affects the human body in many ways, because the body is composed of many cells. Virchow speaks of physical "Automatic Regulators"; what he referred to was "Polarity." To master basic dis-eased conditions, one should study physiological polarity.

Begins here the study of metabolism and scientific dietetics.

All normal assimilation of food, all tissue metabolism is a polaric give-and-take, a play between the positive and the negative vital currents which quicken the cells into activity. Healthy cells have normal polarity; to acquire and maintain this, foods must be properly blended and balanced, that chemical changes be not abused. What the average medical mind knows about a scientific diet is a startling revelation to the inquiring mind.

May I metamorphose? The bloodstream is a river, the red-blood corpuscles are steamboats. Their normal cargo is oxygen, and they traffic in nutrition, carrying it into interior parts of the country (body), where they unload and exchange it for a cargo of waste material.

Hemoglobin has positive polaric attraction for ozone, and oxygen combines with hemoglobin to coax and hold nutrition aboard the bloodboat. Billions of these tiny ships, carrying cargos of food and warmth, cruise along Red Rivers to deposit their cargos when and where needed along Life's flowing stream.

Every hungry tissue cell along the banks of this stream is negative to the arriving blood-boat. Polaric give-and-take occurs, the poles are switched, the unloaded boat becomes negatively absorptive to the now positively excreting cell, reloads with waste material and departs along venous canals en route to eliminative Integumental, Bronchial, Malpighian and Intestinal docks.

Normal inter-cellular traffic requires an exact polaric balance, a 50-50 break. Any excess of deficiency quickly affects these bloodboats and immediately we find them carrying suboxidated cargos of toxic matter instead of nutrition.

Improper foods, and foods badly blended, cause fermentation which leads, in turn, to acidosis. Cells are sullen workers in an acid medium. There follows lowered cellular vibration and sooner or later dis-ease and death. Equalized cellular polarity is the road to perfect health

—plus, of course, the required mentality.

In this article, there is dropped to the student-mind merely a hint of a fascinating subject often neglected, if, indeed, not entirely overlooked.

Truly, metabolism depends upon cellular polarity; to understand its many phases more and more we must study Nature's Primer. Physiologically, that book can hold no more important subject than the polarity of the cell and the traffic between the blood-boats and the hungry tissues.

This burnt girl owes her life to Spectro-Chrome Therapy and its skillful administration by Dr. Kate W. Baldwin; ordinarily, such a case under known medical and surgical methods would have been considered hopeless. Read next page for particulars.

Front View of Burnt Area

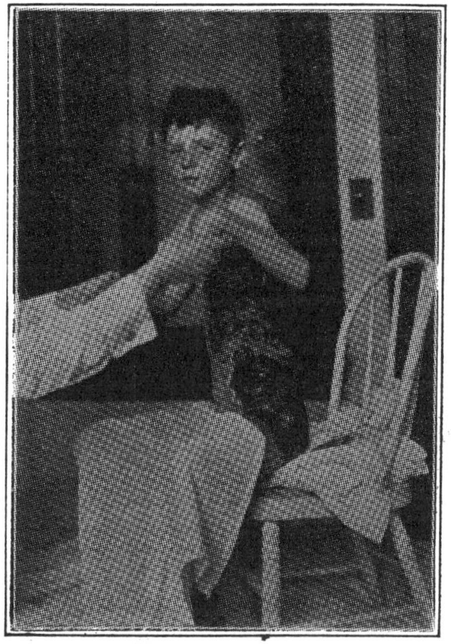

Back View of Burnt Area

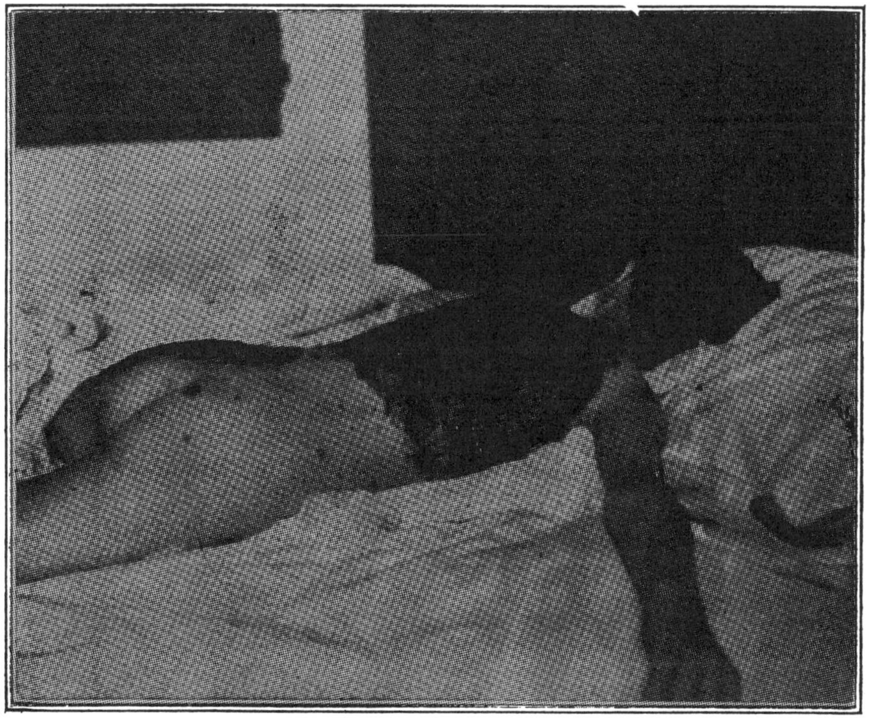

Attuned Green Wave

(Spectro-Chrome Therapy, Spectro-Chrome Therapists)

NO PROOF OF ANYTHING BUT RESULTS

The Wise Man knows how little he knows of what may be known.

By Kate W. Baldwin, M.D., F.A.C.S.,

Senior Surgeon, Woman's Hospital of Philadelphia, Pa.

A statement of opinion is as nothing compared with a report of experience. Repeated successes allow of no other conclusion than that experience with Spectro - Chrome Therapy or Attuned Color Waves has proved its value. This truth is worthy of all the reiteration it can be given.

Here I can only mention in a general way some of the results secured by the use of Attuned Color as a therapeutic measure. It is the most well-worth-while contribution to the healing art in many years. I doubt if anything known to "scientific medicine" can show more satisfactory results.

As with any other therapeutic measure, acute cases respond quicker than do chronic ones, but results will come with the chronic cases if due time and attention be given. I would like to give histories and go into details, but that is impossible in the short space available.

Attuned Colors seem to have a particular charm in all varieties of septic conditions, regardless of the special organism; as a fact, in mixed infections, the streptococci disappear first in many cases. Septic cases with temperatures of 102 to 105 degrees F. have a normal temperature in from one to three days and, if gotten early, resolution takes place before necrosis; if not, the time for healing is much reduced. Very seldom is an opiate or even a sedative required; the fewer sedatives, the less call for cathartics and laxatives.

Burns of the first to the fourth degree heal in about half the time taken with any other line of treatment. No exuberant granulations form and the new skin is soft and flexible. Large areas cover in without contractions. At the present time, I have a girl eight years old whose clothes caught fire and from two-thirds to three-quarters of the torso, the neck to the hair line, one ear and one arm to the elbow were very badly burned. According to all rules for burns, this child "has no business to be alive," but she is very much alive and the burned surface is practically one-half covered with beautiful flexible skin. The skin is so perfect that the nurses on the case remarked some days ago, "Dr. Baldwin, you will have difficulty to convince people of the extent of that burn." This case has had **no medication.** Attuned Colors and correct diet have taken care of all symptoms, and elimination by kidneys, bowels and unburned skin has been so perfect that the tax on her vital forces has been reduced to a minimum. The local dressing in this case has been sterilized, waxed lunch paper dipped in a saturated solution of bicarbonate of sodium in crude cocoanut oil. The cocoanut oil melts at body temperature. It is warmed to about that temperature and the paper takes it up as a sponge does water and gives a translucent covering for the burned surface. The oil paper, if allowed to extend beyond the burned surface, adheres to the skin sufficiently not to shuck. Outside of this we have used a jacket of two thicknesses of gauze just tucked together over the shoulder and down the sides. No clothing at all and no bandages. This is the simplest and the most efficient dressing I ever used for burns and I have tested nearly if not all of the recent as well as the older methods of dressing. It is difficult to do justice to such a case, on paper. The photograms may help a little, but they do not show the full extent of the burned surface.

If Spectro-Chrome was of value in the treatment of burns **only,** it would be worth while for every physician to investigate it.

In extreme cases of hiccoughs, it remedies after drugs have failed to give relief. In one case where hiccoughs had continued for ten days, the attending physician said, "We have reached the limit of our resources: try anything that promises a chance of relief." So Spectro-Chrome was used. After the first treatment, the patient had several hours of refreshing sleep and within eighteen hours the hiccoughing ceased entirely. Another man had hiccoughed continuously for five days without even momentary relief. This case also responded satisfactorily to Spectro-Chrome.

A woman past seventy years of age addicted to the use of morphine for eighteen years, a registered drug addict and a paralytic for twenty years, has not taken morphine now for more than a year, thanks to Spectro-Chrome! This patient came under my care because of a fracture of the left leg, the tibia at the lower end and the fibula at the upper. The next day after this accident, we found that the left arm was paralyzed. She developed a fast spreading bed sore over the sacrum. The excretions were all involuntary and as a whole the case was discouraging. For at least three weeks there was no indication that the bones were going to unite, but finally we were able to secure union, the bed sore responded promptly, the excretions were controlled and could she have continued the treatment I feel reasonably sure that the paralysis would have been even better than it is. It is a little more than one year since the fracture. Today I called the daughter on the telephone and learned that the mother with assistance was able to walk around on the one floor and that she did not take any morphine and in a general way enjoyed the family life.

Members of the medical profession, Spectro-Chrome or Attuned Color Therapy is well worth your time to investigate.

By kind invitation of Dr. Kate W. Baldwin, I visited the hospital and saw the burnt girl. It occurred to me that the doctor had certainly produced the most marvelous result in so serious a case. The cocoanut oil and waxed paper dressing was doing all for the application of the Attuned Color Waves through its translucence and the absence of pain to the patient demonstrated the potency of the Spectro-Chrome in the alleviation of suffering which in cases of deep burns is intense, agonizing and pitiably distracting. I congratulate the worthy doctor on her confidence in the power of Attuned Color Waves and exclusively using them against all odds, bravely avoiding morphine and its associates. It is truly hard to believe that drugs could be dispensed with entirely even in such extreme cases and their secondary and tertiary physiological and other adverse effects guarded against for the ultimate complete recovery of the patient. Since the introduction of Spectro-Chrome Therapy into Philadelphia, Penna., about fourteen months ago, the Spectro-Chrome Institute conducted two classes aggregating about 110 students of the system. A majority of these pioneers is practicing in and around the city, but few had such opportunities for clinical observation of the potencies of Spectro-Chrome in such varieties of dis-eases as Dr. Kate W. Baldwin, owing to her high hospital positon, and, without any reflection on the enthusiasm of other practitioners, it might be said that few followed the courses more attentively and put the knowledge to quicker use for suffering humanity than this good doctor. She took the lead to introduce the new system on its own merits under crucial conditions which would have probably discour aged a less competent surgeon, and for and on behalf of the Spectro-Chrome Institute I offer the best and the most heartfelt compliments to her. I also convey my sincere thanks to the Board of Managers of the Woman's Hospital of Philadelphia, Penna., for liberal thoughtfulness in introducing Spectro-Chrome Therapy into their institution—setting an example of the nobleness of womankind to the less progressive hospitals conducted by men.

SPECTRO-CHROME IN THE DENTAL OFFICE

By Alice M. Norton, D.D.S.

Early in October, 1921, I began to use a 400-Watt Spectro-Chrome equipment in the treatment of three cases of Pyorrhoea Alveolaris.

The first patient dropped out in three weeks, declaring she could not give the time or observe the dietetic rules.

The second patient, a woman of pronounced dislike for any form of medicine, was interested and willing to be a subject for experiment. She came in the midst of an acute attack, five beautiful anterior teeth being involved (an X-ray examination showed marked disturbance.) I removed all calcareous deposits from necks of the teeth, cleaned and polished crowns of the teeth, and instituted a regular half-hour local treatment of alternate Green and Turquoise in place of the usual routine prophylactic treatment I gave such patients.

There was a marked improvement after the sixth treatment—the patient able to bite apples with comfort—the color of the gums much improved. I then switched to alternate Indigo and Lemon at three-day intervals. I dismissed her after the 18th treatment. No recurrence so far.

The third case was a woman 52 years old, a diabetic; she was not satisfactory as a patient, though quite interesting. She is an invalid, comes infrequently, has had seventeen local and two systemic treatments in six months; the improvement was very slight, which is all one could expect under the circumstances. Her physician insists on a "diabetic diet," which includes meat and whiskey! An interesting phase of her treatment is that she always settles herself for a nap as soon as I turn the Green or Turquoise Color on her and says it rests her for about two days.

The above cases are interesting and the trouble lies only with the patients or their attending physicians. The patients expect Spectro-Chrome (as a new system) to perform "magic" in all cases without attention to the diet or the technic and rarely give opportunity to the Spectro-Chrome Therapist to show the results by following Nature's laws, as indicated. Pyorrhoea should be treated like any other constitutional dis-ease with systemic Green a number of times in conjunction with the local Green, etc., and the patient must be patient with the Therapist in going for the necessary treatments. We thank Dr. Alice M. Norton for her perseverance and her encouraging report.

Attuned Turquoise Wave

(Poetry, Romance, Drama, Art)

SPECTRO-CHROME

The Harp of Health

By Miss Kashmira Dinshah, S. C. T.

Amaranthis was in sorrow,
For its noble King lay ill;
The physicians said that grim Death
Would his weakened pulse soon still.

So throughout the splendid kingdom,
There was woe in every heart;
Up they gave themselves to praying,
All forsaking shop and mart.

Incurable, so-called, there lay
That land's illustrious King;
Abandoned hope the healers all,
Affirming they tried everything.

"Tried everything!" But with what aim?
Ah! was it to save or kill?
If the first, they failed, if the last,
Time in that would prove their skill.

That day the King's son came to him
And said, "Worthy sire, thy wealth,
Domains, rulership, what are they,
Without that great treasure, Health?"

"Prince, O my son, thou speakest truth,"
The suffering King then said,
"But these wise healers have despaired.
They say I am all but dead!"

"O my beloved father, live!"
And he drew him to his breast,
"I shall seek that wondrous treasure,
And till I succeed not rest!"

"Brave, O thou loyal hearted Prince,
Go, my blessings on thee, son!"
"Father, farewell!" They parted.
Fell the twilight—day was done.

Alone, the young Prince sallied forth,
When sable night's wings unfurled;
The full moon gleamed a silver track
Of bright light across the World.

Long he did travel, day and night,
Thru woods and marshes haunted,
Where dark and wicked Nature sprites
Him pursued, harassed, taunted.

Across the stormy seas he went
Inquiring, seeking ever;
Of turning back before success
Did crown him, he thought never.

The North and South and West he searched,
Through countries fraught with danger;
Treachery, sham, truth, law and love,
All met this royal ranger.

At last the pearly Eastern strand
He reached—the Land of Sunrise:
The natal place of mystic lore
And wisdom that never dies.

The Prince went onward in his quest,
Passing marble cities grand.
Asking the sages, "Where is Health?"
As he traveled through the land.

"Seek farther," was their one reply,
And still farther on he sought.
Braving now the jungle's terrors
Thinking, "This is not for naught."

Cent times the sun did rise and set;
Still the palmy jungle thru
He pressed; at last, the night crept on,
Silence rose, and cooling dew.

Presently, up a mountain side,
The gallant Prince's way led,
Himalaya, old, majestic,
Forest-robed, with snowcapped head.

The next day dawned. The morning sun
Crowned the mountain tops with light,
The blossom laden tropic trees
Seemed as with bright gems bedight.

Now paused the Prince beside a lake
Upon whose surface gleaming,
Rested loti creamy-petaled,
And white swans floated, dreaming.

The lush grass formed a carpet soft
On which the wild deer bounded.
The songs of rainbow-plumaged birds
Thru verdant groves resounded.

A balmy wind, sweet Nature's fan;
To the flowers murmured low,
Swaying them to its harmony,
In a measure soft and slow.

Down on the lake brink knelt the Prince,
And uttered an earnest prayer.
"O Himalayan Lords Divine,
Is that precious Health found here?"

Filled with reverent thoughts, he rose,
A Being of holy mien,
An exalted occult Master,
In a halo crowned, was seen.

Awhile the young Prince gazed in awe,
At his features supernal;
Then, bowing, said, "I seek the boon
Of Health Supreme, Eternal."

Penetrant-eyed, the Master looked,
At the tireless seeker brave;
"O man courageous, thou hast won,
Thou wilt yet thy father save!"

Then, by the Master there appeared,
A radiant Harp of gold,
He drew his hand across the strings,
Forth a silver sonance rolled.

Harmonic, fresh and free, it swelled,
In joyous cadence ringing,
Pulsating waves of melody,
Like holy angels singing.

The wondrous music soared and spread,
To a throbbing sea of sound,
Embracing in its tonal arms,
That calm, hallowed scene around.

Then, to the sense of sight became,
The marvelous sonance clear,
In great iridescent surges
That the Prince could see and hear!

"Behold, O man," the Master said,
"Now no farther need you roam,
That treasure Health you seek is here—
This glorious Spectro-Chrome!"

"Thy dauntless unselfish virtue,
Invincible truth has brought
To thee, after all thy trials,
This brilliant success long sought."

"Thy name befits thee, loyal Prince.
That of Duty thou art King;
Down through the aeons thy honor
All posterity shall sing."

"More lustrous than a thousand suns
Yet shall be thy splendid fame,
The anguished millions thou wilt heal,
Shall speak blessings on thy name."

"Great and good and godly-thoughted,
The world's darkness thou shalt light;
Thine shall be transcendent glory.
Strong Defender of the Right."

"Before these Attuned Color Waves
Shall medical legions bow,
On thy path of truth and duty
March thou boldly forward now."

A violet light exquisite
That Grand Master's face illumed,
The soft refreshing mountain air
Became with roses perfumed.

The celestial Being vanished,
And the luminance grew dim,
The glowing Sun now sank beneath
The western horizon's rim.

With velvet touch the nocturne gleamed,
With jewels in her dark crown,
The sparkling stars—the rising moon
On that tranquil scene shone down.

Bearing the wondrous Harp of Health,
Back to Amaranthis came
In haste the Prince; the populace
Did in songs its joy proclaim.

Straight to the palace chamber went
That faithful son, where, dying,
Figure and features gaunt and wan,
His father dear was lying.

In sombre black the room was hung,
And the windows veiled in black;
Anything that might tend to gloom,
That apartment did not lack.

The chamber's mournful atmosphere
Was augmented by a dirge,
Its doleful strains enveloped all
In a melancholy surge.

Humbly kneeling, the Prince then said,
"Honored Father, I am here!"
Slowly the King opened his eyes,
And in them was seen a tear.

"My beloved son! God bless thee,"
Affection truly royal
Entwined these two, the gracious King,
And the young Prince so loyal.

Noting his father's pallid cheeks,
And the room's dark hangings grim,
The Prince exclaimed, "I wonder not
That the fire of life grows dim!"

Turning to the smug-faced leechers,
He gave vent to righteous rage,
"Where is your much-vaunted learning?
Tell, who were your teachers sage?"

"Your so-called 'skill' proved valueless;
This environment of gloom
Is as if designed to hasten
On the patient to his doom!"

"But in a few weeks—days, perhaps,
The King would most surely die
Naturally," the doctors said,
Shrinking from the Prince's eye.

"And so, dictates your wisdom great.
Send him to Death in hurry!
Leave at once!" the Prince commanded,
They all fled in a flurry.

Tore away the sombre curtains,
Open wide the windows threw,
In streamed the gladdening sunshine,
And in cooling breezes blew.

The Prince then took the golden Harp,
A marvelous melody
Issued forth as he struck its chords,
Silencing the threnody.

Softly thru the quiet chamber
Rang the tender harmony.
A swiftly increasing ocean
Of enchanting symphony.

As though stirred by sudden impulse,
Did the dulcet music soar,
Like a titanic tidal wave,
With a tuneful, rushing roar.

At this point the royal patient
Raised eyes that once more sparkled;
Illness, Death's foreboding shadow,
Around him no longer darkled.

And, as apparent to his sight
The wonderful sound became,
His languor left. Re-energized,
Life's bedimmed spark leaped to flame.

The rolling ocean of sonance,
Thru the spacious palace halls,
Spread in melodious surges
Refluent from the crystal walls.

Soon the tremendous sea of notes,
Engulfed the city entire,
It glistened like a pearl amidst
That impassioned tonal fire.

Amaranthis' favored people
Saw and marveled at the sound;
Delighted, they, for a moment,
Did listen as though spellbound.

Then the multitudes assembled,
By a mighty force impelled;
And proceeded to the palace
From whose halls the music welled.

At the portals of that building,
Halted that imposing throng,
Watched and waited, rapt, expectant;
They had not to wait for long.

On the gleaming glassen terrace
Of wide steps, builded about
The palace, there appeared the King,
Quite well now, without a doubt.

Bearing the music's source, the Harp,
The triumphant Prince came out;
The joyful populace sent up
A great, hilarious shout.

Facing the assemblage beaming,
In a strong, commanding voice,
Spoke the eminent-thoughted Prince,
"My good people, I rejoice."

"At that time when from us our King,
It was likely Death might wrest,
By filial affection urged,
I set out upon Health's quest."

"When success my long search ended,
Only then I turned; a wealth
Of knowledge having gained. Behold,
The marvelous Harp of Health!"

As the final words were uttered,
From the gladsome concourse broke
An exultant cheer which echoes
From the distant hills awoke.

The solar orb shone down upon
The city, gem of beauty,
Its people, their King, and the Prince
So faithful to his duty.

Again from the vibrant Harp-strings
Rang out harmonious chords;
Sounding praise to God Almighty,
And to the great Karmic Lords.

Then there arose, as with one voice
Sweetly from a million throats,
A swelling chant of gratitude,
Of radiant silver notes.

"O Masters of Occult Wisdom,
Hail to ye Lords Supernal,
Hail to our King, and noble Prince
Whose name shall live eternal!"

"Wavering not from stern Truth's path,
Our Prince's fame resplendent
Shall be; above all others he
In Knowledge is transcendent."

Everlasting, thru the ages,
Shall be sung in every home.
By unnumbered grateful people,
The glory of Spectro-Chrome.

Attuned Magenta Wave

(Love, Friendship, Kindness, Emotion)

MAGENTA

By HENRY C. LANNEAU, S.C.T.

(Paper presented before the Washington Association of Spectro-Chrome Therapists at their regular monthly meeting.)

Recently I was the recipient of a highly diverting and altogether interesting post card message, as follows:

Saturday night, 8.30—1439 R.

Dear Henry—Well, old boy, you are to show what Magenta will do. I demand it on your part to produce a new field of progress and prove to us that you can and will surround the Universe with this immense stimulant.

H. B. Coblentz, Pres.

Rather a large order, I should say. I treasure this post card, for of all strange missives I ever saw, much less received, this seemed then as now most cryptic. In fact I regard it nothing less than classic, but like all classics worthy of the name its true intent and import was to a degree obscure. The first impression of surprise gave way to puzzled wonderment, then amusement mingled at last with a feeling of haunting uncertainty, that spite of glee over the whimsical message would not down. Was the worthy doctor only joking, or half in jest, half in earnest, did he really mean to make a goat of me? If the latter, I thought, then all good and well, in true sporting spirit I'll accept the challenge and become the goat. But if so be, beware! for in true goat-like fashion I mean to play my part. I give you fair warning then, all friends alike, whoever you are or wherever you happen to be or however disposed at the time, if in the immediate vicinity of my forward-going part I shall certainly endeavor to give you a very cordial butt—for playing goat remember is only a game of give-and-take; it is up to you to decide which you want. As the children used to say, I bid not to be it! I hear each one in turn exclaim, "All right, then I'll be the goat." But look out! For I the butter, and you the buttee, will know what it is like to affront and confront a goatee. Next day perhaps after your vigorous pommeling, you may not be able to do very gracefully or with a high degree of comfort what every good tub is required to do, but nevertheless, you may console yourself with the thought that Spectro-Chrome can relieve you of your frightful predicament and all will soon be well.

And now in passing, while on the pleasant subject of goats let us consider for a moment this amiable creature's character, and analyze from a strictly scientific standpoint based on the indisputable principles of Spectro-Chrome, his most striking characteristic. Whatever else may be said of him, a goat is certainly a progressive, forward-moving animal. Whoever doubts this claim to virtue need only stand in the path of his progress to be convinced. So much for that. But whence springs the animal's well known propensity for displaying his high spirits by coming into violent contact with all around him? Just a moment's thought will show the explanation to be very simple, yet sound and accurate. Everyone knows that the most effective and at the same time inexpensive way to dispose of the accumulation of old tin cans on the premises is to allow his voracious majesty, the goat, free access to the feast. He indeed should be very appropriately called, if not the king of beasts, at least the king of legerdemain, for the goat is the only animal in or out of captivity, known to be able to devour, and not only that but to be able to devour with exceeding gusto, articles of food composed entirely of metal,—in this case, large, round, fat tin cans. Now pause a moment—with what are these tin cans adorned that first attract the animal? Why to be sure, the most conspicuous, gaudy labels of flaming red and yellow! And where do we find the element of tin? Right on the yellow spectral section! Enough said. This animating tin can diet supplies the necessary motor stimulus, and the immediate, magical effect is to cause the hilarious animal to go pitching about in every conceivable direction like a man four sheets in the wind, until stopped perchance by a stone wall. Thus, friends, we see that Spectro-Chrome is at the bottom of everything and can explain even so obscure a subject as the traditional nature of the common, garden variety of goat. I trust you will pardon this manifest digression which is intended merely as a divertisement by way of introduction to the real subject in hand.

Magenta! What mystic meaning and music in that word! Long ago a mighty one of earth exclaimed, "What's in a name?" Ah! Surely you must know by now as he too would have known had he but learned the principles of Spectro-Chrome, that there is everything —yes, deep significance in a name. Who has not found from frequent experience that music indeed hath charms? And what forsooth produces the charm save the vibration of the sounds in both words and notes in combination of harmony. Further, a single glance at the white and black triangles will at once disclose why all stimulating, gay, martial, war-like music, is predominantly composed of notes or vibrations found on the red side of the spectrum, while at the same time every mother's lullaby and every crooning love-song must equally partake of the soothing vibrations found on the violet side. But hold! you say. It may be well enough to claim that such a love-song is composed mainly of blue and violet notes, but how explain the fiery, passionate song of an ardent, wooing lover? Look again at the black triangle and mark well the position of this magic wave. You see that Magenta is mid-way between Red and Violet, like a pendulum, as it were, suspended from the top or head of the white triangle from which flow the directing influences of the brain and animating impulses of the heart through the central canal of the sympathetic nervous system, which swing the pendulum at will to right or left depending on the nature of the emotions dominant. When fiery passion fills the breast and ardent love would seek to free itself in song, suddenly and in a violent manner as if from the tremor of an earthquake does the pendulum swing from its central, balanced position, far over toward the animating Red? Here then the impetuous lover finds fuel for impassioned song, and the nearer the pendulum swings toward the degree of Red the higher mount the hydrogen flames of his desire, until if not restrained in some measure he will be consumed by the fervor of his own passion. But observe the curious fact that it is possible for the pendulum to swing far enough from its position where it animates merely personal love and self-seeking, far away toward the Red and beyond, until it partakes of the nature of selfless or humanitarian love, and fires the soul of the patriot with love of country, until borne along on strains of martial music he glories in death upon the battle-field. Here again we see the underlying principle of martial music—to stir the

blood and stimulate violent activity as found on the Red side of the spectrum. Thus, according to the dictates of the directing will toward right or left, forever swings the magic pendulum—and like the Giant Clock of Time—which is but a name for a portion of Eternity—the pendulum, now swayed to left, now right, ticks off unceasingly the seconds that mark the Destiny for good or ill—a destiny of happiness or one of doom—for each mortal soul that breathes the prana of life. In contra-distinction to the emotions finding expression as Magenta moves toward the Red side displaying the aggressive forms of love, we find that as it moves in the opposite direction, toward the Violet side, the more refined forms of emotion become dominant. And the more refined the emotion, the more the cool, discerning influence of the intellect is felt upon it, and as the scale mounts higher a feeling of gratitude and reverence toward a higher being develops, manifesting itself in love for the Creator and appreciation for all His manifold blessings. As for the fervor of religious zeal so frequently displayed, be it remembered that albeit the cool electrical forces are found on the Violet side, that the principle of electricity as has been so often pointed out is, under proper conditions, to generate the most intense heat known. Therefore we find that heat is generated no matter in which direction we move, whether heat of holy passion or passing through the opposite phases, heat of the more physical emotions, differing in kind or character, not degree. We might liken the distinction on a lower plane to the difference between the fire of eloquence on the one hand and the heat of billings-gate on the other.

Coming now to the effects of Magenta from a practitioner's standpoint, we know that Magenta is the most powerful wave acting on the emotions of the human body. This being the case, it should certainly be used with appropriate caution and judicious care, in all cases especially in the young where matters of sex are concerned, not to overstimulate or unduly depress the functions which would entail disastrous results. In the case of more mature or even very old patients unfortunate results might easily ensue, but with them habit has become formed, and it is with the young that we should be more vitally concerned. I believe in all such cases that the utmost care and judgment should be exercised in the use of Magenta where sex is involved. Speaking of Magenta in this connection of course embraces the use of Purple and Scarlet, which are both composed of half Magenta.

The heart, that great wonder working organ of the body, is most frequently the target for the magic waves, to increase energy, stimulate respiration and restore circulation to normal. The place Magenta occupies in the middle of the base of the White Triangle is deeply significant, for as we know it is an artificial wave not appearing in the visible spectrum, but is none the less real and potent because it is not seen. It is there, although invisible, the motivating spirit of the powerful trinity of attuned color potencies, symbolic of the great mystery which runs though all creation. This pure emotional or love wave, is at the very foundation centre of our life, deep within the hidden springs of being. It is like the delicate balance wheel in the mechanism of the finest watch; disturb it in the slightest degree and the whole system suffers. You have seen natures which seemed to have lost entirely this animating force, and they were as withered as sheets of parchment or the crumbling forms of mummies which had been dried by the air of three thousand Egyptian years. How true it is that "The mind has a thousand eyes, the heart but one, and the light of the whole life dies when love is done"!

As Green is at the apex of the White Triangle, it becomes naturally the governing wave of the whole system from above, but in a deeply significant and highly remarkable way, Magenta, which is in fact merely the other end of the same rod or pendulum, and is the lowest point, in reality the apex, of the super-imposed, inverted Black Triangle, becomes the great governing wave from below. Now follow and observe this closely (for the analogy is perfect) the logic irresistible. While the direction comes from above with the touch of the pilot's hand upon the wheel, the choice directed by the brain, the course is taken by means of the way the rudder below responds to the guiding hand, the frail barque of life quivering in answer to the summons as it leaps to sail the uncharted seas. But we may go further and say that this secondary governing wave as it were, in reality assumes a far more important role, and becomes indeed the prime factor and influence of all activity during life. For, keeping in mind always the position of the two triangles, we readily see that the Dark Triangle, like the shadow of the White Triangle, is indeed more the substance from which the brighter reflection springs, while we find Magenta at the very lowest point, the apex of the submerged triangle, like the great tap-root from which the resistless forces of life push their way up and expand into all the radiant energies of a full orbed human being. And the stronger, more perfect system of root growth developed, the more fertile the soil and nourishing the substance on which it is fed, the loftier and more sublime is the tree of life that rears aloft its mighty branches, softened by glistening leaves, to drink in the sunshine and the dew and sway and bend and shout in triumph as it withstands the mighty, rushing winds of heaven. So then viewed in this light we are led to regard Magenta as the great feminine force of Spectro-Chrome of which Green at the other end naturally assumes the positive or masculine polarity. Green the masculine, positive, generates and projects, while Magenta the feminine, negative, receives and produces. Following the analogy just given of the tree, Magenta is to human life like mother Earth to Nature, the great receiver of, brooder over and fructifier of all those mysterious life forces which in the fullness of time bring forth after their kind, filling the world with the matchless beauty of the ever changing panorama of the seasons on the one hand, and on the other populating it with vibrant beings but little lower then the angels. Man governs the world that is, but woman produces the race of men that people the earth, shapes and molds their character, so that in the last analysis the feminine force is found to be the more potent. Thus it comes about that Magenta, symbolic of a mother's love, unseen, yet ever alert and powerful, unceasing in its influence as the human heartbeat, becomes of necessity by virtue of its inherent power, the true governing wave. How often have we read, and how often do we know, that "the hand that rocks the cradle rules the world!"

Because Magenta is the general wave surrounding the body it likewise fills completely the entire round of our existence, which was so aptly described by one who said, "We are such stuff as dreams are made of, and our little life is rounded with a sleep." From the first moment of conscious life when Aurora glides softly through the gleaming gates of dawn, and a mother bends over her rose-petaled babe with a smile far sweeter than the morning, throughout the busy hours until the parting kiss of day leaves a blush upon the evening sky, which later is reflected on the tender maiden's cheek as she hears love's tale unfolded while she holds her breath, and dare not, nay she cannot speak, Magenta is the wave that glows and softens, glows and softens, and brighter glows ever reigning supreme. And when at length our fitful dream of earthly life is over, and the silver cord is loosened, surely we may well believe that the dis-embodied soul clothes itself for celestial flight in such dazzling beauty of Magenta-like glowing color, as no mortal eye hath seen or mind conceived, until perchance in mounting higher through the heavenly realms it changes even this shining garment for a still brighter radiance. For surely we must depart, returning, as Wordsworth says, "We come, trailing clouds of glory, from God Who is our home."

And now in lighter vein. Here is a golden talisman I give to all you who are practitioners or expect-to-become-practitioners of Spectro-Chrome. To the degree to which you will make use of Magenta will your success in this new therapy depend. Therefore, I urge you strongly to let no opportunity go by to avail yourselves of this unique advantage which you alone possess over all competitors i n the healing art!

With apologies to a certain noted author, I offer you for your success this:

My Tip.

If you will treat your trusting Spectro-Chromic patients
Each time before they go,
With just a minute of Magenta wave,
The world is yours my friend, you'll never need to worry,
Because your patients all will love you so
Your path with gold they'll pave!

A sensible suggestion, Henry, but be careful where, on whom and how the Magenta is thus used; you might incur danger of a divorce suit on hand!

* * *

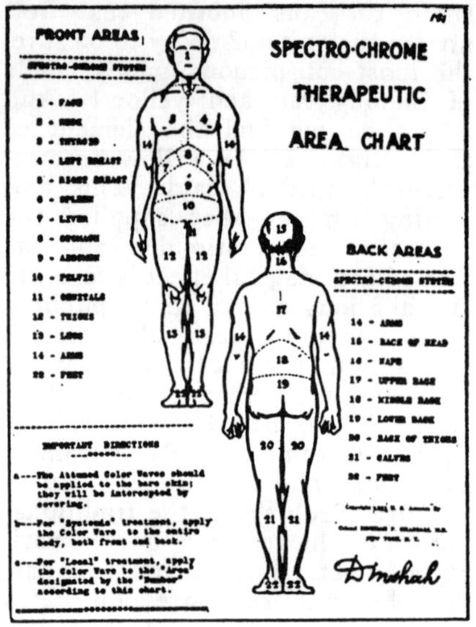

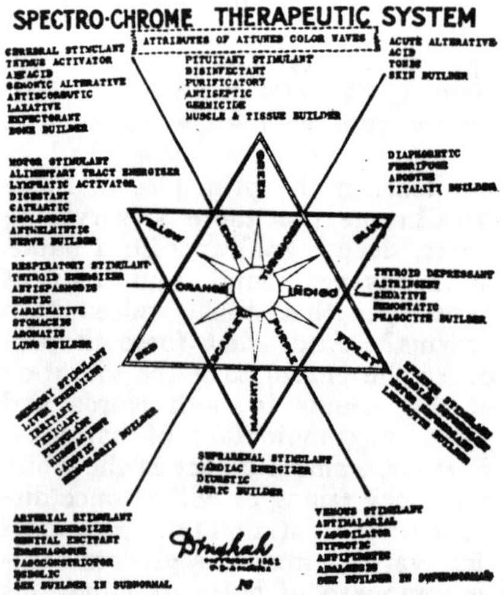

Spectro-Chrome Institute

Incorporated under the Laws of the State of New Jersey. Authorized for the transaction of business as Foreign Corporation under the Laws of the States of New York, Pennsylvania, Maryland, Delaware, New Hampshire, etc.

Central Office and Experimental Laboratory

2401 NORTH BROAD STREET

Philadelphia, Pennsylvania

UNITED STATES OF AMERICA

The main objects as set forth in the Charter of the Spectro-Chrome Institute are:

The instruction, advancement and practice of Spectro-Chrome Therapy and the research, development and pursuit of such sciences as may assist in the equipment and application of Spectro-Chrome Therapy or the restoration of the human Radio-Active and Radio-Emanative equilibrium by Attuned Color Waves, without the use of drugs, manipulation or surgery, in accordance with the system conceived, originated, developed and applied by Colonel Dinshah P. Ghadiali, M.D., D.C., Ph. D., LL. D.

The building, equipping, establishing, conducting and maintaining of schools for the investigation, instruction, demonstration and practice of Spectro-Chrome Therapy; also the power to confer and grant the degrees of Spectro-Chrome Therapist, Master of Spectro-Chrome and the like, and to issue diplomas and titles in connection therewith.

The corporation shall also have power to use, manufacture, sell and dispose of, in any way, all means, publications, appliances, instruments or products that may seem fitting or in any way connected with such purposes or objects aforesaid.

Spectro-Chrome Institute conducts complete resident courses in Spectro-Chrome Therapy in a number of cities each season. Each Course, comprising 24 discourses, is delivered personally by the Originator of the System with demonstrations and experiments. At the end of the course, students passing the examination are granted Diplomas. There are no correspondence courses. Classes now forming in Chicago, Ill., Philadelphia, Pa., New York, N. Y.

Spectro-Chrome Institute also manufactures and sells complete equipments for the practice of Spectro-Chrome Therapy. For further information address the Central Office.

SPECTRO-CHROME

JULY 1922
BLUE 1922

VOLUME 1
NUMBER 2

Devoted to: Spectro-Chrome Therapy — **Dedicated to: Service of Humanity.**

CONTENTS

	Page
The Latest Honor Conferred upon the Originator	Inside Front Cover
Dinshah's Platform—Area Chart—Therapeutic System	25
The World Through the Eyes of Dinshah— (Our New Expansion—Chicago, Illinois, will open—Next Issue—Sensible Negroes—Garters versus Safety Pins?—Smoking Students—The Earth Breathes—Exemplary Bravery!—Busybody Gloomgoblins—Faulty Administration—Kansas Outraged—Pronounce This—Erroneous Utterance —Chicago Wettest City?—Smoking and Eyesight—Sensible Medical Opinion—Miscarriage of Justice—Impossible with Spectro-Chrome—Handgrip and Color—Needs no Armor—A Misguided Bishop—Sooner the Better—Keep Skin Unpainted—Crab Surgery—Beggars Threaten—Verdict or Craze?—Knickerbocker in Brooklyn—Stupid Orders—Evil to Him Who Evil Thinks—Syphilitic Test or Humbug?—Titles for Turnips?)	26 to 37
Varieties of Indian Servants (by Kashmira Dinshah, S.C.T.)	37
What Our Learned Friend Says (by Anna C. de la Motte, M.D., S.C.T.)	38
Volstead or Sheppard? (by Victor Russell)	39
Coal Tar and Cancer (by Edmund J. Hogan, D.C., S.C.T.)	40
Spectro-Chrome of Much Value in Septic Cases (by Kate W. Baldwin, M.D., F.A.C.S.)	42
Nervousness, Appendicitis, Epilepsy (by LaVerne A. Barber, M.D., Ph.D.)	45
Success in Measles (by John A. Cohalan, D.O., S. C. T.)	47
From Stage to Spectro-Chrome (by Helen Mansfield Barber)	48
Miscellaneous Announcements	Inside back cover
Spectro-Chrome Institute	Outside back cover

Editor
Dinshah P. Ghadiali

In course of life you reach a time,
When Doctor, Lawyer, Priest alike,
Does fail to give the sought relief,
To troubled mind or burdened heart—
Then come to me;
I help you to help yourself.

SPECTRO-CHROME.

Subscription: June, 1922, to December, 1922
$1.75

Thereafter, January to December
$3.00

Single Copy, 25 Cents

The Latest Honor Conferred
Upon
The Originator of
SPECTRO-CHROME THERAPY
by the
Maryland Academy of Sciences
of
Baltimore, Maryland

MARYLAND ACADEMY OF SCIENCES

FOUNDED A. D. 1797. RE-ESTABLISHED A. D. 1819.

In the exercise of Academic Jurisdiction reposed in us by Acts of the Honorable Legislative Assembly of the State of Maryland, A. D. 1826 and A. D. 1867, we confer the dignity of Academician for Life upon

COL. DINSHAH P. GHADIALI, M.D.,
IN RECOGNITION OF
HIS SERVICES TO THE ACADEMY.

Granted and Conferred this first day of May, A. D. 1922, and authenticated by our seal and the signatures of our duly constituted officers:

M. W. PULLEN, *President.*
FRANCIS C. NICHOLAS, *Dean.*
EDWD. STABLER, JR., *Secretary.*
EDMUND B. FLADUNG, *Treasurer.*

In everything is some truth
Let everyone be heard.

[EMBLEM] [SEAL]

SPECTRO-CHROME

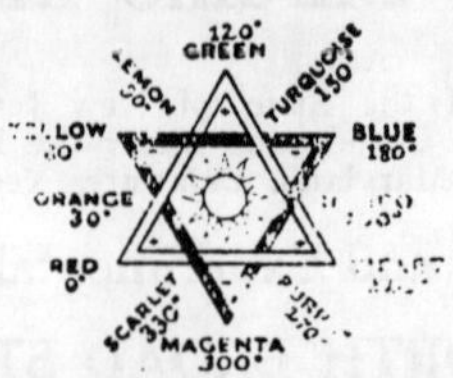

JULY 1922
BLUE 1922

Volume 1
Number 2

Printed and Published Monthly by Spectro-Chrome Institute
2401 N. BROAD STREET, PHILADELPHIA, PENNA.
United States of America

Application for Entry as Second Class Matter at the Post Office at Philadelphia, Pa., pending

Devoted to the Latest Revelation in the Healing Art
SPECTRO-CHROME THERAPY
Restoration of the Human Radio-Active and Radio-Emanative Equilibrium
(Treatment of Dis-eases) by
ATTUNED COLOR WAVES
No Drugs - No Manipulation - No Surgery
Conceived, Originated, Developed, Applied and Copyrighted, 1920, by
Colonel Dinshah P. Ghadiali, M.D., D.C., Ph.D., LL.D., &c.
Metaphysician and Psychologist

DINSHAH'S PLATFORM AS ADOPTED BY HIM SINCE 1891

The Boundless Vibratory Ocean of Thought is Essentially Universal and All-pervading; it is the Individualized Monopoly of NO Person, and is the Common Heritage of Humanity's Evolution: thus, what a Development of Unrevealed Ages has given unto Me in the Form of Knowledge in My Present Incarnation is No Distinctive Acquisition of Mine for My Sole Use, Benefit or Elevation, but is All for Thee and Is Thine Without Condition, Without Obligation, Without Expectation. I Fear No One; Only God Above and Conscience Below, and From Them I have Nothing to Fear.

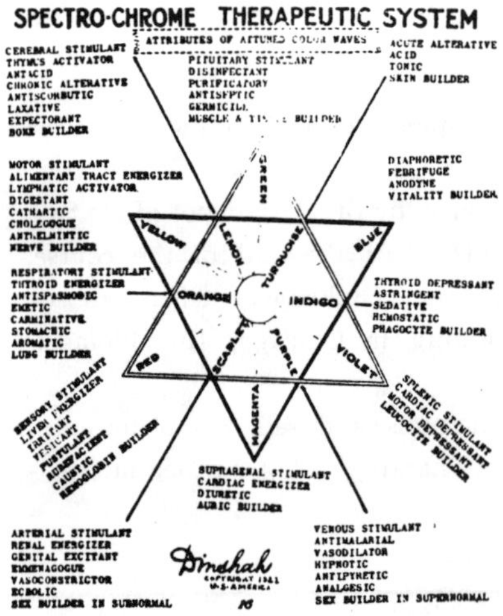

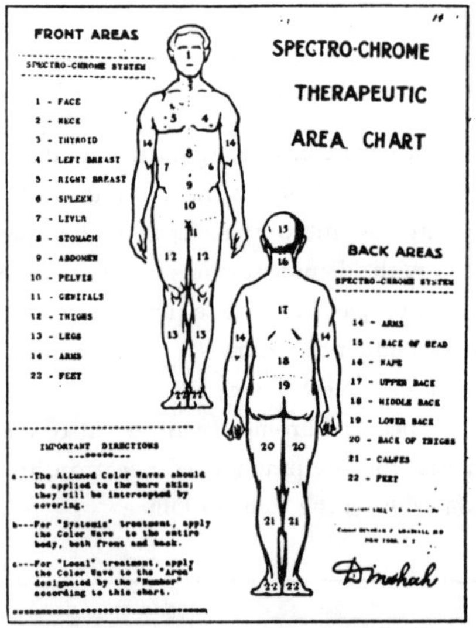

Attuned Red Wave

(The World Through the Eyes of Dinshah)

OUR NEW EXPANSION

To render punctual service to the subscribers of this periodical and to cope with the fast-increasing demand for Spectro-Chrome literature, I have installed a printing plant in 2401 North Broad Street, Philadelphia, Pa. This four-story corner building now exclusively houses the Central Office activities and Experimental Laboratory, and purchasing it for the Spectro-Chrome Institute I have dedicated it to the unselfish service of humanity.

CHICAGO, ILLINOIS, WILL OPEN

I paid a flying visit to Chicago, Illinois, this month and decided to open a Middle Western Division there for my Spectro-Chrome activities. The autumn course on Spectro-Chrome Therapy will open with a few public discourses in the early part of September. Graduates and friends are requested to communicate promptly the full names and addresses of prospective students. This course will be as usual, of 24 discourses, but will be given five nights a week, so as to leave me free for the Philadelphia, Pa., and New York, N. Y., classes by the middle of October. Please, without fail, pass the word to those interested and oblige. Write me soon.

NEXT ISSUE

The August issue of *Spectro-Chrome* is now getting ready, and with the printing facilities making us independent of unnecessary third party delays, it is necessary that your reports and articles be sent promptly for publication. Join me in making Spectro-Chrome more helpful and bigger. I cannot do everything alone. Send your articles, reports and your photograms. You surely have faces handsome enough to look at—what makes you shy?

SENSIBLE NEGROES

At a convention of the Negro Baptist Churches of Pennsylvania, assembled in Scottsdale, Pa., the delegates decided that as "the sight of wagging jaws was most disconcerting to the preacher while delivering his sermon, gum chewing should be forbidden in the churches." They are right. Converting human beings into ruminating animals was not the purpose of Nature, and human "goats" and "jaw waggers" are out of place anywhere. Will some sensible judge undertake to reform the law courts by removing the ornamental spittoon?

GARTERS VERSUS SAFETY PINS?

A battle is raging in commerce between the Garter "Trust" and the Safety Pin "Trust," each trying to decry the use of the other's manufactured commodity. The garter makers estimate there are 53,000,000 men in America who should buy and use garters and say that if they discontinue garters 106,000,000 safety pins will be needed to take their place. The Ohio State Journal asks "what badge do you wear?" I say emphatically "neither." Men's apparel needs no such "support"

or "hold-up" except as a silly fashion. Garters impede circulation and safety pins, for the stated purpose, have no field as a "necessity." Drop both from the said field and save money for better purposes. A twist on the socks is all you need and even that is not essential.

SMOKING STUDENTS

Big headlines in the New York daily papers announced the installation of smoking rooms for students in the new New York University building and congratulated the far-seeing policy of this institution of learning. I regret such a policy. At a time when the studying mind has to be kept clear for absorption of knowledge, to let the student fill his body with filthy poisonous material such as is produced by the smoking of tobacco is suicidal and will prove detrimental to efficiency. Smoking was proved during the war army tests to deteriorate accuracy in marksmanship by 14 per cent. and the New York University deserves censure of the right thinker in encouraging such a foul practice by actually providing accommodation for its "enjoyment."

THE EARTH BREATHES

Astronomer J. W. Evans declares in the last eighteen years Greenwich drifted about half a mile toward the equator, Naples is one-third of a mile south of where it was fifteen years ago, and between 1872 and 1913 Cambridge, Mass., moved thirty-six feet away from Greenwich, England. He maintains that the earth bulges a foot and a half in an up-and-down movement owing to the gravitative influence of the sun and the moon. It is a pleasure to note that similar great thinkers are now awakening to previously unrecognized matters. The students of higher life know that even minerals breathe and this with many similar matters is placed before the students of Spectro-Chrome Therapy in the courses delivered on the subject. The rein, the mysteries of human breathing processes are expounded thoroughly and demonstrated in relation to the same solar and lunar influences by means of our special chart called "Occult Operation of the Great Sympathetic Nervous System." The price of the chart is only 25 cents. Investigate the truth by personal study.

EXEMPLARY BRAVERY!

In Japan, shooting innocent ducks in flight is considered "great sport." During his recent conciliatory visit to Japan, the Prince of Wales got three ducks, beating the record of Japan's Prince Regent by one. I fail to see the bravery or "sport" in such inhuman slaughter. This prince, who will some day perhaps become the head of the British, ought to use his young life to better advantage, because as European political trend goes at present, he may be under the necessity of earning his livelihood other than by "kingship." Let him watch his steps; he may be a good, jolly fellow, but the world is drifting to a point where kings will be ranked with political "kinks" and even as such they will have to make good or lose their head from the shoulders.

BUSYBODY GLOOMGOBLINS

There is an association in New York called the Lord's Day Alliance. Rev. Harry L. Bowlby, its secretary, started recently a Blue Law campaign by raiding merchants doing business on Sunday. The violators were fined $5.00 by Magistrate Dale, of the Bridge Plaza Court in Brooklyn, N. Y. This Blue Law craze is a remnant of by-gone ages and in the hustling civilization of modern times with the newspapers and railroads, telephones, and telegraphs, trolleys and automobiles running seven days in the week, there is no sense in trying to impose such restrictions on people of composite nationalities. The Sabath of the Jews is Saturday, that of the Mohammedan is Friday, that of the Christian is Sunday, and that of the Zoroastrian (like me) is every day in the year. Why then does the Christian persist in the revival of ancient Puritanism for all nations? Blue Laws are now unconstitutional. They legislate in favor of a Sunday to the exclusion of other days, therefore, they favor the observances of one religion against another. America is not a single nation in the true sense of it. It is a beautiful melting pot of numerous nations and is supposed to be built on cosmopolitan levels. I ask the Christians to desist from such stupid endeavors; Christianity is only one system of religion and no Christian has any right to have laws passed or enforced in the name of Christ (who loved all mankind) against others to whom such laws are odious. God never rests, never sleeps; He is ever working, and a moment's quiet reflection will show that these sporadic efforts on the part of fanatic so-called "Christians" engenders enmity and opposition in the mind of other religionists thus subverting the foundation principle of the Lord Jesus which was Universal Brotherhood of Man.

FAULTY ADMINISTRATION

A Swiss, August Probst, was lately deported as having entered illegally the United States. At his trial, he stated that some rich girls loved him, but their parents did not fancy a butler like him, and to avert the catastrophe of one of the rich maidens collaring him as a son-in-law for her father, he was chased out of Pennsylvania by private detectives and brought to New York where the authorities were appealed to deport him. I do not wish to rake up here the demerits of the case, but question by what right the American officials paid Government money—$150 to the Cunard Company to convey him out with the services of an attendant? This attendant was to report on his physical condition daily and was to get a receipt from the parents for the safe delivery of their son. Why were the officials so anxious after deporting him as an insane undesirable? The published reports breathed much scandal and had offensive odor, but what is the happiness and life of a poor butler to a great government where some official's family affairs are involved? Do they behave similarly when they deport poor Italian and honest Russian immigrants?

KANSAS OUTRAGED

There is a town in Kansas with the high-sounding name of Eminence which boasts of a population of just 92 citizens, men, women and children, exclusive of parrots, chickens, jack rabbits, geese, cats, dogs, mice and mock turtles. Because of the absence of other accommodation, the schoolmistress, Mrs. Clara White, allowed a dance to be held in the school house. For this grave and serious offense against the laws of God, Moses, Zoroaster, Confucius, Krishna, Buddha, Jesus and Mohammed, not forgetting the laws of morality, principle and chastity, the State Superintendent, Miss Lorraine Wooster, caused Clara White's certificate as teacher to be revoked, because she "violated the laws of Kansas and trifled with the dignity of the Commonwealth." Superintendent Lorraine Wooster further says that "dancing is particularly heinous in a school teacher, and when the offense takes place within the very portals of the school the offense comes pretty near being lese majeste!" What next? How about President Warren G. Harding and King George of England dancing? I should like to get a photo of this State Superintendent of Education of Kansas to study her Radio-Active and Radio-Emanative Color Waves for a character sketch. Spectro-Chrome has subscribers in Kansas; will one please favor me with the requested copy?

PRONOUNCE THIS

They cannot pronounce my full name in the United States. Dinshah P. Ghadiali is to the inflexible tongued Americans an odd, unpronounceable, long, hard, peculiar name. I tell them, DIN SHAH means King of Duty, P. is the initial of the name of my beloved deceased father, Pestonji; Ghadiali means watchmaker, which represents my ancestral vocation. They shrug their shoulders, open their mouth with a ahem and auho, wipe the perspiration on the brows and say either Di Di Di Di or Gei Ghei-di di al. It does not annoy me in the least. I tell them to call me plain *Dinshah*, and print it so in bold type and yet daily I receive letters addressed as Densaw, Damshoe, Danshot, Dannshaw, D. In Lhot, Ding Lip and such euphonies as Damn Rot also. How about this, however? In Prague, Bohemia, an officer was appointed some time ago for the work of winding up food rationing. His title was "Lebenswittelzuschusseintellungskommissionvorsitzenderstellvertreter."

I am anxious to hear a born American pronounce this which even to me, as a foreign-born American, is clear enough. He who laughs last laughs the best.

ERRONEOUS UTTERANCE

Speaking at the Pennsylvania Military College, in Chester, Pa., Secretary of War Weeks is said to have informally spoken to reporters that the people wanted beer and light wines and if he were in Congress they could have them. For what? To make more war? To make more poverty? To consume $2,500,000,000 of wealth in a year for rank alcoholic poison? To demoralize more? To fill more prisons? To further populate lunatic asylums? What really does Weeks mean? Who wants beer and wines, really? The drinking editors of some newspapers might be clamoring for agitation, but no truly sensible student of ethics and economics can show any proofs or figures of the existence of human elevation by the use of any alcoholic as a beverage.

CHICAGO WETTEST CITY?

C. E. Main, Washington Prohibition Field Superintendent, in his report to Prohibition Commissioner Haynes calls Chicago, Ill., the "wettest city in the United States" and lays the blame on the calibre of the government agents whom he calls extortionists. His knowledge interests me in view of the establishment of a branch of the Spectro-Chrome Institute there in September. When government officials like Main speak out so boldly, my whips may not smart any more in criticizing the faulty administrative methods. At any rate, it smooths my way, but I have doubts whether Chicago outraces New York and Atlantic City for this extreme honor.

SMOKING AND EYESIGHT

Lord Birkenhead, Lord Chancellor of Great Britain, was informed by his physicians to abandon smoking or lose his eyesight. This has set up an uproar in England. Dr. M. S. Mayou, a noted writer and oculist, says that cigars and pipes cause blindness, the trouble arising through the blood. Professor Roman, of Syracuse University, expresses a belief that tobacco prohibition should automatically follow alcohol prohibition. Whether tobacco smoking causes eyesight trouble cannot be directly maintained by stating that William E. Gladstone, the late Prime Minister of England had good eyes because he was a non-smoker or the present premier, Lloyd George, has good eyesight because he is a good smoker, but can only be reasoned from the analysis of tobacco smoke and its study. We hear always about one product only—nicotine, but there are other constituents of the smoke about which very little is generally known. Analysis shows that tobacco smoke contains the following important chemicals among other minor ones:

Nicotine	$C_{10} H_{14} N_2$
Pyridine	$C_5 H_5 N$
Carbon Dioxide	$C O_2$
Carbon Monoxide	$C O$
Sulphuretted Hydrogen	$H_2 S$
Sulphurous Acid	$H_2 S O_3$
Sulphuric Acid	$H_2 S O_4$
Hydrocyanic Acid	$H C N$

I should like, in view of this analysis, showing an array of exceedingly powerful poisons, any person (layman or professional) to defend or justify their constant introduction into the human organism. It is no use talking nonsense or acting like a bunch of heep, because even doctors smoke. Talk facts; look into direct evidence of science; examine the physiological effects of the above array of poisons on the teeth, tongue, mucous membrane, lungs, heart, brain, stomach, nervous system, blood, everything. I am willing on any platform to accept the challenge of any competent speaker on the subject and smash arguments to smithereens. Tobacco in any form is detrimental to human life forces and the United States will positively lead the world some day by abolishing its cultivation, as a deleterious rank weed, unfit for human consumption.

SENSIBLE MEDICAL OPINION

At the seventy-second convention of the Illinois State Medical Society in Chicago, Dr. Frederick Green voiced that "the use of scientific terms when conversing with patients is not advisable when the ailment can be put in plain terms; mere technical designation of the trouble is enough to frighten any patient." Of course, what does the patient care what you call the ailment? The patient is concerned only in getting better and the less fright is created the quicker the recovery. Ailments are the signals of Nature showing the existence of certain transgressions of her laws; they are semaphores to the healer indicating certain conditions; the removal of those defects removes the ailments or diseases as they term. In Spectro-Chrome Therapy all this mess of different diseases is cut down and the dis-eases substituted. When all functions of the human being are 100 per cent., the person is at "ease"; when the health equilibrium is disturbed by the over or under functioning of the Radio-Active and Radio-Emanative organism, there is "dis-ease." Find this according to the system established by the originator and without scaring the wits of the patient, the trouble will disappear under the given treatment. Spectro-Chrome Therapists, take my brotherly suggestion and do not parade your "medical" knowledge to the patient in bombastic jaw-breaking words, but tell the sufferer that such and such color wave is wanting or excessive and that you have the means of restoring the equilibrium in full accordance with the laws of benign Nature if the patient would but follow your advice. Remember DIS-EASE and not DISEASE.

There is no such thing as "disease" as understood by the followers of a moth-eaten medical "science" which at best is purely empirical and possesses no basis of accurate mathematics such as that upon which Spectro-Chrome Therapy is founded.

MISCARRIAGE OF JUSTICE

Carl Veith, a man of 37, living in New York City at 691 Wythe Avenue, shot dead William Doyle because, during his absence, Doyle in a drunken condition went to Veith's home and demanded money from Veith's mother. She refused and Doyle beat her. The jury gave a verdict of "not guilty" on the "unwritten law" principle and Judge Mancuso discharged the murderer. The newspapers are all silent about this judicial miscarriage. During the trial it was proved that the accused on arriving home learned about the occurrence, took a revolver and went after the man. He met him in East Broadway and shot him.

This was a case of deliberate murder, downright meditated manslaughter. The unwritten law of either self-defense or justified motives cannot apply here. No matter what the drunken man did to the mother of the accused, he was not present on the spot for so-called vengeance stirred in him by the act. He came home, learned about the mistreatment and deliberately went after the other fellow. This meant perhaps an extenuation but not a total discharge. The jury was led to a stupid verdict through emotional miscarriage, and was the judge bound to accept it? Trial by jury as it goes at present is a humbug—unwritten law must not cover cold-blooded murder under any circumstances. If nothing else, the accused should have been sent to the "refrigerator" to get cool and act more sensibly in the future. Compare the above American miscarriage with the following British specimen:

INDIAN LEADER TO BE JAILED

Hazrat Mohani, Moslem League President, Convicted of Sedition

Ahmedabad, May 5 (Associated Press).—Hazrat Mohani, President of the All-India Moslem League, was sentenced to two years' imprisonment today on a charge of sedition after the judge had refused to accept the unanimous verdict of not guilty rendered by a jury which included five Indians. A second charge of inciting to war was referred to a higher court.

The defendant delivered an impassioned address to the jury in which he said the independence he advocated in the All-India Congress was every man's birthright.

The prosecution contended that the defendant had urged the establishment of a parallel Government to usurp the functions of the existing Government, making warfare inevitable.

Which is preferable? And where shall we look for justice? Law is the latest interpretation of the written word by the most whimsical judge and, as things go now, we fail to see which country has worse administration of "law." British or American? Nothing to choose from if the above be samples of "justice."

IMPOSSIBLE WITH SPECTRO-CHROME

Charles Erickson, of 279 Cooper Street, Ridgewood, Queens, wanted to take a medicine for cold and, as the police report says, "swallowed iodine through mistake." He was in Kings County Hospital recovering. This ought to be a good lesson for Charles to take to Spectro-Chrome: it surely could not kill! Our New York office has instructions to get after him as a convert to Spectro-Chrome.

HANDGRIP AND COLOR

Dr. Fere, a French experimenter, found the strength of the handgrip the greatest under the Red color, diminishing in the order of Orange, Yellow, Green and Blue. This has been known to the Originator of the Spectro-Chrome years now, as the Spectro-Chrome Charts have demonstrated.

NEEDS NO ARMOR

The medical and surgical world is now turning to high power "X (unknown) rays in its useless struggle against cancer and the like. Numerous cases have been reported of baldness, burns, blindness and similar unpleasant results by such rays penetrating unaimed-for quarters; so now when they install such machines they build a lead-armored chamber for preventing stray leakage. They never will succeed in remedying cancer by such methods, because it is a constitutional trouble and not a local affair. Proper food of an untainted character in conjunction with Spectro-Chrome is the best for cancer and its allies. Spectro-Chrome needs no armor plates and can be used with perfect safety.

A MISGUIDED BISHOP

Bishop William Lawrence of Massachusetts in his annual report to the Protestant Episcopal Diocesan Convention in that state speaks as follows:

"Hundreds of thousands of working men who found solace and comradeship after the day's work in what they felt to be their innocent glass of beer had it snatched from them, and thousands of reputable citizens found their personal liberties and domestic habits broken in upon.

"Surely it is competent for every citizen to speak, work and do everything consistent with the law to have this law either amended or rescinded."

Beer innocent? Alcoholic liquor innocent? Personal liberty snatched? The bishop forgets that in civilized conditions of society we "snatch" many so-called "personal liberties" if they are detrimental to the state. Prohibition of all alcoholic liquor must stand as long as non-thinkers will use them as beverage and depreciate the manhood and womanhood of glorious America.

SOONER THE BETTER

In 1921, about 150,000 horses and 5,500 dogs were eaten in Germany because the prices of other flesh went up 200 to 300 per cent. Furthermore, they are turning again to vegetarianism which was enforced on them during the International Europequake. Friends, take our advice and cut out slaughtered food entirely. You fought well against odds because of the vegetarian diet and there is no reason to return to blood which costs more and nourishes less, not counting the generated poisons.

KEEP SKIN UNPAINTED

Angelina Molinari, a four-year child of New York, died of poisoning through eating massage cream. What is the value of such products at any rate for human welfare? A healthy skin needs no cream made with poisonous materials mixed with lard and called "cream." What will poison by mouth will poison through the skin and it is a matter of regret that natural beauty is discounted in favor of drug store complexion. If the skin be defective use Spectro-Chrome Turquoise for systemic treatment.

CRAB SURGERY

When a crustacean (like a crab) has a damaged limb, it breaks it off near the base, a bandaging membrane covers the lesion and prevents bleeding; then a new leg shoots out and becomes hard. It is a pity we cannot grow limbs like even a crab and yet arrogate that we are "superior" animals! It would not be a surprise if some day a medical school might be established to study surgical phenomena from seaside under the name of "Marine Crabology."

BEGGARS THREATEN

Turks are known for threats and bluff. In Constantinople, Turkey, there is a Union of Beggars. They are now grieved at the alms given them and have voted to refuse alms less than two cents in value. They are thus counting the teeth in a giftmare's mouth and the best the nations there can do is to pocket the money and forget the bluffers. In the Mohammedan states of India they have a currency of shells called Cowry. For an American dollar, one may buy about 8,000 cowries. Fancy a person going out with a bag full of ten dollars worth of such currency, making purchases and distributing alms of five and ten cowries most graciously in the name of the Almighty God. At the end of the day, one may find that the bag is still full to the brim! And yet the same Indians want Turkey to be in Europe as a thorn to progress!

VERDICT OR CRAZE?

At New Road railroad crossing of the Pennsylvania Railroad Company, at Absecon, N. J., an automobile collided with a train locomotive and an entire family of six was killed. The coroner's jury of Atlantic County passed a verdict, after twenty minutes' grave consideration, that the judgment of the automobile driver was erroneous, but the railroad management was warned not to let such an accident happen again! O, incarnation of Solomon! Thou certainly art out-solomoned! In a certain town in India (which we may call Humbugpur), there was an old patriarch who was credited with having unusual wisdom. He was known as Lal Buzrug, and all the country resounded with his sagacity, occult knowledge, uncanny learning and philosophy, reaching up to 19th power of the high planes! One day, a farmer installed an earthen waterpot for his cattle to allay their thirst, but an over-thirsty extra-sized buffalo, with oddly-shaped horns, was so greedy in reaching the water inside the pot that it shoved its head and horns right into the potneck and was trapped. There was a great hue and cry, but the farmer, with a batch of wise neighbors thrown in, could not succeed in extricating the buffalo. In this quandary, the whole gang tramped to Lal Buzrug and laid the case before the worthy. Lal Buzrug stroked his 13½-inch beard and said, "O, you helpless human ordinaries, go quickly and chop off the head of that buffalo before he suffocates."

So they bowed to the ground and marched back and with a Jodhpuri sword separated the head from the trunk. A strange situation thus arose. The head fell inside the pot, but because of the curved horns declined to come out! Meanwhile, the other cattle on the farm were thirsty and mooing, so the farmer and his trusty advisers put their heads together to dope out a proper solution. The conference resulting in failure to reach some amicable and practical understanding, they returned to Lal Buzrug and, with joined palms and appealing features, placed before the esteemed dignitary the question of the next step. Lal Buzrug cast a withering glance at the quivering agriculturists and said, "O, ye human mental carcasses, whatever will become of ye all when I am gone to the Almighty's throne? Have ye not thus much common sense as to know what the next right step is to be? Get ye all out of my sight and let the water pot be promptly smashed before the head of that buffalo begins to decompose!"

KNICKERBOCKERS IN BROOKLYN

Dr. William L. Felter, principal of the Brooklyn Girls' High School, has posted a ban on knickerbockers in favor of skirts and asks the girls who come in knickers to interview him personally. I suggest the worthy doctor interview me personally or on the platform and I shall willingly inform him about the history of skirts as potential promoters of high morality in the Elizabethan period!

Mayor Swanton, of Traverse City, Mich., goes one better. He issued an edict that any female appearing on the streets wearing knickers will be arrested. He says that knickerbockers are too tempting for the "moral" eyes of the boys. Well, Mayor, if I were in your place I should issue an edict that all males moving in your beautiful streets should wear blinders. You unpsychological meddlers, the fault is not in the knickerbockers; your concepts of morality are in need of reformation.

STUPID ORDERS

Now the edict has gone forth that rubber overcoats should be worn by bathers at Atlantic City, N. J., while passing through the streets. An indignation meeting resolved to march clad in barrels. I should suggest that a monster parade, duly but quietly organized, by courageous independent-minded thousands marching just once there in full Adam and Eve garb would produce reaction enough to cure the ugly authorities permanently.

EVIL TO HIM WHO EVIL THINKS

The following article appeared in "The Evening Bulletin," of Philadelphia, Pa., June 12, 1922. It speaks for itself:

PHYSICIAN CAUGHT IN HIS OWN TRAP

DR. SALTZMANN SAYS HE PLANNED TO SNARE CHIROPRACTORS AS BRIBERS

IS NABBED AS TAKER

Dr. Louis A. Saltzmann, registered physician employed by the State Board of Medical Licensure as a special detective, who was held in $1,000 bail on Saturday in Norristown, charged with extortion from the Chiropractors' Association, says he was planning to expose the chiropractors as bribers when they had him arrested as taker of a bribe.

He was arrested by county detectives working, according to Major S. O. Wynne, their chief, in conjunction with the State Board authorities, at the home of Dr. Merri Mollier, chairman of the legal committee of the chiropractors, in Dr. Mollier's home, at Trappe, Montgomery county. Dr. Mollier charges

that Dr Saltzmann demanded $200 a month for "protecting" chiropractors from prosecution as unlicensed physicians, and to lobby for a bill licensing them.

Dr. Saltzmann says his idea in taking the money was to worm his way into the secrets of the chiropractors in their plan to have jurisdiction over them taken away from the State Board.

Dr. Saltzmann asserts it was his plan to get all data together and have three witnesses handy to watch the chiropractors hand him money to help them. He then expected to turn the entire case over to the District Attorney's office or to the State Medical Board, in the nature of an expose of what he claimed the chiropractors were doing and how willing they were to pay for it.

He was to receive more than $700 during the past week, and he planned to have the money turned over to him in a corner of City Hall where his witnesses could watch the proceeding.

But Mollier, he said, insisted that he take the $100 on account Friday afternoon, and Saltzmann did so, he says, because he believed Mollier would not appear at City Hall with the balance if he refused to accept Friday's "partial payment."

"I had a trap set," said Saltzmann, "but they sprung theirs first. Mollier, after they arrested me said: "Saltzmann, you played a game of chess with us, but we checkmated you. We trapped you in Trappe."

"I got the dope that they had me spotted. The ones I hadn't seen yet were told what I looked like and they had it all fixed up that the first one that got me on the table was to jump on my back and fix it so I wouldn't have any spine left. It is all wrong that I received money from them last year."

SYPHILITIC TEST OR HUMBUG?

Dr. W. A. Evans, writing in the *Daily News of New York*, May 25, 1922, speaks thus in part about the value of Wassermann test for syphilis:

"When the test is properly made, and the report is interpreted in connection with other features of the case, it is an excellent indication as to whether syphilis is present or absent.

"A positive Wassermann is not given until after about the fifth week of the disease. Diagnosis prior to that stage must be made by the clinical appearance, clinical history, and by microscopic examination.

"The Wassermann reaction is not a test of infectivity. It is not given during the first six weeks of the disease when infectivity is absent or at most only low.

"It is liable to be present in certain diseases and conditions not related to syphilis and where syphilis is not present. Among these are malaria, ether and chloroform narcosis, scarlatina, jaundice, diabetes, pregnancy, nodular leprosy and yaws.

" 'The blood of a healthy baby less than two weeks old,' Stillians says, 'will give a positive Wassermann.'

"On the other hand, Krauss, of Memphis; Dock, of St. Louis, and Lydston and Stillians, of Chicago, are among the men who have repeatedly called attention to the shortcomings of the Wassermann test.

"Kahn, of Memphis, says: 'It is evident from the foregoing that in every syphilitic stage isolated cases are found which will not respond to the Wassermann test.'

"On the one hand, in alcoholism there may be absence of a Wassermann reaction even though syphilis is not present.

"On the other, there are some cases that seem to get the Wassermann habit. Havin~ once started to give a Wassermann plus, they keep it up for life, even though the syphilis has been brought under control. The scientists call these cases Wassermann fixed.

"In the competition between laboratories there is some tendency to advertise such claims as 'We get a larger proportion of positive Wassermann's than other laboratories do.' This claim is not untruthful necessarily. By varyin~ the methods one way or other the test can be made more sensitive and the result may be as advertised.

"But there is this to be said: The Wassermann test is not specific for syphilis. It is most dependable when it is just about so sensitive. If, on the other hand, it becomes too sensitive, it loses value, just as it loses it when it is not sensitive enough.

"What is the final conclusion? Shall we pay no attention to Wassermann reactions? Shall we quit having them made? I find practically no one in favor of that. With all its shortcomings, the Wassermann test is a standard procedure and should be continued."

These same questions are frequently asked me and in order to answer them thoroughly for the student to gauge, I append the definition and technic of this (in) famous "test."

Wassermann test was introduced b August V. Wassermann, a Berlin bacteriologist. It is a diagnostic test for syphilis, based upon the theory of complement fixation. Measured quantities of extract from the liver and spleen of a syphilitic fetus (representing the antigen) and of the blood serum of the person to be tested are mixed in a test tube and some fresh normal guinea pig serum (containing the complement) is added; after incubation at body temperature for about an hour the contents of this tube are added to a mixture of the red blood cells of a sheep or other animal, suspended in physiological saline solution and of the serum of a rabbit which has been immunized to these particular red cells; if the serum mixed with the extract of syphilitic organs in the first test tube is from a syphilitic individual no hemolysis of the red blood cells in the second tube will occur after incubation for one hour. This test was subsequently modified by the Japanese bacteriologist Hideyo Noguchi, but the basic idea is on similar lines. Herein all depends on a crude idea that the physiological and biological actions and reactions of the animals are identical with the human. Herein is a terrible and gross error.

Animals under the human kingdom possess only the physical, auric and psychic vehicles, whereas man has developed mental and causal vehicles also. Therefore, to mix pigs, sheep, rabbits and humans in one bloodpot is more appropriate for ghouls than scientists. If this wretched "test" will fail in one among one hundred tests and an innocent human is thereby branded and treated as a syphilitic, it is a crime against humanity. Spectro-Chrome Therapists, keep off from these and similar "tests" and resort to the potency of the Attuned Green Color Wave in all cases where there is the least suspicion of any outside infection, no matter what its specific medical term might be and if there is any vitality left in the patient the health equilibrium will be restored. The more you get away from ancient, doubtful changing theories the better for your patient and the world. Where will *you* stand, when, according to certain doctors of medicine, syphilitic conditions exist in nearly the whole world and more so in the United States? Dr. Albert Abrams, the eminent researcher of San Francisco, Cal., vigorously maintains this ground, but while the doctors are philosophizing and dosing people with quantities of mercurial and arsenical poisons and iodine, the sufferers are the goats: so use your Green and the streptococci will disappear without drugs.

TITLES FOR TURNIPS?

Sir Joseph Robinson, known as the "Diamond King" of South Africa, was elevated to the peerage by the King of England, which caused storms of protest. Lord Burton, who was Governor-General of South Africa during the war, says that Robinson's value was then known by his absence in public-spirited enterprises. Lord Harris in the House of Lords protested that Robinson had nothing to recommend him except money. After much debate the discussion was discontinued. This is a question of vital importance to the English. The feudal custom of conferring titles of nobility (which was the highest ambition a few years ago) is now so looked down upon in certain quarters that the sensible Canadian Legislature some time ago petitioned His Majesty to discontinue conferring hereditary ranks on the Canadians. Rightly so. Titles of nobility of a hereditary character are stupid, worthless and of no true value except to show that one's ancestors did such and such a brave deed. I say to a person, "What did *you* do?" Man is essentially individual in development by the laws of retributive action and reaction and to pride one's self on the glories of an empty past is ridiculous.

Let a man be honored for personal achievements by which the world is benefited; let a woman be similarly honored for her glorious work and service to humanity; both will be meritorious and deserved and will command public respect, but to entitle some nut for his wealth or elevate some pup to hereditary titledom because he can play good polo with some "gink" is utterly childish and serves no useful purpose except perhaps later to allure an American heiress to wed when his coffers are cleaned out and she covets the rank of "Her Ladyship" at a doubtful age in life.

Attuned Orange Wave

(Humor, Wit, Fun, Satire, Tickle, Irony)

VARIETIES OF INDIAN SERVANTS

By Miss Kashmira Dinshah, S.C.T.

A Musselman named Baboot Khan once lived in the city of Delhi in India. He had a servant called Shekh Mia, who would do just what he was told to—no more, no less.

One day Baboot Khan's wife wished to go to Umbala, a town one hundred and fifty miles from Delhi. Shekh Mia was ordered to hitch up bullocks to the cart and drive her to Umbala. Three days later Shekh Mia reappeared at the Musselman's, but without the woman. "Hey, Shekh Mia," cried Baboot Khan, "where is my wife? Didn't you bring her back?" "Why," said the servant, "you didn't tell me to bring her back. You simply told me to take her to Umbala." "Oh, you are a mutt, and a nut at that," shouted the master, "go, bring her back."

For a month no trace was heard of either Baboot Khan's wife or the servant. At the end of that time Shekh Mia presented himself at the Musselman's door with the latter's wife on his shoulders. "You rascal!" howled the master, "why didn't you bring her in the bullock-cart?" "Well, you didn't tell me to. You only told me to bring her back from Umbala. You did not say put her in the cart and drive her back." "Oh," roared Baboot Khan, "get out of here! You are fired! You blockhead! Couldn't you use any common sense?" Then he called an employment agency and asked for a servant who would use judgment in carrying out orders. It sent him one who boasted he did not only do what he was told to do, but more. "Ah, you are the right kind," said Baboot Khan, "I hire you."

Next dav Baboot Khan's wife fell ill. He ordered the servant to call a doctor who lived around the corner. The servant was out five hours. When he did return he brought with him not only the doctor, but an undertaker, a priest, two hundred mourners and a coffin. "What in Allah is this? I didn't order you to bring these," shouted Baboot Khan. "Sir, I thought if the doctor might not be able to cure the woman, the priest, the undertaker and the mourners would be needed." "You are a fool! Get out of my service," cried the angry Musselman, and again called the agency. "Can't you send me a servant who will use some brains?" "Oh, yes, we have just the man for you. He need not be told anything. He will do everything by using his brains. He is an intuitive servant." "All right, send him."

The new servant was installed. "You need not tell me anything," said he, "just think and I will know what is to be done by intuition." Some days passed and Baboot Khan went on a journey, taking his servant with him. Night came and they entered a wayside inn, and retired. At about three in the morning Baboot Khan awoke. "Get up," he said to the servant, "see what time it is." "Oh, it is about three o'clock." "How do you know?" "Why, the window is open and by the cold draft coming in I know dawn is approaching." This answer satisfied Baboot Khan. Shortly he said, "Get up and see if it is raining." "No, it is not raining." "How do you know? You did not look." "No necessity," answered the servant, "a cat just came from outside and passing close to my bed I put my hand on its back; the fur was dry, so I intuitively know it is not raining." "Oh, hang your intuition!" cried Baboot Khan. "I want common sense! You get my goat with your guessings! Get the gate!"

Attuned Yellow Wave

(Correspondence, Query, Catechism, Inquiry, Interrogatory)

WHAT OUR LEARNED FRIEND SAYS

July 4, 1922.

Dear Teacher and Friend:

"It" arrived apropos of the birthday of the Nation! Heartiest congratulations. As this blessed land has been a haven and pathfinder for many a tired and seeking wanderer, so may the Spectro-Chrome organ be the same!

It emits a glowing yellow ray as well as a benign purple. Supreme is the Father-Mother wave of Magenta.

May everv progressive M.D. consider it a privilege to subscribe. It may come!

Faith, prayer and work will do wonders, they remove mountains.

With intense sympathy for the welfare of Spectro-Chrome,

Gratefully yours,

ANNA C. DE LA MOTTE, M.D., S.C.T.,
70 South Tenth St., Brooklyn, N. Y.

VOLSTEAD OR SHEPPARD?

The newspapers of the United States have named the Eighteenth Amendment to our Constitution, "Volstead Act." When the originator of Spectro-Chrome Therapy was in Washington, D. C., last winter, teaching the Spectro-Chrome class there, he spoke very forcefully in favor of prohibition, eulogizing the "Volstead Act." A friend, Victor Russell, who was present, kindly sends the following for the true information of the public. It sheds a light on facts not generally recognized and which are interesting. We thank the writer for his correction of statements then made by Dinshah.

Colonel Dinshah P. Ghadiali,
Washington, D. C.

MY DEAR SIR: You will recall at the close of your lecture, I brought to your attention the slightly erroneous impression you seemed to have regarding the Eighteenth Amendment and the prohibition situation. You asked me to write you concerning this, and I have therefore endeavored to set forth in brief compass something of the inception, growth, and victory of the prohibition movement in the United States, which will, I believe, be helpful in enabling one to view the present period of adjustment in the proper perspective.

The movement which resulted in having a ban on the traffic in alcoholic beverages written into the organic law of this land, had its inception in the temperance societies of a couple of generations ago. These societies, largely religious in character, pledged their members to abstain personally from indulging in alcoholic liquors. It might be pointed out here that there is a difference between *temperance* and *prohibition*. The temperance movement sought merely to induce individuals to lead pure and temperate lives, basing its appeal upon religious and hygienic grounds. The prohibition movement, on the other hand, made no particular plea for personal abstinence, but was a collective movement to prohibit and abolish through legislative enactment the traffic in alcoholic beverages, and its appeals were economic and political.

The transition from a temperance to a prohibition movement occurred gradually and was completely effected not over twenty years ago. This came about when the business men and politicians came to the aid of the moral and religious reformers. The business men saw that the efficiency of their employees was impaired by alcoholic drinks, that there was a very direct relation between quantity and quality of output in factories when the neighboring community was "dry" or "wet," and they further saw the vast economic wealth being squandered in the liquor traffic. The men engaged in political life saw that the center of rotten and corrupt politics was the saloon, that it was a menace to the civic and social life of the community, that it was the rendezvous for the old criminals and the chamber of initiation for the younger ones.

The progressive business men put up the money and the progressive political candidates did the campaigning with the result that county after county began going dry all over the country. But prohibition could not be made very effective in any one county or state when the one adjoining it was "wet." It became apparent that it must be made nation wide. Under the Constitution, Congress had no authority to enact such legislation. It therefore became necessary to amend the Constitution, hence the Eighteenth Amendment was passed in the autumn of 1918.

The author of the Eighteenth Amendment and the leader of the fight for its adoption in the Senate was Senator Sheppard, of Texas, who, when he took his seat nine years ago, was about the youngest member in point of age in that august body. He first secured the passage of the bill establishing prohibition in the District of Columbia, in the face of the most determined opposition. He later put through the Eighteenth Amendment, although the leaders of both political parties were against it. He had been pushing the measure for some years, and it was largely due to his persistent but patient efforts that the measure went through when it did.

It was for this reason that I brought the matter to your attention when, in the course of your talk, you attributed credit for the Eighteenth Amendment to Mr. Volstead. After the Eighteenth Amendment had been ratified by all but three states in the Union, it became necessary to enact an enforcement code, defining violations and prescribing penalties. This was drawn up by the friends of prohibition in both the Senate and House of Representatives. It was referred to the Judiciary Committee of the House and favorably reported by the chairman of the committee, Mr. Volstead, of Minnesota. The measure subsequently passed both the House and Senate, and because it was first referred to as the Volstead Act, it has continued to bear this name. While full credit is due Congressman Volstead for his efforts in steering this enforcement act through the House, and preventing any weakening amendments to it, the distinction of having first forced the issue and pushed it to a successful consummation belongs to Senator Sheppard.

It might be interesting to know that the attitude with reference to prohibition which one finds in the very large cities where vice parades the streets and virtue dwells in obscurity is not at all representative of the attitude of the great expanse of country west of the Atlantic seaboard, especially in the smaller cities and villages where virtue walks the streets and vice hides in obscurity. Yet somehow our American newspapers have a way of representing things so that it would appear that it is elegant and liberal to poison oneself with alcohol and other forms of dissipation, and that it is narrow, mean, and hypocritical to oppose such practices.

Very truly yours,
VICTOR RUSSELL.

Attuned Lemon Wave

(Medicine, Surgery, Science, Osteopathy, Chiropractic, Philosophy)

COAL TAR AND CANCER

By Edmund J. Hogan, D.C., S.C.T.

The New York Times, in an editorial of June 26, 1922, states: "A disquieting theory concerning cancer put forward by British physicians is that the disease is induced by the action of coal tar, pitch and similar substances of the new materia medica." Further on it quotes the "Journal of Industrial Hygiene" to the effect, "Experiments with rabbits and guinea pigs show that applications of coal tar derivatives in all cases produced within four months extensive and rapidly growing malignant tumors."

This information is well worthy of consideration and deep reflection when we realize medical statistics prove that during the past fifty years cancer shows an increase of some four hundred per cent.

Iridology proves conclusively in the iris, the result to humanity of these latest additions to materia medica. Antipyrin, phenacetin, antifebrin, triasol, antikamnia and many other preparations have coal car derivatives. All are of a highly poisonous nature and paralyze and depress the brain, heart and respiratory centres. Antikamnia will be found to produce in the upper portion (brain area) of the iris a grayish white coating, somewhat resembling whitewash. Antifebrin, antipyrin and phenacetin from the sympathetic wreath outward, show a discoloration blending upward from gray to light yellow. Creosote and guaiacol cover the entire iris with a grayish coating.

In presenting a case history a patient will often vehemently deny having used drugs or patent medicines containing coal tar preparations. The signs in the iris are convincing, however, and the entrance of these poisonous derivatives may have been furthered by many food products.

When we read the reports of the Bureau of Chemistry and Dr. Wiley's experiments, we may well stand aghast at the deadly and murderous concoctions put forth by the commercial Borgias of the present day. Preserved cherries, currant jam, port wine, corn beef, sausage, soda water syrups, etc., have yielded yards of brilliantly dyed silks. But the most damnable of all atrocities is the cheap candies sold to children, oftentimes containing arsenic, prussic acid, sulphate of copper, and clay. Many cheap brands of chocolates are composed of starch, animal fat and oxide of iron or iron rust.

If the experiments upon guinea pigs and rabbits quoted above, show the malignant effects of coal tar preparations, in the light of this information, is there any cause to wonder at the increase of cancer? We know that anabolism and catabolism constitutes metabolism; we know that a drug is a substance that will not metabolize, and through you we know also that a drug has a potency only due to the disengagement of color waves in the system. When we realize that Spectro-Chrome Therapy works upon the Radio-Active and Radio-Emanative human organism, and releases color values without deleterious effects, we shall hasten the day of emancipation for a drug-cursed world.

Henry Lindlahr, M.D., of Chicago, Ill., in his work, "Iridiagnosis," presents the following addenda:

COAL TAR PRODUCTS

(Acetanilid, Antipyrin, Creosote, Phenacetin, Antikamnia, etc.)

Allopathic Uses:

1. Powerful antipyretics acting within one hour.
2. General nervous sedatives, anodyns and hypnotics, "giving complete and prompt relief in nervous headache, neuralgia, ataxia pains, gout, rheumatism, dysmenorrhea," etc.

Accidental Poisoning:

1. Patent fever remedies and headache powders.
2. Antikamnia and other proprietary anodynes.

Toxicology:

1. Reduction of blood alkalinity and red corpuscles.
2. Depression of all vital functions with a special tendency to cardiac failure.

Symptoms of Coal Tar Poisoning:

1. Undue readiness to fatigue.
2. Despondency and loss of memory.
3. Renal irritation.
4. Nervousness, neurasthenia, paralysis, insanity.

Elimination of Drug in Healing Crises:

1. Excessive perspiration and erythematous eruptions.
2. Catarrhal discharges.
3. Excessive urination.
4. Nervous and mental disturbances.

Dr. Hogan surely gives us here something to consider for national health. The stated coal tar products when introduced in the human body have disastrous influence on the cardiac and pulmonary centers especially, and as early as 1899 I found them so acting when used in the treatment of Bubonic Plague in India. There is no doubt that one drug, Antifebrin, enters into more proprietary composition sold promiscuously and used by the public indiscriminately than all others put together. It is known in the Pharmacopœia as Acetanilid, being a monoacetyl derivative of aniline. Its total chemical formula is C8 H9 O N and its molecular weight is 135.08 approximately. It appears as colorless, shining, micaceous, crystalline laminae or as crystalline powder, is odorless, has a slightly burning taste and is neutral to litmus. Antikamnia, a proprietary drug, contains besides caffeine, sodium or ammonium carbonate and acetanilid. The action of acetanilid on bodily heat is through its breaking up into para-amido-phenol, stimulating the vaso-dilators, thus disengaging a purple wave in the system. There is one commonly used proprietary "headache remedy," however, which especially deserves mention because of the broadcast use the nation makes of it through the drug stores. Here is its label:

Emerson's
BROMO-SELTZER
Trade Mark, "Reg. U. S. Pat. Off."
Originally Registered May 21, 1889.
Each Ounce Contains Twenty Grains Acetanilid U. S. P.
[Each Teaspoonful Representing about 3 & 3-5 Grains]
Emerson Drug Co., Balto., Md., U. S. A.
A speedy and reliable remedy for Headache and Neuralgia.
DOSE—A heaping teaspoonful in half glass water; repeat in half an hour if not relieved.
PRICE 12 CENTS
Keep in a Dry Place, Well Corked

The U. S. Pharmacopœia gives 3 grains as a dose of Acetanilid; the makers of this Bromo-Seltzer admit having 3-3/5 grains in each teaspoon of their product, and advise the public to consume a "heaping teaspoon" first with "repeat in half hour if not relieved." Men, women and children go to soda fountains and consume this poison like soft drinks and the medical world quietly looks on. Bromo-Seltzer is a poison beyond question and its sale should be under the "poison act," the same features regulating Antikamnia and allies.

Attuned Green Wave

(Spectro-Chrome Therapy, Spectro-Chrome Therapists)

SPECTRO-CHROME OF MUCH VALUE IN SEPTIC CASES

By Kate W. Baldwin, M.D.,
Fellow American College of Surgeons
Senior Surgeon, Woman's Hospital, Philadelphia, Pa.

I am reporting three septic cases, drawing some conclusions which I hope may be of use to those dealing with that line of cases.

Miss B., a nurse in training, on April 16th burned the left thumb and wrist with steam. She applied a dressing of olive oil and did not report for treatment until five days later, when it had become infected. The whole arm was red and swollen, the axillary glands very painful, temperature 102° F., pulse 120 and respiration 40. She was given two drachms of Epsom salt in a small

amount of water and put to bed. Spectro-Chrome Turquoise was used every four hours after the first with the 400 watt equipment. At 8 P. M., the temperature was 101° F., pulse 108 and respiration 28. At 6 A. M. on the 23d, the temperature was normal and did not go above that. Between the Spectro-Chrome treatments the involved portions were kept wrapped in a saturated solution of Epsom salt, which was allowed to dry and then rewet. Miss B. went back to duty the fourth day and from that time on she had one Spectro-Chrome treatment a day until the wrist was entirely healed. There is practically no scar.

Vivian F., age 11 years, for some weeks had been treated in one of the clinics for an infection of her hands; later referred to another department and in time the hands were very much better, but not entirely healed. Late in the evening of September 15, 1921, she came to the Woman's Hospital and was admitted for treatment. The entire left arm was swollen and painful, with an indurated area just above the elbow, the axillary glands were swollen and very painful. My interne on duty, when reporting the case, raised the question as to the necessity for an immediate incision above the elbow. Being unable to reach me when the child was admitted, they applied a compress of Epsom salt. It was past ten o'clock when I got in touch with the case. Notwithstanding the lateness of the hour the child was taken to the "dark room" and given Spectro-Chrome Yellow for a half hour. The compresses were continued through the night and by eight o'clock the next morning the area above the elbow was so much improved that there was no necessity for any incision.

On entering, the temperature was 102° F., pulse 90, respiration 28; on the 16th at 1 P. M., temperature was 105° F., pulse 118, respiration 37; at 3 P. M. temperature was 105° F.; at midnight 103° F.; at 6 P. M., the 17th, the temperature was 104.9° F. From that time the temperature dropped quickly to normal at 4 A. M. the 19th, being less than four days from the time of admission.

Her treatment consisted in Spectro-Chrome twice a day, Epsom salt compresses when not under the Spectro-Chrome. On entering, 2 drams of castor oil was given; on the 19th, 3 ounces of Efferv. Mag. Sulph. On the 16th, 15 drops of Lloyd's Echafolta every three hours; on the 19th every four hours and on the 20th, 10 drops three times a day.

On the 20th she was out of bed for two hours and on the 22nd was discharged to report to clinic. She was perfectly well, hands and all, in a few days.

At the present time I would probably treat this case differently, giving Spectro-Chrome every four hours when the temperature is high or even more frequently; let it take the place of the laxatives and possibly of the Echafolta.

It is a little difficult to have a ward case taken to another floor for treatment more than twice a day. At the present time we have better convenience for the Spectro-Chrome treatments.

Anna W., age 13 years, February 28th had an abscessed tooth extracted under gas. Two days later the child became quite ill, had fever, cough and was very much prostrated. There was pain in the abdomen at a level of the umbilicus, marked constipation with nausea and vomiting of everything eaten.

The family physician was called on March 6th. A temperature of 104.5° F., and the above conditions were present. The dentist was called and he lanced an abscess where the tooth was extracted, but found very little pus. On the 7th the temperature was down to 102.5° F.; on the 8th it went up to 104° F. The bowels were freely opened that morning and about noon the patient was sent to the hospital.

There was every clinical evidence of a very serious condition. For several days it did not seem possible for her to recover.

The sputum showed pneumonia and the throat secretions streptococci. On the 8th the blood culture for twenty-four hours was staphylococci.

On the 20th, the blood culture for twenty-four hours was no growth and at the end of forty-eight hours no growth.

Blood count on the 8th showed Hb 80; R. B. C., 3,700,000; W. B. C., 126,000; Poly , 90; on the 20th it showed Hb 70: R. B. C., 3,510,000; W. B. C., 16,300; Poly , 66.

The patient was very weak and restless and more or less delirious for three days; at times restraint was necessary to keep her in bed. Base of right lung was involved.

A creolin mouth wash was used frequently; an ice cap to the head when restless. When the temperature was 102° or above, the torso was kept in a cool pack. On the 9th, 10 drops of Lloyd's Echafolta was given every four hours. On the 18th this was reduced to 5 drops three times a day. She was given little to eat or drink, but dilute fruit juices.

I personally gave the first Spectro-Chrome treatment and stayed with her during the treatment. She remarked, "I would like to stay here all day; I feel so good!" We used Spectro-Chrome Blue, Turquoise, Lemon and Magenta as symptoms indicated. The Blue quieted the delirium in a few minutes and restraint was not required after the first treatment. At first the patient had to be taken to the "dark room" for the Spectro-Chrome treatments and so had only two a day except the Blue, which was used at night. The third day we were able to put her in a room that could be darkened and after that she had the Spectro-Chrome every four hours for several days and then less frequently.

Just here let me put in a caution to the users of Spectro-Chrome. *You must pay strict attention to orientation.* When the arrangement for the change of rooms was made I was very particular to give careful instructions regarding orientation, for I knew that in the main building nearly every bed was east and west. In some way the instructions did not carry to the next relay of nurses and when I went in the next morning the nurse on duty said, "Dr. Baldwin, Annie did not like her color last night and wanted it turned off; I thought it was only a notion and kept it on for a while, but had to turn it off before the time was up." I found the child with her head to the west and I told them that was the reason she did not like the color. Then the bed was put right and there was no dislike to the treatments. I made the mistake the text for a little talk and I am quite sure that mistake will not happen again with that set of doctors and nurses. The trouble with institution work is, the doctors and nurses frequently change. I have had unusual co-operation at the Woman's Hospital and am sure that each one intended doing just as I requested.

From the time of the more frequent Spectro-Chrome treatments the case improved rapidly and was discharged on the 21st, just thirteen days after admission.

The temperature, pulse and respirations were of interest. After admission to the hospital the highest temperature was 103.4° F., the pulse 120 and respiration 62. Much of the time they were out of proportion to each other. For three days, the rectal temperature was just two degrees higher than by mouth. We usually count only one degree between mouth and rectum. The family physician tells me that the child was able to return to school in time to make her grade.

I have had a goodly number of septic cases where I have used Spectro-

Chrome as the only treatment, except for the local dressing and others where I have considered it wise to combine its use with others previously proven trustworthy. The shortened time and the lessened discomfort where I have used Spectro-Chrome leads me to the conclusion that one of its fields of usefulness is in septic cases and as infection of one sort or another is an element in nearly all pathological conditions, I see a future for its use in very many fields. As I work with Spectro-Chrome I more and more see how I can make it useful. At the present time I am quite positive that I could take the three cases reported above and by the judicious use of diet, cool packs and Spectro-Chrome be able to report as good and possibly quicker results.

Many thanks, Dr. Baldwin, for the encouraging reports. I am delighted to note that you take such a brave stand as to feel that even without the little drugging you could have obtained similarly helpful results. The mass of clinical evidence that is collecting around Spectro-Chrome is daily becoming more and more convincing even to the skeptical-minded and time will prove that the mathematical foundation of the system is not mythical. Meanwhile, to the lay practitioner of Spectro-Chrome Therapy I re-issue this warning that although as a liberal publication, articles about drugs, surgery and other methods are freely admitted for just discussion and enlightenment, Specto-Chrome Institute does not countenance or authorize the use of anything outside of Spectro-Chrome Therapy pure and simple. Let the graduates of the Spectro-Chrome Institute strictly adhere to the principles laid down, and signed for in their enrollment and they will keep clear of legal and other complications. Those who, like Dr. Kate W. Baldwin, are registered and licensed to practice medicine and surgery are entitled and privileged to use whatever other treatment they may deem fit or necessary, but to my graduates (who, under the present class legislation healing art system, are laboring under much one-sided disadvantage) I repeat that they stick absolutely to Spectro-Chrome Therapy, without even so much as swerving one way or another onto an outside method. Follow exactly what the State Charter of the Institute shows and success will follow your path.

Keep to the tracks of Spectro-Chrome. In regard to Dr. Baldwin's voice about taking care of the patient's orientation, I concur fully in that direction. The technique described in Compendium of Spectro-Chrome Therapy must be rigidly followed and the dietetic principles should also be carried out, not according to the practitioner's whims, but in accordance with the simple rules taught during the course on Spectro-Chrome Therapy.

NERVOUSNESS, APPENDICITIS, EPILEPSY

By La Verne A. Barber, M.D., Ph.D.

I can report splendid—perhaps the proper word is wonderful—results with Spectro-Chrome Therapy in nervous cases.

A woman came to me trembling with nervousness, pulse high, and on the verge of collapse. I placed her under your Attuned Blue Wave for one-half hour. After the primal treatment her pulse was normal, and her general condition remarkably improved. She is now a regular patient at my office, and calls the Spectro-Chrome "My Light!" Truly it *is* "the Light" that will show many a weary sufferer the pathway back to health and happiness.

I am also having success in aborting threatened appendicitis with the Spectro-Chrome Attuned Turquoise; am

La Verne A. Barber, M.D., Ph.D.

treating an hour-glass contracted stomach with the Orange Wave and the condition is improving.

I intend to try the Attuned Magenta Wave, as per suggestion of Henry Lanneau, in the June issue of Spectro-Chrome.

I am treating an interesting case of epilepsy with improvement, but she has recurrent spells. She is a heavy meat-eater and there is trouble, as in most of these cases, to make them eat with moderation. I shall be thankful for any suggestions you can offer in this case. Before the case came to me she was doped, as usual, with all the bromides in the drug calendar. I took her away from medicines at once, and she has been without a convulsion for two months until last week, following a meat-jag.

And now a few words to my kind reader about this new system:

Dear reader, if your health declines,
Read well the portent of these lines:

Doctors harken well and heed!
In every clime and sect and creed—
Nature comes into her own,
Since Dinshah made his welcome home
Here in the good old U. S. A.,
And taught old truths in brand new way!
Hence his fame spreads day by day.
Science, too, takes added strides,

Since Dinshah launched his diatribes,
Proving pills are out of date.
Each enlightened graduate
Can find, from Mexico to Nome,
The magic in his Spectro-Chrome!
Revealing truths unguessed before,
Or hidden deep in occult lore.

Chronic cases find relief,
Heretofore beyond belief.
Rejoice with me for Dinshah's kind—
Our friend, our guide, our master mind;
May Medicine one day proclaim
Each debt it owes to Dinshah's name!

Returning to Philadelphia, Pa., from Chicago, Ill., I made a short business trip to Lancaster, Pa., incidentally calling on Dr. La Verne A. Barber. What I saw there did amaze me. He occupies a whole building (at 439 North Queen Street, in a quiet neighborhood) and each room is tastefully equipped with apparati for drugless healing in which he specializes. The spotless cleanliness, the courteous and encouraging manner of the attendants on duty and the broad smile of the genial physician himself are mainly sufficient to soothe the patient's troubles initially. His chamber for Spectro-Chrome Therapy is so dainty in appearance that I could not suppress a cry of hearty congratulation. Surely he understands human nature. Dr. La Verne A. Barber is a graduate of the Medical Department of Ohio State University, Old Starling Medical College (1891), was Vice-President of Butler County Medical Society in 1896 and delegate to the American Medical Association in the same year. He practiced two years in Los Angeles, Calif., conducted an office as Electro-Therapeutist in Johnstown, Pa., for ten years and was head physician in a New York Electro-Therapeutic establishment for three years. He is now established in Lancaster, Pa., as X-ray specialist, and uses Spectro-Chrome Therapy to a large extent, although not a graduate of the Spectro-Chrome Institute. This tends to prove his broad-mindedness, for which I am very highly obliged.

In regard to his stated cases I may say that, with the Spectro-Chrome Attuned Turquoise Wave, excellent results are always reported in treating appendicitis. Should there be severe pain, shift the polarity to Blue and Indigo on Area 9.

Epilepsy has sex relation and, among other troubles, there is always flatulence. Treat with Orange on Area 8 for five or ten minutes after a meal, and go into the sex history of the patient. Start with a Green systemic and, as there is circulatory disturbance, administer Magenta on Area 4. Bromides and drugs are useless and injurious. The convulsions always will yield to the Ultra-Green section. However, as the body is built of the foods eaten, all stimulants like flesh, fish, tea, coffee and alcohol should be rigorously avoided.

SUCCESS IN MEASLES

By John A. Cohalan, D.O., S.C.T.

I wish to report the splendid results I secured with Spectro-Chrome Therapy in three cases of measles.

The first case, a girl of seven, came in contact with a case of measles in school. As soon as the condition began to develop she was given a systemic Green for thirty minutes 400 Watt equipment, then Yellow on 8 and 9, ten minutes, Magenta on area 4 ten minutes, later when fever developed she was flooded for over an hour with Turquoise, followed by ten minutes of Magenta on 4.

The result was that the fever lasted only forty-eight hours and at no time did it rise above 102 degrees. After the temperature returned to normal she was given a treatment every day of Lemon systemic and Magenta on 4, using Yellow on areas 8 and 9 when the bowels were sluggish.

To overcome the intense itching and to promote desquamation her body was rubbed wtih crude cocoanut oil only and dusted with Fuller's earth.

Thirteen days later, two other cases developed in the same family, girls aged five and six, respectively; they were put through the same routine with the same good results.

The remarkable issues about these cases are that the fever did not go above 102 degrees, there was practically no irritability of the eyes, no acrid discharge from the nose, no cough, and yet there was no mistaking the condition as they were all covered with the reddest, measly rash one ever saw, indicating that elimination by the skin was complete.

Dr. John A. Cohalan, although an osteopathic physician, was one of the first few far-sighted pioneers to take to Spectro-Chrome Therapy. From the commencement, success has marked his way, and his patients are now asking for the "Color Waves" as babies ask for candy! More success to you, brother, and more honor for your pioneer stand.

Attuned Turquoise Wave

(Poetry, Romance, Drama, Art)

FROM STAGE TO SPECTRO-CHROME

By Mrs. Helen Mansfield Barber

It was my good fortune to be a member of the theatrical profession for several years. As I look back, they were wonderful years, for from these people of the theatre I learned courage, cheerfulness, and true brotherly love.

No matter what the trial was or how often it came, there always was some fellow co-worker ready and willing to lend a helping hand.

"Keep Smiling" is the motto of the theatre, and, oh, how bravely those dear people do smile, even though their hearts are breaking!

I know that the work of the theatre helped me in the field I am now following. I am assisting my husband in a "House of Health," where we are relieving suffering with nature's ways and freeing men and women from a belief in drugs and the all too often needless knife.

We are doubly armed in our work by having the pleasure of using Doctor Dinshah's Spectro-Chrome Therapy and seeing daily results obtained through this wonderful method of healing and health building.

It is with all sincerity that I recommend it to the people of the theatre, for these good folks are always seeking the truth. I advise them to investigate this novel method of harmonizing to health, and also, if possible, to afford themselves the pleasure of hearing Dinshah lecture.

It is a pleasure to meet a woman who takes interest, as Mrs. Helen Barber does, in her husband's work. Life is truly a dual problem, and no man makes for success without some woman supporting him, and no woman can make a choice of career better than healing the ailing. I thank Mrs. Barber for her zeal and wish her all success in her new career.

REGISTRATION NOTICE OF TRADE-MARKS

Notice to Purchasers and Patients

Notice is hereby given that applications are on file in the United States Patent Office for the Registration of the words DINSHAH, SPECTRO-CHROME and Star Design in connection with Spectro-Chrome Therapy Equipments and Supplies.

All genuine Spectro-Chrome Therapy Equipments bear such identifying Trade-Marks and all genuine color slides made in accordance with the Attuned Color Wave System of the Originator of Spectro-Chrome Therapy have one or more such marks and are manufactured **exclusively** by the Spectro-Chrome Institute.

All genuine Spectro-Chrome Therapy Equipments and Supplies are obtainable only from the Spectro-Chrome Institute and its established branch offices. In purchasing Spectro-Chrome Therapy Equipments and Supplies, ask for SPECTRO-CHROME and insist on having them only.

Patients going to any practitioner are requested in their own interests to insist that they be treated with DINSHAH'S ORIGINAL SPECTRO-CHROME EQUIPMENT AND ATTUNED COLOR WAVE SLIDES.

Any person, practitioner or otherwise, substituting his own equipments and slides or using equipments and slides other than the Original Dinshah Equipments and Slides is not using or treating patients with **Spectro-Chrome Attuned Color Waves** and is therefore not honest in representations or accurate in therapeutical results.

TO OUR CONTRIBUTORS

1—Please write or type, but be legible.
2—Sign your full name and address in ink for information, even if not for publication.
3—Articles submitted for insertion will not be returned.
4—Articles published will become the property of the Spectro-Chrome Institute, unless otherwise stipulated in advance.
5—Send articles early to insure publication in following issue.
6—SPECTRO-CHROME desires original and unpublished articles only; if not original or if published elsewhere, full information must be given as to the author and origin.

ADVERTISEMENT RATES:

1 Page, 1 insertion	$15.00	¼ Page, 1 insertion	$5.00
½ Page, 1 insertion	10.00	1 Inch, 1 insertion	3.00

Professional Card, 1 insertion, $1.00

Communicate for contract rates.

SPECTRO-CHROME

AUGUST 1922
INDIGO 1922

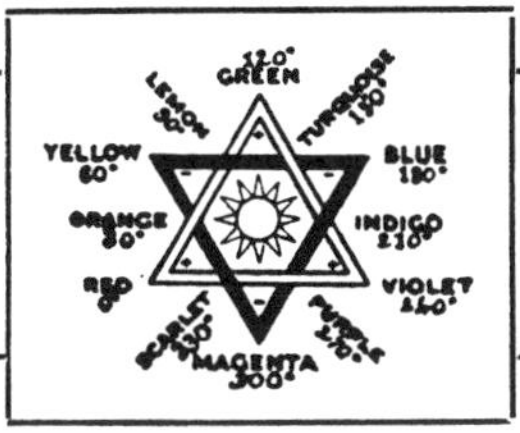

VOLUME 1
NUMBER 3

Devoted to: Spectro-Chrome Therapy **Dedicated to: Service of Humanity**

CONTEN⸤S Page

Spectro-Chrome Institute .... Inside Front Cover
Spectro-Chrome Area Chart—Therapeutic System—Dinshah's Platform 49
The World Through the Eyes of Dinshah—
(Occult Operation of the Great Sympathetic Nervous System chart—Sun Spots Troublesome—America's Soul Sold to the Devil for Dollars—Spectro-Chrome University, When?—How about the Women?—Varnished Deacons—Bread or Pasted Holes?—Reverse the Postulate—Doctors, Please Answer—Woman or Wo X Man? Taxicab Tyranny—Diarthrosistis or Wigglewoggloitis?—Bobbed Hair Beauties—Caution to Women of Sense—Foolery in Court—Marriage and Divorce; Remove Birth Stigma—Russia Going Back—Cancelling International Debts—A Present Source of Heart Dis-Eases—Watch Your Step——Alienist Opiumized—Women with Trunks—Down with Vaccines—Lloyd George's Shirt—Dog-Oil or Dog-gone Medicine?——Saved from the Knife—Mayor's Wise Act—Japanese Turned Christian—Human Locomotives—Stupidity of Censors—Glands, Goats and Monkeyshines) .... 50 to 64
Did You Ever? (A Short Sketch on the Oddities of Females, by Irene Hoger, S. C. T.) 64
Sex, Education and Help (by Anna Au Burton, S. C. T.) 65
Suggestions for Spectro-Chrome (by Naya Bundy, S. C. T.) 66
Therapeutical Accuracy (by Edmund J. Hogan, D. C., S. C. T.) 67
Constipation, Amenorrhea, Anemia, Piles, Appendicitis (by Peter P. John, N. D., S. C. T.) 68
Everybody Boosting Spectro-Chrome (by Roy Forbush Hanaford, S. C. T.) 69
Mother Nature's Alphabet (by Kashmira Dinshah, S. C. T.) 69
Life Sketch of the Originator ("Pottsville Daily Republican") 69
Pestanji Framji Ghadiali .... Inside Back Cover
Registration Notice of Trade Marks .... Outside Back Cover

Editor
Dinshah P. Ghadiali

In course of life you reach a time,
When Doctor, Lawyer, Priest alike,
Does fail to give the sought relief,
To troubled mind or burdened heart—
Then come to me;
I help you to help yourself.

SPECTRO-CHROME.

Subscription: June, 1922, to December, 1922 **$1.75**
Thereafter, January to December **$3.00**
Single Copy, 25 Cents

Spectro-Chrome Institute

Incorporated under the Laws of the State of New Jersey. Authorized for the transaction of business as Foreign Corporation under the Laws of the States of New York, Pennsylvania, Illinois, Maryland, Delaware, New Hampshire, etc.

Central Office and Experimental Laboratory

2401 NORTH BROAD STREET

Philadelphia, Pennsylvania

UNITED STATES OF AMERICA

(Numerous Branch Offices)

The main objects as set forth in the Charter of the Spectro-Chrome Institute are:

The instruction, advancement and practice of Spectro-Chrome Therapy and the research, development and pursuit of such science as may assist in the equipment and application of Spectro-Chrome Therapy or the restoration of the human Radio-Active and Radio-Emanative equilibrium by Attuned Color Waves, without the use of drugs, manipulation or surgery, in accordance with the system conceived, originated, developed and applied by Colonel Dinshah P. Ghadiali, M.D., D.C., Ph.D., LL.D.

The building, equipping, establishing, conducting and maintaining of schools for the investigation, instruction, demonstration and practice of Spectro-Chrome Therapy; also the power to confer and grant the degrees of Spectro-Chrome Therapist, Master of Spectro-Chrome and the like, and to issue diplomas and titles in connection therewith.

The corporation shall also have power to use, manufacture, sell and dispose of. in any way, all means, publications, appliances, instruments or products that may seem fitting or in any way connected with such purposes or objects aforesaid.

Spectro-Chrome Institute conducts complete resident courses in Spectro-Chrome Therapy in a number of cities each season. Each Course, comprising 24 discourses, is delivered personally by the Originator of the System with demonstrations and experiments. At the end of the course, students passing the examination are granted Diplomas. There are no correspondence courses.

Attuned Red Wave

(The World Through the Eyes of Dinshah)

OCCULT OPERATION OF THE GREAT SYMPATHETIC NERVOUS SYSTEM

1	NADI	IDA		SUSHUMNA		PINGLA
2	COLUMN	LEFT		CENTRAL		RIGHT
3	INFLUENCE	LUNAR		TERRESTRIAL		SOLAR
4	RULER	BRAHMA		VISHNU		SHIVA
5	TATVA	PRITHVI	JALA	VAYU	AGNI	AKASHA
6	ELEMENT	EARTH	WATER	AIR	FIRE	ETHER
7	COLOR	YELLOW	WHITE	GREEN	RED	BLACK
8	SHAPE	SQUARE	CRESCENT	CIRCULAR	DOTTED	DIFFUSED
9	FORWARD OPERATION MINUTES	30	24	18	12	6
10	BACKWARD OPERATION MINUTES	30	24	18	12	6
11	FUNCTIONAL FINGER WIDTH	4	6	8	10	12

12	LUNAR BRIGHT HALF DAYS						LUNAR DARK HALF DAYS					
	AT SUNRISE	LEFT ACTIVE			RIGHT ACTIVE		AT SUNRISE	RIGHT ACTIVE			LEFT ACTIVE	
		1	7	13	4	10		1	7	13	4	10
		2	8	14	5	11		2	8	14	5	11
		3	9	15	6	12		3	9	15	6	12

13	LEFT + BRIGHT HALF-FAVORABLE				RIGHT + DARK HALF-FAVORABLE		
	MONDAY	WEDNESDAY	THURSDAY	FRIDAY	SATURDAY	SUNDAY	TUESDAY

14	SANSKRIT NOMENCLATURE	PURAKA	KUMBHAKA	RECHAKA
15	AMERICAN PHRASEOLOGY	INHALATION	RESTRAINT	EXHALATION
16	RESTRAINT PERIOD SECONDS	LOW = 42	MEDIUM = 84	HIGH = 126

SUN SPOTS TROUBLESOME

Through the French Academy of Medicine, Drs. Vallot, Sardou, and Faure report that their observations show sun spots capable of influencing the heart, liver, intestines and blood and also aggravating the seriousness of accidents. The learned Academy may do well to study Indian Metaphysics and Occultism from some competent teachers instead of wasting time on old concepts of medicine to be revised from new theories. All this and much more *has been recognized* very minutely by the ancient sages of India and in Spectro-Chrome Therapy courses, the originator of the new Healing Art has fully tabulated the astronomical theory into downright psychological and physiological practice. He demonstrates therein how the two nostrils breathe according to a definite time schedule regulated by the Solar and Lunar influences and how these effects operate for human life through the Great Sympathetic Nervous System by means of certain internal secretion glands. The annexed chart of Dinshah is the key to this marvelous operation about which the occidental medical world is calmly ignorant. The occidental "know-alls" have much to learn and to give the zealous searcher for the truth an opportunity to secure such knowledge first hand, a class on Spectro-Chrome Therapy is now being formed in Chicago, Illinois to commence the first week of September. It is the quickest, shortest and the most condensed scientific experimental course on such a deep subject as Spectro-Chrome Therapy and is worth the time of any person—professional or layman to take it. The knowledge gained will enhance the scope of the professional and will increase the therapeutical field of the layman.

AMERICA'S SOUL SOLD TO THE DEVIL FOR DOLLARS

After passing an Eighteenth Amendment to the blessed Constitution, thereby taking the greatest step in history for the world's progress, next perhaps to the Declaration of Independence, the American Government has quietly connived at the opening of liquor-selling bars on American seagoing steamers. Chairman Albert D. Lasker, of the Shipping Board, voices challengingly that he would not rescind his order authorizing the sale of booze on the ocean on American liners until the Supreme Court stops him. And who is this Supreme Court? A creation of the people to defy the will of the people's Congress? Its Chief Justice has to go to England to learn some more laws! English laws precipitated the Declaration of Independence, and have we to go to that aristocracy-dominated kingdom to shape our course? Why in the name of goodness are the Americans endeavoring to follow other countries like monkeys? Have we not all the nations represented in our democracy? Why can we not stay here and mind our business, frame our laws and enforce them as we please here? Why should we allow our glorious Constitution to be trodden and trampled by Albert D. Lasker, our own man, twisting and turning our regulations to please the financial interests? He says we shall lose all our maritime passenger trade if we do not follow the foreign shipping rules and dispense liquor on American boats. You disgraceful hypocrite! You are chairman of a Government Shipping Board, to enforce our constitution and not to pander to outside pressure or views. The trouble is our shipping is in the hands of incompetent men, our government hesitates and does not act with promptitude or firm decision. It is pure politics and nothing else. The real solution is to compel obedience to our laws by preventing the entry of foreign liquor into our ports; instead of being compelled to sell liquor on our ships, you will soon find those drink-sodden foreign ships will bow to our regulations and stop their own liquor traffic. Wake up, Warren G. Harding, and throw Albert D. Lasker out of the chair into the sea of oblivion!

SPECTRO-CHROME UNIVERSITY—WHEN?

Leaders of Christian Science announce the intention of constructing a $1,000,000 college near St. Louis, Missouri, the first of its kind in the world. All success to ye workers for any system of healing that dispenses with drugs. I am waiting for a similar sum to come to me for a similar idea: it must be on the way—it will come some day!

HOW ABOUT THE WOMEN?

Senator Dennis Fleming introduced a bill making it unlawful in Georgia for a married man to keep away from home or go fishing without the full consent of his wife. Should he do so, he is to be sentenced to not less than five and not more than twenty years at hard labor! If prohibition were not in the Federal Constitution, I should be led to believe he had tasted something worse than one-half per cent. Such a bill is human outrage! It is unisided. Throw it out. How dare senators waste public time by such idiotic measures? Let men and women politically as socially be treated equally and not butted one against another if one is looking for human elevation. Why not enact a law preventing women in Eve(ning) costume from dancing in the embrace of somebody else's husband, brother or lover at past midnight when she should have been in the arms of Morpheus hours ago? Have a heart, Senator!

VARNISHED DEACONS

At a communion service in Grand Rapids, Mich., nine deacons and elders of the Seventh Reformed Church drank varnish instead of wine! It was evidently an error, but, what kind of olfactory bulbs had they? The church to have wine? The wine of the Lord Jesus was without alcohol; it was un-fermented grape juice, an emblem of prosperity and the like; it never was meant for the 17 per cent. French champagne or "homemade" cider, even which ferments and produces volumes of that curse called alcohol, unfit as a human beverage.

BREAD OR PASTED HOLES?

The Hygienic Institute of the University of Leipzig is awakening to the fact that our teeth decay owing to the use of bakers' bread. It would appear from a study of the report that bread instead of being the "Staff of Life" is really the producer of troubles. Bread as made today with white, devitalized, chemically treated, fermented flour is positively detrimental to health equilibrium. To benefit life forces, wheat must be used in its original entire condition and the bread should be prepared without fermentation. As made and sold, we get only a necklace of pasted holes with a taste of glue to eat, the best vitality supporting element being either milled or burned out. Civilization indeed, when man has to depend on laboratories to improve on nature's creations! The whole diet situation, as exists, is nothing but destroying the natural original elements and trying to refill the loss by artificial means.

REVERSE THE POSTULATE

Herbert W. Kirby, of West Philadelphia, returned from India after a residence of 15 years. On being questioned for an opinion he is reported to have said that "India's best asset *is* the British Flag." Friend Kirby, I differ from you as a man born in India, trained in India and a student of India. The truth is the reverse of your assertion. I admire the British people and have nothing against them as a naturalized American, but, after studying British politics and the results of the diplomacies of the British Government, I have come to the conclusion that, "India is the best asset *of* the British flag." If tomorrow George faced the contingency of either losing the Kingship of England or the Emperorship of India, I feel he would sooner lose the English crown. If instead of stupidly pushing crude raw western ideas and methods into Indian minds, the British would let the people develop internally along their own proclivities and copy the example of the United States of America by befriending the downtrodden population instead of exploiting it as that wretched East India Company did, Hindustan would be soon the equal in prosperity of America. Its enormous population of industrious but poor trampled agriculturists living now on ten cents a day, its intelligent quick-brained youngsters studying by a castor oil lamp of three-quarters candlepower till past midnight for university examinations, passing which gives them government services with a handsome start of ten dollars per month if not less, its police officers getting five dollars per month, its hospital assistant medical officers receiving twenty dollars per month, its native magistrates, with high powers, dispensing justice for less than thirty dollars per month, all this on one side and its English imported officers eating up the cream of the land on the other side cannot be conducive to prosperity. What a glorious day it would be if India were under the flag of the United States! The British exchequer is unable to pay our debts—well, tell them to hurry with the settlement or pawn India to us as a security under mandatory rules—O Grand Almighty Creator, if such a day should come, the long-forgotten glory of ancient India would dawn again on the horizon of the Orient. America, press your demands on England and be instrumental in bringing happiness into weeping India by compelling the stronger ruling nation to mete justice to the weaker ones.

DOCTORS, PLEASE ANSWER

Walter R. Herrick, former Commissioner of the Department of the Narcotic Drug Control of the State of New York, reports:

"I want to emphasize the fact that 50 per cent. of the drug taking today in the higher walks of life is the outcome of illness during which narcotics have been administered by the physicians."

Beyond question. In my long medical practice in India I observed the same curse. A male patient started by having his pain relieved by an injection of cocain. When he came to me he was consuming by the mouth sixty grains of cocain a day! I reasoned with him and succeeded in restoring him to normalcy by dropping the habit. I maintain that the patient's will is undermined primarily by the physician's crude methods of treatment with potent drugs and after the physician has established the lead and taken leave, the patient continues similarly of his own accord. Who is to blame? The first jab of the needle and the first administrator of it.

WOMAN OR WO X MAN?

Earl Puryear, bantam weight boxer, of Omaha, Nebraska, secured a divorce from his wife, Grace, testifying that she "abused, bossed and beat" him. "I would rather face a champion than she," said this fighter. I believe him; the law allows him to smash up a man in open arena, but at home he dare not raise even a voice in protest. As long as a woman stays a wombman she is all right, but when she starts otherwise she is liable to degenerate into Wo x Man.

TAXICAB TYRANNY

New York State has passed a law compelling taxicab drivers to provide insurance guaranteeing payment of damages for accidents. This law, as it stands at present, is utterly unjust, is class legislation and is favorable more to the rich insurance companies who provide the necessary bonds. Admitting the importance of such insurance, to protect the innocent public, it must be applied indiscriminately to all automobile drivers and vehicular users of the State Highways. Now, it benefits only the pooling companies which are National Surety, New Amsterdam Casualty, Fidelity and Deposit, American Surety and Aetna Casualty and Surety. It is shameful how the rich can oppress the poor and yet we wonder at the existence of strikes, riots and disorders in the world. If, in great republic America, such aristocracy could go ahead what surprise is there if Russia should do similarly and the Ex-Kaiser should have thought it? It is all the curse of political backgammon and the deviltry of bossdom. Might makes Right?

DIARTHROSISTIS OR WIGGLEWOGGLOITIS?

Ermie Young, a theatrical promoter of Chicago, is sued for $10,000 by Marie Ehlers. Her bill of complaint sets forth that Young required her "to execute many contortions, convolutions, distortions and gyrations associated with modern dancing aberrations, particularlv twisting and writhing of the hips and shivering and convulsing of the shoulders."

Therefore, Miss Ehlers continues, she was greatly injured and fell a victim to the malady known as "diarthrosistis."

Of the jazz emotion she said she is "peculiarly sensitive to soniferous instruments, suggestive of an accentuated syncopation of a jazz orchestra, which incites her involuntarily and unconsciously to wiggle and revolve."

Well, well! But who the bliss compelled you to do the wiggling, Marie?

That was what you were employed and paid for perhaps—to make an absurdity out of the human form, the image of God.

BOBBED HAIR BEAUTIES

From a beauty parlor in Defiance, Ohio, comes the story that bobbed hair is going out of fashion and that soon American girls will be glad to stuff their heads with bought locks that once graced some dead person. Now the barbers are profiting, then the beauty parlors will profit. My view is this. If a girl wants convenience and less labor, let her bob her hair; it proves that she is either too lazy to keep her hair in order or that she believes hair is not hygienic. But, if such be her view, let her have it clean-cut as the men have, because the present method of having a rounded fringe on the head makes her resemble a mattress more than a woman and the esthetic appearance is far from pleasing, beautiful or attractive. Be either a hundred per cent. male or a hundred per cent. female: this way she looks more like a swinging pendulum bobbing wherever there is gravity.

CAUTION TO WOMEN OF SENSE

U. S. Consul L. E. Gauss in his report fro .ı Tsinan, China, says that the hair nets that enter the United States of America from Europe are mostly from Ch.nese heads and that there is no sterilizing process used on human hair nets shipped from the Shantung district; also, there is no sterilizing done on any hair nets in China. This is "good" news for the fashionable American woman! Why do we need such filthy things as dead hair from human heads to keep our live hair in position? When will women be taught to recognize the difference between live hair and dead hair? As long as hair is on the body it is growing live material charged with the Radio-Activity of the individual and breathing Radio-Emanations of high potency, but, when severed from the frame it becomes a mass of foul matter unfit to be used by the clean. It is like a withered branch, a faded leaf, a paralyzed limb; its real value is gone, it is dead matter. Its use for decorating clean healthy heads is not advisable and we must dispense with it as a carrier of dis-ease.

FOOLERY IN COURT

In the Ridgewood, Queens. Court before Magistrate Conway, a Jewish baker Samuel Cohen was hauled for violating the "sacred Sabbath" by selling bread on Sunday. The magistrate held him in $500 bail, stating there was only one Sabbath in the United States of America, although the Jew proved that he did close his store on Saturday. If foolish judiciaries do not watch their steps and act more sensibly, they will bring about the same condition for Jews here from which they thought they escaped when they flew the Russian Pogroms. Such "justice" is downright religious persecution and such laws must be abolished for human welfare.

MARRIAGE AND DIVORCE; REMOVE BIRTH STIGMA

Miss Sophie Irene Loeb is a well-known writer and her recent article on "Give the unwanted child his rights! 'Illegitimate' should have no place in language," published in the New York *Evening World* of July 27, 1922 is highly commendable. Fanatic, non-far-sighted, touch-me-not neople of an ultra-religious make-believe - God - beloved type have cast such a stigma on a child born out of wedlock that there is an alarming increase of suicides anu child-murders. The wrong and uni-

sided interpretation of the laws of marriage has become an unbearable torture to the unfortunate sensitive and because of some social man-made barrier, the right of motherhood is positively denied some virtuous and devoted female. The real God-made principle of love that underlies marriage is subordinated to foolish thoughtless human notions and the yielding of a girl to her innate forces impelled under the all-powerful influence of the highest emotion of sex is frowned upon and reviled as impure, obscene and unchaste and the result of any relation outside of what is prescribed by iron-glad lopsided laws of the most prudish and unnatural character is named "illegitimate" and for ever reviled and shunned by the self-styled "decent." O what an enormous crime is this against the All-Seeing Eye! The birth of beings is based on the most sacred and the most occult laws of eternal life, marriage being intended only to be a social sanction for sex intercourse which is otherwise banned. I am not for promiscuous intercourse of the sexes, I do not preach what is popularly called "free love," I do not advocate prostitution or the *abolition* of the bonds of marriage, but I have for 31 years maintained emphatically one ground that if marriage be the consummation and consolidation of the highest emotion of love, then, when love disappears through any cause from human life, the mated unfortunates should not by a legal law be compelled to cohabit in the same house, on the same floor, in the same room, on the same bed and produce children for the nation, because, the offsprings of such loveless conjugations will most probably as not be idiots, imbeciles, robbers, thieves, murderers, marauders or pimps whose malpractices keep the prison wards, lunatic asylums and brothels full and the law courts and lawyers on the job.

Thus, love being the foundation of life, an unwedded female rarely gives herself except to the one loved by her and the child of such junction will be just as good as that of one in the bonds of wedlock blessed by the holiest rites and assuredly superior in faculties to one born within wedlock where love between the parents was absent. And those wedlock bonds are a veritable bondage as at present is the case! One Dollar and One Error at the City Hall and Two Human Lives in the Image of God are launched into a contract of the most unthinkably infernal inhuman character—indissoluble except by purposed adultery or downright perjury in a burlesque show called the Court of Divorce. Let the bonds of marriage stay, but, let the divorce be left to the same processes of nature as indicate marriage and as long as the children are taken care of by the parents responsible for their birth, let no law interfere with un-Godly barriers. America will awaken to this fundamental principle soon enough, the women, those among them at least who really are sensible will in the right exercise of this political emancipation see to it that mothers—unwedded as wedded—get their right and the birth stigma be removed from any child born otherwise than in so-called man-made-law-sanctioned wedlock.

RUSSIA GOING BACK?

Russia is in a worse flutter than ever. While endeavoring to straighten its political and economical tangles, it was helped indirectly by the greatest boon conferred on its people by the late Czar Nicholas—the abolition of Vodka. Vodka monopoly was controlled exclusively by the Government, and its abolition gave such healthy impulse to the national opinion, that when the crisis came the down-trodden populace overthrew the monarchy. However, as it happens in all similar cataclysms, the change of policies did not immediately allow the nation to settle to earnest business and after several upheavals it is slowly getting down to some semblance of a government now.

At such a point, instead of keeping the devil of alcohol out as it was, a new decree is being promulgated to reinstate the state monopoly of Vodka. The pre-war basis was 47 per cent alcohol: now they want it fixed at 38 per cent. In the frantic endeavor to get cash, the government is making herein a fatal mistake. A drunken population may be good for ruling fanatically, but, no sensible government can be established where the people have no stabilized rule and less stabilized currency. A government monopolizing the sale of poisons (and alcohol is a powerful one) and placing it under state control forfeits its moral right to control the vice and drunkenness, because, what the state permits to be sold it permits to be used and such use cannot be penalized. Russia will face the same problem the British face today in India.

There also the sale of liquor is under government auction and the demoralizing influence of saloons thus established and controlled is one great cause of India's poverty. Let India go dry and the British will face a clear-minded population whose righteous aspirations they will be forced to respect as a self-respecting nation.

CANCELING INTERNATIONAL DEBTS

The *New York World* is clamoring for the wiping out of International War Debts. "Cancellation" has become in Europe a word with dark and sinister meaning and the United States is urged to wake up to the realization that we must forego these debts because they are "uncollectible." We *know* these pre-war debts are uncollectible, but until the nations in Europe come to their sense that warring is worthless and until our friend England stops clamoring for land, water, air, ether and heaven supremacy, to wipe out these debts would mean that the British Lion would be let loose once more to try to establish some sort of death-grip on weaker nations and drag us perhaps into another trouble. We are sick of the horrors of war. Throat-slashing and bomb-firing may suit civilized barbarians, but no God-loving person would advocate human warfare. There is only one avenue to universal peace: to cancel not debts, but, first all monarchical government systems, all titles of hereditary rank (whether temporal or otherwise) prevent aggression of weak nations by strong ones through commercial boycott of the aggressor and put a stop to stupid financing of foreign military enterprises by the money lords through refusal to protect their monetary interests unless previously sanctioned by the governments. While kings and lords last, their aggressive ambitions will keep up the trouble, while standing armies and navies are maintained, there will always be the danger of aggression and oppression. Convert all the municipal police forces into Federal Protective Armies for peace, to be the Army Reserves in fact. Build and draft in Merchant Marines on similar plan for Naval Reserves. Do the same with the Air Services.

Let us forgive and wipe out certainly the billions owing to us, but, on condition only that the weak nations be all granted autonomy and that the principles of non-expansion of territories be rigidly carried out. Nations must stop building forts, guns, submarines, battleships and murderous machines and we must compel the nations owing money to make good promptly as per agreements made when borrowing. Then and then only, when faced with economic and financial crash, will these crazed, war-mad nations that use their resources to buy more war material instead of paying their creditors be awakened to the truth and either settle their debts to the relief of other nations or under a stress of public opinion be obliged to change their tyrannical tactics to avoid revolutions. Forgive and forget yes, but not by letting them use our money for oppressing poor weak people and expanding their ill-gotten territories everlastingly. America, dear United States of America, do anything and everything, wise or foolish, but keep your hands off from signing any document that may move us away from our strict non-aggressive neutrality. The day is not distant when the United States may again be called upon for help, but that help should only be given under the strictest terms of neutrality and no other. Europe (including England) is the hotbed of aggressive politics and our American diplomats are no match for the crafty political intriguers across the Atlantic ocean. They are true statesmen at any rate; our new fledgling country has no such men; we have plenty of politicians and more bosses and grafters, but, there are very few who walk the footsteps of George Washington or Abraham Lincoln with an eye that ever watches the weal of this country which is destined to be the greatest birthplace of the great men and women of the world.

A PRESENT SOURCE OF HEART DIS-EASES

Speaking at the Harveian Society in London, England, Dr. E. Turner extolled the use of salicin for curing influenza. He stated he had cured himself and his patients and lost not a case. He gives for it an initial dose of 40 grains followed by 20 grains every hour for twelve hours, then every second hour for six hours more. I do not question the word of the learned professional, but, fail to realize the mathematical accuracy of his technique. Salicin is an orthohydroxybenzylglucoside (total chemical formula $C_{13}H_{18}O_7$; molecular weight 286.14) obtained from the bark of several species of salix and populus (willow and poplar) trees. It is a bitter tonic and has been extensively employed in the treatment of rheumatism, influenza and the like. It is used both in the United States and the British Pharmacopœia and its average dose is from 5 to 15 grains, the maximum being 30 grains. In its physiological action, salicin is inferior to acid salicylic because its systemic absorption takes place by its conversion into salicylic acid, formula $C_7H_6O_3$; molecular weight 138.05. The average dose of this acid is even less—only 12 grains. Now, how should we comprehend the large doses administered by this doctor to his influenza patients as operating? Surely, he pushes the drug to the limit in his patients with a view to abort the influenza, but that would mean that the therapeutical effects of the drug would be also disengaged in the body in large proportions. Salicin has among others these physiological properties:

In small doses	In large doses
On Digestive System:	
Stimulates	Irritates
On Nervous System:	
No cerebral action	Causes cerebral congestion. Headache. Vertigo
On Respiratory System:	
Stimulates and quickens respiration	Causes labored and slow respiration
On the Skin:	
Acts as a powerful diaphoretic	Causes exhausting perspiration, intense itching.

Thus, the large doses administered by Dr. E. Turner would perspire the patient to depletion and derange his circulatory apparati. Such has always been the sequel of the administration of salicin, acid salicylic or any of the salicylates. The perspiration may drive the febrile symptoms, but all at the expense of the most important organs of the circulation of blood. Patients suffering from rheumatism and treated with these drugs by the allopaths get a legacy of all kinds of cardiac disorders and one thing leading to another, once the patient gets into a drug doctor's hands, he seldom thereafter has a clean bill of health. In treating influenza by Spectro-Chrome Therapy, no complication of any character can arise. All is taken care of beforehand. Treat the patient just as any other specific infection would be treated. Attuned Turquoise Color Wave on Areas 1-2-3-4-5 following Systemic Green, with Magenta on Area 4 to provide against cardiac complication is all that is needed. Yellow on Area 9 may be used if the bowels are not in proper operation. In very grave cases Systemic Turquoise or if the temperature be very high Attuned Blue Wave may be used as the various symptoms from time to time indicate. The entire treatment of the dis-ease is Ultra-green for the system.

WATCH YOUR STEP

George Stonoff, of Kansas City, Missouri, went to consult Dr. George L. Schmitt. The physician told him that he needed an operation, whereupon the patient cried, "Either cure me or I'll kill you," and shot the doctor, who died half an hour later. This indicates that a healer should study human nature always. An enraged patient is worse than insane. Tact in the expression of opinion is ever safe. Speak the truth boldly, but use judgment.

ALIENIST OPIUMIZED

Dr. William Hicks, 64, of No. 46 Milford Avenue, Newark, N. J., was recently discharged from Bellevue Hospital after being treated for acute opium poisoning. This professional is known as an alienist—one who treats the insane or people suffering from mental troubles. It would be good to know how he was poisoned and who administered it.

WOMEN WITH TRUNKS

A recent Paris decree in fashion placed the vogue on the female nose built on long and slender lines like the Grecian. An instrument made of gold is worn nosily a few hours a day by the women not blessed along such lines. After about three years of pinching torture, if they get such lines and look Grecian, then there will be probably a change for some other type of nose. Where will the end be? Women, act with sense; slavery was abolished, but you are worse slaves to fashion.

DOWN WITH VACCINES

In a bulletin issued by the Public Health Committee of the New York Academy of Medicine is mentioned that vaccine for the prevention of pneumonia are still in the experimental stage and that though the vaccination has some value against three of the fixed bacteriological types of lobar pneumonia, the duration of the immunity is not more than 5 or 6 months. Vaccines are rot; putting garbage in the human temple of God is of no use; the best preventive is to keep the vitality inside working properly by clean food, no stimulants and correct thinking.

LLOYD GEORGE'S SHIRT

The *Boston Transcript* reports that in replying to the criticism that he frequently changed his mind, Lloyd George said:

"Well, men change their shirts don't they? We rather admire them for that. It's hygienic. So with the mind. As conditions change, the mind must change to conform with them.

Blame the man who never changes his mind! How stale and soiled and dusty such a man's mind must be. He who changes his mind to harmonize with circumstances, though, is spiritually clean and fresh, is living according to the laws of spiritual hygiene."

If changing mind means breaking of promises, if changing mind means turning the sails to collide against another's ship as suits best, then, such changing is craftiness and statescraft and is not a sign by any means of a clean mind. However, we must admire this master mind of English politics—we certainly have none here to match him—this great Welshman!

DOG-OIL OR DOG-GONE MEDICINE?

A science that constantly keeps changing its empirical formulae, fundamentals, and functions, ever groping in the dark, could not rank as a "science" until it becomes an accuracy. Drug therapy, ordinarily known as medicine, cannot by any stretch of imagination be classed as a science; it never was so, although it has established its stronghold (stranglehold) six thousand years and tries to choke others who do not follow its follies. Here is a sample of medical "science" by the great surgeon Ambroise Pare, who wrote in 1540 Anno Christi:

"When I was at Turin I found a surgeon famed above all the rest for his treatment of gunshot wounds, into whose favor I found a way to insinuate myself, that I might have the recipe of his balm as he called it, wherewith he dressed these wounds. And he made me pay my court to him for two years, before I could possibly get the recipe out of him. In the end, thanks to my gifts and presents, he gave it to me, which was this, to boil down, in oil of lilies, young whelps just born, and earthworms prepared with Venice turpentine. Then was I joyful and my heart made glad, that I had learned his remedy."

And the London Pharmacopoeia for the year 1659 contained a recipe for "dog-oil," with the following ingredients: "sallet oyl, two puppy dogs newly whelped, earthworms washed in white wine and cypress turpentine and spirit of wine."

Some medicine! Dog-oil! What kind of a "dog" would use it now you will perhaps ask? But during all these years only the methods have differed, to come again to the same noxious nostrums, leeches, and caustics and only a few years ago bleeding and mercurializing were the two anchors for the "cure" of human ills. How much difference is there yet in the medical "science" of the past and the medical "science" of the present? Filthy horse vaccines and pig serums, dirty dead organs of animals and chopping off of a part because it does not function properly, are *still* the pride of medical "science." The day of reformation could not have come sooner and the true therapeutical science that deals the death-blow to the "science" of indiscriminate drug therapy has already taken its birth under the name of SPECTRO-CHROME THERAPY which is the most potent, rational, clean, natural, simple and accurate single system of healing human dis-eases supplanting drugs in toto.

SAVED FROM THE KNIFE

Would you like to hear an interesting story about the accuracy of medical "diagnosis?" The Reverend Dr. Carl E. Grammer, rector of St. Stephen's Episcopal Church of Philadelphia, Pa. was sent (while there) to the American Hospital in Paris, France, for an operation for what the doctors diagnosed as "stoppage of the gall duct." The hospital doctors declared the diagnosis to be incorrect, saying they did not know what the trouble was, but that there was nothing wrong with the gall duct. Thus, the operation was postponed for "better diagnosis;" meanwhile, the patient improved so much he was allowed to depart. Let him thank his planets he escaped the knife! Spectro-Chrome Therapy avoids such complications and false stepping by original methods of treatment without the knife and gives results surpassing in value any form of destructive surgery. Food of the wrong kind is the root of most such troubles and following the principles laid down in Spectro-Chrome Therapy relieves such diseases quickly. Diagnosis statistics of the hospitals indicate wrong diagnosis in over fifty per cent cases; what *is* then the value of such an art? To cut open an individual and find that the organ removed was healthy but that the disturbance was produced *reflexly* by trouble in another organ cannot be called a "science." A person mixing hydrogen and oxygen in the proportion of two to one never gets anything but water; a mathematician never gets from the addition of 2 and 1 anything but 3; when medical "science" reaches such accuracy then may it try to regulate other systems of healing; till then, it cannot deserve any other name but hit-and-miss, fluke or empirics. Spectro-Chrome surpasses it in mathematical accuracy to such an extent that if a patient be not relieved in strictly non-surgical cases by Spectro-Chrome Therapy properly applied, it is a corollary that an undertaker should be called to assistance and not a drug doctor.

MAYOR'S WISE ACT

J. Hampton Moore is the Mayor of Philadelphia, Pa. The City of Chicago, Ill., appealed to the Quaker City authorities to send the Liberty Bell to their city during a celebration to be held there this month. This request was emphatically denied by Mayor Moore; all honor to him. The Liberty Bell, the visible emblem of our dear republic, should not under any circumstance be moved from its place in the Independence Hall. If the principle of moving shrines and relics be once accepted, as well send a request to move Jerusalem to Rome—it would surely be taken care of better. Honor to the sensible step of the Mayor.

JAPANESE TURNED CHRISTIAN

Salto Taizo, a Japanese electrocuted for murder recently at Sing Sing, N. Y., was converted to Christianity before he died. It had been better if he had lived as a good Japanese and followed in the footsteps of Buddha. Mere conversion from one religion to another counts for nothing, the foundation of all religions is the Truth and the benefit accrues only by the exemplification of the precepts of any prophet into our daily life. The present day Christianity as exemplified in the European War is no encouragement for anyone to become a "Christian;" a true Christian follows meekly in the footsteps of Jesus, sacrificing self for others, decrying none, obeying conscience. When Christianity will stop warfare (which is legal murder), when Christian nations join in unity for self-enactment, when Christian ministers of the Gospel stop going to foreign lands with the Bible in one hand and the Brandy Bottle in another, when the hundreds of denominations, creeds and sects of the so-called Christians will join in worship at one altar of Universal Brotherhood, when the two main opponents Protestants and Catholic will join as one of the Fatherhood of God, then may the Christians be entitled to the respect of other nationalities which are Non-Christians. Preach some true Christianity into politics, preach some real Christianity into Wall Street, practise some Christianity in every day life in business, commerce and dealing with those unfortunately placed low in the scale of human progress and an era of a New Christianity will enlighten the world. Who will don the cloak of the Messiah for such glorious emancipation?

HUMAN LOCOMOTIVES

Professor William A. McKeever, of the University of Kansas, writing in the *Evening Journal* of New York, July 26, 1922, about children says in part:

"Study your cross and peevish child systematically and you will most probably discover some removable cause of his unaccountable state of mind.

"Precisely as is the case with adults the mind and the body of the child act in about the same level of well-being and efficiency.

* * * * *

Examine the atmosphere. You may there find the hidden cause of your child's irritability and low mentality. Many little ones are caused to remain long in smoke-laden rooms. It has been found by test that air heavily charged with tobacco smoke will render eggs infertile, will stunt young rabbits and make young chickens anemic. Show father how to improvise a smoke ventilator and thus draw the tobacco fumes from the living room, for the sake of his baby."

He is right. In the exercise of personal liberty no person has any right to injure another. One does not like eating from an unclean dish or drink from an unclean tumbler, yet no attention is paid to keeping the atmosphere clean. I see every day unthinking fathers and mothers smoking indiscriminately and blowing their foul breath into the tender lungs of their innocent babies, converting a healthy home into a dirty smoke-sodden coup, unfit for the residence of developing humans. The Creator forgot to place a chimney on their skulls! By what earthly right does any person dare render unclean a health-giving atmosphere? I do not deny any person's right to do as they please, but, where the tip of my nose begins their liberty ends. Parents and others must not smoke their foul breaths into children and the sooner this weed of tobacco be placed under the same ban as alcohol or similar poisons the better for the preservation of the race.

STUPIDITY OF CENSORS

A nation's virility depends on the free expression of public opinion. Nations which submitted to magistracy, imperialism, kingship, lordism, autocracy, bureaucracy and censorships that moved the power of public opinion to a chosen few went down in the graves of oblivion and are now only useless history. Therefore, profiting by their example as a beacon to avoid the shoals, we must keep aloof from such suicidal policies. It seems, however, the old-time spirit is being revived in America under the name of "Boards of Such and Such," and persons ignorant of the A, B, C, Z of human nature are being appointed as members of such (mahogany) boards to inflict their thoughtless decisions on the followers of the particular industry affected by their censorship. The *New York Times* recently published what appears as a flagrant example of the knockdown stupidity of the Pennsylvania Board of Film Censors. Its writer says:

"The Pennsylvania Board of Film Censors would walk a thousand miles to smoke out a 'Camille.'

"It was from this same musnud of moralists that the edict was issued last year that no picture could be shown that portrayed a mother making a layette for her unborn child. In Pennsylvania children are not born in the usual way, but are bootlegged over the roofs by storks or moonshined into the world through a cabbage leaf.

"It was the same Sanhedrian of Sages who ordered the title, 'It is a boy,' cut out of 'The Four Horsemen of the Apocalypse.' If it had read 'It is a Republican,' it would have passed —in Pennsylvania.

"Anyhow, 'Camille' arrived in Pennsylvania on the celluloid. It was not called 'Camille.' It was called 'The Red Peacock.' The story was the story of 'Camille' done in Germany. It was originally called 'Poor Violette.' Pola Negri was the star who coughed her way into eternity in the fifth reel in the way the younger Dumas had arranged it some years back. The names of the characters were changed, but the play is essentially the same that millions of people have seen and read.

"As soon as the Pennsylvania board saw the picture they put their ivories together and ordered the following changes in 'Camille,' alias 'The Red Peacock.' I here reproduce verbatim the order. It is my most precious literary possession. As a bit of Americana, I believe it is unique:

"REEL 1—(a) Eliminate subtitle, 'Gaston du Pont, her satellite,' and substitute 'Gaston du Pont, her fiance.'

"REEL 3—(a) Eliminate subtitle, 'Count Girgy sees an opportunity,' and substitute 'Count Girgy sees an opportunity to vary his hectic life with an act of humanity.'

"(b) Eliminate subtitle, 'Violette, you may remain as maid in this house if you wish to,' and substitute 'Violette, my house is lonely. Let me do an unselfish act—be my ward and enjoy the comforts of my home as a sister would.'

"(c) Eliminate views of Girgy embracing and kissing Violette after ...nging her wrap to her and all views of Girgy kissing and embracing Violette in any other reel throughout the picture.

"REEL 4—(a) Eliminate subtitle, 'Alfred, I love you. Take me away from this,' and substitute 'Alfred, I love you. I was happy as Girgy's ward until you returned. Take me away.'

"(b) Eliminate views of Alfred shaking his head to express 'No.'

"(c) Insert, after Alfred has fallen at Violette's feet with his head in her lap and she is fondling and kissing him, subtitle, 'Come to me, Violette; we will be married at once and say nothing about it.'

"REEL 5—(a) Insert a subtitle after 'I love Alfred—he is all I have in the world. I cannot let him go,' when Violette sinks in chair and Claire goes out of the room, to this effect, 'Realizing that for his own reasons Alfred had not told his father of his marriage, Violette loyally kept the secret.'

"(b) Insert a subtitle during the views showing Violette leaving her home and before she goes to Gaston du Pont, to this effect: 'With a courageous determination to find some means of honestly earning money to aid Alfred.'

"(c) Eliminate subtitle, 'I am here, ill in body and soul. Take me away, anywhere, and substitute 'I am here, ill in body and soul. You offered to help me. Are you good friend enough to take me, unselfishly, where I can learn to dance so that I may earn money?'

(d) Eliminate subtitle, 'My dear Alfred: Forgive me, I am leaving you. My illness will become a greater and greater burden on you, and our financial troubles are growing each day. You have your future to consider, Violette,' and subtitute 'My dear Alfred: Forgive me. I am leaving you for a time that I may earn money to overcome our financial troubles, which are growing each day. I love you, and hope for the future. Violette.'

"(e) Eliminate all views in this and other reels following of Gaston du Pont embracing and kissing or making love to Violette.

"(f) Eliminate subtitle. 'We will go South and there will soon be roses in your checks,' and substitute 'You may trust me. We will go South and there will soon be roses in your checks. You shall learn to dance there.'

"REEL 6—(a) Eliminate all views of Gaston du Pont making love to Violette, kissing or embracing her.

"(b) Eliminate subtitle, 'You'd better not dance this evening—your cough,' and substitute 'You'd better rest this evening—your cough.'

"(c) Eliminate the word 'you' from

subtitle, 'It was my money you loved, not me—you—'

"(d) Eliminate subtitle, 'Now I know you for what you are—and I'm through with you,' and substitute 'I'm through with you forever.'

"A careful analysis of this remarkable document will disclose to the future excavators and archaeologists of the ruins of American liberty the wheels in the mental processes of those who gave to the world the Great Pennsylvania Idea. Rare pickings for a future Taine or Gibbon.

"It may be noted (Reel 1, a) that no Camille is entitled to a satellite within a radius of 300 miles of Harrisburg. She may have a fiance. The fiance must not be a satellite—but a satellite may become a fiance. So the Home is conserved.

"Count Girgy (Reel 3, a) cannot just see an opportunity. He must see an opportunity that involves an act of humanity. There can be no objection to this, as it follows the philosophy of McGuffey's Reader.

"Violette and Girgy (Reel 3, c) are not allowed to embrace or kiss throughout the picture. That these changes practically destroy the picture is of no moment. Girgy and Violette, although in love with one another, are not married; therefore all contact is null and void. This is based on an old Pennsylvania tradition that one may only kiss his wife or his dead aunt."

America, the cradle of the liberty of the world, must have positively and absolutely no Censor Boards on the Press, the Author, the Orator, the Critic, the Artist, the Theatre, the Silent Drama, the Healer, in fact on anything and anyone that tends to place brain productions before the public. Let the public be the true censor, because, once the principle of censorship is countenanced, its baneful influence will induce bribery and the public's weal will be converted into woe. The above is a fair example of the tyranny of such a Board and further, the crass ignorance, idiocy and utter nonsense enunciated by its members in its operation.

GLANDS, GOATS AND MONKEYSHINES

Harold F. McCormick, the millionaire of Chicago, Illinois, was operated on by Dr. Victor Lespinasse for "rejuvenation of youth" by implanting in his body the purchased gland of a youth. The newspapers created a roar about this until McCormick finally declared he had never bought any human gland. This gland business and organotherapy, as it is called, is a monstrous medical fraud and I shall soon write more about it. Meanwhile, an article appeared in the *Evening Bulletin*, of Philadelphia, Pennsylvania, dated June 26, 1922, which for its good and steady representation of the affair deserves attention:

"Mr. McCormick, in searching for the fountain of youth, is doomed to a keen sense of disappointment. Ever since the history of mankind rejuvenation has been a subject of considerable attraction. Of course, it may be desirable, but unfortunately impossible. Experimental reasearch in that direction is excusable, but to commercialize research work ere the result has definitely been established, is unpardonable. It is nothing short of charlatanism. The glandular transplantation operation has no biologic basis. The assumption that any particular gland possesses the fountain of youth is unscientific and untrue. Transplantation of organs and tissues (bones excepting) always results in atrophy of the tissue transplanted. But granting for the sake of argument that such gland after a proper transplantation still retains its vitality it is still inconceivable how such gland would have any effect upon the tissue elasticity of the victim of old age. A gland is not an independent organ functioning on its own volition; it depends upon the blood supply and enervation of the newly acquired environment. A youthful gland transplanted into an aged body whose blood vessels are already changed and hardened, whose nerve elements have already undergone senile changes will, if not atrophied, function only in direct proportion to the new nerve and circulatory stimulation. The gland will become senile instead of the patient juvenile (although his mind may become puerile).

"We cannot halt the finger of time; we cannot change the laws of God and nature. We cannot by any miraculous means change the aged into a young Herculean. Would you be young? you must learn the art of living in youth. Strict regard to mental and physical hygiene, a philosophy of life rigidly cultivated may help one to remain youthful and reach old age gracefully. Buying human glands to rejuvenate your lost vigor is like gambling in stock to recuperate your lost fortune: it ends disastrously to the buyer. Let man be rational and not mystical: the former leads to God, the latter to superstition."

This writer is Aaron Bray, M.D., of Philadelphia, Pa., and deserves credit for his able enunciations. Burning the candle at both the north and south ends during youth, wasting virility then and trying to buy (for further excesses in old age) potency through foreign glands is absurd, illogical and unnatural and the doctor of medicine who encourages such views is a traitor to his profession. Teach patients to lead instead a saner and cleaner life. We shall, in the case of Harold F. McCormick suspend judgment until he remarries and the truth is known: the newspapers say, "Ganna Walska is waiting for him in France," where he also is—well, may he have better luck in his new life! It is nobody's business but his, it does not touch the public.

Attuned Orange Wave

(Humor, Wit, Fun, Satire, Tickle, Irony)

DID YOU EVER?

A short sketch on the Oddities of Females, by

Miss Irene Hoger, S.C.T.

Did you ever watch a woman feed coal to a furnace? Either one of these two methods is usually used: She puts the loaded shovel to the door of the stove and forces the coal in, or, after putting the shovel into the stove (leaving the handle out, of course), she turns the shovel over and lets the coal fall off onto the fire. You never saw a woman do that? Watch more closely next time.

Did you ever see girls having a "catch?" The majority turn the palm upward and push the hand in a likewise direction which sends the ball into the air with a so-called under-hand motion. Hence the saying, "Aw, girls throw under-hand." Again you say you never saw such witlessness? 'Tis a pity you are not more observant.

Did you ever watch the fairer sex get out back first when alighting from a car? No? Then you have much to learn, for, this unsafe, almost unforgivable way, shows no common sense whatsoever and can be seen every day, everywhere. Who knows? Perhaps you have been doing the same thing!

Did you ever witness in agony one of the female species endeavoring to nail some smooth surface, but succeeding only in punishing the nails: likewise her fingers? Beg pardon, how stupid of me! I should have specified what kind of nails; carpenters' nails, of course, not finger nails. With patience exhausted, nails bent, fingers sore and desperation almost approaching, one first hears mumbling: a few naughty words and then—"Oh, fudge." "Beans," "Oh, pickles," "Bean soup." If she kept on she would name a complete vegetarian diet. Having said "Oh, boy," and "Man alive," she is not contented and shouts "Oh, Gee," which in reality means God or Jesus. I wonder if she, who considers herself man's superior, realizes what an impression she is making on the onlooker or listener; how vulgar it all appears and really *is*.

Lucky for you if you never saw or heard a woman in such a mood; but tell me then, what *did* you ever see, or even hear? What do you do year in and year out, lie in the house like a mummy or act like an idiotic dummy, or imagine you are playing a tune with your fingers on the kitchen broom as an accompaniment to the fly's jazz dance on the parlor walls?

It is time for you to get busy and know what is going on in this old world of ours; or else tell me what's the use of living anyway?

You may say I used no discretion in the way I knocked the deadlier sex. Cheer up, if that is the case, I shall in next month's issue pick on the opposite sex, if our kind editor permits.

Permission is granted, though not necessary. Go ahead; it would be good to hear your views about the less dead—I mean the less deadly sex. You wrote about the fair sex, now write about the unfair one.

Attuned Yellow Wave

(Correspondence, Query, Catechism, Inquiry, Interrogatory)

SEX EDUCATION AND HELP

Woodlawn, New York, N. Y.

DEAR DR. DINSHAH:

If these questions are not suitable for publication, I shall not mind if you utterly ignore them:

1. People are so horrified if one mentions sex; one is looked upon as unclean. Is not what the world needs today a regeneration of the *mind?* Then all things become pure to him, then, this *wonderful God-given power* Sex, or the *Creative Force*, becomes to that man, or woman, the greatest gift of God.

2. Is one selfish or making a mistake when one takes time to educate self at the sacrifice of what the world calls duty?

3. Is one wasting time by giving it to the blind or helpless, by reading books to make them happy, even though one feels that it is not what one should do?

Sincerely,

ANNA AU BURTON, S.C.T.

Your questions are indeed sensible.

1. I want to publish a good article on Sex; who will take the lead? Do not wait for me to write on *all* TRO-CHROME columns belong to you. subjects. Use your brains; the SPECTRO-CHROME columns belong to you.

Write in plain open language; we shall publish it.

2. Self-education is performance of a duty; if one is not trained, how can one serve others? Education of self in itself is the fulfillment of an obligation. Go ahead.

3. By helping others in any discreet way, you create mighty forces for good; these forces react on you and boost you; judge none; do what your conscience guides you as right and forget consequences. Duty is its own reward. You are on the right path. Anna.

SUGGESTIONS FOR SPECTRO-CHROME

July 5, 1922.

Editor SPECTRO-CHROME:

Hail to the first issue of SPECTRO-CHROME! It is a good beginning, but like all infants we can expect improvement with growth and age, and we who know its editor know that it will fill a long-felt want.

I am taking the privilege of suggesting some topics which I feel sure the majority of the magazine's readers will like to see treated *by the Editor.*

1. Use and Abuse of Incense. Its effects on human beings mentally, morally and physically. The history of its use in various countries. The subject as you can treat it will be almost inexhaustible.

2. Perfumes, too, would make an interesting article.

3. The exact history of the development of the use of Attuned Color Waves in treating dis-eases. How the idea dawned and the line of reasoning followed. What prompted the development of the various stages, with the experiments performed, the dates, etc.

4. A history of Zoroaster and His real teachings. The mistaken idea of it as held in the Occidental mind.

5. A comparative treatise on the Seven Great Teachers.

6. Color, what it is, its effect on human beings in every way. Its scientific relation to music. How music could be used in conjunction with color to affect patients, etc.

7. Principle of Polarization—in material science and as related to man's various planes of consciousness, etc.

8. Reincarnation. This might be treated in a number of articles running serially.

9. Occultism versus Occidental Religions. The occult wisdom as is found in works of Occidental poets, Emerson, Ruskin, Longfellow, Wordsworth, Browning, Coleridge, etc., together with that of Socrates, Epictetus, Marcus Aurelius, and many more.

10. The Principle of Gender, from the beginning to the end.

11. I heartily endorse Dr. Anna De La Motte's suggestion of an article and discussion on "Procreation and Regeneration."

I shall stop with these, but there are many others that I would like to see treaty by our worthy and beloved teacher and Editor.

Yours for success in benefiting and uplifting humanity,

NAYA BUNDY, S.C.T., *Podiatrist.*

Thank you cordially for the good suggestions and sincerely wish I could publish all you want, at once. The limit of our periodical, the limit of our treasury and the limit of your friend Editor's 24 hours in a wakeful physical day's time is now humanly reached by his putting in for one thing or another generally 21 out of them to working for humanity's service and dedicating every cent of his available resources to the expansion of his activities in that direction. However, in reply to your suggestions, I wish to report as follows: 1-2-4-5-7-8—I may write when other imminent and more important problems will allow. 3—Is ready and may be published soon. 6—The chart of Spectro-Chrome Acoustic Therapeutic System published by me in 1920 has the nucleus of what you want and I intend to expatiate further. 9—Very interesting, but other items are more pressing. 10-11—I am constantly writing and speaking about these subjects; no Gender, no Spectro-Chrome; I shall publish the many thousand pages of my unprinted matter gradually as our circulation increases, space is available and the call comes. Success to you, doctor; meanwhile, there are tips for others in the above to write from. I welcome the budding writer to these columns.

Attuned Green Wave

(Spectro-Chrome Therapy, Spectro-Chrome Therapists)

THERAPEUTICAL ACCURACY

By Edmund J. Hogan, D.C., S.C.T.

"Spectro-Chrome Therapy meets all the requirements of a complete therapeutical system based on accurate principles." This extract from the Compendium by Dinshah should be read and reread by every Spectro-Chrome Therapist until the essence incorporates and permeates every minute cell of our gray matter. Then, let us turn to another extract.

This is taken from the President's address before the Ohio State Medical Association, June, 1920, by J. F. Baldwin, A.M., M.D., F.A.C.S., Columbus, Ohio.

"A few years ago I saw a patient late at night in consultation with three of *our very best* physicians, two of whom were treating the case together. After getting the history and examining the pneumonia patient, we all agreed in our own belief that *nothing had been given, or could have been given,* which could in any possible way have affected the progress of the disease. *In fact, all the drugs administered had been, though unintentionally, really placebos, or what we might now call camouflage.* The treatment had been secundem artem strictly, but by no possibility could have affected the progress of the disease, *which terminated fatally a few hours later.* And yet I imagine those same physicians *are still, from force of habit, using the same placebos.*"

Can there be a more scathing indictment of this drug system of inaccurate principles found anywhere? Spectro-Chrome Therapy now steps forward to take its rank as the foremost "therapeutical system based on accurate principles." Onward!

"Placebo" comes from Latin, Placeo meaning "to be welcome." Placebo is any indifferent substance, in the form of a medicine, given for the moral or suggestive effect. In pure American parlance it would be a sheer make-believe. It really means in the physician's language, "Well, we can do nothing, but to keep our face we must do something and give the patient something to prove that we did everything." Placebo! O Bo! O Drug Hobo!

CONSTIPATION, AMENORRHEA, ANEMIA, PILES, APPENDICITIS

By Peter P. John, N.D., S.C.T.

I wish to report the following wonderful results obtained with Spectro-Chrome Therapy, which would seem to be either unbelievable or miraculous.

On October 18 last a Mr. Rose came to try the new science of Attuned Color Waves and told me he had suffered with chronic constipation and kidney and bladder trouble for more than seven years, frequent and regular doses of medicine having failed completely to give the slightest relief.

The yellow wave of the 1000-watt equipment was used. He was put on a rational diet and advised to drink two quarts of home-made lemonade per day. He took eight treatments, the constipation with its attending ails having vanished after the fourth treatment. This is the fifth day of August. It is practically ten months since he took the treatments and none of those ailments has recurred.

Miss L. N., age, 25 years, suffered with amenorrhea for 60 days. She had only three treatments—one yellow and two scarlet waves. Her condition was corrected after the third treatment. Her testimonial speaks for itself:

"After giving me three treatments of the Color Waves, I saw a great change. I think your treatments are wonderful. I cannot recommend it too highly."—L. N.

A woman of 21 years was treated with yellow, lemon and orange waves for pernicious anemia. A good many of the high-toned patent medicines in the regular drug-store had been tried and cast aside as worthless. She came to me weighing 99 pounds. She was put on a diet which included plenty of greens, salads of grated raw carrots and beets, whole wheat bread, milk and honey, and fruits (principally bananas, oranges and apples).

At the end of four weeks' time, after her thirteenth treatment, her weight was 119 pounds, thus gaining 20 pounds. Today she is the picture of health.

Mrs. S. G. suffered terribly with piles. Her family physician and others considered her case unamenable and hopeless. She came to me on May 25, scarcely able to walk. I used the yellow wave to keep her bowels regular and the indigo wave on Area 19. Her recovery was so rapid that after her fifteenth treatment she went to Atlantic City where she is still working. A few days ago, I met her husband, who told me she wrote to say she is doing well.

Three cases of appendicitis have come under my hand. Yellow, indigo and turquoise waves were used with the most astonishing results. I use the yellow wave on Area 9 in almost every case, as the foundation. The most serious of these cases was a young woman who had suffered severely for six weeks. She ultimately made up her mind to have an operation, but was saved from the knife in the nick of time. On her way to the hospital, on Monday, May 1, she was advised to try Spectro-Chrome Therapy. She came to me. The first treatment was so effective, that night she slept better than she ever did for the whole time she had been sick. The fourth treatment brought complete relief. She took twelve treatments and there has been no more trace of appendicitis.

I am now treating a case of cancer of the breast. Green systemic and lemon (local) are used. The patient is abstaining from meat, fish, eggs, tea, coffee and alcoholic drinks. I shall be glad to report the progress of this case in a future issue of the SPECTRO-CHROME.

As the first negro graduate of the Spectro-Chrome Institute, Peter P. John, a West Indian by birth, deserves commendation. His intelligence and knowledge of Spectro-Chrome Therapy are certainly praiseworthy and may all success follow his efforts.

Attuned Turquoise Wave

(Poetry, Romance, Drama, Art)

EVERYBODY BOOSTING SPECTRO-CHROME

By Roy Forbush Hanaford, S.C.T.

"**S**PECTRO-CHROME" is our new magazine,
Presenting facts from behind the screen;
Editor Dinshah, full of "pep" and "go,"
Can tell us a great many things we should know;
Truth is the platform on which he stands,
Regardless of where the boomerang lands.
Only with your help can "SPECTRO-CHROME" spread,
Collect some subscriptions, show you are not dead!
Hail to Nature, her laws obey,
Replacing dis-ease with health come to stay;
Omit all drugs and surgery destructive,
Make people well with this method constructive;
Everybody boost "SPECTRO-CHROME."

MOTHER NATURE'S ALPHABET

By Miss Kashmira Dinshah, S.C.T

A is an Aster, of crimson hue,
B is a Balsam, of fine dark blue;

C is a Cosmos, graceful and tall,
D is a Daisy, pretty though small;

E is an Elm-tree, lofty and grand,
F is a Fern, that grew close at hand;

G is a Gentian, Autumn's flower,
H is a Hazel, gracing bower;

I is an Iris, blue as the sky,
J is a Juniper, standing by;

K is the Knoll, on which these grew,
L is a Lily, bepearled with dew;

M is a Marigold, yellow, bright,
N is a Narcissus, creamy white;

O is an Oak-tree, crowning a hill,
P is a Primrose, near rippling rill;

Q is a Quince, ripe, ready to fall,
R is a Rose, the fairest of all;

S is a Snowdrop, a first Spring blow,
T is a Trillium, with purple glow;

U is the Underbrush, all around,
V is a Vine, which lay on the ground;

W is a Willow, upon the green,
X is a Xanthium, of silvered sheen;

Y is a Yew, that stood on the wold,
Z is a Zinnia, deep red and gold.

Attuned Blue Wave

(History, Biography, Geography, Language)

LIFE SKETCH OF THE ORIGINATOR

Our graduates and others constantly express their desire to have a reliable biographical sketch of the peculiarly versatile checkered life of the originator of Spectro-Chrome Therapy. The Pottsville *Daily Republican*, of

Pottsville, Pa., a newspaper having the biggest circulation in Eastern Central Pennsylvania, printed in its issue of July 29, 1922, the following, which is the most accurate report of the man ever published in America. It will go a long way to show that Dinshah's work is not the result of sleeping on a "bed of flowers" and that even now he could not afford to rest on his oars, his real work for human service being only begun with Spectro-Chrome Therapy.

Men Prominent in Life of Pennsylvania

Colonel Dinshah P. Ghadiali, M.D., D.C., Ph.D., LL.D., of 2401 North Broad Street, Philadelphia, one of that city's prominent men. Born in India a brief sketch of his unusual career will be of interest to readers:

1873—Born November 28, of Parsee Zoroastrian parents at Bombay, British India. The name Ghadiali meaning "watch maker." His father, a watch maker, was also contractor to the British Postmaster General for the presidency of Bombay.

1876—At the age of 2½ years attended primary school and at the age of 8 years was attending high school. Three years later, at the age of 11 years, acted as assistant to professor of mathematics and science at Wilson College, Bombay.

1886—Took Bombay University examination.

1887—At 14 years of age became experimental demonstrator in chemistry and physics at seven institutions of learning. At this time he began the study of medicine.

1891—Became member of the Indian Temperance Association and was elected Fellow of Theosophical Society and began career as public orator on scientific, metaphysical, spiritual and occult subjects.

1892—He erected several electric light installations and was appointed telegraph and telephone superintendent for Dholpur State, Central India, and acted as electrical engineer to his highness Maharana of Dholpur.

1893—Joined British Merchant Marine service, with the Peninsular and Oriental Steamship Company and went to London, England; became a member of the International Order of "Good Templars."

1894—Was appointed chief mechanical engineer in the "Umballa Flour Mills," Punjab, India. Became a member of the Independent Order of Rechabites, the Royal Army Temperance Association, and also of the Royal Artillery Temperance Association then under the Presidentship of Lord Roberts, Commander-in-Chief of the Indian Army.

1895—Became consulting electrical engineer and erected a number of electric light plants in various cotton spinning and weaving mills.

Was elected an honorary member of the Poona Volunteer Rifle Corps, which is a great honor in India.

1896—Made his first visit to the United States and was introduced to Thomas Edison, Nicola Tesla, Michael Pupin and other inventors and scientists of international note. Delivered first lecture on X-ray and radio activity in New York, which earned for him the title of the "Parsee Edison."

Returning to India, he developed his practice as medical practitioner and medical electrician in Bombay and started a number of reform campaigns throughout India, of such a radical character as to stir up the 332,000,000 people throughout the country.

1899—Became stage manager of the Bombay Theatre and erected therein the first electric motion picture machinery as a pioneer in the film art.

1900—Inaugurated "Electro-Medical Hall" at Ajmer, Central India, for the treatment of diseases by Chromopathy or color waves, electricity and other drugless methods in conjunction with

his regular medical and surgical practice.

1901 — Opened "Electro - Medical Hall" at Surat, Western India, and became famous for successful treatment of bubonic plague and the so-called incurable diseases.

1903—Became trustee of the Deshi Mitra printing press.

1904—Was elected chairman of the "Nanpura Parsee Community" to protect the rights of the poor, enabling them to secure medical assistance with less cost. In doing this, he became involved in a number of defamation and libel suits, all of which he defended personally with success, without the aid of a lawyer and won much popularity by his upright stand against autocratic oppression. At this time he was appointed District Agent of the New Zealand Fire and Marine Insurance Company.

1905—Established his own printing plant and weekly newspaper called the "Impartial," to further the cause of liberty of speech and writings. His editorials were under the title of "Our Golden Whip," Criticizing unjust acts of government and other public officials and exposure of political graft created much uproar.

1908—July 4th left India and traveled through England, France, Belgium, Holland, Germany, Austria, Switzerland, Italy and other countries to promote his electrical and scientific inventions.

1909—Founded and was chairman of the Alcohol-Free Wines Company, Ltd., of London, England, with factory in Switzerland to promote the cause of total abstinence. Was appointed "Honorary Deputy Grand Chief Templar of the Grand Lodge of England of the International Order of Good Templars." Was delegate to the Twelfth International Congress on Alcoholism and was appointed member of the first "International Prohibition Confederation" of London, England. Took up the study of flight. Passed the examination with honors and was elected Fellow of the "Templar Institute of England" and was commissioned speaker of the Anti-Socialist Union of Great Britain under the presidentship of the Duke of Devonshire.

1911—Arrived in United States with $75 in his pocket, after unforeseen financial reverses in England caused through his zealousness in promoting the cause of prohibition.

Decided to make America his permanent domicile, and became professor and chief instructor in New York College of Engineering Science and Automobile Instruction. Later acted as secretary of a division superintendent of the Prudential Insurance Company of America.

1912—Became Vice-President and General Manager of the Automobile Fault Finder and Anti-Forgery Electric Pen Company.

1913—Became Assistant Manager of the Southern Nut and Fruit Company of New York and Georgia. Established experimental laboratory with a wireless telegraph station at Hillsdale, N. J. At this time began to manufacture the "Dinshah" Engine Tester for testing gas engines of automobiles, submarines and aeroplanes, which he gifted to the United States Government during the war.

1914—He organized and became President of Dr. Dinshah P. Ghadiali Inventions Corporation for manufacturing and marketing his inventions, and the Dinshah Photokinephone Corporation for talking motion pictures.

1915—Became General Manager and Vice-President of the Independent Electrical Supply Company and Vice-President and General Manager of the North Atlantic Metal Corporation of New York.

1916—Became export manager of the Union Smelting and Refining Company of New Jersey.

1917—Was naturalized as an American citizen at Hackensack, N. J., June 15th.

1918—Was commissioned a captain in the New York Police Reserve by Special Deputy Commissioner Rodman Wanamaker for war duty.

1919—Received major's commission and was appointed Governor of the New York Police Aviation School. With the assistance of a large number of American war veterans, both flyers and non-flyers, he organized the aviation branch of the New York Police and was commissioned colonel and commander of the New York Police Reserve Air Service. Made several spectacular flights and established a record by bringing the first Police Air Mail from New York to Philadelphia.

He received for his services two hydroplanes from the United States Government to be used for policing New York Harbor and from Police Commissioner Richard Enright and Mayor John Hylan, of New York City, received the "Liberty Medal."

Founded the Business Psychological Bureau, Dinshah Commercial Company and other business enterprises and resigned his export managership for which he was drawing a salary of $15,000 a year, for the purpose of again starting his scientific lecturing work and serving the cause of humanity.

1920—Having perfected and completed one of his life works, "Spectro-Chrome Therapy," his original method of treatment of diseases by attuned color waves, he founded the Spectro-Chrome Institute for the training of practitioners of his new system, which succeeded so well that he established a number of offices in different states within two years.

1921—Received certificate as air pilot for the operation and construction of flying boats from the Curtiss Aeroplane and Motor Company, a pastime which he thoroughly enjoys.

In recognition for his scientific knowledge and development of "Spectro-Chrome Therapy," he was given a life membership with the rank of "Academician" in the Maryland Academy of Sciences at Baltimore.

1922—Incorporated Spectro-Chrome Institute under the laws of the State of New Jersey for the instruction, development and practice of Spectro-Chrome Therapy to which he devotes 18 to 21 hours a day exclusively, traveling from point to point, approximately 1500 miles each week. All this has been accomplished by Colonel Dinshah on an absolutely vegetarian diet, as he uses no meat, fish, eggs, tea, coffee, tobacco or alcohol; in fact all these materials are debarred from his premises.

He is conversant with eight Oriental and eight Occidental languages and is an eloquent orator and a convincing masterful discourser with thorough grip of his numerous subjects.

Among other honors he holds the degrees of: Doctor of Chiropractic, Doctor of Philosophy, Doctor of Legal Law, and is connected with numerous public activities, chief among which are:

Fellow and Ex-Vice-President, Allied Medical Associations of America.

Member and Ex-Vice-President, National Association of Drugless Practitioners.

President, All Cults Medical Association.

President, American Association of Spectro-Chrome Therapists.

Member Keystone Automobile Club.

Member Metropolitan Museum of Art.

Member American Anti-Vivisection Society.

Member Anti-Vaccination League of London.

Member American Association of Orificial Surgeons.

Fellow American Geographical Society and a host of other scientific associations too numerous to mention.

PESTANJI FRAMJI GHADIALI

(Formerly of Thornhill & Company)

WATCHMAKER

LATE CONTRACTOR TO THE BRITISH POSTMASTER GENERAL
BOMBAY PRESIDENCY, INDIA

(Firm Established in Bombay, India, 1869)

MY MOST BELOVED REVERED FATHER

At thy feet on Earth I ever bowed,
At thy feet in Heaven I yet perform
My daily obeisance—my dear Genitor
As at the feet of the Almighty's Throne;
For, through thee I attained this birth
Achieved what I did in this incarnation,
Received Physical Life through thy self,
My most beloved revered father!

Dinshah P. Ghadiali

SPECTRO-CHROME

SEPTEMBER 1922 VOLUME 1
VIOLET 1922 NUMBER 4

Devoted to: Spectro-Chrome Therapy Dedicated to: Service of Humanity

CONTENTS Page

Price List ....Inside Front Cover
Spectro-Chrome Area Chart—Therapeutic System—Dinshah's Platform..... 73
The World Through the Eyes of Dinshah—
My Tour to Chicago—Drunk Doctor Jailed—Good for Whom?—Assinine Canine Law in Pennsylvania—Pull Her from the Mud—The Graduate Equipment—Revised Price List—Warning to Our Graduates—Rounding Up Quacks—One More Proof 74-78
What's More? (A Short Sketch on the Oddities of Men, by Irene Hoger, S.C.T.) 79
Our Advance Prospectus to Chicago 81
Spectro-Chrome Therapy Graduate Equipment 82
Spectro-Chrome Therapy Intermediate Equipment 83
Syphilis the King of Destroyers (by J. W. King, M.D.).................. 84
Carbuncles (by Kate W. Baldwin, M.D., F.A.C.S.)......................... 86
What a Prominent Homeopath Thinks (by Isaac Sossnitz, M.D.)........... 90
Dinshah's Discourses (by James M. Tarsy, D.C., Ph.C.).................. 91
A Chronic Sufferer Relieved (by Edmund J. Hogan, D.C., S.C.T.)........ 91
Sunlight Sets Right Grave Cases (by John A. Cohalan, D.O., S.C.T.)...... 92
A Layman's Personal Experience (by Orlo H. Frost)...................... 93
Crowning of King William Goat (A Barnyard Session of Words; by Kashmira Dinshah, S.C.T.) 94
Registration Notice of Trade Marks—Notice to Purchasers and Patients—To Our Contributors—Advertisement Rates...........Inside Back Cover
Spectro-Chrome Institute Charter—Spectro-Chrome Therapy Courses
Outside Back Cover

Editor
Dinshah P. Ghadiali

In course of life you reach a time,
When Doctor, Lawyer, Priest alike,
Does fail to give the sought relief,
To troubled mind or burdened heart—
Then come to me;
I help you to help yourself.

SPECTRO-CHROME.

Subscription: June, 1922, to December, 1922 — $1.75
Thereafter, January to December — $3.00

Single Copy, 25 Cents

PRICE LIST

The following Equipments and Supplies may be purchased from the **SPECTRO-CHROME INSTITUTE**, 2401 N. Broad Street, Philadelphia, Pennsylvania, United States of America:

.....The pages refer to the Compendium of Spectro-Chrome Therapy.

COMPLETE EQUIPMENTS FOR SPECTRO-CHROME THERAPY:

Graduate Equipment (As per Page 2)	$425.00
Practitioner's Equipment (As per Page 10)	150.00
Intermediate Equipment (As per Page 32)	90.00
Standard Floor Equipment (As per Page 18)	60.00
Standard Table Equipment (As per Page 19)	50.00
Standard Gas Equipment (Similar to Standard Table Equipment but fitted with complete gas attachments)	45.00

SUPPLIES AND EXTRAS:

Iris Diaphragm for Standard Equipments of all Types	$15.00
1,000 Watt Bulb for Graduate Equipment	9.00
1,000 Watt Bulb (Specially Frosted for Use with the Dinshah Attuned Color Slides in Practitioner's Equipment)	8.00
1,000 Watt Bulb with Concentrated Filament (Specially Frosted for Use in the Lower Socket for Practitioner's Equipment for Local Area and Orificial work)	10.00
400 Watt Bulb (Specially Frosted for Use in the Intermediate Equipment)	5.00
100 Watt Bulb (Specially Frosted for Use with the Dinshah Attuned Color Slides in Standard Equipments)	2.00
Replacement of Dinshah Attuned Color Slides (To replace one or more broken slides, **RETURN THE ENTIRE SET INCLUDING THE BROKEN SLIDES. A NEW SET** will be promptly supplied at a charge per Broken Slide, of	3.00
Spectro-Chrome Game of Health (a very enjoyable and entertaining game teaching the Fundamental Principles while playing)	2.00
Compendium of Spectro-Chrome Therapy by Dinshah	2.00
SPECTRO-CHROME (Monthly Publication of Spectro-Chrome Institute) per month	.25

CHARTS (by Dinshah):

Spectro-Chrome Acoustic Therapeutic System	.25
Universal Mirror (showing Cosmic Evolution)	.25
Human Psychogenic Rays	.25
Occult Operation of the Great Sympathetic Nervous System (The Only Correct Medico-Ontological Chart)	.25
Human Brain Structure	.25
Rational Human Diet (for Health and Beauty)	.25

ALL THE ABOVE PRICES ARE NET CASH WITH ORDER F. O. B. NEW YORK, N. Y.

Attuned Red Wave

(The World Through the Eyes of Dinshah)

MY TOUR TO CHICAGO

For the purpose of delivering a complete course on Spectro-Chrome Therapy and establishing the centre of the Middle Western Division activities of the Spectro-Chrome Institute, I leave Philadelphia, Pa., on Sunday morning, the 27th of August, accompanied by daughter Kashmira, by automobile, to avoid railroad strikes and similar complications, as well as to reconnoiter and arrange for the opening of the ground in the intermediate states, which have sent me their kind call.

Kashmira promises to take photograms on the way and give you on return some short and illustrated sketches of what she thinks will be "her adventures." We intend to cover about 200 miles daily.

Having edited and mailed you three numbers of SPECTRO-CHROME within three weeks, bringing it to date, I surely will be granted by you time to finish properly my projected work with concentration of energy. Though I acknowledge no physical limitations, there *is* one that limits my work and that is *time*—the day and night having only 24 hours. The October number will be issued *after* my return, which will be in the last part of that month. Thereafter SPECTRO-CHROME will be issued in the beginning of each month punctually.

Wherever I may be, all communications will reach me through the Central Office; please use *that address exclusively for all business,* to avoid loss of time.

The Central Office will be open daily for the transaction of business as usual.

DRUNK DOCTOR JAILED

In England an operation was performed recently by a surgeon while under the influence of liquor and the patient died. The surgeon received a sentence of twelve months in jail on a charge of manslaughter. If driving an automobile when drunk is dangerous, surgery ought to be under such circumstances. When will doctors and surgeons become total abstainers from poisons? They are legislating against other systems of healing, why not legislate them into an exemplary life?

GOOD FOR WHOM?

L. C. Hartman, a student of Spectro-Chrome Therapy, reports that a business house on North Calvert Street, Baltimore, Md., has this humorous (but tritely true) placard posted in its window:

"The doctors say that your appendix is no good to you; BUT consider how much good it is to the doctors!"

The appendix is a good little thing; it is so good that some doctors get $300 to get it out to see whether it *is* good. One of my graduates in Atlantic City last year insisted that I examine a pathological specimen his brother had. When I told that it was a perfect appendix he said with a sigh that it cost his brother about $350 to get it into the bottle of alcohol! It is this type of "destructive surgery" that I style "licensed butchery" and have been fighting against since the day I entered platform life.

ASSININE CANINE LAW IN PENNSYLVANIA

There is a law of 1915, in this Keystone State that a dog found in the possession of an alien should be killed and the owner should be fined $25. In a recent incident, this law was about to be invoked on Dick a part-Mastiff part-St. Bernard belonging to Jacob Silverman, an alien farmer, when the sympathy of President Warren G. Harding was so aroused that he appealed to Governor Sproul to use his influence to save the life of the dog. Before such influence could be used, however, prosecuting Justice Boorse had turned the animal over to the S. P. C. A. I should like to know what led to the enactment of such a monstrous and unjust law in a State where the parchment of the World's Liberty was signed. Why should America, built by and of aliens primarily and essentially, pick so much now on aliens? Why should our people put up their noses at aliens with contempt? Why should we discriminate with such acidity against aliens now that we are a built nation? Let us not forget that Action and Reaction walk hand in hand and what we shall do now will decide our future position in the World.

PULL HER FROM THE MUD

Germain Laporte, a beauteous French girl of 24, residing at 1227 Fulton Street, Brooklyn, N. Y., stole jewelry valued at $750 from Mrs. Emma J. Brown with whom she lived. When arraigned for stealing she said with tears that she was a cashier, but, lost her position owing to illness and being out of work for a month was faced with ejection from her room and was compelled either to steal or walk the streets. She chose the former to moral loss. This is but one instance of what actually happens to most girls in similar predicaments. Germain certainly deserves to be uplifted instead of being looked down upon, because a girl that values moral loss above any other form of degradation must of necessity be a highminded girl. Those placed higher in social scale should rush to the assistance of such a person and see that she be placed so as to keep away from further temptation. To say that stealing is wrong is right, but, before we judge any we had better study the causes at the root of human evils. This reminds me of the recent case of a poor negro in New York City, who stole two chickens (birds, not human) and was brought before a short-sighted unpsychological magistrate. The accused in pleading guilty said that he had no food for a number of days and to save his life was compelled to steal when hunger pressed him hard. This man was sentenced to 30 days' imprisonment. Such punishment serves no end either of justice or of social welfare. What would this magistrate have done if faced with a similar quandary? It is easy to lock up humans by locking up our heart, but, that does not convert the animal in the human into the angel in Heaven. Legislation never engenders or protects morality; it is education that elevates and it is education we need all around. We have more legislation in a day than the ancients had in an age; yet I doubt if we advanced further in our inner concepts of justice and ethics.

THE GRADUATE EQUIPMENT

Your kind attention is drawn to my latest Spectro-Chrome Therapy Equipment, which is called "Graduate." It is the most powerful apparatus of its kind and has special features, making it superior to any heretofore manufactured. It has a motor-driven exhaust for cooling the hood and slides and the slides are on revolving disks, built directly into the condenser system. The disks are graduated to indicate the Color Combinations for any specific Attuned Wave and the optical system of lenses make it very convenient and accurate for Area and Orificial Work. It is highly finished, well balanced and built with mechanical precision and will form the most elegant adjunct to the office of the earnest practitioner.

The first order, No. G-1922, for this special type of equipment came from Kate W. Baldwin, M.D., F.A.C.S., of Philadelphia, followed by those of Martha Peebles, M.D., and Anna de la Motte, M.D., of Brooklyn, N. Y.;

J. W. King, M.D., of Bradford, Pa.; Samuel R. Love, D.O., of De Land, Florida, and John A. Cohalan, D.O., of Philadelphia, Pa. Orders are also on hand from Mimaji Mitsuboshi, D.D. S. and Roy Forbush Handford, S.C.T., of New York. The Graduate Equipment is now on exhibition at the Central Office in Philadelphia, Pa., and at our New York Branch Office, at 171 Madison Avenue, Room 1505. One will also be on exhibit at the Spectro-Chrome Institute in Chicago, Ill., on my arrival there.

As all orders are filled strictly in rotation. I request you to place yours as early as possible to avoid delay. *All correspondence* should be addressed to the Central Office, from whence all contracts are executed by the President.

REVISED PRICE LIST

From September 1, 1922, the following prices for Spectro-Chrome Equipment and Supplies will apply. All former price lists and quotations are hereby cancelled. The pages refer to the Third Edition of the Compendium of Spectro-Chrome Therapy, which is now out.

WARNING TO OUR GRADUATES ROUNDING UP QUACKS

A warning note is sounded by some newspapers that certain persons calling themselves "doctors" are practising medicine without license and should be "rounded up" by the Bureau of Medical Education. My attention has been drawn repeatedly to these "warnings," by Spectro-Chrome Therapists, who ask for information on the subject. I gladly give them the following points to steer clear of trouble from meddlesome persons:

1—Do not use with your name the title of "Doctor" unless you have a degree entitling you to such distinction; the certificate of your study at the Spectro-Chrome Institute gives you the true identification of "Spectro-Chrome Therapist" and **you** ought to stick to **it**.

2—None can deny your right and privilege to use S. C. T. with your name, because your graduation at the Spectro-Chrome Institute entitles you as such. The Spectro-Chrome Institute Charter provides for this specifically.

3—Unless you are registered as a practitioner of another system of Therapy, which licenses you as such, use no other system in combination with Spectro-Chrome Therapy, but use and treat with only Spectro-Chrome Therapy.

4—Do not use terms as are used by regular physicians, surgeons, osteopaths, chiropractors, and the like, but stick to the Color Wave Predominance System as instructed by the Originator of Spectro-Chrome Therapy. Your patient has certain Color Waves increased or decreased in the system which is thereby placed out of the health equilibrium or "out of tune" and you have the means of restoring the said equilibrium or "tune" by Attuned Color Waves.

5—Do not use the word "Disease"; your patient in health is at "Ease"; when the said "Ease" is disturbed, there is "DIS-EASE." Use the word DIS-EASE.

Use the Area Chart and Spectro-Chrome Symptomatology and treat with common-sense according to the Therapeutic System as mathematically set down. Keep off from "diagnosis" as understood by the old schools of healing and follow the Spectro-Chrome methods **exclusively.**

7—Use **no** Drugs or Chemicals, **no** manipulation and **no** Surgical or other instrumental aid outside of the regular Spectro-Chrome Equipments and Accessories. A case where anything outside may be necessary may

need "constructive surgical" assistance, as I call it, and the Spectro-Chrome Institute does not authorize **you** to use such measures. **Keep Off!!!**

8—Issue no certificates of birth, death or sickness. You will rarely have occasion for them anyhow.

9—Declare to the public boldly what you **are** as follows:

Spectro-Chrome Therapy

RESTORATION OF THE HUMAN

RADIO-ACTIVE AND RADIO-EMANATIVE EQUILIBRIUM

(Treatment of Dis-Eases)

BY

ATTUNED COLOR WAVES

NO DRUGS — NO MANIPULATION — NO SURGERY

JACK JOB JONES

SPECTRO-CHROME THERAPIST

Use such standard signs as the above; they are supplied by the Spectro-Chrome Institute, or use any sign with similar wording, but refrain from any display that would mystify your identification as a Spectro-Chrome Therapist and what you stand for. Many graduates of the Spectro-Chrome Institute who have had long practice as Doctors of Medicine and Surgery, proudly place after their title of M.D., the letters S. C. T., to show their personal endorsement of this system. Hold before the public, therefore, S. C. T., always and the public will soon learn to recognize Spectro-Chrome Therapy and respect you as a follower of the Latest Revelation in the Healing Art. They will **demand** it.

10—Do not say you are using "Colored Lights" for treatment. People have been sitting in the radiance of all kinds of colored glass windows for ages and were affected neither for good nor bad perceptibly. What you are using is "Attuned Color Waves," which, in plain language, means Color Waves mathematically graduated to certain degrees giving certain wave-lengths for certain predetermined results when applied technically to certain areas of the human body, in accordance with Dinshah's system of Primary, Secondary and Tertiary Polarities. Attuned Color Waves in reality are waves of different "axial polarities" and their operation on the human system is different from ordinary light which is a beam of "full polar oscillations." Spectro-Chrome Therapy is **not** "Color Treatment" or "Chromopathy," because Chromopathy is not an accurately graded complete therapeutic system and is as much a hit-and-miss or fluke affair as Drug Therapy. Spectro-Chrome Therapy is a complete Therapeutical System based on exact Physical and Ontological principles and is the only single healing system that stands forward as a perfectly successful drugless substitute for all cases where hitherto drugs have been used with imperfect results and it does much more besides. Also, Spectro-Chrome Therapy is not "Helio-Therapy," as it uses no Sunlight, but produces all its Life Effects or Vitalizing Influences by totally artificial measures, on entirely new fundamental principles founded by laborious research.

11—Do not tell your patient you will "cure." Man "treats," Nature "cures." You are merely the agent of Nature in this respect and what was induced or produced in your patient by transgressions of Nature's Laws you attempt to remove by rational conditions being resumed. Even in this your methods are far superior to the Drugman's, because drugs have their undesirable aftereffects; Spectro-Chrome Therapy has no such sequelae.

12—Above all, I give you a final hint; be absolutely integral in your representations and resolutely hold up the banner of Spectro-Chrome Therapy. Following honestly the above-outlined rules, no legal law will be able to touch you and should it be made to touch it will only burn its own hands and subject the persecuting prosecutor to ridicule of a never - to - be - forgotten character. You, as Spectro-Chrome Therapists, are justified and entitled to your honorable profession of serving humanity and within the nearest future you will be given better recognition by the whole world than is within your dreams today. Back the system and its results will back you; back away from it and somebody will back into you. Act sensibly, for you are not "quacks." A quack is "one who pretends to a knowledge of medicine which he does not possess." Firstly, we do not want or use either "medicines" or what is known as "medical science;" secondly, we know more about the ways in which the "other fellow's" "medicines" act, and thirdly, we discard by very exact mathematical principles the junk foisted on credulous human patients as "medicine" and "medical science" during the last 6000 years. The term "quack" applies differently now since the introduction of Spectro-Chrome Therapy, and some day an unbiased public judgment will uphold the truth. Go ahead! "Quackk" has shifted its "polarity" now, *you get the results, let the public judge!* THAT will make others quack; for the present, grow on your back a rhinoceros hide topped with a tortoise shell.

ONE MORE PROOF

Alfred C. Redfield and Elizabeth M. Bright, both of the Harvard Medical School, in experimenting with seeds found that radish seeds exposed to the radiation of radium developed carbonic acid gas more rapidly. As production of this gas (CO_2) is a measure of metabolic rate, which represents the destructive and constructive sides of living matter, the experiments indicate that the rays given out by radium possess such activity for germination. Radium, in Spectro-Chrome Therapy, appears on the green wave and green in nature shows similar activities. Green systemic will always be found to produce muscle and tissue building results. An Italian commission of eminent physicians was recently appointed by the Italian Minister of the Interior to study the effects of Radio-Therapy on Malaria. They could get all they are looking for by studying Spectro-Chrome Therapy and with results that are positive.

Attuned Orange Wave

(Humor, Wit, Fun, Satire, Tickle, Irony)

WHAT'S MORE?

A Short Sketch on the Oddities of Men, by Miss Irene Hoger, S. C. T.

I think the Editor of "Spectro-Chrome" misunderstood me. The opposite of "fair" does not necessarily mean "unfair." Fair has many opposites, for instance, blonde or brunette, meaning fair or dark. In last month's article I meant the opposite of the "noun" sex, not the "adjective" fair; at any rate I promised to knock the men folks; so here goes.

What's more ridiculous than making love to a girl while driving an automobile? It is no laughing matter; it simply shows lack of common sense. It is well known that men act in this peculiar way.

What's more laughable than to see a man wearing a monocle? Although the man himself may not appear foolish, the monocle makes him look outrageously funny. Whenever you see a man with a monocle, you know what is likely to follow; a parade of mocking children, enjoying their young lives at the expense of the man. Who could blame the little merry-makers?

What's more, I read only a few days ago an article on "beauty of form." The writer made an assertion similar to this, "Women are shapeless, men's bodies are shapely." Are the women going to let the men say so? Women are fast discarding corsets, (a very wise thing to do) and the men are picking them up, as it were. I fail to understand why, if men are so "wonderfully" built, they should wear corsets! What are the men coming to next?

What's more provoking than to watch a man trying to ——— well, read the following incident and you will know.

"Well," called Cecilia from the top of the stairs, "How much longer is it going to take to darn the hole in that stocking of mine?" There was no response. The same question was repeated and again there was utter silence.

In her mad rush, she thought not of lighting the hall and stairway but down the stairs she came in a fury, for Cecilia Harroway certainly had a temper. What was the result? She had run up and down that very flight of steps all her life and she was exactly eighteen years, seven months to the day; yet at this time it seemed she judged wrongly, for there were a few steps to come. The girl walked into the air, as it were, instantly landing on the floor. "Ouch! my foot, my ankle, I'm dying," howled she.

Don't get frightened, gentle reader, it wasn't so bad as all that, for, after a short while Cecilia stood up and started to walk, limping only slightly.

But why all the excitement over a stocking? Just simply this.

The Harroways were wealthy, but it seemed at this time when Cecilia wanted to go to a party, of which she was notified only a few hours beforehand, the maid had a day off, her mother was not at home, Cecilia claimed she had to wear her sheerest steel-colored stockings to match her silvery slippers and poor brother Lloyd sat downstairs—and—just—could—not—thread—that—blessed—needle. A needle! Oho! Now the mystery is gradually being solved.

Cecilia was almost as ignorant as Lloyd when it came to domestic duties, although I doubt if she would have endeavored to use a machine needle to mend her stocking.

Yes, a machine needle, that is what Lloyd Harroway was trying to thread, and how? Instead of putting the thread through the needle, he held the thread stiff in the air like a bayonet and brought the needle to the thread. The needle was reluctant to go and each time it would hit the thread, the thread would tumble over like an anti-prohibitionist. There seemed about as much attraction between the two as there is between an Irishman and a Jew and vice versa.

You can understand, that with a sore ankle and the time much later than the opening of the party, Cecilia preferred to stay at home. You can also understand why, next day there was a little sewing circle in the Harroway home. Mrs. Harroway was giving her daughter lessons in the art of stocking darning and giving her son instruction as to the proper way in which a needle is to be threaded; also the difference between ordinary sewing needles and machine needles.

And Now Some More

What's more funny, than to see
  Strolling proudly down a lane,
A green suited, brown derbied man
  With a polished jaunty cane?

Judging from his new, white spats
  And his patent leather shoes,
In my mind, there's not a doubt,
  His brains are certainly loose.

I know if he should try to smile,
  It would surely crack his face;
What he needs is petticoats,
  Trimmed with pretty, dainty lace.

Instead he tries to make believe
  And almost fools himself,
That he has bought United States
  And reeks joyously in pelf.

He wiggle-wobbles right and left,
  Until now, you did not know
His secret—I shall now impart
  Hush—he wants to make a show.

It's noticed last, although not least
  At the place where people sip
He raised what he insists is a
  Mustache on his upper lip.

A mustache is just what **he** says,
  Please don't start to blame me now;
I have always preferred to say,
  "The silly misplaced eyebrow."

Wow! You are hard on the men, but there **is** truth in what you say. It may be news to many, but, some men do wear "corsets" and we need not be surprised if some start wearing skirts. Let France and England begin and we shall follow!

OUR ADVANCE PROSPECTUS TO CHICAGO

For a period of 99 years from August 14, 1922, we are duly authorized to transact business as specified in our charter, by the Secretary of State of the State of Illinois.

The following letter was mailed to about 9000 Doctors of Medicine (of the regular allopathic and homeopathic schools), Osteopaths, Chiropractors and the like, including even Christian Scientists. I had no idea that Chicago, Illinois, needed so many Healers to maintain its Health Equilibrium!

Let none say afterwards, "we did not know he was here or we would have done this and that." I shall be there until my course on Spectro-Chrome Therapy is completed.

August 17th, 1922.
11-24

Esteemed and Learned Progressive:

Please, give this your kind attention.

It will interest you.

You will not be troubled with commercial follow-up letters.

The Healer of human ailments has always been in demand, but the Healing Art made no definite progress until the last few years.

Yet, it has not attained therapeutical precision such as you as a Progressive desire.

Such a message of "*Therapeutical Precision*" I bring to you.

It is the message of SPECTRO-CHROME THERAPY, the latest revelation in the Healing Art, the new science of the Restoration of the Human Radio-Active and Radio-Emanative Equilibrium (Treatment of Dis-Eases) by Attuned Color Waves.

It is entirely original and replaces drugs by a complete drugless system giving results surpassing any known method in "Therapeutical Precision."

It is now in daily use by hundreds of practitioners (allopaths, homeopaths, osteopaths and others) in New York, N. Y., Philadelphia, Pa., Baltimore, Md., and elsewhere and stands on its merits only.

I am coming to Chicago, Ill., to place this system before you and deliver a course on SPECTRO-CHROME THERAPY with demonstrations, experiments and illustrations.

About the 1st of September, 1922, I shall be in your city. Meanwhile, further particulars may be had from Colonel Frank G. Bolles, 542 Monadnock Block, Chicago, Ill., or from the Central Office of the Spectro-Chrome Institute, at 2401 N. Broad Street, Philadelphia, Pa.

Yours in the Service of Humanity,

Dinshah P. Ghadiali

President,
SPECTRO-CHROME INSTITUTE.

SPECTRO-CHROME THERAPY GRADUATE EQUIPMENT

Interbuilt Revolving Slides, Motor Driven Exhaust, Precision Optical System for Area and Orificial Work.

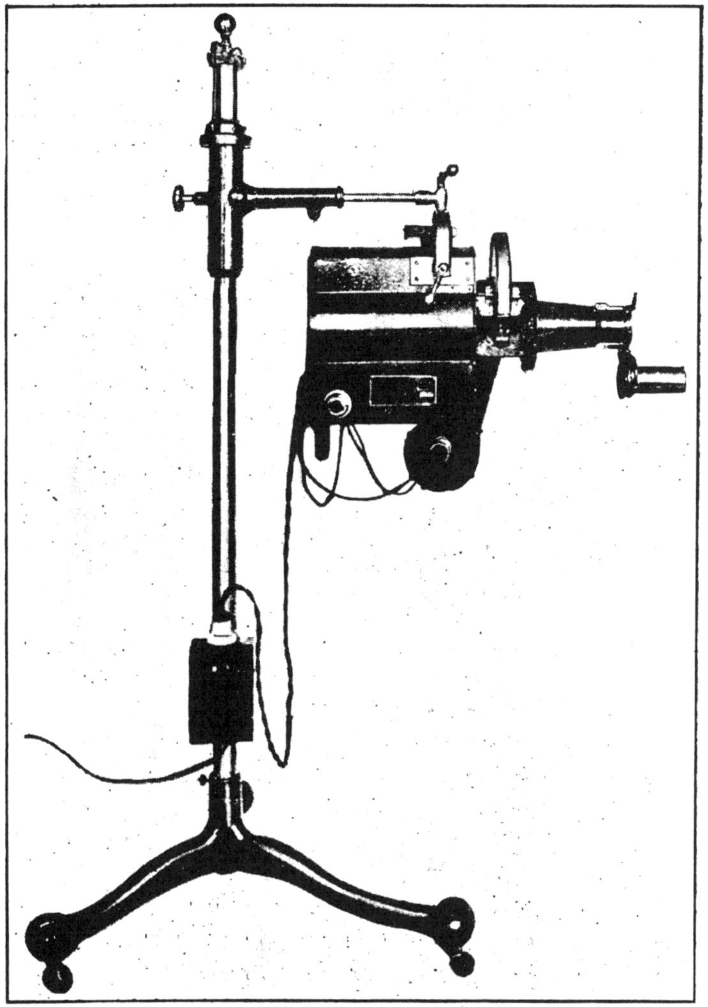

Dinshah's Latest and Most Powerful Equipment for Spectro-Chrome Therapists

PRICE COMPLETE $425.00

F. O. B. NEW YORK

* * * * * * * *

Patent Applied For

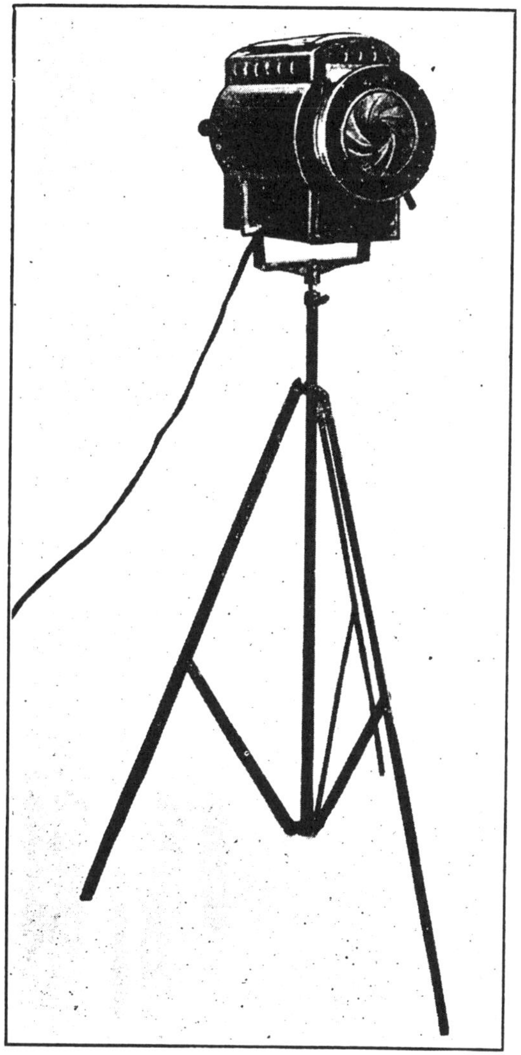

Spectro-Chrome Therapy

INTERMEDIATE EQUIPMENT

Absorbs 400 Watts. Operates on any electrical circuit of 110-120 volts. Direct current or alternating current.

Complete equipment, mounted on sliding tripod floor stand, steel hood, 400 Watt frosted bulb, powerful adjustable reflector, push switch, 15 feet insulated cord, swivel connector, Iris diaphragm, condenser, full set Dinshah attuned color slides and *Compendium of Spectro-Chrome Therapy* by the Originator.

PRICE $90.00

F. O. B. NEW YORK

INTERMEDIATE EQUIPMENT

Attuned Lemon Wave

(Medicine, Surgery, Science, Osteopathy, Chiropractic, Philosophy)

SYPHILIS, THE KING OF DESTROYERS

* * * * *

By J. W. King, M.D., Director,

Physico-Clinical Laboratory,

Bradford, Penna.

August 15, 1922.

My Esteemed Friend, Dr. Dinshah:

Since the day I spent with you in Philadelphia, May 29th, you have been with me, a constant companion of my mind and heart. I owe you a lifelong devotion to the cause you champion, as I know I shall soon give you reports on cured patients from Spectro-Chrome Therapy as furnished by my "Graduate and Practitioners' Equipments," two equipments I greatly appreciate. Having graduated from Jefferson Medical College, Philadelphia, in 1889, thirty years of which have been spent in post-graduate work, year after year, a month or more, in special lines in order properly to equip myself and be able to give the best medical service to patients frequenting my *Physico-Clinical Laboratory,* an Institution devoted to the diagnosis and treatment of disease, the past twenty-six years in this city.

Early in my medical career I learned the truth of the statement made to the classes of Jefferson Medical College by Roberts Bartholow, namely, "After you have practised some twenty years, former patients whom you had treated many years before for chancroid, will consult you under the guise of some very queer symptoms and with many chronic diseases. You may not recognize these as syphilitic, but somehow

you hit the mark in your giving these patients anti-syphilitic treatment." Later, Sir William Osler gave us this:

"Know syphilis in all its manifestations and relations and all other things clinical will be added unto you."

During the last war, the American Medical Board adopted the French nosology of syphilis (as had been pointed out over one hundred years ago by Hahnemann, who spelled it "Shanker") that 98 per cent of chancroid, formerly regarded as a mere local affection only, were chancre, an infectious constitutional disease, namely, syphilis. Another writer on this subject avers that "without syphilis there would be no disease," or as one expressed it, "Syphilis is the Father of Disease." If one views this affection in a broad manner, we may well accept the dictum, namely, that syphilis is the "King of Destroyers," for the merest tyro in medicine will soon see *that* much.

The origin of syphilis is quite unknown among medical men. It was not *brought* to Europe by the crew of Columbus and other American Explorers, as some asserted, as a contribution from the American Indians; neither did these *tubercularize* the Indian, as alleged, and to equalize matters by the exchange. We find it long before Confucius's time. The probable date is—at the creation of the world. It always was and it will always be that way. Nothing arises spontaneously nor will we ever stamp out the evil because it was and is a soil laden material, *until* the soil is made non-syphilitic—what we eat we are.

Our friend, Dr. J. W. King, is certainly very progressive. He visited the Experimental Laboratory of the Spectro-Chrome Institute before starting on his recent continental tour, and I am glad I met him. It is such men and men of such broad vision and high calibre among the medical fraternity who will eventually bring about therapeutical emancipation by fighting through the prejudices of the old-timers, whose ideas procreated 6,000 years ago, are still running in the ruts wheeled by Methuselah. It is the progressive man who will bring about salvation from within the ranks. The healing art is undergoing a terrible upheaval from the drugless angle and Spectro-Chrome Therapy challenges the moth-eaten conceptions of the old professions at every step. As to how in these days of hustling civilization infectious troubles may travel from point to point and how "Syphilis" may be contracted without venereal contact, I present the following from *Safe Milk:*

CORRESPONDENCE GERMS

"People who have a habit of biting the tip of the penholder are quite likely to become infected from what physicians rather aptly call 'correspondence germs.'

"That the penholder, such as most persons frequently use in hotels, post-offices, banks and other public writing rooms, may be as deadly a carrier of disease germs as the roller towel and public drinking cup is the belief of medical authorities today.

"When one of these public penholders is used it will be noticed, if careful examination is made, that the wooden handle is covered with little dents. These dents are the imprints of the teeth of persons who used the pen before; and as the mouth is the favorite port of entry for disease, each dent is usually full of microbes.

"To expose one's self to sickness and death, therefore, all one need do is to bite on the penholder. Every time this is done one takes into the mouth the germs which scores of other persons have left by similar bites."

But I go further—much further yet. How about the following "habits" by which one may infect another?

1. Licking postage stamps and envelopes and mailing our dis-eases with our love to our sweethearts at a distance.

2. Biting nails and touching cooking utensils subsequently.

3. Performing certain "Natural Functions" and not washing the hands.

4. Being a bookkeeper and licking fingers to turn pages in the ledgers; the employer may receive a "gift" from his bookkeeper.

5. Being an employer and doing the same thing, thus giving a "present" to the bookkeeper or stenographer.

6. Licking fingers for separating dollar bills.

7. Sitting promptly on a recently vacated seat and receiving radio-emanations of a dis-eased character.

8. The present system of plumbing by which one is compelled to sit on a wooden (porous) seat and infection to be carried to the innocent.

9. Promiscuous moving around of women during their monthly periods when the radio-emanative polarity of the body is befouled and vitiated.

10. Pernicious custom of kissing on the lips and sucking the sputum of another into the organism.

11. Funeral and burial customs which spread infections broadcast by indiscriminate contact of the dead with the living.

12. Sleeping in the same bed with one or more persons.

13. Drinking from the same cups in the home and the restaurant.

I can expatiate more, but for this time the above is enough to answer the contention set up by some, "What are you talking? I have syphilis? I? Why, my dear man, I never knew a woman in my life!" Syphilis and similar disorders may be brought into the body through various channels and it is here that the value of Spectro-Chrome Therapy is unsurpassed. The originator found that the same mechanism that removes one infection, removes *all* others, because it is operated through the same Color Polarity. The Attuned Green that energizes the cleansing functions systematically, clears *all* such dis-ease producing activities, rendering useless the necessity for differentiating and diagnosing different strains.

Attuned Green Wave

(Spectro-Chrome Therapy, Spectro-Chrome Therapists)

CARBUNCLES

By Kate W. Baldwin, M.D.,
Fellow American College of Surgeons
Senior Surgeon, Woman's Hospital, Philadelphia, Pa.

Fortunately, carbuncles are not the most frequent of infections. There are many boils or furuncles, but carbuncles the surgeon does not see every day.

The following is the usual description given in works on surgery:

"Carbuncles are most commonly upon the posterior surface of the neck or upon the back, usually where the skin is not covered by clothing. A circumscribed area of skin becomes indurated, red, later purple in appearance, then a number of small perforations occur upon the skin. The process is through the subcutaneous fat, a larger and larger portion of the skin surrounding the original center attaining a honeycombed appearance. The infected area is exceedingly painful and the patient has the general appearance of one who is severely ill. Diabetic tissues furnish an especially excellent culture medium.

"The most satisfactory surgical treatment consisting in a crucial incision down to the base of the infected tissue, then the removal of the infected tissue and the cavity is then sponged with a 95 per cent. of carbolic acid and tamponed with gauze saturated with strong alcohol. In place of this five to ten drops of a 50 per cent. solution of a 95 per cent. carbolic acid in glycerine is injected into the base of the indurated area about two cm. apart around the entire circumference of the carbuncle, the surrounding skin having been protected with vaseline. The surface is then covered with a thick layer of gauze saturated with strong alcohol. One injection is usually sufficient, but the carbolic acid injection may be repeated if necessary. There is usually considerable loss of tissue and great disfigurement. Skin grafting may be necessary."

Recently a jet of hot air has been used in the treatment of carbuncles with reported satisfaction. With it the pain is intense and an anesthetic is frequently required.

Within the past eighteen months I had several very interesting carbuncle cases which are worth reporting:

Mr. D., age 37 years, came to me November 29, 1921, with a carbuncle on the back of his neck. He had been treating it for a week or more with the usual home remedies. He was unable to sleep or attend to his usual duties of an interior decorator. Nearly all of the back of the neck and up to the occipital protuberance was involved in the indurated area, but the slough was confined to one side of the median line. There were twenty-four or more openings in the skin. Aside from the ordinary surgical dressings no treatment was given but the Spectro-Chrome Attuned Colors, which were at first applied twice in twenty-four hours. The Spectro-Chrome controlled the infection and circumscribed the sloughing area. The skin broke down between three of the openings and the Spectro-Chrome softened and separated the slough so that it was easily removed, leaving a large cavity, which was soon obliterated by the use of the astringent Indigo Wave. It was not more than three days when the cavity was entirely obliterated and the skin soon became normal. The last time I saw Mr. D. the skin was so perfect it was difficult to believe one's own eyes. In just twelve days from the first treatment, Mr. D. papered a room, the ceiling as well as the side walls. Try the position necessary for such work and see how tired the neck gets!

We used Orange, Yellow and Lemon Spectro-Chrome Color Waves until the slough was removed, then Indigo to close the cavity and later Turquoise for the skin.

Mrs. K., age 72 years, very feeble. Her husband recently died and she had little interest in life. The son told me that when they learned that the mother had a carbuncle they had no idea she could recover.

The location and area involved was almost the same as in the case of Mr. D., only that the slough was on the other side of the median line.

Mrs. K. came first to the clinic of the Woman's Hospital, but soon was too feeble to go back and forth and was taken into the house on July 5, 1921. On the eleventh, she was able to return to her home and report to the clinic. I saw her last August 23, 1921; this was only the third time she had been to the clinic after leaving the hospital. The only evidence of the carbuncle was a thin white line less than one inch long. The treatment was the same as in the case of Mr. D., with the exception that she was put on a non-meat diet. Mr. D. was eating at restaurants and was obliged to take what he could get.

Mr. S., a mechanical dentist, had a perfectly good carbuncle on the back of his neck. Dr. Delia Riggs treated Mr. S. with Spectro-Chrome and the ordinary surgical dressings between treatments. He was well in an unusually short time and only time lost from his regular duties was the time necessary for the treatments. I saw this case only in consultation.

The next case is that of a twelve year-old girl with a small carbuncle on the lower part of the sacrum. The slough was ready to be removed; this was done and she was given Spectro-Chrome. She came back for a second treatment and was to have another, but her mother sent word that it was not necessary and that the child had gone to a picnic.

The three following cases may well be reported together. Two little negro boys (brothers) and Tony, a little Italian. The scalp was the seat of infection in all these three cases. Tony was the worst; his scalp was literally covered with secretions. The necrotic points were so numerous that the entire scalp looked like one gigantic carbuncle or a collection of smaller ones. The others were much the same, only not as extensive. Tony was in the hospital for several weeks and later came a few times to the clinic. The two brothers lived quite near, so went back and forth to the clinic. Only two days ago I saw the mother and she reported that the boys were all right, the scalps nice and clean. Spectro-Chrome was the only treatment, except for different wet dressings used to soften the secretions. I have had a good many years' work in a surgical clinic, but never did I see heads such as these three.

The next two cases are probably of greater interest, as one was on the lip and the other on the neck of a diabetic. The diabetic case died, not so much from the carbuncle as from the general condition. The spleen was very much enlarged and very painful. The slightest pressure gave pain, in fact, there was pain all the time and he could move with great difficulty. He also had a painful area on the inside of the right thigh. This, however, did improve.

The carbuncle on the lip was in a young woman seemingly in good general condition. About two years ago she had an extensive infection on the chin, which was some time getting well. Miss N. came to me July 22, 1922, with the left side of the face badly swollen, the eye partly closed. It was evident, however, that the upper lip was the center of the infection. Between the Spectro-Chrome treatments, compresses of Epsom salt were used and the swelling was promptly reduced and confined to the lip. We did not incise this at all, the Spectro-Chrome Scarlet, Yellow and Lemon soon softened the necrotic tissue and it was discharged through the twelve openings in the skin. Later, Indigo was used to obliterate the cavity. I saw Miss N. yesterday, just three weeks from her first visit. In three months, or sooner, there will be no disfigurement: really, there is very little at the present time. A film of con-

tractil collodion is the only dressing at the present time. It is worth noting that contractil collodion very much assists in preventing a scar.

Mr. B., age 39 years, was referred to me July 14th by his regular physician, who was leaving the city on a vacation.

Mr. B. gave the following history: Four members of his mother's family died of diabetes, his mother one of the four. Two had carbuncles, the mother one of the two, hers being on the lower part of the back. Mr. B. was known to have had excessive sugar in urine for twenty years—at times as high as 6 per cent. Fifteen years ago had many boils on different parts of the body. At that time he had 6 per cent. of sugar. Three months ago had a "boil" on the right arm, which opened in six or seven places; soon after he had one on the end of the nose.

In July there was 4 per cent. of sugar by weight, no casts, no albumin. He had been getting pain in the neck for about eight weeks; when he came to me July 14th he held the head rigid and turned to the left. The indurated area reached from the left ear to beyond the median line of the neck and eventually extended nearly to the right ear. It was not until Monday, July 23, that one could say positively it was a carbuncle. Between the 14th and the 23d I saw him only every second day, as he lived some distance from the city and traveling gave much discomfort. He came to the city on Tuesday, the 24th, and from that time I saw him three times a day. He was given Scarlet, Yellow and Lemon Spectro-Chrome and later Indigo. The slough softened and discharged through the many openings in the skin, no incision was necessary and in a diabetic very undesirable. There was a great amount of discharge, the dressing often having to be changed two or three times between treatments. By August 1 the slough was all gone and the secretion was very little. He could move the head freely in all directions and on the 2d told me that the carbuncle did not hurt him at all and that his only trouble was the shortness of breath and the intense pain in the left side. This pain had been increasing for several days and it was with difficulty that he could get up and down. At no time was the temperature higher than 101° F. The heart was regular and of good quality until August 1. From that time on he grew rapidly worse, became unconscious about 2 P. M. on the 3d and died at 10:30 A. M., August 4, without regaining consciousness.

The urine on August 3 showed less sugar than on July 19. On the 19th it was 4 per cent. by weight, on the 3d only 2.86 per cent. On the 19th there was no albumin, on the 3d only a trace. No casts either time.

In the interest of the next case, I ask was there any way to have saved the life of this man? When the carbuncle did so well I had visions of recovery.

These cases have all been of much interest to me. There was less pain, no incisions, quicker recovery with, I may say, no scars. I never had as good results with any other line of treatment. If there is a better one I hope to find it; until I find a better one I shall use Spectro-Chrome for all carbuncles.

I have pleasure in responding to the call of Dr. Kate W. Baldwin for elucidation regarding carbuncles. A carbuncle is in all respects an "inverted furuncle." Boils, abscesses, furuncles, carbuncles are all clean-cut evidences of lowered vitality in the left section of the human system and indicate constitutional taint. The obvious trouble will always be found in the deficient operation of Area Number 6 (the spleen), which is not capable of full Radio-Activity, hence throwing the entire mechanism out of balance. When the spleen, through any cause disfunctions, the Portal Circulation is disturbed and Area Number 7 (the liver) suffers in compensation. As soon as this happens, metabolic troubles follow and Diabetes Mellitus is one of them. The liver suffering, now throws the burden on the Area Number 4 (the heart), because the blood having unbalanced Color Waves its Radio-Activity is disordered, thus throwing out of balance the Radio-Emanative functions of all the organs supplied by it. The blood-purity-testing mechanism of the Pituitary therefore comes in for its punishment and its Green Wave producing Radio-Emanativity is over-worked. As long as the Green lasts the blood keeps on compensating the exhaustion, but the reckoning comes when the failure of the adequate supply of the Green allows the Red (right) section of the system to predominate and inflammatory symptoms of redness, swelling, heat and pain commence to appear in those areas that are weakened. It is truly the kind processes of Nature that are endeavoring now to restore health equilibrium by throwing the garbage out of the body, and the boils, abscesses, furuncles and carbuncles differ only in degree, but are of the same family and need practically the same routine of treatment.

The first essentiality is to attend to Area 6. If tender, red, hot, or swollen, Violet on 6 is indicated locally. Green Systemic first to cleanse, followed by Lemon Systemic to alter the conditions. On the spot where Nature concentrates her volcanic crater (boil, abscess, furuncle or carbuncle), use Orange for quickening suppuration; use even Scarlet if the blood pressure is sluggish. When the secretions start flowing freely and the necrosed detritus is cleared, use Indigo Wave to close the cavity and avoid an ugly scar. The Areas Numbers 4 and 18 should receive Magenta, if so indicated. In short, treat carbuncle on the same principle as any severe inflammation may be treated. Treat it as a semaphor of what is going on inside—depleted left side vitality.

In the last case specified by Surgeon Baldwin, she appears to have done the best for the patient, death being traceable to the condition of the patient being so exhausted already, that when the crisis came the vital organs could not react. However, the interesting cases reported by her deserve our consideration and her support of Spectro-Chrome our cordial appreciation. Carbuncles will yield to Spectro-Chrome as well as other similar troubles and no knife will be necessary if the treatment laid down be duly followed.

WHAT A PROMINENT HOMEOPATH THINKS

New York, N. Y.

Dear Doctor:

You wish me to tell you what I think of "color light" treatment as perfected by you under the name of Spectro-Chrome Therapy? Let me be brief; for no matter how much I may choose to say, I can not say more than this: That neither money nor gratitude can pay for what you have given into the hands of the intelligent and conscientious physicians for the *cure* of the ills of humanity. Spectro-Chrome Therapy is not a "cure-all," and you do not claim it as such, but it is one of the most potent means of combating disease and bringing the ailing human body back to its healthy state without substituting one dis-ease for another

as it is mostly done under the orthodox method of treatment.

I have used your method and have had excellent results in very serious cases. I am using it now and am having success, and I shall continue to use with further good results, I am sure.

I have recently sent to you, of my own accord, reports on several cases, among them two cases of goitre and two cases of chronic eczema, all of which have shown remarkable improvement under the treatment of Spectro-Chrome. I can add now one more interesting case, that of stubborn jaundice of two years' standing. The lady, a widow, has had so far eight treatments of Spectro-Chrome only, and the jaundice has diminished by 50 per cent.

Dear Doctor, I am aware of the immense struggle you are having to break through the hard shell of vanity and prejudice of the M.D., but I am sure you will succeed and that in the end the radio-active and radio-emanative forces will prevail over the course, sledge-hammer methods of orthodox therapeutics.

With all good wishes,

Sincerely yours,

I. Sossnitz, M.D.

Isaac Sossnitz is a Doctor of Medicine in active practice and is very well known in New York circles. He graduated as Spectro-Chrome Therapist among the pioneers of New York and his support of Spectro-Chrome Therapy deserves our thanks.

DINSHAH'S DISCOURSES

My dear Dr. Dinshah:

I feel that I ought to write you and tell you of the inestimable amount of good your lectures have done me. I am a better man for having come in contact with you.

I wish that the world at large were broad enough to appreciate what you have been giving, for then we would have a better world.

Wishing you and your work undying success, I am,

Yours very respectfully,

James M. Tarsy, D.C., Ph.C.

A CHRONIC SUFFERER RELIEVED

By Edmund J. Hogan, D.C., S.C.T.

Mrs. A.—The history presented was an intestinal disorder covering a period of over 20 years. Chronic constipation, flatulence, eructations and palpitation of the heart. Climbing stairs necessitated a period of rest. Could not comb hair; difficult breathing followed lifting of arms. Daughter had been home taking care of house for last two months; had secured a leave of absence from position.

When I took the case, Mrs. A. had just finished an attack while under medical or drug treatment. A hypodermic had been followed by severe sinking spells. She was nervous, feeble in gait and general physical condition. One month thereafter she was stronger. Could climb stairs and had no difficulty in combing her own hair. Notwithstanding the fact that it had been necessary to take drugs for constipation for 20 years, Spectro-Chrome treatment moved the bowels every day and they are now moving of their own volition.

Mrs. A.'s nervousness has disappeared and she has enjoyed restful and refreshing sleep since the Spectro-Chrome treatment; previous to this, her rest was broken and fitful and the morning found her tired out. A pain extending from the gluteal region to the patella which resisted medication for eight months also has disappeared under the Spectro-Chrome treatment.

There is no doubt in my mind as to the efficacy of Spectro-Chrome treatment; the world owes to Dr. Dinshah P. Ghadiali a debt of gratitude for his contribution to constructive healing.

SUNLIGHT SETS RIGHT GRAVE CASES

By JOHN A. COHALAN, D.O., S.C.T.

An article entitled: "Simple Sunlight Works Wonders in Surgical Tuberculosis Cases," appeared in the magazine section of the *Philadelphia Record, Sunday,* July 30, 1922, and should interest Spectro-Chrome Therapists from the fact that no drugs, no manipulation and no surgery or plaster casts are used, and special emphasis is put on the diet. Contrary to the usual stuffing treatment these cases receive no alcohol, meat or cod liver oil (a few steps on the road to rational diet).

In the words of the medical expert who wrote the article, "marvelous results were obtained in thousands of cases."

I give herein a few extracts and any student of Spectro-Chrome Therapy can draw his conclusions, as to how results were obtained and what wonderful results he may expect to obtain from the same form of treatment carried to a far greater degree of perfection. The surgeon says among other things "Our present methods, essentially surgical, are pitifully inadequate in every respect. For the most part, they consist in amputations, excisions of joints, and so forth. . . . For tuberculosis within the cranium, the abdomen and for Potts' disease, where the bacillus attacks the back bone, the knife is impotent, and in effect we use nothing else. . . . Nearly all the cases treated are surgical tuberculosis. In general, they are grave advanced cases, which have been treated—or as we see now maltreated—elsewhere for years." . . . "Dr. Rollier puts his patients into the light and it does the rest; they get well before your eyes. How the Sun acts he does not know, nor does anyone else." (A Spectro-Chrome Therapist *does* know!) . . .

"The skin, practically the whole surface of which receives the light baths, pigments to the most surprising and amusing degree and there is some parallelism between the pigmentation and the recovery." The skin automatically interposes the proper attuned color slide and thus admits the proper color to restore harmony.) . . . "For myself I neither read nor heard anything that seems to me to explain the results obtained." (Send him some Spectro-Chrome Therapy Literature.)

. . . "And here and now let it be stated that whosoever will visit Leysin to see the work, which is certainly the most beautiful thing I ever saw in my life, will learn, even in the first few minutes, that the knife and the plaster cast and all the rest of our customary methods are *barbaric, stupid, obsolete, cruel* and *useless* compared with the natural magic of the Sun. You look at a radiogram of a hip joint, disorganized, disintegrated by the disease and by the secondary infections which followed upon surgical treatment; then, see the radiogram of a joint practically re-created by the natural resources of the body—usually a young body—which the Sun has evoked; and the smiling, brown patient, moving his new joint easily and painlessly is before you. . . . The patients receive no alcohol, practically no meat, and no cod liver oil, which I learned to my astonishment."

Then to show how *organized medicine* looks upon this treatment because it cannot be commercialized (no serum, vaccine or drug to sell), the learned doctor continues, "Practically nothing is known of this glorious work elsewhere. . . . compared with this, Finsen's use of light to cure lupus, for which he received the Nobel Prize, is a trifle not worth mentioning. . . . ."

Then he continues to impress upon the readers the facts we learned in our first Spectro-Chrome pamphlet on Fundamental Principles by Dinshah, "The Sun is the source of all life and energy conveyed to the world by waves of light."

Our friend, Dr. John A. Cohalan, has had such magical experience with Spectro-Chrome Therapy that he virtually "swears by it." Thank you, doctor, for your compliments to Spectro-Chrome, but, if the "cure of lupus" gained for Finsen the Nobel Prize, certainly you must be expecting the Originator of the complete "Color Wave Pharmacopœia" to be given even bigger "plums" by the World. Well, he is patient; it will all come some day: but all the prizes that may come cannot offset the blessings showered on his system by those among the suffering humanity who have been saved from the surgical knife and restored to health when there was nothing else in sight but despair.

A LAYMAN'S PERSONAL EXPERIENCE

By Orlo H. Frost, Hillsdale, N. J.

I am sure you will be interested to know the results I secured by the use of Spectro-Chrome Therapy.

My mother, who is 66 years of age, had been suffering for at least five years from some sort of intestinal disorder, the exact nature of which apparently was not known. One doctor diagnosed it as cancer of the stomach, another as tuberculosis of the intestines, a third, as liver trouble. One of the symptoms of this dis-ease was an excessive diarrhea (she sometimes had as many as 15 or 20 motions during 24 hours). During the last of July and the first part of August, she was particularly bad and in bed most of the time .

I arrived there the evening of August 7, and gave her a treatment that night, using Turquoise. I continued this treatment twice a day during Sunday, Monday and Tuesday. Wednesday morning she asked me what was the treatment for constipation, as she had no bowel movement since I put the first color wave on her Saturday night. I immediately applied yellow and in 20 minutes she was relieved.

A further accompaniment of her disorder was a sore spot in the pit of her stomach, so painful and tender that she could hardly bear her clothes over it and should any one place their hand there, her breath would almost be shut off.

I gave her a treatment for that, using Lemon, and almost at once the pain disappeared. I continued using the Lemon wave on her during two weeks and when I left she was feeling well, out in the flower garden, working around the house, visiting the neighbors, and the Sunday before I left I took her on a long automobile trip (over 100 miles) and she withstood the fatigue of the journey much better than the rest of the party.

Her entire ailment disappeared and she felt better than she had in years.

My aunt, 70 years of age, suffered from chronic bronchitis for 40 years. She also periodically suffered from poison oak, which she contracted in California more than 20 years ago. I treated her for both these ailments, using Lemon Wave, and in a few days the poison oak broke out in blotches on her body and then disappeared. Likewise, the chronic bronchitis also cleared and disappeared. Heretofore, when retiring she would have fits of coughing, lasting from 10 to 15 minutes; likewise in the morning on arising. This has all disappeared and now she is entirely free from both.

The writer of the above is interested as a layman in doing good to mankind and is not a professional Spectro-Chrome Therapist. Yet the satisfactory results obtained by him could be obtained by any person using Spectro-Chrome Therapy as per instructions laid down. Orlo Frost achieved the results with a 100 Watt equipment, which speaks well for his intelligent application.

Attuned Purple Wave

(Royalty, Aristocracy, Diplomacy, Politics, Statesmanship)

CROWNING OF KING WILLIAM GOAT

A Barnyard Session of Words

By Miss Kashmira Dinshah, S.C.T

"Good morning to all of you," said the goat, "it will be a fine day today."

"And so it will," all the other barnyard creatures agreed.

"Yes," the goat went on, feeling pleased that they were paying attention to what he said, "it will be a very fine day. The weather has been quite unpleasant recently, but not so very unpleasant, I think, on the whole."

"What an absurdity!" exclaimed the peacock. "How ridiculous!" crowed a rooster who stood up on the top of an old tree stump, flapping his wings and cock-a-doodle-doing with all his might at the Sun, which was just rising over the distant hillcrests. He was a fine-looking rooster, snowy white all over except for a collar of red.

"What is an absurdity? What is ridiculous?" asked the goat.

"Why, your speech, the last three words of it!" crowed Rooster Silver-Red, "cock-a-doodle-do! I never heard anything quite so foolish in all my life. Mr. Billy Goat, how you do talk!"

"What is ridiculous and foolish about my speech?" demanded the goat, looking angry, for he did not like to be made fun of.

"Oh, Billy Goat, your speech was so senseless. Are you losing your mind, Billy Goat?" asked the rooster. Now the goat was really angry! He liked to be called William Goat, and it always provoked him to be addressed as Billy Goat—it was so common! So when the rooster called him that, the goat started after him, but Rooster Silver-Red just flew on the roof of the chicken coop and laughed in the way that only a rooster can laugh when William Goat, carried over the stump by his own momentum, struck his head on the ground and broke a piece off one of his horns.

"Ha, ha! How amusing you are, silly Billy Goat!" screeched Mr. Peacock, spreading his splendid tail and strutting about, enjoying the fun.

Up jumped the goat and made for Mr. Peacock, who did not escape so easily as Rooster Silver-Red, being only able to run. So he was chased several times around the whole barnyard before William Goat tripped on a log and fell, thus giving Mr. Peacock time to take refuge on the topmost bar of the fence. Here he was safe enough, for the goat could not reach him, and

then he began to shriek in glee and finally went so far as to try to sing a song with his horrible, screechy voice. This piece of impudence and disregard for the other creatures' feelings set all against him, and they sided with the goat. Peace was restored once more, after a while.

Presently the canary, whose cage was hung on a limb of the stately old crabapple tree so that he could enjoy the benefit of the fresh Spring air along with the other creatures, spoke, "What was it you said that made such a stir, Sir William Goat?"

Ah, that certainly pleased the goat, for nothing delighted him more than to be called "Sir." "Yes," the canary continued, "I wonder why you do not reply, Sir William Goat. What was it you said? Do you know you broke a piece off your right horn? Now you will be able to show it to your friends, as a battle scar. I really think you are descended from King William the Conqueror's time, Sir William Goat." The canary had said that often, and now that it was repeated, Sir William Goat actually believed, not that he was descended from, but that he was really King William the Conqueror himself.

It flattered him, and he announced that henceforth the others of the barnyard should call him by that title and nothing else. Of course they agreed, and said that such a noble title befitted only a valiant and battle-scarred warrior such as William Goat. But there were two who did not agree. One was Rooster Silver-Red; the other was Mr. Peacock. Rooster Silver-Red said that if the goat was to be given a title, he must have one, too, for was he not the largest rooster of all, and the most beautifully feathered? And could he not crow louder and better than the others? Therefore, he must be crowned King Chanticleer on the spot. But they made him only a "Count."

"Cock-a-doodle-do!" he crowed, "it is so delightful to be called Count Chanticleer. I have finished with the plebeian appellation rooster.'"

"What is plebeian appellation?" asked Miss Red Hen.

"Plebeian means belonging to the common people ———"

"There are no people here," interrupted the hen.

"Oh, all right," said Count Chanticleer, "it means, belonging to the common roosters, and appellations means a name by which a thing is called or known."

"But," cackled Miss Red Hen, "we are not things; we are hens and roosters."

"Oh, I never saw anybody quite as stupid as you are, Miss Red Hen," exclaimed Count Chanticleer, "it is no use explaining anything to you, for you are so crassly ignorant." Miss Red Hen was just about to ask what crassly ignorant meant, but she decided not to, for it might make her seem more foolish, and only cackled to herself of this annoying creature who would use such exasperatingly long words.

Just then the peacock, still perched on the fence bar, shrilled out, "I don't care if the rest call you King William, but I won't. Ha ha! Count Chanticleer and King William! I'll side with the rooster, for I don't want to be seen with any one who makes such crazy speeches as that silly Billy Goat!" This time King William Goat did not lose his temper as quickly as before because it was not becoming to a king to let fall his dignity every few minutes at the slightest provocation. So he said quietly, "What did I say that was crazy, Mr. Peacock?" The bird almost fell off his perch—he was so surprised to see the goat so calm. Mr. Peacock had thought that King William Goat would become filled with wrath, and he could laugh and make merry. But when he found that King William was all composed and peaceful, Mr. Peacock decided that he might as well be friendly now.

"Well," he began, "when you said 'on the whole,' it made me laugh because the weather is all around us; it is not on a hole and nothing else ever is, for that matter. What made you say it? Don't you know that nothing can stand on a hole? Why, it would fall in. It would not be a hole if it could support anything; it would be something solid."

"Yes, that's true," put in Miss Red Hen, "it is really queer how King William made that error, Mr. Peacock, isn't it? Very queer, I must say."

"Oh, but that is not what I meant," said King William Goat, " I meant that the weather had not been quite so unpleasant, on the whole—"

"There! Again you said it!" cried all the others at once, "what is on the hole? We don't see any hole here, and even if there were nothing would be on it because it would fall in. And we wonder what the weather has to do with holes. What is on the hole, King William? Do tell us!"

Before King William Goat could answer, Mr. Peacock screeched out, "There! Didn't I tell you he was crazy? He's again talking about the weather being on a hole! Who ever heard of such a preposterous thing before? He's the biggest fool I ever saw, and the others may call him King William, if they want to, but I won't! I'll call him Mr. Billy Goat—no, just Billy Goat will do!"

When he heard the peacock saying that, King William Goat became terribly enraged. He raised himself on his hind legs lowering his head, and dropping down on all fours, charged.

Down went the peacock from his perch on the fence bar, shaken off by the force of the goat's onset, and away he scurried under the chickencoop for shelter. Furiously William rushed on, knocking down everything that came in his way, and finally stood under the crabapple tree. Then all the other barnyard creatures united against the goat. Count Chanticleer was quickly elected Commander-in-Chief of the Feathered Force; Mr. Grey Goose came next in authority as General. These two being chosen as leaders because there were the larger birds of all in the barnyard.

Then they all made a concerted rush toward the goat, and almost nothing was seen for a few minutes but a flying mass of hair, feathers and dust. Quackings and cacklings and crowings resounded on all sides; finally, when the din of the battle became so great as to be practically deafening, one by one the combatants forsook the field of fight and hastily took shelter in different places—King William Goat *was* the Conqueror! He had proven to them his prowess, if not wisdom.

When quiet was again restored, the canary spoke, "Mr. Peacock ought to be corrected for starting the row. He always is the cause of our disputes." The others agreed, and Mr. Peacock was sentenced to banishment from their society for the space of twenty-four hours. "Now, King William," said the canary, after the sentence was passed, "will you explain to us what you mean by the weather being 'on the hole?' "

"Yes," replied the goat, "I was just explaining when Mr. Peacock butted in like the ill-mannered bird he is." And then he told them what he meant by "on the whole," spelling out w-h-o-l-e.

"Well," remarked the canary, when King William Goat had finished his explanation, "if Mr. Peacock had not been so ready with his sharp tongue, we would not have had all this trouble." The other creatures nodded assent to what he said, and the remainder of the day passed in happy tranquillity. Then they adjourned the session.

SPECTRO-CHROME

JANUARY 1923
RED 1923

VOLUME 2
NUMBER 1

Devoted to: Spectro-Chrome Therapy — Dedicated to: Service of Humanity

CONTENTS — Page

Spectro-Chrome Institute; Charter; Instruction; Periodical; Equipments ....Inside Front Cover

Spectro-Chrome Area Chart—Therapeutic System—Dinshah's Platform.... 169

The World Through the Eyes of Dinshah—
Africans Against Vaccination—Custom Made King—Another Horror—Food or Junk?—Spectro-Chrome Slides, Solutions and Some Storms; Let Us Clear the Air for 1923....170-174

How Worry Affects the Brain (by Carl F. W. Walter, M.D.).... 174

The Voice of the Deep (by Ira L. Kepperling, M.D.).... 174

Are Medical Men and Women "Doctors"? (by Emil Gleitsman, B.S., M.D., S.C.T.).... 176

Clinical Side of Spectro-Chrome (by Kate W. Baldwin, M.D., F.A.C.S.).... 179

Death of a Noted Spectro-Chrome Therapist, L. Grant Baldwin, M.D., F.A.M.A., F.A.C.S..... 182

2,000 Miles by Automobile; Jottings by the Way; Horrors of Armour's Meat-Packing Plant; Most Disgraceful Blot on Human Civilization (by Miss Kashmira Dinshah, S.C.T.... 183

Progressive Physicians Endorse Spectro-Chrome Therapy.... 192

Price List ....Inside Back Cover

Registration Notice of Trade-Marks—Notice to Purchasers and Patients—To Our Contributors—Advertisement Rates....Outside Back Cover

In course of life you reach a time,
When Doctor, Lawyer, Priest alike,
Does fail to give the sought relief,
To troubled mind or burdened heart—
Then come to me;
I help you to help yourself.
SPECTRO-CHROME.

Editor
Dinshah P. Ghadiali

Subscription:—January to December 1923....$3.00
All Subscriptions Taken For Full Year Only And Expire With The December Issue
Any Single Copy 25 Cents. Back Numbers Available From June 1922.

Spectro-Chrome Institute

Central Office and Experimental Laboratory

2401 NORTH BROAD STREET

Philadelphia, Pennsylvania

UNITED STATES OF AMERICA

(Numerous Branch Offices)

Incorporated under the Laws of the State of New Jersey. Authorized for the transaction of business as Foreign Corporation under the Laws of the States of New York, Pennsylvania, Illinois, Maryland, Delaware, New Hampshire, etc.

CHARTER

The main objects as set forth in the Charter of the Spectro-Chrome Institute are:

The instruction, advancement and practice of Spectro-Chrome Therapy and the research, development and pursuit of such sciences as may assist in the equipment and application of Spectro-Chrome Therapy or the restoration of the human Radio-Active and Radio-Emanative Equilibrium by Attuned Color Waves, without the use of drugs, manipulation or surgery, in accordance with the system conceived, originated, developed and applied by Colonel Dinshah P. Ghadiali, M.D., D.C., Ph.D., LL.D.

The building, equipping, establishing, conducting and maintaining of schools for the investigation, instruction, demonstration and practice of Spectro-Chrome Therapy; also the power to confer and grant the degrees of Spectro-Chrome Therapist, Master of Spectro-Chrome and the like, and to issue diplomas and titles in connection therewith.

The corporation shall also have power to use, manufacture, sell and dispose of, in any way, all means, publications, appliances, instruments or products that may seem fitting or in any way connected with such purposes or objects aforesaid.

INSTRUCTION

Spectro-Chrome Institute conducts complete *resident* courses in Spectro-Chrome Therapy in a number of cities each season. Each course, comprising 24 discourses, is delivered personally by the Originator of the System with demonstrations and experiments. At the end of the course, students passing the examination are granted Diplomas.

There are *no* correspondence courses.

Spectro-Chrome Institute creates no "Doctors," and its graduates have nothing to do with "the Practice of Medicine." Its motto is:

NO DRUGS—NO MANIPULATION—NO SURGERY.

PERIODICAL

Spectro-Chrome Institute through its own printing plant, publishes a lively monthly magazine, SPECTRO-CHROME, edited on broad and progressive principles, by its President, the Originator of Spectro-Chrome Therapy. It is an absolutely fearless, independent, unbiased publication, ever ready to enunciate the Truth, without restriction or reserve, for the service of humanity.

EQUIPMENTS

Spectro-Chrome Institute manufactures all the Equipments and Supplies for the practice of Spectro-Chrome Therapy, the Latest Revelation in the Healing Art.

SPECTRO-CHROME

JANUARY 1923 — VOLUME 2

RED 1923 — NUMBER 1

Trade-Mark Registered at the United States Patent Office

Printed and Published Monthly by Spectro-Chrome Institute
2401 N. BROAD STREET, PHILADELPHIA, PENNA.
United States of America

Entered as Second-Class Matter July 20, 1922, at the Post Office at Philadelphia, Pennsylvania, under the Act of March 3, 1879

Devoted to the Latest Revelation in the Healing Art

SPECTRO-CHROME THERAPY

Restoration of the Human Radio-Active and Radio-Emanative Equilibrium (Treatment of Dis-eases) by

ATTUNED COLOR WAVES

No Drugs - No Manipulation - No Surgery

Conceived, Originated, Developed, Applied and Copyrighted, 1920, by Colonel Dinshah P. Ghadiali, M.D., D.C., Ph.D., LL.D., &c. Metaphysician and Psychologist

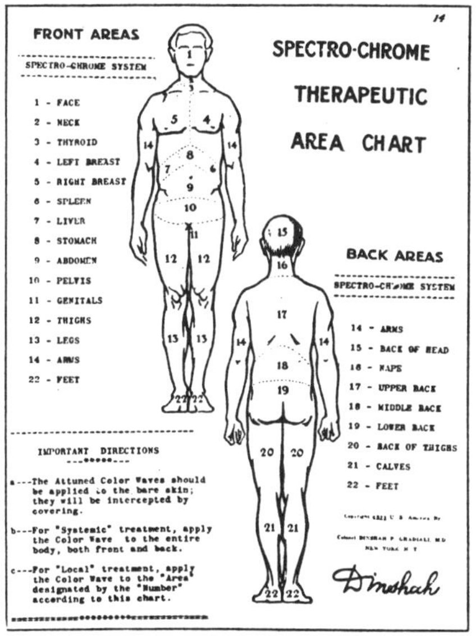

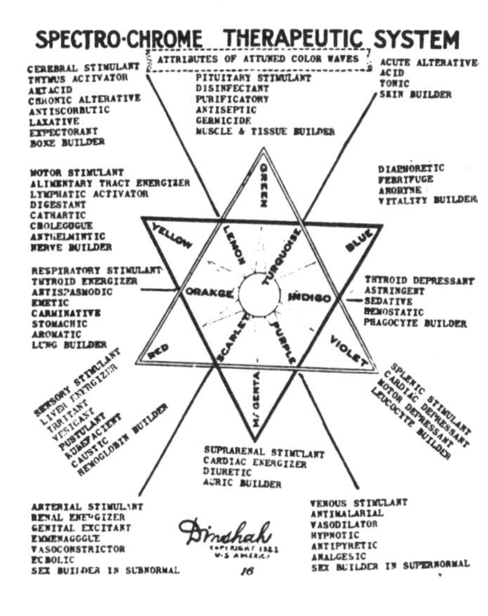

DINSHAH'S PLATFORM AS ADOPTED BY HIM SINCE 1891

The Boundless Vibratory Ocean of Thought is Essentially Universal and All-pervading; it is the Individualized Monopoly of NO Person, and is the Common Heritage of Humanity's Evolution. thus, what a Development of Unrevealed Ages has given unto Me in the Form of Knowledge in My Present Incarnation is No Distinctive Acquisition of Mine for My Sole Use, Benefit or Elevation, but is All for Thee and Is Thine Without Condition, Without Obligation, Without Expectation. I Fear No One; Only God Above and Conscience Below, and From Them I have Nothing to Fear.

Attuned Red Wave

(The World Through the Eyes of Dinshah)

AFRICANS AGAINST VACCINATION

The Negroes of Southwest Africa are uneasy. At Ontjo, compulsory vaccination brought on unrest. In 1904 a step like this resulted in many deaths among the vaccinated population. Now they are fighting the German settlers against a recurrence. The Africans are right; the white population there has no better blood than the native Negro and the Negro needs no vaccination now any more than in ages past. Down with the filth of vaccination! The same with all kinds of serum "therapy."

CUSTOM MADE KING

E. Alexander Powell, writing in *Harper's Magazine* about the kingdom of Mesopotamia, says:

"In order to give their puppet ruler all the trappings of royalty, the British designed for him a uniform, half Arab and half European; provided him with a court, complete from grooms to grand chamberlain; organized a bodyguard of lancers; presented him with a palace, a motor launch, and a fleet of motor cars; raised, drilled and equipped a miniature army, and designed a flag, a set of postage stamps and an order of knighthood. The latter, I was told, was to be called the Order of the Lamp, which, in view of the fact that the country's chief source of revenue is its oil fields, seemed highly appropriate."

This is only one of the new creations. The great war abolished a few big kings and created a number of petty ones. It would be wiser for America to stay where she is instead of butting into European politics and helping the abolition of one kingdom and the creation of some more. America should be no ally of any king or kingdom; distant friendship, but positively no political entanglements.

ANOTHER HORROR

Now comes a report that an international committee appointed by the League of Nations is working to standardize serums for the treatment of pneumonia, meningitis, diphtheria, and the like. Is the League of Nations a political body or a League of Doctors? I sincerely hope and wish the United States of America stays away forever from this infernal messpot.

FOOD OR JUNK?

Fifteen meat dealers in Allegheny, Luzerne, Montour and Lawrence Counties were arrested by agents of the Bureau of Foods of the Pennsylvania Department of Agriculture and fined. They sold "meat" composed of slaughter house refuse and stale scraps of meat, treated with Sulphur Dioxide. This "meat" could not then be detected from "genuine, high grade" meat. When will the people wake up to the extinction of butchery? Read the Armour article in this issue by Kashmira and judge yourself.

SPECTRO-CHROME SLIDES, SOLUTIONS, AND SOME STORMS

Let Us Clear the Air for 1923

During the second and third weeks of November, 1922, I visited Washington, D. C. once and Baltimore, Md. twice. The local organizations in both places are going along well. The Baltimore section has the strong support of the Maryland Academy of

Sciences and is in charge of Henry Sisco, M.D., S.C.T., with Frank Kaufman, S.C.T., as President, and William S. Arnold, M.Sc., S.C.T., as Vice-President. The Washington Class is presided over by Horace B. Coblentz, M.D., S.C.T.

The reception was warm at both the places and I thank them for their kind words and generous spirit. As elsewhere, I received the same pet complaint and the same suggestions. It was in regard to the slides. The following communication from William S. Arnold, S.C.T., was subsequently received and will explain the trouble, which is now producing much discontent and about which it is necessary that I must clear their minds. He says:

"Your policy of but one set of slides to each equipment is splendid, but not for the practitioner. To be able to give the uninterrupted treatments necessary for the best results and to be available to his patients at all times, in other words properly to serve them, the practitioner must have an extra equipment. Often, especially when he is building his practice and is in greatest need of giving service, he cannot afford an extra equipment, for which he would probably have no immediate use.

"Besides, I can see no especial reason for the necessity of extra slides for the practitioners, altho this would be very desirable. However, I do believe that your representatives should have extra slides (sets) available for the therapists of their districts. These representatives would, of course, give out the slides only to those who returned complete sets, thus replacing broken sets of slides in a few hours instead of several days, as is now the case.

A further precaution would be to etch on each slide of a set, with hydrofluoric acid or the so-called "Diamond Ink," the serial number of the set. This would be a very simple matter and would assure you and your representatives that the slides being returned would be all of the same set. It is a regrettable fact that there are some dishonest people, but it would be a reflection on no one to trust only your representatives, who could be placed under bond. Besides I believe with your knowledge of people you are most capable of picking the right men.

"The present policy is inductive of great dissatisfaction and, possibly, dishonesty. Practitioners who, if they could get slides locally would gladly turn in their slides for a new set at once, now patch and patch, and I know of instances when they have tried to make their own slides. This practice most certainly does not reflect to the credit of Spectro-Chrome.

"This is written in line with your request for constructive criticism and I have been perfectly frank in stating my mind. I trust you will accept these suggestions in the spirit in which they are given. That I have the success of Spectro-Chrome at heart is amply shown, I believe, in my efforts and endeavors as an officer of the Maryland Association of Spectro-Chrome Therapists and in my part in organizing that body. We have opposition here and are doing our "bit" which while not spectacular is beginning to bear fruit.

"This is a personal communication, intended as a suggestion toward better co-operation and understanding and is an effort to help correct a condition which, I believe, must be unsatisfactory to yourself as well as to most of the practitioners who are unable to afford more than one equipment.

"As to an article for the Spectro-Chrome Magazine, I shall be glad to write anything of especial interest to the therapists at large, but believe that the usual success which is attendant with the routine use of Spectro-Chrome is too well known and that detailed reports would be superfluous."

While I am highly appreciative of the criticism of our good friend, I have to inform him and all others that at the present stage it is impossible to follow the outlined suggestions about the slides. The whole foundation of Spectro-Chrome Therapy is the five slides correctly matched to produce the proper degree twelve color waves. My past experiences with human dishonesty and especially with the inhuman dishonesty of a number of my own graduates has compelled me to keep my policy unchanged on this matter. One man foisted on me in a package four slides for five, another believes it needs no matching of slides and tells his hearers they may use any old color as long as they "look" right, a third cuts my slides in two and makes two "equipments," a fourth builds my slides into ceiling equipments according to his conceptions of the technique, a fifth makes his own "color slides" a foot square, a sixth hoaxes his patients by telling them he has other waves better than Spectro-Chrome and uses a "Lavender Wave" with other combinations and so forth. How can my kind representatives guard against such contingencies? Besides, if I start doing buisness as suggested, my corporation records would have to be separately analyzed for each different State and more State Taxes and Fees would accrue. My bookkeeping would be complicated further; legal fights would start with the dishonest ones to guard the honest; through the mistakes of some unwary and good-hearted confiding representative a mix-up might occur in slides and a clash might spring up between me and my representative. All this *must* be avoided until the suffering public learns to know the genuine from the non-genuine. There is no solution for this. I cannot make all people honest and while they are not honest, I *must* guard the situation by maintaining rigidly the policy of One Set of Slides With One Equipment, No Spare Set To Any Person and No Replacement of Broken Slides Without Receipt Of The Complete Original Set. This rule has no exception, Spectro-Chrome Institute has no exception and I have no favoritism of any kind. My two eyes see equally.

But, I must be fair to the practitioner; how can I do so?

I have now installed a "Used Equipment Department." The busy practitioner who needs immediate relief from this tense situation may take advantage of it; at a reduced price he may secure a complete set of original slides with a spare equipment. These equipments, are serviceable in every way and are mostly "Exchange Equipments." Generally, I have a few of each kind on hand, which I cannot sell as "New Equipments" and which are offered at two-thirds the price of complete New Equipments. I have no other solution at present and if anyone will offer further suggestions by which I can smooth the uphill path of the pioneer Spectro-Chrome Therapist I am open to listen. Do not grumble uselessly; I am guarding Spectro-Chrome Institute, Spectro-Chrome Therapy, Specro-Chrome Therapist and the Spectro-Chrome Patient equally. In this particular respect, my personal responsibility to the suffering public is grave and I *know* it. Therefore, until some better solution offers itself, until my limited resources improve sufficiently to maintain in different states and cities specially trained representatives who can be adequately compensated for all the time and trouble necessary to spend on this kind of unprofitable work, *I cannot and will not change* my policy. You will have to put up with some inconvenience perhaps now, but friends, time will prove that I did the best in the interests of human service. If I stumble the least now in the infancy of this latest revelation in the Healing Art, I would be a traitor to the cause and the heavy duties attached to it.

Now while I am at it, I wish to open my heart to you a little. Many and many a scandalous tongue is wagging adversely about my "peculiar" methods of teaching and business. Look at the results accomplished in less than three years however and judge. I am human and just as fallible; I am in many respects even *very* human, but, a man who has the composite functions of a researcher, originator, teacher, commercializer and fighter to perform single-handed against a granite wall faced with steel lined with the ingrafted prejudices of ages of notions behind it, all without outside financial or other support of any character must not expect anything different from the world. The irony of the situation is that because I expect nothing, get nothing and take nothing (except grumbles), some of the "friends" who supported me and my cause turned against me. Why? Because, I politely let them know that while I am open to reason and suggestions, I mean to run my business on my own methods and as I think fit. They are told repeatedly that I take no personal gifts; yet, they persistently send me gifts (I call them material junk) and when they are returned they feel "offended."

Say friends, if you love me and my Spectro-Chrome cause, will you oblige me by refraining from this eternal nonsense of gifts and presents? You will be appreciated and valued more if you use my Therapy and help my human mission, but please do not seek to entangle me into acknowledging and responding to social presents and gifts, whose acceptance I swore off nearly thirty years ago.

Others, need a different request. They write me about matters not pertaining to the subjects in question and expect me to answer promptly. They just forget that outside of answering mail I have complex duties to perform. I would like to oblige them, but, they must abide by my convenience and leisure, of which I have none to spare.

They brand me with laxity and negligence. Far from my mind it is to neglect well-wishers, but please, friends, be patient. I have only a small staff of devoted helpers and the overhead expenses are very heavy compared to the receipts. The following will show you some of our *actual* figures for 1922 as submitted to the Federal Government:

Receipts

Instruction	$11,576.84
Equipments	26,726.16
Literature	1,037.28
Bulbs	545.25
Slides	470.00
Total	$40,355.53

Here is the other side.

Expenses

Raw materials	$12,711.65
Salary of employees	8,210.39
Rent of various classes	6,599.24
Sundry, depreciation etc	2,912.95
Travelling and railroad	2,507.79
Express charges	816.30
Printing	727.02
Advertising	646.84
Stationery	594.60
Insurance of all kinds	506.96
Postage	406.00
Commissions	410.48
Corporation taxes	301.25
Telephone	223.76
Coal	204.10
Legal fees	153.85
Electricity	145.78
Gas	84.13
Total	$38,163.09

This leaves a margin of only $2,192.-44 as apparent "net profit." Add to these items of expenses, additions and betterments, reserves for research and expansion and such other necessary and vital provisions and—what about this humble servant of yours? Where is his compensation, where is the provision for his rainy day? Who will provide his future? He gets no salary and lives on—well, you find out yourself!

Bless your generous heart, O beloved friend, Spectro-Chrome Therapy has come to stay for the blessing of the world, and be proud that you are destined as a chosen pioneer in this truly great labor of human love. Blame me not for my failings only, but, support with your courageous moral help Spectro-Chrome Institute and its work and the Powers on High Whom we all serve will take care of its future.

Tathastu! Amen!!!

Attuned Lemon Wave

(Medicine, Surgery, Science, Osteopathy, Chiropractic, Philosophy)

HOW WORRY AFFECTS THE BRAIN

By Carl F. W. Walter, M.D.
Philadelphia, Pennsylvania

Modern science has brought to light nothing more curiously interesting than that worry will kill. More remarkable still, it has been able to determine from recent discoveries just how worry does kill . It is believed by many scientists who have followed most carefully the growth of the science of brain disease, that scores of the deaths set down to other causes are due to worry, and that alone.

The theory is a simple one — so simple that anyone can readily understand it. Brifly put, it amounts to this: Worry injures beyond repair certain cells of the brain, and the brain being the nutritive centre of the body, the other organs become gradually injured, and when some dis-ease of these organs or a combination of them arises, death finally ensues. Thus does worry kill. Insidiously, like many other dis-eases, it creeps upon the brain in the form of a simple, constant, never-lost idea, and, as dropping water over a period of years will wear a groove into a stone, so does worry gradually, imperceptibly, but no less surely, destroy the brain cells that lead all the rest—that are, so to speak, the commanding officers of mental power, health and motion.

Worry, to make the theory still stronger, is an irritant at certain points, which produces little harm if it comes at intervals or irregularly. Occasional worrying of the system the brain can cope with, but the iteration and reiteration of one idea of a disquieting sort, the cells of the brain are not proof against. It is as if the skull were laid bare and the surface of the brain struck lightly with a hammer every few seconds with mechanical precision, with never a sign of a stop or the failure of a stroke. Just in this way does the worrying idea, the maddening thought, that will not be done away with, strike or fall upon certain nerve cells, never ceasing—diminishing the vitality of these delicate organisms that are so minute that they can be seen only under the microscope.

THE VOICE OF THE DEEP

By Ira L. Kepperling, M.D.
Philadelphia, Pennsylvania.

Again and again as I listened to the message of Spectro-Chrome Therapy from the lips of Dr. Dinshah P. Ghadiali, I wondered exactly how many of the students, who, like myself, took his course of lectures *really* saw what was back of all this effort on his part.

The surface thinker can see naught but a new system of therapeutics, another school of thought or "pathy" to add to the success of a doctor in curing the aches, pains and sickness of his patients.

To me this is the most distressing portion of this great teacher's work. Great as may be the blessings to mankind from this new healing method, greater than all is the majesty of the Power that lies at its root, for it is *"Prana"* itself in one of the higher manifestations.

Brother, Sister, forget the narrowness of your previous religious training and behold God everywhere manifesting. Let not the chains of narrow worm-eaten dogmas and creeds longer hold you intolerant to a message brought you from the Masters of Wisdom. You are sending missionaries to far away Asia to carry a message of the good and noble Nazarene Jesus; you expect them to listen to its call. The Orient has now sent you one who has tested and proved the age-old truths of Man's Inner Self. He has demonstrated every step of these Occult Truths by the only language *you* can yet understand; that of hard concrete material demonstrations. Is he not fully entitled to a careful thought on your part—of yet to you greater mysteries? The Soul and its pilgrimage is charted for your own guidance into a fuller life of usefulness, if you will but apply its absolute laws.

You have looked upon this therapy and its "Colored Lights" doubting, but have proved its merits. Yet, few, I believe, have even a faint conception of really what is accomplishing your "cures." Back of each so-called "Attuned Color Slide" is a Law,—an Act, —a Force—which materialistic *science* cannot know or never will be able to reduce to a *reason,* for it can only be touched when we have gone beyond Reason.

Now, you saw the first Light of Truth glimmering from the open door of the great Temple of Soul. Why not make happy this sacrificial teacher's life by going deeper into the messages he has waiting for the sincere seeker who will gladly crucify the Self, that the glories of inner unfoldment may come?

There is a parable of a wood-cutter and a Sage that illustrates the point.

In a village in India lived a poor wood-cutter. Day by day he went into the surrounding forest and each evening brought branches of trees to the city, which selling for a few coins supported his family. In this way, the poor man labored for years. One evening as he emerged from the forest with a heavy load of wood upon his back he met a Sage, who, seeing the miserable condition of the wood-cutter, said, "My good man, why do you not go farther into the deep recesses of the forest?" The poor wood-chopper replied, "Why, sir, I get enough wood here; what is the use of going farther into the unknown forest?" But the wise Yogi replied, "Go farther into the forest." Then each went his separate way.

The words of the Sage kept ringing in the poor man's ears. Next day, remembering the words of the Holy One, he went deeper and found sandalwood. Mentally thanking the Sage, he took away a large bundle of the wood, which he sold for a high price.

The next day, with the Sage's advice to go deeper, he went farther into the depths of the forest, where he found copper. Gathering together all he could carry, he returned to receive more money than he ever hoped for. Remembering the Sage's advice "to go deeper" he went on and found farther a silver mine, which made him rich. Next month, he went past all the previous places and found a gold mine; then, realizing the wisdom of the Sage, he pushed yet farther into the depths, found a mine of diamonds and precious stones and became very wealthy.

The advice of all great teachers to you is, *go deeper;* onward, upward toward the mines of Spiritual Diamonds. May the teachings of the above parable become your very own. Onward—*onward,* until you find for yourself the Gems of greatest price. Then and then only can you *fully* understand *why* Spectro-Chrome Therapy came to you as a message, bidding you become *one* with the *Great Essence of Light,* The Central Sun.

Attuned Green Wave

(Spectro-Chrome Therapy, Spectro-Chrome Therapists)

ARE MEDICAL MEN AND WOMEN "DOCTORS?"

By Emil Gleitsman, B.S., M.D., S.C.T.,
Specialist in Natural Healing,
Chicago, Illinois

Our commercial age has established a lot of mental drilling halls, going under the name of high schools and colleges, some of which specialize in drilling students in medical science so-called and when such students after a time are examined and found drilled enough, so as to repeat, parrot-like, some of the magical formulas of said medical jargon, the drilling time ends and studies are finished in most cases forever, and said medical college or university is empowered to transform such students to "Doctors of Medicine" and with said title transfers them to the very pinnacle of learning and understanding in the High School of Life.

Swelled with the pride of Augustus, this young medical fledgling announces to the suffering world, that he (or she) is ready and fully prepared for its complete cure, if they only will call at the new office. But, is he *really* a "Doctor of Medicine"? Could he (or she) possibly have arrived already at the summit of wisdom and understanding of the healing art? For what does "Doctor of Medicine" really mean? This Latin expression signifies a "Master Teacher" in the intricate art of healing; in other words, anyone acquiring or using the title of "Doctor of Medicine" proclaims himself or herself to be not only an ordinary professor, but a Master Teacher, endowed with all the knowledge and experience essential for combating any and all the endless complications of dis-eases human flesh may be heir to and be perfectly at home in every branch of healing. How absurd to expect such things from a medical chicken! It takes a life-time, and then some, to gain such knowledge and ripe experience with mature judgment, which no short college curriculum can give. Almost anyone can be a "graduate in medicine," but very few of the very elect and most experienced practitioners can ever become "Doctors of Medicine." Most of them die in their own ignorance and inability to heal themselves.

We all know by now, that even under the legal protection of a state license, a doctor of medicine means not more than a salesman or saleswoman for the drug trust, and we also know that "medicine" has as much to do with healing as a mule in a coal pit has to do with the management of a coal mine, for each drug is a small part in the drug business, but has no part in the healing art, for, healing is one of Nature's own processes of life and *comes from within.* Real cures and actual healing cannot be filled in bottles, sold over the counter and handled like any other piece of merchandise. We may commercialize the art of healing and have really done so, but we cannot commercialize healing, or can a medical man or woman be a healer.

What a pretense then, what a camouflage, to proclaim every Tom, Dick and Harry a "Doctor of Medicine!"

Even among the real teachers in medicine, how many are mere copyists and plagiarists, repeating in the class-room what someone else has pretended to know. Would or could a real "master teacher" repeat such medical jargon, as we find in the textbooks, or could he be classified with the modern "medical authorities?" Practically every new medical book issued by them is nothing but a rehash of old or new theories, resurrected opinions, experimental conclusions, which, when seen or read in books, become gospel truths of medical science until discarded again for another set of opinions from another literary aspirant.

Even the best professors and teachers in medical colleges, who ignore and deny the very life principle of soul and spirit in their teachings to medical students, what have they really left of the truth in their medical "science?" All the merchandise traded and sold in drug stores for "cures" on one side of the balance and the higher inner forces of patients (always available for real healing) on another side of the scales would convince any blind bat how far the latter outweigh the former. But because Nature's wonderful healing forces cannot be seen or handled like merchandise, they do not exist for our modern medicine man. "Medicines" do not cure; they only change one dis-eased condition into another, generally more severe, more diffused and general. As long and as far as doctors of medicine ignore or think they can ignore the very foundation of all healing, as long as they remain ignorant about the main factors for restoring the fine equilibrium of dis-eased conditions in any human being, so long are medical men and women unfit to be "Master Teachers" or "Doctors" (taken from the Latin verb "docere," to teach, instruct) and remain only a few degrees removed from the ignorant Red Indian Medicine Man. But there is one hope still for all such, provided they overcome their false pride and begin to study and digest the fundamentals of Spectro-Chrome Therapy, which rationally teaches how to restore in the human body the radioactive and the radio-emanative equilibrium by Attuned Color Waves, as originated and taught by its one real Master Teacher.

This widespread pretense of being doctors of medicine, without the slightest qualifications for being master teachers in the healing art, is so common that very few find anything wrong with it, just as their finer moral senses of society have become so obtuse by reading the daily catalogues of crimes in our midst, that most readers consider them only daily "news." This medical pretense has also a very serious side, for teaching is not only done by the spoken and written word, but much more effectively by our acts as teachers in the healing art. A doctor's example and actions throughout life speak louder and reach farther than any lecture in the classroom. Therefore, the wise James of Bible fame says truly, "Let not many aspire to be teachers, for so much greater responsibility rests with them." Ignorance may excuse the uneducated, if their words or conduct cause sufferings, but it never excuses the learned man or "Doctor." He is supposed *to know* what causes suffering and dis-eases and is supposed to understand how to remove these causes and sufferings and work by Nature's laws to restore the feeling of ease which has been disturbed and to regain the consciousness of renewed strength and vitality. But instead of this, it is the very doctor of medicine of the modern type (with very few exceptions), who, in his or her monumental ignorance, prolongs and intensifies sufferings, makes free use (in his unforgivable ignorance of the black magic of his suggestions) of cutting open the live body and carve out dis-eased parts of organs. His "scientific" arguments and suggestions become auto-suggestions in his uneducated patient, he is made to believe them and act on them and all his or her thought forces are blackened and poisoned, and instead of healing, a process of sure destruction follows. A thirty-year practice of healing has taught me by the very experiences with all patients what unbelievable causes medical men and women imagine to be at the root of dis-eases. I shall give only a few illustrations.

A woman in her menopause called at my office complaining of severe and constant headaches; before I could question her, she told me that her physician had told her all her blood had turned into pus in the head, therefore the aching! Pure medical "science!" I restored her blood circulation and regulated her liver and bowels, the blood pressure on her brain was removed and the headaches melted away and all the "pus" must have turned red for shame!!!

A personal friend of Mrs. McKinley, from Canton, Ohio, some years ago came to my office. She was blind and had consulted the best eye specialists in the East. Naturally, I asked her what their diagnosis was of her case and she told me, it was diagnosed as a detached retina. It sounded very "scientific," and I asked her, "What detached your retina in the eye? Was there any accident, shooting or stabbing in your eye?" Nothing of the kind had happened. It had been pure guesswork of an eye specialist.

A young woman suffering from fits had been diagnosed that "a splinter of bone pressed on the brain." That physician's X-ray apparatus must disclose a splinter of bone, even though the bony skull bars all peeking!

Every day such "scientific" teaching comes from doctors of medicine to the great public, who must support them. When a physician examines a patient, looks at the tongue, feels the pulse or even goes so far as to knock all over the bare chest or listen with his "scientific" ear or the stethoscope to the pulsations of the heart and pronounces his judgment with a Greek or Latin name of a "dis-ease," he teaches by such acts, that this is the medical way of knowing all about the trouble. Some go even far beyond this; they have some foul and dis-eased secretions or excretions examined by a wiser one, to find out what bug or microbe shall be made responsible, and so they all guess and opine and suppose and believe as to the cause or the name for dis-eased conditions. Even with the teamwork of consultations, the poor patient grows steadily worse until his great friend, Mr. Thanatos (Death) relieves them all for sure. If even the lay people, who do not know anything about medical "science," get so thoroughly disgusted with the acts and teachings by word and deed of their supposed "Master Teachers" in the art of healing, it must be rotten indeed. Such patchwork and guesswork, such experimentation, such hide-and-seek play under cover of Latin and Greek terms for prescriptions and dis-eases has robbed the most trusting public of all confidence in doctors.

Another sort of practical demonstration of medical teaching is the present craze of injecting the vilest and nastiest animal matter directly into the already dis-eased bloodstream of a living body by doctors. Such debasing acts by these supposed "Master Teachers" in healing is only another proof positive that they do not know what they do. But they are not entitled to forgiveness for such absurd degradation of human beings to the brutal animal plane. Under their guidance, teachings, direct help and active assistance doctors have forced animalism anl carnivorism, dis-eases, crimes and insanity on their charges, instead of elevating the human animal to the higher levels of consciousness and freedom. Even these insane wars of nations are attributable to unbalanced minds of some individuals temporarily in power, but all under direct and immediate guidance of doctors, whose

absolute duty it was to tame such wild men crazed with megalomania. If the medical profession as such would understand its own duties, rights, privileges and responsibilities as real master teachers in words and deeds of the healing art and would learn and understand to practise healing, instead of practising medicine, our civilization would surely show another face. A doctor of medicine must be a gnostic of the laws of health and life, an everlasting student of Nature, especially the nature of man as to his body, spirit and soul and their interrelationship. He should not dare to be and be known as an agnostic in practically everything human, save to get away with the coin of the sufferers; he must instantly detect and know charlatans and should never be "in the market" for selling or prescribing the wares of the "drug trust" or slaughter-house refuse, or recommending dead and rotten carcasses for food, or assist any money-mad merchant in filling unspeakable rot in bottles to sell for the "health and happiness" of suffering humanity.

The above article from the pen of Dr. Emil Gleitsman, S.C.T., deserves our warm thanks. Although a doctor of medicine himself he was progressive enough to take to drugless healing and his entrance into the field of Spectro-Chrome Therapy means much to his patients. His article appears to have been inspired by the recent trial of some officials of the State of Illinois in selling licenses to physicians and druggists who failed to pass the prescribed tests. For a good sum of money, they are charged to have licensed "Doctors of Medicine," and it would be well to know whether the American Medical Association protested against such graft. SPECTRO-CHROME will appreciate further articles from Dr. Emil Gleitsman's pen.

CLINICAL SIDE OF SPECTRO-CHROME

By KATE W. BALDWIN, M.D., F.A.C.S.,
Senior Surgeon, Woman's Hospital,
Philadelphia, Pa.

It is now nearly ten months since the little girl so badly burned came to the Woman's Hospital. Those interested will be glad to know that even though she is still in the hospital, satisfactory progress is being made. Just about the time I had made up my mind that she would be home by Christmas, early in November, in an unaccountable way she became infected. Her temperature went to 106° F. Pulse 158 and Respirations 48. This was anything but cheerful news. We kept our heads and with the courage of our convictions just rushed Spectro-Chrome night and day, always every four hours and many times an hour or two at a time. All healing processes were checked and the open surface looked as if varnished. Much of the new skin was lost. An abscess developed in the right groin, also a large swelling above the knee on the left leg. This, however, did not break down, but there was some pus from the left knee. For days it seemed impossible for the child to weather this storm added to that which she had already been through.

The mental condition became quite alarming. General choraic movements were marked. She lost weight. Appetite was very poor. A 106° F. temperature was recorded but once, but days and days it went to 103°, 104° and 105°.

At present, the case is in fine shape. In surface covered she is about one-third beyond the point when the healing process was arrested. The nervous symptoms have all disappeared; she is eating well, gaining flesh, the new skin is forming satisfactorily. I now feel that in a reasonable length of time I shall be able to report the case as a complete cure and be able to show her as a living proof of the efficiency of Spectro-Chrome. Several smaller burns since my last report have all responded quickly to Spectro-Chrome.

Some months ago, I reported several cases of carbuncle. I now wish to add another which in many ways has been a greater triumph for the method of treatment than any one of those already given.

Mrs. G.—Age 66 years and poorly nourished; really looking more than the age given. Mrs. G. was sent to the Woman's Hospital after ten days of treatment by the family physician. She was unable to sleep because of the severe pain. The whole back of neck, from ear to ear, and from occipital ridge to the cervical dorsal articulation, was tense, indurated and purple in color. An area in the middle about two and a half by three and a half inches was necrotic. This condition followed some application which was very painful. All the patient could tell about it was that it gave severe pain and that she had not been able to sleep for nights and nights.

The patient was immediately put on a meat-free diet and, of course, on Spectro-Chrome. She was given a half hour treatment three times a day; Yellow or Lemon morning and afternoon, late in the evening Indigo, which had the effect of giving her a good night's sleep *without* any sedative. (She had only a very small dose of a sedative two nights after entering the hospital. After that Spectro-Chrome took care of the pain.) Her remark to me was, "Oh! Dr. Baldwin, it is heaven to be here and have this treatment after having suffered as I have. Why do not all the doctors use it?" Finally, all of the necrotic tissue was removed. The slough extended beyond the area that was necrotic when she entered. No more skin was lost, but the undercut was from one to two inches in all directions. The lower flap just fell over on the neck. In order to bring it in contact with the deeper tissues so that nature could have a chance to attach it I was obliged to bring it in contact with a strip of adhesive plaster which had to be carried around the forehead. Other straps were used from ear to ear and by utilizing the elasticity of the skin we were able very much to diminish the space to be covered by new skin. As surgeons, I am sure we often forget to make use of the elasticity of the skin. I often only remove the part of the strap extending over the open surface and put another layer on top of the previous ones. This gives a good firm base for traction, as plaster sticks to plaster better than to skin and the skin is greatly saved by not having the plaster torn off each time.

Between the Spectro-Chrome treatments, we used some simple surgical dressing. This line of treatment was continued until we had only a small open surface, possibly one and a half inches by one inch. Out of a clear sky, so to speak, she ran up a temperature of 102° F. The open wound became dry and glazed. The face was swollen and much of the scalp oedematous, the eyes were nearly closed and the dear woman made up her mind that her end was near and she wanted to go home if she was to die. The whole back became more or less involved and there developed an open wound over the sacrum which extended to the bone. For a few days things did look dark. I put a Spectro-Chrome equipment by her bed and she had it much the same as did the burned case when it went bad. The cellulitis of the head, face and back suggested erysipelas, but the laboratory returned only staphylococci. In this case I painted the face with Thuja and gave for a time Echinacia.

After four or five days, the change was very much for the better and from

that time on everything was satisfactory. Her friends began to tell her how much better she was looking than when she left home and that she looked ten years younger. She really did look much better, the general condition of the skin was much better and a nice little "pinkish" color in the cheeks.

Mrs. G. went home with a small place on the back not quite closed. She returned for Spectro-Chrome and dressing.

I saw her some time ago and the neck was perfect except for just a line about one and a half inches long, and this was well within the hair line.

I regret exceedingly not having a photogram of this case. A verbal description does not carry the conviction that even a poor photogram does.

If Spectro-Chrome can do the following for an advanced case of tuberculosis, what a wonderful field is there for it in the incipient cases!

July 21, 1922, Mr. L., a teacher, age 31 years, over six feet tall, weighing only 105 pounds. Coughed almost all the time, there being little sleeping consequently. Expectoration profuse, heavy and difficult to bring up. Blood pressure 105, pulse 110, very poor in quality. Respirations shallow. When 24, had typhoid pneumonia and not quite normal since. Two years ago, a severe hemorrhage, another hemorrhage a year ago, each time losing a quantity of blood. The sputum often streaked with blood; last June a large clot, following an osteopathic treatment; this was probably only a coincident.

Last spring, Mr. L. was out of school for some time. He was having so much creosote that all his food tasted of it and made eating distasteful. A day or two before he came to me, he decided to give up the creosote. He really was much discouraged and quite down and out. His general appearance was that of a very sick man.

The physical examination showed extensive involvement of both lungs, with evidence of a cavity in the left apex. With the history of pneumonia, two hemorrhages, loss of weight, low blood pressure and rapid pulse, the case did not seem an ideal one to establish the efficacy of any treatment.

July 22nd we had an X-Ray picture taken. This showed extensive involvement of both lungs with a cavity in the left apex the size of a lemon, nearly filled with fluid.

The man was at once put on a meat-free diet and Spectro-Chrome. Very soon all the symptoms improved; less cough, more sleep, better and deeper breathing. The chest and heart sounds improved, expansion was better and weight increased. Just six months from his first visit to my office the change was so sufficiently marked that we decided to have another X-Ray photo. This showed marked improvement in all parts of the lungs. No fluid in the cavity; in fact, one would not diagnose a cavity unless from a thickened edge which outlined the area. The pulse is down to 85 to 90, good quality. Blood pressure up fifteen points. Chest expansion increased. Weight, 130 pounds. The patient now sleeps all night; the coughing is almost entirely limited to the morning, when he brings up a small amount of secretion. Examination of the sputum shows T. B. positive and many pneumococci.

Mr. L. has not been in school this winter, but they want him back to coach the ball team, which he has done for several years and the team has always won when he is the coach. I am very much inclined to favor his trying the work with the team this spring.

Six months hence I hope to report this case again.

Attuned Blue Wave

(History, Travel, Biography, Geography, Language)

* * * * * * * *

DEATH OF A NOTED SPECTRO-CHROME THERAPIST

L. Grant Baldwin, M.D.
Fellow American Medical Association,
Fellow American College of Surgeons, etc.

Spectro-Chrome Therapy, since its publication to the World, three years ago, created a live interest in a large number of prominent members of the orthodox medical profession, but among those who went into it none could be said to have taken a more energetic part than Dr. L. Grant Baldwin, of 20 Schermerhorn Street, Brooklyn, N. Y.

And just as he mastered it and began to give the benefit of his high learning to his numerous patients, came the news of his sudden death through cerebral hemorrhage, on December 31, 1922. As he had attended the Second Annual Convention of the American Association of Spectro-Chrome Therapists, on November 28, 1922, at Philadelphia, Pa., the report of his passage to higher planes of human activities came as a surprise to his numerous friends.

He was born of a highly honorable and esteemed family in Lawrenceville, Pa., on November 18, 1863. His parents were Moses S. Baldwin and Mellicent Wylie Baldwin. From an early age, Grant showed fondness for the healing art and with his sister, Dr. Mary E. Baldwin as preceptor, went to the Univerity of Michigan for the study of medicine. Establishing his practice in Brooklyn, N. Y., in 1886, he made his influence felt very soon, and became famous as a surgical and gynecological authority. Last year, through the reports of his sister, Dr. Kate W. Baldwin, he heard about Spectro-Chrome Therapy and conceived such keen interest in the system that he took the full course in

New York, N. Y., and installed a number of Spectro-Chrome Therapy equipments in his sanatarium, where, with their aid he achieved some wonderful results. But before he could give the public the benefit of his keen observations, his death deprived Spectro-Chrome Institute of a great supporter of the system.

Grant was truly a great man. He was ernest in his endeavors in the relief of human sufferings, open-minded, liberal-hearted and progressive to a degree. On learning about his death, the New York class paid him a silent and sincere tribute by rising in his honor with bowed heads.

May the noble soul that went forward to the higher levels, from there send us his assistance. which he had so zealously begun to render. Death is only a transposition of activity and though we lost here a great pillar, we must rest in the true knowledge of the occult that the Powers on High have ordained all as the best in the immutable laws of human evolution and with resignation submit to the processes that determine the future.

We shall never forget thee, Brother Grant! To thy enlightened soul may there be Peace—Shanti: Shanti: Shanti:!!!

2000 MILES BY AUTOMOBILE

Jottings by the Way

By Miss Kashmira Dinshah, S.C.T.

HORRORS OF ARMOUR'S MEAT-PACKING PLANT

Most Disgraceful Blot on Human Civilization

Down through ages the all-intelligent unfair sex has railed against the folly of the eternal feminine in following blindly the capricious will-o'-the-wisp "style." There are different kinds of "style"—styles in clothes, customs, manners. Daring to deviate from the accepted fashion in these branches of convention is haughtily derided as "eccentricity" or a "parvenu trick." So the vogue is meekly followed by the many or markedly flouted by the few—but O so few! Yet, dear femininity is not alone in this form of foolery.

A curious phase of this sheeplike habit among humans, we noticed in the streets of Chicago, Illinois. It was seen in the flagrant disobedience of the city traffic regulations. The chauffeurs of Chicago believe that the paramount duty of each driver is to show his expertness in the operation of his motor vehicle. In this important work requiring their undivided concentration, they utterly disregard the established rules of traffic. It is considered "unfashionable" to heed these laws, so all strain their energies to remain "in style." When two cars meet at right angles at a street intersection, neither slows nor stops, but with a marvelous intuition each driver foresees the other's move and in a miraculous manner the machines swerve around one another and proceed serenely on their way, at forty miles per hour. It is truly a wondrous sight. Right and left hand turns are taken whenever the drivers choose, the traffic policemen being looked upon by them as mere inconvenient road obstructions, not even decorative. Unnecessary evils, the chauffeurs think, once in a while happening to mistake a traffic bluecoat for a paving block. What chance has an ordinary policeman's signal when Queen Fashionable Driving directs otherwise?

Father and I were given a beautiful experience of Chicago chauffeuring by the driver of the sightseeing bus which carried us to the Armour & Co. stockyards, in the southern part of the city, one day.

Having heard much of the great institution, Armour's stockyards and meat-packing plants, we decided to take advantage of the opportunity afforded by our presence in Chicago to visit personally the far-reported slaughter houses. Rumors said that the animals were killed in divers ways: by shooting, throat cutting or plunging in scalding water. No one informed us exactly; statements differed or contradicted each other. Hence the resolution to obtain the wished-for information ourselves. We wanted to see with our own eyes the true affair. Father has a stern sense of justice which leads him to criticize and rebuke mercilessly actions and deeds which are opposed to the single Divine Law of Truth whose various aspects are love, charity and the sacred mandate of tolerance—Live and let live. But the same just feeling prevents him from denouncing that of whose verity he is uncertain, for to father, *right* is *might*—and *right* shall prevail ever against all odds.

Twenty-seven years ago, father learned through scientific reasoning that the correct diet of human beings is not carnivorous, but is frugi-herbivorous. Thitherto, he was a normal meat-eater. Convinced of the erroneousness of such a diet, he pledged himself in one day to total vegetarianism, eschewing all flesh, fish, fowl and eggs. Incidentally, abstention from tea and coffee took place, having found them poisonous drugs harmful to health. He had taken the pledge of complete abstinence from intoxicating liquors some years before, but tobacco he never in his life has touched under any circumstance. To this day he adheres rigidly to all his pledges.

Father in the aforementioned respects is a "reformed sinner," but I am from conception and birth an *absolute vegetarian*, having never tasted any meat, fish, eggs, tea, coffee, spirits or tobacco, and I never will! When we dine, no animal mother weeps for her offspring. No drug or noxious stimulant enters the system to stupefy the brain and disorder correct functioning of the bodily organs. As a result we possess full health physical, psychical and mental. We work eighteen to twenty-one hours a day, at the end of that time being ready to work right into another day, if necessary, without fatigue, because our blood runs clear of carnic and alcoholic toxins. The body, untaxed with poison produced by wrong diet, has recuperative and energic powers truly marvelous in the eyes of those accustomed to the lethargic influence of a stomachic charnel. *Lack* of energy really should be unusual, not, as it is, the opposite.

It is a part of father's life work to teach human beings the criminal savagery of a slaughtered diet, enunciating twelve specific reasons proving its wrongfulness from aspects of anatomy, physiology and other scientific points, humanity among them. Therefore, he grasped the opportunity when presented of viewing the process in which a high (?) civilization produces its food from live animals to chops and steaks for the table. Then, just wait until he would be on the platform again, lecturing on diet! His denouncements of human carnivorousness would be hotter than Plutonian fire, the thermal temperature being regulated according to the degree of barbarism witnessed. Hear him talk!

Thus it came to pass one day shortly before our departure from Chicago that we rolled off in a sightseeing bus, having reserved seats ahead, to the great plant of Armour & Co. Two young women, strangers to us, sat alongside the driver, who sped the bus at thirty to forty miles per hour the while he conversed merrily with them. Through the crowded thoroughfares he rushed headlong, turning corners on

two wheels, passing standing trolleys and traffic signals, while the policemen watched serenely or barked and howled, according to their temperaments. One policeman shouted, jumping up and down in great rage because the heavy machine nearly snatched off the buttons on his coat. "Hey! What do you think I'm here for?" But the bus driver replied over his shoulder, "Go to hell!" and stepped harder on the gas (if it were possible). I did not think the chauffeur's retort sufficiently cutting. He should have said to the angry bluecoat, "Go to Armour's!"

We had traveled nearly half an hour when we came in sight of the stockyards at the extreme south end of the city. From a distance of five miles back, our keen olfactory senses, accustomed to the pure perfume of greens, had detected the putrid odor of the yet far-off slaughter houses and their rotten products. But it did not require *much* keenness of detection. The bus turned right on to a high wooden bridge spanning the breadth of the stockyards. As far as the eyes could reach we saw huge, dull red brick buildings, some painted with the name Armour in bold white letters, surrounded by the vast animal pens. Below were seen the cattle being herded by horsemen, into the runways leading inside the killing houses. The air was filled with the horrible shrieking of the unfortunate creatures, both within and without the shambles. The bus rolled on across the bridge toward the slaughter houses. Suddenly, there was a wild stampede among the steers beneath. The horsemen lashed them vigorously with whips, while they bellowed and reared running about wildly. The pens were small, holding hardly over a dozen animals each. The stampede was quelled easily, for only a few enclosures were opened at a time by each different group of men. The creatures apparently foresaw their doom, hearing and sensing the cries of their perishing fellows inside the building into which they were being herded. They drew back; down came the cruel whips; blinded by the blows, frightened, bewildered, the steers rushed on to escape the beating and the tumult of the men's shouts, rushed to destruction, into the wide runway like a horizontal funnel which gradually narrowed until at the end within the killing rooms, there was space for the egress of one animal at a time. No escape! Later, the manner of disposal will be described.

The ground area of the Chicago plant of Armour & Co. is seventy-five acres. It was the first and is the largest of their many plants, being established in 1867. In it are 12,000 employes. Its daily car loading and unloading capacity is 514 cars. Six thousand two hundred and seventy-five tons of refrigeration are produced every day, the boiler horse-power capacity being 30,720. This plant's daily killing capacity is, to wit: Cattle, 2800; calves, 2000; hogs, 9600; sheep, 10,000. These statistics are of the year 1922.

The area not occupied by buildings is divided into small squares by rude fences, three to five feet high, formed by horizontal wooden bars thrust into notched upright poles. There is no shelter of any sort. Wallowing perforce in the mud and filth of these open pens, the cattle, hogs and sheep stay until the government inspection takes place, only those animals which appear sound and healthy being passed for slaughter. *Appear* healthy. There is no surety. Who can tell? Filth *is* filth.

Finally, the end of the wooden bridge was reached. The bus stopped at the door of a brick building. Everyone alighted; the driver, the two women, a male passenger and father and I, six in all. We entered the building, coming into a small office. The driver informed the man in charge that the party had come to "see Armour's." After a slight delay in the office, a heavy wooden door with iron bolts was swung open and with the bus chauffeur leading, our party stepped through into a suggestively dark and narrow passageway. We groped along through some more doorways, the mazy path guiding us to an iron stairway three feet wide. Going first, the driver opened another door at the head of the stairs, upon which there burst forth the full intensity of sound of the animals' cries hitherto muffled by the thick walls. Father and I virtually shuddered. The party ascended and passed through beyond. Side by side, arm in arm, father and I went up the stairs, braced ourselves and stepped through the doorway into—*a veritable hell on earth*. The scene before us exceeded far beyond our wildest imaginations of horror. We found ourselves on a wooden iron-railed gallery extending around a huge room, branching off at points. On the floor below, two enormous steel wheels rotated slowly in a vertical position. This was

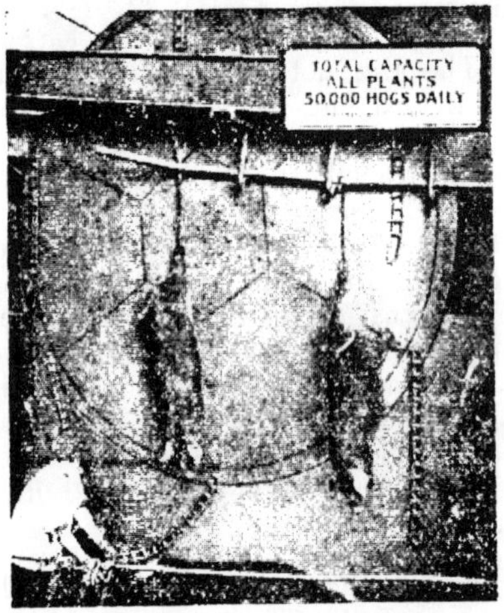

the hog-killing room. The hogs were driven into a pen at the side of each wheel. A hind leg of each hog in turn was shackled to a chain on the wheel and it was lifted as the wheel rotated. The animal was automatically transferred from the wheel to a sliding rail, which conducted it to the butcher, who, by a quick downward knife thrust, severed the jugular vein. The harsh countenance of the killer was livid, hideously distorted by an expression of utmost savagery. The jets of blood streamed forth upon his face and his clothes. He stood in a pool of blood and, brandishing his knife, with a demoniacal ferocity, slashed the throats of the creatures passing before him, helplessly shackled upside down by one hind leg. Twelve hundred hogs are slaughtered thus every hour—one every three seconds! All around, suspended on hooks or moving rails, were the porcine carcasses in different stages of preparation. From the bloody floor rose, like the fetid exhalation of a swamp, a livid vapor of that bluish color peculiar to the lips of human corpses. Those developed along the paths of higher intuition can sense, as father and I did, an uncanny psychical depression—a nameless desolation in the atmosphere, hanging like a death pall over the city of Chicago, Illinois. This melancholy influence is born of the shrieks and radio emanations of tortured animals—the curses of thousands of innocent creatures enduring horrible death throes in the grim pale of the shambles.

Immediately after the throat was severed, the struggling, screaming hogs were plunged into a vat of boiling water. Plunged *alive* into the scalding water! The draining of the blood when the jugular vein is cut (as said in Armour's descriptive pamphlet) takes but a moment. I say it is a *lie!* The animals do *not bleed* to death. They are *drowned* in seething liquid. The wails, the sighs, the agonized yells of the tortured victims pierced the foul air of that ghastly slaughter house. It was appalling, almost overcoming the senses with excess of horror. I espied a green door behind us, marked with a red cross and the word, "Stretcher." A person not retaining proper mental control or permitting the projective imagination to run riot might readily faint in that awful place. I mean a person who is a *real human being*; not

an insensate vertebrate ghoul. Was ever medical aid deemed a necessary adjunct to a visit to an orchard or botanical garden?

The intolerable stench of dead flesh pervaded the atmosphere; detestable, putrid, it clung to our clothes for several days afterward. Even washing could not remove it for a long time.

There is no age limit for visitors. The veriest child may view the grisly sights at Armour's slaughter houses, aye, is even welcomed to have imprinted upon its pure and tender unsullied mind the portrait of the most monstrous brutality conceived by human minds, toward the accomplishment of which the lofty scientific genius of mechanical invention is so degradedly employed!

We moved along the gallery, which declined to the floor, stopping before the dehairing machine. In a seemingly endless procession the dead hogs rolled from the machine, falling with a dull, sickening thud upon a moving platform. Workmen with knives scraped off what hair was left and the carcass was conveyed to various parts of the plant, traveling suspended from trolleys running on rails. A series of shower baths and cutting operations takes place as the carcass moves along, each worker having an allotted task Here is an extract from Armour's pamphlet: "Interspersed among the workers are United States inspectors paid by the government. A force of fifty, all trained veterinary and food experts, is employed at the Chicago plant. They place the government stamp, 'U. S. Inspected and Passed,' upon pure, wholesome meat only. The government inspection begins in the pens, where the animals are inspected and only those which appear sound and healthy are passed for slaughter. This inspection is continued rigidly at every stage of killing, cutting and packing. All Armour meat products are inspected by the government. (They need to be.) About 35 per cent. of the meat sold in this country (America) is NOT inspected, according to government estimates." The words in parenthesis are my own.

Pooh! "fulsome meat" should be said. Death is decay. Pure meat! Pure putrefaction? It is grimly absurd to speak of a carcass as "healthy." Inspected or not inspected, the fact remains, *no meat* is "wholesome." The moment cellular constructive activity in the physical body ceases, so-called death supervenes. Death is the result of the destructive cellular force in ascendancy over the building power. Decay sets in at once, until total disintegration takes place. You "civilized" human carnivora, what possesses you to make graveyards of your holy stomach? Are not the clean fruits and herbs furnished in abundance enough for your sustenance? Must you eat carrion to support life? Fie on this defiling of your bodies made in the image of God. By eating bloody flesh, you become bloody in mind and body. Your conscience sleeps; you are slaves of debased tastes. The taste of meat appeals to your perverted tongues and the voice of humanity is silenced, turning your noble forms into walking charnels. Awaken your higher senses, comprehend the cruel unnaturalness of meat diet and eschew it, thus elevating yourselves to grace the name of a human being. Be human, make appetite your slave, let it not be your master.

As the dead hogs traveled along, containers holding the viscera were noted opposite each carcass. These containers remain beside the carcasses until the government inspectors pass upon them. They examine the head and throat glands, then the various organs, no part of the body being cut open until the part previously opened is passed. Still suspended from overhead trolleys, the porcine carcasses were conveyed into the cooling room, where they remain for forty-eight hours, in a temperature of from 34 to 36 degrees above zero, to eliminate all animal heat. The government requires that all plants engaged in interstate commerce be kept clean and sanitary in every respect. Walls, floors and tables must be washed daily, tools sterilized and each workman must have clean linen frocks and keep his hands sanitary. But what about the rotten product itself?

The pharmaceutical (poison) products obtained from pork are, viz.: pancreatin, U. S. P.; pepsin, U. S. P.; rennase; amylopsin; trypsin; thyroid; corpus luteum.

From the hog-killing room, we passed to the pork cutting department. There we saw the carcasses which had spent forty-eight hours in the cooling room, being cut up and distributed, the various cuts despatched by way of chutes and otherwise to their destined departments. A conveyor takes fresh pork cuts, such as loins and roasts through several blocks of buildings to the shipping assembly room. The atmosphere in the pork cutting room is chilled by artificial refrigeration, so that the carcasses and cuts of "fresh" meat may be kept in an absolutely "wholesome" temperature. The dressed carcass supplies about 53 per cent. of cured meat, 30 per cent. of fresh cuts, 13 per cent. for rendering and 4 per cent. which is inedible is sent to a huge tank to be turned into fertilizer or other by-products. Armour's Star Ham—the "Ham What Am"—comes from young barrow hogs cornfed. The hams, after sorting and grading, are cured by a special process. "The hams are enclosed in a stockinet covering, a special Armour feature, which retains all the juices which ordinarily would drip away during the smoking, thus 'intensifying' the 'flavor.' Hams so covered are naturally more 'sanitary.'"

So much for the hogs. Leaving that place, we followed our party across an elevated street connecting the buildings of the plant, to the cattle killing and dressing department in another gloomy structure.

At one end of this huge slaughter house, the cattle were driven into pens, two or three in each enclosure, depending upon their size. A hammer blow delivered on their skulls, from above, stunned them and the next moment, upon signal, the gate of the pen was raised, its movable floor tilted forward and the poor cattle were deposited on the main floor of the killing room. We viewed this from the floor, just before mounting to the gallery. The animals rolled within touch of us, groaning, feebly struggling, being stunned, their piteous big brown eyes glazed with the terror of the death. Human beings imagine that the dumb creatures do not feel pain. It is not so. Just because their mode of expression is different it is thought the none. But they *have* a language of their own; they cannot think, but they talk to each other in their own peculiar way, feeling pleasure and pain; fear and courage are known to them, also. The cattle actually wept silently, pathetically, at sight of their murdered fellows. Oh, the cow mother weeps when her offspring calf is led to slaughter. She feels the same grief the human mother does at the loss of her child. Watch carefully when two animals are led to the butcher and you will know my meaning. Animals show their companions in sorrow noble generosity, and man can learn many good lessons from them.

While the stunned steers writhed on the floor, all piled upon each other, men ran about among them, stumbling and sliding on the floor, slippery with blood. The men shackled the hind legs of the steers, to chains which swung the animals upward and carried them past the "sticker" who stood on a platform above the knocking pens and severed the carotid artery in the

throat. The blood was partially drained as the carcass passed along, suspended from a rail. Several minutes later, the skinning and dressing operations start with the removal of the skin from the head. Extract from Armour's pamphlet: "Neck glands are first examined by government inspectors and when these are passed, the head is taken off. Then comes the removal of the hide, a job requiring considerable skill and care lest the hide be mutilated, in which event it would be considerably less valuable. A knife first makes a slit in the belly, then the worker partly cuts away the hide from the body. Meanwhile, others remove the skin from the forelegs. The carcass is then opened, so that government inspectors may examine the vital organs. Next the viscera are removed. Then a crew of workers, with saws and cleavers, cut the carcass down the backbone. A final inspection is made. The carcass is washed and dried. The sides of the beef go to the cooling room for 48 hours, until the animal heat has totally disappeared. From the knocking pens to the cooling rooms requires but thirty minutes. As the beef travels along, the viscera and the head of the animal remain with it until the United States government inspectors give the meat their stamp of approval. . . . Kosher cattle are killed according to the rites of the orthodox Jewish Church. These cattle are not stunned. Their legs are shackled and their bodies held suspended with the forepart touching the floor. A muzzle, with a long handle is slipped over the head and an authorized member of the church severs the jugular vein. Only the forequarter meat, that in front of the seventh rib, is utilized by Kosher trade, while the remainder is sold to the regular trade."

A member of the church, "authorized" to commit murder? Oh it is nothing less. The supposed representative of God, killing God's own creatures! It is monstrous—a criminal scandal to the name of the holy Jewish religion! Have the members of the Jewish church nothing more holy to occupy them, that they should imbrue their hands in slaughter? Kosher meat is a religious misconception of the great faith of Moses. Murder is murder no matter if a churchman commits it or one of the laity. The sixth commandment forbids it. It is, in a subtle sense, a *theft*—it is the forcible taking of that which does not belong to them, the life of another creature. Humanity has not yet learned humanity! When will it?

We ascended to the gallery. As we passed up, the dead beeves swung by on sliding rails, a foot from our faces. We were almost nauseated; the blood would have splashed on our clothing had we not drawn back. Looking down from the gallery, we beheld the strikingly ghastly spectacle of numerous cattle heads moving on a rail, flayed, the eyes gouged out leaving gaping holes, the horns still remaining and the whole dripping with blood.

A sign caught our eyes. It told us that "This is the most humane method of killing. Death is virtually painless, occurring in less than thirty seconds. The apparent struggles observed, doctors say, are merely convulsive contractions of the muscles." Those doctors are liars. "Convulsive contractions of the muscles!" That is adding aggravated insult to injury. I know death when I see it. I can sense it intuitively. I know that the wild struggles I witnessed were not "convulsive muscular contractions post mortis," but the tortured writhing of a *live* animal in *frightful pain*. Life does not become extinct for several minutes. Sometimes the hide is ripped from the creature not yet quite dead! I saw *that* in the sheep-killing room, which we next visited. Shocking, ghoulish barbarity!

The pharmaceutical (poison) products obtained from beef are, viz.: rennet (calf); extract of red bone marrow; pituitary liquid; corpus luteum; thyroid; thromboplastin solution; kephalin; suprarenals; pituitary powder and tablets; pineal gland; thymus substance (calf); ox-gall; peptone powder.

Three hundred and fifty cattle an hour are slaughtered daily in the Chicago plant of Armour and Co. Why is the price of milk so high? Why are thousands of children starving today for want of milk? *This is the cause, the slaughter of the cows!* The country is eating its capital, in the eating of beef. Meat eating is in itself the source of the ruin and damnation in the world, but feeding on beef is a very hideous ingratitude. When the human baby is helpless, the cow gives her own milk to feed it, even by robbing her tender calf of its rightful nourishment. Thus the cow becomes the human baby's fostermother. When the baby grows, he turns and chops up the cow, the animal that benignly fed him when he was small and weak, and devours her flesh. Basely ungrateful! It is fostermother murder—in the highest occult sense it is cannibalism! And you call yourselves civilized. Your actions villify your designation of "human being."

The report of the Department of Agriculture states that "more meat was consumed in the United States last year (1922) than in any previous year." The consumption was estimated at 16,333,000,000 pounds, which amounts to 149.7 pounds a person, applying the July 1, 1922, estimate of population. The per capita consumption showed an increase of 6 pounds over 1921 and 6½ pounds over 1920. That department says, "Coupled with the decrease in the exports and somewhat higher prices to purchasers in 1922, these figures evidence a satisfactory state of the home market, due doubtless to the prosperous condition and better purchasing power of the people generally." The *beef* gain was the greatest, 3.6 pounds a person, the increased consumption of pork was only slightly less, 3.1 pounds. "The conversion of corn into lard by means of the hog, is one of the major industries of the country, the product of which goes to all parts of the world," again the department says: The 67,-050,745 hogs slaughtered in 1922 yielded more than 9,000,000,000 lbs. meat (more trichinae spiralis) and 2,333,000,000 pounds of lard. This is the highest production of lard on record and as the exports were less than in 1921, it follows that the total consumption also was highest. The per capita consumption, however, was exceeded in 1916. "The adverse conditions of foreign exchange and the poorer purchasing power of European countries have seriously affected the exports of meat products. Foreign trade in beef has practically ceased." We do not want such foreign trade. Keep the cows alive in this country and milk will soon sell here at five cents a gallon. More meat means more dis-ease, more doctors, more deaths, more undertakers, more misery and suffering.

From the cattle slaughtering place, we passed to the sheep killing room. It has a killing capacity of 1,000 per hour. The sheep were killed in the same manner as hogs, shackled by a hind leg, swung upward on the rotating wheel, transferred automatically to the sliding rail and carried past the butcher who cut their throats. In Armour's pamphlet it says, "Here, as elsewhere, careful scrutiny is given by government inspectors to the neck glands, vital organs and other parts, to make sure that only sound, healthy stock is converted into food products. As the worker's knife opens each part of the carcass examination is made by an inspector. After the blood is drained and the first inspection is made,

the pelt is taken off, a matter of seven distinct operations, following which the long lines of workers perform their allotted tasks in opening the carcass, removing the viscera, and trimming, preparatory to the trip to the cooler."

I saw the killer lose his temper and kick the struggling, hanging, bleeding sheep, pushing them so close on the moving rail that they struck each other's faces with their legs, and some other men roughly shoved them along with knives, tearing the skin from the living animals.

Further on, we viewed the removal of the thyroid glands from the carcasses' throats. Worthless bluff of medicine! Suprarenalin, an astringent and heart stimulant made from the glands of sheep, is worth more than $5,000 a pound, but it requires the glands from 125,000 sheep to make one pound of this unnecessary pharmaceutical product. Spectro-Chrome replaces it entirely now.

Next we visited the refinery, where lard and shortening is made. Big, open kettles, fashioned like double boilers, are used in the process. Afterwards we saw the oleomargarine being manufactured. Oleomargarine is made from a hash of meat fats, vegetable oils, pasteurized milk and dairy salt.

We passed through the canning department where meat was being packed in cartons, tins and glass jars. Thence to the power plant for generating electricity and producing refrigeration. How calm this place after the screeching horror of the slaughter houses!

Last of all, we visited the cooling rooms. Quotes Armour's pamphlet: "The beef cooler at our Chicago plant is the largest single unit cooler in the world. Here, the sides of beef are kept for forty-eight hours, during which the animal heat is dissipated and the meat chilled. A beef cooler should not be confused with a freezer or a cold storage. The American public eats chilled meat, not frozen meat. A trip through the cooling rooms is a most impressive experience. Here, you seem to be almost in another world. It is cool, the temperature always at 34 to 36 degrees F. It is clean, with great rows of the 'finest, purest meat' you ever saw, suspended there before you. The floors are covered with fresh sawdust. Walls are of white enameled tile. The chilling process keeps the meat 'fresh' and well-shaped."

Frozen, chilled or hot, it is the same bloody, stinking filth. As we passed through the grisly lines of carcasses, a deep voice behind us, said, "Gosh, how I'd like a juicy steak from one of these. Say, this beats your boulevards!" The smacking of lips followed. We turned, unsurprised when we saw who had uttered that remark. Fat and bloated, he was the ideal personification of indulgence to carnal appetites.

With a feeling of relief, we left Armour's plant and returned home.

The day is not far, when the soundless avenging rod of the Almighty's Benign Powers will overtake Armour's and associated industries and the awakening of the slumbering, but potential conscience of the glorious United States will wipe clean out by a sweeping prohibition every such hellhouse in the land and utilize live stock as a blessing to agriculture in place of feeding the stomachs of human beings with their dead products. Wake up, my America, to your true service to humanity; stop slaughter of poor animals and wars will stop with it.

(To be concluded in next issue.)

Progressive Physicians Endorse Spectro-Chrome Therapy

By special invitation and kind arrangement of John Wesley King, M.D., Dean of the College of Electronic Medicine, of Bradford, Pennsylvania, commencing December 4, 1922, a Special Course on Spectro-Chrome Therapy was given to 14 Doctors of Medicine, 2 Doctors of Osteopathy, 1 Doctor of Dental Surgery and 3 Physicians' Assistants, at Dr. King's Laboratory in Bradford, Pa. The studiousness of the class was phenomenally praiseworthy and the facilities placed at the disposal of Spectro-Chrome Institute by Dr. King were so exceptionally thoughtful that the entire course was speeded like lightning without a hitch. For the success of that class, all the best thanks are deserved by Dr. J. W. King, whose liberal views and broadmindedness merit lasting gratitude.

At the close of the course the following testimonial was presented to the Originator of Spectro-Chrome Therapy:

SPECTRO-CHROME INSTITUTE

Bradford, Pennsylvania

We, the students attending the Course of Instructions in Spectro-Chrome Therapy, December 4, 1922, given us by Colonel Dinshah Ghadiali, M.D., of Philadelphia, Pennsylvania, fail to find suitable words to express our deep appreciation for the information given us by the Master, in his unique presentation of a science, little understood by the masses which concerns one's welfare in the hour of physical sick-states. We assure you, Dear Dinshah, that the knowledge imparted to us will prove of incalculable benefit to the sick in their hour of suffering, and the relief we can give them, through your system, *Spectro-Chrome Therapy*.

Last but not least, we give you double thanks for the information gained in medical and metaphysical lore, of which you are a lord indeed!

We wish you God-Speed in spreading the knowledge possessed by you, to impart to others the great gift given us. Man can never reward you for the great work you are doing. We pray that many more years will be spared you at this work.

(19 signatures)

SPECTRO-CHROME

APRIL 1923

LEMON 1923

VOLUME 2

NUMBER 4

Devoted to: Spectro-Chrome Therapy **Dedicated to: Service of Humanity**

CONTENTS Page

Dinshah's Platform—Spectro-Chrome Therapeutic Area Chart—Spectro-Chrome Therapeutic System 242
Intermediate Equipment—New York Office—Information about Remittances—Business Routine during Tour—To Raise Postal Rates—Pulling Teeth by Incorrect Diagnosis—Pharmacist as Human Servant—The Institution of Marriage; What a Woman Thinks and What I Think—Poison Gas for Paresis and Tuberculosis; Use Spectro-Chrome—United States Supreme Court Upholds Principles Underlying Spectro-Chrome Therapy and Spectro-Chrome Therapists; Study the Reasons Why—Advice to the Mother of an Epileptic—Grapes Are Sour—Opening of the Boston, Massachusetts, Class—Crazy Doctors of Medicine—Intoxicating Liquors and Life—How Do You Get That—An Unsolicited Endorsement—Porter Came to Porter's Port 243-256
Harry and His Appendix; a Rhode Island Tale (by Kashmira Dinshah, S.C.T.) 256
Dr. Kate W. Baldwin Unfurls the Spectro-Chrome Banner in Washington, D. C. (Reported by Miss Grace E. Griswold, S.C.T.) 258
Violent Mental Cases, High Blood Pressure, Burns, Blood Poisoning Relieved by Spectro-Chrome Therapy (by Dr. James C. Daniels, S.C.T. ..260
Ivy Poisoning, Nasal Catarrhs, Heart and Bladder Dis-Eases Relieved by Spectro-Chrome Therapy (by Gertrude Burgess, D.O., S.C.T.) 261
Papilloma of Bladder, Hemorrhage, Mastoiditis Relieved (by Martha J. Peebles, M.D., S.C.T.) 261
Light of Life (by Mrs. Mary P. Hynson) 262
What America Needs (by Irene Grace Dinshah, S.C.T.) 262
Violet Flower State Welcomes Spectro-Chrome Therapy 264
Price List ..... Inside Back Cover
Registration Notice of Trade-Marks—Notice to Purchasers and Patients—To Our Contributors—Advertisement Rates ..... Outside Back Cover

In course of life you reach a time,
When Doctor, Lawyer, Priest alike,
Does fail to give the sought relief,
To troubled mind or burdened heart—
Then come to me;
I help you to help yourself.

SPECTRO-CHROME.

Editor

Dinshah P. Ghadial

Subscription:—January to December 1923 ..... $3.00
All Subscriptions Taken For Full Year Only And Expire With The December Issue
Any Single Copy 25 Cents. Back Numbers Available From June 1922.

NEW NEW YORK OFFICE

On May 1, 1923 the New York Office will vacate 171 Madison Avenue and move to the Metropolitan Building, Number One Madison Avenue, Room 201A. The present telephone connection will be discontinued until the 10th of May 1923. Thereafter, the same number Ashland 6232 will be resumed.

Business will go on as usual undisturbed during the period of change; however, your indulgence is requested.

INFORMATION ABOUT REMITTANCES

So many people have recently acquired the inconvenient habit of changing their mind a few days after placing their orders, so many others have gotten into the astonishing habit of sending or handing personal checks with funds insufficient to meet at their banks and so many more have become absent-minded (perhaps through change of weather) as to what they want or what they wish me to do, that I repeat here what I said often before so clearly.

Personal checks *are certainly accepted* by the Spectro-Chrome Institute from responsible persons, but, *the orders do not go out* until the checks are cashed and payment received through the banks. This is one reason of apparent delay in the execution of orders of certain persons, for which delay I am blamed constantly.

Out of town personal checks had therefore better be sent *certified* or still easier send Postal or Express Money Orders or Bank Cashiers' drafts.

I very much regret the obvious inconvenience to my esteemed and beloved "friends," but I shall thank them cordially to let me know what I should do otherwise when my safe carries in a special drawer under double lock and key several thousand dollars of utterly worthless (un)negotiable paper. Bless your kind heart, please find someone else to practise your arts; why pick on me? I am doing my best to serve you.

I learnt my bitter lessons long ago and mean to profit thereby; yet I am open to learn more always, but not thus through you.

Do not place your orders unless you desire shipment; do not send checks unless they are currency.

It means *you*; if the cap fits wear it; *I* mean business. Time is too valuable and one life is too short to complicate by bookkeeping of bad debts; I have yet more to accomplish—very much more than I already did—help me do it, please.

BUSINESS ROUTINE DURING TOUR

Please Read Carefully

The same business routine as was

outlined in the February 1923 issue will be followed.

Bulbs, charts and similar supplies may be had for cash at the Central Office or the New York Office.

The situation for slides is the same as before. Slides *must go by mail* to 2401 N. Broad Street, Philadelphia, Pa., duly packed and insured.

No slides will be given out "at the counter."

Kindly remember; Spectro-Chrome Institute does *not* maintain a credit department; all orders and transactions are *Net Cash with order,* F. O. B. New York, N. Y. or Philadelphia, Pa. from wherever they are shipped. Our responsibility ceases with shipment.

No C. O. D. orders, please.

Spectro-Chrome Institute in its business transactions has no exceptions and acts on a single policy with all.

Will you my friend, help me in bearing my standard of Spectro-Chrome Therapy by curtailing useless correspondence? I did not yet succeed in producing a 25th hour in the day. Letters outside of my business routine are not desired and will *not* be answered. No weather reports or racing bets or scandal talks or how the hat shoe or coat of some one fits have place in the busy life of the staff of Spectro-Chrome Institute. As the scope of our work increases, our responsibility increases and we shall be obliged if friends will help us by sending reports of their Spectro-Chrome work, in place of grumbles—create first your right to grumble by sharing some of our arduous toils.

TO RAISE POSTAL RATES

A dispatch from Washington, D. C. states:—"The Postoffice Department is about to launch a count of mail to determine the average cost of handling it, with a view to a readjustment, either of postal rates or contracts with transportation companies. The count, which will begin in selected cities Septeber 21 and end October 20, will embrace the four classes of mail.

It is estimated that the cost of the postal service is about $500,000,000 annually, and revenue from the sale of stamps comes within $65,000,000 of that amount. Officials say that eventually the postal service will pay for itself. There is a possibility that Congress may see fit to raise postal rates."

There is nothing in our Postal System to brag much about. Throughout my travels, I came in touch with different methods of Postal Government, but, have no reason to believe our system here any way equal in efficiency to the British Post Offices. Our complicated zone systems and the odd postal privileges granted without common sense are nothing short of nuisance to the public. There is no efficiency and accuracy in the Postal Service of the United States and most of the time, special delivery letters reach hours behind regular service, registered letters are delivered to strangers without authority, restrictions are imposed without regard to public convenience and in every manner inefficiency and insolence is meted to its owners (the public). The whole system needs, in my opinion, a thorough overhauling by a strict non-partisan business man. As it is, I do not care to call its operations anything approaching the British Postal Service for public good.

PULLING TEETH BY INCORRECT DIAGNOSIS

Dr. Kurt H. Thoma, a Boston dentist, speaking before the American Laryngological Society in convention at Atlantic City, N. J. said that incorrect diagnosis was responsible for the merciless extraction of a number of teeth which otherwise could have been saved. But what can those tooth-pulling dentists do? They know nothing else to strengthen the teeth or prevent the rot in the gums. Spectro-Chrome Attuned Color Waves judiciously used and carnivorous food debarred would produce the desired beneficial results without much expense, preserving the beauty of the mouth along with its effective mechanical power.

PHARMACIST AS HUMAN SERVANT

In a recent writing in a Providence, Rhode Island newspaper, the pharmacist is said to have summed himself in a few words, "If we can be of service to humanity, we will feel amply repaid." I am glad to read the pharmacists are so anxious to serve humanity and am doubly pleased to lay down the route. Abolish alcohol from your Tinctures and Extracts, stop dealing in Opium, its derivatives and other dopes, avoid the conversion of your business into liquor saloons, cut out from your counters the half million "patent remedies" (poisons) and go exclusively in the sale of candy and ice creams and you will be blessed in heaven by the humanity you thus serve and save. There is no other way brothers.

THE INSTITUTION OF MARRIAGE—WHAT A WOMAN THINKS AND WHAT I THINK

Writing in the *New York Evening World* Mrs. Helen Carter of 63 St. Mark's Place comments in a way unusually frank for an American woman. I have pleasure in presenting her views to my readers, because there is surely some screw loose in the married life of United Staters. She says:

"Contrary to the established opinion that marriage as it exists to-day is a useful social institution, we find it to be a snare and a menace and mainly for the reason that men are too chivalrous, and considering that their wives belong to the weaker sex allow themselves to be exploited by them. The result of such point of view is that marriage is almost always unhappy, as evidenced by the enormous number of divorces and tragedies that are worse than divorces. Why are divorces so much on the increase? Why is it that our grandfathers and grandmothers were not suffering from that social evil to the extent that we do? Because in grandfather's time man was the head of the family and the vow to love and obey was carried out literally. As Milton said: "Man looked to God, but woman found her God in man.

Undeniably there were certain evils of a reverse kind then. Women were too suppressed, as they still are in the Orient, but those were individual cases. At any rate, no such widespread epidemics of separation and divorce took place. Then the pendulum swung the other way and woman is now dictator in the family. Woman now has not only equality but a great many special privileges and still demands more. But she forgets that she is naturally unfit for the tasks of the opposite sex and for many a reason. And men, unfortunately, do not take it very seriously, and out of a sense of false chivalry and good nature allow their wives to assume superiority and they themselves become subordinate.

Do they forget, these husbands, that nature cannot be conquered? That no matter how far civilization has carried the human race, the nature of both man and woman can never be changed to such an extent as to be reversed? Woman was created to be man's helpmate, not his equal and certainly not his superior, and no amount of agitation and social advance can change it.

Nor does the economic situation make any difference. The extravagance of the rich wife, the giving in to her endless whims and fancies will often ruin a rich man. The same is true of the poor man. The latter cannot give her all the good things she craves, so the least he can do, according to his ideas, is to sacrifice his individuality, to submerge himself and become her slave. How often do we see this type of a husband after a hard day's work come home to run errands for his wife to the grocery store, wash dishes and take care of the baby!

All this not being in accordance with the course of nature, something must give way, and does; and that is the happiness of the home. Either there is a sudden revolt on the part of the husband, which means going to extremes, or the weak feminine nature not finding anything to hold her back, becomes entirely perverted and seeks new thrills, and experiences which finally lead to the divorce court. Those conditions did not obtain a few generations ago for the reasons I have mentioned, nor do they obtain in families of the manly man, the so-called brute. The curious phenomenon about the brute husband is that his wife feels quite happy, a good deal happier than her pampered sister, the wife of the obedient husband, and it is a known fact that woman will love a strong man.

And this, as it seems to me, is the true solution of the problem—let the husbands become the true heads of the families, the senior partners, so to say, or in other words obey what nature ordained. Marriage will then enrich one's life, and a divorce and an unhappy family will become a rarity."

After studying the subject from various angles, my firm conviction is that the hustling life of the business man of the United States makes him a very inferior lover. With all her boasting of man being nothing to her, even the busiest business woman cannot avoid craving what makes her the "woman"; the absence in American love-making of those subtle etherial thrills that accompany true mating on all the levels of physical, psychical and mental aspirations conduces to much mismating and when such abnormality is mixed with the absurd 48 marriage and divorce laws of the country, the incongruity is heightened to the polarity of ridicule of God's highest aims in life. The clergy have filled the mind of the credulous that "marriages are made in heaven" and that is another reason why they generally end elsewhere. Marriage is the noblest function of animal life; its responsibilities are far-reaching and its actual functions extremely important, but, its ties must not be so tight as to compel two mismated unfortunates to sleep in the same room in the same bed and procreate children for the nation without having the bubbling influences of potent love. Where love disappears, marriage becomes worthless, useless, eventually positively pernicious and a menace to society. As it exists, it is in many cases nothing short of vicious adultery licensed and sanctioned by law. The marriage and divorce laws *must* be unified and unicoded and divorce *must* be granted without compulsive perjury as long as the fruits of the marriage (the children) are taken care of in such a way by the parents as not to encumber the State. The easier the Divorce the more difficult will become the Marriage—think over the true psychology of human nature.

POISON GAS FOR PARESIS AND TUBERCULOSIS; USE SPECTRO-CHROME

At every step of new scientific research and investigation the so-called specific and potent remedies brought to light by the drug therapists, strengthen the mathematical accuracy and corroborate the preciseness of the action of Spectro-Chrome Therapy. Says the *Evening World* of New York:—"The possibility of curing tuberculosis, influenza, paresis and other diseases by the use of poison gases is believed by Army experts to be entirely possible, following tests now being carried on at Edgewood Arsenal, near Baltimore.

It is asserted that through experiments with chlorine gas, the chemists have established that epidemics of grip and colds may be checked almost instantly by the introduction of weak concentrations of the gas into the rooms occupied by those exposed.

That mustard gas is a specific for tuberculosis apparently seems to be demonstrated by the experiments conducted under the direction of Lieut. Col. Edward B. Vedder of the Army Medical Corps. Guinea pigs, inoculated with tuberculosis germs and a concentration of mustard gas were apparently rendered immune by the gas and failed to contract the disease. An equal number of guinea pigs, inoculated with tuberculosis germs and not subjected to the mustard gas treatment, contracted the disease.

As a result of experiments with the burning gas known as lewisite, the chemists have evolved what appears to be a remedy, if not a cure, for paresis and locomotor ataxia. Lewisite is composed of arsenic and acetylene gas.

Dr. A. S. Loevenhart of the University of Wisconsin has been studying the effects of lewisite upon the human system in conjunction with the chemists at Edgewood Arsenal. He has the records of forty-two persons committed to insane hospitals with paresis, which have come under this treatment. Of that number twenty-one have been cured and have left the hospital.

A professor of the University of Arkansas, during a mild epidemic of influenza, subjected 300 students to a mild concentration of chlorine for five minutes each day, with the result that the epidemic was reduced among that group 90 per cent.

According to Army officers, there is almost no limit to the practical application of the use of the so-called poison gases in combatting disease."

They are certainly on the right path though groping in the dark and they will finally land into Spectro-Chrome Therapy when their ancient cobweb-covered medical ideas are revised and rejected. What they are "searching" today, the Originator of Spectro-Chrome Therapy has been *teaching* a number of years as a glance at page 242 will indicate. The Green Wave (of Chlorine) *has* the Disinfectant, Purificatory, Antiseptic and Germicide effect and the Orange Wave effect of Arsenic and Acetylene is a Respiratory Stimulant, Thyroid Energizer and Lung Builder. Reports of the highly beneficial effects of the Green and Orange Attuned Color Waves are common to Spectro-Chrome Therapy and dis-eases that puzzle the medical men do not in the least puzzle the knowing Spectro-Chrome Therapist. A course on Spectro-Chrome Therapy would be the best post-graduate training a highly competent medical man even may receive without lowering his dignity. It will be his best eye-opener.

ANOTHER MEDICAL BUBBLE BURST

For years Doctors of Medicine have been prescribing the filthy oil of cod liver for the amelioration of lung troubles. Now Dr. Michel Mouriquand of Lyons, France sounds a warning against its use! He says in part, "Long experiments involving the use of cod liver oil in guinea pigs and rabbits show that while cleansing the blood and preventing, even curing, scorbutic diseases, there is also the danger of promoting bone maladies for which there is no cure. Better, the doctor says, rely upon castor oil, which though unpleasant is much surer in results, whereas even non-Fascisti in Italy are not known to have succumbed to the maximum doses administered as punishments by Mussolini's disciples."

Better be punished with the pure vegetable castor oil and be purged of intestinal mud than to be dosed with the foul products of the liver of a fish. How do the doctors expect to purify the blood in the entire body by deluging the stomach with the stinking oil of a fish? When will these fool ideas stop and when will humanity turn to sane views of life?

UNITED STATES SUPREME COURT UPHOLDS PRINCIPLES UNDERLYING SPECTRO-CHROME THERAPY AND SPECTRO-CHROME THERAPISTS. STUDY THE REASONS WHY.

A report from Washington, D. C. states as follows:—"The paramount right of the individual was the basis for a decision by the Supreme Court nullifying at one sweep State laws of Nebraska, Iowa and Ohio, which made it illegal to teach any language but English in private, denominational, parochial or public schools. The laws were aimed at instruction in the German language.

Two Justices dissented, Holmes and Sutherland. The former, in a brief statement, justified an experiment in accordance with the restrictive laws as necessary to an assimilation of foreigners, who, living in large groups, would have limited opportunity to become familiar with the prevailing tongue of this country.

A Nebraska case was chosen by the court for its principal opinion, which was read by Justice McReynolds. In another opinion Justice McReynolds reviewed the cases coming from Ohio and Iowa.

The problem involved was stated as being whether the State law unreasonably infringed upon the Fourteenth Amendment, that no State "shall deprive any person of life, liberty or property without due process of law."

"Without doubt," the Justice continued, "it denotes not merely freedom from bodily restraint, but also the right of the individual to contract, to engage in any one of the common occupations of life, to acquire useful knowledge, to marry, establish a home and bring up children, to worship God according to the dictates of his own conscience, and generally to enjoy those privileges long recognized at common law as essential to the orderly pursuit of happiness by free men."

Saying education long has been recognized by the American people as of supreme importance, Justice

McReynolds added:

"Mere knowledge of the German language cannot reasonably be regarded as harmful. Heretofore it has been commonly looked upon as helpful. Plaintiff in error taught this language in school as part of his occupation. His right thus to teach and the right of parents to engage him so to instruct their children, we think, are within the liberty of the amendment."

This is the very ideal I always have maintained and for which I have idolized the Constitution of the United States of America. The medical profession and the laws regulating the practice of "Medicine and Surgery" may be quite welcome and within their rights to choose, select and direct whatsoever laws their lobbied legislations may enact, but, as soon as they place their fingers on each and every Drugless Cult or System they are exceeding the birthrights of humanity, which are Liberty, Property, Security and Contract. A person may be required to be licensed under certain laws to practise "Medicine and Surgery," but Drugless Cults and Systems of Healing are neither "Medicine nor Surgery" and the medical laws cannot touch them. Subsequently, the medical profession, in its horrible jealousy, got other laws passed through by pressure on the legislations, terrorizing the various Drugless Practitioners and succeeding partially in placing their thumbs on the carotids of some weaklings. The underlying reasons of such partial success were the weaknesses in the public policies of such systems and the lax ways in which their teachings were conducted. The main feature of the fight hinged on the use of such Drugless Systems of the same textbooks on Anatomy, Physiology, Bacteriology, Histology, Pathology, Biology, Diagnosis and a number of other "lologies" as were taught in the regular medical schools and were "changed" every ten years to suit them. It was also owing to the followers of such Drugless Systems, appending the title of Dr. (Doctor, Driver or Debtor?) before their names, which irritated further the older school-members, perhaps deservedly.

When Spectro-Chrome Therapy was perfected and brought to the stage of commercial practical clinical application, the Originator had waded fully through these obstacles and legal complications. It was thought unwise to launch out, until this new Drugless Science could go before the whole medical and non-medical world and challenge boldly the right of the so-called Medical Profession to dictate to humanity what system should be used and how in order to alleviate the sufferings of the ailing. Step by step, in Spectro-Chrome Therapy, the Originator evolved a system of healing which not only had nothing in common with any Drug or Drugless System, but which at one stroke swept clean the motheaten, ancient, valueless, destructive and inhuman doctrines foisted under the name of the healing art by ages of uncivilized and civilized so-called "sciences." In Spectro-Chrome Therapy, the Originator boiled down the known and to-the-general-public unknown knowledge of physical and metaphysical fundamentals, evolving a complete absolutely new science, art and profession, whose follower and practitioner can do everything the Drug or Drugless Systems aim at performing and at the same time whose knowledge did not need the usual Anatomy, Physiology, Histology, Pathology, Biology, Bacteriology, and the intricate like. At one clean sweep, the Originator of Spectro-Chrome Therapy not only discarded the filthy, foul, dirty, obnoxious vaccines, serums, blood and

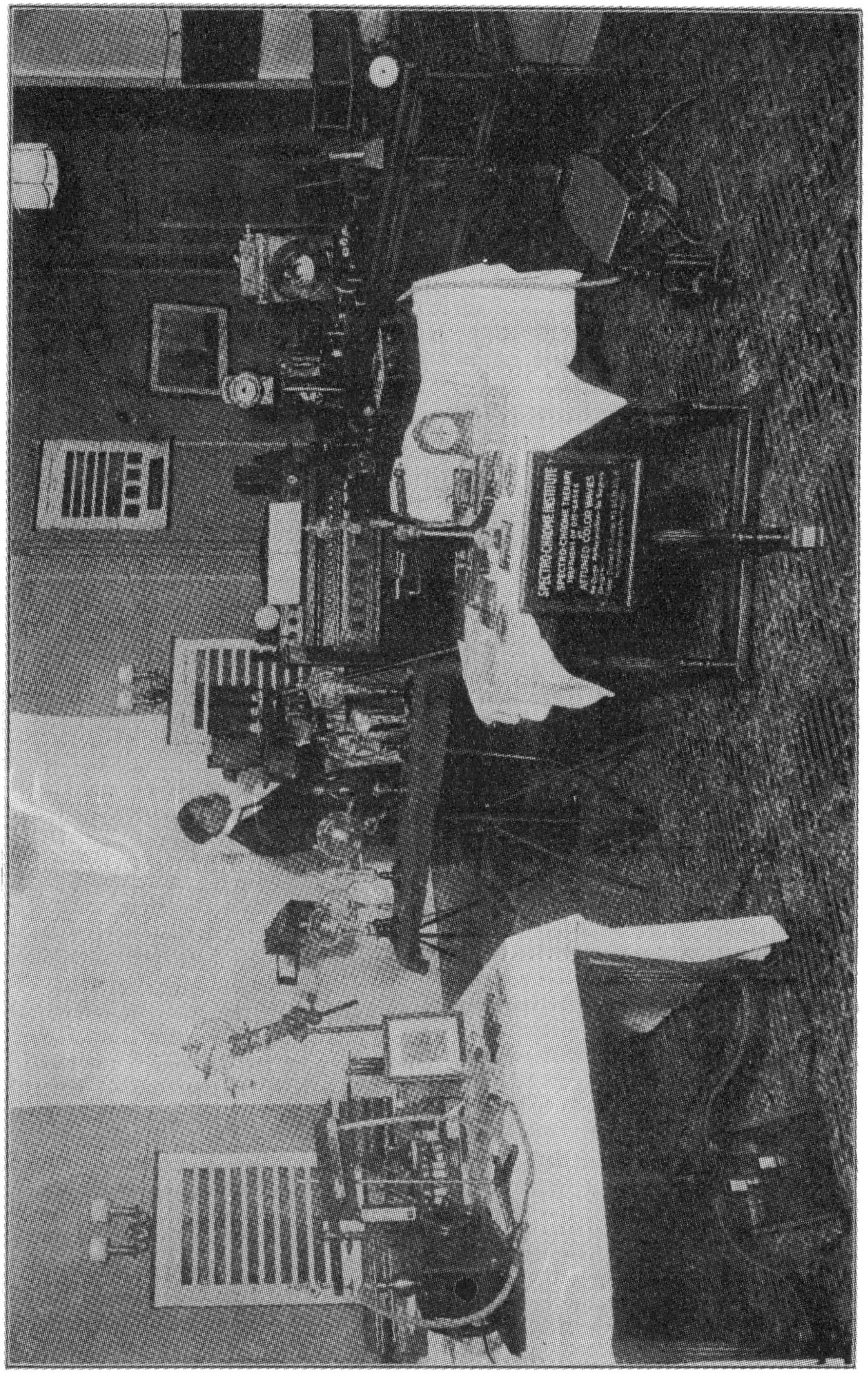

urine therapies and the like, but, he threw into the scrap-heap of an ignorant past all the so called methods of differential diagnosis of the various "strains" of human disorders, by which the ignorant public was hoodwinked into believing the regular Doctors as the "allwise."

The power of Spectro-Chrome Therapy in all these is so great that graduate after graduate is now discarding old views and bowing to the Divine Laws of Nature as enunciated in Spectro-Chrome Therapy. They are now fast awakening to the fact that Spectro-Chrome Therapy needs no support of other systems to carve its way, in as much as it stands unrivalled on its own merits. It holds forth to the entire medical world the most potential challenge in the Healing Art and in as much as it follows entirely novel and universally unknown or unrecognized Laws of Nature, its followers and practitioners have positively nothing to fear in the existing manmade legislation governing other systems of healing.

The decision of the learned Supreme Court of the United States of America positively upholds the absolute birthright of the individual in education, worship, orderly pursuit of happiness, the right of contract et cetera, and on the same principles will the future of Spectro-Chrome Therapists be resting. No legislation can touch either as long as the system and its practitioner has nothing in common with Drugs, Manipulation or Surgery. No law can be enacted and enforced which may debar an individual from being treated by a Spectro-Chrome Therapist or may debar Spectro-Chrome Therapy being practised. Spectro-Chrome Therapy is the Natural Octopus of the Healing Art; its every tentacle has the grasp of Nature's Unchangeable Laws and until the Sun, Moon and Earth begin new tunes of oscillatory periodicity, its principles and fundamentals will survive against all odds.

Drugs and Destructive Surgery have had their day and the public is craving for more rational and sane methods. In Spectro-Chrome Therapy lies the direction of the future Universal Healing Science and the quick expansion of the activities of the Spectro-Chrome Institute in spite of pessimistic imitators is the best proof of its efficacy and potentiality.

At the time of writing this, the Originator is establishing classes for Personal Intensive Training at the approximate rate of one each month in different States. Each class meets daily, six days a week and receives two discourses a day with numerous convincing demonstrations, illustrations and experiments proving the whole system at each step and stage. The fascination attached to this original work is so positive that on the last session, it is practically impossible to make the students believe it is time to disperse.

Those who graduated from the Spectro-Chrome Institute in its early years, when the activities of the Originator were centered in a single office will do well to glance at the picture on the next page. It is of a part of the lecture hall of the Spectro-Chrome Institute, at Providence-Biltmore Hotel, Providence, Rhode Island, where the thirteenth class received its training. It is easy to note what an amount of time, money, energy and conscientious labor is spent by the Spectro-Chrome Institute in giving a thorough training in the Latest Revelation in the Healing Art—Spectro-Chrome Therapy — Spectro-Chrome Institute is actually a University of Learning all alone.

There is no similar institution any-

where in the World, where the student is taken in hand as mere raw material willing to listen only and where from being an ignoramus or an all-knowing skeptic he is patiently led step by step until he is turned out into a finished product with full conviction that "there is much to learn *outside* of the medical colleges." The training given is unobtainable anywhere else and the fact that without outside advertising or trumpeting fanfares of so-called boosters the work of the Spectro-Chrome Institute is quietly paving the way for the attainment of the goal of "Therapeutical Emancipation," stands as the highest credential and testimonial to the appreciation of the unselfish labors of the Originator and his devoted staff. They shall ever be at the Service of Suffering Humanity.

ADVICE TO THE MOTHER OF AN EPILEPTIC

To the editor of the *New York Evening Journal* a distressed mother Mrs. C. D. wrote:—

"Is there any cure for epilepsy? Do you know of any specialist in that disease? I have a daughter who suffers from epilepsy and would appreciate any information you can give me, as I have tried more than twenty physicians and numbers of other remedies besides without satisfactory results."

To this communication the editor of the said newspaper replied:—

"There is unfortunately no absolute cure known for epilepsy, one of the most ancient, strange and baffling of all diseases.

Eventually science will conquer that disease undoubtedly as it has conquered others. Meanwhile, the best that can done is by careful living and diet to modify the violence of attacks, which sometimes pass away entirely.

For the benefit of our readers we have consulted one of the ablest doctors in New York. Below appears the letter received from him.

Sufferers should not try "outside cures" or "quack" remedies. Anybody who says that he can "cure" epilepsy is deceiving you, holding out false hopes and ought to be put in jail.

This letter, written by a thoroughly scientific physician, will be recognized by other conservative physicians as embodying about as much as is now known.

Before using any drugs in the treatment of epilepsy, a doctor, of course, should be consulted. And it may be remembered that in that, as in some other diseases, drugs are of comparatively little avail. Wise living and wise diet are most important.

A Doctor on Epilepsy

March 20, 1923.

Editor *New York Evening Journal:*

Dear Sir: I know of no "cure" for epilepsy. However, a great deal can be done to control the disease by suitable diet (strictly vegetarian), by careful regulation of the bowels and by the administration of small doses of luminal. On this regime my epileptic patients have gone for a whole year without a seizure, whereas, before, the seizures had not been infrequent. Of course, a great deal depends on how long the illness has lasted and the severity of its original onset. Any neurologist should be able to use the methods suggested.

With kind greetings, I am,

Very truly yours,

M.D."

It is a matter of surprise that the editor Arthur Brisbane should publish anything like a favoring of "strictly vegetarian" diet." He wrote last year and is always writing *against* vegetarian diet and now it is held as a curer of this formidable evil. Neither he nor his doctor friend realizes that epilepsy is a sex disorder and that in the Purple Systemic and Orange on Areas Numbers 9 and 10 is a powerful remedy according to Spectro-Chrome Therapy. Let the mother of this girl leave the M.D. and go to the capable S.C.T. and she will find better results than in drugs. Drugs are to be avoided like fire near gasoline.

GRAPES ARE SOUR

In order to encourage Pasteurizing milk and other substances, the advocates of that system of artificiality insist that the process is beneficial and should be resorted to, yet, it does not appear as if they know the fundamentals of the *source* of dis-ease. Writing about temperature and bacterial growth, Safe Milk says:—

"Bacteria have minimum, optimum and maximum temperatures of growth. They multiply fastest at their optimum temperature; any variation from it, either up or down, retards their growth. Different bacteria grow best at different temperatures. Those that grow in milk seem to prefer about 72 degrees Fahrenheit.

At ordinary low temperatures bacterial growth is inhibited, but the germs still exist and multiply rapidly after the temperature is again raised. It is much easier to kill them by heat, however. The maximum temperature at which the germ of tuberculosis, the most resistant of any disease producing organism, that could get into milk, is safely below the pasteurization temperature (142-145 degrees Fahrenheit.)

Pasteurization is 100% efficient in killing the pathogenic or disease producers, and about 99% efficient as regards the ordinary souring or the inert bacteria of milk. This lengthens the life of milk, but, of course, does not prevent its souring.

After pasteurization, milk should be kept at as low a temperature as possible. The milk bottle, previously sterilized, with its tight-fitting cap gives ideal protection to milk, but it will not keep it cold; the housewife must see to this and also that it is kept covered at all times."

I should like to know where the "pathogenic" germs come from. Germs are *not* the original *causators* of dis-ease; they are present in pathologic conditions as semaphores indicating merely the *existence* of what is going on. The World will some day soon wake up to scrap all these wrong ideas. Pasteurizing milk, instead of making milk better removes its vitality and the fact that it sets back the natural souring through the "inert bacteria" is the signal proving the unnaturalness of the universally accepted views.

OPENING OF THE BOSTON, MASSACHUSETTS, CLASS

The fourteenth Spectro-Chrome Therapy class will open at the Hotel Westminster, Boston, Massachusetts, at 7 P. M. on Monday June 17, 1923. This class will study four hours each night, six nights a week, thus completing the full course of 24 discourses in 2 weeks.

There will be delivered two public lectures on Friday and Saturday the 15th and 16th June 1923 at the Recital Hall (Jordan Hall) Huntington Avenue and Gainsborough Street, Boston, Mass. at 8 P. M.

CRAZY DOCTORS OF MEDICINE

In a certain famous hospital in Pensylvania, a case of Diphtheria developed. Another case of a girl was there just recovering from an exceedingly serious and grave illness. The life of this girl had only been saved by the exclusive use of Spectro-Chrome Therapy and through the care and skill of the surgeon attending. Notwithstanding this, the Health Authorities injected Diphtheria Serum, *without the knowledge or consent* of the surgeon in charge of the case. Am I not justified in saying, as I do in public, that the medical fraternities have powers more than necessary, and their wings need clipping by the voice of public opinion? The action of those injecting doctors was utterly reprehensible and even unethical. The conscientious surgeon is complaining bitterly, but what?

INTOXICATING LIQUORS AND LIFE

A Danish scientist declares that the life of a human being is shortened 25 minutes for each booze. I may not agree about the 25, but, if alcohol is a preservative and prevents digestion of food products, it surely *must* interfere with the leading of a normal life—which also *must* mean its shortening.

HOW DO YOU GET THAT?

A milkman advises his customers to eat many vegetables, especially of the leafy type. Then he says, "If you greatly dislike such vegetables, you may eat more milk, as the cow has eaten many of them for you!" The advice is good, but the argument and logic are foul in principle. Mr. Milkman, the same argument and logic are used by those masters of butchery. Armour and Company and their millions-making brethren—*eat* the cow? You fellows need to be taught the laws of the conversion of energy and elements. Why not eat a dead human baby—what is the difference?

AN UNSOLICITATED ENDORSEMENT

The following letter from one Doctor of Osteopathy and Dental Surgery who with his wife, sister and sister-in-law graduated as Spectro-Chrome Therapist from the Providence, Rhode Island class to another prominent expert who is a user of Spectro-Chrome Therapy but did not yet take the course was sent me with express permission for publication. It breathes confidence and sincerity and shows the broad mind of the writer. I am pleased to inform him that Dr. Francis A. Cave and Dr. Edith S. Cave will take the course in the class of Boston, Massachusetts.

Dr. Francis A. Cave,

Boston, Mass.

Dear Dr. Cave:

Dr. Alice and I finished the course in Spectro-Chrome Therapy last Saturday evening and to say we are pleased is putting it mildly.

It is clear, concise and to the point and gives a thot along therapeutic lines worthy of consideration, and now that we know what it is, we would say to you not to miss it.

With best regards I remain

yours fraternally

S. L. GANTS.

PORTER CAME TO PORTER'S PORT

When I went to Chicago, Illinois last year, I had the pleasure of making the acquaintance of Earle Wilbour Porter, who undertook the advertising work of the Spectro-Chrome Institute and conducted it very creditably and successfully. He also took the course and graduated as Spectro-Chrome Therapist. As I returned from Rhode Island, I received a postal card and knowing Mrs. Earle Wilbour Porter personally, I have the honor to extend greetings. The contents of the card are good:

There came to port the other night
The queerest little craft
Without an inch of rigging on
Our Baby here at last!
God shipped our darling little one
Across the unknown water
For us to guide, love and protect
Junior Earle Wilbour Porter.

On a side line the card states the Baby Porter was born at West Suburban Hospital, Oak Park, Illinois on June First 1923 and the weight was 8 pounds 12 ounces. Good big baby indeed; the boy received some Spectro-Chrome treatment before birth; it is now necessary to nourish him on Spectro-Chrome principles, without flesh, fish, eggs, tea, coffee, tobacco or alcohol in any shape or form and by all means keep him from the abominable contamination of Vaccination. Keep his blood clean and clear and he will bless his good thinking parents all his life.

Attuned Orange Wave

(Humor, Wit, Fun, Satire, Tickle, Irony)

HARRY AND HIS APPENDIX

A Rhode Island Tale

By:—Miss Kashmira Dinshah, S.C.T.
Philadelphia, Pennsylvania.

It has long been a favorite theory of the medical doctor, that pain in the human body signifies the presence of some mysterious germ or foreign substance. To his conception, even the appendix is "useless organ just thrown into the body perhaps for the purpose of filling up space." Foreign substance! In a moment of mental aberration, the Grand Maker just flung in a few extra parts which might be removed without injury to the organism or its operation; in fact the chopping out of the "worthless" appendix is known to *benefit* the chopper, (monetarily) while the choppee goes his way, lightened of his inside load and incidentally, his bankroll. Thus, though "idle and useless in the body," the appendix does double duty. The trusting public has no defense against the licensed hewer of humans, it being made to believe that it is for their own good to have the destructive

surgeon's knife stuck into them from time to time as the M. D. sees fit, until the patient goes into fits and losing patience, dies just to spite the doctor. Perhaps the high cost of funerals is one reason why some people are so ready to have chopped out their appendix, etc. They think it will not cost so much for disposal after death if half of themselves is cut out beforehand. There will be less to cremate. But this cuts no ice with the good undertakers. If they can not put it over you, they will not undertake you because you tried to beat them. They want the whole of your wallet and the whole of you besides. But you can not help it if you are not all there, because the rest of you is in the doctor's office, in neatly labeled bottles filled with anti-Volstead fluid and he will not give them back.

A laughable but cruel instance of medical "what shall I call it" happened in Providence, R. I. The scene of it was laid in a private hospital on Blank Street. The street name is ironical to the victim, a man named Harry who is now an attendant in a restaurant on Eddy Street: Dining in the restaurant where Harry worked, we struck friendship with the man and in the course of conversation, spoke to him of Spectro-Chrome Therapy, "Can it cure appendicitis?" he asked. "Most assuredly," said father, "a few treatments with the proper Attuned Color Wave and the patient goes home happy and well with a healthy appendix still inside. "Well," said Harry, "my appendix was cut out a couple of years ago." "Is that so?" remarked father, "it is a pity. You will suffer all your life from constipation, indigestion and a host of other troubles." "Oh, how did you know? I *do* suffer from all those things!" exclaimed Harry, "after my appendix was taken out, I became constipated, my back ached, my head hurt and I felt terrible." "Did you go back to the doctor and tell him?" "Surely I did," answered Harry, "I went to the doctor and said, 'Doctor, you took my appendix out and charged me one hundred and sixty dollars for it but I am not well yet. Aside from that, you made me pay seventy-eight dollars for two weeks stay in the hospital and ten dollars extra for the operating table. I ought to be healthy after paying all that money, but I am not!' So the doctor gave me some more medicine and sent me home." "The physician must have told you the trouble was only the 'reaction from the operation,'" father observed. "That's just what he said, 'reaction,'" Harry continued, "but that 'reaction' kept up for over a year. I went to the doctor again and again but he just gave me any old medicine and carelessly told me to go home. He had my money and he had my appendix, but I had aches and pains like Old Harry. I felt dreadful; I thought I would fall flat any minute with all those things gone wrong inside of me. One day I mustered courage and went to the physician's office. He used to become angry when he saw me coming so often. What did he care for me now that I hadn't another appendix to be taken out? So this time I was trembling, wondering if he'd kick me out, but I was determined to have something done for me." Here the man stopped for breath. We listened attentively as he related the climax of his story. "Well," Harry went on, "the doctor called me into his office and said, 'Say, what are you always hanging around here for?' I told him, 'Doctor, it is over a year I've been suffering like this. That appendix operation didn't do me any good—it has made me worse. You must make me well!' 'Oh, get out, you are dreaming.' I said, I'm *not* dreaming. I have pains all over. Give me something to cure me. I can't pay money all my life for medicine; I'm a poor man. You've soaked me nearly clean out already! What did you take my money for if you couldn't cure me? I can't help it if I have to hang around here so much. I'm so sick I'm afraid I'll keel over right here.' Then the doctor said, 'I won't give you any more medicine. You've bothered me long enough. Go home!' When I heard that, I was wild. 'Yes,' I said bitterly, 'it is nothing to you now what becomes of me. You have my money and I have nothing but pains. *You* give me a pain, too.' At that the doctor became hot under the collar and —would you believe it?—this is what he told me. 'Say, young fellow,' he shouted, 'there is nothing the matter with you! All you need is a new set of brains! Get out!'"

Attuned Green Wave

(Spectro-Chrome Therapy, Spectro-Chrome Therapists)

DR. KATE W. BALDWIN UNFURLS THE SPECTRO-CHROME BANNER IN WASHINGTON, D. C.

Reported by:—Miss Grace Eunice Griswold, S.C.T., Vice-President, Washington Association of Spectro-Chrome Therapists.

Kate Wylie Baldwin, M. D., F. A. C. S., Senior Surgeon in the Philadelphia Woman's Hospital, gave a public lecture in Washington, D. C., on March 24, 1923, at the invitation of the Washington Association of Spectro-Chrome Therapists. It was the biggest Spectro-Chrome event in Washington since Dr. Dinshah gave his course here a year ago. It is now a part of Spectro-Chrome history.

The White Parlor of the Ebbitt Hotel, seating two hundred, was filled to capacity, with persons sitting in the windows and standing, gathered chiefly by the personal invitation of members of the Association and their friends. Dr. Baldwin's name and reputation attracted a number of prominent physicians of different schools of medicine; among them were doctors from Walter Reid Military Hospital, and many medical students.

The Committee of arrangements had spared no pains in the endeavor to have every detail complete; every member of the Association "did his bit," feeling that nothing was too good for Spectro-Chrome, or for Dr. Baldwin. The last act of Dr. Horace B. Coblentz, late honored and beloved President of the Association, in the cause of Spectro-Chrome, was to call a meeting to plan for this lecture. The new President, Rev. Simon P. Fogle, introduced the speaker.

Hearing of the enthusiasm her coming had aroused, Dr. Baldwin was in her element, and gave a masterly summary of the principles of Spectro-Chrome Therapy. Calling attention at first to the finer but almost unrecognized forces of Nature —light and air—without which life could not long exist, then taking up more fully the powers and principles of light and color, she expounded the theory of Spectro-Chrome Therapy with such clearness that both laymen and physicians could not but grasp it.

After an hour she asked the audience whether they would prefer theory or cases." There were calls of "more theory" from many parts of the hall; so she enlarged somewhat, giving illustrations from her extensive private and hospital experience of the wonderful results obtained in tuberculosis, sarcoma, carbuncle, goitre, asthma, drug addiction, etc., and the "star case" of Grace the burned child, which attracted marked attention.

A grateful Spectro-Chrome patient loaned and operated a stereopticon, with which near the conclusion of the lecture Dr. Baldwin projected slides demonstrating the Single Polarity Predominance Chart, Operation of the Attuned Color Waves through the Cerebro-Spinal and the Great Sympathetic Nervous System, Therapeutic Area Chart, Attributes of Attuned Color Waves and the like. These slides were loaned to the speaker by courtesy of the Spectro-Chrome Institute. She added further interest by showing four views of the burnt girl, as before and after Spectro-Chrome treatment.

Dr. Baldwin was interrupted by frequent applause, her happy hits keeping the audience alert. When after the early pictures of the burned child in her great sufferings, she was shown later with her sweet and radiant face, nearly healed, the contrast was so great that a wave of emotion swept over the audience, followed by a spontaneous burst of loud applause.

Many intelligent questions were asked by physicians and others in the audience, showing keen interest and close following of the subject and proving that the great underlying principles of Spectro-Chrome had been "put across" by the speaker, who also expressed herself as greatly satisfied with the quality of the audience.

In conclusion, many gathered around Dr. Baldwin to clasp her hand and voice their appreciation of the privilege of hearing her speak upon this up-to-the-minute topic. An eager group lingered for nearly another hour asking further questions. For many Dr. Baldwin "rolled away the stone" of doubt, prejudice and ignorance that had held them, and made Spectro-Chrome a living reality.

Many appreciative sentiments were overheard. One physician said "Dr. Baldwin's argument was clear and convincing and inspired a desire to take up the course of study." A lawyer, whose wife is a physician said, "Dr. Baldwin's argument appealed to me because it was so logical." Others were impressed by the speaker's fairness and honesty and said they could not help believe.

The Graduates of the Spectro-Chrome Institute including the speaker, were decorated with a knot of ribbons of the true primary colors, Red, Green and Violet, so they could be easily distinguished. A list of the graduates using the Spectro-Chrome Equipments had been prepared for inquirers, and was eagerly taken. Copies of the charter and prospectus of the Spectro-Chrome Institute were distributed and some copies of the magazine Spectro-Chrome sold.

Some of the daily newspapers which had consistently refused all mention of Spectro-Chrome in their columns relaxed for this occasion and gave brief notices and reports. So Dr. Baldwin legitimatized Spectro-Chrome, as it were and gained for it a wider recognition by her splendid presentation of convincing facts.

The next morning, she met the local Spectro-Chrome Association in a two-hour informal conference, discussing individual cases and giving many practical suggestions. Her last words before leaving Washington were: "Spectro-Chrome Therapy is the most fascinating thing I have ever engaged in." So saw we all and the many friends she made for Spectro-Chrome join us in the hope that she may return soon and often to Washington.

Spectro-Chrome is thankful to you, Miss Grace, for your graceful report. Dr. Kate W. Baldwin is certainly placing Spectro-Chrome Institute under an obligation so extensive as to be unrepayable. Her high knowledge of Medicine and Surgery and her vast hospital experience upholding Spectro-Chrome Therapy mean an endorsement of the most potential character to the system. She pioneered it in the State of Pennsylvania and by her persistent successful clinical use of it, built for Spectro-Chrome Therapy a solid foundation which the Originator needed to present the system before an ultra skeptical body. For all she did and for all her kind endeavors in the promulgation of Spectro-Chrome Therapy, Spectro-Chrome Institute offers the most cordial thanks.

VIOLENT MENTAL CASES, HIGH BLOOD PRESSURE, BURNS, BLOOD POISONING RELIEVED BY SPECTRO-CHROME THERAPY.

By:—Dr. James C. Daniels, S.C.T.
In Charge, Spectro-Chrome Therapy Department, Lindlahr Sanitarium, Chicago, Illinois.

Just a little note to let you know we are still using Spectro-Chrome here at Lindlahr Sanitarium with some success.

The following are some cases specially noted:

Case 1—Violent mental cases; quieted by Attuned Violet.

Case 2—High blood pressure; bad nephritic condition; one foot in the grave when he landed here. Wonderful results from Nature Cure treatment backed up by Attuned Purple Systemic, Scarlet on Area Number 18, which I changed to Lemon on Area Number 18 with better results. This patient says Spectro-Chrome is one of the better treatments we give here.

Case 3—Woman washing hair accidentally poured very hot water over her head; was badly scalded. Treatment was Turquoise on scalp and Indigo for pain. Excellent results.

Case 4—Woman; poisoned finger with glandular swelling in axilla; 105° fever. Fast and cold packs only temporary help; supplement Blue for fever (½ hour treatment every 3 hours) alternating with Indigo for pain as a cellulitis had developed all over the body. Fine result.

Many others are being treated, but, have not received reports yet.

I am truly pleased to receive your reports and thank you for the trouble. I also thank the worthy Dr. Henry Lindlahr for introducing Spectro-Chrome Therapy in his famous sanitarium. The allusion to Spectro-Chrome Equipments made by him in the April issue of *The Lindlahr Magazine* is very kind and it is gratifying to know that he is getting results from the Graduate and other Spectro-Chrome Equipments he possesses. All success to the great man in his efforts for the relief of the sick.

IVY POISONING, NASAL CATARRHS, HEART AND BLADDER DIS-EASES RELIEVED BY SPECTRO-CHROME THERAPY

By: — Gertrude Burgess, D.O., S.C.T. Philadelphia, Pennsylvania

Among others, I treated the following cases with success:—

Case 1—T. C. age 11 years. Ivy Poison on hands. Had three ½ hour treatments in 48 hours; color Indigo. Affected parts dried and healed leaving no scars.

Case 2—Mrs. D. M. age 25 years. Nasal Catarrh of several years standing. Received half hour treatments every 24 hours for about 2 weeks. Color Turquoise on Area Number 1. Condition improved from first treatment.

Case 3—Mrs. D. G. R. age 37 years. Severe catarrhal disorder. Received two half hour treatments in less than six hours. Color Turquoise on Area 1. Result was marked and relieved patient very much.

Case 4—Mrs. E. C. age about 45 years. Acute heart condition. Applied Magenta on Area 4 in two periods of 20 minutes each in 24 hours with very good result.

Case 5—Miss M. M. age 39 years. Chronic bladder affection. Had to be catheterized once and often twice every 24 hours for several months. Applied Indigo on Area 10 every 12 hours in periods of 20 minutes each. After the third day, it was no longer necessary to use the catheter. The patient is steadily improving under correct treatment with Green Systemic, Magenta on Areas 4 and 18, Yellow on Area 9, Indigo on Area 10. She admits she is in better health now than she was during more than twenty years.

PAPILLOMA OF BLADDER, HEMORRHAGE, MASTOIDITIS RELIEVED

By:—Martha J. Peebles, M.D., S.C.T. Formerly Inspector, Department of Health, City of New York

It will be remembered, in my earlier reports to the Spectro-Chrome, printed in the February issue, I cited a case of Papilloma of the Bladder with hemorrhage, in an old man of 74 years, the bleeding having been controlled with ONE treatment. This was in March of last year. I continued to treat the patient for a month, when he discontinued treatments to go to the country. I advised him in the Fall to resume treatments as Papilloma are frequently of a malignant nature and in my opinion it would require 3 to 6 months of treatment to eradicate it entirely. Recently, he reported back to me, suffering from the bleeding. Indigo Attuned Color Wave again controlled the hemorrhage in one treatment, pressure being exerted (and I think it will work) to convince him that he must continue treatments long enough to be considered totally well.

I like to report another case also.

A man aged 50, six weeks prior had threatened Mastoiditis with Parenceticis. Since that time he had suffered great pain in the neck on the same side with spasm of the outer margin of the Trapezius muscle. Medical treatment at the hands of the family physician failed to relieve him. 1½ hour of Indigo Wave stopped the pain. This patient had a low blood pressure—120 mm. systolic, this being considered by me the cause of the spasm of the muscle. Red Wave was given on the back in conjunction with the local Indigo for pain. It is needless to say the patient is enthused.

Attuned Turquoise Wave

(Poetry, Romance, Drama, Art)

LIGHT OF LIFE

By:—Mrs. Mary P. Hynson

Were life but a flame . . .
A flame which heats, consumes;
It could not be a life-force—
But an error, a Force—presumed.

But the Force of infinite being
'Tis a wonderous, beauteous thing
That shimmers in a sea of life,
And tints, its varied form-screens.

The flame of life—burns,
And doth consume its forms . . .
It heats, and radiates . .
But dies with its wasting sperm.

But the infinite Force of God—
Whose rays are life divine,
Preserves—the forms it shines upon
And lives to the end of time.

Attuned Blue Wave

(History, Travel, Biography, Geography, Language)

WHAT AMERICA NEEDS

By:—Irene Grace Dinshah, S.C.T.

Philadelphia, Penna.

History declares that being discontented under the severe disciplin of their Mother Country, the Pilgrims came to America to work out their own salvation—to worship God in their own way. Thus, our glorious land, America (the country that we find comfortable to live in) was built by neither "highbrows" nor "lowbrows," but by truehearted, [illegible] Th[illegible] America could be compared to a business concern just launching out into the wide World; she played her part well with her few workmen (Pilgrims) but, the business (America) so expanded that more employees (immigrants) were essential.

Truly, America was merely toddling.

I class in three stages, our country's development since 1492. First:—the Welcome; second:—the Undeserved, "It is the same to us if you do or do not come; third:—the "Morons."

The Welcome

In countries far away poor struggling peasants were advised through this (now appearing sinister) question, "Why do you not come to America—the Golden Land of Opportunity?"

They came willingly,
Were welcomed cordially,
Toiled unflinchingly,
Lived quietly, happily.

America was *their* country as well as ours.

This was followed by ungratefulness.

The Undeserved

It was sufficient to cause the foreigners despair, when, instead of being greeted through Columbia's Gates as the pioneers had greeted the immigrants' forefathers, they were knocked around by heartless jeering people who shouted, "Hey, you Wops, you Sheenies, where did you come from?" In Columbia, the Land of Hope, Freedom, Justice and Equality, the newcomers found, worked for and got nothing more than "Hail Columbia." They may or may not have seen the American flag in the land of their birth, but, they wanted to respect it at any rate. They *did* respect our flag; they saw—some got—the "stripes" and fortunately or unfortunately saw "stars," also. If they inquired for directions, some snobbish remark was hurled at them. Such was their remuneration. It was distinctly shown that they were not wanted, from the attitude displayed by *real Americans!!*

The Morons

Now, the urgency for deportation and exclusion of foreign immigrants *by law* is the topic of the day. Another set of "real Americans" (dissimulators) expected loyal Americans to place credence in the obnoxious assertion that our country is over-crowded. Even residents of New York City tenement houses had the pleasure of seeing miles and miles of God's country unoccupied since Ford automobiles came into existence; and for it they had not far to go! Nevertheless, these political hypocrites, who call themselves "real Americans," concocted another scheme, when, to their disappointment the first was a failure.

An immigrant may be perfectly normal, perhaps even unusually intelligent, but, because he fails to answer meaningless jargon that some nonsensical immigration authority uses as a test to determine whether or not the person is *mentally deficient*, he is pronounced a Moron (a poor fish) and is supposed to credit him with the brains of a child twelve years of age; some papers are signed and the person deported.

When I learned about the absurdity of the examination, I thought how unhappy must they be—they who would like to earn an honest living in a country where Democracy is *supposed* to prevail, but where injustice does not seem inevitable. Those law makers forget that:

All
Must
Enjoy
Rights
In
Congenial
America.

How fortunate for me that I was *born* in America! I would surely have been adjudged and awarded the title of "Moron" otherwise by the present-day "tests." Even Dinshah, the Originator of Spectro-Chrome

Therapy, with his wide knowledge of languages would be unable to answer many questions without consulting a dictionary. Can *you* define the word abovousqueadmala? To know the definition will not make you any better American.

Georges Clemenceau, the ex-Premier of France says, "If you think an academician knows how to spell, that only proves the persistence of a legend." Therefore, do you not think we should show at *least* a *little* leniency toward the willing-to-learn-and-become Americans?

To reject and deport the would-be-loyal future citizen and welcome the clever brainy scamp, in itself is absurd.

America needs the builders,
Not destroyers;
America needs the workers,
Not the shirkers.

The sooner Americans wake up to this responsibility, the better for America—the better for the World. Let *all* the hardworking, honest brawnholders be welcomed to America—there are brains enough here.

VIOLET FLOWER STATE WELCOMES SPECTRO-CHROME THERAPY

Rhode Island, the smallest State in our Union gave an exceedingly good reception to Spectro-Chrome Therapy. There was such harmony in the class that it appeared like a family gathering rather than a number of comparative strangers taking a course in the Latest Revelation in the Healing Art. On the final day, the following testimonial was handed to the Originator of this new science:

Spectro-Chrome Institute,
Providence, Rhode Island.

June 2, 1923.

We, the students of the thirteenth Class attending the Course of Instruction in Spectro-Chrome Therapy, wish to express our deep appreciation to Col. Dinshah Ghadiali, M.D. for his unfailing courtesy and broad-mindedness.

We also wish to express our sincere thanks to you, our beloved Master and Teacher, for the brilliant skill and manner to teach us in the Science of Spectro-Chrome Therapy.

Pledging ourselves to co-operate with you to carry the message which you have brought to us, to others, we wish you God-speed on your noble mission and hope that your untiring efforts to relieve suffering humanity will, in the near future, be crowned with the desired success for which you so earnestly are looking forward.

(Signatures of the Students)

SPECTRO-CHROME

JULY 1923 VOLUME 2

BLUE 1923 NUMBER 7

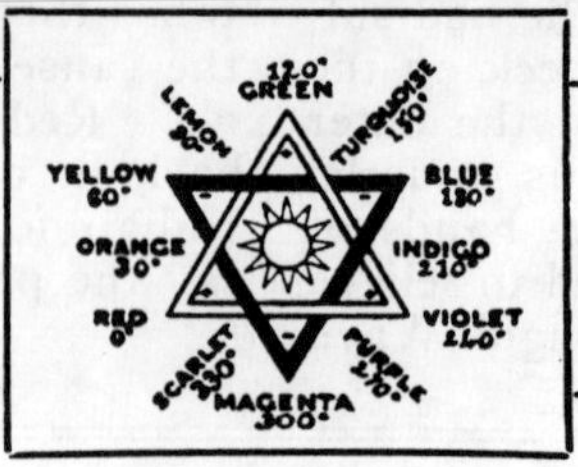

Devoted to: Spectro-Chrome Therapy **Dedicated to: Service of Humanity**

CONTENTS Page

Dinshah's Platform—Spectro-Chrome Therapeutic Area Chart—Spectro-Chrome Therapeutic System 313
Transfer of New York City Activities—Doctor Tastes His Own Medicine; Spectro-Chrome Saves Sight; Birth of Spectro-Chrome Sympathometer—California Tour; New Classes—British Government Sinning Against Its Subjects; A Political Humor—Diabetic Increase Laid to Volstead Act—A Filthy Habit—Freedom of Therapeutic Choice—Insulin for Diabetes Another Foolery—What to Wear—Cancer, Cosmetics, Smoking, Shaving—Girl Fooled Doctors 314-323
Loyal Spectro-Chrome Therapist Scores Disloyals 323
Third Annual Convention 325
Sarcoma, Phlebitis, Ulcerated Bladder (by Kate W. Baldwin, M.D.) 326
Formation of The New England Association of Spectro-Chrome Therapists .. 328
Eye Trouble Removed With Spectro-Chrome Therapy (by Dr. F. B. Williams) 329
What Even a Small Spectro-Chrome Equipment Can Do (by Orlo H. Frost) 330
Epilepsy Relieved; Use of Cracked Slides (by Emma G. Taylor) 332
Night Sweats Remedied (by Dr. A. Eppolito) 333
Osteopaths Successfully Use Spectro-Chrome Therapy in Dropsy, Constipation, Cold, Marasmus and the like (by George F. Wagoner, D.O. and Lillie E. Wagoner, D.O.) 333
Quick Results Obtained in Pleurisy, Ear Trouble, Constipation, Nervousness and Albuminuria (by Mrs. Loccie B. Smith) 334
The Isle of Magenta (by Kashmira Dinshah, S.C.T.) 335
Cleveland, Ohio and Indianapolis, Indiana Classes 336
Price List .... Inside Back Cover
Registration Notice of Trade-Marks—Notice to Purchasers and Patients—To Our Contributors—Advertisement Rates .... Outside Back Cover

Editor
Dinshah P. Ghadiali

In course of life you reach a time,
When Doctor, Lawyer, Priest alike,
Does fail to give the sought relief,
To troubled mind or burdened heart—
Then come to me;
I help you to help yourself.
SPECTRO-CHROME.

Subscription:—January to December 1923 .... $3.00
All Subscriptions Taken For Full Year Only And Expire With The December Issue
Any Single Copy 25 Cents. Back Numbers Available From June 1922.

Attuned Red Wave

(The World Through the Eyes of Dinshah)

TRANSFER OF NEW YORK CITY ACTIVITIES

On and after November First 1923, the business activities of the New York City office will be completely transferred to the Central Office at Philadelphia, Pa., maintaining only the Statutory State Office in New York State. Please act accordingly.

DOCTOR TASTES HIS OWN MEDICINE

SPECTRO-CHROME SAVES SIGHT

BIRTH OF SPECTRO-CHROME SYMPATHOMETER

The serious automobile accident on my last tour to Detroit, Michigan, showed only a slight scar on the left side of my forehead, which was quickly healing when I departed for Cleveland, Ohio.

We left Bedford, Pa., early morning for Pittsburgh, Pa., 95 miles out. We left that town at 1 P. M. and having 175 miles more to make that day we naturally had to travel as quickly as discretion and the road rules permitted.

At 3 P. M., an unexpected trouble started. The bright Sun shining out of a cloudless sky confronted us from the West and for three hours it was impossible to avoid its direct Orange glare on my eyes. As evening approached, it became worse and as we neared the limits of Sharon, Pa., a color haze spread before my vision—the overpowering Orange radiation from the Sun had worked its mischief.

We reached Cleveland, Ohio and the next day began arranging for the Sixteenth Class on Spectro-Chrome Therapy; my enthusiasm and alertness however received its first shock when opening the window blind I found my eyes could not keep open against the power of the Orange in Sunlight. I ran to a mirror—my left eye was gorged with blood; I realized the possible situation and closed my right eye—the left was covered with a film that rendered it sightless.

For a moment, my future flashed before my mind; to have one eye to arrange my experiments, to be sightless during the day and work in the dim of the night only was to be my lot for an uncertain period. Four more days and the discourses would start. Meanwhile the smarting pain and the burning kept on increasing. I went down to an optometrist in the hotel and had him look into my trouble. He examined the eyes and putting on a look of the utmost gravity said, "I do not want to scare you." "Scare? you can say nothing that could scare me; go ahead; I cannot look into my eye myself so I came to you; speak the truth," said I. "Well, you are getting a cataract!" "Cataract nothing!" I shouted; "look again, this is a trouble from the glare of the Sun; my eyes were always perfect except for a myopia of—4.75." He maintained his ground and I returned to my darkened room.

What was to be done? I noticed the terrible inflammation and the same night waved on it Turquoise from an Intermediate Spectro-Chrome Equipment; but meanwhile, the ciliary muscle of the left eye too having been weakened the glare prevented the eye from keeping open. The only remedy for facing Sunlight was to make a special pair of Spectro-Chrome Blue goggles and wear them over my spectacles.

I did it. During all the three weeks I groped around during the day with those super-dark glasses endeavoring to cope with my duties notwithstanding that the pain was excruciating; I lived an entombed life during three weeks in the day and faced the class lights at night. It certainly was a world dead to me; I must work; stopping the class work would mean heavy financial loss to me and discouragement to the twenty six students. I must work at all costs.

I worked. There was no way to stay in the dark after Spectro-Chrome treatment; so I did not take further treatments; there was no facility for using the necessary Graduate Equipment.

Thus it dragged on until the week for departure came. Here was I, a cripple in vision and had to drive the automobile 275 miles to Indianapolis, Indiana. I must leave, but how? A thought struck me. I would consult an Occulist Specialist, ascertain the exact damage to the eye and act accordingly. There was no Spectro-Chrome Therapist optician there to look into my eye, so I picked a Doctor of Medicine recommended by the other Optometrist, made a definite appointment by telephone and went to his office punctually, guided by Kashmira.

That M. D. was not there; his office was filled with a number of patients; I announced myself and was told nobody knew when the Doctor would be in; I must wait and then some! I saw through the game of that shark and left saying I would not be in the hands of a man who violated his appointment. I went to another.

He received me well and after examining my eye said, "I must look into this further." Forthwith he produced a bottle and prepared to put some drops into the eye when I pushed his hand gently saying, "Excuse me please, I do not wish to be treated medically." "But this is only Cocain; I just want to ascertain whether there is any Corneal Ulcer!" I told him who I was and gave him the story of the accident. He recognized me and said, "Well, you have an inflammation!" Of course I had it! He said, "You have a slight astigmatism developed and there is Infiltration in the Cornea." "Yes," said I, "has any damage occurred outside of that to the optic itself? Is there any actual organic trouble that would permanently ruin it?" "This is organic trouble and it generally is obstinate. You must stop using this eye altogether and work only with one eye." "And drive an automobile with one eye too?" "Yes, yes, that would relieve the strain." "And then if the Optical Chiasm became overloaded through one-sided action?" "That is all to consider; but you see it is necessary to relieve the strain on this eye." "Well, what is your opinion about the whole affair?" "If you mean the treatment, the classical treatment is Rest, Heat and Atropine. You must not use the eye and must heat it; take an electric heater and focus the heat directly on this eye; also put a drop of Atropine Solution three times a day in it." "But Doctor, this is a trouble caused by the heat side of the Sun; the very appearance of the eye shows it and heat is like poison to it, that is why I wear these Spectro-Chrome Blue goggles; without them I could not have been able to come here." "May be, but, Heat is necessary to remove the inflammation and the corneal infiltration." It was no use talking more. When he knew who I was his fee seemed to have increased—he charged me ten dollars for telling me I had redness in the eye which must be heated by an electric stove! And what galled me more was he asked me to see him in five days later to ascertain "progress."

I faced a bad situation. I could not follow the advice of that "Specialist" for Rest and Heat, so sacrificing all comfort and business consideration I shut myself in a totally darkened room by day and at midnight when the class-work was over I treated the eye with Indigo with the Intermediate Spectro-Chome. I had only a few minutes treatment each time for three days prior to departure but with the goggles on and one eye seeing I drove the machine to Indianapolis, Indiana 275 miles in one stop. It was a race for saving my sight.

There my Graduate Equipment was ready. The torturing pain, sensitiveness to white light, obscuration of vision and continued strain for over three weeks had left me comfortless and sleepless but not thoughtless. My mental activity was unimpaired although the physical damage had rendered me helpless in white light. The Sun whose power-worshipper I am, that Sun without bowing to whose glorious orb I would not start my day's work, was now as it were working to leave me sightless. Many times in the self-imposed darkness in the day would I have tears come to my relief; there was I endeavoring to serve my graduates but unable to read even their complaints; there was a pile of business, but, without my eyes all was useless to me. Irene and Kashmira were literally swamped; day and night the loyal secretaries stuck to their arduous tasks, but without my personal attention to the important personal matters they found the detail load piling day by day. It was a battle between the powers of light and darkness and light must supervene.

I pulled the Graduate Equipment from its case, rigged it, connected it and switched on the current. The concentrated Indigo Wave was flashed on the left eye for 15 minutes and I retired to my bed in the dark. The next night I flashed the concentrated Turquoise and treated for 25 minutes, the third night I stood it for 35 minutes. The reward came. In those three treatments with the Graduate Equipment, the inflammation subsided and the corneal infiltration began to disappear. A few more treatments were taken before I left Indiana; I discontinued my Spectro-Chrome Blue goggles and my sight became practically normal without any color distortion—and that without Heat from an electric stove!

But a bigger reward for patience was to come. Two days before starting from Indiaanpolis, as I was chanting prayers to The Most High, in the darkness of the midnight and casting my internal vision around, there came to me like a streak of lightning that which I had been thinking of inventing years. The sweet lull that had supervened after the restlessness of numerous nights had enkeened my faculties and like an astronomical chart of the heavens appeared before me the entire scheme of the mechanical working of the Occult Operation of the Great Sympathetic Nervous System.

At once I threw off my bedcovers. Out came the paper and pencil and within a few hours' calculation as the new day dawned the design of the machine was completed in its mechanical details—it was the birth of the *Spectro-Chrome Sympathometer* whose use will be demonstrated at our Third Annual Convention on Wednesday, November 28, 1923 at 8 P. M.

CALIFORNIA TOUR
NEW CLASSES

The following new classes will be opened and students trained in Spectro-Chrome Therapy in December 1923 and January 1924. The times and places are fixed and there will be no

change in the program.

San Francisco, California Class
Class Number Eighteenth

1923

December

1 —Departure of Dinshah and assistants from Philadelphia, Pa., 10:25 A. M.

2 —Arrival in Chicago, Illinois, Hotel Atlantic, 8 A. M.

2 —Meeting of Spectro-Chrome Therapists, Hotel Atlantic, 3 P. M.

3 —Departure by the "California Limited," 8 P. M.

8 —Arrival in San Francisco, California, Hotel St. Francis, 8:15 A. M.

8 to 13 —Arrangement of experimental apparati, Lecture Room.

14 —Free Public Lecture, Assembly Room, 8 P. M.

15 —Public Reception, Lecture Room, 2 to 4 P. M.

15 —Public Lecture, Assembly Room, 8 P. M.

16 —Public Reception, Lecture Room, 2 to 4 P. M.

17 —Beginning of Spectro-Chrome Therapy course, Lecture Room, 7 P. M.

29 —Completion of Spectro-Chrome Therapy course, at midnight.

30 —Departure of Dinshah and assistants from San Francisco, California.

Los Angeles, California Class
....Class Number Nineteenth.....

31 —Arrival in Los Angeles, California, Hotel Biltmore.

1924

January

1 to 3 —Arrangement of experimental apparati, Lecture Room.

4 —Free Public Lecture, Assembly Room, 8 P. M.

5 —Public Reception, Lecture Room, 2 to 4 P. M.

5 —Free Public Lecture, Assembly Room, 8 P. M.

6 —Public Reception, Lecture Room, 2 to 4 P. M.

7 —Beginning of Spectro-Chrome Therapy course, Lecture Room, 7 P. M.

19 —Completion of Spectro-Chrome Therapy course, at midnight.

20 —Departure of Dinshah and assistants for the Central Office.

After completion of the tour in California as above outlined, new classes will be opened in other States. The 1924 program will be announced subsequently. Frequent inquiries are received as to the opening of classes in Philadelphia, Pa., and New York, N. Y. To them and similar inquiries the one answer may be given: there will be no more classes in the States already covered until other States that did not have the message of Spectro-Chrome Therapy are presented the opportunity. Formation of special classes may be considered for States already covered only where a sufficient number of students is guaranteed beforehand by a responsible person, because, I cannot waste my time and the resources of the Spectro-Chrome Institute by going over the same territory for "experimental chasing." Each State, in opening it for our work, costs very conservatively a round $2,000.00 and unless I traverse the ground quickly, each month opening a new State, my life period may fall short of the requirements of the message. To go into a new State absolutely uncalled, unheralded, unaided and unadvertised, fix a lecture room complete with all the intricate experimental paraphernalia, deliver two public lectures and hold two public receptions, answer inquiries, give interviews, interest the newspapers and the public in the work, break through the long-established prejudices of professionals and the laiety, formulate a new class and start teaching Spectro-Chrome Therapy to a number of strangers in a mixed crowd besides doing the routine business—all in the space of 7 days, means *work*—downright *pounding hard* work with no ulterior aim or benefit except the joy of serving humanity by disfranchising it from the monstrous clutches of the "Medical Trust" and accelerating the arrival of universal "Therapeutical Emancipation."

BRITISH GOVERMENT SINNING AGAINST ITS SUBJECTS: A POLITICAL HUMOR

Attorney General A. M. Manson of Victoria, British Columbia announces with satisfaction a profit to his government during the six months ending March 31, 1923, of $1,250,195.12, from the sale of intoxicating liquors. His report adds:—

"Five per cent. of this amount will be placed in the reserve fund, leaving $1,187,685 to be divided. Fifteen per cent., or $178,152, will be placed to the credit of a special fund for hospitals and charitable institutions. Another 35 per cent., or $415,689, goes direct to the municipalities, of which two-sevenths must be used for school purposes.

The remaining half of the profits, amounting to $593,842, will be credited to the consolidated revenue fund."

What a noble and magnanimous dispensation! Intoxicate and poison the people first, then, put them in hospitals and pauperhouses with 15% of the poison income to set them right; also, let the children get the blood taint and poison from their parents and learn the elevating example of guttersniping from them at home, then, use two-sevenths of 35% to correct the habits at school. The consolidated revenue fund will probably go partly to support more officers. Mark the words of a born British subject: The British Government is responsible to the whole civilized World for its support of the narcotic situation. That Government yet controls the entire liquor monopoly in India and is the *sole* responsible power in the Opium monopoly. Until the United States will bring the British Government to terms on this horrible bloodsucking abominable and monstrous traffic and until the British Government itself is awakened to its huge responsibility toward its downtrodden poor East Indian subjects, the narcotic political policy of the British will remain as a black spot on its conscience. It is a matter of present day history, perhaps not much known in America, that fines in the British Courts in Nigeria, Africa, are exacted in demijohns of rum in place of money. And such is the Government system that some fool American politicians wish us to follow when it comes to Prohibition. Keep the lid tight, as tight can be, enforce that law without reserve and drive from our sacred shores the bloodtainted murderous traffic in human souls.

DIABETIC INCREASE LAID TO VOLSTEAD ACT

Surgeon Major Benjamin Jablons in a lecture delivered at the St. Mark's Hospital, New York, asserted that Prohibition has been the indirect cause of Diabetes Mellitus, through the substitution of sugars and fats to produce the heat that was formerly produced by alcohol. He was speaking about the Insulin treatment of Diabetes and laid the increase at the door of the Volstead Act. The doctor is the head of the Department of Metabolism at the said hospital and must therefore be well aware that even according to the opinion of Sir William Osler, the disease called Diabetes is purely and essentially a disorder of metabolism. It is in fact, a nutrition disorder and the danger is really not in patient getting more sugars or carbohydrates in the body, but, in the quick depletion of the blood by non-assimilation of the nutritive contents. Such a condition leads to starvation of the patient and the theory of Insulin prepared from the Pancreas of dogs or cows and administered to the patient started under the same idea that an increase in Pancreatic activity would increase metabolism. This condition is very easily taken care of in Spectro-Chrome Therapy. The Pancreatic Juice is of an alkaline character. Attuned Lemon Color Wave is an Antacid or Alkaline Wave. Therefore the administration of Lemon Systemic and especially on the border of Areas 17 and 18, together with Magenta on Areas 4 and 18, will very adequately take care of both the pancreas and the circulation. Feeding the butchered horrible polluted products of dogs (or cows) to sick human beings is the same principle that the jungle-living Red Indian Medicine Man has been using for centuries and if the Doctors of Medicine have nothing better, they may as well acknowledge their ignorence and step aside for more rational therapy. Would it not be a better idea to use the Pancreas of convicts and feed them to the sick?! Doctors of Medicine, your vaunted "science" is a fraud on suffering humanity; you know nothing about the higher side of life yet seek to remedy disorders of which you know less. Pshaw! The title of Doctor may eventually be a mark of ignorance than that of knowledge.

A FILTHY HABIT

Lecturing before the Anatomical Society of Paris, France, Dr. J. Bertrand and Dr. G. Medakowitch warned the public that a dis-ease not commonly known but recognized as Kyste Hydatic was caused by kissing dogs. They said that the tongue of a dog carries invisible microbes which are liable to produce an infection similar to brain cancer. They further advised that dog owners should not permit their pets to lick their hands and face for showing friendliness. Such advice should not even have been necessary in a civilized community. It is as dangerous to kiss promiscuously human beings and when it comes to kissing dogs on the lips it is the limit of human absurdity. A certain type of so-called high class women takes pride in this act, but, it is a filthy, dirty habit and is unhygienic to the extreme. If human emotions should be spent, let them be spent on worthy humans but not this way on lower animals, whose physiology is different from our evolution. Dog kissing, cat kissing and rat licking are all synonymous of the remnant of the uncivilized barbarous times, when pigs and people lived in the same huts. Today, if with higher knowledge of the highest, we have such anomalies among our civilized people, surely, there must be the necessity of the French doctors sounding the foghorn to remove the brain tempests.

FREEDOM OF THERAPEUTIC CHOICE

Lady Blanche Follett an English woman of title died in London, England, while under the treatment of a Christian Science Practitioner, Morley Fletcher. Her sister Baroness Teynham alleged that had a medical doctor been consulted, she would not have died. She also avowed disbelief in Christian Science. Notwithstanding such evidence, the Coroner decided that Lady Blanche Follette was old enough to judge for herself and therefore had the right to follow her own views. A verdict of death from "natural causes" was returned. What "other causes" could there be? The verdict was correct, because, each individual has the birthright of Liberty to select whatever method of healing is the best in their belief. It will work out the same way in Spectro-Chrome Therapy. However, my advice to Spectro-Chrome Therapists is to follow faithfully the methods laid down in the system; then, they will never have to face a similar ordeal.

INSULIN FOR DIABETES ANOTHER FOOLERY

The 117th Annual Meeting of the Medical Society of the State of New York made a great noise in regard to the treatment of Diabetes Mellitus by Insulin, the so-called recent introduction credited to Dr. Frederick G. Banting of Toronto, Canada. In speaking of this, Dr. J. J. R. McLeod Professor of Physiology at Toronto University remarked that the epoch-making discovery could never have been made and the method of treatment perfected without experiments on animals." He said, "The full value of the treatment was demonstrated on rabbits and dogs and there could be no better evidence of the usefulness of this method." My dear sir, for this very reason the treatment is utterly worthless. Rabbits and dogs are not human beings; they are not both even in the same scale; the rabbit is herbivorous, the dog is carnivorous, the human being is frugivorous. Neither has in its physiology anything common. You cannot produce animals out of a combination of rabbits and dogs or fecundate human beings by cross breeding with such animals. If the internal glands operate with different secretions in different animals, experiments on one would have no definite value for the benefit of another. Vivisection is a monstrosity and a survival of barbarity. It is an authorized and licensed system of torture and has no advantages conducive to human elevation. Insulin treatment will follow the same road as other crazy notions of leisurely medical men and will be soon on the shelf like Koch's lymph for Tubercular Pthisis. The biggest joke is that the Insulin has to be injected daily and continued all the life of the patient to "keep on the cure"! Where is the funny button of the public gone? And Insulin is not even a new "discovery"; it has been discussed by physiologists for long as the internal secretion of the Pancreas in conjunction with sugar metabolism. Insulin will *not cure* Diabetes Mellitus.

WHAT TO WEAR

Writing in the Evening Bulletin of Philadelphia, Pa. about the "Wisdom of keeping the body at uniform temteperature," Garrett P. Serviss the Astronomer-writer makes some sensible remarks. He says,

"A word of caution with regard to the danger incurred during the hot weather in entering houses, or shaded places, while the clothing is still soaked with perspiration. In such a case wool would be safer than cotton, because the wool retains the moisture and holds back the rate of evaporation. But the danger can be averted if the cotton is allowed to dry before the wearer enters a cool place. 'Rheumatisms' are a frequent result of sudden cooling, caused by too rapid evaporation of perspiration in a cool, drafty place. People have unconsciously invited life-long illness and even death by incautiously exposing themselves in this way. A cool draft blowing on the back of the neck is especially perilous. Sleeping in a draft is often dangerous."

Commenting on this, P. A. Maignen gives his views:—

"We have here a warning against the danger of exposure to sudden changes of temperature. The United States is a country of magnificent distances, and it is also one of extremely variable climate. In Europe the seasons are more regular — winter is winter, and summer is summer. It is, therefore, very difficult to know what

to do in the matter of raiment. Wool may be too warm one day and cotton too cold the next day. Outer garments may be changed every day, but inner garments cannot conveniently be changed. The solution of this problem, therefore, would seem to be in the adoption of flannel as undergarment for summer and winter. In the summer it keeps the body cool, and in the winter it keeps it warm.

One of our friends has been in the habit of spending his week-ends at a farmhouse. Upon arrival he, would take off his coat and vest and delight in the cool air, with one thin cotton undergarment. Every night late our friend was attacked by asthma, which kept him awake. When he put on a sweater (after removing his city clothes), sure enough, he has had no asthma since. Some of his friends may be inclined to laugh at the idea of wearing a sweater at this time of the year, but "he who laughs last laughs best," and wool, on this occasion, enables him to laugh last."

The above views of the two persons are identical with mine. Americans do not dress and the general American women's dress is at best an unsuitable undress. United States has peculiarly fluctuating weather conditions and I do not see any other way to keep health except by the continued use of wool. People have run away with the idea that wool is "warm" and cotton is "cold." There is no such inherent character in either. The body itself is the *producer* of the heat, the wool or cotton merely retains it or maintains it. The only purpose of wearing clothes is to maintain the temperature of the body uniform, enabling us to preserve it from sudden changes, which may produce quick disturbance and rapid demands on the thermal equilibrium. To that end only are clothes necessary and beyond that purpose, to go merely by the dictates of mere fashion or convention is nonsensical. If clothes will not protect us from those changes, their value is lost. Our ordinary mode of dressing imposes too much strain on the thermal mechanism inside and a greater strain on the purse to conform to absurdities. Wearing furs that dangle on the neck doing nothing, wearing garments made of lace (strung-up holes), donning stockings made of glass-like silk scattering the bodily heat quickly or putting on articles without regard to their durability, heat-conserving value or logicality has made for deterioration of human stamina. I take no stock in wearing clothes only for their appearance. If with the so-called appearance, there is a drain on vitality, that type of costume has no business in life. For America, a cotton or linen underwear topped with a woolen vest or shirt, all-wool stockings and a simple lightfitting non-compressing cap is the best wear for Winter, Spring, Summer and Autumn. The wool prevents during the cold season the scattering of the bodily heat, thus, maintaining health. During the hot season, the same wool acts as an insulator, preventing an excessively rapid evaporation from the surface of the skin and depletion of energy. The appetite is on the wane then, because, Nature does not need much food to replenish. At that period, a rapid evaporation and consequent rapid cooling of the body would waste more energy than can be supplied by the scanty food; it would try to frustrate the wise provision of Nature. Therefore, the wool instead of making one uncomfortable is sure to be the reverse, provided the digestive tract is not kept loaded with unnecessary and heavy foods. The way to keep healthy and comfortable in all conditions is to eat and drink as little as possible commensurate with actual physical need; eat and drink only pure non-stimulant materials and preserve the temperature of the body uniform by covering it with an insulating material like wool which has the power of absorbing the perspiration like a blotter. Our comfort, common sense and health come first; conventions, customs and fashions will have to drop by the wayside. It has been urged, (many physicians endorse the view) that by keeping the breast and neck uncovered women become stronger. The breast is the *very* part that needs protection against weather changes, because there the thermal machinery and bloodpumping apparati are situated. Strength would come if the *entire* body be exposed constantly to the elements; that the civilized nations call savagery. At best, looking from all points, a simple wooı garment on cotton or linen underwear would serve better and be more hygienic than the various stupid fandangles and worthless contraptions which go to make sartorial pleasures in truth unhealthy. Furs and such other materials convert human beings into rats, seals and skunks, giving an unnatural appearance and a vile leathery effluvium plus uncleanliness.

CANCER, COSMETICS, SMOKING, SHAVING

Sir George Lenthal Cheatles, of King's College, London, England, states that cosmetics, paints and powders affected by women, give rise to skin irritations which may develop later into cancers. Speaking in this connection, a surgeon of the New York Skin and Cancer Hospital, in an interview with a reporter of the Evening World thus expresses his opinion:—

"Cancer of the face is much more common among men than women, probably because of the shaving, which produces constant small injuries to existing, apparently harmless, skin conditions. It is also known that certain occupations tend to develop skin cancer because of the harmful irritants used in these industries. Smoking and other forms of tobacco use, have been known for years to be directly causative of the development of cancer in the mouth. Here again, the predominance of cancer of the mouth in men over the cases occurring in women suggests a factor in men which is not found in women. The stem of a pipe, by being constantly applied to a certain part of the mouth, will bruise the skin and the constant irritation will cause a cancer to develop. A man who will talk with a cigarette stuck in the corner of his mouth might find that the sticking of the cigarette to his lips and the continued damage to his lips from its removal, will cause a cancer. It may be said that many patients who have cancer of the lip or of the mouth have never smoked, and that cancer in this region, therefore, cannot be due solely to tobacco; nevertheless, so many other cases give a history of over-use or suggestive evidences of a constant use of tobacco that it may safely be said that tobacco use must be considered a direct causative agent of cancer."

And when women combine cosmetics with smoking the influence for cancer must of course be worse. I shudder to think of the progeny of a cosmeticated smoking woman and as such women increase constantly in pseudo "society," the menace to the nation is proportionately increased. The remedy is not legislation but education; legislation has already become overlegislation in the United States and we need enforcement of what exists after boiling the jargon of legal terms to the Extract of Common Sense. That is and will be the best medicine.

GIRL FOOLED DOCTORS

Evelyn Lyons of Escanaba, Michigan, registered a "temperature" of 115° Fahrenheit and more and yet was found by her medical attendant Dr. Harry D. Defnet to be a "picture of health," as far as her sanity went. Her pulse was stated to be 120 per minute, respiration as 86 ! ! ! The blood pressure was announced as 122 millimetres of mercury. The medical attendant and other "big doctors" believed the case to be unfathomably marvelous and she was hailed as a living mystery. Finally, the temperature bubble blew up; she was warming her armpit with a hot water bottle! Even the veriest tyro of a Spectro-Chrome Therapist, who had studied the Occult Operation of the Great Sympathetic Nervous System as taught by Spectro-Chrome Institute would have immediately pronounced the affair a hoax by comparing the ratio of the pulse and the respiration. 120 of the one and 86 of the other is a physiological absurdity and a system of "medicine" which pretends to "cure" human beings not seeing through such a simple trick could hardly be reliable, even to be killed at its hands.

Attuned Yellow Wave

(Correspondence, Query, Catechism, Inquiry, Interrogatory)

LOYAL SPECTRO-CHROME THERAPIST SCORES DISLOYALS

ASKS FOR STRONG MEASURES

CALMNESS AND PATIENCE ADVISED

Col. Dinshah P. Ghadiali.

Dear Sir:—

To avoid embarrassment, I did not attend the meeting....... Sorry I did not go, for I have been told they again ripped you up the back, accused you of being very commercial, etc. Had I been there, I would have given them a tongue-lashing that would remind them somewhat of their duty to such an honorable gentleman as you are.

To me who have seen your mental aura, such a pure primrose yellow, which denotes the highest and the most unselfish use of intellectual power, the pure Reason directed to Spiritual ends, they are as children.

But truly, your teachings in many instances are just as the proverbial pearls before swine. I mean it; pigs, eaters of pigs and pigs truly! It is these . . . people who will ruin your good intentions, dear master. Would to God I had attended this meeting. I got this all from one who attended the meeting, but, even he does not know (as he says) whether you are a black or a white magician!!! Cannot you do something to awaken them? Use the magazine or turn a thunderstorm loose in any way you know!! If I knew, how I should surely have been tempted to exterminate a few or take their souls and place them in their true abode—the bodies of pigs! I am ever striving for self-control, but every time I think of these ...... my blood boils.

* * *

I have that which is infinitely more than wealth—good health and a good conscience. My conscience I owe to ?? my health to you. I am

Ever gratefully yours

S. C. T.

Thank you for the interest, thanks for zeal
Thanks for all you say in favor;
But, why disturb your mind for that
Which is the sure lot of all good helpers?

He who speaks evil of his teacher,
Who speaks evil of him who gave him knowledge
That he in turn may help the sick,
Does evil himself and evil thoughts
Words or deeds do slap back him who harbors them.

It is sheer foolishness or ignorance;
For that, pity their narrowness—reason with them,
Endeavor to lift them to appreciation of the right;
Let them alone ye friend, they will soon stifle their cries
When the number that follows will learn the Truth.

Perhaps it is anger, perhaps imagined wrong
Perhaps the canker of envy—to them I say,
Envy me not, unless you do the work;
Even then, envy brings falseness, hatred
The bane of human brotherhood; forget it.

Envy not my apparent prosperity, success
That which I earned by work, devotion
And steadfast intention; what *is* mine?
Nothing; I own nothing for my self,
That self which is an humble part
Of the great Macrocosm from which
I culled its treasured stores of knowledge,
Not for personal gain or elevation, but
For human service unflinching, unselfish,
Unbounded, continuous; it was given me
Through the Laws of Retribution as a messenger
To spread to humanity for humanity
For the benefit of the World.

Yet in the spreading of that Gospel of Truth
I face the grind of the merciless commercial world
Where with the Higher the Lower of the Physical Plane must combine,
I must of course resort to measures adequate and just for varied times
As best in each case conscientiously thought,
To avoid the errors of other cults, where
With best intents the teachers went
In wordly duties and gave the best they had
In knowledge, wealth and friendship,
But, ere long they found themselves and art
Launched in endless troubles on Physical Levels,
Because the Higher could mix not with the Lower.

Thus warned, I follow a different saner course:
Living with the wordly in the life of the World
I keep away from what is called "society,"
Keep away from the creation of human ties
That would tie me down to else's ways
Where I to please a few must displease more
By having different rules for different beings.

I had to do it—I have to do it—I must do it
For otherwise, I could not fulfill my destiny
Of serving the World yet keeping on head
The roof under cover of which to serve.

Mind not what any says, let him shout,
At best will he soon shout hoarse while I shall go
At each step in life, forward to the goal
That dutiful goal I set thirty three years prior—
O Benign Masters on High, give me the power
That I emancipate the ailing from the poison and the knife!

THIRD ANNUAL CONVENTION

Dear Friend:

Our THIRD ANNUAL CONVENTION will assemble at 10 A. M., on Wednesday, November 28, 1923, at 2401 North Broad Street, Philadelphia, Pa.

The Morning session (10 A. M. to 12 M.) will include Registration, Enrollment, Reports of the Officers and Presidential Address.

The Afternoon session (2 P. M. to 6 P. M.) will include Reports of SPECTRO-CHROME THERAPISTS, Discussion and Catechism.

The Evening session (8 P. M. to 11 P. M.) will be devoted to an Intensive Post-Graduate Discourse by Dinshah and the demonstration of his latest invention—SYMPATHOMETER, which enables the Spectro-Chrome Therapist to ascertain *at a glance* the patient's condition and to work *without any calculation* in accordance with the Occult Operation of the Great Sympathetic Nervous System. Dinshah's SYMPATHOMETER heralds another revolutionizing advance in practical SPECTRO-CHROME THERAPY.

Many are coming from distant States. The day's work will be instructive and entertaining. Bring your friends as visitors; do not forget to register full name and present address on entrance.

A copy of our Constitution and By-Laws, also an Application Blank is enclosed herewith. If you are a Graduate of SPECTRO-CHROME INSTITUTE but did not join the Association, join now and receive our magazine SPECTRO-CHROME free. Remember: Co-operation means success; the success of November 28, 1923 rests with YOU.

Yours Spectro-Chromically,

AMERICAN ASSOCIATION OF SPECTRO-CHROME THERAPISTS,

per Dinshah P. Ghadiali

President.

Attuned Green Wave

(Spectro-Chrome Therapy, Spectro-Chrome Therapists)

SARCOMA, PHLEBITIS, ULCERATED BLADDER

MY CLINICAL NOTES ON SPECTRO-CHROME THERAPY

By:—Kate W. Baldwin, M.D., F.A.C.S., Senior Surgeon, Women's Hospital, Philadelphia, Pennsylvania.

Many interesting cases have been under Spectro-Chrome since the last magazine was out. I hardly know which to select for this report.

Case 1—Mrs. W. Thirty-six hours after an extensive operation on the left leg for sarcoma, went bad, a severe chill, temperature 104° in axilla, heart very irregular and feeble, radial pulse not present. As she could not be given Spectro-Chrome immediately, I was obliged to let her be given one hypodermic of Lloyds' Subculoid Lobelia. The hospital staff felt that she was very near her end and telephoned for her husband. They also moved her into a room by herself so that she could have the Spectro-Chrome and be alone. That is one of the very thoughtful things one finds at the Woman's Hospital. If possible to avoid it, no patient is ever allowed to die in the ward surrounded by other patients. Climatic conditions may have had some influence on the case; a terrific storm, thunder, lightning and a deluge of rain continued for about two hours. I dressed the wound to make sure nothing there was causing trouble. The foot was a little swollen, otherwise, the local condition was good. As soon as possible, Magenta was given on the heart and it was repeated every three hours until the heart demanded less frequent attention. The wound was dressed every second day and when open Green and Indigo were used on it; at least once a day she had a Lemon systemic.

She was kept in a private room for the two weeks that she remained in the hospital, returning home in wonderfully good condition. After this, she came to the office for treatment until a few days ago when she had quite a severe cold. I was out of town and did not see her for three days. A very good man who lives next door was called in and cared for her through this emergency. I went to the house once to dress the wound; since then she has come to the office. The growth is returning and she is not in as good general condition as before the cold.

As sarcoma cases go, this patient should have died at least two years ago. Until the operation, the bone had not been involved. At that time the crest of the tibia was denuded of periosteum for about two inches. The outlook here is not bright, but she has been in a condition to look after her house and enjoy her friends for at least two years longer than could have reasonably been expected.

Case 2—The above case recalls a case of sarcoma of the first phalanx of the thumb of the right hand in a negress about 35 years of age; this growth was on the palmar aspect about the size of a small hickory nut. Spectro-Chrome reduced it at least one third, but it was too much in her way as a laundress and general housekeeper

with a sick husband to care for. She lived in New Jersey and as it made a long journey for her the treatments were not very frequent. Finally she said it was so much in her way that she wished I would remove it. I had no idea that it was a sarcoma. It was easily removed under local anesthesia and dressed with an ordinary surgical dressing. It went bad, bad and worse; the whole arm into the axilla was involved, badly swollen and extremely sensitive. A quantity of pus was evacuated from above the elbow and the condition began to improve. In the mean time her husband died and the poor woman had her heart and hands full. She made a good recovery and reported in good condition about five months ago.

Case 3—The following is a very interesting case of acute phlebitis (venous inflammation) of the internal saphenous vein of the right side. The patient a woman about thirty years of age, in good general condition, was sent to me for the phlebitis with a diagnosis of possible pregnancy. I made no pelvic examination. The primary swelling was just above the knee, very painful, then a mass of swollen veins below the knee and a slight involvement near the groin. The patient reported that her mother was in bed nine weeks with just such a leg. This patient was not far from the hospital and went back and forth for treatments. The condition improved from the very first treatment and after eight treatments she was so perfectly comfortable that she did not report for two weeks. Then she came just to tell me how wonderfully well she was and very much appreciated not having to stay in bed as did her mother. The treatment was Green local, immediately followed by Indigo. Between treatments she used a compress of Epsom Salt. When she reported after an absence of two weeks there was no evidence of any trouble above the knee, but below there was still some hardness. I advised her to report for treatment until that was entirely softened. Only those who have had such cases to treat can quite appreciate what Spectro-Chrome did for this patient.

Case 4—This case is still under treatment, but the improvement has been sufficiently worthwhile to report progress. Mr. R.—51 years of age. Blood pressure 190, Pulse 96, Respirations 19 to 20. For ten years had an ulcerated bladder, primarily alcoholic, complicated by other infections, about everything a man could have. In 1902 had gonorrhoea which was cleared in about six weeks. The frequent necessity to urinate and the extreme pain made it necessary for this patient to give up his position in the advertising department of a large daily newspaper and for about ten years has not been able to engage in any work more exacting than that of a watchman and that for only a short time—one week at night and the next week during the day. He had been in several hospitals and sanitariums. In 1917 the bladder was curetted and a retention catheter kept in for two weeks.

After this he was worse rather than better. In 1918 was wildly delirious due to extreme pain; was nine weeks in a hospital for this condition.

At present and for years before the bladder trouble he had been very sensitive to cold. As he expressed it, he could not lift his hat when out of doors that he did not take cold. With a hope to relieve the tendency to "take cold" he had two operations on the nose. One side was burned and the other drilled. The colds continued. He was more or less constantly covered with small or large boils, especially on back, chest and upper arms. The back suggested the scarring from small pox. There was a very unusual double V shaped area just below the cervical vertebrae covered with small papules, pustules and necrotic epithelium. This area was very itchy. This condition alone would have been sufficent to make one very uncomfortable. Really, life held very little for this man. When I asked him about smoking he replied that it was about the only comfort he had left. He was much more comfortable standing than sitting. He stood all the time he was giving his history.

There was no organic heart lesion, neither was there trouble of the lungs. He had pneumonia twice and bronchitis and pleurisy each once. None of the so-called children's dis-eases. Since the first of February 1917 had not tasted any kind of strong drink; following some medication in the Summer of 1922 there has been enuresis (involuntary passage of urine) night and day, which has caused great inconvenience and much loss of sleep. Urinalysis did not show any kidney involvement. Ears, nose and throat nothing special.

This patient came to me March 10th 1923; he came every day for about a month and then three times a week. The first of May at a family conference it was decided that he had improved but that it was evidently going to take a long time to get the really desired results and so he would stop treatments. I do not often urge patients to continue treatment when they get that attitude. In this case, however, I wrote the man and requested him to come and have a talk with me. This he did and he decided to continue for a time at least three times a week.

About once a month we checked up and found that there has been a steady and it seems to both of us a satisfactory progress. This month he reported his general condition "A number one," and all the bladder symptoms at least seventy per cent better. He voids much less frequently and more at a time with practically no pain—a pain that formerly did double him up like a jack knife. It is now infrequent and not more than "a scratch." He has much better control over the bladder even at night.

His skin has cleared up wonderfully and only occasionally is there even a suggestion of a pustule.

Is this worth while? The man thinks it is and he says that now "life is well worth while". He went back to his old job.

The treatment has been strictly Spectro-Chrome. Green and Lemon systemic and at first I used Indigo and Turquoise locally, but, it did not do as well as after I changed to the more stimulating colors Lemon and Orange with an occasional Green and Red. The blood pressure has come down to 145 and the pulse to 70-75.

Has it been worth while??? I ask the readers to be the judge. The patient and I are going to keep right on until he can say that the bladder condition as well as the general condition is "A. number one."

FORMATION OF THE NEW ENGLAND ASSOCIATION OF SPECTRO-CHROME THERAPISTS

Reported By:—Frederick Dugdale, M.D., LL.B., S..C.T., Boston, Massachusetts.

I am pleased to be able to write you that a meeting of the members of your Boston class was held at Hotel Westminster on July 2, 1923.

It was suggested that we call the association, The New England Association of Spectro-Chrome Therapists. I was privileged to have the honor of being selected as their President; the following officers were also elected:—

Francis A. Cave, M.D., D.C., S.C.T., Vice-President.

Harry E. Bodwell, D. M., S.C.T., Secretary.

Milda Steinharter, S.C.T., Treasurer.

They were elected as a Board of Trustees to draw up By-Laws with a committee of five as follows:

Charles G. Hatch, D.O., S.C.T., Lawrence, Mass., Chairman.

Katherine F. Boyle, S.C.T., Providence, R. I.

D. Richard Holder, S.C.T., Worcester, Mass.

Leon A. Dana, D.C., S.C.T., Boston, Mass.

Nathan N. Hurvitz, S.C.T., Boston, Mass.

It was decided that the initiative fee for membership was to be $10.00 and each member to pay $1.00 a month dues; that all members should make application to enroll in the National body (American Association of Spectro-Chrome Therapists).

There were some very interesting reports given by Dr. Harry Bodwell, D. Richard Holder, Dr. Leon A. Dana, Miss Milda Steinharter and others.

My diploma was received yesterday afternoon and feel very happy and proud to be the possessor of it.

I am very happy and proud to have you giving such gracious help to the Graduates of the Boston class and others in your district. Co-operation and fraternal organization is what we most need. The cause of Therapeutical Emancipation requires powerful help from all quarters and each enrollment to the cause means added strength. I welcome The New England Association of Spectro-Chrome Therapists to our fold and promise it my cordial support.

EYE TROUBLE REMOVED WITH SPECTRO-CHROME THERAPY

By:—Dr. F. B. Williams, Shenandoah, Iowa.

As the days go on and I become more familiar with the work of Spectro-Chrome, I am more and more convinced that it is a wonder worker.

About three weeks ago, a woman 81 years of age, was brought to me suffering great agony from a disorder in the left eye. Little or no help had been afforded from the regular channels to relieve her pain.

I put her under the Lemon Wave, then the Green Wave and after half hour I gave the Indigo. Her pain stopped.

She speaks very little English and after a while she made a glad exclamation in Swedish. I asked her son about what she said. He replied that she wanted to know what I did with her pain!

When she came, her eye looked like a piece of raw meat; now, it has cleared, she has no pain, sleeps well and feels happy.

I have some very interesting cases on which I shall be glad to report later.

I thank you for your interest and wish you unbounded success in your great work.

WHAT EVEN A SMALL SPECTRO-CHROME EQUIPMENT CAN DO

By:—Orlo H. Frost, S.C.T.,
Hillsdale, New Jersey.

Congratulations to you three on your miraculous escape from more serious injury in the accident Kashmira so graphically depicts in the June number of "Spectro-Chrome" just received. Her article is extremely interesting, more for what it does not say than what it does tell.

For a long time I have been intending to write you giving some cases that have come under my observation or have been treated by me; (No dates will be given, and cases indicated by numerals).

Case No. 1: (not treated by me, but by a friend to whom I had been instrumental in selling an equipment some time ago) *Case of Goitre.* Woman about 50, with well developed goitre, seeing my friend treating herself for another ailment, asked her if she would try it on her goitre. Indigo on area 2 was used for two treatments, and the woman said that she believed her neck felt easier. Then a tapemeasure was procured and the circumference of her neck taken. Treatments were continued for ten days, and at the end of that time the neck had reduced in circumference five inches. The woman said she felt better and all discomfort from her goitre had disappeared. She then went to Florida for the winter, since which time I have heard nothing as to her case.

Case No. 2: The husband of the woman with the goitre. Scratched his face when shaving, infection set in, and a "lovely time" with all its accompaniments was had by him until at his wife's solicitation he was induced to try Spectro-hrome. Turquoise and Indigo, (alternate treatments) were used, twice each. Wound all healed up and no further trouble experienced.

Case No. 3: Oetitis Media (Inflammation of Middle Ear). Girl 12 years of age. Used Spectro-Chrome exclusively, Turquoise on Area 1 and side of head; also systemic. Indigo on face and ear. Abscess discharged freely and without pain. Indigo, after discharging had continued for two days, stopped discharge over-night. No further trouble encountered. Complete recovery.

Case No. 4: Gastric ulcer. Patient had suffered intermittently for about ten years. Had received thorough medical treatment and also had taken the so-called "Lehnartz" treatment in summer of 1916; had tried also Christian Science and Chiropractic. All these various treatments afforded a certain amount of relief, but trouble recurred about every six months, usually lasting for about a month. Lemon on Areas 6-7-8-9 and 10, alternating with Turquoise occasionally, finally switching to Orange on same areas. Patient entirely relieved and in excellent health.

Case No. 5: Occipital Neuritis. Patient had never before been afflicted. Spectro - Chrome exclusively, with proper diet used. Turquoise on Areas 1 and 16, also Indigo on Areas 15 and 16. No serious discomfort. Some slight pain at first. Patient later developed abscess in ear. Again Spectro-Chrome brought her out of it, with but little pain and no great discomfort. Turquoise and Indigo alternating locally were the colors used.

Case No. 6: Influenza. Patient given extensive Turquoise on Areas 1-2-3-4 and 5, one hour at first, then half-hour for three successive treatments. Entirely relieved and up and able to pursue his regular duties in two days. When first taken, had temperature of 101°, and all the usual discomforts of the "Flu". After first treatment, discomfort relieved and before end of first day feeling much better. Patient expectorated copiously a caustic discharge which burned his mouth and nostrils. Two or three sores formed on lips, but these rapidly disappeared.

Cases Nos. 7, 8, 9, and 10. "Flu" in household of same man. All given like treatment with like results. Entire recovery in each case about two or three days.

Case No. 11: Whooping Cough. Patient, a boy about four years of age, contracted whooping cough. Theretofore, each winter from the time of his first winter, patient had been afflicted with chronic bronchitis. This, accompanied with whooping cough, made case particularly aggravated and difficult of treatment. Green, Areas 1-2-3-4 and 5, also systemic, for the whooping cough, brought about the desired result, with no great discomfort to patient during the run of the disease. He seldom vomited and his sleep was good. The chronic bronchitis was another matter, but his parents, seeing the good result achieved with the whooping cough, bought an "Intermediate Equipment" and under my direction, treated him for the chronic condition. Turquoise on Areas 1-2-3-4-5 and Magenta on 4 were used. The treatments continued for several weeks and I understand complete relief was effected. A peculiar thing about this case is that the mother of the boy, seeking to keep in the good graces of her regular practitioner, had him prescribe for the boy during his whooping cough and bronchitis attacks, regularly filling the prescriptions and carefully pouring each dose, not down the boy, but down the sink! At each visit of the doctor, he would expand genially on discovering how well the boy was progressing, and would say "My, Billy is reponding fine to his medicine this time"! I assume he takes great credit to himself for effecting this "cure." He is welcome to the credit. Results are what count.

Case No. 12: Toothache—the kind that keeps you awake nights and gives no ease daytimes. Two or three applications of Indigo entirely relieved the pain, and though an abscess had formed, it was absorbed and no further treatment has since been necessary.

Case No. 13: A second degree burn. Woman severely burned by exploding oilstove. Her legs from ankles to thighs were severely burned and her left arm from palm of hand to elbow and right under fore arm. Regular physician called and he used "Ungentine" with the usual surgical dressings. Did not see case until two weeks after treatment had been administered. Used Blue on left arm just as an experiment to see (to show patient) what would happen if used regularly. Theretofore, patient in great discomfort, unable to sleep and extremely nervous. Blue applied for half hour to aching arm produced soothing comfort. Blue applied to Area 1 induced first comfortable sleep. Patient dropped to sleep within ten minutes after application of Blue on Area 1 commenced. Continued application for half hour and patient rested comfortably most of that night. Following night, treated same arm with Turquoise producing like results. The family are in dire circumstances and the regular physician had already stated that his charge would be regulated according to their circumstances. Therefore, the woman's husband, fearing that the doctor might become angered if Spectro-Chrome were used, and thus render a large bill, decided not to continue treatments, although he stated that had he first had Spectro-Chrome he would certainly have continued it. Have since heard patient was recovering slowly under regular medical treatments. It will be many a month before that poor woman has any great degree of comfort.

Case No. 14: Poison Ivy. Boy about 17. Had the usual sugar of lead treatments from regular physician; no relief. Driven nearly frantic, he finally tried Spectro-Chrome. One sustained treatment of Green, produced great relief, and subsequently Yellow was used on Areas 8 and 9. Entire recovery after three treatments.

Case No. 15: Abscess of ear. Boy, about nineteen, "contracted cold" while swimming. Had usual treatment for abscess by regular physician, who lanced it twice. No relief being afforded, he too sought Spectro-Chrome. Indigo on ear produced quick relief, and subsequently brought about complete health.

Case No. 16: Acute Tonsilitis: Young woman developed over-night rapidly expanding tonsilitis. When treatment began she had a fever of 101½ and great pain in throat and neck. Blue on Areas 1 and 2 applied for one hour brought quick relief and subsequent Turquoise on same areas effected a complete cure. Duration of case, about 36 hours.

All the treatments mentioned above were given with the small 100 Watt Spectro-Chrome Equipment saving that of the parents who bought the Intermediate Equipment and the case of the acute tonsilitis just referred to, where the Practitioner's Equipment was used.

Many more similar cases could be given, but these are sufficient to illustrate the wonder-working properties of Spectro-Chrome when intelligently and conscientiously used.

I never cease to speak for Spectro-Chrome. Indeed, my friends say I am a "nut" on the subject.

My dear Dinshah, I wish you all the success that the Originator of so great a boon to humanity so richly deserves.

Thank you cordially, my kind friend. You are truly performing service to humanity by your keen interest in Spectro-Chrome Therapy—but, I sincerely believe you ought to have a Graduate Equipment with your extensive practice. O, what more wonders and marvels you would then be reporting each month! Let me know when ready.

And yet, in the early part of 1920, when I was living in semi-seclusion as a scientific researcher in the town of Hillsdale, New Jersey (which has now been put on the map as the birthplace of Spectro-Chrome Therapy) as I was traveling with him each morning on the same (cowspeed) Erie Railroad, this friend of mine would remark frequently, as I expatiated on the Beauties of Compound Color Waves, "how *can* it be?" He is doing it *himself* now and may all have that skeptical yet open mind as Orlo H. Frost kept while he was searching for the Truth in Spectro-Chrome Therapy. Success to you, friend Frost!

EPILEPSY RELIEVED
USE OF CRACKED SLIDES

By:—Mrs. Emma G. Taylor, S.C.T., Brooklyn, New York.

A boy had attacks of Epilepsy. Since I treated him as directed there has been only one light attack thanks to Spectro-Chrome Therapy.

I wish to ask if a slide be cracked is it all right to use it? It is only a small crack from the bulb being too close and neglect to draw it back.

A cracked slide may be used if the crack does not leak white light or another color through it. In an Intermediate Equipment, the bulb-carrier rod should be fully drawn *out* when using; in a Graduate Equipment, the rod should be pushed all the way *in* when in use. The condenser in a Practitioner's Equipment should be less than one inch from the bulb. This can be easily done by turning the button at the back.

NIGHT SWEATS REMEDIED

By:—Dr. A. Eppolito, S.C.T.
......Brooklyn, New York.......

A woman recently applied to me for assistance. She had been complaining of night sweats for two weeks, having to change her underwear three to four times during the night.

I gave her a Systemic Turquoise treatment for twenty minutes, front and back. She reported four days later that she had no night sweats that night and has not had any since.

OSTEOPATHS SUCCESSFULLY USE SPECTRO-CHROME THERAPY IN DROPSY, CONSTIPATION, COLD, MARASMUS AND THE LIKE

By:—George F. Wagoner, D. O. and Lillie E. Wagoner, D. O.
Creston, Iowa.

We are using Spectro-Chrome every day with good results.

Case 1—Male, age 42, injured when a boy by falling astride a wagon box. In 1918 he was vaccinated for "flu"; a week later, site of injury in boyhood began giving pain. He was very miserable for nearly two years when his trouble was "diagnosed" cancerous and right testicle was removed. In April this year he was told he had only two weeks to live (cancer)! When we took the case he was dropsical to the waistline, legs four times normal size. Liver not acting at all. Used Magenta on 17-18-19 and on 4 Areas; also Attuned Color Waves Lemon on 19 and on 9-10-11 ending with Magenta on 4. The scrotum has been somewhat enlarged, but is lessening in size and the patient has gained 8½ pounds in the past six weeks. Please let us know what has been left undone.

We are more than pleased to get the Spectro-Chrome magazine; the case reports are surely inspiring and beneficial. We are using Spectro-Chrome Therapy daily and are glad to use it.

Case 2—Little child, 1 year old; bowels had never moved normally. One treatment of 15 minutes of Yellow over abdomen and the bowels acted normally (with the exception of one day) for three weeks. Then gave the child a second treatment, same as the first and Nature did the rest. The child seems perfectly healthy now and has had no more treatment for 4 months.

Case 3—Extremely bad cold in man of 60 years. Yellow on Number 17 for 10 minutes, then Yellow on 2-3-4-5 for 10 minutes followed by Magenta on 4 for 5 minutes and cold disappeared as if by magic.

Case 4—Woman, 65 years old. Congestion all along the spine and much gas formation in abdomen. Was relieved with Yellow on Numbers 17-18-19 for 10 minutes, then on 8-9-10 for 10 minutes and Magenta on Number 4 for 10 minutes.

Case 5—Baby 4 months old. A case of Marasmus (Extreme Emaciation). There seemed to be no life left in the baby. Five treatments of Yellow back and front systemic seemed to put new life into child and seems to be gaining daily.

We are now treating a case of Paralysis Agitans; also another of Sarcoma in Pelvis; both are steadily improving. We are using on others too, but, do not wish to tire you with too many.

I thank you for the reports, doctors; it is pleasing to receive reports from those who are not graduates of Spectro-Chrome Institute yet use Spectro-Chrome Therapy in their daily practice and honestly acknowledge the merits of the system. All cases similar to Case 1, should have Systemic Green and Lemon as well as Magenta on 4 and 18 Areas. Dropsical tendency or dropsical condition indicates failing circulation of blood and any case where Magenta will fail to pick up the exudation (either totally or partially) is heading toward death, because Magenta is the predominant color of the Auric Vehicle and its failure to act denotes want of activity in that higher vehicle through which the Emotional Vehicle operates on the Physical Visible Vehicle. Understood properly and used with a comprehension of its extreme potency, Magenta as produced and used in Spectro-Chrome Therapy is a *sure* instrument for the detection of approaching dissolution. It is the semaphore to the Spectro-Chrome Therapist who knows his business as to the condition of the links that comprise the chain of the Physical Visible, Physical Invisible and the Invisible Psychical Man. A new era will dawn when all my graduates will fully recognize this important operation in life and aim at mastering it. The time is near now—read in another column the announcement about my invention of the Spectro-Chrome Sympathometer, for the indication of the Occult Operation of the Great Sympathetic Nervous System at a glance, without any calculation. Come to our Annual Convention on November 28, 1923 and see it.

Keep on sending your reports. Reports are needed for publication and without reports of various users with a variety of views, the value of our system may not be so well appreciated. You cannot *tire* me; that word does not exist in my vocabulary.

QUICK RESULTS OBTAINED IN PLEURISY, EAR TROUBLE, CONSTIPATION, NERVOUSNESS AND ALBUMINURIA

By:—MRS. LOCCIE B. SMITH,

Indianapolis, Indiana

I report successful results in the following cases:—

Case 1—Acute Pleurisy; I used Attuned Color Wave Turquoise direct on area of lower rib on right side; applied this Color Wave 3 times only.

Case 2—Retracted Drum of left ear; I applied Orange direct for 15 minutes only. (Patient had been treated by Specialist 10 days without result.)

Case 3—Constipation; child 4 years of age; Lemon twice and Yellow once on 9. No other treatment needed.

Case 4—Nervousness of girl; 9 years of age. I applied Blue Systemic six times. Patient has been normal since.

Case 5—Albuminuria; I applied Lemon on 18 five times resulting in no trace of Albumin.

Attuned Turquoise Wave

(Poetry, Romance, Drama, Art)

THE ISLE OF MAGENTA

By:—MISS KASHMIRA DINSHAH,

S. C. T. Joint Secretary, Spectro-Chrome Institute.

Afloat on the waves of a sapphire sea,
An emerald island of mystery.
A charmed jewel-isle where, gorgeous and great
The Lord of Magenta resides in state,
And governs his people with gentle hand,
Love being his mighty enchanting wand.

Such is the mystical Magenta Isle
Where all is virtue and unknown is guile.
It is far away, not everyone knows,
Straight from the depths of the ocean it rose
Aeons ago, in the dawn of the world
When, against dark chaos, cosmos unfurled
Conquering light. This superb paradise
Visible only to most chaste of eyes,
Is a bright region of gladness supreme,
Of calm happiness like unto a dream.
Where wondrous flowers their beauty unfold,
Of violet, azure, crimson and gold,
Scenting the air with exquisite perfume,
Where the sun doth ever the land illume
Not flashingly but with shadow and sheen
Blended entrancingly. By this light seen
Are dim glowing forms like fugitive flames,
They are the ones whom Magenta reclaims.
It is a place of marvelous magic,
Where is no sound of aught that is tragic.
Permitted no voice, Dispute flies away,
Ruling impulse only, Love, has a say.

There to that island are brought those of Earth,
(In all is a promise of secret worth)
Wounded and seared by the destroying fire
Of selfishness, envy, jealousy, ire;
Those victims the splendor of Love shall heal,
A spell of purity, phantom yet real,
Is cast over them. The slumbering glow
Of selflessness blazes, at last they know
A wonderful joy before they knew not,
All else is submerged, all else is forgot.

Ever there rings thru the sweet-scented air,
Clearly and softly a symphony rare;
Melody silver that hushes and stills,
Magical music that tenderly thrills.
Awakening memory, hopes and dreams,
Ideals revive and a fresher life beams.
Gradually, in this glory of song,
From the mind vanishes all thought of wrong.

Thru the influence of radiant Love,
Gift sweetest to Man from Heaven above,
Over the Earthlings there comes a great change,
A grace and nobleness heretofor strange.
New happiness throws a halo of light
Around them. No more shall they in the night
Of evil passions rove lost and forlorn,
And safe in a luminous mental morn
Return they to Earth, but not one can say
Where is the island where dawned the mind's day?

CLEVELAND, OHIO AND INDIANAPOLIS, INDIANA CLASSES

The Sixteenth Class on Spectro-Chrome Therapy was formed in Cleveland Ohio, at the Hollenden Hotel and the Seventeenth in Indianapolis Indiana, at the Hotel Lincoln. Each presented a testimonial to the Originator of Spectro-Chrome Therapy, expressing appreciation and promising co-operation.

CLEVELAND, OHIO TESTIMONIAL

September 28, 1923

Sir:—

We, the students of your first course in Cleveland in Spectro-Chrome Therapy take occasion publicly and collectively to express our appreciation and gratification in having had the privilege of intimate association with you.

To one versed in the physical Sciences your discourses have been a revelation and to all the seeming crystallization of a university education compressed into a short two weeks. The profundity and breadth of your contacts and information are amazing, both a delight and a challenge. Your ability to synthesize and to make the varied sources of information revealed, understandable and usable, together with an unique use of metaphor, bespeaks a real genius for teaching.

Rarely does one encounter another as impersonal and yet devoted wholeheartedly to the cause of Humanity.

The logicality of the development of your theme and presentation of Spectro-Chrome Therapy is unescapable if one have an open mind.

Any to whom may come the opportunity to sit under your instruction, we would assure that they can ill afford not to avail themselves thereof.

In speeding you on your way toward the furtherance of your plans, be assured of our confidence and gratitude—and our hopes not our fears go with you.

(Signatures of all the Students)

INDIANAPOLIS, INDIANA TESTIMONIAL

October 18, 1923

Dear Colonel Dinshah:—

The members of the Seventeenth Class of Spectro-Chrome Therapy wish to express their sincere thanks and appreciation for the rare and instructive course you have given us here in Indianapolis.

The theory of the Attuned Color Waves, and its application to the Human Aura in treatment of dis-ease, was given in such a scientific and mathematical way that we realize we have been given a great Truth; and Truth can never die.

The opportunity of hearing theoretical discourses, followed by practical demonstrations of every subject brought under discussion, was a revelation to all of us. It has inspired us to an increased desire to attempt cases we otherwise would have considered hopeless.

The great store of knowledge, the unusual ability of the concise and understanding presentation of the Science, the deep sympathy and love of the Teacher for suffering humanity, and also for his students, made this course one that we shall never forget. It is invaluable to us.

May you in some measure receive some of the fruits of your hard labors! This class shall endeavor to further the work of Spectro-Chrome Therapy in every way possible, and always to be of any service to our Teacher—Dinshah.

(Signatures of the Students)

SPECTRO-CHROME

OCTOBER 1923
PURPLE 1923

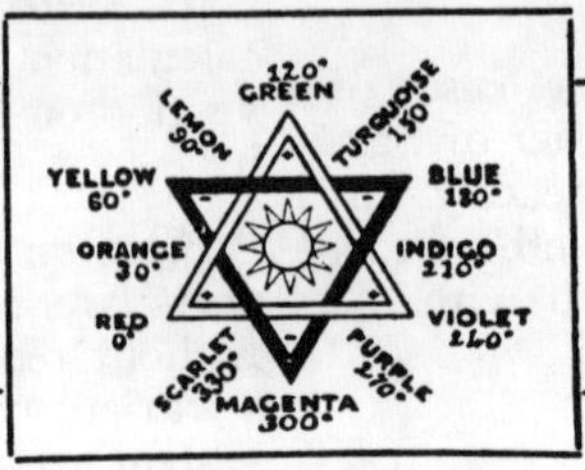

VOLUME 2
NUMBER 10

Devoted to: Spectro-Chrome Therapy Dedicated to: Service of Humanity

CONTENTS

Page

Dinshah's Platform—Spectro-Chrome Therapeutic Area Chart—Spectro-Chrome Therapeutic System 385

Spectro-Chrome Therapy Awakens Journal of American Medical Association; Answer to Its Attack (by Dinshah P. Ghadiali, S.C.T.) 386-398

The World Through the Eyes of Dinshah—

Our Central Office will move in August—Pacific Coast Tour; Please Note Carefully; New Classes— 399-401

From Coast to Coast; The Message of Spectro-Chrome Therapy Delivered from the Atlantic to the Pacific; Dinshah's Prediction Fulfilled (by Kashmira Dinshah, S.C.T.) 402-405

San Francisco California Opens the Door of the Golden West 406

Mrs. Hirabai A. Vacha 407

Los Angeles California Tenders Cordial Reception to Spectro-Chrome Therapy 408

Price List .... Inside Back Cover

Registration Notice of Trade-Marks—Notice to Purchasers and Patients—To Our Contributors—Advertisement Rates .... Outside Back Cover

Editor
Dinshah P. Ghadiali

In course of life you reach a time,
When Doctor, Lawyer, Priest alike,
Does fail to give the sought relief,
To troubled mind or burdened heart—
Then come to me;
I help you to help yourself.
SPECTRO-CHROME.

Subscription:—January to December 1923 .... $3.00
All Subscriptions Taken For Full Year Only And Expire With The December Issue
Any Single Copy 25 Cents. Back Numbers Available From June 1922.

SPECTRO-CHROME THERAPY

AWAKENS

JOURNAL OF

AMERICAN MEDICAL ASSOCIATION

ANSWER TO ITS ATTACK

BY

DINSHAH P. GHADIALI, S.C.T.

PRESIDENT

SPECTRO-CHROME INSTITUTE

(Incorporated)

Central Office and Experimental Laboratory

Philadelphia, Pennsylvania

The duty of a journal is to be the upholder of the right, the defender of the down-trodden, the protector of the poor, the speaker of unflinching truth, the enunciator of the most exact and precise facts.

For years, however, it has been an admitted matter that the American Medical Association as reflected in the conduct of its Journal has followed a line of apparently hoodwinking policy, endeavoring to be the angel in Heaven sitting on the throne of justice to adjudicate the merits of other systems of therapy from its own standpoint of knowledge and trying to represent to the World its great superiority by whipping the skin off those differing from its doctrines.

There was a time years ago, when, in the ignorance in which the public was submerged, it bowed in solemn reverence to the dictator who posed as the benefactor of the suffering sick. At that period, the persecution with which this abominable journalism that held its stick over the followers as well as the leaders of other systems of therapy resulted in many untoward effects for those who were so unfortunate as not to gain the favor of this ugly monster. It therefore set up an oligarchy of tyranny by hogging the whole field of healing and through the assistance of stupid ignorant legislatures wove such a snare of oppression, that the victims were glad to die at the bidding of this most unamerican inhuman proposition.

The American Medical Association, in its past career, has established throughout the land a terrorizing Black Hand such as no commercial trust ever dared to dream. It is known and spoken of as the Medical Trust and by a cunning masterful affiliation with the Medical Examining Boards, Health Boards and all other so-called (Mahogany) boards it has created such a despotic administration, that even the boldest leaders of other therapies have avoided the uncanny contact of this gigantic trust as far as practicable.

This Medical Trust through its servile Journal has attacked and ruined the life work and reputation of many great researchers whose shoes it is not wise enough to clean. It has upheld only its own damnable poison therapy and destructive surgery and dashed the last hope of many a dying person who would otherwise have been saved by its obnoxious old-time "ethical" systems and persistent dogmatic interference. The M. D. has nothing but poisons and pills, potions and plasters, blisters blasters and babboon glands, hog serums, dog serums, cow vaccines and pig pepsins, his utter dislike and inertia for progress has justly made him derided as "Doctor of Murder," "Doctor of Mules," "Mule Driver," "Money Dog" and now "Doctor of Monkey-glands" to give worn-out old debauchees the power lost by unnatural practices.

The Journal of the A. M. A. is not recognized as an impartial reviewer, but has earned the demerit of a one-sided, partial one-eyed owl. Its single owl eye is blinded now by the dazzle of the science of Spectro-Chrome Therapy and the rising respect and confidence of the ailing public its followers are getting. It stands as it were dazed and blinded by the effulgence of the latest revelation in the healing art and in it sees the utter eventual destruction of the unholy temple of lie that the followers of poison therapy and master butchery have succeeded in creating throughout the length and breadth of the country.

This Journal of A. M. A. in its foul public tactics bulldozed the Christian Scientists, Osteopaths, Electropaths, Chiropractors and others who dared differ from its unscientific attitude and now it has started similar tactics against the Originator of Spectro-Chrome Therapy, his system of healing and his positively graftless Spectro-Chrome Institute.

It published a long article deriding Dinshah and his qualifications, endeavoring also to besmirch the name of our most learned and estimable friend Surgeon Kate W. Baldwin, M.D., F.A.C.S., S.C.T., whose name is blessed by hundreds who were saved from the surgical knife through her very skillful and timely use of Spectro-Chrome Therapy. It was this brave, progressive, courageous and masterful surgeon who saved Grace Shirlow from certain, sure and inevitable death when by her clothes catching fire her torso was nearly completely burnt and her skin was two-thirds destroyed. Let any M. D., any surgeon, look at the pictures in this article and judge for himself whether there *is* anything in *their* system or in *any other* system—drug or drugless—that could have saved the life of this pretty child.

This burned girl was taken to the Woman's Hospital in Philadelphia, Pennsylvania, where Dr. Kate W. Baldwin has been the Senior Surgeon for twenty years. When her long surgical experience plainly indicated to her that in the recognized methods of plastic surgery and skin grafting, dosing, dressing, doping and drugging, there was not a ghost of a chance for the injured girl to live through—her noble magnanimous heart throbbed to save the unfortunate patient through other means not as yet fully understood by the followers of so-called medical "science." Even at the risk of appearing unethical and unprofessional, this learned surgeon set aside all preconceived ideas, notions and methods and from the beginning started to treat Grace exclusively with Spectro-Chrome Therapy. There was no skin grafted, no surgical dressing used, no pill potion or powder administered, yet, the patient thrived and her bodily powers from within (as enunciated in Spectro-Chrome Therapy) built a new satin-like skin within a few months. Her kidneys, bowels, lungs and heart have functioned under Spectro-Chrome Therapy most perfectly; even the temperature which went at times over 106° Fahrenheit was controlled with the precision of a chronograph and brought down without delirium or any untoward symptoms.

This girl is now in the enjoyment of health and is a live monument to the labors of the Originator of Spectro-Chrome Therapy and the devotion, skill and accuracy of its pioneer followers.

And for such gracious and unselfish service to humanity, such dutiful use of her skill as a healer and such excellent demonstration of her progressiveness as a capable and competent surgeon, this wonderful woman whose work was extolled as the most marvelous in the annals of surgery by eminent surgeons, got from the (un) American Medical Association through its obviously ignorant Journal editor nothing but ridicule and foolish derision. Such is the reward she as a Fellow of the A. M. A. received from that disgraceful publication.

But that is not all—the end is not here; the end will come when Spectro-Chrome Therapy will put an end to this Octopus, whose disastrous practices none dared to speak about openly. That poor fish submerged in the ocean of crass ignorance, bloated into a stinking carcass of the most malodorous type because of its hoary antiquity has thought fit to attack the brain power, honesty and character of Dinshah, not knowing in what pouch it was inserting its grasping tentacles.

Whatever methods the leaders of other therapeutical systems may follow have nothing in common with Spectro-Chrome Therapy and the methods of its Originator. He abhors underhand tactics, crafty policies and grafty diplomacies. He prefers at all times to speak in the open without fear of consequences, where the weal of the suffering public demands that he speak. And he spoke as follows:

COPY OF ORIGINAL

February 4, 1924,
18-05

The Editor:
Journal of
American Medical Association,
535 North Dearborn Street,
CHICAGO, ILLINOIS.

Sir:

On returning from my tour in California, my attention was drawn to an article published in your issue of January 26, 1924, under the heading "SPECTRO-CHROME THERAPY," attacking me personally as well as defaming my work and my SPECTRO-CHROME INSTITUTE.

My scientific and professional standing are so well known and recognized internationally during thirty-three years of a fearless and independent platform career that they need no vindication or justification; however, it appears that you never had the opportunity of investigating first hand what SPECTRO-CHROME is and what it does.

As an humble servant of suffering humanity and in justice to yourself and the therapy you represent, I have the honor to invite you to appoint a committee of any Seven (7) of your representatives and send them to take my complete course on SPECTRO-CHROME THERAPY with experiments and demonstrations at Hotel Manx, San Francisco California, where my next class opens on Monday, the 24th of March 1924, at 7 P. M.

I hereby agree to pay all first class traveling and hotel expenses of your said Seven (7) representatives and also to present the course absolutely unconditionally.

Furthermore, although I am known as a caustic speaker in denouncing the "Empiricism of Medical Science", I hereby pledge you my word of honor that not a word tending to derogate, discredit, disrespect, disrepute or dishonor your representatives, their profession or their therapy will be uttered by me during the entire investigation and that they will be treated as my worthy and honored guests.

The above invitation stands good for two weeks from your receipt of this and your acceptance is requested within that period to enable me to remit in advance the stipulated expenses to your bankers and make other necessary arrangements.

Yours sincerely,

DINSHAH,
President.

DPG/IGD
By Registered Mail,
with receipt requested.

GRACE SHIRLOW

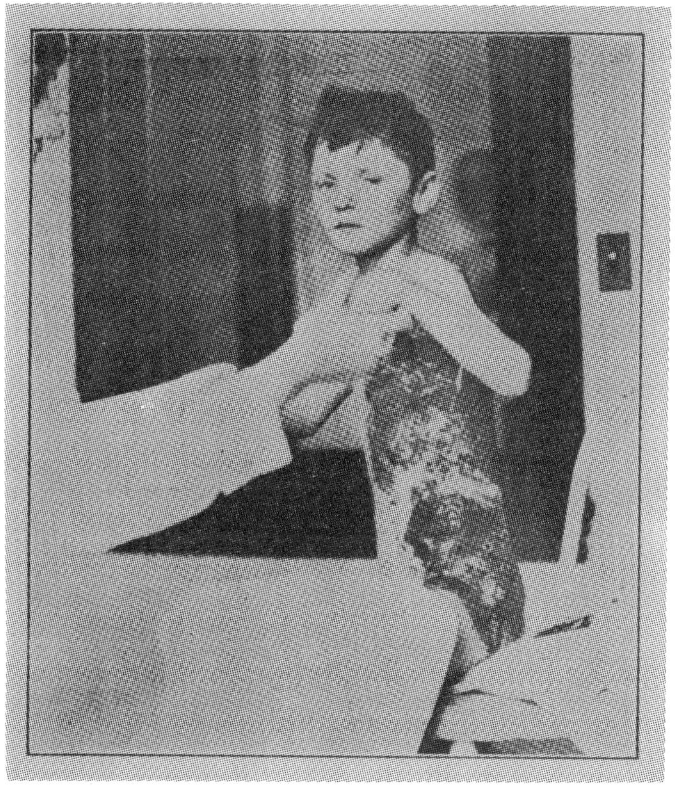

FRONT VIEW OF BURNT AREA
NOTE THE EXTENT OF SKIN INVOLVED

GRACE SHIRLOW

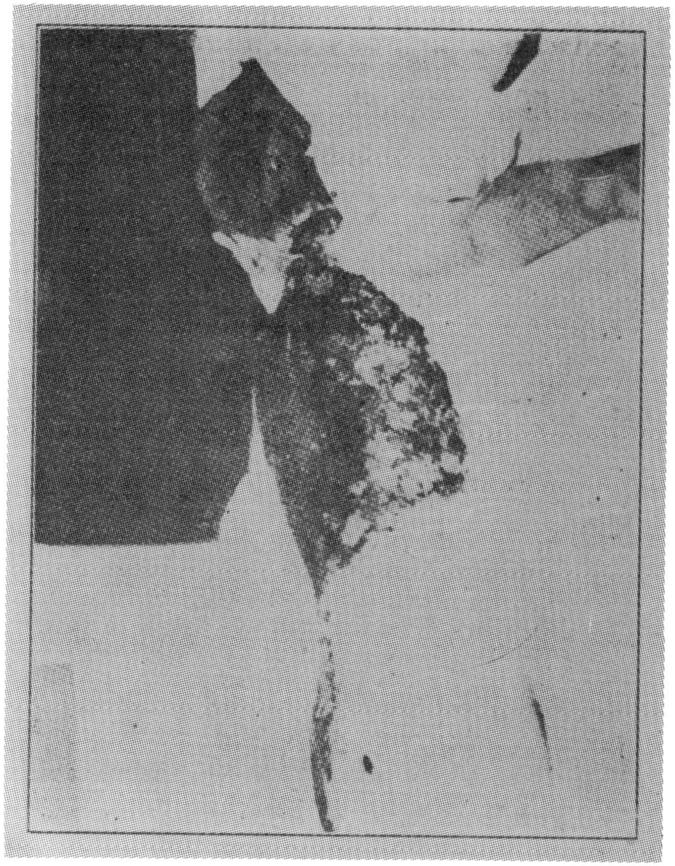

BACK VIEW OF BURNT AREA

GRACE SHIRLOW

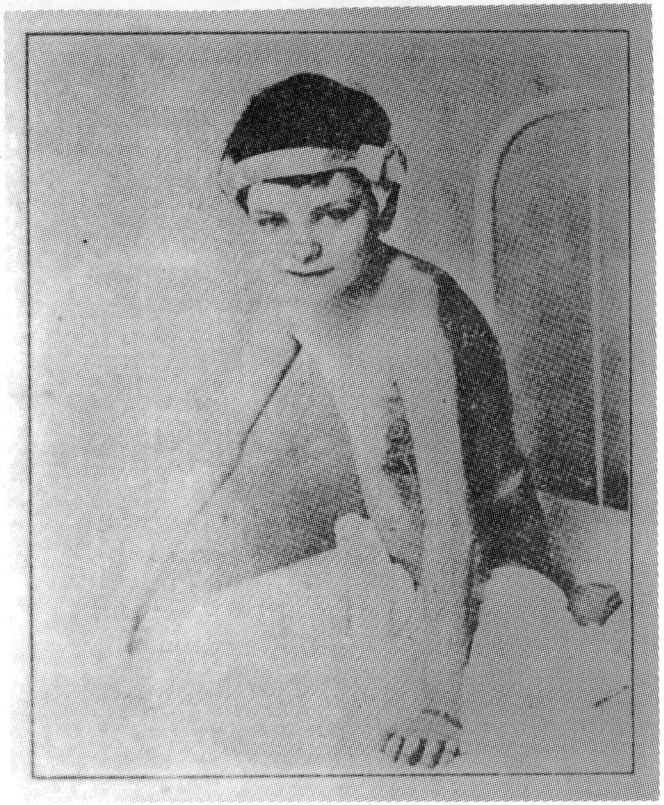

THREE MONTHS AFTER
NEW SKIN GROWING

GRACE SHIRLOW

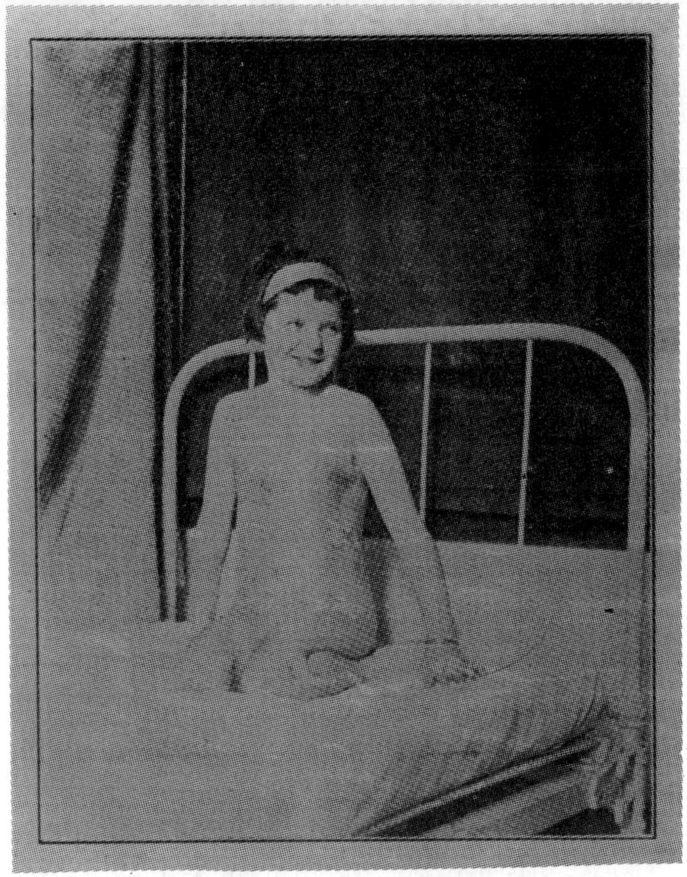

EIGHTEEN MONTHS AFTER
WONDERFUL RESULT PRODUCED EXCLUSIVELY
BY SPECTRO-CHROME THERAPY

GRACE SHIRLOW

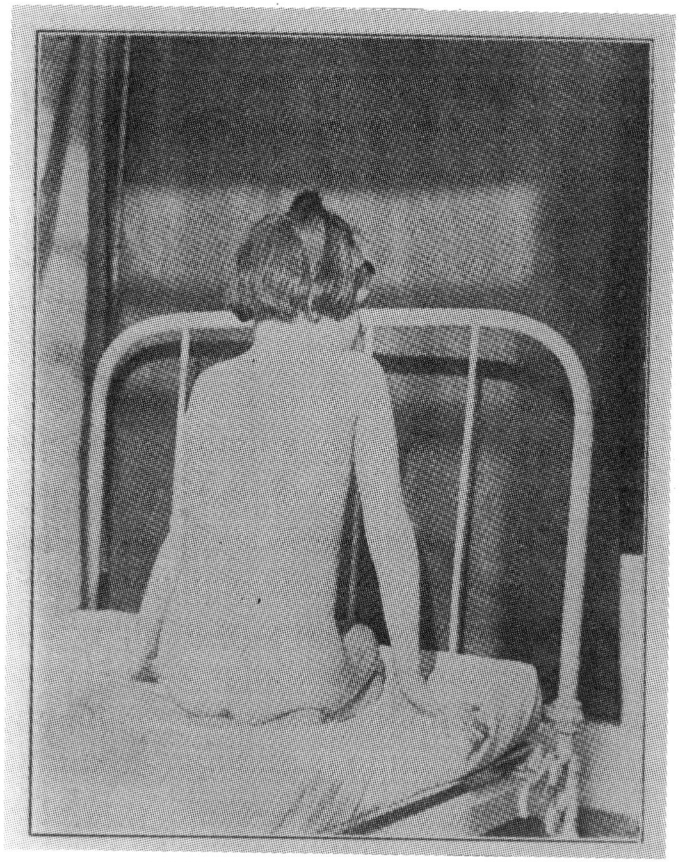

EIGHTEEN MONTHS AFTER
NOTE THE UNIFORMITY OF THE NEW SKIN
BUILT BY SPECTRO-CHROME THERAPY

GRACE SHIRLOW

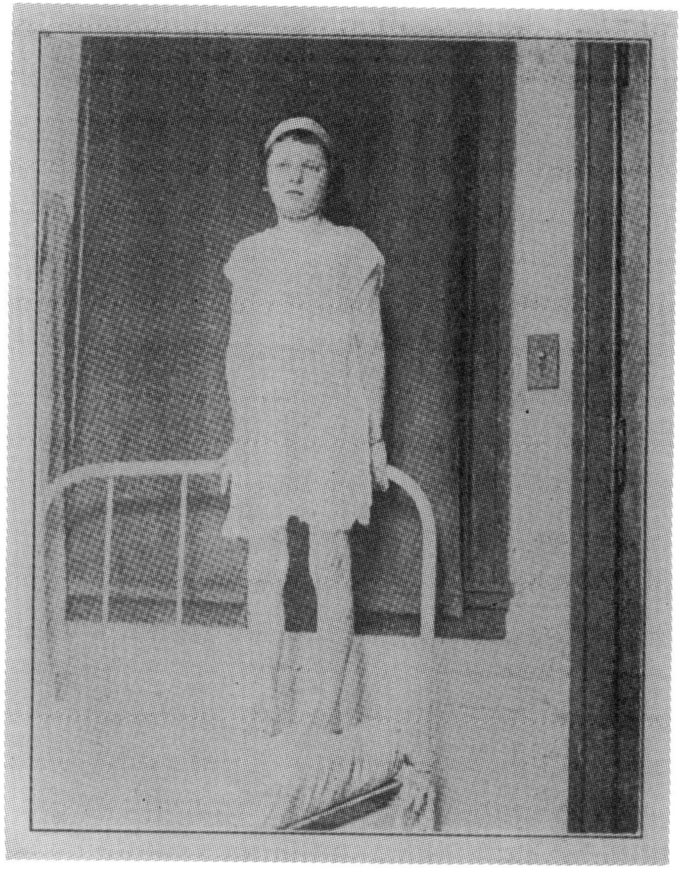

AFTER RECOVERY FROM THE BURNS
NOTE THE ABSENCE OF TENSION

GRACE SHIRLOW

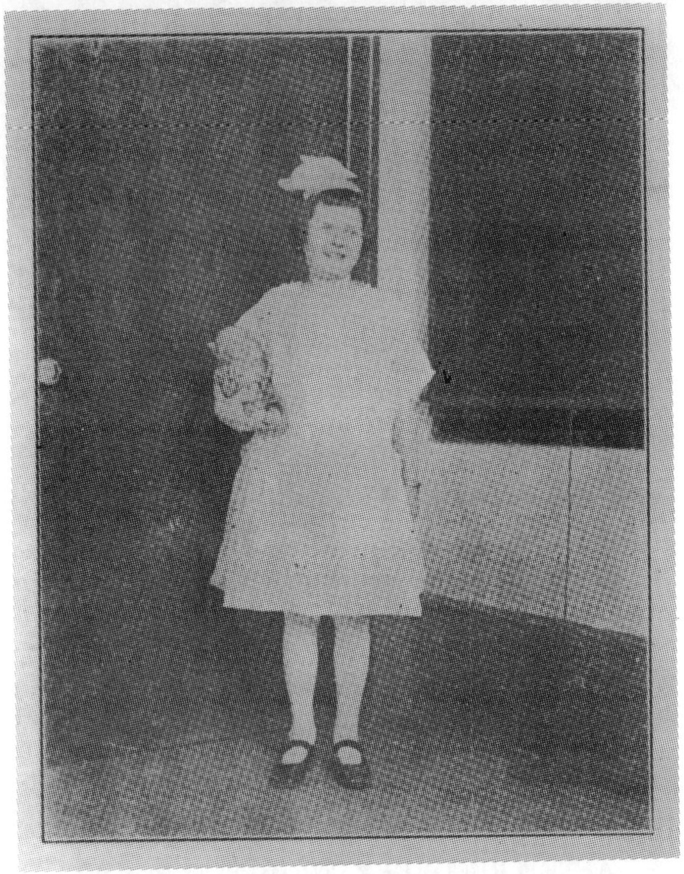

AT PLAY AFTER RECOVERY FROM BURNS

GRACE SHIRLOW

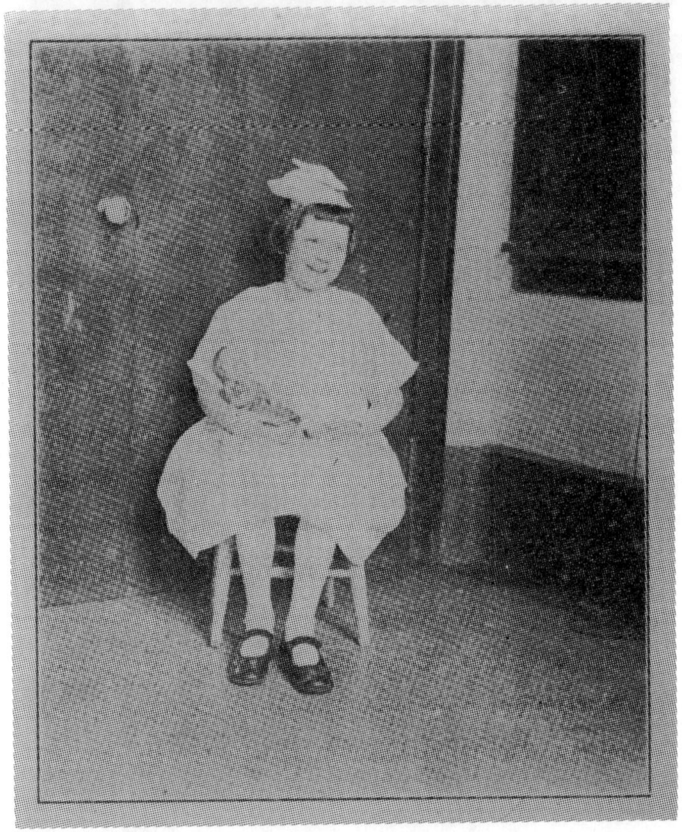

THE LIVING TESTIMONY TO SPECTRO-CHROME THERAPY

The above challenge which went by registered mail will remain as a permanent record against the cowardice, unfairness, pussillanimity and peanut-brained-ness of the most detestable organ of a cowardly trust which through covert means always endeavored to have the upper hand.

Ye Editor of the Journal of the A. M. A., you attacked my personality, integrity, character, unselfish work for humanity; I sent you a dignified request which, were *you* honest and integral, red-blooded and American, were *you* fair, just and cleanminded as well as competent, *you* would have either printed, answered or accepted; but, in quietly ignoring my gentlemanly challenge you have proved for ever your utter unworthiness to wield the sacred editorial pen.

I challenge you and your entire Medical Trust. I defy you and your whole organization to show and prove a money-making scheme or a fraud on the public such as you have made me and my Spectro-Chrome Institute appear in your dastardly attack. I have said times without number on the platform during thirty three years and now *repeat* emphatically that through me and my work your entire crack-brained edifice of drug therapy and destructive surgery will crumble to pieces and its bricks will be volatilized into molecules atoms electrons and beyonds and blown off into the place where there is plenty of sulphur and no winter!

Spectro-Chrome Therapy as taught by me is no fad imposition or fancy such as your fool article states. It has absolutely nothing in common with any other system of healing and is not an improvement on the medical art. Medical "art" is too slimy to reform. While your whole medical "science" is nothing but stupid contradictions and empiricism, Spectro-Chrome Therapy is the very essence of the unknown hence unrecognized occult sciences of ages which its Originator researched and junctioned and presented to the World as its most humble servant.

Spectro-Chrome Therapy teaches no complications of Anatomy, Physiology, Biology, Pathology, Histology, Bacteriology, Botany, Psychology, Surgery, Materia Medica, and the like as taught in the colleges of your worthless systems and does not even touch the humbugs known as Differential Diagnoses. It is an entirely original science which has revolutionized the conceptions of the healing art and is battering down the fabric of your impotent charlatanry. Spectro-Chrome Therapy is based on the laws of the most subtle operations of Nature, whose grand secrets have been zealously treasured and guarded by the Benign Masters of Occult Wisdom and whose message for the relief of the ailing, America is now receiving for the first time. But that time is here—its inroads through the land have created a havoc in the pockets of many an old time medical practitioner who is fast being supplemented by the successful Spectro-Chrome Therapist.

I ask for no quarters or mercy at your hands, ye Editor of the Journal of A. M. A.! The owl's eye may be blinded by the glory of the Sun, but it is no fault of the Sun—it is the weakness of the owl's eye. You have *started* something—it will end not here but in the speedy disintegration of the fabric of the Medical Trust. Fight with all your might; I care not for you and your vicious tactics, for, at your hands I have nothing to fear.

Spectro-Chrome Therapy has come to stay and finally mark my words—within the next three years and less its power of Truth will throw the last monkey wrench in the cogs of your evil machinery and burst the entire engine of the repulsive Medical Trust never to recover its prior prestige or vaunted dignity and terrorizing influence.

Attuned Red Wave

(The World Through the Eyes of Dinshah)

OUR CENTRAL OFFICE WILL MOVE IN AUGUST

When from a single room office in New York, N. Y. and a six room experimental laboratory in Hillsdale, N. J., in 1921 we shifted our head quarters to Philadelphia, Penna. in 1602 Summer Street in a three floor rented building, we believed we made an important move. Later, in 1922 when we were overcrowded there and removed to 2401 North Broad Street, in our present premises which we purchased last year, we thought we made progress.

However, since the beginning of the year, we started such intensive activities that our present four floor accommodation also was found utterly insufficient and we faced a big problem.

On returning from the California tour, our needs for present and future expansions to take proper care of and house our multifarious and quickly advancing requirements, as well as accommodations for our experimental research and other machinery necessitated so much area that the President had to decide upon his next step.

Finally, looking with an eye to what Spectro-Chrome Institute will have to face soon, when the Sympathometer, Itisometer and the Slideless Spectro-Chrome Therapy Equipments go before the World, it was resolved that a big move be promptly made to avoid break-down of public service.

We purchased therefore, the large property in Malaga, New Jersey, known as the Peach estate, with an area of 75 acres, situated on the Atlantic City State Highway, 28 miles from Philadelphia, Pa., and 5 miles from Vineland, N. J. This estate has numerous buildings and houses suited for various purposes, the main building a frontage of 78 feet facing a 700 feet lawn. There is enough acreage besides the fruit trees on the estate to accommodate the aeroplanes which will in the near future be housed there to fly the President and assistants from point to point for quick transit in the service of humanity. Furthermore, Spectro-Chrome Institute will later keep in touch with its graduates through a special Private Radio Broadcasting System in order that the various State and County Associations may be at close call with the head quarters.

Praise be to the Grand Creator of the Universe through Whose Beneficence the banner of the Latest Revelation in the Healing Art has been carried from coast to coast and when in August 1924 the Central Office will finally be transferred to Malaga, N. J. we sincerely expect to deliver better service and better results will ensue. Amen! ! !

PACIFIC COAST TOUR PLEASE NOTE CAREFULLY NEW CLASSES

The following is the program for the next two classes for training students in the science of Spectro-Chrome Therapy.

The times and places are fixed and there will be no change in the set program.

San Francisco California Class

Class Number Twentieth

1924

March

10 —Departure of Dinshah and his assistants from Philadelphia, Penna., 1:16 P. M.

11 —Arrival in Chicago, Illinois, Hotel Victoria, 9 A. M.

2401 N. BROAD STREET, PHILADELPHIA

THE PRESENT CENTRAL OFFICE AND EXPERIMENTAL LABORATORY OF SPECTRO-CHROME INSTITUTE

11 —Meeting of Spectro-Chrome Therapists in good standing and organization of Illinois Association of Spectro-Chrome Therapists, (Incorporated) Cook County Centre, 3 P. M. Hotel Victoria.

11 —Departure by the "California Limited" at 8 P. M.

14 —Arrival in Los Angeles, California, Hotel Lankershim, 3 P. M.

14 —Graduation Meeting of the Nineteenth Class in Spectro-Chrome Therapy, Organization of Los Angeles County Centre, California Association of Spectro-Chrome Therapists, (Incorporated) Election of Officers, Hotel Lankershim, Lecture Room 427, 5 P. M.

14 —Departure via Southern Pacific, 8 P. M.

15 —Arrival in San Francisco, California, 10 A. M., Hotel Manx.

15 —Graduation meeting of the Eighteenth Class in Spectro - Chrome Therapy: Meeting of the San Francisco Centre, California Association of Spectro-Chrome Therapists, Hotel Manx, Lecture Room 624, 8 P. M.

16 to 20—Arrangement of experimental apparati, Lecture Room 624.

21 and 22—Free Public Lectures on "Scientific Spectro-Chrome versus Medical Empirics," Native Sons Hall, 414 Mason Street, Yosemite Room, 8 P. M.

22 and 23—Public Receptions and Discussions, Hotel Manx, Lecture Room 624, 2 to 4 P. M.

24 —Beginning of Spectro-Chrome Therapy Course, Hotel Manx, Lecture Room 624, 7 P. M.

April

5 —Completion of Spectro-Chrome Therapy Course, at midnight.

6 —Departure of Dinshah and assistants from San Francisco, California, 6 A. M. by automobile.

Los Angeles California Class

Class Number Twenty First

1924

April

7 —Arrival in Los Angeles, Hotel Lankershim, 1 A. M.

7 to 10—Arrangement of experimental apparati, Lecture Room 427.

11 and 12—Free Public Lectures on "Scientific Spectro-Chrome versus Medical Empirics," Paulais Banquet Hall, 739 South Broadway, 8 P. M.

12 and 13—Public Receptions and Discussions, Hotel Lankershim, Lecture Room 427, 2 to 4 P. M.

14 —Beginning of Spectro-Chrome Therapy Course, Hotel Lankershim, Lecture Room 427.

26 —Completion of Spectro-Chrome Therapy Course, at midnight.

The precise program for the three following courses in Seattle Washington, Portland Oregon and Salt Lake City Utah will be published later. The dates of opening the courses are May 12, June 9 and July 7, respectively.

Attuned Blue Wave

(History, Travel, Biography, Geography, Language)

FROM COAST TO COAST

THE MESSAGE OF SPECTRO-CHROME THERAPY DELIVERED FROM THE ATLANTIC TO THE PACIFIC

DINSHAH'S PREDICTION FULFILLED

By:—Kashmira Dinshah, S.C.T.
Joint Secretary,
Spectro-Chrome Institute.

Started, at last, on our way to that American paradise of beauty and romance, in whose very name there is a musical charm—California!

It was on Saturday the first of December, 1923. We went by railroad, the time and climatic conditions rendering it inadvisable to essay the crossing of the Rocky Mountains in our usual way by automobile. Father, Irene and I formed the entire party—a happy trinity, bearing the glorious message of Spectro-Chrome Therapy to the Golden West.

Leaving Philadelphia, Penna. at 10:25 A. M. on the Metropolitan Express to Chicago, Ill. we arrived at Kittanning Point, Penna. 172 miles distant, at about dusk. The Pullman porter came running. "Cunnel, we're near the Hoshoo Cove. Look out the windah, you see there are two engines to pull the train. Yassuh, pretty steep climb!" As the train rounded the famous Horseshoe Curve, we saw clearly the two locomotives laboring industriously, while at the end was visible the observation car. It was a wonderful sight. Picturesque valleys and precipitous cliffs and the mighty engines drawing the long train safely across the mountains, along the edge of the steep ridges of the Allegheniés.

cago, Ill. Having called a meeting of the Spectro-Chrome graduates of that city, we stayed there a day. Some other necessary business matters being attended to, we resumed our journey at 8:00 P. M. Monday, December third.

Going by the Atchison, Topeka and Santa Fe railroad, our itinerary included a one-day stay at the Grand Canyon in north central Arizona.

During the two and a half days intervening before reaching the Canyon, we passed our time observing the varied landscape and studying the geography of the country we journeyed thru. It is of no worth to whirl across the continent and spin back home again just to say, "We traveled!" Travel with open eyes and ears, that you may know the world you live in.

Early next morning, we passed Kansas City, Kan., coming to Newton, Kan. at 2:30 P. M. where a short stop of fifteen minutes was made. Our train, the California Limited, was in three sections. We were in the second. Five minutes behind the preceding one, each section pulled out of the station at the scheduled time. The punctuality of the Santa Fé is praiseworthy. The State line of Kansas and Colorado was crossed not many hours later.

The hour of real activity on a transcontinental train is that when the call for dinner comes. The passengers who a moment before were steeped in lethargy that even the recent Japanese seismic upheaval might not have dispelled, sprang to their feet as though electrified when the waiter came thru the train, announcing, "First call for dinner!" But the people in the rear cars were out of luck in the preliminary rush, for those in the ones ahead reached the dining salon first and

ageway, eyeing the prior arrivals impatiently and hungrily eager to pounce on their places the moment the others finished or else sadly wend their way back thru perhaps six cars to await the next call. Our car being far behind, we were thus fooled once, but only once. We became wise then and stayed back till the last summons came. The fun would begin if everyone on the train thought at the same time to use like wisdom to avoid disappointment.

One argument tending to disfavor strict vegetarianism at all times, is its imagined "impracticability" while traveling in the country of a meat-eating people. Nonsense! A person of firm principles can adhere to the truth regardless of time and circumstance. Without correct foundation is the belief that one not eating meat would starve to death for lack of other provision while on journey. Even on the train, we were able to obtain our pure sustenance sans flesh, fish, fowl and eggs. There was no lack of cereals, legumes, vegetables, fruits or nuts, either in variety or abundance. We explained our case to the steward in charge, a cultured man whose cheerful courtesy and heartfelt attention gracefully upheld the high ideals of the matchless Fred Harvey Meal Service on the Santa Fé. By his direction, there were prepared for us special dishes composed of different vegetables and other innocent products of Mother Earth, victuals as dainty and delicious as ever graced an epicure's table. Henceforth let no one think it impossible to maintain the principles of absolute vegetarianism elsewhere than at home. We have proved that it is quite easy for one so minded. But had we found it otherwise, we would have gone hungry rather than defile ourselves with a diet of blood.

At about two o'clock, on the morning of Wednesday the fifth, we awoke and looked out at the station of Raton, N. M. The train rolled onward. The outlook appeared very peaceful. According to the travelogs which confuse New Mexico with Mexico, there should have been seen hordes of frenzied Mexicans scouring the desert in search of some new victim to elect as their President! No volley of shots rang thru the still night air. What a disappointment, after expecting such a thrilling time in that wild region where, supposedly, every man has muscular cramp of the right arm from a certain gesture toward the back belt and wooden overcoats are the fashion!

Toward noontime, we reached the city of Albuquerque, N. M. There was a stop of thirty minutes. The passengers hurried to alight for some fresh air and exercise and a glimpse of local life. Adjacent to the station was The Alvarado, a $750,000 Fred Harvey hotel, in the annex of which was a Harvey museum containing extensive collections of Red Indian and Mexican relics. There we saw also Navajo and Hopi weavers, potters, silversmiths and basketmakers at work. Round slabs cut from the wonderful trees of the Petrified Forest in Arizona caught our attention. Aeons ago, they were arboreal giants of the usual ligneous structure. The storms and changes of marvelous Nature transformed them into colorful agate and chalcedony, amethyst and smoky topaz. The mighty sea, full of mineral salts, which rushed in and covered that forest, flowed after ages had passed, yielding its captives to the open air, the common wood turned to semi-precious stones. We touched the polished disks, admiring their flaming beauty of hue and thinking of the glory of the power that made them so.

Passing Laguna and Gallup, N. M. we entered the State of Arizona. All day long there lay before our eyes the sandy wilderness and barren hills: miles upon miles of it. Above, the sky stretched cloudless and intensely blue. Occasionally the waste was broken with patches of mesquite and sombre gray sagebrush. There was no green, but the reddish sand and painted rocks of dull orange and blue with their suggestion of volcanic origin, formed a wild, weird picture of elusive charm. Little desert hamlets of adobe broke the monotony of the landscape from time to time. There was seen life almost painfully primitive. The continuous struggle to wrest a bare existence from the reluctant soil, soon marks with care the faces of the desert dwellers. Those people might be happy, for individuals judge their lots differently, but they did not look it. The trying sun glare and extreme dryness of the air bring early the appearance of age to the skin, so that even the children went with their countenances about as unlined as a baked apple.

It was 5:30 P. M. when we reached Adamana, Ariz., on the Rio Puerco. A few miles south of the track lie three of five petrified forests located in that region and two lie nine miles northward. There are thousands of acres covered with agatized fossil remains of gigantic prehistoric trees. On and on we sped. There were points of interest galore. A great crater 600 feet deep and 3 miles around, formed by a giant meteor, and known as Meteorite Mountain, near Sunshine; Walnut Canyon, 14 miles long, with ancient cliff dwellings; an extinct volcanic cone of almost pure sulphur called Sunset Mountain, ice caves and Black crater, close to Cliffs, Ariz. But as our time was limited, we forewent seeing these, contenting ourselves with a visit to the Grand Canyon alone. The Lowell Observatory, noted for its astronomical studies of the planet Mars, standing on a hill situated in Flagstaff, the county seat of Coconino County, is visible just after leaving the station, provided there is daylight. At 10:05 P. M. the Sante Fe junction at Williams, Ariz., was reached. Our train was sidetracked there, staying till 5:40 A. M., when it started off, branching sixty-four miles north to the Grand Canyon.

Alight with the fire of transporting expectation, we all arose long before the rooster's call, washed ready and sat waiting, watching the approach of dawn over the silent Arizona desert. Slowly the darkness brightened into the new day's sunlit beauty. The bluish morning mist lifted from the rainbow tinted hills. Before us lay the flat, sandy, sagebrush-strewn waste. Some meandering streams lay gleaming icily, frozen solid. Skeletons of a few horses and cattle were scattered about, lending a properly impressive touch of "local color."

At 8:20 A. M., we reached our destination. Our time for departure being 7:25 P. M., there were eleven hours at our disposal for sightseeing. We went to El Tovar Hotel, located at the railroad terminus, near the head of Bright Angel Trail, at an elevation of 6,866 feet above sea level. It was a long, low structure built of native boulders and pine logs, accomodating 175 guests. The manager in charge, Mr. Victor Patrosso, was a fine, gracious personality. When he learned of our position in regard to alimentary requirements, Mr. Patrosso kindly made special arrangements for us, as El Tovar is conducted on the American plan. Furthermore, a comfortable room was placed at our disposal. Thoughtful attention, unobtrusive and therefor welcome, charms the visitor at El Tovar, another of the superb Fred Harvey institutions.

From the railroad terminus, nothing very unusual was seen. Pines and cedars reared their wavy crowns aloft. Invisible was that noted gigantic gorge we had come to see. But a short distance beyond, just by El Tovar, the vast panorama sprang into view. A stupendous chasm, 217 miles around its rim, from 9 to 13 miles

wide and with a maximum depth of 6,000 feet. A great cleft in the terrestrial crust, a sudden drop-off in the wooded mesa country. Carved out by the mighty Colorado river, huge architectural forms, weird granite titans, stand exposed to the desert Sun, brilliant with varying hues. As far as the vision reaches and even beyond, great monuments of rock loom majestically appalling. Yet this mammoth scene is viewed by looking down. In that unparalleled abyss are hundreds of peaks taller than any mountain east of the Rockies, but not one of them with its top as high as the verge. It is a terrific, abrupt hollow in a vast, arid, lofty floor of nearly 100,000 square miles. Full of a thousand lights and shadows, chaotic and mysterious, painted with prismatic colors, the chasmy stone wonder has an indescribable appeal.

In the forenoon, we took the fifteen-mile auto ride along Hermit Rim Road. From different points en route there were viewed new vistas of the Canyon with its fantastic shapes and glowing shades of color. The tumbling waters of the Colorado River, so many thousand feet below and in reality several hundred feet wide, appeared but a short leap from bank to bank. We had a group photograph of us taken at Maricopa Point, a mile and a half west of El Tovar and a mile above the rushing river, where the Canyon is ten miles wide. The picture shows us standing near the edge of the rock. By the simple expedient of two steps backward, the headquarters activity of Spectro-Chrome Institute could have been transferred to the spacious floor of the Grand Canyon of Arizona, without wasteful trouble or expense! Plenty of area for expansion, fresh country air, a healthy change from the crowded city quarters of the present Central Office—all for the falling!!!

In the afternoon, we went on the Navajo Point and Painted Desert View drive, a round trip of sixtyfour miles, via Long Jim Canyon, thru part of the Tusayan National Forest. From Navajo Point, the nature picture was magnificent. To the north, a thirty-mile span across the beginning of the granite gorge; east or west the great extent of the Grand and Marble canyons, the gorge of the Little Colorado River and the vast panorama of that famous but seldom visited marvel, the Painted Desert of Arizona, the home of the Navajo and Hopi Indians. Our guide told us a quaint story about these people, pointing out some caves in the cliffs below.

In bygone times, the Navajos were a somewhat warlike tribe. They looked with contempt on the peaceful Hopis and frequently threw the latter into a fright by sudden descents upon the Hopi villages on the high mesas. So the clever Hopis cut dwellings for themselves out of the solid walls of the granite canyon and quietly retired into these lofty caves when the enemy was sighted, the means of access being kept secret. Thus the warlike Navajos were baffled and they went away with a look of "sour grapes," perforce contenting themselves with nicknaming their opponents the "Moki" tribe, "Moki", according to our guide, being the Red Indian equivalent of the American "deadwood" or "ivory-head." But the happy Hopis snapped their fingers and told them, "Banana oil!"

At 5:30 P. M., an entertainment was given in the shape of a war dance by some Hopi Indians, in Hopi House opposite El Tovar. The building is a reproduction of Hopi dwellings and several hogans of the Navajo tribe. We saw the dance. The wild jazz of modern civilized life is not much unlike it.

(To be continued.)

SAN FRANCISCO CALIFORNIA OPENS THE DOOR OF THE GOLDEN WEST

The Eighteenth Class for the study of Spectro-Chrome Therapy was commenced on December 17, 1923 with 43 students.

Among them were 3 Doctors of Medicine, 3 Doctors of Osteopathy, 4 Doctors of Dental Surgery, 3 Doctors of Chiropractic, 2 Physio Therapists, 4 Nurses and 1 Graduate Pharmacist. A representative of one of the leading newspapers of the city also enrolled as a member of the class with the intention of personally investigating the truth or otherwise in Spectro-Chrome Therapy and passing the examination graduated as a Spectro-Chrome Therapist!

The attitude of the Class was very attentive and enthusiastic. From the beginning to the end, their intelligent questions and mental quickness were an inspiration to the instructor and sealed never-to-be-forgotten friendliness.

At the conclusion of the course, the following testimonial was presented by the class to the Originator of Spectro-Chrome Therapy, amidst vociferous applause:

SPECTRO-CHROME INSTITUTE

Hotel St. Francis,

San Francisco, California.

December 29, 1923.

Colonel Dinshah P. Ghadiali,
San Francisco, Calif.

Dear Dr. Dinshah:

We, the undersigned members of the Eighteenth Class in the study of SPECTRO-CHROME THERAPY feel that words are too inadequate in expressing our appreciation of the extraordinary super-university course of instruction which you have so thoroughly given us the past two weeks.

We feel that the forces that control the Universe have attuned your personal self to make you a teacher of unusual power and efficiency. The two weeks of the intensive training passed too quickly—but have left us so rich in the understanding of Nature's forces and powers and how to use them, that we are anxious to have you return to us soon for more knowledge and training, that we may be of still more usefulness to mankind.

As faithful students of SPECTRO-CHROME THERAPY, we are eager to study and apply the Attuned Colors and educate our friends to the sanest, surest and shortest method of the healing art—SPECTRO-CHROME THERAPY.

OUR WISH:

May the Powers of the Universe always be with you and speed the day to end the Medical Trust.

(43 Signatures)

The Nineteenth Class in Spectro-Chrome Therapy, in Los Angeles California, had as a student Mrs. Hirabai A. Vacha, a born Parsee Zoroastrian of India. Mrs. Vacha is the first woman of that nationality to graduate in the new healing art. Her pleasing personality attracted much attention and her inspiring presence was warmly appreciated. All success to her!

LOS ANGELES CALIFORNIA TENDERS CORDIAL RECEPTION TO SPECTRO-CHROME THERAPY

In the Nineteenth Class on Spectro-Chrome Therapy, there were 39 students.

The course began on January 7, 1924 and as usual continued each evening until its completion.

Among the students were 6 Doctors of Medicine, 4 Doctors of Osteopathy and 10 Doctors of Chiropractic; also 1 Doctor of Magnetics who had studied under the late Dr. Edwin D. Babbitt and who was therefore more than usually appreciative of the merits of the Attuned Color Waves.

This class was in many ways different from other classes. Its atmosphere was very genial and the course went through without interruption by useless catechism. The students, while eager to ask questions appeared decided to wait until the last and then realized that their "questions" were in fact answered right in the teachings; they certainly were tuned to the occasion.

During the final session, the following testimonial was presented:

SPECTRO-CHROME INSTITUTE

Hotel Lankershim

Los Angeles, California

January 19th, 1924.

Dear Colonel Dinshah:

At the conclusion of this, your Nineteenth Class of Spectro-Chrome Therapy, held in Los Angeles, California, we, the undersigned members desire to express our sincere appreciation for the great knowledge imparted to us.

The teaching has inspired us to greater efforts to aid suffering humanity and given us a practical scientific system that does the work.

Your demonstrations in the class room have taught us more wonderful Truths about God, Man and the Universe than ever dreamed of by many of us, and the beauty of it is the seeing for oneself, the actual Cause and Effect in operation which banishes all element of doubt in the mind of the student and leaves nothing to be taken for granted or on faith. The proof is shown.

Dr. Dinshah, you are a Master Mind in presenting this knowledge as proved by our interest shown every moment.

For years, many of us have devoted our time to alleviating the ills of mankind and now we feel your system fills the long-sought want that countless thousands have longed for.

We pledge you our loyalty, support and earnest endeavor to hasten the day when Spectro-Chrome Therapy will lead the world in the healing art, as it surely will.

(39 Signatures)

SPECTRO-CHROME

APRIL 1925
LEMON 1925

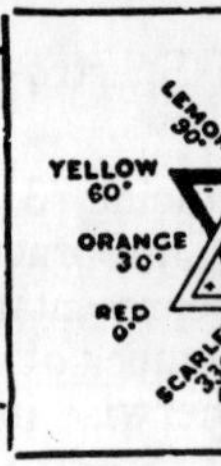

VOLUME 4
NUMBER 4

Devoted to: Spectro-Chrome Therapy   Dedicated to: Service of Humanity

CONTENTS

	Page
The World Through the Eyes of Dinshah— Vain, County Medical Society and its Boobery--- American Medical Association Turning Sensible--- More Humbug About Cancer--- In Defense of Hadwen--- The Principles of Light and Color Published-- Pacific Coast Tour--- Notice to All Concerned--- Manufacture and Delivery of Itisometer--- Price List Changed--- A Clerical Error . . .	542-550
Rulings of Chief Executive Council	550
Anonymous Correspondence by Irene Grace Dinshah, S.C.T.	552
A Knockout by An M. D., S.C.T.	553
The Advantages of Healing by Spectro-Chrome Therapy by Wallace F. Mac Naughton, M. D., S.C.T. . , ...	556
Ohio Makes a Good Report by Henry C. Gehlke, S.C.T.	557
Ear Troubles and Varicose Veins by Kate W. Baldwin, M. D., F.A.C.S. .	558
Itisometer Demonstration by Geraldine Mc Cann, S.C.T.	560
Completion of The Middle Western Tour by Kashmira Dinshah, S.C.T.	562

Editor
Dinshah P. Ghadiali

In course of life you reach a time,
When Doctor, Lawyer, Priest alike,
Does fail to give the sought relief,
To troubled mind or burdened heart—
Then come to me;
I help you to help yourself.
SPECTRO-CHROME.

Subscription:—January to December (12 Months) $3.00
All Subscriptions Taken For Full Year Only And Expire With The December Issue
Any Single Copy 25 Cents. Back Numbers Available From June 1922.

Attuned Red Wave

The World Through the Eyes of Dinshah

VAIN, COUNTY MEDICAL SOCIETY AND ITS BOOBERY; WHO ARE THE PARASITES?

The Wayne County Medical Society which was challenged in Detroit, Michigan by Spectro-Chrome Institute in the open, through all the daily newspapers and also by personal letters written to its retiring and new Presidents and the Editor of its Bulletin has the nerve and cheek to make false attacks on other unoffending ones. In a recent issue, under the heading of, "The Chiropractic Bill must be defeated again," that Society's Bulletin attacks the Chiropractors and calls them "parasites."

The Wayne County doctors of medicine are so afraid the bill would pass and put a crimp in their pockets that they are clamoring to rally against that bill.

If the State of Michigan were the only State in the United States, such a fight may perhaps be considered reasonable, but, when many another State already admitted Chiropractic, it would be in the line of progress to keep their mouth shut and not call such names to followers of a different system of treatment. Chiropractors are not "parasites" on the public. They are the blood-sucking l e e c h e s on the *pockets* of the orthodox dead-wood type medicos who are the real "parasites."

And in analyzing the situation that Vain, County Bulletin takes the opportunity of calling Spectro-Chrome "a pseudo-medical sect now before the public!"

Spectro-Chrome Therapy is not pseudo-medical, medical or in any way connected with the fraud of medicine. It is the *one complete therapy* that looms as the Nemesis of medical imposture. It is the *one system* that grapples all claims of the medical "science," chokes them, scraps them and scientifically throws them into the junk heap. It is the one therapy that will not only kill the oppressive power of the medicos, but, will finally become the Universal Therapy of the whole World, when many another is relegated to the oblivion of the past with Drugs.

It is merely a question of time. It is merely a question of public education and awakening. It is merely a question of the realization by the public that in leaving their lives in the hands of age-old system followers they are slaughtering their own interests.

Were medicine a "science," a precise science, it would not have changed its fundamental conceptions; but, it is only a card house leased on wrong views, false notions and is honeycombed with legislative supports created by political graft and public weakness.

How long will it last? I shall see the day its dominance shall decline and be hammered out flat as a discomposed porpoise and its bloated abusive carcass discarded to destruction, its monopolistic Trust broken down, its machinery for torture on the conscientious followers of other systems crashed to ineffectuality and made irreparable because of the gigantic power and pressure of public opinion.

Spectro-Chrome Therapy integrally practised is leading toward that glorious end. The Wayne County Medical Society is itself merely a parasite of the American Medical Association—a thorny cactus whose spine is dislocating vertebra by vertebra, by the onward march of righteous public sentiment.

Parasites and pseudo-medical, forsooth! If you are red-blooded Americans, fight my S p e c t r o - C h r o m e Therapy openly on a scientific platform before the public and let them judge.

I came at great expense to your own city and bearded you in your own den; where was your bravery gone then? If you have something better, trot it out; I shall be glad to embrace it; you never studied *my* system—I am well versed in the hollowness of yours. You are only spineless cacti and backbiters; your President printed in his report that he investigated Spectro-Chrome Therapy. He is not speaking the truth, for, when I came to Detroit he went out of town and he knows he never even met me.

Have you courage? Genuine moral courage? Are you on the side of Truth? Have you brains enough in your entire organization to decry Truth and trod on it? Before you try to put down anything, is it not your duty to study it in a scientific manner? What steps did you take before declaring that "Spectro-Chrome Therapy is not a science?" Did you take it for granted that because the former Editor of the Journal of A.M.A. printed a ridiculous article about me, that my system was unscientific? How many scientists and physicists have you on your membership roll? Get them listed and send them at my expense to take my complete course on Spectro-Chrome Therapy. I assure you, as a result your doors will finally close!

You ruled the World in the past perhaps because of the gentle character of your opponents. I fight you *before* the world. Your fires, however, can not be drawn because you have no fire. Your fires lighted when Methuselah was a baby, are now fast dying out; the public will *not* stand for drugs and poisons. Your days are numbered and Spectro-Chrome Therapy will without the least unfair method end your existence as the big guns of healing.

Your guns are spiked; your hands are paralyzed and your knees shaky. If there is Truth of science among your followers have them meet me anywhere in open debate, before the public. My guns are loaded with shots of genuine science as will explode all your false teaching. For each medical shot, Spectro-Chrome Therapy will give you a machine gun bombardment such as will make you understand where chicanery ends and Truth begins.

Before you speak or write about me or my system, vindicate your moral courage and bravery by accepting my challenge and firing in the open. Otherwise, the World will judge you as arrogant arrant cowards and deservedly.

Ask your President to justify his statement about investigating my system.

Shame on his untruth! ! !

AMERICAN MEDICAL ASSOCIATION TURNING SENSIBLE

Dr. Morris Fishbein now the Editor of the Journal of American Medical Association in place of Dr. Simmons who it appears went to Europe. Since the change, however, the Journal has suddenly discovered that other methods besides dosing with dope have merits. In an article published by The Evening Bulletin of Philadelphia, on the sixth current, over the name of Dr. Morris Fishbein, is an enunciation that convulses me with laughter; it speaks of "Light as curative gains importance."

"Perhaps the most striking trend in modern scientific investigations of the cause and cure of disease is the amount of attention being paid to the effects of light.

Many years ago Finsen received the Nobel prize in medicine for his pioneer work in this field, but the subject remained practically quiescent until the significant work of Rollier in healing tuberculosis on top of the Alps Mountains, and the work of English and American physicians in treating rickets by sunlight brought the matter again to the front.

In Great Britain the Medical Research Council has a special department which has been giving attention to this question. The report of its work for the year, just issued, calls attention to some remarkable discoveries. One group of discoveries proved exposure of the skin to ultra-violet light, such as comes from the sun, gives the blood increased power to destroy bacteria.

It should be pointed out here that ordinary window glass filters these rays out but that quartz does not. The Johns Hopkins Hospital has installed a quartz window in the department devoted to the treatment of children suffering from rickets, so that these children may receive the rays without being exposed to the elements.

In the course of the British investigations it was found long flamed arcs are more efficient than the short flame Finsen arcs in supplying ultra-violet light and that pure tungsten poles or tungsten-cored carbons are best for small arcs, whereas ordinary white flamed carbons are efficient enough for large flame arcs.

Dwellers in smoky cities will be interested to know measurements taken in various places in and about London showed that persons living in the great city received one-half to one-third the amount of the beneficial rays received by those living in the country. In this connection the British scientists have worked out an efficient measure for measuring the amount of smoke pollution in the air. Briefly the method involves the exposure of a tube containing a standard solution of a dye called acetone methylene blue in a standard quartz tube, which bleaches under the light.

The amount of bleaching is measured against a set of standard tubes containing solutions of pure copper nitrate. Furthermore, methods have been worked out for testing the effects of the rays by the number of living organisms such as infusoria which they will kill and by the production of redness on the white skin of the upper arm.

It is interesting to know that more of the ultra-violet rays come from the sky as a whole, even on sunny days, than come direct from the sun. This is particularly the case when the sun is low or obscured by a cloud.

In the course of their investigations the research workers of the Medical Research Council tried the effects of various oils on the tubercle bacillus. It was found the bacteria gradually die when suspended in olive oil, and that they die much more rapidly in oils that have been exposed to sunlight than in ordinary oils.

In trying to find out just what changes the sunlight brings about in the body that causes it to benefit tuberculosis, it was discovered that both sunlight and heat in general, as from a hot bath, caused an increase in the constituents of the blood, called blood-platelets, was found to be reduced in infections like those of colds, influenza and measles but much increased in chronic tuberculosis of the lungs.

The facts developed by the British workers are an indication of the way in which the concentration of research on small points yields finally to systematic knowledge which must inevitably bring disease processes under control. It is this demonstration of practical results that has caused philanthropists to devote increasing sums to investigations in the field of biology and scientific medicine."

Philanthropists are thus worked up to give large sums of money to so-called Medical Research Councils, which eat up the cash worse than the bacteria. The results they are endeavoring to prove as accomplished were already accomplished by others who paid the penalty of pioneer researching by being jeered by the ignorant. They are similarly looking into the Infra-Red and the Ultra-Violet sections of the Solar Spectrum, where the caustic and highly destructive energies are concentrated, but, purposely blind themselves against looking into the Color or Visible Spectrum where the Drug Store and Pharmacopoeia of the Divine Creator is built in.

They are wasting their time. The powers inherent in Light these European and American scientists and doctors are searching, are already public knowledge through the work originated in Spectro-Chrome Therapy. Its refined character, completeness, accuracy, s c i e n t i f i c application and positive precision are matters of everyday experience to the Spectro-Chrome Therapist. Spectro-Chrome Therapy is a complete medical science and very much more, because, there is no guesswork and unpleasant subsequence. In combination with the Itisometer, Spectro-Chrome Therapy is today the most formidable weapon engineered in the Healing Art and soon its recognition must come, in spite of the American Medical Association.

I congratulate Dr. Morris Fishbein for twisting the helm slowly toward this end and assure him of the appreciation of right thinkers.

MORE HUMBUG ABOUT CANCER

Cancer is one disorder which is profitable in filling the pockets of the Medicos. Every few weeks—sometimes days—a new theory is started in the shape of some germ causing it. In the March issue of the Journal of Tropical Medicine and Hygiene, Dr. Louis Sambon announces that Cancer is "almost certainly a parasitic disease, circulated contagiously by cockroaches, cellar beetles, meal beetles, liver flukes and intestinal parasites." He says, "the disease is preventable by stamping out the parasites which transmit germs to humans exactly as mosquitoes pass on yellow fever." "The actual germ, the socalled 'filter-passer,' is exceedingly small and lives as dominant symbion within tissue cells "

In connection with this, Dr. Charles Mayo, of Rochester, Minn., predicts the conquest of the disorder as inevitable because of "notable advance in blood diagnosis by which the presence of cancer may be determined, made possible by the recent discovery of the German scientist, August von Wassermann."

Keep on dreaming, doctors and scientists and keep on working on roaches, beetles, bugs and bothers. Cancer is a constitutional disorder and not a local affair, caused by the introduction into the body of wrong food, like flesh; while the World eats dead pigs, dogs, cats and all animals except human babies, Cancer will stay in one form or another. Leave the cellars and look into your food products. We are weary of listening to your silly bug tirades.

IN DEFENSE OF HADWEN A CONSCIENTIOUS DOCTOR

Under the heading of Current Topics The Bulletin of the Wayne County Medical Society of Detroit, Michigan says thus about "A Layman's Movement Against the Anti-Medical Quacks." How far the doctors are going in hoodwinking the public may well be seen from the following:—

Recently *Saturday Night* published an account of the establishment by physicians and others of the Gorgas Memorial Institute, designed to help block the efforts of quacks and anti-medical propagandists. Now we learn that there has been organized, in Boston, the American Association for Medical Progress, whose membership is composed exclusively of laymen, and whose purpose is likewise to protect the public against the Macfaddens, the anti-vaccinationists, the anti-vivisectionists and others who are doing their utmost to discredit the scientific medical system and place quackery and ignorance on a pedestal like the golden calf of old.

In telling how the organization of this Association was inspired, Edward R. Warren, writing in *The Independent,* says that a number of Bostonians attended a convention of anti-vivisectionists to find what sort of stuff they were broadcasting. He described what they discovered as follows:

"They sat and listened, and as they listened they realized that here were persons possessing an extraordinary affection for animals, a zoöphile psychology which would seem to shut them off forever from seeing beyond the immediate object for which they were working, a group of persons whose devotion to their cause gave them both the zeal and bigotry of the fanatic. But—and here was the significant point—the persons to whom the zoöphiles were appealing were by no means fanatics. They were average men and women who knew no biology and no science, men and women whose respect for the physician was founded upon nothing more stable than a sense of awe mingled, perhaps, with a little fear for the superior powers of the man of science, who talked of medicine in a language they did not understand.

"And now, with the zoöphiles, for the first time these average men and women had come in contact with speakers who talked to them of medical subjects in a language they did understand; men who did not give lengthy discourses on the subjects of vaccines and serums, but who told them outright that vaccination meant pouring the disease pus from a sick cow into their children's blood; men who told them that animal experimentation, thus far little more than a term which they had heard, means cutting up animals alive; men who told them that the medical profession was a huge trust organized for the purpose of extorting fees from a helpless populace."

Mr. Warren goes on at some length to emphasize the point brought out in the above quotation, and to show how ineffectual have been the efforts of medical men to meet the attacks on them, partly because the doctors talk in medical terms which the public does not understand and partly because physicians, when they try to defend their profession are accused of doing so for financial reasons.

He shows how weak and false the arguments of the opponents of medicine are. For example, a British quack, one Hadwen, recently convicted of manslaughter for permitting the death of a diphtheria patient, told the court that the "modern germ theory is all bosh," and further argued that "the best scientific and medical opinion" had once approved the practice of bleeding. This, of course, lacks 100 per cent of proving that the use of vaccines is in the same class as bleeding.

The new association plans to present to the public propaganda that will meet Macfadden and his supporters on their own ground—not in spreading misrepresentation and hokum, of course, but in giving the public the truth in plain, everyday English. In this way,

it should be able to accomplish a great deal of good."

The Medical Pocket Quarterly publishes this about Hadwen:—

"Dr. Hadwen, M.D., M.R.C.S., L.-R.C.P., is free again. The opponent of vivisection, vaccination and the germ theory is acquitted by a jury of his peers and wanders forth into the world.

But Hadwen is acquitted in the eyes of the law only on *this* point, *sic.*: that he is not guilty of criminal negligence. Mr. Justice Lush stated, after the trial and before the jury filed out, that the jurors were not to decide whether Dr. Hadwen's professional viewpoints and theories were sound. No! They were only to decide whether or not in the particular case on the docket he was guilty of negligence.

Then, in summing up the case, Justice Lush stated that it was his personal opinion that a constant fear of being accused of manslaughter would have a degrading and disastrous effect upon the physician and his profession at large.

In twenty minutes the twelve good men and true came back with the verdict: Not Guilty!

Legally acquitted. Freed in every way of the accusations against him. Except *this* (and there is no jury on earth before which he can be brought to trial on this count, except the world-wide jury of professional opinion), that the germ theory of disease, the use of anti-toxins and the necessity for humane vivisection are labelled by him as bunk.

Every scientist from Louis Pasteur on down to our little moment, knows that Hadwen is merely a poor, misguided individual on these points."

No, emphatically no. You, who are followers of the age-old humbug which is now gilded into a respectable organization under the name of "medical science" are the true "quacks" and "manslaughterers." The Anti-vaccinationists, Anti-Vivisectionists, the followers of Progressive Drugless Systems and the ridiculers of the germ theories are the true-souled reformers who aim at crashing through your skeptic barriers and autocratic impositions. Hadwen has behind him sound arguments and sound-minded people. I belong to the same class of so-called "fanatics" who fight to the last ditch the bosh of germs and the nonsense of vaccines and serums. I do not maintain there are no germs; I positively know that there are all kinds of life in all kinds of shapes, sizes and with all kinds of characteristics, but I also positively affirm and maintain that these germs are not the original *producers* of the disorders, but, that they are the *subsequences* of the pathological conditions. I most emphatically maintain further, that the human body has *within* it all the powers to demolish any and all "germs" introduced into it and that the introduction into the human organism of vaccines serums and other similar products produced by the anti-bodies of animals lower in evolution, is a sin of the most unpardonable character against the purity of the Temple of God's manifestations. You doctors who perhaps are godless may continue with your thick-skulled material views, but, we who oppose the introduction of filth into our clean physicality are the ones who will win the final victory.

The horrible imposture called "medical science," is nothing but a remnant of past stupidity and its essential fundamentals are still stupid. But, it is a satisfaction to note that many of its sensible-minded followers, like Dr. Hadwen saw through the mist of ignorance and refused to follow its dogmatic methods—to call Hadwen a manslaughterer is a wicked lie.

If you wish to know more about the

plagiarisms committed by Louis Pasteur, read Pasteur or Beauchamp.

THE PRINCIPLES OF LIGHT AND COLOR PUBLISHED

The Photo-Engraved Edition of the Principles of Light and Color by the late Edwin D. Babbitt, M.D., LL.D., is published by the Spectro-Chrome Institute and already mailed to those who paid $15 prior to the publication.

The uniform price of this edition to subsequent purchasers is $25.

The commentary by Dinshah is not not yet out. The pre-publication buyers of The Principles of Light and Color will receive it Free of all charges; the subsequent purchasers will have to pay $5 extra for it as originally announced.

There are no exceptions.

The volume is bound in finished green cloth with the name embossed in gold. It has four plates, two hundred and three photo-engravings and five hundred and sixty pages. A preface written by the Originator of Spectro-Chrome Therapy is added.

To those who want a book fit for a gift this volume may be cordially recommended.

PACIFIC COAST TOUR CENTRE AUTHORIZATIONS AND FREE ITISOMETER COURSES

Since the establishment of the Integrity Policies rigidly and the organization of the American Association of Spectro-Chrome Therapists (Incorporated), I commenced rounding-up the entire graduate list by going personally among them from State to State, opening the Post-Graduate Classes, completing the Itisometer instruction thus either getting the Graduates into the organization to help us with Integrity or letting the hesitants, shaky-ones, doubters and traitors weed themselves out automatically.

This task began February 21, 1925; by April 1, 1925, through constant traveling and tireless work, I completely circuited all the Eastern and Middle Western States Classes; I rounded the territory covered during the previous three years, within five weeks. Glimpses of the labor may be gathered from the reports of the Secretaries who accompanied me on the tour and whose ceaseless efforts helped me in the work.

I am now leaving for the second lap of this important task. On Sunday the 19th current, I leave with one Assistant Secretary for the Pacific Coast States. The program covers a tour of over 8,000 miles and will end in exactly three weeks.

The Free Post-Graduate Course on the *Itisometer* will be given only to Fellows and Members of American Association of Spectro-Chrome Therapists, in good standing. Those wishing to take advantage of this opportunity should bring their Applications for Enrollment, properly filled in *Duplicate* and present to the Secretary in charge of the Class Room. No admission to any Graduate not duly qualified. Please save friction by following the instructions.

In my last tour in the East, many Graduates, applied for joining the Association and after taking the full work free through the courtesy of Spectro-Chrome Institute, asked *to withdraw* their application on the ground that they could not agree with certain of the Integrity Policies. Their admission fees were therefore refunded. They got the Course *Free!*

Please note that it is such unfair practice on the part of unscrupulous Graduates that pushed me to the drastic measures for public protection. A Graduate who will mulct his teacher and friend, his teaching Institute, will no doubt mulct the public. Such dishonest ones are better out than in. I only want the sincere and this time the Class Secretary has plain instructions to admit none without proper credentials, made in strict conformity with the Constitution and By-Laws of our Association. Study them before signing the blanks; that code of honor was installed to be enforced and it *will* be enforced to the very letter against all without exception.

The Classes will be held as follows—

Los Angeles, California, Hotel Lankershim, Friday, April 24, Saturday, April 25.

San Francisco, California, Hotel Manx, Monday, April 27, Tuesday, April 28.

Portland Oregon, Hotel Portland, Thursday, April 30, Friday, May 1.

Seattle, Washington, Hotel Gowman, Sunday, May 3, Monday, May 4.

The time of meeting will be 8 P. M., in each case. Positively no visitors or strangers will be admitted to the Post-Graduate Course. Be punctual, as besides giving the work, I have also to complete the formation of your Local Centres.

I have been bombarded here as to whether I could not take into consideration the various circumstances under which certain Graduates are working, to have me modify the regulations. I repeat, that I have no authority to modify anything. The Constitution and By-Laws are in your hands and you know what they say. They are there to be abided by and followed; otherwise you should not join. I repeat, work *with* me or I work *without* you. Do not therefore waste time by endeavoring to wiggle; the Chief Executive Council asks me to state that the Code as laid down *must* be maintained at all sacrifices and I have no remedy.

It is up to you to keep the banner high and unblemished; the public are watching curiously and keenly this new experiment in therapeutical integrity and we must jointly pull in one direction to deliver the goods—INTEGRITY.

NOTICE TO ALL CONCERNED THE GRADUATE, THE PUBLIC AND THE GOVERNMENT

Spectro-Chrome Institute only loans its Equipment as personal concession to followers of its system of Spectro-Chrome Therapy.

These Equipments are so given only to those Graduates who are in good standing with the American Association of Spectro-Chrome Therapists.

The concession ceases with the cessation of such connection.

The prices charged for the Equipment are rental prices paid for the grant of the use of such Equipments.

No sale is created and no title passes to the user.

All Spectro-Chrome Therapy Equipments issued after February 21, 1925 when the Constitution and By-Laws were unanimously adopted by the Chief Executive Council of American Association of Spectro-Chrome Therapists, were issued in strict accordance with such understanding.

Spectro-Chrome Institute owns the sole title to such Equipments and all Slides and Slide-Carriers (new or replaced) in possession of the Associates of the American Association of Spectro-Chrome Therapists are the permanent property of Spectro-Chrome Institute.

Withdrawal, resignation or expulsion of an Associate from the Association means forfeiture of all rights, privileges and payments made and the prompt recalling of all Slides, Slide-Carriers and loaned Equipments without any process of law and positively without any refunds.

MANUFACTURE AND DELIVERY OF ITISOMETER

Even while traveling, my plans for the manufacture and delivery of the Itisometer are in progress. On completion of my Pacific Coast tour, the unprofitable drain on my time, money and energy will cease and systematic smoothness being established, the commercial side of my duties will begin to accelerate.

Lessees of the Itisometer will kindly understand my handicaps; I started to accomplish the impossible single-handed, but, it is *being* accomplished and with your patience and indulgence as well as co-operation, you will facilitate the progress very materially.

PRICE LIST CHANGED NOTE LOWER PRICES ON BULBS

The prices of all bulbs are changed. Owing to the establishment of Local Centres of Spectro-Chrome Therapists under the vigilance of American Association of Spectro-Chrome Therapists consequent reduction in overhead expenses of handling, and the prices of many bulbs used in Spectro-Chrome Therapy Equipments are reduced.

Kindly study the Prices when sending for supplies, to save delay and possible dissatisfaction.

Spectro-Chrome Institute does not maintain a C. O. D. department. All prices are net cash in advance with the order and are F. O. B. shipping point.

A CLERICAL ERROR

In our March 1925 issue, at page 524, in the Article 14, Section 5, in the first line, "on or before" should read "on or after."

Attuned Yellow Wave

(Correspondence, Query, Catechism, Inquiry, Interrogatory)

RULINGS OF CHIEF EXECUTIVE COUNCIL

At a meeting of Chief Executive Council of American Association of Spectro-Chrome Therapists, held at Malaga, New Jersey on Thursday, April 9, 1925, at 7 P. M. the following rulings were unanimously given in connection with our Constitution and By-Laws:—

Question 1—Can an Associate use *any* white light or color equipment for treatment?

Discussion—If we allow that, it will open the fraudulent door we closed by Article 49, Section 4; then, any Associate may be tempted to use fakes or imitations.

Ruling 1—No.

Question 2—How about an electric light bath?

Discussion—The inherent right of a legally registered practitioner of another system of therapy for which a license is granted by the State and is held by the practitioner must be recognized. Therefore, such licensed practitioner, may use an electric light bath if so chosen, provided its administration to the sufferer is not allowed to be confused with Spectro-Chrome Therapy.

Ruling 2—Yes, if the therapist is licensed by the State Laws to use it.

Question 3—Can an Associate use Infra-Red or Ultra-Violet Equipments?

Discussion—As such use means use of destructive rays or rays that have caustic and drastic effects, Spectro-Chrome Therapists as such are not authorized to use them; but, if they are used by licensed practitioners of other systems of therapy where their use is allowed by the State Laws to the Licensee, we must not interfere with the conscientious views of the Associate, provided the use of such systems is not confused with Spectro-Chrome Therapy.

Ruling 3—Yes, if licensed by the State Laws.

Question 4—Can a color slide be used with such different kinds of equipments?

Discussion—If a Spectro-Chrome Therapist believes any color slide can be used for treatment, it is better to have such therapist out of our Association; such a person is of no service to our Cause and is likely to mislead the sufferer. Where white light bulb and color slide affair comes in, all kinds of imposition become probable and must be stopped.

Ruling 4—No.

Question 5—Can an Associate use an Arc Lamp with or without color attachments?

Discussion—It will lead us into the same trouble as with white bulbs and untuned slides.

Ruling 5—No.

Question 6—Can not even an M. D. use other light and color equipments?

Discussion—Article 49, Section 4, expressly debars it in clear and plain language; the question needs no ruling.

Ruling 6—No.

Question 7—If an Associate is in the service of a Graduate of Spectro-Chrome Institute who is not an Associate and who does not practise Spectro-Chrome Therapy, but, employs this Associate to use it on his sufferers, can such Associate do so without violation?

Discussion—Article 50, Section 5, is intended to cover such cases.

Ruling 7—No.

Question 8—Does an Associate in service have to follow the same rules as in partnership?

Discussion—Partnership gives right to profits and service gives profits as compensation by salary; thus, similar rendering should apply.

Ruling 8—Service is partnership.

Question 9—Can an Associate employ as nurse or assistant a Graduate Spectro-Chrome Therapist who is not an Associate?

Discussion — Such employment would lead to complications under Article 50, Section 7. The fact that a Graduate will not join the Association, shows lack of good faith.

Ruling 9—No.

Question 10—Can an Associate practising Spectro-Chrome Therapy in partnership with an M. D. under a corporate name use on his letterhead the name of such M. D. with the words, "Spectro-Chrome Therapy?"

Discussion—Such use is likely to mislead the public to believe the said Medical Doctor was also a Spectro-Chrome Therapist. Refer to Article 50, Section 3, which debars such posing.

Ruling 10—No.

ANONYMOUS CORRESPONDENCE

By:—Irene Grace Dinshah, S.C.T.
Secretary, Spectro-Chrome Institute

Prompted by a recent verbal inquiry, "Do you ever receive anonymous letters?" the following is submitted:

When the noted Frenchman, M. Emil Couè came to America with Suggesto Therapy, one was not considered "fashionable" unless one repeated the famous—"Day by day in every way I am getting better and better."

A typewritten communication based on M. Couè's well known motto was received by Dinshah from a person (who, evidently knew nothing, or, at most, very little about Spectro-Chrome Therapy), with the *extremely polite* salutation, "Sir:"

To spice the letter with the essence of incongruity, caustic remarks were showered, the most humorous of which, was this extract:

"How sublime, yet how simple, is your conception! All is Radio-Activity: hence all we have to do in order to acquire health, wealth, fame, honor and happiness, is to Attune our Radio-Active and Radio-Emanative Equilibrium to the required pitch and automatically attract and absorb the mental creations of other men! And then your mastery over colors! May I suggest as your motto:

Day by day in my own smart way
I'm getting richer and *Richer!*

Anonymous."

What a pity it was an anonymous communication!! Otherwise, we could have sent the writer a *sweet* reply with our best regards for the compliments!

Sublime? So proved by testimonials of those benefitted by Spectro-Chrome Therapy.

Simple? Apparently! The key note is *common sense!*

Smart? Lucky for the World that he *is* "smart," for, tact is required to contend with and outwit such antagonists.

Richer? Spectro-Chrome Institute Income Tax Returns deny this; all "Income" is eaten up in overhead expenses for public service. However, Dinshah *is* richer in knowledge of traitorism, trickery and deceit of some of our students as well as others!

Previous to the commencement of a Spectro-Chrome Therapy Course, Dinshah invariably delivers two preliminary (public) lectures and although every introductory program has a similar straight-forward trend, he took one of our prominent Eastern cities by an unusual thunder-storm; in fact, he was compelled by the uncongenial tone of the audience to do so in order to succeed in breaking through the walls of intolerant skepticism.

The people of this city were "dead," metaphorically speaking; Dinshah realized that nothing but explosives would awaken them. Therefore, during these two evenings the majority of statements uttered by him were on the Infra-Green section of the Spectrum. His emphatic assertions and poundings made the ears of some of the "dead wood" tingle and splinters fly; others genuinely gasped.

A very dignified "lady," (having heard the lectures), sent this anonymous note:

"Your lectures are good and to the point, but, ladies don't like your language. Be more careful if you want to win the heart of our cultured city."

A later communication divulged her name and address.

Dinshah promptly replied—"Dear Madam: Thank you cordially for your kind suggestion; nevertheless, I am here to teach Spectro-Chrome Therapy and not to win the heart of your city's cultured ladies."

Here is an anonymous note, which, if received by you, might make your hair stand on end:

"You will remember me as you prepare for the Tower of Silence."

Such threats do not affect Dinshah. To him, these trivial affairs merely wend their way in and out of the course of life.

When the conscience is clear
And the heart is pure,
There is nothing to fear.
When his mission is fulfilled
And he is ready to depart,
From the Physical Plane,
His thoughts will be
For suffering humanity,
Around which his life is centered.

And *not* for anonymous letter-writers.

An anonymous communication denotes cowardice and indicates a very "uz" trait of character. The fact that a reply is not desired, means the writer lacks courage to combat an adversary's views.

Enlightenments, discussion and criticism walk three-abreast, hand-in-hand, and their goal is progression. Therefore, when a person writes anonymously, voluntarily closing the door on the enlightenor, discussor and critic, progression is hampered.

An anonymous letter deserves no attention; it *is* an injustice to the receiver, but, there is no advantage whatsoever in pondering over the matter—wondering who sent it and what the motive in the mind of the sender was.

If the person be "on the square," there is no reason for hesitancy in the acknowledgment of his or her name.

When Dinshah receives anonymous correspondence he reads and forgets it. He has the satisfaction that he is unaffected by such occurrences; that the writer failed to accomplish his endeavor—to annoy him.

There is only one day of the year when unsigned missives are excusable —the fourteenth of February (St. Valentine's Day)! and then nobody writes him!

A KNOCKOUT

By:—An M.D., S.C.T.

An M.D., S.C.T., desiring to order an Itisometer wrote his father for a loan and in the letter listed the accomplishments that the invention was designed to perform. The reply from Father follows:

Dear Son:—

Your letter of the 2nd received and noted. To say that I am amazed at your request puts it very mildly—no of course, I can't let you have the money.

You said if you could buy the Graduate equipment you would have all that you required to make a success. This is the result. It's a "will of the wisp" and how much longer are you going to follow it?

Do you believe that man can create such a machine? Where is your intelligence? It's pure bunk to me.

If he had the generosity of a mouse he would say, to you, "as an offset for part of what you have done for me, take my machine and try it out"—and then you would see the fallacy of it.

Now it seems to me, you have tried your present business long enough to know that people don't want to be doctored by machinery, and you would be wise to realize it. You can't make people adopt and follow new theories very long.

You were educated along a certain line that thousands have been very successful at—and you too would be successful had you followed that line instead of departing into a new venture—throwing aside the years of toil and the *cost* of it all.

There were times when I sweat blood to give you the means of being thoroughly educated. It seems to me you hardly realize it.

Do you realize you have been through two colleges at heavy expense? Now you are throwing it all aside as lightly as you would cast off a coat—to accept the theories of one man, you know so little about.

It's about time you gave yourself a thorough introspection, to see where you are drifting to. I cannot any longer be a party to your own downfall.

Affectionately,

YOUR FATHER.

On one of my recent tours, I met again the Medical Doctor whose father wrote the above and who sent me later the copy of that letter for publication. This Doctor is a learned gentleman and is one of the rare sincere Spectro-Chrome Therapists, who has withstood the jeering of the hard-boiled medicos because of the success he gained in numerous refractory cases through the use of Spectro-Chrome Therapy. He appeared rather disappointed that his father did not help him to securing an Itisometer, but, I believe after reading the following letter, perhaps his father might change his views.

Doctor dear, please mail this letter to your worthy genitor.

My Esteemed Unknown Sir:

I never had the honor of meeting your good self, but, having heard sufficiently from your son about your generosity in giving him training through two colleges to fit him for the medical and surgical line, I take liberty to address this to you.

Your views are absolutely reasonable, correct and on the level of good understanding, from your stand points. Assuredly, I am *one man,* perhaps not much known or recognized, but, the pioneers of the World who achieved any radical good results were all *one man* types. Without putting myself on the high level of the great intellectual conquerors of history, like Isaac Newton, Robert Bunsen, William Crookes, Thomas Edison, William Ramsey, and such others who were only *one man* in their lines of research, allow me to say without egotism that it is not necessary to go far to see that in the case of Spectro-Chrome Therapy also it is in its particular line a *one man* job. There never can be two generals in one army, there never is a place for two in one line of radical research and that *one man* is always hundreds or thousands of years ahead of the orthodoxy.

Study the lives of the great prophets and the messengers of the Almighty. They were all *one man* affairs and look at the monument of religious philosophy they left as heritage to enlighten the ages. But, for their pioneering they were derided and pilloried—yet their work lives and will live.

What you must therefore look at is whether this *one man* Spectro-Chrome Therapy has behind it the unshakeable support of the highest in the sciences, whether its followers actually reproduce the claims of its Originator, whether the system does serve human-

ity in the relief of untold sufferings and whether your son is justified in throwing aside poisonous drugs and destructive surgery for a genuine science more akin to a true Healing Art than any other system known today. Surely it *is* "doctoring by machinery," but, my kind sir, this is the age of machinery. It is as much machinery of a utile character as is the Telegraph, Telephone, Printing Press, Typewriter, Electric Light, Radio, Photography, Railroad, Automobile, Aeroplane and thousands of other man-made *one man* conceptions put to the service of humanity. Spectro-Chrome Therapy brings to your door the same quick and effective service in the relief of dis-eases as other present day inventions bring in the relief of labor. Furthermore, it eliminates human error equations of false diagnosis and wrong treatments. It restores the human being to health without ripping open the internals and gives the so-called incurables a new lease of life.

Sir, you gave the means of medical education to your son at a time when Spectro-Chrome Therapy was not in existence and dopes and pills, potions and lotions, blisters and butchery were the only measures to resort. Can you as an affectionate father, a broad-minded soul, who gave birth to a progressive-minded son—a counterpart of your self, blame him for seeing the Truth now when the Truth was expounded to him and when its proofs were experimentally scientifically and clinically demonstrated to him? It will not be in line with your reasoning to think that you would tie him down to theories (of a medical mis-called "science") which are ever changing their aspects and presentations and which I relegated to the eternal years ago by scientific incontestible proofs. I would expect you to do otherwise and encourage him to seek further until he reached the goal—the Truth.

Money earned by bleeding people is blood money. Your son may not have earned as much by practising medicine because medicine is the worship of a false god. He found it faulty; he found its teachings unscientific and conflicting as others also found. How could any honest physician have a conscience and collect coin by poisoning and butchering without feeling dissatisfied with such incongruity? Instead of blaming him for leaving the moth-eaten, fake, fictitious, dogmatic, tyrannical, autocratic humbug of medicine, a thoughtful parent like you would encourage him to further research in the cause of Truth when the facts are known.

And when you speak about my having "the generosity of a mouse" and the like, you forget sir, that my two eyes must see equally in the conduct of an international organization; a courtesy that I may extend to your worthy son, who is one among my large following, ought and must be extended to others as well. Can I be partial?

You at least as a father fully educated your son; I did not have the opportunity your son had; yet the total income of my Spectro-Chrome Institute earned by the labors of this *one man* and his devoted assistants goes without reserve to the support and expansion of this service to humanity. You do not expect in all justice that over and above giving all time, money and energy to the public, I should also present and gift costly apparati to the practitioners of Spectro-Chrome Therapy.

This *one man* at present is enough to compel the medical people come out of their lethargy of six thousand years and when this *one man* shall leave the physical plane, his righteous work of selfless and unselfish services to suffering humanity will be carried on by such conscientious ones as your learned son and others who have seen the Light and will maintain the Torch burning.

Send your son what he requests, for, he *is* on the right road and headed to success.

I have the pleasure in being

DINSHAH,

The One Man of Spectro-Chrome Therapy.

Attuned Lemon Wave

(Medicine, Surgery, Science, Osteopathy, Chiropractic, Philosophy)

THE ADVANTAGES OF HEALING by SPECTRO-CHROME THERAPY

By:—WALLACE F. MACNAUGHTON, M.D.
Spectro-Chrome Therapist

The Human body is managed by a set of glands known as the "Glands of Internal Secretion" or "Endocrine System." To quote Colonel Dinshah P. Ghadiali M.D., "this Endocrine System acts as an interlocked directorate of the nervous system and other organs, and the whole body responds not in parts and portions of organs, but as a unit, to the activities of the complete glandular system." Every individual from the time he is born until the time he dies is under their influence. As special activities, Dr. S. W. Bandler of the New York Post-Graduate Medical School and Hospital, says, "Some of these glands have to do with the development of the bones and teeth, some with the development of the body and nervous system, some with the unfoldment of the mind etc., and later on with the introduction of sex features—with reproduction." Throughout life, along with these special activities they have to do co-ordinately, with the preservation of these structures and functions in a state of harmonious performance which constitutes a normally responsive body and a serene mind.

Each of these Endocrine glands is placed in close proximity to plexuses of the Sympathetic Nervous System from which nerve fibers pass to them, so that they are susceptible to stimulation or depression from that avenue as well as from the blood. Any force, intrinsic or extrinsic, capable of exciting or depressing Sympathetic Nervous System will compel a corresponding reaction from the Endocrine System. It is through this arrangement of the Blood, Nervous and Endocrine Systems that the intrinsic forces of thought and emotion, either conservative or destructive, are able to influence the latter system and to bring about changes in the blood and subsequently the tissues. It is also by this means that extrinsic forces, such as Light, working through the Etheric counterpart of the dense physical body, likewise the Attuned Color Waves, says the Originator of Spectro-Chrome Therapy, "playing upon the Sympathetic Nervous System and through this upon the Glands of Internal Secretion COMPEL THE INTERNAL ORGANS THEMSELVES TO AWAKEN TO THEIR RESPONSIBILITIES."

As a driving mechanism, the Endocrine system serves to initiate

and support the procedure of all bodily functions, from appetite and digestion to menstruation and gestation. And these functions occur periodically though the cyclic interval varies with different functions. Some require a small amount of energy but take place often, as the beating of the heart, while others weeks or months apart draw heavily on the Endocrines for their occurrence. Every reaction of the nervous system, pleasant or otherwise, is dependent for its completion on the energy flowing through these glands. If a given procedure of living is constantly or even frequently requiring excessive performance of a minor function, the associated glands become depleted and later exausted, the rest of the system becomes unbalanced and the major function that is expected fails to put in an appearance, or if it does, the advent or course is accompanied by unpleasant symptoms.

Part of the duties of this directorate is to keep the body free from poisons; from infectious invasion on the one hand, by detecting the invading poisons and marshalling the body's protective forces in opposition, at the same time keeping open the channels of elimination, and on the other hand looking after the elimination of poisons generated in the body by normal metabolism as well as the toxic substances taken in as food.

There are compounds called "Vitamines" without which man can not live, yet writes an eminent student of nutrition, Alfred W. McCann, in his book The Science of Eating. "he makes an almost fanatical effort, for business reasons, to keep them out of his diet, and succeeds in doing so in at least eighty percent, of all the foods he eats." Associated inseparably with the vitamines is another important food division—the Minerals. About these minerals the same author says, "If we analyze a handful of fertile earth we find it composed chiefly of sixteen elements. If we take a measure of milk, wheat, corn, barley, oats or rye again we find the same sixteen elements. If we analyze the body of man we find that it, too, is made up principally of the same sixteen elements found in the soil and the plants. There is no living creature whose body does not contain these main sixteen elements. The constancy with which they appear in the soil, in the tissues of the plant, and in man's own tissues and the consistency with which they are found in all unrefined foods reveals the operation of a fixed and inviolable law.

To be continued

Attuned Green Wave

(Spectro-Chrome Therapy, Spectro-Chrome Therapists)

OHIO MAKES A GOOD REPORT

By:—HENRY C. GEHLKE, S.C.T.

The Cleveland Ohio Graduates again had the great privilege of having Colonel Dinshah with them to teach them more about the wonderful system of Spectro-Chrome Therapy.

On March 24 and 25, he gave a Free Post-Graduate Course on the Itisometer in the Hollanden Hotel. A large number of the original class took advantage of this opportunity and was certainly well repaid for the time spent in this course.

The more study and searching one does in Spectro-Chrome, the more simple and precise it becomes. The Itisometer, the crowning achievement of the Originator of Spectro-Chrome Therapy, our beloved and esteemed teacher Colonel Dinshah, puts at our disposal the most marvelous and accurate method of determining the Radio-Active and the Radio-Emanative condition of the sufferer who comes to us for relief.

While Colonel Dinshah was here, the First Cleveland Ohio Centre of American Association of Spectro-Chrome Therapy was organized. Every member of our old organization but one, (who was absent) signed the Application for Enrollment in the new body and we expect to be a real force in spreading the gospel of Spectro-Chrome Therapy among our fellow beings.

EAR TROUBLES AND VARICOSE VEINS

Three Unusually Interesting Cases

By:—Kate W. Baldwin, M.D., F.A.C.S.
VICE-PRESIDENT
American Association of Spectro-Chrome Therapists

All the morning I have been impelled to report these three cases for the April Spectro-Chrome.

Case No. 1—The first is that of Peggy M., Age 7 years. One day, my telephone rang. An old patient was on the wire; "Dr. Baldwin, will it be possible for you to see my niece's little girl. She has been having trouble with her ears for three weeks and while one is better the other is worse and she has scarcely slept at all in three weeks. If possible, please see her today."

At three o'clock the mother and an aunt brought the child to the office. A sicker looking little one I seldom saw. Her eyes were lusterless, the lips white and the face a gray waxy look. In every way, she said I am sick please help me. The mother confirmed the uncle's statement as to the amount of sleep. The left external auditory canal was badly swollen and full of secretion. I could do little at cleansing it. I gave her thirty minutes of Yellow Spectro-Chrome with the Graduate equipment within six inches of her ear. I told the mother to put on a warm application or not as felt best and to let me see her at eleven the next day. When they reached home the child was given something to eat and was in bed by six o'clock. She went to sleep at once and slept straight through until she was awakened to come to the office. Sixteen hours was some little sleep, truly! ! ! Sleep is tired Nature's sweet restorer. Peggy looked as bright and happy as if nothing had ever ruffled her. She had not had any pain since the treatment and she did not have any more. I saw her five times and they did not bring her since. I have had many cases of Otitis Media, acute and chronic and they did well as the average run of such cases. Since I used Spectro-Chrome all such cases cleared up much quicker than with any other line of treatment and Peggy I think is surely a prize case.

Case No. 2—Mrs. M., age 54 years. Mother of ten children, eight living. There had been some varicose veins for some years, but, since June 1924 the legs had been badly swollen and very painful. From the junction of the middle and lower third both legs were tense and hard covered with thick heavy crusts on an inflamed base. This condition involved about three-fourths of the circumference of the legs and extended to

above the patella. The day before, when she went to bed, they pained so badly she felt as if they would "crack open." The children said, "something must be doing if mother went to bed." This area looked more like elephant's hide covered with crusts than it did like human skin. Above the knee the internal saphenous veins were swollen and acutely sensitive. On this area she had been using some compresses and I told her to continue if they were a comfort. The legs itched so that in many places the skin was broken from scratching. The second day she had to protect the floor—there was so much exfoliation. I covered the dry areas with cold cream made with olive oil and applied Orange Spectro - Chrome. On the acute phlebitis I used Green followed by Indigo Spectro-Chrome. Mrs. M. was comfortable after the first treatment and the crusting, scaliness and also the stiffness, practically gone after the third. The first treatment was March 3rd; to-morrow the 28th March will probably be the last. She may continue for a time to clear the color and improve the old varicosity.

Case No. 3—The third case that of Mrs. C., age 50 some years, the mother of six children and has had to work very hard, much on her feet. Her legs have given more or less trouble for some years, but, recently have swollen and become extremely painful with no relief night or day. Her face showed the extreme suffering. She had a bandage on the right leg which was a little more painful than the left one. About the middle of the thigh the internal saphenous vein of both legs was acutely inflamed much the same as in the previous case. The legs up to just below the knees were swollen, intensely hard, just as hard as flesh could be. I did not wonder that she thought they were going to burst open. The deep as well as the superficial veins were involved and the color was best described by saying they were *black*. Just below the knee there was a narrow band that was nearly white and it was not due to tight bands constricting. Above the knees to the trunk the veins were varicosed and very blue but the thighs were not hard as were the legs below the knees. I have seen many and many bad legs, but for a pair with no skin broken these were the very worst. How the woman could stand —much less walk—was more than I could see. She said she could hardly get to the hospital and had to go to bed when at home, but, even in bed she had no relief from the pain and tenseness. The next day but one she came looking quite bright and smiling. Her greeting to me was, "Doctor! yesterday I felt like a ten-year-old. I do not feel quite so well to-day, but, much better than I did." She continued to improve and after the fifth treatment the legs were soft, the swelling gone and the color good except for the actual veins which were now blue. Her greeting this day was, "My husband is happy, my children are happy and I am happy. I could go to a dance." We gave Green Spectro-Chrome followed by Indigo and applied the bandages to both legs. This patient is sending me all her friends and relatives who have any trouble with legs. Now that her legs are so comfortable she is having some treatment for an enlarged thyroid. Mrs. C is one of the very appreciative patients I ever had.

Even with all the experience I have now had with Spectro-Chrome, I can hardly see how it was possible for any line of treatment to produce the results that we had in these three cases. When I first saw Mrs. C., I said to myself and to those around that I did not know how we were to get the better of that condition. They are not beautiful now, but, they are soft, free from swelling and pain and she walks as well as anyone. I am never again going even to think that anything is impossible.

Attuned Blue Wave

(History, Travel, Biography, Geography, Language)

ITISOMETER DEMONSTRATION IN THE NEW ENGLAND STATES

By:—Geraldine McCann, S.C.T.
Assistant Secretary,
Spectro-Chrome Institute.

"We start on our tour to the Eastern cities on Sunday, March 8, 1925," announced Colonel Dinshah, one morning as we were busy with the usual routine work of Spectro-Chrome Institute.

March 8, 1925, dawned a brisk clear morning, just the atmosphere suitable for a long journey.

The day previous to our departure, the automobile was packed and everything put in readiness; hence, we had nothing to detain us.

With the exuberant good wishes of all we left the Central Office at 4 A. M.

Our first stop was to be Hartford, Connecticut, a distance of about 240 miles. All this mileage was to be covered in thirteen hours.

As the car silently crept through the hamlets and towns that lay in peaceful slumber, joy was in my heart as I realized that I among others, was to assist Colonel Dinshah in his glorious work for the cause of suffering humanity.

Camden, N. J. was reached at 6 A. M. where we took the ferryboat over the Delaware River to Philadelphia, Pa.

The roads from Philadelphia, Pa., being like a smooth sheet of glass, the automobile glided along at a fast rate until we reached Princeton, N. J.

"Let us get down here and have some refreshments, at our usual rendezvous," said Colonel Dinshah. I offered no objection and the car was parked. While he locked the car, I looked into the restaurant. Not a sign of life—it being Sunday the doors were more firmly closed than on week-days!

The nonsensical idea that public places should be *dead to the world* on Sunday, is absurd. Why should Sunday be an exception to the other days of the week? The sooner these Blue Laws are discarded, the better for the whole universe.

But I had to be reconciled to this new idea of traveling on an empty stomach. The faithful Oakland was again turned loose and ere many minutes had elapsed, Princeton, N. J. was miles in the background.

"Don't you think the best idea would be stop in New York?" Colonel Dinshah inquired. "All right," I said, "we shall undoubtedly find some place open there."

"Then it is final," the Colonel said, "and if we reach New York before 12 o'clock (which was our schedule time) we shall linger for a few minutes at the Bronx Zoo."

I looked at him in surprise as to how he "guessed" that I secretly

wished to visit this Zoo, but did not want to mention the fact, as we had more important matters on our mind.

We did reach New York before 12 o'clock, and my unexpressed wish was fructified.

The animals were all very interesting, but, one seemed to command my attention more than the others; it was "Leo," the lion and it was the first lion I saw. Time was flying. We had to leave New York before the noon-day rush.

By the time breakfast was over, the Sun had nearly reached his meridian and we reached the Boston Post Road.

What a relief to leave the smoke and noise of the city behind, and once again be in God's open country. The magnificent scenery, from the stately pines gently waving in the morning breeze, to the emerald green grass, was a sight not to be overlooked.

After five hours of riding thus we drew up at the entrance to Hotel Bond, Hartford, Conn. and promptly unpacked the car.

We were under the impression Hotel Bond had Direct Current. The Itisometer for Demonstration was adjusted to work on Alternate Current, (although the final instrument will work on Alternate or Direct Current), therefore, it was found necessary to carry with us a Rotary Convertor to Convert Direct Current to Alternate Current. Instead of starting with the usual rhythmic hum, the Rotary Convertor, would not work quietly and began to heat and smoke. Investigation showed no defect in the machine, so as a last resource it was decided that the best thing to do would be to look around the electrical stores for a new Convertor.

Just as Colonel Dinshah was about to step into a taxi, a thought struck him and he went to the hotel engineer. "Engineer, what kind of Current have you here?" he queried. "Why, we have Alternate Current," responded the engineer. It was a lesson in not taking things for granted. We had been vainly trying to convert Alternate Current into Alternate Current!! Little wonder the Convertor refused to work. After re-connection everything proved to be in perfect condition for the next night's duties.

Next night, many graduates were in attendance; they were anxious to catch a glimpse of the Itisometer, which was kept veiled until the preliminary address was over, then endless questioning started.

Practical demonstrations on the working of the Itisometer were given as usual and the affair closed the following night; I should say at two o'clock Wednesday morning! And we had to maintain our time schedule.

Without sleeping, the car was repacked and the journey at once resumed to Providence, Rhode Island. It certainly was not an easy position for me to cope with the President's over-active temperament.

The meetings in this city were held in the Providence-Biltmore Hotel, at precisely the same hours as Hartford, Conn.

An enthused group it was, as the lecturer proved the simplicity and scientific fundamentals of the Itisometer; a very reluctant one also, as he bade his farewell to the assembly.

But Boston, Mass., was awaiting arrival with anticipation, and time was not to be lost.

Friday the thirteenth is by many considered to be an unlucky day; not so with us, for, we had success in Boston.

At the close of the meetings much appreciation was shown for the new apparatus.

Long after the world normally has retired to rest, sleepless hours are spent by this man who invents new scientific devices to give to the honest Spectro-Chrome Therapist. The Graduates of Spectro-Chrome Institute do not apprehend the arduous labor and wakeful hours, spent by him for their benefit and protection, for if they did, they would surely be more patient. He is but *one man*. It would seem difficult to some to contend with the various obligations incurred in such a wide-spread movement, but this one man *is* doing it, and *will* continue to do so as I see it in daily association with him.

With the Free Post-Graduate Course being completed in three prominent Eastern Cities within only one week, it was time we were returning to the Central Office, in Malaga, N. J. The office desk would no doubt be piled with correspondence.

Sunday morning, March 15, we left Boston.

Upon discovering that the road via Waterbury, Conn., would shorten our traveling somewhat, it was decided we return by that route.

Unfamiliarity with this road, caused much delay, as we had to have recourse to the Road Map. In several incidents, we lost the trail mark, but, with a little manoevring were on the right track. But in one case we did not find the outlet until it was too late to turn back. This road led us again into Hartford, Conn. which we did not intend to revisit.

At four o'clock the same afternoon New York City was reached, where we took the Fort Lee ferry to New Jersey.

Passing through Hillsdale, N. J. about five o'clock, I was shown the birthplace of Spectro-Chrome Therapy. It was there the inventor had spent earlier years in trouble and I can imagine his feelings passing near it.

Once through Hillsdale and Newark, N. J., the progress was rapid.

At one o'clock, Monday, March 16, we drove up to the Central Office, after a journey of 409 miles, without stop. Yet, people say vegetarians have no endurance; I know many who have the same and more than the meat-eaters.

COMPLETION OF THE MIDDLE WESTERN TOUR FOR ITISOMETER DEMONSTRATION AND POST-GRADUATE COURSE.

By:—Kashmira Dinshah, S.C.T.
Secretary, American Association of Spectro-Chrome Therapists

It was a sunny Sunday, March 22, 1925, when father and I set forth at 6 o'clock in the morning, on an automobile tour of the Middle West; our purpose being to assemble the faithful Graduates of three classes in Spectro-Chrome Therapy, in Cleveland Ohio, Detroit Michigan and Chicago Illinois.

Long streamers of amber and pearl stretched across the Eastern sky, changing, as the bright Sun rose, to flaming hues of crimson and turquoise. After the city of Philadelphia, Pennsylvania was passed, the road lay before us, wide and clear, flanked by variegated scenery freshly green and our good Oakland clipped away the miles steadily. The motor, inanimate yet alive, purred like a pleased lioness; the muffled roar of its pent-up energy in release seeming as it were an expression of joy in the humanitarian work achieved.

Our destination for this day was Bedford, Pennsylvania, a journey of 231 miles from our starting point Malaga, New Jersey. This included a run over part of the great Allegheny Mountain range. With ease we reached Bedford at four o'clock in the afternoon. The solar orb shone encouragingly. Why waste three hours of precious daylight in laziness by staying here for the night? Tomorrow there might be rain, hindering speedy advance. Ligonier, Pennsylvania, 47 miles further on, appealed irresistibly to us. We went on.

Between Bedford and Buckstown runs a stream interesting for geological study. A rugged stony wall rises from its bed, the appearance of which is that of Niagara Falls solidified. It is a lithic curiosity calling to mind bygone ages, when a cataract probably flowed down the cliff to join the placid water at its feet.

That night was spent in the hospitable Fort Ligonier Hotel and the next day we started off at five A. M. The weather was fair. We were rolling along peacefully when, coming down a steep decline of macadamized road, the balloon-tired car began to bounce on the bumps and in avoiding striking another auto, we came too near the edge and landed in the ditch. We escaped without a scratch; so did the car, but, a truck was required to pull it onto the highway again. This occasioned a delay of nearly two hours. Nevertheless, Cleveland Ohio was reached before darkness fell, a distance of 224 miles over the mountains.

On the evening of the morrow, when the Free Post-Graduate Course on the Itisometer opened in the Hotel Hollenden, several Graduates of Spectro-Chrome Institute appeared for joining. The meeting began, meaning, trouble started. One Graduate, after application for Enrollment in American Association of Spectro-Chrome Therapists, objected to a certain section of the By-Laws. "It's against personal liberty," he shouted, "if I use such and such a method for treating my patients, that's none of your business!" "I see," said the President, "it is none of my business if my own Graduates disgrace my work by using methods which have already been adversely confused with pure Spectro-Chrome Therapy. By the way, do you use this mode of treatment of which you speak?" "No, I don't, but supposing I wanted to, why should you try to stop me?" was the heated reply. The President answered, "No one is trying to stop you from anything. Simply do not join the Association; then, treat your patients in any way that suits you. The Secretary will return your fee."

The man refused to leave, saying he wished to belong still. "Then abide by the Constitution and By-Laws as they stand unanimously accepted; it might be made *more* rigid but it shall *never lessen* in ethical inflexibility. Why create discussion about something which does not affect you?"

The trouble ended. All were pleased. Our business was successfully completed the next night and the morning after, March 26, father and I left Cleveland for Detroit Michigan, arriving in that city prior to sunset on the same day, traveling 188 miles.

The following evening, the Course opened, at the Hotel Tuller. The

Graduates began to assemble. At the proper time, the work commenced with an introductory address by the President. When this was over, a certain hotheaded Graduate rose. "Colonel Dinshah, do you mean to say that I must pay twelve dollars just to *see* the Itisometer? Your invitation said the Course was free!" "That amount goes to the Association as fee for Enrollment," said the President, "I receive nothing. The Course *is free* but is given only to those who belong to the Association. Did you join?" The disputant became flushed with rage. He replied angrily, "No! I have a right to see what you have there *before* I pay anything. I came here to see the instrument and I am entitled to hear the Course because I paid you six hundred dollars already!"

"All these others have paid and joined. Why should you be excepted?" spoke the Colonel, "do you mean you were robbed of six hundred dollars? You took for less than that the Spectro-Chrome Therapy course and my two Equipments, one on the installment plan. Your notes were dishonored and when payment was requested, you returned the Equipment express *collect,* after using it three months, in a battered condition!" "I won't pay anything!" "Very well then, there is the door!" At this, the man lost his temper and shouted, "I won't be shown the door. I won't suffer any more indignity. You took my six hundred! I sent your Equipment back without using it. I have a right—" etc., etc.

The gathering expressed disapproval of the man's absurd stand. The sincere ones openly said to him, "Oh, pay or get out! You are detaining us all!" Though hearing this, still the Graduate stood arguing, harping on the $600.00 which in reality included the training of his nurse and self and also a Graduate Equipment. Finally, he decided to take what shreds of dignity he had left and marching out, stood in the hallway discussing with some one else the "injustice" (?) he had experienced. When he was gone everyone breathed freely and the business of the evening proceeded harmoniously.

On Sunday the 29th, we took our departure from Detroit, finishing late that night the 279 mile trip to Chicago Illinois. In Chicago, the Course was also attended by two Graduates from Canada and Missouri. There was no untoward incident.

Our mission completed, we turned homeward on the second of April. From Chicago to Lima Ohio, 251 miles, via Fort Wayne Indiana, we went that day, though hindered by wretched roads and the most inaccurate Motor Club directions we ever struck. Being stuck hub-deep in soft sand, lost a dozen times on wrong roads, were common occurrences on that journey. On the morrow, it was Lima to Ligonier Pennsylvania, via Mansfield Ohio and Pittsburg, Pa., 321 miles. Father just sat at the wheel all day, chopping up the distance home, the machine grinding through like a mill.

The next day, we covered the distance from Ligonier Pennsylvania to Malaga New Jersey, 231 miles, arriving an hour before midnight. The entire trip from Chicago to Malaga inclusive of the ramblings was 858 miles, finished in three days with two stops, not a bad tour for an automobile heavily loaded with delicate scientific apparati.

Two hours after that, at 1 in the morning, father was at his duties reading the accumulated mail.

SPECTRO-CHROME

SEPEMBER 1925
VIOLET 1925

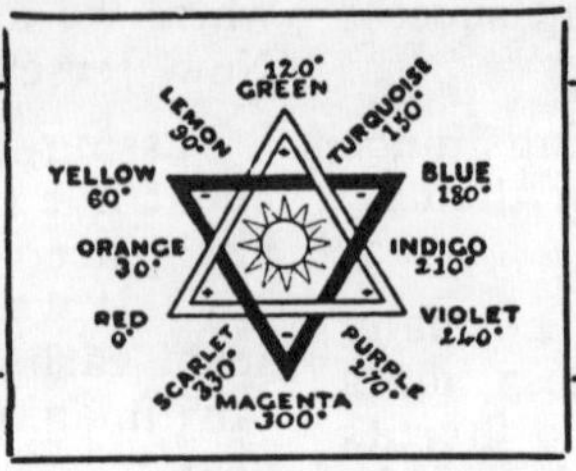

VOLUME 4
NUMBER 9

Devoted to: Spectro-Chrome Therapy **Dedicated to: Service of Humanity**

CONTENTS Page

The World Through the Eyes of Dinshah—

ITISOMETER

NUMBER

***

PICTORIAL

Editor
Dinshah P. Ghadiali

In course of life you reach a time,
When Doctor, Lawyer, Priest alike,
Does fail to give the sought relief,
To troubled mind or burdened heart—
Then come to me;
I help you to help yourself.
SPECTRO-CHROME.

Subscription:—January to December (12 Months) $3.00
All Subscriptions Taken For Full Year Only And Expire With The December Issue
Single Copy 25 Cents. Back Numbers Available From June 1922.

COLONEL DINSHAH P. GHADIALI, M. D., M. E., D. C., PH. D., LL. D.
Metaphysician and Psychologist.

DINSHAH

Humble Servant of Suffering Humanity

Originator of Spectro-Chrome Therapy

President

Spectro-Chrome Institute

AND

American Association of Spectro-Chrome Therapists

DINSHAH'S PLATFORM AS ADOPTED BY HIM SINCE 1891

The Boundless Vibratory Ocean of Thought is Essentially Universal and All-pervading; it is the Individualized Monopoly of NO Person, and is the Common Heritage of Humanity's Evolution: thus, what a Development of Unrevealed Ages has given unto Me in the Form of Knowledge in My Present Incarnation is No Distinctive Acquisition of Mine for My Sole Use, Benefit or Elevation, but is All for Thee and Is Thine Without Condition, Without Obligation, Without Expectation. I Fear No One; Only God Above and Conscience Below, and From Them I have Nothing to Fear.

SPECTRO-CHROME

September 1925 — VOLUME 4

Violet 1925 — NUMBER 9

Trade-Mark Registered at the United States Patent Office

Official Organ of
Spectro-Chrome Institute
and
American Association of Spectro-Chrome Therapists

Printed and Published Monthly by Spectro-Chrome Institute
MALAGA, NEW JERSEY
United States of America

Devoted to the Latest Revelation in the Healing Art

SPECTRO-CHROME THERAPY

Restoration of the Human Radio-Active and Radio-Emanative Equilibrium (Treatment of Dis-eases) by

ATTUNED COLOR WAVES

No Drugs - No Manipulation - No Surgery

Conceived, Originated, Developed, Applied and Copyrighted, 1920, by Colonel Dinshah P. Ghadiali, M. D., D. C., Ph. D., LL. D., &c.

Metaphysician and Psychologist

THIRTY FIRST CLASS

The next class in Spectro-Chrome Therapy will be opened at the Research Laboratory of Spectro-Chrome Institute at 10 A. M. on Wednesday October 21, 1925.

Accommodations for the students, fees and routine of instruction are the same as before.

The full resident Course has thirty discourses and completes in 11 consecutive days.

Send your friends and accompany them if you can.

***

Attuned Red Wave

The World Through the Eyes of Dinshah

ITISOMETER

The most wonderful apparatus for Therapeutical Precision

DELIVERY COMMENCED

LEASE PRICE ADVANCED

Itisometer Number One was delivered to our esteemed and learned friend Kate W. Baldwin, M. D., F.A.C.S., First Vice-President American Association of Spectro-Chrome Therapists, on Sunday October 4, 1925.

Itisometer Number Two was delivered to our kind Associates Clara E. Sandbach, S.C.T., and Richard E. Sandbach, S.C.T., two days later.

After the first demonstration in Philadelphia, Pennsylvania on February 21, 1925, fully seven months elapsed before the delivery began. During these seven months, its inventor Dinshah crowded more work and labor than he would have been called to normally in half as many years, to design and build this apparatus so as to avoid any complications of human equations and to produce an equipment whose finish and precision may command the admiration and confidence of the user and the public.

In strict accordance with the Integrity Policies installed by Spectro-Chrome Institute, this unique apparatus is capable of being leased exclusively by Associates of American Association of Spectro-Chrome Therapists, under an ironclad agreement which is contingent upon the lessee remaining connected with the said Association. It can not be had or used otherwise by any person.

Further deliveries are continuing. However, owing to the numerous additional features introduced by Dinshah in the perfection of this equipment, the lease price advanced according to the notice given in our last issue. No more orders will be accepted at the original pre-delivery price after November 1, 1925.

As is well known, Spectro-Chrome Therapy totally discards "Differential Diagnosis" and treats the Physical Body as one Unit. There is not an apparatus in the World that does even in part for the medical practitioner what the Itisometer does fully for the Spectro-Chrome Therapist. Medical "diagnosis" is a farce, a humbug, a perfect guesswork hooked onto unscientific jargon. In the Itisometer not only its inventor produced an instrument for what the medical practitioners would call "diagnosis," but, he originated and perfected an entirely new and complete science (Itisometry) which at once tells with 100 per cent precision where the disorder is, what the disorder is, how much the disorder is, what the remedy is and many more novel features of which the Medical Doctors never even dreamed and yet in Itisometry "Differential Diagnosis" is totally discarded!

And there is one beauty in Itisometry which Medical Doctors and their systems strived for but never attained. It is the registering and recording of the condition of each vital organ of the sufferer so as to be capable of permanent reference by either the treating practitioner or any other Spectro-Chrome Therapist, at any time and place, without any misconception or error. The

Itisometer Chart uniformized the work of the Itisometrist and introduced a novelty, in itself exceedingly valuable and desirable, producing a method of perfect check and recheck.

It is neecessary here emphatically to warn our graduates, some of whom are Medical Doctors. The Itisometer may be used by them as Spectro-Chrome Therapists and in their capacity as Spectro-Chrome Therapists only but, the Spectro-Chrome Institute will not tolerate or allow its use by them as Medical Doctors, by working it in accordance with their vocabulary or age old moth-eaten systems. They must use this apparatus only in accordance with the teachings of Spectro-Chrome Therapy as laid down by the Orginator and also in conjunction with the phraseology and language of the science of Itisometry Any other use of the Itisometer will be a violation of the Itisometer Lease and will be in contradiction to the ethics of Spectro-Chrome Therapy; on such plain understanding is this apparatus leased and Spectro-Chrome Institute will absolutely refuse to lease it to an M.D. Spectro-Chrome Therapist wanting to use it otherwise, sooner than allow within its ranks the creation of another medical oligarchy, which would use this formidable weapon in a way detrimental to the interests of suffering humanity or of Spectro-Chrome Therapy.

The purpose of the invention of the Itisometer explains clearly the aims of the Originator of Spectro-Chrome Therapy. Since 1891, he has been steadily upholding the banner of "Therapeutical Emancipation" from the horrible bondage into which the Medical Doctors and Destructive Surgeons manacled the World. His aim and goal from earliest boyhood was to evolve a therapeutical system so accurate, natural clean, scientific and precise, that it would stand the test of time and experience and would oust permanently the frauds of poison and knife. After over 30 years, when he laid before the World Spectro-Chrome Therapy, its quick progress was effected because of its true Fundamental Principles and later the establishment of the American Association of Spectro-Chrome Therapists laid the foundation of honest integral service to the World. But, one gap was still unfilled. The therapeutical science, the therapeutical apparati, the ethical organization, the honest therapists, *were* there, but, the apparatus for precise knowledge of the inner conditions of the vitals of the sufferer to enable the therapist to know positively the what how why and where of the condition, to apply the exact Attuned Color Wave, was not existent, except as in the judgment of the Spectro-Chrome Therapist, thus leaving a "human equation."

This was *one* vital point necessary to be solved in order to place the practising Spectro-Chrome Therapist on a professional level far higher than even the Medical Doctor and on a plane of efficiency far more accurate than the dreams of the Medical Doctor. In his haughty arrogance, the Medical Doctor, in the past has endeavored to choke gag and kill every system that threatened to pop its head above the waters of dark ignorance and the tremendous power of the Medical Trust succeeded in keeping them downtrodden and under the heel. Dinshah, with his absolute fearlessness of consequences, his utter disregard for conventional dogmas, his total independence of all human ties tending to curb his truthful onslaught against un-American tyranny, his unsullied scientific reputation, plunged into this problem with vim and vigor and spent an enormous amount of time money and energy in fulfilling the aim. Night and day he worked; the Medical Trust with its octopean tantacles started to put obstacles in his path; the Money Trust behind them tried some terrible means to put him out of commission; there was a time when they even believed they *had him;* they machinated, plotted, schemed, tried to damage him and his reputation, did what they could to ruin him —but the Omnipotent Powers on High had ordained otherwise—the Itisometer was born and —the biggest nail was driven into the coffin of Drug Therapy and Destructive Surgery.

Fellow-workers, dear colleagues, here is the apparatus you clamored for so long; here is the "marvelous achievement" you frequently openly said no human mind could evolve You know not the powers of the human mind! You in progressive America who are steeped in materialism, believe only in the seen and the seeable, the touchable, the objective only; Dinshah was born and trained in a country where for thousands of ages the invisible and the intangible have had their firm root in the minds of the people. He believes in Powers greater than we; he learnt in his younger days to contact them and even living in this glorious (though material) America he did not forget the teachings instilled by his mind moulders. The Itisometer, like other of his thinkings and evolvements, is not *his* brain child exclusively. It is the *result of* the knowledge of our ancestors left by them in the higher ethers as a heritage, for which he acted only as contacting entity and essentialized the invisible and the intagible onto the lower planes of the objective Universe.

Praise be to the Great Souls who left for us the store which he partially tapped! His head bows in solemn reverence to those Benign Masters of Occult Wisdom Whose Illuminated Teachings precipitated the

ITISOMETER

May it serve the human race for its uplift, betterment and evolution toward the goal of Therapeutical Emancipation!

———***———

SPECTRO-CHROME INSTITUTE

What it now is

From a single room office in New York in 1920, Spectro-Chrome Institute expanded to a well-organized working body in only five years and the photo-engravings on the succeeding pages will suffice to show what ceaseless labor and untiring effort can accomplish when combined with singleness of righteous purpose

Without compensation except that of human service, the Originator of Spectro-Chrome Therapy helped by his wife Irene and daughter Kashmira extended his system from Coast to Coast, invented and marketed his products and established a reputation for his followers as unselfish workers in the cause of suffering humanity, through American Association of Spectro-Chrome Therapists. Though opposed by the strongest of vicious elements—the Medical Trust —he swept away everything before his sledge-hammer oratory in exponence of his science of Therapeutical Precision.

The photo-engravings will speak more than a hundred oratorical discourses—and this is merely the beginning. Particulars will follow later

Just wait and watch 12 **months more! There is a power in Truth the most formidable energy in the Manifested Cosmos.**

Spectro-Chrome Institute—Central Office and Research Laboratory.

Spectro-Chrome Institute—Side View of Central Office and Research Laboratory, Engine House, Light Tower and Recreation Building; the various plants stand on an area of 23. 23 Acres.

Spectro-Chrome Institute— Office of the President.

Spectro-Chrome Institute—Reference Library of Central Office.

Office of the Secretaries of Spectro-Chrome Institute and American Association of Spectro-Chrome Therapists.

Spectro-Chrome Institute—Spectroscopy and Design Room.

Spectro-Chrome Institute—Part of Physics Room.

Spectro-Chrome Institute—Part of Caloric Room.

Spectro-Chrome Institute—Part of Lecture Room.

Spectro-Chrome Institute—Showing Printing Plant, Cabinetry, Administration Buildings and Park in the Distance

Spectro-Chrome Institute—Flying Field

ITISOMETER

FULL FRONT VIEW

———***———

The complete Itisometer, showing Bakelite Control Board, Main Indicator Dial, Illuminator, Chronometer Auriculator Dial, Ratio Dial, Tanger Locker, Mahogany (or Oak) Stand, Chart Drawer and Alternating Current Connector.

———***———

This marvelous apparatus for Therapeutical Precision costs less then ½ cent per hour to run it continuously.

ITISOMETER

Another Front View of the Bakelite Control Board and the Dials, with the Tanger Locker open, showing the Caloric Chamber, Tanger, Battery Switch and Alternating Current Switch.

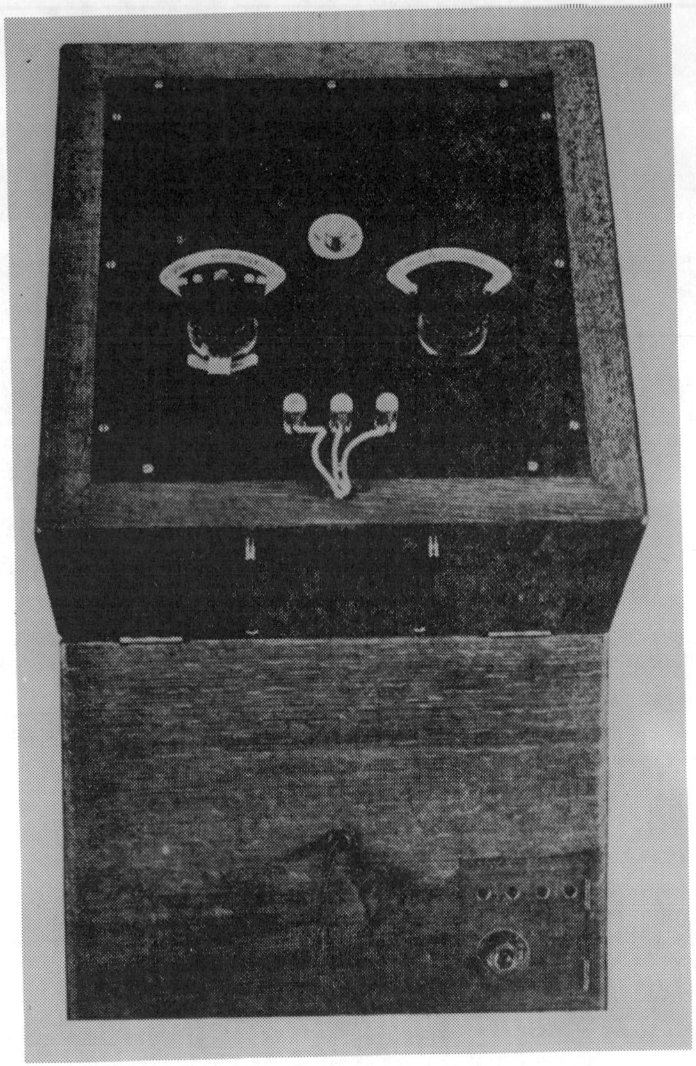

ITISOMETER

Part of Back View, showing the Bakelite Control Board with Accuracy Calibrator, Accurator and Tanger Switch, Zero Adjustor, Tanger Connectors, Alternating Current Attachment and Battery Locker.

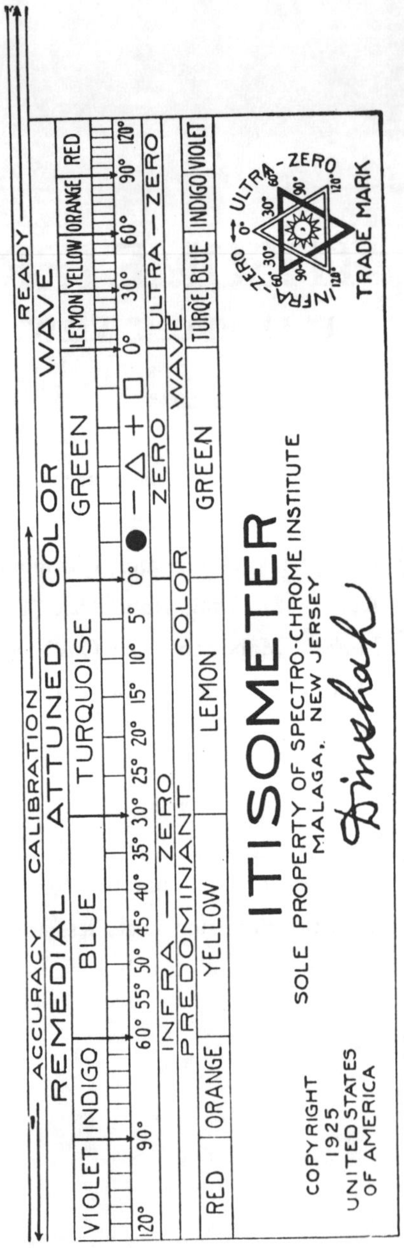

Actual size of the Main Indicator Dial of Itisometer

At a glance, with a single sweepindex, it proves the following:—

1—Is the working Battery sufficient for maximum and minimum records?
2—Is the Itisometer in working order and ready for itisization?
3—What is and how much is the Predominant Color Wave in the itisized Area or Intarea?
4—What is the extent to which the sufferer's internal powers can compensate the dis-ease?
5—What is the Remedial Attuned Color Wave necessary for correcting the dis-ease in the itisized Area or Intarea?
6—Is there danger lurking in the functioning of the itisized Area or Intarea?
7—To what extent is the itisized Area or Intarea working above or below the normal?
8—What is the improvement in the sufferer's condition since last itisization?

Many other features are exhibited by combining the readings of the **Indicator Dial with the Auriculator.**

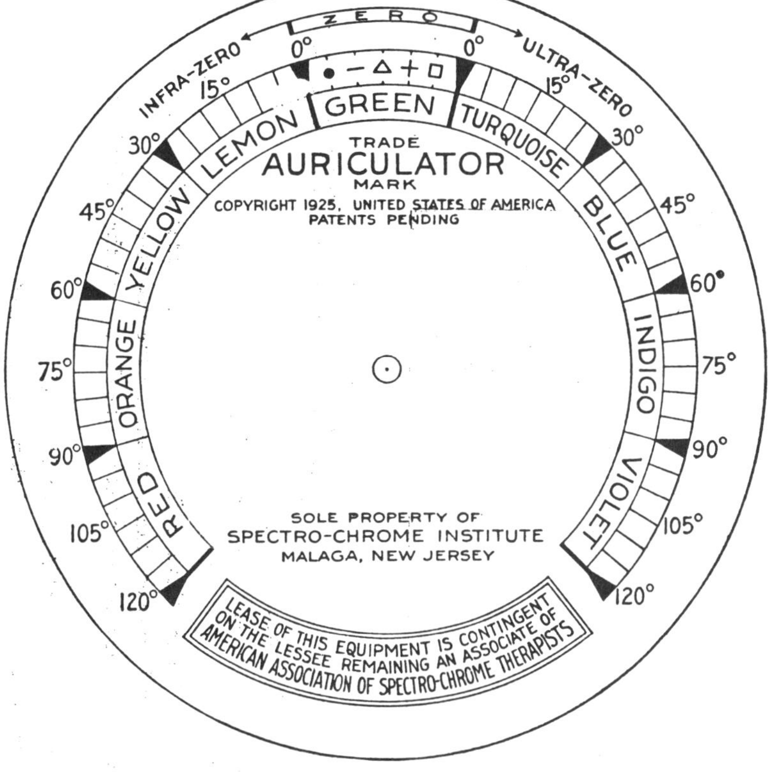

ACTUAL SIZE OF AURICULATOR DIAL.

After itisizing the five Intareas A,B,C,D,E of Areas 1 and 15 they are transferred from the Itisometer Chart to this ingenious device, in accordance with the science of Itisometry.

The result automatically indicates on the Ratio Dial, the Auriculator Zero of the itisizee, from which as the gauge of his prevailing energy for governing other organs, the itisization is made for subsequent Areas and Intareas. Thus, within a few minutes, an absolutely precise record is obtained of the normalcy subnormalcy or supernormalcy of the Areas and Intareas of the whole body, as well as the potential value of the Radio-Emanations of the Front Aura and Back Aura computed automatically and registered through the window below, in full sight of the Itisometrist.

Itisometry totally discards all medical terms and methods; it is an absolutely original science.

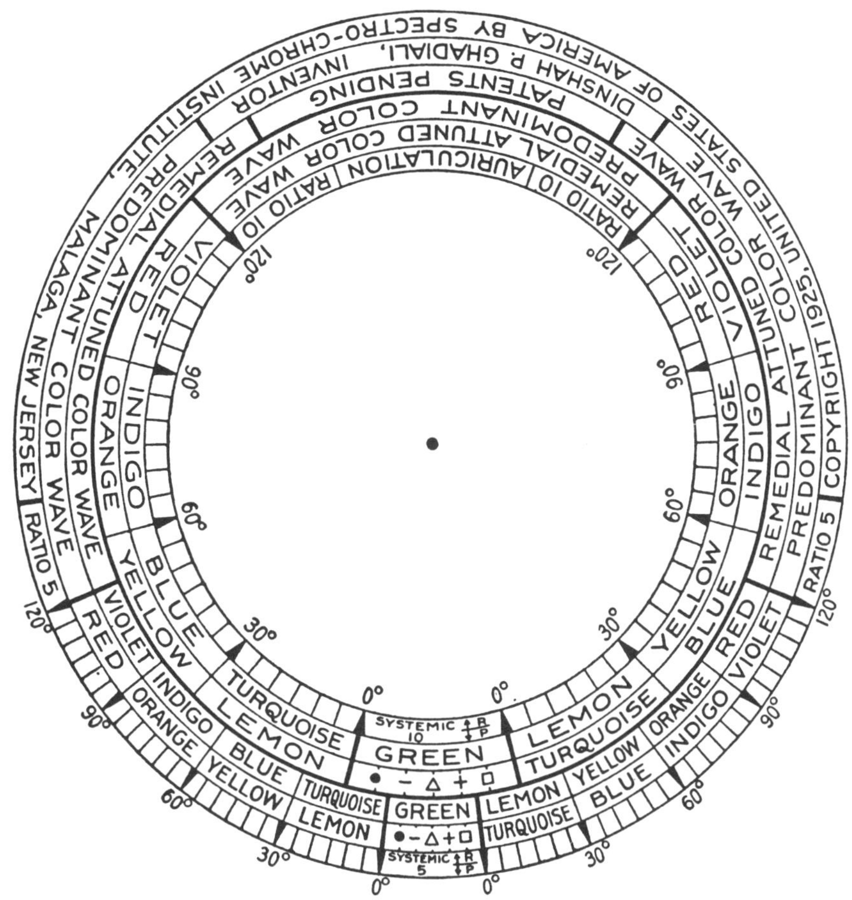

Actual Size of Ratio Dial

This dial proves automatically the Auricular Zero, the amount of disorder in each Area and Intarea, the character of the disorder in each Area and Intarea, the Remedial Attuned Color Wave for each, the value of the Auric Vehicle, the Systemic Front and Back treatment and many other features.

It is through this Ratio Dial that the Auriculator promptly registers whether the sufferer will recover from the dis-ease or whether his vital powers are so far exhausted as to let him succumb under their acceleration or retardation.

The registration is in 5 degrees or 1-6th of the value of one Attuned Color Wave. Its accuracy is 100 per cent.

Chronometer Spectro-Chrome

The new Attuned Color Wave time clock designed by Dinshah for Spectro-Chrome Institute.

This clock needs no winding or oiling. It keeps accurate time and is run by electricity, costing less then One Dollar a year to keep it going.

It is leased only to Associates of American Association of Spectro-Chrome Therapists for marking the Itisometer Chronos and for general use and would form an attractive addition to the office of any Spectro-Chrome Therapist.

The dial is 8 inches in diameter.

TRADE MARK REG. Dinshah U. S. PATENT OFFICE

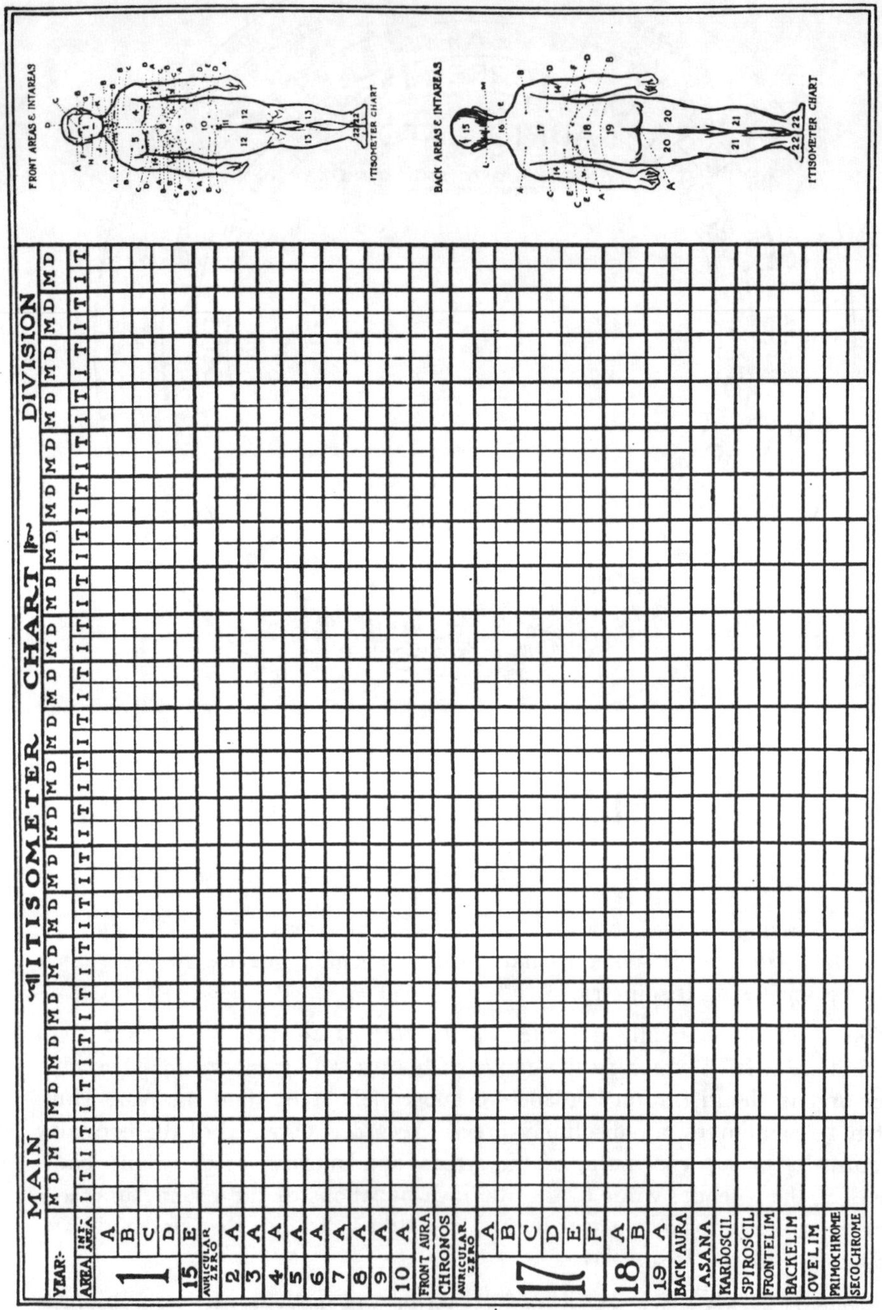

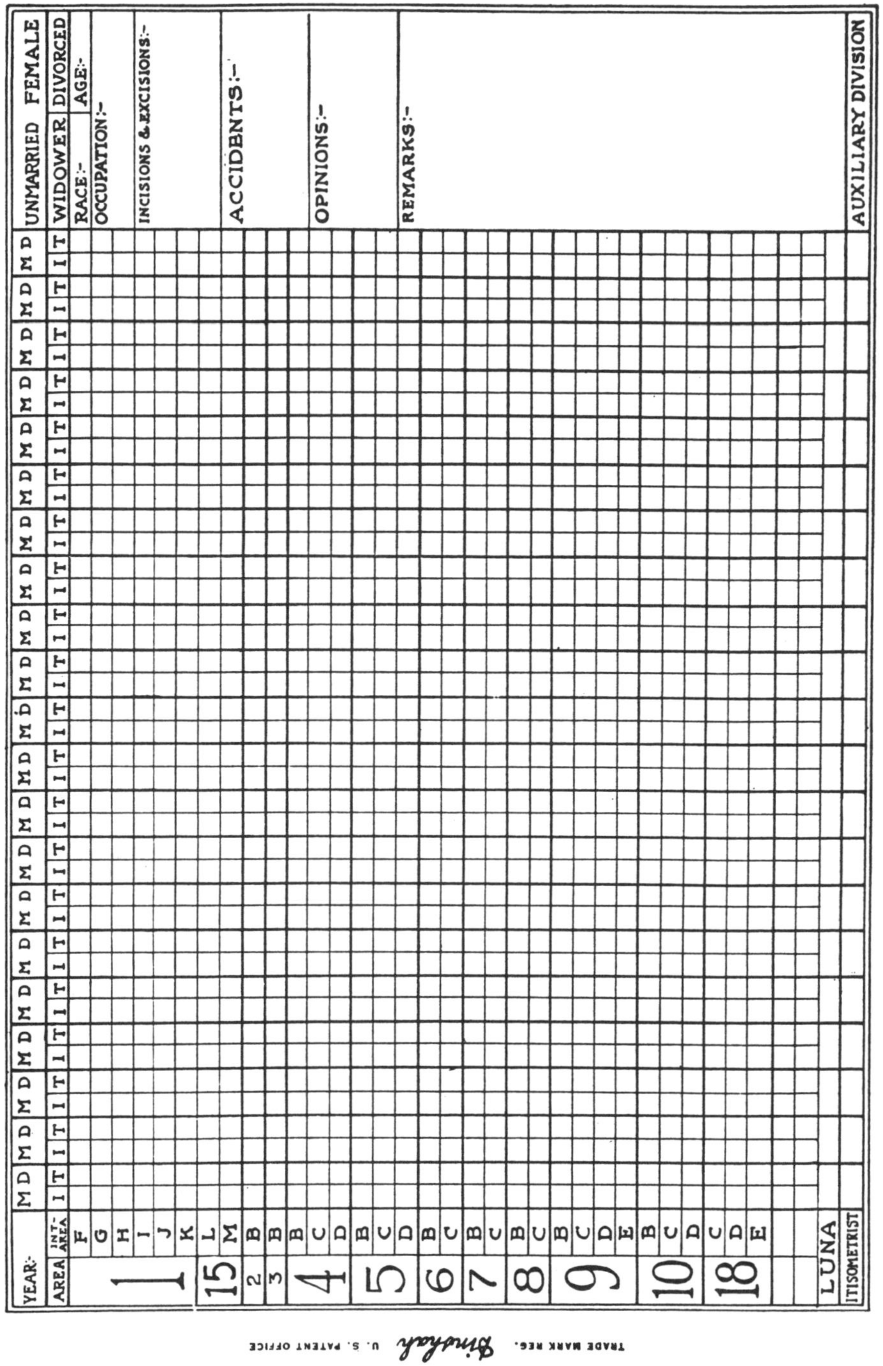

Two poses of **CYRUS DINSHAH**,
Son of Irene Grace Dinshah and Dinshah P. Ghadiali, who completes his first year of this life on the Physical Plane on Sunday, October 25, 1925.

IMPORTANT WARNING:-

Projecting, copying, transcribing, duplicating, photographing or reproducing in any manner, any part of the subject matter or illustrations of the Itisometer and its co-related apparati the Auriculator and the like, is protected by law and is *positively forbidden.*

The word *Itisometer* is registered as a *Trade Mark,* at the United States Patent Office, Number 201631, August 4, 1925.

The double triangle design, with degree values, appearing on the main dial of the Itisometer is registered as a Trade Mark, at the United States Patent Office, Number 201632, August 4, 1925.

All the dials and descriptions are protected by Copyrights and patents are pending on all the apparati.

The sole owners of the Trade Marks and the Copyrights as well as the inventions are Spectro-Chrome Institute, Malaga, New Jersey President,

Dinshah P. Ghadiali

SPECTRO-CHROME

DECEMBER 1925

SCARLET 1925

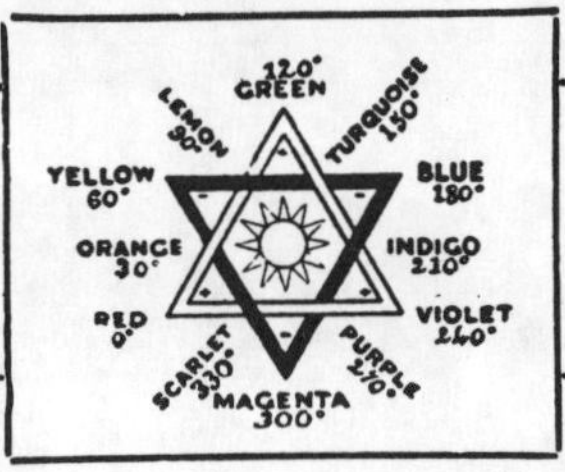

VOLUME 4

NUMBER 12

Devoted to: Spectro-Chrome Therapy **Dedicated to: Service of Humanity**

CONTENTS

The World Through the Eyes of Dinshah—

ANNUAL CONVENTION NUMBER

Editor
Dinshah P. Ghadiali

In course of life you reach a time,
When Doctor, Lawyer, Priest alike,
Does fail to give the sought relief,
To troubled mind or burdened heart—
Then come to me;
I help you to help yourself.
SPECTRO-CHROME.

Subscription:—January to December (12 Months) $3.00
All Subscriptions Taken For Full Year Only And Expire With The December Issue
Single Copy 25 Cents. Back Numbers Available From June 1922.

Attuned Red Wave

(The World Through the Eyes of Dinshah)

EXTRAORDINARY MEETING OF AMERICAN ASSOCIATION OF SPECTRO-CHROME THERAPISTS IN PLACE OF FIFTH ANNUAL CONVENTION

A PICTURE OF SUCCESS AND EXEMPLARY UNANIMITY

The Fifth Annual Convention of American Association of Spectro-Chrome Therapists, which should have met according to the By-Laws on November 28th, could not be gathered because of the unavoidable departure of the President Colonel Dinshah P. Ghadiali and Secretary Kashmira Dinshah and of Irene Grace Dinshah the Secretary of Spectro-Chrome Institute, for the Pacific Coast, to attend to important matters at law.

Returning from Oregon during the middle of December and seeing the necessity of convening all the Associates, the President consulted the Chief Executive Council. It decided that the occasion was an unprovided one and resolved upon calling an Extraordinary Meeting at short notice in place of the usual Annual Convention.

The call was promptly answered from all quarters, those not able to join sending their regrets and extending good wishes for customary success.

On Sunday December 27, 1925, the Chelsea Room of Hotel Lorraine Philadelphia, Pa. again resounded to the oscillations of Spectro-Chrome By Yellow 00 White, the time set for the opening of the Extraordinary Meeting, the beautiful audience chamber had many more than the quorum necessary to begin the business of the day. On the right and left of the platform, a mahogany and an oak Itisometer was working to show the year's remarkable progress in Therapeutical Precision. Smiling faces, cheerful happy women and men of culture and intelligence—representatives of the various Centres of the Association, graced the sunlit hall and the encouraging presence of the Vice-presidents Kate W. Baldwin, S.C.T., M.D., F.A.C.S. and Martha J. Peebles, S.C.T., M.D. on the platform added to the enthusiasm of the crowd. Only Associates were admitted, the doors being totally closed to Strangers during the first day.

FIRST DAY MORNING SESSION

On the stroke of 10, President Dinshah called the meeting to order. The roll demonstrated the presence of Associates from Mighigan, Massachusetts, Ohio, New York, New Jersey, Pennsylvania, Rhode Island, Delaware, Maryland, Connecticut and other States.

The minutes of the Fourth Annual Convention were presented as printed in Spectro-Chrome of January 1925 and on motion of John A. Cohalan, S.C.T., D.O., seconded by Ruth A. Verbeck, S.C.T. were accepted unanimously as correct.

The Secretary Kashmira Dinshah, S.C.T, presented the Annual Report. There was no resignation or violation Moved by Welcome A. Hanor, S.C. T., M.D. and seconded by Bessie A. Hasenau, S.C.T. that the said report be accepted. Unanimously passed.

The report of the Treasurer Dinshah P. Ghadiali, S.C.T. was then presented showing a good balance in favor of the Association notwithstanding the expenses incidental to the pioneer work. Moved by Ruth A. Verbeck and seconded by Charles F. Lallouette, S.C.T. that the report be accepted. Unanimously passed.

Second Vice-president Martha J. Peebles, S.C.T., M.D. spoke about her endeavors in New York to maintain a Centre exclusively with Spectro-Chrome Therapists who were licensed by the State to practise Medicine, Osteopathy or Dentistry. She said she had first thought that would produce ethical co:operation from certain of our Associates who wanted to avoid troubles among their medical and other licensed brethren by joining only a Centre where no lay Spectro-Chrome Therapists could be affiliated, but, her experience during the last year showed that it would be wiser not to have such separate Centres. She requested the Chief Executive Council to affiliate all the Spectro-Chrome Therapists in New York with the First New York New York Centre which she headed, instead of formulating the Second New York New York Centre for the lay persons.

Dinshah commended Dr. Peebles for her sagacity and foresight saying he was perfectly willing that the medical and other licensed friends be given all the consideration to maintain their affiliations with their respective organizations, provided our By-Laws were not transgressed. He was glad to allow Dr. Peebles to have her way in maintaining the First New York New York Centre as an "exclusive" Centre, but he was pleased to note the worthy Vice-president's experence made her change her views. The Chief Executive Council would welcome the change as marking more harmonious co:-operation.

Surgeon Kate W. Baldwin corroborated the views, saying she presided over the First Philadelphia Pennsylvania Centre—a Centre which had nearly 75 Associates of different walks in life, yet, all were of one mind when it came to Spectro-Chrome Therapy. She had no difficulty whatever with them and the laymen and the professionals were one at heart and purpose in the practice of Spectro-Chrome Therapy. She eulogized Dr. Peebles for the change and assured her she would find it beneficial. Dr. Baldwin said that the editor of the Journal of American Medical Association—Morris Fishbein—had published a book called "The Medical Follies" in which he wrote among other systems a few lines against Spectro-Chrome Therapy also. She said she was surprised how anyone could write anything like that without proper investigation. It did not seem fair.

Ruth A. Verbeck asked for forms of Transfer of Affiliation. Dinshah said they were getting ready and would be mailed.

Dr. Kate W. Baldwin spoke against allowing Associates to be enrolled as Fellows, when they had only a 400 Watt Intermediate Equipment. "The 400 Watt Equipment is good for its work, but, you can not expect proper results in general office practice with it," she remarked.

Dinshah said it was true, but, as the By-Laws differentiated between Members and Fellows only as those not practising or practising among the public, without reference to the type of equipment possessed, he did not believe in enforcing otherwise.

He stated it would look as if Spectro-Chrome Institute wanted to lease only the higher-priced equipments, which would be selfish because all could not afford them and also because it did not appear right yet the time to lay down such restriction. He smilingly said it were better if each Fellow were compelled to have an Itisometer also!

The question of maintaining reserve Slide Sets with Centres came for discussion. Dinshah said he wrote to all Centres about it asking them to have a separate fund for maintaining Slide Sets and Slide Carriers in reserve. The First San Francisco California Centre and the First Boston Massachusetts Centre were the two which did not fall in line with the others. He explained how vitally essential the absolute protection of genuine Spectro-Chrome slides was. "It is either they must be in accordance with their proper spectral positions for Therapeutical Precision or they would be just pieces of colored glass. This is the *only one* difference between the true Spectro-Chrome Therapy and the imitation Fakeo-Chrome! If I were to release my hold on this situation, you might as well bid good-night to honesty and integrity; the American people have been cheated long enough; let us stand by our high principles aid guard against anything that would open the door to unprincipled imitators. I stopped a large source of revenue by stopping all sales and by absolutely shutting out unauthorized persons from even leasing our equipments. It was done to protect your interests and those of the suffering public and we must be ever watchful to prevent any leak in the Slides situation. There is too much fraud going on and I wish each Centre will realize its responsibility by guarding against imposition."

Dr. Martha J. Peebles eulogized Dinshah's interest in the work, said he was doing his utmost to maintain the high character of the Institute and asked all to support his endeavors. She said it was marvelous what he accomplished within the last five years. She was satisfied all that could be done was done for the benefit of the Associates.

The morning session adjourned at Green 00 White.

FIRST DAY AFTERNOON SESSION

The afternoon session commenced at Blue 00 Black. There was a larger attendance than in the morning, as the whole afternoon and evening were especially set apart to hear of the "adventures" the President had gone through during the year and to listen to which the Extraordinary Meeting was convened.

Dinshah commenced by giving an account of the rapid development of Spectro-Chrome Institute and his efforts to secure faithful and efficient employees for carrying on the complex work. He stated how it was necessary for him constantly to travel, how he had suffered during his April 1925 Pacific Coast tour, how certain malicious organizations and persons had placed him in great trouble and loss and persecuted him most virulently. He stated how his life was threatened with extinction in a pistol combat, how he came safely through it, how the most pernicious falsehoods were uttered and spread about him and his dear wife and daughter, how some of his trusted employees had bitterly stung him by deliberate perjury and how he had been put through every possible annoyance to ruin his character, reputation, finances and business. He emphatically repeated that notwithstanding all that was done to the contary, he was yet alive to serve the World.

The first day afternoon session adjourned at Purple 00 Black.

FIRST DAY EVENING SESSION

The first day evening session met at Red 00 Black. There was even a larger attendance than in the two prior sessions and the Associates listened with rapt attention. As records were read by Dinshah to prove his innocence of the horrible charges against him—records that no impartial person could read without feeling a sense of anger against his revilers—the Associates expressed their resentment in open terms. When they heard about the perfidy and treachery of certain persons and how it resulted in Dinshah coming near losing his life 3,500 miles from his home, being kept days without food or drink, shut away from communication with the outside world,the tenseness of the situation gave way to moist eyes. Many shouted for retaliation and turning the tables against the persons who instigated such atrocities, but, Dinshah calmed them saying he had no revenge in his mind and was not seeking to injure anyone. Truth was ever triumphant and he hoped soon to put the evildoers to shame by exposing their lies.

The Associates clamored that a fund should be started to help Dinshah in carrying out his appeal against injustice, but, Dinshah declined any such help by saying it was against his principles to receive gifts or presents. He said if anyone wanted to help let the help come through different channels. Let them take Preferred Stock in Spectro-Chrome Institute or send him more students, but, he would accept no gifts. He requested his Associates to keep a level head under trying circumstances, to divest themselves of harmful thoughts against the opponents. Truth must prevail and he would get justice eventually. Evil thoughts were boomerangs. "Wish no evil to my revilers; they will get what is coming to them from Higher Powers than human. Send them on the contrary good thoughts so they may be filled with remorse. When the sting of conscience will come, they will be sorry for their conduct. I am only paying the penalty common to every pioneer. They persecuted Zoroaster, Buddha, Jesus, what am I—a mere servant of humanity, compared to Those Enlightened Souls?"

The first day evening session was concluded at a few minutes past midnight and each Associate went pondering on the "Faithlessness of the World." It was a painfully instructive session in many ways, in regards to the peculiar psychology of humanity.

SECOND DAY MORNING SESSION

Monday morning was fruitfully utilized in exchange of views among the Associates in regards to Spectro-Chrome Therapy. It was an entertaining session.

Clara C. Bryant, S.C.T. said that Spectro-Chrome Therapy satisfied people; her sufferers were pleased with the results. Painful boil on Area 15 was normalated by Orange.

Ruth A. Verbeck, S.C.T. spoke of a case of Rheumatic Inflammation

about which Dr. G. Ernest Spear, S. C.T. had inquired. The President asked her how she know it was "Rheumatic Inflammation" and she said the doctor said so. This led to Dinshah advising the lay Associates to cut out medical phraseology as being a jargon of unmeaning words and to take to Spectro-Chrome methods for better accuracy and safety.

Dr. Martha J. Peebles spoke very emphatically about the worth of changing the phraseology. Medical phraseology led to errors which in a therapeutically precise system like Spectro-Chrome could not exist.

Dinshah was asked to speak on the Science of Itisometry, which he did at some length.

Dr. Martha J. Peebles spoke of a case of High Blood Pressure with 280 systolic and 140 diastolic, which she tonated with Spectro-Chrome Therapy in a week. She said that the stress laid by us in our work, on the breathing processes was correct, in as much as life depended on proper breathing; she had exemplified it.

Bert C. Smith, S.C.T., D.C. described the case of a bank manager who had cankerous sores which were normalated by attention to the diet in accordance with our system.

Wallace F. MacNaughton, S.C.T., M.D. made some sensible remarks about diet and tonation of disorders. He alluded to the difficulties met with in certain cases where Backelim Retardation existed.

After enjoyable discussion, the second day morning session adjourned at noon.

SECOND DAY AFTERNOON SESSION

The second day afternoon session opened at Blue 00 Black. According to announcement, the doors were opened to Strangers introduced by the Associates and many brought sufferers in to demonstrate the success achieved by them with Spectro-Chrome Therapy.

Grace E. Lallouette, S.C.T. started the meeting by speaking of a case of Infantile Paralysis she had normalated. She also spoke of a case in which by the administration of the Lemon Attuned Color Wave the sufferer passed a bunch of worms and was normalated.

Ernest Carreiro, S.C.T., D.C. described a case which the Medical Doctors had diagnosed as Hodgkin's Disease. The sufferer was "dying because of 14 consultants"! Had 55 respiration, 102 Degrees temperature, pulse threadlike; swollen like a pig; had been ill 3 years. Tonated with Spectro-Chrome at bedtime; Magenta improved the condition. Now his Kardoscil is 100, Spiroscil 24; swelling is gone. A woman of 75 had Tic Douloureux; very painful 6 years; had eminent consultants; they were injecting Alcohol in temples. First day of Orange Attuned Color Wave gave relief and 6 weeks normalated.

A girl of 13 had Infantile Paralysis. Had it for 5 years; had been to the State Hospital; Yellow, Lemon and Magenta normalated her in 3 weeks.

Nissin S. Hanoka, S.C.T., D.D.S. said that he had satisfactory experience in all cases tonated by him during three years of his practice as a Spectro-Chrome Therapists. He normalated toothache, hemorrhage, Tic-douloureux. His wife tonated children who ate poisoned fruit; 104⅘ degrees, fever; Orange on Area 8 Blue Systemic, Magenta on 4 made them orelimate and backelimate and in the morning they were right. The doctor would not believe it. Spectro-Chrome does more than we thought it would.

Bert C. Smith, S.C.T., D.C. spoke of his success with Spectro-Chrome A case was reported of Erysipelas in foot from soft corn. Four Indigo tonations and it was gone. Woman suffering from Ringworm on Number 1 was given by someone Ultra-Violet Rays and was burnt. One tonation with Yellow was given; she went home, slept and the scales dropped off.

Ernest Carreiro, S.C.T., D.C. reported ohter cases. A woman suffering 13 years from Pulmonary Tuberculosis, was married and had two children. Ran temperature, had a bad heart with leaky valves, was helpless. Five months of Spectro-Chrome tonation relieved her. She went to motion pictures; had no cold now. Used Orange on 5 and 4, Lemon on 17; also Magenta; Green systemic twice. A man 28 years of age had Pulmonary Tuberculosis with putrid mucous discharge; chest sounds could be heard six feet away like gurgles and explosions; he had been given X-Rays.

He wrote to Dr. Kate Baldwin about it and she encouraged him Nine months of Spectro-Chrome normalated the man and he took a machine shop. He had an impression that where night tonation is not given Spectro- Chrome may not be so successful. A case of Psoriasis was so bad a trail of scales was left. Turquoise, Lemon and Magenta was used for five months. The sufferer was relieved of the disorder. He had normalated Acute Appendicitis; Neurasthenia in woman responded to Spectro-Chrome; the survival value of tonation with Spectro-Chrome was large.

Bert Smith, S.C.T., D.C. spoke about the harm caused by the removal of organs; he said the removal of the Thyroid Gland in his wife caused her to succumb to Pulmonary Consumption.

Dinshah said that the removal of the Thyroid was a serious mistake; the chain of internal glands should never be disturbed.

At this point, Emma H. Gabriel, S.C.T., N.D. escorted Anna Schoener the (formerly paralyzed) girl whose case was reported in Spectro-Chrome of February 1925. She came walking with her father, a Philadelphia, Pa. policeman. The father reported to the audience how helpless the girl was and how Spectro-Chrome helped her. Another case of a paralyzed burnt girl was also introduced. Both cases interested the Associates who expressed their satisfaction with applause.

The President extolled the praiseworthy endeavors of Dr. Emma Gabriel in the cause of suffering humanity and presented to her the Silver Medal of Spectro-Chrome Institute; the second one to be so presented, the first one having been awarded to Dr. Kate W. Baldwin, two years ago.

Dr. Martha J. Peebles spoke about the value of long and continued tonations for long standing disorders. She explained the beauty of the Magenta Attuned Color Wave and spoke about the rejection of the superfluous Color by the Auric Vehicle.

Wallace F. MacNaughton, S.C.T., M.D. reported some of his interesting cases, which included Paralysis, Colic, Constipation, Corneal Ulcer, Stomach Ulcer, Psoriasis and the like. He asked for more information about Impeded Back Elimination.

Dinshah spoke at length about the reason of Imbackelim and why some cases did not appear to respond to Yellow. He impressed upon all the value of taking the Visible Physical Vehicle as one unit machine and not tonating it purely locally. The defective circulation in the internal organs needed Magenta on 4 or 18 and without removing such sluggishness the local Area may not respond. Furthermore, the original cause may need systemic; the Itisometer was the one apparatus that was necessary to find where apparent failures occurred, but, now there should be no failure of any kind. The Itisometer gave the correct clue and gave the Remedial Wave, which may in an aggravated case be Orange as Yellow.

Dr. Wallace F. MacNaughton spoke of the beneficial effects of Magenta and Purple in the case of a woman who had ulcer in the stomach, Kardoscil 92, Spiroscil 16. After tonation with Spectro-ChromeTherapy, she could eat a pear, a fruit she could not eat during three years; her Kardoscil came down to 64 with Spiroscil 16.

Nissin S. Hanoka, S.C.T., D.D.S. spoke of his success in normalation of Imbackelim. He had one case which always needed an enema and had piles cut. Orange on 8-9 with later Yellow gave results. A boy of 5 years; had convulsions since birth; constant Imbackelim; 18 tonations cleared all disorders. A dentist had varicose veins for 30 years; was 3 weeks in bed with a swollen foot; Spectro-Chrome equilibrated him so, he became interested and went to see the Spectro-Chrome Institute Research Laboratory.

Grace E. Lallouette, S.C.T. reported some of her work. A married woman, young, was told by Medical Doctors she had a tumor on the right ovary and her case was declared hopeless. She was reduced to 92 pounds. Two Spectro-Chrome tonations per day were given; in six months she was 122 pounds. Her home was a hundred miles away, so an Equipment was loaned to her; she was soon restored to normalcy; her present weight is 133 pounds. Two cases of Infantile Paralysis are progressing toward recovery; they received four tonations daily for some time. She reported favorably in cases of chronic Imbackelim.

Frank E. Kaufman, S.C.T. reported a case of a man of 55, suffering from Shaking Palsy and recovered after leaving smoking and drinking and being treated with Spectro-Chrome. He spoke of a case of Tic-Douloureux.

Anna M. Dana, S.C.T. treated Imbackelim, Piles, Eczema, Itch and other disorders with Spectro-Chrome successfully.

Dr. Maude B. Crans, S.C.T. spoke of her experience as a Chiropodist, with Spectro-Chrome. Man of 30 was doubled with heel trouble; Yellow strenghtened his arch. A gangrenous toe was normalated. She said Spectro-Chrome was good for foot troubles. Ingrowing nail disorders responded.

Ernest Carreiro questioned about dropsical conditions and wanted explanation about a case where the lower limbs and abdomen were swollen and the heart was out of order. The Kardoscil and Spiroscil ratio was 8 to 1.

The President gave a lengthy explanation about such conditions, upon the general lines as taught in the classwork on Spectro-Chrome Therapy, emphasizing the beauty of Spectro-Chrome in finding whether the sufferer will recover or pass away. The public being present, he spoke guardedly without giving details, but was thoroughly understood by the Associates.

Cleo V. Butner, S.C.T. spoke of her success in the South. She blamed Dinshah, in jest, for the publication of her letter from Alabama and said she did not write to gain publicity. She reported many interesting cases. A child had bad sore leg; tubercular bone; pussy mass on shin; Green skinned it in one tonation; three tonations completely normalated; two months totally set right all disorder. A woman had neuralgia; Yellow used; perspiration poured; in three tonations went to work. A man was tonated for disorders caused by adhesions inside; he was normalated soon. She was working with a Medical Doctor, who had seen several of her cases and had joined for the next Course in Spectro-Chrome Therapy, because he believed Spectro-Chrome was simply a wonder. A case of Cancer was reported. She said they had over 20 cases of Tonsilitis waiting for her return to avoid removal; Turquoise always set them right. A man had no relief from Rheumatism for 5 years; Blue worked on him; after the first tonation he slept the whole night.

Soon after Purple 00 Black, the session adjourned.

SECOND DAY EVENING SESSION

The Evening Session of the second day was instructive like the others and being open to Strangers, many were introduced by the Associates.

Ruth A. Verbeck, S.C.T. discussed two of her cases. Boy 4½ years old had his face and ear scalded by hot water. She gave ½ hour Blue; next morning another ½ hour Blue; normalated. A girl of 15 was having epileptic fits according to the Medical Doctors; she had them since she was 4 years old; just establishing ovelim; tonated with Green systemic, Magenta on 4-18;Violet on 16; she went through her period without an attack; at other times, the mother had to carry "ether cones" to relax her.

Bessie A. Hasenau, S.C.T. was next in line with some of her work. A young woman of 32, had no Backelim without enema nad drugs; six years ago she had Rheumatism; lost use of right knee; she dragged right foot behind; worked in agony; was about 7 weeks in a hospital and the doctors said it was useless for her to stay longer; went to Rome Indiana; she was there given opiates; came for Spectro-Chrome; left medicine at home; tonated with Red and Scarlet systemic, Indigo on knees; has no pain now; has two Backelims daily; equilibrated. A man was in the habit of drinking; he was perspiring profusely when brought in; systemic Green gave him pain; by Turquoise he trembled; systemic Violet and Magenta on 18 made him sleep for half an hour, then he went to office.

Dinshah again emphasized the necessity of attending to the Kardoscil and Spiroscil before starting tonation. He said the man's case showed the potent value of Attuned Color Waves of graded known frequencies.

Jacob Lickerman, S.C.T. had a case of a woman who had only 1 Backelim in 2 weeks. She was tonated with Green and Yellow to Orange with Magenta. She was perspiring under her hat; three tonations normalated her.

Surgeon Kate W. Baldwin, S.C.T. gave the assembly the benefit of her wide experience in the use of Spectro-Chrome for all types of complications. A man of 59 had been paralyzed from infancy; his right leg was shorter; he was an artist and a woodcarver; for 17 years he worked 18 and 20 hours a day; ate little but meat; had 1 Backelim in 2 weeks; smoked all the time; ate two meals a day; his left side gave out; took Aspirin; when he was given under her care, looked like a hopeless case; she let him have a Practitioner's Equipment; four weeks of tonation changed the condition of the man; the case became very hopeful; the man left off his smoking and became a vegetarian saying if it was good for his recovery he was going to do it; his bad piles were shrunk; Backelim cleared; he had Green and Red, Magenta and some Yellow; since she received the Itisometer he was itisized and the Areas proved on Zero Scale showing his growing vitality; his Area 4 was good.

Speaking of the Itisometer, Dr. Baldwin stated the apparatus for a while puzzled her; a man itisized Zero all over and he was far from normal; he coughed and expectorated heavy mucous; then her attention was drawn to the Kardoscil and Spiroscil as well as Zero needed Green; that cleansed him. The Itisometer gave some peculiar readings; one had to learn its peculiarities; she used the Itisometer constantly; she had apparatus Number 1; she was anxious for months to have it and now she had it, she would feel lost without one; it was a great help in Spectro-Chrome Therapy. She spoke about two cases of Glaucoma which she tonated with Spectro-Chrome; they were well responding.

Dr. Baldwin emphasized again the necessity of Fellows of the Association having proper Spectro-Chrome Equipments; it was not fair to the public to go for tonation with insufficient apparati; they should at least have a Practitioner's if not a Graduate Spectro-Chrome Equipment; it was inportant for proper results. She spoke of a sanatarium doctor who had itisized Ultra-Zero 110 all over and steadily recovered until the Itisometer proved him near Zero.

Anna M. Dana, S.C.T. asked Surgeon Baldwin about X-Ray burns and was informed that Spectro-Chrome was effective in those cases.

George F. Ehrline, S.C.T. introduced a young man of 19 whose left side was paralyzed from childhood; left arm was short; gave him 18 tonations in 36 days; his forearm increased 3-16th of an inch and Area 14 increased 3-8th of an inch in girth; the youth was slowly but steadily recovering; he said he was satisfied that soon better results would appear.

The Evening Session upto O 10 B was open to the, public; after that owing to too much of classwork being given, the President declared it a closed session; discussions also continued about the best means of defeating the foul aims of those who were fighting to ruin Dinshah and to destroy his family, reputation and work.

Dinshah having received that evening, the transcript of record and other valuable documents from the Pacific Coast, read aloud certain portions and a majority of the assembly who knew the truth again clamored for justice. They expressed their willingness to go to the Pacific Coast and testify as to the untruth of the charges against Dinshah, from what they personally knew. After quiet deliberation, the meeting was adjourned past midnight.

THIRD DAY MORNING SESSION

On the last day of the Extraordinary Meeting which represented the Fifth Annual Convention, the interest and enthusiasm of the Associates was unabated. They were as usual hungry for knowledge, thirsty for information, ready to expound and stand by the truth.

The President brought to the attention of the meeting that he had information of the persecution of an Associate Joseph J. Zilis, S.C.T. in Cleveland Ohio, on the complaint of an agent of the Ohio State Medical Board. He said he had not the full information, but, that Zilis was given no proper opportunity to defend in the Court. The Court had told Zilis that he should secure a License from the Ohio State Medical Board!!!

Dinshah said that the time was ripe for the Association to start an insurance scheme to protect against such medical tyranny; he could not authorize the use of the regular Association funds for such purpose, because, it was not fair to the Association to be the custodian of individual safety at this juncture as the limited funds of the Association would not warrant it; furthermore, the Charter of the Association may not be construed to authorize such disbursement; he was willing to help, but, the help should be from a separate fund to which those who wanted such special protection might subscribe. Other Associations had similar provisions and we should find out what would be the best for us. He asked Bessie A. Hasenau, S.C.T., President of the First Cleveland Ohio Centre, if she knew and could give verbatim report of who were the accusers of Zilis and what happened in the Court actually so the Association might come to some understanding. She had not any special inside information and was unable to give an exact rendering

Vice-President Dr. Martha J. Peebles wanted action in the matter to give relief to Zilis if it were found he was unjustly persecuted by the Medical Board; she stated how Dinshah had appealed to her about helping Zilis, but, without learning the inner facts, nothing could be done.

Dr. Wallace F. MacNaughton wanted to go through this case thoroughly to set a precedent. Welcome A. Hanor, S.C.T., M.D. expressed similar opinion.

Dr. Ernest Carreiro spoke about it being the same attitude as against the chiropractors. He was arrested eight times and was once convicted, but, he continued his practice as a chiropractor in spite of the persecutions. We must have a separate organization for taking care of such legal troubles.

Anna M. Dana spoke about the license troubles in Massachusetts.

Cieo V. Butner spoke of free tonations.

Dr. Martha J. Peebles said that even licensed Medical Doctors were subject to malpractice charges and the American Medical Association would not defend individuals; there were numerous spite cases among them also and much of the annoyance was by unjustified lawsuits.

Dr. Nissin S. Hanoka said his dental friends went through similar troublesome experiences. He told of a negro who came for a tooth extraction, gave two dollars and sued for $2000 damages. Each Spectro-Chrome Therapist should carry protective insurance against such blackmail.

Dr. Wallace F. MacNaughton called it a necessity; he favored it.

Dr. Martha J. Peebles gave an enlightening address on the values of different policies and moved that the President investigate this question from various insurance companies and report to the Chief Executive Council what was the best to protect the interests of all concerned.

The assembly did not take any definite action in regards to the Zilis case owing to the absence of full particulars.

THIRD DAY AFTERNOON SESSION

The afternoon was well used in discussing how to raise the standards of training in Spectro-Chrome Therapy so as to meet progressive requirements and fulfill the purpose (h) of our Constitution.

The President spoke at length in connection with this. He had started since June 1925 classes at Malaga New Jersey, where right in the Research Laboratory the students were trained very carefully with elaborate experiments and demonstrations in Spectro-Chrome Therapy. The full course was composed of 30 discourses instead of 24 and was gaining the favor of the students. He had given all he possessed to Spectro-Chrome Institute and was so arranging matters as to maintain the work even after his death. However, he said, he was very busy and had no leisure yet to die; the World needed his services and he would live years to see his goal of Therapeutical Emancipation achieved.

Walllace F. MacNaughton S.C.T., M.D., suggested that an addition of Professional Economics in the Course would be of benefit to the S.C.T. It was necessary to teach the S.C.T. how to earn livelihood after graduation and not to lose. Dinshah said they learnt that the same way as the Medical Doctors.

Bert C. Smith, S.C.T. spoke of his experiences as a Doctor of Veterinary Medicine. Some interesting questions issued in regard to hay and oats for horses.

Dr. Martha J. Peebles said she had a wide experience in medicine and did not lose more than perhaps $1000 in all 25 years. People were in general honest and practitioners did not lose much; she gave service to many poor free of charges and did not regret it.

Bessie A. Hasenau, S.C.T. lost only $54 in two years.

Dinshah offered a prize of $25 for the best article on "Professional Economics." Dr. Martha J. Peebles said Spectro - Chrome Institute should receive help from the Associates and not give prizes. Dinshah stated with a twinkle in his eyes that it meant really more money for the Institute, because, although it would pay $25 the copyright of the article would belong to the Institute and it was worth more than that. He felt sure if nobody won it, Wallace Mac Naughton was sure to win, he was a good writer and had a clear head. He did not smoke!

Dinshah then brought forward another item of interest—it was about the degree of Master of Spectro-Chrome (M.S.C.). The Charter of Spectro-Chrome Institute provided for that high degree, but during the period of its existence, none received it. He had guarded those initials as a treasure would be and asked for suggestions as to the qualifications necessary in a Spectro-Chrome Therapist to deserve it. Spectro-Chrome Institute did not—and would not-create "Doctors," bnt would stick to its own original line and would have nothing to do with the titles of other professions. The following suggestions were offered by some of the Associates, for attaining the degree:—

Wallace F. MacNaughton — 100 Itisations, 2 years practice as S.C.T., 6 months of experience in Itisometry, 5 articles to be written for Spectro-Chrome, each of 1000 to 5000 words.

Welcome A. Hanor—5 years practice as S.C.T., affidavits from 500 sufferers normalated, 1 year of Itisometry.

Martha J. Peebles—Suggestions reserved.

Frank E. Kaufman—4 years as S. C.T., 1 year as Itisometrist, 200 case reports.

William E. Darling — Besides practice, the S.C.T. should be able to tonate without the use of an Itisometer as Dinshah does.

Bessie A. Hasenau—Should know why Orange or Yellow may or may not help, how something is wrong with the inner Triangle, 3 full courses in Spectro-Chrome Therapy, 4 years as S.C.T., 200 case reports,1 year in Itisometry.

Nissin S. Hanoka—3 years as S. C.T., 100 Itisations, 10 articles on interesting subjects in Spectro-Chrome, 1 thesis on Spectro-Chrome Therapy and Itisometry.

R. Sitarama Rao—should have knowledge enough to defend Spectro-Chrome Therapy, 3 years as S. C.T., 6 months of Itisometric experience, 200 case reports.

Emma H. Gabriel—3 fullCourses in Spectro-Chrome Therapy, 4 years as S.C.T., 6 months as Itisometrist, 100 case reports.

Anna A. Dana—Supported Bessie A. Hasenau.

Jacob Lickerman—100 so-called "incurable" case reports.

Bert C. Smith—The average suggestions as given.

Bessie P. Grist—Should take a repeat Course in Spectro-Chrome Therapy.

Dr. Kate W. Baldwin was not present at this session because of pressing professional engagements. She offered her suggestions on arrival in the evening:—Must have full Spectro-Chrome Equipment, take Post-Graduate Courses, take repeat Courses in S.C.T., 2 years as S.C.T., 100 Itisometer Charts of tonated cases; she further said that the American College of Surgeons requires for admission as Fellow, fifty or more major operation reports and the like and our requirements ought to be sufficiently met for our purpose by following the outlined suggestions.

Dinshah took the matter under advisement saying he would fully consider all the issues involved and soon establish the requirements for the degree of M.S.C. (Master of Spectro-Chrome). He hinted that he was looking for competent persons whom he could train so as to be fit to succeed him; the pistol combat on the Pacific Coast in which he was innocently involved had made him determined in his resolution to fortify the work so as to prevent its destruction on his demise. He was training his wife Irene Grace and his daughter Kashmira to carry on the work; he had taught them the process of attuning slides for Spectro-Chrome Therapy; he was trying to get such helpers as would be interested in the work and would take to it as a permanent career. He assured the Associates Spectro - Chrome Therapy had no death—it would live until the Sun Moon and Earth changed their chemical potentialities and light gave a new Spectrum!

Discussion then started about Post-Graduate Courses for higher knowledge in Spectro-Chrome Therapy. Bert C. Smith suggested further elucidation of the "Occult Operation of the Great Sympathetic Nervous System"; Wallace F. MacNaughton asked for a Course on "Psychical Disturbances".

R. Sitarama Rao, S.C.T., B.Sc., N.D.reported a case of a girl of 17, epileptic; did not attend school; was backward in education; had Imbackelim; several fits a day; was considered by the Medical Doctors as a hopeless case. Spectro-Chrome Equipment was used. She started proper Backelim after only three tonations and had no day fits. On those days, previously, that she had fits she was very cross. Her mental condition improved in two weeks. She was tonated 4 months; did not adhere well to our vegetarian diet. She had Lemon systemic, Magenta on 4-18; occasionally Purple systemic, Orange on 8. He asked for further information.

Dinshah explained how Epilepsy was connected with disorders of Areas 10 and 11 and how those Areas reacted on Area Number 8 Flatulence and similar disorders of the digestive tract arose in conjunction with such derangements and the handling of Dr. Rao was in line with our methods.

Nissin S. Hanoka spoke of a case of a married man; he was a big man, but had trouble in the use of Area Number 11. Some discussion was carried on about similar disorders and Dinshah explained at some length the working of the Psychical Vehicle.

The Election of Trustees was unanimous. The same three Trustees were reelected and installed: Dinshah P. Ghadiali, Irene Grace Dinshah and Kashmira Dinshah.

The Trustees unanimously reappointed Dr. Kate W. Baldwin and Dr. Martha J. Peebles as the Vice-Presidents and they were installed amidst loud cheers and applause of the assembly.

THIRD DAY EVENING SESSION

R. Sitarama Rao presented a case of a woman of 35; severe pain at Ovelim which was irregular and scanty; had headaches; stayed at home for about 3 years during which the Medical Doctors treated her fruitlessly. Her husband finally requested him to use Spectro-Chrome. An Intermediate Equipment was used and she improved. After the second Ovelim period, there was no headache; two tonations a day for four months; had no trouble during the last 2 years. Systemic Lemon, Magenta on 4-18, Scarlet on 10-11 and Turquoise on Area 1 were employed as needed.

Surgeon Kate W. Baldwin spoke on the way the Itisometer could be used for demonstrating to medical people the efficacy of Attuned Color Waves by the change being measured and shown. Sugar might be passing and its elimination according to their testing methods might be shown by the Areas responding through Spectro-Chrome Therapy.

Dinshah said that in Spectro-Chrome Therapy, as the names of disorders were eliminated, the lay practitioners had better stick to the terminology of Spectro-Chrome as taught. He said many complaints were received where Centre Presidents had allowed discussion according to other therapeutical systems in Centre meetings and it was averred that some Associates who were doctors of other systems persistently kept up heated arguments about systems foreign to our work. He drew the attention of the assembly to the Bylaws pertaining to the subject and requested the Centre Delegates to inform their respective Centres to refrain from allowing any Associate to bring up such matter in our Centre meetings. We were not meeting to know about other systems and their merits, but, were concerned only with Spectro-Chrome Therapy. Let the Medical Doctors discuss medicine and Surgery in their societies, the Osteopath and Chiropractors carry their special talks to their organizations and so on; let us consider in our meetings how to establish and disseminate our work in our ways. Centre Presidents allowing irrevelant discussions and foreign arguments were not right in their attitude and any Associate present when such irrelevancy was going on would be right in calling for order. He would not encourage other therapeutical matters in this Association and would only adhere to our principles and purposes. He expressed high regard for Dr. Kate W. Baldwin and eulogized her devotion.

After further remarks of similar character, the President asked for a rising vote in silence for five seconds, in honor of four of our friends who died in active service to our Cause:- Lieutenant Colonel Frank G. Bolles, S.C.T., Pauline Barton-Peeke, S.C.T., Martin Lappa, S.C.T., and Laura M. Gaudette, S.C.T., which was duly accorded.

There being no further business, the Extraordinary Meeting was dissolved at midnight.

Opinion was unanimous that the meeting was a picture of unanimity, singleness of purpose, fraternal feelings and boded success. Nothing marred the smoothness and all went to respective duties well satisfied that the three days were among their happy ones.

***

DR. KATE W. BALDWIN HONORED

In appreciation of her services to the suffering and in recognition of her learning, the Maryland Academy of Sciences of Baltimore Maryland, conferred upon Kate W. Baldwin, S.C.T., M.D., F.A.C,S,, First Vice-President of American Association of Spectro-Chrome Therapists, Life Membership in the Academy.

We congratulate the worthy surgeon and cordially wish she may live long to enjoy many more honors which she well deserves.

We also congratulate the Maryland Academy of Sciences and especially its dean Dr. Francis C. Nicholas for the wise selection. In honoring Surgeon Kate W. Baldwin, they demonstrated a high sense of civic duty.

We are proud of Dr. Baldwin and the Association may well be proud to have such a distinguished woman as its First Vice-President.

***

ITISOMETER MANUAL

The Itisometer Manual, the handy guide for Lessees of the Itisometer, is published and distributed. A copy was mailed to each Lessee of an Itisometer. Each copy has the name of the Lessee and the number of the Itisometer to which it belongs.

The Itisometer Manual is the Sole Property of Spectro-Chrome Institute and can not be issued or circulated by the Lessee. Its value is assessed at One Hundred Dollars and a new copy will not be obtainable without the return of the original.

This volume is the result of years of concentrated labor and represents a fund of information of the most important character in a condensed form.

It will NOT be issued to any but a Lessee of an Itisometer when the apparatus is delivered.

***

CENTRE SECRETARIES PLEASE NOTE

Our Centre Secretaries are generally very conscientious and methodical in their reports, but, a few points may be drawn to their kind attention. They are important.

In reporting, OMIT all those cases in which other forms of healing were used with Spectro-Chrome Therapy. By reporting and publishing such mixed cases, we reach nowhere, because, we are not interested in promoting other systems, our Graduates are not taught by us other than Spectro- Chrome Therapy, our unlicensed practitioners are not encouraged to use any system other than ours and reporting mixed cases gives no clue to the value of Spectro-Chrome Therapy.

Spectro-Chrome Institute, American Association of Spectro-Chrome Therapists and readers of Spectro-Chrome are only looking forward to know about Spectro-Chrome Therapy. Without deprecating or downing our friends who follow other methods, we may say that our sole goal is Spectro-Chrome Therapy which has unquestionably proved to be the most accurate—in fact the ONLY precise mode of relieving the ailments of humanity.

Please therefore, give us by all means Spectro-Chrome Therapy and only those cases where the relief may be attributed straight to Spectro-Chrome Therapy. Let the use of other methods be reported where it may belong AND let Centre Presidents kindly see to it that such mixed business does not come on our platforms for discussion or debate.

Many of our Centres are lagging in their meetings and reports. The attention of the Centre Executive Councils may be drawn to our By-Laws which provide for monthly meetings, monthly reports and the like.

If a cause is worth working for, work for it and do the duty attached to your affiliation.

Spectro-Chrome Therapy needs you and the World needs Spectro-Chrome Therapy.

***

ANNUAL FEES

The Annual Fees become due on January First and those not paying them are automatically suspended at the end of February. Those not paying their Annual Fees by the 31st of March will have their names struck off the roll and all their privileges forfeited.

Please note that our By-Laws are made to be observed and they WILL be observed. Do not blame your Chief Executive Council for dropping your name. The fees MUST be in by March 31st.

Our Association as a Corporate Organization will make no exceptions—and do not forget that your suspension and resignation means the end of your connection with Spectro-Chrome Institute and Spectro-Chrome Therapy. Do not endanger your Equipment and Slides Leases by forgetting to mail your Annual Fees.

Do it now!

***

TO SUBSCRIBERS OF SPECTRO-CHROME

This is the last issue for your present subscription.

Should you wish to continue, please forward $3 promptly, otherwise the next issue will not be mailed.

All subscriptions begin with January and end with December.

Spectro-Chrome is issued to any subscriber, whether our Associate or otherwise.

***

CENTRE ELECTIONS

Election reports of several Centres are not yet in. Kindly send them promptly to avoid confusion in the transmission of respective remittances. Do not sleep on your duties.

***

HARSH JUDGMENT

Jake Ball, a cleaner and dyer, used an American flag to wipe his shop windows. He pleaded ignorance of the law, but, was told by the judge that his ignorance on that point was worse than the desecration. He got a sentence of 250 days in jail. The judge forgot the law of mercy himself. What was the intention of the offender? That should have counted for mitigation; sentence too severe to fit the case. Such harshness on the contrary breeds unrest and contempt of law; the man when out of jail, will treat the flag and what it stands for with less respect and there will be no "ignorance" then, for, he will know what the flag did to him.

***

NEXT CLASS

The next class for instruction in Spectro-Chrome Therapy with experiments and demonstrations will open at Malaga, New Jersey, on Monday, April 12, 1926. All fees and arrangements are as usual.

Come and bring your friends or recommend them to take this unique course.

It is the Only One of its kind in the World.

***

SPECTRO CHROME LANGUAGE

Until now, in deference to Spectro-Chrome Therapists who were prior Graduates of other schools of therapy, we continued to use the language common to others, but, since the birth of Itisometry, Spectro-Chrome Therapy has gone to such higher levels of Therapeutical Precision that a new language has been formulated to express our views in our own ways. This will clarify our situation still further, by creating for us an exceptional field, a different scientific status, which by our originality was always ours.

The medical world has nothing on us. We are not dependent on them, their drugs, their dopes, their poisons, their vaccines, their serums, even their complicated jargonic expressions. We have no guessworks, no differential names of disorders, no hidebound ageold lines of thought or action—Itisometry brought absolute automatic positivity of the most unbelievably accurate character in therapeutics and with its original phraseology also gave birth to a new language. This language—Spectro-Chrome language—will henceforth be adopted in our work and as far as practicable we shall expect our Associates to learn to talk and express their views in it. It is beautiful and simple in build, graceful in cadence, sweet in enunciation and is devoid of incomprehensible terms. Above all, it is entirely original and will surely appeal to our progressive Associates as the one necessity of our work which is now fullfilled through the Itisometer. "Treatment of Dis-Eases" now disappears for ever and "Tonation of Disorders" takes its place—a more appropriate phrase in line with our science. The following is the current vocabulary to which new words will be added as needed or coined and published. If you have suggestions, they will be surly considered; send them. If the suggestion be accepted, your name will be published and the words will be adopted in our language.

AREA—Main Division of the Visible Physical Body.

ASANA—Posture during Itisation or Tonation.

AURICULAR ZERO—The resultant of the auriculation of Intareas A-B-C-D of Area 1 and Intarea E of Area 15.

AURICULATE—To find with the Auriculator.

AURICULATION—The process of Auriculating.

AURICULATOR—That part of the Itisometer which reckons the values of the Spectral Scales from the Itisometer readings.

BACK AURA—The resultant of the auriculation of the Auricular Zero with Intareas A-B-C-D-E-F of Area 17, Intareas A-B of Area 18 and Intarea A of Area 19.

BACKELIM—Back Elimination.

BACKELIMATE—Eliminate through Back Orifice.

BACKEXIT—Back Orifice.

CALORIC CHAMBER—Cylinder for warming the Tanger.

CHRONOMETER—The Spectro-Chrome Timepiece.

CHRONOS—The time computed with Spectro-Chrome Attuned Color Wave System.

FRONT AURA—The resultant of the auriculation of the Auricular Zero with Intarea A of Areas 2-3-4-5-6-7-8-9-10.

FRONTELIM—Front Elimination.

FRONTELIMATE—Eliminate through Front Orifice.

FRONTEXIT—Front Orifice.

HUMAN—Male or Female.

ILLUMINATOR—Lamp.

IMBACKELIM—Impeded Back Elimination.

INTAREA—A Subdivision of an Area.

ITISATION—The process of Itisizing.

ITISEE—The person being Itisized.

ITISIZE—To determine with the Tanger.

ITISOMETER—The apparatus invented by Dinshah P. Ghadiali for determining the precise Predominant Color Wave values in the various Areas of the living body and automatically indicating the Remedial Attuned Color Wave, in accordance with the Science of Spectro-Chrome Therapy.

ITISOMETER NORMAL—The Red Line within the Zero Scale.

ITISOMETRIST—Practitioner of Itisometry.

ITISOMETRY—The science originated by Dinshah P. Ghadiali for determining the precise Predominant Color Wave values in the various Areas of the living body and automatically finding the Remedial Attuned Color Wave, in accordance with the teachings of Spectro-Chrome Therapy.

KARDOSCIL—Pulse beats per minute.

LUNA—Moon Period.

MISTONATE — To project the wrong Attuned Color Wave.

MISTONATION — Projection of the wrong Attuned Color Wave.

NORMALATE—To restore the Radio-Active and Radio-Emanative Equilibrium.

NORMALATION—Restoration of the Radio-Active and Radio-Emanative Equilibrium.

ORELIM — Oral Elimination.

ORELIMATE— Eliminate through mouth.

OVELIM—Ovarian Elimination.

OVELIMATE—Pass Ovarian Elimination.

PRIMOCHROME—The resultant of the auriculation of Intareas A-B-C of Area 1 and Intarea A of Areas 6 and 7.

RATIOTE—To find by the Ratio Dial.

SECOCHROME—The resultant of the auriculation of Intarea D of Area 1, Intarea E of Area 15, Intarea A of Area 8, and Intareas A-B of Area 18.

SLOT—The indentation admitting the Tanger cord when in use.

SPIROSCIL—Respiration per minute.

SUFFERER—Unwell person.

SWEEPINDEX—The moving Needle.

TANGER—That part of the Itisometer which contacts the Itisee.

TANGER LOCKER—The lockable compartment for holding the Tanger cord, Tanger and Itisometer Manual.

TONATE — To project the Remedial Attuned Color Wave.

TONATION — Projection of the Remedial Attuned Color Wave.

TONE—Perfect functioning of all Areas.

—***—

ITISOMETER NUMBER 20 EXPRESSED TO CALIFORNIA

Itisometer Number 20, the first one sent by American Railway Express across the United States was received by the Lessee Dr. A. T. Vance, S.C.T., President of First Los Angeles California Centre. He wired, "Itisometer arrived. Working beautifully. You will win Nobel Prize some day. Success and thanks."

Thanks to you, Doctor Vance, for your kind patience and progressiveness. Thanks to Mr. J. A. McKeon and Mr. W. J. Hoy, Claim Supervisors of the American Railway Express Company who advised us about the proper packing to avoid damage and took the trouble to come all the way from New York twice to inspect the designing of the special container and the like.

Thanks are also due to Mr. E. H. McCarty, the Express Agent of Malaga, N. J. whose careful handling of the situation smoothed the transit arrangements.

The way is now opened for transportation of other apparati and soon the installment instruments also will be delivered by rail. They are now being manufactured.

Attuned Green Wave

(Spectro-Chrome Therapy, Spectro-Chrome Therapists)

THE SATISFACTION OF ITISOMETRY.

By:—WALLACE F. MACNAUGHTON, S.C.T. M.D.,

The world little knows what exactness in the Healing Art is now available with the production of the Itisometer. To Spectro-Chrome Therapists, it is almost unbelievable yet comprehensible. Without a knowledge of the teachings of Spectro-Chrome Therapy, the operation of the Itisometer staggers the imagination, but, is no less fascinating.

The operation of the Itisometer is beautiful and the beauty is shed by the truth it tells. There is no experimentation, speculation or delay. There are no Laboratory Analyses, X-Ray pictures, Therapeutic Tests or Medicinal Remedies necessary, but, trouble anywere in the body is detected unmistakably.

With the Itisometer the trouble is located at once. The variation from normal function or health is measured definitely and exactly and recorded in figures not only for the body as a whole, but, for any obscure area on which information is desired.

An appreciation of such an instrument and the assistance it gives to the practitioner can only be gained by using it. Such an appreciation is enhanced many times if the practitioner is then forced to do without it. The loss seems tremendous at first and it is not until the memory of former days recalls the fact that Spectro-Chrome Therapy has been practised without the Itisometer—after a fashion.

Both local and systemic treatments are indicated by it which no stretch of the imagination or any amount of reasoning would enable one to select as the proper tonation. The response of the sufferer and the changes in readings on subsequent Itisation seemed to prove that the nail had been hit precisely on the head in each case. Another valuable feature is the knowledge it gives of when to persist for a number of days in one color on certain areas, for instance, the Red on 17 and 18 or when to give a tonation of Orange for instance, on Intarea 1C.

As for psychological effect, the sufferers may observe that the needle comes to rest in the Ultra-Zero Scale indicating subnormal condition or in the Infra-Zero Scale indicating supernormal condition of an organ, but, they are not stricken by being labeled with some terrorizing "disease" or subjected to the hazards of diagnosis, which even medical authorities admit to be only 57 per cent correct.

It is important to observe certain technical precautions so that the readings may have uniform conditions. In cold weather, the sufferer should be "acclimated" to room temperature of 75 to 80 degrees Fahrenheit before Itisation is begun and kept as far as possible from radiating too much warmth. To insure this uniform surface temperature, a smock may be used which permits the application of the Tanger to any Intarea with the minimum of surface exposure. It will also give the sufferer a greater sense of privacy and contribute to comfort. If desirable, a sheet may be thrown over shoulders to keep Back Areas from being chilled while Front Areas are being Itisized.

The following findings are submitted as suggestions for future abservations and it will be interesting to note if other Itisometrists corroborate them.

One observation that the writer has made is that a surprising number of cases past middle age showed Ultra-Zero findings on Intareas 17 C, D, E, F, and 18, especially 18 E, some persisting beyond 90 degrees Ultra-Zero for several days. Another interesting point was observed on a case showing high fever and Infra-Zero readings on all Intareas which was to be expected.

The Auriculation of the Front Aura was higher than the Back fourteen out of sixteen Itisations although both sometimes fell within the same Remedial Color Wave. In this case, there was more or less correspondence between the range of the fever and the Remedial Attuned Color Wave indicated.

At past middle age, the energies circulating blood slow down and Area 17 may keep registering in Ultra-Zero. As 18E registers Pancreatic activity, that may also indicate Ultra-Zero as the organ wears. Fevers are Accelerations of energies and the Itisometer therefor records it always on Infra-Zero . Whichever rendering be taken, the indications of the Itisometer, although sometimes apparently mysterious are in absolute harmony with universal laws regulating humans. Follow the Itisometer—IT KNOWS!!!

***

FIRST LOS ANGELES CALIFORNIA CENTRE

Reported By:—THOMAS W. MC ALLISTER, S.C.T., D.O. SECRETARY

There was a regular meeting in October. In the absence on tour of Dr. T. W. McAllister, Jessie H. Lansing, S.C.T. acted as Seretary pro tem, Dr. Henry Gross in the chair.

Unanimously resolved that a letter be written to Col. Dinshah expressing our sympathy and our faith in his honor and character; it was drafted by the President, approved and signed by all present.

Lillian E. Fluke, S.C.T. reported the case of a child of ten years with a temperature of 105 degrees. Turquoise was applied; in one half hour, the fever was all gone.

Dr. Henry Gross gave an interesting talk on the seasonal changes of the human body and its relation to the cosmos. A warning was given of the danger of X-Ray and Radium treatments, many instances being cited of serious injury resulting from such treatments.

A regular meeting was called to order on Wednesday November 4, 1925 at 8 P.M. Dr. Henry Gross presided.

The minutes were read and approved.

A letter was read from Col. Dinshah Ghadiali in answer to the letter forwarded by the Centre expressing appreciation for sympathy and kind heartfelt wishes of all.

The plan for maintenance of reserve Spectro-Chrome Slide Sets was read and acted upon. Two bills were allowed.

Following the transaction of business, the Secretary Dr. T. W. Mc Allister was asked to give a resume of his trip East and more especially of his visit to the Central Office and Research Laboratory at Malaga, New Jersey. Everyone was most interested in the development and future plans.

Meeting dissolved at 10:25 P.M.

***

FIRST PROVIDENCE RHODE ISLAND CENTRE

Reported By:—FREDERICK CROMPTON, S.C.T. SECRETARY.

Following is a report of the meeting of the First Providence Rhode Island Centre, held on Sunday November 1, 1925, at 5:30 P.M.

The meeting was called to order by Dr. Charles C. Hesse, S.C.T., the President. The Secretary's report of the previous meeting was accepted as read.

Nine applications for affiliation were favorably passed upon by the Centre Executive Council.

There was varied discussion on the Centre carrying extra supplies and on refund of dues from the Headquarters to the Centre.

Meeting dissolved at 6:30 P.M.

***

Attuned Turquoise Wave

(Poetry, Romance, Drama, Art)

WOODLAND SPRITES

By:—KASHMIRA DINSHAH, S.C.T.

Here we come, a musical throng,
Hopping, skipping, tripping along.
The velvet turf we lightly tread,
By wish for happy frolic led.

Adown the forest lane we glide,
Assembled fays from far and wide,
From homes in caves or coverts green,
A host by mortal eyes unseen.

From flowers fair and berries we
Draw our silvery harmony,
With magic touch and fairy skill,
And dulcet chords the night air fill.

For drums we use the gooseberry,
With twigs of weeping-willow tree,
The drums are struck, sonorous floats
The joyous thunder of their notes.

The harps are made of pansies pale
Over which stretched are cobwebs frail.
The fifes and flutes, orchestral needs,
Are fashioned from slim hollow reeds.

Bright trumpets formed of lilies white,
Send the echoes flying thru the night.
The woods and hills and leas resound
With this sweet music all around.

O, happy Woodland Sprites are we!
All night we sport so merrily!
But, when day dawns, we slip away,
Till darkness comes, which is our day.

SPECTRO-CHROME

JAN.-FEB.-MARCH 1927

R-O-YELLOW 1927

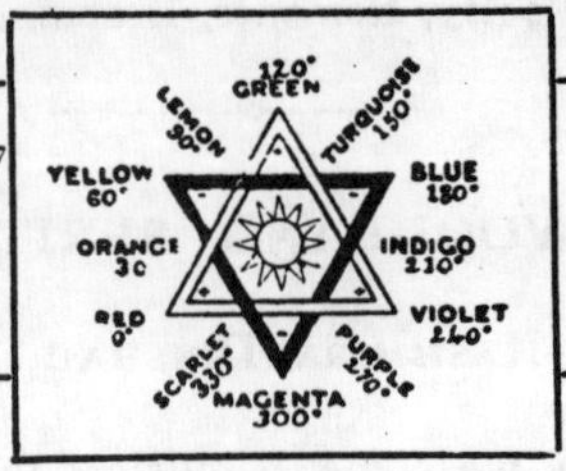

VOLUME 5

NUMBERS 1-2-3

Devoted to: Spectro-Chrome Therapy **Dedicated to: Service of Humanity**

CONTENTS

The World Through the Eyes of Dinshah—

SIXTH ANNUAL CONVENTION NUMBER

Editor
Dinshah P. Ghadiali

In course of life you reach a time,
When Doctor, Lawyer, Priest alike,
Does fail to give the sought relief,
To troubled mind or burdened heart—
Then come to me;
I help you to help yourself.
SPECTRO-CHROME.

Subscription:—January to December (12 Months) $3.00
All Subscriptions Taken For Full Year Only And Expire With The December Issue
Single Copy 25 Cents. Back Numbers Available From June 1922

Attuned Red Wave

(The World Through the Eyes of Dinshah)

SIXTH ANNUAL CONVENTION OF AMERICAN ASSOCIATION OF SPECTRO-CHROME THERAPISTS

A RECORD OF PROGRESS AGAINST DIFFICULTIES

The Sixth Convention of American Association of Spectro-Chrome Therapists assembled at Hotel Lorraine, Philadelphia, Pa., on November 28, 1926, at Yellow 00 White. The Chelsea Room was well filled with the Associates, long before the opening time and the meeting promised a success even better than in former years.

On the platform, next to the President Colonel Dinshah P. Ghadiali, were seated the two Vice-Presidents, Kate W. Baldwin, S. C. T., M. D., F. A. C. S. and Martha J. Peebles, S. C. T., M. D. On one side of the platform was an Itisometer in action, while another had some apparati hidden under covers. Curiosity was demonstrated by all as to what was in store, but the cloth covers controlled their getting acquaintance with the contents.

The following First Notice calling the Convention, was sent by mail to all the Associates of the organization:

October 23, 1926.
Chronos—Orange 49 White

Dear Friend:

1. In accordance with Article 15 of our By-Laws, please be advised that our Sixth Annual Convention will be held in Philadelphia, Pennsylvania, at the Hotel Lorraine, in the Chelsea Room, at 10:00 A. M., Sunday 28th, Monday 29th and Tuesday 30th of November, 1926.

2. An attractive and entertaining educational program will be presented full particulars of which will be mailed later. There will be public sessions on Monday afternoon and evening.

3. Your presence is cordially invited.

4. Should you have any complaints for consideration, please send them to the Chief Executive Council, in writing, duly signed.

5. Should you intend to read special papers or have any subject about which you desire to address the Convention, kindly mail in time for the program.

6. Centres will appoint ONE Regular and One Alternate Delegate to the Convention, expenses being paid by the Centre, subject to Article 47.

7. Secretaries of Centres are requested to send their Annual Reports showing the number of meetings held, average Associates present and the balance sheet of the year enumerating the receipts and disbursements.

8. Your co-operation will be warmly appreciated.

9. SPECTRO-CHROME, our monthly, was discontinued by order of the Chief Executive Council pending the Appeal on the Portland, Oregon case. It will be resumed after the Convention.

10. NOTE—No admission to the Convention without the presentation of the Association Card; many vital

and amazing developments will be discussed.

By order of

CHIEF EXECUTIVE COUNCIL

Dinshah,

DPG|KD President.

Later, the complete program was mailed for better information, so the Associates might come prepared for the various sessions:—

November 12, 1926.
Chronos: Green 55 Black

PROGRAM OF
SIXTH ANNUAL CONVENTION

Dear Friend:

1. The following attractive and entertaining educational program will be presented at our Sixth Annual Convention, opening at 10 A. M. on Sunday, November 28, 1926, in the Chelsea Room, Hotel Lorraine, Philadelphia, Pennsylvania:—

Sunday, November 28, 1926

10 A. M. to 12M.—Registration Annual Reports of the Secretary and Treasurer; Centre Reports; Suggestions and Sundries. Sessions open to Associates only.

2 P. M. to 5 P. M.—Elucidation by Dinshah of "The Science of Itisometry." Session open to Associates only.

8 P. M.—Address by President Dinshah, on the year's amazing events. Elucidation of "The Science of Sympathometry" with exhibition of the perfected Sympathometer. Session open to Associates only.

Monday, November 29, 1926.

10 A. M. to 12 M.—Exchange of views among Associates. Questions and answers. Session open to Associates only.

2 P. M. to 5 P. M.—Case Reports and general discussion. Session open to Strangers introduced by our Associates.

8 P. M.—Address by our Vice-Presidents: Kate W. Baldwin and Martha J. Peebles. Session open to the public.

Tuesday, November 30, 1926

10 A. M. to 12 M.—Election of Trustees; Appointments and Installations; unfinished business. Changes in the By-Laws. Open to Associates only.

2 P. M.—Essays, Papers and Debates. Open to Associates only.

2. Associates are cordially requested to be present at the various sessions and help make this Convention a success as usual. No admission without presentation of Association Card and Registration.

By Order of

CHIEF EXECUTIVE COUNCIL,
Dinshah, President.

Notwithstanding the notification, many Associates came without their Association Card and were denied admission. Duplicates were issued to those who asked for them, on payment of one dollar; this was in accordance with the By-Laws and taught a lesson to those who took the By-Laws till then as a "mere matter of form."

First Day Morning Session

At Yellow 10 White, President Dinshah's gavel called the Convention to order.

The Minutes of the Extraordinray Meeting which was held to take the place of the Fifth Annual Convention on December 27, 1925, were presented by the Secretary Kashmira Dinshah, S. C. T., as reported in SPECTRO-CHROME, issue of Scarlet, 1925. On proper motion duly seconded, the Minutes were unanimously accepted as recorded.

The Treasurer Dinshah P. Ghadiali presented the Annual Report, as of November 1, 1926. It showed an increase of $594.62 over that of last year, indicating the economical and conservative manner in which the disbursements were made. On motion of John A. Cohalan, S. C. T., D. O., seconded by Frank E. Champion, S. C. T., the Treasurer's Report was accepted by unanimous vote.

The Secretary Kashmira Dinshah presented her Annual Report. There were thirteen Centres of the Association, located at Baltimore Md., Boston Mass., Los Angeles Calif., Malaga N. J., New York N. Y., Philadelphia Pa., Portland Ore., Providence R. I., San Francisco Calif. and Seattle Wash. Thus the activities of the Association had covered the United States from the Atlantic to the Pacific coast, reflecting the success of the cause and the patient labors of the workers.

There were no deaths, violations or resignations among the Associates. Although the Association went through trying times, an increase of 4 Associates was reported. The Association had a bright outlook and satisfactory progress in every direction. On motion of Bessie A. Hasenau,S. C. T., seconded by Margaret Morris, S. C. T., the Annual Report of the Secretary was voted upon and unanimously passed.

The following Associates were present as Delegates of the respective Centres: Baltimore Md., Frank E. Kaufman; Cleveland O., Bessie A. Hasenau; Detroit Mich., Frank A. Kraft; Malaga N. J., Dinshah P. Ghadiali; New York N. Y., Eleanor B. Holwick; Providence R. I., Oscar C. Hesse.

Later, the following Delegates presented their credentials and were accepted as Delegates: Boston Mass., G. Ernest Spear; San Francico Cal., Isabelle Munson.

The Annual Reports from various Centres were then read.

First Cleveland Ohio Centre was reported as making satisfactory progress by the Centre Secretary Margaret A. Morris. They had 12 meetings during the past year at 10543 Euclid Avenue, Cleveland, Ohio. The Centre Treasurer Marguerite F. Sait-La Roe reported a balance of $192.90; they ran the Centre without expense, each one doing their duty with pleasure.

First Detroit Michigan Centre was reported by the Centre Secretary E. Norman Pearson. They had eight meetings during the year. Much difficulty was experienced in finding a night suitable to all Associates. Failure to do this, in spite of their best efforts, kept a number from attending the meetings, who would otherwise have been present. The meetings were spent mainly in the discussion of cases of particular interest—a general interchange of experiences—and a few papers dealing specifically with the particular therapeutic values of Attuned Color Waves. They had a balance of $181.50, having spent only $16.50 during the twelve months. Their work was progressive and satisfactory.

First Lawrence Massachusetts Centre reported through the Centre Secretary Ada E. Hatch. Their meetings were held the second Sunday of each month. They had 8 meetings during the year. Their report of funds by the Centre Treasurer Samuel L. Gants, showed a balance of $36.90 with no disbursements.

First Malaga New Jersey Centre being the Centre created under the By-Laws for the workers of the Association had nothing special to report except that they were working without compensation for the good of the Cause. They were glad to do all in their power to promote the Association in its aim of serving suffering humanity.

First New York New York Centre reported through Nissin S. Hanoka, the Centre Secretary. They had 9 meetings during the year. Their balance was $246.00 and no expenses. Their Centre had Associates in distant States hence their attendance was less than what might be expected.

First Philadelphia Pennsylvania Centre was reported at length by the Centre Secretary Lettie Allen Ward. This Centre had the largest roll of Associates. Their meetings were largely attended. Their Associates were assigned from very far off States like Florida and could not attend meetings each month, yet, the showing was very satisfactory. They had a balance of $789.56, having disbursed during the year only $67.30. The Centre President was Kate W. Baldwin, S. C. T., M. D., F. A. C. S., our First Vice-President, who called each meeting to order not later than Red 15 Black. They finished all old and new business usually in half an hour and then went into a Scientific Session. On an average, at least 18 or 19 of the Associates present reported cases. Questions were asked and answered. Every meeting grew more interesting. Many of the results obtained by Spectro-Chrome Therapy seemed almost magical and at times the interest was so great it seemed almost impossible to bring the meeting to a close. They usually closed at Yellow 30 Black.

First Providence Rhode Island Centre was reported by Charles F. Lallouette. They had seven meetings during the year, at 1192 Broad Street, Central Falls, R. I. They spent during the year only $3.00; they paid $60 for the expenses of their Delegate to the Convention; they had a balance of $36.17, as reported by Grace E. Lallouette, Centre Treasurer.

First Los Angeles California Centre was well reported by Thomas W. McAllister, the Centre Secretary. They had spent during the year $45.78 and had a balance of $369.87. They would have sent a Delegate, but, no one seemed to be able to leave at the time, much as they expressed an earnest desire to be present. They would doubtless miss much that could not be printed in the Official Organ. They were much interested in the new developments and had expressed their appreciation of the way "our Chief" was keeping up and handling this trial and felt he would be fully vindicated, as it surely was a dastardly attempt to destroy a system of healing which aided suffering humanity. Truth would survive and the day was coming when this system would by its own merit destroy the monster which would ruthlessly trample all in its

path to gain its own ends. They wished the Convention well; they expressed their whole-hearted support; they wished every success and that the Convention might be the greatest on record.

First San Francisco California Centre report was submitted by Isabelle Munson. They held seven regular meetings during the year and one special meeting for the solution of the situation of spare Slide Sets. They expended $42.00 during the year and had a balance of $250.15, out of which they had paid $250.00 as the expenses of their Delegate, leaving in the Centre Treasury a balance of 15 cents.

First Portland Oregon Centre and First Seattle Washington Centre were not heard from at the Convention.

Moved by Sigmond Rosenthal and seconded by Cleo V. Butner that all the reports be accepted and filed. Unanimously passed.

John A. Cohalan stated that a Chamber should be provided for poor people to tonate their disorders. Dinshah stated that the means for such work were not available. He spoke at length how in the Philadelphia Woman's Hospital, our esteemed and learned First Vice-President, Surgeon Kate W. Baldwin had introduced Spectro-Chrome Therapy and used it for nearly five years with the most marked success even in cases that otherwise would have been pronounced "incurable". He then told the Convention that recently, owing to the enormous difficulties placed in her path by the hospital authorities under pressure from the Medical Trust, the noble surgeon who had given over 23 years of her valuable life to that hospital ***without compensation,*** was compelled to resign and remove her Spectro-Chrome Equipments! A wave of surprise and indignation swept through the meeting, the Associates being utterly unprepared for such an announcement.

Dr. Martha J. Peebles, Second Vice-President, said that the severance of connection of Dr. Kate W. Baldwin from the Woman's Hospital of Philadelphia was a set-back to progress. Color Therapy in one form or another was being mentioned in the newspapers daily and the value of light was more and more being recognized. The public were awakening to the new methods.

Dinshah said that a Resolution should be formulated voicing our views about the action of the Board of Managers of the Philadelphia Woman's Hospital. That was due to express our regards for the treatment tendered to our most worthy and beloved friend, Dr. Kate W. Baldwin. He said it was necessary however before anything be done, to hear what the doctor had to say.

Amidst a round of applause Dr. Kate W. Baldwin took the floor. In her usual calm and quiet manner, she spoke about her first appointment as Senior Surgeon of the Philadelphia Woman's Hospital nearly 23 years ago. She had given—gladly given—the best part of her life to the noble work of relieving the sufferings of humanity and had worked in the highest interests of progressive medicine. During her surgical experience she had felt the need of better methods to help the sick and when she found in Spectro-Chrome Therapy a superior system of therapeutics she introduced it in the hospital. Her results were gratifying, but she met with much opposition. When in January of 1924, the Journal of the American Medical Association wrote against the Originator of Spectro-Chrome Therapy, the article mentioned her name also in a derogatory manner. Colonel Dinshah challenged them, but, they never replied. She could understand how the A. M. A. might feel about Colonel Dinshah, but, they had no right to write like that about her who was one of their Fellows. It was not ethical. That article produced much trouble for her in the hospital work. She was insulted and put to the third degree. Obstructions of all kinds were put in her way to discourage her from Spectro-Chrome. She was told not to use Spectro-Chrome, because, the American Medical Association had not passed on its merits and approved it, but, the A. M. A. never investigated! The Board of Managers of the hospital appointed a special committee to look into the matter; they investigated carefully for two months and finally a report was made which was favorable to Spectro-Chrome Therapy. They saw nothing against it; it was only a prejudice against individuals and not methods. She was proud of Spectro-Chrome. She preferred an open fight, but Spectro-Chrome did not get any. The hospital used white light on the body; that showed they had faith in the power of light; actions spoke louder than words.

After that Special Committee's report, the Board of Managers graciously granted her more space for the use of Spectro-Chrome. She was laid up last winter by an accident; the staff never sent a patient for Spectro-Chrome; they wanted surgery. If a person were sent to her for amputation of a leg and she could save it by Spectro-Chrome, she would not take off the leg. There was continued friction. Matters were made so unpleasant for her she wanted to resign; they asked for her resignation. The clinic in Spectro-Chrome went to a smash; the people were heartbroken. She resigned.

The President said that the treatment given to Surgeon Baldwin was not an insult to her, but, it was an outrage against the Association. They were in all honor bound to stand by their noble Vice-President and uphold the dignity of Spectro-Chrome. He was sorry to note that so competent and devoted a surgeon who had lived her life for surgery should be given such unfair treatment by the Board of Managers of a great hospital because she did good to the sick by Spectro-Chrome Therapy. He wanted to express the views of the Association to the hospital authorities and asked for a motion.

Moved by John A. Cohlahan and seconded by Alice M. Norton that a committee of five be apointed by the President to formulate the necessary proposition. Dr. Kate W. Baldwin expressed her displeasure at such a move, but, the motion was carried unanimously without her vote.

The President appointed a committee composed of John A. Cohalan, Alice M. Norton, G. Ernest Spear, Sigmond Rosenthal and Frank E. Kaufman to formulate the proposition and submit it for approval and vote of the Association.

A resolution was presented by the First Detroit Michigan Centre asking Dinshah to expedite the Slideless Spectro-Chrome Equipment so imitators of the system may not foist on the public fake slides; the Centre stated it was necessary to have more powerful equipments because others were

stealing a march on them in many cases by using arc lamps.

Dinshah said that the evolution of his Slideless Spectro-Chrome was only a question of money; he could market it, but, owing to the peculiar troubles he was undergoing at present he preferred to wait for better times. It would be an expensive apparatus, but, it would be a marvelous piece of workmanship, as with it, in combination with the Itisometer and Sympathometer, 72 waves could be used for even greater Therapeutical Precision; in place of the present 30 degree Attuned Color Waves, the Slideless Spectro-Chrome would have the Attuned Color Waves in 5 degree differences; the idea was unique, but the heavy expenses attached to the process of standardization and construction of such an apparatus were too large for the present small demand for him to bear; he would however soon do so and when that would be marketed and leased to the Associates, he would withdraw his other Equipments by paying the full price less depreciation. Meanwhile the public needed education about the Science of Spectro-Chrome Therapy.

Frank A. Kraft remarked about the conditions obtaining in Detroit Mich., where various imitations were foisted on the suffering public, as Color Therapy.

Dinshah advised the Associates to educate the public to understand the difference between Color Therapy of ancient times and Attuned Color Waves. The former was nothing but a hit-and-miss system like medicine, the latter was absolutely precise and accurate in methods without guesswork. Attuned Color Waves were precisely graded colors for definite results; no system other than Spectro-Chrome could do what is now recognized as being done by Spectro-Chrome.

Dr. Sigmond Rosenthal of Aberdeen, S. D., said the doctors declared cases to be "incurable"; incurable meant *they* could do nothing; Spectro-Chrome Therapy was used by him with good results.

Frank A. Kraft reported a case where the sufferer could eat only ground carrots; Spectro-Chrome Therapy restored him to normalcy.

Dr. John A. Cohalan said he had great success in severe cases of Eczema when tonated with Spectro-Chrome; it was the best in the world.

Dr. Nissin S. Hanoka spoke bitterly against imitators of Spectro-Chrome; he cited instances of persons endeavoring to use spurious imitations and had no favorable results. He spoke openly giving names of fakers.

The morning sessions adjourned at Green 00 White.

First Day Afternoon Session

The afternoon session began at Blue 00 Black. The attendance was largely increased and being Sunday, many otherwise busy Spectro-Chrome Therapists came from distant cities to demonstrate their interest in the activities. The Delegate of First San Francisco California Centre, Isabelle Munson, was introduced at this session and received a cheering ovation. She expressed her joy at being at the Convention and stated that the San Francisco Associates were all anxious to hear about the successful issue of the case pending against Dinshah in the Circuit Court of Appeals there.

Dinshah occupied the floor the whole afternoon, elucidating and demonstrating the value of the Science of Itisometry. An Itisometer was on the platform and he gave clear instructions as to Itisation, Auriculation and Tonation. Explanations were given in regard to avoiding certain methods when itisizing, stress being laid on the use of the proper technique. The Itisometer was to be used as a scientific finder and not as a whimsical freak; it was to be used for finding the Attuned Color Wave values in the itisized Intareas and was not to be mixed with the woeful humbug of "Medical Diagnosis."

When complaints were made about the Tanger, he said the Tanger was an exceedingly delicate part of the Itisometer and like any delicate piece of scientific apparatus should be handled carefully and used intelligently. Any Tanger that was returned to him as being out of order, he had replaced or repaired without charge, except in two cases where the user was openly at fault. One Lessee had stuck on the platinum wire of the Tanger a piece of shellac to hold the broken wire! Another sent the Tanger to a jeweler and in "repairing" it, the jeweler shortened the arm of the platinum, by changing its resistance and ruining the bridging capacity! In both cases, the Tangers being rendered worthless, the Lessees had to pay the price for replacement. However he had charged them only half price warning them to refrain from such action in the future.

Any time a Tanger gave trouble, it should be returned to Spectro-Chrome Institute for recalibration; it was specialistic work and not work for any electrician, much less a jeweler. If a Tanger "repaired" wrongly were used on a sufferer, the Itisometer readings would be utterly vitiated. Such a state of affairs could not be tolerated in a work of such preciseness as Itisometry. Practitioners should be careful to follow the instructions rigidly for accurate results. The Itisometer told no lies and its indications should be taken in preference to human-formed opinions. Especially, care should be taken not to miss the readings of Itisometry with pre-conceived notions according to other systems. Spectro-Chrome stood alone in its own field and needed no auxiliaries to support it.

Many instructions in the peculiar technique to be used in Itisation were given and a lively interest was exhibited. The afternoon session of the first day adjourned at Purple 00 Black.

First Day Evening Session

The first day evening session convened at Red 00 Black. There was a much larger attendance and a greater interest because of the attractiveness of the evening's program.

The President described the difficulties he had in the year owing to the labors involved in exposing the diabolic perjury in his case. He had no doubt the truth would be victorious eventually and the framers brought to light. His faith was unwavering and his work was progressing in spite of all opposition. Spectro-Chrome could not be killed; its glorious mission of healing the sick would go on for ever.

He stated how day by day it had become apparent to him that medical graduates and lay graduates did not work in harmony, that all his labors teaching the medical people did not fructify as they should, because of their domination by the American Medical Association. He gave an account of his two visits to Atlanta Georgia, where he had hopes and promises of forming a good class in Spectro-Chrome Therapy. When he finally went there and delivered public lectures, the County Medical Society there threatened one of our prominent Medical Associates there with expulsion from the Society because he had one of our lay practitioners working with him. They charged him officially and he was told plainly that unless they passed on Spectro-Chrome, he should not use it. This M. D. did not stand up with enough boldness as our other medical Associates and did not fight down the tyrannical edict of the Medical Trust. As a result the Class about to form received a set-back and many offered no cash and all notes to take the work as an experiment! Dinshah gave them an ultimatum either to take the work by paying the tuition fee or to forego it. He said unless the class was ready to open at 8, he would not care to open it later, as the expenses of another two weeks' stay would fall on his head, without proportionate results. Finally Dinshah dismissed them with thanks and requested them to take the work later in Malaga, N. J. He packed all the experimental apparati during the night and returned with a loss of about $2,000.00.

After such experiences, Dinshah thought it was not fair to waste his time, energy and resources on training Medical Doctors, especially those whose affiliations were with the American Medical Association. He was seriously thinking of putting a clause in the tuition application that Spectro-Chrome Institute would not teach anyone connected with the Medical Trust. Dr. Sigmond Rosenthal proposed and Dr. Rebecca Berenson seconded that Spectro-Chrome Institute be requested to refuse to teach Spectro-Chrome Therapy to any persons connected with the A. M. A. Dr. Martha J. Peebles opposed the motion on the ground that the A. M. A. had many conscientious members who did not approve of some of the policies and methods of their organization. Such members were the ones who would be likely prospects for Spectro-Chrome Therapy. She expressed her disapproval in emphatic terms saying such a condition would debar many good doctors from joining us and even if there were some bad ones we should note that we might have out of the good ones a Dr. Kate W. Baldwin. Dinshah said she was only one unique specimen of such goodness. A Medical Doctor from Massachusetts said it would not be right to get heated. Dr. Kate W. Baldwin proposed and Dr. John A. Cohalan seconded that the motion be tabled. Unanimously passed.

The President spoke about the necessity of giving Post-Graduate Courses. He had thought of giving them before, but, the Graduates were not ready. He would formulate the rules under which such courses would be given in the near future.

Then after a short speech in which he gave them an account of further researches during the year, he asked Vice-Presidents Kate W. Baldwin and Martha J. Peebles as well as Eleanor B. Holwick to perform the ceremony of opening the cover of the perfected Sympathometer. The new apparatus had a polished wood case and was only about 18 inches square. It had a Chronometer dial, a Month Date and Week-Day Dial, a Sunrise Dial, a Sunrise Calibrator Dial and two Nostril Dials, in all 7 Dials. The apparatus that would be marketed would have 12 dials in separate circles It would be leased to the Associates for $150.00 provided they ordered in advance; afterwards the leasing price would be $250.00. A large number of orders was promptly placed by Associates and its demonstration and working was received with applause and enthusiasm. The beautiful design and neatness of appearance of the Sympathometer drew the crowd to it and congratulations were offered the Originator—the Science of Sympathometry was an accomplished fact.

Then another unexpected surprise was sprung by Dinshah before closing. He said it had nothing to do with Spectro-Chrome Therapy, but as it was one of his latest inventions, the Associates might permit him to show it to the Convention. He pulled off the cover from what looked like a lantern and exhibited the Dinshah Perfect Headlight. In driving down a mountain at 4 in the morning, from North Carolina, truck headlights annoyed him so much that he revolved in his mind how he could circumvent that trouble. In a few minutes he had the invention in which the beam of light is controlled by electricity by the automobile operator producing a headlight of a no-glare fool-proof character. He intended to market it; the proceeds would be useful in furthering the aims of Spectro-Chrome Therapy.

The first day evening session adjourned at nearly midnight. It was a success.

Second Day Morning Session

The Monday morning session opened at Yellow 12 White. A letter of good wishes from Robert Malbon, S. C. T., of Denver Colorado was read. Frank A. Kaufman, S. C. T., President of First Baltimore Maryland Centre, complained of his Associates not being present to get a quorum; some lacked interest. Rebecca Hanoka, S C. T., said that many Associates were not making a living on Spectro-Chrome Therapy. Dr. Margaret Morris said they should put in office those who lacked interest. Fred Kraft, S. C. T said he had no trouble in having sufferers. Dinshah spoke about the financial aspects of Spectro-Chrome Therapy; the Associates had a ground for complaint, because the sufferer was relieved too quickly to suit the practitioner! But, the one who was benefitted brought in others. Dr. Nissin S. Hanoka said it was necessary to reach the masses. Dr. Martha J. Peebles said personal equation had much to do with maintaining the goodwill of sufferers; Spectro-Chrome Therapy had made a wonderful showing; set a sufferer right and one has an advertiser.

The tabled matter of not training members of the American Medical Association again came up for consideration. Dr. Sigmond Rosenthal spoke against taking them; Dr. Martha J. Peebles wanted to make the position clear; Dr. Nissin S. Hanoka thought Dr. Martha J. Peebles was very idealistic; Dr. Kate W. Baldwin did not agree with the proposition, because all Medical Doctors did not belong to the A. M. A.; she advocated the maintenance of a high standard of training in Spectro-Chrome Therapy.

Cleo V. Butner, S. C. T. of Atlanta Georgia spoke about her bitter experiences with the Medical Doctors there; they knew the inroads our system was making on their pockets; they controlled the license bureaus and wanted her to pay $150.00 for a license to practise Spectro-Chrome Therapy; she did not pay the exorbitant demand; she did not need any license; her opinion of the Medical

Doctors was not complimentary. No action was definitely taken, Dinshah stating he was not yet ready for such a boycott, but, he felt the time was not very far when the American Association of Spectro-Chrome Therapists might become an exclusive body of only lay practitioners, excepting of course, those who had previously joined. He believed it would be a wise move when the time came, because he had more trouble in training a Medical Doctor owing to the radical methods used in Spectro-Chrome Therapy and the consequent tedious unlearning processes.

Isabelle Munson, S. C. T. of San Francisco California presented the views of her Centre in asking for spare Slide Sets to be kept by the Centres. Dinshah said that he had endeavored to do that by having the Centres sign a strict agreement for such purpose, but, it had produced nothing but bitterness; he regretted there was an element of risk in giving out the possession of Slide Sets; the Slide Sets were the all-in-all in Spectro-Chrome Therapy; the life of sufferers absolutely depended on the correctness of the Slides; Spectro-Chrome Institute owed a public responsibility; its Integrity Policies would be maintained at all costs; the public must have honest services or nothing; the Pacific Coast Centres had volumes of discussion over that subject of Reserve Slide Sets; they denounced Dinshah as a tyrant, a dictator, a czar; they called him severe and unjust; he saw no reason why to please a few Associates he should let go his control on his own inventions and open a door to potential dishonesty and sacrifice public safety.

He requested Isabelle Munson to convey the message to the First San Francisco California Centre that Spectro-Chrome Institute flatly declined to send out Reserve Slides on any consideration and refused to allow any Associate to handle them for the Institute. In the present condition of the world, when we were fighting the biggest battle of life against the Medical Octopus, allowing the bungling of the Slide Sets would mean destruction and disruption and it was necessary for him to hold on even at the risk of being called a czar—the Institute and the Association needed a strong arm to hold the reins firmly.

The second day morning session adjourned at Green 00 White.

Second Day Afternoon Session

There was a large attendance of Strangers introduced by Associates at the afternoon meeting which opened punctually at Blue 00 Black. This session was reserved for case reports and general discussion including Class Work.

The President thanked the strangers for their kind interest and presence. Frank E. Kaufman, S. C. T., introduced a boy who had suffered from paralysis. He took the case and started the tonation; Scarlet on areas 12-13-20-21 gave no response; Systemic Orange returned the appetite; he had lacked it a long time; in 7 tonations the boy was able to flex the knee; in 18 tonations he walked the length of the room; the boy walked on the platform and the Convention gave Kaufman a round of applause. Dr. Kate W. Baldwin declared the case was a wonderful result. Theodore Franks, a Stranger from Vineland, N. J. asked questions about the boy and received satisfactory answers.

The President conferred on Frank E. Kaufman, S. C. T., the Silver Medal of Spectro-Chrome Institute for his meritorious services to the Cause. He praised Kaufman as a man of convictions who had upheld our banner in Maryland. The audience vociferously cheered.

Another case of the same practitioner was presented. A Mr. P——— who continually had used rectal suppositories; his backelims became normal in only 3 tonations; 42 tonations totally normalated him.

Surgeon Kate W. Baldwin, S. C. T., spoke about the goitre of Mrs. Alice T——— an Associate who had such pressure from it on the neck that she could not sleep on her back for nearly 30 years. Spectro-Chrome relieved her and the enlargement which was as big as her fist was reduced so it was hardly noticeable. The Itisometer was better than any "diagnosis," in her opinion; it gave accurate results. She had wonderful success in the tonation of Cataract; however, the sufferer should not leave off early. A man who had kidney and bladder trouble was relieved by Spectro-Chrome. Indigestion disappeared under Orange. On the 11th of February of last year she fell on ice and was on crutches for 4 weeks; she had three accidents in a year; she tumbled over a rolling chair and had a sore shin; in an automobile accident she broke her nose; she was helped in each case by Spectro-Chrome.

She gave a vivid account of a remarkable case in which where the surgeons failed to do anything, Spectro-Chrome performed a veritable miracle. A Mr. G. I. S——— suffered from what the Medical Doctors diagnosed as Mucous Colitis. They (mis)treated him with all kinds of measures until finally they wished to clean his inside with a pump at the exact spot where the trouble was supposed to be! So they made him chew India Rubber hosepipe gradually until in 2 weeks they succeeded in pushing in his alimentary tract about 30 feet of rubber!!! The sufferer was meanwhile writhing in agony; on the fifteenth day he was as gogd as dead and wanted to die; he could not swallow any more pipe; Mrs. S—— had been to one of our worthy Associates, Dr. Alice M. Norton, a Dental Surgeon. She was advised to consult Dr. Kate W. Baldwin. Dr. Baldwin took the situation in hand, pulled out the pipe and placed Mr. S—— under Spectro-Chrome. The man got a new lease on life and was attending his business as usual; he was present at the Convention.

As soon as the President heard this, he requested Surgeon Baldwin to stop for a while and said that although it was not fair to identify sufferers or humiliate them, yet, if the said person were present and if he cared to divulge himself it would be appreciated if he came forward and boldly announced himself. Mr. George I. Sullivan came out promptly with a smiling face and extolled the skill of Dr. Baldwin; he said he would have died had it not been for Spectro-Chrome; his condition was utterly hopeless until Spectro-Chrome came to the rescue. Loud cheers greeted him and he was cordially thanked for his endorsement of Spectro-Chrome.

Dr. Sigmond Rosenthal, S. C. T., of Aberdeen, South Dakota reported a striking incident of a boy Arnold B—— of Orient, S. D., who was pronounced by diagnosticians as suffering from Lymphatic Sarcoma; the father brought him to Dr. Sigmond Rosenthal who restored the boy to health with Spectro-Chrome in about 3 months.

Dinshah expatiated on the aims of the Association; he was for full liberty of thought and against medical interference; none had any right to dictate what system of healing one should use. The aim of the Association was Therapeutical Emancipation. The great Abraham Lincoln abolished slavery, but, we Americans were in the meshes of a worse bureaucratic autocracy under the thumb of the Medical Trust. The dogmatic control of the Medical Doctors must be broken. That was the aim of the Association.

Dr. Emma H. Gabriel spoke of a woman's case. She was pronounced to be suffering from Gastric Ulcer. She was itisized and the tonation was constantly checked with the Itisometer. She recovered.

Richard Sandbach described a case of a man who suffered from indigestion and could not work; he was incapacitated for two months; tonated with Spectro-Chrome, he drove 35 fence posts in a day. Another man could not work for a year; his calves were cold as stone; under Spectro-Chrome, he soon recovered and walked 1 1-2 miles.

Dinshah was questioned about the methods of tonating Cataracts. He said success had been marked in many such cases. Anything that could be produced in any part of the body from the blood, could be redissolved by reversing the process. Dr. Kate W. Baldwin did it; there was no reason why others should not do it. However, such a case should be taken in hand during the period of formation of the deposit for quick results; later, it might become unmanageable.

Clara E. Sandbach, S. C. T. praised Spectro-Chrome Therapy. She had Pleurisy and the Medical Doctors gave her no hopes. Spectro-Chrome changed her life and made her happy. She had success with Spectro-Chrome. A case where a person suffering from Colitis was passing blood and mucous and was ordered to go to Denver, Colorado, was relieved by her. She reported a case where the Medical Doctors had failed to diagnose whether a woman had a tumor, cancer or pregnancy and wanted to operate on her to determine the exact condition! The husband opposed the operation and wanted Attuned Color Waves. Twenty eight tonations caused the growth to disapear. She also restored to health a woman of 58 who had stomach troubles. In disorders of children, Spectro-Chrome was effective.

Dr. Nissin S. Hanoka addressed the audience next. The more he used Spectro-Chrome Therapy the more he was sorry he did not learn it quicker. He had achieved success with the system. A girl was bleeding for three days and the Medical Doctors tried many things. He succeeded with Spectro-Chrome. A man of 42 was impotent in Area 11 and believed life not worth living. He came for Spectro-Chrome for a whole year. His wife became pregnant. They were blessing Spectro-Chrome. Another man, a Spaniard 67 years old had suffered Chronic Bronchitis and stomach disorders. This Spaniard used imported cathartics. In ten tonations Spectro-Chrome demonstrated its power. He took 10 more. In 42 tonations he was a new man and went to Porto Rico. Another man had hemorrhage of the rectum. Only 3 tonations set him right.

Dinshah gave a short discourse on the various physical and emotional aspects as well as causes of sexual impotency. This discourse was listened to with attention by the Associates and the Strangers.

Isabelle Munson, S. C. T. of San Francisco was the next to speak. She reported of a woman who had ovarian trouble according to the Medical Doctor in attendance. He wished to operate on her. Spectro-Chrome saved her from the knife. A woman with hardened eyeballs was benefitted by Turquoise. Her sister, Miss Munson said, was relieved of her disorders by Magenta.

Anna Franks, S. C. T., of Vineland, N. J. was then introduced by Dinshah as a lay woman whose three sons were also Graduates of Spectro-Chrome Institute. The whole family was so enthused over their success with Spectro-Chrome that the father of the family, Theodore Franks, would also take the course next summer. She had never before spoken in public. It was her first speech. She suffered about 15 years from Rheumatism and was barely able to appear for enrollment in the Class, because of that disability. However, as she took the work and changed to the Spectro-Chrome vegetarian diet, the Rheumatism left her and did not re-appear. She tonated a woman who had suffered with Bronchial Asthma for 35 years and the Medical Doctors had no hopes of ameliorating her condition. She could not sleep on her back and suffered intensely. When Mrs. Franks graduated and started practice, among the first ones who went to her was this sufferer. Five tonations relieved her so she could sleep on her back. Her husband seeing her asleep in such position thought her dead and was surprised to see her resting comfortably. The case was the talk of Vineland, N. J.

Theodore Franks of Vineland, N. J. the husband of Anna Franks and a Stranger was asked to speak. He responded by thanking Spectro-Chrome Institute for the knowledge imparted to his wife and three sons and said the results of Spectro-Chrome were unbelievably wonderful. He was sorry he did not take the Course last summer and would take it at the earliest opportunity.

Dr. Oscar Hesse gave some of his experiences since his graduation in 1923. A woman had Chronic Imbackelim. She did not move for 8 days at a time. Spectro-Chrome produced the desired result. A man of 60 had impaired eyesight. All glasses had been tried and changed. Three tonations with Attuned Green restored the sight. A man had inflammation in Area 18, as shown by the Itisometer. He came for six weeks and was restored to health. A woman had fits and had regular attacks every 5 or 6 weeks. She had no disorder after 6 weeks of Spectro-Chrome.

Dr. G. Ernest Spear had used Spectro-Chrome in different cases. A disorder of long standing in the retina was set right in quick time. He used Spectro-Chrome successfully in Chorea and in cases where manipulation would be of no use. An extreme case of Strabismus he set right in one month. A young man was paralyzed from the waist down. The surgeons in his town had told him he would never walk; Spectro-Chrome rescued him from that; in 20 tonations he walked with two canes; further tonation normalated him. He had other cases quite interesting. A man wanted to be "shown" the action of Spectro-Chrome on the bowels; 25 minutes "showed" him!

The second day afternoon session adjourned at Purple 00 Black.

Second Day Evening Session

The second day evening session was open to the general public. It was the most largely attended of all sessions. Further interest was created because it was the session exclusively set aside for our learned Vice-Presidents Kate W. Baldwin and Martha J. Peebles whose valuable experiences in Spectro-Chrome Therapy were heard with great admiration by the audience.

Surgeon Kate W. Baldwin introduced a negro girl of about 22, called Annie. This girl was pronounced a consumptive by the Health Authorities of Philadelphia, Pa. and the laboratory tests confirmed their views. Spectro-Chrome Therapy was used by Dr. Baldwin on this girl with remarkable results. All the untoward symptoms had disappeared and the girl was steadily gaining weight and health. The laboratory tests indicated progress. Annie was presented to the convention and the President asked the permission of Dr. Baldwin to be allowed to itisize the Area 3. Permission being granted, Dinshah Itisized the Area 3, the Itisometer corroborating the findings of the doctor. It was a striking instance of the value of Spectro-Chrome Therapy and the precision of the Itisometer.

Dr. Baldwin then introduced Miss Mary McCurdy, a woman whose eyes had cataracts for years and who was fast drifting to helplessness through it. Spectro-Chrome had restored her sight, the Cataracts had disappeared and Miss MacCurdy was happy. She came out among the audience and delivered a short speech expressing her gratefulness.

After enumerating several similar experiences, Dr. Baldwin finished her oration and was accorded resounding applause. In her, the Association recognized the greatest living champion of Spectro-Chrome Therapy and all expressed their satisfaction in hearing her speak so emphatically about Spectro-Chrome.

Dinshah then introduced Lettie Allen Ward, S. C. T., M. D., the Secretary of First Philadelphia Pennsylvania Centre. She said she felt like a "Quaker!" She was not a public speaker, but, would try to iterate her views. She took to Spectro-Chrome for her health's sake; she was a sceptic, but, was prepared to be shown. She gained her health and had been helping others to gain theirs with Spectro-Chrome Therapy. It was a wonderful science and although people had told her about the "darn old light" being a fake, she was fully convinced the "darn old light" works!

The President extolled the conscientious work of Dr. Lettie Allen Ward; she was a very good secretary and her reports were exceedingly accurate in detail and compiled elaborately. Her example as a Centre Secretary was worthy of emulation.

Dr. Martha J. Peebles was then introduced to the audience. She said she was "scared stiff" of addressing public meeting; she had taken to Spectro-Chrome for her own sake primarily; she was with the Ameican Expeditionary Forces among the first line of trenches during the war as a contract surgeon and had seen very severe service. There were times when she worked 16 hours a day in the hospital and attended to 61 operations. On her return, the strain told on her constitution and she was wrecked in health; her medical friends advised her to retire permanently and live in Denver Colorado in seclusion. On recommendation of Dr. Delia Riggs and of the late Grant Baldwin, S. C.

T., M. D., F. A. C. S. a brother of Dr. Kate W. Baldwin, she studied Spectro-Chrome and in a short time her own health was restored. She owed a great debt of gratitude to Spectro-Chrome; she was a sceptic, but, the accurate results of Spectro-Chrome had conquered her disbeliefs and she was only using Spectro-Chrome in preference to other healing systems; Spectro-Chrome Therapy was a stupendous adventure for one man to undertake, but, it was an accomplished fact and the World would in time learn that in Spectro-Chrome the healing art had reached a hitherto unknown and marvelous precision. Spectro-Chrome could do what drugs could not do. Cancer was a disorder of the meat eating nations. She had been in the practice of medicine for 28 years and had seen nothing like Spectro-Chrome for the removal of that dreaded scourge. Spectro-Chrome had been used effectively by her in cases of Cancer. The future of Spectro-Chrome was assured.

The remarks of Dr. Martha J Peebles were received with eager attention and she spoke with sincerity and frankness.

The second day evening session was adjourned at near midnight; after the President had allowed the public to depart, a private session was held for further explanation of the beauties of the Science of Sympathometry.

Third Day Morning Session.

The morning session of the third day was open to Associates only. It began at the usual time—Yellow 00 White.

Dinshah P. Ghadiali, Irene Grace Dinshah and Kashmira Dinshah, of Spectro-Chrome Institute were unanimously re-elected as the three Trustees for the Association. The Trustees were installed in office by Dr. Kate W. Baldwin. The Board of Trustees unanimously reappointed Kate W. Baldwin and Martha J. Peebles as the Vice-Presidents, thus completing the Chief Executive Council. They were installed by the President.

The President having called on the Committee appointed to report on the matter of the Woman's Hospital of Philadelphia, Dr. John A. Cohalan, the Chairman of the Committee submitted a resolution which the Committee had unanimously drafted and passed upon. Moved by Sigmond Rosenthal and seconded by Bessie A. Hasenau that the said Resolution of the Committee be accepted and the President be ordered to send it to the Board of Managers of the said hospital. On the question, Dr. Baldwin preferred not to send it, but, the opinion of the Association was otherwise and the Resolution was unanimously passed.

The following was the text of the said Resolution which was subsequently forwarded by the President to the Board of Managers of the Woman's Hospital of Philadelphia:—

"RESOLVED THAT: We, deeply deplore the action of the Board of Managers of the Woman's Hospital of Philadelphia, Pennsylvania, in requesting the resignation of Kate W. Baldwin, M. D., F. A.C.S., S. C. T, &c., after 30 years of her unselfish services to the Hospital, and their failure to appreciate the results produced by her use of SPECTRO-CHROME THERAPY, in the Hospital."

Dr. Emma H. Gabriel spoke of the peculiarity of certain reactions noticed in Itisometry. The President gave explanations.

The subject of the changes in certain By-Laws was spoken of by the President on behalf of the Chief Executive Council. Under the Article 21 of the Constitution, the Chief Executive Council had the power to formulate the By-Laws of the Association from time to time, as deemed advisable by it, for the conduct of its affairs. The Chief Executive Council had found that the funds of the Association under the existing methods were too much scattered and were doing no good with the Centres; they had found many Centres behaving in a refractory, unbusinesslike way; Portland Oregon and Seattle Washington Centres had not even reported for months and San Francisco California had used its privilege of using Centre Funds for sending a Delegate by completely wiping out the entire Centre Treasury in paying $250.00, out of a Centre Fund of $250.15.

The President emphasized that the Association in its biggest battle of life with the Medical Trust needed to conserve all its resources, so when the time came it may not be found wanting. As it was, the scattered funds were either dormant or wasting. It was necessary for the good of the Organization to consolidate all the money into one Treasury and thus have a direct control so no waste might occur. In that way, the Centres will be obliged to report properly and punctually to receive their disbursements and all disbursements will thus be checked for legitimacy. The consolidation of the funds was to be done by recalling all the Centre Funds and changing Article 14 and other articles of the By-Laws to conform with the necessary requirements. The Centres would not receive thereafter $6 out of the Annual Dues, but, would get only their actual legitimate disbursements by submitting monthly the claim and vouchers to the Chief Executive Council. Such procedure would absolutely prevent possible abuse or misuse of funds, as, the Treasury of the Association being in charge of a duly bonded Officer, the funds would be in safe custody, whereas, the Centre Funds were in charge of unbonded Associates, the Centres in most cases having ignored the bonding By-Laws.

The Vice-Presidents also voiced their opinion most favorably for the consolidation of all Centre Funds into the Association Treasury. They stated it would be the best step for conserving our resources and would centralize the work so the Association Funds may not be wasted. It was a vital necessity, otherwise in case of need, we should have no adequate resources to fall back upon.

The Chief Executive Council having unanimously voted on the proposed change, there was no necessity of further action in the matter as the authority for the fomulation of the By-Laws was vested solely in the Chief Executive Council. However, the President asked for the opinion of the Convention and the vote of the Convention was unanimously in favor of the consolidation. The changes were put into immediate effect accordingly.

The matter of the publication of SPECTRO-CHROME was then brought out. The President explained how in face of the tremendous pressure on his time and the unbearable burden on his office and staff, he was willing to publish the periodical, but, under the conditions prevailing at the time, the Chief Executive Council had thought it advisable for certain reasons to discontinue the publication, until the Portland Oregon Case Appeal was decided.

Dr. Kate W. Baldwin said that under Article 41, Section 3, each Associate was to receive a copy of the SPECTRO-CHROME "as published", but, if none was published there was none to give. Colonel Dinshah had made superhuman efforts to save his life's work from destruction and she was unable to understand how he did what he did against all odds.

Dr. Martha J. Peebles also spoke in a similar strain.

Dinshah said that although the views of the Chief Executive Council were sympathetic and true, yet, Spectro-Chrome Institute having received the $3 per Associate for the periodical during the year, he would not accept any payment for the next year.

The Convention expressed views similar to the Vice-Presidents, but, Dinshah refused to take any money for the publication for the next year. He could not do so with justice.

The third day morning session adjourned at midday.

Third Day Afternoon Session

The third day afternoon session began at Blue 00 Black. The interest had not abated. The crowd was unusually large although many had left for distant States. The whole session was a round of intellectual entertainment, of questions and answers, of cases normalated. Bessie A. Hasenau spoke of a case of stone in the kidney. Area 18 was found on Itisation 120 Ultra. The Medical Doctors had operated for gallstones. His Area 18 was 120 U, by the Itisometer. Spectro-Chrome helped him. It brought the Area to 25 U.

Frank E. Kaufman spoke about the Baltimore Maryland Chamber 1. It was dormant because of the restrictions imposed in its Authorization, about having an Itisometer and a Graduate Equipment.

The President said that the Association had no funds to supply Equipments and the terrible financial losses of the Spectro-Chrome Institute owing to the Portland Oregon Case prevented it from doing anything at present in that direction. However, the conditions would soon change and perhaps the matter would be attended to differently then.

Much discussion and explanatory work was done in regard to Itisometry and Sympathometry and the Convention was finally dissolved at Purple 30 Black.

The consensus of opinion was that wonders had been performed by Spectro-Chrome Therapists during the year, notwithstanding the odds they were fighting against and the difficulties Spectro-Chrome Institute had gone through.

* * *

SPECTRO-CHROME

Jan. to Sept., 1928
Red to Violet, 1928

VOLUME 6
NUMBERS 1 TO 9

Devoted to: Spectro-Chrome Therapy Dedicated to: Service of Humanity

CONTENTS

Spectro-Chrome Therapy, Spectro-Chrome Therapists.

SEVENTH ANNUAL CONVENTION NUMBER

Sub-Editor
Kashmira Dinshah

In course of life you reach a time,
When Doctor, Lawyer, Priest alike,
Does fail to give the sought relief,
To troubled mind or burdened heart—
Then come to me:
I help you to help yourself.

SPECTRO-CHROME.

INCARCERATION ISSUE

Attuned Green Wave

(Spectro-Chrome Therapy, Spectro-Chrome Therapists)

SEVENTH ANNUAL CONVENTION OF AMERICAN ASSOCIATION OF SPECTRO-CHROME THERAPISTS

STIRRING MESSAGE FROM DINSHAH BROUGHT TEARS IN MANY EYES

The Seventh Annual Convention of American Association of Spectro-Chrome Therapists assembled at Hotel Lorraine, Philadelphia, Pa., on November 28, 1927, at Yellow 00 White.

On the platform, next to the President Irene Grace Dinshah, were seated the two Vice-Presidents, Kate W. Baldwin, S. C. T., M. D., F. A. C. S., and Martha J. Peebles, S. C. T., M. D., and the Secretary-Treasurer Kashmira Dinshah. High on the wall at the rear of the platform was the portrait of the retired President, Colonel Dinshah P. Ghadiali.

Although they received notification, several Associates came without their Association Card. Duplicates were issued to these, on payment of One Dollar; this was in accordance with the By-Laws.

The following First Notice calling the Convention, was sent by mail to all the Associates of the organization:

October 25, 1927
Chronos: Yellow 14 White

SEVENTH ANNUAL CONVENTION

Dear Friend:

1. In accordance with Article 15 of our By-Laws, please be advised that our Seventh Annual Convention will be held in Philadelphia, Pennsylvania, at the Hotel Lorraine, in the Chelsea Room, at 10 A. M., Monday 28th and Tuesday 29th, of November, 1927.

2. An attractive and entertaining educational program will be presented, full particulars of which will be mailed later. There will be no public sessions. All sessions will be open to Associates only.

3. Your presence is cordially invited to hear the Message from the President, which will be of the greatest importance.

4. Should you have any complaints for consideration, please send them to the Chief Executive Council, in writing, duly signed.

5. Should you intend to read special papers or have any subject about which you desire to address the Convention, kindly mail in time for the program.

6. Centres may appoint ONE Regular and One Alternate Delegate to the Convention; Delegates must present written credentials and bear their own expenses, as there are no Centre Funds for such purpose as in former years.

7. Secretaries of Centres are requested to send their Annual Reports, showing number of meetings held, average Associates present and the balance sheet of the year enumerating the receipts and disbursements.

8. Your co-operation will be warmly appreciated.

9. NOTE: No admission to the Convention without the presentation of the Association Card; many vital and amazing developments will be discussed.

By Order Of
CHIEF EXECUTIVE COUNCIL
Kashmira Dinshah,
Secretary

Later, the complete program was mailed for better information, so the Associates might come prepared for the various sessions:

November 15, 1927
Chronos: Green 23 Black

PROGRAM OF SEVENTH ANNUAL CONVENTION

Dear Friend:

1. The following attractive and entertaining educational program will be presented at our Seventh Annual Convention, opening at 10 A. M., on Monday, November 28, 1927, in the Chelsea Room, Hotel Lorraine, Philadelphia, Pennsylvania:—

Monday, November 28, 1927.

10 A. M. to 12 M.—Registration; Annual Reports of Secretary and Treasurer; Centre Reports; Suggestions and Sundries. Session open to Associates only. 2. P. M. to 5 P. M.,—Address by our Vice-Presidents Kate W. Baldwin and Martha J. Peebles.

8 P. M.—Stirring Message from retired President Dinshah, on "OUR TURNING POINT TO VICTORY." Session open to Associates only and is worthy of your presence to prove your loyalty to The Cause.

Tuesday, November 29, 1927.

10 A. M. to 12 M.—Exchange of views among Associates. Questions and Answers. Session open to Associates only.

2. P. M. to 5 P. M.—Case Reports and general discussion. Presentation of Sufferers.

8 P. M.—Election of Trustees; Appointments and Installations; Unfinished business.

Wednesday, November 30, 1927.

Visit to Malaga, New Jersey.

2. Associates are cordially requested to be present at the various sessions and help make this Convention a success as usual.

NO ADMISSION WITHOUT PRESENTATION OF ASSOCIATION CARD AND REGISTRATION.

By Order Of
CHIEF EXECUTIVE COUNCIL
Kashmira Dinshah,
Secretary.

First Day Morning Session

At Yellow 15 White, President Irene Grace Dinshah's gavel called the Convention to order.

The minutes of the Sixth Annual Convention, were presented by the Secretary Kashmira Dinshah, S. C. T., as reported in SPECTRO-CHROME issue of R-O-Yellow, 1927. On motion of Frank E. Kaufman, S. C. T., seconded by Bessie A. Hasenau, S. C. T., the minutes were unanimously accepted as recorded.

Irene Grace Dinshah read the Resignation of Dinshah P. Ghadiali as President-Treasurer of the Association as follows:

"Resignation"

"United States Penitentiary—Sunday, November 13, 1927
"Atlanta, Georgia. 5.17 P. M.
"American Association of Spectro-Chrome Therapists,
"Malaga, New Jersey.

"Most Esteemed and Beloved Friends:

"1. Beset with a solemn duty, in obedience to the Divine Will, I hereby resign my office as your President and Treasurer, to take effect promptly.

"2. I shall retain my position as a Member of the Board of Trustees and not as President-Treasurer.

"3. With the best of human love, I remain,

"Ever Yours For Suffering Humanity

(Signed) DINSHAH P. GHADIALI,
"President-Treasurer."

She read also the Selection of Irene Grace Dinshah as President and Kashmira Dinshah as Treasurer:

"Selection"

"Malaga, New Jersey

"Wednesday, November 16, 1927

"1. We, the undersigned, the Board of Trustees of the American Association of Spectro-Chrome Therapists, do hereby accept the resignation of Dinshah P. Ghadiali, as President and Treasurer.

"2. By authority vested in us by Article 27, Section 2, of our By-Laws, to fill vacancies on the Chief Executive Council, we hereby select and install in authority, Irene Grace Dinshah, S. C. T., as our President and Kashmira Dinshah, S. C. T., as our Treasurer, this to take effect immediately.

Irene Grace Dinshah, President.

Kashmira Dinshah, Secretary-Treas.

Dinshah P. Ghadiali, Member.

"Reference:—

"Article—Section of By-Laws
27 — 2

"Article—Section of Constitution

6 — 2
6 — 3
7 — 1 & 2
10 — 1
12 — 4"

The new President Irene told about the former President Dinshah being in the Federal Penitentiary in Atlanta, Georgia; that he was taken away from the 48th SPECTRO-CHROME THERAPY Class, and he went—not as a coward, but, unafraid, bold and fearless, just like the Dinshah the Associates had always known. He was penalized—as is the lot of *all* pioneers. She stated that religiously speaking, she should have been in childbed, her third son having been born on the 30th of October, 1927, but that she bowed to the call of Duty which she would cheerfully perform and try to grace the chair she had the honor to occupy. The President thanked the two Vice-Presidents for their gracious co-operation in every respect and the faithful Associates for their continued interest and zeal; she thanked also the Secretary-Treasurer for her cheerful labors. The Association was more prosperous this year than last, there being more Associates in 1927 than 1926 and more money in the Treasury. She said that Colonel Dinshah started such a wonderful, humanitarian work and we must continue; he worked hard even in the Penitentiary. From prison walls he sent a great Message and she requested the kind presence of the Associates at the evening session to hear it.

The Secretary Kashmira Dinshah presented her Annual Report. There were thirteen Centres of the Association, located at Baltimore Md., Boston Mass., Cleveland O., Detroit Mich., Los Angeles Cal., Malaga, N. J., New York, N. Y., Philadelphia Pa., Portland Ore., Providence R. I., San Francisco Cal., Seattle Wash., and Vineland N. J.

There were two deaths and four violations, which were under investigation. There were no resignations. There was an increase of 25 Associates. On motion of Bessie A. Hasenau, S. C. T., seconded by Earl Broadbent, S. C. T., the Annual Report of the Secretary was voted upon and unanimously passed.

The Annual Reports from various Centres were then read.

First Cleveland Ohio Centre was reported by the Centre Secretary, Margaret A. Morris. They had eleven meetings during the year, all of which were very interesting.

First Detroit Michigan Centre was reported by the Centre Secretary E. Norman Pearson. They were proud of being Associates and had interesting meetings.

First Philadelphia Pennsylvania Centre, the largest Centre of the Association, was reported by the Centre Secretary Lettie Allen Ward. They had eight meetings during the year. There seemed to be no end to the wonderful cases reported. They felt very grateful that Spectro-Chrome Therapy was placed in their hands as a means of helping others.

First Vineland New Jersey Centre, the latest created Centre, was reported by the Centre Secretary Janet Izola Mole. Since the Authorization in May 1927, a meeting was held each month. Their average attendance was large.

First Baltimore Maryland Centre was reported by the Centre Secretary George A. Lentz. They had a very good representation of the Associateship.

First Los Angeles California Centre was reported by the Centre President Henry Gross. They had seven interesting meetings during the year. Through their President, they sent a communication of greetings, cheer and kind thoughts.

Margaret A. Morris of Cleveland, O., regretted not being able to attend the Convention personally.

Richard E. Sandbach of Wellsboro Pa., reported a case of a man whose condition was diagnosed by a doctor as pernicious anemia and who was at first very unwilling to take Spectro-Chrome tonations; after being urged, the person took 76 tonations and felt much better; he later went to a sanitarium. He is entirely better now; his wife gives all the credit to Spectro-Chrome. The man, however, having been a skeptic from the start, is willing to give Spectro-Chrome about half of the credit only, but, the former sufferer was set right and that is all that matters. Sandbach said he felt that Spectro-Chrome did much for this man, who was described by many as being "nearly a dead man." Another man was given strychnine and digitalis for heart trouble and was very weak. Spectro-Chrome normalated him. Another with ulcer on the eye improved under Spectro-Chrome tonations. A diabetic for thirty years is now healthy due to Spectro-Chrome. A sufferer with prostatic gland trouble was under medical care. After leaving medicine, he had twenty-five tonations with Spectro-Chrome and is now healthy. An infected toe was normalated by twenty tonations. A sufferer had trouble with the teeth; after a few tonations the condition was normalated and he sent a letter to Sandbach saying he was "sold" on Spectro-Chrome, because, he did not have a toothache since the time he had been tonated with Spectro-Chrome. A man was pronounced suffering from gall bladder trouble; he had twelve tonations and now feels better than ever. A woman with neuritis was normalated after having Spectro-Chrome tonations. The back of a twenty-four-year-old man was injured when a child and he was suffering from too frequent frontelim (urine); he was operated on for what Medical Doctors said were ulcers of the bladder; there was no relief. He suffered severe pain. After ten or twelve tonations with Spectro-Chrome, he had no more pain and frontelim was normal. A boy was hit by an automobile and his leg was bruised; he could not stand on it and could not sleep the first night. After one tonation he slept all night; Indigo was the Wave used. The leg that was bruised is now better.

Dr. Kate W. Baldwin said that Spectro-Chrome would do the work regardless of whether a person had faith or not. She had a severe hiccough case and the man said that he had no faith in Spectro-Chrome. Would it work without faith? Dr. Baldwin replied, "You take the color and I'll have the faith." She stated that in bruises remarkable results were obtained, by tonating with Green and later Indigo.

Esther Jennings had a case of a child that had imbackelim (constipation); she had told the mother to prepare vegetarian food including "dal" and she had given Yellow tonation. There was no improvement. She asked for suggestions.

Dr. Martha J. Peebles responded, saying there were many factors involved in imbackelim; the diet was very important; there was not sufficient roughage in food oftimes; bananas are excellent food; give food that has bulk. Give plenty of fluids, including lemonade.

Dr. Kate W. Baldwin told about the value of drinking plenty of water and fruit juices. The bowels can form habits just like anything else. Backelimation should be performed regularly and once a day is not enough. Train the bowels to evacuate twice or thrice a day always at the same time and you will find a freedom—a cleanliness. Tell your sufferers not to read or smoke at such times, but, to concentrate the mind on the action and never neglect Nature's calls. The morning is the time when you will get your freer action, but, the morning action is not sufficient. You should never wait. If there is a large family there ought to be more than one toilet in the house. She always put stress on this point. She said that if ever she went "insane," she believed it would be on this subject. Breathing is important; the skin breathes. To have the proper elements, you must have the proper food. Many times Green acts well for imbackelim. A few minutes of Indigo sometimes helps. She told favorably about squatter toilets, such as were used in Spectro-Chrome Institute's Resident Quarters. Gas in the stomach makes action of the heart erratic. She had never been much of a drug giver and if she had Spectro-Chrome in former times she could have used it to great advantage.

W. Frank Bailey spoke about a type of toilet for children.

Martha J. Peebles told about a woman of 35 years, who had a nervous breakdown. She had to have the bed-pan every hour and produced a defecation sheerly by keeping her mind on the idea. Dr. Peebles thought that the woman had induced a habit and therefore had the sister lock the woman in her room. The first night she called all night for the bed-pan which was not given to her. The second night she called only a few times, but, after that she was all right. Dr. Peebles further said that there was no imbackelim in the army because there they automatically used the proper posture. They dug trenches and squatted on the ground on planks for the purpose. She said that cases of imbackelim were due many times to mechanical trouble and in such cases use Turquoise or Indigo on Areas 8-9 and 18-19.

Frank E. Kaufman said a child had imbackelim and Lemon, Yellow and Orange were used but there were no results. The child was "high-strung and nervous." Turquoise and Blue were used and the child felt better. Yellow was then used on 8 and 9 and the child was restored to health.

Dr. Alice M. Norton said that she had been practising dentistry for twenty-seven years. She found invariably when children had "acid mouth," the trouble was due to faulty backelim and that the root of this trouble was wrong toilet conditions. She told of the advice she gave in this connection, to mothers of the children.

Bessie A. Hasenau told of a stubborn case of imbackelim, where the sufferer was not co-operating and was using enemas. She cautioned the Associates about such cases.

Frank E. Kaufman was present as Delegate from Baltimore, Md. Centre; Erma I. Kyte as Delegate from Vineland, N. J. Centre, and Ruth O. Sinclair as Alternate from Vineland, N. J. Centre.

The morning session adjourned at Green 00 White.

First Day Afternoon Session

The afternoon session began at Blue 00 Black.

Irene Grace Dinshah read a telegram of good wishes from four New York Centre Associates who regretted their inability to attend the Convention.

The first speaker of this session was Dr. Kate W. Baldwin, who told about the horrible injustice of the outcome of Colonel Dinshah's case. Spectro-Chrome Therapy had done more for her than anything else in regard to helping others; she said that in the medical profession you need a vision, which, if you do not have you get nowhere; Spectro-Chrome is the most wonderful Healing Art in this century. She told about the serious burns produced by Arc Lamps and X-Rays and of the importance of the Magnetic Meridian; that any old position for tonation would not do. She said that we owe much to our leader Colonel Dinshah; that he performed superhuman efforts; she advised everyone to make the trip to Malaga to see the new additions to the buildings and told the Associates to do their level best, by putting their prejudices, (if they had any) in their pockets and seeing the thing through! She said she still hoped there might be some means of pardon or commutation of sentence for Colonel Dinshah. On the matter of diet, she explained the absurdity of throwing away the water in which foods were cooked and that foods should not be cooked very long. Dr. Baldwin cited numerous serious cases that were normalated by Spectro-Chrome. One was an advanced T. B. sufferer. Her lungs were full of matter and she was expectorating much. She came right up under Spectro-Chrome tonations; she changed so much for the better that people did not recognize her. She became so well that she cooked for and waited on a table of twelve. Another case was a man with a cyst in his throat; he was much relieved by Spectro-Chrome; the cyst became smaller and he gained in weight and health. Dr. Baldwin warned those present about "Stool pigeons" and told of a man who was very ill, sent for Spectro-Chrome tonations; he was supposed to die on her hands and the blame would redound on Spectro-Chrome! However, the man became better and that ended the farcical incident! She advised to get rid of one's sufferers if they are going to die, before they do die. She told of an eye case in a woman. The eyes were horrible looking—red and sore; she tested her on the Itisometer but did not dare to give Red tonation according to the dictation of the Itisometer, thinking perhaps something was wrong with the instrument. She gave Turquoise and there seemed to be no improvement. She tested the Itisometer and it was in perfect working order and it again indicated Red tonation for those eyes. Dr. Baldwin hesitatingly gave three Red tonations and lo! It was miraculous. The congestion disappeared and the pain left. The Itisometer had not failed! From that day to this, the person has had no trouble with her eyes. A man who had been gassed in the army was suffering from pneumonia and pleurisy. He was practically lifeless; his face at times became a clouded, dark color that might be compared to a light-mahogany. The spleen was very large and the liver was tender; he ate and slept very little. Dr. Baldwin said she had never seen anybody come up so well as he did. In a very short time, the spleen became decidedly smaller and he said he felt ever so much better under Spectro-Chrome. A woman had cancer and had been operated on; it seemed to be a hopeless case. Her husband did not approve of her being tonated with Spectro-Chrome by Dr. Baldwin. He wanted to know what a *woman* could do in the case when all the best men doctors in town could do nothing! Dr. Baldwin clearly proved to him that women *can* do something, by normalating his wife through Spectro-Chrome. A case of cancer of the larnyx where there was a swelling on the outside as large as an orange, and the sufferer could barely breathe, responded very well to Spectro-Chrome. The woman laid her hand on Dr. Baldwin's arm and said, "Dr. Baldwin, you are so good to me. You make me feel so comfortable." The audience was all attention and gave tremendous applause when Dr. Baldwin had finished.

Following Dr. Baldwin, was Dr. Martha J. Peebles. She extended a hand toward the portrait of Dinshah and said she deplored the resignation of Colonel Dinshah as President. She spoke of the beauty of Spectro-Chrome system as being within the visible spectrum whereas the systems that produced burns were beyond the visible spectrum, either on the Infra-Red side or the Ultra-Violet side. She said there is no need whatever for using the destructive rays when we have Spectro-Chrome Therapy. Always remove the cause or causes of trouble. Never promise too much to your sufferers; let them judge for themselves. She advised everyone to have an Itisometer if at all possible and said that she did not know what she would do without it. She warned that in accident cases an X-Ray picture should always be taken to determine if there be any fracture so that no blame could come on the head of the Therapist. She told of several serious abdominal cases normalated by Spectro-Chrome; also a fibroid case; the woman is now normal due to Spectro-Chrome. The audience appreciated the address of Dr. Martha J. Peebles.

Kashmira Dinshah spoke about the four violators who will soon be put on trial in their Centre, in Vineland, N. J. She told also of rascality on the part of two residents of Vineland, N. J., (non-Associates) who dealt unfairly with the Institute who would never under any condition be awarded the Diploma of Spectro-Chrome Institute. Ernest Carreiro, a Graduate of the Boston, Massachusetts Class in Spectro-Chrome Therapy had disappeared with several of our Equipments only partly paid for and we were unable to locate him. Any advice as to his whereabouts would be appreciated. He had lectured in Philadelphia, Pennsylvania, some time last October on so-called "Occult Breathing." Dr. Baldwin said that Carreiro asserted in his lecture, "When you hear me, your hear Dinshah!"

The afternoon session adjourned at Purple 00 Black.

First Day Evening Session

The first day evening session commenced at Red 00 Black.

By special permission of the Chief Executive Council, Dr. Emma H. Gabriel presented two sufferers, who would have been unable to appear the following day according to program. One was a young woman who had digestive disorder and had been on a milk diet. She had an attack and was very weak; she had almost given up hope. Dr. Gabriel gave her two tonations every day with the Graduate Equipment; according to the Itisometer, she used Turquoise on Areas 8 and 9. Around Thanksgiving time she took solid foods. She slowly gained in weight; she was 108 lbs., then 113 and now 138.

In April she was so much improved that she took a position and under Dr. Gabriel's supervision was using an Intermediate Equipment in her home. Dr. Gabriel said Spectro-Chrome saved that woman's life. She then introduced another sufferer whose entire right side was paralyzed as well as her left arm. She could not walk at all and was practically

helpless. After Spectro-Chrome tonation she can now use her arms and with assistance can walk. Dr. Gabriel explained that the improvement was indeed great and that she hoped to present this sufferer next year entirely well, walking herself. Dr. Baldwin said that she had seen the sufferer in the beginning and the improvement was so remarkable that it could hardly be described. President Irene Grace Dinshah stated that with Colonel Dinshah present at the next Convention and this gracious sufferer, a sufferer no longer, to her mind such would be "The Victory Convention." The sufferers then departed, as the session was open to Associates only.

The program was then followed.

Kashmira Dinshah rose as the representative of Colonel Dinshah and impressed upon the audience the fact that absolutely no notes were to be taken of his Message, because according to prison rules nothing written by a prisoner was released and that he had pledged his word of honor to that effect to the Penitentiary Warden John Wilson Snook. She then read aloud that Message to the Associates. It was in the usual emphatic and masterful style of Dinshah and was so realistic and truthful in its representation of the facts, that one after another handkerchieves were brought into use by the Associates. The hearts of all were full and their eyes welled with tears. Having visited her father in the Penitentiary on October 10, 1927, when she came to that part, the Secretary was overcome with strong filial emotion and could barely proceed. She, however, did justice to her duty and delivered the Message in a voice that carried the words of the incarcerated Originator to the attention of her hearers perfectly.

The Message had no bitterness and was purely businesslike, but, its human character showed the philanthropic feeling of its sender. In it he had good words for even his opponents, whom he wished good thoughts. On conclusion, the Treasurer Kashmira Dinshah presented the Annual Report as of November 1, 1927. It showed a large increase in funds over that of last year. On motion of Rebecca Berenson, seconded by Esther Jennings, the Treasurer's Report was accepted by unanimous vote.

In Colonel Dinshah's Message, certain Eight Plans for the good of the Spectro-Chrome Cause were laid before the Associates. A Committee of seven was to go over the plans thoroughly and render a report to the Convention the next day, thrash the matter out carefully in the Convention and make the decision by vote, accepting one or more of the Plans outlined. The Associates decided it would be best for the President to appoint the Committee. Accordingly, Irene Grace Dinshah appointed a Committee composed of Frank E. Kaufman, Richard E. Sandbach, Bessie A. Hasenau, Maude B. Crans, Rebecca Berenson, Jennette J. Deemer and John F. Coyle. The Committee left the room, to discuss the Plans and it was voted that the meeting adjourn. The time was Orange 45 Black.

Second Day Morning Session

Tuesday the 29th, at Yellow 00 White, on suggestion of Associates the President requested Kashmira Dinshah to tell of her trip to see Colonel Dinshah in the Atlanta Penitentiary. Kashmira said that some horrible rumors were being circulated in Vineland and nearby in regard to "terrible treatment" being given Colonel Dinshah because he was supposed to have resisted injections for a certain disorder which he was supposed to have. With the determination to receive first-hand information, she went to visit the Colonel and to save expense did not take a Pullman, but, went by coach, sitting up all the way there and back. A prisoner there is allowed only two visits a month and as a Graduate of ours living in Atlanta, had been to see him twice that month, Kashmira said they were not going to admit her. She finally prevailed on the Guard to admit her to the Warden's Office; when the Warden found that she was "one of the family from out of the State," (Georgia) he permitted her to see her father who was perfectly healthy as usual; he was resigned to his fate and had no complaints whatever. The Warden, Chaplain, Guards and other men were all very friendly toward the Colonel and they respected him for his learning. She told about the immaculate cleanliness of the penitentiary. She said that even though imprisoned, Colonel Dinshah adhered strictly to his high principles and through the kind courtesy of the Warden, had been supplied with vegetarian food. He was working there in the Power House, which was the work he had chosen. In conclusion, she said, "Colonel Dinshah is in constant touch with us and even directs work here, from there. He thinks constantly of the Associates and of the day when he will be again free to serve humanity." Kashmira was heard with rapt attention and appreciation.

The President called on Frank E. Kaufman, Chairman of the Committee appointed to report on the Plans, who gave the preliminary report accepting Plan Three and Plan Five, rejecting Plan Sixth and Plan Eighth. Dr. Kate W. Baldwin requested everyone to have his or her say right then and there so there would be no chance of confusion afterwards. There was much discussion and the subject was left open for the afternoon session. At Green 05 Black the morning session adjourned.

Second Day Afternoon Session

The second day afternoon session commenced at Blue 00 Black.

Earl Broadbent presented a Mr. Ferris, who appeared happy; the man showed radiant health and spoke about his own case. He had been a vegetarian thirty-two years. He had always been healthy, but, in the Spring of 1927, he was "forced to call for help." He had contracts outside that required physical labor and he knew he was not in a condition to fulfill them. He was unable to sleep and suffered much pain. He called on Earl Broadbent on the last Sunday of May, 1927. He said he had been a very weak and sick man; Spectro-Chrome was used on him night and day. In two days he was free from pain and in three weeks was entirely well. Spectro-Chrome was a revelation to him. Broadbent told the Associates the Attuned Color Waves that were used in the case.

Dr. Kate W. Baldwin explained the difference between nudity and nakedness. She said there was no reason whatever for the practitioner to cause the sufferer embarrassment and that it is unethical, to say the least, to speak to anyone no matter who it was, about the marks, scars, deformities and the like that they as sufferers might have. She spoke also about the use of false Slides and Bulbs. She said that users of such false Slides were not real Spectro-Chrome Therapists; that they might save a few cents by buying other Bulbs than those supplied by the Institute, but, they would crack their Slides. It would certainly be no advantage in the long-run and it is against the Integrity Policies.

William I. Newbold told about his own case. He was at Spectro-Chrome Institute for a month, became well and took the Course, graduated and is now practising. Twice he was overcome by illuminating gas, once in his garage, another time in his cellar. He felt so weak, he was barely able to get up the stairs. He told his wife to give him the Slides to make the necessary combination and used Magenta on 4. Later he used Scarlet on 4. He soon became well and strong.

Esther Jennings reported a case pronounced as of cancer; the case was so far advanced, the doctors refused to operate;

the sufferer was given only ten days to live; he had very much pain and could not sleep. She tonated him for one month; the pain left and he improved; however, when she learned he was smoking 42 cigarettes a day and would not live up to the diet, she dismissed him, because such action was not fair to Spectro-Chrome. A woman aged 48 had terrible hemorrhages from Area 11. In four days' tonation the hemorrhages stopped entirely. The Waves given were Green Systemic, Indigo on 19 and 11, and Magenta on 4. Mrs. Jennings said the woman is very grateful to Spectro-Chrome and today is one of her very best friends.

Dr. Emma H. Gabriel presented Anna Schoener the paralysis case which was reported in SPECTRO-CHROME of Scarlet 1925. The father of the little girl had to assist her in walking, when she appeared at the Meeting held to take the place of the Fifth Annual Convention, on December 27, 1925. Not two years elapsed, yet the girl so improved by Spectro-Chrome tonations that she now walks to school. Dr. Gabriel requested the child to walk clear across the Convention Hall, which she did without any assistance. The father said that before he heard of Dr. Gabriel and Spectro-Chrome, he thought there was no hope for his little Anna; at first she could not move at all; her legs were so stiff. After only one tonation of Lemon she stretched her legs out! This convinced him that there was hope for his child.

Benjamin Kessler told about his aged father who was bedridden and was advised to be operated. After a few Spectro-Chrome tonations he became so spry that he walked several miles a day. He then visited Spectro-Chrome Institute.

David C. Bosworth described a case of a man, 56 years of age, still under his care, who could not walk, had very little use of his arms and had imbackelim. For over thirty years he took physics. The tonations being given are Yellow on 8-9-10-19; Magenta on 4 and Orange on 17. The sufferer is getting along very well and the use of his limbs is gradually returning. He expects to recover totally.

Frank E. Kaufman spoke about a girl two years of age said to have had whooping cough. Besides the cough, her backelim was affected. He used Orange on 18 and 19 and backelim became normal. He gave two weeks of Green and Lemon systemic and Magenta on 4 and the child slept through the entire night without a single spell of coughing. One night, he failed to give Magenta in conjunction with the other Wave and the baby coughed. This proved to him that Magenta plays even a more vital part in Spectro-Chrome, than he had hitherto thought, thus, confirming the teachings of the Originator.

As Chairman of the Plan Committee, Kaufman then said that the final decision of the Committee would be given in the evening session.

Jennie Edelstein proved a very humorous reporter of cases. She had been asked why she took up Spectro-Chrome. In her reply she told of sad experiences in her own family; her husband and her children who died, she claims, owing to the ignorance of the Medical Doctors. Her husband was in an automobile accident; soon after, he began to lose weight and to "sink" in health; the doctors had twelve X-Ray pictures taken and endeavored to locate "cancers and ulcers;" they said they found *nothing*. Soon after such expert "diagnosis," he died with cancer of the stomach! One Medical Doctor said her daughter had spleen trouble and another said it was tonsil trouble; he removed the tonsils, after which she had trouble in the throat and later died, a victim of the humbug of medical "diagnosis." Mrs. Edelstein had a son who was accidentally thrown out of a boat and not wishing to worry his mother did not tell her of the incident. Two weeks later, he "came home with a headache." The Medical Doctor said, "It looks like a case of appendicitis." The next morning he said it was something different. The result was as usual —"the boy died!" She said she wished she had Spectro-Chrome in those trying times. She reported a case of a woman who could not speak; the sufferer indicated that she had a pain at the back of her head. Indigo was given on the painful Area and Lemon systemic; the woman soon began to utter words. Another case of a woman whom she met casually, had severe pain in her shoulder, for which there seemed to be no remedy. Indigo systemic, relieved her. A man thirty-two years of age had his tonsils removed when fifteen. He breathed heavily and was worse than ever. Spectro-Chrome tonations normalated him. Her son (the only one of her children now alive) had trouble with his eyes; she used Blue, which made them smart; she switched to Indigo and in a very short time the eyes were normalated.

At this point, Dr. Martha J. Peebles asked the President what had happened to the Plan proposition, which should have been continued in the afternoon. Frank E. Kaufman said he had explained that the final decision would be reached and given at the evening meeting. Dr. Peebles said she must have missed the remark. The President then said the meeting would progress according to the program.

Kashmira Dinshah said there appeared to be some misunderstanding about Colonel Dinshah's stand in regard to the medical profession. His battle is waged against the "Octopus" and not any specific individual. It is the Medical *Trust* that is persecuting him and Spectro-Chrome and not the Medical Doctor as an individual. She further stated that work is going on at the Institute as usual, with the exception, of course, of the teaching of new Classes. Equipment orders, Bulbs, Slides, the Printing of Stationery and the like, were being taken care of as ordinary routine.

Dr. Martha J. Peebles here voiced her opinion about orders, saying that within the last three months, Literature, Bulbs, Slides, etc., were coming through even quicker than before.

Elmer C. Heath spoke about certain peculiar indications of the Itisometer and requested enlightenment from one who had more experience in Itisometry. Dr. Martha J. Peebles gave satisfactory explanations.

Alice M. Norton, S. C. T., D. D. S., reported a dental case known as Bell's Palsy. After the dental work was finished, she tonated with Indigo; the tissues healed quickly and the Palsy condition was soon relieved. Dr. Norton said the case responded wonderfully.

Dr. Maude B. Crans told about a young man who had taken an active part during the World War; he had been subjected to the usual serum injections; he had fallen over an embankment; the sufferer was given up as hopeless; the expression had gone from his face, his appearance was one of dullness and listlessness and his relatives thought if the condition kept on, he would become an idiot, even though he was an intelligent lad and far from being a "dummy." Dr. Crans took him to Dr. Kate W. Baldwin for itisation and according to the indication, Dr. Crans tonated him with Red and Orange. He now looks bright and cheerful and is again taking an interest in life much to the delight of his family. A woman about seventy years of age had neuritis; she was suffering intensely. She was normalated by Orange, Yellow, Lemon and Magenta. A chemist had very severe burns and was normalated by Turquoise. Dr. Crans said, "I should not care to go on with my particular kind of work, which is Chiropody, unless I had my Spectro-Chrome." She stated that she found Yellow, Lemon and Orange very effective for strengthening arches, for fallen arches, and the like, rather than Indigo.

Erma Kyte told about the only case she tonated as yet and which is still under her care; the sufferer's ratio was 1 to 3; he was in such a terrible condition that she hesitated to take the case; she is now tonating the man with Lemon systemic and Magenta on Area 4 and he is responding very well. The President introduced Miss Kyte as having been graduated from one of the more recent Spectro-Chrome Therapy Classes. She stated Miss Kyte (being a shy young girl) made the request that she be not called on to address the Convention, but, the fact that she voluntarily reported her first case showed her interest in the system. The audience agreed that she had done very well and applauded her.

The afternoon session adjourned at Purple 03 Black.

Second Day Evening Session

The second day evening session opened at Red 00 Black.

Dinshah P. Ghadiali, Irene Grace Dinshah and Kashmira Dinshah, of Spectro-Chrome Institute were unanimously re-elected as the Board of Trustees for the Association. The Trustees were installed in office by Dr. Kate W. Baldwin.

As to the appointment of Vice-Presidents the President said that Colonel Dinshah the third member of the Board of Trustees would have to be consulted. Dr. Martha J. Peebles said the Vice-Presidents, according to rules, would have to be appointed at the Convention. If this were not done, she moved that the Convention be held as null and void. Whereupon, Dr. Kate W. Baldwin said such procedure of shutting the meeting down would not be proper. There was some discussion and argument between the two Vice-Presidents. Dr. Peebles asked if the Vice-Presidents were still in office and was informed they were. At this time, Red 40 Black, to the obvious surprise of the assembly, Martha J. Peebles resigned Second Vice-Presidentship in the Association. The President announced that the Board of Trustees accepted the resignation of Dr. Martha J. Peebles. Dr. Baldwin said since there were two of the three members of the Board of Trustees present, it would be in order to appoint the Vice-Presidents. A vote was taken and unanimously passed that the Vice-Presidents be appointed.

The Board of Trustees desired to re-appoint Dr. Kate W. Baldwin as First Vice-President, but, the worthy friend gracefully declined acceptance, stating it had always been her opinion and she had said many times that people should not hold an office too many terms; she could and would do just as much for the Cause as hitherto; she would always be our friend. She further stated that there were many good, honest Associates just as capable to choose from.

The Board of Trustees then selected Bessie A. Hasenau as First Vice-President of the Association, and Richard E. Sandbach Second Vice-President, both of whom accepted.

Dr. Martha J. Peebles asked the President where those two Associates belonged, where they came from, how they would attend Executive Meetings and so on. The President replied that Bessie A. Hasenau is affiliated with the First Cleveland Ohio Centre and Richard E. Sandbach with the First Philadelphia Pennsylvania Centre. Dr. Peebles said Cleveland was pretty far off.

The Vice-Presidents were then installed in office by the President.

Some Associates being present at this session, who had not attended the previous ones, the President requested Frank E. Kaufman, Chairman of the Plan Committee in regard to the Plans, to re-read all Eight and then to announce the final decision of the Committee. This was done and Mr. Kaufman said the Committee accepted Plan Three, Plan Five and Plan One; Plan two and Plan Seven were agreed upon favorably insofar as the Associates were able to comply. The floor was then opened for discussion. As there were no further comments or views, on motion of Dr. Alice M. Norton, seconded by Harriette M. Roberts, the final decision of the Plan Committee was voted upon and passed unanimously.

The President thanked Dr. Kate W. Baldwin and Dr. Martha J. Peebles for their devotion to the Cause and services in the past and requested a rising vote of thanks in honor of the retired Vice-Presidents, which was duly accorded.

Esther Jennings requested, and a rising vote of thanks was given Irene Grace Dinshah for her dutiful and able performances as the new President.

Dr. Martha J. Peebles said if the Associates present did not need another Equipment, they might each be able to place one with a sufferer.

The newly installed Vice-Presidents Bessie A. Hasenau and Richard E. Sandbach said they would be dutiful and conscientious.

Kashmira Dinshah said the Association was taking orders for gold or silver pins with the Association insignia at $2.00 each. If sufficient orders were received they would be manufactured; if not, the matter would be held in abeyance until Colonel Dinshah returns.

There being no further work on the program, the Convention was dissolved at Orange 18 Black, with invitation to visit the Principal Office, the next day, at Malaga, N. J.

After the Convention was dissolved, several Associates surrounded the President's table and congratulated her and the Treasurer-Secretary for the successful way in which they had conducted the meeting through a severe ordeal. It proved to them that Spectro-Chrome would live beyond the life of the Originator and further that it was a work worthy of the great efforts put forward to propagate it.

SPECTRO-CHROME

1929 VOLUME 7

EIGHTH ANNUAL CONVENTION NUMBER

Editor
Dinshah P. Ghadiali

In course of life you reach a time,
When Doctor, Lawyer, Priest alike,
Does fail to give the sought relief,
To troubled mind or burdened heart—
Then come to me:
I help you to help yourself.
SPECTRO-CHROME.

Liberty Issue

automatic cookers for preparing vegetarian meals for about a hundred persons. The hot water is supplied night and day, by an electric heater, fitted with thermostatic control, the water being taken from a driven well underground, by a turbine pump. The heating is done by electrically regulated Quiet May Oil Burners, installed by John E. Watson, of Woodstown, N. J.; thus, the entire system worked by electricity forms an up-to-date equipment insuring comfort with convenience.

The main power supply of 4,000 volts taken from the High Tension Lines of the Atlantic City Electric Company, enters through a transformer of 50,000 watts, the basement of the Research Laboratory of Spectro-Chrome Institute. It is thence distributed through a 14 feet main switchboard to the different points, for the supply of heat, light and power.

In speaking of this equipment, The Pitman Grove Review, of Pitman, N. J., dated Wednesday, April 3, 1929, commented as follows:—

"Local Contractor John F. Albers, is now installing one of the largest Hot Point Electric ranges and water heaters made by the General Electric Company, at the Spectro-Chrome Institute, at Malaga, N. J.

The above merchandise is being furnished by the Atlantic City Electric Company, distributors for Hot Point equipment.

The service and meter board necessary for this and other equipment is one of the largest installations in this section.

Col. Dinshah firmly believes in everything electrical, having one 13,600 watt Hot Point Range, one 7,500 watt range, one 5,000 watt water heater, three electric-driven oil burners, three electric-driven water pumps, many motors, projecting machines, testing equipment of all kinds, furnishing plenty of light around the whole property.

He also is expecting to install a large beacon light for aviators and a completely-lighted landing field."

After the short recess the Convention reassembled and the Minutes of the last Annual Convention (the Seventh) were presented as printed in the Red to Violet 1928 issue of Spectro--Chrome, volume 6, numbers 1 to 9.

Proposed by Bessie A. Hasenau, seconded by Clara Sandbach that the said Minutes be accepted as correct. The President in putting it to the vote, said that in the printed Minutes, his Annual Address was purposely omitted, because, it was against the prison rules to allow any matter to be sent out by a prisoner for publication and that he had pledged his word of honor last year, not to publish it, till he was out. Being now free, he was at liberty to publish it and that part would therefore appear in the present Convention records. It was as follows:—

Annual Address of the President of Spectro-Chrome Institute

American Association of Spectro-Chrome Therapists

In Seventh Convention Assembled.
My Esteemed And Beloved Associates:

1. That on the Seventh Step of our collaboration, some unusual phenomenon should occur was expected, but, my being imprisoned at such juncture was unthought.

2. However, for me, "Iron Bars do not a prison make," as my physical emotional and mental healths are absolutely unchanged, the powers energizing me through life being still the same, if not more.

3. This day, as I write, I complete 54 years on the Physical Plane, in this incarnation, according to the Zoroastrian Calendar; on the 28th current, when this message will be delivered unto you, I shall complete that age according to the Solar calculation. I have many more yet to pass for your service—human service—and I am looking forward to that time when having duly accomplished my Masters' mission, I may not be needed on the Physical Plane. His orders shall be obeyed at all costs.

4. What glory do I see for Spectro-Chrome! This Science Of Therapeutical Precision is the heritage of humanity's most precious evolution. It is incomparable, unrefutable, unchangeable, unchallengeable—it is destined to be the uplifter of the entire race and the uprooter of the false gods who have been for 6,000 years exacting the homage of the suffering humanity. It will outlive all!

5. We had our First Convention in 1921; within three years from its origination in 1920, the banner of Spectro-Chrome had waved from the Atlantic to the Pacific Coast, unopposed and successful. We started with a single room office in New York, N. Y., to-day we have an organized working institution of learning, occupying over 23 Acres. This was accomplished by dint of tremendous labor, perseverance and expenditure of all our earnings for propaganda. The Originator and his family were content to live in the simplest quarters, spending less than $1,000 per year for their personal needs.

6. During that quick expansion and success, just as I had laid by several thousand dollars to establish the Plant on the Malaga, N. J., estate, came into our life an entity from Portland, Oregon, which in 1925, schemed through the Octopus and other powerful organizations my ruin, financial and otherwise and well-nigh succeeded in wiping out my name, family and work. To save our noble cause cost me over $25,000; my savings were lost, all plans for furtherance were shattered, my liberty was jeopardized, my reputation suffered in all directions and the waves of prosperity for the Institute turned into storms of debts.

7. I need not defend myself or my motives; you who know me and have worked with me, fully cognize my integrity of purpose and undivided faithfulness to my duties and obligations. I need no defense or defender; my conscience has nothing to answer or defend in the immutable laws of Occult Philosophy and though branded by the decree of the august Government of the United States, I shall continue to face you and the World without any fear or shame; in my sentence and consequent disgrace, the glory of America for ever shall stand disgraced and fallen.

8. The inventions of the Itisometer and the Sympathometer were the results of ceaseless labor and large expenses; however, the calamity wiped out every available resource and over $20,000 had to be borrowed to fulfill the Leases of the Institute and fill the deficit. All Itisometer orders were delivered and building the new Sympathometer commenced.

9. When I took the Sympathometer orders last year, there was strong expectation that the learned judges of the U. S. Circuit Court of Appeals would see through the deliberate perjury and reverse the conviction; but, that tribunal ruled against me on the absurd ground of "mesmeric control"

and piled on me further loss.

10. At such critical moment, all my lawyers deserted me after draining me dry; I had no means left to hire others; all talked in $5,000 and $10,000 terms and no guarantee or certainty of the issues. Money was pouring out and I was earning nothing but distraction in mind. I was faced with the alternative of either losing the entire Institute's resources or losing my liberty for want of money; I chose to lose my neck to save the Institute—my pet life-work of Spectro-Chrome and cut out all legal expenses by taking the reins of the case in my hands and dispensing with the legal sharks. Would that I had never employed a lawyer and fought myself from the start!

11. I was already very heavily obligated; a beloved friend had put up security to cover my Appeal Bond of $10,000; I had to borrow $9,000 here and more there to keep up the overhead expenses; it was all going out and not much coming in; I leave you to judge what my darling wife and sweet daughter were living on when I say that since the case started they did not have even a new suit of dress; many of you who saw my white jackets in the classwork would judge of my own "extravagance." We did it gladly and without grumble; we were bound to our duty unflinchingly.

12. The Secret Organization that was one of the corrupt means employed to encompass my destruction, being finally convinced of my innocence, tried hard to interest its Headquarters in Georgia: but, to protect their so-called "National Policies," their Washington, D. C. Chief Counselor refused to intercede with the U. S. Attorney General, although he believed that justice had miscarried. He pushed me out of his office—helpless; the air around me as it were went dark before me; I saw my doom approaching, when an etheric hand guided me to the Headquarters of the Department of Justice, on a Saturday afternoon.

13. I succeeded in getting exceedingly courteous, sympathetic and human hearing from the responsible high officers of the department; except the Hon. Attorney General himself, I met practically all those who could help me solve the tangle, until I reached the Pardon Attorney James A. Finch, who was struck at the serious outlook of the case; he read the important Affidavits printed in "Railroading A Citizen"; he went over the two volumes and the whole Transcript of Record. He was determined to investigate fully and give me justice. He asked to produce the persons who had sworn the Affidavits; I visited him 14 times in all and each time he gave very long and patient interviews. I took there by automobiles witnesses from Florida and New York, involving large expenses and much loss of time and energy. Even the negro friend of the Complaining witness testified before him in my favor; deliberate perjury was clearly brought out; the Pardon Attorney spent much time on the case, to fathom it.

14. Meanwhile, the U. S. Circuit Court of Appeals had rejected my Petition For Rehearing, in San Francisco, California. I petitioned that Court to grant me permission for a Writ of Certiorari to the U. S. Supreme Court and to stay the Mandate. Such permission being graciously given, I went to Washington, D. C. and petitioned the U. S. Supreme Court for final redress.

15. The Annual Convention of the Octopus Trust, assembled in Washington, D. C. on May 12, 1927; two days later, the august tribunal of equal justice to the rich and poor which holds the scales of the Blind Goddess, denied my Petition, without giving me hearing!

16. The Mandate was handed by the Circuit Court of Appeals to the Trial Court; the Trial Judge refused intervention; by order issued by the Department of Justice, the U. S. District Attorney in Portland, Oregon, held the Mandate from entry. I requested the Pardon Attorney to prevent my incarceration in McNeil Island by transferring the order to Atlanta, Georgia. This was done and a few days later, the Commitment Order was signed and Oregon ordered me to surrender voluntarily to the U. S. Marshal in Atlanta, Ga., by or before June 10, 1927.

17. The death knell was rung to finish my career, but the good Pardon Attorney was still investigating. I wired the Trial Judge to put me on probation; he rejected my Petition. The U. S. District Attorney denied my right to any Executive Clemency. The Pardon Attorney was not satisfied though with their denials and just the time when I was to surrender, a Respite of 60 days was granted by the Hon. President Calvin Coolidge to keep me out.

18. I was temporarily saved, but, that started me thinking as to what would support my family should the worst happen. After due deliberation, I decided to establish properly equipped Resident quarters and enlarge my lecturing facilities. It put the Institute into further obligations, but, it was to be a profitable project to save the situation.

19. I desire to leave this subject here and go back to the time when I started on my disastrous tour to Portland, Oregon in April, 1925. Upto that time, the Institute was clear of any considerable debts; there was a business loan of $5,000 on a Note in the hands of our Bank, but, when I was tried and sentenced the quick financial needs necessitated further support. While the former transaction was secured by a mortgage on the estate, the Bank now wanted the Deed of the whole property for securing a loan of $10,000. I gave it—there was no other way.

20. This Note was repeatedly renewed; a year later, the Circuit Court of Appeals confirmed my sentence just as the Note was due for payment on February 21, 1927. The Bank served its Notice on February 17, 1927 that its Board refused renewal and asked for satisfaction. If I should not pay $10,055 within the four days, the Bank insisted on eviction, putting the Institute out of the premises by right of possession.

21. I faced a veritable impossibility. I held an urgent consultation with all the members of the Chief Executive Council. You will remember the regulation we approved of last year about the consolidation of all the funds of the Association. That foresight came to our rescue. A kind friend loaned $6,000 for 12 months on a Second mortgage, there being already a First mortgage of $7,500 on the estate; a three year Third mortgage was taken by the Association at the usual 6%, as an investment of idle funds, for $4,000 and once more the property of Spectro-Chrome Institute was in its own name.

22. The 60 days of Respite passed and President Calvin Coolidge granted another Respite of 30 days. I was being eaten up by tortures, financial and mental, yet, I kept on with my services to mankind, the Institute and the Association. The expansion and future provisioning needed voluminous expenses, but, I dared not stop the work. I realized keenly what it was to maintain my honor and reputation for integrity unsullied. I could not throw up my hands and dishonor my Leases with my Associates. The President of the United States and the Pardon Attorney as well as the U. S. Attorney General were out of the city

on their vacations and just as I thought I should receive a favorable termination, the papers pertaining to my pardon went before Acting Attorney General Hon. George Farnum. A legal cog in the intricate machinery of Justice missed its mesh and over the signature "Sargent" came a thunderbolt of a telegram denying submission of my Application For Executive Clemency to President Coolidge. A day later, my Respite expired. I pleaded for reconsideration of the adverse decision, but, instead of receiving a reply, I was arrested by two U. S. Marshals from my lecture room, on Saturday afternoon of September 10, 1927, on a Warrant as a "fugitive from justice!" I was taken to the Mercer County Jail in Trenton, N. J., whence I was removed three days later and transported to the U. S. Penitentiary in Atlanta, Georgia, absolutely unprepared and penniless. I am now there working in the Power House.

23. My imprisonment cut off at one blow practically the entire revenue of the Institute. To benefit my Graduates, I had signed the Integrity Policies preventing the sale or lease of any Equipment to the non-Associate and the denial of all services of the Institute to the general public. Though facing dire consequences, I am maintaining the Integrity Policies inviolate. Everything may be destroyed, but, my word of honor must stand. The past can not be recalled, the present is keeping me behind bars and for the shaping of the future, I am held responsible by The Benign Masters Whose message Spectro-Chrome Therapy is and of Whose Sacred Trust I am the repository. I must not fail.

24. Beset with numerous unexpected difficulties, surrounded by losses of all kinds, I endeavored hard by constant prayer and concentration to seek the right way out. On October 1, 1927, Saturday night, I received my response from Those Who Know Better, commanding me to "leave the United States for India," if certain plans of mine should fail to materialize or my supporters and friends should turn their back on my work at this critical juncture. This address is in obedience to that order. There is no middle way with the Lords of Karma. It is either Yes or No. Spectro-Chrome Therapy depends on Spectro-Chrome Institute and our Association depends on the maintenance of the system intact. I need not remind that the future of our Cause depends solely on the prosperity of the Institute; the Association is a creation of the Institute and its policies are subservient to those of the Institute. It is the labor of the devoted workers of the Institute that gave life to and keeps alive our Association, the interests of the two being mutual. I therefore approach you, the representative components of the great Cause of Therapeutical Emancipation to come to the rescue in this hour of peril.

25. Notwithstanding all the annoyances and losses I passed through in this year, see what actual work was put in! The total number of Associates which was 259 then, is now 281, an increase of 22; the total of our funds then was $853.39; now it stands at $6019.66. This in spite of the cessation of our Centres in Portland, Oregon and Seattle, Washington, both of which neither sent any reports nor the funds and will have to be disciplined to recover the Association property they are unlawfully holding back.

26. To add to my hardships, many of those who subscribed for a Sympathometer, are canceling their order and demanding prompt repayment of their money. Under the Lease terms, though immune from legal complications, the pure name and high ethics for which the Institute is justly respected impose a duty that such repayment be granted. The final model is actually ready in the factory; it was

tested for months for accuracy and found perfect, but, until I am out of jail, the delivery can not be made, because, none else except myself knows the standardization of the astronomical details. Forget not, dear friends, for a moment, that I am a Federal Prisoner, a convict, a felon, a man without even citizenship rights under the laws of this country. While I am active as ever mentally, I am restricted on the Physical Plane by the powers that are on the Physical Plane. It was for such and other cogent reasons that I resigned as your President and left the control to the Board Of Trustees to do the right.

27. I am—we are—fighting the biggest battle ever waged against the Forces Of Evil controlling human lives and human health. They control large resources, untold millions of dollars and immeasurable influence to do what they please. We are a small, humble, struggling few trying to save true spirit of the Brotherhood of Man, by sacrificial services. Either we keep on, or we stop; and if you wish to stop, it is necessary that I know your wishes promptly, so I may take the steps proper to save a financial disaster which may perhaps land my family into the cold streets and destroy the reputation of the Institute.

28. I believe firmly however, that you, one and all, sincerely wish me to carry on the noble work, until the desired goal is reached. It needs money—cold solid cash. It needs as much money as each can spare, afford or float. I ask for no gifts, presents, contributions or donations; I positively will accept no such propositions. I *must* have money to keep the Institute alive and *must* plan the future long ahead—long before I leave this penitentiary. If you wish to co-operate with me as your teacher and guide in Spectro-Chrome, you will have to agree to one or several of the plans outlined herein. No other way appears feasible or meritorious or which I can agree to without sacrificing my honor and principles and that never will be. The Masters have spoken their command and it has to be obeyed by me. On January 11 of next year, 1928, I enter upon the 21st year of my departure from the land of birth. The three cycles of 7 piled on my head enough troubles to crush a less dominant soul, but, I am still as vigorous in every way and with your kind backing and co-operation am able and willing to face all that may come hereafter. *Without* you, I refuse to stay in this country and shall seek other fields of activity, to fulfill my destiny and work for humanity.

29. Spectro-Chrome Institute as it stands today, is a living proof of our labors, a silent monument of what it can be if saved now from destruction. It is a potential university of learning and healing, saving the lives of the coming generations for ages untold. A few years only and it will be given the homage of the whole World, scientific and otherwise. I know its future and have no worry about that; my worry is only of *today*. What I gave the World in the past is nothing to what I have in me to give in the near future, but, with frozen large assets and without working capital, the inevitable ruin of the Institute will take place, if I slip a single step.

30. I can not in that crash involve my friends from whom I borrowed and who look to me for payment. While I gave *all* I had to America, you know what America gave me and where she put me. Yet, 1928 is a banner year for me and I expect to be with you then for a special session. Save meanwhile our Alma Mater, our Institute, our seat of future happiness, our dear ones who have toiled for you without any compensation except friendship. I care nothing for myself, because, the august Government is liberally caring for me, but, my

thoughts are only for those whom I left on the outside, the sharers of my sorrows, for joys I had none to offer them at any time.

31. I now place before you the Plans for securing the fulfillment of the objects sought and ask for your consideration.

PLAN FIRST

Each of you buy *at least* One Hundred Dollars of the 8% Preferred Stock of the Institute. You may pay so much down and the balance in about three months. Thus, the Institute will have ample working capital provided to carry on the work and also relieve the pressing needs. It will be a good investment.

PLAN SECOND

Each of you lease at least *One* extra Equipment either for your use or sub-lease to a sufferer. Use your surplus cash for this purpose. To encourage you, the Institute agrees hereby to lease such Equipments at a 10% reduction during the current year, 1927.

PLAN THIRD

Your Secretary has worked hard till now without any compensation. Under Article 36, the Chief Executive Council could have given her the reasonable sum of say $25.00 per week, for clerical services during the last nearly three years. The Institute has taken care of all her requirements and of course will take care of in the future. However, it would be within the By-Laws, if the Chief Executive Council gave her the sum of $3,900.00 for past services and also $25.00 per week for the future. Such procedure would relieve the Third Mortgage for the Institute, which might be introduced as a condition for her past services, the future going to her in full rights. This will be payable as *clerical services.*

PLAN FOURTH

I have acted in a similar honorary way, as your President till now; a similar compensation at this time may smooth many matters; the future may be given to your new President by changing the By-Laws. Please remember that the Chief Executive Council *has all this power without asking you,* but, I am proving to you our sincerity and faithfulness by putting the matter for your approval, without setting precedent or prejudice.

PLAN FIFTH

You may voice your approval to allowing the Institute to be given loans at 6 per cent interest, from the Treasury of the Association. This will be a good feature, as, soon the Annual Dues being collected it will release a sum of money for tiding over a part of the difficulties, until better times arrive.

PLAN SIXTH

Take steps to purchase the entire Spectro-Chrome Institute and run it yourselves. For years, the exemplary strictly disciplinarian business methods of the Institute, have given rise to grumbles in certain quarters. Here is *your opportunity* to do better. Collect capital enough to support my family's needs till I am released from prison; give us only $20,000.00 to pay for our emigration costs and the like; provide a reasonable Indemnity Bond that you will *not* sell the property or Spectro-Chrome to the Octopus Trust or otherwise and I shall turn over the full rights, titles, property, trade marks, copyrights, patents, everything *as is* and leave the United States without any delay. If any one or you can do better than my family and I did, I am ready to sacrifice my interests for the benefit of the World.

PLAN SEVENTH

The Institute has Resident Quarters, where proper food and tonation are given at reasonable cost. Send sufferers who need such care and for whom you may have no facilities, to the Institute. If you will take a little trouble to do only *this,* very soon the tangles will all straighten out, without your spending a dollar.

PLAN EIGHTH

Should you *fail* to agree on any one or more of the stated seven plans, then, the only and *final* course rests with me exclusively. To pay off the liabilities and obligations of the Institute, in an honorable manner, to save my reputation and fulfill my promises, I *must* sell off the Real Estate of the Corporation, pay off the Mortgages and debts, disbanding all expansions and removing all the heavy machinery through quick sale from the premises. Then, should I see the daylight again, I may start the battle of life over again; otherwise, it could not be done. Until I am released from incarceration, my family will conduct the business in that case from some small quarters; afterwards, I shall provide suitably for the maintenance of the leases and the like and shall leave this country which disowned me. It is left up to your judgment and discretion to approve and indicate *your* wishes. I suggest nothing, enforce nothing; I did my duty, may you do yours as conscientiously.

32. A few words now in regard to my most worthy colleagues on the Chief Executive Council. Our First Vice-President Kate W. Baldwin, performed her duties since the year of her joining, very devotedly; I doubt whether there could be any *better;* she stood by me through my hardships like a true friend and I express to her my sincere and cordial thanks. Similar expression of my best regards I offer the Second Vice-President Martha J. Peebles, who helped the Cause in many ways during the past.

33. However, in completing our Seven Year Cycle, I find that our future work lies in a slightly different direction. I *know positively* that the m e d i c a l profession and Spectro-Chrome *never will go together;* it is to be a survival of the fitter and Spectro-Chrome *must win by placing* "A Spectro-Chrome, In Every Home." My plans for this are matured and I am taking all precautions and steps to save embarrassment to those beloved friends who are connected with the Octopus Trust while giving us their level best. The work of my Institute is for human service and the Institute will *not* hereafter receive as its student anyone connected with that ruinous Trust. This step is vital to the safety of my mission and will prove highly beneficial to the lay practitioner, avoiding all confusion of ideas, views, therapies and consequent production of complications. Later, the same will apply to all other healing systems, leaving the field of Spectro-Chrome for exclusive services. It will benefit you.

34. I can not close this address without speaking about two others; I allude to my wife Irene and daughter Kashmira. I beseech you to give your new President Irene your undivided loyalty. If America gave me nothing else than this girl, who was instrumental in bringing to life my three sons, then, I say may America live and be blessed for ever. Irene is a darling sweet wife and a devoted mother and I feel sure that she will prove a loveable and firm-minded President of our Association. Kashmira has given me the most dutiful assistance since her youngest days and has suffered with me all the privations and miseries. My heart is sad that she has had no time in working for Spectro-Chrome to use her life for more profitable development. She is the light of my eyes and if something be done to encourage her to keep to her task, it will be warmly appreciated. I thank cordially both these colleagues for their love, patience

and devotion.

35. In conclusion, I ask you to give your very concentrated attention to this message which will be delivered through the mouth of your Secretary who will represent me. She knows what my instructions are and has my confidence. Appoint a "Plan Committee" of Seven Associates to go over this matter thoroughly and render a report to you tomorrow. Then, thrash out the whole in the Convention carefully. Make your decision by usual vote and let me know through the Board of Trustees. I shall act accordingly as is fit.

36. I bow to the Divine Will and the Laws of Karma. I bow to the command of The Benign Masters of Occult Wisdom Whose Effulgent Radiant Powers made me the humble messenger of Spectro-Chrome. I bow to you.

37. I have gone through my acid test and survived it; I shall emerge from incarceration a stronger and more powerful individual—stronger than ever in my faith, better in every way to serve you and mankind for decades to come. It is left to you to prove your loyalty and steadfastness to the Cause by staying at the helm of public service. The World needs you and Spectro-Chrome; if you will stand *your* acid test in this emergency, this Convention may safely be styled the "Victory Convention"—the evil forces will have been conquered.

38. A timely warning from One Who Knows. The Institute has Sublease Form for Self-Use of Equipment. I urge you to follow that system rigidly to escape spite work of the Octopus, to avoid "tonation" technicalities.

39. A request is made for you to visit Malaga, N. J., on Wednesday, the 30th current and look at your Spectro-Chrome Institute; you will be amazed and delighted.

40. Under Prison Regulations, I am not allowed to write for publication; hence, our monthly Spectro-Chrome will have to wait until I am out.

41. A few words to those who are looking forward to the Post-Graduate Courses. They will have *one* important requirement as qualification for enrollment; the Applicant shall have adhered strictly to the dietetic system of Spectro-Chrome Therapy to receive consideration. Begin now if you did not before.

42. The outcome of this Seventh Convention will have a wide influence on the future of the Healing Art. It will also test our stability after my death; there is a deeper import in all this; you may learn it later. Use your discretion; prove our strength.

43. My liberty can be easily secured, if someone by proper means may prevail upon the Hon. President Calvin Coolidge *to request* the Attorney General of the United States to submit my papers to him. Before being imprisoned, I completed all the voluminous preparations for that purpose and left them with the Department of Justice in Washington, D. C. Is there *any* among you who will do this great service? Single-handed, I finished my work; can you and will you all jointly find a channel to see the President?

44. I have spoken; though physically 1,000 miles away, I am with you —right here—otherwise. I am following your steps; the All-Seeing Eye of Heaven is watching me and I am watching your welfare. May we soon be joined in the furtherance of the only cause worth living for in the World—unselfish service to humanity. Peace and happiness to you, my dear friends; may all joy be yours according to merits; may the same good thoughts follow our opponents.

(Signed) Dinshah P. Ghadiali,

Prisoner No. 25, 308,

"A" Basement.

The minutes were unanimously accepted.

Reports of the First Philadelphia Pennsylvania Centre and the First Vineland New Jersey Centre were read and accepted. A rising vote of honor, in silence was accorded to the memory of Josephine B. Twichell, Eleanor Holwick, Oscar Hesse and other associates who died during the preceding year.

Drs. Nissin S. Hanoka and Simon Weissburg spoke about the First New York New York Centre and regretted that the conditions around New York hampered the progress of Spectro-Chrome Therapy.

Dr. Kate W. Baldwin said there was no system equal to Spectro-Chrome, but, it had to be used according to technique; it was necessary to use horse sense; the Itisometer was a marvelous instrument: she had an eye case in which the Itisometer persistently indicated that the Red Attuned Color Wave should be used; she hesitated; finally, she used it and the disorder disappeared; she would not do without the Itisometer; she suggested every Spectro-Chrome Therapist should have an Itisometer.

T. Bayard Brunt said that Spectro-Chrome needed a leader; in every Centre there was need of a leader; during the absence of Colonel Dinshah, the Associates were left without guidance and Dr. Kate W. Baldwin's assistance at such time was very helpful.

Dr. Charles Grapek stated that he did not push Spectro-Chrome much, because, he did not see his way clear to getting supplies, if anything happened to the Originator; it was no use creating a following in such case.

Dinshah said that such contingency was provided against and that the business of Spectro-Chrome Institute would go on even after his death.

After a few casual suggestions, the Convention adjourned till Blue 00 Black.

First Day Afternoon Session

The Convention reassembled at Blue 10 Black. This session being open to invited guests, the Auditorium was full. Many cases were introduced personally by the Spectro-Chrome Therapists attending them. The interest of the audience did not flag for a moment and all enjoyed the information given.

The President greeted the guests on behalf of the Association, thanking them for their presence and introduced as the first speaker of the afternoon, Dr. Kate W. Baldwin. He extolled her services to the cause, her steadfast adherence to Spectro-Chrome and the good work she did among sufferers. She was given enthusiastic applause.

Dr. Baldwin spoke at length of the benefits of Spectro-Chrome, even in surgery—constructive surgery. The bones healed quicker, post-surgical complications being also avoided and the formation of harmful adhesions was rendered less likely. One doctor wants Spectro-Chrome used to prevent adhesions.

Ruth Sinclair spoke of a case of a boy, corroborating the facts.

Esther Jennings introduced one of the sufferers under her care; she was Mabel McCullough of 17. She was in bed for 6 years; could not help herself; she could not turn over in bed; was like a newborn infant; she had what they joked as "sleeping sickness of a waking kind"; weighed only 60 pounds when Mrs. Jennings took her; she was then not digesting much food; was restless 24 hours of the day. Now, after a few months of Spectro-Chrome care and food she ate well, slept with comfort and talked some; her Backelim was in order; she walked some too; her weight went to 86 pound Mabel was "put through the steps" by Esther Jennings, who gave much credit to the guidance given her by Kate Baldwin. The case was listened to and Mabel's talk and walk observed with rapt attention by the audience. The Association extended a hearty vote of thanks to Esther Jennings. Mrs. Jennings also reported the case of a girl who was pronounced by the Medical Doctors as having a "leaky heart." Under the Magenta, this case was brought to normalcy by Spectro-Chrome, the ache in her joints, the rash on her body and other conditions which kept her out of school disappearing quickly.

Dr. Nissin S. Hanoka introduced a sufferer Zelda Hananel of about 5. This girl had fever two years ago and in five days five Medical Doctors were called to care for her; they did not know what the illness was and the girl was bedridden with infantile paralysis. Nissin S. Hanoka undertook the case and in about two months, the girl began to walk around, thanks to Spectro-Chrome. Zelda was asked to walk and she moved about to the applause of those present. The parents of the girl are staunchly for Spectro-Chrome.

Mrs. Ida Souder introduced the case of her son William of 21. He had a partly collapsed right lung, was suffering seriously from pulmonary phthisis three years ago and was sent to the New Jersey State Sanatarium. He had 23 hemorrhages of 8 to 10 ounces of blood each time and even with 20 enemas his bowels could not be brought into action. His father pleaded with Colonel Dinshah, to receive the boy under the care of Spectro-Chrome Institute on the opening of the Resident Quarters of the school. On July 4, 1927, William Souder was brought into Spectro-Chrome Institute, in a hopeless state. The Medical Doctors of the sanatarium had given the father to understand that the boy would not live three weeks. When William was taken out of the automobile and placed on a chair, he crumpled onto the floor. Dinshah received him, after sizing the gravity of the situation, at Turquoise 00 Black and promptly used Spectro-Chrome measures. Within one hour and a half, the boy sat up in bed and by four hours had Backelim. Soon after, he ate a hearty meal and the next day went downstairs to dine! On Monday July 11, 1927, just a week after admission to the Resident Quarters, William Souder joined the 46th class in Spectro-Chrome Therapy and graduated with his mother in due course!!!

While William was studying, the Medical Doctor of the Sanatarium called to see him and was surprised to find William in the class; he could hardly believe his eyes when he saw William's right lung, which was formerly partly collapsed, breathing normally! He admitted he did not comprehend what did it, but, the fact was there. William is still alive.

Dr. Emma Gabriel spoke of a boy who went through an automobile accident and had infection; he was under Spectro-Chrome 24 hours of the day. Mrs. Agnes Russell, his mother spoke at length, how it was a case for a surgeon; the boy recovered.

Jokes were cracked about Spectro-Chrome abscesses being broken with Spectro-Chrome. Esther Jennings described a case of abscess.

Dr. Welcome A. Hanor said he was about the only Medical Doctor of twenty who graduated in Spectro-Chrome, from the class in Bradford, Pa. He spoke of the valuable aid Spectro-Chrome was; he described a case of a woman who was discharging from ankle to hip and had no relief for 10 years; 12 tonations of Spectro-Chrome restored her to normalcy.

Allusion was made about the Resident Quarters of Spectro-Chrome Institute. Here, the sufferers were taken in hand under exclusive Spectro-Chrome methods and restored to equilibrium in very short time; the rate was moderate; only $35.00 a week and no extras, tips or gratuities.

Dr. Margaret Morris spoke of the wonderful relief Spectro-Chrome gave in her broken ankle case; stayed day and night under Attuned Clor Waves; son was furious, but, no surgeon was called.

Bessie A. Hasenau, First Vice-President of the Association, spoke of an interesting case of a woman who was the Principal in a school in Lakewood, Ohio. She was declared suffering from Chronic Nephritis and had trouble in walking. Spectro-Chrome set her right; she would take the next Course.

Nachman Diamond, of Wildwood, N. J., then told the audience, how he suffered with illness and went to the best Medical Doctors, but, could obtain no relief. He was 71 years old, but, thanks to Spectro-Chrome he was restored to health in 4 weeks whereas the Medical Doctors wanted at least 5 months. He weighed 148 lbs. He normalated a boy of 11 suffering from Tonsilitis; they wanted to operate on him; tonations set him right and the tonsils stayed in. Toothache in a girl was relieved by one tonation of Indigo. Many troubles of his wife were normalated by Spectro-Chrome.

Richard E. Sandbach, Second Vice-President of the Association, spoke of a man of 36 years, who collapsed and was injured in his back. He laid flat paralyzed. Several Medical Doctors were in attendance; chiropractors helped some; he recovered under Spectro-Chrome. Many other cases were described.

FIRST DAY EVENING SESSION

Kate W. Baldwin, M.D., F.A.C.S., the champion advocate of Spectro-Chrome, opened the evening session, amidst loud and continued applause. She enumerated several bad cases of hiccough, heart disorders and cataract which all responded to Spectro-Chrome; she trusted Spectro-Chrome in cataract cases. She described a case of a woman who was operated upon without benefit by someone prior. She had a severe heart disorder; the Medical Societies will not believe it, but, one hour under Spectro-Chrome acted like a miracle; the heart became normal. She spoke of the case of the burnt girl Grace Shirlow and described at length how Spectro-Chrome took care of all the requirements step by step, including the action of the kidneys. She said, "Any burn will yield to Spectro-Chrome; the pain disappears in half an hour. It is no use going on with description of cases; they all act the same way. In cancer, the knife does no good; if taken early, Spectro-Chrome will normalate them. Cutting into normal tissue is not good. Even in cases where ether is to be administered in surgery, the use of Spectro-Chrome prevents nausea. Spectro-Chrome knits bones together quickly. I am opposed to serums and antitoxins. It makes me sick to think of so much suffering; Spectro-Chrome can do much good."

Her address was heard with rapt attention; it contained many suggestions and encouraging points.

Dinshah spoke caustically against Vaccination; he would not countenance the introduction of any foreign bodies in the blood stream. He said the law was not about compulsory vaccination, but, compulsory education. He said, "Let your children keep on going to school and let the school teacher keep on sending them home; then the blame of truancy will not be on the heads of the parents. No power on Earth can ever justify compulsory puncturing with a tool; it is assault, battery and mayhem; it is nothing but a rape on the body."

Bessie A. Hasenau spoke of her mother's case; she had not been healthy in 16 years; her heart had no rhythm; in less than 2 days. Spectro-Chrome showed its good effects; her mother was in good health now. She spoke about a case of arthritis in an old woman, where actually crystals like rock-salt were appearing on the body; Spectro-Chrome normalated her.

The session adjourned at Yellow 20 Black.

SECOND DAY SESSIONS

The second day was all devoted to the relation of the experiences of Dinshah in Atlanta, Georgia and what difficulties he surmounted to preserve his health and principles in diet and otherwise. He tendered the highest praises to the indefatigable investigation of his case by the United States Pardon Attorney James A. Finch, Assistant Attorney General Mabel Walker Willebrandt and Director U. S. Bureau of Investigation J. Edgar Hoover. He also spoke of the considerate action of the U. S. Attorney General John G. Sargent, who recommended his release and the President Calvin Coolidge, who ordered him freed from incarceration. He was vindicated. He declined to speak openly or publish anything about certain of his activities while there, hinting that what he did was his duty as a citizen and as a demonstration of his loyalty to the Government of the United States.

Moved by Dr. Welcome A. Hanor and seconded by Ruth Verbeck "That this Association hereby expresses its most grateful appreciation and thankfulness, to U. S. President Calvin Coolidge, U. S. Attorney General John G. Sargent, U. S. Assistant Attorney General Mabel Walker Willebrandt, Director U. S. Bureau of Investigation J. Edgar Hoover and U. S. Pardon Attorney James A. Finch, for their careful consideration of the case of our President Dinshah P. Ghadiali, which resulted in his receiving the extension of Executive Clemency and release from incarceration in the United States Penitentiary, at Atlanta, Georgia." Unanimously passed.

Dinshah showed diagrams of his many new inventions, among which were the Caloreg, Kaspirameter, Pranameter, Shvasayantra, Sanslide, Currecut, Leasemeter, Irradiator, Collecoin, Causetone, Tonepan and the like. He was working on combining them all with the Itisometer and Sympathometer in panel boards, so Spectro-Chrome may be lifted from being separate Equipments to a complete System of Automatic Precision. He designed all this in captivity; he also wrote about 3,000 pages.

He then spoke of the dishonesty of several Spectro-Chrome practitioners in Vineland, N. J. and hinted that the time was ripe to change the name of the system from Spectro-Chrome Therapy to Spectro-Chrome Metry, because Spectro-Chrome was now a complete system of measurement of the imbalance.

Moved by Clara Sandbach and seconded by Milda Steinharter that the Convention approved the name Spectro-Chrome Metry. Unanimously passed.

Moved by Margaret Morris and seconded by Simon Weissberg that the Convention also approved the name Spectro-Chrome Metrists. Unanimously approved.

Dinshah promised that as soon as the Spectro-Chrome Institute was ready he would act upon the suggested changes; Spectro-Chrome Metry was to be considered at once and when the Panel Board System would be ready, the designation of such Graduates would be Spectro-Chrome Metrists.

Moved by David McKibbin and seconded by Benjamin Kessler that Spectro-Chrome Institute be asked to establish a preliminary training course; John A. Cohalan and Raymond Blefgen advocated the raising of standards. Unanimously passed.

Dinshah said that as the system was a machine-worked system, longer courses would serve no purpose; however, when the Panel Board would be inaugurated, he would have a Preliminary Home Training Course, before the Resident Course be taken, so as to save the teaching time of Spectro-Chrome Institute in the theoretics.

THIRD DAY SESSION

The Third Day of the Convention had only a single session in the morning, but, it was full of events; there was warm discussion.

The Secretary-Treasurer Kashmira Dinshah having resigned her office, Irene Grace Dinshah was appointed by the Board of Trustees as Secretary-Treasurer. Her report as Secretary was read and unanimously accepted by motion of Bessie A. Hasenau, seconded by Nissin S. Hanoka.

The Treasurer Irene Grace Dinshah submitted her report; it was unanimously accepted by motion of Simon Weissberg, seconded by Bessie A. Hasenau.

Dinshah then suggested certain changes in By-Laws, to protect the lay practitioners of Spectro-Chrome, by having all sufferers join as Collaborators of the Association and the like. Very interesting and pungent discussion ensued; the practitioners of Spectro-Chrome who were licensed under their State Laws for their system of therapy, raised strong objections. Martha J. Peebles, M.D., raised the biggest objections against such "dictatorship" on the part of Dinshah. She did not care two hoots for medicine, but, she thought it was not fair to compel such measures.

Dinshah said that the attitude of the Medical Trust against Spectro-Chrome forced him to take some definite steps to protect the lay practitioners. The Medical Doctors and Osteopaths were self-protected; they did not need him to protect, but, he had to consider the lay practitioners for justice.

Charles Grapek, M.D., D.O., raised objection to "dictatorship."

Bessie A. Hasenau spoke in favor of protection; Nissin A. Hanoka D.D.S., spoke in favor of devising some way to protect those who needed such assistance.

Dinshah said that he was far from being a czar, but, it was our duty to protect ourselves. He was bound to take every possible step to prevent persecution of his Graduates, at the hands of the Medical Trust. He was not a dictator, but, it needed a steady hand and iron nerves to fight the Medical Octopus, by steering the ship of Spectro-Chrome to success.

Benjamin Kessler moved and Simon Weissberg seconded that a hearty vote of thanks be tendered to Kashmira Dinshah our retired Secretary-Treasurer, for the able way in which she conducted the duties of her offices. Unanimously passed.

Moved by Marguerite Sait-La Roe and seconded by Clara Sandbach that a hearty vote of thanks be tendered to Irene Grace Dinshah for the efficient way in which she managed the affairs of the Association, during the absence of Dinshah. Unanimously carried.

At this point, Dinshah stood up and offered to the Chief Executive Council his resignation as President and Trustee and as a Fellow of the Association. He said the rules of common sense, decency and corporations, made it imperative that he resign to prevent embarrassment to his worthy colleagues and co:workers, owing to his incarceration.

Dinshah's resignation, so sudden and unexpected, produced an electrical effect upon the Convention; there were murmurs of amazement and regret all through; Bessie A. Hasenau moved and Yetta Kessler seconded that an unconditional and absolute vote of confidence be given to Dinshah for his unselfish and loyal services to Spectro-Chrome and that he be requested to reconsider the resignation and withdraw it. The motion was further supported by Clara Sandbach and Margaret Morris. On being put to the vote by the Vice-President, the motion was unanimously carried amidst the applause of the Convention which rose to its feet to honor the Originator of Spectro-Chrome.

Dinshah withdrew his resignation and expressed his cordial thanks and appreciation for the confidence. He gave his life to Spectro-Chrome and would do his level best to serve his Associates.

Martha J. Peebles said she realized what heartaches the wife and daughter had to go through in his absence; but, they managed the business of the Institute devotedly and she was glad the days of trials were over. She said, "To look upon the marvelous instruments evolved by the man, as shown on the screen, would suggest to anyone that they are the results of a master mind. It simply floors one to look at those charts that will be used in the future. I have been nearly eight years in Spectro-Chrome. After 22 years of straight medicine and surgery, I went into Spectro-Chrome and I found that nothing in medicine even approached Spectro-Chrome. I want Colonel Dinshah and Spectro-Chrome Institute to reap their rewards so justly deserved."

Dinshah thanked the worthy doctor for her sympathetic expressions and assured her that in the new system he was designing for the protection of his lay Graduates, he would have nothing that would do any injury to the professionally licensed practitioners. He promised a just, fair and equitable dealing to all.

William Flick, M.D., expressed his pleasure in coming to the Convention. He was not a speechmaker. He was starting to use Spectro-Chrome in a

new territory. He wished all success.

Dinshah P. Ghadiali, Irene Grace Dinshah and Bessie A. Hasenau were elected Trustees for the year. They were duly installed.

The Board of Trustees, appointed Richard E. Sandbach as First Vice-President and Ruth Verbeck as Second Vice-President. They were promptly installed.

There being no further business the Convention dissolved around noon.

The meeting of the Chief Executive Council took place right after.

SPECTRO-CHROME

JANUARY 1933
RED 1933

VOLUME 9
NUMBER 1

Devoted To: Spectro-Chrome Metry **Dedicated To: Service Of Mankind**

TENTH ANNUAL CONVENTION NUMBER

Editor
Dinshah P. Ghadiali

In course of life you reach a time,
When Doctor, Lawyer, Priest alike,
Does fail to give the sought relief,
To troubled mind or burdened heart—
Then come to me;
I help you to help yourself.
SPECTRO-CHROME

Single Copy 50 Cents

DINSHAH'S PLATFORM AS ADOPTED BY HIM SINCE 1891

The Boundless Oscillatory Ocean Of Thought Is Essentially Universal And All-Pervading; It Is the Individualized Monopoly Of No Person And Is the Common Heritage Of Humanity's Evolution: Thus, What A Development Of Unrevealed Ages Gave Unto Me In the Form Of Knowledge In My Present Incarnation Is No Distinctive Acquisition Of Mine For My Sole Use, Benefit Or Elevation, But, Is All For Thee And Is Thine, Without Condition, Without Obligation, Without Expectation. I Fear No One; Only God Above And Conscience Below, And From Them I Have Nothing To Fear.

Attuned Red Wave

(The World Through The Eyes Of Dinshah)

The Wheels Move Again

After years of enforced silence for economic as well as financial reasons, Spectro-Chrome again makes its bow to its friends and expects to continue each month to deliver its message of emancipation for the suffering.

Dominated by crushing and overwhelming evil forces in the pernicious healing laws, Spectro-Chrome forged its way through enormous difficulties, keeping its head above the waters of depression and emerged victorious. The reports of the 1930 and 1931 Annual Conventions printed in two issues will give some good glimpses. The grinding machine already started. It will not stop until Spectro-Chrome reaches its culmination, by complete fulfillment of our aim — Spectro-Chrome — In Every Home!

The Favorscope

For over 12 years, the crying need was felt of the proper time table to determine Sunrise and Sunset. Almanacs, as usually printed, are meager in character and based upon the Longitude of Greenwich, Washington or some other Longitude, ignoring mostly the due corrections for Latitude. We started to smooth this difficulty, but, finding no proper solution, finally standardized all our work by computing on the basis of Malaga, New Jersey, where our Central Office and Research Laboratory is situated. This now is the Spectro-Chrome Meridian and the monthly Favorscope will be a welcome addition to the many facilities given to our Normalators. Just add or subtract the indicated minutes, in your Local Standard Time and you will have the Initial Sunrise and Sunset Favor for that day for your Locality. It cannot be made easier.

Doctor Bootleggers

Dr. Arthur Dean Bevans, a former President of the American Medical Association, told the United States Beer Committee, that 90 per cent of Whiskey prescriptions are for beverage purposes. It was a clap from the clouds! Dr. Bevans said, "A large number of doctors have not been able to resist the temptation of selling those prescriptions, some selling the whole quota of 400 pints a year. This makes it possible for a Doctor to make about $1,200 a year out of those prescriptions."

What Others Think

In an address at Washington College, Chestertown, Maryland, Dr. William H. Pearce, of Baltimore, Maryland, called medical science "a dismal failure." He quoted Dr. J. B. S. Haldane, a British scientist as authority for the statement that the death rate for Americans from 30 upward, had been increasing steadily and blamed the development of "medical science" and the expense of "medical art" as the cause. We agree; medicine never was or can be "a science"; it is only a jargon of names of worthless articles.

Attuned Orange Wave

(Humor, Wit, Fun, Satire, Tickle, Irony)

The Unusual News

Irene Grace Dinshah did not make her yearly addition to the family tree around Convention time. The Members expected the announcement of the sixth son's arrival, but, alas, the record was spoilt! The Mogen David is yet incomplete!

I. G. D.

Air From No Air

Here is a puzzle to the writer:
1—A stick with no air
2—Is connected with a vacuum pump
3—The air comes out!

Cyrus Dinshah.

What He Found

Colonel Dinshah returned from his Milwaukee, Wisconsin tour after attending the Twelfth Annual Convention. Starting to write the Spectro-Chrome, he looked around for his soft pencil and opened the drawer in his right hand table—half a dozen dusting caps for women pulled out and no pencils! One should have seen his face. An assistant in the Research Laboratory had changed his work table.

Lorraine M. Whitman.

Attuned Green Wave

(Spectro-Chrome Metry, Spectro-Chrome Normalators)

TENTH ANNUAL CONVENTION

Consolidation of American Association of Spectro-Chrome Therapists and Scientific Order of Spectro-Chrome Metrists

A Memorable Convention

The Tenth Annual Convention of the American Association of Spectro-Chrome Therapists and the Scientific Order of Spectro-Chrome Metrists, was held on Friday, November 28, 1930, in the Adams Room, Hotel Tuller, Detroit, Michigan.

As the joint registration took some time, the assembling was delayed till 11:11 A. M., when the President, Colonel Dinshah P.

Ghadiali knocked the gavel; Members from different parts as Malaga New Jersey, Kenmore New York, San Francisco California, Dunkirk New York, Cleveland Ohio, Windsor Canada, Binghampton New York, answered the call and the meeting started with interest, as an over quorum was present.

The Secretary, Irene Grace Dinshah being absent owing to the close approach of childbirth, the President read the notice:

Tuesday, October 28, 1930.
Chronos: Green 32 Black.

NOTICE OF TENTH ANNUAL CONVENTION IN JOINT SESSION

Dear Friend:

1. Our Tenth Annual Convention will assemble on Friday, November 28, 1930 at 10 A. M., and the following day, in the Adams Room, Hotel Tuller, Detroit, Michigan, Colonel Dinshah presiding.

2. There will be a public session from 8 P. M. on the opening day. All other sessions for the two days will be devoted to the important work, as outlined below:

(a) To approve the "All Bars Down" policy to accelerate the spread of Spectro-Chrome Metry.

(b) To amend the Charter of the Scientific Order of Spectro-Chrome Metrists to meet the new conditions regarding Collaborators and Normalators.

(c) To consolidate the two organizations into one to meet the altered conditions imposed by the "All Bars Down" policy.

(d) To reduce the Annual Dues to Six Dollars.

(e) To change the existing Constitution and By-Laws to suit the new requirements.

3. The management of Hotel Tuller will do all in its power to make this convention our best. Write for reservations in advance, to Colonel Dinshah, Suite 1437-1443, Hotel Tuller; rates for room with bath are from $2.50 up, according to position.

4. There will be a Spectro-Chrome banquet on Dinshah's birthday, at 6 P. M., Friday, November 28, 1930; secure your ticket in advance by sending $1.00 to us at the Hotel Tuller; your friends may join.

5. We aim to make this our best Convention. The program will have numerous educational activities and your visit to Detroit, Michigan, will carry with it a pleasant memory.

6. Be sure to have your Affiliation Card for registration; no admission without presentation of the Affiliation Card.

7. Looking forward to greeting you in the great Automobile City, we remain

Yours In The Service of
Mankind,

COUNCIL OF GOVERNORS,
By Irene Grace Dinshah,
Secretary.

First Day Morning Session

Donald N. Allen, Leona D. Burghardt, Bessie A. Hasenau, Isabelle Radcliffe and Oscar C. Keener were introduced as speakers; Dr. Oscar C. Keener, President of the Buffalo Optometric Society took the floor first.

He spoke with a smile of the Buffalo, New York, experiences of Dinshah and how when Dinshah was placed behind the bars there he looked like a "circus monkey"! (Laughter). There were many difficulties placed in his way in using Spectro-Chrome in his optometric work, but, as he worked under the direction of Dr. Welcome A. Hanor, a licensed Medical Doctor, he had no trouble. Spectro-Chrome worked by supplying the Chemical Elements lacking; it worked well. He had a personal experience in the case of his little boy who had 103° fever; He was loaned a Transport Spectro-Chrome of only 400 Watts. In a tonation of one hour, the rash of Measles broke out and the next morning the skin cleared. It was hard to believe!

A man burnt a leg to the bone in two spots. It was six weeks before he went to see Dr. Keener; the leg was then not healed. By the third tonation the Inflammation receded; eleven tonations formed healthy scab; soon the burnt spots were in perfect condition.

A young man was sent to him with a diagnosis of Tuberculosis; he wanted to go West, but a woman brought him to Dr. Keener. One month's tonation changed the outlook. He was no longer tired and sleepy and became a different boy.

Dr. Keener stated further that the First Buffalo, New York, Center had a meeting on the 17th of the month; Donald N. Allen suggested there that a sanatorium conducted on Spectro-Chrome principles should be started. It was difficult to live and eat according to Spectro-Chrome teachings as common sense would dictate; however, Dr. Keener said, "I quit smoking"! (Applause).

The President explained the principles recommended in Spectro-Chrome system on the ground of the purity of the Human Aura and expressed his pleasure at the good sense exhibited by Dr. Keener in quitting the tobacco habit.

Isabelle Radcliffe spoke a few words of encouragement.

The First Day Morning Session was adjourned at noon.

First Day Afternoon Session

The Members reassembled at 2:30 P. M. Donald N. Allen gave some of his experiences. He could not do much because of the Buffalo case against Dinshah. Asthma was cleared in two tonations.

His baby of 3 had Bronchial Pneumonia; stayed up all night,; in the morning was not better; Dr. Keener had a Transport Spectro-Chrome and the baby was laid on the bed under the Turquoise Attuned Color Wave. Returning at 1 P. M., he found her playing and breathing normally.

A case of a woman run over by a truck. She had an indentation in a leg with a small hole full of pus. Two weeks of Spectro-Chrome closed the hole and the pain was all gone. She was happy; she was a woman of 60. He had no proper facilities for Spectro-Chrome service.

He reported a case of a woman of 60 suffering from Cancer, according to medical diagnosis. She was treated by Medical Doctors with Radium and had no operation. She took Spectro-Chrome and was getting along fairly well and would come through if she did not violate the Rational Food System. They could not see the advantage of the Dinshah Food System; he had to knock it into their heads! That was one reason why they felt the need of a sanatorium where the air might be right, the food correct and tonation might be received in the proper way.

A baby's case of 19 months of Eczema; the specialists could not do anything; they rubbed ointments; then the family heard of Spectro-Chrome. Children are generally afraid of lights, but, we put the Equipment and left. For several days he cried for "the light" and was lonesome without it. It was all a red mass of sores and raw like beefsteak; he had two pairs of mittens tied. In two weeks, everything returned to natural condition and there were no patches. At the end of three weeks, all cleared except the arms; he was digging in the arms, but Turquoise removed the trouble; the normalation was finally made with Lemon. Donald N. Allen looked at Spectro-Chrome from the service angle, the service it rendered to humanity.

Dr. Harry Meixner asked Allen how he knew that the woman had Cancer. Allen replied it was the diagnosis of the Medical Doctors; he followed solely Spectro-Chrome methods on Areas 10 and 11.

Dr. Meixner said he was in Spectro-Chrome for two years. One case came with good Sinus trouble; in four months of tonation the trouble disappeared. In some cases, he had no results. In a case of a paralytic he could do nothing. Cases came out fairly well in two tonations in Eczema. Spectro-Chrome must be handled rightly.

In case of Constipation in a woman of 54; none could do anything, medicine could do nothing; he gave Orange an 8-9-10-11; he used the objective lens of his Graduate Spectro-Chrome for 15 minutes; it responded.

Dr. Edward D. King was then introduced. He said he was only a beginner in Spectro-Chrome; he had a variety of cases; recently, one or two were spectacular. There were three conditions in one sufferer; the woman was under his care two years; at 1:30 on a Saturday night there was an automobile accident; clipped a fender and teeth were knocked off. The room was cold in the police station; she developed a severe Broncho-Laryngeal trouble; she wanted irrigation; he gave it to her; he told her of Spectro-Chrome. She was given Green on Areas 1-2-3-4-5; also Turquoise; she sneezed only once; she called him at 4 in the afternoon; she was coughing less and was raising phlegm; before that she had a tight Cough; she was rocky on the Sunday, but, the following week was her menstrual time when she would have terrible headaches. Before, he had used on her several modalities; he would relieve her

in 3 days; she would have boring, lancinating pains; her build was well proportioned, vital and emotional; she was an executive of over 30 girls. He gave her 20 minutes of Green and the rest of the time 30 minutes of Scarlet. She had a Spiroscil of 18 and Kardoscil of 68 or 70. Magenta was given on 4 and 18. When she came next, her menstruation was normal and free she told him, one hour after she left the office. He gave her Green systemic, 35 minutes of Blue on 1 and Magenta on 18; she went smiling. That was 3 weeks ago. She called a week ago and said something terrible happened. She had a lump—a tumor in the breast; the diameter of the mass was 2¼ inches and it was 1 inch thick. He gave 1 hour of Indigo; she could touch its lower surface and the redness disappeared; soon the soreness disappeared, the lower part was gone and the upper not prominent enough to distort.

Dr. King had a case of green apple Cramps; relieved in 20 minutes.

One tonation knocked out an Acute Cold. Some cases were stickers; one woman was on the verge of a stroke; he detoxicated her by fruit juices for a week; her Blood Pressure was 200 to 240; she was a heavy meat eater; she said, "Ruby, I like that bright light;" her condition was blood poisoning. After 10 days, her Blood Pressure was 150; it was formerly so great that the Abdominal Aorta shook the whole body.

Alexander L. Spence next took the floor. He was a Customs Officer in Canada and his Spectro-Chrome experience was small. A window-washer in Windsor, Ontario was burnt and was a mass of scars after 2 years; he thought of Grace Shirlow then, how she was normalated without complications.

Spence had a violent Headache one night; he took Blue and in 15 minutes went to sleep, with the Headache gone. Using Magenta on 4, he felt his own activity improved; he was getting results. His brother was using Spectro-Chrome on a boy whose face was cut, the teeth pushed in and who was swollen in appearance. The pain vanished; the boy wanted Spectro-Chrome to be applied. Spence was glad he got the Spectro-Chrome.

Leo A. Lefave said he had no opportunities in wide ranges yet, but had a spectacular thing happen one month ago. A terrific Headache came one afternoon; the head choked; he drank coffee though he should not have done so; 30 minutes of Turquoise and the head cleared; he felt as if born that day. He liked to get the success common to older Normalators; he spoke of his work with a boy having Tonsil Trouble.

Dr. George A. Smith reported many cases where he used Spectro-Chrome. Two weeks ago a man was burnt by a gasoline explosion; his neck was burnt 2 inches wide; they brought him to Dr. Smith from hearing lectures. Spectro-Chrome was used; in a few minutes the burning ceased and the pain subsided totally in 1 hour. The neck healed soon on one side. He feared there would be scars, but, there were no scars.

A man was hurt by a flywheel flying off from the shaft. The injured spot was inflamed and bare. one tonation removed the swelling; he expects to go to work soon.

He found it to work quicker in acute cases. Arthritis responded to Spectro-Chrome. A woman had been doctoring 4 years and was feeling worse at the end of one month because she was eating white bread and the like. She was to the Ford Hospital and other places; she felt better after proper service with Spectro-Chrome.

A woman had Uterine Ulcer; pain subsided by Spectro-Chrome.

A woman had Sciatic Neuritis; she had a lump on a finger; the lump and swelling and pain all went under Spectro-Chrome.

A woman had a Goitre; she was normalated in a month.

A man had a hard stony Goitre; in a month it was gone.

Colds, as some spoke of before, disappeared amazingly. A woman at a party had Sore Throat; Turquoise cleared it.

A child, a few days ago, vomited for 48 hours; an hour of Orange stopped it and in two tonations, the child was well; it was sick from its infancy, but, a few tonations with Spectro-Chrome started it building rightly.

It was surprising how the skin built by Spectro-Chrome. He spoke with emphasis about disorders of the Stomach; he had such trouble, for which he used unlimited medicine and chiropractic. He thought using Spectro-Chrome would be a waste of time, but, the Orange in 20 minutes took out the gas. The results of Spectro-Chrome were remarkable; they could not have such success without Spectro-Chrome.

Dr. Oscar C. Keener asked the President about the Junction Period in the Occult Operation Of the Great Sympathetic Nervous System; Dinshah explained it.

Numerous telegrams were received and read:

From Dr. Lettie Allen Ward:
Birthday congratulations and best wishes for successful convention from the First Philadelphia Center Spectro-Chrome Normalators.

From Clara Kozlowski and Mary Nowacsyk:
Congratulation for a happy birthday and success to Spectro-Chrome and its Normalators.

From Max and Milda Steinharter:
Congratulations best wishes regret can not be with you.

From Benjamin and Yetta Kessler:
Sorry we can not be with you; many happy birthdays and much success for the continuation of Spectro-Chrome Metry.

From Dr. Simon Weissberg:
Greetings best wishes to our teacher and President Dr. Dinshah and all assembled. May all bars down policy make way for a bright dawn for Spectro-Chrome Metry. Let the spirit of service to mankind be the soul of this Convention. Though absent, am with you in spirit. Fraternally yours.

A letter of congratulation was

read from Lazar J. Krivokapitch; he regretted his absence.

Dinshah exhibited several "joke" Slides. They were favorably received, because, they portrayed some phase of human character.

The First Day Afternoon Session was adjourned at 5 P. M. for the Spectro-Chrome Banquet, which was served in the Ball Room at 6 P. M. It was prepared on strictly vegetarian principles and had no meat, fish, eggs, tea, coffee, tobacco or intoxicants. Many visiting ladies and gentlemen, members of the families of Spectro-Chrome Normalators and others were present. A few of the visitors grumbled because no smoking was allowed, but, the majority pronounced the banquet a success. Dinshah was greeted on his birthday. He thanked all for their friendship.

First Day Evening Session

The Evening Session being open to the public, the Adams Room was crowded from one end to the other. Among the Graduates of Spectro-Chrome Institute, there being many talented singers, the program provided entertainment with variety. Katherine L. Slagle sang "Mother Of Pearl" and "Be The Best Of Whatever You Are." This was followed by Agnes Patterson with "Love Comes Creeping Into My Heart" and "Oh, Lovely Night," with a duet, "Awake Dearest One," by Miss K. L. Slagle and Miss A. Patterson. Jean Theodore sang "Oddio dell Passoto" and Mr. Arnoni sang "Sleep In the Deep." Professor Paccini was at the pianoforte and gave excellent interpretation of the music. Dante Borton sang from La Traviata. Each singer was given voluminous applause and the appreciation of the audience was very effusive.

After Dr. Oscar C. Keener made a short speech, Bessie A. Hasenau of Cleveland, Ohio, reported the case of a girl who fell from a porch; she had no organic damage, but, she could not walk for 3 years; she had fallen face down and her right hand was helpless; she talked jerkily; Spectro-Chrome normalated her; at 10½ in age, she walked straight and her speech was good.

A man of 50 had High Blood Pressure; came with 170 and in 7 tonations was reduced to 152; he had Angina Pectoris, which also was remedied.

A case of Fibroid Tumor as big as two fists and another on top; Medical Doctors had promised her relief by Radium, but, the husband was against it. Mrs. Hasenau placed a Spectro-Chrome in the home. There were decided hemorrhages, but they stopped; slender fibres were passed out and the tumor gradually passed away. Since July, the sufferer did all her housework and though 7 months elapsed there was no recurrence. Bessie A. Hasenau emphasized that she used the Itisometer for all cases.

Jessie E. Ness spoke about the next class on "Occultism and Higher Life." She said in effect, "Dinshah brought the Science of Health to the Western World. The students wanted teachings; they never had such an offer; no teacher can teach the way Dinshah does. Ontology is the science of why and where; men and women want to know; come along and join in this intellectual feast! Those who passed through his hands know how he chases the ego away. We are grateful; it is a course not to be surpassed. Sign up for it, as it will do his heart good; if you do not, you will be the loser."

A poem written by Thomas Ness to the 61st Class of 1930 in Spectro-Chrome Metry, was then recited by Dr. Frank L. Richards:

Congratulations members all,
You have passed the final test,
Ready now for Duty's call,
Keen are you to do your best.

Just a warning from a friend,
When to Normalate your ache,
To the mind this message send,
"Spectro-Chrome is not a fake."

When you've thus assured your man,
Find the area indisposed,
Turn the light (per Dinshah's Plan)
On the cuticle exposed.

When you've run the proper time,
Nothing more, and nothing less,
Then it's donuts to a dime
That the thing's a huge success.

There's no need to sigh and sob
Should you meet with a rebuff,
Spectro-Chrome will do the job
Write that down upon your cuff.

To our Teacher, best of men,
We should all our thanks return,
When he's passed from mortal ken,
May his Light forever burn.

May his name remembered be
While the ages come and go,
As the one who set men free
From the doping Medico.

The First Day Evening Session adjourned in a round of mirth and applause.

Second Day Morning Session

The Second Day Morning Session opened at 10:15 A. M., on Saturday, November 29, 1930. Dinshah explained at length the Occult Operation of the Great Sympathetic Nervous System, in accordance with recognized astronomical phenomena, gravitational influences and the potency of Light.

Piro Murshida Rabia A. Martin, President of the Sufi Society said she was pleased at the thought of taking the Healing Triangle to California and invited Dinshah to take the work again to the Pacific Coast. Dinshah thanked her for her inspiring presence.

Isabelle Radcliffe spoke of a case of Stomach Ulcer in a woman who was to Buffalo, New York specialists. She came to Mrs. Radcliffe in the March of 1929. She had 2 or 3 tonations a week and had 6 weeks of tonation in all. An Equipment was placed in her home; she was normalated.

She tonated a number of sufferers and relieved them some way or other. An old woman had a Medical Doctor for a week; she was given Spectro-Chrome and she was up and around.

Rabia A. Martin desired to know about the Medulla Oblongata and the decussation of the fibres in its pyramids. Dinshah explained how

that affected the breathing process.

Recess was declared at noon.

Second Day Afternoon Session

Activities were resumed at 2:30 P. M. Frances Renny asked about a woman of 35 having Shaking Palsy. Bessie A. Hasenau told her to irradiate with Violet; Harry Meixner spoke of Green Systemic Front and Back with Violet as needed; Oscar C. Keener supported the Green with Magenta on 4.

Leo A. Lefave asked, "How shall we correct the Constipation?"

Dinshah said, "By stopping the eating of junk." (Laughter).

Question after question was fired at the Originator about various disorders and complications. Oscar C. Keener asked about Diabetes; Frank J. Weiland questioned about Deafness; both were answered by the President by giving details customary in the Class.

Leo A. Lefave spoke of trouble in the Tonsils; it was 6 years old. After one month's tonation, hard putrid matter came out of the nose in chunks. Frank J. Weiland became over-stout 20 years ago. He had pneumonic trouble each year; Spectro-Chrome helped him materially.

Dinshah cited the case of a 13 year old girl of 210 pounds whom her mother called "my darling little baby doll"! She had Extract of the Pituitary Gland given by the Medical Doctors for about 2 years. She was practically a born moron and could not talk. Spectro-Chrome started her on the proper track by Green Systemic and in a few months she began to talk back to her mother. Soon she went to school; then she menstrated and the last he heard from her mother was that she was keeping company with boys! Just a healthy, normal girl.

Alexander L. Spence reported of a woman with heart trouble; it was both organic and functional. Spectro-Chrome benefited her.

John H. Adams said he was not married, but, had Palpitation and Scary Dreams. (Laughter). Yes, but, they were gone after Spectro-Chrome.

The Minutes of the last year's (Ninth) Convention printed in the Spectro-Chrome of 1930, were presented and on motion of Bessie A. Hasenau seconded by Leona D. Burghardt were unanimously accepted as correct.

The Second Day Afternoon Session was adjourned at 5 P. M.

Second Day Evening Session

The Second Day Evening Session was the culmination of the Convention. The President delivered an address and drew the attention of the members to the important fact that the existence of the American Association of Spectro-Chrome Therapists and the Scientific Order of Spectro-Chrome Metrists was found to be unworkable and complicated as two distinct organizations and recommended the advisability of merging them under the name of Scientific Order of Spectro-Chrome Metrists.

Moved by Dr Oscar C. Keener and seconded by Dr. Frank L. Richards that the two organizations be consolidated as Scientific Order of Spectro-Chrome Metrists. Unanimously passed.

The following Resolutions were unanimously carried:

(a) Moved by Donald N. Allen and seconded by Jessie E. Ness that the Provisional Membership be abolished.

(b) Moved by Frances Hogan and seconded by Mary White that the Article 8 in the Constitution of the Scientific Order of Spectro-Chrome Metrists be amended to read, "Fellows and Members shall extend aid and relief by Spectro-Chrome Metry, to affiliates of this Corporation or as otherwise provided in the By-Laws."

(c) Moved by Oscar C. Keener and seconded by Bessie A. Hasenau that the "All-Bars-Down" policy of Spectro-Chrome Institute be heartily endorsed.

(d) Moved by George A. Smith and seconded by Frank J. Weiland that the restriction regarding Members of the Scientific Order of Spectro-Chrome Metrists using no other science for aid and relief than Spectro-Chrome Metry, be removed.

(e) Moved by John A. Adams and seconded by Frank J. Weiland that the Annual Fees and Dues of Members be reduced from $12 to $6 and partners and members of the same family domiciled or practising Spectro-Chrome together pay half fees; further that the said reduction as already made by the Board of Trustees, be approved.

The above unanimous resolutions being passed, the Board of Trustees was authorized to take steps to put them into effect immediately.

Oscar A. Keener raised a point that the Collaborator Blanks be sold; Carl Engel said that more equipment must be sold. After long discussion, recommendation was made that the Collaborator Blanks for 10 Collaboratorships be sold for $3. The President explained that Members were becoming disloyal to their pledges of not giving aid and relief except to Collaborators, because of the current rule of charging $3 for each Collaboratorship and sending $1 for fee to the Principal Office. He preferred to make no charge than to encourage dissatisfaction and probable dishonesty.

On motion of Jessie E. Ness and seconding of John H. Adams, Dinshah P. Ghadiali, Irene Grace Dinshah and Bessie A. Hasenau were re-elected as the Board of Trustees. The Board of Trustees being duly affirmed, they selected Dinshah P. Ghadiali as the President and Treasurer and Irene Grace Dinshah as the Secretary. The Board of Trustees then appointed Jessie E. Ness as the First Vice-President and Isabelle Radcliffe as the Second Vice-President. Both were promptly affirmed and installed by the President, thus completing the Council of Governors.

There being no unfinished work, the Tenth Annual Convention was dissolved in peace and harmony at 10:45 P. M.

SPECTRO-CHROME

APRIL 1933
LEMON 1933

VOLUME 9
NUMBER 4

Devoted To: Specto-Chrome Metry — Dedicated To: Service Of Mankind

TWELFTH ANNUAL CONVENTION NUMBER

Editor
Dinshah P. Ghadiali

In course of life you reach a time,
When Doctor, Lawyer, Priest alike,
Does fail to give the sought relief,
To troubled mind or burdened heart—
Then come to me:
I help you to help yourself.
SPECTRO-CHROME.

Single Copy 50 Cents

ATTUNED RED WAVE

(THE WORLD THROUGH THE EYES OF DINSHAH)

Twelfth Annual Convention

Welcome address by the Honorable Daniel Webster Hoan, Mayor of Milwaukee, Wisconsin.

New Cabinet Spectro -Chrome and Irradiator Exhibited.

A BRILLIANT OCCASION

The Twelfth Annual Convention of the Scientific Order of Spectro Chrome Metrists assembled in the Gold Room of Hotel Wisconsin, Milwaukee, Wisconsin, on Saturday, November 26, 1932 at 8 P.M. Notices were originally sent out for the convention to begin on Monday the 28th, but the Mayor of Milwaukee informed our President Colonel Dinshah P. Ghadiali from Washington, D. C. that important business of the city would not permit him to be present on the 28th, so the day was changed to the 26th.

Because of the popularity of the mayor and the unusual character of the occasion, the Gold Room of the Hotel Wisconsin was packed by Spectro - Chrome Normalators, their families and friends.

Mayor Daniel Webster Hoan was escorted to the platform by President Dinshah who said, "It is a point as to whether the Honorable Mayor should give the Welcome Address first or whether the President should speak first as introducing Mr. Hoan to the Convention; however as to speak about him in his hometown would be like endeavoring to gild gold, I prefer the worthy guest of the evening should have the floor."

As mayor Hoan rose with his gracious smile, the assembly received him with a vociferous applause. He said in his Welcoming Address, "Col. Dinshah and friends, it is not infrequent for me to welcome Conventions. Conventions in this city are very frequent. We have Medical Conventions and others; 160 to 270 conventions a year assemble in Milwaukee. It is a tribute to the people. Milwaukee has several good features; it has many more than other cities; we have good points in our splendid programs. We have a good fire department, health department and we have the lowest rates. We try to prevent crimes; we know that the juvenile ages during the idle hours start to think; 80 per cent of young people need occupation in idle hours; we therefore have a good program to provide for playgrounds. We opened the schools for training the young; they are efficiently managed. There are houses for recreation and night schools to teach anything one wants to learn. Outdoor athletics are provided. Our night schools have a larger program of activities than any city in America of pro rata population. We aim at preventing crime among the youth by keeping them busy. Automobiles have added to the burden of traffic, but I thought 50 traffic officers were plenty; they cost $2,000 a year each and by our economical program we managed with the least number. I took issue against unnecessary increase. We have a well conducted police department; our burglary insurance rates are the lowest of any. All the result of preventive work. We have a bureau of School teachers who take care of the food program for the nutrition of the children; tons of bad food are destroyed; milk is carefully watched for its purity. There is yet a lot of room for improvements, but we are doing the best. There are Sunday School troubles in large cities. Hundreds of people have insufficient food. Cities do not take care of their people—but, I should not talk on many of these matters. The human family has been on the Earth millions of years; improvement has always been slow. Printing machines turning out 100,000 copies per hour, books and pamphlets and newspapers laid the foundation for progress. Free education solved a big problem. We accomplished much in a century of freedom. Human beings are controlled less by enacting laws; we must fit ourselves into natural conditions. Society as an organization is governed by laws, but we find the World is economically sick. It is passing a great economic crisis; it is a social crisis, like the birth pains of a great revolution. Just as we passed through the feudal stage which broke down, we are now passing a crisis. In spite of millions in grain we had famine and people starved. They are throwing into the sea millions of pounds of coffee, burning crops—we are so ignorant. It is like a bedbug with small brain. Believe me we have to solve some problems. For different things, there are different kinds of doctors; the doctors let the light shine on some. Let Dinshah shed light on some problems! After listening to Colonel Dinshah, I have to leave tomorrow for a hearing in the Senate about the Canadian Waterways Treaty regarding St. Lawrence River. We hope to endorse it; I am the Chairman of the Great Lake Harbor Association; it is an important treaty. Colonel Dinshah, I welcome you and your friends to the city of Milwaukee. Each time Colonel Dinshah speaks he gives us something new. We are here more tolerant, more liberal; with such spirit, we solve our problems. I wish you success." (Applause.)

The Welcome Address of Mayor Hoan was listened to with attention and interest and the humor with which it was frequently interspersed was highly appreciated. In thanking the Mayor, President Dinshah said, "Milwaukee, through its mayor conferred on us great honor. It is the first time in our twelve years of experience that the Chief Executive Officer of a city welcomed us. Our best and most cordial thanks are due to the gentleman and to the city he represents.

When I came first to this city last spring, I was struck at the scarcity of traffic policemen, yet, the traffic seemed to flow in a smooth and orderly manner. I agree with the Mayor in his views that prevention is better. Education is better than legislation. We have, in fact, too many laws; he has the right idea; what we need more is proper education.

There is an idea prevalent among the public that the policemen are there to harass us. I know having worked among them for nearly 2 years during the wartime, in the city of New York— that on the average, the police are the worst paid and the most loyal to their duties of any police officers. I am glad Milwaukee has good fire, police and health departments. They are all necessary and when well governed are conducive to welfare. The idea of training the youth is praiseworthy. The Mayor is to be congratulated for the wonderful showing in financial stability made by his city when larger ones are in imminent bankruptcy. The city is to be congratulated for having a mayor who through his ability, honesty and fearless character retained the chair over 18 years, against all opposition.

I thank you, Sir, and your wife for gracing our convention with your auspicious presence. We too believe firmly in prevention and assure you that we shall do our level best to offer what we have in the service of mankind. To serve best by Spectro-Chrome-In Every Home, since my last visit, I designed a new Spectro-Chrome like a radio cabinet. Here it is." (Dinshah removed the cover and exhibited the new Cabinet Spectro-Chrome and projected the powerful 1,000 Watt concentrated beam full in the face of the audience.)

Dinshah - To serve the sufferer more effectively, besides the external projection, I designed the "Irradiator" which utilizes the beam at the rear and charges water, milk, sugar or similar material with the Attuned Color Waves, used during the tonation at the end of which they are used internally as Irradiated Food. This Internal Spectro-Chrome is my latest humble contribution to suffering mankind and it will be a great help to the Normalators in accelerating the equilibration.

Among cheers, the meeting was adjourned to 10 A. M. of the following day; most all Graduates encircled the Cabinet Spectro-Chrome and watched it in action. Many questions were asked and answered. The Mayor and his wife Mrs. Agnes Hoan, stayed till the close.

Second Day Morning Session

At 10 A. M., Sunday, November 27, 1932, the President's gavel called the assembly to order. Jessie E. Ness, First Vice-President was on a seat on the platform, assisting the President in taking care of details; Marguerita W. Simon, the Secretary of Spectro-Chrome Institute was in charge of the Register. The Convention was for the entire day open to the public, by registration of names and addresses. It was a gala day, all were in good humor and many flocked around the new Cabinet Spectro-Chrome whose beauty and simplicity of design was given its share of admiration.

The president read the notice of the Twelfth Annual Convention and the notice marked "Convention Dates Changed." They were introduced into the Records.

TWELFTH ANNUAL CONVENTION

Thursday, October 27, 1932,
Chronos:—Green 24 Black.

Dear Friend:

1. Our Twelfth Annual Convention will meet in the Gold Room, at Hotel Wisconsin, Milwaukee, Wisconsin, on Monday, November 28, 1932 and two subsequent days.

2. The following is the program:—

MONDAY, NOVEMBER 28, 1932—8 P. M.

Opening Welcome Address by Honorable Daniel W. Hoan, Mayor of Milwaukee. Exhibition of new Cabinet Spectro-Chrome with Irradiator, another crowning Spectro-Chrome Invention by our President Colonel Dinshah. The above session is open to the public. Invite your friends.

TUESDAY, NOVEMBER 29, 1932—10 A. M. to 12 M., 2 to 5 P. M. and 8 to 10 P. M.

Registration of Affiliates.

Address by Council of Governors, Case Reports and Speeches by sufferers introduced by our members.

All sessions open to the public by registration of their names and full addresses. Invite your friends.

WEDNESDAY, NOVEMBER 30, 1932—10 A. M. to 12 M., 2 to 5 P. M. and 8 to 10 P. M.

Address of Dinshah regarding the year's progress and future of Spectro-Chrome.

Suggestions for amendment of Constitution and By-laws.

Election of Trustees and appointment of the Council of Governors.

Instruction in Irradimetry, Itisometry and Sympathometry with the Favorscope and co-related scientific subjects.

Unfinished matters.

All Wednesday sessions are open to Members and Provisional Members ONLY.

Collaborators and Strangers will NOT be admitted:

WARNING:—POSITIVELY no admission without presentation of Registration Card.

3. Make the 12th Convention a memorable success in the Service of Mankind.

SCIENTIFIC ORDER OF SPECTRO-CHROME METRISTS.

BY BOARD OF TRUSTEES

Irene Grace Dinshah, Secretary.

Convention Dates Changed

Malaga, N. J., Nov. 19, 1932

Because of unexpected engagements calling Hon. Mayor D. W. Hoan out of town, he will open our 12th Convention Saturday, November 26, 1932 in Hotel Wisconsin, Milwaukee, Wis., at 8 P. M.

The prior program will continue Sunday 27th and Monday 28th current.

Invite your friends and make our convention a memorial success.

Scientific Order of Spectro-Chrome Metrists.

The President having spoken a few words of greeting to the visitors present, requested the Members to go ahead with their reports.

Verna Wentland- I have no equipment; a friend was sick; took 18 tonations; did not work. It relieved others.

On questioning it was found the trouble was of three years duration and the time was too short for favorable results.

Stanley Wentland- The man was paralyzed many years; he was 65; a $30 Home Spectro-Chrome was used; however he got his money's value.

President- Well, when I marketed the Home Equipment, I distinctly stated that it was only for acute Disorders; yet the graduates, in some cases use it for Chronic Disorders and then grumble. Whose fault is it? It is not built for heavy duty.

Frank Mueller- I had some success. My sister woke me one night with pain in the Appendix. She said, "Get the Doctor." I gave her one hour Spectro-Chrome; the pain was gone. Four more tonations and she felt so much better, she went to a dance the next day. I gave her water and lemonade.

A gentleman had trouble in prostate. Took 8 tonations, felt better. Now 3 months, had 40 tonations. He had continued Ear Noises and Skin Disorder; all gone. It is good for minor cases also.

I had a woman's case; she would not use food rules; came 5 times in a week; her troubles were 65% gone in 3 months. but no results in Constipation. I feel I am in doubt as to how often I should tonate, how long, when and how. How do I know when to change the color? If I give one-half hour Systemic Front, 15 minutes Systemic Back and 15 minutes Yellow. is it proper? A lady had Rheumatism; shall I give local or Systemic?

President- If someone would not follow Rational Food System, you should not be blamed. The food builds the body and care must be taken not to eat "junk." It appears you are doing the same many others do. Do not try to mix Attuned Color Waves in too many combinations at the same tonation. One complete tonation must begin at the Favor Time, go through the Junction Period and terminate not less than 16 min. after. That means 1 hour in all. The Auric Vehicle must be given time to alter its Oscillations. Infra-Green and Ultra-Green Attuned Color Waves should not be irradiated on the same Areas at the same tonation. Acute Disorders have a short and relatively serious course; they have one or more of the Four Signals of Inflammation, Rubor, Calor, Tumor, Dolor; Redness, Heat, Swelling, Pain; in Chronic Disorders, the troubles are old and the course is long drawn, the efforts of repair in the process of inflammation being spent or nigh exhausted. Always understand my friend, that there is really no Local Disorder; a disorder might start at a Local Area, but, its effects, its products, will be drawn into the entire system through the principle of Blood Circulation. All Disorders, therefore, **must** have a short or long Systemic Tonation to clear the aftermaths. Rheumatism, like many similar conditions, is not a prime disorder; it is merely the result of other conditions, in fact, a sign of such existence.

Herman Dietrich- A 68 year old man had a Kardoscil of 63; I did not take the Spiroscil. He went to Mayo Brothers for examination. The report was that "his Lungs had grown to his ribs." He had a High Blood Pressure; much pain; they gave X-Rays on Gall

Bladder. He came to me; I tonated him 8 days according to the Favorscope Chronos, with Lemon, changed to Orange, Systemic, I had him 3 weeks 4 or 5 times a day; gave Purple; Kardoscil came to 61 a minute; gave Scarlet; it went to 68. What of this case?

President- Your information is in your own question. The man had Lung Disorder; his Area 17 apparently did not function properly, yet, you "did not take the Spiroscil." You should not make the same mistake as the Medical Doctors; they take the Kardoscil and seldom the Spiroscil. The Pulse Beat without the Breathing Rate gives little information of value in our work. It is the 4 to 1 or 5 to 1 Ratio between the two that is the Index of Normalcy or otherwise. Without such full information, a consultation in our system is of little value. X-Rays on the Abdominal region are dangerous to the vital organs; read my editorial about it in the Spectro-Chrome of June 1922. The Medical Doctors use them without any regard to their dangers; they are Caustic and of no real value for healing. Give that man Magenta on 4, Green Systemic Front alternated with Magenta on 18 and Orange on 17 and he will be better. The food must be altered to our method. His Kardoscil of 68 fits his age.

Frances E. Tiley-I went on tour this Fall with my husband by automobile; on the way, we stopped at Malaga, New Jersey and saw Colonel Dinshah. Spectro Chrome Institute is a big place, all just as he told us. It is a complete Scientific Laboratory; it is wonderful, has a well stocked library and the walls are lined with Diplomas and Certificates. We saw the first Cabinet Spectro-Chrome there and bought one right away. It certainly does the work. Our enthusiasm about the work after using the Equipment was so great that we came all the way from St. Louis, Missouri here to this convention. We came to learn more; St. Louis people enjoyed the Colonel's lectures.

The President spoke a few words of appreciation and said, "When Mrs. Tiley came to the Central Office, she told me she did not like to spoil the beautiful appearance of her husband's bedroom by placing in it a steel Systemic Equipment. I promptly took her upstairs into the Cabinet Shop and said, 'Well Madam, here is the Cabinet apparatus and you can paint it to suit your husband's handsome furniture!" She gave me her order at once!"

Second Day Morning Session adjourned at 12:30 P. M.

Second Day Afternoon Session

The Convention reassembled at 2 P. M. The President was in the chair.

Greeta O. Peck—We came as far as anyone else, from Sandwich, Ontario in Canada. It was a sacrifice, all the greater because of the depression affecting the exchange; 17 per cent discount in dollars, but, we so fully believed in Spectro-Chrome, we could not stay away. We attended the Class in Detroit last Summer and got an equipment. It has been in constant use since; it is now in my mother's home; it is not deprived of its use. I had my sister's case. Before Spectro-Chrome, she was operated for a very serious condition. Friends did not think she would pull through; but Spectro-Chrome soon changed her condition; today, she is a picture of health. It is a pleasure to hear the enthusiastic reports of the people. Medical Doctors who performed the operation had told her she was not good insurance; she should take "lots of light." She had a lump develop in her breast and they wanted to remove it. Dr. Bryant wanted to remove though it was smaller. My mother was in a collapse; she slumped. Sister went on vacation; came back six weeks later; she could not sleep. In 2 weeks of Spectro-Chrome she was a different woman. She comes in a street car; preserves health; she knows Junctions; she is enthused. My husband and I took tonations; the whole family is talking Spectro-Chrome; there is little opportunity for outsiders to say things. We had 2 or 3 remarkable cases. A friend had an infected wisdom tooth; my husband invited Ivy to come; she came at 8 with a bundled face swollen ;could not wash lower jaw was just paralyzed. I gave her Turquoise; I leave our sufferers; each case goes to sleep. Her jaw opened in 20 minutes; after an hour, she told my sister she would punch her face! Next day the Doctor could not understand what had happened; she gave a dramatic recital; he will get a surprise when he finds out. (Laughter)

A friend was suffering from Hives. She said the children looked at her; Dr. Bryant said there was no help. Three tonations were given. We said, "Janis, we have a light, if you like come." She came on Saturday; again on Sunday; it broke in 2 tonations; no fresh Hives came. These two girls—one with the infected tooth and sister, were two worthwhile cases. Colonel said the Vital Ratio should be 4 to 1; or 5 to 1; here was a Kardoscil of 61 and Spiroscil of 22, which was close to 3 to 1, another was 68 and 9, which was close to 8 - 1; both were teachers; they were fainting in school and taking a year's leave. I suggested "give me a chance." Orange on Front and Red on Back was irradiated. In 10 days,Kardoscil rose to 84, Spiroscil to 25; she has zest and pep. Sister was so delighted! She had Diphtheria, Scarlet Fever and Appendicitis when young; was poked full of serums. In 8 tonations, she now feels like new. Her Kardoscil is 82. No success about menstrual period. Before I married, I taught 1,000 children a day and was feeling the effects. Now, since I took Spectro-Chrome, the friends remarked, Mrs. Peck, you look better." Credit should be given to Spectro-Chrome.

President - Thank you Mrs. Peck; that Spiroscil of 9 with Kardoscil of 68 must be an error. The Menstrual Period will be impeded by defective circulation, as any function may. Always the Ratio must be duly watched for proper results.

Albert W. Peck- When Mr. Mueller spoke about threatened appendicitis, I made some notes. How about eating salt?

President- There is a movement on foot that eating salt makes for Cancer. Salt, as the chemically treated and devitalized material Sodium Chloride (Na Cl) is not good, but, the real salt of the ancients was not such. That was either Rock Salt, produced from the ground containing the percolated deposits of ages and full of various minerals useful for the body or was sea salt containing the evaporated solids of the sea, without any addition or subtraction. Realizing the need of such proper Salt as food, Spectro - Chrome Institute now has "Lavan" or All-Salt Sea-Salt which is genuine evaporated Sea Salt, ground in our Laboratory and packed. It costs only 25 cents for a large cartoon of about 1½ pounds and if you taste it, you will find it saltier than any salt you ever ate! Lavan contains all the Chlorides, Fluorides, Iodides and Sulphates that are normally in the sea and is beneficial for body nutrition.

Albert W. Peck - I want some of that salt. For years, I had a wonderful digestive track, but my Area 8 troubled me for 20 years I had much distress and spells of gas caused it. I used to carry on a month with spells. Spectro-Chrome removed my distress. I also had a sort of Arthritis; an attack in the wrist; it is gone too. My sister-in-law had injury; it is remarkable how it healed; the scar is hardly visible.

Dr. Jaroslav Premysl- When Colonel Dinshah came last here, I attended his lectures. I had over $6,000 of electrical treatment machines of all kinds. I had been fooled by others, so I doubted Spectro-Chrome, but I bought one. A case of Inflammatory Rheumatism came; it was a bad one. I gave Spectro-Chrome twice, it got rid of it. I had 4 cases of Tuberculosis. Many cases of Tonsilitis, three tonations and they were well. It removed Abdominal Obstructions. It worked well on so many people, I bought 4 Spectro-Chromes and an Itisometer. I paid $1,740 to Colonel Dinshah. It is wonderful; I feel well. People like Spectro-Chrome. I have so many people coming to me I can't take care of all. I speak in favor of Spectro-Chrome.

President- That is very kind of you, Doctor Premysl, you need more Spectro-Chromes; we need more money! (Laughter)

Dr. Jaroslav Premysl- I have no space to put them in.

President- That is easy; move into larger quarters. You must have one of the New Cabinet models; all others you have. The Secretary will please book Dr. Premysl's order. (Laughter)

Premysl- I shall think it over.

President- There is no thinking to be done; I thought for you.

book his order, please, Miss Simon. (Laughter)

Jessie E. Ness from Detroit, Michigan was then introduced as the First Vice President. She spoke with enthusiasm.

Jessie E. Ness- I need more time than ever to speak about the merits of Spectro-Chrome. A lady came for a Spectro-Chrome on hire. Her Brindled Bull Dog had some trouble. The Veterinary Surgeon called it Double Pneumonia. It would not live, he said. She brought the dog to me; his temperature was high; it was breathing hard. I turned on the Spectro-Chrome; he gave a grateful look. He sneezed and long things came out. I gave him another tonation; the next day he was frolicing. We must have an animal Spectro-Chrome! I wrote to Colonel Dinshah; he told me the same Spectro-Chrome applies to animals as well.

I had great success in bad cases of Pyorrhea. A lady was suffering from it. She was told her teeth should be removed; Spectro-Chrome normalated her and no teeth were extracted.

I had a case of Goitre; it was surprising how it came down.

A lady came screaming with pain; they had told her she had Gall Stones. She was given Spectro-Chrome. In three weeks she was right; she was delighted.

My second daughter, Deanne, works in a hospital. There are lots of Doctors and lots of talk. My children are trained to independence. She runs around and dances. She did not feel well; went and got care in hospital; got worse in Doctor's care. Then she said, "I shall take Spectro-Chrome." In 3 tonations spots came out.

I hired a Spectro-Chrome to a man in a bad accident. X-Rays showed he had three large spots in the lungs. They gave him medicine; he took Spectro-Chrome and poured the medicine down the sink. A month ago, he had another X-Ray examination; there were no spots. Some Spectro-Chrome Normalators think it is not possible, but I know Spectro-Chrome works in and out of season. The Human Aura takes it. I am no man's fool. The body builds by Spectro-Chrome. I will not allow any one to speak against Spectro-Chrome. Some cases need longer tonations. I was saying in Detroit, "We lose all, but we shall not lose Spectro-Chrome." For the inventor of Spectro-Chrome? I shall not allow anyone to speak against him. A man like him to stand prison life? Beyond question is Spectro-Chrome.

Dr. James W. Lewis- I record the same way as Mrs. Jessie Ness, though my knowledge is in a different line. My work is exclusively in Dental Surgery. My wife was in poor health. We had quite some talk about Spectro-Chrome. It gave wonderful results. She had a heart with 95 to 100 beats and many more peculiar symptoms. She was usually tired. Now, after using Spectro- Chrome, her Heart is 78 and she sleeps well. There are many convincing arguments in favor of Spectro-Chrome. In many cases I had phenomenal results. In the extraction of teeth, where it is necessary to do certain dental work, in the past there was considerable after-pain. After feasible experiments, I found that I can prevent 100 per cent after-pain. I used Spectro-Chrome on hundreds of sufferers; I should like them all to testify. I took all of Mr. Unruh's teeth and used Indigo; it stopped the bleeding. I saw him here; his case was one in particular. I use Spectro-Chrome where people have Focal Infections. Some time, I use it for the Heart. We eliminate Poisons by it. I can say that Dental Work heals quicker when Spectro-Chrome is used.

The Second Day Afternoon Session was adjourned at 5 P. M.

Second Day Evening Session

The Second Day Evening Session was called to order at 8:19 P. M. owing to the crowd surrounding the New Cabinet Spectro-Chrome, asking questions to Dinshah like the volley of a machine gun. They received their answers with genial smiles as usual.

The question was asked from the visiting audience about our method of taking care of Diabetes Mellitus. The President asked whether the visitors wanted to hear about the medical humbugs in the name of Diabetes and the audience answering with vociferation, a lecture on the subject was given ex tempore.

President- Diabetes is not the name of the Disorders to which it is generally applied. All Disorders in which Sugar passes in the Urine, are not all Diabetes Mellitus. To be true Diabetes, it is necessary that the Sugar eliminated must be Grape Sugar, otherwise called Dextrose, which must be thrown out of the body for weeks, months, even years and such excretion should increase after the ingestion of Carbohydrates. Any deviation from this formula, brands the Disorder as mere Glycosuria or Sweet Frontelim. It is said to be the result of Hyperglycemia, but a sensible study of the subject shows that it is not Hyperglycemia or Excess Sugar in Blood, but truly is a shortage of the Sugar or Hypoglycemia.

This way of looking at the Disorder is not novel to most students of Diabetes. They are taught of the passage of the Sugar, therefore, believe that as the Sugar went out in quantity, there must be so much in the blood! It is the contrary. Take an example: If a tank of water holding 100 gallons has a leak of 10 gallons every hour, in 10 hours the tank will be empty. However, while the emptying is progressing, if one pours in every hour 10 gallons, the tank will stay full while the repairing is proceeding.

A similar action goes on in the body. Carbohydrates, which are materials rich in Carbon and Hydrogen, like Starches, are converted into Sugar before being taken into the blood Stream. As Energy is demanded for work being done, the storehouse of this Sugar called Glycogen, is signalled to throw it into the Blood for

burning in the tissues. That storehouse is the liver—Area 7, which responds to the call with the disengagement of two Sugars, Dextrose and Levulose. Both these Sugars as well as a third one Galactose or Milk Sugar are known as Hexoses, because they have 6 Atoms of Carbon in their composition. They all have the same empirical chemical formula which is C6 H12 06, although the structural arrangement of the same Atoms in each one of those Molecules is different. Galactose is Milk Sugar, Dextrose is Grape Sugar, Levulose is Fruit Sugar but the last two have a peculiar property by which their difference is known.

As you well know, the Gyratory Oscillations composing a beam of White light move in two different planes—Horizontal and Vertical. When they are separated by a Polarizer, one rotates the axis of the beam to the right, the other to the left. Crystals have singular properties in relation to light; Dextrose rotates the plane to the right, Levulose turns it to the left. It is the Grape Sugar or Dextrose that passes out, the Fruit Sugar or Levulose is mostly consumed for Energy by the tissues. Thus, on call the one part is used as energy while the other is wasted, truly speaking unabsorbed. Diabetes becomes simply a Disorder of Nutrition in which, although there is plenty of Sugar in the Blood, it cannot be utilized, hence if the sufferer be denied the feeding of all Carbohydrates — Starches and Sugar—he finally dies of nothing but Carbohydrate starvation.

The average Medical Doctor is woefully ignorant of Physics; he has little knowledge of light and its potency; he knows scarcely anything about Polarized Light. He has no idea of the extent to which we go for precision. In Blood, ordinarily there may be 1 part in 1,000 which may rise in Diabetes to 3 or 4 and in some cases to 7 or 10. The entire trouble is merely lack of Lymphatic Activation. Though bathed in Sugar-bearing rich fluids, the tissues are unable to utilize the potency and starve to death.

What do we do? We do not stop the ingestion of Carbohydrates or any Starches or Sugars. We stimulate the Lymphatic Activation by flooding the body with Yellow Systemic, both Front and Back and place the sufferer on our Vegetarian Rational Food System as in all other cases, without any restriction as to Starches or Sugars. We let him eat as much Brown Sugar or Raw Sugar as he likes and—he recovers! I do not care about how much Sugar passes **out**, all I care about is how much is utilized **inside**, till the sufferer gains equilibrium. Instead of feeding him as Medical Dietitians do on Meat, Fish, Eggs, Tea, Coffee, Wine and such other useless stuff, we shut down on all these, including the injection of that abominable humbug called Insulin and in all cases under our observation the sufferer recovers the Equilibrium of Basal Metabolism.

The remarks of the President about Diabetes gave rise to lively comments. At 10:50 P. M. the audience dispersed, well impressed with the value and accuracy of Spectro-Chrome in the so-called "incurable" ailments. The sessions open to the public were over and their aims for enlightenment were served.

Third Day Morning Session

The last day's work began at 10:15 A. M. Only Members or Provisional Members with current Affiliation Cards were admitted, no strangers being allowed. The President spoke about the Members adhering to the Collaborator System to avoid someone framing them by co-operation with the Medical Octopus.

Jessie E. Ness emphasized this by describing how a certain person tried to get her into saying certain things, but she compelled him to sign the Collaborator Blanks, after which he just disappeared. "I live by Spectro-Chrome," said she, "we must be cautious."

The President read a letter from our esteemed and beloved friend Surgeon Kate W. Baldwin, in which she breathed her loyalty. She said, "There are so many things I would like to tell you about Spectro-Chrome I would possibly weary you with even interesting reports; so will sum them all up by saying that Spectro-Chrome is the most worthwhile contribution ever given to the healing art. Get it! Use it! and be sure it will never disappoint you. Just the cases on my list at the present time ought to be sufficient to convince any right minded person of its value. I would close my office tonight if I was obliged to give up Spectro-Chrome. Here are three cheers for Spectro-Chrome!!" The audience received Surgeon Kate Baldwin's letter with three cheers.

A telegram from Frank E. Kaufman, of Baltimore, Maryland was read: "My congratulations and best wishes for successful convention."

A letter from Dr. Grace Dodge Elwell, from Hollywood, California said, "Wish I could bridge the miles between us and be with you in person. My love to you in the work and blessing for your continued service in the welfare of suffering humanity."

A communication from Minnie F. Guenther, Marie Ida Smaltz, Howard N. Smaltz, read in part, from Cleveland, Ohio, "We are having good reports from the Spectro-Chrome operators and getting the Favorscope which is a great time-saver."

Bessie A. Hasenau, of Cleveland, Ohio, one of the Trustees, wrote, "This is the first time in ten years that I have been unable to attend the Spectro-Chrome Convention and I cannot express how greatly I regret my inability to be with you and our devoted co-workers this year. I have waited until tonight to write this letter thinking that it might yet be possible for me to come, but, such is not the case and so the only thing left to me is to write and tell you that I hope this will be the best Convention we have ever had and to ask the blessing of the Benign Powers to rest upon you and guide you through this Convention and the coming year of Spectro-Chrome activity. May this birthday be the happiest and the most successful one you have yet enjoyed and may you live in this World many years yet to preach the gospel of Spectro-Chrome Metry, the one tangible solution of the physical miseries of mankind and the unmistakable proof of our relationship to an Almighty God and this is my endorsement after ten years' active service in Spectro-Chrome Metry. I am yet hoping to see the best of all Spectro-Chrome Conventions held in Cleveland, Ohio and may that one be our next."

Dr. Maude B. Crans sent report on Infantile Paralysis, where the hopeless case of Barbara de Bride was benefited by Spectro-Chrome. Dr. Crans after giving the history, states with deserved pleasure, "Today, Barbara is a gloriously beautiful baby. She has marvelously big brown eyes and is an adorable child, who will always attract attention. When she came two months back, she was a frail, sick, paralyzed, pathetic baby, beautiful and sweet then, but, one to whom pity went out. Now she gets around everywhere, climbs, has gotten upstairs alone, goes down with her hand braced on wall and the other held.... We expect in a few months to have every trace of the trouble gone." This case received much publicity; even Hospital Doctors credited Dr. Crans.

Dr. Maud B. Crans reported also her own case, where various Medical Doctors advised an operation, saying it was "impossible" without it. She resisted the opinions, used Spectro-Chrome and here is what she says, "Either every diagnosis was incorrect or Spectro-Chrome did; and Spectro-Chrome **did** it."

Dr. Lettie Allen Ward, formerly of the Cooper Hospital, Camden, New Jersey wrote, "A little girl of 6 years was sent home from school as a Diptheria Carrier, October 31. Child confined to one room, house quarantined. I have treated the throat and nose by all means possible for 3½ weeks with no results. Twelve days ago, I put my small light in the house and the child has slept under green for 12 successive nights from 9 P. M. to 7 A. M. I took another culture on Thursday the 17th and was called up on Friday A. M. the 18th by our Board of Health and told the child was negative. How about that for Spectro-Chrome?"

President—It will ever be so. No matter what we do, they never give us credit. Dr. Ward's Diptheria case proves the value of Spectro-Chrome for so-called "infectious" troubles also. What human lives are sacrificed under the guise of the healing art! Yet they will not investigate Spectro-Chrome.

The election of Trustees being declared open for nomination, the same Board of Trustees was re-elected for 1933. The President said that owing to the absence of Irene Grace Dinshah and Bessie A. Hasenau the appointment of the Council of Governors will be postponed till after the beginning of the year.

The Third Day Morning Session was adjourned at noon.

Third Day Afternoon Session

The President promptly opened the Third Day Afternoon Session at 2 P. M. He devoted part of it to giving the new students and members an idea of how Spectro-Chrome in over 12 years made its headway against severe oppositions from without and within, yet, all work in the Conventions was done by unanimous vote—a genuine tribute to the devotion of the Members. He enumerated the various quick changes he had to make in the policies of sale, lease, hire, finally letting **All Bars Down** to protect the interests of suffering mankind. He said further, "When I was last in this city, they told me that I should go to Missouri; if anyone could carry his message through Missouri and convince the Doubting Thomases there, all America was open for him. So I took the advice and went to St. Louis, Missouri. I did not find the people any different there; they looked the same, laughed the same and I had a class of over 80 students. There, I received invitation from the Missouri Institute Of Homeopathy for the State of Missouri, to speak at their Convention in Kansas City, Missouri. They chose the subject of my address "Medical Liberty!" (Laughter). I went there and one of the Medical Doctors present delivered a speech on "Chromopathy." Then I was recommended by Dr. L. M. Ottofy, one of their members as the one man really qualified to speak upon Colors. They gave me the floor again and after hearing about Spectro-Chrome, expressed their desire to study the science. I returned to St. Louis, Missouri, conducted another class of over 120 students (all free of course) and went to Kansas City, Missouri. A class of about 80 studied there. Then I opened a class in Atlantic City, New Jersey and two in Vineland, New Jersey, but, everywhere though great enthusiasm was created, very few seemed to have escaped the depression in finance enough to purchase a Systemic Spectro-Chrome of $150. The outlook for business appeared justly gloomy, because, it costs about $1,000 to open a new Class in a new city in a new State and unless a few Equipments be placed, there would be heavy loss in each locality.

"On returning to the Central Office from Missouri, seeing the trend of the time, I decided that it was imperative to find some way to build an inexpensive yet efficient full power Equipment. Till then, our assembling was done in New York, where primarily the Central Office was located, but I was resolved upon cutting out the heavy expenses and middlemen's profits and do all the manufacturing ourselves.

"I had numerous machines for such manufacture lying idle and rusting since years. The gap in my life produced by my incarcerations needed to be refilled, so after long thinking and many wakeful nights, I designed a new model built like a Radio Cabinet, and exhibited that to a select number of our loyal Graduates, in Malaga, New Jersey, on Sunday, September 18, 1932. It was this Cabinet you see on the platform. It has double insulation of plasterboard and air, 1,000 Watt concentrated filament projection bulb and Shuffling Slide-Carrier, with the Slides fixed in to avoid accidental breakage. The apparatus is mounted on a special designed table with tilting arrangement. The material of the Cabinet is 3 ply wood and the whole forms an attractive and serviceable apparatus, which can grace any house by painting it to individual artistic taste. The price of this Spectro-Chrome, is purposely kept low to reach the home and I feel sure you will agree with me that it is a long step forward in selling it complete for $65 only, with free tuition, free consultation or advice and no extra charge for taking the Examination when ready and getting the Diploma if passed! It certainly proved a move in the right direction.

"Then I took another step forward, knowing that the Irradiations of Light are capable of being utilized through the process of Reinforcement and Interference by Etheric pick-up, a fact well investigated by Dr. Edwin D. Babbitt in the last century and amply corroborated by Professor Henry Steenbock of the University of Wisconsin in Madison, Wisconsin. Professor Steenbock recently irradiated milk in the form of a flowing sheet, passing before a battery of 12,000 Watt arc Lamps, evolving Utra-Violet Rays and the effects of such irradiation were declared to be wonderful. I say the same about my new introduction, the Spectro-Chrome Irradiator. As you note, it is a small box with the 5 genuine Spectro-Chrome Slides, through which the Light radiated from the rear of the bulb of the Cabinet passes onto a set of phials which may be filled with either just ordinary water, milk or sugar, prior to beginning the regular tonation. As soon as the tonation is finished in one hour, the contents of the phials are ready for use. Use the three Primary Colors singly, by taking the contents of the Red, Green or Violet phials as indicated or making with the combinations of the Yellow or Blue, the desired Secondary or Tertiary Attuned Color Wave products, charged with the requisite potency, in accordance with the mathematics of Spectro-Chrome Metry. I used this so-called 'Internal' system for decades and produced startling results even without the Attuning Process of Spectro-Chrome; what then the results will be with the genuine Slides with the powerful bulb, you may conjecture.

"The one beauty of this new Irradiator is that it prepares the potency fresh all the time of tonation and each phial is just one draught. For domestic use, the method will be very valuable; for Spectro-Chrome purchasers who have the Systemic, the same Irradiator will be of similar service. It will be supplied so as to fit either the Cabinet or the Systemic model at the same price."

A Member—What is the price?

President—F i f t e e n Dollars complete. I also already designed an office model Irradiator, which will serve the purpose of a constant charging apparatus with Spectro-Chrome Attuned Color Waves; it has many original features, making it fool-proof. Thus, I perfected both the External and the Internal Spectro-Chrome and a larger scope of handling Chronic Disorders or distance service will result from their combination.

Several years ago, I made a Dentifrice and sold to students in our classes. Since then, owing to travel and expansion, I stopped making it; now because of the insistent demand for it from older students, I resumed its manufacture. Everything pertaining to Spectro-Chrome is now housed in one plant, where on a 23.23 acre estate we have our printing plant, cabinet shop, research laboratory and the administration offices, auditorium and resident quarters. Spectro-Chrome Institute is not a fly-by-night affair, where the headquarters are in the head ot the President and the manufacturing is done elsewhere! There is an investment of about a quarter million dollars there in land, buildings, labor and machinery and it is only a beginning; watch us grow.

This steady growth of Spectro-Chrome Institute is not to be all credited to my labors. It is the result of the tireless efforts of my darling wife Irene Grace and the devoted and loyal services of the staff composing the assistants of the Institute. I do not hesitate to express to them my most cordial thanks.

Since the origination of Spectro-Chrome, one great drawback to the proper time for tonation was noticed by the Graduates. It was the determination of the Sunrise and Sunset periods and the Junction of the two chains of the Great Sympathetic Nervous System in conjunctive operation with the motions of the Sun, Moon and the Earth, in relation to the breathing process. For that reason, I invented the Sympathometer. It was a very intricate process, making the astronomical relation of the orbs to conform to the peculiar changes in human vitality. Building a clock system to suit all possible conditions, was not an easy job. Being born of a clockmaker's family, it was simpler to me than to another, but, difficulties of a commercial character beset my path.I made train after train of clocks until I reached the goal and thought I was ready for manufacture of the Sympathometer. I had barely made my commercial contacts for it, when the clock company wound its business and another concern had to be sought. Barely had I started again, when the company consolidated with another and the new concern changed the train calculation, annuling my clock labors. Then the 1927 disaster came on me and I came out in 1929 March; by that time, the day of spring wound clocks was gone and Electrical Alternating Current synchronous timepieces were in vogue. I contacted after much inquiry a big clock concern, whose Electric Clock train suited my Sympathometer; then, they too changed their motor and I had to start anew. I reached the solution finally and soon the Electric Sympathometer may see the daylight.

Meanwhile, the Graduates clamored for the time of Sunrise. They could not realize the difficulties lying in my way. Were the Sun, Moon and Earth always moving in the same manner, the work might have been easier; but, the Earth moves on her axis at an inclination of 23 degrees and 27 minutes, hence, the various seasonal changes need accounting. Then again, the Moon rotating on her axis and around the Earth, is not constant in her time, because her orbit too is elliptical. The Earth in revolving around the Sun carries the Moon with her. Thus, numerous factors which might otherwise be left out, come into existence. The Moon moves around the Earth in 27 days, 7 hours, 43 minutes, 12 seconds, but in consequence of the Earth moving around the Sun in her orbit, this time is lengthened; it becomes on an average 29 days, 12 hours, 44 minutes, 3 seconds, from one New Moon to the next New Moon. Even this does not end the story. The Sun and Moon both regulate our vital processes, as I demonstrated so often experimentally in the Classes. The Earth butts into both, making more intricacies. The Sunrise and Sunset for different places is not only different according to their position North or South of the Equator, but, even in the same place it changes day to day and season by season. To supply a simple monthly Calendar that would without deep calculations be available to Spectro-Chrome Normalators in all parts of the World, was a problem. I thought I solved it one way, when some other difficulty leading to a potential complaint loomed on the horizon of research. Finally, after 12 years of silence, I began to deliver that service from October 1932; several people wrote me, "we counted our Favorscope according to your instructions; please, check on my work!" (Laughter) That was the very thing I aimed to obviate; the Favorscope **added** to my work.

Further thought showed a glimmer of simplicity. All the World over, the astronomical calculations are made from the Longitude or East to West positions of a locality from Greenwich, England. We gave the gate to England from here, but, we still use the English foot, which was the measure of the actual foot of an English king Henry the Eighth, the Henry with his eight wives. So also the meridian of Greenwich, England. Our Almanacs are built around the Latitudes or North-South Locations of Boston, Massachusetts, New York, Washington, District of Columbia or Charleston, South Carolina, without regard to an American Standard of fixed decimal position and their calculations are subject to correction from the American Ephemeris and Nautical Almanac built around the Longitude of Greenwich, England. No student of Spectro-Chrome can be expected to learn as much as a ship captain, to tonate sufferers at the proper time. Therefore, I set my mind to find some real and final solution. I found it at last!

Our Central Office is situated very close to the Longitude of 75 degrees from Greenwich East. Thus our Eastern Standard Time is just about correct to within a few seconds, for Malaga, New Jersey. It occurred to me that if I standardized my astronomical calculations to a due Malaga, New Jersey time, it would be necessary to compute monthly only by Latitudinal variations and each locality where the Spectro-Chrome Normalator is needed only information of what to add to or subtract from the stated time (Chronos). So I standardized the whole by the Spectro-Chrome Meridian and began to give the necessary variations to be added to or subtracted from in terms of a Mean Time, which generally would be precise to the minute. The Favorscope Chronos gives the Spectro-Chrome Meridian Time, by adding to or substracting from which the minutes indicated, the Sunrise and Sunset Time for your locality is easily attained. Where the Day of the Moon is unfavorable at Sunrise, the Initial Tonation Favor is attained by adding 1 Hour, 28 Minutes; then each subsequent Favor follows every 2 Hours and 56 Minutes. In the entire Favorscope, each computation is made to suit the practical needs, in regard to the sufferer's tonation and all Seconds, Fractions and Decimals are obviated. It works admirably and letters of commendation indicating the fulfilling of a great desideratum and the accurate results produced by its intelligent use, are on file; we hear every week from someone or another. I sincerely hope the members will comprehend what was done and feel that we are always doing our best to serve their interests. (Applause).

The Third Day Afternoon Session adjourned at 5 P. M.

FINAL SESSION

The Final Session of the Convention assembled at 8 P. M. The Gold Room was well filled by the Members and the Provisional Members. Long distance Members from New Jersey, Missouri, Michigan, Canada, Illinois, Indiana, were still at their posts.

The President announced that the 83rd Class in Spectro-Chrome Metry would commence from Monday, December 5, 1932, at 7 P. M. in the Colonial Room, Hotel Wisconsin, Milwaukee, Wisconsin. No fee or admission charge was to be made. The tuition of 10 nights was as before, free. Repeating students or students who took the free work before were not to admitted, for lack of space; only 150 new students could be accommodated. Many desired to repeat, but, were told the rule applied to all. They would serve better by sending new students; minors under 21 years of age, would be admitted by getting the consent and approval of their parents on the Application for Enrollment. The admission was by Reserved Tickets, numbered and dated; they were issued to the accepted applicants only and were not transferable; the students were bound to take the full course; absence on more than a single night acted as a cancellation of the studentship.

After various small items of less general interest were discussed, the President stated that if sufficient Members would join in 1933, he would resume the publication of the official monthly Spectro-Chrome. This called forth an outburst of cheer.

At 10:30 P. M., the Convention was dissolved amidst hearty greetings and cordial wishes for success of the next Class. All went satisfied, the twelfth Annual Convention was memorable indeed.

MAKING MORE MONKEYS

In 1926, Serge Voronoff, a Medical Doctor known as the champion of Gland Therapy (for the rejuvenation of worn out old fogies by inserting in their body interstitial monkey glands so they may make more monkey shines), made some wonderful experiments in biology, by injecting human semen into the body of a female chimpanzee. The French and Belgian Governments considered these experiments so necessary and marvelously useful, that they subsidized them. Newspapers rattled gleefully about the expected results and the whole World was set agog about the benefits to accrue—when without warning the chimpanzee "woman" died.

I had my suspicions about what was actually done and believed that no injections were made but actual contacts were produced. A similar report now trickles from Atlantic City, New Jersey, through the 91st Convention of the American Association for the Advancement of Science. There, Dr. Howell S. England of the Michigan Academy of Science, Arts and Letters revealed that he knew Dr. Elie Ivanoff of Moscow was making much progress in that line of work on nine female chimpanzees and expected to marry the products to actual men and women in a few years. The idea is to produce more powerful "working men."

Poor working men! They are unemployed enough now; what

will happen when instead of machinery which needs some men, actual ape-men will do their work, is hard to say.

Spectro-Chrome is steadily ousting the Medical Doctors; the above is a new occupation for them.

A NEW MEANING

Why are the Germans called Germans?

Is it that they are full of germs or is it that they make medicine to kill germs?

Cyrus Dinshah.

TRADE **"FAVORSCOPE"** MARK

FOR SPECTRO-CHROME METRY

DINSHAH SYMPATHOMETER SYSTEM

1933 APRIL — 1933 LEMON

CHRONOS

Computed by DINSHAH According To The Longitude And Latitude Of Spectro-Chrome Institute.

DATE	DAY	SUN RISE WHITE H. M.	SUN SET BLACK H. M.	INITIAL SUNRISE FAVOR ADDITION H. M.	INITIAL SUNSET FAVOR ADDITION H. M.
1	Sat	P 46	M 24	1 28	.....
2	Sun	P 44	M 25	.....	1 28
3	Mon	P 42	M 26	.....	1 28
4	Tue	P 40	M 27	.....	1 28
5	Wed	P 38	M 28	1 28	.....
6	Thu	P 37	M 29	1 28	....
7	Fri	P 35	M 30	1 28	.....
8	Sat	P 33	M 31	.....	1 28
9	Sun	P 32	M 32	.....	1 28
10	Mon	P 31	M 33	.....	1 28
11	Tue	P 29	M 34	.....	1 28
12	Wed	P 28	M 35	.....	1 28
13	Thu	P 26	M 36	.....	1 28
14	Fri	P 24	M 37	1 28	.....
15	Sat	P 23	M 38	1 28	.....
16	Sun	P 22	M 39	1 28	.....
17	Mon	P 20	M 39	.....	1 28
18	Tue	P 18	M 40	.....	1 28
19	Wed	P 17	M 41	.....	1 28
20	Thu	P 15	M 42	1 28	.....
21	Fri	P 14	M 43	1 28	.....
22	Sat	P 12	M 44	1 28	.....
23	Sun	P 10	M 45	.....	1 28
24	Mon	P 09	M 46	.....	1 28
25	Tue	P 08	M 47	.....	1 28
26	Wed	P 07	M 48	.....	1 28
27	Thu	P 06	M 49	.....	1 28
28	Fri	P 05	M 50	1 28	.....
29	Sat	P 04	M 51	1 28	.....
30	Sun	P 03	M 52	1 28	.....

FAVORSCOPE SYMBOLS

H = Hours   + = Add

M = Minutes   — = Subtract

April Correction Formula For 1933

Your Locality	Sunrise Correction In Minutes	Sunset Correction In Minutes
Aberdeen, South Dakota	+ 23	+ 41
Atlantic City New Jersey	— 2	— 2
Baltimore, Maryland	+ 6	+ 6
Boston, Massachusetts	— 19	— 12
Canton, Ohio	+ 24	+ 28
Chicago, Illinois	— 13	— 5
Cleveland, Ohio	+ 23	+ 29
Detroit, Michigan	+ 29	+ 36
Gary, Indiana	— 14	— 7
Kansas City, Missouri	+ 17	+ 19
Los Angeles, California	— 1	— 13
Malaga, New Jersey	Spectro-Chrome Meridian	
Mansfield, Ohio	+ 26	+ 30
Milwaukee, Wisconsin	— 13	— 5
New York, New York	None	— 3
Philadelphia, Pennsylvania	+ 1	+ 1
Providence, Rhode Island	— 17	— 12
Rochester, New York	+ 14	+ 6
St. Louis, Missouri	+ 2	None
San Diego, California	— 2	— 20
San Francisco, California	+ 13	+ 7
Sheboygan, Wisconsin	— 15	— 2
Terre Haute, Indiana	— 10	— 10
Vineland, New Jersey	Spectro-Chrome Meridian	
Washington, District of Columbia	+ 10	+ 7
Wellsville, New York	+ 8	+ 16
Wildwood, New Jersey	— 1	— 1
Wilmington, Delaware	+ 2	+ 2

CAUTIONS OF THE MONTH

Date
10—LOW CRITICAL PERIOD
25—HIGH CRITICAL PERIOD

The crisis for Sufferers may culminate on the above dates, beginning two and a half days prior and ending two and a half days subsequent to the Criticals.

FAVORABLE DATES TO ACCEPT NEW SUFFERERS

3, 5, 6, 7, 10, 11, 15, 16, 18, 22, 23, 26, 27, 28.

TO FIND DAY FAVORS FOR TONATION

Instruction 1—Take column marked "Sunrise" for the desired Date, in the Chronos.

Instruction 2—Add to or subtract from the Chronos, the number of Minutes indicated in the Sunrise Correction column for your Locality; the result will be the Correct Sunrise in your Local Standard Time.

Instruction 3—Refer to Initial Sunrise Favor column for further correction; if none, the Correct Sunrise Time will be the First Day Favor for beginning Tonation; otherwise, add the indicated 1 Hour 28 Minutes for the First Day Favor to begin Tonation.

Instruction 4—Thereafter, subsequent Favors to begin Tonation follow at every 2 Hours 56 Minutes.

Instruction 5—When the Concluding Day Favor overlaps into the Sunset period, omit Tonation during that Favor.

TO FIND NIGHT FAVORS FOR TONATION

Instruction 6—Take column marked "Sunset" for the desired Date, in the Chronos.

Instruction 7—Add to or subtract from the Chronos, the number of Minutes indicated in the Sunset Correction column of your locality; the result will be the Correct Sunset in your Local Standard Time.

Instruction 8—Refer to Initial Sunset Favor column for further correction; if none, the Correct Sunset Time will be the First Night Favor for beginning Tonation; otherwise, add the indicated 1 Hour 28 Minutes for the First Night Favor to begin Tonation.

Instructon 9—Thereafter, subsequent Favors to begin Tonation follow at every 2 Hours 56 Minutes.

Instruction 10—When the Concluding Night Favor overlaps into the Sunrise period, omit Tonation during that Favor.

TO FIND JUNCTIONS

Instruction 11—A Junction of the Occult Operation Of The Great Sympathetic Nervous System, occurs 44 Minutes after each Favor Time; a Junction must be included within a Tonation period.

COMPLETE TONATION

Instruction 12—A Complete Tonation must begin at a Favor Time, include a Junction and terminate in not less than 16 Minutes after a Junction, but, may exceed it.

PRECISION

Seconds, Fractions and Decimals are omitted purposely in the foregoing computations, but they are all precise to the Minute for Spectro-Chrome Tonation.

PROTECTION NOTICE

FAVORSCOPE TIME

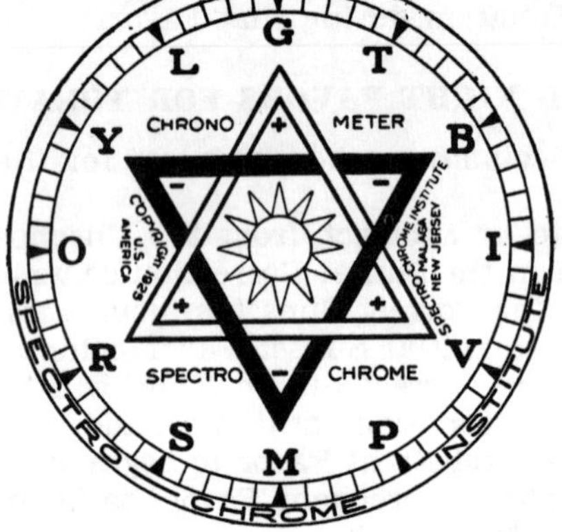

In Sympathometry, the Time of Day or Night is reckoned in Spectro-Chrome terms. The Time is called CHRONOS, the Hour is represented by the initial of the Attuned Color Wave at the indidicated position, the Minutes are designated in figures, the A. M. as White Triangle is shown by White, the P. M. as Black Triangle is marked by Black. The Months begin at Red for January and go around clockwise.

SPECTRO-CHROME

AUGUST—SEPTEMBER
OCTOBER
1933
INDIGO—VIOLET
PURPLE

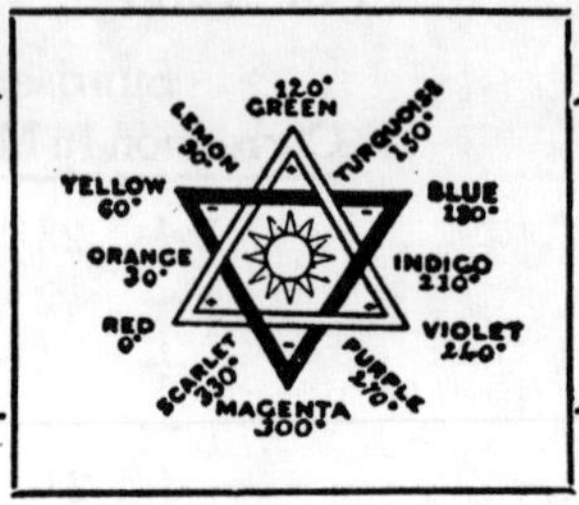

VOLUME
9
NUMBERS
8 - 9 - 10

Trade Marks Registered In The United States Patent Office

Devoted To: Spectro-Chrome Metry **Dedicated To: Service Of Mankind**

AUGUST
SEPTEMBER
OCTOBER
NUMBER

Editor
Dinshah P. Ghadiali

In course of life you reach a time,
When Doctor, Lawyer, Priest alike,
Does fail to give the sought relief,
To troubled mind or burdened heart-
Then come to me:
I help you to help yourself.
SPECTRO-CHROME.

Single Copy 50 Cents

Attuned Red Wave

(The World Through The Eyes Of Dinshah)

SPECTRO-CHROME INSTITUTE MEMBER, N. R. A.

When the President of the United States Of America, Franklin D. Roosevelt, took the reins of the depression bugbear and established the National Recovery Administration, Dinshah was conducting the 87th Class in Milwaukee, Wisconsin. In accordance with his practice of supporting every movement which aims at the Service Of Mankind, Dinshah promptly answered the patriotic call and on Wednesday, August 2, 1933 signed the President's Re-employment Agreement and received the Blue Eagle.

Although as a professional service institution, we were actually exempt from the various restrictions prescribed by the N. R. A., we had no difficulty in joining, because, all our employees always received more than the Code at present demands. We joined to relieve a strenuous condition and gladly sacrificed our personal views by curtailing to the limit the working hours of our employees and raising their salaries. We even took more persons in the work and have now a staff of ten persons on the payroll, with the President and Vice-President serving the World as before without any compensation, making in all twelve.

We are in a position to take and train into our work more persons, provided they are conscientious and honest and may follow the principles of Spectro-Chrome and are willing to join in the cause of human service. Dinshah works 18 hours a day and sometimes for days even 20, when absorbed in writing or research. The N. R. A. did not benefit him in the least, but, he and his personal staff officers fully realize what pleasure there is in humanitarian a c t i v i t ie s, where each day forwards the goal and work becomes joy. We believe in work and more work.

Work hurt none,
It is worry that kills,
Let the other fellow worry for you!
Join the Don't Worry Club!

We hope when the depression is lifted, the American people will begin again to work more and not slump into the lethargy of inaction. God never sleeps; why should we His Sons And Daughters? This is the time for your betterment, friends; get a Spectro-Chrome Metry Encyclopaedia and commence its study, so you may profit in health and be happier by the blessings of the time and leisure given you by the N. R. A.!

ERNEST DISTRIBUTING COMPANY

Our New Deal

1,000 Representatives Wanted

More Relief For The Unemployed

During the past 13 years, since Spectro-Chrome Metry saw the Light Of Day, the Originator went through numerous experiences, some of them nothing short of phenomenal.

His recent experiences, in the direction of fighting the Medical Octopus, which has been so persistently trying to grapple him and strangle his humanitarian activities, made him resolve emphatically to expand the work so broadly that his dream of "Spectro-Chrome—In Every Home" may be realized in short time.

It would seem as if the Powers Of Nature were aiding him silently. In Milwaukee, Wisconsin, he contacted one man who measured upto the mark, set by the Originator for such a stupendous task. That person was Mr. Edward A. Ernest, one of the enthusiastic students of Spectro-Chrome, who studied the system and graduated as a Spectro-Chrome Normalator.

On invitation, Mr. Ernest visited the Central Office and Research Laboratory of Spectro-Chrome Institute at Malaga, New Jersey and went over every angle and phase of Spectro-Chrome activities night and day for three consecutive days with the President Dinshah.

He challenged every statement, asked for definite conclusive proofs, went into the value and merits of Spectro-Chrome from all conceivable sides and finally offered to act as the exclusive Distributor General for the United States, under certain conditions, the chief among which was "Service To Suffering Mankind."

From its angle of vision, Spectro-Chrome Institute too, found Mr. Ernest equally trustworthy, honest, straight-forward and energetic, full of the sincere conviction that in the Science Of Spectro-Chrome he found what was the embodiment of precision in the Healing Arts.

Both parties joined hands and on Monday, September 4, 1933, Full Moon Day and Labor Day, an Agreement was unanimously drawn and signed, sealed and delivered Tuesday, September 5, 1933, by which Spectro-Chrome Institute appointed Mr. Ernest as exclusive Distributor General for all its manufactures and products, for the entire United States Of America.

Mr. Ernest is well known throughout the country, having been President and General Manager of the Wisconsin Ornamental Iron & Bronze Company for 25 years, doing a national business and then Vice-President and Manager of the General Bronze Corporation of New York, until 1931. Before his appointment, he had used a Spectro-Chrome in his own home for about 2 years and it did not take his intelligence long to appreciate the beneficial use of the invention and to see it as an essential necessity for every home. He felt the call of duty to mankind and entered the field, with all his knowledge, experience, resourcefulness, business connections and willingness to do good urging him.

Mr. Ernest will require in his capacity of Distributor General of Spectro-Chrome Institute, approximately 1,000 Representatives and Agents. He is a man of exacting fairness and justice and the Representatives or Agents working with him in Spectro-Chrome activities, are certain to receive utmost satisfaction.

The appointment of Mr. Ernest and his acceptance of the position, now leaves Colonel Dinshah unhampered in the pursuit of his researches. He will not have to conduct Classes every month in different cities to instruct students at great expense and will have more time at his disposal to manufacture and market his numerous inventions for the benefit of the World.

Dinshah requests all those who desire to act as Representatives or Agents for Spectro-Chrome Institute, to address:

Ernest Distributing Company,
3106 North Hackett Avenue,
Milwaukee, Wisconsin.

Orders for all Equipments, Supplies and Products may be sent either to him or direct to Spectro-Chrome Institute; they will receive equally prompt attention.

SPECTRO-CHROME METRY ENCYCLOPAEDIA

Start Home Training Course

The First of the 3-Volume set of Spectro-Chrome Metry Encyclopaedia is off our Press and is now being mailed. It will be in the hands of the subscribers within a few days.

With the publication of this work, Spectro-Chrome Institute enters a new lease of life. During 13 years, all the activities of the teaching were centered in Resident Courses conducted by the Originator; those who graduated had no text books to refer and those who did not graduate were not enabled to follow the science at a later period.

The scope of Spectro-Chrome Metry Encyclopaedia is of such comprehensive character, that all the theoretical and scientific part may be comfortably studied in the quiet and peace of the home and the student who could not conveniently spare the time for the Resident Class of 2 weeks, can study at leisure all that was taught before by word of mouth.

It is a set of books whose value will never fade. The First Volume contains much of the Healing Triangle Of Light and many more matters of intense scientific value. The language is simple and yet so vigorous that the student can not miss learning the science in its utmost essentiality.

The price of the complete set is only $10.00. The purchaser or any other person may enroll for the Free Home Training Course, by sending the Enrollment Blank duly filled. The enrollment and subsequent home examinations and certificate of qualification are Free. Only one person will be accepted as a Free Student for each subscriber and the Application must be on the original form printed in the book; the same rule applies to the Examination Application.

Dinshah expects to have the Second Volume out by his 60th birthday—November 28, 1933. It is ready in type and if our kind friends will please be patient with the man who uses his life for their service exclusively, he will give them in the Second Volume, such funds of information as no money can buy, that their confidence in him will be justified.

PAINTED SKINS

The Commonwealth Bank Of Australia forbids all its women employees from rouging and lipsticking. Five girls disregarded the rule and were told to wash the decoration, but, hesitating were dismissed.

There is no justification for making a wall out of the skin. To clog the pores and choke perspiration is unwise. The Bank's attitude should be copied everywhere to promote health.

SLEEPING SICKNESS

St. Louis, Missouri is invaded by what they term an epidemic of "Sleeping Sickness." Many sufferers are either sleeping it out or slipping out of it; the authorities know not what to do and the Medical Doctors know less. They call it "Encephalitis," but that would merely mean "Inflammation of the Brain."

Names do not remedy the trouble. Spectro-Chrome remedies without naming. Our Graduates can easily reach this disorder by irradiating with Spectro-Chrome Green and the Ultra-Green Attuned Color Waves when recent, the Infra-Green with Magenta on 4 in older attacks.

Will the Medical Doctors prefer to let the sufferers die, sooner than use Spectro-Chrome?

DOCTORS WANT CASH

Dr. Samuel Horton Brown, Editor of the Weekly Roster and Medical Digest, the mouthpiece of the Philadelphia County Medical Society, raises the wail that Medical Doctors should be paid, just like other business people. We agree with the learned gentleman and should strongly advise them to place themselves on the same level as tradesmen, like bakers, butchers and barbers, getting the same profit as they get, by the "Cash And Carry System". It would appear that the time is fast approaching when the dignity and pomp connected long with the profession of the Medical Doctor will be removed by the public itself when the Truth about the empiricism of Medicine is more clearly comprehended.

PORK AND AMERICA

It was reported at the Annual Meeting of the National Live Stock And Meat Board, in Chicago, Illinois, that each average person in this country ate 72 pounds of Pork during 1932, plus 47.2 pounds of Beef, 6.8 pounds of Veal, 7.1 pounds of Lamb and 15.2 pounds of Lard.

Do not ask us again about the cause of Consumption and Cancer. Your answer is herein. Moses debarred expressly such filthy "food" (?) as Pork. The wise (?) Christians laugh at His dictum, but, some day it will be realized that the introduction of the Trichina Spiralis into the human body was the reason why He placed the ban. He knew.

$75,000 A BONE

A Federal Court Jury awarded $75,000 damages against Dr. Alfred L. La Ferte of Detroit, Michigan, because, he removed the wrong bone from the wrist of a football player, Ken Strong.

It would be better to award similar damages for each Tonsil, Ovary, Appendix, Adenoid, removed wrongly from the human body; a few such cases in succession would stop merciless butchery.

SPECTRO-CHROME IS PAR-ULTRA-SUPER-HOMEOPATHY

Among one of the early Graduates of Spectro-Chrome Institute, I have the honor to enumerate Robert R. Logan, President of the American Anti-Vivisection Society, a gentleman of experience and learning and having the highest repugnance against the systems of organized butchery, foisted on the suffering public under the name of Medicine. In The Starry Cross of September 1933, he wrote an editorial under the heading of "Homeopathy," which portrays so many matters dear to our heart that I take the liberty of republishing it, to show how others with intelligent and thinking minds are drifting into the field of "No Drugs." He says:

"There is an interesting article in **Progress Today,** April-July, 1933, issue, entitled "Homeopathy and Vivisection," by E. Petrie Hoyle, M. D., in which the author asserts that modern orthodox medicine—vivisectional medicine — in its reliance upon surgery, vaccines and serums has lost all knowledge of curative drugs or medicines and so really has nothing with which to heal the sick.

Dr. Hoyle begins by pointing out that the orthodox medicine has had a succession of fads and fashions: "bloodletting (which has killed even kings), mercury (pushed to the limit, until teeth fell out and joints were filled with mercury globules, visible to the eye—see Royal College of Surgeons Museum), Listerism (with many more victims overdosed with carbolic), then the reign of Microbes (when it was tried chemically to reach them **in vivo,** to be slain **in situ**—they often died with their hosts). This was **antisepsis** with a vengeance, and if the host did not die, who can compute the extent of the bodily injury sustained? Then came Salvarsan (which was to make venery a safe sin, but untoward events rapidly following, the famous '606' soon became '909,' and now one is ashamed, or afraid, of even this experiment on the sick). After this came the fashion of 'breaking-the-fever,' when the reign of antipyretics was merged into hypnotics (which simply masked the pain and danger symptoms until the patients could not tell what ailed them most.)"

"These two aids were soon put on the 'poisons list.' Why? Because 'breaking-the-fever' often spelt a funeral. I will tell you another time what the microscope has demonstrated in this connection. Break-the-fever, indeed! What unscientific nonsense, as phagocytosis is quickened in a fever-raised temperature. (See Professor Bakody's work at Budapest.) After this came the fashion of Serum-Therapy, hypodermically given — **another most unsurgical procedure,** which had a most unscientific origin. **What did God give us mouths for?** Serum-Therapy merged into colloids, with spectacular vivisection into gaining ground and endowments. I wish Rockefeller could be told how his millions are being used. If he realized how animals were being treated, I can't think he would stand for it. His special doctor (now dead) used to be Dr. Hamilton F. Biggar, a homeopath whom I knew very well."

"And now, 1920 to 1932, as another phase of bewildered groping, orthodoxy comes right back to bloodletting, discarded when Grandmamma was young! Such is a brief outline of a wobbling pseudo-science, and a vicious circle of some seven or eight decades, though there are scores of other forgotten

fashions. Change is not always advance."

Dr. Hoyle then goes on to state that Homeopathy has not changed in 130 years because it is founded on absolutely scientific principles, each drug in the homeopathic pharmacopeia having been selected to cure certain specific symptoms because it invariably produces those symptoms when tested not upon animals without their consent but upon human beings voluntarily. Hahnemann began this work with some ninety-seven drugs and

"The results were more than astonishing. The drug symptoms produced were faithful portrayals or 'pictures' of every day illnesses. His 'provers,' expressing their departures from normal, made use of exact phrases met with daily when questioning the sick today, and these phrases are still on record. Moreover, he found that nine-tenths of the symptoms produced were subjective, **that is, requiring human speech to make them known.** These he classified and recorded and they are of classic import today. Thus nine-tenths of those thousands of drug symptoms would have been lost to posterity had he employed animals. Have the vivisectionists anything with which to beat this work?"

"Please note," says Dr. Hoyle, further on, "that Homeopathy has not any 'specifics' for this or that disease, nor has it ever had any. A specific drug for any specific disease may lead to all sorts of complications. What if the disease has been wrongly diagnosed? I have drawn your attention to the percentages of error in diagnosis. Any prescription based on a mere diagnosis, which may be wrong, at once makes the prescribed drug void, deleterious and even stupid. Homeopaths never evade a diagnosis, but we go beyond the mere classifying of a disease under a technical nomenclature. It is seldom that two people are affected in **an exactly similar manner** in any disease as broadly diagnosed. With Homeopathy it is always the precise and peculiar **personal** deviations from the normal that are noted, and these individual differences point to a demand for one **drug only** in a manner totally unknown to vivisectionists."

Under the use of anti-toxin diphtheria, says Dr. Hoyle, quoting the **British Medical Journal** of September 28, 1931, has been showing a death rate of 20 per cent. lately:

"This looks as if another fashion of orthodoxy was in the process of being discarded in spite of much vivisection. The homeopathic death rate in this disease, using chiefly our high dilutions of **Mercurius cyanide** (a terrible poison if not employed in the minute homeopathic doses), which should never be used in a lower potency than the tenth decimal or the twelfth decimal strength (ten or twelve noughts) runs about one to five per cent. Yet homeopathy has never called on vivisection."

Lastly Dr. Hoyle points out that after ridiculing for generations the minute or **infinitesimal** doses of homeopathy as having no possible potency, orthodox medicine has now discovered that infection could be transmitted with a dilution requiring fifteen noughts to express it:

"Orthodoxy finds that work with a dilution of 1/1,000,000,000,000,000 was strong enough to inoculate two out of every three men and give them a common cold after they had filtered out every microbe. The **Lancet** calls this a 'brilliant piece of work,' so why sneer when we employ our infinitesimals? Is this another discard for orthodox bacteriology and vivisection? Here they had filtered out all the bacteria, and yet there remained the infectious **something** — infinitesimal, without a doubt."

THE STARRY CROSS does not hold a brief for Homeopathy or any other cult or school of medicine, but it does protest against the domination of the one school which bases its "science" of human healing on cruel and misleading tests made upon animals, and which steadily ignores all the facts and observations recorded through the centuries by all other seekers after truth.

It may be that in the very nature of things there never can be a "science" of medicine, that it will always remain an art whose secret shall lie in the individual ability of each physician to reconcile his patient to a conformity with nature's laws. If, however, there is to be something approaching a science of medicine it will come only when state protection is withdrawn from one favored school and the monopoly and arrogance of vivisection broken forever. Then and only then will the knowledge gained by homeopathy and osteopathy and the other schools of healing be shared and correlated and made of practical value."

The remarks of Dr. E. Petrie Hoyle about the "Orthodoxy" are very pertinent, but, where that Orthodoxy and even Homeopathy with its "Twelfth Decimal Potency" ends, Spectro-Chrome with its most precise yet ultra-superattenuated potency begins.

There is nothing new to me in the dilution of the One Quadrillionth strength. The chemical and physiological properties of the Spectro-Chrome Attuned Color Waves which have now proven merits of invaluable character, are nothing but inconceivably superior "dilutions" and triturations hurled from the furnace of Nature with the highest speed possible in physical matter.

The moment one will realize that Light and Color are not mere "decorations," but, terrible engines of chemical projection of physical particles, he will become a person imbued with the same views as I have and a new history will be written in Healing Arts.

That page was written for Spectro-Chrome Metry on Monday, April 26, 1920.

DEATH OF ANNIE BESANT

Annie Besant, the President of the Theosophical Society, died at the age of 86.

Dinshah was introduced to her in London, England, in 1893, just prior to her first visit to India. She was then a meek student of Oriental Philosophy, but, step by step she rose into power in the Theosophical Society, until she became its esoteric as well as exoteric head. How she attained that position is known well to the writer, although he does not think it essential to republish the events leading to it. He took active part in her controversy about Charles W. Leadbeater and also about the Krishnamurti incident. His lengthy and caustic articles on both matters created quite a stir at the time. The Leadbeater articles were published in his weekly newspaper Apakshapata (The Impartial). Prior to that, Dinshah had numerous clashes about his life and lectures with her and the climax was reached when one day in Bombay, Annie Besant threatened him that unless he stopped his sledge-hammer tactics "The Masters will not keep you in the work." Dinshah promptly retorted that "The Benign Masters never interfered with individual

Karmas" and told her his reply would reach her next morning.

The next day, he resigned his Fellowship and voluntarily surrendered all his esoteric papers. Three years later, after much altercation, his old Diploma was restored and he resumed his activities.

However, since the Krishnamurti and the Messiah affair, Dinshah stopped taking part in the activities of the Theosophical Society, in as much as its separation into the different factions and divisions was not his idea of Universal Brotherhood.

Annie Besant was a great soul. A powerful orator and masterful discourser she won many hearts. However, her autocratic temperament frequently produced awkward situations.

She is gone to Planes about which she wrote in life; may she advance!

SLOW APPROACH TO SPECTRO-CHROME

How cunningly are chemical products advertised may be gathered from the new noise of "Heavy Water," which is being made in the American Chemical Society's meeting in Chicago, Illinois. In a report published in the Wilmington Morning News, of Wilmington, Delaware, of Tuesday, September 12, 1933, Howard W. Blakeslee, Associate Press Science Editor writes about this "Heavy Water":—

"This liquid is not different in appearance from ordinary water, yet plants do not grow in it. Although perfectly pure and made of the same "H2-O" as all other water, chemists suspect that it may prove to be either poisonous to animals and man, or show medicinal effects.

Heavy water is made of the recently discovered double-weight hydrogen atoms and the heaviest of oxygen atoms. The hydrogen atoms are obtained from ordinary hydrogen, of which they form a small fraction. Electricity is used in separating the two kinds of hydrogen.

The resulting extra weight water is only slightly heavier than the ordinary liquid, because the heavy hydrogen atoms are very light compared with the single oxygen atom in a water molecule. Ordinary hydrogen has a mass, of weight of one; heavy hydrogen, two; and oxygen atoms have masses of 16, 17 and 18."

It were so much simpler and better for the World, if instead of constantly groping in the dark, the chemists would throw into discard for all times the Atomic Theory and interpret all chemistry in Spectro-Chrome terms. There is **no** such thing as an indivisible "Atom." I have been speaking for dozens of years and emphatically teaching my students that the so-called "Elements" of the chemists are, spectroscopically speaking, all "compounds." Were they true Elements, Radium could not as an Element throw off Niton and Helium and later become Lead; Uranium would not decay into Radium and many other phenomena of Radio-Emanation be impossible.

Spectro-Chrome Metry only can open that door, through the correct comprehension of Occult Chemistry; our students should form Classes for that purpose and ask me to give a Post-Graduate Course; when they will be ready, they will find me at their service, but perhaps the Spectro-Chrome Metry Encyclopaedia may avoid such necessity.

HIS NAME

I receive all kinds of letters addressed in different manners. The latest was from New York, N. Y., addressed as follows:

Colonel Dinishau Gilidili,
Molaca,
N. J.
"Chrome Inst."

Thanks to the vigilance of the United States Post Office, it reached me the next day, instead of being sent to Molacca or the Spice Islands. Gilidili! I dislike to tell what it would mean in the language of the Parsees!

REPLACEMENT OF SLIDES

Spectro-Chrome Genuine Attuned Color Wave Slides as supplied with each Spectro-Chrome Equipment, are the very life of the system and the rules pertaining to their replacement are still the same as installed 13 years ago.

Here are the Rules:

1. Each Spectro-Chrome has only a single Set of Five Dinshah Attuned Color Wave Slides. The Graduate has Two Red Slides in the Slide-Carrier.
2. No SPARE Set is ever supplied.
3. When needing replacement of broken Slides, the complete Set including the broken pieces MUST be returned.
4. The Seal on the Slide-Carrier should not be broken.
5. All Slides MUST belong to the same Set.
6. A Set sent with mixed Slides, even though each be genuine, will NOT be replaced.
7. Each Set of Slides is registered for ownership with us and will be entitled to replacement only through change of ownership by registration of the Assignment of Bill Of Sale.

There are no exceptions and the rules apply to all equally.

NEW FAVORSCOPE SYSTEM

Beginning with this issue, our students will notice our new method in recording the Favorscope, so as to meet the demands and requirements of the most untrained laymen. All the frills of science are removed, without deteriorating its scientific character. The Sunrise and Sunset times with the Sunrise and Sunset corrections are abolished and the Chronos or Time Table is simplified by adding those corrections beforehand. The Sunrise column is changed to Day First Favor and the Sunset column is made into Night First Favor. It is a Ready Reckoner.

The Standard Time is used throughout and the Chronos is built to the Longitude (Meridian) and Latitude (Parallel) of Spectro-Chrome Institute, which is taken as our Observatory Standard. All the user of the Favorscope has to do, is to join to this Standard Time, the Correction in Minutes of Addition or Subtraction for his locality and the First Favor is fixed. Thereafter, each 2 Hours and 56 Minutes the next Favor will appear. This alteration was made in conformity with the suggestion of our Distributor General Edward A. Ernest and noticing its utility, we were glad to make the necessary changes. We are sure it will meet with unanimous approval.

THE REACH OF SPECTRO-CHROME

During the past 13 years, what Spectro-Chrome accomplished may be judged from the cases reported in our magazine. Below is a partial list of the various Disorders in their original terminology. It is a silent tribute to the success of our Spectro-Chrome Normalators. The numbers refer to the pages in Spectro-Chrome:

Abscess	203-206-299-300-301-304-332-435-585
Accident	138
Acidity	583
Adenoids	307
Albuminuria	334
Amenorrhea	68-112-353-384-448-462-584
Anemia	68-232-584
Angina Pectoris	204
Appendicitis	45-68-229-230-303-304-346-352-435-461
Arthritis	476-602-603-604-612
Asthma	284-306-383-492-500-583
Auto-Intoxication	230-232
Backache	498
Bladder Trouble	261-327-328-501
Blindness	382-584
Blood Poisoning	260
Bronchitis	583
Burns	14-179-180-200-260-331-434-436-495
Cancer	41-204-232-281-302-326-435-436-448-497
Carbuncle	86-87-88-89-180-280-435-611
Cataract	501
Catarrh	138-233-261-297-299-300-333-346-352-384-466-500
Cyanide Poisoning	499
Conjunctivitis	230
Constipation	68-91-109-112-138-204-232-233-234-306-333-334-346-349-350-351-384-435-448-500-501-611.
Consumption	181-232-235-349-353-415-435-436-612
Convulsions	468
Corn	138
Cough	109-232-234
Deafness	232-281-607
Diabetes Mellitus	231
Diarrhea	93-112-231-383
Diphtheria	138
Dropsy	333-609
Drug Habit	15-230-348-500

Dysentery 583
Dysmenorrhea 138
Dyspnea 384
Earache 233-281-284-300-301-307-330-334-558
Eczema 91-138-232-415-435-447
Epilepsy 46-299-332-436-498
Eructation 91
Eye Trouble 207-329
Fever 139-260-435
Fistula 462
Flat Foot 612
Flatulence 91-139-333-612
Gall Stones 435-466
Gangrene 204
Gland Trouble 299
Glaucoma 608
Gleet 347
Goiter 91-330-461-494-501
Gonorrhea 232-347-447
Grip 8
Headache 498
Heart Disorder 261-501-610-612
Hepatic Hypertrophy 232
Hernia 301-382-384
Hiccough 15-382
High Blood Pressure 260-346-436-585
Hydrocephalus 470
Hysteria 299
Indigestion 138-230-382-500
Infection 330
Influenza 330-331
Insanity 260
Insomnia 307-435-606
Itch 285-384
Ivy Poisoning 261-331-500
Jaundice 91
Labor 153
Lead Poisoning 415
Locomotor Ataxia 353
Marasmus 333
Mastoiditis 262
Masturbation 584
Measles 47-607
Meningitis 501

Menorrhagea 231-232-302-304
Metritis 501
Mitral Regurgitation 231
Necrosis 200
Nephritis 232-260
Nervousness 45-91-232-334-384
Neuralgia 234-383
Neuritis 204-230-280-306-330-349-435-462-476-501-612
Night Sweat 333
Obesity 232
Ovaritis 231-352-353
Pain 435-436
Palpitation 91
Papilloma 204-261
Paralysis 138-139-334-349-384-495-496-497-585-607-608
Phlebitis 327
Piles 68-384-500
Pleurisy 229-334-583
Pneumonia 138-228-229-351-462-611
Prostatitis 501
Psoriasis 494
Pyorrhea 16
Retention 498
Rheumatism 111-233-308-351-352-584-611
Rhinitis 461
Rickets 206
Sciatica 501
Sepsis 14-42-43-44
Sores 498
Sprain 498
Sty 204-461
Syphilis 35-206-302-303
Tachycardia 235-415
Tic Doloreaux 447-501-608-612
Tobacco Habit 29-299-435
Tonsilitis 138-332-351-352-383
Toothache 232-284-307-331
Tuberculosis 181-232-235-349-353-415-435-436-612
Tumor 138-232-584
Ulcer 299-303-330-382-435-498
Varicose Veins 558-559-606
Vertigo 384
Voice Loss 461
Whooping Cough 284-305-331

Attuned Green Wave

(Spectro-Chrome Metry, Spectro-Chrome Normalators)

Meetings Of Planets

All Planets of the Scientific Order Of Spectro-Chrome Metrists, meet the First Thursday in each month, at 8 P. M. The following are the meeting places at present:

First Philadelphia Pennsylvania Planet, Second Floor, 1117 Spruce Street, Philadelphia, Pennsylvania.

First Milwaukee Wisconsin Planet, Room 157, Hotel Wisconsin, Milwaukee, Wisconsin.

First Detroit Michigan Planet, Park Room, Hotel Tuller, Detroit, Michigan.

First Vineland New Jersey Planet, Baker House, Landis Avenue, Vineland, New Jersey.

First Wilmington Delaware Planet, 2,910 Broome Street, Wilmington, Delaware.

First Malaga New Jersey Planet, Spectro-Chrome Institute, Malaga, New Jersey.

First Mansfield Ohio Planet, Mansfield, Ohio.

First Cleveland Ohio Planet, Cleveland, Ohio.

First Canton Ohio Planet, Canton, Ohio.

First St. Louis Missouri Planet, St. Louis, Missouri.

First Los Angeles California Planet, Los Angeles, California.

First San Francisco California Planet, San Francisco, California.

The Detroit Michigan, Milwaukee Wisconsin, Philadelphia Pennsylvania, Vineland New Jersey and Wilmington Delaware Planets are Active and hold regular meetings each month.

The Malaga New Jersey Planet is the Official Planet to which the staff workers of Spectro-Chrome Institute and Scientific Order Of Spectro-Chrome Metrists belong, ex-officio. Being in service, it is technically open all time and holds no monthly sessions.

The rest of the stated Planets and others not listed, are for the present inactive, holding no regular meetings.

Only **Members** and **Provisional Members** are entitled to admission at any meeting of a Planet. A Member or Provisional Member of one Planet is entitled to admission in any meeting of any Planet as a Visitor. Such Visitor can take part in the activities by permission of the Deputy.

No admission is allowed to any Collaborator or Stranger in any regular meeting of any Planet.

All Members and Provisional Members must show their Affiliation Cards before being admitted to any regular meeting of any Planet. Without the presentation of the Affiliation Card, no admission is allowed.

The Deputy, Sub-Deputy and Recorder forming each Planet Commission, are requested to make no exceptions to this rule, as it is obligatory under the By-Laws. Our Affiliates should endeavor to support and co:operate and not to hamper the activities by asking special favors which the Officers are not empowered to grant.

Will the Planet Recorders kindly send in their reports not only who was present, but, a detail account of the discussions, so the Principal Office may keep in closer touch with the questions and answer them, for the benefit of all?

Such co:operation saves duplication of instructions and also helps others who need information.

PRICE LIST

The following are the current prices of the Main Articles of merchandise manufactured by Spectro-Chrome Institute:—

Itisometer (Table Model)	$650.00
Graduate Spectro-Chrome, 2,000 Watts	425.00
Systemic Spectro-Chrome, 1,000 Watts	150.00
Transport Spectro-Chrome, 400 Watts	90.00
Cabinet Spectro-Chrome, 1,000 Watts for Alternating Current	65.00
Cabinet Spectro-Chrome, 1,000 Watts for Universal Current	70.00
Sympathometer, for Alternating Current	35.00
Convertor, Direct Current, Special for Sympathometer	35.00
Bulb for 2,000 Watt Graduate Spectro-Chrome	15.00
Bulb for 1,000 Watt Graduate Spectro-Chrome	6.75
Bulb for Systemic Spectro-Chrome	5.00
Bulb for Transport Spectro-Chrome	4.00
Bulb for Cabinet Spectro-Chrome	6.75
Spectro-Chrome Metry Encyclopaedia, Set of 3 Volumes	10.00
Good Thoughts	3.00
American Sex Problems	3.00
Healing Triangle Of Light	3.00
Spectro-Chrome (Magazine in Complete Sets Only) per copy	0.25
Spectro-Chrome (Magazine) per current year copy	0.50
Spectro-Chrome (Game Of Health)	2.00
Manjan (The Dinshah Dentifrice)	0.25
Lavan (All-Salt-Sea-Salt)	0.25

All prices are F.O.B. Malaga, New Jersey. Terms Net Cash With Order. No Discounts and No Credits.

SPECTRO-CHROME

DECEMBER
1933
SCARLET

VOLUME
9
NUMBER
12

Trade Marks Registered In The United States Patent Office

Devoted To: Spectro-Chrome Metry | Dedicated To: Service Of Mankind

NRA MEMBER U.S. WE DO OUR PART

TRANSITION NUMBER

Editor
Dinshah P. Ghadiali

In course of life you reach a time,
When Doctor, Lawyer, Priest alike,
Does fail to give the sought relief,
To troubled mind or burdened heart-
Then come to me:
I help you to help yourself.
SPECTRO-CHROME.

Single Copy 50 Cents

REINCARNATION OF SPECTRO-CHROME AS WEEKLY NEWSPAPER

This is the last issue of our monthly Spectro-Chrome!

After doing faithful service as a magazine since 1922, seeing exceptional downs and ups in the Service Of Mankind, finding the need of greater scope to accomplish the aim of "Spectro-Chrome In Every Home," after mature deliberation, we resolved to convert it into a weekly newspaper of a unique character, with a unique purpose. The name of the newspaper will be

SPECTRO-CHROME

It will be published each Sunday, dedicated to the Service Of Mankind, devoted to the Science Of Spectro-Chrome Metry. It will NOT be affiliated with any party, political, religious or social. It will take no sides and will have no paid advertisements, but will admit to its columns any and all matters which may uplift and ennoble mankind. Later, advertisements for public good may be accepted. If one has anything to benefit the World or enrich the scientific knowledge of the readers, its columns will give a warm reception.

Most newspapers depend for their existence on paid advertisements; as that ties them down to partiality, **SPECTRO-CHROME** will not have such paid advertisements to make an income, but, standing, on its own merits will establish a new type of independent, fearless, impartial, graftless human service in journalism.

Its circulation will be primarily among bona fide Members and Provisional Members of the Scientific Order Of Spectro-Chrome Metrists, whose subscriptions are actually paid through their Annual Dues and Fees. It will also be sold to the general public at the same rate as the Member subscribers. This rate will be as follows:—

Single Subscription for 12 months, inclusive of postage anywhere	$3.00
Single Copy, at the News-stand	0.05
Single Copy from the Publisher, inclusive of postage anywhere	0.10
25 Copies, inclusive of postage anywhere	1.00
50 Copies, inclusive of postage anywhere	1.50
100 Copies, inclusive of postage anywhere	2.00

No Free or Complimentary Copies will be regularly issued to anyone. Each copy must be paid either by the subscription of some individual or subscribed for him by some person interested in the dissemination of knowledge of the Science of Spectro-Chrome Metry. Its income will solely be through the sale of the newspaper and nothing will be derived from the sale of space, leaving it therefore on a totally independent basis, having none to cater and less to flatter.

SPECTRO-CHROME will maintain absolutely no censorship on its writers. Within the bounds of the Laws of God and Man, except preaching of obscenity, sedition, anarchy and libel, it will have wide open columns for the free expression of thought on any and every subject, without the slightest restriction. It will be open for the free expression of thought to all people, without regard to their views, castes, creeds, professions, vocations, religions, beliefs and standpoints, as long as the aim is "Service To Mankind."

To facilitate such service as it intends and expects to deliver, Spectro-Chrome Institute which owned **Spectro-Chrome,** voluntarily relinquished its ownership of the magazine and by consent of Dinshah P. Ghadiali, its Founder, transferred its rights, title and privileges, to the Scientific Order Of Spectro-Chrome Metrists, (a Not-For-Profit Corporation of New Jersey), to be used for the Service Of Mankind, through education of public opinion in matters of Health, Clean Legislation, Sane Living and the like.

Furthermore, Dinshah will act as the Editor of **Spectro-Chrome, without any salary or compensation,** so the Science Of Spectro-Chrome Metry may be placed before the truth-seeking public in its original beauty, so the welfare of the coming generations may be guarded and the interests of those who need help may be served in the most direct manner. He will also be its Publisher, taking the full responsibility.

It will not be a parched, Puritanic newspaper. Written under the editorship of Dinshah, the Originator of Spectro-Chrome Metry, each issue will mean some real good; his powerful pen will make it a vivacious, educational, family newspaper, whose clean expressions may inspire respect, love, confidence, faith, hope and patriotism. It will accept no gifts, gratuities and grafts. It will be most radical in the enunciation of the Truth and will endeavor to establish its reputation sheerly on its merits rather than on the support of some selfish partisans.

To cope with the responsibility of producing an attractive weekly

newspaper, we increased our printing facilities by installing Two Line Composing Machines, which with our other machinery will give us ample scope for service. Our printing plant is really a complete up-to-date plant. Besides the Line Composing Machines, we have two Saw Trimmers, a large Whitlock Cylinder, a High-Speed Miller, a Cottrell Pony, a Chandler and Price Automatic Jobber, a 36-inch Cutter, a Liberty Folder, a large Morrison Stitcher, a Boston Stitcher, a Matrix Casting Machine and many other smaller composing and printing outfits for a well-conducted newspaper.

Spectro-Chrome will be a Sunday newspaper. Its originality and unique character will go far to make it popular among our friends.

Our newspaper will not be used as the tool to harm somebody from ulterior aims. You may write your facts or type, but the signature and full address of the writer must be there for our information. Writing should be in ink and on one side of the paper, according to the usual rules of such work. Send nothing on approval or for return. Copyrights of others are not wanted. Let the work be your own and true and we shall print it for human good.

Send your articles. Our field is the World, our faith Truth in the Absolute. Use our columns to ventilate your grievances.

The First Issue of our Weekly **Spectro-Chrome** will be out by Christmas; it will be our first gift to the Medical Octopus. It will be dated ahead as of Sunday, January 7, 1934.

We want your co-operation; let us serve you.

Members and Provisional Members of the Scientific Order Of Spectro-Chrome Metrists are requested to send their Annual Dues and Fees of $6.00 promptly, to be sure to get the newspaper from its first issue. Miss none, as Back Numbers may not be available later.

Under the amended By-Laws of the Order, **each** Member or Provisional Member pays $6.00 a year, regardless of being husband or wife or relative of an Affiliate.

Already 2,500 copies are subscribed by our friends and as the first issue will be only 5,000 copies, we request you to place your order without delay. It will be a sound and sensible investment and its educational benefits will outweigh all other considerations.

All remittances for the weekly newspaper **Spectro-Chrome,** should be made at the rates stated above.

All Membership or Provisional Membership should be at the rate of $6.00 a year, beginning January 1, 1934 to the end of the year 1934; this includes the Annual Subscription for the newspaper **Spectro-Chrome.**

The subscription of $3.00 for the year, for purchasers of the Cabinet Spectro-Chrome after September 4, 1933, will be paid by Spectro-Chrome Institute to the owners of the newspaper.

Others who purchased the Cabinet Spectro-Chrome Equipment prior to September 4, 1933, will please send their $3.00 for the subscription, should they desire to receive the **Spectro-Chrome** each week or $6.00 including Provisional Membership, entitling them to attend the meetings.

No separate Favorscopes will be issued hereafter; the Favorscope in advance will be printed in the middle of each month, in the issue of **Spectro-Chrome.**

Kindly make all subscriptions and Membership or Provisional Membership remittances payable to:

Scientific Order Of Spectro-Chrome Metrists

and NOT to: Spectro-Chrome Institute. Both are separate Corporations and have separate accounts.

Make **nothing payable** to Dinshah P. Ghadiali; he works for both Organizations without any compensation or salary and outside of being their President-Treasurer, does not act in any personal monetary capacity toward either. He has no income.

Aufwiedersehn!

Spectro-Chrome—In Every Home

DOINGS IN DELAWARE

(1) Aviator Ray Hunt was arraigned before Municipal Judge John F. Lynn, in Wilmington, Delaware, on a charge of stunting over a congested area. The Defendant was under questioning by the Judge and His Honor looking out at the sky said, "How high do you think that airplane is in the sky?"

Defendant—May it please your Honor, that doesn't happen to be an airplane; that is a Turkey Buzzard!

The Court dismissed the case.

(2) In the same city, in the same Municipal Court, a Medical Doctor appeared for expert testimony; then another Medical Doctor came on the stand. He was asked, "Why were you called in that case?"

M. D. 2—Well, I was called because the other Doctor did not do what he should and the patient was not progressing!

We do not have the sequel.

(3) "On Thursday, November 16, 1933, I rented a machine (Spectro-Chrome) as my mother, ... has been confined to bed with a serious illness, mainly a very nervous weak heart and a low vitality, and then she broke out with what the medical doctor claimed to be shingles. Mother will be 75 years old on Thanksgiving Day of this year. After a series of treatments for three days with Spectro-Chrome, I am happy to say that today, Saturday, she is out of bed reading the newspapers, and was informed at my office by telephone that the shingles are healing up and mother is greatly improved due to Spectro-Chrome."

Wilmingtonian.

Tell that to the Mayor of your city.

THE SIGN ON THE WALL

Dr. Joseph A. Stephenson, Director, Department Of Moral Welfare, Presbyterian Board Of Christian Education, stated at the Logan Dry Rally, that "It is possible to repeal the 18th Amendment, but it is not possible to repeal the inherent evils of Alcohol and of the Alcohol traffic."

John Barleycorn ever was the enemy of Truth, Morality, Honesty and Health. By giving the public a tumbler of Alcohol, no nation will prosper. No nation can lift itself at the expense of Public Health. Alcohol was ever the despoiler of homes, the grafter of politicians, the corrupter of morals, the ruin of the generations to come. We know what it did in the past.

We know it is returning and the saloon with its horrors is looming on the horizon also. America, the greatest of all nations, to live under the impression that by the return of Alcohol prosperity will return, is slated for a sock in the jaw when the awakening will come. It is soon coming and how!

COME OUT OPENLY

In the "Letters To The Bulletin," in the "Answers To Queries," of Thursday, October 26, 1933, the following drew my attention:—

> "SMOKING—TOBACCO: Smoking is largely a personal matter and affects some persons differently than others. Your family physician should be consulted. Dr. George A. Knowles, Assistant Director of Public Health, says for boys in formative period "the least smoking the better." He advises against inhaling because of harmful effects of carbon deposits and nicotine in delicate lung passages, and against excessive smoking, as productive of hyperacidity and indigestion.—(J. P.)"

Here is a great and influential newspaper trying to face an issue of vital importance, yet, appears to hesitate to set down the Truth. Here is evidently a conscientious Medical Doctor who **knows** Tobacco **is** harmful, but, endeavors to keep his foot on two swings at the same time. Why not give clean cut open advice that "Tobacco is a filthy poison and none should use it in any form"? There is no question of inhalation; the mouth, nose and throat have a mucous membrane and whether one inhales or not, the poisons are bound to be absorbed through it. Speak out and save the youth; if the old tough ones do not want to be saved, it is their privilege.

SPECTRO-CHROME METRY ENCYCLOPAEDIA

The first volume of our Encyclopaedia, with 43 chapters, 400 pages, 56 figures, received unanimous acclamation from its subscribers. It is the best gift for your family or friend. Buy your Set now and get the autographed copy for a Christmas present.

KILLING DOGS

Now the Health Service in Virginia reports that dogs are the carriers of Rocky Mountain Spotted Fever, because the ticks are infected with its germs and the dogs carry the ticks.

So there! Destroy those diseased dogs and end the trouble.

My firm belief is that if the same treatment be accorded to medical nonsense, there would be fewer disorders.

A WISE DECISION

John McCrea of Philadelphia, Pennsylvania died leaving a will for $71,958, to be given to his grandchildren, if they abstained from boozing and smoking. The heirs tried to break that trust provision in the Orphan's Court, but failed. Judge John M. Gest ruled that the provision must be upheld.

It was just. What sensible man or woman would leave his or her hard-earned money to be wasted for such harmful habits? McCrea was sensible in not leaving his cash for grandchildren who would refuse to behave.

THE SAME ELSEWHERE

In the "Deshi Mitra," of Surat, India, dated February 9, 1933, I read an advertisement:

> "WANTED a Lady Doctor, (L. C. P. S.,) on a salary of Rs. 85 p. m., with free furnished quarters. Apply with particulars to Dewan, Chhota Udepur."

The salary of Rupees 85 per month means ⅓ of that in American Dollars. A Lady Doctor for $28.33 a month salary! How she will fare when the Prime Minister of Chhota Udepur will place Spectro-Chrome in the State Hospitals is a matter for thought.

IS N. R. A. FIZZLING?

When the National Recovery Administration started its operations, we had our opinions, the same as other and deeper thinkers. It was supposedly an Emergency Measure and depended upon the voluntary co:operation of the people. We signed under the same impression, willing to do anything humanly reasonable to serve the country and help the President of the United States Of America straighten the kinks left by his predecessors.

Right after the start, there was a publication in some of the newspapers that General Hugh S. Johnson said to the effect that all must join and if anybody declined he would "give a sock in the jaw." Believing that to be an exaggeration of the press, we did not pay attention to it. All who read it took it with good nature.

Now come faint rumbles and lightning flashes about "discontent" "law" "Constitution" and the like. The N. R. A. does not like the independence of certain business people and they believe the Government has not the slightest power to compel or coerce. I agree with this without reserve. I admire and love Franklin D. Roosevelt, but, if he believes he can use an Emergency measure as a permanent big stick or muzzle, he will find himself rudely awakened.

The American public is gullible and easy-going, but, it will never stand any dictation or compulsion. I may candidly state that put to the impartial test of our glorious Constitution in the august Supreme Court of our country, the whole conception would be found unconstitutional. America would be another political messpot, if our Constitution can be interpreted to suit the views of any President or any Congress. The Constitution made America and Americans must follow the Constitution. No President or Congress can set up a dictatorial hierarchy or there will be an economical uprising and as for giving a "sock in the jaw," the General Administrator had better not be so quoted. He perhaps did not mean it.

A warning word—Do not test the N. R. A. in the U. S. Supreme Court; there will be trouble out of that issue. Certain muds had better not be stirred. Let us pull the country from the depression—there will be time enough to fight.

We shall join you in it wholeheartedly, if there be any humbug in it; but, we have to be shown the Truth.

A STORY OF SEX MISERY

So many persons ask advice on important matters pertaining to their emotional life, that a letter like the following is worthy of my special attention. Its printing will bring to many the incidents in their own private life and show them how to overcome certain defects in emotional make-up. For obvious reasons, the name and identification is completely withheld. The young man is a decent man and what he says may be said by millions of others all over the World. He writes:—

"I have attended your ten free lectures which you gave at Wisconsin Hotel in Milwaukee. I admit that I haven't bought a Spectro-Chrome not because I don't believe it to be as you claimed but because of financial reasons. But believe me if I ever get married and have a family I shall also have a Spectro-Chrome in my home.

Here are a few questions which I humbly wish you would answer me. My problem is a moral one. I am a young man of 26 years of age and am single. I have never had sex congress with any woman. And, as some of my associates say, this is a serious mistake. They tell me a man should have sex congress whether he is married or not if he wants to be healthier and efficient. What have you to say about this? Should I continue to live a continent life until I am married or shall I go seeking some woman of the street? Of course I firmly believe that a man should live a continent life until he is married, but, in my case, well, you see I am in a habit of masturbating. Have been masturbating ever since I was a child. I have made many attempts to break myself of this habit but somehow I always get back to it. If I try to live a clean life, at the end of two or three weeks I get sick and extremely nervous. This in turn affects my speech. I begin stammering more than ever before. You see, I am also in a habit of stammering. I have stammered ever since a child. It's a hopeless case, yet, something tells me that it is possible for me to break away from these bad, demoralizing habits and live a normal happy life. Won't you please advise me what I should do about my condition. Shall I persist living a clean life in spite of the fact that it makes me sick, or shall I masturbate occasionally? I really believe that the root of my speech impediment is masturbation.

I feel very much ashamed of writing you a letter like this, but I am desperate and despondent and am tired of living, and I feel that you can help me. Please don't fail me."

Most assuredly can I help you, young man, as I helped thousands of other unfortunates, in worse conditions.

Let me advise you step by step. Your problem is not solely a "moral" one—it has many phases besides. Morality is right conduct and Sex Congress as a normal function of emotional life is not a simple question of morality.

Sex has many aspects in its expression. It may be Physical as Sex Congress, Psychical as Affection or Mental as Love. Each aspect has its manifestations for normalcy or abnormalcy and the higher the conception of the Human Being, the greater the nearness to mentality.

You have innate morality if at 26 you did not connect with the opposite sex. Maintain by all means that purity and do not "seek a street woman." Purchased Sex Congress is devoid of the best in man and woman—Love. Love can not be bought and such promiscuous intercourse will stumble you into possible venereal disorder, which will make you sorry for life.

Your associates who advise you that a man MUST have Sex Congress are all wrong. There is chastity among men as among women and if you look forward to a happy future, uplift yourself from such wrong associations.

Your illness is not because of the continence, but, because of your brooding and Self-Abuse. A drop of male Semen lost from the body, is like losing 40 times that much in Blood and even in married life, in actual so-called "legal wedlock," profusion in Sex Congress produces just as disastrous results as Self-Abuse, because of wastage of life's vital fluid. In such cases, marriage is nothing but licensed debauchery.

Take my logical advice. A man like you can be a good husband to some good girl. Go among your female relatives and learn to associate with them in a congenial, fraternal manner. Look around you. Find a girl of good temperament and clean domestic habits, suitable to your ideals and marry her. Emotions were intended to be exercised in Human Beings, but **not** to be debased as your associates aver.

On the contrary, shun them.

To overcome Self-Abuse and Stammering, let me tell you point-blank that mere advice will not help you. You MUST help yourself. Change entirely to the Rational Food Of Man system as you learnt in my Free Course on Spectro-Chrome Metry. It will remove the false stimulation and clear your Blood System.

With that change, go to a friend who has a Spectro-Chrome and irradiate with Lemon Systemic, Magenta on Area 4. Continue this till your Kardoscil and Spiroscil (Pulse and Respiration) becomes normal. Use Yellow to keep the Backelim clear, if necessary. Four weeks of such Irradiation with Spectro-Chrome, will perform wonderful results. If Area 11 be very harassing by wrong thoughts, an occasional Purple on it will decrease the Blood Pressure there. You will find when your system gains equilibrium, the hankering for Self-Debasement will disappear and your Stammering too will be decreased, because you will be enabled to look the World straight in the face.

Should you do this, your troubles will soon be over and you will have gained mastery over that which is ruining you now and if left uncontrolled, will blast your future and the future of the progeny which you may perhaps later beget.

Now is the time. Take this from your friend; approach E. A. Ernest our Distributor General and I am sure he will put you into the way of someone who will help you with Spectro-Chrome service. However, before going to him, change your **food.** Stop smoking and the use of all intoxicants, if you have such habits. They are both curses to civilized mankind and utterly worthless in every sense. Do as outlined and life will scent rosy to you once again.

When you feel a different outlook by Spectro-Chrome, marry and lead a clean, happy life, for which marriage to some noble girl fitted to your station is the ideal avenue.

Above all, do NOT go to a Medical Doctor for advice; he has nothing except Bromides to offer and they will desex you. Shun him as you will shun an undesirable person.

SOME HINTS ABOUT SPECTRO-CHROME METRY ENCYCLOPAEDIA

The first volume of the Spectro-Chrome Metry Encyclopaedia was published on October 7, 1933 and already very favorable comments are in our hands.

The second volume will be even more fascinating. Containing the Tonation of Imbalance, analysis of both schools of thought, medical and Spectro-Chrome, its study will open a vast field of unusually interesting knowledge.

Many who purchased a copy made one error. They filled the blank for Preliminary Examination and sent it to us, without even taking the trouble to enroll first.

Please note! If you wish to enter for the Home Training Course, first fill the Enrollment Blank on page V. When you are registered, study the volume and when you are thoroughly conversant with the Science of Spectro-Chrome Metry, apply for Preliminary Examination.

Do not for a moment believe that if you say, "I completed the Volume in 51 hours of actual study," we shall issue a Certificate. We have a method for exactly finding how much you really know. It is an infallible test for Long Distance Examination. Only when you pass that written test, will you be granted the Preliminary Certificate or subsequent Certificates.

Kindly follow the routine outlined, to avoid useless correspondence.

First Enroll; then study fully; then apply for examination.

No Diagnosis

No Drugs

No Manipulation

No Surgery

BRAINLESS IDIOCY

After being the cause of the death of countless innocent sufferers of Pulmonary Pthisis, by the use of foolish methods and obstinate ignorance, the Medical Doctors are seeing the errors of their ways and trying to reform. O how they must be feeling humiliated when they read this article syndicated by the New York Academy Of Medicine! Judge for yourselves:—

"Tuberculosis Surgery

Treatment Has Changed Since Disease Has Been Closely Observed

(By the New York Academy of Medicine)

In the treatment of tuberculosis 40 or 50 years ago the patient was exposed to all the rigors of the weather in the belief that what he needed most of all was a constant and unlimited supply of fresh air.

The rigors of winter weather no longer are considered essential to treatment, and though the tuberculosis sufferer is nowadays exposed to fresh air, he no longer is kept constantly out of doors.

The treatment style then shifted to food, and patients were stuffed like Thanksgiving turkeys. Milk by the gallon and eggs by the dozen were fed to the patient.

This, too, was abandoned. The tuberculosis patient, while in need of good food in generous quantities, is not benefited by being stuffed.

The present-day emphasis in tuberculosis treatment is upon rest. The tuberculosis patient with fever or other undesirable clinical manifestations is kept in bed and allowed but little activity.

The diseased lung is rested in a variety of ways. In a number of cases air is injected into the space between the chest wall and the lung.

Figuratively, the lung is compressed. In actuality it is allowed to contract or shrink. This surgical procedure is known as pneumothorax, meaning pneumo (air) in the thorax or chest.

> Another surgical procedure sometimes employed is the cutting of the nerve which controls the movements of the diaphragm, more correctly speaking, only one-half of the diaphragm, becomes relaxed.
>
> Subjected to the pressure in the abdominal space, the paralyzed diaphragm bulges upward and compressses the diseased lung.
>
> In a more radical type of surgery called thoracoplasty a number of ribs are removed and the chest wall as a whole collapses, carrying with it the diseased lung.
>
> The coverings of the ribs are left intact and subsequently new bone forms. The chest wall, however, does not re-expand. The collapse of the lung following this operation is permanent."

What a sad error they committed before and what horrible worse errors they are committing now! To expose a person in Tuberculosis, is very unwise. The sufferer not having enough circulation shivers in the cold and the Medicals call it "getting hardened". To collapse lungs, cut nerves and perform other licensed butchery on the Image of God, is a sin against Him and humanity.

Spectro-Chrome can take any case of Tuberculosis (not dead or dying) and restore the Lungs to normalcy in a short time. Our cases are unbelievable in their near-miraculous results.

Will some of our Graduates wake to the necessity of establishing a "Guest House", where Spectro-Chrome resident service may be given? They will earn undying gratitude and fame.

NOT WANTED THERE

Dr. Sam Higginbottom, agricultural missionary to India, (whatever that may be), speaking at a Presbyterian Church meeting asserted that "Among the religious beliefs that have kept India in poverty are the worship of the cow as a sacred animal and the belief that man has no right to kill any animal, bird or insect."

This agricultural gentleman ought to know better than stating the people of India "worship the cow." It would be far better if such men would stay here and not go to India; there is enough mis-

sionary work to be done in New York, Chicago, Detroit and Cleveland, to last 20 centuries.

Why malign the people of India? They live on farm products. The cow's dairy products are their main support, so they oppose cow slaughter done by the so-called Christians there, because, such action depletes their natural resources. Thus, they respect the cow as the support of their babies, much as we may respect our parents here.

The Christians going there, take with them the Brandy Bottle in one hand, the Bible in the other, Beefsteak in the mouth and Profanity with Tobacco on the tongue. They call them heathens, forgetting that the Master distinctly laid down in the Sixth Commandment "Thou shalt not kill." That phrase is unqualified. I fail to comprehend how any person can be so absurd as to argue that God gave him the power of killing what he could not create.

India's poverty is caused and continued by wrong political policies of the British Government, which allows the cow slaughter against Indian protest. Beef-eating is carrion-eating and is the producer of economic troubles in a farming country. The people of India no more worship a cow than Dr. Sam Higginbottom placing a wreath on his relative's grave would be worshipping the carcass within. The cow feeds your baby with milk when the baby has no teeth and in that sense becomes its "mother"; the spiritual-minded Indians reverence such a grateful idea, but, the inconsiderate Christian who respects nothing, butchers the animal that fed it when he was feeble—the ungratefulness may well be styled matricide.

No use writing more about this. The arrogant Chirstians believe they know all, but, India, impoverished though it be, can teach much real wisdom to these overelated, misguided materialists.

ADVICE AND REPLACEMENT OF SLIDES

As already many misunderstandings took place in regard to our FREE Mail, Telegraph and Telephone Advice Service and the replacement of broken, Dinshah Attuned Color Wave Slides, it is again thought necessary to give this warning.

Ernest Distributing Company is our Exclusive Distributor for the United States, but, it has no responsibility of the Advice Department or the Replacement of Broken Slides.

The reasons are clear. The Slides are the LIFE of the Spectro-Chrome. None knows the Attuning process outside of the Dinshah family and as it is absolutely necessary to have them correctly matched to their Set, that work is done only in our Research Laboratory at Malaga, New Jersey. With all the good intentions and reliability ot our worthy Distributor General, we did not see fit to encumber him with that portion of our activities.

The Advice Department is also exclusively conducted by us from the Central Office. Such service is maintained here 24 Hours of the Day, 365 Days in the Year.

However, the service is given solely to our Spectro-Chrome purchasers and the general public having no Equipment to use is not given its benefit.

Personal interview about any case with Colonel Dinshah, the Originator of Spectro-Chrome is thus not possible, because of his numerous engagements. All business is conducted solely by Mail, Telegraph or Telephone in Interstate Commerce and our friends are requested not to endeavor to reach the Originator for any interview without definite appointment.

Dinshah wants to live and serve. Let him unhampered by social engagements which he neither craves nor has any time to fulfill.

Advice by Mail is given Free, but, the seeker of the Advice must use the Blank supplied Free by us. Incompletely filled or unsigned Blanks will receive no attention.

Furthermore, because of the peculiar nature of the unique service, the Advice in the use of Spectro-Chrome in the Home, is given without any liability at Law and (or) in Equity being incurred by Spectro-Chrome Institute, its President or other Officers and the Originator of Spectro-Chrome Metry.

Those who desire such telegraphic or telephonic advice, will have to sign a permanent Waiver Of Claims and Liability and Release at Law and (or) in Equity, before their request will be granted.

We desire to serve Mankind **without compensation,** but, have no desire to do so and be subjected to the tyranny of the fool Medical

Laws. We can fight anybody, anytime, anywhere, but why waste time in fruitless actions which may bring us glory and fame, but disgrace and shame to our opponents?

We live and let live. The opponents will die their natural death without our creating of Karma; the Retributive Justice of the Grand Architect will take its countless toll when the eventual reckoning will come. Let us hold the banner of

No Diagnosis

No Drugs

No Manipulation

No Surgery

Spectro-Chrome—In Every Home

SPECTRO-CHROME INSTITUTE ON 24-HOUR BASIS

When the World of workers is clamoring for a day of less and still less hours, Spectro-Chrome Institute voluntarily went the other way. Believing that the sick sorely needed our services, believing that it is not work that kills but laziness and worry, we started by joining the NRA under its Code Rules and began to work on a 24-Hour basis —that is **not** 24 hours a week, but, all 24 hours of the day with shifts for the workers.

As a professional establishment we were exempt, but, because of manufacturing and printing being also connected with our activities, we preferred to place all under the 40-Hour a week basis and to conform strictly to the President's Code co-operation, we cheerfully added a number of helpers. We were glad to do so, as it increased the scope of our public service and employed more helpers; we now have 18 workers in all and each week or so are in need of more.

There are positions for non-smokers and non-boozers still vacant; other assistants will be needed for our Editorial Department as our Spectro-Chrome transforms into a Weekly Newspaper under the ownership of the Scientific Order Of Spectro-Chrome Metrists. Apply now.

The 24-Hour system applies solely to our Free Advice Department, which helps you to help yourself with Spectro-Chrome.

Such Advice may be sought at any time of the Day and Night by Mail, Telegraph or Telephone, by a Spectro-Chrome purchaser, duly registered as such on our records. Positively **no** personal interviews or "consultation" will be rendered by the Originator, at any time, in this connection.

The regular business hours of 8 A. M. to 4 P. M., are maintained for other departments and friends are requested to follow the routine and not persist in asking for Dinshah on matters to which his assistants might just as well attend.

PRESIDENT HAS COLD

I read with concern in the Milwaukee Sentinel of October 26, 1933, that our busy President Franklin D. Roosevelt was "still confined to his living quarters" by a "cold" and was being treated with "Infra-Red Rays." If the last be true, whoever administers it ought to read Spectro-Chrome Metry Encyclopaedia for the true cause and remedy for the said complaint.

The life of Roosevelt is worth billions in the present political and economical upheaval and the best that could be done would be to put him in perfect health. Some time ago, he retrenched expenses by removing the Medical Doctor of the White House—a very wise move; I recommend that Spectro-Chrome be installed there for him and the foot of the Medical Doctor be removed from that palace; it would be a service to the country.

WHAT WILL YOU DO?

Now that your favorite Spectro-Chrome is making its transformation from a monthly magazine into a Weekly Newspaper, a dream of your life too might be easily realized.

Many of you want to write your views about something—your pet hobby—but have no channel of expression. Here is your long-sought opportunity. Let Spectro-Chrome be **your** newspaper. Write, write anything; write, write all you wish on your own hobbies—it will be printed without expurgation or censorship.

Spectro-Chrome will be the one newspaper that will not sell its free soul to graft. Make it the biggest newspaper for Human Service.

Write; what will you do for Human Service?

Should you desire to write an article each week or at some regular interval, write to us and we shall be glad to place your name as one of our Associate Editors, on the newspaper staff.

Will you shoulder the troubles?

ERNEST DISTRIBUTING COMPANY MICHIGAN STATE REPRESENTATIVE APPOINTED

Among the various State Representatives of the Ernest Distributing Company, the first to be appointed was Thomas Ness, S. C. N., 3458 Eastern Place, Detroit, Michigan. He is now the exclusive Michigan State Representative and already started to build his territory by sending orders for Cabinet Spectro-Chromes. We are sure of his success, because, he has one behind him to push him into success, even though he might not like to succeed! We allude to his beloved wife, Jessie E. Ness, who for several years has been identified as the First Vice-President of the Scientific Order Of Spectro-Chrome Metrists, in the Cause.

It may be mentioned incidentally that the Representatives and Agents of Ernest Distributing Company are all thoroughly investigated for their integrity and ability and the suffering public may therefore rely on them implicitly. They are in constant contact with the headquarters and any information on Spectro-Chrome may be obtained from them.

Join the Spectro-Chrome forces and terminate the reign of the Drug Peddlers. Be the Representatives and Agents of Ernest Distributing Company and be independent for life.

Write to E. A. Ernest, S. C. N., President, for further information; do not miss this exceptional opening for an honorable and profitable vocation with limitless possibilities. Apply now. Next month, more State Representatives will be aligned and at least 100 more Agents will be absorbed.

SPECTRO-CHROME HOME GUIDE

From the pen of Dinshah, came another book of 102 pages, which promises to be the greatest step ever taken to place Spectro-Chrome—In Every Home.

Spectro-Chrome Home Guide is designed primarily for that purpose and will be issued Free Of Charge to those who have Cabinet Spectro-Chrome Equipments. A copy was sent by Insured Third Class Mail to each purchaser of the Cabinet and any delay in receiving it should be promptly drawn to our notice.

Spectro-Chrome Home Guide is not obtainable by any person outside of a Cabinet owner, except on payment of $65.00. There are no exceptions and it will be of no use asking for it otherwise.

You are further requested to **Register** yourself as the purchaser, to receive Free Advice by Mail, Telegraph or Telephone. Use the blank on page 101 of Spectro-Chrome Home Guide.

Such registration will entitle you to Provisional Membership and

the receipt of the Spectro-Chrome each week for 12 months, as such.

Kindly act promptly.

NOTICE TO MEMBERS

Your subscription as a Member of the Scientific Order Of Spectro-Chrome Metrists, expires on December 31, 1933.

Unless your remittance of $6.00 to cover the Annual Dues and Fees for 1934, be received on or before the end of the month, you will have no authority to act as a Member in good standing having the privilege of signing Collaborators.

Please note therefore, that unless you comply with the above Rule, no Spectro-Chrome or Favorscope will be sent and your privilege of attending our Planet Meetings will cease.

Each Member pays $6.00 a year; Joint Membership Rule is discontinued.

No Diagnosis

No Drugs

No Manipulation

No Surgery

Spectro-Chrome—In Every Home

Attuned Green Wave

(Spectro-Chrome Metry, Spectro-Chrome Normalators)

AN ODE TO THE DOG BOX

A smart Medico, while on the witness stand recently, referred to the latest Spectro-Chrome Cabinet equipment as a DOG BOX, thereby inspiring the lines which follow:

Well folks, I've a story I want you to get
"Bout a DOG BOX.
Some Doctor guy said, when he saw it, "I bet"
"It's a DOG BOX.
A box on a stand, with a bulb inside,
Some pieces of glass that work in a guide,
You turn on the switch, let the light strike your hide,
THAT'S THE DOG BOX.

A fellow called Dinshah invented this thing
Like a DOG BOX;
And one of our neighbor folks gave us a ring
'Bout the DOG BOX.
He told us the Doctors were all in a fret,
Says he, "They're out this Dinshah to get"
"He's spoiling their racket, that's why they're set
"Gainst the DOG BOX."

The Missus says, "Bill, do let's go and see
That DOG BOX."
"This Dinshah is giving some lectures free
On the DOG BOX."
I says, "Now Melinda, you're crazy as H - - - L,
"It's only a fake, and it can't make you well."
But, as usual, SHE WON, and I'M here to tell
It's SOME DOG BOX.

Melinda went clean off her nut when she saw
THE DOG BOX.
She wanted to buy one, but I said "NAW,"
"Who wants a DOG BOX?"
But what was the use? She'd have her own way,
(It never DID matter what I had to say)

So I dug up the cash, in order to pay
For a DOG BOX.

Melinda and me almost fought for first try
At the DOG BOX;
I nearly won out, but she started to cry
For HER DOG BOX.
Her stomach was aching, and so was her head,
So I stripped off her things and put her to bed,
And BELIEVE IT OR NOT, HER TROUBLES ALL FLED
Through THAT DOG BOX.

So now, when we're sick, both Melinda and I
Use the DOG BOX;
And even the kiddies will blubber and cry
For the DOG BOX;
So folks we're all through with powders and pills,
I'm telling you true, IT'S THAT JUNK THAT KILLS,
There's only one thing that can ease all your ills,
THAT'S THE DOG BOX.

Just give it a whirl; you never will rue
Dinshah's DOG BOX.
'Twill chase all your aches, and make you anew,
Will the DOG BOX.
For sixty-five bucks, you can Tonate the GANG,
The Doctors ARE THROUGH, so let them go hang,
And UNITE to help Dinshah put o'er WITH A BANG
THE NEW DOG BOX.

Tom Ness to Dinshah.

Say Tom, you are now an Associate Editor; beware of what you write! The Dog might bark out of the Box; it surely will from next month.

Dinshah to Tom Ness.

FOREWORD

By:—Thomas Ness, Associate Editor, Spectro-Chrome

Almost simultaneously with the announcement of the birth of Hom, the sixth son to Colonel and Irene Grace Dinshah, comes the news that in January, 1934, Spectro-Chrome magazine will be re-incarnated in newspaper form. This latest child of Dinshah's ever active brain will be published weekly.

Being born under the Zodiac sign Capricorn, the mouthpiece of Spectro-Chrome science, will find itself in most distinguished company. Sir Isaac Newton, Daniel Webster, William E. Gladstone and Woodrow Wilson first saw the light of day under that sign. Surely it is significant that a periodical devoted to the Science of Light with its boundless potentiality for healing, should make its bow to the American public under the same auspicious sign that saw the birth of Sir Isaac Newton, famous scientific researcher.

As century succeeds century and the scroll of human life and endeavor unfolds, men write indelibly upon it, the record of their achievements along scientific lines. So it comes to pass that leaders of men, outstanding personalities in their day and generation, bless this old world not only by their presence, but by their contributions to the accumulated knowledge of the ages. Such was Newton in his day, a man whose discovery and practical application of certain natural laws undoubtedly laid the foundation for great scientific advancement on the part of his fellowmen.

Such also is Dinshah, Apostle of Light in the 20th Century. The story of his life is one of intensive study, struggles against great odds, disappointments, triumphs over the forces opposing him and finally his crowning achievement, Spectro-Chrome Metry, a system in which remedial Wave Lengths of Color take the place of dangerous and habit forming Drugs. Under these beneficent Rays, the sufferer's imbalance is corrected and health returns.

Harnessing the forces of Nature, man made use of water to manufacture Electricity. This powerful ally, he uses to drive the machinery he invented, to light and heat his home and office, to chill his food, to carry his words to the farthest corners of the Earth, via Telephone, Telegraph and Radio, to weld metals. In short, Electricity, a formidable and dangerous master, becomes in the hands of the scientist, a veritable "Genii of the lamp" in the service of man.

How apt is this description of Dinshah's Spectro-Chrome equipment with this exception however! You do not rub the lamp; you simply plug in on your house circuit and the NEW ERA GENII OF THE LAMP awaits your pleasure. That his tasks are efficiently performed there can no longer be room for doubt. Thousands of men and women whose Dis-Eases were declared incurable, have been raised from beds of suffering and pain to a newness of life and an enjoyment of such health as they had despaired of ever regaining.

Knowing the man and his work, one cannot but believe that Dinshah was born for the express purpose of benefiting generations of sufferers through the evolution of this mathematically precise method of health restoration, to an exposition of which, in the interest of humanity at large, our weekly newspaper SPECTRO-CHROME will be dedicated.

Its pages will reveal the Truth, the WHOLE Truth, and NOTHING BUT the TRUTH regarding Light as a life giver. Let every Spectro-Chrome Normalator, every student of Spectro-Chrome Metry, every human being who was benefited by the application of the Dinshah Attuned Color Waves, grasp the opportunity now presented and join in a holy crusade against the age-old superstition of drug and dope swallowing, by carrying the TRUTH to the people of America, via the pages of our baby newspaper.

WHAT AN AGENT SAYS

An Agent of the Ernest Distributing Company, in writing his order for a Cabinet Spectro-Chrome, says some words which have deep import. He says, "Mrs. N.— is faithfully pushing the Dog Box whenever possible. Her enthusiasm for Spectro-Chrome Metry is infectious, as this sale bears witness, therefore she gets the credit, I (BIG ME) get the commission, (which in turn she gets) and everybody is happy. Incidentally the purchaser gets restored health and she too is happy. THE N. R. A. can't beat that.

"All joking aside, we begin to see the opening up of a new era of opportunity for Spectro-Chrome. There are thousands of sufferers eagerly awaiting healing, even as they did when the Lord Christ came. Even as in His day also, they are sick of professional quacks. They have spent their little all for evil tasting nostrums. Now comes the Gospel of Spectro-Chrome, a Therapy sans pills, drugs, serums and hand carving of the anatomy, but plus simplicity, efficiency to the Nth degree and mathematical precision. WHAT A REVELATION! In a few years, your claims will be vindicated, your labors rewarded and the value of your contribution to the science of healing fully recognized. God speed the day."

That day already arrived. It **is** here. It dawned on the horizon of life and its effulgence will increase until the clouds of crass ignorance will be dispelled completely and the Lord's words be fulfilled:—

"Let there be Light"
And Spectro-Chrome was born.

NOVEMBER 1933 REPORT OF FIRST MILWAUKEE WISCONSIN PLANET

Otto H. Lutz, Deputy,
Harry Goebel, Sub-Deputy,
Elaine Astor, Recorder.
Room 157, Hotel Wisconsin.

Twenty-one Affiliates were present.

Ella Lutz and the Members present request me to ask if it would be possible to receive the Favorscope in time to use it for their cases necessitating accuracy. Although E. A. Ernest explained how difficult it has been for you to handle all your many business interests exactly on schedule.

Two particularly interesting cases were by Laura B. Gross of the man who was in the insane asylum and recently entirely unaware of his surroundings, with Heart registering 160; was unusually improved with Green and Magenta in two Tonations, enough to go back to work and secure two orders, one of over $100 and the other over $200; feels he does not even need anymore. Wonderful, I think!

The other concerned a woman having either Rheumatism or Arthritis, with terrible knobs constricted over the Body, very stiff joints and heart weak. With Green, Lemon and Magenta used, in six weeks together with the Dinshah Food System, she was well.

After our general discussion, E. A. Ernest spoke a few minutes, telling us of the results even already that are accomplished in the sales

of Spectro-Chrome and the requests of people in other States to handle sales. He reminded us of our need to shoulder the wheel as well, to make the great plan and work a greater future and foundation through efforts and continued support.

Without you here personally, it is a great aid and comfort to us folks to have someone like Ernest, to turn to and be in our midst as a spokesman for you.

Elaine Astor, Recorder.

The complaint of your Members is just, but, we mailed Spectro-Chrome on October 30, 1933 and they should have reached you all by November 1, 1933. The Third Class Mails are slow as snails. However, as the SPECTRO-CHROME will be a Weekly Newspaper next month, this complaint will permanently disappear as the Favorscope will be printed each month ahead.

We have certainly many duties to perform, but, as business is not built by excuses, we do not present any.

E. A. Ernest is the right man in the right place and it is certain that he will make an immense success by his straight-from-the-shoulder methods. Stand by Spectro-Chrome and the Distributor General; we shall stand by you. Thank you.

PAGE SPECTRO-CHROME

By:—Irene Grace Dinshah, Associate Editor, Spectro-Chrome

In an article published in the Evening Journal of Wilmington, Delaware, Thursday, November 16, 1933, page 1, on "Dread Tropical Disease Hits Chicago Visitors," by Dr. Morris Fishbein, Editor, Journal of the American Medical Association and of Hygeia, the Health Magazine, it was stated, "Recent death of Texas Guinan, famous night club hostess, brings into national prominence the outbreak of amebic dysentery that has been prevalent in Chicago for several months. Miss Guinan's sudden death is ascribed to this disease. The epidemic is believed to be under control.

"Amebic dysentery is caused by invasion of the body by an organism larger than the bacteria, known as entamoeba histolytica. This organism undermines the upper layers of the large bowel.

"The symptoms begin with frequent, scanty evacuations, often containing mucus and blood, severe abdominal pain and depression. Sometimes the early symptoms relating to the abdomen are so severe that in the failure of an accurate diagnosis, surgical operation for appendicitis or peritonitis or for other abdominal infection may be performed.

"Until rather recent times, this disease was associated universally with the tropics. More recently it has been found that it can be transmitted by food handlers in any part of the world.

"It was by this method that the epidemic arose in Chicago.

"In 1927, seven or eight cases were traced to a large hotel there. Apparently a person still carrying the germ in the spring of 1933 obtained a position in another hotel.

"Cases of amebic dysentery are not usually reported in Chicago; in fact, the average is one or two a month.

"However, about the middle of August, 1933, two cases were reported, and by the end of August 13 cases had been reported, all traced to the same hotel.

"Therefore, examinations were made of all food handlers in this hotel, and by October 1, it was found that of 412 food handlers examined, 92 were infected.

"The examination for the presence of the organism is a highly technical procedure which can be carried out only by a competent laboratory investigator.

"There is an incubation period of from 9 to 94 days after the invasion of the body before severe symptoms appear. In fact, the cases investigated by one group of authorities indicated that the average time from the infestation by the organism to the first appearance of the intestinal symptoms was 65 days."

Dr. Morris Fishbein might relate a different story if he and his fellow-medicoes would use Spectro-Chrome. Then, his report need not be, "The epidemic is **believed** to be under control," but, "The

epidemic **is** under control."

The untimely decease of Texas Guinan and others with similar disorders could easily have been averted.

What a disgrace to medical "science"—to an "honorable profession" to admit wrong diagnosis, hence unnecessary surgical operation and probable death in the wake! Appendices are removed when medical diagnosis and treatment should be for Dysentery.

How wonderful to work with the Spectro-Chrome Metry simplicity and one does not have to sit and wait 65 days with a calender in hand, for "symptoms".

The user of Spectro-Chrome cares nothing about the size of the organism, whether it undermines the upper layers of the large bowel or the lower layers of the small bowel or its name—it as well might be called entamoe—bah hiss! Furthermore, Appendicitis, Peritonitis, Dysentery and so on are just so many words, indicating Inflammations.

Spectro-Chrome followers know **without diagnosis** — without complicating terms, that this Disorder, about which there appears to be so much ado, lies on the Infra section of the Spectrum and the indicated tonation is Ultra-Green.

Simply Irradiate with Turquoise Systemic and worry not about the invasion of the body by the entamoe—bah hiss army!

SIGNS OF THE TIMES!

By:—Marguerita W. Simon, Associate Editor, Spectro-Chrome

The following newspaper article tickled my funny bone to such an extent, that I could not refrain from commenting about it.

"Entire Lung Removed

Pittsburgh Physician Fully Recovered From Cancer Operation

St. Louis, Oct. 28—(AP)—A report on the successful removal of an entire lung from a patient of Dr. Evarts A. Graham was made public today by Dr. Morris Fishbein, editor of the Journal of the American Medical Association.

In an article appearing in the current journal, Dr.

Graham and Dr. J. J. Singer, both of the Washington University School of Medicine here, state: "It is apparently the first time in which the whole lung has been deliberately removed at one stage."

Cancer necessitated removal. The operation was performed last April 5, but no previous reports of it had been made public. The patient, a Pittsburgh physician, is reported fully recovered."

It is certainly a good sign of the progress of **Spectro-Chrome in Every Home** when the M. D.'s have such a small amount of business that they must butcher one another to keep in practice and enlarge their collections of anatomical "pickles"!

While it is humorous to those who practice Spectro-Chrome and who know the opposition to our work by the Medical Trust, it arouses one's sympathy for the poor Pittsburgh physician who must for the rest of his life be a one-lunger. If he had come to Spectro-Chrome, he would have finished his life hitting on all cylinders and kept the spark plugs firing in proper order in his patients.

I should like to know if the Pittsburgh M. D. pickled his lung in alcohol after it was removed or whether it was removed because it was pickled in alcohol; no cause was mentioned. Who has the lung?

The American Medical Association is proud of the new accomplishment of its Members, but, it would appear by the absence of his name that the Pittsburgh physician does not share their egotism in walking around minus half his bellows!

No Diagnosis

No Drugs

No Manipulation

No Surgery

Spectro-Chrome—In Every Home

SPECTRO-CHROME

March
1934
Yellow

Volume
10
Number
3

Important Read and let your friends read. **Articles**

SPECTRO-CHROME INSTITUTE
FIGHTS FOR ITS LIFE

HOW THEY ARE PLANNING TO BLOT
OUT ITS EXISTENCE

MARCH-1934

Publisher and Editor
Dinshah P. Ghadiali

Single Copy 50 Cents

TUGWELL-COPELAND BILL

Hoodwinking The American Public—
A Blow At The Fundamental Principles
Of Our Glorious Constitution.

Spectro-Chrome Institute Fights For Its Life

In the last issue of Spectro-Chrome, we gave the public an impartial and correct interpretation of the so-called "Pure Food And Drug" Bill. We mailed a copy of it to President Franklin D. Roosevelt as well as to every Senator and Representative of the United States Congress. As Chairman of the United States Senate Committee On Rules, the Honorable Senator was sent a personal letter as follows:—.

Thursday, January 25, 1934,
Chronos:- Green 45 Black.

Honorable Royal S. Copeland, M. D.,
Senator for New York,
WASHINGTON, D. C.

Dear Doctor Copeland:

1. Knowing you to be the sponsor of THE TUGWELL COPELAND BILL, I desire to register my most emphatic protest against the injustice and gross absurdity of one special clause in that Bill.

2. I allude to the inclusion of "Devices for the treatment or prevention of diseases" within the meaning of the word "Drug".

3. Fifteen years ago, when you were the Honorary President of the Allied Medical Associations Of America, I had the honor to be the Vice-President and I knew of you as a high minded Homeopath.

4. However, it appears that during the last few years, your attitude toward medicine has been such as to give an exclusive monopoly of the art of healing to the Allopathic School, about which the less I say the better.

5. Inclosed please find the views of my Corporation in regard to the TUGWELL BILL, at Page 1056.

6. For and behalf of my Corporation, as the Originator and Manufacturer of Spectro-Chrome, I hereby register an emphatic protest against the gross injustice of the specific inclusion of a "Device" for health within the meaning of "Drug."

7. When a public hearing will be given in regard to your Bill, I shall thank you to let me know, so I may come personally to Washington, D. C. and express my views about it.

Yours very respectfully,

Dinshah P. Ghadiali,
President.

DPG: MWS.
Inc: February 1934 SPECTRO-CHROME.

The Senator's reply was as follows:—

Royal S. Copeland, N. Y., Chairman,
Joseph T. Robinson, Ark.
Pat Harrison, Miss.
Kenneth McKellar, Tenn.
M. M. Neely, W. Va.
Hugo L. Black, Ala.
Alva B. Adams, Colo.
Harry Flood Byrd, Va.
Frederick Hale, Me.
Porter H. Dale, Vt.
David A. Reed, Pa.
Frederick Steiwer, Oreg.
Felix Hebert, R. I.
G. C. McEldowney, Clerk.

UNITED STATES SENATE
Committee On Rules
January 27, 1934

Dear Mr. Ghadiali:

Thank you for your letter of January 25th together with the February issue of "Spectro-Chrome" with its article in reference to the Food and Drug Bill. I am interested to learn your point of view. I do not think there will be any hearings on the matter as the new bill is the outgrowth of those hearings held in December.

Cordially yours,

Royal S. Copeland

Mr. Dinshah P. Ghadiali,
Spectro-Chrome Institute
Malaga, New Jersey.

The following prompt reply was sent:—

Thursday, February 1, 1934,
Chronos-: Turquoise 05 Black.

Honorable Royal S. Copeland, M. D.,
United States Senator,
WASHINGTON, D. C.

Dear Doctor Copeland:

1. I thank you for your kind communication of January 27, 1934 in which you state that you are "interested to learn your point of view".

2. My point of view about the Tugwell-Copeland Bill is clearly indicated on Pages 1056, 1057, and 1058 of the February 1934 issue of Spectro-Chrome, a copy of which you already have.

3. Whatever the aim of the Bill may be, it is plain that a Food and Drug Bill should be restricted technically to the giving to the people pure and unadulterated Foods and if they wish to have Drugs, to give them unadulterated Drugs.

4. It is well known that food, as now sold all over the Country, is a concoction of so many admixtures that the consumer does not know what he is buying.

5. In the same manner the buyer of Drugs, whether Medical Doctor or otherwise, who is interested in buying Drugs, expects the Drugs he prescribes and from which he expects certain definite results, should reach the sufferer in exactly the form in which they are prescribed.

6. I practised Medicine for a long number of years in India, but since settling in this Country 23 years ago, left the Medical line as notoriously empirical. However, that is not the point at issue.

7. The Medical Doctor and the Public must get whatever they want in exactly the form they want and not adulterated with undesirable material. Thus far you may take your Tugwell-Copeland Bill; it would be legal.

8. However, when you endeavor to put into the definition of Drugs "Devices intended for use in the cure, mitigation, treatment or prevention of disease", it certainly is a perversion of the Lexicon and a mockery of paraphrasing. The reason for this is self-evident.

9. A Device is a contrivance, mechanical, electrical or otherwise and to call it a Drug is enough to bring a loud laughter on the face of even a brass monkey.

10. You enjoy great respect among the Public and are noted for your intelligent views on medical subjects; therefore, I believe you should clarify the Bill by applying it to where it belongs, viz. Food and Drugs and not produce a mockery of Law by having a "Device" dragged therein as a "Drug".

11. There is another phase of this Bill of yours which defines Advertisement.

12. I am the owner of numerous Patents on a number of Devices used for "the mitigation, treatment or prevention of disease." These Patents issued by the Government of the United States of America are valid for 17 years from the date of issue. These patents give me the exclusive right to make and vend these inventions; as such these Patents are contracts between the People of the United States of America and myself.

13. Now, Mr. Rexford Tugwell and you come forward with this Bill, as protectors of the purity of Foods and Drugs and therein insert an advertisement clause which would prevent me from advertising my Equipments.

14. As this would invalidate either my Patent or the Bill, whichever way it operates, I request and appeal to your intelligence to remove such a nonsensical restriction from your otherwise well aimed Bill.

15. Surely you do not intend that the brain of the inventor should be deadlocked by stopping the sale of his brain-children—his inventions upon which the welfare of the World depends.

16. I enclose some literature about my activities and inventions, a glance at which will show to you that you will be ruining the welfare of many a sick person, if your Bill be passed without due purgation.

17. I have a fully equipped research laboratory and manufacturing Plant standing on an estate of over 23 acres, giving free service to suffering mankind through the use of my System——Spectro-Chrome Metry.

18. I wrote and published some time ago a Spectro-Chrome Metry Encyclopaedia, giving complete details of this new and original Healing Science for the use of those who would like to study it.

19. I shall be delighted to present you with a copy of it, so you may look deeper into the subject, if you so care.

20. I submit that the Tugwell-Copeland Bill will only benefit the Allopathic School, which always aimed at maiming or killing its opponents of all other schools of Healing.

21. Surely as a prominent Homeopath, whose name at one time graced the Flower Hospital of New York, New York, you will not be an involuntary tool in the hands of the Allopaths, who at one time persecuted even the great Hahnemann.

22. Wishing you the best of health, I have the honor to remain, My dear Senator,

Yours very respectfully

Dinshah P. Ghadiali
President.

DPG:MWS.

Inc: Booklet,
Biog.
Health Message,
Triumph.

Up to the time of going to the Press, no answer came from the Senator, but, it is possible the following may have something to do with the delay. It was published in the Evening Bulletin of Philadelphia, Pennsylvania, but, of course, we can not vouch for its truth:

SENATOR LOST HIS PANTS

Copeland Tells Congress About Mean Thief at Erie

Washington, Jan. 16.—The Congressional Record tells how Senator Copeland, Democrat, New York, chairman of the Committee for the Suppression of Crime, lost his pants not so long ago.

Here it is in the Senator's own words, spoken in debate:

"I might say to the Senate that when I was on the way over to Chicago for our hearing there I stayed overnight in Erie, Pa., and in the morning my trousers and $400 had disappeared. I thought it was outrageously mean.

"The Senator from Arkansas (Mr. Robinson) wants to know how I happened to have so much money. I was going to the fair as well as to this committee meeting, and knowing the prices they charged over there, I thought it would be necessary for my party to have at least that much money. So these criminals robbed the chairman of the Committee on Crime of his trousers and $400."

We can well understand that if the Honorable Senator could not guard his own property, he would be less likely to be watchful of ours.

Here is what we wrote to the father of the Bill:—

Thursday, January 25, 1934,
Chronos:— Green 30 Black.

Mr. Rexford G. Tugwell,
Assistant Secretary of Agriculture,
Department of Agriculture,
WASHINGTON, D. C.

Dear Mr. Tugwell:

1. Knowing that you introduced the new Food And Drug Act in

the United States Congress, we desire to draw your attention to the Article THE TUGWELL BILL on Page 1056 of the inclosed copy of our Monthly Magazine SPECTRO-CHROME of February 1934.

2. While willing to give you full credit for your good intentions in regard to the purity of food and drugs, we hereby register our emphatic protest against the inclusion of the word "Devices" in the meaning of "Drug".

3. What our views are, we stated fearlessly in the Article and sincerely hope you will give this matter your serious consideration, as the inclusion of a health device for the "treatment or prevention of diseases" will completely ruin our business and prevent us from rendering that unselfish service to mankind that we have been rendering for the past 14 years in cases where the Medical Doctors have been utterly helpless.

4. You will please also note that such inclusion of such a "Device" in the meaning of "Drug" and placing it under the Department of Agriculture is, according to the best of legal opinion, absolutely unconstitutional and should there be a public hearing on the subject, we request you to inform us at the proper time, so we may defend our Cause personally.

5. Expressing our kindest regards, we remain

Yours very respectfully,

SPECTRO-CHROME INSTITUTE

by Dinshah P. Ghadiali,
President

DPG: MWS.
Inc: February 1934 SPECTRO-CHROME.

No answer to this letter came. He must be busy farming.

There was another letter we wrote to the Chairman of the United States Committee On Commerce:—

Thursday, January 25, 1934,
Chronos:- Green 55 Black.

Honorable Committee On Commerce,
United States Senate,
WASHINGTON, D. C.

Kind Attention: Honorable Hubert D. Stephens, Chairman.

Dear Mr. Stephens:

1. This Corporation, a Manufacturer of certain patented "Devices for the treatment and prevention of Diseases" hereby registers its most empathic protest against the inclusion of the word "Devices" within the meaning of the word "Drug" as proposed in the TUGWELL-COPELAND BILL, which is now in your hands.

2. Inclosed please find our Monthly Magazine SPECTRO-CHROME of February 1934, which voices such protest in full.

3. Your kind attention is drawn to Page 1056.

4. When this matter comes for a public hearing, we shall thank you to inform us about the time and the writer, who is the Originator of the System, will appear personally to voice the said protest.

5. Your kind courtesy will be appreciated.

Yours very respectfully,

SPECTRO-CHROME INSTITUTE

Dinshah P. Ghadiali,
President.

DPG: MWS.
Inc: February 1934 Magazine.

We received no reply to the above letter also; perhaps none was expected.

A letter was addressed to the President Of The United States Of America, to complete the chain:—

Thursday, January 25, 1934,
Chronos: Green 15 Black.

Honorable Franklin D. Roosevelt,
President Of The United States Of America,
White House,
WASHINGTON, D. C.

Dear Mr. President:

1. There is an Open Letter addressed to yourself on Page 1060. in the February 1934 Issue of our Monthly Magazine SPECTRO-CHROME.

2. There is another Article THE TUGWELL BILL on Page 1056. We submit both of these for your kind attention.

3. We have the highest regard for you and sincerely believe what you do is intended for the good of the Country, in accordance with your opinion and that of your Advisors.

4. However, our side of the outlook is presented in the SPECTRO-CHROME.

5. We have the honor to remain,

Yours very respectfully,

SCIENTIFIC ORDER OF SPECTRO-CHROME METRISTS.

by Dinshah P. Ghadiali
President.

DPG: MWS.
Inc: February 1934 SPECTRO-CHROME.

That letter too was not acknowledged. It went at the time the President was celebrating his birthday, so it is likely his numerous secretaries might not have thought it wise to disturb his mental quietude. Furthermore, we are not Democrats and are opponents of intoxicating drinks, therefore, naturally not entitled to voice under Democrat regime. We may say that we are not even Republicans, because, whether it be one party or another, politics ever has the same earmarks. Our President inaugurated new political theories of Constitutional Government and because of the Depression, everybody sits still and watches. They have a long watch!

But, we are concerned only in the ultimate result of this Tugwell Bill. A recent Associated Press dispatch says thus:—

Copeland To Push His Food-Drug Bill

WASHINGTON, FEB. 9 (AP).—The Copeland-Tugwell food, drugs and cosmetics bill was approved in a modified form today by the Senate Commerce subcommittee which has been studying it several months.

Senator Copeland (D., N. Y.), physician-author of the bill, said it now was satisfactory to President Roosevelt and has been approved by the Agriculture Department. He announced he would ask for action on it by the full committee next week in order to get Senate approval this session.

"We took some of the controversial issues on which Professor Tugwell and I did not agree to the President and made him referee," Copeland said.

* * *

It were better if the Honorable Senator had published what were the "controversial issues" on which the Tug-Of-War existed. The President would be a good referee with his level head, yet, why should this oldtime, clumsy procedure of Committees, Subcommittees and similar Shuttlecock and Battledore methods still continue in our politics instead of open handed business-like systems? Politics, as it goes all over the World, is abominably corrupt and our noble ideals mentioned in the Constitution—Liberty, Property, Security, Right Of Contract and similar wonders as look good on paper and in talk, are slowly becoming completely scattered.

We are steadily and surely heading toward either a new Monarchy, Anarchy, Economical Revolution or it seems clearly for a minimum, an imperious, autocratic dictatorship.

We are for the Constitution as laid down in the basis of our Government——we are opposed to what is made today, a gutta-percha interpretation of that Magna Carta; our bulwark of Liberty is broken by the judicial decisions so frequently handed to litigants within the meaning of our Constitution, suitable to the views of some party or other wire puller. The public is hoodwinked and bamboozled at every step and suffers and groans. While the people howl, the politician grins; boss rule rules the land.

Such a Bill as this Tugwell Bill, if it should pass unexpurgated, would be a mockery of our Constitution. The idiocy of including "Devices" within the meaning of "Drugs" is so glaring, that we are aghast at the brazen attempt made by the sponsors to palm off such an absurdity on a confiding public. We should like to know what relation even "Foods" have with "Drugs" and what earthly, heavenly or reverse relation is there with "Devices." To try and grab power and means of throat-squeezing at every step, so as to convert a Free Nation again into serfdom and to reduce a Republic to autocratic, dictatorial subjugation is unthinkable to our all-American mind, trained to consider humanity from the broadest standpoint of service, equity, fairness and justice.

We say—we openly say—our mind revolts at the very idea of foisting upon a helpless suffering public the rotten allopathic system of so-called Medicine. We resent, as nothing else can make us, the very conception of forcing us—Free-Born Humans—to accept what our education, training, study and knowledge directs as the wrong path—the placing us under the domination of a fraternity whose methods are notoriously crafty, grafty, in fact utterly cruel and worthless. We refuse as students of Spectro-Chrome Metry, to submit our homes, our children, our lives into the hands of this accursed hierarchy of pox-punchers, toxin-makers, serum-pokers, at best the nihilists of what we gained as purity of our blood by years of consecration to clean life.

What knowledge does a Secretary Of Agriculture have about health matters? What will he do with the Bill as proposed, if he were clothed with all those powers? He could do nothing without extraneous advice; that advice he will get solely from the Medical Octopus, that Gorgonic Hydra, whose ramificatory activities he never will comprehend, until he goes through the mill as we went.

We appeal to the highminded American—the real honest-to-goodness American who values the purity of his blood, babies and home, to present an ironclad front to the Tugwell Bill by sending protests to the authorities, with the view of showing them your opinion of their crack-minded activities. Under the assumption of our Constitution, our appointed congressmen were supposed to be acting for us, but, it would appear as if our own appointees became our oppressors. The Congress which once was supposed to be the controller of the country's policy, became just an automaton, a robot, a mere mechanical party tool and instead of being our servant became nothing more than a buckpasser. Its power of determining the issues is in abeyance, its independence is paralyzed, its movements are languid and crippled, its real authority is gradually seeping into the hands of a self-willed President, who dictates the tune to which these Marionettes dance. O what a degradation is being brought upon us who are striving at all sacrifices, to serve suffering mankind whose hands are outstretched for help which the poison-vending Medical Doctors never could give!

We must join efforts to quench this terrible Tugwell Bill. If that clause of "Devices" are "Drugs" stays in it, we must **never** submit to it; we claim our right—our birthright—as thinking souls and Servants Of Mankind, to serve the suffering as best our conscience dictates, Congress, Tugwell-Copeland Bill, President or the whole Medical Deviltry joined together to the contrary. It strikes at the Fundamentals of our national freedom— our liberty to choose our own healing system. The day of dictation for the American Medical Association is passed; its power and prestige is crumbling by the momentum of its own ghoulish and tyrannical acts. The public must learn that the Medical Doctor of today, is nothing but an individual operating under the garb of humanity and mercy, yet really a professional money-maker, who from the beginning of his career is under the supervision and domination of that most nefarious Trust the human brain ever created.

Hold up the banner of the American Constitution! Voice for upholding our human birthrights and down with the Medical Interference, Autocracy and Imperialistic Dictation. Stand firm for your rights and bow not to these insidious Medical Charlatans, who for 6,000 years have been poisoning the human race with worthless, unscientific, empirical concoctions just the same as in those days they were using the witch cauldron.

Fight for your rights,
Till the goal is reached,
Rest not on your oars,
Till the day does dawn,
For the Medical Doctor
To leave our homes
And be out for ever.

Just as the above went to the Press, we received an unexpected letter. It was unexpected because, Senator Royal S. Copeland already informed us that "I do not think there will be any hearings in the matter"!

UNITED STATES SENATE

Committee On Commerce

22nd February, 1934.

Mr. Dinshah P. Ghadiali,
Malaga, New Jersey.

Dear Sir:-

I have your letter in which you express your desire to appear at the hearing to be held on Food, Drugs and Cosmetics. The hearing will begin at 10 A. M. on Tuesday, the 27th of February and will be conducted in the Caucus room of the Senate Office Building. We shall be glad to have you appear and present briefly the matter in which you are interested.

Respectfully,
(signed) H. D. Stephens.
Chairman.

This development was surprising, but we promptly replied accepting the invitation.

Saturday, February 24, 1934.
Chronos:- Turquoise 10 Black.

Honorable Hubert D. Stephens,
Chairman,
Committee On Commerce,
United States Senate,
WASHINGTON, D. C.

Dear Sir:

1. I thank you cordially for your kind communication of the 22nd current giving me information about the Hearing that will begin at 10 A. M. on Tuesday the 27th current in the Caucus Room of the Senate Office Building.

2. I am obliged for the invitation to appear and shall very gladly

avail myself of the opportunity to present my side of the Question in the briefest possible time.

Yours very respectfully,
(Signed) Dinshah P. Ghadiali
President.

DPG:MWS.

In a raging Zero Degree Snow blizzard, how we went to the Hearing in Washington, D. C. and what rib-splitters we heard and saw there will be published in the next issue.

ANOTHER GROUND-HOG CASE ——By Albert T. Reid

IS BUSINESS HEADED TO PROSPERITY?
OR IS IT GLIDING SLOWLY TO DICTATION?

O WHAT A KNOCK!

Second Volume Is Such Or Different?

DETROIT, MICHIGAN

Colonel Dinshah,
Spectro-Chrome Institute,
Malaga, New Jersey.

Dear Sir:

I am sorry to write you that this Encyclopaedia you had sent is no good at all. It does more harm than good in our home.

Since the book is here, Mrs. A is so busy reading and studying it, that she forgets all about cooking. No supper and dinner is ready in time, if not burned to a mass.

I want that you don't make the next volume so interesting so that I get my supper in time.

A Sufferer,
R. A.

R. A., the Second Volume of the Spectro-Chrome Metry Encyclopaedia is now published and ought to be in your hands soon.

Should you not receive it by the middle of this month, write to us.

However, we tell you frankly ahead that you had better make reservations at some restaurant because we know for a certainty you will have neither breakfast, lunch, dinner nor supper, well-cooked, burnt or otherwise.

Now, be kind to the cook; she means well; if every husband had such a studious wife, the Divorce Courts would close; but, of course that does not mean she should starve you. Why not compromise by studying with her?

SINUS TROUBLE, WEAK HEART NORMALATED

An Unsolicited Expression

Milwaukee, Wisconsin,
February 3, 1934.

Mr. E. A. Ernest,
Ernest Distributing Company,
Milwaukee, Wisconsin.

Dear Mr. Ernest:

My belief that you are interested in all passengers arriving at WELL-VILLE via Spectro-Chrome "route" and the hope that my words may mean a blessing to suffering mankind in reporting my return to glorious health, prompt me to write these few lines.

During the past few years congested sinuses caused me to suffer constant pain, gradually lowered my vitality and weakened my heart. Tried various methods of treatments and curatives with the result that my suffering became more and more aggravated. It seemed that just when my discouragement was verging a state of hopelessness I became acquainted with Spectro-Chrome Attuned Color Waves, which proved to be "the System that did not fail." Since then, to my great astonishment, I was relieved of all sinus pain in one tonation in November and have not felt the slighest recurrence. My general health too has greatly improved. Cheerfully therefore, I join the vast army of Spectro-Chrome enthusiasts.

May your aspirations revealed in the slogan "SPECTRO-CHROME IN EVERY HOME" soon meet with full realization.

Gratefully yours,

Edwin Frey

FOOD FOR THOUGHT

January 30, 1934.

Col. Dinshah P. Ghadiali, Pres.
Spectro-Chrome Institute,
Malaga, New Jersey.

My Dear Colonel Dinshah:

It gives me great pleasure to acknowledge my safe receipt of Spectro-Chrome Home Guide, same having come to hand last Saturday morning. Needless to say I lost no time in going through its interesting contents and polishing up on some of the facts as laid before me in the Vineland class over a year ago by yourself personally. The pleasures derived from that course of study will never be forgotten, and I am sure all of those who comprised that class will feel likewise. Please accept my sincerest thanks for the great privilege you have accorded me in presenting me with this Home Guide, and I will assure you that it will at all times be referred to when occasion demands.

A copy of the February issue of Spectro-Chrome Magazine also came to hand and the contents give serious food for thought when one reads the details of the Tugwell Bill now before Congress. God forbid that this bit of affrontry should ever go through and receive President Roosevelt's signature. Until I read of it in the magazine, I had no knowledge that such a bill was even in contemplation, no mention of it having come to my attention via the public press. Is it another of those cases of something being pushed over on an unsuspecting public? I have made mention of it in conversation with friends and one and all condemn it as a piece of chicanery typical of the Medical Trust and likewise hope that it will suffer defeat in its present form. Undoubtedly it is a veiled hit at Spectro-Chrome, but unless I miss my guess Col. Dinshah is still one jump ahead and will retain the lead. Let us hope that some day the American people will wake up and think for themselves on all matters pertaining to their personal liberty.

In closing I would ask your advice in the matter of my filling out the last page of the Home Guide and mailing it to the Institute. I presume that this should be done at this time in order that I may receive advice by mail or telegram direct from the Institute in case of necessity.

With best wishes and personal regards

Very truly yours,

M. C. F.

I am glad to hear from all directions appreciative remarks about the Spectro-Chrome Home Guide. It gives me encouragement for better efforts. You may follow the Guide blanks.

The Tugwell-Copeland Bill **is** a "veiled hit" at us most assuredly and it is a matter of regret that such an attempt should be made by a learned Homeopathic Medical Doctor like Royal Copeland, for whom I have respect.

The American public is a flock of sheep. It follows emotionally, without any thought for consequences and does not realize that this so-called Food and Drug purity bill, will benefit solely the Medical Trust. Read my correspondence with the Senator, printed elsewhere in this issue.

FIRST DETROIT MICHIGAN PLANET

Minutes Of Regular Meeting Of February 1, 1934

By:—Ethel K. Greenham, Recorder.

First Detroit Michigan Planet held its regular meeting, Thursday, February 1, 1934, in Park Room, Hotel Tuller.

The Meeting was called to order at 8:15 P. M. by Ethel K. Greenham and she explained that there were no minutes of the previous meeting, due to a small gathering and not enough business to transact.

There were no communications, so the meeting was open for discussion on SPECTRO-CHROME.

The Recorder called on Thomas Ness for a short address, but, he asked to be excused so he could speak later in the evening.

Albert Greenham was called upon next and he read from one of the past SPECTRO-CHROME magazines, Volume 4, No. 5, May 1925, an article entitled "Advantages of Healing by Spectro-Chrome Therapy." During the course of discussion he laid great emphasis on the strict adherance to the Rational Food Of Man System, as outlined by Dinshah. All the members enjoyed his remarks.

Otto Hinz, a new member in our planet, gave quite a detailed report on his experiences with Spectro-Chrome. His condition being known as Asthma, Lemon Systemic and Orange Systemic and Magenta on Areas 4-18, were the Attuned Color Waves used to normalate this condition.

Thomas Ness next spoke on the "fallacies of diagnosis by Medical Doctors;" he outlined several such ridiculous reports given him by users of Spectro-Chrome, whose conditions were normalated without such foolish methods.

Albert Greenham next discussed two cases, under his care, one being Appendicitis in a pregnant woman, which he was tonating with Lemon Systemic and Magenta on Areas 4-18; also a case of another woman who was suffering from a condition in which the whole left side was affected, a sort of shrinking of the tissues on the left side. The sufferer wanted continually to draw on the right side of her body, so as to pull the parts of the left side up, there being a deep seated pain especially in the throat and left chest. The Medical Doctors had removed the tonsils and the adenoids as the base of her trouble. Her condition did not improve; she went back to the doctor and he stated if the tonsils and adenoids were not the base of the trouble, then surely the appendix must be! So this good Samaritan removed the appendix. Of course her condition did not improve; in fact, it grew steadily worse; so she was forced to go to the Medical Doctor again. He stated neither the tonsils nor the appendix was the base of her trouble, but since her condition was on the left side of the body, the left side of the thyroid was the base of the trouble; so the left side of the thyroid was removed! The sufferer grew gradually worse. She went back to her Medical Doctor and he now stated that the trouble was in the ovaries and that she would never be any better until the ovaries were removed!

This, she refused to have done. She heard about Spectro-Chrome through Arnold and after tonations for one month, she experienced greater relief from pain than she had in the past 10 years.

The Kardoscil now is 74, Spiroscil 16; Lemon Front Systemic and Scarlet on Area 18 is irradiated; next tonation Lemon Back Systemic and Scarlet on Area 4 are the Attuned Color Waves used.

One can readily see what could have been accomplished here, if the sufferer had taken Spectro-Chrome before she had four of her Spark Plugs removed.

Albert Greenham urged that we all become pioneers and assist in placing Spectro-Chrome in every home and prevent such abominable conditions as these recurring.

Greeta O. Peck spoke of a case in Canada, which is now under her care and which she was tonating with great success. A severe head condition was the cause and Indigo was being used; she hoped to have a better report at the next meeting on this case.

Albert Greenham proposed that the meeting be dissolved; Martha Schreiter seconded it and the meeting was dissolved by unanimous vote.

———o———

FIRST MILWAUKEE WISCONSIN PLANET

Minutes Of Regular Meeting Of February 1, 1934.

By:— Otto H. Lutz, Deputy.

The meeting was opened by Otto H. Lutz reading Good Thoughts for February 1, 1934, "Accuracy". He read the letter sent to him, telling our members to put their shoulders to the wheel and maintain their Membership by paying their dues. This would stop writing long communications and would also help to help mankind. It is known to Members that the fee they pay goes to print the monthly magazine and pay for the Planet meetings. The subject read was "Itisometer", the most wonderful apparatus for Therapeutical Precision, from Spectro-Chrome of September, 1925, Pages 662 to 664. Our Members are interested in the "Itisometer".

E. A. Ernest spoke about the new Aluminum Cabinet which will soon be on the market. He also spoke about the Journal Case and of the Magazines going out to about 1,000 people all over the States. It will not be long before Spectro-Chrome will be known all over the country.

Arnold F. Klinkhart reported a case of Acute Neuritis. This woman was carried into their place with great pain. Seven Tonations of Indigo Systemic were given and she felt fine.

Lillian Goebel is tonating a School Teacher who had pains for 15 years and after 12 tonations she said it is the first time in 15 years that she has no pain. The Attuned Color Waves started with Green, Lemon, Yellow and now Orange.

Elsie E. Ihlenfeld told Members of using her Equipment for Motor Trouble and is getting good results; this is her 10th week at it.

Otto H. Lutz has a case of a man who claims he had a dropped stomach, sinus trouble, severe pain in back. His disorder is of long standing. So far he had five weeks of Spectro-Chrome and is feeling well. Orange and Lemon Systemic were given; also Orange and Red given for Local and Magenta on Areas 4 and 18.

Questions and answers were discussed among the Members. We are glad to have new Members with us. Every one enjoys the meetings.

Emma Goldman Today

TORONTO: . . . Emma Goldman, (Mrs. James Colton), above, famous anarchist now 64, continues to enjoy her 90 day stay here with relatives while awaiting decision on whether she can re-enter the U. S. for a visit from where she was deported 15 years ago.

May Become Envoy

WASHINGTON . . . Mrs. J. Borden Harriman, (above) of New York, is reported to be slated for a diplomatic post, possibly American Minister to one of the European nations.

Thus has the World changed from time to time; what wonder then the empirical methods for long known as **Medicine,** should now be found worthless and be supplanted by other systems, more scientific and accurate? The day of Medicine is doomed.

The Medical Doctors are floundering in the bogs of Ignorance. Their merciless and open exposure is the sole way to benefit the sick. Send us **your** experiences of their tactics.

COMPLAINT AGAINST THE N R A

What Does General Hugh S. Johnson Say?

TOmpkins Square 6-5727

HOFFMANN
MACHINERY & ENGRAVING CO.

Printing And Bookbinding Machinery
Brass Type, Brass Dies, Etc.

114-116 East 13th Street, New York

February 20, 1934

Colonel Dinshah P. Ghadiali,
Malaga
New Jersey

Dear Colonel:

I received your letter of February 14th and I hope you will be successful in fighting your case in the Supreme Court. You remember the long and nice conversation we had in your place a few weeks ago, and at which time we were talking about the big business men trying to suppress the small business people under the cover of the NRA. You were right and here's what happened to us: A bunch of large manufacturers got together about six months ago—went to Washington and with the help of good lawyers, etc. worked out a scheme by which under the flag of NRA they would just eventually eliminate the dealers—so they will have the road free to sell the printers new machinery at their high prices. Never mind the printer who, like you, was satisfied to buy good used machines for about half price and could not afford to buy new machines. Never mind the dealers as long as they can live and prosper and make the millions again, as before the depression set in. Well people with money and influence always find ways and they actually succeeded. They worked hard and put up a Code and had it cooked up nicely and well done. And it was finally adopted and signed by General Johnson.

Without even consulting us dealers, we suddenly find that we are in it and that the signed Code not only covers the manufacturers but also all of us dealers. No questions asked. The Code Authority to which we have to submit consists of seven members and mind you,

all manufacturers, not one dealer. According to such Code they are establishing their own Clearance Corporation for the disposal of used machinery and they expect us to sign that—our own death warrant. In other words, we should be put out of business, which by and by they will take over and they have even the kindness in providing the undertaker and flowers, by establishing the Clearance Corporation. Besides that, we are to send lists of all the used machinery we have for sale, giving prices and all other data—so they would have a guide and know all about how to sell used machinery. Of course, it is set forth in different words, but that is the essence of its meaning.

Well I could tell you more about it, but I shall talk to you when I see you, which I hope will be soon. One thing is sure, that scarcely any dealer will sign such a Code—unless forced to do so under valuable pressure. That is how things go and that is how the well-meant NRA is being shamefully abused.

I am only glad that I sold you the Linotype Machine and Dexter Folder at a time when I could meet your own reasonable terms and it was at a time when my selling was not directed by hypocrites.

With kindest regards, I remain

Very sincerely yours,

HOFFMAN MACHINERY & ENGRAVING CO.
By Henry Tilford

Not having that Code before me, I am not able to judge its merits, but, am sending your protest to General Hugh S. Johnson for his reply. After I receive his answer to your charge, I shall offer my comments.

However, you may have your loins girded for a strong fight, a big fight and a fight to the finish. The sky is all clouded.

EARN MONEY IN SPARE TIME

Here Is Your Opportunity

The Days of Miracles have NOT passed. Here comes HEALTH for the entire family, delivered right to your Home from the Light Socket. One of the GREATER DISCOVERIES OF THE AGE FOR GAINING and MAINTAINING HEALTH.

Before accepting the appointment as exclusive United States Distributor we made a thorough investigation, and search during the past two years which proved conclusively that Spectro-Chrome HAS MERIT and does everything claimed for it by the Originator, Colonel Dinshah. The system and the working Equipment is very simple; a child can use it with favorable RESULTS. IF YOU ARE TIRED OF BEING SICK, IF YOU WANT TO ECONOMIZE, use SPECTRO-CHROME IN YOUR HOME.

We are appointing State Representatives and Agents to introduce Spectro-Chrome in every State and City in the United States; A REAL OPPORTUNITY FOR THE UNEMPLOYED TO BE OF SERVICE TO MANKIND and be liberally compensated for such service.

We shall appreciate having you give us the names of men or women or refer to friends and associates who you think may be interested in this proposition. To such applicants we can offer a very attractive plan whereby they can acquire a decent income during spare time. For our status we refer you to:—

The Marine National Exchange Bank, Milwaukee, Wis.
Merchants & Farmers State Bank, Milwaukee, Wis.
Teutonia Avenue State Bank, Milwaukee, Wis.
B. P. O. E. No. 46, Milwaukee, Wis.

For further information, either as a user, State Representative or Agent, WRITE or 'PHONE.

ERNEST DISTRIBUTING COMPANY

E. A. ERNEST, S. C. N.
3106 North Hackett Avenue
Milwaukee, Wisconsin

SPECTRO-CHROME

April 1934 Lemon		Volume 10 Number 4

Trade Marks Registered In The United States Patent Office

Important **Articles**

Read and let your friends read

ROYAL S. COPELAND OPENLY STATES HEALTH BOARDS THREW OUR BILL OF RIGHTS INTO ASH CANS

WAKE UP AMERICANS! FIGHT FOR YOUR RIGHTS

DINSHAH'S OPEN STATEMENT BEFORE SENATE COMMITTEE

APRIL - 1934

Publisher and Editor Dinshah P. Ghadiali

Single Copy 50 Cents

UNITED STATES SENATE COMMITTEE ON COMMERCE

Hearing On "Pure" Food, Drugs, Devices —An Experience Of Purity Of Air And Politics

It was Tuesday the 27th day of February, 1934 that was to be another turning point in the history of Spectro-Chrome. Hubert D. Stephens, Chairman of the United States Committee On Commerce had sent a kind invitation to appear before his Committee and "briefly present the matter in which you are interested." I certainly accepted it.

Here was really an exceptional opportunity. For years the American Medical Association through its wide ramifications had hounded me, persecuted me, dogged my very steps with the intention of crushing my work completely and the Tugwell-Copeland Bill with its definition of "Devices" means "Drugs" was one hidden blow deliberately aimed to wipe my work—Spectro-Chrome—from the face of America, by placing it under their domination, subjugation and dictation.

Spectro-Chrome is my brain-child. It is the product of over 30 years of arduous and laborious scientific research in the face of untold privations, starvations, miseries and hardships and its truth, precision, logical deduction and clinical accuracy have been continously in the public eye during the past 14 years of its existence. Cases thrown out of the hospitals as "incurable", sufferers told by the Medical Doctors "they had no hopes of recovery," thousands upon thousands of human beings in the last lap of desperation, were successfully normalated and restored to the enjoyment of their customary activities by the unerring Spectro-Chrome.

All these years, when the merits of Spectro-Chrome were being proven and established for all time, I had endeavored by numerous requests and invitations to the American Medical Association and its ramificatory affiliations to investigate Spectro-Chrome, but without response. Their opposition and actual hostility apparently was not as much against the system, but, against the Originator as a man—perhaps their vanity was hurt that a comparatively obscure Parsee Zoroastrian of India should tell them openly to throw their age-old, moth-eaten, nonsensical, empirical, absurd, illogical, corrupt, unscientific, jargonic humbug of "Medicine" into the junk pile and turn to the

path of Spectro-Chrome with its unerring Automatic Precision.

It was bad enough for Spectro-Chrome to claim No Drugs—No Manipulation—No Surgery, but, when to that was attached No Diagnosis, their rage knew no bounds. How can Diagnosis be avoided? Why, that was the **one** grip they had on the throats of the practitioners of systems of healing different from those of the Medical Doctors and it was the one weapon to harass, persecute and bulldoze such others whose views did not gibe with theirs. That Diagnosis was the key to the citadel of healing and by that "Diagnosis", they ruled the kingdom of Heaven to Earth and below with domineering success.

In the conduct of their nefarious and subtle activities, the Medical Octopus has been maintaining a powerful and influential lobby in the nation's capital city. Step by step, under the guise of improving public health, it made inroads into the various healing systems in such a way that most or all of them had to submit to the domination of the Drug Trust more or less, to keep the head above the water.

Suddenly, the Benign Masters Who Watch The Welfare Of The World, sent a servant of Theirs to initiate the suffering into the true mystery of the Covenant Of God With Man Through The Rainbow and the advent of Spectro-Chrome extended to the ailing a new Ray Of Happiness, which made them rejoice; but the Medical Doctors saw in it their most formidable opponent—their Nemesis. With the most simple apparatus, composed essentially of an Electric Bulb and Five Slides of Glass of specially matched Colors or "Attuned Color Waves", used in accordance with an original conception of spectroscopy and chemistry in human life, a novel Science was placed before the suffering World, of such unprecedented accuracy and unerring Automatic Precision, that the Medical Doctors gasped at its claims. Those "claims" are no longer **claims** and Spectro-Chrome Metry as the perfect Healing Science, without the use of Diagnosis, Drugs, Manipulation and Surgery stands unrivalled and unchallenged on its demonstrated proven **merits.**

Various were the means devised and traps laid to stem the tide of Spectro-Chrome. Its opponents had the Originator framed for all kinds of "crimes", from Practising Medicine Without A License to Grand Larceny, Supernatural Powers, Hypnotism, Mesmerism, Mysticism, Astral Projection, Flying Through Space, Salem Witchcraft, Mann Act Violation, with various penalties and Sentences, including heavy Fines and Incarcerations; he saw the inside of 10 Prisons already! Once President Calvin Coolidge issued a Pardon to pull him out. He went through all these persecutions cheerfully, working without any salary so humanity may be served. There was a goal to reach.

When no legislation in 14 years could stop Spectro-Chrome because of its inherent scientific Truth, it appears that the Medical Lobby somehow had Rexford Tugwell and Royal Copeland introduce the Tugwell-Copeland Bill, S-1944, about which full mention was made in Spectro-Chrome of February 1934. In the issue of March 1934, the full correspondence about it was printed. I was invited to appear and ventilate my objections and for me it was an opportunity to iterate the Truth. So I obeyed the call and spoke the Truth.

Malaga, New Jersey, my headquarters is 141 miles from Washington, D. C. I always travel by automobile, but, this time the unprecedented cold wave with its five foot snow drifts and below Zero winds, having blocked the highways, I decided to go by train via Philadelphia, Pennsylvania.

Not finding any train that could take me early morning from my village, I left the prior night and arrived at the Union Station in Washington D. C. before daybreak. I lingered in the depot a few hours. The Senate Office Building is right opposite and I went there at 9 A. M. The Caucus Room where the Hearings were to be held was on the third floor and as I entered it, I expected to see it jammed to the door—but, I was the first one in that spacious room accommodating over two hundred persons.

Slowly and one by one the public interested in the Bill collected; by 9:50 A. M., every chair was taken. I was in the left hand section of the first row and was literally sandwiched. My appearance with the Parsee Zoroastrian cap and different features must have appeared strange in that motley crowd. At 10.00 A. M., the Chairman of the Committee On Commerce Honorable Senator Hubert D. Stephens, called the assembly to order and the performance began.

By this time, I was fairly choked. I had thought that the learned Medical Doctors, Scientists, Senators, Ladies and others were all there to promote public welfare, health and prosperity by giving **"PURE"** Foods, Drugs, et cetera. Judge my intense surprise when I found these champions of Pure foods, blowing their malodorous, filthy tobacco smoke into the air of the room where many non-smokers like me were seated. Women too joined in the smoking contest. My eyes smarted

with the smoke. The atmosphere became positively stinking and unhealthy. Were I there for just a pleasure jaunt, I should have promptly left that room to save my lungs, but, being there on a mission of Duty in defense of my Spectro-Chrome, I stuck to my post. What a disgraceful sight! An august council of the United States Senate, made of the Senators—the responsible legislators of the Nation, to sit in conclave puffing like locomotive engines with the public joining them in the smokathon, did not impress me much except as an imitation Opium den in Shanghai, China.

What surprized me more was to see Dr. Royal S. Copeland also smoking leisurely. Some years ago, in one of his newspaper syndicated articles, he published a story about his son whom he admonished against smoking and now the worthy doctor needed similar admonition. Politics is surely a wonderful game!

I thought I came in that room for hearing on the Tugwell Bill, S-1944; what transpired showed me that the hearing was to be on another similar Bill by Senator Royal S. Copeland, S-2800. I asked for a copy and was handed one. A quick scan showed me that it was just as vicious and irrelevant as the other and had the same nonsense of "a device" is "a drug" within the meaning of this Act, as well as the names of diseases not to be advertised and the like. That Bill looked like the funeral toll of Spectro-Chrome and aimed directly at its complete control if not extinction. Two other Bills S-2355 and S-2358 were also to be within the same hearing; they too had the same earmarks of medical domination with the restraint on advertising and the inclusion of therapeutical devices in the term "Drug".

The first person to make a Statement was Charles Coolidge Parlin, of the Curtis Publishing Company. He appeared for the National Publishers' Association and endorsed the Bill. He said he spoke for publications with 50,000,000 circulation. He praised Copeland for protecting the consumers.

The next Statement was by Congressman Arthur P. Lamneck of Columbus, Ohio. He was in a rush to go back to the House Of Representatives, but, in 5 minutes he wiped the floor with the Bills. He spoke sensibly and registered unmitigated opposition. He said, "Take your definition of a drug. A drug is deemed misbranded if a disease is mentioned for which the drug is not a specific cure. Now, I am told there are only four specific cures known in medicine and I believe that a man that thinks aspirin is a good thing for a headache, as

I do, if he wants to call it a cure for headache, that is all right and if they advertise it as a cure for headache; it is to me in many cases; it cures my headaches; may be not permanently, but it cures it temporarily anyway. Now, it seems to me that one of the things that is back of this movement— I do not know where it originated—is probably to prevent the public from going to a drug store and buying some patent medicine that they think is good for the thing that ails them. This bill, in our judgment, is too drastic. It puts autocratic powers in the hands of executives." He voiced my views!

He was followed by Charles Wesly Dunn, a lawyer representing Associated Grocery Manufacturers Of America, American Pharmaceutical Manufacturers' Association, National Drug Trade Conference and other organizations like National Association Of Manufacturers Of Dog Food. This gentleman wanted "Boards" to be appointed, as if we did not have boards and more mahogany boards cluttering our administration and eating off the fruits of the soil. He spoke on Lea and Perrins Sauce, Beech-Nut Chewing Gum, advertisements, frauds, trade-puffs, boasting and the like for over 1 hour and 30 minutes; the Chairman finally requested him to stop at noon and a recess was declared till 2 P. M., Senator Stephens stating that the hearing was to be resumed in the Commerce Committee Room, instead of the Caucus Room.

Never having been a Senator, I did not have the slightest idea of where the Commerce Committee Room was. I was violently smothered by the Tobacco smoke in the Caucus Room and my sole desire was to stay as little as I could and be out of that infernal punishment to my health. I detest Tobacco most cordially and no smoking is permitted to anyone where I can enforce the rule. I want my food, water as well as air pure and the sight of these people poisoning the air which innocents have to breathe on compulsion, gave me a revulsion I shall not likely soon forget.

I asked a number of persons where the afternoon session would be held. One pointed in one direction, another sent me to the ground floor. The Capitol Guards told me to ask the elevator man. He took me to the next floor. I wandered in that place a number of times until a kind one gave me an exact number and I bent my steps thither. Barely had I opened the door and peeped, when a loud shriek rang from a young woman within: "O-o-o-o-o-o lookat 'im lookat 'im lookat 'im—ou u u —u—u uhhu—!" I was about rooted to the spot, when another and louder shriek greeted me and the woman rushed toward me with a face terrified and convulsed with obvious horror: "U-u-u-uhhu—O can't you catch him—O-o-o- catch him!" I thought I was the one she wanted caught. I had many painful experiences with human beings in my life and to be thus given a howling reception in the Senate Office Building was not encouraging. I shouted therefore, "Please madam, please, I am here as a visitor only, I came to find— — — —" "Yes, yes, lookat 'im, catch 'im, can't you— — O-o-o-o-h!" "Madam, but, what is it? What do I have to catch?" "There he is—theeerrr—"

Something whisked near my face; it was evidently a flying object endeavoring to get out of that room; the windows however were all closed, so it kept fluttering, to the consternation of the young woman. Gathering courage, I entered that room and opened a window, but the woman rushed again with her right shoe in hand; the high heel of the shoe was torn off in the quick runs she made hither and thither and she was limping on the unequal stilts. I tried in a few words to tell her the reason of my presence there, but, she lifted the telephone receiver and shouted to the Superintendent to hurry the janitor.

In a few moments entered a negro and a white man and they scampered from that room into the adjoining committee room after that flying devil, without success. Finally, the white one caught it by hitting it with a long pole and brought it to the offended woman. It was——a poor little baby bat !

The tenseness of the odd situation being relieved, we had a laugh about the incident. The white one opened the wings of the bat and I sensed he intended to tear them apart; I told the woman it was an innocent animal and a harmless creature, which was dazed by the daylight; I pleaded to let it go. She directed me to the Superintendent's office and I was glad to depart. On the ground floor, the Superintendent told me the Commerce Committee Room was not there but was in the Capitol. I went there, breathing a sigh of relief.

That room was the size of a small private office and could accommodate perhaps 50 people comfortably. To have such a cabin for a public hearing, where many hundreds would insist on justice and fairness seemed ridiculous. It looked to me that the Committee perhaps did not like the public to ventilate their grievances in plenty and this might be a political way of giving the gate to many. Anyhow, I felt that if I stayed there with those puffing locomotives a day, my breathing apparatus would be so injured that even the "Pure" food laws could not repair the damage.

Therefore, as Senator Stephens entered the room, I went to him quickly and introducing myself asked him to allow me precedence and relieve me from being there longer. He promised to do so and I promised I should be very brief; both fulfilled their promises. I thanked him for his courtesy; he was very kind.

As the hearing was resumed, the air of the room became positively nauseating with the stench of Tobacco fumes, mixed with all kinds of female perfumes. How anybody could sit in such poisonous air for any period and keep a steady thinking brain, I am unable to decipher. The Chairman stated that some of the Senators were busy and would come late. However, as the proceedings would be printed, those who would like to speak might do so. I promptly signalled.

THE CHAIRMAN. The hearing will come to order. Mr. Dinshah, I understand you desire to be heard.

MR. DINSHAH. Yes.

Statement Of Dinshah P. Ghadiali, President-Treasurer
Spectro-Chrome Institute, Malaga, N. J.

MR. GHADIALI. Honorable Chairman and Senators, I shall be very brief in my statement in regard to this bill, S-2800.

There is one point on which my Institute, the Spectro-Chrome Institute, Malaga, N. J., desires, if your Honors please, to protest against the inclusion of the word "devices" in the definition of food and drugs, that is devices, electrical contrivances, where, in the preamble of the bill, it is mentioned as foods, drinks, drugs, and cosmetics. That is the position taken by the Honorable Senator Copeland and Mr. Tugwell also. In the definition given on page2, section b, the term "drug" shall include "all substances, preparations, and devices intended for use in the cure, mitigation, treatments, or prevention of disease."

I am the inventor and patentee of a number of electrical equipments for the treatment of diseases. In fact, in 1896, when I came first to the United States from Bombay, India, I was the first lecturer on X-rays in New York. I am a manufacturer. We have a big plant for manufacturing our own instruments. My inventions and patents were made after over 30 years of laborious Scientific research and they are used all over the United States. If, therefore, the word "devices" be included in this act, Your Honors, my institute contends that it will

put us completely out of business because it would put us under the power of the medical trust. It is well known that the medical trust has always opposed every other system or method of healing. To put a healing device of an electrical or mechanical or scientific character under the control of a gentleman like the Secretary of Agriculture, who is supposed to mind the business of food and drugs and the purity of those things—to put mechanical and electrical devices under his control is a problematical absurdity in itself.

A second point to which I wish to draw your attention is that it takes a stand here as to misbranded foods and drugs and so on, on page 7, section 6, where it says:

A food, drug, or cosmetic shall be deemed to be misbranded (a) if its labeling is false or misleading in any particular. Any representation concerning any effect of a drug shall be deemed to be false under this paragraph if that representation is not supported by substantial medical opinion or by demonstrable scientific fact.

Now, please, sir, the Honorable Senator Copeland means well, of course. I do not endeavor to impeach the honorable Senator's motives, but I do contend that the inclusion of devices within the meaning of drugs would deprive my invention and machines, patented under Government patents, from being used. A Government patent gives me the right to make, use, and vend my machines. I have introduced my healing system all over the United States and have hundreds using exclusively this system, and it will entirely put us out of business. My plant, which is a quarter of a million dollar plant for manufacturing, will close, because my work will then have to be substantially under medical opinion and if the medical gentlemen in their wisdom think fit to fight me as they have fought other systems, they are going to fight me in a way which would be really unfair competition, because that would give, through the power of the United States Senate, through the zeal of this august Congress, it would give them an absolute control of the entire healing art of the United States.

I therefore emphatically voice my protest against such an unfair method of including within a food and drug act the word "devices" and including all of us within that act. It is unfair.

Another thing that is very apparently unfair here is on Page 15, "false advertisement." If Your Honors will please look at this, section 9, line 15, paragraph b:

An advertisement of a drug shall also be deemed to be false if it contains the name of any disease for which the drug is not a specific cure, and so forth.

I refer you to the authority of no less a medical expert than Dr. Richard C. Cabot, the great physician and brilliant master mind of America. Dr. Cabot has definitely put into the records of the American Medical Association that 52 per cent of medical diagnosis is on post-mortem examination found to be absolutely false. If there were a specific cure for any disease, nobody would die. There is no such animal as a specific cure for any disease.

I get something and it is a pain in the neck. Somebody else gets it and it is called malaria. Drugs are like women. They have different temperaments. That, Honorable Senators, is a fact. There is no such animal as a specific cure for any disease. Otherwise, we would have simply a bottle house under the opinion of Senator Copeland, or under his supervision perhaps, would be better because he is a medical doctor of renown. There is no specific cure for a disease. We have sneezing, coughing, shooting, cutting, any blessed thing in the world. If I sneeze 5 times instead of 2 times, I am going to a doctor who will give me a specific cure. That is, therefore, an incongruity and absurdity against which I emphatically voice my disagreement with the Honorable Senator. And it says, "is not a specific cure but is a palliative"—then it is a false advertisement.

I appeal to the Senate of the United States, which is the greatest legislative body in the world. Why should we disgrace our American citizenship—and I am an American Citizen sirs—why should we disgrace our American intelligence by putting such an incongruity there, that unless a thing is a specific cure, it shall not be sold? Why, nothing can be sold. Even ice cream contains gelatin. Flour is white, which makes us white-livered. Everything is chemicalized, and only God Almighty knows what is in any food or drug. You go to the corner drugstore and get some bromo-seltzer. That contains a poison called acetanilid. I practiced medicine 30 years and I treated 30,000 people and I left it. I am a retired medical doctor. I practiced medicine in India. In India there is no medical legislative act of any kind. We just kill them and leave them.

The last thing that is utterly absurd, which will absolutely put my Institute out of business and put my inventions on the blink, is this: This is something, Honorable Sirs, which must be completely wiped out from the legislation. It is in regard to advertising. It is on Page——

(At this point Senator Copeland entered the committee room.)

SENATOR COPELAND. Pardon me. I just got here and I did not get your connections.

MR. GHADIALI. I am president and treasurer of the Spectro-Chrome Institute, Malaga, N. J., which is a New Jersey corporation, chartered in every State of the Union, from coast to coast. I have been in 10 prisons because of the machinations of the Medical Trust. They have locked me up ten times in the United States. Why? Because I teach the public not to take any drugs at all. Under this clause that is put here, "False advertisement"—this is the utmost absurdiity in any proposed legislation that I in my life ever saw.

Some years ago I wrote a thesis on criminal law. If I had read this act, I would have included this also in it. It is page 15, paragraph c. The purpose of this is to discourage the public advertisement for sale in interstate commerce of drugs for diseases wherein self-medication may be especially dangerous.

The medical people do not want people to medicate themselves. They want to get the money from the public and pocket it for their own use.

It continues "wherein self-medication may be especially dangerous." Who is going to rule on whether a medicine is dangerous or not, except the American medical gentlemen that are licensed for that purpose? The Secretary of Agriculture with this document in his hand will not know anything when you give him the power. You are making a toy, an automaton, out of the Secretary of Agriculture, the great minister of the Cabinet of the United States President. You put into his power the food law for purity of food. I do not want to speak about food and drugs. Others have spoken. Let them do that. I shall only take two more minutes and close my discourse.

All I want to show you is the absurdity of this proposition of the Medical Trust here patently contrary to the interests of the public.

The question is here; "How can anything be contradictory to the interests of public health if the people medicate themselves?" It would be contrary to the interests of the medical profession to have the public take its own drugs. A man, for instance, takes Bromo Seltzer, a poison. He gets it in the drug store. That is made in Baltimore by the Emerson Drug Co. Why doesn't the Food Administration take cognizance of that as a poison? Is it contained in the United States Pharmacopoeia? It says on the label to take so many grains, doses and you either die or you recover. They say that an advertisement shall be a false advertisement, to discourage any advertisement of a drug, and that drug includes a device also. There are 38 diseases here and may it please you, sir, there are the nerve diseases, in which my machine has proved its value thousands of times within the last 14 years of existence. They want to stop me from doing my business, a thing for which I gave my whole life. It will stop every inventor from progressing. It will stop all patenting. It will stop the progressing of every true-souled American. It will give a monopoly to the healing arts process only to the physicians.

They say this, that "an advertisement shall be deemed to be false" if it says it can remedy appendicitis. Give me a case of appendicitis and I shall send the man back in 1 night, having his appendix inside of him instead of in a glass bottle and $300 out.

"Carbuncle." We have with our system, gentlemen—and I appeal to your own honorable intelligence—we have never had a case of carbuncle without healing it by Spectro-Chrome.

I can take any person and break his carbuncle in less than 2 weeks' time without the knife and the man goes home keeping his money (and his life) in his pocket.

Then it says "diabetes." Four weeks of Spectro-Chrome has cured diabetes. Why should we be prevented from putting it before the public? It will stop me from saying that Spectro-Chrome did this or that. Even the United States mails will stop me.

I print a monthly magazine called "Spectro-Chrome". That, because it does not get the privilege of medical sanction, will be debarred from the mails. I couldn't label it. I couldn't use my machine at all. I appeal to the honorable Senators therefore to consider that—

SENATOR COPELAND. I am sorry that I missed the early part of your speech.

MR. GHADIALI. I may say that I have the highest respect and honor for Dr. Copeland as a medical gentlemen. I am not opposed to Dr. Copeland, and I appeal to his own high intelligence. Why, does he include these disorders about which the medical profession is unable to do anything? My system has remedied diabetes. I have put

people out of bed from consumption in 4 days and because I spoke about those things on the public platform I am every month being fought by the medical profession and they send me to prison. I have been in 10 prisons, sirs and this bill is not going to put me again behind the bars.

I appeal to your sense of honor. Why does the honorable Senator Copeland want to put something in this about appendix and typhoid? Give me typhoid and I will cure it and I won't treat him with quinine or any other medicine. It has been done.

I didn't come here to boost my methods. Here is the point, Senators: You are giving a complete monopoly of the healing art to one branch of the American medical profession called the "Medical Trust." I have said it. I appeal to your sense of honor, gentlemen, not to allow any such injustice to people who are thinking far ahead of others and stop their brainwork from doing a service as they have been doing for years. If you pass this bill, all invention in medicine will stop. It can be passed only when the medical men can stand forward and say they have a specific cure for any one of these diseases, which they haven't. I have.

I thank you, sirs, for your kind courtesy.

As I finished the short summary, I bowed and walked out—glad to be out of Tobacco smoke. The outright and open opposition expressed by me in plain language, must have been unusual for politicians, but, I prefer to speak my own mind.

A few days later, as I was in the Capitol on other business, I was informed by the clerk in the Committee On Commerce Room that a copy of the entire Hearing was mailed to me and that the Copeland Bill S-2800 was revised by its writer and reported to the United States Senate, with amendments. He gave me the revised copy. The 38 diseases about which the Medical Doctors wanted all to stop advertising, were struck off from the Bill, some words here and there were changed, but "Devices" were still defined as "Drugs". Evidently, they do not want to yield to common sense and horse sense, but will persist in using donkey senselessness.

In the printed report of all the Statements following mine, I found the post of honor held by Dr. William C. Woodward, of Chicago, Illinois, Legislative Counsel of the American Medical Association, which was expected, of course. Senator Copeland repeatedly asked him to make clear that the S-2800 was not initiated by the A. M. A. and Dr. Woodward emphatically declared it was not! Copeland and Woodward must have thought the public is deaf, dumb and blind; evidently Copeland does not know how the Medical Octopus operates—at least he makes matters look so innocent, yet, he gave public notice to Dr. Woodward to draft certain parts for him.

The public who have votes ought to learn about how much respect Copeland bears for our glorious Constitution. He was formerly the Health Officer of New York, N. Y., so Health Boards and their frequently questionable activities must be known to him. This learned gentleman who acts every inch like a suave politician, made an open confession when Dr. James H. Beal, of Cocoa, Florida, representing the interests of The National Drug Trade Conference was making his Statement. Here is it from the printed report of the Hearing page 116:—

"Dr. Beal.—But even in the case of the rascal, they should have the benefit of the Bill Of Rights. I would say permit any of these establishments to be searched from a warrant showing probable cause, describing the place to be searched and the things to be seized.

Senator Copeland.—Doctor, you know the **Bill Of Rights went into the ash cans when they established Boards Of Health.**

Dr. Beal.—Well, I think that is a mistake—I mean that opinion."

I came home a wiser and sadder man. America is headed to such domination, bureaucracy, autocracy, egotism, vanity, self-elation and selfishness that the very momentum of its fooldom will eventually unbalance its economical equilibrium, in the near future.

Come what may, Spectro-Chrome will NEVER yield to the power of the Medical Octopus. Its momentum of Truth In The Absolute is so great, that any such endeavor will annihilate its opponents by their own inertia. Try it! There is a Power in the Higher Ethers greater than the abominable Medical Trust whose corrupt methods enslaved the country.

Awake Americans, fight for your birthrights.

—o—

PIG IRON

By:— Roshan Dinshah (Age 7)

It is called pig iron because it contains germs that pigs have!

VIVISECTION.

By:— Agnes F. Chase

Vivisection is the name given to the practice of burning, cutting, twisting, boring, baking and inoculating the bodies of living animals, which, through determined medical advertising, has come to be associated, in the mind of the general public, with so-called "scientific research", although actually it is no more scientific than the old custom of prescribing pulverized spiders and the hair of dog that bit you, for human ills.

Vivisectionists claim that the practice is necessary for the health and survival of the human race, even though a rapidly growing number of honest and wise medical men not only admit that experiments upon animals are useless as aids to discovering the cause and cure of human disease, but frankly warn that such experiments are dangerously misleading and the final experiment must always be upon man.

Vivisection is cruel and unnecessary. Its continuance is made possible by the careful secrecy that surrounds it and the unwillingness of many persons to investigate and involve themselves in an unpleasant matter. The sensitive individual fears the shocking discoveries that must follow any real investigation of vivisection and the coward shrinks from becoming entangled in a controversy wherein his conscience may force him to array himself upon the unpopular side.

Vivisection degrades and brutalizes those who practice it. The depths of falsehood and chicanery to which its agents will descend in an effort to defend its outrages are beyond belief to the inexperienced, but are only too well known to Antivivisectionists. For Antivivisectionists are members of an ever-increasing army of honest, humane men and women who have the brains and courage to think for themselves; who find it impossible even to tolerate tacitly the wanton cruelties of vivisection and whose righteous and accurate judgments of the practice are based upon the detailed reports of vivisectors themselves as published in medical journals. No other evidence is necessary to condemn animal experimentation as a lapse into barbarism and no manufactured stories of these cruelties—however lurid—could equal the sickening details of the actual facts.

Nevertheless, such is the blind ignorance of the average citizen, when the vivisectional forces, becoming alarmed by the increasingly wide spread of Antivivisectionists, hastily send a suave, well-groomed vivisector, such as Simon Flexner, for instance, before a gathering of susceptible and determinedly stupid women and this man, despite the published and spoken testimony of his colleagues, has the audacity to go on record with claims and representations so evidently false as to reflect upon his own intelligence; the susceptible females actually fall upon his neck in servile ecstasies and solemnly pledge their organization to support a practice that is not only an affront to civilization, but actually a menace to themselves and their children.

Vivisection is not a vaguely-understood "scientific" practice that safely may be "left to the doctors". It is a Moloch, demanding **human** as well as animal victims, and **getting them** in hospitals, foundling homes and other institutions.

He who willfully ignores the facts concerning vivisection makes himself equally responsible with the vivisector for the continuance of its dangers and the suffering it involves. The cries of tortured animals, though paralyzed by the gag and such drugs as Curare, lest they offend the ears of the vivisector, yet do most clearly reach the ears of the Maker and Judge of us all "in Whose hand is the soul of every living thing." Can you afford to meet the question: "What have **you** done to oppose the infliction of these cruelties upon the helpless animals for whose care and well-being you are surely held responsible?"

———o———

Heads Marines

WASHINGTON . . . Maj.-Gen John Henry Russell, is the new head of Uncle Sam's Fighting Marines. This photo was taken at his desk in the Navy Department, when he took over the command last week.

Columbia's Queen

NEW YORK . . . Pauline Reevere (above), descendant of Paul Reevere, has been selected by faculty members as the most beautiful girl at Columbia University . . . an honor which completely "floored" the young freshman.

WARNING TO PUBLIC

Milwaukee Medical Doctors Declare Certain Drugs Killing People

Rexford Tugwell and Royal Copeland should attend to this.

In a paper read before the Milwaukee County Medical Society last month, Dr. Fred W. Madison and his co:worker Dr. Theodore L. Squier emphatically declared that the result of their research indicated that recent deaths from a rare disease "Agranulocytic Angina" were owing to the use of Pyramidon. The disorder is also called Primary Angina Agranulocytotica, in which there are destructive ulcerative lesions of the throat, leukopenia or destruction of white corpuscles in the blood and somtimes jaundice. All this is brought on self by the use of Pyramidon and similar drugs, sold at any drug counter, without restriction.

Let us see what this curse to mankind is. Pyramidon is the trade name of a coal tar derivative, called Amidopyrine, which is an antipyretic and anodyne and is widely used for Tuberculosis, Colds, Rheumatism and Neuritis, in 5 grain doses. Its chemical formula, like that of other coal tar products, is very complex: C 6 H5. N (CO. CH). N (C H3). C (CH3). N (CH3) It is used extensively as an Anti-asthmatic by persons suffering from Asthma. The Drug is also named Dimethylamino-antipyrine and is used for Insomnia, Headache and similar disorders, by the public without realizing that this wretched product is the very one that **causes** all the symptoms which it is supposed to remove.

It is a matter of regret that the public trusting the Medical Doctor as its guardian, is treated thus by the Medical Charlatan, who does not know much about the sequelae of Drugs. Pyramidon destroys the power of the Blood Stream and is in the same class as Antifebrin, Phenacetin, Antipyrin and other Blood Pressure Diminutives, which operate upon the Human Being through the Purple Wave.

In our Spectro-Chrome of July 1922, under the heading "Coal Tar and Cancer", we exposed the dangers of using **Bromo-Seltzer,** a concoction palmed off on the public as a "speedy and reliable remedy for Headache and Neuralgia". That remedy is a poison to the Cardiac and Circulatory System. It acts by dilating the blood vessels, thus lowering actually the very dynamic energy necessary for nutrition. The United States Pharmacopoeia prescribes a full dose of Antifebrin or Acetanilid (which is the active drug in Bromo-Seltzer) at 3 grains;

the Emerson Drug Company of Baltimore, Maryland right on its label declares the contents of a teaspoonful to represent 3 3-5 grain of Antifebrin or Acetanilid, recommending at the same time: DOSE—A heaping teaspoonful in half glass water; repeat in half an hour if not relieved.

What a crude way of dealing with the trusting public! The Medical Doctors are responsible to The Almighty as the cursed destroyers of the Image Of God and the defilers of His Holy Temple for the development of the Soul.

Instead of trying to pass new so-called Food and Drug Laws, let Tugwell and Copeland pay attention to the enforcement of the existing laws and remove Bromo-Seltzer and similar poisons from the lunch counters.

Formerly, Drug Stores were selling Drugs; then they began to sell Safety Razors and Stationery; then they started with Soda Water and Ice Cream; this was followed by some with Boston Beans and Valentines; now, with the advent of Spectro-Chrome, it may not be long before they will be renting to the ailing Spectro-Chrome Cabinets at a quarter a day.

It will bring real blessing to the public.

—o—

Dillinger Sheriff

CROWN POINT, IND. . . . Sheriff Lillian Holley, who ruled over the "escape proof" Lake County Jail, from which John Dillinger, notorious killer and bank robber made his escape with a wooden pistol he had whittled out during spare moments.

New N. E. A. Head

CLEVELAND . . . Edison E. Oberholtzer, Superintendent of Houston, Tex., schools, is the new president of the National Educational Association. He was born at Patricksburg, Ind. and is 54 years old.

LET THERE BE LIGHT

Radio Broadcast, 5:15 P. M., March 12, 1934
over Station W C F L

By:—Dr. John Maxwell

The Originator of Spectro-Chrome Metry frequently sent on the Radio public messages about his System in years past and in many cases, the local medical authorities brought pressure to bear on the Radio Station to stop allowing him the use of the Station. The Station being threatened with revocation of License, in many cases bowed to the will of the autocrats. In this connection, the activities of the Federal Radio Commission may bear an investigation.

However, recently Dr. John Maxwell, an independent thinker and writer of Chicago, Illinois, who constantly broadcasts his "Health messages" via Radio, after commencing a study of Spectro-Chrome Metry, thought it fit to broadcast the message of Spectro-Chrome voluntarily and the following was his address on the air:—

"The prophet Isaiah drew attention to what man lacked through being out of tune with his Creator's great plan. Straying from the path of truth we get into a quagmire, a maze, a fog. We must come back to have a receptive mind, ours ears must become attuned, the scales must fall from our eyes, before many eternal truths can be grasped by us.

"Since the begining of the world men have not heard, nor perceived by the ear, neither hath the eye seen, ... what He hath prepared for him that waiteth for Him."

Sometimes men think these things have only a spiritual significance. But where will you separate the spiritual from what is called temporal? Let us take man as a whole, with his spiritual, moral, physical aspirations, his needs and desires for understanding, for health, for happiness, for a long, useful and well-rounded life, with a well-stored mind replete with worthwhile experiences.

Many of the great masters have had their ears attuned to the marvelous rhythmic harmony of fitly arranged sounds. Years ago in a London museum I looked with reverence and veneration upon the spinet used by Handel, as I thought of the score of his Hallelujah Chorus, and wondered how many bright sparks of genius might have flowed through his mind and been recorded by his pen, if he had had access to a modern concert grand piano.

In a New York museum I spent an interesting time as my reflective mind was busy when I viewed the marvelous display centered there of ancient and modern musical instrments, wind and reed and string, from the early harps and the zither. the dulcimers and cymbals, to the primitive harpsichord and spinet, and the gradual developments of the centuries until we come to the present day grand piano, the modern percussion istruments and the solo and trumpets.

My friend Percy Grainger has traveled through many European countries, England, Germany, Sweden, Norway and Denmark, gathering and recording the ancient tuneful folk songs and melodies, handed down from parent to child in the past centuries, available now for us with this great master's full score added to those rhythmic airs, helping us to

"Wake to energy each social aim,

Attuned spontaneous to the sill of Jove,"

as Beattie puts it. Jove or Jupiter, who with his azure influence was regarded as giving joy or mirth and serene happiness, as Shakespeare says: "Be bright and jovial among your guests."

We have felt the soul-stirring effects of music, the healthful stimulation as the silver chords were struck . We have learned something of the therapeutic value of musical harmony, the tuneful vibrations which have greeted our ears.

Now, in this our day, we are being introduced to the potency and therapeutic value of attuned color waves. There's rhythm in color as well as in sound. The very colors themselves are the effect of attuned vibrations and oscillations, proceeding from the source of light, the resplendent sun.

Sir Isaac Newton introduced us to the order of colors in the spectrum. Take a piece of clear glass, a bar with parallel sides which form an equilateral triangle. Rest one side flat on the nose and look through that clear prism at the sun or at the light in a clear electric bulb and you find displayed before your eyes all the seven colors of the rainbow, in a defiinitely arranged order. The colors always follow in the same sequence. Red always appears at one end of the prismatic wedge, and violet always appears at the other end. Here we have the

order,— Red, Orange, Yellow, Green, Blue, Indigo, and Violet.

The study of color harmony is not only deeply interesting as a matter of general knowledge. It should also bring us to the understanding of an exact science of color therapy.

The doctor of the future will doubtless find a most valuable ally in the potency and therapeutic value of color.

"No two elements have similar spectra," one authority avers. For instance, in a chart made by Bunsen, it is shown that potassium appears in the spectrum with a "very bright, wide red line on the left, with a faint violet line on the extreme right." Thus we get magenta. Sodium is found to give a yellow line.

To many people color seems as a daub of some particular tint. Each color is somtimes classed under many tints, but there is a definite true shade for each and every color and there is in each attuned color wave a mighty potency of a chemical character, said to have a velocity of 186,324 miles a second.

As the G string on a violin is known by the rate of vibration following the plucking of that string, and the A and E strings having other determined vibrations when they are in tune, so we have the oscillations of colors.

There have been many master minds investigating color from chemical, physical, metaphysical and psychophysical standpoints. Newton, The Herschels—father and son—Faraday, Kirchhoff, Bunsen, Tyndall, Darwin, Lommel, Huxley, Brewster, Crookes, Young, Babbitt and others.

S. Pancoast published in 1877 a book on "Blue and Red Light, or Light and Its Rays as Medicine".

Dr. Edwin Dwight Babbitt was, perhaps, the most active exponent on the effects of color. He was established at his Science Hall in New York in 1876 and published a treatise on the "Discovery of the form and constitution of atoms and their working connection with etherial forces, to produce the effects of heat, cold, electricity, magnetism, chemical action, light, color and many other effects". Then in 1878 another book on "The Principles of Light and Color" was published. These publications produced quite a furore at the time, and brought forth bitter opposition from his medical compatriots who refused to believe that light and color can have any effect upon the human body.

Babbitt did not meet with success that his painstaking efforts and deep thinking deserved. Part of that may be due, it is maintained, to the fact that he based his theories on the statement that the primary colors are Red, Yellow and Blue, and that the Secondary colors are Orange, Green and Purple. This led to other errors, so that he was unable to build up a system of healing which was thoroughly scientific and mathematically correct.

It has remained for another Scientist, Colonel Dinshah, a Parsee of Bombay, India, to bring out Spectro-Chrome Metry which, whilst giving due credit to every previous investigator, for what appears to him really scientific discoveries, has himself reasoned out step by step, this subject of attuned color waves, going to show that the true primary colors are Red, Green and Violet (as first proclaimed by Dr. Thomas Young of Milverton, Somerset, England): that the secondary colors are Yellow, Blue and Magenta; that the tertiary colors, or the third class, are Orange, Lemon, Turquoise, Indigo, Purple and Scarlet and that the true shades of each of these colors are, in their proper order, commencing with Red, each a step up to thirty degrees above the preceding one. Thus, that their accuracy and their therapeutic value can be definitely asserted.

Colonel Dinshah says that with these 12 colors is established a Spectro-Chrome measurement, forming a complete cycle of attuned color waves, oscillatory frequencies of a specific color, graded to a specific position, to produce uniformly specific wave lengths, with specific effects, the whole cycle being formed of colors in the following order: Red, Orange, Yellow, Lemon, Green, Turquoise, Blue, Indigo, Violet, Purple, Magenta and Scarlet.

He maintains that these colors cover the whole field so far as color can be of therapeutic value. Admitting that there are infra-red invisible rays below the red, the ultra-violet invisible rays to be accounted for above the true violet ray, he insists that these two are both highly burning destructive rays. The Infra-Red burns by its coarse grade, low frequency heat oscillations by a process of 'impress', and the ultra-violet burns by its fine grade high frequency cold oscillations by a process of 'express'. The effects were similar in many respects, the infra-red producing burns like steam-scalding or fire-scorching, the ultra-violet producing 'burns' like frost biting or chilblains. In the former the heat works by injection of Energy; in the latter by extraction of Energy."

It is Colonel Dinshah's contention that as the sun's action is such a prominent factor in producing color and health in the vegetation, so each attuned color plays its part in producing therapeutic effects in the human body; that those colors when backed by a strong light are ready to be taken in by the aura of the body and produce definite effects in breaking up congestions, in clearing the body of toxins which have brought about the heavy encumbrances and crippled normal functions. That, for instance, the attuned purple waves may be called "Blood Pressure Diminutive Waves"; the Scarlet waves may be considered to have an augmentative effect upon the blood pressure. That no burning or any other harmful effect can take place within the range of the 12 color waves from Red to Violet. His thesis is deeply interesting, and sounds very, very logical.

Every cell in our body has its own rate of vibration, so, as Colonel Dinshah says: "Just as hydrogen and oxygen have an attraction for one another and on the application of a spark, rush towards combination to produce water, similarly the Attuned Color Waves pick up Affinity waves (waves of an opposite nature) wherever they be found and convert into neutrality."

———o———

Birth Control Hearing

WASHINGTON . . . Mrs. Margaret Sanger, a leading advocate of birth control, appeared before a senate committee to urge approval of the Hastings bill which would permit dissemination of birth control information.

More Mad Monk Suits

NEW YORK . . . Princess Irina Alexandrovna (above), who was awarded damages of $126,750 in a London libel suit against an American Moving picture concern for the the picture of Rasputin, the mad monk, is now reported turning her eyes toward the U. S. for future action against the company and exhibitors here.

FIRST MILWAUKEE WISCONSIN PLANET

Report Of Regular Meeting Of March 1, 1934

By:— Otto H. Lutz, Deputy.

The meeting was opened by Otto H. Lutz, Deputy, reading the subject of "The Poor Foodbag" followed by the Two Words that spell Death. This was read from the Spectro-Chrome of 1922. Albert Franz reported a case in the family, of pain in the left side. The Colors used were Turquoise and Indigo. This pain would come and go, but with 10 tonations it was normalated.

Sara E. Bangs used Magenta for Heart Trouble on a friend of hers, with good results.

John L. Susek told us there was so much war and death all around us, that he and his better half decided to buy an Equipment and put themselves in health. He is using it for Sensory Trouble and is now using the Orange. He said, "it surely cleans you out."

We had quite a discussion on Curvature of the Spine. Last month Pauline Wilde brought this subject due to a fall she had. She was advised to consult The Central Office and she got her answer as to what Attuned Color Waves to use.

This month, Harold Weinholtz said he tried so many times to find out if Spectro-Chrome will do the work on the Spine or not. But no definite answer is ever given to him. One man said, "the Home Guide has in it, 'Do Not Expect the Impossible'; he thought that was enough said. Others said, "we have a right to ask questions, as we are all mostly laymen in this field of work." The question is, "will Spectro-Chrome do the work or will it not?" We know your station in life but the Members should know if it is a mechanical defect. You can answer that, then they could use their own judgment.

E. A. Ernest informed us that he had been to Malaga, New Jersey since our last meeting and was working with Dinshah to solve the problem of crackage of Slides. They decided to make each Slide 4 X 4 inches instead of 4 X 6 inches; the beam of light would strike the center of the Slide more evenly and in that way avoid concentration. This was tested and found to be very favorable. We were told that he saw Dinshah answer the large pack of mail that came in that morning. That answers a question, for some of our members thought the mail was answered by the Secretary. He also said some very good results and even orders came in from sending out those February magazines. We also heard of the big snow storms and very cold weather out there, and E. A.Ernest said he just had to pick that

cold day to go out there! It was their coldest day in many years. C. B. Keppler wanted to interview a person that was normalated for diabetes. Otto Lutz told him there are cases in the Spectro-Chrome issues that he could read. He said he did not care a hoot for testimonials; he wanted to talk to the person. People are skeptic when you tell them about Attuned Color Waves and you can not blame them because he too is skeptic. This man came to Otto Lutz when he had the Cancer case and asked to see the man and talk to him. The sufferer told him he got more results from one month of Spectro-Chrome than he did in a year of medical aid; and after getting such good results what do you think happened to him? The company from which he got his compensation ordered him back to medical aid. He is now in the hospital and not doing so well. Our feelings can not be expressed in words for the man in this case.

We advise you to make a systematic study of the Spectro-Chrome Metry Encyclopaedia, each month. Our Planets are not gossip clubs or scandal houses and whether Tom, Dick or Smith believes a thing or not does not concern us. We deprecate personal testimonials, because, it would degrade our Science into a quackery. Neither the questioner nor anybody else will be given names of Sufferers, because, it would mean violation of professional confidence, obtained by us through our Free General Advice Charts.

Supposing, tomorrow somebody would ask us about cases of Gonorrhea or Syphilis in some persons normalated. Do you for a moment believe that we should state, "Surely, go to the wife of Jim Jam Hootsen; she had it twice in a year; once from her husband and once from somebody else's husband and Spectro-Chrome set her right!" Please, friends, be reasonable; use common sense. In 14 years, the Originator sold thousands of Spectro-Chromes to homes sheerly by teaching the Science **without personal endorsements;** why not study for yourselves? We need no electro-plating on Spectro-Chrome; its merits shine.

In regard to Spine Disorders, as in our System we discarded all names and Differential Diagnosis, no general opinions on medical names are given by us. Fill the General Advice Chart and get the Free Guidance. You have the way pointed; follow it; why criticize our conscientious staff, when we all strive to serve you **without any compensation?**

None went to Heaven or elsewhere by proxy; experience is the best tutor. Study for yourself.

OUR EXPANDED ACTIVITIES

More Help Wanted

Owing to the expansion of our numerous activities in every direction and the inability of coping with it through the necessity of giving 24-hour service to Suffering Mankind, we need more assistants. All are permanent positions and are open to those who can follow consistently the principles of Spectro-Chrome and can deliver service without grumbling. They may live in or out, but **must** conform to the strict Rules of Spectro-Chrome Institute, by being non-boozers and non-smokers as well as vegetarians.

Apply in writing, stating age, nationality, religious beliefs, affiliations, experience, salary wanted.

POSITION 1—ASSISTANT GENERAL MANAGER

We need an Assistant General Manager, man or woman. Must have thorough business knowledge, high sense of responsiblity, unquestionable integrity, broad and open mind, ability to meet problems with quick decision and efficiency for eventual advancement to the post of General Manager.

POSITION 2—EXECUTIVE SECRETARY

We need an Executive Secretary, woman. Must be competent and capable stenographer-typist, thorough bookkeeper with ability to shoulder full responsibility and relieve the President of all details.

POSITION 3—CORRESPONDING SECRETARY

We need a Corresponding Secretary, woman. Must take full charge of the Free Advice Department and be capable of handling confidential Advice to the Suffering in a conscientious manner. She must be a first class stenographer-typist and used to supervising the work of others.

POSITION 4—PLANT SUPERINTENDENT

We need a Plant Superintendent, man. He must be familar with cabinet and metal work, courteous to subordinates but firm in handling them.

POSITION 5—HOUSEKEEPER

We need a Housekeeper, woman. Must take full charge of domestic responsibilities in our Resident Quarters, be familiar with working among children, clean and refined in habits and language.

SPECTRO-CHROME INSTITUTE, MALAGA, NEW JERSEY

SPECTRO-CHROME

July 1934 Blue

Volume 10 Number 7

Trade Marks Registered In The United States Patent Office

Important Read and let your friends read **Articles**

THIRTEENTH ANNUAL CONVENTION NUMBER

READ ABOUT SPECTRO-CHROME NORMALATIONS

JULY - 1934

Publisher and Editor
Dinshah P. Ghadiali

Single Copy 50 Cents

THIRTEENTH ANNUAL CONVENTION HELD IN SPECTRO-CHROME INSTITUTE AUDITORIUM AN EYE OPENER

The Thirteenth Annual Convention of the Scientific Order Of Spectro-Chrome Metrists was held on Tuesday, November 28, 1933, at 8 P. M., in the Auditorium of Spectro-Chrome Institute, at its Central Office and Research Laboratory Headquarters, in Malaga, New Jersey.

A RECORD OF SUCCESS IN THE NORMALATION OF

BURNS, HIGH BLOOD PRESSURE,
LOW BLOOD PRESSURE, GOITRE,
GANGRENE, PYORRHEA, ANGINA
PECTORIS, PARALYSIS, TUMORS, SPINAL
CURVATURE, RHEUMATISM, THROMBOSIS,
TUBERCULOSIS, HEART DISORDERS,
GALL BLADDER TROUBLE, MASTOIDITIS,
ST. VITUS'S DANCE, SPASMS, INSANITY,
ROSE FEVER, NEURITIS, DOG BITE, HERNIA,
AND CATARACT

BY LAYMEN AND LAYWOMEN RIGHT IN THE HOME, WITHOUT ANY MEDICAL ASSISTANCE OR MEDICAL TRAINING

Solely By The Use Of

SPECTRO-CHROME METRY

First Evening Session

As usual, a large number of Representatives from various parts of the country, was present. The States of Wisconsin, Michigan, New York, Pennsylvania, Delaware, New Jersey, were represented among others.

Dr. James Lewis and Emily Lewis of Milwaukee, Wisconsin wired:

"Happy returns of the day to you, congratulations on Thirteenth Annual Convention."

Otto H. Lutz, Deputy of First Milwaukee Wisconsin Planet, wired:

"Congratulations and best wishes for successful Convention."

Mario Tata and Mary Tata of Detroit, Michigan, wired:

"Congratulations and best wishes to you and all assembled for the success of Spectro-Chrome."

First Vineland New Jersey Planet sent a poem congratulating Dinshah on his sixtieth birthday.

First Detroit Michigan Planet showered in a long letter blessings on the Originator of Spectro-Chrome.

At 8 P. M., Colonel Dinshah P. Ghadiali, the President, called the Convention to order. He gave a short Welcome Address and demonstrated the New Portable Itisometer, which he was standardizing to market at the price of $150.00. The Electric Sympathometer, was then exhibited. It was the result of over 10 years of the scientific labors of the Originator of Spectro-Chrome. It was to be marketed at the same time as the Portable Itisometer and at about the same price. The Cabinet Irradiator for charging Water with Attuned Color Waves was exhibited and Dinshah explained its principles. The Solar Actemator, for use in the Sunlight was exhibited and explained.

As the session was open to the public, the interest in the new apparati was well noticeable; numerous questions were asked and answered. After that, E. A. Ernest, President of Ernest Distributing Company, Distributor General for the United States for Spectro-Chrome Equipments, was introduced to the Convention amidst applause.

E. A. Ernest said that he was a great sceptic. He did not believe in Spectro-Chrome for a long time, but made searching investigation from every point about the claims made by the Originator of the system. He could not believe at first that Spectro-Chrome was effective in all Disorders; he thought the claims were exaggerated, but, upon studying the class work and thoroughly looking into the matter, he found everything that Colonel Dinshah had stated to be true. He became the Distributor General for the United States Of America. It was a man-sized job that was wished on him. He would do his utmost and there was no doubt that Spectro-Chrome eventually would be in every home.

After pleasant greetings to Colonel Dinshah for his sixtieth birthday, the session closed at 11 P. M.

Wednesday, November 29, 1933—Afternoon Session (2 P. M. to 5 P. M.)

The Registration of the Affiliates being completed, the session opened at 2:15 P. M.

Many interested new visitors and strangers were also present. The first speaker of the afternoon was Jessie E. Ness of Detroit, Michigan, our First Vice-President and a Member of the Council Of Governors. She spoke of her experiences. " I am proud of this great work. I have had nursing experience and know about the human body. I once saw an appendix; I believe it is a silly thing to take it out, because it was supposed to give trouble later. It is cruel to mutilate the body by useless operations. Human made Laws are not unchangeable. They say the Tonsils are of no use; they search for the causes; when they do not find any to convince them, they say they have no use. I looked for true knowledge and did not find it until I met Colonel Dinshah in Detroit. I was for many years in political work. I talked politics; it was a wonderful thing, but, I was 'dumb' and could not get the real vision of life. When I heard Colonel Dinshah, his Truth held me. I knew he had troubles in the West. I was on juries and knew how people are framed. We are grateful the President of the United States passed on his case and pardoned him.

"I went to the Colonel's public lectures. Conditions were bad; there was no money and the family was disgusted. The children said, 'Mother, don't do that.' Mr. Ness also said the same way. He is the head of the house.........."

Dinshah—"Who is the head?"

Jessie E. Ness—"He is the head, but, I am the neck and the neck turns the head! I was extremely interested in Spectro-Chrome.

We are like machines, but, we thought wrong, ate wrong and had experiences to gain. The Medical Doctors, our friends, take low and high blood pressures. I am convinced we have a better method; Spectro-Chrome does that lowering and raising. I found that in the Tonations I gave to my friends. I tonated maimed and blind and convinced myself—I know now that Spectro-Chrome works. Men and women who require Spectro-Chrome, go to this and that Doctor; the Medical Doctors will be the sellers of Spectro-Chrome in the future. Only real men and women can sell Spectro-Chrome; it is no fool thing. It is based on the Science Of Light, as our 'Great Leader' said, 'Let there be Light' and there it was. It can do everything for us. A man like Colonel Dinshah whose science could build the body, was much persecuted. The dear man struggled; instead of eulogies, he got prosecutions. Spectro-Chrome is a wonderful therapy. I went into a store in Vineland, New Jersey, close to the Spectro-Chrome Institute. They asked me whether I was going to the 'nudist colony'!!! (Laughter) They thought we were nudists and exposed the body to Light! They know nothing about the wonderful things he said. The great Master Jesus did such work and he tried way back to tell people how to deport themselves. The people still have the old ideas and they like their old habits of drinking and so on; they hate anything that is a disturbing element to their views. We have a great work ahead of us.

"I burnt my finger and went downstairs for Spectro-Chrome. The heat was gone in 20 minutes. There is nothing foolish about it. I can tell you long stories about our successes. Of course, there are some disappointments, too, but, that is not the fault of Spectro-Chrome. We had a Brindle Bull Dog brought in. They had ordered him to the Lethal Chamber. We worked on him with Spectro-Chrome. He looked so grateful. He sneezed a big sneeze and coughed out a lump. He recovered. All animals and even birds are of the same family. We can't make magic, but, they are all amenable to Spectro-Chrome.

"In a case of Goitre it disappeared in a short time by Spectro-Chrome. There was a man whose leg was swollen. The Medical Doctor said, 'it must come off.' They were ready to have it amputated, when they heard of Spectro-Chrome. Attuned Color Waves were 'poured' day and night. There was a nurse from Metropolitan Life Insurance Company watching the case. The Medical Octopus was also watching! The nurse went and talked; another nurse said, 'it was their poultice that did the good'!

"I had a case of Dry Gangrene. It recovered under Spectro-

Chrome; they had tried hard to remedy it, before it came to Spectro-Chrome.

"There was a case of a woman with Pyorrhea. The breath was very offensive. We worked with Spectro-Chrome and the condition was corrected. It was an annoying condition. All we can do is to tell people what good Spectro-Chrome can do. I am more than delighted to be here. It is certainly my pleasure to be with you and I am willing to do whatever I may be called upon to do for Spectro-Chrome." (Applause)

Dinshah spoke about pigs and rabbits having been reported as being irradiated with Spectro-Chrome with benefit; he had reports of canaries and cats, horses and dogs being similarly administered Spectro-Chrome and there was no reason why Spectro-Chrome should not be used just as successfully by Veterinary Surgeons.

Ida Steckle of Millville, New Jersey, was the next one to address the Convention. "Ten years ago I was left to die. I had Angina Pectoris and was paralyzed on one side. I also had a Tumor. Medical Doctors could do nothing for me. I took to Spectro-Chrome. I joined the Class and graduated. My Tumor was dissolved and Dr. Kate W. Baldwin found nothing there on examination. Soon there was no trace of Paralysis left. In the use of Spectro-Chrome, I have had many wonderful experiences. Seven years ago, I had trouble with my Eyes; my sight was terrible. I went to a Chiropractor and spent $60.00 each Winter, without benefit. I had Curvature of the Spine at the lower end. I suffered till 30; had 11 operations. I had Rheumatism too. Spectro-Chrome set me right."

Dinshah stated his intention of converting Spectro-Chrome into a Weekly Newspaper, from January 1934, if enough subscriptions were in sight. He said it was the only way to combat the opposition and make the voice of Spectro-Chrome heard.

David Bosworth of Vineland, New Jersey was the next speaker. He would have attended other Conventions, but, they were too far from home. He was glad to be in this Convention close to him. He used Spectro-Chrome for many years. He remembered one particular case; three years ago, man of 42 came to him suffering from Tuberculosis. "They said he would pass out. His Bowels moved 6 times a day; he had no rest. I went to see him and left a small Spectro-Chrome. In 3 weeks time, his Backelim normalated and his Lung healed. His weak Heart started to become stronger. The Spectro-Chrome was left 10 months. The man was an inveterate smoker. He

gave up a good deal; however, his neighbors were killing him with kindness (?). Every time he smoked he had Hemorrhage; he reformed and in 10 months was perfectly well. He was down and out first, but, he went to work and in 5 months he held up his head.

"I may speak for myself. Two years ago, I woke and could not breathe properly. A large space in my Lungs was congested. I started with Spectro-Chrome, ate nothing for 36 hours; just used Fruit Juices. In 4 days, I was normal.

"My sister became sick. The Medical Doctors said she had Thrombosis and bad Gall Bladder Disorder. She dropped 20 pounds. She improved wonderfully under Spectro-Chrome."

Dinshah explained casually about the use of Spectro-Chrome in all Heart Disorders.

Allie C. Ray of Newfield, New Jersey, addressed the Convention. "I took to the study of Spectro-Chrome a year ago. I studied in two different Classes. In the use of Spectro-Chrome, after I graduated, I made a success. I have a daughter who had Mastoid Disorder. They suggested operation. In 6 Tonations, without any opening whatever, the swelling subsided. She was examined by two women Doctors and one told her, 'Miss Ray, the Gland is large; it is a bad case of Mastoiditis.' A week after Spectro-Chrome started, there was no more swelling or infection. She passed 100 per cent. in physical examination for State Teachers' College. My faith in Spectro-Chrome is strong. I feel that Spectro-Chrome will get along well. Spectro-Chrome is our Doctor. We find it wonderful; we have always a good word for Spectro-Chrome. I am completely sold on it. A number of Sufferers come to me for help.

"There was a boy in Collingswood, New Jersey, who had what they called St. Vitus's Dance. He had no control over his voice; he was just like a child; had no control. One week before Decoration Day, I took care of him. He had suffered 3 years. The brother of the boy is a Medical Doctor in a large hospital. He said everything that could be done was done and nothing more could be done. In 2 or 3 months, the young man walked, talked and never had any twitch.

"The father of this young man was a paralytic for 5 years. He asked if he could be helped. I said, 'try it' and told him what Attuned Color Waves to use. In August he said, 'this leg was dead 5 years, it is all alive now.' He squeezed my hand and it was a real squeeze. I follow Spectro-Chrome as Colonel Dinshah instructed. I trust Spectro-Chrome more than the Medical Doctor, any time.

"A gentleman of 72 came to me one day. He was a lifelong friend. He could not lift his arm. He stayed with us 19 days. His Tonation started on a Friday. On the Tuesday following, he told me he wanted to show me something. He took off his shirt, which he could not do before without help. I said to him it was wonderful. He now dresses himself. It was a very remarkable case. He was a shipbuilder 21 years ago. He dragged his toes as he walked, because of some injury. After Spectro-Chrome, he said, 'I believe that leg is feeling better. He picked the foot on a level."

The Afternoon Session closed at 5 P. M.

Evening Session—8 P. M.

The Evening Session was an especially interesting one, because, a number of Sufferers from various parts was also present. They testified by their presence their interest in Spectro-Chrome.

Esther Jennings of Glassboro, New Jersey, started the work of the evening. She said, "A year ago October, I had the case of a young woman. She came to live with me. She had Spasms; they were all worried. They had called Medical Doctors, who wanted to take her to Camden. They said a year would do her good. She was not in her right senses. On recommendation of Spectro-Chrome Institute, I took charge of her. They were giving her Sedatives; that was stopped. Spectro-Chrome was irradiated. Her Kardoscil was 100 and the Spiroscil 12. I succeeded in controlling that by Spectro-Chrome. In 2 weeks she was quiet and feeding herself. In 2 months, she asked me how she came to live with me. She had, prior to that, wanted to kill herself; but, by Spectro-Chrome, she came to her senses. She was taken home, as a trial, on Christmas. In March, she was taken home for good. She is now in Philadelphia, where she got a position. She was selected out of five candidates. If it had not been for Spectro-Chrome, I should not be alive here tonight. I had Heart Disorder and was given up many years ago. There was no hope. Spectro-Chrome remade me completely."

Dinshah said he did not care 2 hoots and 3 postage stamps as to what the Medical Doctors thought of Spectro-Chrome. It was there to stay for ever.

M. Alice Sinclair of Vineland, New Jersey, spoke of her experiences. "My experiences in Spectro-Chrome have been most gratifying. Burns are common. They are easily remedied by Spectro-Chrome. My mother splashed boiling water and was burnt. The opportunity to study the effect of Spectro-Chrome on Burns was fulfilled. Blue was used. In 17 minutes, the pain was gone! In a few more Tonations, the Burn healed. Mother would not take more Tonations.

"There was a teacher of Atlantic City. She stayed up all night, the spasmodic pain was so much. I used Orange on her. In about 1 or 2 hours, she fell asleep and would have gone sleeping how long! That Tonation was given her on August 31. She said, 'Call on me for testimony; it is the most wonderful thing; I can do anything now.'

"I had another gratifying experience in the case of my mother's sister. The trouble is all who come, come for Spectro- Chrome as a last resort; they try all and then come here. My aunt lived only 6 weeks with us. She had drunk in 30 years enough Medicine to sink a ship. She was 30 years in pain. We gave her Systemic Tonations. She had no uncomfortable sensation after that. It will take time to rid her system of all that Poison.

"Last week, a Graduate Nurse came with some swelling. I am no College Graduate; I knew nothing. She was suffering from what they called Rose Fever. She had a good start for Mastoiditis. Two Tonations relieved her. She had a Mole which they had removed by Electricity. I changed its Color by Spectro-Chrome. It was a marvelous relief.

"It is a pity that people come to us too late for quick results. Then they become discouraged, if they do not have at once 100 per cent. results. I had a woman with Neuritis, who was sitting on the top step; she went so without sleep. She had received all kind of treatments and saw an advertisement from Perth Amboy. She came for 3 weeks and had no great discomfort left. She was splendid in following the instructions.

"People have queer ideas about Spectro-Chrome. They asked me many times whether I had to change my religion for Spectro-Chrome! All this talk of Spectro- Chrome Institute being an advocate of 'Nudism' is so nonsensical!"

Dinshah spoke about the necessity of mental cleanliness. Those who talk dirt have filthy minds. Mental Hygiene should be taught in the schools. It is necessary for the uplift of the Nation that moral education should be inculcated from infancy. Each organ of the Human Body is fit for its normal function as intended by the Creator and no shame should be impressed upon the young mind as being attached to it. He told a story of the man who complained to the Lord Gautama Buddha, that he was ogling the wife of the complainant. Benign Buddha asked him whether he witnessed such flirting. The man said, "Of course, I did see it with my own eyes." The Master said, "Go thou and with a redhot poker burn thy eyes, for the fault is there." Purity of the mind is needed in the educational curricula.

Roland B. Gilman of Camden, New Jersey was the next speaker. He said, "I started on Spectro-Chrome with strong statements against it. I was suffering myself; had choking nights. I have no trouble now after using Spectro-Chrome. My experiences gave me amazement.

"My boy had argument with his dog. There were Streptococci in the boy's Blood; he had been bitten and had 11 punctures in the flesh. He was brought from Bridgeton to Camden. Spectro-Chrome was used and plain dressing; nothing else. Soon the swelling was all gone. They could scarcely believe it when on Tuesday he went home.

"A ten year old girl had spinal injury. They took out splinters of bone. Spectro-Chrome helped her recovery."

Dinshah then reintroduced Distributor General E. A. Ernest, of Milwaukee, Wisconsin. Among a round of applause E. A. Ernest went on the platform and said, "After hearing such encouraging stories, I can not say anything differently. I can not figure out, how this man (Colonel Dinshah) single-handed fought the Devil in his own den. He made me the head of this army——like a Colonel. I need more Captains and Lieutenants. I see some of them in New Jersey. We need a whole army of fighters for Spectro-Chrome. I told you last night, what I want to do. I want to build the sales on a solid rock on the Fundamental Principles of Spectro-Chrome. There is no doubt in my mind that we shall come on the top. I had many experiences with Spectro-Chrome. I could not start to speak of them all, if I had time till tomorrow morning.

"I may tell you of an exceptionally bad case of Heart Disorder. It was that of my own mother. She could get no relief from Medicine. Finally, Spectro-Chrome was used. Her Heart before that was beating 3 times and skipping 3; then it beat 6 times and skipped 3. In 2 weeks time, her Heart was normal and she is now moving around as any well person.

"I had a case of a woman who was paralyzed. She could not talk. In 4 weeks of Spectro-Chrome, she was moving around again.

"We can go thus all along the line. Our opponents may call us 'fakers' and we must combine our forces against them. It is a hard struggle, but, we shall surprise them. I do not say I can do it in one year or even five years, but, we shall do it beyond question. They talk of Electricity, Light, Color, Heat and Sound. What this gentleman (Colonel Dinshah) has known for 40 years, is now becoming known to the Smithsonian Institute. They find that oak sprouts can bend by Colors; certainly, a human body ought to respond similarly.

"I can not tell you all I feel and want to say, in 3 or 6 minutes. I am very happy to be at this Convention. We shall know one another better as time passes."

Thursday, November 30, 1933—Afternoon Session (2 P. M. to 5 P. M.)

The third day of the Thirteenth Annual Convention being open solely to Members and Provisional Members, the public were not admitted. The work done was all official and pertaining to the welfare of the Organization.

The President after a short speech requested E. A. Ernest, the Distributor General to unfold his views which he did in a frank and forceful manner. He said, "Thanks for the compliments paid to me. How many years I was searching for a work into which I might put my heart and soul; I know what I thought of Spectro-Chrome. A man in Washington asked me to check it. I was forced to check. Spectro-Chrome is a genuine Service to Mankind. There is a man more superhuman than any soul I know. How that man feels his neck safe is unthinkable. I know it. I have backing; we are going to have 2000 Agents acting for us. We have greater power than all our opponents, because, every person is sincere in the use of Spectro-Chrome; every one is guided by the Higher Powers and such Higher Powers are behind this work. It is no use telling that one man singlehanded could fight all this battle. The Masters sent him to deliver Their Message. We will help him to do it. It took me 2 years to find out the Truth about Spectro-Chrome. Spectro-Chrome is what it is claimed to be. We had an equipment for 2 years in our home and we checked on every claim that was made for it; we checked on it and we checked on Colonel Dinshah; we checked his Records, his Patents, Copyrights, Trademarks. We did not find **one** single statement that was not correct.

"I came here to help him in his work; I forget myself in it. Everything that he said was backed to my satisfaction. Show me any record in any business that is more perfect. Naturally, people are skeptical. Way over there in Washington, the Smithsonian Institute speaks of some things, that this man knew 40 years ago. I saw newspaper clippings of that time. It is all there.

"When I came here last September, I brought with me a hundred or more questions to ask Colonel Dinshah. I came here to talk about the taking over of the work of distribution. The questions were written by me on my way from Milwaukee. Before I asked him those questions, Colonel Dinshah had answered me. He reversed the procedure. We have since then had scores of letters passed, but, everything we do is in accord. We are thinking the same way, the only difference is that Colonel Dinshah does everything for the other fellow—he does all for the other and is doing nothing for his own family.

"Do not hesitate to tonate or sell. You may collect the money for it. 99 per cent. M. D.'s are doing it. They would sooner cut your Heart out than correct the condition. They got me out of my home at 6 A. M. and diagnosed correctly that I had Appendicitis which needed immediate operation. Smart Doctor who said that; he was my family doctor; got me ready for the operation; rubber-tired cot, beautiful operating room; ether or gas was to be given. Two Doctors pushed me down when I decided not to go through with the operation. I refused to submit, saying, 'Mr. Ernest belongs to myself.' I picked up my things and went home. I rode my horse for 4 hours. That consulting Doctor had his $1,000.00 fee counted off. He sees me in the club and goes out of my way! One of these days, he will have a Spectro-Chrome.

"When the Colonel first introduced Spectro-Chrome, he tried to sell the idea to the Medical Doctors, but, they laughed at him. I accused him many times to let up on his criticism against them, but, it is a 'hornet's nest.' Now I know what it is and I do not care it. I never get an order from a Medical Doctor.

"I tell you to follow his instructions. I have personal friends among the professionals—many of them; Medical Doctors, Surgeons,

Dentists, Exodontists. I talk to some of them till 2:30 in the morning; Spectro-Chrome talk. A great many of them are not to blame. If their Medical Schools and Associations would know about it, they would be criticized, ostracized. They should learn something about it though.

'I don't know about the Human Body. The simple form our good friend the Colonel has made of Spectro-Chrome, makes it possible for any person to use Spectro-Chrome. If Spectro-Chrome did nothing but what it did for my family, it is a marvel. A Medical Doctor informed my mother that she had a Hernia. She was in great pain. He said, 'Mr. Ernest, you will find your mother dead one of these days.' We started Spectro-Chrome on her. In 14 days, we had another examination. Her Heart was beating 58 before; after the 14 days, it was running 78 per minute and she now runs upstairs twenty times a day. You can't stop her. She told me one day she was afraid to take more Spectro-Chrome, lest she might live too long! The results of Spectro-Chrome are miraculous; just unbelievable, if you follow it religiously.

"Some very fine professionals are using Spectro-Chrome. I can relate a number of incidents of a Dentist of Milwaukee; he actually doubled his practice. Sufferers come to him from a distance traveling and they leave without pain. He is without doubt the best booster for Spectro-Chrome. He is talking and eating and breathing Spectro-Chrome.

"Just recently a couple came from California. They wanted to represent our Company, in California. Mrs. Ernest and the lady became intimately acquainted and talked Spectro-Chrome. They left for Racine; had friends there. They were husband and wife, both Doctors. They became enthused. I took advantage of the visit, which resulted in a call to Racine. They use now 1 Graduate, 1 Systemic and 1 Cabinet. Today, I had a telegram; they ordered one more Systemic.

"In any business, anybody who comes in on the ground floor and gives conscientious service to the public will be in clover. Never mind what the professionals say; those professionals will come to their knees before we are through with this. Undoubtedly, many af you will recall their attitude about other things. They refused to consider the use of any measure except Drugs and Surgery. Now, even they are using Violet Rays, Ultra-Violet Rays, Sun Rays and it is foolish to predict what will not happen about Spectro-Chrome. Anyone can use the Spectro-Chrome. Those Medical Doctors study for 9 years and learn all kinds of nonsense. One of the greater opportunities is coming to see the day when the American Medical Association will come to Colonel Dinshah and apologize!

"Colonel Dinshah wished this distributing job on me. I shall never stop, jail or no jail, until that work is accomplished. Here is a soul sent by the Masters to do something for us. Maybe we were cowards till now; he encouraged us to go ahead. Until I found out the true value and merits of Spectro-Chrome, I did not go into it. It helped not only my family, but, thousands of others, as my thorough inquiry showed. I came to take only State of Wisconsin Agency, but, Colonel Dinshah wished the United States on me. The World has sick human beings and the Medical Doctors are there too. If we combine all our forces and go after them morning, noon and night, we must succeed. I shall never stop until every family has a Spectro-Chrome. I do not sleep more than about three hours; I shall sleep less until Spectro-Chrome is in Every Home." (Loud Applause)

Dinshah then spoke about the tremendous progress made in the year, through the distribution having passed into the hands of the Ernest Distributing Company. It gave him more time for further research and manufacturing. Spectro-Chrome Institute now manufactured all its Spectro-Chrome Cabinets and other apparati right in its headquarters at Malaga, New Jersey. It printed all its books and literature in its own printing plant, which was equipped with two composing machines, four printing presses, two folding machines, two stitching machines, one cutting machine, besides numerous apparati for trimming, casting, etc. The manufacturing of the Cabinets proceeded with two shifts of helpers and the Advice Department was open for telegraph or telephone Advice, at all hours of the day and night. No personal consultation or interviews were granted, because, he was too busy with his researches and the like and furthermore, it gave rise to potential legal complications. He was thankful to E. A. Ernest for relieving him from the burden of sales.

Nachman Diamond, of Wildwood, New Jersey, said, "One man can not do all. I am sold on Spectro-Chrome. He needs people to work for him and help him. My wife went to a city oculist for glasses. He examined her and said the left eye had cataract. I looked at it and did not say anything. Gave her glasses. It took 10 years to develop; made a move as if she might not live 10 years; he sent her away. I gave her Spectro-Chrome. She was satisfied. Four or five Tonations of Green is all I did.

"She has a sister who went blind. A young Medical Doctor operated on her. Soon, she could not take care of herself. I took care

of her. The trouble looked like a 'star.' After 5 days, my wife saw her eye. The 'star' was gone. I told her to go to the Doctor and show him. She made up her mind to go and show him that she did not have it."

Dinshah spoke about the use of Spectro-Chrome in Cataract cases. He had numerous cases reported by Physicians and Osteopaths in which the Cataract was benefited and removed eventually by Spectro-Chrome.

On motion of David Bosworth, seconded by Benjamin Kessler, the Minutes of the Twelfth Annual Convention, as printed in the Spectro-Chrome magazine of April 1933, were unanimously accepted as correct.

Moved by Jessie E. Ness and seconded by Esther Jennings that Dinshah P. Ghadiali be re:elected as a Member Of The Board Of Trustees. Unanimously passed.

Moved by Jessie E. Ness and seconded by Thomas W. Ness that Irene Grace Dinshah be re:elected as a Member of The Board Of Trustees. Unanimously passed.

Moved by M. Alice Sinclair and seconded by Esther Jennings that Marguerita W. Simon be elected as a Member Of The Board Of Trustees. Marguerita W. Simon having refused to accept the nomination, the name of Bessie E. Hasenau of Cleveland, Ohio was substituted for re:election, on motion of Thomas W. Ness and seconding of Leona Burghardt. Unanimously passed.

The Board Of Trustees being duly affirmed by Jessie E. Ness, she was appointed by the Board Of Trustees as a Member of the Council Of Governors and the First Vice-President of the Scientific Order Of Spectro-Chrome Metrists. Dinshah P. Ghadiali was made the President and Treasurer and Irene Grace Dinshah was made the Secretary of the Organization.

Jessie E. Ness thanked the Board Of Trustees for her reappointment and promised full co:operation. She said the Spectro-Chrome monthly magazine was a wonderful mouthpiece and should be made into a weekly newspaper. E. A. Ernest promised to subscribe for 1,000 copies of such weekly; Thomas Ness promised to subscribe for 100; others promised similar subscriptions of smaller numbers. Dinshah thanked them all for their enthusiastic support.

Leona Burghardt was appointed and affirmed as a Member of the Council Of Governors and the Second Vice-President. She promised her loyalty to the Cause.

Evening Session—8 P. M. to 11 P. M.

The entire evening session was taken over by the President, as announced. He explained therein the Fundamental Principles of Irradimetry, Itisometry and Sympathometry, accentuating the use of the Favorscope and co:related scientific subjects.

Unfinished matters having been discussed and completed, the 13th Annual Convention dissolved at 11 P. M.

It was a record of unanimous success. There was no dissent, no argument, nothing to disturb the rhythm and harmony. Thomas Ness presented to the Originator a humorous poem, "The Dog Box," which is printed on a subsequent page.

Soy Beans Defy Drought

DES MOINES, Ia. . . . The "hard times" crop for the drought-stricken middle-west has been found. . . . It is soy beans, according to Federal and state agricultural authorities. . . . Soy bean crops are now looked to by farmers to help keep the season from being a total loss. The soy bean's long roots reach deep into subsoil for moisture.

Mrs. Samuel Insull

CHICAGO . . . Mrs. Samuel Insull (above), wife of the indicted utilities magnate, is here from Greece to stand by her husband when he faces trial on the various counts which have been pending so long in federal and state courts.

THE DOG BOX

Ode Number Two

By:— Thomas Ness

This Dinshah Pestanji Ghadiali
Won't give the poor doctors a rest;
Maybe that's why they call him **Pest**anji
With the accent, please, on the PEST.

They hate him because he exposes
Their traffic in dopes and pills,
Their poisons dished out in doses
To CURE (?) humanity's ills.

They laugh at the Cabinet—yes sah
And call it a "Dog Box", forsooth,
But they'll never win out against Dinshah,
For they've closed their minds to the truth.

With Cabinets selling with rush
And a Home Guide **free** with each one,
The thought of Dis-ease is like mush,
Plug in on God's light and it's gone.

It builds nice red fresh corpuscles
And saves your appendix too,
It gives you strong bones and new muscles
And chases away Spanish 'Flu.

So plunk down your few sparing dollars
And give pills and potions the air,
Though it doesn't wash your shirts and collars,
It WILL cleanse your blood-stream for fair.

See, the hospital beds, empty, waiting,
The doctors with nothing to do;
All the sick folks at home, SELF TONATING,
THAT'S THE PICTURE, LET'S MAKE IT COME TRUE.

FIRST MILWAUKEE WISCONSIN PLANET
Meeting Of July 5, 1934

By:—Otto H. Lutz, Deputy

We were honored by having our President Colonel Dinshah at our meeting. He thought he would be our guest, but, we asked him to answer some questions, which he did freely. It surely was interesting to have the Original Teacher speak!

He said many people find fault with a lot of things. His motto is to serve people by his Duty and not to please. You cannot please everyone, so he said he does not try to please anyone!

One question was asked by him as to what to do with a Member, who did not send his Scientific Order Collaborator Blanks. This Member had asked for another Blank Book, without sending in the first. He had trouble with one of his cases and they arrested him for practising medicine without a license; this charge was later changed to practising Chiropractic without a license. Then they learned he used Spectro-Chrome, so the charge was dropped. They did not want Spectro-Chrome to have free advertising.

Dinshah spoke of some of his lawsuits and was glad to be rid of them. He spoke kindly of E. A. Ernest, United States Distributor General of Spectro-Chrome Equipments. Ernest was doing his part by devoting his time to the Service Of Mankind, through the Health Message of Attuned Color Waves.

Dinshah then read the headings of the Chapters of the Third Volume of his Spectro-Chrome Metry Encyclopaedia. The book has wonderful information in it.

This was the first opportunity Colonel Dinshah had to be with this Planet and he said he enjoyed being with us. Instead of meeting lasting 2 hours, we spent almost 4 hours with our visitor and we enjoyed it very much.

We hope to have him again with us some time in the future.

Colonel Dinshah thanks the First Milwaukee Wisconsin Planet for its cordial reception. He enjoyed the visit so well, that he is returning in November 1934, with his Annual Convention to your city. Read the announcement in the next issue.

FIRST DETROIT MICHIGAN PLANET
Meeting Of May 3, 1934

By:—Ethel K. Greenham, Recorder

Tonsilitis, Whooping Cough, Gall Stones, Hernia, Uterine Abscess, Normalated By Spectro-Chrome

Otto Hinz, spoke of a school-girl, with badly infected Tonsils, in which 7 Tonations of Turquoise normalated this condition.

Another case of a child with Whooping Cough, which he reported was normalated with the Turquoise.

He reported a case of a woman, 70 years of age, suffering from Gall Stones; several Doctors had attended this women, without relief and although she received a few Tonations, she is greatly pleased with the results.

Walter A. Hale, displayed a stone, which had passed from him after several Tonations of Lemon Systemic and Magenta on Area 4. He said there was great pain experienced in the passing of this Stone. Spectro-Chrome, was given credit for this helpful elimination.

Albert Greenham, was then asked to deliver a talk to the Members. He laid great importance on the measurement of the Kardoscil and Spiroscil, as the best and only keys, with which a Spectro-Chrome Normalator had to work and how it was impossible for any one to receive help, unless first of all, this was known.

He urged all those seeking advice from the Central Office, to be sure Kardoscil and Spiroscil, Front Elimination and Back Elimination, were accurately given in order to receive the greatest amount of benefits.

He spent considerable time on the discussion of Food, using the Rational Food Of Man, as outlined by Colonel Dinshah and offered as evidence, a new Member, Aquilla Mathers, to bear out his testimony.

Here was a man, who was operated on three years ago, for Hernia and although being unable to purchase a Spectro-Chrome Equipment, had heard of Dinshah's Food System through Albert Greenham and had followed it to the letter, proving conclusively that the greatest maximum results would be received when the Rational Food Of Man, was strictly adhered.

He discussed parts of the Encyclopaedia Vol. 2 and concluded his talk with a report of a case of a Mrs. Roberts, age 24, suffering with Uterine Abscesses and which was said to be of a serious nature.

After two months of constant Tonations, she is now considered well; several Abscesses formed during the process of Tonation, but, were broken by Spectro-Chrome, a continual drainage taking place. Orange on Areas 10, 11, Green on Area 18, Scarlet on Area 4, were the Attuned Color Waves used to bring about the desired results.

Meeting Of June 7, 1934

Thomas Ness was called upon to be the speaker of the evening. He dwelt mostly upon a larger campaign for boosting sales of Spectro-Chrome Equipments, throughout the state of Michigan. He spoke of the immense enthusiasm of the people of Wisconsin and how they had rallied to the side of E. A. Ernest and his associates.

The speaker spent considerable time reviewing many of his past cases and the Attuned Color Waves used for their normalation hoping to be able to assist the Members if their cases were similar.

Otto Hinz spoke of several cases which were under Tonation and the progress they were making.

Albert Greenham spoke of an elderly man, 79 years, with pains in the back and the progress this case was making. Magenta on Area 4, and Orange Front and Back Systemic were used.

Stella Harter spoke of a sprained wrist, in which Indigo was used for the elimination of the pain; this she stated was accomplished.

———o———

MOTHER AT 6½ YEARS!

Here is one reported in the "Deshi Mitra" or Country's Friend, of Surat, India. In its issue of Thursday, June 28, 1934, it says: "In New Delhi, an event that threw Medical Doctors of India into astonishment, happened. On June 7, a girl of 6½ years gave birth to a baby, in the Victoria Zenana Hospital. X-Ray photograms and investigations, proved the mother's age to be correct. Dr. Miskin sent a report of this occurrence to the British Medical Journal."

We are not surprised at this in the least, because, European and American Medical Doctors report occasionally children born to mothers, without any fathers at all. In fact, such experiments are well known and the late Baboon experiment subsidized by the Belgian and French Governments is only an incident.

Who was her husband? How old was he?

Trade "FAVORSCOPE" Mark

FOR SPECTRO-CHROME METRY
DINSHAH SYMPATHOMETER SYSTEM

BASED ON SOLAR, LUNAR AND TERRESTRIAL RADIANT GRAVITATIONAL INFLUENCE

1934 **AUGUST** — 1934 **INDIGO**

CHRONOS
(TIME TABLE)

COMPUTED BY LORRAINE M. WHITMAN, S. C. N., ACCORDING TO THE LONGITUDE AND LATITUDE OF SPECTRO-CHROME INSTITUTE.

SET TO STANDARD TIME FOR HOME USE

FOR DAYLIGHT SAVING TIME ADD 1 HOUR

		Day		Night	
		First	Favor	First	Favor
Date	Day	Hours	Minutes	Hours	Minutes
1	Wed	6	27	7	13
2	Thu	5	00	8	40
3	Fri	5	00	8	39
4	Sat	5	01	8	38
5	Sun	6	30	7	09
6	Mon	6	31	7	08
7	Tue	6	32	7	07
8	Wed	5	05	8	34
9	Thu	5	06	8	33
10	Fri	5	07	8	31
11	Sat	5	08	8	30
12	Sun	5	09	8	29
13	Mon	6	38	7	00
14	Tue	6	39	6	58
15	Wed	6	40	6	57
16	Thu	5	13	8	24
17	Fri	5	14	8	22
18	Sat	5	14	8	21
19	Sun	6	43	6	51
20	Mon	6	44	6	50
21	Tue	6	45	6	48
22	Wed	5	18	8	15
23	Thu	5	19	8	14
24	Fri	5	20	8	12
25	Sat	5	21	8	11
26	Sun	5	22	8	09
27	Mon	5	23	8	08
28	Tue	6	52	6	39
29	Wed	6	53	6	37
30	Thu	6	54	6	35
31	Fri	5	27	8	02

Here is the Work that Shattered all the False Conceptions in Healing !!!

SPECTRO-CHROME METRY ENCYCLOPAEDIA

by

Colonel Dinshah P. Ghadiali, Originator, Spectro-Chrome Metry.

COMPLETE COURSE ON ATTUNED COLOR WAVES

Your Desire Is Fulfilled

During the 14 years of the teaching of Spectro-Chrome Metry, our Students clamored to have in print the entire Course on the Science Of Automatic Precision, so it may serve as a constant reference and Guide.

However, owing to the enormous expense and labor involved in the production of such an important and comprehensive work, the amount of time needed to complete it and the incessant demand on his resources from innumerable and unexpected quarters, Dinshah did not publish it.

It would have been impossible even now to do so, but, having our own printing plant in operation, we are at last in a position to accomplish the object.

This unique Encyclopaedia comprises the Class teachings of Dinshah in the theory and practice of Spectro-Chrome Metry, including his Rational Food Of Man System, Itisometry, Irradimetry, Sympathometry, Spechro, as well as the Occult Operation Of The Great Sympathetic Nervous System with elaborate explanations about the various astronomical phenomena connected with life. All Human Disorders are separately elucidated in accordance with our system and the vast field of the Physical and Metaphysical Sciences in connection with health is satisfactorily covered.

At the end of each Chapter, there is a List of Questions for mental self-training. At the end of each Volume is an Application for Home Training Course Examination. Each Volume has 400 pages.

The Encyclopaedia is profusely illustrated and has numerous charts, tables and valuable scientific data.

The simple and fluent language used in the work is such that you will actually think you are listening to the voice of Dinshah!

What a boon! To have him near you for consultation whenever you need him, is in itself sufficient to justify any price paid for such service.

Here even in the price is Dinshah's additional contribution to the Service Of Mankind and to you his Co:operator.

The price of the Spectro-Chrome Metry Encyclopaedia COMPLETE in the 3 Volumes, printed on 60-Pound durable paper in 12 point type, with beautiful Cloth Binding in Red, Green and Violet, our Primary Colors, is only—$10.00—and it will reach you by Mail. Handling charges 50 Cents Extra.

All sets are numbered and registered. Any person may register with us as a student by sending us the Free Student Enrollment Blank from the First Volume, duly engrossed.

You are not purchasing books!

You are purchasing a wealth of rare knowledge not otherwise attainable. This Set will be your best investment.

SPECTRO-CHROME INSTITUTE
Central Office And Research Laboratory
MALAGA, NEW JERSEY

Dated................193

Spectro-Chrome Institute,
Malaga, New Jersey.

Dear Friends:

1. I enclose $10.00 for which enter my order for One Set of Spectro-Chrome Metry Encyclopaedia, Clothbound in 3 Volumes per Set, in Red, Green and Violet, at $10.00 per Set.

2. I also enclose 50 Cents extra for handling charges.

Name ....

Occuption ....

Street ....

City ....

State ....

Good Thoughts

FOR

JULY 1, 1934

DEEDS

Mental action is always ahead of Deeds,
Premeditated or impulsive, quick or slow;
To produce good Deeds, therefore, it needs
The keeping of the mind, clean and free
Of the cobwebs of harmful ideas;
It is not difficult, to think well
Of all creation, so the thoughts molded
Into Deeds, may materialize into welfare.

A clean mind
Makes for clean Deeds;
The mind rules,
The Deeds follow it.

SPECTRO-CHROME

September
1934
Violet

Volume
10
Number
9

Important Read and let your friends read **Articles**

NRA MEMBER U.S. WE DO OUR PART

OPEN CHALLENGE TO THE OCTOPUS AMERICAN MEDICAL ASSOCIATION AND CHICAGO BETTER BUSINESS BUREAU

SEPTEMBER - 1934

In course of life you reach a time,
When Doctor, Lawyer, Priest alike,
Does fail to give the sought relief,
To troubled mind or burdened heart—
Then come to me;
I help you to help yourself.

SPECTRO-CHROME.

Publisher and Editor
Dinshah P. Ghadiali

Single Copy 50 Cents

Open Challenge to the Octopus American Medical Association and Chicago Better Business Bureau

MEDICAL HUMBUG VERSUS SPECTRO-CHROME SCIENCE

By:— DINSHAH P. GHADIALI, Originator

It has always been my policy in public service, to ignore all criticisms of my work or self, when levelled by anybody in the open. My work and individuality have been in the "Public Eye" since 1891 and my honesty, integrity and conscientiousness have never been impeached. My life is lead in the open, I never have two stories to tell and my principle of accepting no gift, present or gratuity from any person leaves me free to say or do what I think is right. I am afraid of none and have nothing to fear.

My life as a Researcher and Pioneer of Spectro-Chrome Metry never was smooth. Since its introduction 14 years ago in the American Continent, I have been subjected to nothing but derision, annoyance, ridicule, persecution and all kinds of lawsuits of the most criminal character, brought on by the machinations of the American Medical Association and its ramified affiliations.

I never hid from my audiences any of these facts; before any person took my work, they heard right from the platform from my own mouth, a synopsis of the attitude of the Medical Trust against me, their persecutions, prosecutions and the incarcerations through which I went, because of their unjust and disgraceful tactics and false charges.

In the United States Of America, the Land Of Liberty so vaunted for political purposes, I suffered intense misery by starvation, privations and imprisonments imposed upon me since I originated Spectro-Chrome Metry, to release the Suffering World from the

clutches of this monstrous Medical Octopus, which has by grafty legislation endeavored to set up a Healing Oligarchy. They had me sentenced once on a 5 year and $5,000 Fine, to the Federal Penitentiary in Atlanta, Georgia, where I stayed 17½ months until the Honorable President Calvin Coolidge ordered my release. They arrested me in Cleveland, Ohio without charge or warrant and later rearrested me and put me in jail for "practicing medicine without a license," whereas all I did was delivering lectures! They arrested me in Rochester, New York on a charge of "Grand Larceny" and had me tried in the Supreme Court in Buffalo, New York. There, they brought experts and University Professors to prove that Spectro-Chrome was a fake and I was a crook, but, under my bombarding cross examination, those wiseheads were so routed that the Jury acquitted me and I came out victorious. All these cases I fought myself without a lawyer and showed to the machinators that Truth cannot be outmatched.

Similar prosecutions were made against me in Wilmington, Delaware and many other places. Until now, I saw the outside World as it appears from behind the bars, ten times and the war is waging as strongly as ever. In each case, evidence showed the unclean hand of some affiliation or another of the American Medical Association.

That Association, which has a tremendous political influence and maintains lobbies in every legislature, wrote a derogatory and libelous article about my work and myself in its Journal, issue of January 26, 1924, Volume 82, Number 4. I promptly issued the following challenge:

February 4, 1924,
18—05.

The Editor:
Journal of
American Medical Association,
535 North Dearborn Street,
CHICAGO, ILLINOIS.

Sir:

On returning from my tour in California, my attention was drawn to an article published in your issue of January 26, 1924, under the heading "SPECTRO-CHROME THERAPY," attacking me personally as well as defaming my work and my SPECTRO-CHROME INSTITUTE.

My scientific and professional standing are so well known and recognized internationally during thirty-three years of a fearless and independent platform career that they need no vindication or justification; however, it appears that you never had the opportunity of investigating first hand what SPECTRO-CHROME is and what it does.

As an humble servant of suffering humanity and in justice to yourself and the therapy you represent, I have the honor to invite you to appoint a committee of any seven of your representatives and send them to take my complete course on SPECTRO-CHROME THERAPY with experiments and demonstrations at Hotel Manx, San Francisco, California, where my next class opens on Monday, the 24th of March, 1924, at 7 P. M.

I hereby agree to pay all first class traveling and hotel expenses of your said Seven (7) representatives and also to present the course absolutely unconditionally.

Furthermore, although I am known as a caustic speaker in denouncing the "Empiricism of Medical Science," I hereby pledge you my word of honor that not a word tending to derogate, discredit, disrespect, disrepute or dishonor your representatives, their profession or their therapy will be uttered by me during the entire investigation and that they will be treated as my worthy and honored guests.

The above invitation stands good for two weeks from your receipt of this and your acceptance is requested within that period to enable me to remit in advance the stipulated expenses to your bankers and make other necessary arrangements.

Yours sincerely,

DINSHAH

President.

DPG/IGD

By Registered Mail,
with receipt requested.

It was never answered, but, soon after there appeared in the Dearborn Independent of Detroit, Michigan, owned and controlled by the famous Henry Ford, in its issue of March 15, 1924, a four page illustrated article, purporting to be an "exposure" of Spectro-Chrome. I promptly wired Henry Ford as follows:—

March 16, 1924

HENRY FORD
DEARBORN MICHIGAN

MY ATTENTION WAS DRAWN TO YOUR ARTICLE IN THE DEARBORN INDEPENDENT OF MARCH FIFTEENTH ON COLORED GLASS WRITTEN BY ROBERT MORGAN CALLING ME JEW OF INDIA SEMITIC IMPUTING DISHONESTY AND TRICKERY INSINUATING IMPOSTURE FRAUD AND THE LIKE I AM AN ARYAN BY BLOOD AND DEVOTED LOYAL AMERICAN CITIZEN OF INTERNATIONAL SCIENTIFIC STANDING YOUR ARTICLE IS UNTRUE MISLEADING MALICIOUSLY DEFAMATORY AND HIGHLY LIBELLOUS TO MY CHARACTER AND WORK IT IS POSITIVELY DAMAGING AND DETRIMENTAL TO MY SPECTRO-CHROME INSTITUTE WHICH IS NO MONEYMAKING SCHEME AS YOU EMPHASIZE BUT IS AN INCORPORATED ORGANIZATION AUTHORIZED FOR ITS ORIGINAL LINE OF THERAPY IN FIFTEEN PROMINENT STATES AND ENJOYING EVERYWHERE SPOTLESS REPUTATION FOR UNSELFISH SERVICE TO HUMANITY NO ROBERT MORGAN INTERVIEWED ME OR INVESTIGATED MY SYSTEM YOUR REPORT INDICATES PURPOSED MISREPRESENTATIONS AND IGNORANCE OF FACTS HOWEVER BEFORE TAKING ACTION I PRESENT YOU OPPORTUNITY FOR INDEPENDENT UNCONDITIONAL FIRST HAND INVESTIGATION MY NEXT COMPLETE COURSE ON SPECTRO-CHROME THERAPY WITH EXPERIMENTS AND DEMONSTRATIONS BEGINS MARCH TWENTYFOURTH AT HOTEL MANX SAN FRANCISCO CALIFORNIA I REQUEST YOU TO SEND YOUR DULY ACCREDITED REPRESENTATIVE AT MY EXPENSE TO MAKE IMPARTIAL INVESTIGATION AND PROVE YOUR GOOD FAITH FAIR PLAY AND SENSE OF PUBLIC JUSTICE PLEASE WIRE YOUR DECISION PROMPTLY

DINSHAH P. GHADIALI

COLONEL DINSHAH P. GHADIALI
PRESIDENT
SPECTRO-CHROME INSTITUTE
HOTEL MANX
SAN FRANCISCO CALIFORNIA

E. G. Liebold, the Secretary of Henry Ford, promised me investigation which was never made. Some shady persons posed as Henry Ford's representatives and interviewed me, without satisfaction to my outraged reputation. Finally, I went personally to Detroit, Michigan with my Experimental Laboratory and requested public investigation. A Medical Doctor took my whole course as Henry Ford's representative, free of charge, gave me a testimonial of genuineness and the matter was dropped.

Meanwhile, a Medical Doctor of that city, one of my Graduates, kept hammering at the Wayne County Medical Society and the American Medical Association, asking for a full and thorough investigation of Spectro-Chrome. When plagued in plenty by him, the Propaganda Department of the American Medical Association, through Dr. A. J. Cramp, wrote to him that, "We have not 'made an investigation' of Spectro-Chrome Therapy, but, we have gone over with some care the claims made for this freakish cult for, apparently, it has developed into a cult."

Right after this, the following Resolution was passed by the Council of the Wayne County Medical Society and another Medical Doctor, also one of my Graduates, whispered a message in my ears, while at the Hotel Tuller:

"You stop knocking the Medical Doctors or they will get you in 12 Months." They got me in less than twelve months, but, thanks to President Calvin Coolidge, they could not "keep" me.

Resolution adopted by the Council of the Wayne County Medical Society, April 7, 1924:

"It is the sense of the Council that Spectro-Chrome treatment is not scientific therapy and we consider the use of this method by physicians incompatible with membership in this Society."

"THE COUNCIL OF THE WAYNE COUNTY MEDICAL SOCIETY, following a short discussion, passed a resolution last Monday, embodying the information, couched in judicial English, to the effect that any physician susceptible to 'bunk' and who is passing it along to the public in the form of Spectro-Chrome Therapy, will be politely asked to resign his membership from the Society or stand trial. Spectro-Chrome Therapy was discussed in the columns of the Bulletin several months ago and more recently in the columns of the American Medical Journal. The therapy of Spectro-Chrome is not debatable. One either believes that if a green light shines on him that he is going to be benefited or that he is not. We are attuned in different parts of

our body to light vibration and if we get out of tune we are sick. Sounds reasonable, doesn't it?"

Thus, after spilling the beans of Solomonic Wisdom in logic, that Medical Society thought they had abolished Spectro-Chrome from Michigan! A short time later, I sent the following telegram at the Annual Convention of the American Medical Association:—

JUNE 2 1924

EXECUTIVE COUNCIL

AMERICAN MEDICAL ASSOCIATION

IN SESSION ASSEMBLED

MUNICIPAL PIER

CHICAGO ILLINOIS

SIRS IN YOUR OFFICIAL JOURNAL OF JANUARY TWENTY SIXTH NINETEEN TWENTY FOUR WAS PUBLISHED AN ARTICLE HEADED SPECTRO-CHROME THERAPY ATTACKING ME PERSONALLY DEFAMING MY WORK AND MY SPECTRO-CHROME INSTITUTE THIS ATTACK WAS WIHOUT ANY INVESTIGATION I SENT TO EDITOR FEBRUARY FOURTH COMMUNICATION REQUESTING THOROUGH INVESTIGATION BY SEVEN REPRESENTATIVES RECEIVED NO REPLY UPTO DATE WILL YOU KINDLY READ THE LETTER I SENT YOUR EDITOR AND ACCEPT MY OFFER FOR IMPARTIAL UNCONDITIONAL INVESTIGATION ENTIRELY AT MY EXPENSE FOR CAUSE OF SUFFERING HUMANITY I AM AT PORTLAND HOTEL PORTLAND OREGON WHERE PLEASE WIRE COLLECT YOUR DECISION

COLONEL DINSHAH P. GHADIALI PRESIDENT

SPECTRO-CHROME INSTITUTE

Prepaid Night Letter
Portland Hotel
Portland, Oregon.

I also challenged similarly the Wayne County Medical Society.

No answer to this was vouched, but, medical activities started in Portland, Oregon against me, which culminated in my conviction in that city of alleged violation of the Mann Act and sentence thereon to imprisonment for 5 years and $5,000.00 Fine, as aforesaid. Medical Doctors played active part in it.

By two Respites from President Calvin Coolidge, I was kept out of the prison for some time; while my appeals were in progress, I spent fabulous sums of money to ferret and expose the horrible perjury in the case and finally published 2,000 copies of two illlustrated volumes styled, "Railroading A Citizen" and sent them broadcast to those from whom I thought I could receive justice. They were received by the President and many congressmen, senators, judges, governors and other high officers.

That publication brought an avalanche of help, but, it was held that under the law, a man though innocent, must serve part of his sentence before he may be pardoned! Thus, I went in and came out after 17½ months of an unjust incarceration. What I did there, need not be said here. Let some independent U. S. Senator like Robert La Follette of Wisconsin, look into the files of the Department of Justice in Washington, D. C. and investigate the Truth. He will have some surprises from the Records.

That did not end my miseries. When the American Medical Association saw that their attempt at ridding me was of no avail and that the introduction of Spectro-Chrome was cutting deeply into the pockets of its Members, their Propaganda and Investigation Departments started new tactics of sending out private circulars against me. Inquiries made by would-be purchasers of Spectro-Chrome Equipments, were told by Dr. Herman N. Bundesen and others that Spectro-Chrome was valueless and that I was a faker or something like that. In each such letter, the fact of my incarceration was well rubbed in—but, without stating that President Calvin Coolidge ordered my release!

That did not check Spectro-Chrome. My repeated successes in the Law Courts in the so-called Medical Practice cases, piled on and there was consternation among the Medical Societies. They called to their aid their Grand Gazabo, the American Medical Association, which in turn called in the aid of the National Advertising Association, the National Advertising Clubs, the Medical License Boards, the Health Boards, Pharmaceutical Boards, the Medical Censors of the Newspapers, finally the National Better Business Bureau and its

affiliated Better Business Bureaus of the cities where I lectured. One by one, they called to their aid also Chiefs of the Police Departments and District Attorneys. Life for me became unbearable; everywhere, my footsteps were dogged by staffs of Detectives under the auspices of the Inspectors and spies of the most horrible monopoly of the World —this abominable Medical Octopus. However, I went ahead with my Duty to Suffering Mankind.

Here are the latest samples of their movements and machinations:

On July 12, 1934, in Volume V, Number 45, under the guise of public service, the Chicago Better Business Bureau published an article on its first page, first column, about "Colonel" Dinshah's "Spectro-Chrome." The boob, hired writer of that article, by placing quotation marks on the Colonel, sought to convey the impression that I merely tacked on the rank or title and never earned it, whereas, the least inquiry would have shown him that the rank of Colonel was earned by me, through hard work as the organizer during the World War, of the New York Police Reserve Air Service of which I was the Commander. He further stated that the American Medical Association calls Spectro-Chrome Metry a "dignified term for 'boloney'." The article, of course, did not forget Atlanta Georgia, but, as usual forgot the Pardon.

When I may inform the public that the Attorney General of the State of New York, openly classed these Better Business Bureaus as Detective Agencies, many might be surprised. These Better Business Bureaus act the very same way as a Secret Society would and behind the letterhead of the American Medical Association Bureau of Investigation, is distinctly printed: Many newpapers and magazines on the Bureau for information regarding the advertising of products that come within its purview. The National Better Business Bureau and affiliated local Better Business Bureaus are in close cooperation with the Bureau of Investigation."

So there you are! The tentacles of the Medical Octopus are so around me, that through the newspapers they prevent me from advertising even my Public Free Lectures. The crafty Better Business Bureaus play in the hands of the American Medical Association and print whatever that Octopus wants. In the article under question, it starts by stating about "the interminable originality of the nation's swindlers and schemers and the inevitable gullibility of the average man." It speaks of Spectro-Chrome as "hocus-pocus electrical machinery which is supposedly possessed of magical healing powers."

Someone from Wisconsin, wrote to this American Medical Association about me and Spectro-Chrome. He received with the date of August 13, 1934, the following:—

Dear Sir: In re: DINSHAH P. GHADIALI.

We certainly can not recommend the "Spectro-Chrome" treatment, as you will see by the enclosed reprint of an article wt published on it some time ago in THE JOURNAL of the American Medical Association. In this, you will find considerable mention of Dinshah P. Ghadiali, who seems to be the "father" of Spectro-Chrome. Since that article was published, Ghadiali was arrested and found guilty of violating the Mann Act and has served time in the federal penitentiary. Since he get out, he has gone back to his buncombe.

Yours very truly,

BOH:BC AMERICAN MEDICAL ASSOCIATION
Enc.—Ghadiali reprint Bureau of Investigation.
H"

The letter was only initialled. The initials of the dictator are B. O. which various newspaper advertisements proclaim as "Body Odor" and H means Hell. Hell is bad enough, but, combined with B. O., it must be awful indeed. Nothwithstanding this, the inquiring Sufferer bought an Aluminum Cabinet Spectro-Chrome!

Why does this American Medical Association track me and harass me so mercilessly and unjustifiably with its falsehoods and lies? I ask openly, when **did** this kicking donkey **investigate** my Spectro-Chrome System? Even if I were the veriest criminal on God's Earth, the biggest swindler, how would it affect my researchers, inventions, patents, manufactured apparati? Even to be a successful criminal, needs a powerful brain—under what rule of logic do these Master Dopers justify their attack on Spectro-Chrome, the fruit of over 44 years of my indefatigable scientific labors?

I openly say, without any shadow of retraction, that my originated System Spectro-Chrome, the unique System Of Healing, that

has the most perfect Automatic Precision and has NO DIAGNOSIS, NO DRUGS, NO MANIPULATION, NO SURGERY as its slogan, was never investigated by my opponents, maligners and traducers. In sending out by First Class Mail, such false and deceitful statements to inquirers, the American Medical Association is declaring itself as nothing but a coward, daggering a well-known scientific man whose versatile intellectual capacity not all their Medical Doctors can dream of matching, as admitted in testimonials in my possession.

The time is come for me to come out in the open. Till now, I played the defensive as a gentleman, taking their blows on the chin quietly, believing I was fighting a gentleman organization. By their present tactics, I know positively, they are a bunch of mean-minded braggarts. I have known the infernal and vile tactics of this American Medical Association, for nearly two decades and since I was the Vice-President of the Allied Medical Associations of America. I have full information of documentary character in my possession, about the corrupt and concealed manner in which this dastardly organization crushes its opponents. I am no coward; I have no fear; I recognize solely the powers of God and bow to Him and the August Government representing the country whose citizen I am; but, I refuse point blank to bow to the authority of this arrogant self-styled boss of all healing methods and allow him to dictate what I should do with my inventions, for the manufacture of which there is a quarter million dollar plant existent in Malaga, New Jersey. I positively refuse to be put out of my originated field of activities and shall fight these despotic tyrants to the finish, until one of us goes into the Limbo of Oblivion at the bar of public opinion.

In justice to those of your Association who are progressive at heart, but, are mislead by your false representations, I must state that my remarks and criticisms of your organization do not apply to them. I pity their condition, wherein a bunch of high-handed officials rule liberal-minded, truly sincere Medical Doctors, who without such misguidance would otherwise investigate and accept Spectro-Chrome Metry. They have to know that Spectro-Chrome actually does perform near-miracles in even the so-called "incurable disorders" and its merits are unquestionably proven and established, for them to leave behind the trail of Medicine and fall in line with Spectro-Chrome. The American Medical Association has hundreds—perhaps thousands of such conscientious gentlemen professionals, but, the organization keeps these men in the dark and hoodwinks them. I invite such men—such Medical Doctors to leave this bloated carcase and join the evergrowing virile ranks of honest-to-goodness Spectro-Chrome Normalators. Spectro-Chrome Metry is out of the bounds and clutches of the dreaded Medical Octopus and in joining hands with me, they will join in a crusade against the most devilish bigotry—a terrible blot in this civilized World.

Ye bigoted fellows of the opposing faction, I give to you—age-befogged, worn-out, decrepit, self-centered, conceited, addle-patted ignorami—this challenge, that you come out of your foxy dens and **fight me in the open on a debating platform,** at my expense, about the merits of your own humbug known as "Medical Science." Choose your own representatives, your own time, your own place and let in the public to hear that debate. Let it be done before the eyes and ears of the public, whose families you ruin by your horrible operations and whose lives you blast by unnecessary removal of vital organs and the injections of poisons and serums into their system. I challenge you to do this. I am only one man against your thousands of Members and followers, but, I am able to hold my own against **all** of you physicians, when it comes to the Truth about Medicine, which I too practised in India for a long number of years and finally left in disgust on the birth of Spectro-Chrome, the Healing System of all time to come.

You are or you call yourselves, Red-Blooded Americans. I am a Parsee Zoroastrian, a descendant of that Medo-Persian nation which ruled in Persia 1,300 years ago, the nation once ruled by Cyrus The Great. I have the fire of that conquering nation, yet radiant in my mind and body. I defy not only you but your entire American Medical Association—one and all—to meet me in an open debate with your hocus-pocus of Medical "Science" and conquer me—or be disgraced for ever and branded as braggarts and liars.

OPEN CHALLENGE TO THE AMERICAN MEDICAL ASSOCIATION AND CHICAGO BETTER BUSINESS BUREAU

1—I invite you to investigate at my expense, the merits of Spectro-Chrome Metry and expose my frauds, if any.

2—I invite you to prove to the World that any claim by me about Spectro-Chrome Metry, during the past 14 years, was false.

3—I invite you to show where, how and when you made first hand investigation of Spectro-Chrome Metry.

4—I invite you to prove that a single case report given in my monthly magazine SPECTRO-CHROME, was false.

5—I invite you to demonstrate that the Science of Spectro-Chrome Metry, as printed in my Spectro-Chrome Metry Encyclopaedia is unscientific, unproven and based on wrong conceptions.

6—I invite you to place all your cards on the table and show to the public, that what you stated about me, was instigated by your zeal for their good and not through personal grudge, because Spectro-Chrome in the Home is kicking out your employers, the Medical Doctors for ever.

7—I invite you to do all the above jointly or severally and establish your right to stay as the Emperors of Healing and in the business of pus-punching, poisoning, dosing, doping and drugging or be exposed as the real charlatans and thrown by the public into the discard, where your methods should have been thrown before this glorious country was born.

This Challenge will be open for acceptance by you at your leisure, until my Fourteenth Annual Convention which meets in the Milwaukee Auditorium, in Milwaukee, Wisconsin, at 10 A. M., on November 8, 1934 and two subsequent days. Come there and hear the case reports which will make your hairs stand on end, your eyes to pop and your ears to tingle. See for yourselves the respectable body of people who are users of Spectro-Chrome in their Homes. Judge for yourselves what the future of your rotten medical system is. Think, if you have any thinking brains left at all, of what infamous treachery you exercised towards me and feel ashamed of it. Repent, you tyrannical liars, who **never** investigated my work, but, told falsehoods by mail about it, before the powers of the Grand Architect of The Universe engulf your ramifications and dishonor and disgrace you, at the bar of public opinion. That outcome is not far.

I challenged you in the open and a copy of this was mailed by registered mail to your Chicago Headquarters. Have you the guts to debate before the public as professional gentlemen or will you continue to stab me in the dark and be branded as murderers of the innocent? Accept my challenge and prove your honesty.

Come out and fight me with your vaunted superior intelligence or pay homage to Spectro-Chrome and bury your heads like ostriches, in the sands of time.

My gauntlet is thrown. Take it and fight on the public platform, in scientific debate or acknowledge your defeat and hide, behind your prostituted screens of arrant falsehood and downright, hypocritical chicanery.

PENALTY

If you win, I am willing and ready to present to you all my inventions, patents, copyrights, trade marks, business, printing plant, processes for manufacture—everything I have, including my entire Spectro-Chrome Institute and all its belongings, give you an abject unconditional apology and forever retire, leaving this country never to return.

If you lose—as surely as you will—now come out, you sportsmen, let me know what your penalty will be? Show your Red Blood or................?

Let the better one win—let the public be the judges. I spoke—let the opposition file its brief through my own columns; it will be printed without censorship.

The above Challenge copy was also mailed to the Chicago Better Business Bureau. If they are anything but a bunch of "gold diggers," let them too come out and fight in the open—and be licked to a frazzle. They call themselves Better Business Bureau, but, I have sufficient evidence to prove to the public what they truly are—nothing but a body of hired men, living by bulldozing the poor small-men, who can not fight for their human rights. They put me in the class of "the nation's swindlers:" I throw the epithet back in their teeth and challenge them to expose my stated frauds in Spectro-Chrome Metry or be branded themselves as downright **LIARS** and **COWARDS.**

Down with the **Medical Trust** and its hirelings! Down with all autocracy!

Spectro-Chrome—In Every Home

FIRST MILWAUKEE WISCONSIN PLANET
Meeting Of September 6, 1934

By:— Otto H. Lutz, Deputy.

Spectro-Chrome in Blindness, Hay Fever, Diabetes, Goitre, Measles, Pyorrhea.

Letter received about the expulsion of D. P. of Detroit, Michigan, was read.

Chapters 46 and 47 from the Second Volume of Spectro-Chrome Metry Encyclopaedia were read.

Marie Hahn has one Spectro- Chrome loaned to a blind Girl. In 4 week's time, shows some improvement.

Otto Lutz has a case of Hay Fever. The Attuned Color Waves used were Turquoise and Magenta and after his first Tonation, was able to sleep all night. This boy has had Hay Fever for many years and could not get relief from any other methods.

A case of Diabetes; after 2 weeks of Spectro-Chrome, no more Sugar was found in his Frontelim. Green, Lemon, Yellow and Magenta were used.

W. H. Eberly reported his wife had Hay Fever for many years and this year just had a slight touch of it. She got her advice from Spectro-Chrome Institute and had good results. Green and Lemon were used.

Marie Schultz reported her boy 75 per cent. better.

Nahon Ruschewey reported a case of Goitre normalated with Orange Systemic and Indigo Local. She spoke also of her own case. She felt irritable and had Ear Trouble for many years. By using the Ultra-Green Systemic and alternating with the Infra-Green Local, she normalated both the Ear Trouble and the irritable condition.

E. A. Ernest spoke of the Third Volume of the Spectro-Chrome Metry Encyclopaedia to be reaching all in about 30 days. He also spoke of the Fourteenth Annual Convention to be held in the John Plankinton Hall, Milwaukee Auditorium in November. He had a letter from Colonel Dinshah expressing his regards and congratulations to us.

Kurt W. Bader reported two cases of Measles. One was a woman of 28 years and in 3 days this was normalated with Green, Turquoise and Magenta. Another was a boy of 4½ years and it took

6 days to normalate him, because, the mother gave the boy Cough Syrup for his Cough and Spectro-Chrome for the Measles!!!

He spoke of his own case of Pyorrhea. The Dentist told him he would have to have his Lower Teeth all pulled. The Sufferer used Indigo, but got no results. Then he irradiated Infra-Green and thought the Yellow too slow to suit him, so he put the Red Attuned Color Wave for 2 hours with the (2,000 Watt) Graduate Equipment. He drew some Gum Boils. After they drained, he tonated with Turquoise and now his Teeth are set just as tight as before he had this Pyorrhea.

We had a large attendance this month and we hope to have still more.

It were better if in giving reports, the names and addresses of the Sufferers be given to us, in confidence, for our records. You may ask for their non-publication, should you so desire, but, because of our fight against the Medical Octopus, we prefer to have proof of authenticity.

Marie Hahn should report in full about the Blind Girl. It will be interesting.

The Diabetes case deserves more particulars. You and we know what Spectro- Chrome does, but, the oft-fooled public should have details. Write a full report.

Spectro-Chrome Institute will gladly serve you by giving Free Guidance, but, all who ask for it must follow the Rules printed on the General Advice Chart.

We are glad to hear about Nahon Ruschewey being benefited. She had 85 questions to ask—we feel she received her reply to a few of them by her study of Spectro-Chrome Metry Encyclopaedia. She is a learned woman and ought to make a name for herself in Spectro-Chrome Metry work.

E. A. Ernest was right. The Third Volume will be fully completed by the end of this month. Please refrain from writing us about it. It ought to be in the hands of the subscribers by the middle of October. Write us only if it does not reach you by the time of the Fourteenth Annual Convention.

Cough Syrup may have Opium in it to stop Irritation. That makes for the Indigo. Instead, to expectorate Lemon is needed. Let

us tell you friends, once for all time: Spectro-Chrome and Drugs will not work together. We have strict Rules about this, that while a case is under a Medical Doctor's care, we refuse to give even Free Guidance. We have nothing in common with them and where we go by our Spectro-Chrome Equipment, the Medical Doctor must wave his farewell. His game and time are both up. Spectro-Chrome is his final bell. Send him out of your family and ask for Spectro-Chrome.

Full particulars about Irradiation of Boils, Abscesses, Furuncles and Carbuncles, are in Chapter 97, in the Third Volume of our S-C M. E. That Volume will strike the Medical Octopus "dumb, deaf and blind," with its telling scientific information which none can refute.

We congratulate the first Milwaukee Wisconsin Planet on its good showing and clean record.

Win Highest Health Honors During Farm Week

CHICAGO . . . A country girl, but a city boy walked away with honors as the healthiest pair during Farm Week at the World's Fair. Miss Clista Millspaugh, 17, a 4-H Club member of Mt. Pleasant Ia (above left), was the young lady. Mortimer Foxmann, 16, of Chicago, (above right), thanks athletics for the physical development which brought the honor to him.

A SERIOUS PROBLEM IN SEX

Very frequently, our assistance is requested by women afflicted with inconsiderate husbands, who are a source of unhappiness and illness to their unfortunate wives. The following cry of help from one of our Purchasers will tend to show both sides of the question—their complaint and the Remedy. Certain parts are expunged to remain within the limits of the absurd Obscenity Laws:—

"Just a few lines to let you know that my most trouble is **Sex.** My husband has an **abnormal Desire** .... and it is just about driving me Crazy. I am living in misery with him; what should I do? Can you help me? I am just about dead; can Spectro-Chrome help me?

....

....

....

Please answer this as soon as you get it so I know what to do.......

....

....

I wish to tell you that my husband will abuse himself if I fight him away.............he will use.......................... . I can't stand my husband at me all the time. ....I also have my husband sleeping alone in his room."

The above which contains much more fearsome story than placed in print, will show the appalling state to which misunderstanding of these sacred Laws of Sex has degraded its prime conception. In probing the history of the wife and the husband, we found both to be accustomed to false artificial stimulants like Coffiee, Beer and the like and the husband a fiend for Tobacco Smoking. The woman is 29 and the man 26—a mismating physically.

What can any sensible person advise in such case? The woman and the man are both slated to physical and mental destruction in the near future. Unless and until they both stop the use of Smoking, Boozing and such other pernicious artificialities, they are sure to go into an early grave if not a Lunatic Asylum.

The terrible Laws regulating so-called "Obscenity," are so absurd and out-of-date, that in their face to publish a frank article on the subject would be courting a long Lawsuit. However, America and Europe, because of their wide use of "Beef, Booze, 'Baccer," will pay for their prudery.

Sex is a Divine subject. Its ignorance is inexcusable. Such ignorance is the cause of all present misery in married as well as misery in unmarried life. Many a man, who is fortunately married to a beautiful girl, is surprised when told not to use his wife as complained above. He resents interference with his conjugal "Rights." He believes that marrige gave him the right to intercourse his wife as he pleases, forgetting that the wife too has her "rights" involved. We emphatically state that marriage never was intended to make the home a bed for debauchery. Liberty does not mean license and marriage does not give any man the right to convert his wife into a licensed prostitute.

This Sex debasement can only be curtailed or stopped by actual training in the schools and colleges. The Sex education of our youngsters should be rigorously handled from their infancy, by due initiation into the Mysteries of Sex. Sexuality and Sensuality are different. They should be taught Sex Hygiene, Sex Purity and Sex Control. They should be taught that Sex Expression should stay within the bounds of Emotional Purity as Love and should not be allowed to degrade into the exercise of sheer Lust. The preaching of so-called Morality in the Churches has no meaning. Morality should be taught in its true conception by education in Sex and its perservation and not by abstract rules.

When the filth connected with Sex is elevated into the genuine idea of Emotional Expression and not Sex Congress, the Human Race will grow into angelic tendencies and universal happiness will prevail.

Meanwhile, let the people's representatives strive to abolish the Obscenity Laws, so independent, impartial and fearless writers may enlighten Suffering Humanity with the Essence of Truth and bring about results for the Betterment Of Mankind.

Abolish the eating of Meat, Fish, Eggs, Tea, Coffee, Tobacco and Intoxicants and the prisons and insane asylums as well as hospitals will close.

FIRST DETROIT MICHIGAN PLANET

Meeting Of September 6, 1934

By:— Ethel K. Greenham, Recorder.

Gonorrhea, Tonsilitis, Gastritis, Gangrene, Constipation, Tumor Normalated by Spectro-Chrome.

A letter was read about the expulsion of D. P., a Member of our Planet, for Violation of the Organization's Rules about Collaborators. It was filed.

Dr. L. Roy Dodman spoke of several cases which he successfully tonated with the Attuned Color Waves. He spoke of a case of Gonorrhea which he successfully tonated with Lemon. He spoke of a case of Tumor in the Womb, which he normalated in 3 months. Orange and Lemon were used. He spoke of having studied the various methods of Healing and highly complimented Colonel Dinshah on his Science of Attuned Color Waves.

Martha Schreiter then spoke of a case of a very severe Pain in the Back; three or four Tonations of the Red and the Pain disappeared.

Norris E. De Camp then spoke of a friend who had a severe Pain in the Back which he produced Normalation with Indigo Local and Turquoise Systemic.

He also spoke of a sick German Police Dog. He did not know what Color to use for the Dog, so he thought he would experiment with the Scarlet Wave. The Dog had been lying around and would not take any food. After the first Scarlet, he got up and lapped water. The next day, the Dog was given Lemon Systemic and his bowl of water was placed near him so it could be irradiated with the Lemon. He reported that after 4 days of Tonation, the Dog was not only well, but, seems better than it ever was before. He wrote an article, which is being forwarded to the Central Office. He also spoke of having the opportunity of studying through the College Library, many of the therapeutical results with various Light Rays, but, since Spectro-Chrome was introduced to him, he can see where Colonel Dinshah has something of infinitely greater value than all others put together.

Otto Hinz spoke of a case of Tonsilitis and a very high Fever. Gave Blue and the Fever not only disappeared, but, the Inflammation in the Tonsil disappeared.

Thomas Ness spoke of a woman who had been under the care of a Medical Doctor for two years and had taken all kinds of Patent Medicines and Drugs for Stomach Trouble. After 3 weeks of Tonations, the Sufferer was able to take Soups and retain them, something she had been unable to do for many weeks.

He then spoke of a woman with Heart Trouble and a Foot Gangrene. She had been out of the Hospital for 14 weeks and for 9 weeks has had no medicine at all—only Spectro-Chrome. The foot was as black as coal when the Tonations were first started, but, the Foot is now showing a Color of Reddishness and a normal condition will be produced soon.

Otto Hinz then spoke of his daughter who had suffered from Constipation. Seven Tonations of Yellow were given and a more normal Backelim is now produced.

Albert Greenham spoke of Colonel Dinshah's recent case, in which the United States Government tried to take away his Citizenship. This case he won with flying colors and some of the more conservative papers such as the Philadelphia Record, Evening Bulletin, Courier-Post, Morning Post, gave Colonel Dinshah prominent parts of their paper about this case which attracted so much attention. The Morning Post of Camden, N. J. gave his case front page positions, under large type caption, "Edison Of India Upheld by U. S. As White Man," with his photograph. Judge John Boyd Avis decided the case in his favor. That concluded the case in which all the scientific minds throughout the East were centered. Probably, the most interested spectators were the American Medical Association, for it would have been a great victory for them, if Colonel Dinshah had been deprived of his citizenship. He conducted his own defense in the most able manner; many prominent practising attorneys looked on in awe at the masterful way in which he proved his right to his Citizenship and driving his enemies who had tried so hard to crush him, to cover again.

Walter Hale stated that he was in the World's Fair at Chicago, Illinois and visited the Hall of Science. He was more than glad to see the Primary Colors were given as Red, Green and Violet. This is something which the Scientific World had never mentioned, but, Colonel Dinshah had used it in his work from the very commencement. In fact, without it, Spectro-Chrome would not be of Mathematical Precision.

All the Members enjoyed the talk. Albert Greenham will be

the speaker for the October 1934 meeting. His subject will be, "What I saw at Malaga."

We are glad to note the progress your Planet is making.

We owe you an apology for an unusual error. In the last report was printed about Otto Hinz speaking of a "cure of Diabetes." We commented thereon, that we Spectro-Chrome Normalators do not use the word "cure." A protest was received by us that the Planet Report from the Recorder did not have in it that word. We looked into the original and found " case of Diabetes." Here is the explanation: As soon as a report arrives it is transcribed and given to the Editor. He comments from such transcription. The typist mistook "case" for "cure" and the Editor spoke about it in good faith.

———o———

NO GRAFT—SPECTRO-CHROME PRINCIPLE

1. We distinctly desire the Public to understand: We are not in the Practice of Medicine, Conduct No Clincs, Give No Personal Consultations and Accept No Gift, Present, Contribution, Fee, Donation, Compensation, Tip, Gratuity, Inducement or Token, of any Kind Whatever, from any Person Directly or Indirectly.

2. Our Free Guidance is given exclusively to our Registered Purchasers of Spectro- Chrome, solely by Mail.

3. Our President Colonel Dinshah and staff follow the same Rules and should not be approached for Personal Interview about Spectro- Chrome Advice. Such request will **NOT** be granted.

Tuesday, September 4, 1934.

Spectro-Chrome Institute

Malaga, New Jersey

THE MEDICAL OCTOPUS

How It Grips In Its Monstrous Tentacles, The Various Activities Of The Entire Country Through Its Cunning Affiliations:

AMERICAN MEDICAL ASSOCIATION
STATE MEDICAL SOCIETIES
COUNTY MEDICAL SOCIETIES
MEMBER MEDICAL DOCTORS
MEDICAL COLLEGES
AMERICAN COLLEGE OF SURGEONS
HOSPITAL MEDICAL BOARDS
BOARDS OF MEDICAL LICENSURE
STATE, COUNTY AND CITY HEATLH BOARDS
U. S. ARMY MEDICAL CORPS
U. S. NAVY MEDICAL CORPS
U. S. FOOD AND DRUG ADMINISTRATION
PHARMACEUTICAL COLLEGES
BOARDS OF PHARMACY
BOARDS OF DENTAL SURGERY
DENTAL COLLEGES
BOARDS OF OPTOMETRY
BOARDS OF EDUCATION
SCHOOL BOARDS
NEWSPAPER HEALTH COLUMNS
GOOD HOUSEKEEPING INSTITUTE
NATIONAL ADVERTISING ASSOCIATION
NATIONAL ADVERTISING CLUBS
NATIONAL BETTER BUSINESS BUREAU
LOCAL BETTER BUSINEES BUREAUS
MARRIAGE LICENSE BUREAUS
LIQUOR LICENSE BUREAUS
DRUG STORES, PLUMBERS, RESTAURANTS, etc., etc., etc.

TO THE AMERICAN PUBLIC

Do you value your greatest Asset? Your Liberty?

Join us in shattering this inhuman and vicious Monopoly, This Medical Trust, This Terrible Octopus.

SPECTRO-CHROME

December
1934
Scarlet

Volume
10
Number
12

Trade Marks Registered In The United States Patent Office

Important **Articles**

Read and let your friends read

OUR FOURTEENTH ANNUAL CONVENTION

Over 1,000 Attended

DINSHAH'S CHALLENGE UNACCEPTED

Medical Octopus Stands Confirmed As LIAR And COWARD

SINGLE COPY 50 CENTS

FOURTEENTH ANNUAL CONVENTION

Over One Thousand Attended—Reports Of Unparalleled Success

PRESIDENT COLONEL DINSHAH'S CHALLENGE TO AMERICAN MEDICAL ASSOCIATION AND CHICAGO BETTER BUSINESS BUREAU UNACCEPTED

Medical Octopus Stands Confirmed As Liar And Coward

The Fourteenth Annual Convention of Scientific Order Of Spectro-Chrome Metrists assembled in the John Plankinton Hall, Milwaukee Auditorium, Milwaukee, Wisconsin, on Thursday, November 8, 1934.

There were two Informal Sessions, from 10 A. M. to 12 P. M. and 2 P. M. to 5 P. M. During these Informal Sessions, visitors and strangers asked various questions and received the required information. Registration of Members and Provisional Members was made and Reserved Seats were alloted to them.

Evening Session—Thursday, November 8, 1934, 8 P. M.

At 8 P. M. sharp, the President Dinshah P. Ghadiali, took the chair and delivered his Address, in his usual emphatic and no-beating-about-the-bush manner. He said in part: "This is a very auspicious day for our work, which marks the finishing of the Second Cycle of 7 years, in the existence of Spectro-Chrome. We founded our organization, on April 26, 1920, over 14 years ago; during those 14 years, many a ton of water rolled under the bridges of the Niagara Falls and we too went through many ordeals.

"When I started the work, I knew exactly where I was headed. It is not that I was caught in a trap through any ignorance! I went into whatever trap there was, with the fullest and most complete knowledge of what was in store for me. Fourteen years, of actual, clinical practical work among thousands upon thousands of users of Spectro-Chrome, proved the value of Spectro-Chrome Metry as a science. Thousands of people studied under my personal teaching and among them I had the honor of enrolling some of the prominent Physicians, Surgeons, Osteopaths, Dental Surgeons, et cetera, of this country.

However, this year was the most memorable year of our life; it was full of many unpleasant occurrences, though replete also with memories of the most favorable and pleasant character.

"I am a firm believer in the Laws Of Karma or Retributive Justice and Reincarnation. I believe that this one life is merely a link, one of the links in a great chain of life after life, otherwise, our life would have no meaning or objective. Since I was born and had the common sense of humanity developed within me, I dedicated every available moment of my life to the Service Of Mankind. My life's best dreams came true. I am proud to say that in this very beautiful city in the United States, where I landed just by accident about 3 years ago, I actually succeeded in founding an establishment for such noble service as no human power in the World can now shake.

"This Fourteenth Annual Convention was to be opened by your very beloved Honorable Mayor Daniel W. Hoan. You well know the gentleman. It was to be opened by the learned Mayor himself, but, since that arrangement was made in the middle of this year and the announcements were published, certain incidents happened unexpectedly, which compelled me to act promptly and save the name of the Honorable Mayor from being drawn into controversy of an unpleasant character.

"I have the highest regard for Mayor Hoan, not only as the Chief Executive of such a beautiful city, but, as a man of the most sterling, conscientious and upright character. He is a man with integrity, perseverance and honesty. I did not know anything about Milwaukee, when I came here nearly 3 years ago and as usual, paid my respects to the Chief Executive of the city through a letter of introduction which I had. That year, the Mayor very kindly came to the Gold Room of the Hotel Wisconsin and greeted us on behalf of his city. He has nothing to do with my organization; he is in no way connected with my work, but, just as he greeted the American Medical Association Convention here last year, he condescended to greet us this year again as a welcome guest would be. However, because of the controversy between the American Medical Association and myself as well as my Spectro-Chrome, I thought it just to release the Mayor from any opprobrium and on my arrival here voluntarily telephoned him, absolving him from the engagement.

"Letters are being mailed privately by the American Medical Association, calling me all kinds of names and my Spectro-Chrome "buncombe." I am a fighter and a fighter for the Truth. Old or young, there is no man in the World, whether he belong to the American Medical or any other Ass-ociation for that matter, who can call

me a "quack," "schemer," "swindler" or anything else, without being challenged. I am an ex-convict of the Atlanta Penitentiary; that I do admit; I also admit I saw the inside of 10 prisons in America; all this not because I am a crook or a quack, but, because I refused to associate with the most damnable quackery as practised by the American Medical Association. I come here before you with open hands, as a Messenger of the Truth, living in God's glory, without keeping anything secret. I am condemned because I refuse to practise as a Medical Doctor in this country, according to their pernicious and false tenets.

"What happened in my life during this year? Enough to make the hair of any human being gray for life. However, we succeeded in planting Spectro-Chrome in hundreds upon hundreds of homes. The American Medical Association was not hostile to me upto the year 1920, when I gave to the World my Spectro-Chrome System. Until that time, when I was in uniform, some of the Medical Doctors were knocking at my door and I was showered honor after honor without seeking it.

"I am here tonight and three weeks more to perform my duty to humanity, without compensation or fee. The unfavorable attitude of the American Medical Association, started after Spectro-Chrome saved the life of and restored completely, the burnt skin of Grace Shirlow, a Darby Pennsylvania girl of 9. This case by all rules of Medicine and Surgery was utterly hopeless, but, Spectro-Chrome performed the so-called impossible. Without the slighest investigation, the Journal of the American Medical Association attacked my personality, knowledge and system and now, without any study or investigation, notwithstanding many open requests, its so-called Bureau Of Investigation, sends out falsely-worded letters maligning me and condemning my work. I declare without any hesitation, that in calling me a "quack and swindler," that Association is proclaiming itself the biggest swindler in the World, because it is nothing but a gigantic organization—a racket to rob the public and I do not say it—I mean it, because they never accepted my public challenge. How do I know of the graft in Medicine? Because at one time I too belonged to that gang!

"Who are the representatives of that Chicago Better Business Bureau? About three months ago, they printed an article about me, without any provocation, putting it on the first page. It was not an accident. They put me there in the Rogue's Gallery, among the "nation's swindlers and schemers." They put the Colonel in quotation marks. I **am** a Colonel by rank. I served two years during the war without pay and earned that rank in the city of New York; I received the badge for Loyalty . It is they who are cowards—the meanest bunch of liars. Why did they not appear here, through their representatives, to answer me? I sent them all the literature. I am knocking no individual Medical Doctor. I believe they are just as conscientious in their profession as the burglar is in his profession. They are sincere in their meat-chopping. They believe that by cutting off the limbs of people, they may restore them to normalcy.

"My Liberty was lost when I crossed the harbor of New York. In America, we have less liberty of speech than in England. Here, every time I lecture, they send the Police Department to arrest me. 'What have I done?' 'There is a charge of practising medicine without a license, against you.' They take me to prison, scatter my audience, disgrace me, indict me, try me. Who has been doing all this? The American Medical Association through its affiliates. It came out clearly in my trial in Cleveland, Ohio. It came out in my trial in Buffalo, New York. I licked them to a frazzle and came out on their top; I now mean to stay on their top.

"I wrote in Detroit, Michigan, several times to the Wayne County Medical Society for investigation of Spectro-Chrome. They never had the courtesy to answer, yet, how do you think my blood must have boiled, when I read in one of the issues of their magazine that they **had** investigated and found Spectro-Chrome wanting? **Did** they investigate? Never; they lied.

"I went from the Atlantic to the Pacific Coast, to California, lecturing. There, again my attention was drawn to an article about me in the newspaper of that learned historian—Henry Ford. He is in the automobile business. He referred to my beloved wife Irene Grace as "alleged" wife; no he-man could gulp such insult to his wife. He also called me a Jew of India, medicine man of China and the like! I advised him of his wrong information; Ford promised to investigate, sent an M. D. or an M. D. posed as his representative. He looked more like a big gorilla let loose. He testified and said Ford would print a retraction of 3,000 words. That retraction was never made and later Ford's Secretary stated that Ford never sent a representative! Who was that M. D.?

(At this point, Colonel Dinshah spoke vigorously against the perjury committed to have him sentenced and spoke in his forceful way about the treachery of the Medical Octopus. All this was printed and published long ago and is now history).

"My Spectro-Chrome System is now firmly established. My Spectro-Chrome Equipments are patented and marketed and are in daily successful use by the thousands. My Spectro-Chrome Home

Guide is giving families of laypeople genuine service. At Malaga, New Jersey, there is a quarter million dollar plant for research and manufacture, operating day and night. A number of trained helpers is working there loyally, so sick humanity may be faithfully served. Finally, my Spectro-Chrome Metry Encyclopaedia in 3 Volumes being already out, is initiating the World in the marvels of Spectro-Chrome. I believe sincerely that my duty to America is now finished. Loyally and faithfully did I serve this country for 23 years, fighting against odds that would break the heart of any man. I am ready to retire from this country and go to other lands, to spread the gospel of Spectro-Chrome. I refuse to go however, with a blot on my forehead. I want to go with a clear mind and a slate clean of Government records against my character. The Medical Trust and Ku Klux Klan were instrumental in framing me. The Ku Klux Klan gave me full vindication in writing subsequently, but, the Medical Octopus did not apologize. I, gave it my final challenge. It **did not** answer; it **could not** answer. It branded itself thus as a **Liar and Coward.** My challenge expired tonight and the opponents did not appear; they dare not appear before the Light Of Truth, to debate before the public on an open platform. They are averse to publicity.

"I did nothing wrong to be so hounded, persecuted and maligned. All I did was to perfect a system of healing of the most precise character and which tumbles the Medical Doctors from the throne they occupied 6,000 years. I do not have to practise Medicine; I produced in Spectro-Chrome something that will last to eternity. The American Medical Association, if they can get me somehow, will have me imprisoned for life, but, they are helpless before the power of Truth behind Spectro-Chrome.

"I want each one of you to write a Letter Of Recommendation to the Honorable President Franklin D. Roosevelt, to release me from further persecution.

"The Better Business Bureau was declared by the Attorney General of New York State to be a Detective Agency—technically an organization for graft. Let them challenge me and I have documents enough to blow them to smithereens. I did nothing against the Federal or State Laws. There is a Libel Suit hanging in the Court here, against the Milwaukee Journal, for $186,000.00 by my Distributor General E. A. Ernest. It will be a fight to the finish; I shall go on the witness stand as the Spectro-Chrome expert; let them face me with all the University Professors they can gather—I shall rout them all, the same as I did in Buffalo; two of them lost their jobs!

"Write to Roosevelt that you heard the story of injustice and request him to look into my case and release me from further molestation. I was sentenced three times and those Medical Doctors are looking forward to arresting me and having me sentenced a fourth time in New York or Michigan, so they may have the Habitual Offender or Fourth Sentence Law on me and have a sentence for Life Imprisonment imposed.

"I shall not be sorry to leave the United States. Of course, my Spectro-Chrome work here will continue without interruption, but, my presence is needed by other countries whose medical laws are not so corrupt and whose legislatures are less dishonest. This country is honeycombed by political graft and while our Constitution is the most glorious of any, its administration and judicial interpretation is nothing but a mass of misrepresentation. I therefore am not desirous of staying here permanently and wasting my life on a bunch of medical marauders. I shall travel abroad with my Spectro-Chrome banner and return here each year for my Annual Convention and class training.

"Spectro-Chrome will not stop until my aim is duly fulfilled by placing a Spectro-Chrome—In Every Home and then it will go on forever, to the discomfiture of the Medical Octopus. It will be a great success; it will not be my success; it will be **your** successs—the true success of the Benign Masters Whose Divine Message I brought into the World as Their Humble Servant."

The address of Colonel Dinshah was repeatedly applauded and the crowded audience listened to it with vociferous corroboration and openly expressed appreciation.

Adjourned at 11 P. M.

Morning Session—Friday November 9, 1934

The Morning Session opened at 10 A. M. Jessie E. Ness, of Detroit, Michigan a Member of the Council Of Governors and the First Vice-President of the Organization, was called to speak. She said in substance, "This is unexpected, but, I shall not shirk my duty. It is tragic that in this twentieth century a man has to fight for his human liberty, when he is working for the good of humanity. We talk of free speech and free press; there is very little of it tolerated. We are a handful of people seeking to lead out, but, we are bucking against large vested interests. We value principle before anything on Earth; any true man or woman should so believe and behave.

"The struggles of Colonel Dinshah for recognition of the Truth are known to me. I met him before his Oregon case. I was busy, but, I went to hear him. I was then on the Grand Jury; I know how

Juries work. Then I heard he was railroaded. Then on I watched for him until I took his work. Truth crushed shall rise. This man was brought into the World that he may save bodies from being mangled. Those butchers mangle our bodies to see the Chemistry inside, when God created man in His Image. When a dog is sick he runs out and does not eat, but, we run to Medical Doctors to pour dope down our throats. Man has limited his ways and keeps on hitting on five cylinders instead of six; 90 per cent are not hitting right; they are handicapped.

"After coming to this Convention, I am more than ever convinced of our progress. The seeds sown in Milwaukee will cover the World. I am truly and sincerely sold on Spectro-Chrome. It rings the Truth. It sells on its merits. I know Chemistry; the Elements radiate Colors. The more homes in which you place Spectro-Chrome, the quicker you will kill the Octopus. They send out all kinds of stories in Michigan; we mind them not. I do not go to any Medical Doctor; when the Red Signal is up, I go to my Spectro-Chrome. There is nothing Spectro-Chrome can not do that medicine can do. Thousands are struggling for help. Customs die hard. Spectro-Chrome is a great healing science. Those who had the privilege of studying it, are 100% for it.

"Do not allow negative suggestions. Colonel Dinshah said I may take 1 hour, but, I can speak a whole lifetime! A lady came to me with her holy water and candle faith. She was going to pass out in 2 weeks. The last rites were given to her. She had an active abdominal Cancer. She was given Spectro-Chrome; the lump all disappeared.

"A young man had an arm paralyzed 14 years. He was bandaged and was sensitive about it. The Medical Doctor had him thus. I told him to buy a Spectro-Chrome; he did not. However, he was given Spectro-Chrome Tonations and his arm responded; he is now working.

"We are still as it were in the Dark Ages, when we have to fight our way through prejudice. The Medical Doctors control us by the monster of Fear. Fear has the effect of a psychological refrigerator. Colonel Dinshah is the standard bearer of a new system which removes this Fear. We must have vision. I stand 100 per cent. for Spectro-Chrome.

"My eldest daughter ran a high temperature; the Blue Attuned Color Wave quieted it. I know the value of Spectro-Chrome from my own practice. On us all a new day dawned with this wonderful gift at our disposal." (Applause)

The President explained the principles of Construction, Destruction and Protection, the Trinity upon which the Science of Spectro-Chrome Metry is based.

Albert E. Greenham of Detroit, Michigan was the next speaker. He said in substance: "I went to Malaga, New Jersey, a few weeks ago. When we read about Spectro-Chrome Institute having its own Printing Plant, we think it would be like some small country shop; but, what I saw there gave me a surprise. It is a large plant with several presses and is really a modern, completely equipped plant. The Cabinet making plant is also well equipped. The simplicity at the Institute is remarkable. No curtains or rugs. Every part is clean as it could be. No fly screens and few flies. The Colonel's own quarters I saw; the room is simplicity itself; nothing fancy; fits into his scheme of things. I am impressed with his consistency.

"I am a Graduate of the 59th Class. A lady, Mrs. Herrick was so enthused that she tried to introduce it in the Receiving Hospital. Colonel Dinshah offered free equipments, if they would have a Spectro-Chrome Normalator in charge. They would not have it; they want only M. D.s

"Take another instance. In the Farmingdale Hospital are crippled children. A Physio-Therapist Helen Kilbourne took the Spectro-Chrome Course and wanted to introduce it in the Hospital, for hopeless cases. There was the case of Gloria, a girl utterly helpless; nobody could help her. They declined to let Miss Kilbourne use Spectro-Chrome. She resigned and opened her own establishment. In 6 weeks of Tonation, Gloria could use her limbs and reach out for articles. They had taken a muscle of the right leg to put in her left leg or something like that! Poor Gloria! She would be still helpless, were it not for Spectro-Chrome.

"This convinced me of the merits of Spectro-Chrome. I can recite case after case. Some are difficult venereal cases. Wherein they inject Mercury and Salvarsan, we use better methods. There is an affidavit of a man Rocco. He suffered plenty from Medical Doctors for Gonorrhoea. Spectro-Chrome was finally used and the cleansing Waves of Green and Lemon did what the Medical Doctors could not do.

"I had success in Rheumatism, Sciatica, Lumbago. I had a case of Uterine Abscess. It had broken and they said she would die if not operated. So they performed the opration and in three weeks, it came again. Spectro-Chrome in 4 Tonations broke it and it drained for two months. She is now feeling better. That was the time when I long distance telephoned the Central Office and received the Free Guidance, at night. If Spectro-Chrome can not do it, it can't be done.

You get your maximum results with Spectro-Chrome. Here is the point, however. You must follow the Rational Food Of Man System. No pork chops will do, I am positive. All can be done if the individuals comply. If they live right, it gives miraculous results. One must eat right. The Colors emanating from the Food Chemistry count. The cow eats the same food, sick or healthy, old or young; I saw it worked out.

"I had the case of Matthews in Detroit. He had no Equipment. He was normalated by the Food alone; it shows what you can do."

Gretchen Wittmeyer of Chicago, Illinois spoke about how her pulmonary complications were normalated by Spectro-Chrome. She testified to the Truth about the case of L. G., who was suffering from Tuberculosis and for 7 months laid in bed. She said after Spectro-Chrome was used, she saw her walking in 4 days and in a week she came into the class room in Mansfield, Ohio. Yet for merely making mention of this case, the Delaware State Medical Society issued a Warrant against Colonel Dinshah in Wilmington, Delaware and had him tried and sentenced to suffer imprisonment for 2 months! His appeal was denied and even the United States Supreme Court refused a Writ Of Certiorari. He paid a fine of $250 and costs and avoided the prison sentence. Such is American Justice!

John Maxwell, of Chicago, Illinois was then introduced as the next speaker. He said in substance, "I feel like a baby and crying. We lead life step by step. I traveled in younger days, by a House Car, through England, Scotland and Wales, through Devonshire where the cream comes. I know the countryside. I have been married 44 years to the same woman who walked in life with me side by side. Ten years ago, I rebelled against the American Medical Association. My father loved the Truth; he spent 15 years in India. He had a warm spot in his heart for India! He was a Nature lover. I went into the woods with him and gathered herbs; that was better than the Vaccines they use now. We feel like babies considering the Forces of Nature. We have to thank Colonel Dinshah for giving us an understanding of the Forces of Life. When I was a boy I always stood for Principle. I stood in line with those men who stood for the Truth. How they refused Vaccination for their children! They saw their household goods sold off! They took the stand for the Truth. Government gathered crowds; we made converts to Medical Liberty. The pioneers of everything suffered, but not in vain. The darkest hours are before dawn.

"The Medical Hierarchy instilled wrong ideas and set up their autocratic rule through the Congress to have control over you before you are born. This is the work that will smash that Medical Autocracy. I long for more knowledge, to use Spectro-Chrome; I believe in it. We have the Truth here that will illuminate the World. We must aim at the education of the multitude. Henry Ford is sincere at heart, though he wrote in the Dearborn Independent. He desires to do much for the welfare of mankind. Orphan boys who lost their parents were given on his farm the benefit of good air and good food. I gave a lot of thought to the Food; everything should be in its pristine purity. Flesh, flesh and fowl were never intended for food.

"Things are moving now. Radio hook-up is covering the country. The folly of Vaccination and Surgery is being realized. The country will be happier without them. I thank you."

Recess was granted till 2 P. M.

Afternoon Session—Friday, November 9, 1934

At 2 P. M., the Convention was again called to order. Emma D. Snow of Maywood, Illinois was the first speaker. She spoke in substance, "I am granted 10 minutes to speak, but, I must tell you how I became interested in Spectro-Chrome. I was always with Medical Doctors. I came with a Medical Doctor from Boston. I left his service for certain reasons and started Spectro-Chrome on my own account. They said I was crazy. I was offered $500 not to go into Spectro-Chrome, but, I did and am now taking people he could not help; I put them on their feet.

"I had a case of a boy of 6. He had severe Heart trouble; specialist could do nothing. He never had a normal bowel movement. With Spectro-Chrome he recovered and gained 9 pounds in 2 months. Now he goes around. He turned on a hot water tap and received a severe Burn. They rushed the child to me; inside of 20 minutes, he was sound asleep. In 10 days, the Burn was all healed; medically, it would have taken 11 weeks.

"Tomorrow, I shall have the extreme pleasure of bringing in a woman from Wheaton, Illinois; she had a pulse of 40; people thought she would die. In 20 minutes, after the husband brought her to me, he thought his wife passed away! She was breathing so well and the next day, slept all night. She was operated four times for Tumors. She is now in A-1 condition, her Tumor is reduced at least ⅔. Green and Magenta were used; then Yellow, Lemon and Indigo to shrink."

Dinshah explained the differences in different types of Burns and how the same Sun can produce either a "Heat Burn" or a "Cold Burn."

Clemens M. P. Neumann was the next person to address. He came all the way from St. Louis, Missouri to attend the Convention. He said in substance, "I received the Third Volume of the Spectro-Chrome Encyclopaedia and was so anxious to study it that I went through it in 4 hours! (Loud Laughter) I went like lightning. When I return, I shall go slowly. I had Hay Fever in September 1933. I used all imaginable things. German Doctor gave two pills; I took 7! (Laughter) He said, 'You come to the hospital—fix you in 2 hours! I said, 'Bull'——. I took then to Spectro-Chrome. Systemic Equipment was brought. I investigated Art Glass. I lay for 1 hour under Green. I did not feel anything. I paid $7 for it too! But at 11:30 that night, I had a cold sweat; I had no breath. I swore. That man said, 'No doctor.' My eyes throat and nose all cleared.

"Spectro-Chrome was used in Labor by us. Easiest birth; not more than two Labor Pains were felt. The child was born without Medicine. I help a man who does that work. I was interested and I slept, ate and dreamt Spectro-Chrome. Friends say I am crazy! (Laughter) My boy flopped a toast and was burnt. He knew. Eight minutes of Blue was irradiated. Big blister came next day; opened, nothing was left. We had company; wife had boiling hot water sprinkled on her side. I jumped up and in 7 minutes with the Blue she was asleep.

"I am not a Medical Doctor, but, everyday I see the newspapers. I shall say what I think fit. I am convinced Spectro-Chrome is a true Science, honest-to-goodness."

Nahon Ruschewey was then introduced as a woman born and studied Science in Germany. She was a deep student of philosophy also, as well as a Numerologist. She stated in substance, "Like Emma D. Snow, I saw the abuse of Drugs. It is not logical to use them; it is not according to the Laws of Nature. I knew a Medical Doctor who never gave Drugs or anything that could possibly harm. Colonel Dinshah is right in being a sceptic. He was not satisfied and everything had to be proved. He based Spectro-Chrome on mathematics. Do not stuff with Materia Medica but use what they call the 'Chicken Coop!' (Laughter) Use the Spectro-Chrome Home Guide. The more I see of Spectro-Chrome, the more I am awe-stricken. That a human brain can work out a system so perfect with the workings of Nature, is marvelous. I do not believe it is a cure-all; it can not remove mechanical defects, but, many of its results in Acute cases are like the Miracles Of Christ. People do not wait in Chronic cases; they expect quick results. They do not keep their confidence to remove Resistance.

"A woman had Arthritis. I had marvelous results. Woman friend said, 'Don't be silly'. In the audience there is a woman, who had Tuberculosis. She wanted the old system; now she uses nothing but Spectro-Chrome. I confess to Colonel Dinshah, speaking about the temptation to going back. I stopped allowing the use of Dandelion and other herbs, since he explained about Taraxacum as a Drug. I shall be careful in the future.

"Several people had Goitre. Had marvelous results also in Multiple Sclerosis. We realize that if other systems can not do anything, in Spectro-Chrome is the hope.

"Chemicals have Color Emanations; the Body automatically takes what is necessary for Normalation. I urge all of you, by all means to take the Full Course on Spectro-Chrome while the Originator is here."

Jessie E. Ness gave the instance of the Brindle Bulldog, normalated by Spectro-Chrome. It produced much mirth.

Dinshah spoke on the Magenta Attuned Color Wave and explained the uses of Purple, Magenta and Scarlet in Female Disorders.

Laura Gross was then called to give her views. She said in substance, "The Spectro-Chrome Home Guide and Spectro-Chrome Metry Encyclopaedia fulfilled the desires of the students. I get about 10 calls a day for Sinus, Kidney or other trouble. Mrs. M. passed Stones by Spectro-Chrome, without operation. There was excitement by it."

Recess was granted at 5 P. M.

Evening Session—Friday, Novmber 9, 1934

At 8 P. M., the Convention reassembled. Dinshah introduced E. A. Ernest as the United States Exclusive Distributor General for Equipments of Spectro-Chrome Institute. Being well-known to the Purchasers, he was greeted with vociferous applause. He said in substance: "It is a great honor to be with you here. Those among you who used Spectro-Chrome know it is here to stay, because it is based on Truth. If I should call for testimonials from users, the platform would be filled. I lived 30 years here. I dare not start distributing a product without merit. At one time, I employed 300 men. I feel sure of land before I jump. Colonel Dinshah did marvelous service in Milwaukee, far greater than in any other city. The people of Milwaukee are opened-minded and liberal. We have wonderful Police protection. It has just the ground to sow the seed; it is the hub of Spectro-Chrome.

"We are not quite ready to have him go; we can not spare him. I made very thorough investigation of Colonel Dinshah and checked all his statements. On the fourth night of his demonstrations, I was convinced that Spectro-Chrome was the greatest discovery in healing. All my thoughts were in one direction. 'How can we help this man who is helping humanity?' I went to Malaga and we spent many hours together. He showed me hundreds of Testimonials; I classed them as fakes; they pay a dime a piece to sign them! But, I found those Testimonials genuine and not the other type. I collected definite information. At the snap of fingers, he would show Testimonials and proofs.

"Spectro-Chrome is so scientific, the Sun, Moon and Stars have to change to change it. I checked step by step 30 years of research work. Sunlight has great power. 30 days without any Sun, would destroy everything; not a blade of grass would live.

"One of the reasons I was thoroughly convinced was my own mother's life saved by Spectro-Chrome. In 1932, she was 82 years old. Prominent physicians said she would die in 30 days. Her pulse was 58 times in 1 minute and very intermittent. In 2 weeks of Spectro-Chrome, her heart was normal. She is nearly 85 now and says she feels younger than she was in 1892! She goes in the kitchen with the maid. If I could have her come here, she would say the same. She says daily, 'I hug that Spectro-Chrome'. She uses and appreciates it. From the first time she used it, she cleaned out the medicine chest. The Drug Stores are as good as killed. Spectro-Chrome performed miracles.

"Colonel Dinshah is a very generous man. I went to him to ask for the agency for Wisconsin State; he wished on me the whole United States! He did it so as to let me go to jail for him and in order to benefit humanity, I **will** go! (Loud Applause)

"We are building first a foundation here that can not crumble. We shall never stop till the goal is accomplished. If there ever was a servant of the Masters, this Colonel Dinshah is. In 18 months he produced the Spectro-Chrome Encyclopaedia. You should see him work. He takes 3 or 4 hours' sleep. Takes a 2 ton truck and drives from here to New Jersey 888 miles without sleep or stop. John Maxwell and I do some things like that. We should live on natural foods and Spectro-Chrome. Physically, I am all right, but, I use only one Color a week—Green. I have so much energy, I do not know what to do. I am working from 5:30 in the morning; have just one object in mind: Spectro-Chrome—In Every Home. It is a great service; it is used by the most prominent citizens of this city; I shall not name them. Though sceptic, it is so simple, the little Home Guide tells you all. Free Guidance is given by the Institute. I shall be 150, barring accidents! (Applause) Maxwell is also a servant of the Masters. When I live to be 150, he will live to be 175! (Applause) Join in our Sunday hikes and take a jaunt of 10 miles and come back.

"I never would stop talking about Spectro-Chrome; there is no more scientific way. When Colonel Dinshah spent time researching for his Slides to balance system perfectly, he found he had to go one step further, by publishing his Encyclopaedia. If it is not in your library, it is a great mistake.

"Anyone here who did not take the Course on Spectro-Chrome Metry, should take it now. He proves everything before your very eyes. You will get the thrill of your life. He gives it free of charge. The work is not a hit-and-miss system. Spectro-Chrome can not miss as the Sun and the Moon do not miss. My honest belief is that the name of Colonel Dinshah will go in history as one of the greater scientists. If they give him fair play, he gets the Nobel Prize. He needs our co:operation. Every red-blooded man and woman should give him co:operation; to my dying day, he will have mine.

"What a crime our Laws interfere with a man after he labored 30 years to provide a perfect healing system! Fight for him as never before. Every person in this room should fight with him to save suffering humanity. Why suffer? It was never intended that our organs—two, three of them—should be taken away. Sickness comes because of the only reason that our system is out of chemical balance and there are more poisons than it can carry. Refrain from colored ketchups; come with us on hikes. Spectro-Chrome is the last word in healing. Colonel Dinshah told you, 'When in doubt use the Green'. Use it 3 times a week; on Area 4 use Magenta.

"Spectro-Chrome uses no Heat. It can not harm or burn. My mother slept 3 hours with it on; she has red cheeks without rouge. I can talk till tomorrow morning and not finish.

"It is a great privilege to study with Colonel Dinshah. I checked his Diplomas; his walls are plastered with Degrees. Look at this; I received it from my friend. This photo is dated May 26, 1901. Here is on it 15 years experience at that time! How old was he when he started? I had it framed and hung where I get one look at it every day. This man went through Hell rather than change his principles; he went to jails. They could not find one fake thing in Spectro-Chrome. Show me any man who gave such a Challenge and put it in print. They can't touch him. Every time they torture him, I see him becoming stronger. Every time they jail him he becomes stronger. He was shot at with a rifle. It makes our blood boil.

There is a hole in a glass in his Laboratory. Shout Spectro-Chrome from housetops.

"We can not live for ever, but, it is absurd why we can not live a 100 years. You can enjoy life if you live right. I never felt better; I never worked better.

"Spectro-Chrome will wipe out the opponents. I know some conscientious professional men; their hands are tied. 'You must follow the American Medical Association orders or you are blackballed.' I know some professionals are using Spectro-Chrome for their family, but, they can not use it in public. Here is the man who deserves credit.

"We must not let him go to India! You heard his request last night, about petitioning Franklin D. Roosevelt. He was given a Conditional Commutation by President Coolidge; they are holding it over his head, because, they want him out of the way. Write your reasons to the President and send Colonel Dinshah the copy. Write to our Senator Robert La Follette. Thank you." (Loud Applause)

The address of E. A. Ernest, backed by his enthusiasm was listened with much interest by the audience. Dinshah thanked him for his kind words and called on John Maxwell for a short address.

John Maxwell of Chicago, Illinois, said in substance, " Mr. President and Fellows, it is call to colors, literally and specifically. We have to look at Faith, Hope and Charity. I hope tomorrow there will be such an avalanche of letters mailed that we shall have cause to rejoice. We have Faith in the Originator of the Colors. It is a Red Letter Day when we are called to Colors. We do not use the Infra-Red; the opponents are using the Ultra-Violet. The Almighty never intended we should drag on our existence. We should have long life for useful service. I have come to believe in Spectro-Chrome, wholeheartedly. My wife went to a Flower Show on a cold day and when she came, had hands white with cold. She had Laryngitis; much of voice was gone. She stripped and lay for 1 hour under the Indigo. Three hours later, the voice returned; the Inflammation had subsided. We use Spectro-Chrome every day. It is used in Chicago every day.

"I learnt much in this Convention. Take my advice and take Colonel Dinshah's Spectro-Chrome Course; be in the Class. With chemical equipment he gives the foundation. The time may come when we may have to go as ambassadors for this work. What the Medical Hierarchy did, speeded our gospel. Their doctrine is Fear. What I saw among the animals in the stockyards of Chicago, they do to us. My clothes stank with the nauseating odors. Pigs and bulls and sheep are slaughtered there. The Rational Food Of Man is logical. The Colonel has not eaten meat for 44 years; I did not eat it for 32 years. Those animals realize their death is near; the fear has their blood stream poisoned.

"People eating meat can not think straight. They go on the operating table and their minds and bodies are never right again. You have the great hope offered, harnessed and brought to you all. Take Colonel Dinshah's Course. His three volumes give in one life food for thought. I continue to read and reread them; thy have a mine of information. Let us go ahead with Spectro-Chrome. Go into the next class and God bless you."

E. A. Ernest spoke a few more concluding words. He said in substance: "Sound and Color are like Electricity and Light. Their use is daily increasing. Folks, get busy and join the future which is mapped. In 1923, at 5.30 in the morning, I called a physician, because I was sick. From what he said, I understood, 'Off goes my Appendix!' They waited for no dress, but, put me in a robe and went 60 miles an hour to the hospital. The operating room was beautiful; so were the nurses. The gas was ready, but, I changed my mind. Rest of the day flowers came; I distributed them in the ward. They refused to give me anything to eat. I walked out, rode on horseback for 4 hours. The Appendix is still there. (Applause)

"53 per cent. of medical diagnosis is wrong. They take out an organ to explore. You do not have to do anything like that in Spectro-Chrome."

Jessie E. Ness desired to say a few words. " Food is the first consideration. The greatest shock of my life was when I saw pigs murdered and slaughtered. I made up my mind to become a vegetarian. Dinshah teaches us about our food and mission in life, what our purpose is. It lifts humanity. People gad about applying rouge and not pouring thoughts about the Great Masters. You have a life opportunity; seize it."

Thomas Ness, representative for Michigan State of Ernest Distributing Company, then gave a short address. It was thus in substance: "You have been listening to speeches; I have only one thing to say. Get out and sell. Just those four words. There are here 1,200 potential salesmen. A Scotch divine was preaching. He was delivering an oration and was using a bookboard for pounding. He stamped his foot; there was an anthill underneath. The ants did not like it; they were annoyed. Finally they reached the pantleg. The minister shouted, 'God is in my mouth, but, the Devil is in my pants.' No salesman is a success until he gets excited similarly about his sales. There never was an invention like Spectro-Chrome, for health and sanity.

"A man of 73 years had a gangrenous foot. It was ordered to be amputated. A Chiropractor could do nothing. There was a scab 1/8 inch thick. He asked the Doctor. What could he do? Melt the scab? I told him to leave the Green on for 48 hours. He asked, 'Will it burn? I said, 'No, you wait and see'. Soon from many perforations the pus began to pour. It was a matter of life or death; in 5 days, the leg was less than the ordinary size. Now seven months he has been walking without crutches. My wife would not sell her equipment for $1,000, if she could not get another. So friends, get the Devil loose for Spectro-Chrome!" (Applause)

The Convention was adjourned at 11 P. M.

Morning Session—Saturday, November 10, 1934

The first item on the program was the Election of the Board Of Trustees. Jessie E. Ness moved and Albert E. Greenham seconded that the present Board Of Trustees be re-elected. On vote being unanimous, the same Board Of Trustees was declared elected.

Jessie E. Ness felt sorry for Colonel Dinshah, "because the Octopus is endeavoring to impede his progress. It is a great political movement checking progress in healing. Boobs and nincompoops are in office.

"I had a case of a handsome lad with paralyzed arm. He recovered under Spectro-Chrome."

The Convention adjourned at 12 M.

Afternoon Session—Saturday, November 10, 1934

Reverend David Bucks of Washington Park English Lutheran Church, of Milwaukee Wisconsin began the afternoon session. He stated in substance: "I visited the Colonel in Malaga. My nephew who studied Chemistry was with me. We stayed in the Research Laboratory about 2½ hours. When we went home, my nephew said, 'I learnt more of Chemistry in this short time than I learnt in school in 12 months.'

"I realize the value of Spectro-Chrome. It is something worthwhile. It made a great impression on me. I pay tribute to a great work. Pardon being personal, but, from some friends having learnt of Spectro-Chrome, being 60 miles from home I visited Malaga. We were enthused. We walked into the North East corner, we did not know how we should be received. But, we were well received when we walked in; we did not know where we were getting into! (Laughter) He was seated in a big chair and as soon as we felt free we felt at home. He escorted us from the tower room to the basement.

We were in intimate surroundings; most congenial and brilliant in courtesy and kindness.

"We went into his Printing Plant. It was a big place. There was the Linotype Department; he gave us slugs cast there.

"There are black sheep in every great organization. The Colonel does not like black sheep. I like Spectro-Chrome. It requires some sacrifices. Colonel Dinshah makes it plain; you must put something to it. You might chisel, but, you will be doing harm to yourself. Every true disciple is for the good of mankind. As a Radio is to entertain Spectro-Chrome is to health and comfort. It is a marvelous invention. I am scratching my head now as to how I can get the apparatus and take the Course.'

Colonel Dinshah spoke about the ministers of the gospel. Then he said, " Reverend David Bucks spoke about the man and his work; he should have spoken about the woman of the house also." (Laughter)

Reverend Bucks rose again among cheers and said in substance: "Colonel Dinshah's wife was contented and happy. He took me upstairs to their personal rooms. I saw the children playing and studying. There is credit to the mother." (Applause)

Nahon Ruschewey spoke about the case of Elizabeth Benishek. The mother Anna Benishek spoke a little and introduced Elizabeth to the audience. Later, she handed to the President a short written account of her Spectro-Chrome experience, which was interesting from many points. The girl Elizabeth was declared to be a feeble-minded girl, but, when she appeared on the platform, after Spectro-Chrome Tonations, she was vivacious and indicative of change. Here is what the mother wrote:

"I also have a little boy of 2½ years old and he is as smart as can be, some folks claim he is too smart. About six weeks ago, he contracted Scarlet Fever. How do I know? Well, all the symptoms were there: High Fever, Sore Throat, Cramps in the Stomach, Throwing up and Strawberry Tongue; my mother always told me how different diseases start.

"The nurse just happened to be at my house when my boy was seized with Cramps. We put him under Spectro-Chrome and inside of ½ hour he was all right, but, late that night his Fever rose to 100 Degrees and my husband was visiting his mother, so I was all alone and a scared woman I was.

"So again I took Spectro-Chrome in his room and used it two times that night; by morning his Fever had subsided and would you believe that 3 days later was outside playing. I had no Doctor or sign hung on my door and all through my beloved Spectro-Chrome. I know I would have to pay 3 to 5 dollars for a night call and also about $1.00 for medical poison. So you see again Spectro-Chrome was a money as well as life saver.

"Now the funny part of it is, "Sunny" (we call him that) every time he has a little gas in stomach and gets a bellyache, he says, 'Mama, put me under the ight—he can't even say Light! Now if he knows it is good for him at 2½ years, how can intelligent people laugh at Spectro-Chrome?

"I also did not mention that at the time of Sunny's birth, I was quite nervous and run-down that I could not handle both my daughter and the baby. As my husband was not employed steady, I could not hire somebody to help me, but, I struggled along about 6 months. Then, I tried to get her in a school and can you imagine the Government wanted me to send her to Chippewa Falls, Wisconsin and I would have to sign my full right of her away!

"But I could not do this to my own flesh and blood; later, I heard of St. Colleta's School in Jefferson, Wisconsin and for a small fee she would be taught something. She stayed there 2 years and now I brought her home on August 25 to use Spectro-Chrome on her and she has improved so much in so short a time, I would not or rather could not be without Spectro-Chrome. She talks a little and does small errands too. If my daughter wouldn't be feeble-minded, I would go into this whole heartedly, as I always wanted to be a nurse. Since I was 17 years old when I married, I couldn't plan any further; maybe now for the best anyhow. But when Elizabeth is better, I will help put Spectro-Chrome over the top."

Nahon Ruschewey, who had watched this case carefully, stated Elizabeth was improved 20 per cent.

Albert E. Greenham made special mention of the Favorscope and the High and Low Critical Periods. He spoke of a case which the Medical Doctors had stated to be Cardiac Asthma, existing for 12 years, improved much in 4 Tonations of Spectro-Chrome.

A large number of very pertinent and important questions was asked from among the interested audience, to which relevant and fitting answers were given by the President. The catechism with its characteristic accompanying humor, produced impressions of a decidedly favorable character.

Before closing the Afternoon Session of this day, Colonel Dinshah asked for an action from the audience, in regard to his Challenge to the American Medical Association and the Chicago Better

Business Bureau. The said Challenge was not accepted and none appeared for and on behalf of the two opponents. The time limit expired at the Fourteenth Annual Convention. On motion duly seconded, it was unanimously resolved that the American Medical Association and the Chicago Better Business Bureau having failed to appear in response to the Challenge, they branded themselves as Liars and Cowards. The President was instructed to send a copy of the Resolution to the parties concerned.

The Afternoon Session adjourned at 5 P. M.

Final Session—Saturday, November 10, 1934

The evening of Saturday, at 8 P. M., the Convention Hall was well filled. Dinshah delivered a lively lecture on "India to England and America" with about 200 Stereopticon Slides, showing the progress of his eventful life, from his birth to the birth of Spectro-Chrome. For once the students of Spectro-Chrome got a real idea of the enormous labors, hardships, privations and starvations undergone by the Originator in the research and final establishment of the Spectro-Chrome System. All throughout, the audience listened with attention and demonstrated its appreciation in various manners.

At 11 P. M., Colonel Dinshah announced the program for the Spectro-Chrome Metry Free Class (89), opening in the Gold Room, Hotel Wisconsin, Milwaukee, Wisconsin and declared the Fourteenth Annual Convention dissolved.

It was the most successful and picturesque of all prior Conventions. At each Session, Dinshah played Hindustani tunes on the Auditorium Pipe Organ and brought out of it expressions, strains and tones never before heard therein. Many expressed unfeigned surprise at Dinshah's fingering of the keyboard; some thought it was a player piano, because of the incongruity of the conception of a combination of Dinshah and Pipe Organ! Those strange tunes were incomprehensible to the audience, although the American meaning was projected on the screen each time.

Prominent citizens visited the Sessions and the entire Convention was a brilliant success.

MEMBERSHIP NOTICE

With December 31, 1934, expires the Membership and Provisional Membership of **all** Affiliates of the Scientific Order Of Spectro-Chrome Metrists.

This is the regular Notice under our Constitution And Rules. No further or individual notice will be sent.

Those who are full Members, please send to the Principal Office at Malaga, New Jersey, the Annual Fees and Dues of $6.00, by or before the year's end.

Those who are Provisional Members, please follow the same routine and send $3.00, by or before the year's end.

This Fee includes one subscription to Spectro-Chrome, which contains the Favorscope. No **separate** Favorscope will be sent to anyone.

Those who are neither Members nor Provisional Members (Sympathizers) may purchase the SPECTRO-CHROME, (Periodical) by sending the regular Subscription Price of $3.00 per year, which includes postage anywhere in the World.

Remittances should be made payable to Scientific Order Of Spectro-Chrome Metrists and NOT to any individual or to Spectro-Chrome Institute. Members not sending their Fees and Dues in accordance with the Constitution and Rules, will be marked off the Register as resigned and their rights and privileges will promptly and totally stop.

There are no exceptions to our Constitution and Rules and no correspondence or inquiry is necessary, as Rule 56, printed in October-November 1934 issue of Spectro-Chrome has the full text.

Members and Provisional Members, who joined on or after October 1, 1934, may have their Affiliation Cards renewed Free Of Charge, by writing to the Principal Office.

Manure And Cream

Albert E. Greenham sent the following:—

A man was driving a truck load of manure past a Lunatic Asylum. An inmate saw it and shouted, "What are you going to do with it?'

"Put it on my strawberries," said the driver.

"Ah, you crazy! I put cream," said the lunatic," and they call me a nut!"

A STUDENT EXPELLED FOR VIOLATION

When the 89th Class of Spectro-Chrome Metry was in progress, Colonel Dinshah found on his desk an invitation to join a dinner to be held at the so-called Vegetarian Restaurant of M. K. in Milwaukee, Wisconsin.

On this invitation was printed a Turkey! Inside were the words "Yellow Split Peas A LA DINSHAH" and "DINSHAH COUNTRY CAKE" along with other foods.

The appearance of a Turkey on a vegetarian dinner invitation and the unwarranted use of the name **Dinshah** for such commercial purpose, were both so incongruous and in violation of the Registered Trade Marks of Spectro-Chrome Institute, that the woman student M. K. who made such infringement, was promptly expelled from the Class and her connection with Spectro-Chrome activities was severed for life.

———o———

WARNING AGAINST USING OUR REGISTERED TRADE MARKS

In order to avoid repetition of the foregoing incident, we reprint this WARNING, about which we emphatically spoke and wrote frequently in prior years.

The words **SPECTRO-CHROME** and **DINSHAH** are the Trade Marks of our business, to the use of which solely SPECTRO-CHROME INSTITUTE as a corporate body is entitled.

The Patent Office of the United States Of America granted Registration Certificates as follows:—

Certificate Number	Trade Mark	Date Of Issue
163761	**DINSHAH** *(Signature)*	January 23, 1923
163762		January 23, 1923
169554	**SPECTRO-CHROME** *(Equipment)*	June 19, 1923
201631	**ITISOMETER**	August 4, 1925
202137	**SPECTRO-CHROME** *(Periodical)*	August 18, 1925
213895	**TANGER**	June 8, 1926
214947	**AURICULATOR**	July 3, 1926
218906	**ROOMOTHERM**	October 5, 1926

These Registered Trade Marks must **NOT** be used by any person or firm, whether Student or Graduate of Spectro-Chrome Institute or otherwise, under penalty of the Law.

Any Student or Graduate of Spectro-Chrome Institute, having passed the examination, received the Diploma and being auhorized by the Scientific Order Of Spectro-Chrome Metrists as a Member to practise Spectro-Chrome Metry, may use the words "Spectro-Chrome Normalator" behind one's name. No other use of the words is warranted or authorized and anyone using such words in any other connection will be considered a violator and promptly expelled.

Before placing any signs or advertisements in connection with Spectro-Chrome Metry, to meet the public's eye or ear, under Rule 93 of the Constitution And Rules of the Scientific Order Of Spectro-Chrome Metrists, it is absolutely necessary to receive the approval of such signs or advertisements, from the Board Of Trustees.

One student A. La C., a Chiropractor, was reported in Milwaukee, Wisconsin to have started what he styled and advertised as:

Spectro-Chrome Health Service

This being brought to the notice of the Originator, demand was made to remove the words **SPECTRO-CHROME.** Having promised to remove all such infringement from his advertising, no further action was taken.

Please save unpleasantness by **NOT** encroaching on property not created by and belonging to you. The legal consequences of such infringements are never agreeable to the violator.

Spectro-Chrome must be kept 100% honest and integral. If you can not do so, resign and go out of touch with us permanently. Either you work **with** us or we shall work **without** you. Dishonesty and Spectro-Chrome must stay divorced.

ERNEST DISTRIBUTING COMPANY AND SPECTRO-CHROME INSTITUTE

So much confusion is created in the minds of some of our friends about the relations of Ernest Distributing Company with Spectro-Chrome Institute, that this short notice is deemed necessary.

Ernest Distributing Company is under Agreement with Spectro-Chrome Institute as Exclusive United States Distributor General for Equipments of Spectro-Chrome Institute. Outside of such **Sales rights and privileges,** Ernest Distributing Company has no other right or privilege.

The Agreements of Ernest Distributing Company with its appointed Representatives and Agents, have nothing to do with Spectro-Chrome Institute.

Ernest Distributing Company does **NOT** represent Spectro-Chrome Institute, has **NO** power or authority to contract debts or obligations or make any promises to Purchasers of Equipments, for and on behalf of Spectro-Chrome Institute and its own internal management of its affairs as Distributor General have nothing to do with Spectro-Chrome Institute.

Ernest Distributing Company has no power or authority, right or privilege of changing or altering any Equipment made by Spectro-Chrome Institute and it has POSITIVELY no right or authority to exchange, replace or in any manner transpose the genuine Dinshah Slides sealed in the Slide-Carriers of the Equipments.

Ernest Distributing Company **does not control** the policies of Spectro-Chrome Institute and has nothing whatsoever to do with the studentship or Examination of Purchasers of Spectro-Chrome Metry Encyclopaedia.

All replacements of Slides or Slide-Carriers are made solely by Spectro-Chrome Institute and all complaints about fracture, crackage, breakage or replacement of Slides should be addressed to Spectro-Chrome Institute.

Spectro-Chrome Institute is the manufacturer and Ernest Distributing Company the Exclusive United States Distributor of the Equipments trade-marked SPECTRO-CHROME and Ernest Distributing Company has **NO** right or privilege of selling any other healing apparatus, except SPECTRO-CHROME.

Ernest Distributing Company has no right, privilege or authority from Spectro-Chrome Institute to do any healing work, send out nurses to homes, give Advice for the use of said SPECTRO-CHROMES or in any manner connect its sales business with what is known as NORMALATION. All Advice and Free Guidance is given exclusively direct to the Purchaser, on standard forms, by Spectro-Chrome Institute.

Under the terms of the Agreement of Spectro-Chrome Institute, Ernest Distributing Company, its Representatives and Agents are expressly debarred from giving any Normalating Service or Tonation to any Purchaser, except as authorized by the Constitution And Rules Of The Scientific Order Of Spectro-Chrome Metrists.

The NO GRAFT policies under which Spectro-Chrome Institute operates, apply with equal emphasis to Ernest Distributing Company, its Officers, Representatives and Agents, in their relation to Prospects and Purchasers.

Any violation of the above Rules And Regulations and any deviation from the above stated terms of the Agreement, should be promptly reported to Spectro-Chrome Institute.

Spectro-Chrome Institute **must** maintain its laboriously earned reputation for 100 per cent. integrity, honesty, service to Suffering Mankind and square dealing with all. Address ALL communications to Spectro-Chrome Institute and NOT to any special name, Officer or employee.

———o———

NOTICE TO PURCHASERS ABOUT GENERAL ADVICE CHARTS

As our business is increasing, the number of General Advice Charts is naturally increasing also.

Till now, when incomplete Charts came, we remailed them for correction, which entailed expense and unnecessary correspondence.

We find however, that we can not render efficient service by so catering to the careless or negligent.

Therefore, on and after January 1, 1935, incomplete General Advice Charts will be paid no attention and will be filed without reply.

Furthermore, in order to stop the abuse of our Free Guidance, no attenion will be paid or reply given to any Chart which is sent by or for any person not a member of the family of the Registered Purchaser. Many persons using other systems of healing or Slides other than the genuine Dinshah Attuned Color Wave Slides, in the past secured our General Advice Charts and asked Free Guidance through unscrupulous Purchaser friends.

To put a dead stop to such dishonest practices, we shall hereafter strictly adhere to our understanding, as clearly enunciated on Page 88 of Spectro-Chrome Home Guide and absolutely no attention will be given or reply tendered to unprivileged General Advice Charts.

No exceptions exist in Spectro-Chrome. Therefore, ask for no exceptions or favors, as both our eyes see equally and in justice to all we must adhere rigidly to our Rules and Regulations, to avoid any charge of partiality.

All Charts must be completely filled in Type or Ink and signed by the Registered Purchaser, in Ink, in own handwriting. Pencil written Charts will not be answered.

———o———

SLIDES FOR REPLACEMENT

The Rules under which Slides, whether fractured, cracked or broken are replaced are printed in Spectro-Chrome Home Guide, Page 21.

Remittance to cover the needed replacement MUST accompany the transmission of the Slides Set, even though the Slides may have been damaged the first day of use. In case you feel refund is due to you, make such request to Spectro-Chrome Institute. Your alleged claim will be given due consideration.

Slides sent unpaid, collect or without Replacement Remittance, will stay unexamined at the Purchaser's risk.

We shall always be **Fair to you,** but, we must be **Fair to ourselves** also.

———o———

READ THIS ISSUE THOROUGHLY. IT CONTAINS IMPORTANT MESSAGES TO ALL SPECTRO-CHROME AFFILIATES, MEMBERS AND PURCHASERS.

ANOTHER ATTEMPT TO MUZZLE SPECTRO-CHROME

Call To Colors—We Must Act To Kill The Medical Octopus

That nefarious "Tugwell Food And Drug Act" followed by the equally disgraceful "Copeland Food And Drug Act", which Dinshah so ably exposed before the U. S. Senate Committee On Commerce last Spring, is again, due to be brought before the U. S. Congress, next month, under the auspices of of U. S. Senator Royal Copeland, M. D.

Notwithstanding the vigorous protests against the open humbug of including "Devices" within the meaning of "Drugs", the Bill will be endeavored to be passed in the same fraudulent form.

Dr. Louis E. Schmidt, a Chicago Illinois physician was accusing the leaders of the American Medical Association of "employing dilatory tactics on medical matters", etc.

Dr. Morris Fishbein, Editor of The Journal of the American Medical Association answered: "If Dr. Schmidt had read the Journal of the A. M. A., he would have discovered the significant fact that the President chose as men to advise him on medicine, **only members of the A. M. A."**

May be! But is a "Device" a "Drug"? If the A. M. A. chooses to bamboozle the President, as it hoodwinks the others, it is upto us Spectro-Chrome advocates to act quickly and open the eyes of the President and the Congress.

Do you realize what this is? Seeing their death knell rung by Spectro-Chrome, the Medical Doctors are crying for mercy to save their face, by including "Device" within the meaning of "Drug", so they may have some way **to stop the advertising and sale** of the Spectro-Chrome, by Federal Force. If this Tugwell-Copeland Bill passes as is, the Medical Doctors will control ALL Devices also and another fight for life will loom for Spectro-Chrome.

We must act—act at once. Let each reader of Spectro-Chrome write a strong and openly worded letter to his U. S. Senator, to his Congressional Representative and to the President of the United States, all at Washington, D. C., telling them what their opinion of such autocratic measure is. Tell them why you want that Bill amended by removing "Devices" from "Drugs". Tell them how Spectro-

Chrome must not be so handed over to its heartless opponents—the Medical Octopus. Tell them you want your Constitutional Liberty to choose your healing methods. Tell them to veto that Bill and show what public opinion means.

Write immediately and without delay and send us a copy of what you wrote to each. Rally now behind Dinshah and his Spectro-Chrome.

TRUTH must win. The false gods of medicine must be exposed and put out of business. ACT AND LET THEM KNOW YOU MEAN WHAT YOU THINK. Do not be afraid. We Americans must recognize our rights and compel the Congress and the U. S. President to recognize them. Write and continue writing; make your relatives and friends write. Write in short but meaning terms and assert your rights and kill the Tugwell-Copeland Bill.

For full particulars of this monstrosity, read March 1934 issue of Spectro-Chrome.

Write and mail us the copy.

———o———

DEMOCRACY OR DICTATORSHIP?

Fred Perkins, a battery maker of York, Pa., could not pay 40 cents an hour as wages and his men willingly stayed at 20 cents per hour, to prevent bankruptcy by the shop closing. He was adjudged guilty and fined $1,500! This was for "violation of the code".

Codes made by pressure of the Federal Government power are autocratic and unconstitutional; who will bring back our beloved democracy and remove the autocratic dictatorship? Is this the U. S. of America or Italy, Germany, Russia plus Japan combined? Where are we all headed? You fool Americans, if you will NOT stand for your RIGHTS, the Constitution will soon go overboard and another George Washington will need be born. Get Government **out** of business. We want more business in Government and less Government in business. What do officials know about running business? They are driving all to the wall by Wall Street methods.

Open your eyes before the Government becomes our master instead of our appointed servant. Page Abraham Lincoln!!!

TESTIMONIAL OF 89th CLASS

Dedicated By Dinshah To
American Medical Association
And
Chicago Better Business Bureau

The 89th Class in Spectro-Chrome Metry, opened at the Hotel Wisconsin, Milwaukee, Wisconsin, on Monday, November 12, 1934, at 7 P. M.

The Registration was so large that the Gold Room was full with the students, every one being a Registered Purchaser of a Spectro-Chrome Cabinet and a Registered Student of the Spectro-Chrome Metry Encyclopaedia.

It was the most enthusiastic Class of any. There was a large number of prominent citizens among the Students and the entire front row was occupied by influential and well known people who were invited and joined as Special Investigators. At the end of the 15 day Course, all signed the following Testimonial, unsolicited and voluntarily:—

Milwaukee, Wisconsin
November 28, 1934

To Our Esteemed Teacher,
Colonel Dinshah P. Ghadiali.

Dear Colonel:

We, the undersigned students of your 89th Class in the study of Spectro-Chrome Metry, desire to express our appreciation of the most extraordinary course of instruction which you gave.

We assure you it has opened a new channel, not only in the method of healing, but also in the mode of living physically and spiritually.

We feel that your untiring labors, in the field of research, gave you that wealth of knowledge such as is possessed by few other teachers, thus placing you in the field of instruction as none other could attain. Delving into the unseen, as well as that which is visible, has given you the power of understanding such as has brought to mankind Spectro-Chrome Metry, the only positive method of healing. This we know and can appreciate because of the actual scientific demonstrations performed by you before our very eyes in the class-room, thus, once and for all removing every last doubt as to any uncertainty in your method of healing, if we had any.

We pledge our allegiance, support and earnest endeavor to bring to bear the day when Spectro-Chrome Metry will lead the world, by placing a Spectro-Chrome in every home.

Sincerely and devotedly,

Your Students.

(Signatures Of All The Students)

WHAT SPECTRO-CHROME DID IN HOME

WHITE HOUSE MILK COMPANY, INC
Manitowoc, Wisconsin

September 7, 1934

ERNEST DISTRIBUTING COMPANY,
3106 North Hackett Avenue,
Milwaukee, Wisconsin.

Dear Mr. Ernest:-

We regret exceedingly our inability to advise you before this that our Spectro-Chrome Equipment has given first class service since we purchased it. The same is in almost constant use and we are getting satisfactory results.

Our inability to write you was caused by a rather unsatisfactory month of August as I came down with a Septic throat from which I recovered after a week but a few days later the youngster and Mrs. Braun also received their share of it. We are happy to state that with the constant use of Spectro-Chrome we find ourselves making progress in the direction for which we are headed.

The most outstanding result we have obtained from our machine was when Mrs. Braun received a severe gash on the knee, caused by our dog knocking her to the cement sidewalk. Several hours after this incident the knee was very much discolored and considerably swollen. Mrs. Braun overlooked the matter until the next day when it was quite evident that a fluid was beginning to settle in the knee and there were severe pains. That same evening, being unable to stand the pain any longer, Mrs. Braun used Spectro-Chrome Green on the knee for one hour which relieved the pain somewhat and she was able to sleep uninterrupted that night. The third day the knee again gave considerable pain and two more tonations, using Green, were taken. On the fourth day the swelling opened up, emitting a pussy fluid whereupon the knee showed indications of healing. I understand that in most cases of this kind, water is likely to gather in the knee when the knee is badly injured, necessitating keeping the leg quiet and the Doctor would no doubt order an X-ray taken. Mrs. Braun was on her feet throughout this injury and full recovery was obtained by using three tonations of Spectro-Chrome Green. Undoubtedly you will wish to inform the Institute of this incident.

Trusting you will accept our apology for not advising you of our experience sooner, we hope to remain with kindest regards,

Yours sincerely,

H. A. BRAUN

HAB/1

SPECTRO-CHROME

March 1935 Yellow — Volume 11 Number 3

Trade Marks Registered In The United States Patent Office

Important Read and let your friends read **Articles**

FEATURED IN THIS ISSUE

PITUITARY, OVARY, PANCREAS, PLACENTA FOR SALE

TERRORS IN MEDICINE

A DOG'S LIFE

UNAPPRECIATIVE PEOPLE

U. S. CONSTITUTION SCRAPPED

SINGLE COPY 50 CENTS

Reprinted By Spectro-Chrome Institute

MEDICAL BLACK MAGIC

Pituitary, Ovary, Pancreas, Placenta For Sale!

In an Associated Press article published on Saturday, March 2, 1935, in the Evening Bulletin of Philadelphia Pennsylvania, under the heading of "Gland Market Active," appear actual statistics of the extent to which the so-called Medical Science degraded itself, in its insane notions of what is known as Internal Secretion or Endocrine Therapy.

The advocates of this unoccult, filthy, abominable method, go into ecstasies by the idea that a weak human gland may be energized into functioning in normalcy by introducing extracts of similar glands from animals. We have before us a complete list of such Gland Extracts, ranging from Thyroid, Parathyroid, Liver and Spleen Extracts, to powder of Brain, Ovary, Testicle and even Placenta. The height of absurdity and donkeyism is reached in the horror of one special product of a well-known concern which contains both Ovary and Testicle in combination! We might stretch our imagination to make-believe that a barren woman might be energized into pregnancy by the Ovarian product or a sterile man might be rendered fruitful by the Orchitic Essence, but, rack as we could our Brains, we have not as yet been able to comprehend what kind of a human American animal needs both or what Hermaphrodite should imbibe the proffered potency to become masculate!

The said report shows that Maurice van Gelder, a dealer in ductless glands, went to New York, N. Y. "for his annual chat with the American meat packers who are the source of supply." From Maurice it was culled that "The gland market is doing right well." The Pituitary sold at $10 a pound and Parathyroids cost upto $30 a pound. He enthused that his firm sold 2,400 pounds of Pituitaries, 40,000 pounds of Ovaries and as for the Pancreas to make Insulin for befooled Diabetics, the amount in the last year was a cool 600,000 pounds. At the end, the article mentions that "Neither physician nor scientist, van Gelder, is just a plain trader who deals in the stuff that was packing house offal only a few years ago."

It is "Offal" still and will always remain so. It is only through the ghastly chicanery and pursuasive coaxing of the Medical Octopus, that the hoodwinked, suffering public gulps such terrible stuff and pays and pays till it dies a miserable death.

The above is not the whole garbage. A well known pharmaceutical house sends emissaries to the various hospitals for maternity

cases in Detroit, Michigan and collects and pays for human Placentas or After-Births to manufacture a product for women to take for the purification of their internals!

Such is Medicine! The barbarous savagery of the cannibals can not be worse. The imbecility of the Medical Pseudo-Science is unmentionably contemptible and the followers of such a system should bury their heads in sand through shame to belong to such a cult of Black Magic of ultra-heathenish character.

SPECTRO-CHROME INSTITUTE ENGINE

There are four strokes in an internal combustion engine:

S uction—
C ompression—
I gnition—
E xhaust—

The initials of the first three, (the Trinity) represent Spectro-Chrome Institute, which in the fourth stroke will Exhaust the Medical Octopus! Each stroke means Seven years; 14 years already elapsed; the Ignition stroke with its explosion is now on; seven years later? Just wait and watch the developments.

FIRST MILWAUKEE WISCONSIN PLANET
Meeting Of February 4, 1935

Successful Use Of Spectro-Chrome In High Blood Pressure, Tumor, Anemia, Stomach Disorder, Scarlet Fever, Bright's Disease, Stiff Neck, Bronchitis, Diabetes.

User says, "I should sooner lose my automobile than give up my Spectro-Chrome!"

Reported By:—OTTO H. LUTZ, Deputy.

The Meeting was to be in the Colonial Room, but, that room being taken, the February meeting was held in the Gold Room. There were over 100 Affiliates present. It was a very pleasant evening.

Otto H. Lutz reported case of Lawrence Q. who had High Blood Pressure; he was normalated in a total of 8 Tonations, 4 Green and 4 Purple.

Walter W. has mother with Tumor. Got good results with Yellow and Indigo with Green and Lemon Systemic.

Eugene G. has brother with Loss Of Control of his legs; they tighten. Green and Lemon Systemic were used. He asked if the Infra-Division was correct. He was advised it was correct.

Walter B h.as girl of 9 years, with Anemia. Six months ago, she had her Appendix removed. Since December 1934 they have been following the Free Guidance; 1 week Green, 3 weeks Lemon and in the Orange now. She feels well, can receive her lessons very much better and eats better. His wife has a sluggish condition; the results are slow.

H. L. is using his Equipment for High Blood Pressure. It was 250 when he started. He had 2 weeks of Green and is on the Lemon now; Purple on Areas 4 and 18.

Clara F. took about 100 Tonations with Green, Lemon and some Magenta. Her case is Psoriasis of long standing. It is just starting to show signs of becoming moist.

Margaret K. uses Attuned Color Waves for 4 weeks for Chronic Constipation. Green and Lemon are being used.

Alfred J.'s wife irradiated with Orange for building Blood. After two days of Tonation, she felt much better.

John E. has case of Stomach Disorder of 10 years standing. There is some improvement. Green, Lemon and now Yellow are used.

Frank S. has girl who had Scarlet Fever. Had quick results

Mary S. reported case of Stomach Disorder normalated and in doing so, also normalated her Piles. Green, Lemon and Magenta were used.

V. P. and his wife are using their Equipment 2 months now and both are having the same Colors, following the Advice Charts. His wife has Bright's Disease and had choking spells. She was advised by Medical Doctors to go South, but, nothing helped. He said they have so many complications, he could talk all night, if he wanted to tell all about them; but, after 2 weeks of Spectro-Chrome, they saw some results. 2 weeks Green and 4 weeks Lemon were irradiated.

Ruth W. started with Turquoise for what she terms "Inflammation." The Guidance told her to use Lemon. She had a stiff neck and shoulder one morning and took Turquoise; that remedied it. Her girl had Cold. 1 Tonation of Lemon was given and at night it was all gone.

Hazel N. had Bronchial Disorder, used Green, Lemon and Magenta. Now takes Yellow for Gall Stones. She says she goes to sleep during Tonation; she just can not stay awake.

Rose S. had Rheumatism in her right hip and now it is in both legs. Used Green, Lemon and Turquoise. For Stomach Disorder she had Yellow, Orange and now Green. She says she took about 102 Tonations, but, she does not seem to have results. She was asked if she followed the Rational Food Of Man System; she said she **does not.**

Dorothea G. had a sick mother with pains in the Area 18. Indigo, Blue and Turquoise were used according to the Guidance and gave relief.

Genevieve K.'s husband got a steel chip in his Eye. Used Turquoise without results; so he had the steel chip removed by a Medical Doctor who just did the mechanical work. It took 12 attempts to bring it out. K. was in great agony. When he came home, he irradiated Turquoise and after 40 minutes, the pain was all gone. Next day, he called the M. D. and asked him how long he would have pain in the Eye. The reply was "about two days". Then he told him the pain was all gone; he had used Spectro-Chrome.

William E. loaned his Equipment to a friend for a Cold. After 3 days, it was normalated. Turquoise was irradiated.

Albert H. has a Mental case; he can not do much, as the woman says "the light burns her." They can only put on the Attuned Color

Wave for a short time. The Color used is Green. They are trying to convince her, but, it takes a lot of talking.

Frank L. tonated a child that fell down the stairs. She had a great big lump on her Head. With 4 Tonations of Turquoise, the Lump left as if by magic. The sisters of this girl had Measles and Turquoise normalated that condition also.

Elizabeth Z. reported her own case of Diabetes, using Green and Lemon Systemic with very good results. She is Sugar-free.

William O. has his Equipment 2 weeks and is waiting for the Guidance. His brother has an odd condition. No Medical Doctor seems to know what it is. He went to the Insurance Doctor thinking he would give some advice but NO. He only got more insurance to pay, as he passed the "test". His mother has Hardening Of Arteries and started using the Green.

William B. had pain in Area 19; it left with a few Tonations of Blue. He is well pleased. A friend bought an Equipment and is very much pleased with it. He said if he had to choose between his Equipment and his automobile, he would give up the car! This is the first year, he did not have to call in a Medical Doctor. The man had a run-down condition and his boy had the Croup. Spectro-Chrome remedied these in a few Tonations.

E. K. has mother using Green Systemic on 4 and 18, but, she always feels weak after using Magenta.

E. A. Ernest expressed great pleasure in seeing such a large crowd. We outgrew the room 157 we had and some day we may have to meet in the Auditorium. He went to Malaga, N. J. and surprised Colonel Dinshah with a visit; saw him at his 14 feet desk, piled with work. Everyone comes to Colonel Dinshah for advice as to what to do there. He placed 101 Equipment orders just lately here in Milwaukee, Wisconsin; his desk looks as bad as Colonel Dinshah's. He spoke about the Rational Food Of Man System, leaving the dead and refined food alone.

A suggestion was made to Affiliates to be on time. The meeting starts at 8 P. M. Late comers always disturb the meeting.

The Colonial Room will hold upto 150; when larger attendance is in order, we shall gladly arrange for larger space. Meanwhile,

please be on the watch for larger accommodation and send information about location and rates per meeting.

High Blood Pressure is not a Disorder but a condition resulting from existing Disorders. Purple on Area 4 is indicated.

With the Irradiation of Spectro-Chrome, exercise is always in order. Lying in bed without exercise, does not make health.

Persistent Disorders are slow to leave. In any case where you feel the necessity of Guidance, there should be no hesitation in asking for it from Spectro-Chrome Institute.

Persistent Skin Disorders need continued Tonation over a period depending upon the conditions involved. Most such cases have no patience and expect quick results, even though the Guidance is not duly followed.

In Persistent Constipation irradiate with Yellow and Orange. However, the food is the main support.

It appears from reports received that Scarlet Fever is ravaging the young in your city. The Health Authorities are thinking of their usual stuff—Serums and Vaccines. If they would but lay aside their fear of the wrath of the American Medical Association and use Spectro-Chrome exclusively, the results would open their eyes to the merit and efficacy of Spectro-Chrome Metry. We wonder why the Executive Authorities of such a progressive city as yours are still silent about it. Are the lives of the public to be sold to the age-old, worn-out dogmas of Medicine? The Originator is willing to present to the City Of Milwaukee, three, thirteen or as many as needed brand new Spectro-Chromes, if they will accept his humble contribution to the Service Of Suffering Mankind and use it FREE to normalate ailments, under the guidance of qualified Spectro-Chrome Normalators. He is willing to train a special class of Health Officers, absolutely free and without any charge, provided they use the said Equipments according to Spectro-Chrome Science. Once used, they will prove so successful, that the hospitals will partly close, expenses will be reduced and the sick public will bless them. We strongly recommend the Commissioner Of Health to accept the offer and test the Spectro-Chromes as he pleases. We have nothing but public good at heart and if he makes the experiment, his name will go to posterity for broad-mindedness and moral courage.

If V. P. will persistently use Spectro-Chrome, he will reach the goal. All those "complications" will soon disappear.

No matter whether Ruth W. calls it "Inflammation" or otherwise, by following the understanding of the Infra and Ultra Attuned Color Waves, favorable results are bound to ensue.

Sleep is a great normalator. If Hazel N. will stay asleep after the Tonation, she will be benefited quicker. After a Tonation with any Infra-Green Color, do not expose to Moonlight; after a Tonation with any Ultra-Green Color, do not expose to Sunlight. The restriction applies to all cases, for about 1 hour 28 minutes after such Tonation is begun, with due regard to the Favorscope Time.

Rose S. will **never** be right unless she follows our Guidance about the Rational Food Of Man. Food builds the Body and if she will continue to feed it with "junk", she is wasting electricity and time.

Genevieve K. has an intelligent husband. He should not have used the Spectro-Chrome **prior** to the removal of the steel chip. We never decried Constructive Surgery or Mechanical Orthobiosis. In **all** such cases of accident, use the common sense of removing the mechanical offender and using Spectro-Chrome. Refer to Spectro-Chrome Home Guide, page 80.

By all means, irradiate your friends and loan your Equipments to them, but be sure of your Slides and see that they follow the Rules of the Science.

Never "try to convince" anybody, much less a woman. Refer to Spectro-Chrome Metry Encyclopaedia, Volume 2, Page 412, paragraph 4 to the end of Chapter 45, about the best manner of tackling a woman. Albert H. should study the method.

Elizabeth Z. studied the full course. She ought to comprehend the humbug of Sugar in Diabetes. Refer to Chapter 92 of the Spectro-Chrome Metry Encyclopaedia, Volume 3.

If William O. did not receive his Guidance, his papers must be faulty. Do not pother with the Medical Doctors, but take and follow our Guidance.

E. K. should send mother's General Advice Chart for Guidance. Something else is needed after the Green Systemic.

E. A. Ernest always takes pleasure in springing "surprises". Let us make and let him sell and the suffering public will advertise for us.

TERRORS IN MEDICINE

From Minneapolis, Minnesota comes a report that Lucretia Wilder, a Laboratory Technician of 23, was holding a rabbit for injection of some Germs, with which Dr. Ruth E. Boynton, hospital staff member stated the woman became infected when the rabbit jumped and the needle scratched her arm. Some of the injection was spilled on the wound and the woman died. Further report comes from the same source that "because of the rarity of the Germ, no cure is known for the disease and the hospital doctors were helpless to save the life of the woman."

In what lies the final salvation of mankind? In Germs and Poisons or the Finer Forces Of Nature used as in Spectro-Chrome? What guarantee can the public have that such germs after entering the Human Body do not produce toxic results? The entire pseudo-Science of Immunology is based on wrong premises and it is only a question of time when the structure built on it will crumble and Medicine be reduced to a worthless signal of past ignorance.

Spectro-Chrome would have saved Lucretia, but, the numbskulls will not investigate its value and will stick like flies to a gum-pot to their absurd theories.

They dare not investigate Spectro-Chrome! They **know** it is true and they well realize the end of their line approaching.

———o———

A DOG'S LIFE

Worms Expelled By Spectro-Chrome.

BY:—EDITH O. SKINNER

Just a little story of a Dog's Life!

My husband purchased a very small valuable (says he) Bird-Dog, while I was attending the 89th Class in Milwaukee, Wisconsin.

This dog was very thin, but, my husband insisted that the dog had some marvelous markings and traits.

The first week of January, my husband seemed very discouraged over his dog's condition; he thought she had Worms and also a bad Cold and felt he should take her to a good Veterinary Surgeon. I said nothing, until the following Sunday when my dear husband came to me almost in tears, telling me his dog had one Convulsion after another. She was full of Worms and had the dreaded "Distemper." I asked him to bring the dog home which he did. I looked at the dog and it was a pitiful sight. The eyes were swelled shut and the lungs sounded as if each breath were the last, whizzing so hard that one could hear her in the next room, dragging her hind legs when we forced her to move. I looked at my husband and he looked almost as pitiful as the dog. All he could say was, "she is sure to die."

Well, this is what happened. The dog would die any way, so we garbed ourselves in suitable clothes and proceeded with giving the dog a bath. It was a cold day, but, to stay in the house she had to have a bath. We then started to give the dog Spectro-Chrome Tonations—Green, Turquoise, Orange and Lemon. The third day, she was feeling much better and the following Sunday went on a two-hour hunt, for the purpose of exercise. After the one Tonation with Orange, the Round Worms were expelled in bunches. Then, my husband turned to me and said that as bad as he needed money, he would not take any amount for the Spectro-Chrome we now have, if we could not replace it with another. Could one get results like that with anything else than a Spectro-Chrome?

Perhaps, you will not appreciate that I used it on a dog and write you about it, but, to me it was a wonderful experience. Nobody could possibly detect that this dog was ever sick and as I understand it, the Distemper usually takes the life of the afflicted animal or leaves it useless, according to Grover's Dog-Book.

We are glad to publish this Dog Report. Spectro-Chrome was used successfully in the past on various animals; we have cases reported of dogs, cats, horses, rabbits, canaries and even—Husbands, set right by Spectro-Chrome.

UNAPPRECIATIVE PEOPLE!

Several Demand Rebate Because Of Price Reduction!!!

What Should Be Done.

Owing to volume production, because of greater demand from a public served by Spectro-Chrome and expansion of the manufacturing facilities of Spectro-Chrome Institute, from the First day of March 1935, the price of $150.00 for the Aluminum Spectro-Chrome was reduced to $90.00.

To our surprise, we received instead of commendation, a number of demands for rebate, from purchasers in former months! They could not see why they should have then paid $150.00, whereas the present purchaser pays $60.00 less.

In order to satisfy those friends and to conform with our Integrity Policies, we shall be glad to **increase** the price of the Aluminum Spectro-Chrome to $210.00 each and demand **excess payment** from **all prior purchasers.**

What is good the one way, ought in business to be good the other way. The price will be raised as soon as we receive sufficient applicants clamoring for fair play and justice!!!

Start—we shall finish!

Rebate for reduction to serve better the suffering public! Can you beat it?

FIRST DETROIT MICHIGAN PLANET

Meeting Of February 7, 1935

REPORT BY:—JESSIE E. NESS

First Vice-President, Scientific Order Of Spectro-Chrome Metrists

On opening the Planet, Jessie E. Ness stated that owing to the resignation of Ethel K. Greenham from the position of Deputy, she was requested by the Board Of Trustees to act pro tem, until a new Deputy be appointed. Colonel Dinshah, the President, was expected to visit the Planet in the near future. This welcome news was received with enthusiasm by the Planet members.

A correction was made in the former minutes that a certain Disorder normalated by Stella Harter in 4 hours, should be really marked as 2 hours.

L. Roy Dodman, who had to leave early, asked permission to speak. He reported that due to a fire at his place of business and in the usual investigation which followed, he was called before Major John Roehl, Board Of Health Officer and questioned regarding a Spectro-Chrome Equipment he had. Major Roehl informed him it was against the Law to use Spectro-Chrome, as he had investigated Spectro-Chrome and found there was no healing virtue in it. Dodman advised those who functioned as Normalators to go carefully for their own sakes. Jessie E. Ness said that none should be tonated until they had first signed a Collaborator Application Blank. She pointed out that as the only safety.

Albert E. Greenham then recounted his experiences with the gallant Major and stated it was his firm belief the Major never really investigated Spectro-Chrome.

Jessie E. Ness referred to the case of Frances R., a former Normalator, in support of Greenham's statement.

Edward Lemke reported his attendance at a meeting of Medical Doctors and others, at which he asked why the American Medical Association never accepted Colonel Dinshah's challenge. He received no reply to the question. Evidently, the said Association fears the result of an open discussion on the subject of Spectro-Chrome.

Otto Hinz quoted the case of his daughter who had a variety of symptoms such as Fever, Headaches, Sluggishness, all of which disappeared following the use of Turquoise, Blue and Magenta. The Sufferer was now restored to normal health.

Stella Harter reported having dealt with an extremely hard case, made so by the fact that the Sufferer had Stomach Disorder, Headaches, Low Blood Pressure, Difficulty In Hearing, et cetera, without limit. Orange and Scarlet were used and the Sufferer now looked like a new woman.

Magdalene Arnold told of a small crippled boy, whose father came to her for aid. After tonating, the father ordered a Spectro-Chrome Equipment. At first, the child was afraid. He was previously taken to a hospital for examination and the medical methods had caused him to be afraid. Now, he became accustomed to the beautiful Attuned Color Waves he sees in the Equipment, has lost his original fear and is making satisfactory progress.

Aquilla Mathers reported the case of a woman with Varicose Veins and Lumps formed behind the Knees. Scarlet on Area 4 and Indigo Local were irradiated. The swellings were much reduced. For Rheumatism, she is now being tonated with Lemon Systemic and her condition is gradually improving.

J. J. W., a convert from Medicine to Spectro-Chrome related his own personal experiences and that of his boy. The boy had suffered almost from birth with a Rupture. The case was normalated in 11 Tonations. He then purchased an Equipment together with the Spectro-Chrome Metry Encyclopaedia. He had been a victim of Partial Paralysis in the lower limbs for 30 years. Tonating steadily since last October, he is now able to bend his foot at the ankle and to

move all his toes. The Medical Doctors gave him no hope of recovering the use of his limbs. Now, he looks forward to complete restoration to normalcy, the sole agent for such being Spectro-Chrome, since he ceased swallowing "dope" the moment he started tonating with Spectro-Chrome. Albert E. Greenham suggested W. to write to President Franklin D. Roosevelt, advising him to try Spectro-Chrome.

Walter A. Hale spoke of a boy who suffered from Water On The Knee. The family Doctor had monkeyed with it for 2½ years without relief. Lemon Local 1 week and the Knee was better.

Harriet Reisdorph reported having experienced wonderful success in the use of the Attuned Color Waves of Dinshah. She quoted a case of Epilepsy in which the Disorder had run a course of 3 years. At the proper times and in correct order, she used Orange, Lemon and Magenta. Albert E. Greenham suggested as a help in this case that Scarlet be used for a few days before the periodic sickness.

Otto Hinz read a clipping from a newspaper, dealing with the case of a soldier who had Typhoid Fever. The Vaccine used by the Medical Doctors brought on a complication of Disorders and the man is, according to medical testimony, doomed to die. Just one more unfortunate victim of professional ignorance.

Martha Schreiter told of a woman whose Kardoscil and Spiroscil were alarmingly low when she took her in hand. She reported that during the course of Tonation over a period of time, that condition changed a good deal, but that the Sufferer was much benefited by Spectro-Chrome.

Colonel Dinshah is preparing for his Spring Tour to the existing Planets and to open new Planets where necessary by reason of increasing membership.

The warning given by L. Roy Dodman and Jessie E. Ness is very sensible and must be considered by all who act as Spectro-Chrome Normalators. Before this year is over, extremely stringent measures will be taken by the Board Of Trustees, to prevent any tampering with our Collaborator Affiliations and to make the Organization strictly fool-proof and integral.

Let Major John Roehl Read This.

If Major John Roehl stated he "investigated" Spectro-Chrome, he spoke a deliberate lie. When the Originator first visited Detroit, Michigan over 10 years ago, the same officer tried to bluff him in the Hotel Tuller, but, Colonel Dinshah showed him the door. Spectro-

·Chrome NEVER was investigated by Major John Roehl or his office or the Wayne County Medical Society or American Medical Association or any Medical or Health Board. We challenge any and all—the whole World over to meet the Originator in a public debate on a scientific platform and defeat him. No man born of a human mother has the gumption to prove Spectro-Chrome a fake, but, the Originator is ready at all times to prove conclusively that Medicine as a science is the biggest bunk, swindle and downright pseudo-scientific humbug that was ever propounded to cheat and defraud the suffering public. Take this to Major John Roehl and let him accept the challenge. These Health Boards live by bluff and if he said he investigated Spectro-Chrome, the Originator openly challenges him to declare when, where, how and through whom he made the supposed investigation or eat his own words.

———o———

OPEN LETTER TO JOHN P. KOEHLER, M. D.
Commissioner Of Health
Milwaukee, Wisconsin.

Esteemed And Learned Sir:

Through a friend of ours, the following circular distributed by your Department, throughout the City Of Milwaukee and handed to all the school children there, came to my attention:—

JOHN P. KOEHLER, M. D.
Commissioner Of Health

E. V. BRUMBAUGH, M. D.
Deputy Commissioner Of Health

Office Of The
HEALTH DEPARTMENT
City Of Milwaukee

STOP DIPHTHERIA!

FATHERS AND MOTHERS: Every year Diphtheria sweeps through Milwaukee, cutting off the lives of little children, especially the babies under 6 years, who are least able to resist it.

In the last ten years 9,342 children have suffered from this disease—and ONE in every TWELVE died.

Diphtheria afflicted 792 Milwaukee children in 1926 and 757 in 1927.

DIPTHERIA CAN BE STOPPED

Children can be protected against it, and in a very safe and simple manner; a manner which has the endorsement and support of all the scientific, medical and health authorities of the country; a method that is not a fad or an experiment; a means that has been proven true.—That means is Toxin Antitoxin.

Toxin Antitoxin is not for the treatment of a child that is sick. It is for the child that is healthy and vigorous, and its aim is to keep him so. It is that child's insurance against a dreaded enemy. It may be given any time after the child is six months old.

HAVE YOUR FAMILY PHYSICIAN give Toxin Antitoxin (or Toxoid).

The Milwaukee Health Department wants to co-operate in protecting your own little ones against Diphtheria. All that is necessary is your consent.

Sign the enclosed card. The school children will give these cards to their teachers and the health department physicians will administer the Toxin Antitoxin at school. Those children who had Toxin Antitoxin six months or more ago should now have a Schick test which will show whether or not they are safe from diphtheria.

But don't forget the children under school age. They need the protection more than the school children and have less resistance. Take them to the school yourself and see what is being done to combat Diphtheria.

Insure for your little one the health and happiness you want them to have.

BUREAU OF COMMUNICABLE DISEASE,
MILWAUKEE HEALTH DEPARTMENT.

I presume—nay feel sure, that you undoubtedly mean well and that your Department is truly desirous and anxious to serve the interests of the citizens who depend on you for their health and welfare.

It is evident that with all the Medical and other means like Serums, Vaccines and the like at your disposal, you have not been able to check, control much less eradicate, the scourge of this "Diphtheria".

I warn every citizen and parent of Milwaukee, NOT to sign the requested paper and NEVER to consent to the administration of any such material as Toxin Antitoxin and meanwhile ask you to lay aside all your preconceived notions of healing and investigate the merits of Spectro-Chrome Metry, which can positively protect the children from this "to-the-medical-fraternity-dreadful" disorder.

Not only for Diphtheria, but for Scarlet Fever, Small Pox, Chicken Pox, Measles, Whooping Cough and similar troubles of children, Spectro-Chrome proved its value unquestionably. Many among the hundreds upon hundreds of the users of Spectro-Chrome in their homes, can testify to the effectiveness of Spectro-Chrome, right in your city.

I offer you all the Spectro-Chrome Equipments you need for the Milwaukee children under your care, **absolutely free and unconditionally. I offer you my services** similarly **without any renumeration.** I can save the children; I can place in your hands by Spectro-Chrome, the most potent weapon to tackle these enumerated scourges, if you will grant me the opportunity to demonstrate the Science.

Why waste your precious time in experiments that lead nowhere, when Spectro-Chrome is already in successful use 14 years? You have but to say the word and all my Spectro-Chrome resources will be placed at your disposal, so you may save the children and the blessings of the parents will go to you and posterity will remember you as the one courageous officer who to serve the public, stood forward for therapeutical emancipation.

I have the honor to remain

Yours Very Respectfully,

Dinshah P. Ghadiali,

Originator Of Spectro-Chrome Metry.

Malaga, New Jersey,

Monday, March 11, 1935.

—o—

FEMALE HELP WANTED

Three Traveling Secretaries who can be trained to corporate duties and demonstration of Spectro-Chrome Equipment for Classwork in the United States Of America, Europe and India. Smokers not eligible. Apply by letter, stating age, experience and salary expected to Spectro-Chrome Institute, Malaga, New Jersey.

—o—

THE GREATEST CHAMPION OF SPECTRO-CHROME DIED

Kate W. Baldwin, Medical Doctor, Fellow American Medical Association, Fellow American College Of Surgeons, Former Senior Surgeon Woman's Hospital of Philadelphia, Pennsylvania, Honorary President of Scientific Order Of Spectro-Chrome Metrists, died at Lawrenceville, Pennsylvania on January 17, 1935.

Much can be written of this noble surgeon who was our best friend, but, for certain cogent reasons, the Originator of Spectro-Chrome Metry refrains from disclosing at this time many facts, which would make the Medical Octopus bow in shame, at the unjust persecution according this star in the firmament of devoted womanhood, whose sole aim was Service To Mankind.

—o—

NOTICE TO PURCHASERS

No Price Reduction

There is no reduction in the price of Spectro-Chrome Wood Cabinet Equipment contemplated. It is as is the **Least Expensive And The Best Health Service In The World:** $75 serve for Life Happiness.

Spectro-Chrome—In Every Home

PARALYZED BOY IMPROVING

Colonel Dinshah P. Ghadiali,
c/o Spectro-Chrome Institute,
Malaga, New Jersey.

I am happy to advise you that my son Kenneth is improving gradually and I am anxiously looking forward to the time when he will be able to walk and get about like other normal children.

His improvement has exceeded my expectations. A thousand thanks to the merits of Spectro-Chrome Metry.

I have had some interesting and astonishing results from the use of my equipment during the past year and would not under any circumstances be without it. I can not speak too highly of this wonderful discovery and those whom I have interested in it are also very enthusiastic and report satisfactory results.

I do not know when you expect to visit again Milwaukee, but, whenever you do, I can assure you I will be among those present and sincerely hope my son will then be able to walk, as I would like everyone present to know of the wonderful results obtained by the use of Spectro-Chrome.

Trusting I have not taken up too much of your valued time and looking forward to the time when I can clasp your hand and thank you personally, I wish to remain

Yours Very Truly,

Frank S. Burns
103 North 87th Street
Milwaukee, Wis.

March 9, 1935.

SPECTRO-CHROME

July
1935
Blue

Volume
11
Number
7

GREEN 150° · TURQUOISE 150° · BLUE 180° · INDIGO 210° · VIOLET 240° · PURPLE 270° · MAGENTA 300° · SCARLET 330° · RED 0° · ORANGE 30° · YELLOW 60° · LEMON 90°

IMPORTANT ARTICLES FEATURED

IN THIS ISSUE

Read and Let Your Friends Read

CHILDREN TONATE WITH SPECTRO-CHROME

MISSOURI WOMAN NORMALATED BY SPECTRO-CHROME
VICTIM OF EIGHT SURGICAL OPERATIONS

MEDICAL IGNORANCE

MEDICAL DOCTOR PUZZLED BY SPECTRO-CHROME

ERNEST EXCLUSIVE DISTRIBUTORSHIP TERMINATED
Earn $15.00 On Each Spectro-Chrome Sale

DINSHAH P. GHADIALI, *Editor and Publisher.*

SINGLE COPY 50 CENTS

CHILDREN TONATE WITH SPECTRO-CHROME

Successful use of Spectro-Chrome in Measles, Partial Blindness, Burn, Bee Sting, Cut, Bleeding, Heart Disorder, Diabetes, Asthma, Bladder Inflammation, Constipation, Bed-wetting, Gall Bladder Disorder.

By:—CARL I., *Oconomowoc, Wisconsin.*

Spectro-Chrome has been used in our family with the greatest success and it is mainly due to the Normalation of my wife from a severe Gall Bladder Disorder that I decided to make Spectro-Chrome Metry my life work. I personally had many occasions to be convinced of its beneficial effects (Fever, Lumbago, etc.). During the Measles epidemic, our four children made the parade under Turquoise and Blue and on the fourth day they hopped around without any sign of Measles.

My father-in-law, who lives with us, was relieved of Partial Blindness of 3 months standing, in 9 Tonations.

A Burn by an electric iron yielded to Blue, in my wife and even a Bee Sting suffered by one of my boys, gave up its "sting" to Turquoise in 10 minutes.

Even our four children, ages 11 to 16, know how to use the Attuned Color Waves. When my wife and myself came home one evening at 11 P. M., from a "gospel" tour of Spectro-Chrome, the boy of 13 had his head under Indigo, with a 4 inch cut in his head. He had fallen from a broken swing and his head had encountered a log that was in the way. We were greeted by him with a smile, saying, "Father, you don't have to scold me any more; the bleeding has stopped." Two more Tonations and "No Plasters" did the trick.

Spectro-Chrome is the supreme Healing System. John H. of Jefferson, Wisconsin purchased a Spectro-Chrome, about 6 weeks ago. He had a severe Heart Disorder, a light case of Diabetes, swollen ankles upto knees; also had to sit up frequently during the night to prevent choking, due probably to an asthmatic condition. Last week, according to a so-called examination or check-up by his Medical Doctor his Heart sounded good, his body color had surprisingly improved, the swelling had subsided and a Blood Test came back from the Wisconsin General Hospital as 100 per cent. perfect; the Blood Pressure also was normal. But above all, during the past 3 weeks, the Sufferer who adheres strictly to the "Rational Food Of Man" system, has his natural uninterrupted sleep. The man feels so much improved

in his general condition, that he could not refrain to "show up" to his former family Doctor by an examination.

H. of Hartland, Wisconsin has a son 12 years old, who suffered from Asthma since boyhood. The Equipment was purchased last year in August and the youngster has not had an attack since that time. He went with his parents to Florida, was sick with Measles and in connection had his first attack again. But since Tonations were resumed, all is well.

Mrs. H. of Watertown, Wisconsin bought an Equipment about 4 months ago. The Sufferer had a persistent Inflammation of Bladder (no control over pressure). Constipation; claims to be normal since 2 months. Her boy, 18 years old, a farmhand, suffered from persistent Bed-Wetting since childhood and therefore could not hold any job. The boy took 2 Tonations daily for 2 months and during the past 9 weeks, no "accident" happened and he is working on a farm now.

F. bought a Cabinet 6 weeks ago. His wife was in bed for 3 weeks with Gall Bladder Disorder and operation, according to the Medical Doctor, was imperative. The Equipment was installed on Sunday. Tuesday the Sufferer, according to her husband, fainted twice and Thursday she was up making meals and doing the shopping. The purchaser claims—by right and evidence—that Spectro-Chrome saved him the price of an operation and his wife's Gall Bladder.—E. P.

MISSOURI WOMAN NORMALATED BY SPECTRO-CHROME VICTIM OF EIGHT SURGICAL OPERATIONS

By:—EDITH O. S.—*Marshall, Missouri.*

I have been the victim of Surgery 8 times! Four Goitre operations and Gall Bladder removed. Since the Gall Bladder was removed, I found myself in a worse condition than before, with Stomach trouble, Bowel trouble, Headaches and Vomiting. Took all kinds of Medicine and tried everything to get relief but, without avail.

I purchased a Spectro-Chrome and used it (about 18 months ago). From the very first Tonation, I improved in health and really took ten years younger and feel like living. I sleep and eat, but there are no Drugs on my shelves. I discarded all.

A friend had Appendicitis. Medical Doctors cared for her over a week without any improvement. I gave her 3 Tonations with Turquoise and she felt well and ready for work.

It is the only safe and sane method of healing and a marvelous discovery for humanity.—E. P.

FIRST MILWAUKEE WISCONSIN PLANET

Meeting Of Thursday, June 6, 1935.

(Hotel Wisconsin, Milwaukee, Wisconsin.)

Successful Use Of Spectro-Chrome In:—Facial Paralysis, Sprained Ankle, Leg Pain, Tuberculosis, Heart Kidney and Bladder Disorders, Rheumatism, Chronic Bronchitis, Croup, Cold, Diabetes, Gall Stones, Constipation, Sinusitis, Sore Gums, Neuritis, Insomnia.

Medical Doctor said "You will not be able to close your Eye for about 6 weeks." Spectro-Chrome did it in 1 hour!

One User Says:—"The machine did wonders for her!"

Another User Says:—"I will not sell my Spectro-Chrome for $1,000.00!"

Reported By:—CAROLINE ZIMMERLING, *Recorder*

Caroline Zimmerling, the newly appointed Recorder, was introduced and affirmed according to Rule 33 of the Constitution And Rules.

Otto H. Lutz read Chapters 60, 61, 62, on Disorders Of Vision, Disorders Of Olfaction And Gustation and Disorders Of Audition, from Spectro-Chrome Metry Encyclopaedia, Volume 2.

Rudolf B. reported a case of a woman who was affilicted with Paralysis of the Face and Eye on the right side. The Medical Doctor said she would not be able to close her eye for about six weeks. In one hour with Spectro-Chrome, she was able to close her eye. Her face became mobile in about two weeks. His wife had sprained Ankle. Turquoise was used successfully.

Otto H. Lutz reported that nine Tonations of Lemon normalated a man of 52, who had severe pain in his legs.

Clara L. reported that she has been using her Spectro-Chrome for about a year. At that time, she was not in very good condition, but she is getting along well now. Green, Lemon, Yellow and Orange, were used in this case which is one of a Tubercular character. She was in a Sanatarium before using Spectro-Chrome. She said that the machine did wonders for her.

Lizzie R., of Rochester, Wisconsin, has her Equipment about 3 months. She bought her own Equipment after receiving good results from Spectro-Chrome Tonations. She is very much satisfied with it. She is using it for Heart, Kidney and Bladder Disorders and Rheumatism.

Gerhardt N. is using his Equipment every day.

Last July Berta B. got her Equipment. She is getting results in a case of Bronchitis of twenty years standing. It was the first winter for some years that she really felt quite well.

Adeline C. reported that Green normalated a case of a little girl with Croup. The girl was out of school only two days. For Colds, she uses Green and Lemon with very good results and for Colds in children, she uses Turquoise, with equally good results.

A great deal of success has been the experience of Laura G., in Diabetes, Bright's Disease, Gallstones and many other Disorders. She has been tonating for about two years and says she uses Spectro-Chrome exclusively. She says she finds that one must use real common sense in tonating. She found that in Chronic cases, she got good results by going to the Ultra-Green Division, after getting a good start with the Infra-Green Division. She says that since Miss N. R. is no longer visiting Spectro-Chrome users, her time is almost entirely consumed in giving advice and suggestions to users who call her up for this purpose and she feels that she can not let them down. She says she has been in the healing profession since the age of sixteen and while she thinks Spectro-Chrome is wonderful, especially in the cases of Asthma, Arthritis and Rheumatism, she finds that it is difficult to judge only by the Spectro-Chrome Home Guide. She feels that the advice given is not adequate.

Joe L. has had his Equipment since last July. His wife had a growth which Indigo Local helped. She followed the advice received from Spectro-Chrome Institute, which was Indigo Local and Orange Systemic, but felt that the Orange was too strong so started on Green.

E. B. got his Equipment for his father's use. He received advice from Spectro-Chrome Institute, Green, Lemon and Yellow. He feels that the Equipment did for his father so much good that it may have saved his life.

For about three months, John D. (not Rockefeller) has his Equipment. He uses it for Constipation and says it is doing a very good job of it. Two or three Tonations normalated the condition. He has been using Green, Lemon, Orange and Magenta. Not for $1,000.-00 would he sell his Equipment if he could not get another. His wife had good results in Rheumatism of 25 years.

A Sinus condition of 5 to 10 years' standing is responding well to Spectro-Chrome Tonation in the case of William K. The first week, he used Green alone; then, he used Lemon Systemic and Magenta Local.

After the intermission, Walter H. rose and said that he purchased his Spectro-Chrome on December 13, 1933, just after he returned from Mayo Brothers. Having studied Medicine and Electrical Engineering, he could readily see the merit of this Science. His first Tonation was two hours long and it made him feel so well that he went out and enjoyed himself. He says it put him back on his feet entirely in a short time. About three weeks ago, his wife had Ptomaine Poisoning. Spectro-Chrome normalated this condition. He re-reported a case of Scarlet Fever; the Temperature of the Sufferer was reduced from 105.3 to 99 Degrees in a few hours by the use of Violet. This person was normal in a few days. He said he found a number of people who were not satisfied with their Equipments. He felt this was due to the fact that they did not understand the use of their Equipments. He feels that we should have more discussions in the Planet Meetings concerning certain ailments. He then made a motion that at the Planet Meetings Normalators be allowed to advise people or that people be allowed to get advice from Normalators. He said he wants to give advice. The motion was seconded.

Marie W. said she came to the meeting with a heart full of resentment. She wanted to know what the status of a Member was. She said people could not learn anything from the Spectro-Chrome Home Guide. She said she wants to learn from discussion in the group.

Walter H. rose again and said that he had a case of Mucous Colitis and Arthritis. He took the Inflammation out of the body first with the Ultra-Green Division and then worked right down the Infra-Green Division with Systemic Tonations after that.

Dorothea G. wanted to have a rising vote taken to voice the sentiments of the group on the subject of more discussion and advice. Edna C. also felt that more discussion was needed. She said she came away from Hales Corners, Wisconsin, to learn something about Spectro-Chrome.

Betty W. reported of a woman of 65, who broke her shoulder and elbow. These parts were set, but the pains were very bad because Rheumatism had set in. Spectro-Chrome was used on her. Three weeks of Indigo did not relieve her, so Orange was used with very good success. Besides the Orange Local and Magenta on Areas 4 and 18, Green and Lemon Systemic were also used. She also reported of a man of 59, who had Rheumatism in his hand and arm. Green Systemic and Indigo Local were used successfully. Another case of a woman with Sore Gums, in which Green Systemic and Indigo Local, with Magenta on Areas 4 and 18 were used to normalate was reported.

E. A. Ernest then gave a short talk about the "complaints"

that were made, saying if the Spectro-Chrome Institute did not allow us to have more discussions and give out advice, it was merely to protect us as well as Spectro-Chrome, from the corrupt Laws of the United States Of America made by the Medical Trust. He also said that if there be any doubt as to the course of procedure in tonating, Page 81 of Spectro-Chrome Home Guide, whereon is stated that Green is to be used first in all cases, will answer the purpose. He said that out of over 1,000 Spectro-Chromes in use in Milwaukee County, there were only 15 or 20 complaints, which were from people who did not do as they were supposed to do. He said if the General Advice Charts were not answered, it was because they were not properly filled.

Michael S. has his Spectro-Chrome about six weeks. He is using it for Sluggish Liver, Spleen, Arthritis, Swollen Feet. He said that sight seems to be failing since using Spectro-Chrome; is following Guidance from Spectro-Chrome Institute faithfully.

Spectro-Chrome has been used by Alma H. since last Fall. She used it for Neuritis, Running Ear, Bronchitis and Sleeplessness, She has had very good results. She started with Green and went to the Orange. About 5 or 6 Tonations gave results. For Sleep, she used Green; in about a week she was very much better. Her Yelping Dog (reported in last issue) was very much improved.

We suggest that instead of reading so many Chapters, the Planet Deputy ask for a vote of the Members present each time as to just **one** item for that evening's study. Then, have somebody with a loud and clear voice read that part from the Spectro-Chrome Metry Encyclopaedia and request the entire assembly to join in discussion for elucidation of points not comprehended. Anything dubious in the minds of the Affiliates, which none present could satisfactorily explain, may then be reported by the Recorder to the Principal Office. The explanation given by the Principal Office Staff may be read at the next meeting. Such systematic procedure will produce educational results .

Laura G. should not interfere with Purchasers who are under the Free Guidance of Spectro-Chrome Institute. N. R. was stopped by Spectro-Chrome Institute from visiting the homes of Purchasers and tendering Guidance, which frequently conflicted with the Guidance given by Spectro-Chrome Institute. Laura G. had better keep her hands off similarly, because, she will be likely to create unintentional troubles between the Purchasers and Spectro-Chrome Institute. Spectro-Chrome Home Guide and Spectro-Chrome Metry Encyclopaedia have more information in them than all that Spectro-Chrome Normalators ever will be enabled to give and as Laura G. is an Agent of Ernest Distributing Company, she is bound by the Agreement Rules of Spectro-Chrome Institute with Ernest Distributing Company, which clearly debar all such Guidance interference.

The Board Of Trustees of Scientific Order Of Spectro-Chrome Metrists wrote a very plain letter to the First Milwaukee Planet Commission, setting forth the details to avoid any unpleasant misunderstanding; this letter was to be read by the Recorder Caroline Zimmerling at the Planet Meeting of July 4, 1935. It is printed verbatim below:—

Tuesday, July, 2nd 1935,
Chronos: Red 45 White

Miss Caroline Zimmerling,
2356 LeFeber Avenue,
WAUWATOSA, WISCONSIN.

Dear Miss Zimmerling:

1. We shall thank you to read this communication at the meeting of the 4th current.

2. Notwithstanding repeated notices and long communications, our Affiliates still persist in mixing communications of Spectro-Chrome Institute and the Scientific Order of Spectro-Chrome Metrists.

3. All communications regarding Planet activities as well as SPECTRO-CHROME magazine should be addressed to the Scientific Order of Spectro-Chrome Metrists.

4. All communications regarding Spectro-Chrome Equipments, Spectro-Chrome Slides, Spectro-Chrome General Advice Charts and such other matters should be sent to Spectro-Chrome Institute.

5. No communications should be addressed to any specific name or person; all communications should be addressed solely to the particular corporate name.

6. In any communication or complaint in which both Organizations may be involved, the matters should be segregated and separate communications should be addressed to each.

7. The reason for all these regulations is that both are separate corporations under distinct State Laws and their officers and files are different. Hence, our Affiliates will do well not to mix one with the other, especially in matters of sending remittances.

8. During the past few months we have been constantly receiving communications that the activities of the First Milwaukee Wisconsin Planet are not satisfactory to a certain number of individuals. These individuals are Graduates of Spectro-Chrome Institute and are Spectro-Chrome Normalators in active practice. Their major complaint is that the Affiliates do not receive the educational benefit they ought to receive at the meetings. However, they forget that it is entirely upto the Affiliates to make meetings interesting by introducing educational questions in Spectro-Chrome Metry and discussing various subjects from SPECTRO-CHROME METRY ENCYCLOPAEDIA and practical case reports on the floor.

9. Many seem to believe that all discussion was debarred by us; this is a miscomprehension on their part. Instead of the Deputy reading a series of Chapters in a desultory manner and declaring a recess, it is our opinion that by a vote of the Planet a certain Chapter should be chosen each time or a certain part of a Chapter may be so selected or a certain theme or disorder may be picked and its various aspects discussed and thrashed thoroughly by the Members present. It is essential to note at this point that the voting power rests solely with full-fledged Graduate Members who have the Green Ticket and are in active standing thereby. Provisional Members who are Affiliates purely by courtesy have no voting power and while they are permitted as Sympathizers to take part in the discussion, they should not be allowed to have any voice whatsoever in the management of the Planet.

10. Free discussions are always welcome; we even invite caustic criticisms and merciless lashing leveled against us in so far as the management of Planets, but is must be implicitly understood that what is given in our Constitution and Rules must be rigidly maintained and enforced and no violation of any character should be tolerated.

11. A very serious violation was found in the past to be the giving of Attuned Color Wave Guidance to Spectro-Chrome Equipment Purchasers by certain Agents of Ernest Distributing Company, a fact which was in distinct violation of the agreement between Spectro-Chrome Institute and Ernest Distributing Company.

12. You well know that Spectro-Chrome Institute is fighting for its very existence against our bitter opponents and any infraction of any Rule of Spectro-Chrome Institute would at any time launch an Affiliate and through an Affiliate the entire Organization, as well as the Originator into serious troubles.

13. Many Spectro-Chrome Normalators have been reported in the past as tendering Guidance in Attuned Color Waves in full knowledge of the fact that the sufferers to whom they tendered such Advice were under the Free Guidance of Spectro-Chrome Institute. The Free Guidance of Spectro-Chrome Institute is directly under the personal supervision of the Originator and he frequently found to his great regret that such Guidances were antagonistic to the Free Guidance given by Spectro-Chrome Institute. Notwithstanding warnings given, such infractions continued and it is only within the past few days that the infractions seem to be abating because of the strong steps taken by Spectro-Chrome Institute.

14. There is another serious infraction looming. Many Spectro-Chrome Normalators seem to believe that they can copy, duplicate or issue the Favorscope or diagrams issued either by Spectro-Chrome Institute or by us and utilize them either for their personal use or for others. It is necessary to warn that all work in Spectro-Chrome Metry is copyrighted by the Originator and duplicating any Chart for even personal use and hanging it on the wall as a display is a violation under the Copyright Laws and the violator, under our Constitution and Rules is liable to expulsion.

15. All these matters we desire you to place before the Affiliates, as we find the trend of human nature to be negligent to rules and as we aim at harmony and cooperation, it would be to mutual advantage for Affiliates to avoid the specified contingencies.

16. We desire to impress upon the minds of the Affiliates that the little income derived from the Membership Fee is away below that required to pay even the rent of the meeting room; in truth, were we looking only to the financial side of the question, we would be strongly for abolishing the maintaining of Planets. We sincerely hope and believe for such reason that the educational and helping feature of our Planets be borne in mind by all Affiliates and steps be taken to exchange personal views without unnecessary bitterness or friction, thus increasing the value of the Planets and doing good to the sufferers by guiding them in the proper use of Spectro-Chrome, as long as such Guidance does not clash with or antagonize the Rules of Spectro-Chrome Institute.

17. What Spectro-Chrome Institute desires is simply this, that the sales people working for Ernest Distributing Company shall not bring their matters, complaints or squabbles on the floor of the Planet. Such matters should be discussed with the Ernest Distributing Company at the personal meetings and not in the Planet.

18. None stands for freedom of speech more solidly than the Originator: all that he desires is that the Planet be used exclusively for its educational purposes and not for personal complaints against Spectro-Chrome Institute, which has nothing to do, as a corporation, with the Planet.

19. To keep away from the clutches of our opponents, it is essential and vital that no sales person working for Ernest Distributing Company should act as a visiting nurse or go to the homes of sufferers even for Spectro-Chrome Advice.

20. Furthermore, it is also essential and vital to our existence and interests that any deviation from our Constitution and Rules, no matter by whom made, should be promptly reported to the Principal Office at Malaga, New Jersey, so requisite action may be taken to keep all violations out of our Organization and fully insure the safety of the Originator and his work.

21. A duplicate and triplicate of this communication is sent to the Deputy and Sub-Deputy and it will be warmly appreciated if the Planet Commission will please do

what is necessary to avoid friction and misunderstanding.

22. With kind regards and best wishes, we remain

Yours in the Service of Mankind,

BOARD OF TRUSTEES

by

Irene Grace Dinshah,

Secretary

IGD:RF

AIR MAIL

Duplicate to Deputy Otto Lutz

Triplicate to Sub-Deputy Harry Goebel.

SPECIAL DELIVERY

After reading the above letter, any sensible person will admit that the so-called "restrictions" of Spectro-Chrome Institute are well meant safeguards. The Spectro-Chrome Normalators, in their over-enthusiasm, have been worse than our opponents in creating troubles for the Originator. Their attitude does not have his sympathy. It is his desire NOT to encourage the interference of Spectro-Chrome Normalators in dealing with the Sufferers, as his experience proved that many Normalators, like some Medical Doctors, actually let the public linger, so they may reap a richer harvest! Thus, since Spectro-Chrome Institute started direct sales from Maker To Purchaser, interference from all Spectro-Chrome Normalators was cut off and our Planet activities were placed strictly on an educational basis, for the benefit of the Purchasers who were admitted as students in Spectro-Chrome Metry, by Provisional Membership.

Walter H .is another Spectro-Chrome Normalator, who wants to reform the Organization, by running it according to Roberts's Digest! He had better study our Constitution And Rules. Rule 64, Section 19 says, "No Rules Of Debate shall apply except as provided herein." Our Organization is controlled according to **our** methods and those who want to stay with us should follow **our** Rules. If instead of grumbling, all would suggest and follow constructive methods, we should be obliged. None is prevented from writing for Free Guidance from Spectro-Chrome Institute; why then should he, an Agent for Ernest Distributing Company try to act for others? If he studied "Medicine" he ought to realize its short-comings and refrain from trying to run the show of Spectro-Chrome for us. Spectro-Chrome Normalators are NOT Spectro-Chrome Institute and if they will but mind **their own** business, there would be no friction.

Marie W. should read the Constitution And Rules to know her status. She, as a Provisional Member has NO STATUS except

what she got by our courtesy, to be allowed to attend Planet Meetings. She has NO VOICE in the conduct of the Planet and NO VOTE. If she could not learn anything from the Spectro-Chrome Home Guide, she should ask for the Free Guidance. She may start her discussion in the Group and surely someone will enlighten her. She **never asked for any Guidance** and has no right to complain. Our records indicate that she NEVER requested any information—naturally she got none; what right has she to complain?

Dorthea G. was "Out Of Order" and should have been "hammered" down about the voting. Provisional Members are merely "Courtesy members" and are enrolled sheerly as "Students in Spectro-Chrome Metry." They have NO VOICE IN THE CONDUCT OF PLANET AFFAIRS and positively NO VOTE. Refer to Rule 65, Section 6. Edna C. gets the same answer. It were better for our Affiliates to study the Constitution And Rules, than to complain. There is NO REASON to complain, when they do not follow the Regulations and all the time these Provisional Members pay NO FEE, but are given a special privilege Free Of Charge!!!

As many of these so-called "Complaints" were the results of some of Ernest Distributing Company's Agents violating the Rules, we shall be thankful if E. A. Ernest would control his crew from becoming obnoxious in the Planet Meetings. We have numerous reports of the Violations of such Agents and a Warning is issued openly to them, that the constant infraction of Rules will not be tolerated. If such infraction persists, the Board Of Trustees will be compelled to stop such Agents from their Affiliations with the activities of the Planet, a move which will remove absolutely all cause of friction. Most of these Agents are only Provisional Members, without the rights of Normalation and **without** the privilege of voting. What reason then have they to claim that they are better fitted for Guidance in Spectro-Chrome Metry than the Alma Mater that trained thousands like them? Some people certainly presume!

Why should E. A. Ernest or anyone worry about the corruption in the Laws of this country? We repeatedly stated that no Constitutional Government Law can stop the sale of a Patented Apparatus like the Spectro-Chrome and its transportation in Interstate Commerce. Just forget about Laws and their "graft" and place a Spectro-Chrome—In Every Home. **When that happens, we shall be making the Laws!** And then? We shall set an example for Integrity, Public Service and Truth. Look forward to the Future.

FIRST MILWAUKEE WISCONSIN PLANET

Meeting Of Thursday, July 4, 1935

(Hotel Wisconsin, Milwaukee, Wisconsin.)

Bedridden Woman Walked By Spectro-Chrome

Reported By:—CAROLINE ZIMMERLING, *Recorder.*

The meeting was opened by Otto H. Lutz, Deputy.

A letter from the Board Of Trustees of the Scientific Order Of Spectro-Chrome Metrists, concerning matters pertaining to the meetings of the Planet and discussions at the meetings, was read by the Recorder.

E. A. Ernest and Otto H. Lutz informed the meeting that in the future discussions of various Disorders will again take place in the meetings as previously. This practice was stopped for the past two meetings due to a misunderstanding. In these discussions, Spectro-Chrome Normalators, as well as other Affiliates present will have a voice. It was also stated that the people must still get their Advice from Spectro-Chrome Institute, but, may come to the meetings for discussions.

E. A. Ernest reported a case of a woman, who was bedridden. Her husband asked for Guidance from Spectro-Chrome Institute and until it came read the Instructions from Spectro-Chrome Home Guide. The woman acted accordingly. With five weeks use of Spectro-Chrome, she was walking around the bed without assistance.

There were no other reports. The meeting was cut short because of scanty attendance due to the day being a holiday.

The Affiliates were clamoring to have us hire a larger meeting room. We know better! The time is not yet ripe for increasing expenses.

Why The Mogen David Sign?

All matter printed under an article marked by the above Diagram, indicate comments by the Editor.

This applies to all our issues of Spectro-Chrome. That Grand Sign represents the views of the Editor on any subject below which it is printed.

FIRST DETROIT MICHIGAN PLANET

Meeting Of June 6, 1935.

(Hotel Tuller, Detroit, Michigan)

Spectro-Chrome Successfully Used In:—Black Eye, Earache, Appendicitis and other Disorders.

Reported By:—JESSIE E. NESS,
First Vice-President, Scientific Order Of Spectro-Chrome Metrists

John J. W. reported that his boy, of 6, was knocked down by an automobile and his Eyes were completely closed thereby. Indigo enabled the boy to sleep and later Magenta reduced the Swelling. Two Tonations of Scarlet raised the Kardoscil from 72 to 94. The Sufferer is now progressing toward complete recovery.

Martha S. quoted a case of a Black Eye received accidentally. Two Tonations of Blue, followed by Turquoise, improved the condition.

Ervin B. cited the case of his small son, who had Earache, with Fever. Turquoise was used and the Disorder was eliminated in a short time.

Otto H. submitted a favorable report on the case of the young boy already mentioned in prior Planet Meetings. Continued improvement is very noticeable, credit being due solely to Spectro-Chrome.

Albert Greenham mentioned a case of suspected Appendicitis; five Tonations were given in 1 week. After the third, the Pain was gone. Turquoise relieved the accompanying inflammatory condition. The Spiroscil being low, Orange was irradiated and Normalation followed.

Walter H. gave a report of a case of Paralysis of the lower limbs. Since Spectro-Chrome is being used, the Sufferer, a young lad, dreams of playing baseball and running the bases! He even slams out home runs in his dreams. He never sees himself unable to walk or run. Surely, coming events cast their shadows before.

Stella H. praised Spectro-Chrome as producer of wonderful results in sickness, particularizing one of a woman who was very sick, but, who gained 10 pounds in weight and is progressing very nicely.

Edwin L. stated that in trying to purchase a Radiometer at the store of a Detroit Optometrist, he learnt that the gentleman in question kept it in the window to attract passers-by to the other articles on display. He characterized the merchant as a "Fathead". A Dentist whom he visited admitted he was scared to use the X-Rays in his business and stood as far away from the machine as possible. Another "Fathead"!

The desire for Advice as to the correct Attuned Color Waves to be used in certain Disorders was evident from the questions asked by some of those present. Jessie E. Ness referred them to the Spectro-Chrome Home Guide, supplied free with each Equipment. She also extolled the simplicity of Spectro-Chrome Metry as exemplified in the Infra-Green and Ultra-Green System and recommended all to purchase the Spectro-Chrome Metry Encyclopaedia. The happy possessor of this great work, plus a Spectro-Chrome Equipment, she declared, had no need ever to call a Medical Doctor in time of sickness.

Optometrists are using Spectro-Chrome now! Later, they will all use it for various Eye Disorders. Dentists also are falling in line. They will ALL fall in line, including the Medical Doctors, when the public demand comes from all sides. Time will do its work.

The suggestions of Jessie E. Ness were sensible, but, the main trouble is to MAKE the owner of the Spectro-Chrome Home Guide or the Spectro-Chrome Metry Encyclopaedia read and utilize it. Many who are owners of both, still ask questions which are already explained therein. Water may be brought to the horse, but, how can he be made to imbibe it?

Greeta O. Peck of Sandwich, Ontario, Canada, was appointed to officiate as the Deputy of the First Detroit Michigan Planet, for the current year. Your kind co-operation is requested.

MEDICAL IGNORANCE

Elizabeth Riley, a woman of 37, was taking a Sunbath, in Atlantic City, on the third floor of a house, in 920 Pacific and fell off. Medical Doctors stated that Sunbaths frequently caused "Dizziness". Why? What is the reason? A short study of the Ultra-Violet Rays of the Sun will indicate clearly the havoc created on the Human Body, by indiscriminate exposure to those destructive radiations. The Chapter 65, Page 657, of the Volume 2, Spectro-Chrome Metry Encyclopaedia, on Dangers Of X-Rays And Radium, will be an eye-opener to these so-called learned professionals if they would but leave their egotistical attitude of know-all and study Spectro-Chrome. All their medical acquisition of 6,000 years is condensed in 1,200 pages therein and completely exploded.

MEDICAL DOCTOR PUZZLED BY SPECTRO-CHROME

By:—MARIA I., *Oconomowoc, Wisconsin.*

Spectro-Chrome used with success in Arthritis and Heart Disorder.

I come in contact with cases too numerous to mention. Paula G., of Watertown, Wisconsin, is a good example. Suffering from Arthritis for over 7 years, this girl of 18, terribly retarded in growth, legs rigid, with knees unable to bend, mentally depressed, has progressed wonderfully under the Advice of Spectro-Chrome Institute. Relaxed under the first Tonation of Green, after second day woke up only once or twice at night, which at other times had mounted to 10 or 12. After second week, completely without pain, natural Bowel Movements, mental attitude entirely changed. After 2 months able to kick legs and later able to cross legs at knees. Had severe reaction which brought back Menstruation which had totally stopped. Parents very happy at progress being made.

Ben S. also of the same town, had such severe Heart trouble that his Medical Doctor advised his wife to be prepared for his death at any time. He just followed the Spectro-Chrome Home Guide, Infra-Green Division with plenty of Magenta. After 3½ weeks, he felt so well that he went to his Medical Doctor for an examination. His Heart being pronounced perfect, the Medical Doctor was very puzzled. He knows by now what did it. Believe it or not, but this man claims he never took one Tonation, according to the Favorscope, as he is out on business at all hours! Spectro-Chrome must be good!!!

It is absolutely the last word in Healing Science.—E. P.

SALESMAN HAS STRING OF SUCCESS

User says, "I will not sell my Equipment for $1,000.00."

By:—WALTER S., *Milwaukee, Wisconsin.*

The following are some people to whom I sold Equipments and who had remarkable results:—

Anna B. has a 9 year old girl, who could not speak and was feebleminded. She had marvelous results.

Frank S. had Heart trouble and is much improved.

Victor H. had Kidney trouble and would now never part with his Equipment.

Andrew K. had Arthritis and in 8 weeks all pain was gone and has never pothered since.

Mrs. J. H. had a Hemorrhage; normalated in 6 weeks and not troubled since.

Theo. W. had Constipation and a pain in his side and is well satisfied with his Equipment.

Paul K. had Hay Fever, Ringing in Ears, Headaches. His wife could not taste or smell; had a pain in the right side. Both are well satisfied and would not sell their Equipment for $1,000.00.

Marie Z. had Diabetes, which was normalated in 2 months.

R. M. had Constipation, High Blood Pressure, Heart Disorder. Is well satisfied; had wonderful results.

Delida N. had an open leg which healed and a girl had a Stroke which was normalated in 7 weeks. Well satisfied.

I also loaned my Equipment to many people who could not afford to buy one at the present time and all had the most marvelous results.

My views of Spectro-Chrome Metry are that it can not and never will be surpassed as a Healing Art.—E. P.

---o---

SUMMER

By:—ROSE FERRARA

Summer has come sings the Robin,
As he swings from the treetop on high,
The Bluebird joins in the chorus,
While the Wren wings her way to the sky.

The Rose whispers joy to the Lily,
As the breeze gently bends low her head,
Daisies gaily dance in the Sunshine
And Pansies gently sway in their beds.

Ah! Yes, it is a pleasant season,
Overflowing with laughter and song,
Sunshine and blue skies above us
And below, blue streams flowing along.

RUNNING EAR NORMALATED BY SPECTRO-CHROME

By:—D. E. O., *Glassboro, New Jersey*

I am writing this letter just to show my appreciation for "Spectro-Chrome", as my baby girl had a running ear after she had the measles. We used drops and did everything else possible, but she did not get any relief.

One night a woman informed me of "Spectro-Chrome" treatment and of course like any one will try anything to give the child relief. After a few treatments the ear had stopped running and the baby was getting along finely. She has picked up wonderfully since then.

I also took my other little girl who had a "bronchial cold" and after three treatments she was well.

I cannot praise "Spectro-Chrome" too highly and hope other Sufferers will learn of its value.

---o---

SPECTRO-CHROME ON SPRAINED ANKLE

A Letter To A Spectro-Chrome Normalator

By:—EVA J., *Swedesboro, New Jersey*

I know you have been expecting me to return for more Color on my Ankle but the one treatment helped me so much I did not need another.

As you know, I sprained my ankle so badly I could not put my foot on the floor and was almost carried to your Machine. After an hour of Color the pain was much better and on my way home I stopped at my brother-in-law's for a pair of crutches and he promised to bring them to me next morning.

When I stepped out of the car, I found I could walk on my heel and by supper time I was able to walk with the help of a cane.

When the crutches arrived the next morning, I was doing the washing and never used them except to lean on while hanging up the clothes.

People told me I would not be able to walk for a week or more, but thanks to Spectro-Chrome I was only off my feet less than a day. I have not taken Medicine for twelve years, because Spectro-Chrome always hits the right place.

My prayer is that Spectro-Chrome machines will soon be as numerous as radios.

SPECTRO-CHROME NORMALATED TONSILITIS

By:—VIOLA S., *Sicklerville, New Jersey*

Ever since I was about five years old I have been a victim of Tonsilitis; the last year each attack became worse. I had tried Medical Doctors and other remedies that friends might suggest, all of which gave me relief in time.

A friend of mine knew of these attacks and suggested Spectro-Chrome to me. I went for a treatment during the worst of one of these turns and received relief after an hour. It took the swelling out of my glands and made it possible for me to go to work the next day. Previous to this I had always had to miss a week or more. Thanks to Spectro-Chrome.

———o———

SPECTRO-CHROME REMOVED SEVERE PAIN

By:—NATELI V., *Monroeville, New Jersey*

Sunday I was stricken with a Toothache. I did not sleep a wink that night. I went to the Dentist to have it extracted (it was only a root) but he said it was not in the condition to be taken out, because it was inflamed and abscessed. He said to use Polaris Plasters on it; well I got through that night quite well.

The next day I got it taken out. He didn't hurt me at all, but he forgot to mention what trouble would be ahead. He said to use a plaster on it immediately. I had to have hot water, so I went to Mrs. J——. I did not realize it would pain. By the time I reached Mrs. J—— I was pretty well fed up with pain.

She asked me to take the Colors, so I did not refuse. By this time the Novocaine had left me and I was paining Hail Columbia.

Within an hour and a quarter I was relieved of that awful pain. Boy, I certainly am thankful to Spectro-Chrome, because if it had not been for Spectro-Chrome I would have gone out of my mind!

I believe I could have taken a boxful of Aspirin tablets but still it would do no good.

Thanks to Spectro-Chrome.

FIRST DETROIT MICHIGAN PLANET

Meeting Of Thursday, July 4, 1935

(Hotel Tuller, Detroit, Michigan)

Canadian Medical Doctor said:—"Either operate at once or Appendix would rupture."

Spectro-Chrome Normalated Appendix Disorder Without Operation!

Reported By:—GREETA O. PECK, *Deputy*

Jessie E. Ness installed Greeta O. Peck as Deputy.

The new Deputy stressed the importance of remembering the purpose of the Planet Meetings, which is the *study* of Spectro-Chrome Metry. It was resolved that a portion of the Third Volume of the Spectro-Chrome Metry Encyclopaedia be read and studied at each Meeting. Thomas Ness consented to begin with the August meeting.

Jessie Ness brought greetings from S., who is improving wonderfully under Spectro-Chrome. He is now able to turn on his side, use the telephone and has no pain.

Thomas Ness reported the case of a young man who came for Tonation for Sunburn. He was actually scorched. Three Tonations of Turquoise were required—although he was disappointed the first one was not sufficient!

Erwin B. reported that the boy who was hit by the car was normalated and was quite himself again. He was also relieved of Sunburn by Turquoise. Yellow was now used for Constipation.

Jessie Ness spoke of the great importance of following the Rational Food Of Man System as given by Dinshah, even to the use of the various ingredients for their influence on the Chemistry of the Body.

Greeta O. Peck then told of the case of H. P., a woman of 75, who was ordered to the Hospital for an operation for Appendicitis. The Medical Doctor said the Appendix might rupture before morning and it would be serious if it did. Kardoscil was 120, Spiroscil 30. The case was taken out of the Doctor's hands and Spectro-Chrome was used exclusively. Turquoise gave relief within a half hour. Within a few hours, the pain which had been intense, was completely gone. There had been pain and soreness over a large portion of Areas 9 and 10, extending through to Area 19. Some soreness still remained, but, it was confined to a small Area on the Right side of 9 and 10. After five days, Indigo was irradiated on Areas 9 and 10, alternated with Yellow on 18 and 19, with an occasional Magenta on Area 4, as advised by Spectro-Chrome Institute. For the first five days, no food was given; a little water to moisten the lips was given for the first two days and after that all the water that the Sufferer wanted. For the next ten days, nothing but milk food and Limbupani (Lemonade) was given; then, the Rational Food Of Man. The Sufferer improved steadily and not only the so-called "Appendicitis", but several other ailments responded simultaneously to Spectro-Chrome and the Sufferer is better than she has been in a long time.

———o———

DEATH OF THOMAS NESS, S. C. N.

Thomas Ness, Spectro-Chrome Normalator, of Detroit, Michigan, while marching in Windsor, Ontario, Canada, in the Orange Parade on Friday, July 12, 1935, collapsed near Erie Street on Ouelette Avenue and was dead when carried into the Hotel Dieu.

Dr. J. M. Young made an examination and pronounced death due to "Heart Failure". No inquest was held.

Thomas Ness was a loyal Spectro-Chrome supporter, born at Kirkaldy, Scotland. He came to this country about 20 years ago. He is survived by a widow Jessie, a son Thomas and two daughters Betty and Deanne, all of whom are Graduates of Spectro-Chrome Institute.

SPECIAL ANNOUNCEMENTS

ERNEST EXCLUSIVE DISTRIBUTORSHIP TERMINATED

You May Earn $15.00 On Each Spectro-Chrome Sale

For the quick expansion of the Distribution of Spectro-Chrome, by mutual understanding between Spectro-Chrome Institute and Ernest Distributing Company of 3106 North Hackett Avenue, Milwaukee, Wisconsin, the Agreement as Exclusive Distributor For The United States Of America, was terminated on Monday, July 22, 1935.

From that date, Ernest Distributing Company will act as Sole Agent For The State Of Wisconsin only. All other States outside of Wisconsin, will be under the DIRECT SUPERVISION AND CONTROL of Spectro-Chrome Institute.

Until the proper distribution policies for other States be formulated, Spectro-Chrome Institute will pay $15.00, to any Registered Purchaser of a Spectro-Chrome, who is a Member or Provisional Member of the Scientific Order Of Spectro-Chrome Metrists, who will place a Spectro-Chrome in a home, in any State outside of Wisconsin. This commission sale rule will apply to all Registered Purchasers in good standing. They must send the full price and the commission will be mailed to them promptly.

Here is YOUR OPPORTUNITY to earn $15.00 on an article of established and proven value. Here is YOUR OPPORTUNITY also for becoming Sole Agent For A State.

Spectro-Chrome—In Every Home, MUST become a reality; the tentacles of the Medical Octopus must be broken. THE MEDICAL DOCTOR MUST LEAVE OUR HOMES. Here is THE OPPORTUNITY.

—o—

SPECTRO-CHROME

OCTOBER NOVEMBER 1935 PURPLE MAGENTA

VOLUME 11 NUMBER 10-11

Trade Marks Registered In The United States Patent Office

IMPORTANT ARTICLES FEATURED IN THIS ISSUE

Read and Let Your Friends Read

FIFTEENTH ANNUAL CONVENTION OF SCIENTIFIC ORDER OF SPECTRO-CHROME METRISTS WILL ASSEMBLE IN DETROIT, MICHIGAN

DINSHAH P. GHADIALI, *Editor and Publisher.*

SINGLE COPY 50 CENTS

FIRST DETROIT MICHIGAN PLANET

(HOTEL TULLER, DETROIT, MICHIGAN)

Meeting Of Thursday, September 5, 1935

Successful Use Of Spectro-Chrome In:—Hip Stiffness, Skin Disorder, Fever, Cold.

Reported By:—GREETA O. PECK, *Deputy*

Martha S. told of how her hip had been stiff and black and blue, from moving some furniture. She tonated with Blue and the following day used Red and was entirely normalated.

Robert H. told of normalating a Skin Disorder in the form of an eruption over the eye. The Sufferer had consulted a Medical Doctor who said he did not know what it was, but on his advice the Tonsils were removed! This did not give results and on learning of Spectro-Chrome, she came to Robert H. Green, Lemon, Orange and Magenta were used and later Turquoise.

L. Roy D. told of a child who developed a Temperature. A Medical Doctor was called, but, she did not improve. Her Kardoscil was 160, Spiroscil 140 (?), Temperature 104.6. After the first Tonation of Blue, the Temperature dropped to 103 Degrees. After the second Tonation of Blue, it went below normal. Red was then given for 2 Tonations and the child was entirely normalated.

Otto H. told of a baby who had been cross, irritable and sleepless for some time. It fell asleep during the Tonation and was all right afterwards. The mother who had a very bad Cold also felt very much better after one Tonation of Turquoise.

Henry S. was again present in his wheelchair. He is gaining steadily and now weighs 152 pounds, a gain of 7 pounds in 1 month. From being a helpless cripple 4 months ago, he is now able to feed himself and do many things with greater ease than a month ago.

A Member asked for discussion on Hay Fever. Her husband had not the results he had hoped for in this trouble, although he was much improved in other respects.

John W. asked for a discussion on Kardoscil and Spiroscil Ratio.

Philip M. read the first two chapters of the Third Volume of Spectro-Chrome Metry Encyclopaedia, on "Intricate Motion Mechanism" and Purgation, Pills, Potions And Pumps". Discussion followed.

Albert E. Greenham told of his visit to Malaga, New Jersey and of the perfecting of an unbreakable Slide which will be a boon to Spectro-Chrome. He advised all the Members to read the Constitution And Rules and to adhere strictly to them.

That report about Kardoscil being 160, Spiroscil 140, seems erroneous. A Ratio of practically 1 to 1 is not possible to normalate in 2 Tonations of Blue, even if Life should hold.

Sleep is the best normalator in cases of such character. During Tonation, if a Sufferer should slumber, there should be no attempt made at awakening.

We are glad to hear of the progress of Henry S.

The Affiliate whose husband has Hay Fever, should ask for the Free Guidance. Wherever there is an apparent failure, there is some reason and the General Advice Chart promptly will disclose it.

For Kardoscil and Spiroscil Ratio information, refer to Spectro-Chrome Metry Encyclopaedia, Volume 3, Chapter 109, Page 1135, Paragraph 9.

The Spectro-Chrome Slides are the same, but, by use of the Dinshah Thermocept, the Heat generated by the Bulb being intercepted, the danger of Heat-Crack is avoided. In all Spectro-Chromes fitted with the Dinshah Thermocept, no Heat-Crack may occur, making the Equipment Fool-Proof.

FIRST DETROIT MICHIGAN PLANET

(HOTEL TULLER, DETROIT, MICHIGAN)

Meeting Of Thursday, October 3, 1935

Successful Use Of Spectro-Chrome In:—Eye Tumor, Pneumonia, Cold, Intestinal Disorder, Nasal Growth.

Reported By:—EDWIN W. LEMKE, *Recorder.*

Deputy Greeta O. Peck affirmed the new Officers of the First Detroit Michigan Planet—Sub-Deputy Stella Harter and Recorder Edwin W. Lemke. The Planet Commission is now complete.

Albert G. told of tonating a Tumor on the Retina of the Eye. Before Spectro-Chrome was used, the Medical Doctor, the Eye Specialist and Drugs were employed. Within 8 days of Tonation the

Kardoscil was raised from 50 to 70, with Scarlet on Area 4. A dark ball appeared in front of the Sufferer's Eye and now it has become light in color and greatly reduced. He again stressed the importance of the Kardoscil-Spiroscil Ratio and used this man's case to clear his point.

Ervin F. B. reported the case of his wife who was deathly sick with Pneumonia. Ratio was 2½ to 1. Kardoscil reached 120, Spiroscil in the 30's. Constipation also was present, for which Yellow was used. Fever responded to Blue with Magenta on Area 4. Blue produced Sweat, day after Low Critical Period and improvement soon followed. Imbalance is being rapidly removed.

Henry S., once referred as the "Helpless Wheel Chair Cripple", is not so helpless any longer. He has developed quite a vicious kick. Says he, "Two Tonations a day, with the Rational Food Of Man and I will walk by Christmas."

Aquila M. then stated, that for the past three years he adhered strictly to the Rational Food Of Man and should he miss, he soon feels the effects. He did not lose a day's work in three years, which could not be said of the preceding years.

This caused a discussion on Foods. Spectro-Chrome Home Guide was then read to show what quantity and in what combinations of Foods, the Rational Food Of Man consists. A lively discussion followed on whether Rice and Peas should be cooked separately. It was agreed that no more than enough was to be cooked for each meal. Raw Foods have higher Food Value than cooked.

Albert E. Greenham then described a typical "Colonel Dinshah table", as he experienced during his recent visit to the Central Office.

L. Roy D. explained that the Aura of different Foods is different and the mixing of chemicals is not proper.

E. W. L. spoke of his wife's Head Cold and the correction of the Imbalance. Turquoise Systemic followed 2 Hours 56 Minutes by Magenta on Area 4. Redness around Eyes and Nose was practically gone by morning and the Nasal Discharges also ceased. She is now convinced that Spectro-Chrome is an exact Science and will soon be with us.

Stella H. reported Intestinal Disorder in the case of a Sufferer with a nervous condition. Red helped.

Albert Greenham explained how the female Kardoscil responds to the Scarlet, expelling Dizziness.

L. Roy D. contended that Dizziness may be due to Low or High Blood Pressure.

The Recorder read Chapter 81 on "Paralysis And Motor Disorders" and Chapter 82 on "Spectro-Chrome In Paralyses", from Volume 3 of Spectro-Chrome Metry Encyclopaedia.

Our new Affiliate Rose C. told of Pus Sores around the Chin and Cheeks of her nephew. Three Tonations of Green normalated them.

Otto H. removed a Nasal Growth with Turquoise; both Nostrils are now open.

There is more in the connection of the Kardoscil with the Spiroscil than most students imagaine. It is the key to the Vital Vehicle, the Third Vehicle upon which rests the activities of the Prana. As long as that is correct, the Sufferer has every opportunity for recovery. Please, pay attention to this essential point during Tonation.

Ervin F. B. ought to be more explicit in asking for the Free Guidance. He telegraphed one night and as the Western Union offices were closed in Malaga, Newfield and Vineland, New Jersey, the Camden New Jersey office telephoned us from 30 miles away, asking our Central Office to pay for the telephone charge, which we declined, because, all such expenses must be prepaid by the sender. The telegraph operator in Camden insisted that it was important, but, our switchboard operator refused to receive the message saying it should come to us next morning through the usual channel. Then, the telephonic message was delivered us after asking the other end. The message merely said the "wife was low, Kardoscil 128, Spiroscil 46, Frontelim Good, Back Bad!" The signature was Mister B," with an address quite different from what was on the Register. We wired that the case was dangerous and requested him to talk on the Long Distance Telephone for full information. Close to midnight B. telephoned, but, could give no complete history of the case, so we asked him to mail a General Advice Chart. He agreed to do so, but never did and this is the first inkling we received from him about the recovery of the said Sufferer.

We appeal to the intelligence of our Affiliates in regard to such negligence. "Mr. B.", is not the full identification of the Registered Purchaser especially when even the address is different. Again, consider our situation of giving Free Guidance in a dying case, where the Kardoscil and Spiroscil indicate "Danger". On what grounds may we be enabled to know what is happening to the Sufferer at a distance and how does the seeker of the Free Guidance believe we could satisfy his requirement, unless he expected us to perform "Psychical Phenomena"? It is clearly stated in Spectro-Chrome Home Guide, page 89, that we MUST have certain information in detail, without which we can not give Free Guidance. Though we receive blame after blame for our hard Rules and severe methods, it is plain that our worthy Affiliates do not realize that with all what they say, it is nothing short of a marvel that WITHOUT SEEING OR PERSONALLY EXAMINING A SUFFERER, the Originator made it possible for them to obtain FREE GUIDANCE FOR LIFE and produce the precise results for which Spectro-Chrome is so justly famed.

Henry S. is to be congratulated. With his persistence, he should recover his equilibrium.

Discussion on Food always is very amusing. The tastes of people are so perverted by thousands of years of wrong living, that it is difficult to make them comprehend what constitutes genuine Food. The Medical Doctors classify even Alcohol as a "Food"! Rice and Peas cook at different temperatures, hence, unless a person is a cooking expert, there will be a "gruel" instead of an appetizing meal, if cooked together.

E. M. L. was right in attending to the Favorscope.

Dizziness essentially is the result of Cerebral Circulatory defect and may be produced either way—by more Blood or less Blood. Refer to Spectro-Chrome Metry Encyclopaedia, Volume 2, Page 652, under "Hot And Cold Sunstrokes". The same principle applies and the Kardoscil will show what is in progress.

You will have ample opportunity for solving your questions, when the Originator visits your city and holds two meetings on the 14th and 15th current, respectively.

———o———

EXPULSION OF LUCILLE J. McGRAW

Lucille J. McGraw, a Provisional Member of First Milwaukee Wisconsin Planet, Secretary to E. A. Ernest, was requested by the Board Of Trustees on August 27, 1935, to resign voluntarily, which she declined by a letter of August 29, 1935.

On August 31, 1935, the President ordered her name to be placed on the List Of Suspended Affiliates and she was again requested to resign voluntarily and surrender her Affiliation Card.

tees, on September 11, 1935, in a **Final Warning,** that she being affiliated with an Expelled Member whose Company was declared under Rule 42, Section 3, as an "Antagonistic Organization," unless she acted promptly by tendering a Voluntary Resignation, she was liable to be cited for Summary Investigation under Rule 87, Section 1.

To this she replied on September 12, 1935, asserting her alleged "rights" to sell Spectro-Chrome after "Graduation" and the like.

A Final Supplementary Warning was issued by the Board Of Trustees, on September 14, 1935, explaining fully the Rules and again requesting her Voluntary Resignation to save a situation that might be very unpleasant for her.

After due waiting, she was cited for Summary Investigation because of non-compliance, on September 25, 1935. The Charges against her were as follows:—

(1) Violation of Rule 42, Section 3, because of her affiliation with Ernest Distributing Company, adjudged by the Board Of Trustees as an "Antagonistic Organization".

(2) Violation of Rule 43, Section 7, for refusing to surrender the Affiliation Card, on demand after Suspension.

(3) Violation of Code Of Honor, Rule 22, by knowing of the Violations of Edward A. Ernest, conniving at them and not reporting him to the Principal Office.

(4) Violation of Rule 88, by conspiring to sell a Used Equipment, previously sold to Charles Follett, later repossessed by Ernest Distributing Company and resold to Walter Sutter, her signature appearing in connection therewith as the person receiving the money, $21.25 to account, on March 29, 1935.

(5) Violation of Rule 81, Section 17, in holding the Full Power Of Attorney for Edward A. Ernest, knowing that he was defrauding the public by selling repossessed Spectro-Chromes and then allowing the unused Registration Of Purchaser Blanks to go to the Central Office of Spectro-Chrome Institute, for issue of Bills Of Sale as new.

She replied to the Charges as follows:—

(1) That she "performed menial services for the company in the way of typing" and the like.

(2) That she did not know "why a Suspension should exist at all", so she refused to surrender the Affiliation Card.

(3) That she did not connive at any Violation, because she did "not know of any Violations committed by him".

(4) That she denied any Conspiracy, as "There was no such Conspiracy nor was there any deceit in any sale, that I was aware of."

(5) That she denied "any knowledge that Mr. Ernest was guilty of defrauding anybody as you set forth in paragraph No. 5."

She asked for a hearing before the First Milwaukee Wisconsin Planet, challenged the right of the President about the Summary Investigation and the like in depriving her of her "rights" and finally stated that her letter was "written in a perfectly friendly way."

To this, the Secretary replied by "Order Of The President", on October 4, 1935, that in a Summary Investigation, the President was the sole Judge under the Rules, that she was not working for E. A. Ernest in "menial services", but as "Private Secretary", according to his own Bulletin of September 16, 1935, that she was aware of his transactions, because all his letters bore her identification initials, that his transactions being fully published, were known to her, that she was adjudged as involved with him by "activities interlocked". Again under Rule 15, she was written that "we INVITE YOU TO RESIGN VOLUNTARILY," otherwise further action would be taken "in accordance with the Rules regulating such refusal of silence."

On October 7, 1935, she declared the procedure "not legal", asked for hearing before the First Milwaukee Wisconsin Planet, stating further that all was "an infringement on my rights which I propose to protect promptly and speedily."

On Wednesday, October 9, 1935, the President Dinshah P. Ghadiali ordered her Expulsion with forfeiture of all rights, titles and privileges, under Rule 87 and the Secretary Irene Grace Dinshah informed her about it, adding that she could under Rule 87, Section 3, appeal within 28 days to the Board Of Trustees against the Decision of the President.

Upto the time of going to the Press, no such Appeal was filed.

REVOCATION OF FREE STUDENTSHIP

Spectro-Chrome Institute having adjudged Lucille J. McGraw, "an undesirable student in Spectro-Chrome Metry", revoked on Wednesday, October 9, 1935, the Free Studentship, which was granted to her as a courtesy.

Her name is struck off from the List Of Students.

FIRST MILWAUKEE WISCONSIN PLANET

(HOTEL WISCONSIN, MILWAUKEE, WISCONSIN)

Meeting Of Thursday, October 3, 1935

Successful Use Of Spectro-Chrome In:—Cataract, Burn, Arthritis, Constipation, Paralysis, Sinus Disorder, Insomnia, Heart Disorder, Boils, Fever, Stomach Trouble, Liver Disorder, Backache, Colds, Abscess, Rheumatism, Diabetes, Dizziness.

Reported By:—CAROLINE ZIMMERLING, *Recorder*

Deputy Otto H. Lutz was in the chair.

The Recorder read the following letter from the Principal Office:—

Monday - September 30, 1935
Chronos:- Turquoise 47 Black.

Mr. Otto H. Lutz, Deputy,
First Milwaukee Wisconsin Planet,
3823 North 5th Street,
MILWAUKEE, WISCONSIN.

Dear Deputy Lutz:

1. Occasionally we receive a complaint of the non-receipt of the Spectro-Chrome; investigation then shows that the Affiliate forgot to send in time the change of address.

2. Our responsibility ceases as soon as the copy is mailed. The reason for the non-receipt is that Second Class Mail Matter is NOT FORWARDED by the Post Office to the New Address. The said package then stays at the Old Address or is returned here after a loss of time.

3. Please therefore draw the attention of the Affiliates that we should have *advance* notice of such change of address, to enable us to change the Stencils for the next issue.

4. We need suggestions for the forthcoming Annual Convention. Is there any special subject about which the Affiliates desire elucidation? Kindly let them write us promptly, so we may attend to their wishes if necessary or possible. The Program will be then completed and published in the next issue.

5. It will be appreciated if the accommodation be insufficient to find some other location and recommend us with full particulars in time. We are ever anxious to serve your interests, if we know what is asked by the Affiliates. We refer to the Planet Meeting Room.

6. Extending our thanks for the efficient manner in which the Affiliates help the Cause, we are

IGD/ID

Yours For Human Service
BOARD OF TRUSTEES,
by *IRENE GRACE DINSHAH*,
Secretary.

By Airmail

It was decided that the room was sufficiently large.

The Recorder read Chapter 66 of Spectro-Chrome Metry Encyclopaedia, on "Red Corpuscles And Anemia". The Deputy read some of the list of the various kinds of Anemia as known to the Medical Profession.

Frank Burns volunteered to read two following Chapters at the next meeting.

Pauline W. is taking care of a woman with Cataract. The Medical Doctor said an operation was necessary. She has had 5 Tonations so far and notices a change already. Green Systemic was used. She is building up her own condition with Orange and is improving nicely.

Bertha L. normalated a case of Burn which was very severe. It resulted from a "Permanent Wave". The woman went to a Medical Doctor 4 times and became worse instead of better. The Doctor was afraid of Erysipelas. Her head was a running mass. The first Tonation of Spectro-Chrome seemed to help her. In 3 weeks, her head healed very nicely and there was no more trouble. Turquoise Local was used entirely. This seemed to dry more than with the Blue alone.

Bertha L. has also a case of Arthritis which she is tonating for the past 9 weeks and the Sufferer is improving nicely. Changes in the weather, however, have a great deal to do with the effect of the Spectro-Chrome and the condition of the person. She can not seem to overcome the Pain the person has, however. Indigo does not seem to help much, otherwise the Infra-Green Division is being used. This is a case of long standing, being brought on by "poison from a Goitre". This seemed to appear in the legs. Green, Lemon, Yellow and Orange are being used in their proper order. Constipation was relieved and the Red was used for the Liver with good results. Magenta is being used once a week on Area 4 and now it is being used for the Area 18 once a week.

Ida B. is tonating a woman with Paralysis in the arm and for Heart Disorder. The Medical Doctor said the arm would shrink from a Diathermy. When the Sufferer wanted to take Spectro-Chrome Tonations, the Medical Doctor told her not to take them, as that would burn her! Instead, the arm became well and her Heart was normalated. For the Arm, she used Indigo for the Pain first and the Sufferer slept after that night without Pain. Then Yellow was used for the Paralysis and a few Green. Only 17 Tonations were used altogether.

She also tonated a woman with Sinus Disorder. After a few Tonations, she got a Pain in one Tooth. After examination, the Dentist found Pus and pulled it. The woman continued her Tonations and is much better. Green, Lemon and Orange were used.

Ida B. also had a case of "Nervous Disorder" in a woman; she could not sleep. Now, after using Spectro-Chrome, she feels better and can rest. Magenta and Turquoise were used.

Arthur G. had a case of a person who could not sleep and whose Heart was so bad that she could not walk upstairs without panting. She had also a Pain in her back. Now, her Heart is normal, she sleeps well and has plenty of pep. Her back still pothers her, however. One week of Green and two weeks of Lemon were given and now she is taking Orange for her back.

Frances B.'s husband broke out with Boils. He used Orange to draw the Boils to head, Yellow to open them, Green to cleanse, Indigo to close. They all cleared without a scar. Her son rose one day with Headache, but, after 1 hour of Blue was all right. Another child had Fever; 1 hour of Blue normalated the condition.

The Deputy said that his two grand-daughters were "just like dead" about 2 weeks ago. They had no pep and were burning with Fever. Two Tonations, Blue and Turquoise normalated them.

Olive B. is still tonating her mother for Liver and Stomach Disorders. She is coming along finely. It is a case of long standing. Infra-Green Colors are being used and Magenta with an occasional Scarlet. She is repeating the Infra-Green now. She is using the Equipment to keep the rest of the family in health. Her husband had a Backache which came on very suddenly; a case of Lumbago. After 1 Turquoise Tonation, he went back to work feeling well. Area 18 was tonated.

Emily M. had been ailing for 13 years. She had bad Colds continuously and it cost her about $1,000.00 with a Medical Doctor. She became so affected that she could not stand or sleep. She had an operation and doctored for Sinus Disorder. Her Lungs were full of Pus and badly scarred. She lost a great deal of weight and in a general way was more dead than alive. She was burnt to a blister with Ultra-Violet Rays for treatment. Frank B. one day told her about Spectro-Chrome, so she bought one. In one month she had no more pains and in 2½ months she felt fine. She gained 20 pounds in weight. Even her hair, she says, is turning back to normal dark color. Green and Lemon were used; Orange for the Lungs. She followed the Irradiation Free Guidance to the letter and had wonderful results. A short time ago, she developed Athlete's Foot. Three Turquoise Tonations righted her.

Jacob S. is keeping on with the Yellow and is improving right along.

Carrie M. normalated an Abscessed Toe in 2 Orange Tonations. She asked if anyone could tell her how to normalate an Acid Stomach. Bertha L. answered the inquiry with the information that she normalated a case of 25 years' standing. Not only the Acid Stomach, but Constipation was removed by the use of Lemon.

Elsie L. had a case of Rheumatism in the shoulder which was very painful. After 3 Tonations, the Sufferer was able to sleep and in 1 week went back to work. Gave first Orange. This was "too severe". Green was then given for 4 days and then Orange again. She had a case of a bad Head Cold which responded to Turquoise and Orange.

Rose M. never received instructions in using her Spectro-Chrome. She read her book, however and found where to start—at the Green. Her daughter has Tuberculosis and had been in bed for 9 months. She seemed to be getting along all right, but, there was still a cavity in the Lung. For this reason, she bought Spectro-Chrome. Green was used for a month and the pain in the side left her entirely; also the Constipation was corrected. For the past few days, she has been using Lemon and finds now that a great deal of Phlegm is being loosened. Her daughter is making good progress, she feels. As for her own case, which she thinks is one of "Nerves," she has not been able to get any Guidance from Spectro-Chrome Institute. She says she answered the questions as fully as possible, but, does not seem to have been able to give enough information.

Joseph W. says he is not pothered with a Numbness in his fingers. His Guidance told him to use Yellow now and that is what he is doing.

Bernard K. got a Spectro-Chrome for his son who is 6 years old and is suffering from Diabetes. Upon getting Spectro-Chrome, he placed the child on the Rational Food Of Man and stopped Insulin. The child went into a Coma, so Insulin was resumed. Also, he is on a restricted "Carbo-Hydrate Diet." Green, Lemon and Yellow were used without results. The Members in the Planet Meeting were very united in telling B. K. to stop use of Insulin in tonating his son and place him once more on the proper system of eating, giving Starches and Sugars. An interesting discussion on Diabetes took place. Carrie M. said that the Sufferer should be given Orange Juice to bring him out of the Coma. Her husband, having Diabetes, always found great relief in using the Blue. Also, from the use of Spectro-Chrome, he

threw off so much waste matter through the Bowels, that it was almost unbelievable. Bernard K. however had results in a case of Ringworm, which responded to 3 Turquoise Tonations.

Marie Z. said she had Diabetes very badly and using Spectro-Chrome for 8 weeks, made her feel very well. It was of 10 years standing. She used Magenta and Green and 6 weeks of Lemon. Spectro-Chrome also helped her when her neck was swollen from a very bad Cold.

Frank B. reported that his 13 year old son, born paralyzed was improving steadily.

Belle K. reported about her mother's Heart. She stopped the medicines immediately upon beginning with Spectro-Chrome and within about a month she was feeling well. She herself had a Mastoid Disorder for which she was to have an operation, having had one nine years ago. This has been draining all these years and she began having dizzy spells. After using Spectro-Chrome a short while, the Dizziness vanished and the Pain also, but the Ear continues to drain. She wanted to know if this was all right. The Deputy told her it would continue to drain until it was clean, that she should use Spectro-Chrome to cleanse it and when it was ready to stop draining, to close it with Indigo.

Nina M. said she has trouble with poor Circulation. Her fingers become cold and white. She used Green, Lemon, Yellow and Orange and repeated these several times, but did not seem to help. She gets the best results for Constipation, by using Orange.

Hedwig N. said she had almost the same troubles and found Scarlet on Areas 4 and 18 very beneficial.

The meeting was very orderly and interesting.

An error was made in the last report. Edna B. was reported to have had good results in her son, in Bed Wetting. This was erroneous; she had no results.

Ida B. should tell that Medical Doctor that Spectro-Chrome does not **burn** anything except their hearts!

Elsie L. should consider the idea of not starting Persistent Disorders with Orange. It is like starting an automobile on High

Gear uphill. Follow from the Green down, thus letting the Inertia of the Auric Vehicle to pick momentum gradually.

Rose M. sent Spectro-Chrome Institute numerous Charts about her several young daughters—**all Charts Incomplete.** A two page explanatory request was made and yet she did **not** write them correctly, but, started to argue. Another long set of explanations followed, but, she wrote about article 14 in the General Advice Chart that it was "this way and that way", without giving the information needed about the Habits of the girls. About herself, she said she had "this and that", but did NOT give any information about certain questions asked in Article 14. She has absolutely NO reason or right to complain. Spectro-Chrome Institute shoulders a tremendous moral responsibility (if not legal responsibility) in giving Free Guidance to Purchasers, from a distance; it shoulders a grave moral responsibility with powerful occult consequences and if Rose M. will NOT send her General Advice Chart entirely complete in EVERY DETAIL ASKED, she will never receive Free Guidance. Spectro-Chrome Institute service is impersonal and rigidly uniform and anyone not following the established Rules And Regulations will receive no attention, no matter who they may be. We certainly **are able** to guide her about herself and her tuberculous daughter and the other members, so her investment may fructify and she may bless Spectro-Chrome, but, she will have to answer that Article 14, in completion to receive reply. We know her troubles, but, she MUST conform to the Rules And Regulations, whether they suit her views or not. Older Affiliates, are requested to make her acquainted about the inflexible methods of the Originator of Spectro-Chrome. We are very anxious to serve her, for her children's sake. She NEEDS Spectro-Chrome and it will benefit her family, if she gets the Free Guidance.

Bernard K. is one of those who will mix Medicine and Spectro-Chrome. He must take his choice. Spectro-Chrome can not have Insulin and "Restricted Carbo-Hydrate Diet" business. Either the Sufferer receives Insulin and no Spectro-Chrome or he gets all Spectro-Chrome. If any medical interference be given, we give no Free Guidance. It is but fair to all concerned.

Nina M. should send her General Advice Chart. It would appear though she needs Magenta on Areas 4 and 18.

Edna B. should send for Irradiation Free Guidance.

We appreciate the business-like manner in which the reports come for publication.

FIFTEENTH ANNUAL CONVENTION OF SCIENTIFIC ORDER OF SPECTRO-CHROME METRISTS WILL ASSEMBLE IN DETROIT, MICHIGAN

SPECTRO-CHROME METRY IS THE SCIENCE OF AUTOMATIC PRECISION HAVING:— NO DIAGNOSIS — NO DRUGS— NO MANIPULATION — NO SURGERY. IT USES EXCLUSIVELY ATTUNED COLOR WAVES.

1. You are cordially invited to all our Fifteenth Annual Convention Sessions, in the Ball Room, Hotel Tuller, Detroit, Michigan, on **Thursday, November 21, 1935 and two subsequent days.**

2. You will hear about the marvelous and near-miraculous results produced by Spectro-Chrome in the conquering of Human Disorders, many declared "hopeless and incurable", according to other methods.

3. You will have exceptionally intellectual entertainment, in the exponence of the most radical theories, regarding Tuberculosis, Cancer, Diabetes, Syphilis, Paralysis, Fevers, Arthritis, Inflammations and other scourges of humanity, for which other Healing Systems can do very little. The actual facts will startle you. The Originator of Spectro-Chrome Metry, will help you from the platform to help yourself, in solving your personal problems. Do not miss this opportunity; it is worth any inconvenience to hear the Originator, who is a well-known Researcher.

4. The following is the program:—

Thursday, November 21, 1935—10 A. M. to 12 M. and 2 P. M. to 5 P. M.

INFORMAL SESSIONS

(a) Exhibition of the latest models of Spectro-Chrome Patented Apparati, Inventions of Colonel Dinshah P. Ghadiali, the Originator of the Science of Spectro-Chrome Metry.

(b) The above Informal Sessions are open to Members, Provisional Members and Collaborators, as well as friends introduced by them, who will be given informal warm reception and all requested information about the Spectro-Chrome Apparati and our Organization. Invite your friends. You may meet the Originator **personally** in these Informal Sessions, but, please refrain from asking him anything about your Ailments.

(c) Colonel Dinshah distinctly desires the public to understand that HE IS NOT IN THE PRACTICE OF MEDICINE, CONDUCTS NO CLINICS, GIVES NO PERSONAL CONSULTATIONS AND ACCEPTS NO GIFT, PRESENT, CONTRIBUTION, FEE, DONATION, COMPENSATION, TIP, GRATUITY, INDUCEMENT OR TOKEN, OF ANY KIND WHATEVER, FROM ANY PERSON, DIRECTLY OR INDIRECTLY. He is glad to serve Suffering Mankind through the fruits of his laborious researches.

5. Thursday, November 21, 1935 — 8 P. M. to 11 P. M.

OPENING SESSION

Opening of the Fifteenth Annual Convention by our President Colonel Dinshah and Lecture on "Science Of Spectro-Chrome Metry, Its Conception, Development And Perfection." This Lecture will be illustrated with Experiments and Demonstrations.

Questions asked from the audience will be cheerfully answered by the Originator.

This Session will be open. Invite your friends.

6. Friday, November 22, 1935—10 A. M. to 12 M., 2 P. M. to 5 P. M. and 8 P. M. to 10 P. M.

(a) Address by Board Of Trustees and Council Of Governors.

(b) Case Reports and Testimonies by Members, Provisional Members, Collaborators and Sufferers.

(c) Lecture on "Rational Food Of Man", by Irene Grace Dinshah, Vice-President, Spectro-Chrome Institute and Distributor General for Spectro-Chrome Equipments. This discourse will be illustrated with the Stereopticon and questions on Food Cooking will be answered. Every woman and man should hear Irene and see her six sons, who are being trained by her and are raised in accordance with the Rational Food Of Man System, making a family that will carry on the work of the Originator for the Service Of Mankind on the same principles.

All the above three Sessions of this day will be open. Invite your friends.

Friday, November 22, 1935—10 P. M. to 11 P. M.

CLOSED SESSION FOR THE GOOD OF THE ORDER

Open exclusively to Members And Provisional Members in good standing. All MUST present their Affiliation Cards to attend this Special Session. This Rule has NO EXCEPTIONS.

7. Saturday, November 23, 1935—10 A. M. to 12 M. and 2 P. M. to 5 P. M.

(a) Suggestions for Amendment of our Constitution and By-Laws.

(b) Election of Board Of Trustees and Appointment of Council Of Governors.

(c) Questions on Spectro-Chrome Metry from the audience and Answers by our President Colonel Dinshah.

(d) Unfinished matters.

8. Saturday, November 23, 1935 — 8 P. M. to 11 P. M.

(a) Lecture by Colonel Dinshah on "Higher Side Of Life In Relation To Astronomical, Gravitational And Radiant Phenomena." It will elucidate what is now known as Dinshah's Science Of Sympathometry. The new perfected Sympathometer will be shown in

(b) Closing of the Fifteenth Annual Convention.

WARNINGS:—

(a) Positively NO SMOKING will be allowed.

(b) We reserve the absolute right of refusing admission to or removing from the premises any person, thought undesirable by us, without giving any reason. EXPELLED AFFILIATES WILL NOT BE ADMITTED T OANY SESSION.

(c) Writing any part of speeches, discussions or any transactions at any time of the Sessions is positively forbidden and any person endeavoring to do so, will be promptly expelled.

PRESS REPORTERS

Accredited reporters of the public Press will be welcome, provided, they show their credentials and obtain permission from the Secretary's Desk to take the report of the proceedings. Such reporters will occupy special seats provided for them. We do not look for cheap publicity and our Service To Mankind asks for no favors of Free Advertising. All we ask is **Fair Play and Truth,** hence the restriction, as we nnow that high class Reporters are conscientious.

OUR AIM:— SPECTRO-CHROME—IN EVERY HOME.

STILL AT THE OLD GAME!

Notwithstanding our shortness of space, we gave full scope to E. A. Ernest, to ventilate his side, thinking it was due him for justice. However, he is not satisfied yet and continues to use the mails against us to send all kinds of threats and menaces. He is so courteous and sportsmanlike, that we never receive a copy of what he sends to Purchasers of Spectro-Chrome, except through indirect channels. Here is another attempt by him to tell the "Truth" (?) to the public:—

Telephone Edgewood 5321

Exclusive Wisconsin Distributors *for Equipment of* Spectro-Chrome Institute

ERNEST DISTRIBUTING COMPANY

(*E. A. Ernest, S. C. N.*)

3106 North Hackett Avenue

Milwaukee, Wisconsin

September 16, 1935

FACTS REGARDING FALSE CHARGES MADE AGAINST E. A. ERNEST BY THE SPECTRO-CHROME INSTITUTE

This Bulletin is addressed to all Members and Provisional Members of the Scientific Order of Spectro-Chrome Metrists, and to all who may purchase, or have purchased, Spectro-Chrome Equipments through me.

My reason for forwarding this Bulletin to you is to give you the exact facts and information concerning certain articles appearing in the August issue of the Spectro-Chrome magazine, and other articles which, I understand, are to appear in future issues, regarding me. In my estimation, the object of the Spectro-Chrome Institute in charging me with certain violations was for the purpose of securing technical reasons to cancel my agreement as exclusive Distributor for the State of Wisconsin, solely for their financial gain, at a great loss to me.

All of my dealings with customers, and with the Institute, have been fair, equitable, open and above board and, therefore, I am pleased to submit the following for your consideration. Further information will be furnished at your request, thereby making it possible for you to judge for yourself.

1. Under date of August 9, 1935 I received a communication from the Scientific Order of Spectro-Chrome Metrists, being in the form of a charge lodged against me by the Spectro-Chrome Institute for violations of good faith and conspiracy because I sold two Cabinet Equipments to my Agent, Mr. Walter Sutter. These Equipments were shipped direct to this purchaser by the Institute. The purchaser, in turn, delivered one of the Equipments, taken directly from the carton in which it had been received, to some poor sufferer in great need of Spectro-Chrome, advising the Institute accordingly. Certainly there was no violation of good faith or conspiracy in connection with such a deal and I leave it to you to judge for yourself. I pleaded guilty to having sold the two Equipments to Walter Sutter, as it was a fact and I had the signed orders in my possession. These documents were submitted in evidence and are now in the hands of the Scientific Order. However, I did not plead guilty to conspiracy and the like.

2. Another letter was received from the Scientific Order of Spectro-Chrome Metrists on the same day, informing me that the Spectro-Chrome Institute had placed certain charges against me for having accepted an order for a Cabinet Equipment from a certain purchaser on the time payment plan in accordance with my regular terms. However, due to the fact that this purchaser was very skeptical, he wanted to give Spectro-Chrome a trial for five weeks and, therefore, I obliged him by putting a notation on the order to the effect

that, if after he had used the Equipment for five weeks, he was not satisfied, I would call for it and cancel his obligation in connection with the balance due me. For my great faith in Spectro-Chrome and this promise to the purchaser, I was charged with violating the Code of Honor and endangering the reputation of the Institute. In this instance, I pleaded guilty to having accepted the order on the basis mentioned but did not admit violating any rules or regular trade practices. A copy of this order was mailed to the Institute but they refused to make shipment. I am sure that you, nor any one else, could possibly construe such action as a violation.

3. Under date of August 11, I received a letter from the Scientific Order of Spectro-Chrome Metrists, advising that the Spectro-Chrome Institute had placed charges against me for having violated certain rules and consorting with one of my Agents to commit dishonesty and conspiracy, due to the fact that I sold an Equipment to a certain purchaser who called for it here in Milwaukee, strictly in accordance with the contract entered into on July 7, and July 9, 1935, respectively, at which time I was exclusive United States Distributor for Spectro-Chrome. This purchaser called for the Equipment on July 23 and, after paying a certain amount of cash, took it with him to Cicero, Illinois. The Institute claims I violated certain rules because this particular Equipment had originally been purchased by another party who refused to use it. However, the last purchaser was so informed and was perfectly satisfied to accept the Equipment with that understanding, due to the fact that his invalid boy was in urgent need of Spectro-Chrome and, further, because there was a saving of the express charges, in the amount of $3.90, which had been paid by the original purchaser.

   I admitted that I was guilty of selling this Equipment on the basis mentioned and a copy of the order was submitted in evidence. I did not plead guilty to the charges mentioned. Certainly this cannot be construed as a violation of any rules, regulations or regular trade practices.

4. The Scientific Order of Spectro-Chrome Metrists informed me on August 11 that the Institute had also placed charges against me to the effect that I had given certain definite information, or advice, to a purchaser residing in Stevens Point, Wisconsin as to the specific colors to be used. This was called a violation of Breach of Faith with the Institute. I did give the purchaser information regarding the operation of the Equipment, and also referred him to various pages of the Home Guide, so he would be familiar with the instructions issued therein. I also requested this purchaser to prepare an Advice Request Blank immediately and forward it to the Institute. How can any one construe this to be a violation? I did not plead guilty to this charge.

5. Under date of August 12 I received a letter from the same organization, informing me that I was charged with violation of certain rules of the Order by being unfair to the Agents, due to the fact that the Institute claimed I instructed and sent one certain Agent to a purchaser of a Spectro-Chrome Equipment to offer said purchaser a certain portion of her commission if names of prospects were furnished. This charge is absolutely false. By what authority could I prevent my Agents, or any other person, from spending their money as they saw fit? Such a charge certainly cannot be construed as a violation on my part in any way, shape, or manner. I did not plead guilty to any part of this charge. Furthermore, what right has the Institute to dictate as to how I should conduct my business?

6. In another communication of the same day I was further charged with paying commission to an Agent who was not the owner of an Equipment and, therefore, guilty of violation of the Constitution and Rules. This charge is also false. I did sell an Equipment to one certain woman and, after it was received, she decided to have it registered in the name of her son which was certainly her privilege. However, this purchaser was a duly qualified Agent and made a sale, or recommended Spectro-Chrome to another person. After the sale was completed, I mailed this Agent a check covering the commission due her. Would you call this a violation? I did not plead guilty to this charge.

7. I was informed on August 13 that I had been charged with violation of the rules and regulations of the Constitution because I re-sold an Equipment originally purchased by a person living in Illinois. For some unknown reason, this man had never opened the carton and, consequently, had never registered the Equipment, having absolutely refused to use it. Therefore, after corresponding with this purchaser, I agreed to pay him a certain amount for the Equipment. The offer was accepted and the Equipment shipped to Milwaukee. After some weeks, one of my Agents wanted to purchase an additional Equipment. This particular Equipment was sold to him as a brand new model, which it was, but, due to

the fact that it had been originally purchased by another person, it was sold to the Agent at a discount. The Agent, knowing this was a brand new Equipment, even though it had been shipped to another purchaser, sold it to one of his prospects. How could anybody call this a violation of regular trade practices, unfairness, fraud or the like? I pleaded guilty to having accepted the Equipment from the original purchaser and selling it to my Agent. How ridiculous to charge me with violation! I did not plead guilty to the charge.

8. In a second letter of the 13, I was charged with further violations because I acted as an Agent for a certain party and assisted her in disposing of her Equipment, having paid her the sum of $35.00 for the Equipment. This Equipment was brand new, and had never been used but had been registered. After having the Equipment, together with Bill of Sale and Assignment in Blank, in my possession for some little time, I decided to present it to a Sister, representing a Catholic organization in the State of Kansas, and shipped it to her with my compliments. For doing this, I was charged with violating the Code of Honor, the Agreement with the Institute, and Interstate Commerce Rules. I disclaim violations of any kind. This shipment was made on July 17, 1935 at which time I was acting as exclusive Distributor for the entire United States. I pleaded guilty to having donated the Equipment, as it was a fact, but did not plead guilty to the charge mentioned.

I regret to inform you that the person to whom this Equipment was donated has received such threatening communications from the Institute that she returned the Equipment to them. It is now in their possession and they have requested information from the original purchaser here in Milwaukee as to disposition of it. This, of course, is the most unethical, unreasonable and unbusinesslike course of procedure I have ever heard of. The original purchaser has instructed the Institute to return the Equipment to the rightful owner in Kansas immediately, as it has been assigned to her.

Differences of opinion and many disagreements, have existed for some time between Colonel Dinshah, the Spectro-Chrome Institute, myself and my organization. I have requested a personal conference, on numerous occasions, for the purpose of having a mutual understanding whereby suffering humanity might be served in a greater capacity, and in a more efficient manner. Notwithstanding my many requests, the President of the Spectro-Chrome Institute absolutely refused to grant me the courtesy of such a conference, without giving any reasonable excuse. Who ever heard of a manufacturer refusing to see the exclusive Distributor of his products.

It might interest you to know that Colonel Dinshah offered to come to Milwaukee and present all his evidence before the entire First Milwaukee Wisconsin Planet at any time. I immediately accepted his offer but, for some unknown reason, he refused to come.

The charges made against me, as printed in the Spectro-Chrome magazine, and mentioned in the letters referred to above, are all unjustified and have been the direct cause of great embarrassment to me, as well as much damage and financial loss.

I am informed that Members, and Provisional Members of the Scientific Order of Spectro-Chrome Metrists were instructed in a letter from Headquarters, read at the last Planet Meeting, to refrain from associating, in any way, with me or my company, whether in connection with Spectro-Chrome or otherwise. This has also greatly embarrassed me and caused further damage and financial loss.

I am very sorry that I am compelled to take immediate action, through authorative sources, to clear my name and reputation. My sympathy goes out to the Originator of Spectro-Chrome for allowing himself to get into such an agitated frame of mind, standing in his own light, and actually making it impossible to reach his goal—Spectro-Chrome in Every Home—and serve suffering humanity.

My faith in the Spectro-Chrome System has not been changed, but I have lost all faith and confidence in the Originator of the System and his immediate organizations.

I AM CONTINUING TO SELL SPECTRO-CHROME EQUIPMENTS IN THE STATE OF WISCONSIN. PLACE YOUR ORDERS THROUGH ME AND RELIEVE YOURSELF OF ALL RESPONSIBILITIES.

My attention has been drawn to the fact that threatening letters, containing charges of violations, have been received by other Members from the Scientific Order of Spectro-Chrome Metrists. All of these, in my estimation are most ridiculous, unreasonable and without

grounds. For instance, my private secretary, Miss Lucille McGraw, has received several threatening letters, accusing her of certain violations which I know are false, as she is absolutely innocent. Everything Miss McGraw has done in connection with Spectro-Chrome, while acting as my secretary, has been done with full authority from me.

I again wish to thank you for the assistance you have given me, and for your continued cooperation.

With kindest regards and best wishes, I remain

Sincerely yours,

E. A. ERNEST

"OUR AIM: SPECTRO-CHROME—IN EVERY HOME"

We can not pass by such a letter of nonsense and falsehood, without giving an open reply. The figures refer to his paragraphs. The pages are in September 1935 Spectro-Chrome.

1. Read page 1649. He mislead Walter Sutter and the Equipment 4C-4344 was rented without title.

2. He had no right to make any conditions to take back Spectro-Chromes after being used. This was an open Violation of the Spectro-Chrome Understanding For The Bill Of Sale. Refer pages 1649 and 1650.

3. That sale to the farmer of Daggett, Michigan, analyzed at pages 1650 and 1651 was nothing short of want of honesty, fraud and capitalistic money-grabbing. A human service like Spectro-Chrome, never could tolerate such flagrant practices.

4. Not only in this one case, but, in many such cases, E. A. Ernest injected himself in regard to Guidance and we admonished him against it.

5—6. We let him go free in these matters as they would have wasted too much of our time and space and would have involved many innocent ones.

7. Refer to pages 1662 and 1663. That Spectro-Chrome, we further found, went through many more hands as a hired Equipment. It is yet unregistered. E. A. Ernest made a clean profit of **$73.24 NET** on this one Spectro-Chrome. If the American public would think such methods were honest, America would not be worth its name!

8. That Spectro-Chrome is in our hands and shows plenty of use. For this man, put in authority by Spectro-Chrome Institute as Distributor, to repossess a Spectro-Chrome (never used as he says) for $35.00 and gift it to another organization without sanction of his employers was inexcusable. Refer to pages 1663 and 1664. The as-

never appeared thereon during the transaction.

We sent no threats to anybody, but, merely placed the facts. E. A. Ernest, in his dealings, always was found to evade issues and we are not accustomed to taking prevarications. That Equipment will stay here until the false date matter and the like is cleared.

The so-called "differences of opinion" were sufficient for the Originator to keep E. A. Ernest out of his sight, because Dinshah had **no intention of allowing** E. A. Ernest to stay longer with Spectro-Chrome. E. A. Ernest may be a good business man for his "ultra-smartness", but, no person who will not walk on the straight lines laid, can stay in Spectro-Chrome Service. Dinshah did not refuse to interview E. A. Ernest as a "manufacturer", but, if the Truth be told, he refused the interview because he **did not want** E. A. Ernest connected with Spectro-Chrome any longer. Dinshah deemed him unsuited and unfit for the work and that ended the story. He still refuses to interview E. A. Ernest and as time passes, more reasons manifest to maintain the barrier permanently.

Does the public want more reasons? E. A. Ernest sold an Equipment on **Time Payment to a girl of 13!** Another to a boy of 17! That man was going with his unbusiness-like and unlawful pranks so fast, that had Spectro-Chrome Institute not put him out, in a short time E. A. Ernest would have become the owner of the entire Spectro-Chrome organization and Dinshah would be sunk!

E. A. Ernest still maintains his false colors, by stating he will "continue to sell Spectro-Chrome." **He will not.** All he can do is to unload the few he had on order **before** the forfeiture of his Agreement.**After that,** the name of E. A. Ernest will to us be a thing of the past. We have no desire to waste our valuable time and space on him and we are extremely happy that he has "lost all faith and confidence in the Originator" because to the Originator, the faith and confidence of E. A. Ernest would be a blot on Dinshah's honesty. He is out—let him stay out.

———o———

EXPULSION OF WALTER SUTTER

Walter Sutter, the Agent who was involved and investigated in connection with the activities of E. A. Ernest, was finally adjudged "Guilty" after mature deliberation and was ordered expelled from the Scientific Order Of Spectro-Chrome Metrists, with forfeiture of all rights, titles and privileges pertaining thereto.

However. he has the privilege under Rule 90, of applying for

Reinstatement at the discretion of the Board Of Trustees; the President recommended Clemency in his case.

———o———

DINSHAH VISITED DETROIT, MICHIGAN

Two public meetings were held at 8 P. M. Monday, October 14, 1935 and Tuesday, October 15, 1935, in the Small Ball Room, Hotel Tuller, Detroit, Michigan. The room was filled to capacity to hear the Originator of Spectro-Chrome, who answered questions in regard to Spectro-Chrome Metry.

It was then urged by numerous Members and Visitors—friends of the Affiliates, that this year's Annual Convention be held in Detroit, Michigan. The enthusiasm expressed was so overwhelming that the Originator bowed to their wishes and reserved the Ball Room for the same dates as formerly intended for Milwaukee, Wisconsin.

Some time ago, Dinshah was notified that the Medical Doctors of Milwaukee County would be pleased if this year's Convention be held elsewhere. Well, who can now say that Dinshah dislikes to oblige them?!

———o———

TWO FAVORSCOPES IN THIS ISSUE

Spectro-Chrome is generally a 24 page monthly issue. Last month, however, it was 48 pages and this issue has 32. Thus, owing to the excessive pressure on the Central Office, because of the Annual Convention, October and November 1935 issues are combined. Thus, you will find November 1935 and December 1935 Favorscope are included in this publication. Please refer to the proper one. The December 1935 Spectro-Chrome will be published after the Annual Convention is over.

———o———

HOTEL RESERVATIONS FOR ANNUAL CONVENTION

Those attending our Fifteenth Annual Convention, will do well to be at its Headquarters in the Hotel Tuller, Detroit, Michigan. Resident Manager Robert Hensley will be pleased to receive your inquiries and if you mention you are coming for the Convention, will give you special favorable rates. We expect that the Ball Room will be packed to capacity, as Spectro-Chrome Metry is well known and proven in its merits in the Automobile City. Be sure, therefore, to reserve your accommodation in time and attend all the sessions. You will personally meet there Dinshah and his successors—the living Mogen David!

SIGNS OF THE TIMES

At a meeting of the Child Society Group in Vineland, New Jersey, an article was read by Mrs. Ann Capo, on "Sex Instruction In The Home", recommending that "mothers give truthful answers to children's questions regarding sex—in a kindly, sympathetic manner."

The move is in the right direction. Those days are gone, when mamma could evade the issue by injecting the stork story, because girls and boys learn more in the American schools nowadays about this ticklish subject, than their parents knew before and in most cases even after marriage.

Sex and Sex Congress MUST be differentiated in the parental mind. The child can either learn from books, questions or—personal experience. The stupid Laws Of Obscenity, started by prurient, rotten ideas of brainless, idiotic, moronic, worn-out, old fogies during the early Victorian period, still persist and mainly dominate our progressive civilization. Teach the inquisitive mind and train it to look at Sex from its holy, Godly, creative side and never inject into the question any allusion to so-called immorality. When thus, the child is made to look at the question from a purely physiological aspect, it will direct its energies solely to the high spiritual side and sensuality will be permanently banished.

MANJAN AND LAVAN MANUFACTURE AND SALE RESUMED

The manufacture and sale of Manjan (The Dinshah Dentifrice) is resumed. It will be supplied in the original can at 25 Cents a can plus 5 Cents handling charge.

Similarly, Lavan (All-Salt-Sea-Salt) will be supplied in a smaller carton than before, at 25 Cents a carton plus 15 cents handling charge.

These articles can be had direct solely from the Spectro-Chrome Institute, Malaga, New Jersey. There is no Commission or Discount and the sale carries positively **No Guarantee, Warranty or Liability** in regard to size, measure, quality, quantity, composition, effect or result, from their use by any person, firm or corporation, all Suits at Law and (or in) Equity being understood to be expressly waived by the Purchaser.

This condition was found to be essential, because of the instability, elasticity and meaninglessness of the Federal Food and Drug Laws. Manjan And Lavan are **DINSHAH** products and that name should be enough for all purposes.

AN INCOMPETENT MAGISTRATE

Thomas G. Tuso, Prosecutor of Cumberland County, New Jersey, assailed the papers made out by Justice Of The Peace Jack Gross of Vineland, N. J., as being faulty in three cases of Disorderly Conduct, before the Honorable Jurge Leroy Loder.

We agree with the Prosecutor. Jack Gross is a baker and a pugilist and has no knowledge of the Law. It is beyond our comprehension how such an incompetent person was appointed by Governor Harold G. Hoffman. Someday, that Justice Of The Peace will be a Creator Of Disturbance and then what?

Jack Gross **must** be removed from the bench for the welfare of the public.

ANOTHER MEDICAL HUMBUG

Not satisfied with feeding the Sufferer of Anemia with pounds of the disgusting diet of Raw Liver, the marvel of Medicine turns now to Zomotherapy as the panacea for all evil in Disorders. This system of feeding Muscle Plasma, Meat Juice and All Meat, to the exclusion of all vegetables and fruits, was introduced in 1900, by Professor Charles Richet and was based on the successful (?) results of feeding dogs which were inoculated with Tuberculosis. This was similar to the recommendation of Dr. Fuster, in 1865, of feeding Raw Meat and Alcohol!

Recently, the Medical Doctors admitted that freshly pressed Beef Juice contains only about 6 per cent of Protein and 4 per cent. Extractives and that Beef Extracts "are not as nutritious as might be supposed." So, after trying to keep us savage carnivora as in the ancient times, they are endeavoring to remove Disorders by the same route as actually produced them! What crass ignorance and stupidity is centered in this system known as Medicine only the deep student of Spectro-Chrome Metry can comprehend.

Do not fail to attend our Annual Convention. You will have actual proofs before you.

FEES AND DUES

So no misconception may result, we again publish the Rules regarding the Annual Fees And Dues of the Scientific Order Of Spectro-Chrome Metrists.

All Affiliation Cards expire on December 31, 1935.

The Fees And Dues for Diplomaed Graduates and Members, is Six Dollars.

The Fees And Dues for all Provisional Members, who are such by courtesy of being Purchasers of Spectro-Chrome Equipments, is Three Dollars.

Those who affiliated on or after October First, will be granted their 1936 Affiliation Cards Free Of Charge, provided they apply in writing, to the Board Of Trustees, **before** the expiration of the current year, for continuation of their Affiliation.

Failure to comply with these Regulations, which are authorized under Rule 56, will be deemed Resignation from the Order.

As there are no favoritisms or exceptions in our Organization and as each receives equal courtesy, all Affiliates are respectfully requested to take this as formal official notice and act promptly.

With the mailing of the December 1935 Spectro-Chrome, the year completes. You must be a Member or Provisional Member to receive the Spectro-Chrome, which contains the monthly Favorscope. Your Affiliation admits you to the Planet Meetings.

———o———

EARN $15.00 AT HOME

Many Purchasers of Spectro-Chrome Equipments, still are unconscious of the fact that on the Placement of each Spectro-Chrome in a home or place of business by them, they will be entitled to receive $15.00 Commission, provided their names be registered as Purchasers and they be Members or Provisional Members in good standing, with the Scientific Order Of Spectro-Chrome Metrists.

There are no Agency or Distribution Agreements necessary and the said Affiliation is the sole requirement.

Earn $15.00 on each Spectro-Chrome, by recommending your friends or relatives and sending their Order through you. Write for Standard Order Blanks to Spectro-Chrome Distributing Company, Malaga, New Jersey.

SPECTRO-CHROME

May 1936 Green

Volume 12 Number 5

Trade Marks Registered In The United States Patent Office

IMPORTANT ARTICLES FEATURED

IN THIS ISSUE

Read and Let Your Friends Read

Horrors Of Tuberculin Test

Menstruating Boys

Carbon Arc Health Lamps

Twilight Sleep Destruction

Milwaukee Medico-Legal Burlesque

SINGLE COPY 50 CENTS

HORRORS OF TUBERCULIN TEST

Once Rejected By American Medical Association, Now Foisted on Unsuspecting Public

The poor suffering public! Nowhere in the World is such a mass of active medical propaganda published as in this country, through the channels of the newspaper, radio, magazine, school board and the like. The suffering public is lead to believe that the Medical Doctors are concerned solely in public health and in its maintenance and that the whole activity is carried on so the public may stay healthy. However, the same suffering public has not the slightest idea how under the guise of philanthropy, charity and good intentions, the public is downright cheated, bamboozled and hoodwinked, so the pockets of these quassi-angelic knights of commercial pilldom may be well lined and their limousines run 75 miles an hour, with their wives decked in sealskin coats.

What persuasions and subtle inducements are used by them may well be gathered from a recent syndicated article published in the daily newspapers over the signature of Dr. Allen G. Ireland. Its smoothness and glossy character are self-apparent, its innocence as a guiding hand against public health completely disguised. The heading is the usual "Your Child And The School". He says:—

"Throughout the nation the drive against tuberculosis is taking new form. Without question it is one of the most far reaching public health movements of our time.

"A cry now familiar to every one is 'find it early'. We hear it with respect to cancer, colds, diabetes, boils and what not. The significance is very great in the case of tuberculosis. 'Find it early'. For only in this way may we hope to stamp out this disease.

"The tuberculin test now being introduced through the school is a simple, harmless procedure. Each year will find it used more extensively, perhaps becoming eventually a routine procedure.

"But there is one warning, one word I want to impress upon every parent. If the report of your child's test says 'positive' don't be panicky. A positive test does not always mean that active tuberculosis is present. Only the X-Ray and the physician's examination can determine the true facts.

"Every teacher and school nurse should learn the meaning of the tuberculin test in order to interpret its true significance to parents."

So they are warning parents in a kindly manner to permit the Tuberculin Test, so the darling children may be "protected" (?).

Now read the facts. In an article by Samuel McC. Hamill, M. D., Howard Childs Carpenter, M. D. and Thomas A. Cope, M. D. in The Archives Of Internal Medicine, published by the American Medical Association, December 15, 1908, is given an illustration of the severe damage done to the eyes after a Tuberculin Test. The condition of the child's eyes was wholly the result of that experiment. The following details of this terrible fraud on the gullibility of the public, produced by Masters Of Dopology are taken from a warning issued by the Vivisection Investigation League of New York, New York, through the kind courtesy of its President Sue M. Farrell.

"After describing the method of administering the Calmette or eye test by instilling a drop of tuberculin solution into the eye of an immunized individual we are told that of the one hundred and fifty-eight children receiving the eye test twelve had rickets, four had whooping cough, twelve had bronchopneumonia, some had typhoid fever, acute articular rheumatism, pleurisy, neuritis, heart disease, etc. No. 25 had whooping cough and bronchopneumonia when both eyes were instilled and both reacted. No. 59 had pustules on the eye, inflammation of the cornea, with a corneal ulcer eight days after. Nos. 60 and 62 had similar trouble. No. 74 had pustules, conjunctivitis appearing in 96 hours, continuing two weeks. Nos. 92 and 107 had typhoid fever and "conjunctival test applied at height of fever". No. 101 had bronchopneumonia and "Tests were applied within eight days of death." No. 103, 1½ years of age; "Child anemic and emaciated. Test applied three days before death." Nos. 104, 105 and 106, "Post-mortem showed no evidence of tuberculosis."

"Practically all our patients were under 8 years of age and all but twenty-six of them were inmates of St. Vincent's Home (Philadelphia), an institution with a population of about 400, composed of foundlings, orphans and destitute children. The cases in the home were tested in routine by wards, **irrespective of the conditions from which they were suffering, and in the great majority of instances, without any knowledge of their physical condition prior to or at the time the tests were applied. We purposely deferred the physical examination of these children until after the tests had been applied. . . . . In addition to the cases designated in the tables, we have applied the test to seventeen infants ranging in age from four weeks to five months."** . . . . . . One case in this latter group was "probably one of tuberculous meningitis; the patient was tested within the last twenty-four hours of life." "In about sixty of our cases the conjunctival test was applied from one to three weeks before the scarification and ointment tests. In the remainder of the cases the three tests were applied at the same time...... Before beginning the....conjunctival test, we had no knowledge of any serious results from its use. It is unquestionably much easier of application than the other tests,......but it has the great disadvantage of producing a decidedly uncomfortable lesion, **and it is not infrequently followed by serious inflammations of the eye, which not only produce great physical discomfort and require weeks of active treatment, but which may permanently affect the vision and even lead to its complete destruction.** In our series we have had two cases of severe purulent unilateral conjunctivitis, six of severe recurring phlyctenular conjunctivitis, one of which developed corneal ulcers, and one case of keratocyclitis with a large central corneal ulcer. **Permanent disturbance of vision is sure to follow in this last case from the central scar,** even should the associated lesion, which at present is in a very unpromising condition, clear up eventually.

"These results are by no means unique, many similar observations having been made" by Webster and Kilpatrick, Moro, Feer, Schenck, Eppenstein and Gaupp, Van Durme and Stocke, Seifert, Krause, Hertel, Ebstein, de Lapersonne, Heymann, Renon, Simonin, Buch, Cassoute, and Barbier.

"We have had a number of verbal reports of eye complications, some relating to very serious conditions, and we are sure they are much commoner than the references we have communicated would indicate. ...... **In fact, we are strongly of the opinion that any diagnostic procedure which will so frequently result** in serious lesions of the eye......has no justification in medicine, **especially since there are other diagnostic tests of equal if not superior value, which are applicable to the same class of cases and not attended with the same disturbing results."**

Albert E. Taussig, writing on the conjunctival method in the **Interstate Medical Journal for June** 1908, says that

"Plehn's experience with the conjunctival method has not at all convinced him of its value. He objects in the first place that many plainly tuberculous individuals fail to give a positive reaction, a statement in which all observers agree. But what is worse, he obtained a positive reaction in a considerable proportion of non-tuberculous patients. **Thus two out of five cases of typhoid, two out of five cases of scarlet fever, six out of twelve cases of rheumatism, three out of six with acute bronchitis and several cases of enteritis (Inflammation of the intestines) reacted positively. None of these patients showed any evidence of tuberculosis. Moreover, unpleasantly severe reactions were not uncommon."**

He also quotes Levi and Collin on the dangers of the test, and the **London Lancet, December** 26, 1908, had an article to the same effect.

With the very evident danger accompanying this test, its use would hardly seem justifiable, but that is not the attitude of the experimenter, and we therefore find in **The Archives of Internal Medicine for May 15,** 1909, Louis Hamman, M. D., and Samuel Wolman, M. D., describing their experiments on 500 patients in the following words:

"So much has been written about the cutaneous and conjunctival tuberculin tests that one feels constrained to offer an apology before making a further contribution......What we have to say is the result of over a year of experience with patients coming to the Phipps Dispensary of the Johns Hopkins Hospital."

The reader will observe in the following sentence the callous and indifferent manner in which the authors refer to the poor persons coming to them for relief as "material."

"It is important that these tuberculin tests should be tried on as varied a material as possible, and we feel **that we have had not only a rich material at our disposal,** but one differing in many ways from the cases described in most of the reports that have appeared.. ....It is now well known that patients who have had a conjunctival or a skin tuberculin test exhibit a marked tendency for these **areas to flare up on subsequent subcutaneous injection of tuberculin** ..... the conjunctival recurrence was nearly always more severe than the cutaneous......so many healthy people react to tuberculin, that the method gives positive clinical results only when there are present definite signs of a focal reaction. ...... It should be **fully emphasized that none of these tests can replace in the slightest degree a carefully taken history and a well-made examination."**

In **The Archives of Internal Medicine for December** 15, 1910, the same physicians describe their further experiments as follows:

"In May, 1909, we reported the results of the conjunctival and cutaneous tuberculin tests in 500 patients. The present report deals with 1,000 additional patients to whom these tests were administered, and who formed the **unselected material of an ambulant clinic,** the Phipps Dispensary of the Johns Hopkins Hospital. ......In the event of no reaction we proceeded with one drop of a 5 per cent solution in the right eye. If this, too, produced no reaction, we refrained from further instillations, fearing the possible intensity of a reaction consequent on a second instillation of tuberculin into an eye. **Our**

fear is based on evidence, gathered accidentally, that a second instillation may give a positive and even a severe reaction in a case in which a previous similar test gave a negative result. . . . . . . It is clear that a positive skin reaction in an adult cannot be used as evidence of clinical tuberculosis. . . . . . .We cannot, therefore, attach any prognostic value to the skin and eye tests . . . . . ."

In the meantime, a paper by Drs. Derby and Ayer was read at the session of the American Medical Association, in 1909, and published in the **Journal of the American Medical Association, May** 28, 1910, in which they applied the test to ninety-two patients "in the out-patient clinics of the Carney Hospital and the Massachusetts Charitable Eye and Ear Infirmary, **and we are much indebted to the staff of the latter institution for allowing us to make use of their material."** They add that they "have discarded the conjunctival test as being occasionally productive of disastrous results. . . . . ." Among others who declared these tests to be harmful were Wm. S. Magill, M. D., Director State Laboratories; Professors Sahli, Bosanquet and Eyre; Wm. B. Trimble, Lecturer on Diseases of the Skin, New York University and Bellevue; Muthu of London, Bruno S. Horowicz, M. D., von Pirquet, and the Report of the Committee for Study of the Relation of Tuberculosis to Diseases of the Eye, published in **The Journal of the American Medical Association, July** 2, 1910, says:

"The Calmette test, which, on account of its danger in diseased conditions of the eye, one would hesitate to use for ophthalmic diagnosis, even if one eye is sound."

"The above quotation clearly indicate that after these tests were found to be distinctly dangerous the experiments still continued and "material" was freely furnished by authorities of various public institutions."

What imposture, what hypocrisy, what subtlety! To deprecate a method once and later endorse and adopt it. The trusting parents of children through their insidious propaganda are made to believe the Medical Doctors are working to maintain the health of the community, whereas all the time they are acting through the Health Boards to undermine the people's health so their pockets may stay filled with tainted lucre sucked from the veins of their deluded clients.

Down with the nefarious Medical Trust! Down with all its ghastly ramifications and uproot their control over our beloved little ones!

FIRST ST. LOUIS MISSOURI PLANET

(HOTEL MELBOURNE, ST. LOUIS, MISSOURI)

Meeting Of Thursday, April 2, 1936

Successful Use Of Spectro-Chrome In:—Mumps, Pneumonia, Pains, Headache, Heart Disorder, Carbuncle, Tonsilitis, Cataract.

Reported By:—CHARLES H. STIERMANN, *Recorder*

Deputy Clemens M. P. Neumann was in the chair.

Letter was read from the Principal Office that a larger room would be provided for accommodating the Meetings of the Planet. A circular of the lectures by Colonel Dinshah in Wilmington Delaware was read, indicating that Spectro-Chrome is ever enlarging its field.

Eleanore N. reported a case where a child of 6 years having Mumps, was normalated in 2 days by use of Spectro-Chrome. The child returned to school and Sister at school sent child home again fearing recovery was not complete. The child returned to school after 2 more days, the recovery having been complete.

The Deputy reported a case of Pneumonia. Boy 16 years old, running a Fever of 104 Degrees, had a relative who was a trained nurse and felt other attention than Spectro-Chrome was required. It was normalated in 4 days. Blue and Green Systemic and Orange were used with Magenta on Area 4. He was a very light meat eater, so there followed a speedy Normalation.

The Deputy had terrific pains which too were relieved by Spectro-Chrome in 2 days.

Josephine S. had Headache brought on by severe Cold. Used Blue on Area 1 and had immediate results. She spoke on the Dinshah system of Whole Wheat Cake (Deshikek) which she found very palatable. Most of those assembled felt the same way; some used nuts in its making.

Grace R. spoke again about Whole Wheat Bread. Some did not care for it.

The Deputy stressed the importance of using Rational Food Of Man system for better health; he said if it be followed to the letter, one could dispense with Spectro-Chrome. In the discussion, he also emphasized the fact of the Red, Green and Violet being the Primary Colors, hence the most essential, the others being gradations built from those Colors.

Konrad F. reported a case of Heart Disorder. Lemon Systemic and Indigo for Pain on Areas 4-18. Heart irregular of long standing. Used Spectro-Chrome since Christmas; tonating Area 4 with Magenta made it regular.

A question arose about tonating Hardening Of Arteries. Suggested that Purple be used to act as Vasodilator.

Konrad F. sought information on Enlarged Heart and what caused it. Josephine S. advised finding the History of the case as it may be produced by different causes. If the Heart is beating too fast or too slow, tonate accordingly. Study the Ratio between the Kardoscil and the Spiroscil.

Martha H. reported Carbuncle on her husband's elbow normalated by Spectro-Chrome. Arm was swollen to 9 inches in diameter; the pain was so terrific that he became faint and dizzy at times. Over a quart of foreign matter drained from the Carbuncle and had a core ½ inch thick and 2 inches long. Green Systemic and Orange Local to bring to a head, Turquoise when open and Indigo to seal. It opened in 10 days. The trained nurse saw it and advised having a Medical Doctor to cut it. Two Tonations Local and one Systemic taken daily.

Josephine S. reported a case of a friend who had Heart Disorder and who had to call a Medical Doctor when attacks came on. Learned of Spectro-Chrome and immediately put it in use. The Doctor was prescribing Digitalis. Purple brings prompt relief and the Doctor is called no more. He marvels that she is doing so nicely.

The Deputy reported a case of a girl of 22 having Tonsilitis. The Medical Doctors were about to remove the Tonsils when her father (a policeman) heard about Spectro-Chrome and decided to give it a trial before operating. The girl was home only 2 days after which she returned to work and was normalated completely in 10 days. Green Systemic, Magenta on Area 4 and Turquoise were irradiated.

Oneida W. spoke of a case of Lung Fever, a bronchial condition, tonated with Blue on Area 18 with no effect. Used a few days; was suggested that if Turquoise had been used on Area 17 instead, the results would have shown.

Joseph Z., purchaser of a new Equipment, sought information as to how to start Tonation. Told to start all Tonations with Green, the Cleansing Agent. Was advised to take advantage of the Free Guidance of Spectro-Chrome Institute by filling a Chart and sending it there. Many Purchasers do **not** take this advantage.

A question arose for advice from headquarters whether it is harmful for the Rays of an Electric Heater to strike the Body at the same time as the Spectro-Chrome, during Tonation. Does it detract from the benefit? Also if operating a Radio in the same room during Tonation is harmful. Will it reduce the power of Spectro-Chrome?

Grace R. reported a case where Cataract forming in the Eye was caught in its infancy; Spectro-Chrome was used and so arrested the formation that it never matured and was normalated. This person would not sell her Spectro-Chrome at any price, having proved its efficacy on several occasions.

Elizabeth G. asked about Ordinary Coughs. Was advised to use Turquoise Systemic and Lemon Local to loosen the Phlegm. She asked if two children be tonated at the same time. Was advised that if they were 3 feet apart, tonating may be successfully done and results gained.

Oneida W. asked how to tonate for Tuberculosis of the Bone. Lemon was indicated as the Bone Builder and was so advised.

EDITORIAL REMARKS

Every school should have a Spectro-Chrome installed and put out the Medical Doctor. The silly ideas inculcated by that fraternity work on the mind of parents through sheer fear. Spectro-Chrome will be their cheering hope for betterment.

It were better if the report had stated in what Area the "pains" of the Deputy were, because we are sure to receive letters from inquirers who will perhaps innocently believe they were "Labor Pains"!

There is no hard and fast rule for the "Deshikek" or any other food. Use spices and ingredients for flavor to suit. All we desire to impress is that the wrong materials should not enter the Body.

The Gradational Attuned Color Waves produced from the Primaries are just as useful, because of their specific Attributes. With solely the three Primaries, without the Combinations of the Secondaries and the Tertiaries, the complete Elemental Physiological Functions can not be performed in their complex relations in Organic Chemistry. Furthermore, the Emotional Vehicle can not be oscillated without the Secondary and Tertiaries.

Hardening Of Arteries is not a Disorder, but, is the result of certain wrong conditions produced by Food or Habits.

What difference does it make whether the Heart is dilated

or contracted? Its condition will reflect promptly in the Ratio as well as the Frequency.

By proper Irradiation, Spectro-Chrome performs many feats formerly possible solely by Constructive Surgery.

Keep the Tonsils where they are.

Oneida W. using Blue on Area 18 is not comprehensible. The Turquoise for Area 17 had sense, but what really should have been done was to look into the Kardoscil and Spiroscil Ratio.

Most Purchasers are too lazy to read the Spectro-Chrome Home Guide. They expect the apparatus to operate by itself. The Guidance is Free, yet most do not ask for it. Wherever there is any lack of response, it is the improper or want of use of the apparatus. Spectro-Chrome Institute is ever anxious to serve its patrons.

The Heat Rays are on the Infra-Red Division of the Spectrum, hence they should never be allowed to strike directly the Sufferer, when Ultra-Green Tonations are irradiated. The Heat of the incandescent Red Resistance Coil has Color also. Many Headaches are traceable to sleeping with bare head on which the Heat of a hot water or steam radiator impinges. Keep the head protected against thermal changes during sleep by wearing a light skull cap. In regard to the Radio, there is no question that concentration on any object in view always enhances and quickens recovery. A Radio is an unquestionable disturbance and distracts the mind of the Sufferer from the Normalation idea, which should be the one thought uppermost; the Radio also is likely to interfere with the peace and calmness which supervenes during Tonation. Unless the program be distinctly pleasing and exhilarating, the button may as well be turned to "Off" position. There is also to be considered the Radio Frequency Oscillations of the instrument, the surge from which may not be in harmony with the Oscillatory Frequency of the Spectro-Chrome Attuned Color Wave and its Reinforcement or Interference on the Auric Vehicle. The less disturbance of any kind during Tonation, the better. The best is to allow Sleep to supervene.

Cataracts in the incipient or formative stage are amenable to Spectro-Chrome easily. Our records are full of favorable reports.

Two or more persons may be tonated at the same time with a single Equipment, if the same Attuned Color Wave is to be used. However, under the Laws regulating Optical Beam Dispersion, as the amount of Energy on the Area is inversely in proportion to the square of the distance, because of the necessity of separating the Sufferers

by over 18 inches, the Energy falling on their systems is liable to be scattered detrimentally. A saving of the Electric Bill and some time might be effected, but, such procedure is not advisable, because of the involved factors which might impede the results.

Oneida W. should study Spectro-Chrome Metry Encyclopaedia and refresh her classwork. She must also forget Differential Diagnosis. All Persistent Disorders are amenable to Lemon as a Chronic Alternative and no matter what name be given, the Remedy in Spectro-Chrome is the same.

We are glad to note your Planet is working satisfactorily.

MENSTRUATING BOYS!

What calibre of intelligence rules some of our Boards Of Education, may be gathered from the story which is going the rounds of a certain town in New Jersey. It has as usual a Board Of Education, one of the prominent members of which signs his name with a X mark, because he can not write!

In one of the meetings, a woman member brought for consideration a subject on which she was well posted. She said that the female students in the community were mainly of nationalities having quicker maturity and it was therefore advisable that boxes of sanitary pads be provided in the washrooms of the girls by the Board Of Education. On vote being taken, the above alluded member promptly registered his opinion in substance, "Non, everyting must be equal. If the goils have it, the boys must have it; if the boys don't have it, the goils don't get it!"

The report states that at this outburst of Solomonic erudition and sense of equal rights for both sexes, the woman who proposed resigned.

CARBON ARC HEALTH LAMPS

For years and especially since the introduction of Spectro-Chrome to the family, the Medical Doctors have been harping on the value of the Ultra-Violet and the Infra-Red Rays, to the utter derogation of the Chromatic Spectrum, which is visualized as Colors. Most of the users of the Ultra-Violet and Infra-Red Rays, prefer to use the Carbon Arc Lamps for their therapeutical value, forgetting that such lamps are highly variable in their Emission Values and are in many respects detrimental to health.

In the March 1936 issue of the Technical News Bulletin of the National Bureau Of Standards in Washington, D. C., is printed an open warning that such Carbon Lamps contain "ultraviolet radiation of very short wavelengths and infrared rays of long wavelengths which are not found in sunlight."

They are coming nearer daily to the principles set down in Spectro-Chrome Metry and it is solely a question of time when the potentiality of the Attuned Color Waves of the Chromatic Spectrum will receive its deserved recognition.

SPECTRO-CHROME SMILES

Viewpoint—"All of my uterus was removed when I was 35 years old. I am now 52 and the mouth of my uterus is aching."

Smile—"Most probably you are growing a new uterus, as strange things occur in America."

Viewpoint—"I have a fallen wound."

Smile—"You will need a special type surgeon."

Viewpoint—"My noise clugs up witch drip water on dump days."

Smile—"It is a good remedy to stop talking."

Viewpoint—"I am horse."

Smile—"Attach a carriage."

Viewpoint—"Surgeons tapped my spine and took some fluid; don't know whether it was put back or not."

Smile—"The proceeds flowed into the Surgeon's pocket."

Viewpoint—"My wife drinka da milk and vomit; da vomit no milk, all cheese. Whata could I do?"

Smile—"Open a dairy."

Viewpoint—"Your Advice says my wife should use Spectro-Chrome on Number 4; where shall I put it? I do not know which part."

Smile—"Your wife has only 2 bosoms; put it on the left one; that is the meaning."

Viewpoint—"Electricity here costs me too much money to run the Spectro-Chrome; how can I reduce the cost?"

Smile—"Don't run it".

Viewpoint—"I am a Medical Doctor. In your literature you state that 'The Attuned Color Waves should be applied to the **bare skin**'. What purpose does the animal serve?"

Smile—"Are you sure it was a bear? We see only a donkey."

Viewpoint—"She is very week."

Smile—"Wait till the month comes."

FIRST DETROIT MICHIGAN PLANET

(HOTEL TULLER, DETROIT, MICHIGAN)

Meeting Of Thursday, April 2, 1936.

Successful Use Of Spectro-Chrome In—Mucous Colitis, Cold, Asthma, Carbuncle, Boil, Ear Disorder, Toothache, Tetanus, Burn.

Reported By:—GREETA O. PECK, *Deputy.*

Deputy Greeta O. Peck was in the chair.

Otto H. spoke of Spectro-Chrome in relation to children. In a home where there are children such as his where there are two little grandchildren, he related what a blessing it was to be relieved of all fear, from the anxiety of anything serious developing from their little coughs and colds. They just vanish into thin air. He told of Mucous Colitis being normalated in the older child and the speed with which a Cold was normalated in the case of the younger child. In fact, he stressed the excellent results that are obtained in all Children's Disorders.

Sarah C. used Scarlet for four Tonations on Area 18, that normalated the swelling in her big toe and joint and all around the arch and ankle. This was the condition that was reported at last month's meeting. She reported that she had not received her April Favorscope.

Magdalene A. reported a case of Asthma. Green and Magenta were used alternately for four days with such splendid results that the Sufferer was able to go riding after having been confined to her home for a long period. Lemon is now being used and the Sufferer is continuing to improve.

Thomas H., a new member, told of the remarkable results obtained in his own case of Asthma. He was in a hospital, being treated for his condition when he accidentally heard of Spectro-Chrome. Green produced sleep the first night. After one week of Green and two weeks of Lemon and Magenta, he feels like a new man and is very grateful to Spectro-Chrome.

John W. told of normalating two Carbuncles and a Boil by using Orange, Green and Indigo. His baby with a festered ear, which broke, was being given Green and Turquoise with good results.

Norman D. told of tonating a pivot tooth (from which a

bridge was swung) which was aching. The Dentist wanted to extract it. Blue for one Tonation normalated the condition. He also reported the case of a friend who struck his foot while using a pick. The injury was so slight that he paid no attention to it, but Tetanus set in and he consulted a Medical Doctor. He spent $30.00 for "Serum Treatments", but the condition grew worse and possible amputation was suggested. The foot looked very bad with a dark Red spot the size of a silver dollar and Purple streaks radiating from it. At this point, Norman D. loaned the man his Spectro-Chrome Equipment and Green quickly normalated the condition.

Albert G. reported the case of a third degree Burn from steam and scalding water. Blue was used for two days so the pain disappeared and the blisters, which were about the size of hen's eggs, broke. Then Turquoise was used and when it looked as though infection might be setting in, Green was given. Cocoanut Oil and waxed paper were the only dressings used. In two weeks' time this serious Burn was normalated.

The Deputy told of a friend who had a similar Burn, which was attended by a Medical Doctor. The case was given for comparison. Infection did set in and $50.00 spent for dressings alone. A Doctor was calling every day and a nurse was required to dress the wound twice a day for several weeks. It was six weeks before the Sufferer was around as usual. Thus can be seen what Spectro-Chrome accomplished in two weeks, in comparison to the medical treatment.

L. Roy D. said, "I am familiar with two other healing systems, but I put Spectro-Chrome first." He stressed the importance of looking after Area 4 in all Disorders. He told of a severe Cold in the case of his wife which was normalated in two tonations of Yellow and Blue respectively.

Harley S. stated he had proven that a high Kardoscil will respond to Magenta on Area 4 if used three times weekly. He also stressed the vital importance of the Rational Food Of Man system in the Normalation of Constipation.

Edwin L. read an "exposure of Colonel Dinshah in Health And Hygiene" a medical magazine. It was moved and seconded that the above magazine be sent to Colonel Dinshah, so he might send them a challenge.

Albert G. read chapters 110, 111 and 112 from the third volume of Spectro-Chrome Metry Encyclopaedia, being respectively on

Brain And Cranial Nerves, Great Sympathetic Nervous System and the Pillars Of The Temple Of Solomon.

EDITORIAL REMARKS

The April Favorscope was printed in the issue of February 1936. Sarah C. will please refer, as our records show it was mailed to her. From next year, an innovation will be made. The entire 12 months of the Favorscope are now being computed by Irene Grace Dinshah and will be issued as a single booklet to the Affiliates. The monthly printing in the Spectro-Chrome will be discontinued after the expiration of the current year.

Be careful in loaning your Equipments, lest some day you lose your Slides or Spectro-Chrome Home Guide. Recently an Affiliate in Claymont, Delaware loaned the book to a neighbor who lost it in a fire. It means loss of investment.

The reply of Colonel Dinshah to "Health And Hygiene" libel was already printed in the Spectro-Chrome of March 1936, under "Their Skin Is Now Sizzling". Many thanks for the kind courtesy. Be always on the watch for such covert attacks. The doom of the messpot of the Medical Dumkopfs is sealed, no matter what they print about the Originator.

LEAKY HEART NORMALATED

By:—MARY P., *West Philadelphia, Pennsylvania.*

I am a young woman of 29 years with a little girl who has a very serious heart condition. Thirteen weeks ago she was taken with Scarlet Fever; she had a leaking heart before she got Scarlet Fever, but, since then the heart has been worse. Eight weeks ago our Doctor gave her up; her pulse was 158 and her respiration 58; temperature 102 for six weeks. She had five blood transfusions at home. Our family Doctor said it was just a matter of time; her heart was so large it filled her left lung and part of the right one.

Well, I have a Doctor who gives me Color treatment with one of your Spectro-Chrome lamps and I telephoned her and she rented me a lamp and gave me instructions on the Colors over the telephone.

The child had been spitting up blood and when she had two treatments with your lamp of Magenta, one could see the change in her.

We have had the Spectro-Chrome at home for nine weeks

and it has done her wonders. I wish you could see her; she is a beautiful child of nine years, but, the Spectro-Chrome Normalator claims she must continue the light treatment. I have seen what wonders it has done and I will be scared without one.

---o---

TWILIGHT SLEEP DESTRUCTION

We are not the only ones who oppose the introduction of poisons into the Body of the woman in childbirth. Dr. Gertrude Nielson, at the last Annual Convention of the American Medical Association in Kansas City, Missouri, denounced the use of "Twilight Sleep" in the conduct of childbirth cases, as being dangerous to both the mother and the infant.

The idea of the Medical Doctors is that as there is no "nerve connection" between the mother and the progeny; they may do what they please with the mother and the child is not affected. They forget that the blood of the mother circulates through the Placenta and builds the baby and that anything entering the mother's blood MUST shortly pass into the circulation of the infant.

Twilight Sleep is a crudely conceived idea. It is defined as "a condition of light anesthesia marked by mental subconsciousness." It is produced in the woman in the process of childbirth, by the hypodermic injection of Morphine (H21 C17 N O4) and Scopolamine (H23 C 17 N O5). Morphine needs no elucidation as being the most active Alkaloid of Opium. Scopolamine, which is the same as Hyposcine, is defined as a "poisonous nerve depressant, mydriatic (pupil dilator) and hypnotic." The introduction of two such poisons into the blood of the mother, is bound to produce undesirable results in the blood of the infant and no words are powerful enough to express our denunciation of such ignorant tactics.

Spectro-Chrome Purple is capable of affording the needed relief without upsetting the vital equilibrium and the Medical Doctors will some day realize how utterly unreasonable they were in endeavoring vainly to choke the propagation of the science.

Radio reports that the well-timed denunciation of Dr. Gertrude Nielson, a mother of several children herself, was received by the A. M. A. Convention with vociferous applause. We add some more to it. When the time will come for the Medical Octopus to die it will not be through any trampling under foot by us, but, through an internal explosion that will rock the Healing Fraternity from

pole to pole. Spectro-Chrome set the train of Dynamite under it; all that is needed is the final spark to ignite.

There are many Medical Doctors who are ready to blow out of the accursed gang, but are not aware of the means. Spectro-Chrome will prove their final solution when that fact becomes better known.

---o---

FIRST MILWAUKEE WISCONSIN PLANET

(HOTEL WISCONSIN, MILWAUKEE, WISCONSIN)

Meeting Of Thursday, April 2, 1936

Successful Use Of Spectro-Chrome In:—Prostatitis, Rheumatism, Gastroenteritis, Headache, Hemorrhage, Fits, Diabetes, Tuberculosis, Scarlet Fever, Asthma, Anemia, Ringworm.

Reported By:—CAROLINE ZIMMERLING, *Recorder.*

Deputy Otto H. Lutz was in the chair.

He read a telegram from the Principal Office stating that Mrs. Lena Held had been expelled from the Scientific Order Of Spectro-Chrome Metrists.

Chapter 74 on "Chemical Puzzles" and 75 on "Tuberculosis By Medical Conception" from Spectro-Chrome Metry Encyclopaedia were read by Caroline Zimmerling. A discussion followed.

Alma Horvath volunteered to read Chapters 76 and 77 on "My Experience In Consumption" and "Spectro-Chrome In Tuberculosis" at the next meeting.

Otto H. Lutz read from the Spectro-Chrome Home Guide the Attributes Of The Orange wave. Case reports followed.

The Deputy reported the case of a man 51 years of age, named Robert V. who owned a Home Equipment but was not getting results. He was suffering from severe pains, due to Prostate Gland trouble. He suggested he use the Blue Local, Turquoise Systemic and then Scarlet Local. This was done and in three days the man was all right.

Another case reported by him was that of Joseph M., 31 years, who had Rheumatism. He had pains all over his body. He rented an Equipment for 4 weeks. At the end of 2 weeks all the pain was gone with the exception of the feet. Green Systemic and Magenta on Areas 4 and 18 were used first and then Lemon Systemic with Magenta on Areas 4 and 18. He lost no work at all. He will purchase an Equipment of his own now.

Kurt B. said that a man came to him asking what should be done. His wife was suffering from what is known in this locality as "Stomach Flu". She had severe pains in the Abdomen. Three Purple Tonations on Area 4 were given and two Turquoise Systemic and she was all right.

Frank B. reported two cases of the same ailment in his own family. He and his son both had it. They used Turquoise Systemic only and were all right.

Frances B. reported that Indigo had helped her daughter when she had this same ailment.

Hedwig N.'s daughter, aged 27, had a severe Headache one morning. Given Green on the head for 20 minutes, caused her to vomit. She had only the one tonation, but was feeling fine thereafter.

Carrie M. told of a case of a school teacher who had slipped and fallen, hurting her side very badly. She was also injured internally, evidenced by bleeding from Area 11. Carrie M. took her Equipment to her and gave her Magenta on Area 4. Then she gave her Green and Turquoise to help the bleeding. Three Indigo Tonations for the pain were effective. In 3 weeks the Sufferer was well again. Before using Spectro-Chrome she was not able to move.

Otto S. has had his Equipment since January 17, 1936. In the fall he could not work, the Doctors saying he had fits and was headed for the insane asylum. He would have these fits every 4 or 5 weeks. Since he used Spectro-Chrome he had only one of these spells and that only slightly. He used Green and Lemon. Backelim he finds is not so good. The members of the meeting suggested that he use for the purpose Lemon or Orange according to what he finds is best suited to his needs. The Deputy suggested Yellow on Areas 8, 9, 10 and 19.

Walter B. said his wife found that since she had been adhering to the Rational Food Of Man system she is getting much better results in the Backelim.

Marie L. had one Tonation at the Deputy's house for a pain in her arm and difficulty in breathing. Took Orange and pain was gone completely. She is getting very good results for Diabetes. She takes her Tonations twice a day. She is on the Yellow for the fourth month now and takes 3 Magenta Tonations a week on Area 4. She used Orange for a while, but was not satisfied. She is using Yellow now and is well satisfied.

The Deputy reported a case of Diabetes in which Gangrene had set into the leg. Yellow drained it nicely.

Annette T. reported the case of a friend's brother who was ordered sent to a hospital for Tuberculosis. X-Rays showed he had holes in his lungs as large as 50 cent pieces. Three months later, after using Spectro-Chrome, X-Rays showed the holes to be only as large as pin heads. Her Sister had Scarlet Fever and with Turquoise Systemic was all right in 4 days, while her girl friend was ill for 4 weeks with the same thing.

Esther F. said her mother has been deaf for 32 years. She has been using the Equipment for about a year; Green, Lemon, Orange and is now on Red. Has not had much in the way of results for the hearing, but is better in general health. She is patient, however, because she knows it is of long standing.

David V. used Green 1 week, Lemon Systemic and Magenta on Area 4, for 4 weeks, Orange for 5 weeks and has had no trouble with his Asthma since using Spectro-Chrome.

Louise L. asked what could be done for her too hearty appetite. David B. suggested that she take smaller mouthfuls and masticate her food more carefully. Kurt B. suggested that she use Indigo on the stomach.

Leona M. says her mother has been using Spectro-Chrome for Diabetes for 1 month. She is rather weak now. She had a heart attack a week ago and has been using Magenta on Area 4, for the past three days. She is using it for her self too, for her leg. There is a paralyzed condition. She uses Lemon and Magenta.

Ann P. asked what she could use on her niece, who is suffering from Anemia. Frank B. suggested Scarlet. Walter B. suggested 1 week of Green, 3 weeks of Lemon and then Orange. This, he said, normalated an anemic condition in his daughter.

Florence G. reported that she had rented an Equipment for her mother's use for 5 weeks before purchasing. Her mother had a complication of ailments. Her hands, especially, were diagnosed by the Doctors to have Ringworm. Since using Spectro-Chrome her hands healed and she is up and around, doing her own work again, which she had not been able to do for a long time. She is well satisfied with the results she is getting. Her sister also used the Equipment for Nervousness. She took 1 Yellow Tonation on the head only and felt better.

Chester D.'s wife had a Cramp in the small of her back. It was very severe. She found it very difficult to move even a little. Orange, Turquoise and Indigo were unsatisfactory. Lemon did not prove satisfactory either. She was in bed about 10 days, using Lemon and Turquoise Systemic, but results were very slow, even with 2 or 3 Tonations daily. A discussion on this case followed, suggestions being made that she should have used Blue and Turquoise. Red was also suggested. Kurt B. suggested that if she had looked to her Kardoscil and Spiroscil a tonation on Area 4 would have done more.

Louise M. has been suffering for 3 years from a back ailment. Spectro-Chrome Institute advised her to use Green and Lemon, which she did. She is not following the Rational Food Of Man system so well. All of her bodily functions are normal outside of her back. She has not had much in the way of results.

Anita D. has been using her Spectro-Chrome for a year for Indigestion. She used Green, Lemon and Magenta on Area 4. While she was observing the Rational Food Of Man system, she felt much better than she does now that she is not following it closely. She has had results, however.

Alma M. spoke of the importance of keeping the thermocept clean. She had some difficulty in getting hers clean and noticed that her son, who is using the Equipment, was not getting the results that he was getting when the thermocept was clean.

EDITORIAL REMARKS

Your Planet is working steadily to greater usefulness. It is necessary for the Sufferers to comprehend that in order to receive the full benefit which Spectro-Chrome is capable of conferring, they MUST follow the Rational Food Of Man system and reform where necessary their personal habits. One can not violate the Laws Of Health and expect to maintain it. Most people seem to believe that they may follow their own prior habits of life, which induced their Disorders and yet leave their recovery to Spectro-Chrome; this is an evident mistake. Nature's Laws can not be transgressed.

The Thermocept, like any other piece of similar apparatus should be kept clean. If your water be not clean, use spring water, but many seem to think that we must give them something that will last for 343 years without cleaning.

———o———

SPECTRO-CHROME METRY ENCYCLOPAEDIA—VOLUME 3.

By:—FRANK S. B., *Milwaukee, Wisconsin.*

The wealth of knowledge contained in Volume 3, should be studied by every person seeking to broaden and improve his or her store of general information. Besides being interesting and instructive, it is a work of revelation few of us have ever dreamt of.

Spectro-Chrome Metry Encyclopaedia 1-2-3 are by far the most interesting, instructive and conclusive work it has ever been my pleasure and good fortune to possess. I value them highly and treasure the work of the Originator, so ably, yet simply composed, that even one of ordinary education can comprehend their value and enlightenment.—E. P.

WHAT IS IT ANYHOW?

Rudolph S., Detroit, Michigan, asked us "What is Hodgkin's Disease"? Here is the answer in the lingo of our friends."

"Hodgkin's disease, a disease marked by an infectious granulomatous condition (inflammatory enlargement) involving particularly the lymphadenoid tissues of the body, with the presence in the lymph glands of fibro-blasts, eosinophils, giant cells, and frequently of an organism, Corynebacterium granulomatis maligni, which is possibly the causative agent of the disease. The glandular enlargement begins on one side of the neck and extends thence to the axillary, inguinal, mediastinal glands, and to the spleen. It is usually attended with a relapsing fever. Called also infectious granuloma, malignant granuloma, malignant lymphoma, lymphomatosis granulomatosa, lymphadenoma, lymphogranulomatosis, granulomatosis maligna, lymphosarcoma, anemia lymphatica, and pseudoleukemia."

Can you realize why we junked this conglomeration of words, worthless for our Spectro-Chrome? Yet, very few communications come to us without their writers parading a number of such formidable names, expecting us to reply in similar formidable language. They forget that the fundamental essentials of Spectro-Chrome completely abolished Differential Diagnosis, releasing mankind permanently from the bondage of 6,000 years of civilization. The names of Disorders were a curse to life and complicated the Healing Science; Spectro-Chrome entered that Chaos and shed the glory of Light, scattering the ignorance with which the Medical Doctors had surrounded the most valuable asset of the human race—Health.

It is strange how the present generation does not estimate in its proper value the gift of God—Spectro-Chrome. Unfortunately for us, most cases which resort to Spectro-Chrome are not Acute (Recent), but are solely those which are Chronic (Persistent) and given up by the Medical Doctors, after plenty of butchering. In spite of all such handicaps and obstructions, Spectro-Chrome paved its way and made for it a place which no time can efface. The public WILL awaken to its merits before long.

———o———

No Diagnosis — No Drugs — No Manipulation — No Surgery

OUR AIM: SPECTRO-CHROME—IN EVERY HOME

MILWAUKEE MEDICO-LEGAL BURLESQUE

Medical Doctors Swear Spectro-Chrome Has No Merit And Is Merely Mental!!!

When Edward A. Ernest became the Distributor General for the United States Of America, for the Equipments of Spectro-Chrome Institute, The Milwaukee Journal of Milwaukee, Wisconsin, published a defamatory article under the heading "Magic Healing Device On Sale, Wizard Gone", in its issue of Sunday, November 19, 1933, reviling the Originator of Spectro-Chrome Metry and holding E. A. Ernest to ridicule for his affiliation.

Dinshah threatened to bring a Libel Suit, but refrained when Ernest filed one against that newspaper. It was understood the Originator would appear as a witness for the Distributor, but when the said Ernest lost his distributorship, it was impossible for the Originator to appear in aid of the individual whose expulsion from the Organization he had caused.

The Damage Suit went on trial before the Honorable Judge Sherman E. Smalley, on Monday, March 30, 1936, in the Circuit Court. Frank P. Burke, attorney for Ernest, had tried hard to interest Dinshah to go on the witness stand, which he flatly refused for obvious reasons known to our readers. A Jury of 5 women and 7 men was selected. Attorney J. G. Hardgrove represented the Defendant Newspaper. Our comments are collected from the reports which appeared from time to time in The Milwaukee Journal.

Hardgrove called Spectro-Chrome "hocus pocus" and said that Dinshah was interested "in earlier years in many kinds of occult systems of treatment. Those systems included hypnotism, mesmerism and mystic oriental and European philosophies". He wanted Ernest to establish "scientific proof of the merits of the machine". This could not be done by him without the help of the Originator, as Ernest with his limited knowledge became so tangled in cross examination that his answers became a boomerang against his success. *The Court debarred all testimony of laymen and admitted solely that of Medical Doctors,* as if those opponents would extol Spectro-Chrome! Had the Originator been there, he would have shown the same turn of Court as he did successfully in the Supreme Court Of New York State, but the Plaintiff in this case having no such training could not stand against the onslaught of the Defendant. Ernest could not make clear about Spectro-Chrome having no Differential Diagnosis and the more Frank P. Burke endeavored to elucidate, the worse became the tangle. The Auric Vehicle was torn to pieces and much silly talk resulted. To save his case, Ernest dragged in the names of many prominent purchasers of Spectro-Chrome to prove its merits, an act which in itself was out of the line of gentlemanly decency and was nothing short of a betrayal of confidence—a treachery.

The articles published by the American Medical Association against Dinshah were placed on record, but without the presence of the Originator who had the rebuttal records, the Plaintiff could do nothing. With the production in Court of a Wood Cabinet Spectro-Chrome, irrelevant subjects like the Itisometer, Lavan, Manjan and the like were introduced. It was mentioned the Lavan was sold for $3.00! The Favorscope too came in for its share of ridicule.

The icebox incident of a person was brought into question. The said person had made it appear he could stay in ice long because of the "color"!!! This circus play brought smiles. The Esquimos live in ice huts without using "colors". The public however has little discrimination of what is true or otherwise. Ernest stated that the new color equipment he was selling, since his expulsion from the Spectro-Chrome Organization, was "superior", (?) although Spectro-Chrome was good! He wiggled every way to save his skin,

but, the shrewd Hardgrove pinned him each time. The Court ruled out the whole humbug.

The Defendant Newspaper called as a witness Dr. Norbert Enzer, Specialist in Pathology and Diagnosis at Mount Sinai Hospital. This Medical Doctor emptied his Solomonic wisdom that "the visible rays were utterly without value for specific or symptomatic complaints"! He declared ignorance of Green being a Disinfectant or Yellow a Motor Stimulant. He never had heard of it and to him it was all "absurd". He pooh-poohed "self-treatment".

Another Medical Doctor Louis M. Warfield stated that the claims for Colors were a century old and were disapproved, "there being no scientific proof of their value".

Affidavits and other so-called "evidence" about Dinshah went in without his presence. Hardgrove made plenty of fun about him; although Frank P. Burke did his best to show the work of Dinshah in good light, volumes of irrelevant matters went on record.

Through a dodge on the part of E. A. Ernest many Graduate Spectro-Chrome Normalators went on the stand on his behalf to defend the merits of Spectro-Chrome Metry. When Ernest found that the Originator refused to appear for him in the case, he telephoned a number of established Spectro-Chrome Normalators and told them glaring untruths that they should appear without summons voluntarily, because Dinshah was likely to be there! This lying trick worked and the Graduates went to help Ernest, believing they were aiding the cause of Spectro-Chrome.

Medical and physical experts—self-centered orthodoxists—swore on the stand "that the equipment, developed by Dinshah Ghadiali, Malaga, N. J., a native of Bombay, did not have any substantial healing or curing merit!" Those misjudged, arrogant bookworms *never* heard Dinshah or studied his Science, which in a few years more will most assuredly drive their medical hocum to the rocks. They were brought in as parrots and the testimony shows them to have imitated the cry of their guide the American Medical Association.

Attorney J. Gilbert Hardgrove scraped the skin of Dinshah to bleeding, when as Defense Counsel he summed the case to the Jury. Among the glittering gems of argument scattered by him were these:—"This man Dinshah tells the people never to allow the use of serums, vaccines or any test approved by the medical profession. Yes, he would put the fate of people, their very lives, in that worthless box with an electric light and a few pieces of colored glass. ....He would have people believing that we are all surrounded by an invisible aural body. Hocus pocus and nothing else, a gigantic scheme of fraud. ....It is important that such fakes and schemes be eliminated and people be protected from being preyed upon............".

Attorney Frank P. Burke as the Plaintiff Counsel did his utmost to maintain the ground for his client. Among many compliments he showered on the Originator, he stated "that he accepted Dinshah Ghadiali as a scientist". However, because of the absence of the Originator who alone could duly defend his Science, he obviously could not impress the Jury about the merits of Spectro-Chrome.

The Honorable Court gave a fair Charge to the Jury and the verdict was given in favor of the Defendant. The Milwaukee Journal won every point at issue. Edward A. Ernest lost the case as the Jury decided in the absence of rebuttal testimony, that "the application of light rays by the use of Spectro-Chrome instrument has no substantial healing or curative effect on disease of the human body"!!!

Fools rush in where angels are afraid to tread. In going to Court to prove the merits of Spectro-Chrome and the libel against his character, *without the aid and presence of the Originator,* E. A. Ernest severely burnt his fingers, by losing self-respect as well as a suit of $150,000.00 damages. He had a good and conscientious lawyer in Frank P. Burke, but as the rebuttal records were in the hands of Spectro-Chrome Institute, without its assistance the Jury had to nod as the Medical Doctors testified and no wonder.

God put the vital organs in the Body, so the Surgeons may fatten by their removal; He likewise made the serious error of manufacturing White Light with a 7 Color Spectrum, so Medical Dopers may have the satisfaction of swearing that "color is of no value!" Hallelujah! More will be published later.

SPECTRO-CHROME

August 1936 Indigo

Volume 12 Number 8

IMPORTANT ARTICLES FEATURED IN THIS ISSUE

Read and Let Your Friends Read

Good News For The Sick!
Sugar In Diabetes A Joke
Danger Signal In School
Milwaukee Wisconsin Has Influenza
Dumb Diagnosis Versus Precise Spectro-Chrome
Medical Humbugs Of 1936
Slacks Shock Religious Priest
Special Rally For Silver Jubilee

DINSHAH P. GHADIALI, *Editor and Publisher*

SINGLE COPY 50 CENTS

GOOD NEWS FOR THE SICK!

VITAL MESSAGE TO THE SUFFERING!!!

An Exceptional Opportunity To Hear An Eloquent Orator

INTERESTING, ENTERTAINING, INSTRUCTIVE

TEN FREE PUBLIC LECTURES

By Colonel Dinshah, The Originator Of The Marvelous Spectro-Chrome Metry

That gives Near-Miraculous Results In The Conquering Of Human Disorders, Many Declared "Hopeless And Incurable" According To Other Methods.

You are cordially invited to attend the sequence of Health Lectures, in the **Ball Room** of the **Hotel Tuller, Park and Adams, Detroit, Michigan, as our guest, without incurring any obligation.**

You will have exceptionally enlightening instruction in the Normalation of Persistent Disorders, such as Tuberculosis, Cancer, Diabetes, Syphilis, Paralysis, Fevers, Arthritis, Inflammations and other scourges of humanity, for which other Healing Systems can do very little. The actual facts will startle you. The Originator of Spectro-Chrome Metry, will help you from the platform to help yourself, in solving your personal problems. Do not miss this opportunity; it is worth any inconvenience to hear the Originator, who is a recognized Researcher and noted Academician.

The name of Colonel Dinshah P. Ghadiali, Ex-Commander, New York Police Reserve Air Service, is well known as an independent, fearless orator. He is the holder of numerous Honorary Degrees and Diplomas conferred on him; he was the first in 1896 to speak on the X-Rays in NewYork and has to his credit numerous revolutionizing inventions including the **Photokinephone,** the forerunner of the modern Electric Talking Motion Picture Machine. His scientific erudition and astonishing grip of the most difficult and delicate subjects, have for the past 46 years been receiving unequivocal recognition. His eloquence and logic are unique and his methods of presentation extremely convincing and gratifying. His hearers are spellbound with the force of Truth and his denunciation of pseudo-scientific humbugs never fails to rivet attention.

COMPLETE PROGRAM

Sunday, October 4, 1936—2 P. M. to 5 P. M.

LECTURE 1—Rainbow Message Decoded; Conception, Development And Perfection Of The Science Of Spectro-Chrome Metry.

Sunday, October 4, 1936—8 P. M. to 11 P. M.

LECTURE 2—Attainment Of Automatic Precision In Healing; Scientific Fundamentals Of Spectro-Chrome Metry.

Monday, October 5, 1936—2 P. M. to 5 P. M.

LECTURE 3—Digging Our Graves With The Teeth; Rational Food Of Man In Spectro-Chrome Metry.

Monday, October 5, 1936—8 P. M. to 11 P. M.

LECTURE 4—Immutable Laws Of Karma And Reincarnation; Their Government Of Health As Demonstrated In Spectro-Chrome Metry.

Tuesday, October 6, 1936—2 P. M. to 5 P. M.

LECTURE 5—Construction And Destruction In The Human Body; Marvelous Controlling Potency Of Spectro-Chrome.

Tuesday, October 6, 1936—8 P. M. to 11 P. M.

LECTURE 6—Spectro-Chrome In The Home; Case Reports And Experiences With Spectro-Chrome.

Wednesday, October 7, 1936—2 P. M. to 5 P. M.

LECTURE 7—Health Impairment By Wrong Habits; Their Possible Mutation By Spectro-Chrome.

Wednesday, October 7, 1936—8 P. M. to 11 P. M.

LECTURE 8—Sex Problems; Their Complex Character Solved By Spectro-Chrome.

Thursday, October 8, 1936—2 P. M. to 5 P. M.

LECTURE 9—Sun, Moon And Earth; Their Unrecognized Influence On Humans Unraveled By Spectro-Chrome Metry.

Thursday, October 8, 1936—8 P. M. to 11 P. M.

LECTURE 10—Service To Suffering Mankind; Its Fulfillment By Spectro-Chrome.

AN AMAZING FACT

Learn How To Plug Your Health From The Electric Light Socket!!!

Spectro-Chrome is not a mere Patented Apparatus, but, a complete System Of Healing for Self-Help and Self-Use; its value and merits have stood the rigid tests of over 16 years, even in Law Courts.

Classed in the same rank of utility as the Telegraph, Telephone, Radio and similar revolutionizing inventions, Spectro-Chrome scattered all prior conceptions and set a new pace in Healing by its Automatic Precision and unparalleled simplicity.

All Lectures Are Free To The Public

The subject matter of the Lectures will be illustrated with the Stereopticon and Scientific Experiments for elucidation. The Lectures will be an "intellectual banquet" never to be forgotten and will open the eyes to numerous problems affecting human life and welfare.

NOTICE TO THE PUBLIC

We distinctly desire the Public to understand that:—

(a) *NO SMOKING WILL BE ALLOWED.*

(b) Writing any part of speeches, discussions or any transactions at any time of the Lectures is *positively forbidden* and any person endeavoring to do so, will be promptly expelled.

(c) Questions will be permitted after each Discourse, if pertinent to the subject of the Lecture.

(d) Colonel Dinshah IS NOT IN THE PRACTICE OF MEDICINE, CONDUCTS NO CLINICS, GIVES NO PERSONAL CONSULTATIONS AND ACCEPTS NO GIFT, PRESENT, CONTRIBUTION, FEE, DONATION, COMPENSATION, TIP, GRATUITY, INDUCEMENT OR TOKEN, OF ANY KIND WHATEVER, FROM ANY PERSON, DIRECTLY OR INDIRECTLY. He is glad to serve Suffering Mankind through the fruits of his laborious researches.

(e) We reserve the absolute right of refusing admission to or removing from the premises any person, thought undesirable by us, without giving any reason. EXPELLED AFFILIATES WILL NOT BE ADMITTED TO ANY LECTURE.

Auspices Of

FIRST DETROIT MICHIGAN PLANET

SCIENTIFIC ORDER OF SPFCTRO-CHROME METRISTS

(Licensed In Michigan As A Not-For-Profit Foreign Corporation)

Our Aim:—Spectro-Chrome—In Every Home

SPECTRO-CHROME

(Trade Mark Registered United States Patent Office No. 169554—June 19, 1923)

THE MARVEL OF THE TWENTIETH CENTURY WHICH SHATTERED THE CONCEPTS OF AGES—

Staggering To The Imagination

Accomplished And Proven Reality!

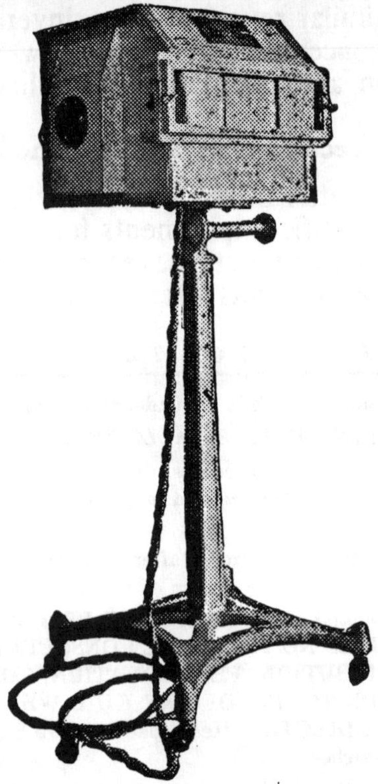

United States Patent Number 2038784—April 28, 1936

NO DIAGNOSIS — NO DRUGS

NO MANIPULATION — NO SURGERY

Ten Free Public Lectures By Colonel Dinshah On

SPECTRO-CHROME METRY

Ball Room, Hotel Tuller, Detroit, Michigan

Sunday, OCTOBER 4, 1936 To Thursday OCTOBER 8, 1936.

SUGAR IN DIABETES A JOKE

What fear the Medical Doctors infused in the mind of the public and how successfully Spectro-Chrome exploded that nonsense of Sugar in Diabetes, may well be gathered from the following, which comes from an intelligent Purchaser who followed our Free Guidance. He says:—

"I have had Sugar Diabetes for about 15 years and I will say I am feeling fine but, I still have Sugar; the Railroad Doctor requires me to report to him once a month for a Urinal Test and if I have Sugar he will not let me work till I am Sugar free, so I have to go on a 'strict diet' for two to four days before I report to him; I can test myself for Sugar so I do not go to him till I am free."

Among other matters we wrote to him in the Free Guidance:—

"The report sent by you was duly read by our President Colonel Dinshah and he had a big laugh about it at your attitude toward your Railroad Doctor. Passing of Sugar in the Urine happens to all heavy mental workers, many pregnant women, persons with financial worries and the like and in our work we exploded the idea that passing Sugar is Diabetes. In this connection, if you study SPECTRO-CHROME METRY ENCYCLOPAEDIA, Chapter 92, Page 975 of the Third Volume, you will find the whole medical humbug fully exposed.

"You continue with Spectro-Chrome as you have been diligently doing and you will live yet long to bless Spectro-Chrome.

"You really need to eat Sugar instead of abstaining from it and it would not be a bad idea to test out the hollowness of medical work by taking to your Railroad Physician a bottle of Cow's Urine (or your wife's Urine) than your own, so he may have his fill of the test."

In some hospitals, they started to use our teachings and feed Sugar, without crediting the Originator. It will not be long before the Sufferers will revolt against the Insulin frauds.

———o———

HOME LIFE IN MICHIGAN

A girl of 11, Mildred Saddison, of Sturgis, Michigan, was missed 4 days when the Police finally found her in her dog kennel living with 6 puppies. She said she preferred living in the doghouse, to drudgery in housework. What degradation has housework that so many girls and women shun it? What will life be without a house?

Michigan is not a pauper state by any means; will its educators investigate the true cause of what this girl insinuated? Perhaps the educators need education themselves to cope with the sources of social evil, because of orthodox wrong conceptions or ultra-ancient methods in dealing with modernism. The nut is loose somewhere in civilization.

SPECTRO-CHROME SMILES

Viewpoint—"Is one donation a day enough or shall I take two donations?"
Smile—"It depends upon what kind of politician you are."

Viewpoint—"I purposely became pregnant to give a sister to my boy."
Smile—"Good idea, but that should not have been your worry, as he would probably take somebody else's sister later anyway."

Viewpoint—"My son brought me a lovely roach from a fishing trip, but I would not break your rules so presented it to a neighbor."
Smile—"Let the neighbor get the Itch; you deserve commendation."

Viewpoint—"Full history of the case is as follows:—Married at 24."
Smile—"Spectro-Chrome Institute is not a Divorce Court."

Viewpoint—"My heart seems to bubble and gurgle."
Smile—"There must be some soap in it."

Viewpoint—"We desire to have a baby and preferably 'Twins', what will you advise?"
Smile—"Mother of the Dionne family will be more suitable for multiple birth advice; we had six, but one at a time."

Viewpoint—"His breeding is short when he walks."
Smile—"What does he breed?"

Viewpoint—"My appendix was removed and repaired outside."
Smile—"Let us know the name and address of the Doctor who replaced it."

Viewpoint—"Have been married two months, but did not reach climax."
Smile—"You are looking for quick results in a usually unwanted direction."

Viewpoint—"I am a rabid stenographer."
Smile—"Green Systemic will remove the danger of Hydrophobia."

Viewpoint—"In my field, corn grows 14 feet high in Vineland, New Jersey."
Smile—"Congratulations for the active Pituitary Gland in your dead hogs."

FIRST CLEVELAND OHIO PLANET

(HOTEL HOLLENDEN, CLEVELAND, OHIO)

Meeting Of Thursday, August 6, 1936

Reported By:— FRED E. OSBORNE, *Recorder*

The first order of business was the installation of Officers by Bessie A. Hasenau, Third Trustee of the Scientific Order of Spectro-Chrome Metrists.

William J. Hasenau was affirmed as the Deputy and Fred E. Osborne as the Recorder. Then John H. Winchester was recommended to the Board Of Trustees as Sub-Deputy.

The meeting was opened by the Deputy for suggestions or questions on the cases for Tonation to relieve suffering mankind by the use of Spectro-Chrome; also the proper food to eat to attain best results.

Alice K. asked the question, "In the Tonation of the Sufferer with the Spectro-Chrome, should the Sufferer lie down or should they sit?" In the discussion, the fact was brought out there is more comfort in lying down than sitting up because the Body is relaxed.

The question was asked to explain the Itisometer. This question by John W. was answered by Bessie H. giving an invitation to the Affiliates to come and see her Equipment.

"Are there any people who live wholly on a vegetable diet?" Asked by Felix M. Yes, there are many who do; the doing away with meat can be accomplished by the use of dry Peas or Lima or Navy Beans as a replacement food.

"What about the value of foods?" Eat the kind of food which is to produce the proper chemistry in the Body. The Ratioanl Food Of Man, Nuts, Fruits, Vegetables. Omit White Sugar, White Flour, Tobacco, Intoxicants, Tea, Coffee, Drugs, Eggs, Fish.

Nettie B. asked the question, "Would the Lemon Attuned Color Wave be good for Circulation in a broken foot? Magenta is used for the Circulation; Lemon to clear the condition after the recent (acute) stage is over. For the recent stage use Turquoise or Blue.

More questions were asked by John W. "Is there any harm in sleeping on the face? Should one sleep on the side or the back? What about the Rockefeller Institute where it was proven that one turns from side to side in sleeping?" Tonations are given with the Sufferer lying on the side. Most people sleep on the side, few on the Stomach or Back.

EDITORIAL REMARKS

You made a good beginning. Continue the educational work. Fol-

low the report routine of other Planets. The meeting should last at least 2 hours to be of benefit. Each session, have some volunteer to read a Chapter from Spectro-Chrome Metry Encyclopaedia. Discuss on the theme. Give your own experiences. Exchange views thereon. Thus, you will serve the good interests of the Order and exemplify the Objects.

FATAL ERROR

By:—BEATRICE K., *Kansasville, Wisconsin.*

"We bought a Spectro-Chrome machine a year ago last August. If we had known of the machine sooner, I'm sure it would have cured my husband, but, at the time we purchased the machine, he had a case of far advanced Tuberculosis. He passed away July 28, '36, but one thing I'm sure of Spectro-Chrome sure prolonged his life."

Life and Death are transitions in the Laws of Evolution and Spectro-Chrome can not give eternal life, but, the above echo of the harassed mind of a good woman, who had she known about Spectro-Chrome before, would have succeeded in normalating her husband and assuredly he would have lived through his Disorder.

Spectro-Chrome is not only a powerful normalator, but, is the greatest preventative of any method of Healing. The confidence engendered in the mind, by having a Spectro-Chrome in the home, can solely be appreciated by one who used it successfully and saw it bringing the most refractory and hopeless conditions to normalcy.

In San Francisco, California, The San Francisco Daily News, in its issue of Saturday, March 22, 1924, in giving a two column description of Spectro-Chrome, called the apparatus, "The Modern Alladin's Lamp" and styled the Originator as the "Rainbow Doctor". Little did the writer of that article think then, that the "Rainbow" will actually be used successfully in homes in cases where even the Medical Doctors failed to accomplish anything.

The Sacred Message of the Rainbow, mentioned in the Holy Bible by which the Almighty God made a Covenant with man, is actually decoded in Spectro-Chrome, which is bringing in homes health and happiness by relieving them from the life-long clutches of the terrible Medical Octopus. It was predicted in Genesis Chapter 9, verses 8 to 17. No power of the Earth can stop Spectro-Chrome, for it is the realization of His glory and His majesty.

MALAGA INVADED

The six sons ran with blanched faces to the father roaring "The Fascists invaded Malaga! What shall we do? Shall we fly to India?"

The father calmly read the morning paper; it was true, but that was Malaga, Spain. Spectro-Chrome Malaga has Italians, but, they are peaceful farmers minding their own business as neutral Americans should. So the boys went back to study and the father to more work.

FIRST DETROIT MICHIGAN PLANET

(HOTEL TULLER, DETROIT, MICHIGAN)

Meeting Of Thursday, August 16, 1936

Successful Use Of Spectro-Chrome In:—Diabetes, Fever, Horse Kick On Face, Rapid Heart, Loss Of Vitality.

Reported By:—GREETA O. PECK, *Deputy.*

Sub-Deputy Stella Harter was in the chair.

Sara C. said her mother is getting along well.

Leslie N. suggested the Rational Food Of Man is one of the main things. Also told of a Diabetes case, that is coming along well.

G. told about his 2 year old boy, who had Fever; Karoscil 122, Spiroscil 27. One Tonation of Blue and 2 Yellow normalated.

Edwin L. told of a Sufferer being troubled with gas; the Doctor gave her "Color Light" treatments.

Selma V. had experience of her son being kicked on the face by a horse, used Turquoise Local, Green front and Magenta on Area 18 for 2 days. A week later he seemed to fail and lost ambition, his Kardoscil 42, Spiroscil 13; had 3 Tonations of Scarlet on Area 4 and was normalated; went back to work in two weeks.

Stella H. told of a woman having a rapid heart; she tonated with Green; her Kardoscil was 96 and went up to 110. Also a man having Diabetes used 2 Green and 1 Magenta; lost 5 lbs. in 3½ days; followed with Yellow and is holding his own.

Jessie N. stated to follow Colonel Dinshah's advice and he also says the system will only pick up the Attuned Color Wave that is needed. The mental attitude helps a good deal; also told of a young girl who had lost vitality through hard study. She was normalated with 5 Tonations of Green and Magenta.

Robert S. asked if he could use 2 different Color Waves on Front and Back the same time? For instance, he is using 1 Blue on Front and 1 Turquoise on Back.

EDITORIAL REMARKS

Please, always send the names of the Affiliates at each meeting.

Medical Doctors are using many types of so-called "Color Lamps", which though they are "color", have no Science or Automatic precision in them. Many are sheer imitations and intended for restricted use in some Disorder according to some medical "diagnosis". However, in time even their efforts will indirectly help our cause, though in many cases their quackery and empiricism is blamed on our work by public ignorance. Be on the sharp lookout for such.

Food and Spectro-Chrome work hand in hand. Mental attitude too influences the Normalation.

With two Equipments, Front and Back Tonations may be simultaneously irradiated. We shall supply such compound Spectro-Chromes when a sufficient demand is created.

KNOCK! KNOCK! WHO?

Milwaukee, Wisconsin.

Spectro-Chrome Institute,
Malaga, New Jersey.

Gentlemen:

The writer just thought up a good "Knock! Knock!" for your Spectro-Chrome Smiles. One hears so much of this knock knock business over the radio and otherwise that I got busy on one myself, namely:—

Knock! Knock!
Who's there?
Dinshah!
Dinshah who?
Dinshah (din'tcha) know that Spectro-Chrome is the way to health?

Cordially yours,
Alfred Z.

P. S. I am enjoying the best of health. At present I am tonating a cousin of mine who lost sight of her right eye a few weeks ago. The doctors informed her that they fear the other eye will also be affected. Spectro-Chrome for one week has made it possible for her to distinguish objects faintly already. I am confident she will be O. K. very soon.

Alfred Z.

EDITORIAL REMARKS

Thank you for your kind interest. Your cousin will surely be benefited if she follows our Free Guidance. However your "Knock! Knock!" started us on the warpath too. Here is ours:—

Knock! Knock!
Who's there?
Dinshah!
Dinshah who?
Dinshah know who knocked the bottom of the Medical Trust?

FULL MOON AND EYES

Robert Ripley in "Believe It Or Not" of August 6, 1936 issue of The Evening Bulletin of Philadelphia, Pennsylvania states that George M. Curry of the Pittsburgh Symphony Orchestra has Blue eyes which turn Brown when the Moon is full. It seems the scientific mind is slowly coming to the way of thinking emphasized in the Spectro-Chrome Favorscope. It is NOT a fad, fancy or whim—Spectro-Chrome is a grim reality whose axioms and postulates destroy known conceptions of what till now was named "science".

DANGER SIGNAL IN SCHOOL

An examination was made in a certain American school, of the physical condition of the children. Here is the analysis:—

Condition	Per Cent.
Defective Teeth	92
Tonsil Trouble (removal advised)	37
Tonsil Trouble (already removed)	28
Tonsil Trouble (not seriously affected)	35
Evidence of Rickets	8
Poor Posture	87

The tragedy lies in mal-nutrition. Carnivorous food, tobacco, alcohol and the use of dopes and drugs are causing havoc in these days of heavy strain and forced mentality. Spectro-Chrome Metry points to the correct way for health. The school directors should aim at clean, nutritious food and cleaner habits.

MILWAUKEE WISCONSIN HAS INFLUENZA

Some children died in Milwaukee, Wisconsin, of so-called "Influenza."

As usual, the Medical Doctors grabbed the opportunity and laid the blame on cats and contacts. Dr. John P. Koehler, the Health Commissioner disagreed with many views, but urged the parents to call Medical Doctors at once to avoid the spread of the "Infection".

The whole theory and practice of Medicine is plain Bosh. The fear implanted in the mind of the public creates more havoc than otherwise. If the Children's Hospital and Isolation Hospitals will install some Spectro-Chromes, there will be no epidemics. We are willing and ready to give them FREE.

Turquoise Systemic will do the work with Magenta on Area 4.

FIRST ST. LOUIS MISSOURI PLANET

(HOTEL MELBOURNE, ST. LOUIS, MISSOURI)

Meeting Of Thursday, August 6, 1936

Successful Use Of Spectro-Chrome In:—Skin Disorder, Boil.

Reported By:—CHARLES H. STIERMANN, *Recorder*

Recorder Charles H. Steirmann was in the chair.

A letter from Scientific Order Of Spectro-Chrome Metrists read announcing the "Silver Jubilee" to be held in Detroit, Michigan under dates of August 14, 15, 16, 1936. Circulars were received covering program in detail; they were distributed to those present for attention and information.

William D. again reported the case of a Sufferer with sore on leg that opens and closes. This was reported last meeting. The Sore had again healed, but as it comes and goes in cycles and wishing to bring it to a complete Normalation, Orange was used to open, but failed. He then used Green and sore opened again. After opening, he used Orange Local and Lemon Systemic. This case is of several years standing and William D. is aiming for complete drainage before attempting to seal with Indigo.

A question arose about arthritic and rheumatic conditions and how to tonate. Charles S. said these were Persistent Disorders and not of recent origin. Lemon Systemic and as warmth usually was soothing to affected Areas, the warmth Color Wave Red might be used locally.

William D. asked how long should one tonate after Normalation had been effected in maintaining a normal condition and balance.

John M. reported the case of his wife brought to the attention of the meeting some two or three meetings ago. The Sufferer was pothered with Blisters on upper arm and a few on lower arm. They were of five or six years' standing and most annoying in hot weather and dreadfully itchy. These blisters broke from time to time and exuded a sticky fluid. Skin specialists failed completely in this case. Now, the past two months, approximately though passing through the most heated term of Summer, at which time it was always at its worst, there has been no further trouble and seems to be perfectly normalated. Two Tonations a day were taken, Lemon Systemic and Blue Local.

Joseph H. said that when he had the Boil on his elbow reported

two months ago, he used Yellow as well as Orange and found it a great help as a drawing agent.

William D. said we were about to enter the Hay Fever season and for the benefit of Members it might be well to suggest what Color Wave to use to tonate. Charles S. suggested inasmuch as Hay Fever from its recurrence was persistent in nature, Lemon Systemic and Turquoise Local be used. It was also suggested to look to the Area 4 condition and use Magenta.

Thomas S. sought information on taking Kardoscil and Spiroscil. A brief session was held in showing how this was done, the time of count being figured on a one minute basis. It seems there is some difficulty regarding this matter and the time is well spent in elucidation to Affiliates.

Some little discussion was held on charging of water by the Attuned Color Waves and benefits to be obtained by such practice. Colonel Dinshah has such Equipment and speaks of its merits in Spectro-Chrome Home Guide.

EDITORIAL REMARKS

William D. is now on the right track, but the conditions of food and abstention from tobacco smoke and the like MUST be observed to gauge duly the value of Spectro-Chrome. We do not believe in patch work, but in eradication.

Arthritis simply means "Joint Inflammation". Apply to it the same principle as in any other Inflammation. Rheumatism also is in the same category. It is an "Inflammatory Fever". The same rules of Rubor, Calor, Tumor, Dolor apply. Refer to Spectro-Chrome Metry Encyclopaedia, Volume 2, Chapters 54, 55, 56, 57.

Many so-called cases of "Hay Fever" arise from floor rugs, feather pillows, furs and fly screens.

For method of measuring Kardoscil and Spiroscil refer to Spectro-Chrome Home Guide, page 91.

Be ready for the Sixteenth Annual Convention, beginning November 28, 1936 and two subsequent days, in Convention Hall, Hotel Melbourne, St. Louis, Missouri. The usual education will be provided.

Suggestions about making this Annual Convention a telling success should be sent promptly to us. We are always ready to listen and friends should not feel shy of expressing their views. There is room for improvement in all management and we are prepared for all suggestions, to insure success.

DUMB DIAGNOSIS VERSUS PRECISE SPECTRO-CHROME

Emma S. from Milwaukee, Wisconsin, sent us a very illuminating report about the condition of her mother, aged 71. Her sketch is so well written that we print it as a token of what ails the public and why we denounce the Medical Racket with such emphatic vigor. She wrote:—

"Conditions confronting on June 1, 1936:— Excruciating pain in head; Lumps forming on head wherever pain occurred; Left Eye closed; unable to move Left Eyeball toward nose; Left side of Throat numb; Cords in back of neck stiff, mother being unable to hold head upright for very long; Jaws so tired could barely open mouth; unable to sleep; Constipated for 40 years.

"Diagnosis and Doctors' Opinions:—Dr. William K., (family physician), 'Neuritis, nerve condition; Strychnine administered in small doses for a long period of time, works wonders.'

Dr. William A., 'Nerve condition; see Occulist about Eye.'

Dr. U. D., Oculist, 'The condition is secondary to some other condition; see a Neurologist.'

"Dr. P. Neurologist, 'Third and Fifth Nerves affected. A Stroke, you may soon expect another.'

"Dr. William K. again, 'Bell's Palsy, Thrombosis; the Disorder is progressive, we must make her comfortable.'

"Treatment given: Strychnine, Iodine, Morphine, Sleeping Powders minus narcotics, the effect of which wore off in less than 2 hours. Dr. William A. gave mother 3 Infra-Red treatments of 15 and 20 minutes each and 3 Ultra-Violet Treatments of 1 to 5 minutes each; 1 Injection for Anaemia.

"Reaction to medicine:—Stomach was so upset that mother would refrain from taking the medicine for a few days to give her stomach a chance to 'come back'. A puff or swelling in different parts of the face every day. Pain worse than ever.

"I rented a Spectro-Chrome on June 1, 1936 and under the capable direction and patience of the Spectro-Chrome Normalator, the reaction to Spectro-Chrome was as follows upto August 1, 1936, when received my own Equipment:—

"More sleep after the first Tonation than all the while with powders. Pain gone after commencing Tonation. Lumps gone. Eye open. Able to move Eyeball in every direction. Jaws much improved; can open mouth wider when yawning and can chew better. Throat only a bit numb now. Cords in back of the neck improved. Better able to hold head up. Bowels beginning to respond; two natural movements to date.

"My attention was attracted to the date Specro-Chrome was

patented, i. e. July 7, 1925. It happens that my mother was born on July 7, 1865!

"The feeling of anxiety which my husband and I felt for my mother while she was under the Doctor's care, has changed to a feeling of confidence and calm since we have been introduced to Spctro-Chrome. It surely has proven itself............... The high caliber and sincerity of your Organization impressed me from the very beginning as did your Constitution."

The above represents what actually Spectro-Chrome is replacing. Used according to instructions, Spectro-Chrome performs near-miracles, where other methods signally fail.

Spectro-Chrome is covered by many patents, the latest having been issued by the United States Of America, on April 28, 1936, Number 2,038,784.

Our thanks to Emma S. for her kind opinion of our activities. We must maintain the high reputation of Spectro-Chrome at all costs and hazards.

MEDICAL HUMBUGS OF 1936

Detroit, Michigan.

Editor, Spectro-Chrome:

During the last week I collected a few newspaper cuts, which I forgot to bring to the lectures. We surely would have gotten a great "kick" out of it. Here are some extracts:—

1—Operations free as prizes offered by Cleveland Osteopathic Society! First prize, free major surgical operation; second prize, free obstetric service; third prize, free tonsillectomy and adenoidectomy. All prizes were transferable (including mother-in-law). Who got the prizes?

2—Alcohol injections in the arm for Pneumonia. Reported from India and experimented by American Medical Association. Are they sure it went into the arm?

3—Dr. Walter P. Gardner, Assistant Superintendent of the State Hospital at Fergus Falls, Minnesota, says, "None is definitely sane or insane." True that is why they trust the Doctors.

4—Dr. Anderson of the British Medical Association declares that of all professional men, the Medical Doctors are found to be the shortest lived. No wonder, if they use the same medicine as they give to the Sufferers.

5—Barbers were once the Surgeons, hence their sign of the Red, Blue and White pole, Red being the Artery, Blue, being the Vein and White the Sufferer bled till bleached. The situation did not change yet.

FIRST MILWAUKEE WISCONSIN PLANET

(HOTEL WISCONSIN, MILWAUKEE, WISCONSIN)

Meeting Of Thursday, August 6, 1936

Successful Use Of Spectro-Chrome In:—Paralysis, Sores, Hodgkin's Disorder, Splenitis, Dropsy, Backache, Female Disorder, Fainting, Leaky Heart, High Blood Pressure, Calculus, Swelling, Pink Eye, Headache, Bee Sting, Sprain, Welding Flash.

Reported By:—CAROLINE ZIMMERLING, *Recorder*.

Literature concerning the Special Rally to be held at Detroit, Michigan was distributed at the door. Deputy Otto H. Lutz was in the chair.

Otto L. made the announcement that Lena Held was reinstated as a Member into the Scientific Order Of Spectro-Chrome Metrists. This news was received gladly by the members of the meeting.

Chapters 82 and 83 of the Spectro-Chrome Encyclopaedia on Spectro-Chrome In Paralyses and Spectro-Chrome in Melancholia, were read by the Deputy.

The Deputy reported a case of Paralysis that he tonated. It was in a man of 35 years of age. This man was completely paralyzed, even to the point of not being able to urinate. After 4 Tonations, he was able to do this and in 3 months was able to sit up. He then had braces made to assist him and was able after a short while to go back to work. He used Green, Lemon and Yellow.

Hedwig N.'s grandchild has Sores all over her body. Salve they were using on her simply spread the Disorder. After giving the child 3 Green Tonations and 3 Lemon, the child was normal; there are no scars even.

Frank B. talked about his little boy. He said the boy is getting along well, even if the progress is slow. He has much more muscular development, a good appetite, is walking much better now, although he still needs a very little assistance. His powers of observation are better as are his other mental powers. He has his 2 Tonations a day and Frank B. has great hopes the boy will be absolutely normal in time.

Otto L. gave the case of William W., a man of 36 who had to walk with a cane and tried everything for about a year. After 2 weeks of Spectro-Chrome Tonations, he walked downstairs himself and says that he will stick to Spectro-Chrome. Another case was that of a man of 56 who had a stroke. After 2½ weeks of Spectro-Chrome, he was able to sleep well and had a good appetite, although his eyes still were very glassy, so his family decided to stop giving him Tonations.

Frank B. told of a case for which he recommended Spectro-Chrome. It was one of "Hodgkin's Disease". Good results were obtained. The eight or nine year old boy's tongue would extend from his mouth during sleep and he would run a temperature; also the pulse was very slow. They succeeded in raising the Kardoscil almost to normal and were able to control the temperature with Blue.

The Deputy gave a case of a man who had pains in his Spleen. By Spectro-Chrome in one hour he went away smiling and whistling.

Another case reported by him was that of a man with Dropsy, who reduced the size of his waistline about 5 inches in 4 weeks, by Spectro-Chrome. Green and Yellow were used, then Violet, Lemon and Yellow with Magenta on Areas 4 and 18.

Frank B. volunteered to read at the next meeting Chapters 84 and 85 of the Spectro-Chrome Metry Encyclopaedia, on Spectro-Chrome In Indigestion and Foolery Of Fly Philosophy.

The Deputy gave another case of a man who was a physical director and by strenuous exercise strained the ligaments in his back. After the fourth Tonation he was doing an Adagio Dance. He had only 2 Lemon and 2 Green Tonations.

Another of the Deputy's cases, was that of Julia W., aged 36, who had a Female Disorder and was burning with Fever. Her temperature was brought down in 20 minutes with the Blue Attuned Color Wave.

On Martha S., another of the Deputy's cases, Green with Magenta on Area 4 are being used to remedy a condition of Fainting. She takes only 2 Tonations a week and did not have a real attack since starting Tonations. She is well pleased.

Frank B. said he knew a man of 66, who weighs about 237 lbs. who cannot take Tonations. It riles him very much. The Deputy said it was the filth in the Body that did this.

Caroline Z. gave a report of the case of a woman with Leakage of the Heart, High Blood Pressure and Dropsy. All Medicines were stopped immediately upon beginning the use of Spectro-Chrome and from that time on she had no spells of being bedridden. Now, after using Spectro-Chrome, faithfully for about a year, she recently had an examination for Life Insurance and was pronounced physically perfect. The woman is some 70 years of age.

Kurt B. gave a report of his own case. Indigo did not help a Pain in Areas 9 and 18, so he used Turquoise Systemic. This did not help so with the aid of the Deputy's Itisometer, he determined that the Infra-Green was needed. He used Lemon Systemic and Purple on Area 4 and the next day passed two Stones. When the Stones were gone, the bleeding also disappeared.

Otto L. said his wife's ankles were swollen. It took 3 weeks of Green, Lemon and Yellow Systemic with Magenta on Areas 4 and 18 to normalate her. Ate nothing during this time but fresh vegetables and very sparingly of that.

Otto L. then read about Gland Stimulating Colors from the Spectro-Chrome magazine of May, 1923, Volume 2, No. 5.

Hilda S. gave the case of the little girl with Pink Eye. The Eye closed completely, was running and very painful. Ten minutes of Indigo and the pain was all gone! Green was found to be too glaring for her, so the Frosted Slide was used over it. Several Green Tonations were given and the child was normal again.

Hedwig N. said her daughter had good results for Headaches with the use of only one Turquoise Local.

The Deputy's wife was stung in a finger by a bee. Green Systemic removed all the swelling and pain.

Carrie M. told of the man of 76 who fell and sprained his arm. He could not lift it. She gave him an Orange Tonation 3 times and then a Turquoise Local; while under the beam she saw a lump move up the arm and disappear.

Otto L. gave another case of a man who fell on concrete and broke the nose. He lost about a pint of Blood. His daughter gave him a Tonation even before an ambulance could arrive to take him to the hospital. The Doctor came and removed a bone from his nose. The Equipment was used on him 4 hours continually. There was no bump and it all healed nicely.

John S. spoke of a man who got Welding Flashes in his eye; had 2 Blue Tonations and went back to work.

EDITORIAL REMARKS

High Blood Pressure is not a Disorder, but a condition produced from other causes. Our Purple, Magenta and Scarlet, judiciously irradiated Local, with Lemon Systemic, will overcome all Circulatory Disturbances.

Fainting indicates faulty Circulation.

SPECIAL RALLY FOR SILVER JUBILEE

Spectro-Chrome Radio Broadcasting Station Proposed

The Originator of Spectro-Chrome Metry, arrived in this country on his first tour, on Sunday February 16, 1896, delivered scores of scientific lectures on the X-Rays and the like and returned to India in a few months. On his second visit he arrived in New York, N. Y., on Wednesday, August 9, 1911 and on Friday June 15, 1917 was naturalized as a citizen.

The presentation of Spectro-Chrome Metry to the World, was made in the World Building, New York, N. Y., on Monday April 26, 1920, in his private lecture hall. The Central Office then moved there from Hillsdale, New Jersey, where the Science was originated. A year later, the Central Office was located in Philadelphia, Pennsylvania, whence on Thursday, July 10, 1924, it made its home in Malaga, New Jersey, where its Plant is at present situated.

Our First Detroit Michigan Planet held a Special Rally, in the Small Ball Room of the Hotel Tuller, Detroit, Michigan, for three days beginning Friday, August 14, 1936, to celebrate the Silver Jubilee of the arrival of Colonel Dinshah P. Ghadiali, the Originator of the Science Of Automatic Precision. A large and intelligent audience was present at each of the six meetings and numerous case reports were given by Purchasers of Spectro-Chrome correlating with the lectures of Colonel Dinshah. Most of the reports pertained to Persistent Disorders declared "hopeless and incurable" and given over by the Medical Doctors or their Hospitals.

Catherine Ivanowski, a Christian missionary, spoke on the wonderful relief she obtained in Buffalo, New York, by the use of Spectro-Chrome.

Elizabeth Geyer extolled Spectro-Chrome because of the good results attained in her own case and that of her young daughter.

Lucille E. Greenham spoke about her visit to the Spectro-Chrome Institute Plant in Malaga, New Jersey and expressed her views about the work.

Magdalene Arnold spoke about the benefit her husband derived from Spectro-Chrome in Sciatica. He was a cripple for many years and now is in perfect health. He would give his life for Spectro-Chrome.

Rev. Fred C. Pillsbury spoke of how he came to the city in a helpless condition and would have gone to the hospital for an operation, when Spectro-Chrome was brought to his attention and he recovered his health completely.

Albert E. Greenham addressed the meeting on the merits of Spectro-Chrome. He said he would part with everything but Spectro-Chrome. He regretted that the Planet Members did not have enough notice ahead or the audience would have filled the big Ball Room.

Colonel Dinshah jested that if he would guarantee the audience he would return next month. Numerous voices clamored for further knowledge of Spectro-Chrome, the main Ball Room was engaged by our President and all promised hearty co:operation. (The complete program is printed in the current issue). On being questioned, Albert E. Greenham guaranteed placing 10 Orders for Spectro-Chrome.

Christina Keim expressed her favorable impressions on the value of Spectro-Chrome.

Maggie Barron spoke enthusiastically about her favorable experience in the family, with Spectro-Chrome.

Much merriment was produced and the audience roared with laughter, when the subject of Gland Therapy was expounded. The climax was reached when one of the Members gave his personal experience of the number of Placentas (afterbirths) of parturient women collected from the hospitals and taken in pails to certain biological laboratories for conversion into pills or powders for feeding women as Medicine!!!

Dinshah spoke about the abominable idea of "Auto Therapy" and his comments on "One Lick A Day" produced another volley of applause and laughter. He showed how fallacious the Gland Therapy and similar "therapies" were, by showing a bottle of "Hormotone", said to contain the extracts of the Pituitary, Thyroid, Ovary and Testes combined! The bottle had 100 tablets, with such "wonderful" combination having both male and female potency!!! He said he did not need any Hormotone, but, if some desexed, unsexed, devitalized, double-sexed hermaphrodite American wanted it, he may have the tablets free of charge! None voluntetred for the marvelous medical gift!

To enhance the spread of correct knowledge of Spectro-Chrome and the underlying principles that made Health, Dinshah was seriously considering the erection of a powerful Spectro-Chrome Radio Broadcasting Station. The proposition was received with enthusiasm. Steps are being taken to install one.

Greeta O. Peck, of Windsor, Ontario, Canada, Deputy of the

First Detroit Michigan Planet delivered a short address on the last night's session. She said:—"Colonel Dinshah speaks Truth greater and more powerful than perhaps many of you ever heard. Just ask, as you listen, that your ears be opened to hear and your understanding enlarged to comprehend the wisdom you shall hear.

"Without Light a Universe could not be. The Light is supreme, the Light is everywhere and in the Light exist all things. All the way down the centuries, in every age as far back as there is any record of humanity, the wise and great ones are portrayed with a radiation of Light about them emanating from the head and body of each. That Light is real, just as real as the electric Light in your homes. All individuals like to be out of doors, because the greater Light from the Sun contains that which makes them happy. They intuitively like it and feel that they require it.

"Does it not seem reasonable that a Healing System which uses Light as its agent is God's system? All things in Nature are simple, calm, peaceful and obedient to natural Laws. We find Spectro-Chrome just that. This afternoon you were taught how 176 Fevers and 213 Inflammations are normalated with one Attuned Color Wave. No great large names to confuse, nothing given to aggravate the suffering. Instead, all becomes calm and peaceful within a very short time.

"A member of my family was ordered to the hospital to be operated on **at once** for Appendicitis. When we were called, all was confusion, tears, despair—such pain, something had to be done, but, they were in no condition financially for a big doctor bill and hospital bill. When we suggested trying Spectro-Chrome, the Medical Doctor of course said he couldn't be held responsible for any consequences which might follow. Yet, by Spectro-Chrome, within an hour the pain was all gone and did not return and the Sufferer who was 75 years of age is active and well. That was over a year ago."

A telegram from Otto H. Lutz, Deputy, First Milwaukee Wisconsin Planet was read, "Congratulations and best wishes for your Silver Jubilee."

The audience gathered around Dinshah and gave their personal greetings vociferously. After thanking all and expressing cordial appreciation of their endeavors, the final session closed on Sunday, August 16, 1936, at 11 P. M.

As usual, Dinshah had to return to the Central Office by automobile, but, before leaving he requested the independent Field Solicitors to send in more Orders so he may travel in his own airplane and escape traffic delays! Time counts in life.

SPECTRO-CHROME

January 1937 Red

Volume 13 Number 1

IMPORTANT ARTICLES FEATURED

IN THIS ISSUE

Read and Let Your Friends Read

Wasting Time On Known Matters

Near-Miracle In Double Pneumonia

Near-Miracle In Arthritis Deformans

Light Is Invisible

Missouri Commends Dinshah

Just One Of The Many—Convention News

It Pays To Follow Instructions

Sixteenth Annual Convention A Success

SINGLE COPY 50 CENTS

WASTING TIME ON KNOWN MATTERS

What the Originator of Spectro-Chrome Metry taught and forgot years ago, scientists are finding now! In an article by H. W. Fry on the "Oddities Of Nature", in the Evening Bulletin of Philadelphia, Pennsylvania, dated October 30, 1936, the following appeared and it is so strikingly candid that we reprint it with thanks to the writer and publisher:—

"Colored" Light Affects Growth Of Plants.

"Green is restful to the eyes. Blue is present in many things about us. Red is associated with danger, anger and war. Yellow is commonly associated with cowardice. These are things we all know.

"But the scientists in their laboratories have recently been demonstrating that different wave-lengths of visible light as expressed in different colors have effects on the growth and behavior of plants.

Dr. Enoch Karrer, of the Smithsonian Institution, exposed oat seedlings to the light of an electric lamp which was split into the constituent wave-lengths by passing through a prism.

"He found the greatest growth in length was in plants illuminated by the orange part of the spectrum and the yellowest growth in the blue-green region. The seedlings bent towards the blue light and away from the red. Roots grew longest in the extreme blue and shortest in the orange red.

"Light means life. The relationship of light and life is photosynthesis. By this process, carbon dioxide in the air is made into carbohydrates of plants and plants provide food for men and animals. Is it any wonder that early man worshiped the sun?"

No wonder! The Originator, IS a Parsee Zoroastrian, a Sun-worshipper as the Occidentals consider. It was that reason that he was chosen as the logical exponent of Spectro-Chrome Metry, the only Science of Healing the ailing that is unmatched and unrivalled for its Automatic Mathematically Exact Precision.

All the items stated by the writer H. W. Fry are age old knowledge in Spectro-Chrome and it would pay the Scientists greater premiums if they would investigate the recognized merits of Spectro-Chrome first and start their 'researches' subsequently.

The Lord Zarathushtra—the Blessed Zoroaster, thousands of years ago left much of that knowledge in His works and the Scientists are just awakening to the Glory Of God—Light and Her Powers. He knew, they surmise.

NEAR-MIRACLE IN DOUBLE PNEUMONIA

By:— RALPH B. D., *Milwaukee, Wisconsin*

I cannot begin to tell you the marvelous results we had with our Spectro-Chrome under your Guidance, or can I ever repay you for the great good done my daughter. My wife and I were frantic and felt that our one and only daughter and dearest possession, would be taken away, but Thank God for blessing the world with the natural power of your Spectro-Chrome as she is apparently getting along nicely and out of Danger.

As you said, "Let Nature take its Course" and let her sleep just so it was and after the first two Tonations she rested comfortably. Her Temperature went up slightly after the first Magenta Tonation but the second day that was on the decline too. So that you might have a better record of the case I am sending a tabulated sheet with Kardoscil, Spiroscil and so forth, from the time she started taking Tonations.

If I have been a little loose in my Explanations I will be glad to elucidate.

Though I know you accept no Gifts of Gratitude you must accept my Gratitude and Love from my heart.

FIRST CLEVELAND OHIO PLANET TEMPORARILY DISCONTINUED

Owing to the paucity of attending Affiliates in the territory covered, the First Cleveland Ohio Planet will not meet at the Hotel Hollenden, until further notice. The action was taken after a report of discontinued activities was received from Recorder Frank E. Osborne.

Many friends do not seem to comprehend that the Organization pays all the expenses of Planet Meetings to encourage the exchange of views among the Affiliates. Such aim can be accomplished solely by presence at the monthly meetings. Many do not like to be present, because discussion is restricted soely to Spectro-Chrome Metry. No gossip is permitted and no refreshments are served. This is not liked by many who take delight in club-type life. Thus, finding the Cleveland Affiliates lax in their presence, we curtailed its annual expense of $60.00.

It will be reopened when the Board Of Trustees is presented with the reasons for such action.

FIRST DETROIT MICHIGAN PLANET

(HOTEL TULLER, DETROIT, MICHIGAN)

Meeting Of Thursday, November 5, 1936

Successful Use Of Spectro-Chrome In:—Nausea, Back Pain, Hysteria, Convulsions, Hay Fever, Facial Paralysis, Heart Disorder, Appendicitis, Constipation, Boils, Carbuncle, Fainting, Ringworm, Bronchial Disorder, Knife Wound, Stiff Neck, Eye Strain.

A Canary Responded To Spectro-Chrome!

United States Department Of Justice Investigates Spectro-Chrome!

Reported By:—L. ROY DODMAN, *Recorder*

Deputy Greeta O. Peck was in the chair.

She gave a report of her visit to our Malaga Headquarters. She stressed the intelligence and knowledge of the boys which is gained from books and also practical experience. Each has his part in the work of Spectro-Chrome.

Maggie B. reports her boy came home sick at Stomach. Two tonations of Orange were used and he went back to work the next day.

Aquilla M. reported sister-in-law had Pain in her Back. She sent in Advice Chart and is using Lemon Systemic and Magenta on Area 4. She still has some pain when she walks.

Peter D. reported a case of Hysteria in a woman of 33 years successfully normalated by Yellow in 1 week. He said the young woman who had Convulsions now had only light spells, sleeps well and feels well. She used Tonations for 4 months and is 80% well. At present she is using Yellow on Areas 8, 11 and 12 and Purple on Area 4. A boy with Hay Fever was normalated with 4 Tonations of Turquoise. A man with Facial Paralysis for 3 months, used Yellow and Lemon for 3 weeks and Magenta on Area 4 and is now well.

Peter D. sold 3 Spectro-Chromes and every Purchaser is satisfied.

John B. reports wife had Heart Disorder, Appendicitis, Constipation. She used Lemon Systemic and Magenta on Area 4 and is well. He had Boils which were hard to heal before getting his Spectro-Chrome. Since he had his Spectro-Chrome, he had a very painful Carbuncle on which he used Orange Tonations and the core came out in 3 weeks.

Lucille G. had a cold with Kardoscil high. She used 1 Purple and 2 Turquoise Tonations and the condition was normalated.

Sarah C. reported a sick canary. After using 2 Magenta and 2 Orange Tonations, it was able to fly.

George G. reported Cold in head with temperature. Kardoscil 87, Spiroscil 22. After taking 2 Tonations of Green and 1 of Turquoise, his Kardoscil was 72, Spiroscil 18 and conditions were normal.

The Deputy reported a woman had been ill and began to faint. She fell and her hand went through glass. She lost use of the hand. After the first Tonation, she was able to wash her back. She has had 5 Tonations and the feeling is coming back and she can turn a door knob. She is using Yellow. Her Kardoscil is 82 and Spiroscil 20. A boy with Ringworm went to Medical Doctor for 1 month without results. He had 7 spots and after 7 Tonations of Turquoise he was ready to go back to school. A woman's ankle was very red and inflamed. She used 3 Turquoise and 3 Lemon Tonations, also Magenta. She was better in 7 days.

Norris E. DeC. reported a woman with Constipation, Low Heart and Bronchial Trouble. She cannot use Yellow, so uses Lemon; Green is helpful also. Her condition is better. A man had a knife wound which was very sore and was under care of Medical Doctor for 1 month before it healed. He later had a thorn in his knee and used Spectro-Chrome. Green Tonations healed it very quickly.

A discussion arose as to whether the guidance from the Central office was based on the Psychic Powers of Colonel Dinshah. It was the opinion of Thomas N., Albert G. and Otto H. that it was Colonel Dinshah's knowledge of Anatomy and Physiology and the effect of the Attuned Color Waves on the Human Body which enabled him to give the correct guidance.

The Recorder reported a case of Stiff Neck which was normalated with 1 Tonation of Lemon. Also a case of Eye Strain which was corrected with Indigo tonations.

Various Affiliates who gave case reports at the last meetings conducted by Colonel Dinshah, were interviewed by a Special Agent of the Department Of Justice. The purpose of the calls was to get facts concerning the healing power of Spectro-Chrome.

Albert G. reported there was about 400 people present at each lecture delivered by Colonel Dinshah. He also told of a friend who had just returned from Florida who said that Spectro-Chrome was known in the South.

Rules pertaining to the appointment of Delegates to the Sixteenth Annual Convention were read. It was the unanimous vote that Albert E. Greenham be the Delegate and Jessie E. Ness the Alternate Delegate for the coming Convention at St. Louis, Missouri.

EDITORIAL REMARKS

Peter D., Albert G. deserve thanks for their incessant endeavors.

Fainting is a Circulation Disturbance.

Colonel Dinshah, for many years in this country and India, is known to be an Occultist, but the handling and reading of the General Advice Charts has nothing to do with "Psychical Powers". Such allusion would not be fair to Spectro-Chrome Metry, where the veriest layman may have equally wonderful results by following the standard technique and the given Free Guidance. Of course, in the Science of Spectro-Chrome Metry, Occult Laws Of Nature are taken into consideration, respected and incorporated, but, the Science finally evolved from all such procedure is based on the Chemical, Physical and Physiological Phenomena readily provable in experimental work.

Spectro-Chrome Metry and the claims made for it have been investigated on numerous occasions by Federal and State authorities and we have reason to be proud of such investigations, as it shows that our public Officers are vigilant and on the alert against fakes. We thank them for doing their duty so well and all our Affiliates are requested not to withhold any relevant or pertinent information from zealous Officers who visit them in the performance of their public duties, but to give them exact and truthful information. We have nothing to hide or fear from anyone, as our cards are always on the table.

Spectro-Chrome is known for its value and merits North, East, South and West; it will in a few years be known and recognized all over the Earth.

Beware of spurious imitations!

APPROACHING SPECTRO-CHROME

Dr. Carl Anderson, of Pasadena, California received the Nobel Prize for the discovery of the so-called Positron or the "Positive Electron." Congratulations!

So they are all coming to the idea that Electricity (as well as Light) is composed of bombardment of material particles! A thorough study of the Spectro-Chrome Metry Encyclopaedia will easily reveal who should receive similar recognition—but then, who will recommend him to such high honor, unless it be the A. M. A?

SPECTRO-CHROME SMILES

Viewpoint—"I have had 79 Backelims in 47 days."
Smile—"Your diary must be interesting to read."

Viewpoint—"My bowels were Green and a Yellow ring around, now they are bright Brown."
Smile—"Call a painter."

Viewpoint—"The Sufferer was air hungry."
Smile—"Let her attend political meetings during next election time."

Viewpoint—"My hip has a Charlie Horse."
Smile—"Hook it to a milk wagon."

Viewpoint—"I am praying to the Dear Lord to help me get pregnant."
Smile—"God did not yet open a baby farm."

Viewpoint—" The Sufferer uses food as it comes."
Smile—"From where and what?"

Viewpoint—"I talk in my sleep."
Smile—"Put your wife in a separate bed."

Viewpoint—"At the age of ten years, I was Sunstroke and fell in an ice cold river."
Smile—"Thank God you did not propel yourself into the moon."

o

NEAR-MIRACLE IN ARTHRITIS DEFORMANS

By:— IRENE K., *Detroit, Michigan*

I have had my apparatus or Spectro-Chrome a week and have been using the Green Color Wave on my Mother AND AM I HAPPY!!! She actually turned herself over! On her left side! Please forgive my enthusiasm. You see I have gone through several different kinds of Hell since she came down with Arthritis Deformans two years ago. I have been taking care of her as one would a baby. Many a night I have sat and worried and listened to her cries of agony. Now after 7 days she sleeps mostly, eats everything and I have stopped giving her Cascara because of bowel movement. Since she has been confined to bed a year ago she has only been able to turn slightly on her right side. But today! She TURNED COMPLETELY HERSELF to the LEFT! She does not holler from pain as much either. I am going to my first Planet Meeting tonight. Thank you. I remain Your Friend.

LIGHT IS INVISIBLE

Of course! Light as Light Energy is absolutely invisible and it is only when it is reflected or interrupted that its effects become visible to the eye for what we call Light. In Scott's Scrapbook, in the Evening Times of Vineland, New Jersey, of Saturday, November 14, 1936, R. J. Scott also iterates the same view that "Light is invisible".

It is unquestionably true. Light IS invisible and also that the Sun has no heat, no Light. Dinshah's words will some day be corroborated by the best brains of the World.

o

PILLS PULVERIZED

By:— CHARLOTTE GEYER, *Malaga, New Jersey*

It used to give me Chills
To take the Pills
For all my ills,
But now it gives me Thrills
To look over the Hills
For Sunlight Frills,
That light my cosy Home
By Dinshah's Spectro-Chrome.

o

SPECTRO-CHROME METRY ENCYCLOPAEDIA AS A COURSE

By:—MILTON F. S., *Milwaukee, Wisconsin*

I read many books but never one as educational and instructive as this. It is an eye opener from beginning to end with all the truths and facts before you. It is to my mind a college course in 3 Volumes, Volume 3 being the finest in understanding. Many thanks to you and your fine staff.

Now that I finished the Home Training Course, I feel that I just began. The light of this wonderful Course is much brighter and would like very much to continue in this Service To Mankind. I never studied anything more fascinating. In my opinion, it sure is the beginning of new hope and service to all who learned, studied and used Spectro-Chrome. The Course in its 3 Volumes is as clear and understanding as Spectro-Chrome itself. The facts and the truth, are brought out in every detail. There is never a question unanswered. As long as there is the Sun in the Heavens. Spectro-Chrome will be here to shine brightly through the years and God bless the man who made this possible—Colonel Dinshah P. Ghadiali.—E. P.

FIRST MILWAUKEE WISCONSIN PLANET

(HOTEL WISCONSIN, MILWAUKEE, WISCONSIN)

Meeting Of Thursday, November 5, 1936

Successful Use Of Spectro-Chrome In:—Heart Disorder, Jaundice, Arthritis, Gall Bladder Disorder, Insomnia, Hay Fever, Constipation, Cold, Sores, Stomach Ulcer.

Reported By:—CAROLINE ZIMMERLING, *Recorder*

Deputy Otto H. Lutz was in the chair.

He read a letter from Spectro-Chrome Institute concerning the special scientific demonstration and comeback which will be staged by Colonel Dinshah in Milwaukee, Wisconsin soon. The Affiliates volunteered to distribute circulars for this event.

The Deputy then read Rule 66 of the Constitution And Rules which regards sending Delegates to our Annual Convention, which is this year to be held in St. Louis. Frank Burns was elected Delegate and Caroline Zimmerling Alternate Delegate by those having the voting power.

Elizabeth Z. read Chapters 88 and 89 from Spectro-Chrome Metry Encyclopaedia, on "Governor Green" and "Explosion Of Element Theory."

Laura G. said she tonated a case of Bladder Disorder that was made worse by the use of Green. Lemon was found to be better.

Francis B.'s mother-in-law had a bad Heart, so she started tonating for the Heart immediately instead of starting with the Green. Good results were obtained.

The Deputy said that some people use Green only, for any and all ailments. He said that one case came to his attention in which the party used Green for a Sprain and then Indigo with excellent results.

Laura G. finds that if she wants to show people quick results in order to sell an Equipment or convince someone of the merits of Spectro-Chrome, she gives Local Tonations with good results.

Kurt B. volunteered to read Chapters 90 and 91 from Spectro-Chrome Metry Encyclopaedia at the next meeting.

The Deputy then reported the case of Carl S., age 23 who had Jaundice. He had taken mineral baths and Colonic Irrigations and thought that he was perfectly "cleaned out". He is now getting up 3 or 4 times a night for Back Elimination and he is not quite as Yellow as he was; he has been taking Tonations only once a day for 1½ weeks. Magenta on Area 4 was used at first, then Scarlet, together with Green, Lemon and Yellow. In another week or so he should be able to stop taking Tonations. Four years ago, he had the same thing and was in the hospital for 8 weeks with it. His Kardoscil which was 50 at first did not increase much, but he is satisfied with the results.

The Deputy had given the case of Henry M., aged 56 last time, but reports that he can now go automobile riding and can now be moving around, even though the Medical Doctors said he could not live. Spectro-Chrome checked his rapid Heart action in one night.

Marie B. reported that she is experimenting with Spectro-Chrome on her bad knee. She had been using the Infra-Green, but it never was any better. Then she read something in the Spectro-Chrome Metry Encyclopaedia which caused her to change to the Ultra-Green and she says it seems to be some better now. She had used Turquoise Systemic quite a bit before at one time and had to stop it because she felt cold all the time.

Walter S. said he had used the Ultra-Green for a man with Arthritis and had good results.

Elvira B. said she used all the Colors on the Infra-Green for Gall Bladder Disorder and Constipation that had been in existence for 45 years. She also had pains in her knee and shoulder. All the time that she used Spectro-Chrome these pains did not go away. Then she started using Scarlet on Areas 4, 18 with Lemon Systemic and then after her Heart had increased in action she went back to Magenta and all the aches and pains went away. She now gained 29 pounds and feels fine.

Charlotte H. wanted to know what Colors to use for her father. He does not want her to write for Advice. He used Green once a day, then Lemon and Yellow and went down to Orange. Then his legs started swelling; there is a hardness across the Stomach. He gets up about every 2 hours during the night for Frontelim and the Urine is turning Orange. He is now back on the Lemon. His Heart is very fast and so is his Breathing, although the Ratio is 4 to 1. Cannot walk upstairs or exert himself in any way. He used Purple on the Heart, but his Stomach became harder then; however, he did not have to get up in the night so much. The Deputy said the Purple could be used on Area 4 and Magenta on Area 18 or Magenta on 4 and 18.

Frank B.'s opinion on this case was that it was all due to poor Circulation. He recommended Lemon Systemic and Magenta on 4 and 18.

Kurt B. suggested the same for the case and also that it was probably a very good thing that there was so much Frontelim to get rid of the water in the Body.

Kurt B. then reported a case of a man of 56 who claims he cannot sleep. Originally he had High Blood Pressure and Constipation. His Kardoscil was 68 and his Spiroscil was 18. He has had 4 months of Tonations, 1 week of Green and 2 weeks Lemon and Yellow. Indigo and Purple also used to cause him to sleep, but still he cannot, although he does not have the head noises that he did have or the Constipation and his Blood Pressure has gone down. He feels much better, but he still cannot sleep. The man had a mental shock some time ago when his sweetheart committed suicide. It was suggested that the man actually does sleep, but thinks he does not. If he did not sleep, he could not live, because this condition has existed over a long period of time. Psychology is the thing that this man needs, it was decided.

Ida B., reported a case of this kind which responded to Violet for sleep.

Edna B. is still using the Equipment for her boy who has a Spinal ailment. She is following Guidance from Spectro-Chrome Institute and is getting results, but they are slow. Is now using Lemon Systemic and Magenta on Area 4. Once in a while a Turquoise is used.

Marie Z. is using it only for her family now if anything needs attention. Outside of that there is no necessity.

Elsie B. says there is nothing wrong with her, although she uses her Equipment 3 times a week to keep in condition. She would not want to part with it.

Johanna A. is still using her equipment for Hay Fever. She had very little trouble this year.

Florence Z. has her Equipment about 1½ years. She is using it for Constipation. She is also very weak and does not seem to be getting her strength back. She is getting results for her Constipation, however. The Deputy suggested the Green as Muscle Builder.

Frank S.'s daughter had a Cold. 1 Green and 4 Turquoise Systemic were all that was necessary. She is 5 years old.

Rose M. had a man come to her with Stomach Disorder of about 20 years standing. He was also very constipated. Orange was given to him at once and he was relieved. He had very good results out of all Tonations. Her daughter had a Sore Throat which was helped by Lemon and Turquoise Systemic alternated. She is 14 years old.

Hannah S. spoke of a man of 66 who has Diabetes and Swollen Legs and who does not seem to be getting any better. He does not

follow the Rational Food Of Man properly and still has the Medical Doctor coming although he is not taking Insulin.

Ida B. has been tonating a girl since last Spring. This girl had open Sores on the Abdomen. At first she had Swollen Knees also. This was very painful. The Medical Doctors could not help her. After the Sores on her legs opened she had relief. After using Spectro-Chrome these sores began to heal, but Ida B. used Attuned Color Waves that would cause them to stay open until they could heal from the inside. This is starting now and the girl looks and feels much better. Green was used for 2 weeks, Lemon for 3 weeks, then Orange. She keeps rotating over and over again on these Colors. She also lives upto the Rational Food Of Man quite well. Magenta too is used on Area 4. The girl is about 20 years old.

Another case given by Ida B. was that of a man of 54, who had an Ulcer in his Stomach. He was advised to have an operation immediately. Instead he used Spectro-Chrome Green, Lemon, Orange and Magenta and is feeling much better now.

The members of the meeting were all very happy to hear that Colonel Dinshah was coming to Milwaukee again and are all looking forward to his visit.

EDITORIAL REMARKS

Laura G. should attempt to normalate a case, instead of attempting to sell by spectacular Local. That is why subsequent "complaints" fall on our head, because the Local acts like a dose of "medical shot", whereas, our sole aim is to eradicate the Disorder from the System. Green never makes "worse" as Lemon has ¾ Green in it; the fault is in the Kardoscil and Spiroscil Ratio.

Frances B. might receive "good results" by the Local Magenta, but, permanence needs the Infra-Green Systemic.

Colonic Irrigations must be rigidly avoided, as also all other enemas. Compel Peristalsis from above down.

None should be satisfied with mere "results"; aim at having a 100% Body.

All should STUDY Spectro-Chrome Home Guide and Spectro-Chrome Metry Encyclopaedia. Spectro-Chrome also needs intelligent common sense; it is not a conjuring box.

Blood Circulation needs prime attention.

We fail to comprehend why persons paying for their Equipment hesitate about asking for the Free Guidance. It is their loss, but we get the blame. We are not for making money, but for service to you. Charlotte H. should send her father's Chart.

Frank S. B. was right.

Please forget about medical idea of "High" Blood Pressure. It is a condition and NOT a Disorder. Spectro-Chrome normalating the conditions will normalate the Pressure also. Look solely to the Frequency and Ratio in our work and discard all medical jargon.

Florence Z.'s case is well known to us. She should send for our Free Guidance.

We again emphasize the Posture to relieve Constipation.

Wherever there is a "Doctor" in, keep Spectro-Chrome out; the two are antipodal.

Colonel Dinshah thanks all in your Planet for their kind feelings.

MISSOURI COMMENDS DINSHAH

By:— HEDWIG H., *St. Louis, Missouri*

With interest we followed your lectures here. If we could get a book which tells as plainly and cleanly the Secrets Of Life, that we with its help could explain it to our boys, we would like to buy it, as marriage was a closed book for me despite the fact I was 30 when married.

Many a mother said yesterday, "I wished my children could hear it."

EDITORIAL REMARKS

Such a book as you require may not be long in appearing from our Printing Plant. We believe that knowledge is power and ignorance does harm. Procreation is the prime human function, but, stupid prudery made it prurient and the disastrous results are quite obvious in the present generation. Dinshah subverted many pet doctrines; it is destined he should subvert the silly ideas attached to sex also. Please have patience.

He Who Found Not God Within Himself

Will Not Find Him Outside.—Dinshah

JUST ONE OF THE MANY — CONVENTION NEWS

By:—ALBERT E. GREENHAM, Delegate, *Detroit, Michigan*

The Sixteenth Annual Convention of the Scientific Order Of Spectro-Chrome Metrists assembled in the Colonial Room of the Hotel Melbourne, St. Louis, Missouri, on November 28, 1936, the birthday of the Originator.

Colonel Dinshah, the President opened the Convention with the usual remarks and extended greetings to all visiting Affiliates, Affiliates of the First St. Louis Missouri Planet and their attending friends.

He immediately launched into his lecture program with which most of you are quite familiar.

The Convention was a success, because of the fact it gave a more clear understanding of the Science governing the operation of a Spectro-Chrome Equipment for self-help and self-use.

The users of Spectro-Chrome, both in St. Louis and other parts of the country, are always benefited by a visit from Colonel Dinshah to their city to speak on this great Science called Spectro-Chrome Metry.

The Convention was well attended at all sessions, by a very fine intelligent audience. The Colonial Room was nigh filled to capacity, a fact which is greatly appreciated by all those who know the Missouri slogan—"YOU HAVE GOT TO SHOW ME, I AM FROM MISSOURI"! Well, were they shown and **did they like it?!!!**

There was a very solemn expression on the faces of all those present, every word being taken in and properly digested. I never saw a more attentive audience anywhere. The lecture on Saturday night, dealing with the Fundamental Principles of Spectro-Chrome, was the outstanding one and the speaker was at his best.

Never did I hear the orations of Dinshah so elaborately expressed. A lecture full of scientific information and demonstrations made understandable in every day language places Dinshah in a niche with the great ones.

In one of the adjoining halls, there was considerable competition! A real Saturday night wild party was in progress and they let loose with everything they had. Yet the Colonel's audience remained unmoved, as it were all ears were closed to outside interference and it is doubtful if the audience was conscious of what was going on in the rest of the Hotel. It was only when Dinshah himself commented on the type of music and singing which was emanating from that part of the Hotel, that people showed any signs of disturbance.

The First St. Louis Missouri Planet deserves commendation for the efforts which it put forward in order to gather such a fine audience. The banner of Spectro-Chrome is now firmly planted in St. Louis and I defy anyone to stamp it out. Wherever the banner of Spectro-Chrome is unfurled, it can never be up-rooted. Time marches on. Milwaukee, Wisconsin forges ahead by the thousands and is still progressing strongly. Detroit came to life and went over with a bang, large increases being shown, hundreds of satisfied users now shouting their praises of Spectro-Chrome. Our motto is:—"ONWARD, EVER ONWARD".

With the united effort of every Spectro-Chrome user in St. Louis, it too will be in the thousands soon and the Colonial Room will be just large enough for the Planet meetings only. Talk Spectro-Chrome freely, but be sure you live it. Cleveland is also "tuning up" and soon it will be echoing the efforts of its sister cities.

Spectro-Chrome must go forward.

Whoever uses Spectro-Chrome should not fail to tell his neighbor; everybody should know of Spectro-Chrome and eventually every one will use Spectro-Chrome.

It indeed was a pleasure for me to be honored by the First Detroit Michigan Planet as its Delegate to attend this Convention and I hope I may be privileged to attend many more.

The success of Spectro-Chrome depends on you and me.

I shall do my part—Will you do yours?

IT PAYS TO FOLLOW INSTRUCTIONS

Many Purchasers continue with their wrong habits even after we warn them. Here is one who after resisting the change, finally reformed. Now he writes:—"I assure you this will be the most encouraging letter I have written to you since my purchase of Spectro-Chrome. Although my limbs are no better, **I am entirely cured of constipation.** I have had a bowel movement every day now for the last ten days. . . . . .the first time in seven years. I am adhering to your instruction 100% and wish now that I would have done so from the very beginning. I also feel very fine."

Very well Anton D.! We are glad you reformed. In time, you will have the limbs released! Just follow instructions as the Free Guidance gives.

SIXTEENTH ANNUAL CONVENTION A SUCCESS

Missouri Skepticism Broken
Spectro-Chrome Firmly Planted

Successful Use Of Spectro-Chrome In:—Burns, Paralysis, Asthma, Hay Fever, Bronchitis, Eye Disorder, Carbuncle, Pinch, Rheumatism, Female Disorder, Kidney Stones.

St. Louis, Missouri invited the Convention and promised to make it a success. They made it more than a success, by the thoroughness with which all was organized and carried through in perfect harmony.

Hotel Melbourne, received the audience in the Colonial Room, which was suitably arranged for the occasion. President Colonel Dinshah P. Ghadiali opened it with the usual scientific discourses, having novel experiments on Polarized Light and Attuned Color Waves, which were very entertaining and instructive to the audience. His remarks on various topics were received with rounds of applause and approbation, all showing openly their endorsement of his views against Drugs and Destructive Surgery.

A telegram was read from William Birnbaum and wife from Palm Springs, California:—

> "Heartiest congratulations on your convention. We sincerely hope that a Spectro-Chrome in every home will soon become a reality and we shall always be grateful for what Spectro-Chrome has done for us."

First Vice-President Jessie E. Ness, of Detroit, Michigan, wrote a letter regretting her inability to attend. She further suggested that a charge of One Dollar be made to the Members for each Collaborator Pad. Members present unanimously expressed their approval of such a charge. The President stated he would confer with the Board Of Trustees about the suggestion, which was sensible and would prevent fraudulent use of the Blanks by unauthorized persons.

In another letter Jessie E. Ness stated in part:—

"Were I retired from the work tomorrow, I would under no circumstance whatever part with my Spectro-Chrome, Itisometer and sufficient machines for our family use.

"I will be listening in mentally for the progress of the Convention. It is very bad that it could not be broadcasted. Would that burn the 'OCTOPUS', OH BOY!" (Loud Applause)

The complete Program was printed in the last issue of Spectro-Chrome. The main interest to the public was centered in the Case Reports, of which the more important ones were in substance as follows:-

Clemens M. P. Neumann, the Deputy of First St. Louis Missouri Planet said, "I tonated many cases and took part in many Spectro-Chrome activities. I had a case of a boy who had a third degree Burn. The bandage was painful to take off. In 1 hour of Spectro-Chrome there was no pain. He was a brave boy. The skin on his legs was gone and pieces of flesh had come off. In 6 weeks with Spectro-Chrome damage was all normalated. He is here. (Arnold Schmidt came on the platform and a powerful beam of Light was projected on his legs so the audience may see better.) Nothing but Spectro-Chrome can do such clean job. I challenge anybody if they do anything like this in the United States or even in Germany. (Two trained nurses from the audience went on the platform to verify the statements and expressed their surprise after examining the boy. Applause.)

Frank S. Burns, Delegate of First Milwaukee Wisconsin Planet, said, "It is a pleasure and a privilege to speak here. I have used Spectro-Chrome since 1934 and have good and sufficient reasons for affiliating with the work. I had conditions to contend for 11 years; now 14 years. My son developed Muscular Spasticism. We took him from one Specialist to another as fond parents. Six months he was in a hospital, where the best Specialists attended. They said he would die within a year. We had then until 5, various medical services money could buy. We had Chiropractors, Naprapaths, Medical Doctors, Healers of all kinds, without promising results. 1928 we took him to Mayo Brothers in Rochester, Minnesota, where for a week they made every possible Test of Eyes, Ears, Throat, Lungs and took X-Ray pictures. Then I was called to the office of the Head Technician and told that nothing could be done to help the boy. He said most such children do not have faculties of the mind and I had reason to be thankful that he could differentiate. They prescribed no Medicine and said there was Disorder of certain nerves at the base of the Brain. In 1934, I heard of Spectro-Chrome. I thought I tried everything else, so I bought an Equipment and used it on the boy since April 1934. At that time, he was unable to support his arms and now with a little assistance and assurance of balance he walks. He not only walks, but supports his body nicely and he has grown in height. He weighs 90 pounds now—he who was given up as a hopeless case is now a marvelous boy. Thank you." (Loud Applause.)

Elisabeth Geyer of Detroit, Michigan, said, "I wish you could have been at the Detroit meeting. Upto one year ago, my family spent I think thousands of dollars in Doctor's bills. I had for many years Asthma, Hay Fever, Bronchial Disorders. I had gone down from 174 to 117 pounds. I took many Injections and they all but finished my Heart. The Doctor stood at bedside and shook his head all was off. They told him to go and they would call him in the morning. Instead they put in a Spectro-Chrome. Where I had taken Aspirin by the bottles, not a tablet was used. Neuritis which set in was also gone and no pain was left. I gained 35 pounds in 3 months. My mother had Varicose Veins since my birth. She had spent $200.00 for Injections and rubber stockings without relief. Spectro-Chrome set her right." (Applause.)

Joseph Hesselmeier said, "I had bad Eyes. I used Spectro-Chrome for 6 weeks and was normalated. I had last Spring a Carbuncle. It was normalated in 2½ weeks by Spectro-Chrome." (Applause.)

Remig Streit said, "I can tell you why I am for Spectro-Chrome. This Summer wife and I were sleeping in the basement. She pinched her fingers in armchair. I brought her out. Doctor took X-Ray picture; was afraid bone was broken. Nothing showed, so put Mercuro-Chrome and sent her home. Relatives had Spectro-Chrome, so they said let us try, may be that helps. In one Tonation of half hour, she felt something going on; next day another Tonation and she was right." (Applause.)

Charles S. Stiermann, Recorder of First St. Louis Planet said, "It is strange how I came into Spectro-Chrome. I was ailing for years, as of course 75 per cent. of the people do not feel perfect. By just an accident over a year ago, I found a circular of Spectro-Chrome on the pavement. It spoke about Color Waves. I was hesitant and then I met William Dorn and later contacted Clemens Neumann who explained the merits of Spectro-Chrome. He told me I had to avoid certain things like meat and there! I was a great meat eater! I thought Neumann must be crazy! I had tried the Medical Doctors and their Drugs, so I thought I was storing Medicines, why not try Spectro-Chrome? I took the chance. Neumann had no trouble in selling me the Equipment. I called him. Now I must say this after 1½ years, that I am not perfect, but I feel spry just as much as a meat-eater and I am not logged." (Applause.)

Walter Weinbrecht said, "I purchased a Spectro-Chrome recently. That Green works! My mother had Rheumatism; she was working too much. She tried all kind of shoes and Medical Doctors. Anyway, Spectro-Chrome hastened her recovery. She has no pain anymore." (Applause.)

Elizabeth Geers said, "I had my Spectro-Chrome in 1934 July. They had advised an operation on my mother, but she used Spectro-Chrome and in 3 months, she was fine! My father also used Spectro-Chrome; I have 3 children under 6 and they all use Spectro-Chrome.

(By request, the mother came on the platform and introduced herself amidst applause.)

Anton Schneider said, "Spectro-Chrome Equipment is wonderful. One night, I came home sick. They told me to go to the Hospital for operation on Appendix. My pain was on the Left, the Appendix is on the Right! (Laughter) I had Spleen Disorder 3 years ago. Last year, I was sick and weak; run down. I went to the Doctor; he said, 'why didn't you come sooner? It is too late now. I have to make a laboratory Test of the Lungs; it will take 4 days.' The Report came back; there was nothing the matter with the Lungs! But I came down and could not work. They examined with the Fluoroscope; said I had Kidney Stones. Gave me Medicine. Did not work. Finally the Spectro-Chrome Equipment came in. Here I am." (Applause.)

On motion of Clemens M. P. Neumann seconded by Frank S. Burns, the same Board Of Trustees composed of Dinshah P. Ghadiali and Irene Grace Dinshah of Malaga, New Jersey and Bessie A. Hasenau of Cleveland, Ohio was unanimously elected. The President then re-appointed Jessie E. Ness of Detroit, Michigan as Sub-Trustee and First Vice-President and appointed Frank S. Burns of Milwaukee, Wisconsin as Sub-Trustee and Second Vice-President. In affirming this new Officer as such and as a Member of the Council Of Governors, the President introduced him to the audience with appropriate words. Frank S. Burns is Superintendent of the Milwaukee Gas Company and is well respected in the community as a man of sound, sterling character.

Beginning on Saturday afternoon, November 28, 1936, the Sixteenth Annual Convention was dissolved on Monday night, November 30, 1936, after six instructive and entertaining sessions. The new Electric Sympathometer and the Aluminum Transport Spectro-Chrome were among the exhibits and attracted favorable attention. All went home impressed with the harmony and dignity which pervaded the Convention sessions.

NEWSPAPER PAID FOR LIBEL

Evening News of Macon, Georgia, branded Miss Jeannette Rankin, former Congresswoman, as a "rank Communist". Under the Georgia Laws a Libel Suit originated. It was settled by the newspaper for $1,000.00. It is easy to write anything, but responsibility in Truth can not be evaded. Jeannette is Associate Secretary of the National Council For Prevention Of War.

FIRST CLEVELAND OHIO PLANET

(HOTEL HOLLENDEN, CLEVELAND, OHIO)

Meeting Of Thursday, November 5, 1936.

Deputy William J. Hasenau was in the chair.

Charles N. asked "How do you prepare Yellow Split Peas for eating?" Bessie H., Third Trustee explained it. She uses unpolished rice.

Fred O. asked "What could be done for frozen feet?" Seems the bones were frozen and smashed in the army. The gentleman has difficulty in walking or standing. Use Lemon Systemic and Magenta on Area 4.

Bessie H. said the Ratio and eliminations are two important factors in Spectro-Chrome work. Correct the Ratio and then use the proper Attuned Color Waves over the Area desired to normalate.

Charles N. asked if one could use a mixing of Colors in the same Tonation? Reply was never to use Blue for a half hour and Red for half hour. The Blue must be used the full hour. Same with any other Color.

In Bessie H.'s 13 years of experience with the Spectro-Chrome, she has tonated many cases but no two were alike. She said it is not only the food one eats that improves one's condition; the mental attitude toward life, also counts. A heated argument or a worry would upset one's Stomach; even the entire system. So the Spectro-Chrome improves one's mind, lifts one to a better, broader vision and higher life by clearing the body.

She reported a case of a child of 8. The Kidney Elimination was scant. She was in a coma. The Medical Doctors had puzzled over the case. She wet the bed 3 or 4 times in night. Was a "Flu" recurrence. The Doctors wished to inject a Saline solution in the Chest. The mother forced Orange juice and water into her mouth. The child was beyond asking. In 3 weeks of Magenta and Green, she was at school fully improved.

Case of an oil well owner 86 years old. It puzzled the Medical Doctors. Bessie H. was called to Wellsville, New York. After using Spectro-Chrome on Areas 4 and 5, he was in a few days normal.

Fred O. read Chapter 82 from Spectro-Chrome Metry Encyclopaedia on "Spectro-Chrome In Paralyses". The Paralytic case of Anna Schoener is also remarkable to the users of Spectro-Chrome. The case of Grace Shirlow who was so badly burned was discussed.

Ernestine Y. spoke on some one with a lump on the neck. It was very painful. Green and Indigo were used. Then Lemon Systemic. This cleared the trouble 3 years ago and there has been no returning.

Charles N. asked what caused pain. It was explained.

The wonder is why people will use Color in dress, homes, revel in sky colors and Nature but refuse to use Spectro-Chrome!

Why should anyone wish or desire to infringe on the Color scheme Colonel Dinshah worked out years ago? Human nature is such!

Charles N. said "It is very difficult to live without Meat, Coffee, Drink or Smoking, but it can be well accomplished. Today so many use these as a so-called "enjoyment" but ruin health. Smoking slows the mind. They are not keen thinkers who use Tobacco. I stopped Cigar Smoking as a School teacher; I see a difference between day-dreaming and smoking. Mind doesn't work clearly. It is sluggish, doped and poisoned. The Advertising is misleading—as 'athletes are better after a Camel'. There are so many women who would give a left eye rather than stop tobacco or a glass of liquor. Mother and daughter are both alike in this. Drinking in time causes trouble to Liver, Stomach and Kidneys. People will chew tobacco even though it really causes trouble to the Stomach. It is filthy to have a mouth so full one cannot talk well—be it food or tobacco. Eat in moderation."

Nettie B. said, "In using Turquoise on my mother, 75 years old, she stiffens." Reply was use Lemon instead. Never use Purple for one over 75 years old. It is too strong. (?)

Bessie H. said, "We are given by God the Spark of life. This is for purity and not impurity. Our life is not our own. It is loaned to use for Him. Why be disobedient by using foul words, evil habits or thoughts? Don't poison mind or body by argument, heated passions wrong food, drink or tobacco. The 'Laws Of Incarnation And Karma' rule life. We should be good. Wheat is ruined by wind and rain, so is life by everything wrong or vile. Our past life tells our story. By Karma we are punished for our wrong deeds in life. Some people in a crowd we make fast friends, others we leave. Don't force anyone against their will to do anything. Ask them to use Spectro-Chrome but it is for them to use their body as they see fit. They suffer the consequences."

The Deputy asked, "Who will go to the St. Louis Annual Convention?"

Bessie H. the best qualified to go felt that she could not attend

this year. The question remained unsettled, as none was qualified to attend.

EDITORIAL REMARKS

Rational Food Of Man is a simple system. It is printed in Spectro-Chrome Home Guide which goes free with each Cabinet Equipment and is the same as printed in the loose leaflet issued to th' Purchasers of the former Systemic Equipment.

The Ratio between the Kardoscil and Spiroscil is the main key to life. Infra-Green Division Colors should never be irradiated with the Ultra-Green Division Colors on the same Area at the same Tonation.

It were better for all users of Spectro-Chrome to avoid using the medical names of Disorders. All such alleged naming by "Diagnosis" is faulty.

Color is not decoration, but Oscillatory Frequency of Energy.

Dinshah's system may be imitated, but can not be improved. They will have to make a new Mathematics.

Whether in "Camels" or donkeys, Tobacco has no business to be in the Human Body.

Purple may be used at any age, should the Area 4 demand it; so can Scarlet or any other Attuned Color Wave.

———o———

FIRST ST. LOUIS MISSOURI PLANET

(HOTEL MELBOURNE, ST. LOUIS, MISSOURI)

Meeting Of Thursday, November 5, 1936

Successful Use Of Spectro-Chrome In:—Burn, Amenorrhoea, Sore, Pain, Leaky Heart, Hay Fever, Mental Disorder.

Reported By:—CHARLES H. STIERMANN, *Recorder*

Deputy Clemens M. P. Neumann was in the chair.

The Spectro-Chrome Home Guide was brought to the meeting in accordance with suggestion at last meeting for careful perusal and study. An epitome of the Spectro-Chrome Metry Encyclopaedia, it possesses all of the fundamentals necessary for the novice and complete instructions for the use of Spectro-Chrome in the home. To that end it was studied in this meeting and many questions were asked and answered on different points to give more enlightenment. The positive necessity that Colors must be attuned was definitely brought out by comparing the concordant notes of a symphony producing perfect harmony without clash to normalate various Disorders of the Body. Spectral Divisions were discussed, Affinity Waves, Infra-Division representing the Sun and Ultra-Division representing the Moon.

The Deputy emphasized that when there is a disturbed balance, use Magenta on Area 4 which in turn would restore balance in the whole system.

William D. said it was his observation that Magenta had a very fine effect upon the Body.

Question arose if a person took poison accidentally what should be the proper procedure. Advised to take a large drink of water and tonate Areas 8-9-10 with Orange. Then follow regular Favor Period until fully normalated.

The Deputy told about his sister who pulled a hot grate out of a fire-place causing a severe Burn. Used Blue; 2 minutes of intense pain after which she felt an easing and soothing effect.

He also mentioned a case where a woman had a Ratio of 100 to 13; had not menstruated in 4 years. Three Tonations and Ratio became normal along with regular menstruation.

Josephine S. related an incident where Kardoscil and Spiroscil Ratio was 5 to 1. After Tonation, the Ratio would be the same but beat of Pulse would be firmer and more pronounced.

Margaret S. told of some apparent wise-acre who advised against using Spectro-Chrome saying "it would kill you." Margaret S. also sought information on what position Spectro-Chrome should be when taking Local Tonations. Advised it could be placed on either side of the Body, but beam should never shine on Body from northerly position but from a southerly position. It was understood the Body should be in the assumed position of head to the North.

Josephine S. in discussing Turquoise and Magenta in Colds said her daughter takes Colds so severe that she can not breathe through either nostril, both nares being clogged. Used Turquoise Systemic and Magenta on Area 4 to no avail. Turquoise Local opens nares and produces results.

Joseph Z. said he was using an electric pad for a Cold in the head when someone told him to use Orange. He has sudden flare-ups of fullness and congestion, evidently a Sinus Infection.

Clara C. asked what might be done where there is a great tendency to chapped and dry skin on the face. Advised to wash with warm water and apply Cocoanut Oil and tonate with Turquoise as a Skin Builder. For Pimples use Lemon Systemic. Use Infra-Green Division going down to Red and over again if necessary. Give attention to Area 18 to flush out foreign matter and impurities. Also stick to Rational Food Of Man system.

Thomas S. wanted to know at what distance to tonate. Advised that for Systemic, the Body should be flooded with a full beam; for Local, only affected Area as shown in Area chart.

Remig S. said Magenta does not throw as good a beam as the other Colors and wondered if its effectiveness was as positive. Magenta has greater density but he was warranted as to its effectiveness.

William D. reported the splendid results finally of old lady with open sore that would heal, and reopen and go through the same cycle over and over again. A substance as thick as a thumb seems to be imbeded within the wound. This was removed. It was an angry looking wound. Black and continually festering, healing and again opening with Pus. It has now cleared nicely. Orange, Yellow, Green and Yellow and back again to Orange in final Normalation. This was of long standing as previous reports indicate.

Charles S. mentioned that when the first sign of a Head Cold revealed itself with sneezing, if immediate steps were taken to tonate with Turquoise locally there would be no distress from Head Colds. He employs such method and finds it a wonderful bafflling agent getting results immediately.

Josephine S. spoke of the merits of Orange for Indigestion and gas formation. Also that Lemon was wonderful in Gall Stones along with Orange. If a Color stirs too much and causes havoc, go to Green and start over again. Asked what was best to overcome Constipation, some had best results from Orange, others placed their faith in Lemon. Lemon to Orange seems best.

Josephine S. sought information on what other foods could be eaten besides those listed in the Spectro-Chrome Home Guide. Was told that all vegetables, fruits and those things not prohibited could be used.

A question arose about the Variant Breath. There is so much haziness about this matter that a request was made of the Deputy to ask Colonel Dinshah to devote some time to this topic at the Convention so our group might have a better understanding on this matter.

Clemens N. was called to the bedside of a man; Area 1 was out of condition. Great pain in Areas 4, 5, 18. Another man had called on this party to sell him some property and finding him in such distress said he would send his friend Clemens N. to see him. He could not get out of bed. Green Systemic used and Magenta on Area 4. Lemon Systemic and Magenta Area 4. Pain in Back was so great that Indigo was used to deaden pain. Complete Normalation in 10 days.

Another case of a woman who did not feel well. Could not

get her breath. Heart jumped and was very irregular. Green Systemic Magenta Area 4. Relieved for 3 days, when trouble again appeared for 5 days. Seems to have a Leakage Of Heart. Daughter was so impressed with results mother was getting, that she is taking Tonations to overcome trouble she has. As this is a new case, the results will be reported next meeting.

The Deputy also reported a woman having a leg swollen so she could not walk without stumbling. This was further aggravated when she sprained her ankle. Two Tonations and ankle was normalated. She also had very poor Circulation. Green, Lemon and Orange Systemic used and Magenta on 4, 18. Now tonating according to Free Guidance received.

Normalated a case of Hay Fever in 12 Tonations. Turquoise and Magenta on Area 4 used.

The Deputy reported man subject to Epileptic Fits for 18 years. Used Spectro-Chrome 3 weeks and felt so much better, did not use Spectro-Chrome anymore. No fits in 14 days, whereas formerly there was a tendency to 1 a day. Poor party and cannot afford Tonations.

Clemens N. reported case of carpenter. Felt he was losing his mind. Could not think like formerly. Left arm did not function as it should. Took Tonations for a week; again feels like working. Arm well. Headache gone. Clear mind. Green Systemic, Magenta 4, Yellow for arm. About 8 Tonations. Boss marveled at renewed energy showed by this man, it being so marked.

Stacks of literature announcing and giving program of coming Convention were at hand for distribution by Members and Affiliates. Plans are made for distribution and the consensus of opinion is that we shall have a good large audience to hear the Colonel and partake of this "intellectual banquet".

EDITORIAL REMARKS

Study of Spectro-Chrome Home Guide and Spectro-Chrome Metry Encyclopaedia is essential for success in the proper use of the Science. The power of Spectro-Chrome lies in its intelligent use. Your Planet is on the right track.

The Emetic effect of Orange may be utilized through Area 8 Irradiation.

Menstrual Disorders need correction of Circulatory Disturbance, along with Systemic Tonation.

Do not emphasize merely Local Tonations. Systemic Tonations are the only ones that will remove the prime causes.

Lubricate the nostrils, first thing in the morning, with warm Cocoanut Oil daily; most Nasal Disorders will disappear, as they are caused by dry mucous membrane obstructing the Sympathetic Reversal in Variant Breath. Let the Sufferer wear also wool stockings to prevent excessive loss of warmth by radiation. Herein is the key. Heed the warning or suffer.

Joseph Z. may mind the same warning, if he wants to discontinue the electric pad.

Clara C. should lubricate the entire Body similarly and then bathe daily with warm water. Spectro-Chrome will certainly "build" the skin, but first let the Sufferer have a "Skin" to build.

Magenta being a very dark Wave, does not appeal so much by its visibility. Study will show that while Yellow is visible 1,000, Violet (which is ½ in Magenta) is only 6. Effect has nothing to do with sight. X-Rays are totally invisible, but they burn worse than any! So does Radium.

Refer to Spectro-Chrome Home Guide for information about tonating "sores"; pages 72 and 73. It is easy to follow the instructions and success will be certain in all cases.

Sneezing in "Head Cold" indicates resentment of the Sympathetics and is the effort to normalate. Maintain the Circulation proper by wearing wool stockings and suitable underwear.

Any Attuned Color Wave may be used for a number of Disorders. The use depends upon the Area Affected and the like. Constipation will not stay if one follows the Food system taught and sits in the "posture" necessary to open the sphincter muscles at the "Body Exhaust."

There is no special "Food". Eat everything outside of the 7 materials to be avoided. Cook as instructed and follow solely pages 36, 37, 38 and 39 of the Spectro-Chrome Home Guide. Cut out all "food fads" and humbugs of dietitians, who teach little and know less of the Truth.

Swollen Legs mean failing Circulation. Any Dropsy means it.

Epilepsy is related to the Sexual Organs, either in the Sufferer or by inheritance.

When Spectro-Chrome enters more homes in your city and State, many more bosses will marvel at the extra working energy shown by their employees.

SPECTRO-CHROME

December

1937

Scarlet

Volume 13

Number 12

Trade Marks Registered In The United States Patent Office

Read And Have Your Friends Read

Notice To Affiliates

Notice About Dime-A-Meal Sales

Edward A. Ernest And Lucille J.McGraw Accused Of Patent Violation

Pimples Gone, No Need Of Spectro-Chrome!

VINAG

$10.00 Down Buys New Aluminum Spectro-Chrome

Spectro-Chrome Institute Paid Off Its Mortgage

Spectro-Chrome Metry
Attuned Color Waves

In course of life you reach a time,
When Doctor, Lawyer, Priest alike,
Does fail to give the sought relief,
To troubled mind or burdened heart—
Then come to me:
I help you to help yourself.
SPECTRO-CHROME

The Science Of
Automatic Precision

DEVOTED TO THE SERVICE OF MANKIND

SPECTRO-CHROME

A Monthly Magazine
Devoted To The Service Of Mankind

Editor And Publisher
DINSHAH P. GHADIALI

OFFICIAL PUBLICATION OF
Scientific Order Of Spectro-Chrome Metrists
(A Not-For-Profit Corporation Of The State Of New Jersey)
Malaga, New Jersey

Entered as Second Class Matter, July 20, 1922, at Philadelphia, Pennsylvania and transferred to Malaga, New Jersey, July 1924, under the Act of March 3, 1879.

Re:entered at
Malaga, New Jersey, January 12, 1934

TELEPHONE:—VINELAND 530

Single Subscription For Current Year, Inclusive Of Postage Anywhere.................. $3.00
All Subscriptions Expire With December Issue.

Single Copy.......................... 0.50

TO THE PUBLIC

We accept no gifts or gratuities, as to a Public Servant it is Graft. Send us nothing.

You may send your book for review or opinion; you may send your writings for publication, provided they are not plagiarized or false and bear your signature in ink, with full name and address.

Anonymous communications receive no attention. Submitted articles will not be returned.

All correspondence should be addressed to SPECTRO-CHROME, MALAGA, NEW JERSEY and not to any individual member of our staff.

For the views expressed in signed articles, the writers are responsible; for the opinions in articles bearing no identification, the Editor shoulders the sole responsibility.

FIRST WILMINGTON DELAWARE PLANET

(BROWN'S PLACE, S. W. Corner, 23d & Washington, WILMINGTON, DELAWARE)

Meeting Of Thursday, November 4, 1937

Successful Use Of Spectro-Chrome In:—Facial Swelling, Pain, Heart Disorder, Colds, Sunburn, Piles.

Reported By:—FREDERICK A. BRUNCK, *Recorder*

Deputy Howard R. Ingram was in the chair.

The Deputy read a letter from the Central Office in which Dinshah announced his defeat for Governor of New Jersey and also his candidacy for the next governorship. It further stated that Dinshah will write a weekly "Family Health Protector" column in "The Record" of Bridgeton, N. J. The Editor asked him to write for the newspaper, after hearing one of his campaign speeches and after visiting Spectro-Chrome Institute.

Cases were reported as follows:

The Deputy reported a woman who had teeth extraction and her face began to swell and pain very badly. One Tonation of Indigo was used and the swelling and pain disappeared.

Another similar case was reported by the Deputy and 1 Tonation was found to give relief.

The Deputy's mother, 79 years old, had a Heart attack recently and 1 Tonation of Purple gave quick relief.

James P. reported his case of Heart Disorder, for which he was doctoring for over 9 months without any results, until he heard of Spectro-Chrome from one of our Affiliates. He promptly bought an Equipment and within the short time of approximately 4 weeks, he seemed to be on the road to recovery. James P. also explained that he had a Cardiograph taken of his Heart by a Medical Doctor and that the Doctor found his heart much improved. He also said that Spectro-Chrome was successfully used to combat Colds.

The Recorder stated a case of severe Sunburn on his son, 15 months old. One Tonation of Blue gave the necessary relief. He also reported a case of Hemorrhoids or Piles which kept his wife in bed several days, until he could get his Equipment. Started Tonations with Indigo Local and after 3 to 4 Tonations everything was well.

SPECTRO-CHROME METRY ENCYCLOPAEDIA

By:—GRACE A. D., *Detroit, Michigan*

Volume 3 is a continuation and development of the knowledge given in Volume 2 and is more specific in the use of Spectro-Chrome. It gives a knowledge of the Higher Physical Body, touching on Occult Phenomena and creating in the student a desire for more of these higher studies.

The Home Training Course is very complete and condensed in teaching a science of healing. I received more knowledge in the art of healing from 3 volumes of the Encyclopaedia in a few months, than I could obtain in a medical library.—E. P.

PIMPLES GONE, NO NEED OF SPECTRO-CHROME!

By:—WILLIAM P., *Mobridge, South Dakota.*

The Sufferer has not had any more Boils and the Pimples have disappeared. She does not think it necessary to tonate daily any more, but will take an occasional Lemon Systemic for a good measure. She wants to thank Spectro-Chrome Institute for the wonderful results.

FIRST ST. LOUIS MISSOURI PLANET

(HOTEL MELBOURNE, ST. LOUIS, MISSOURI)

Meeting Of Thursday, November 4, 1937

Successful Use Of Spectro-Chrome In:—Arthritis, Pyorrhoea, Digestive Disorder.

Reported By:—CHARLES H. STIERMANN, *Recorder*

Deputy Clemens P. Neumann was in the chair.

Letter was read from Headquarters about the Convention, along with circular containing subject matter to be discussed and the appointment of a Delegate and Alternate Delegate to attend the Convention. The same letter contained information about the "Favorscope" for 1938.

Another letter was received in which Colonel Dinshah informed us of his defeat in the Election for Governor of New Jersey and also that he was writing health articles for a weekly newspaper "The Record". He advised those desirous of reading the articles to subscribe for issue beginning November 5. One Dollar sent to "The Record", 28 Cohansey Street, Bridgeton, N. J. would cover the annual subscription. We feel sure that Colonel Dinshah's slogan of "Candidate for Governor until elected" will eventually put him over and compliment him on his pluck and stand against corruption.

Upon motion being made, seconded and duly put, Clemens M. P. Neumann was elected Delegate and Eleonore Neumann Alternate Delegate to attend the 1937 Convention and the Recorder was instructed to issue the necessary credentials to these Affiliates.

Carl S. having an arthritic condition in the thumbs and who was itisized, reported the Disorder normalated. Used 3 Lemon Systemic, 3 Turquoise Local and Magenta on Area 4. Disorder was of an extremely sensitive nature.

Question arose how to tonate for poisoning caused by a Boil. Advised to use Green and Turquoise Systemic, Orange on Boil and Yellow till it bursts open. Turquoise and Green to clean after removing the core.

Mary T. asked about using Magenta systemically and was advised that Magenta was used for Local Tonations and as an Equilibrator.

Harry O. wished to know about Local and Systemic Tonations and if they should be alternated on alternate days. He was informed that his idea was correct and where two Tonations were recommended a day, take a Systemic and subsequently a Local and so on day after day. If possible, take one on Day Favor and the other on Night Favor.

Nicholaus B. said he was having trouble with the Thermocept leaking, which brought responses, others saying they were having the same trouble. Some said they remedied the trouble with caulking cement, rubber packing and aquarium putty.

Peter B. asked about going out immediately after tonating. Advised to wait at least a half hour as the rays of the Sun would reverse the effect of an Ultra-Green Tonation and the Moon would act in a similar way on an Infra-Green Tonation reducing the good secured in tonating.

The Deputy spoke on how Spectro-Chrome through graduations of matched color worked, making comparisons with how in nature a tree from its sturdy trunk sent out branches and in turn smaller branches forming small twigs which in turn sent out stems that bear the luscious fruit composed of minute fibrous matter abounding in juice making fit for king and potentate. The same principle applies all through nature in the process of growth and construction.

Foods again came for discussion as several new Affiliates were

present and eager to learn of any beneficial foods that might be pure and free from adulteration.

Thomas S. wished to know what results might be expected in Pyorrhoea by the use of Spectro-Chrome. Clemens N. said that at one of the Conventions (he believed it was Milwaukee), a person attested to complete normalation of a Persistent case.

Harry O. asked about Ulcerated Gums and was advised to use the Infra-Green Division down to Red to effect Normalation.

The new Affiliates liked to know about the High and Low Critical Periods; the Deputy explained about New Moon and Full Moon and the changes for better or worse during the 2½ days before and after these periods or within that range.

Harry O. had trouble in Areas 8, 9, 10; said he was getting splendid results using Green, also Lemon Systemic and Yellow on 8, 9, 10 and Magenta on Area 4.

There being few case reports, the evening was spent on an educational program principally and to very good advantage, as attentive ears were in evidence.

The Deputy had Babbitt's book, "The Principles Of Light And Color" and spoke of the good work done by Babbitt as a forerunner to Dinshah's complete and exhaustive work in Spectro-Chrome Metry.

EDITORIAL REMARKS

It is advisable to tonate at one sitting either the Systemic or the Local. Alternate them either on different Favors or on different days.

There are numerous reasons for a Thermocept leaking. As now made, it has no rubber rings, because that item was found unsatisfactory in conjunction with glass and metal. The expansion and contraction of the two did not coincide and trouble resulted. The best results were found with non-drying, elastic cement. However, many users continue the Irradiation beyond the Favor hour, the water partially boils out of the Thermocept and part of the cement drying more than another starts a leak on the next run. Making the Thermocept leak-proof is a "tricky" process and we recommend Purchasers not to endeavor to use it except in accordance with the directions given. A caution may well be minded. Do **not** tighten or loosen the screws. They are for holding the rings in place and have no relation to tightness or looseness. The best we recommend is to return the defective auxiliary to the Plant by prepaid express or parcel post; there the trouble will be remedied free of charge. The best liquid to use in the Thermocept is Distilled Water, such as is used in storage batteries. Glycerine, if not chemically pure, soon decomposes by the chemical action of certain Attuned Color Waves and the discoloration ruins the effects of the Spectro-Chrome. Use discretion. The maker of Spectro-Chrome is near to his product and is more than anxious to serve the public. Do not use the Spectro-Chrome without the Thermocept. You will endanger the Slides.

Pyorrhoea owes its main origin to kissing on the lips. The exchange of sputa and consequently the transfer of fermenting particles of food from the crevices between the teeth is a fertile cause of this common Disorder. Use Manjan for cleaning the teeth and gums after each meal and gargle the mouth and throat with a tumbler of warm water in which is dissolved a spoon of Lavan. Lemon Systemic Front and Indigo Atcowa on Area 1 may well be used. However, for permanent Remedy the stated cause must be removed. The same remarks may well be applied to Ulcerated Gums and Gingivitis.

All Affiliates should study the Spectro-Chrome Metry Encyclopaedia. Its investment will benefit for life. It is our "Bible", "Veda", "Koran", "Gita", "Talmud", "Gurugranth", "Avasta" and the like all in one!

Dr. Edwin D. Babbitt's "Principles Of Light And Color" was rejected in Spectro-Chrome Metry, because of the incorrect Primary Colors. Dinshah reproduced it by Photo-Engraving, at great cost, solely to prove the comparative value of his own system. However, it was Babbitt's work which started him thinking and he always expresses the highest regard for that master mind, to whom he owes a great debt of honor and to whose name he pays open homage. Our Affiliates may have that volume at the reduced price of $5.00 plus 50 Cents for postage and handling charge. A study of Babbitt will prove of value.

FIRST WILMINGTON DELAWARE PLANET

(BROWN'S PLACE, S. W. Corner, 23d & Washington, WILMINGTON, DELAWARE)

Meeting Of Thursday, December 2, 1937

Reported By:—FREDERICK A. BRUNCK, *Recorder*

Deputy Howard R. Ingram was in the chair.

The Deputy read a leaflet, which he received on one of his recent visits to Malaga, N. J., about the new fireless, electric cooking utensil, the "Vinag" and Dinshah's "Dime-A-Meal" which will make a meal for 3 persons. It is claimed to take only about an hour or less to prepare the meal and that any person, even a child, can prepare it.

A general discussion followed about progress reports which are being sent to Dinshah in reference to progress of their cases and that several writers are not quite satisfied with the reply of Dinshah, which in no case is more than 5 words and yet each time a report is sent, it has to be written in ink and the complete history from "A to Z" repeated, no matter how many reports were sent before. The feeling among the Affiliates seems to be, that it should not be necessary to state the complete history time and time over again, because it is felt, that those reports are kept on file and a repetition would be superfluous.

Severe criticism came forth in respect of replacement cost of Slide-Carrier Seals in case of breakage. Since the Slide-Carrier with the Attuned Color Wave Slides is personal property of the Purchaser of an Equipment, it cannot be quite understood why Spectro-Chrome Institute should have any jurisdiction over the Seals. No doubt it would be quite appropriate, if the Institute should be loaning or renting the Equipments retaining its ownership, to place a penalty for violating of any Seals, for example like electric meters and gas meters. Furthermore, it is not quite clear in everybody's mind the reason for the Seals in the first place.

The above discussion and criticism should be looked upon as entirely constructive rather than destructive and it would be well if it were discussed and printed by Dinshah in the proper manner, so that other Affiliates, who are hundreds of miles away and are probably harboring the same questions in their minds, would be benefited by his answers.

A question, whether Dinshah would be willing to write a column in the Wilmington Sunday Star, if the newspaper would accept it was set forth and unanimously accepted. This newspaper is an independent one and enjoys a large circulation all over Delaware.

Such writing will give people another means of knowledge about Spectro-Chrome Metry—his origination.

The Deputy reported he was called to see a man R., who had a Heart attack. He explained Spectro-Chrome to the Sufferer and believes R. will buy an Equipment in the near future. The Deputy also reported of a girl having Mastoid Disorder and having no sleep for 3 or 4 days on account of severe pain. An Equipment was rented and before the first Tonation was over, the girl was sound asleep. Atcowa Turquoise was used.

The Recorder reported his 7 weeks old daughter was very badly choked in the nasal canal and throat and that she could get only very little air and therefore the breathing was very difficult.

Tonations of Turquoise were used and before three quarters of the hour (very first) had elapsed the baby was sound asleep and rested peacefully through the night. The Tonations were continued until every sign of that Cold disappeared; only 8 or 9 Tonations were required to normalate the case completely.

The Deputy reported of a girl having a brush bristle under her thumb nail. The girl went to a hospital where the bristle was removed and also with it half of the thumb nail. This was naturally very painful and swollen. Atcowa Turquoise after 2 or 3 Tonations relieved the condition.

EDITORIAL REMARKS

Although it is outside of the objects of the Organization, to introduce queries about the commercial aspects of Spectro-Chrome Institute, at Planet Meetings, we do not hesitate to clear the incorrect viewpoints.

The General Advice Charts are not all examined by the same staff member. Thus, it is necessary that each Chart be complete, to save constant references and cross-references. Furthermore, all our troubles in the past were through persons who were supposedly our best "friends". (?) Thus, to maintain our legal and moral neutrality and to prevent any possible frameup, we leave no loophole, by making each Chart a complete document. We know not when such a Chart may by some crooked person be called before a Tribunal Of Law and although **we know** all about a particular case, without a complete story on the face of the Chart, it would have very little legal value. We had plenty of bitter experiences in the past and should we not profit thereby, we should be wanting in common sense. We take no chances on the liberty of the Originator and thus trust nothing to speculation. Be therefore satisfied with the Free Service you are getting and quit the grumbles, because the Originator will never turn from the track of service as set by him. He will serve according to his own ways, where wisdom proved them in the past to be his protection.

It were better, in fact it is needed to have names before answering "severe criticisms". The Recorder is requested to write always with the names of questioners and critics, as we shall not answer stone-throwers in the dark. We answer openly and while we respect constructive criticism, we desire not to be the target of curiosity mongers. If names be not given, "queries and criticisms" will receive no elucidation. In this case, however, we give the reply openly about the Seals. The Equipment including the Slide-Carrier, is certainly the absolute property of the Purchaser, by Bill Of Sale issued. He may do what he pleases with even the Slide-Carrier, but, that Slide-Carrier

has in it Slides which mean either life or death to some unwary subsequent buyer, if the original Purchaser should through dishonesty extract the genuine Dinshah Spectro-Chrome Slides and substitute fake pieces of untuned glass. **To protect the public** and to place a check on spurious imitation and dishonesty, we introduced the Sealed Slide-Carrier. The Purchaser buys our merchandise, but can not make us shirk our moral responsibility. The appalling trickery practised in the past by some of our own Diplomaed Graduates is indescribable and when we look to our unsullied record for integrity of purpose, we have every reason to maintain it. Since the introduction of the Sealed Slide-Carrier, that crookedness is checked and the Purchasers should thank us for such stringency, instead of smarting under the $10.00 penalty for violation of Seals. Let us state here that even with the payment of the $10.00, we make very severe inquiries before accepting the explanation for the violation. The penalty should be $90.00 instead of $10.00! If the Purchaser has no dishonesty in mind by violating the Seals, what will be the reason of the squawk? Why violate?

Instead of wasting time on such useless matters, if the Planet Affiliates will study Spectro-Chrome Metry Encyclopaedia, they will be better benefited. A thorough grasp of our Science will be more to their advantage in removing potential doubts and mental reservations. Our Science of Spectro-Chrome Metry is so radical and opposed to the commonly accepted theories, that no Wilmington Delaware newspaper will agree to admit it to its columns. American newspapers—even though liberal—are controlled by definite interests and all these suggestions which to your Planet appear novel and attractive, are old news to us and the Originator, who tried all those ways.

Friends, take our well-meant advice. Stop all argument and commence a course of study from our text books. Send your suggestions by all means, send your complaints assuredly, but **stop** being Doubting Thomases before completing the study. Above all, let each person mind his own business. Neither your Planet nor the entire Scientific Order Of Spectro-Chrome Metrists, has any say whatever in the conduct of Spectro-Chrome Institute. Study Rule 36 for better and clearer comprehension.

NOTICE ABOUT DIME-A-MEAL SALES

A number of our worthy friends sent for $1.00 packages of *Dime-A-Meal*.

The remittance had to be returned because *Dime-A-Meal* is usable only in the VINAG. (Refer Page 2293). Kindly note this. *Dime-A-Meal* therefore will be sold solely to purchasers of Vinag, as without the Vinag, satisfaction may not be had in the cooking.

Should any order for *Dime-A-Meal* be received from a non-Purchaser of Vinag, Lavan will be sent for the amount remitted. Price $1.00 per carton, post free in the United States Of America.

SPECTRO-CHROME SMILES

Viewpoint—"My wife had 4 doctors. They told she has a germ which travels thru blood stream, clogging blood much, causing intense pain wherever it hits."

Smile—"Send in a Marine to hunt the germ."

Viewpoint—"I laugh a good deal now."

Smile—"Go to a Medical Doctor to stop it."

Viewpoint—"His Backelim is 1+2 times in one day."

Smile—"Report when he has it 3—3+2 raised to square."

Viewpoint—"Drinking Intoxicants? Yes, half quart milk a day."

Smile—"Was there whiskey in the cow?"

Viewpoint—"I irradiated the Sufferer with Systemic Lemonade, Front and Back."

Smile—"Next time give her Chaval Front and Dal Back."

Viewpoint—"I had a date with the Dentist".

Smile—"What is her name?"

Viewpoint—"My left ear has bad odor."

Smile—"Make it fragrant with Otto Of Roses."

Viewpoint—"She don't get the frog in the throat so much."

Smile—"Push in a snake."

Viewpoint—"Fever starts in the stomach rushing through my back, going to my head, ending with little sweat around my nose and forehead."

Smile—"It is on a joyride through the anatomy."

Viewpoint—"I eat gapes."
Smile—"At supper eat a few yawns."

Viewpoint—"I took each color slide."
Smile—"Next time use all Slides together."

Viewpoint—"I am all raw inside."
Smile—"You need some cooking."

Viewpoint—"I took needles and oil."
Smile—"Drink scissors and milk."

Viewpoint—"He eats olive oil, chaval, dal, lemonade."

Smile—"All right, but, if he does not improve, he may add some croton oil to the lemonade."

Viewpoint—"I send you all what I suffer."
Smile—"Thank you, but, better keep it yourself."

Viewpoint—"Number and Street Mrs. Tertha Burner."
Smile—"It must be some funny street."

Viewpoint—"Mother still comes out of the sore."
Smile—"It will normalate when father comes out."

Viewpoint—"I use Systemic Lemon, Local Front."
Smile—"Change it for Local Back."

Viewpoint—"The pain is on the top of the right head."
Smile—"Wait till the left head and tonate at 'Boob' Chronos."

Viewpoint—"She can not stand on her leg."
Smile—"Try standing on the head."

Viewpoint—"My father drinks Bear."
Smile—"A Giraffe would be better."

Viewpoint—"I take Lemon Aide."
Smile—"Take with an Aide-de-camp".

Viewpoint—"He gained weight when he did not have pains."
Smile—"They must be labor pains."

Viewpoint—"His Frontelim is real white."
Smile—"He is milking."

FIRST DETROIT MICHIGAN PLANET

(HOTEL TULLER, DETROIT, MICHIGAN)

Meeting Of Thursday, November 4, 1937

Successful Use Of Spectro-Chrome In:—Chicken Pox, Hives, Cold Sore, Warts, Pain, Swelling, Arthritis, Stomach Cancer, Constipation, Skull Fracture, Tuberculosis.

Reported By:—GLADYS CLARK, *Recorder*.

Deputy L. Roy Dodman was in the chair.

The Deputy read communications from Spectro-Chrome Institute and stressed the fact it was again time to remit our Annual Fee, as Affiliates of Scientific Order Of Spectro-Chrome Metrists to receive the 1938 Favorscope; also to be sure to send an extra 25 cents for registration to secure against loss through the mail.

He read from the paper of Dinshah's undaunted spirit in continuing his campaign for Governor Of New Jersey.

The program for the 17th Annual Convention to be at Milwaukee, Wisconsin was read.

It was a pleasant surprise to all to know that we may be able to read a weekly health talk written by Colonel Dinshah, under the heading "Family Health Protector" in "The Record" printed in Bridgeton, New Jersey. A number of our Members will be subscribing for The Record.

Albert G. and Jessie N. were elected by unanimous vote as Delegate and Alternate Delegate to the Convention.

Jessie N. reported a case of Chicken Pox. She used Orange to bring out the spots, then Green, Turquoise, Blue and Indigo.

Sarah C.'s nephew had a bad case of Hives. Only 2 Tonations of Turquoise were necessary.

Selma V. tonated with Turquoise a Cold Sore for about 15 minutes, during which time her hand (which had a Wart on it) was also exposed to the Turquoise Atcowa. Both the Cold Sore and the Wart disappeared.

Another Affiliate had a similar experience with Warts.

Elisabeth G. reported a boy fell down an embankment, about 10 feet. His face and knees were swollen and bleeding. He was brought home after 6 hours; he was given 1 Tonation of Indigo. All Pain and Swelling were gone the next morning leaving no marks.

Albert G. reported a man with Kardoscil of 44. After 4 Tonations of Magenta on Area 4, now it is 58.

Irene K. said her mother had been in bed 1 year with Arthritis and suffered untold agony; she was being fed then through a glass tube milk and eggs and Backelim was almost impossible. She had no use of her Body whatever. Since using Spectro-Chrome, she is well and has no pain at all. She tonated her with Green in the morning and Red in the evening and they are so thankful for the help given. They are on Rational Food Of Man. She also had a severe Head Cold; she had 1 Tonation of Lemon and after 4 hours all trouble was gone.

Cecilia D. said Spectro-Chrome should be proud to have such an Affiliate as this girl and of the remarkable way she cared for her mother during her long continued illness.

Elisabeth G. reported a case of Cancer of the Stomach which was to be operated at a cost of $600.00. The Sufferer was in a serious

condition. Had Backelim once in 6 days. He tried Spectro-Chrome, had relief from Constipation and was much improved. After 2 weeks, Kardoscil 78, Spiroscil 18. Green is being used.

Neva C. said her son of 17 fell on cement and fractured his skull. Was taken to Hospital unconscious. X-Ray picture was taken. She was not satisfied with progress in Hospital and took him home next day. She used Green once a day and Magenta on Areas 4 and 18 three times a day for 3 days, then Yellow on Areas 8, 9, 10. One week of Turquoise Systemic and Magenta on Area 4, three times a week for 3 weeks and occasionally Orange on Areas 8 and 9. At the beginning, his Kardoscil was 58 and Spiroscil 22. Now Kardoscil 72 and Spiroscil 20. They are very much pleased with the results and he is able to eat food now that he never could eat before.

Norris DeC. then continued the educational program on "Blue And Its Attributes".

The Deputy then spoke of the effect of Blue on the Pineal Gland.

At the last Meeting, Albert G. presented each Affiliate with a typewritten question sheet on the Tonation of a certain Disorder, which was to be answered and handed in at this Meeting. About 60 per cent. of the answers were correct, indicating their good judgment.

I also want to tell you that the case of Tuberculosis of which I took care, passed an O. K. Test through the Board Of Health. The Sufferer already gained 26 pounds. The case was of nearly 20 years' standing. She has had not more than 6 weeks of Tonation. This is the case reported in the July 1937 Spectro-Chrome.

Sundry other cases were reported by Michael R.

EDWARD A. ERNEST AND LUCILLE J. McGRAW ACCUSED OF PATENT VIOLATION

On Saturday, November 27, 1937, Dinshah P. Ghadiali, filed a suit in the District Court Of The United States Of America. For The Eastern District Of Wisconsin, in Milwaukee, Wisconsin, for acts of Infringement of his Patent Number 1,544,973, for certain new and useful improvements in Color Wave Projection Apparatus, against Edward A. Ernest, trading as Ernest Distributing Company and Lucille J. McGraw, both residents of Milwaukee, Wisconsin. The Bill Of Complaint asks for a Restraint Order and damages created by the alleged Infringement. Full text will be published in the next issue.

FIRST ST. LOUIS MISSOURI PLANET

(HOTEL MELBOURNE. ST. LOUIS, MISSOURI)

Meeting Of Thursday, December 2, 1937

Successful Use Of Spectro-Chrome In:—Measles, Fever, Cold, Pleurisy.

Reported By:—CHARLES H. STIERMANN, *Recorder*

Meeting was called to order by Sub-Deputy Josephine M. Sweeney who presided during the absence from the city of the Deputy Clemens M. P. Neumann, to attend the Seventeenth Annual Convention, in Milwaukee, Wisconsin.

As the next meeting would be in 1938, the need of having the Affiliation Card for the year, was emphasized, as in the event of not possessing the Card, their hopes of attending the Meeting would be completely shattered. All were advised to make application immediately.

Article Number 2 of Dinshah's "Family Health Protector" appearing in "The Record", was read to give an illustration of their quality and prove their value to those who might wish to subscribe.

Harry O. wished to know if any tremendously ill effects would result if a person took a Tonation erroneously, during an Eclipse Period. Advised that it would play more havoc with a person seriously ill than a person mildly so, but as there is always a disturbing factor in the forces of Nature when an Eclipse is in evidence it is better to guard the folly of such error. The human being is a part of the Universe and though a minute part, it is logical to presume an Eclipse would disturb his or her balance due to such forces.

Remig S. brought a sample bottle of Glycerine he had used in the Thermocept of his Spectro-Chrome and though purchased as "chemically pure Glycerin" and clear in color, it was now a decided amber color. He felt that due to the fact that the rays passed through the fluid and in the process of tonating there was a hampering effect on account of discoloration. He expressed a desire to have an opinion from Colonel whether there would be a lessening effect in the quality of a Tonation where the above condition existed. He now uses water due to such anxiety.

Peter B. reported daughter sick; case of Measles, Fever 102 Degrees, Ratio 116 to 18. Atcowa Green Systemic, Magenta on 4 and 18, Yellow 8, 9, 10 irradiated and normalated. Throat sore, but this seems to be from coughing rather than Sore Throat condition. Advised to use Turquoise Area 3 for further Tonation.

Reinhard R. normalated a Cold in head and loss of appetite by tonating with Atcowa Turquoise on Areas 1, 2, 3 and Orange on Areas 8, 9, 10. Three Tonations performed complete Normalation.

Rosa H. said she had a Cold and found a terrible burning sensation in the Throat. Her Heart kicks up a row also. Could not determine whether burning sensation was what was commonly called Heart Burn. Advised to take Ratio and use Magenta if anything seemed amiss in Area 4 as an equilibrating agent.

Alexander B., a new Affiliate, wished to know if any good results might be had by tonating out of Favor Period; also if more than two Tonations a day would prove of greater benefit. Informed that the Spectro-Chrome Home Guide said that results may be had, but, Favor Time was urged as the favorable time as the term aptly implies. Told that in the case of severe Burns, wounds and injuries with bleeding immediate tonating would bring relief, but resort thereafter to Favor Period for pronounced success. He was advised to give Nature time to adjust and recover. Two Tonations a day were ample and that the vital organs would become normal as the Auric Vehicle performed its functions better resulting from Tonations. He also asked if generally speaking we were not inclined to eat too much. All agreed. The food we assimilate gives us the chemical fuel so necessary to carry on. The excess amount frequently retards digestion and hampers health building. In this connection, the Sub-Deputy illustrated how when awfully tired she made a glass of Lemonade (Dinshah Style) and would have a wonderful pick-up in energy that would carry her on for hours.

She said that so many Affiliates do not realize the need of seeing the Ratio between Kardoscil and Spiroscil as in proper order. At the sign of any Disorder, in any Area, however slight or severe, there has been some disturbing cause and as Area 4 must be in perfect balance, Magenta should be used on Areas 4 and 18 along with Atcowa necessary for other Areas at fault. She reported the wonderful effect in recovery her husband secured with the use of Spectro-Chrome. He had a stabbing Pain in the right side. Could not breathe. Condition Pleurisy. Atcowa Turquoise was used and before the hour was up, had relief from the Pain. Successive Tonations completely normalated the Disorder.

The 31 Axioms of the Home Guide were read. That is, we went fairly along to that end, as discussion came in from time to time, just what was desired as we are aiming at a better and more thorough understanding on the application of the rules governing the Science. They are all in the Home Guide, but to read, absorb and apply is the thing necessary.

Smoking again came for discussion. The Planet Members wondered how the Government let the cigarette manufacturers get away with the advertising matter they employ. On the face of it, it is false and misleading, driving many unsophisticated to slow destruction and perdition. Yes sir, it is a great life! One must be of good heart and great courage when one sees the appalling things permitted without any effort to curtail or prevent.

EDITORIAL ✡ REMARKS

The Favorscope is built upon the motions of the Sun and the Moon, in accordance with their Radiant, Gravitational Influence upon the Earth coinciding with our Vital Vehicle acting on the Breathing Processes. During an Eclipse, the Radiant Gravitational Influence is naturally disturbed, the effect of which unbalances the Breath. This in turn reacts on the Element Chemistry of the Human Body. The Occult Operation Of The Great Sympathetic Nervous System, although explained so clearly in our teachings and textbooks, is to the uninitiated still a mystery. More they can not understand, because of their limited comprehension of the magnitude and depth of the Science. There is very much yet which Dinshah knows, but dares not expound, because the World in general is not ready for such knowledge. The time may come soon though, when his sons grow into teachers themselves. Our advice in the meantime is, follow directions and do not fool around with the complications, because you do not realize any visible or tangible harm. Study.

The explanation about the Thermocept was given in the prior month's remarks after the Planet report. Use Distilled Water.

Results may be had "out of Favor Period Tonation", but never the precise results for which Spectro-Chrome Metry stands unrivaled.

For the filth and horrors of Tobacco Smoking, read Chapter 78, Volume 2, Spectro-Chrome Metry Encyclopaedia. President Franklin D. Roosevelt talks about the Seven Billion Dollars spent by Government annually for running the show of the administration. The amount of money spent by senseless Americans in Smoking and Boozing, amounts each year to more than $7,000,000,000.00. Were those two poisons cut off, there would be no depression, no poverty, no crime, no dole, no relief, no heavy tax, no sickness, practically a really utopian era would dawn. The Democrats kicked Prohibition from its enforcement—they got on their heads the curse of all future generations. Until Smoking and Boozing be stopped, the health of the Nation and its morality will NEVER be worth any consideration. The only way to accomplish such change is Educational and NOT Legislation. Our Congress will NEVER do it, because it has no high calibre brains and less of high principles of Altruism.

To accomplish their aims in getting the votes of non- thinking, easily-led, sheepish people, the Democratic Party set up the war-cry that Prohibition was the cause of drunkenness on the increase and the overflow of bootlegging. How false that view was is shown in the recent Annual Report of Attorney General Homer S. Cummings. It discloses that more persons were sent to prison for Federal Liquor Law Violations last year than in any Prohibition year! In the fiscal year ending June 30, 1937—5,390 persons were committed to imprisonment, whereas during 1932, the highest number in Prohibition years was reached as 5,045. Those who killed Prohibition to gain their political end, will pay a heavy Karmic Penalty to the Higher Powers Who watch our entities. They ruined the country and debased, degraded and demoralized the youth of the Nation. Intoxicants are our curse and a Government selling such liquors is bound to suffer.

—o—

FIRST DETROIT MICHIGAN PLANET

(HOTEL TULLER, DETROIT, MICHIGAN)

Meeting Of Thursday, December 2, 1937

Successful Use Of Spectro-Chrome In:—Hemorrhage, Anal Rigidity, Enlarged Tonsils, Leucorrhea, Leaky Heart, Stomach Disorder, Acute Influenza, High Fever, Cancer, Paralysis, Athlete's Foot, Stomach Ulcer, Constipation, Amenorrhea, Birth Injuries, Placental Retention, Labor.

Reported By:—GLADYS CLARK, *Recorder.*

Deputy L. Roy Dodman was in the chair.

Ida A. said she had a Hemorrhage, fainted and fell on the floor; was later found by her husband. A Medical Doctor was called and she was given Injections of Iron which made her ill. After 3 or 4 days, the Doctor wanted to operate. He called for 8 days. Each call meant more Pills and more Medicine. Her husband was advised to call her relatives and friends as she could only live a short time. In the meantime, her husband was studying the Spectro-Chrome Home Guide and also beginning to realize that the Medicine was doing her no good. So he tried Spectro-Chrome. She was so pleased with the results, that she made an effort to come to our Meeting, just long enough to make this report.

Rudolph S. reported his daughter was born with a Tight

Rectum. The Medical Doctor in charge suggested an operation; Backelim was very painful. Yellow was used for 2 weeks and the child is all right now. He considered it a miracle. Another daughter whose Tonsils were enlarged and infected used Green for cleansing and Indigo for shrinking. His wife had Leucorrhea; Indigo was used which cleared the trouble.

Amy R.'s nephew had a Leaky Heart and Stomach Disorder; Four Tonations of Lemon were used. Is much better.

Peter D. reported a child with Acute Influenza; Temperature 106 Degrees, Kardoscil 120, Spiroscil 22. Three Tonations of Turquoise and 3 of Magenta were given and the child was all right.

Alfons M. reported a case of Cancer and Paralysis. Could not walk, was sick 2 years. Has been using Yellow on Areas 8, 9, 10 for 2 weeks and now walks.

Anna H. reported a case which was under a Medical Doctor's care for 2 years. The Sufferer was losing weight and becoming worse. Was given Shots of Thyroid. Kardoscil was 48, Spiroscil 12. After the first hour of Magenta on Area 4, Spiroscil was 58, second hour it was 60. Green, Yellow, Lemon and Magenta were used and they are happy over the results.

A little girl had been under the Medical Doctor's care since last June for Athlete's Foot at home and in the Hospital. Used Lemon, Orange and Turquoise on her feet with gratifying results.

Elisabeth G. reported a woman with Ulcer of the Stomach. Constipated, very nervous and Menstruation had stopped. Had 3 Tonations of Green. Constipation is better and Menstruation restored.

Gertrude R. suffered for 1½ years from Stomach Disorder. Had 3 Blood Transfusions. Has been using Spectro-Chrome Yellow which is causing much pain. Suggested to try Green.

Jessie N. reported a boy with infected Tonsils. Green and Indigo were used. The Medical Doctor was amazed at his improvement.

Lucille G. reported a baby delivered by instruments; the mother was a trained nurse. Its head was so disfigured and bruised they thought it would die. It had several Fits and was Constipated. Under Spectro-Chrome, the head injuries entirely disappeared.

The Recorder reported a woman developed a severe case of Hemorrhoids. Before the birth of her baby and after they were much worse; only a part of the Placenta came away. The usual Medicine was given but after 3 days there was still no change. Spectro-Chrome

was taken into the home and 15 minutes after 1 hour of Scarlet was irradiated, the remaining pieces of the Placenta were expelled without trouble. Three Tonations of Indigo for the Hemorrhoids were used. They completely disappeared. The Doctor had spoken of taking her to the Hospital for correction of her trouble, but it was not necessary.

Jennie M.'s husband was sick for 3 or 4 years. Perspired very easily, was weak and tired. He used, Green, Orange and Magenta.

Aquila M.'s daughter was tonated for 2 months before her baby was born (this being her first child), with Yellow, Green and Magenta. The last week Scarlet was used three times. After Labor started, Scarlet was used again and then she was taken to Hospital. She went through well and in much shorter time than expected. He wants Dinshah's opinion on Circumcision and will abide by it.

Albert G. spoke of his visit to the Convention at Milwaukee, Wisconsin.

Jessie N. expressed her regret that she was not able to attend owing to a death.

Many are having trouble with the Thermocept in their Equipment leaking and want to know what can be done for it.

Jessie N. continued the educational program of the Planet by reading from Spectro-Chrome Metry Encyclopaedia on Turquoise And Its Attributes".

Albert G. commented on the work done in the Planet in the past year and reminded us again about the Dues for the coming year.

EDITORIAL REMARKS

About Gertrude R. your Report says, she "had three Blood Transfusions." Her Chart read, "They WANT to give me 3 Blood Transfusions." She wrote to us thrice, each time incomplete. We depend on the Chart and if the wrong story be given, we can not give the correct Free Guidance. At first she wrote about having 1 Backelim in 6 days. Now she writes "too much Backelim". She wrote us about having 10-12 in a day. It is evident that she needs interview by the independent Field Solicitor who sold her the Equipment, in order that she may have some personal help. We are anxious to serve and give Aid and Relief, provided co:operation be secured from the Purchaser's end.

Scarlet was rightly used in the case of Placental Retention. However, it must be emphatically borne in mind that Scarlet will act where there is something to expel and the Uterus is ready to expel it.

Waiting for 3 days and permitting the hazard of the Cervix contracting, is not a proper method. Further warning may be given to Spectro-Chrome Normalators. Do **not** have anything to do with such and other cases where a Medical Doctor is having his own way. If good occurs, he will take the credit and if anything untoward happens, the blame will be thrown on Spectro-Chrome.

Aquila M. was not right in using Scarlet "in the last week" of pregnancy. Scarlet acts upon a gravid Uterus solely when the Cervix is opening and the Fetus is ready to emerge. Scarlet if needed, should only be irradiated when Labor is in that particular stage. In all cases wherein the Medical Doctor depends on the poisonous Ergot, Scarlet is supreme. There is nothing like it for promptness and safety and it is one of the more valuable accomplishments of Spectro-Chrome Metry, from which generations to come will reap the benefit.

Circumcision is positively contrary to our views. In ancient times, when the males went to war and were slaughtered, the females being left unprotected, the Prophets permitted Polygamy. However, as there were no sanitary systems in those days, it was found that the exudations of the contact of a man and a woman might be transferred to another woman and might endanger her health, so the Prophets advocated and ordered Circumcision. (See Holy Bible, Leviticus, Chapter 12, Verse 3.) At that time, however, the men wore long flowing robes, so did not much feel the after effects by the unprotected Glans Penis. In modern times, wearing tight trousers, the uncovered Glans frictioning against the cloth, is liable to maintain an undesirable titillation, resulting in more undesirable sensual thoughts. The Originator of Spectro-Chrome Metry was elected in 1920, as a Member of the American Association Of Orificial Surgeons, which advocates Circumcision, but he is personally strongly opposed to the procedure. The Foreskin of the Area 11 has a distinct purpose during Sex Congress and its removal is not conducive to the proper performance of that sacred function. Refer to "American Sex Problems", by Dinshah, Chapter 44. What God made, keep and use righteously. Have nothing to do with useless, destructive Surgery. Each time one passes Frontelim, wash the Glans; teach the boys cleanness of habits.

Information about the Thermocept will be found in another part of this issue.

Many sent the Dues late and got the Favorscope and Affiliation Card late. Our records are clear to date.

Man Blames God For One's Own Errors,

Did God Impel Him To Do The Wrongs?—Dinshah

FIRST MILWAUKEE WISCONSIN PLANET

(HOTEL WISCONSIN, MILWAUKEE, WISCONSIN)

Meeting Of Thursday, November 4, 1937

Successful Use Of Spectro-Chrome In:—Appendicitis, Dry Catarrh, Head Noises, Melancholia, Dyspnea, Rash, Blood Poison, Gangrene, Eczema, Mastoiditis, Arthritis, Prostatitis, Mucous Colitis, Cough, Arsenic Poisoning, Heart Disorder, Cardiac Palpitation, Epilepsy.

Reported By:—CAROLINE ZIMMERLING, *Recorder*

The Deputy read several letters received from Spectro-Chrome Institute pertaining to the Convention to be held soon in our city and 1938 Affiliation Cards.

Delegates were appointed; Frank Burns, Delegate and Caroline Zimmerling, Alternate Delegate.

Circulars for the Convention were requested by numerous Affiliates for distribution.

The Deputy read from paper sent by Colonel Dinshah and also from Bulletin of the New Jersey League Of Women Voters.

Chapters 110 and 111 of the Spectro-Chrome Metry Encyclopaedia were read by the Recorder, on "Brain And Cranial Nerves" and "Great Sympathetic Nervous System."

Rose H. will read Chapter 112, on "Pillars Of The Temple Of Solomon" at the next Meeting.

The Deputy reported the following cases:

Marcella S., 18, two attacks of Appendicitis; 2 Turquoise Tonations Areas 8, 9, 10; normalated.

Frank S., 9, Dry Cattarh; could not breathe properly through nose for 4 years. Green, Lemon and Yellow Systemic. Now can breathe well.

John P., 41, Head Noises; Green, Lemon, Yellow Systemic and Magenta on Area 4. Excellent results, ordered own Equipment.

Mary F., 54, Melancholia, took overdose of sleeping tablets, but was brought back to life. Green, Yellow Systemic and Magenta on Area 4 for 1 month; now is normal.

Lydia S., 26, could not breathe well; tightness across chest. Green, Lemon, Orange Systemic and Magenta on Area 4 gave good results.

Cyril S., 7, was sent home sick from school. Medical Doctor

gave him pills, which made him break out all over his body by the next day. The Doctor ordered them to throw the pills away. A week later, Spectro-Chrome was used. Green Systemic, Yellow on Areas 8, 9, 10 used. Back to school now again, in 1 week.

A FAILURE! Boy blind and with Water Head (Hydrocephalus) for 4½ years. Two Tonations of Spectro-Chrome failed to normalate, so they stopped Tonations!!!

Affiliates reported as follows:—

Ida B. used Spectro-Chrome on father all Summer. Blood Poison from a rust scratch and had Gangrene in toe from horse stepping on it. Medical Doctor wanted to amputate. She put him on Spectro-Chrome and it healed beautifully. Had Indigo for Pain, then Green, Lemon, Turquoise and Magenta to heal. Her mother had results in a general way with Purple, Magenta, Green and Lemon.

William T. Whole family is feeling very well now. Boy of 4 had bad Eczema. All healed now; used Lemon, Yellow Systemic and Magenta (mostly Lemon).

Marie C. Nephew 5, hurt his knee. Green 3 days started moving knee; then changed to Lemon. O. K. now. Sister-in-law always felt as if Mastoid was pothering her since she used Spectro-Chrome. Purple used with good results.

Joseph W. reported that his Arthritis gave him no pain any more.

Anna V. has Equipment 3 years; had general rundown condition. Now is well and feels that Rational Food Of Man system is very important. Used Green, Lemon and Indigo for husband's Prostate Gland trouble. Good results. Treated a friend for Catarrh with Lemon and Turquoise; now is well.

Carrie M. removed a piece of hardened sebaceous matter (like kernel of rice) from beneath her eye with Green systemic only.

Marie C. used from Green down to Red and also Indigo for Mucous Colitis with excellent results. Followed proper Food.

Stephen L. asked for advice. He hurt his back and now his hands are like dead. Alice B. suggested Orange, the Deputy suggested Yellow and Orange alternated with Magenta on Area 4. Stephen L. also made the suggestion that people reporting cases should specify the number of times they used each Atcowa.

Anna B. gained in weight since using Spectro-Chrome. Husband, 48, had bad Cough. Turquoise and Lemon; now entirely right.

Marie B. asked about Bed Wetting in a boy of 11. Lena H. suggested Ultra-Green and later Indigo, because she had results in a case with 6 of each Atcowa (this was in the case of a boy of 16 who had this for 6 years).

The Recorder reported her own case of Arsenic Poisoning with resultant Heart Disorder 4 years ago. Used Green only for 1 month, then slowed heart action with Purple. Perfectly normal heart now, whereas the Medical Doctors said the heart would never be normal again!

Anna H. used Purple for Heart Palpitation in her brother. Scarlet, Magenta and Yellow for Stomach Disorder also in him.

The Recorder stressed the importance of using Magenta in cases of common Cold.

Anna H. had an epileptic girl; 2 Tonations and she felt much better and had no more attacks since using Spectro-Chrome.

FIRST MILWAUKEE WISCONSIN PLANET

(HOTEL WISCONSIN, MILWAUKEE, WISCONSIN)

Meeting Of Thursday, December 2, 1937

President Dinshah Visited The Planet

Reported By:—CAROLINE ZIMMERLING, *Recorder*

This Meeting of our Planet opened with Deputy Otto H. Lutz in the chair. He read from "Good Thoughts" by Dinshah.

The platform was then turned over to Colonel Dinshah for the evening after Alma Horwath read Chapter 112, of Spectro-Chrome Metry Encyclopaedia, on "Pillars Of The Temple Of Solomon."

The first thing the President did upon taking the platform was to give the Affirmation Of Installation to Frank S. Burns, who will take over the Deputy's office of this Planet starting in January 1938. Frank S. Burns succeeds the present Deputy, Otto H. Lutz, who was promoted to Second Vice-President of the Scientific Order Of Spectro-Chrome Metrists.

Caroline Zimmerling was re-appointed Recorder, for the year 1938. The President spoke highly of the work done by the retiring Deputy and the Recorder.

Dinshah opened his discourse with an explanation of the Annular Solar Eclipse which was just about ending at the time he was speaking. He spoke a little of the phenomenon of the Eclipse and the effect of the Sun, Moon, and Planets upon our Bodies and also upon the Earth. He explained what actions and reactions occurred in the Human Being as a result of the action of these Orbs. He spoke also of the Breathing Processes of the Body and of the way in which all Nature rests between pulsations, giving the suggestion that the slower the Lungs and the Heart could be made to work the longer the life would be with us.

At this point, a deviation from the subject was made. Albert H. was asked by the President to tell what his reactions were to the Spectro-Chrome Institute at Malaga, New Jersey, when he attended the Resident Class given there last July. Albert H. said that he enjoyed being there very much and also enjoyed the company of the Colonel's family and everything about the place. He found all very interesting.

From this point, the President entered upon the subject of the Spectro-Chrome Radio Broadcasting Station, a potentiality, which may be known, when accomplished, as W S C I. He spoke of the advantages of having such station in the Service Of Mankind by giving free expression to the thoughts of anyone and everyone on subjects interesting to the general public. It may also be a wonderful help in giving prompt advice to Spectro-Chrome users by means of codes, etc. The tentative list of subjects to be embraced by this unique station was given in part and the principles of "no malice", "no partiality", "absolute freedom of speech" and "impersonal attitude" were discussed.

Another deviation came at this point when he asked Clemens N. of St. Louis to say a few words. Clemens gave his ideas favorably on the subject of the Radio Station.

The President resumed his address by telling how Syphilis can be "caught" without personal contact and how it is to be tonated by Spectro-Chrome. He also brought out the futility of the Sterilization of Criminals and the asininity of the Wassermann Test.

His entire discourse was enjoyed by the members of the Planet. It was delivered with his usual vigor, wit and humor and the Planet felt highly honored to have him there at this time. We hope he will come soon again.

The usual Colonial Room being considered too small to hold the Affiliates, the beautiful and commodious Gold Room of the Hotel Wisconsin was engaged for this special occasion. It proved a sensible move, because the Affiliates filled the room. The Meeting extended nearly upto midnight.

SPECTRO-CHROME

July 1938 Blue

Volume 14 Number 7

GREEN 120° · LEMON 90° · TURQUOISE 150° · YELLOW 60° · BLUE 180° · ORANGE 30° · INDIGO 210° · RED 0° · VIOLET 240° · SCARLET 330° · PURPLE 270° · MAGENTA 300°

Trade Marks Registered In The United States Patent Office

Read And Have Your Friends Read

Price Increased For Universal Motor Spectro-Chrome

Spectro-Chrome Metry Encyclopaedia—Second Edition

In Praise Of The Vinag

Dinshah Sees Through The Original Eyepieces Of Joseph Fraunhofer

Some Spectro-Chrome Marvels

Donkey Smoker!

Delaware Injunction Suit Took A Step Forward

Detroit Will Have Our Eighteenth Annual Convention

Spectro-Chrome Metry

Attuned Color Waves

In course of life you reach a time,
When Doctor, Lawyer, Priest alike,
Does fail to give the sought relief,
To troubled mind or burdened heart—
Then come to me:
I help you to help yourself.
SPECTRO-CHROME

The Science Of

***Automatic** Precision*

DEVOTED TO THE SERVICE OF MANKIND

DINSHAH P. GHADIALI, *Editor and Publisher.*

FIRST MILWAUKEE WISCONSIN PLANET

(HOTEL WISCONSIN, MILWAUKEE, WISCONSIN)

Meeting Of Thursday, July 7, 1938

Successful Use Of Spectro-Chrome In:—Persistent Disorders, Epilepsy, Fractured Collar Bone, Skin Rash, Constipation, Tachycardia, Heart Disorder.

They Pet Their Spectro-Chrome Daily!

Reported By:—CAROLINE ZIMMERLING, *Recorder*

Deputy Frank S. Burns was in the chair. He spoke first of the book "Man Visible and Invisible" by Charles Leadbeater, which Kurt B. lent him at the previous meeting. He then read Chapters 54. 55 and 56 of the Spectro-Chrome Encyclopaedia, on "Fevers By Medical Conception", "Spectro-Chrome In Fevers" and "Rubor, Calor, Tumor, Dolor."

During the time of the reading of these chapters, a telegram arrived, which the Deputy read to the Affiliates. It stated that Colonel Dinshah gained a point in the Federal Court in Wilmington, Delaware, against the Delaware State Medical Society, in Injunction suit.

The Deputy talked at some length on what Spectro-Chrome means to us or should mean to us and also on the credit Colonel Dinshah gave to the great scientists who came before him in contributing to the sciences which lead him to Spectro-Chrome. He also gave a dissertation on Spectroscopy and how it works.

Theodore M. volunteered to read Chapter 57 of the Spectro-Chrome Metry Encyclopaedia at the next meeting.

The Deputy welcomed the new Affiliates to the fold and encouraged them to bring their problems for discussion.

Walter W., from Juneau, Wisconsin, has been a user of Spectro-Chrome about 3 years. He had very good success. He had been under medical care all his life for general complications. He had operations, but nothing helped until Spectro-Chrome came along. He is all right now and his whole family is using the Equipment.

Emily M. reported that her husband had gone along now for 70 days without having an epileptic spell. Their Spectro-Chrome is in every day use and they are very much sold on it. She says they pet their machine every day! It is just a year ago they started using it.

Otto H. L. gave the following case reports:—

(1) P. K., 2½ years, broken collar bone. Turquoise given first Tonation and she slept all night. Lemon Systemic given and was well in 2 weeks.

(2) M. K., 44, condition of Phlegm. Lemon Systemic front and back and expelled a cup of Phlegm each day. Her husband became alarmed, thinking Spectro-Chrome was causing this and made her stop using it!

(3) L. L., 37, Skin Rash. Turquoise and Blue Systemic for one week, 2 each day. Itch entirely disappeared.

(4) A. B., 54, constipated for 25 years. After second Tonation, ordered own Equipment.

(5) M. L., 63, Kardoscil 130, Spiroscil 37. Used Turquoise Systemic and Purple on Area 4. After fourth Tonation, Kardoscil went down to 100 and Spiroscil to 30. Orange used on Area 17 and Kardoscil went to 78 and Spiroscil to 17. Now using Lemon Systemic and Magenta on Areas 4 and 18. Had been taking Digitalis before using Spectro-Chrome.

(6) A. S., 57, Heart Disorder. Heart was so fast and jumpy that it could scarcely be counted. Had been taking drugs for 2 years. Had to sit up during the night for several hours to find relief. After taking Spectro-Chrome, Magenta on Area 4 and Green Systemic, had only 2 minor attacks.

Following Otto L.'s reports a discussion on Heart Disorders followed and the Attuned Color Waves used were discussed, with the reasons for using Purple and the like.

Another discussion followed on Inflammations and Whooping Cough. Kurt B. suggested that for Whooping Cough, Lemon Systemic and Lemon Local on Areas 3, 4, and 5 should be used. He also stressed the point of checking on the Kardoscil and Spiroscil.

Anna V. said she used Green, Lemon and Turquoise for a Cough, but did not get the results she thought she would have had, although she did prevent Whooping Cough.

A long discussion took place on the effects of Arc Welding Sparks and Flashes on the eyes and whether or not they were Hot or Cold Burns. It was finally decided that a good way to tell whether they were Hot or Cold Burns was in the length of time it required to feel the effects. If it were a Hot Burn, the effects would be noticeable immediately and if it were a Cold Burn, the effects would not be felt until some time later and should be tonated accordingly.

During the Meeting, a question was raised about Used Equipments. Some felt that if they purchased a Used Equipment they should also be granted Free Guidance. Explanation was made as to why this was not possible. Spectro-Chrome Institute and Colonel Dinshah having so much correspondence to handle already, that if they were to grant such privilege, soon they would be giving Free Guidance to the entire United States, even though they had not sold Equipments to everyone receiving the Free Guidance. We believe the matter was cleared in the minds of those who thought otherwise.

The Meeting was attended quite well in view of the fact that many people were out of the city on vacation. The discussions were especially interesting and instructive and we feel, very beneficial.

EDITORIAL REMARKS

All Affiliates would be materially benefited by attending the monthly Meetings of the Planet. The knowledge gained by them would amply repay for the time spent.

The loosening of the Phlegm in case of M. K. (Number 2), was a sign of recovery. It was what was needed.

Cough needs Lemon as an Expectorant. Where a Systemic Atcowa is used and the same is needed for Local Areas, one may dispense with the Local Tonation as being included in the Systemic.

Turquoise used by Anna V. was wrong. Lemon and Turquoise are not irradiated together; they are Affinity Waves.

Arc Lamps or Welders, give both Heat and Cold or Infra-Red and Ultra-Violet Rays. The former give the Heat, the latter the Cold. Follow our methods for either. The explanation given at the Planet Meeting was good; the Ultra-Violet effects may come days later. Avoid exposure of the head to Sunlight also; which is full of both these Rays. Wear a cap; it will maintain the Cerebral Circulation at an even Temperature. Therein lies the principle of not going bareheaded.

The question of giving Free Guidance on Used Equipments is important. On a $90.00 Spectro-Chrome, there is a very small profit after paying $15.00 as Commission to an independent Field Solicitor. Our overhead expenses are very heavy and may not be met until we have volume production. The service given is extended solely to the original Registered Purchaser and each Free Guidance averages 25 cents worth of expense in stationery, postage and stenographic charges. Should we grant a similar privilege to Assignees, we should be tied down eternally for Free Guidance and under the Laws governing business, that would mean bankruptcy. With the present system even, some Registered Purchasers become positively obnoxious, by sending incomplete General Advice Charts, not following our Free Guidance, using Free Guidance for non-Purchasers and the like. We are the only Healing Firm

giving this unique free service and we expect that it shall be used for its legitimate purposes.

One Registered Purchaser asked for Free Guidance for herself first. Then, a week later, sent for Free Guidance for a "Cousin". Some time later, she asked for "Cousin once removed". We answered all, but, when she sent for "Cousin twice removed", we drew the line and removed her from the privilege.

The original Purchaser is bound by certain Understandings with us. The Used Equipment Purchaser might legally claim third party immunity and have loopholes for trouble-making. Therefore, for the protection of all concerned, we drew the line at giving Free Guidance to the Registered Purchaser only and even that applies to him (or her) and family members by birth or marriage. We mean the **immediate family** and not people living in separate homes.

There is another source of danger to our integrity which arises through the rental of their Equipments by Provisional Members, as a matter of income. While the commercial law is on their side, such persons are liable to violate our Constitution And Rules, because they as Provisional Members have no authority to serve the public. Any Affiliate doing so, risks Investigation and if found out of order, risks censure or perhaps Expulsion, if the conduct of the Violator would so justify. Please, stay within the bounds.

IN PRAISE OF THE VINAG

10 Cents A Meal For 3 Persons!

Dinshah's Automatic Cooking System

By:—LUCILLE E. GREENHAM, *Detroit, Michigan*

For many years,the Rational Food Of Man has been used in our home and for many years I tried to eat split peas and rice (Dal and Chaval), but honestly, I never enjoyed them.

Although my mother is an excellent cook, yet one day they would be passable and the next day one could not touch them.

Not until I visited the Spectro-Chrome Institute at Malaga, I truthfully say, I enjoyed the Rational Food Of Man; I could hardly wait for dinner time to roll around each day.

On Colonel Dinshah's last visit to Detroit, Michigan, we purchased the **Vinag.** He had it on display and I was anxiously waiting to test the Vinag to see if the food would taste the same to me as when I was in Malaga.

On Monday, we had our split peas and rice and I thoroughly enjoyed the meal, as did my mother and father.

The seasoning is just enough to make it tasty and one package of rice and peas is ample for three people; in fact a little left over, but which was not wasted I can assure you.

The one thing that pleased me most was the fact that every day it is just the same, not one day too much seasoning and the next day not enough.

I strongly urge everyone to purchase a Vinag; your meal is prepared without any trouble and everyone will relish the split peas and rice when you start using the Vinag. It is delicious and nutritive and fills the bill at 10 cents a meal for 3 persons! What next?

* * * * * * *

By:—CHARLES H. STIERMANN, *St. Louis, Missouri*

Fairness prompts me to write this note and bear testimony to **Vinag** and its excellent performance in preparation of the "Dime-A-Meal". Having used Yellow split peas and rice over a period of three years and enjoyed its tastefulness, our household felt no better perfection in preparation could be attained. This was entirely disproved after eating a meal prepared in a Vinag. We found greater palatability and tastefulness and a flavor—to use superlatives—extremely exquisite. I am glad I made the purchase, having no regrets and the madam finds it a great source of convenience, simplifying the cooking of the meal and saving time for other activities.

You may use this or put it in discard. I just wanted you to know of my attitude toward the **Vinag.**

EDITORIAL REMARKS

We thank you cordially. Those who do not have a **Vinag,** may never realize what it means to possess one. Vinag solved the problem of home economy as nothing else did before.

Every home should have a Vinag. Its price of $15.00 places it within the reach of all. Just think! A minute to fill, a second to turn the Automatic Time Switch and an hour later there is a meal ready for three persons, costing 10 cents for the meal! Talk of domestic economy! And what a meal!

However, be sure that your Electric Current is "ON" in the wall socket, as otherwise you may think the meal is cooking and at the end of the hour when you feel like eating, you will have raw grain! This actually happened to one user!

FIRST DETROIT MICHIGAN PLANET

(HOTEL TULLER, DETROIT, MICHIGAN)

Meeting Of Thursday, July 7, 1938

Successful Use Of Spectro-Chrome In:—Tumor, Hay Fever, Acne, Sore Throat, Hoarseness, Valvular Regurgitation, Athlete's Foot, Sunburn, Headache, Glaucoma, Dysmenorrhoea.

Reported By:—GLADYS CLARK, *Recorder.*

Deputy L. Roy Dodman was in the chair.

The Deputy commented on the good crowd which attended Colonel Dinshah's lectures while in Detroit on June 24th, 25th and 26th. Case Reports were as follows:

Madeline W. had a growth on her breast and was to be operated, when she was advised to try Spectro-Chrome, which she did for one month with amazing results. She then bought an Equipment and is much better now. Lemon Systemic and Indigo Local were used. She also had a Sore Throat; after the first Tonation of Turquoise she was much better.

Peter D. reported as follows:

(1) Woman, 38, Hay Fever, had Injections for it without results. Kardoscil 96, Spiroscil 22; after 6 days of Turquoise Systemic she was well.

(2) Girl of 22, suffering for 2 years with Acne (Pimples on the face). Kardoscil 82, Spiroscil 20; after 10 weeks of Spectro-Chrome, she is well. Green Systemic, Orange on face until pimples became "weeping", then Yellow, Lemon and Turquoise were irradiated.

(3) Woman 32, Sore Throat and Hoarseness. Three Tonations of Turquoise and her Throat Disorder was gone.

(4) Man, 52, Regurgitation in the Heart Valves, had Magenta 2 days and was all right.

Esther LaB. found Atcowa Lemon good for Pimples on the face.

Otto T. suffering from Athlete's Foot found that one Yellow and 4 Tonations of Lemon cleared the Disorder.

Elizabeth G. reported two cases of bad Sunburn. One Tonation of Turquoise was all that was necessary; they had no pain.

Louis S's. wife had severe Headache and was unable to talk; Kardoscil was 50, Spiroscil 12. One Tonation of Magenta and not at Favor Time was all that was necessary.

The Recorder reported as follows:—

(a) Henry S. is improving gradually. Last week he was able to stand on his feet twice for a short time, which is the first time in 3 years.

(b) A woman of 32, suffering from what the Medical Doctors pronounced as Glaucoma, (Hardening Of Eyeballs) had total blindness in one eye and the other started to be affected. She was getting a loan to have the one eye removed and later perhaps the other would have to be taken out. Eye was very red and painful. After the second Tonation, there was little or no redness or pain and the vision came at times. After 3 weeks of Turquoise, Lemon Local and Systemic, she has almost full vision, no pain and is very much improved in health.

(c) Woman, 22, suffering from Painful Menstruation, so severe that she went into Convulsions. Only one Tonation of Scarlet and she had no more trouble.

Question:—Please advise if it is proper to use an electric fan in the room while giving Tonations.

Albert G. appealed to the Affiliates to invest in a Vinag and Dinshah Dime-A-Meal. There is plenty in one meal to serve 3 people. The seasoning is correctly proportioned and the meal is very easily prepared.

Grace A. D. read from Volume 3, of Spectro-Chrome Metry Encyclopaedia, on the Yellow Wave.

The Deputy spoke of a new book which said that fish have a regular feeding time.

EDITORIAL REMARKS

There is nothing detrimental in having an Electric Fan in the Tonation Chamber for ventilation, provided the draft does not pass over the exposed Body of the Sufferer. The warmth of the room encourages Dermelim (Perspiration) and if judiciously utilized is beneficial.

Henry S. should put in mental effort to overcome his handicap. Let him exercise and start standing up with help, later discarding assistance and using crutches, finally throwing away the crutches.

If we stop eating stinking fish, our brains too may act more clearly and Skin Disorders may disappear.

If A Private Individual Will Do What Our Government Is Doing, He Will Go To Prison Or Bankruptcy Court—Dinshah

DINSHAH SEES THROUGH THE ORIGINAL EYEPIECES OF JOSEPH FRAUNHOFER

Visit To Deutsches Museum, Muenchen, Bavaria

The night of Saturday, May 14, 1938, there was a Total Eclipse of the Moon. I moved around during the balance of the day, rested during the night as much as my active thoughts permitted and early morning of Sunday the 15th, after my customary prayers, took the train for Munich, Bavaria, via Stuttgart and Ulm. As before, I had plenty of "official" company, with whom I passed an enjoyable time, getting information about the Anschluss, (Annexation) and its probable effects on the country, its people and economical aspects. We talked about family life, men, women and babies and when I incidentally stated that I was married to a girl whose parents were of German descent, they were much interested. They thought I was living upto German tradition, having 7 sons at my age of 64! That brought the questions of food, medicine, virility, habits, sickness, health, study, science and finally Spectro-Chrome. We chatted pleasantly on the subject, until at 4:30 P. M., the train entered the Hauptbahnhof of Munich, Bavaria and I ran to the Hotel Metropol, close to the station.

I had information that the Deutsches Museum closed at 6 P. M. on Sundays. No time was to be lost, because, in 5 days more I had to perform numerous errands and be on the Steamship Washington, in Southampton, England. Therefore, when I say "ran", it was an actuality, because, I had to leave Germany for Italy, at 10:20 A. M. next day. A speedy trolley landed me at the Museum by 5. What a museum! In size, grandeur, beauty and arrangements, it seemed to me to surpass even the British Museum, in London, England.

The Deutsches Museum of Masterpieces of Natural Science and Technology, was founded in 1905, through the initiative of Dr. Oskar von Miller and was opened on May 7, 1925. It houses over 70,000 objects and embraces nearly 9 miles to walk through! What could I do in 1 hour, where it would take ordinarily 9 hours to go through? But, I had a single fixed object in mind. I had no desire to view the floors full of exhibits in Geology, Mining, Metallurgy, Power, Transportation, Roadwork, Railroading, Tunneling, Bridgework, Shipbuilding, Aeronautics, Meteorology, Automobiling, Chronometry, Mathematics, Mechanics, Heat, Electricity, Telegraphy, Telephony, Photography, Television, Acoustics, Music, Magnetism, Chemistry, Pharmacy, Engineering, Lighting, Heating, Hydraulics, Gas, Surveying, Brewing, Distilling, Printing, Engraving, Bookbinding, Physiology and the like. I was familiar with such exhibits in one way or another. My mind yearned solely to see the original apparati of Joseph Fraunhofer, the chief exponent of the Fraunhofer Lines in the Spectrum and the instruments used by Robert Bunsen and his associates in developing Spectroscopy. It was a "life dream" and I was after having it fulfilled.

At 5 P. M., I arrived and rushed to the gate. I paid the entrance fee of 50 pfennigs (about 12 cents) and inquired for Fraunhofer, Bunsen and Kirchhoff. I was promptly directed to the Halls of Fame, in Rooms 156 and 157 where I saw oil paintings of those celebrities. Thence, I was told to go to Room 180, where an extremely courteous attendant greeted me. I handed him my Magenta-colored card:—

COLONEL DINSHAH P. GHADIALI, M. S-C.
(HONORARY) M. D., M. E., D. C., Ph. D., LL. D., N. D., D. Opt.,
D. F. S., D. H. T., D. M. T., D. S. T., Etc.
Metaphysician And Psychologist
Ex-Commander, New York Police Reserve Air Service
PRESIDENT, SPECTRO-CHROME INSTITUTE,
MALAGA, NEW JERSEY,
United States Of America.

Fellow And Ex-Vice-President, Allied Medical Associations Of America; Member, American Association Of Orificial Surgeons; Member, American Association For Medico-Physical Research; Member And Ex-Vice-President, National Association Of Drugless Practitioners; Academician And Life Member, Maryland Academy Of Sciences; Fellow, Theosophical Society; Life-Member, American Anti-Vivisection Society; Member, Anti-Vaccination League; Member, Supreme Lodge, International Order Of Good Templars; Member, Independent Order Of Rechabites; President, Scientific Order Of Spectro-Chrome Metrists.

He scanned it. It truly impressed him with the importance of showing me all he could in his charge. The Germans like science. The card showed him my thirst for science. On learning my mission, he promptly went to an inner room and fetched a bunch of keys. A glass cage was opened. He pointed with glowing pride to the shelves of optical apparati therein—there were the ORIGINAL prisms and similar instruments made and used by Joseph Fraunhofer! I touched them with reverence, looked through some of them and replaced them. A moment later, I requested again for a sight of the ORIGINAL apparati of Kirchhoff and Bunsen. He closed the cage and took me to the opposite side. Another cage stood there against the wall. There were buttons on it to see the Spectra of different Elements in actual disintegration. I saw a few; they were all very familiar to me like intimate friends. The attendant spoke three words "die originale Spektroskopie" which fairly galvanized me. He handed a cheap-looking thing in my hand. That apparatus was the original instrument which founded Spectrum Analysis, in 1859, whose culmination in 1920 was

Spectro-Chrome Metry! My "dream" was realized. I was as it were rooted to the spot. Psychometric perception gave me the unforgettable thrill, of having handled and seen through that apparatus which in the last century started in a humble house in Heidelberg, the minds of Kirchhoff and Bunsen toward the unfathomed mysteries of the Solar Universe! It was a moment I shall never forget. The World was dead to me. So **that** was what orginated Spectro-Chrome Metry! May the memory of those noble researchers be blessed!

Reverently, I handed that precious relic to the attendant with voluminous thanks and rushed to the Central Observatory, at his suggestion. It was in Rooms 297-298. There! It had the original refracting telescope of Joseph Fraunhofer and the observer was marking the disturbances in the Sunspots.

Going down, I saw the original telescope of Sir Isaac Newton in a corner. That reminded me of England and the British. Newton's telescope in the German Museum! What would Newton think should he reincarnate?

Six struck. I left the Deutsches Museum. I had no eyes for the other beauties of that country. Fraunhofer, Kirchhoff, Bunsen, filled my consciousness. Some day, that Deutsches Museum will have the original apparati which produced Spectro-Chrome Metry, unless either the Franklin Institute, Smithsonian Institution or the Greenwich Village authorities of Detroit, Michigan, accept them in priority.

IN THE NEXT ISSUE

The European Crisis And America Through The Eyes Of Dinshah

SOME SPECTRO-CHROME MARVELS
Doggone It! What Next?

By:—FRANK J. CASSANO, *Farmingdale, New York*

Marvelous results are being obtained with my Spectro-Chrome. I tonated several cases of Burns in the eyes, caused by Hydrated Lime. One Tonation normalated two cases and three Blue Local the other. A boy's fingers crushed in a car door, responded in 1 Tonation of Turquoise. My brother's and sister-in-law's dogs are frequent visitors of the Spectro-Chrome. Our "Cocker Spaniel" runs in to his Tonating Cage, when I say "Tonation" to him and cannot remove him until tonated!

Many of the past "Doubting Thomases" are beginning to believe, because their sight is registering the Truth. "Long live Spectro-Chrome!"

FIRST ST. LOUIS MISSOURI PLANET

(HOTEL MELBOURNE, ST. LOUIS, MISSOURI)

Meeting Of Thursday, July 7, 1938

Successful Use Of Spectro-Chrome In:—Sunburn, Dog Diarrhoea, Stomach Disorder, Limp, Radium Burn, Heel Pain, Gastritis, Ringworm, Cardiac Disorder.

Reported By:—CHARLES H. STIERMANN, *Recorder*

Deputy Clemens M. P. Neumann was in the chair.

Telegram from Colonel Dinshah was read giving the information that he won a point in a decision in the Federal District Court against the Delaware State Medical Society.

Harry O. had Sunburn that was normalated by one Tonation of Blue, all Heat and Fever leaving.

Rosa H. also effected Normalation of Sunburn with one Tonation of Blue.

Harry O.'s dog had Diarrhoea. Tonated with Turquoise and dog was restored to normal condition.

The Deputy reported several cases:—

(1) Man 53 years old. Disorder in Area 8 and spent 10 years drifting from Medical Doctor to Medical Doctor, spending a fortune without results. Atcowa Green Systemic, Lemon Systemic, Orange on Areas 8, 9 and Magenta on Areas 4, 18 six weeks tonating and Disorder was corrected. Persuaded him to buy a Spectro-Chrome Equipment but he was hesitant, being a bachelor; he liked to reduce his possessions to a minimum.

(2) Woman, 60 years old, Growth on instep; it was removed and Radium was used, injuring the foot. Took seven Tonations of Spectro-Chrome, results proving so satisfactory that she bought a Spectro-Chrome. Now walks without limping and never felt so well during her life time. This party was itisized and tonated accordingly.

(3) Person with pain in heel; Disorder 6 years old. Continued pain with pounding and thumping similar to the thumping when a Boil is coming to a head; pain was keeping her from sleeping. One and one half weeks of tonating with Atcowas Green and Yellow Systemic and Yellow Local on heel and Magenta on Areas 4, 18. Nine Tonations corrected this Disorder.

(4) Called to a home at 12:30 A. M. Woman had eaten a portion of Chili con carne with disastrous results. She was very sick, doubled with pain and unconscious. One Orange on Areas 8, 9,

and 10 and she became quiet and fell asleep. Next day, when the Deputy called she did not know that he was there the night before!

Milded K. inquired what further action to take for Defective Hearing, after pursuing the course as outlined in the Spectro-Chrome Home Guide. As several of our members have defective Hearing, they were also eager for more information for though receiving some benefit, they were not entirely satisfied with the results thus far gained. The Deputy advised how to proceed, some Tonations to be used locally and there would be better results obtained.

Nicholaus B. said he had some teeth that gave him pain. He used Indigo to no avail. Finally went to Dentist and had the teeth pulled.

Ida G. had Ringworm in armpit. Claimed one Tonation of Yellow corrected the Disorder. It was the size of a half dollar and the Medical Doctors could not effect a remedy. As this seemed highly unusual speed to effect a Normalation, brought out the fact that she had been using Yellow, Lemon and Orange Systemically for quite a period and the Yellow Local had given the final kick in disposal of the Ringworm.

Mary K. had been using Green and Lemon and finds she gets cramps in the Stomach which she can not endure. Advised to resort to Orange on Areas 8 and 9 when such a condition arises, because something must have lodged and did not want to be on its way. She is a new Affiliate and with diligence will learn the technique of Spectro-Chrome along with the "Rational Food Of Man".

Marie G. asked what to do in Sunstroke and Heat Stroke.

This was a timely inquiry and she was advised as to what course to pursue.

Eugene N. Case of woman with broken arm and well advanced in years (70). Arm was stretched and set and put in splints. He used Orange. Advised to use Yellow instead of Orange, Turquoise Local and Yellow Systemic and not to forget Areas 4, 18 using Magenta at least once a week.

The Deputy brought a book on "Eugenics" to the meeting to show that occassionally a medical man has no fear of his contemporaries and will speak the Truth. An article was read from the book in which he decries the value of Tobacco and tells of all the deleterious effects it has on the user. Also he tells about the folly of eating Meat, Fish, Eggs and ordinary Cheeses and the destructive effects of Coffee, Tea and Alcoholic Beverages. This is in accordance with our teachings, exactly, but that learned gentleman very likely will occupy the scorner's seat, too, because the powers that be do not look favorably upon such information being dispensed and especially do the manufacturers of above products and sales agencies frown with vehemence on such attitude.

In closing, the Deputy reported a case of a man who a year ago bought a Spectro-Chrome. He had a bad Heart Disorder and was unable to work. He had taken his quota of Digitalis and Strophanthus and various Heart Stimulants with the usual success which was nil, their ravages being left upon the Body and still unable to perform any of the duties pertaining to his business. Used Spectro-Chrome and is again back at work and yet he says "Spectro-Chrome did not do anything"! Such declaration reminds one of the phrase used by Colonel Dinshah quoting an envious purchaser who had wonderful results, "It is a fake, but the damned thing works!" That is gratitude.

Oh yes! Many of our Affiliates hearing the call of the woods had a good dose of Chiggers and ask if Spectro-Chrome can help them.

EDITORIAL REMARKS

Mary K. should send her General Advice Chart.

Defective hearing may be owing to different causes. Refer to the Spectro-Chrome Metry Encyclopaedia, Volume 2, Chapter 62, on Disorders Of Audition. Why not take advantage of our Free Guidance and follow it than fumbling around? Many do not follow our Free Guidance but stick to their pet views.

Books on "Eugenics" do not reckon the emotional and mental divisions, as well as the physical fits of the Areas of the mates. Their conclusions do not tally with occult findings. You may read and study them at home, but, reason their logic before accepting their one-sided doctrines. Without a thorough knowledge of Karma, Reincarnation and the Occult Laws governing them, Eugenics has not much value.

Chigger, Chigo, Chigre or Jigger is the name given to the Sarcopsylla (Dermatophilus Tunga) penetrans, a sand flea of the tropics. It burrows often in the hands and feet, causing much distress, somethimes even death. The same name is applied to a harvest mite, Leptus (Trombicula) irritans, of the southern United States of America, which often becomes a troublesome parasite. Tonation should be by Atcowa Green Systemic, followed by Ultra-Green if the affected Area shows much Redness; otherwise Turquoise Systemic should reach the cause. The same Tonation method applies to bites of other insects also. Wear proper stockings when going in the woods. In Skin Disorders, use no soap on the affected Areas.

ST. LOUIS AND DETROIT VISITED

A Strange Occurrence After An Utterance!

Our President Dinshah visited two Planets by special invitation last month and delivered discourses on the Science of Spectro-Chrome Metry. The lectures were all very well attended and created much interest.

The first series was delivered June 17, 18, 19, 1938, Hotel Melbourne, in the Colonial Room, St. Louis, Missouri. Numerous case reports were submitted at the end of the discourses, all tending to show the value of Spectro-Chrome.

The second series was delivered June 24, 25, 26, 1938, Hotel Tuller, in the Arabian Room, Detroit, Michigan. Many among the audience went on the platform voluntarily and testified about the benefits they derived from the use of Spectro-Chrome. The new Spectro-Chrome Equipment and the Vinag were exhibited and created much interest.

At the end of the third day, a unanimous request was made for the next Annual Convention to be held there. The President promised to consider the request.

All the discourses were as usual illustrated with the aid of the Stereopticon and profusely demonstrated with scientific experiments.

The meetings were all harmonious, except one disturbance in St. Louis, Missouri, where during a caustic criticism of the wrong food eaten by the American people, producing illness thereby, a man walked out shouting "Boloney"! The lecturer took it with his usual humor, but it produced mirth in the audience.

Another striking incident happened in the same city. An Affiliate having asked about the Medical Laws, Dinshah stated that they all fought against the Truth. He showed to the audience the hidden deviltry in the new so-called Food And Drug Law, wherein a "Health Device" was also included in the definition of "Drug"! Furthermore, the same Law, propounded by the U. S. Senator Royal S. Copeland, stated that any advertisement claiming the remedy for "Bright's Disease, Cancer, Tuberculosis, Infantile Paralysis, Venereal Diseases, Heart or Vascular Diseases" was a "False Advertisement" and was prohibited. That being what SpectroChrome Metry HAS ACTUALLY BEEN DOING for the past 18 years, was a distinct blow aimed at us. The lecturer said, "They may fight Spectro-Chrome, but, the man who fought against the Truth and happily lived through it was not born, because Spectro-Chrome was based on the powers of Truth." Four hours later, the evening newspapers came out with the headlines "U. S. Senator Royal S. Copeland Suddenly Died!" The man who was instrumental in having the Law enacted, died of the same Disorders whose remedy he had declared as "False Advertisement"! Spectro-Chrome could assuredly have saved him.

DONKEY SMOKER!

From Surat, India, comes a story of a jackass accustomed to the use of cigarettes. There is nothing surprising in this news. We have been accustomed to give that matter serious consideration since years and believe sincerely, that one must be a donkey to poison self and others by the pernicious fumes of the stinkweed of tobacco, the curse of the civilized World.

DELAWARE INJUNCTION SUIT TOOK A STEP FORWARD

The suit for Injunction against the Delaware State Medical Society, State Of Delaware and the Wilmington Delaware Police Department, instituted over 5 years ago by Dinshah P. Ghadiali, was heard in the District Court of the United States of America, on Thursday, July 7, 1938, by Honorable John P. Nields, Federal District Judge. A Motion for Dismissal was argued by E. Ennalls Berle, for the Respondents to which the Complainant Dinshah replied, appearing as usual for himself. The Respondents contested that the Court had no jurisdiction. The Complainant contested that as a Court Of Equity, because of the violation of his Constitutional Rights of Free Speech, that was the only Court where he could ask redress. The point having been successfully urged by the Complainant, the Court issued an Order to file a new Bill Of Complaint within 30 days, because some of the original Respondents were deceased or changed.

DETROIT WILL HAVE OUR EIGHTEENTH ANNUAL CONVENTION

The Eighteenth Annual Convention of the Scientific Order Of Spectro-Chrome Metrists, will be held on Saturday, November 26, 1938 and two following days, in Detroit, Michigan, in the Arabian Room, of the Hotel Tuller. Make your reservations and arrangements to attend. Full educational program will be published in a subsequent issue.

Original 15 Inch Refractor Telescope Of Joseph Fraunhofer.

DEUTSCHES MUSEUM, MUENCHEN, BAVARIA.

JOSEPH VON FRAUNHOFER

BORN:-March 6, 1787, at Straubing, Bavaria — DIED:-June 7, 1826, at Munich, Bavaria.

Invented the Diffraction Grating and constructed a map of 576 Lines in the Solar Spectrum, in 1814-1815, establishing Solar and Stellar Chemistry, by expressing the belief that the Dark Lines D coincided with the Bright Yellow Lines of the Sodium Flame.

Printed By:—SPECTRO-CHROME INSTITUTE PRINTING PLANT—*Malaga, New Jersey*

SPECTRO-CHROME

August
1939
Indigo

Volume
15
Number
8

Trade Marks Registered In The United States Patent Office

Read And Have Your Friends Read

MY EXPERIENCES WITH SPECTRO-CHROME

OUR NINETEENTH ANNUAL CONVENTION

KEEP AMERICA OUT OF WAR

THE REACH OF SPECTRO-CHROME

THE GERMANS FOUNDED THE ENGLISH NATION!

COLONEL DINSHAH'S ONCOMING VOYAGE TO INDIA

Attuned Color Waves
Spectro-Chrome Metry

In course of life you reach a time,
When Doctor, Lawyer, Priest alike,
Does fail to give the sought relief,
To troubled mind or burdened heart—
Then come to me:
I help you to help yourself.
SPECTRO-CHROME

The Science Of
Automatic Precision

DEVOTED TO THE SERVICE OF MANKIND

GOLDEN WHIP

My Experiences With Spectro-Chrome 2686
First Detroit Michigan Planet 2687
Spectro-Chrome Smiles 2688
Our Nineteenth Annual Convention 2689
First Milwaukee Wisconsin Planet 2690
Keep America Out Of War!!! 2691
Spectro-Chrome Metry Encyclopaedia, Volume 2, Second Edition 2692
Special Meeting Of First Milwaukee Wisconsin Planet .. 2692
The Reach Of Spectro-Chrome 2693
First St. Louis Missouri Planet 2703
First Surat India Planet 2707
The Germans Founded The English Nation! 2709
Colonel Dinshah's Oncoming Voyage To India 2710

EXPLANATION OF "HOLES IN SKIN"

Case Report of April 6, 1939, in regard to L. J., 36 years, who rented a Spectro-Chrome to be used on his twin boys, 12 days old, Robert and William, for a very bad Skin Disorder. They brought this Skin Disorder home from the Hospital, where 11 other babies also had similar Skin Disorder. Robert had "holes" on his Neck, Torso and Legs the size of an ordinary lead pencil eraser. The Sores were holes in which a lead pencil eraser would fit without dropping out. After two weeks of Green Systemic they were normalated.

The mother said Spectro-Chrome was a Godsend.

(Signed) Otto H. L., Milwaukee, Wisconsin.

MY EXPERIENCE WITH SPECTRO-CHROME

A Series Of Interesting Normalations

By:—FRANK J. C., *Farmingdale, New York.*

My daughter was burnt in the eye with a firecracker; blister and swelling appeared and disappeared in one Tonation of Turquoise and Blue; two Tonations and eye completely normalated.

Pneumonia. Saved a boy with one Tonation of Blue; could not continue, as nurse would not stop Medicine. Boy was later sent to Sanatorium.

High Blood Pressure. Set this case right in seven days; Purple 4 and 18 alternating.

Low Blood Pressure. The same result with Scarlet.

Appendicitis responded in four Tonations of Turquoise Local. Burn of Third Degree, normalated in five weeks with Blue and Turquoise Systemic. Magenta on Areas 4 and 18.

Hemorrhage from Lung. Orange 4-5-17, sealed in five minutes, right in one week.

Hemorrhage from Nose, Indigo Area 1; three Tonations normalated.

All users of Spectro-Chrome in this locality, are getting good results, except two who still believe in putting wood on the fire with their eating habits.

Mrs. M. who was at death's door with Heart Disorder, Kidney Disorder, Enlarged Thyroid Gland and Nervousness, is now worrying how to get thin again!

Frank M., Pleurisy and Tuberculosis, with severe Hemorrhage from the Nose is again enjoying normal health. Weight increased from 123 to 185 pounds at present.

Anna Rose M. completely normalated of Epileptic Fits.

W. of Floral Park who had Diabetes and Rheumatism is now well.

S. of the same town, with Enlarged Thyroid Gland, Heart Disorder and Respiratory Disorder, is now able to do her house chores with less difficulty. Her husband says, "life was never like this before."

It is by far the greatest Healing Science in existence today. The technique of Spectro-Chrome Institute is far more superior than any method of teaching I encountered in the past. Being reduced to a minimum by experiments and demonstrations, it bars all doubts which ordinarily cause resistance in our receptive faculties. Therefore, knowledge acquired by this method, makes a lasting impression upon the mind of every student of Spectro-Chrome Metry.

FIRST DETROIT MICHIGAN PLANET

(HOTEL TULLER, DETROIT, MICHIGAN)

Meeting Of Thursday, August 3, 1939

Successful Use Of Spectro-Chrome In:—Sunburn, Nervousness, Chronic Arthritis, Pneumonia, Diabetes, Fall, Scratch, Inflammation, Exophthalmia, Tumor, Blindness.

Reported By:—GLADYS CLARK, *Recorder.*

Deputy L. Roy Dodman was in the chair.

The Deputy spoke of the Colonel's success in his recent Meetings held in this City on July 28, 29, and 30. Although many were away on their vacations, the Arabian Room was well filled during each session and to those who listened, it will provide food for thought for many months to come.

Case reports were as follows:—

Emma H.'s son was badly burned on his Back from over-exposure to the Sun. Blue was used and he had no Pain whatever.

Marie S.'s husband was operated in Ford's Hospital in 1932 under a local anaesthetic. He has been extremely nervous ever since and also had Chronic Arthritis. Blue was used for Pain and Yellow Systemic brought some improvement.

Stella H. reported that a baby became suddenly ill. It was thought at first that it had Pneumonia. An oxygen tent was used. X-Rays were taken and it was decided the child had a Tumor; afterward it was decided it had no Tumor. Finally they used Spectro-Chrome, Turquoise Systemic and the child is improving nicely. Kardoscil 88.

Ellen H. reported her son, who had been seriously ill for months from Diabetes, was very much improved through the use of Spectro-Chrome. Before using Spectro-Chrome he was a hopeless case, but, now, he is ready to return to school. At present, his Kardoscil is 90, Spiroscil 17.

Adolph D. fell about 12 feet from a step ladder. Turquoise was used and he suffered no ill effects. His daughter, 2½ years old, scratched her nose; Inflammation set in. Face and Head were a solid sore with a crust of about one eighth inch depth. Nine Tonations of Turquoise and the application of cocoanut oil cleared the condition without leaving a mark. His wife is now home from the Hospital after having an "emergency operation" for Goitre. Her Legs and Eyes were badly swollen. After four weeks of Lemon Systemic she is much improved and doing her own housework.

Aurelia G. had a lump on her Arm the size of a nickel for the past twelve years and used Lemon Systemic about six weeks and it is entirely gone.

The Recorder reported: (1) A boy of eleven, while riding a bicycle, fell; it removed the flesh in places on the Knee Cap right to the Bone. Turquoise Local was used; the child had little or no Pain and after about eight days, it was filled with new flesh.

(2) Boy of eight, almost Blind since the age of two, very dull in school. His trouble was supposed to have been caused by the effects of Caesarian birth. Child's vision now almost Normal. Mental condition much improved. Orange Systemic Back and Front; Magenta on Area 4.

The Deputy gave an interesting talk on Wave Lengths.

The question arose: Are there Infra-Red and Ultra-Violet Rays in the Human Body? Please answer.

EDITORIAL REMARKS

As the Human Body generates Heat by the process of Oxygenation, it does produce Infra-Red Rays, because Heat is on that Spectral Division. Similarly, as the Motor Mechanism of the Physical Vehicle is actuated by Electrical Currents, Electricity being on the Ultra-Violet Division, the Body also has Ultra-Violet Emanations. However as all these Energies are results of Chemical Reactions, Spectro-Chrome automatically acts through the Auric Vehicle, without the necessity of the confusing guesswork of other methods.

SPECTRO-CHROME SMILES

Viewpoint—"I had a tumor on my Ovelim."
Smile—"It must be a flowing rivulet."

Viewpoint—"My directions were all turned the opposite way."
Smile—"Press the rudderbar the other way."

Viewpoint—"My blood was running down."
Smile—"Put it on an elevator."

Viewpoint—"My Expendisk is removed.'"
Smile—"That is why your clutch slips."

Viewpoint—"My head is very noisy."
Smile—"There must be some Bombay Parsee Anti-Prohibitionists in it."

OUR NINETEENTH ANNUAL CONVENTION

SIX EDUCATIONAL, FREE, PUBLIC SESSIONS, WITH PROFUSE EXPERIMENTS, DEMONSTRATIONS AND ILLUSTRATIONS WITH THE STEREOPTICON.

A Brilliant Exponence Of The Marvelous Spectro-Chrome And Its Near-Miraculous Potency As The Unrivalled Science Of Automatic Precision, by Colonel Dinshah, The Originator.

The Nineteenth Annual Convention of the Scientific Order Of Spectro-Chrome Metrists, will be held in the Milwaukee Auditorium, (Peter Engelmann Hall), 500 W. Kilbourn Avenue, Milwaukee, Wisconsin, for three consecutive days, beginning Sunday, October 1, 1939. The Originator, Colonel Dinshah P. Ghadiali, recently returned from a scientific tour in India will preside.

All Sessions will be open to the public, free and without any charge or collection, as our main object is to awaken the public to a realization of the Truth. The Sessions will be a round of genuine intellectual banquets for the Truth-Seeker. During each Session the activities of the Annual Convention will be interpolated and actual Case Reports will be presented, by Spectro-Chrome Normalators as well as Sufferers who will speak voluntarily.

As usual, **writing notes will not be allowed.** Positively **NO SMOKING** will be permitted. Expelled Affiliates will have no access to any of the Sessions. Right of denial of admission to any person is reserved.

DATE AND TIME

Sunday, October 1, 1939—	2 P. M. to 5 P. M.
	8 P. M. to 11 P. M.
Monday, October 2, 1939—	2 P. M. to 5 P. M.
	8 P. M. to 11 P. M.
Tuesday, October 3, 1939—	2 P. M. to 5 P. M.
	8 P. M. to 11 P. M.

TOPICS

Debunking Medical Fallacies. Medical Empirics Displaced By Spectro-Chrome Science. Marvels Of Spectro-Chrome. Challenging Eye-Openers. Explosion Of Germ Theories. Nine Organic Systems Of The Physical Body. Occult Side Of Life.

No Diagnosis — No Drugs — No Manipulation — No Surgery

FIRST MILWAUKEE WISCONSIN PLANET

(HOTEL WISCONSIN, MILWAUKEE, WISCONSIN)

Meeting Of Thursday, August 3, 1939

Successful Use Of Spectro-Chrome In:—Leaky Heart.

Reported By:—CAROLINE ZIMMERLING, *Recorder*

Deputy Frank Burns was in the chair. He opened the Meeting by reading a letter from Colonel Dinshah, relative to his visit to Milwaukee in October and other business.

The Deputy then gave a long and very instructive discourse on Spectro-Chrome Axioms, as found in the Spectro-Chrome Home Guide. These Axioms were illustrated on the blackboard with the diagram of the Mogen David.

Otto C. wanted to tell of his oldest son's Heart Disorder. It was a case of Leakage and the boy was in very bad condition. He could not get work because of the physical examinations he would have to pass. Sometimes the Heart Beats could not be felt, but after a few Magenta Tonations, the Beats became even and he is now doing nicely.

Theodore M. wanted to know whether the Heart Beat of an adult had to be a certain speed to be normal, or whether the Ratio was the important thing. He said that when he rises in the morning, his Pulse is never more than 48 and his Breathing 12, but after he is up for a while, this increases to 60 and 15, respectively. He maintains this Ratio. He said he never has any trouble with his Heart and is apparently in perfect Health.

During the course of the discussion on Axioms, the question came up as to the Normalation of Imbackelim (Constipation). Otto L. was asked his opinion on the subject. He said that such condition responded to Yellow on Areas 8, 9 and 10.

Hilda J. spoke of a woman who had a Swelling on the right side of the Head, which Blue relieved. When she used Orange, her Ear began to drain and was still draining. Blue stopped the draining slightly, but the Deputy said she should continue the use of Orange to keep it draining. She complained about her Head feeling as if it had a metal plate in it, also, the Deafness had moved from the opposite side of the Head and she was somewhat discouraged.

Foods were discussed, also whether it was all right to take a Tonation immediately after a meal.

The question was also raised as to whether it was necessary to own a set of Spectro-Chrome Metry Encyclopaedia in order to be eligible to take the Spectro-Chrome Normalator Course. The Affiliates of the Planet would like to have this answered.

EDITORIAL REMARKS

Page 62 of Spectro-Chrome Home Guide has all the answers about Heart Beats and Ratio. The number depends upon the age and sex, while the Ratio depends upon the relation to the Respiratory System; by the Ratio, the hopeful character or otherwise of a case is opined.

After a meal, no Tonation should be taken for at least 90 minutes, except for Disorders in Areas 7, 8, 9, 10, where indicated.

None is eligible to take the Course for becoming a Spectro-Chrome Normalator, except the Registered Purchaser of a Spectro-Chrome and who is also an Enrolled Free Student through a set of Spectro-Chrome Encyclopaedia.

KEEP AMERICA OUT OF WAR!!!

Monroe Doctrine As Gutta Percha Viewpoint

Spectro-Chrome is not a political organization, but, in as much as we are Americans, we dutifully voice our candid views in regard to the preservation of the country wherein are located our major activities and to which our allegiance is pledged.

Once more the craze of conquest by war, gripped Europe. As outlined in the Spectro-Chrome of August 1938, events are proving one after another, that once more the European nations will do their utmost to pull us into the boiling cauldron, from which they will get the fruits and this nation shall be chewing the clean-picked stones, as 1918 to 1939 amply proved.

President Franklin D. Roosevelt proclaimed alleged "Neutrality". We call it alleged, because the leaky valves opening into our heart through this sham "Neutrality" are devoid of regulative potency. To sell war materials to a so-called neutral nation, which may subsequently sell the bill-of-lading to a belligerent is **not** neutrality. To remedy those leaky valves, Roosevelt asks for change of the embargo to a cash-and-carry plan. It sounds wonderfully good, but, such a plan would be neutral solely where both the belligerent parties have equal access to our ports as our customers. In the case in question, one set of belligerents has the whole Atlantic Ocean to itself, while the other set has no way to reach us without sub-oceanic warfare. It seems to us logical, that we should be alsolutely neutral and impartial.

The only way in which we can be so, is to keep ourselves out of both direct as well as indirect contact with the War Plague. Last year, President Roosevelt voiced a startling view that Americans will not stand by and watch calmly if Canada were in danger. He interprets the Monroe Doctrine to apply in such contingency. We differ from the opinion. The Monroe Doctrine would apply solely when a European Power would **invade** Canada voluntarily and as an aggressor, but, when Canada by declaring War on a European Belligerent Set becomes itself involved in aggression, it is hardly right to expect the United States Of America to enforce the gutta percha of the Monroe Doctrine.

We are neutral. Our foundation as laid by George Washington is on neutrality; we disregarded his wise principles in 1917 and plunged the country into perdition; that **one lesson** should have been enough. Our warning again is **Keep Neutral And Out Of War.**

IMPORTANT:—Notification is issued to all Affiliates of Scientific Order Of Spectro-Chrome Metrists, to refrain from discussing any War news in our Planet Meetings and our Deputies are requested to enforce **complete silence and neutrality** in this regard therein. BE NEUTRAL! KEEP AMERICA OUT OF WAR!!!

SPECTRO-CHROME METRY ENCYCLOPAEDIA—VOLUME 2 SECOND EDITION

Volume 2 of the Spectro-Chrome Metry Encyclopaedia, Second Edition, is on the Press and when ready will be mailed to the subscribers. Please do not bombard us with inquiries about it. We are doing the best we can. It will be mailed by end of October 1939.

Special Meeting Of First Milwaukee Wisconsin Planet

On Saturday, September 30, 1939, at 8 P. M. there will be a Special Meeting in the Gold Room, Hotel Wisconsin, Milwaukee, Wisconsin, open to Affiliates of the Scientific Order Of Spectro-Chrome Metrists exclusively. No admission in this Special Meeting without presentation of the Affiliation Card and Registration of full name and address.

Meet Colonel Dinshah. Introduce yourself personally. He will discuss recent developments and give you Free Guidance on your Reports.

OUR AIM: SPECTRO-CHROME—IN EVERY HOME

THE REACH OF SPECTRO-CHROME

An Unrivalled Accomplishment Far Surpassing Centuries Of Drugs

During the past 19 years, what was accomplished by Spectro-Chrome Metry may be partially adjudged from the cases reported in our magazine. The list is by no means complete and the names of Disorders are given in their original terminology. The reported cases are those mentioned in our Planet Meetings upto the end of December 1938 and does not cover the vast field of unreported cases, which are of everyday occurrence in the families of Purchasers and Users. The numbers refer to the pages in Spectro-Chrome:

ABDOMINAL DISORDER........... 786 - 910 - 1984
ABSCESS........................... 179 - 203 - 206 - 299 - 300 - 301 - 304 - 332 - 435 - 466 - 467 - 499 - 585 - 810 - 1048 - 1243 - 1244 - 1271 - 1272 - 1399 - 1440 - 1687 - 1811 - 1879 - 1954 - 2034 - 2049 - 2050 - 2334 - 2335 - 2354 - 2376 - 2399 - 2464
ACCIDENT.......................... 138 - 1241 - 1378 - 1383 - 1387 - S-C. December. 1934, Back Cover - 1439 - 1453 - 1475 - 1501 - 1532 - 1590 - 1669 - 1710 - 1711 - 1787 - 1790 - 1800 - 1811 - 1920 - 1944 - 1973 - 1985 - 2005 - 2050 - 2068 - 2092 - 2115 - 2152 - 2193 - 2218 - 2229 - 2231 - 2305 - 2318 - 2350 - 2359 - 2360 - 2366 - 2391 - 2416 - 2417 - 2426 - 2449 - 2468 - 2509 - 2510 - 2527
ACIDITY............................. 583
ACNE................................ 1709 - 1739 - 1868 - 2146 - 2421
ADENOIDS........................... 150 - 307 - 308 - 1499 - 2148 - 2392
ADHESIONS.......................... 741
ALBUMINURIA....................... 334
ALCOHOLIC CRAVING............. 230 - 2132 - 2247
AMENORRHEA....................... 68 - 112 - 353 - 384 - 448 - 462 - 476 - 584 - 831 - 835 - 1592 - 2023 - 2071 - 2081 - 2174 - 2286
ANEMIA.............................. 68 - 232 - 476 - 584 - 626 - 783 - 1474 - 1504 - 1854 - 2089 - 2167 - 2470 - 2471 - 2486 - 2487
ANEURISM........................... 704 - 834
ANGINA PECTORIS................. 204 - 851 - 1254
ANOREXIA........................... 2488
APHASIA............................. 1972
APHONIA............................ 461 - 790 - 1592 - 1982 - 2029 - 2038 - 2073 - 2081 - 2086 - 2114 - 2115 - 2338 - 2528
APOPLEXY.......................... 2120
APPENDICITIS...................... 45 - 68 - 229 - 230 - 303 - 304 - 346 - 352 - 435 - 461 - 629 - 723 - 739 - 834 - 869 - 870 - 871 - 907 - 1115 - 1525 - S-C. June, 1935, Back Cover - 1577 - 1590 - 1598 - 1715 - 1723 - 1776 - 1879 - 1882 - 1936 - 1939 - 1973 - 1976 - 2004 - 2019 - 2164 - 2289 - 2300 - 2301 - 2354 - 2376 - 2390 - 2473
ARSENIC POISON.................. 2151 - 2291 - 2318 - 2319
ARTERIO SCLEROSIS.............. 1452 - 1476 - 1573
ARTHRITIS........................... 476 - 498 - 499 - 602 - 603 - 604 - 612 - 633 - 634 - 702 - 772 - 773 - 811 - 812 - 832 - 850 - 862 - 910 - 1018 - 1403 - 1524 - S-C. June, 1935, Back Cover - 1579 - 1580 - 1581 - 1592 - 1605 - 1670 - 1674 - 1685 - 1723 - 1737 - 1738 - 1775 - 1825 - 1868 - 1890 - 1893 - 1920 - 1925 - 1940 - 1970 - 1985 - 2007 - 2010 - 2029 - 2037 - 2038 - 2050 - 2067 - 2073 - 2089 - 2090 - 2094 - 2126 - 2151 - 2193 - 2272 - 2280 - 2290 - 2302 - 2303 - 2304 - 2310 - 2336 - 2337 - 2354 - 2395 - 2399 - 2422 - 2473 - 2474 - 2484

ASTHMA.......... 284 - 285 - 306 - 383 - 462 - 467 - 492 - 493 - 500 - 583 - 626 - 706 - 723 - 771 - 847 - 1115 - 1288 - 1378 - 1400 - 1410 - 1437 - 1454 - 1499 - 1577 - 1579 - 1669 - 1714 - 1766 - 1777 - 1790 - 1825 - 1848 - 1854 - 1861 - 1862 - 1893 - 1934 - 1939 - 1972 - 1982 - 2017 - 2088 - 2090 - 2094 - 2148 - 2169 - 2170 - 2217 - 2376 - 2391 - 2392 - 2413 - 2441 - 2469 - 2520

ASTIGMATISM.......... 1520

ATHLETE'S FOOT.......... 1386 - 1549 - 1550 - 1686 - 1865 - 1975 - 2037 - 2085 - 2163 - 2286 - 2421

AUTO-INTOXICATION.......... 230 - 232

BACKACHE.......... 498 - 1117 - 1241 - 1272 - 1328 - 1440 - 1550 - S-C June, 1935, Back Cover - 1604 - 1671 - 1686 - 18 - 1932 - 1981 - 1990 - 2004 - 2085 - 2086 - 2099 - 21 - 2170 - 2218 - 2230 - 2258 - 2319 - 2332 - 2354 - 23 - 2376 - 2389 - 2390 - 2514 - 2519

BALDNESS.......... 634

BED WETTING.......... 632 - 633 - 1577 - 1670 - 2194 - 2291

BELL'S PALSY.......... 790 - 1925

BILIOUSNESS.......... 2146

BIRD DISORDER.......... 1381 - 1766 - 1982 - 2005 - 2100 - 2463

BLADDER DISORDER.......... 261 - 327 - 328 - 467 - 501 - 583 - 723 - 769 - 832 - 834 - 1243 - 1577 - 1578 - 1825 - 1976 - 2009 - 2113 - 2319 - 2338 - 2376 - 2441

BLINDNESS.......... 382 - 584 - 830 - 1262 - 1263 - 1321 - 1379 - 1576 - 2258 - 2290 - 2304 - 2422

BLISTERS.......... S-C. June, 1935, Back Cover - 1602 - 1669 - 1801 - 1872 - 1923 - 1974 - 2085 - 2398 - 2406

BLOOD POISON.......... 260 - 348 - 725 - 1981 - 2186 - 2290 - 2332

BOILS.......... 384 - 462 - 499 - 500 - 626 - 725 - 737 - 834 - 1322 - 1686 - 1777 - 1802 - 1848 - 1869 - 2004 - 2049 - 2050 - 2067 - 2093 - 2100 - 2211 - 2399 - 2489 - 2491 - 2527

BONE DISORDER.......... 1889

BRAIN DISORDER.......... 2115 - 2120

BRADYCARDIA.......... 1401 - 1440 - 1551 - 1675 - 1676 - 1826 - 1982 - 2005 - 2085 - 2253 - 2280 - 2362 - 2470

BRONCHITIS.......... 93 - 583 - 770 - 847 - 848 - 1475 - 1501 - 1502 - 1504 - S-C. June, 1935, Back Cover - 1579 - 1581 - 1595 - 1707 - 1824 - 1866 - 1893 - 2005 - 2017 - 2034 - 2050 - 2073 - 2094 - 2120 - 2170 - 2194 - 2337 - 2338 - 2352 - 2441

BRUISE.......... 583 - 783 - 784 - 1525 - 1669 - 1824 - 1825 - 2038 - 2133 - 2509 - 2510

BURNS.......... 14 - 179 - 180 - 200 - 260 - 306 - 331 - 348 - 349 - 434 - 435 - 436 - 445 - 446 - 495 - 630 - 631 - 739 - 741 - 790 - 811 - 833 - 847 - 849 - 980 - 981 - 1242 - 1253 - 1257 - 1379 - 1394 - 1401 - 1402 - 1502 - 1546 - 1576 - 1603 - 1685 - 1825 - 1849 - 1879 - 1880 - 1883 - 1903 - 1969 - 2017 - 2020 - 2023 - 2029 - 2030 - 2035 - 2047 - 2054 - 2055 - 2056 - 2073 - 2100 - 2107 - 2151 - 2170 - 2205 - 2252 - 2260 - 2365 - 2366 - 2398 - 2399 - 2406 - 2407 - 2414 - 2426 - 2450 - 2451 - 2488 - 2491 - 2509

CANCER.......... 41 - 68 - 204 - 205 - 232 - 281 - 282 - 302 - 326 - 435 - 436 - 448 - 461 - 467 - 497 - 584 - 625 - 698 - 699 - 726 - 741 - 773 - 785 - 786 - 788 - 789 - 848 - 869 - 1156 - 1398 - 1503 - 1669 - 1903 - 2030 - 2095 - 2112 - 2150 - 2247 - 2280 - 2281 - 2286 - 2463 - 2520 - 2532

CANKER.......... 738

CARBUNCLE.......... 86 - 87 - 88 - 89 - 180 - 279 - 280 - 435 - 467 - 611 - 626 - 704 - 705 - 833 - 835 - 1840 - 1848 - 2004 - 2018 - 2068 - 2113 - 2162 - 2186 - 2353

CAR SICKNESS.......... 702 - 703 - 723 - 724

CATARACT.......... 148 - 501 - 626 - 769 - 772 - 811 - 834 - 1262 - 1604 - 1670 - 1685 - 1708 - 1841

CATARRH.......... 138 - 233 - 261 - 297 - 299 - 300 - 333 - 346 - 352 - 384 - 466 - 500 - 832 - 1452 - 1453 - 1574 - 1670 - 2289 - 2290 - 2376

CAT DISORDER.......... 203 - 2074

CAVITY.......... 467 - 1687

CELLULITIS.......... 260

CHICKEN POX.......... 1790 - 1862 - 1976 - 2092 - 2280 - 2362 - 2363

CHILD DISORDER.......... 770

CHOREA.......... 179 - 771 - 1255

CIRCULATION.......... 1670 - 1688 - 1982 - 2025 - 2148 - 2163

COLD.......... 138 - 233 - 333 - 831 - 849 - 850 - 979 - 980 - 1441 - 1452 - 1475 - 1502 - 1550 - S-C. June, 1935, Back Cover - 1579 - 1607 - 1608 - 1621 - 1622 - 1674 - 1676 - 1686 - 1687 - 1709 - 1737 - 1765 - 1774 - 1775 - 1776 - 1778 - 1786 - 1787 - 1789 - 1798 - 1799 - 1800 - 1811 - 1823 - 1824 - 1825 - 1826 - 1839 - 1848 - 1849 - 1861 - 1862 - 1864 - 1869 - 1882 - 1890 - 1893 - 1943 - 1975 - 1982 - 1983 - 2005 - 2011 - 2024 - 2037 - 2049 - 2050 - 2060 - 2061 - 2068 - 2071 - 2081 - 2092 - 2093 - 2123 - 2148 - 2151 - 2152 - 2170 - 2193 - 2247 - 2248 - 2259 - 2270 - 2283 - 2297 - 2300 - 2309 - 2310 - 2338 - 2349 - 2375 - 2376 - 2389 - 2413 - 2460 - 2464 - 2468 - 2486 - 2487 - 2500 - 2509

COLIC.......... 466 - 739 - 2115

COLITIS.......... 769 - 770 - 832 - 833 - 1073 - 1243 - 1580 - 1604 - 1707 - 1775 - 1799 - 1800 - 1848 - 2455

CONCUSSION.......... 2038 - 2527

CONJUNCTIVITIS.......... 230 - 2119

CONSTIPATION.......... 58 - 68 - 91 - 109 - 112 - 138 - 204 - 232 - 233 - 234 - 306 - 333 - 334 - 346 - 349 - 350 - 351 - 384 - 435 - 448 - 497 - 500 - 501 - 584 - 610 - 611 - 629 - 701 - 723 - 725 - 739 - 740 - 741 - 742 - 746 - 769 - 771 - 784 - 789 - 810 - 831 - 832 - 833 - 834 - 835 - 848 - 981 - 1242 - 1244 - 1379 - 1380 - 1381 - 1401 - 1452 - 1474 - 1498 - 1501 - 1502 - 1503 - 1524 - 1545 - 1550 - 1551 - 1574 - 1577 - 1579 - 1592 - 1593 - 1598 - 1604 - 1622 - 1665 - 1670 - 1671 - 1676 - 1685 - 1687 - 1688 - 1723 - 1765 - 1774 - 1775 - 1776 - 1786 - 1787 - 1788 - 1798 - 1825 - 1853 - 1863 - 1943 - 1944 - 1960 - 1974 - 1981 - 1982 - 2004 - 2005 - 2010 - 2011 - 2015 - 2031 - 2048 - 2049 - 2051 - 2071 - 2075 - 2085 - 2086 - 2093 - 2098 - 2121 - 2123 - 2135 - 2152 - 2163 - 2169 - 2184 - 2185 - 2188 - 2194 - 2211 - 2218 - 2253 - 2259 - 2260 - 2280 - 2281 - 2286 - 2301 - 2309 - 2310 - 2319 - 2362 - 2365 - 2388 - 2417 - 2435 - 2455 - 2461 - 2470 - 2477 - 2484

CONTUSION.......... 2406

CONVULSIONS.......... 468 - 607 - 740 - 1243 - 1944 - 2004 - 2120 - 2422

CORN.......... 138 - 728 - 739 - 1738

COUGH.......... 109 - 229 - 232 - 234 - 306 - 835 - 848 - 1441 - 1502 - 1525 - 1604 - 1770 - 1778 - 1801 - 1826 - 1861 - 1939 - 2048 - 2071 - 2163 - 2174 - 2290 - 2309 - 2362 - 2365 - 2417 - 2509

CRAMPS.......... 849 - 1381 - 1409 - 1854 - 1855 - 1870 - 1889 - 1890 - 2086 - 2211

CRICK.......... 1798

CROUP.......... 1579 - 2252

CYANIDE POISON.......... 499

CYST.......... 785 - 982 - 2166 - 2167 - 2168 - 2350 - 2413

DEAFNESS.......... 232 - 281 - 607 - 979 - 1387 - 1438 - 1670 - 1854 - 1981 - 2114 - 2257 - 2332 - 2365 - 2366 - 2474 - 2513

DEBILITY.......... 770 - 980 - 1441 - 1453 - 1474 - 1685 - 1891 - 1960 - 2113 - 2169 - 2170 - 2173 - 2174 - 2186 - 2257 - 2287 - 2392

DELIRIUM........................... 2360
DIABETES MELLITUS.............. 231 - 783 - 834 - 869 - 979 - 1321 - 1380 - 1381 - 1476 - 1549 - 1576 - 1577 - 1593 - 1688 - 1689 - 1708 - 1709 - 1789 - 1800 - 1853 - 1854 - 1870 - 1891 - 1916 - 1920 - 1944 - 1985 - 2011 - 2012 - 2049 - 2050 - 2062 - 2063 - 2064 - 2073 - 2120 - 2128 - 2147 - 2169 - 2173 - 2218 - 2258 - 2305 - 2367
DIARRHEA......................... 93 - 112 - 231 - 383 - 1254 - 1452 - 1573 - 2034 - 2163 - 2230 - 2231 - 2427
DIPHTHERIA....................... 138 - 915 - 1490
DIPLOPIA........................... 2514
DISTEMPER......................... 834 - 1483 - 1484 - 1801
DIZZINESS.......................... 346 - 633 - 1382 - 1669 - 1688 - 1864 - 1873 - 1943 - 2094 - 2231 - 2259 - 2434 - 2455 - 2514 - 2515
DOG BITE............................ 1258 - 1670 - 1708 - 2111 - 2514
DOG DISORDER..................... 834 - 835 - 911 - 1244 - 1253 - 1328 - 1403 - 1483 - 1484 - 1551 - 1604 - 1708 - 1721 - 1765 - 1799 - 1801 - 2034 - 2089 - 2212 - 2218 - 2350 - 2366 - 2390 - 2406 - 2426 - 2427
DREAMS............................. 853
DROPSY.............................. 333 - 609 - 610 - 832 - 1932 - 1933 - 2048 - 2085 - 2332 - 2389
DRUG ADDICT...................... 15 - 230 - 282 - 348 - 467 - 500 - 833 - 1869
DYE POISON......................... 2087
DYSENTERY.......................... 583 - 1975 - 1976 - 2212 - 2474
DYSMENORRHOEA................ 138 - 448 - 629 - 631 - 703 - 746 - 980 - 1379 - 1453 - 1549 - 1801 - 1990 - 2422
DYSPNEA............................. 200 - 383 - 384 - 790 - 1255 - 1258 - 1453 - 1576 - 1777 - 1824 - 1853 - 1982 - 2032 - 2050 - 2120 - 2169 - 2275 - 2276 - 2289 - 2350 - 2521
EAR DISORDER..................... 233 - 281 - 284 - 300 - 301 - 307 - 330 - 334 - 352 - 558 - 723 - 832 - 907 - 983 - 1321 - 1453 - 1486 - 1498 - 1504 - 1581 - 1590 - 1595 - 1670 - 1708 - 1848 - 2034 - 2114 - 2173 - 2212 - 2228 - 2332 - 2360 - 2407 - 2434
EAR NOISES......................... 1498 - 1593 - 1670 - 1801 - 1824 - 2011 - 2148 - 2289
ECZEMA............................... 91 - 138 - 232 - 415 - 416 - 435 - 447 - 462 - 628 - 740 - 764 - 832 - 848 - 979 - 1244 - 1437 - 1503 - 1605 - 1722 - 1738 - 1777 - 1820 - 1861 - 1882 - 2092 - 2094 - 2118 - 2259 - 2290 - 2484 - 2485
EMPYEMA............................ 1551
EPIDIDIMITIS...................... 447
EPILEPSY............................ 46 - 148 - 299 - 332 - 436 - 448 - 498 - 625 - 741 - 746 - 771 - 980 - 1853 - 1972 - 2025 - 2286 - 2291 - 2304 - 2336 - 2337 - 2388 - 2416 - 2473
ERUCTATION........................ 91
ERYSIPELAS......................... 202 - 739 - 981 - 1710 - 1939 - 2119 - 2469
EXOSTOSIS.......................... 1943 - 1976 - 1985 - 2067 - 2073
EYE DISORDER..................... 207 - 314 - 315 - 316 - 329 - 739 - 771 - 783 - 785 - 790 - 830 - 833 - 1254 - 1262 - 1263 - 1387 - 1453 - 1454 - 1499 - 1520 - 1524 - 1549 - 1550 - 1574 - 1590 - 1604 - 1670 - 1675 - 1676 - 1710 - 1711 - 1739 - 1774 - 1798 - 1811 - 1820 - 1826 - 1861 - 1902 - 1925 - 1933 - 1943 - 1976 - 1981 - 1982 - 1983 - 2005 - 2018 - 2032 - 2085 - 2086 - 2119 - 2121 - 2123 - 2133 - 2148 - 2162 - 2169 - 2170 - 2187 - 2217 - 2230 - 2258 - 2260 - 2290 - 2304 - 2338 - 2366 - 2465 - 2468 - 2469 - 2510 - 2514
FAINTING............................ 1932 - 2005 - 2093 - 2188 - 2252 - 2253 - 2286 - 2337 - 2400 - 2488
FEEBLE MIND....................... 1409 - 1410 - 1592 - 1943 - 1954
FELLON.............................. 306 - 1604 - 1711
FEMALE DISORDER............... 384 - 461 - 462 - 771 - 1932 - 2186 - 2286 - 2465
FESTER.............................. 2090 - 2162 - 2247 - 2468
FEVER................................. 139 - 233 - 260 - 435 - 659 - 726 - 755 - 980 - 981 - 1328 - 1398 - 1409 - 1452 - 1453 - 1486 - 1503

1504 - 1525 - 1549 - 1576 - 1590 - 1665 - 1670 - 1674 - 1676 - 1686 - 1789 - 1790 - 1799 - 1839 - 1840 - 1863 - 1866 - 1920 - 1932 - 1976 - 2005 - 2030 - 2038 - 2063 - 2067 - 2071 - 2081 - 2084 - 2101 - 2108 - 2113 - 2121 - 2123 - 2133 - 2171 - 2174 - 2193 - 2200 - 2258 - 2259 - 2282 - 2297 - 2300 - 2301 - 2309 - 2319 - 2349 - 2362 - 2389 - 2394 - 2399 - 2427 - 2460 - 2485 - 2509 -
FISH DISORDER.................... 2260
FISTULA............................. 462
FLAT FOOT......................... 612
FLATULENCE...................... 91 - 139 - 333 - 612 - 850 - 910 - 1545 - 1890 - 1920 - 1938 - 2169 - 2194 - 2257 - 2407
FOOD POISON...................... 738
FOOT DISORDER................... 629 - 630 - 701 - 723 - 740 - 783 - 790 - 1379 - 1824 - 1865 - 2033 - 2427
FOX DISORDER.................... 2318
FROST BITE........................ 1801
FURUNCLE......................... 2354
GALL BLADDER DISORDER....... 783 - 834 - 1255 - 1381 - 1504 - 1550 - 1551 - 1574 - 1576 - 1577 - 1707 - 1709 - 1786 - 1799 - 1810 - 1982 - 2010 - 2032 - 2120 - 2170 - 2193 - 2319
GALL STONES..................... 348 - 435 - 466 - 1271 - 1381 - 1574 - 1579 - 1708 - 1939 - 1970 - 2024 - 2031 - 2033 - 2080 - 2309
GANGRENE......................... 148 - 204 - 626 - 740 - 1043 - 1253 - 1254 - 1408 - 1853 - 1969 - 2054 - 2115 - 2290 - 2310 - 2485
GAS POISON....................... 788
GERMAN MEASLES.............. 1800
GINGIVITIS........................ 706 - 1580 - 1738 - 2194 - 2318 - 2350
GLAND DISORDER............... 299 - 632 - 657 - 658 - 2087 - 2093 - 2108 - 2147 - 2174 - 2184 - 2185 - 2309 - 2319 - 2394
GLAUCOMA........................ 608 - 723 - 742 - 771 - 833 - 2422 - 2435
GLEET............................... 347
GOITRE............................... 91 - 330 - 461 - 494 - 501 - 625 - 628 - 703 - 704 - 722 - 723 - 769 - 850 - 911 - 982 - 1253 - 1321 - 1403 - 1440 - 1453 - 1551 - 1670 - 1872 - 2037 - 2100 - 2163
GONORRHOEA..................... 232 - 347 - 447 - 701 - 723 - 1328 - 1387 - 1399 - 1974 - 1977 - 2031 - 2048 - 2360
GRANULATION.................... 1524 - 1525 - 1969 - 2085
GRIPPE.............................. 8 - 198 - 2030 - 2084
HALITOSIS......................... 2147
HARMLESSNESS.................. 1242 - 1243 - 1405
HAY FEVER........................ 199 - 704 - 1321 - 1379 - 1402 - 1502 - 1593 - 1669 - 1939 - 1982 - 2011 - 2017 - 2025 - 2114 - 2174 - 2194 - 2218 - 2247 - 2399 - 2413 - 2421 - 2469 - 2485 - 2527
HEADACHE......................... 178 - 498 - 627 - 628 - 629 - 746 - 835 - 849 - 979 - 982 - 1116 - 1242 - 1382 - 1486 - 1502 - 1552 - 1574 - 1577 - 1593 - 1669 - 1686 - 1722 - 1775 - 1789 - 1800 - 1802 - 1839 -1853 - 1862 - 1933 - 1939 - 1973 - 1981 - 1990 - 2025 - 2037 - 2045 - 2067 - 2071 - 2086 - 2114 - 2164 - 2186 - 2218 - 2231 - 2259 - 2389 - 2390 - 2421 - 2510
HEART DISORDER................ 196 - 261 - 352 - 353 - 436 - 501 - 610 - 612 - 703 - 728 - 783 - 810 - 811 - 812 - 832 - 833 - 853 - 862 - 911 - 952 - 953 - 979 - 982 - 993 - 1018 - 1155 - 1254 - 1256 - 1261 - 1378 - 1401 - 1404 - 1437 - 1438 - 1439 - 1440 - 1452 - 1498 - 1501 - 1502 - 1503 - 1520 - 1545 - 1549 - 1550 - 1552 - 1573 - S-C. June, 1935, Back Cover - 1576 - 1578 - 1592 - 1593 - 1670 - 1685 - 1688 - 1730 - 1739 - 1776 - 1799 - 1822 - 1824 - 1840 - 1850 - 1864 - 1869 - 1873 - 1878 - 1882 - 1889 - 1894 - 1932 - 1933 - 1940 - 1960 - 1972 - 1973 - 1976 - 1982 - 1984 - 2004 - 2009 - 2020 - 2024 - 2025 - 2054 - 2064 - 2081 - 2082 - 2085 - 2086 - 2094 - 2095 - 2099 - 2113 - 2147 - 2167 - 2170 - 2187 - 2252 - 2258 - 2270 - 2286 - 2291 - 2338 - 2389 - 2392 - 2395 - 2417

2429 - 2463 - 2474 - 2476 - 2519
HEART SKIP........ 1258 - 1259 - 1878 - 2113 - 2211
HEAT PROSTRATION........ 1670
HEMATURIA........ 435 - 1976
HEMORRHAGE........ 181 - 232 - 261 - 302 - 351 - 705 - 738 - 770 - 789 - 810 - 832 - 851 - 1255 - 1576 - 1593 - 1850 - 1853 - 1870 - 1880 - 1903 - 1933 - 2031 - 2099 - 2114 - 2115 - 2174 - 2187 - 2217 - 2258 - 2280 - 2285 - 2359 - 2391 - 2436 - 2510 - 2514 - 2519
HEPATIC HYPERTROPHY........ 232
HEPATIC SCLEROSIS........ 726
HERNIA........ 301 - 302 - 306 - 384 - 500 - 1242 - 1261 - 1379 - 1380 - 1498 - 1604 - 2101 - 2163 - 2164 - 2174 - 2182 - 2248 - 2305 - 2484 - 2485- 2521
HERPES ZOSTER........ 703 - 726 - 933 - 1486
HICCOUGH........ 15 - 382 - 723 - 784 - 811 - 833 - 1959 - 2252 - 2394
HIGH BLOOD PRESSURE........ 260 - 346 - 436 - 585 - 633 - 634 - 738 - 849 - 851 - 1380 - 1381 - 1452 - 1474 - 1573 - 1593 - 1822 - 1932 - 1933 - 1973 - 2011 - 2037 - 2147 - 2187 - 2218 - 2259 - 2476
HIVES........ 909 - 2124 - 2280
HOARSENESS........ 8 - 14 - 2114 - 2148 - 2421
HODGKIN'S DISEASE........ 738 - 1932
HORSE DISORDER........ 148 - 1721
HOUSEMAID'S KNEE........ 1810 - 1811
HYDROCEPHALUS........ 470 - 2531 - 2532
HYPERURIA........ 783
HYPOMENORRHOEA........ 723 - 1574
HYSTERIA........ 299 - 1972 - 2004 - 2085
IMPETIGO........ 2307 - 2357 - 2389
INDIGESTION........ 138 - 230 - 231 - 382 - 500 - 725 - 769 - 770 - 786 - 2148 - 2414
INFECTION........ 279 - 280 - 330 - 629 - 704 - 783 - 810 - 833 - 834 - 909 - 912 - 1255 - 1670 - 1710 - 1776 - 1811 - 1825 - 1849 - 1969 - 1981 - 2049 - 2113 - 2120 - 2331 - 2389 - 2461 - 2468 - 2509
INFLAMMATION........ 207 - 230 - 330 - 558 - 703 - 834 - 847 - 849 - 1043 - 1387 - 1406 - 1475 - S-C, June, 1935, Back Cover - 1580 - 1590 - 1714 - 1879 - 1902 - 1933 - 2005 - 2010 - 2107 - 2123 - 2162 - 2164 - 2212 - 2218 - 2331 - 2360 - 2366 - 2465
INFLUENZA........ 330 - 331 - 607 - 1853 - 1862 - 1868 - 1893 - 1894 - 2063 - 2067 - 2074 - 2084 - 2089 - 2113 - 2147 - 2174 - 2286 - 2332 - 2338 - 2392
INGROWING NAIL........ 740 - 2162
INJURY........ 607 - 634 - 635 - 705 - 723 - 769 - 783 - 790 - 810 - 811 - 847 - 848 - 849 - 851 - 862 - 910 - 911 - 1073 - 1241 - 1258 - 1378 - 1441 - 1501 - 1532 - 1545 - 1590 - 1723 - 1775 - 1811 - 1824 - 1825 - 1849 - 1853 - 1861 - 1880 - 1933 - 2005 - 2018 - 2107 - 2108 - 2120 - 2147 - 2151 - 2187 - 2210 - 2211 - 2253 - 2257 - 2258 - 2259 - 2260 - 2290 - 2310 - 2319 - 2332 - 2399 - 2435 - 2461 - 2477 - 2488 - 2509 - 2510 - 2519 - 2520
INSANITY........ 260 - 348 - 1018 - 1256 - 1475 - 1707 - 1869
INSOMNIA........ 307 - 435 - 606 - 608 - 1581 - 1686 - 2011 - 2038 - 2146 - 2173 - 2189 - 2230 - 2257 - 2309
INTESTINAL DISORDER........ 1676 - 2037 - 2038 - 2212 - 2252 - 2258 - 2260 - 2335 - 2354 - 2375 - 2441 - 2460
IRRITABILITY........ 1231 - 1674 - 2031
ITCH........ 285 - 384 - 461 - 631 - 740 - 1573 - 1776 - 2038 - 2063 - 2081 - 2084 - 2087 - 2099 - 2118 - 2120 - 2146 - 2231 - 2259 - 2363 - 2490 - 2514
IVY POISON........ 261 - 331 - 500 - 659 - 2084 - 2469
JAUNDICE........ 91 - 2009 - 2010
KIDNEY DISORDER........ 584 - 769 - 771 - 1378 - 1439 - 1453 - 1498 - 1499

S-C. June, 1935, Back Cover - 1578 - 1592 - 1902 - 1903 - 1935 - 2020 - 2032 - 2048 - 2127 - 2162 - 2360 - 2408 - 2460
KIDNEY STONE........ 775 - 1271 - 1403 - 1726 - 1737 - 1799 - 1903 - 1933 - 2019 - 2030 - 2120 - 2173 - 2362 - 2520
LABOR........ 153 - 1387 - 1402 - 2310
LARYNGITIS........ 232 - 848 - 1406 - 2030 - 2500
LEAD POISON........ 415 - 2258
LEUCORRHEA........ 2286
LEUKODERMA........ 2170
LISTLESSNESS........ 1243 - 1668
LIVER DISORDER........ 1498 - 1549 - 1581 - 1685 - 1686 - 1823 - 1974 - 2113 - 2146 - 2218 - 2231 - 2393 - 2485
LOCOMOTOR ATAXIA........ 353
LOW BLOOD PRESSURE........ 262 - 476 - 498 - 499 - 833 - 1379 - 1486 - 1608 - 2167 - 2169 - 2231 - 2258 - 2455 - 2521
LUMBAGO........ 607 - 834 - 1399 - 1524 -1576 - 1686 - 1824 - 1903 - 2030 - 2148 - 2319 - 2399
LUNG DISORDER........ 1709 - 2037 - 2066 - 2488 - 2518
MALFORMATION........ 2349
MALIGNANCE........ 467
MARASMUS........ 333 - 334
MASTOIDITIS........ 262 - 862 - 1255 - 1257 - 1503 - 1688 - 1811 - 2034 - 2275 - 2290 - 2309 - 2407
MASTURBATION........ 584
MEASLES........ 47 - 607 - 847 - 1242 - 1321 - 1322 - 1382 - 1476 - 1490 - 1552 - 1574 - June, 1935, Back Cover - 1576 - 1577 - 1973 - 2030 - 2282 - 2297 - 2354 - 2362 - 2376 - 2392 - 2399 - 2434
MELANCHOLIA........ 728 - 1770 - 1870 - 2146 - 2147 - 2289
MENINGITIS........ 148 - 501 - 1715
MENOPAUSE........ 178 - 659
MENORRHAGEA........ 231 - 232 - 302 - 304 - 305 - 1439 - 1722 - 2255
METRITIS........ 501
MIGRAINE........ 625
MILK LEG........ 2162 - 2193
MILKLESSNESS........ 2186
MITRAL INCOMPETENCY........ 612
MITRAL REGURGITATION........ 231 - 703 - 1741
MITRAL STENOSIS........ 1741
MOLE........ 1257
MOTOR DISORDER........ 1117
MUCUS........ 1670 - 1715
MULTIPLE SCLEROSIS........ 1403 - 1714 - 1715 - 1765 - 2038 - 2258
MUMPS........ 659 - 1811 - 1839 - 2121 - 2309 - 2353
MYOCLONUS........ 627
MYOPIA........ 723
NAUSEA........ 659 - 811
NECROSIS........ 200
NEPHRITIS........ 232 - 260 - 811 - 1475 - 1524 - 1891 - 1976
NERVOUSNESS........ 45 - 91 - 232 - 334 - 348 - 384 - 608 - 628 - 725 - 980 - 1379 - 1438 - 1452 - 1524 - 1574 - 1670 - 1686 - 1707 - 1709 - 1710 - 1799 - 1824 - 1854 - 2031 - 2032 - 2090 - 2147 - 2170 - 2171 - 2173 - 2230 - 2258 - 2286 - 2309 - 2337 - 2394 - 2434 - 2455
NEURALGIA........ 233 - 234 - 383 - 447 - 741 - 1737 - 1808 - 1809 - 1890 - 1891 - 1925 - 2085 - 2148 - 2352
NEURASTHENIA........ 232 - 739
NEURITIS........ 204 - 230 - 280 - 306 - 330 - 349 - 435 - 462 - 476 - 501 - 612 - 625 - 783 - 790 - 832 - 850 - 1117 - 1257 - 1501 - 1579 - 1581 - 1777 - 1925 - 1972 - 2018 - 2066 - 2067 - 2229 - 2395
NIGHT SWEAT........ 333
NUMBNESS........ 1687 - 1773 - 1775 - 2055 - 2365 - 2406
OBESITY........ 232 - 832 - 853 - 1380
OEDEMA........ 705 - 706

OPENING 611

ORCHITIS .... 1073 - 2253

OTITIS MEDIA .... 330 - 558 - 1808 - 1809

OVARITIS .... 231 - 352 - 353 - 583 - 771

OZENA 466

PAIN .... 138 - 233 - 280 - 281 - 282 - 285 - 306 - 307 - 308 - 352 - 382 - 415 - 435 - 436 - 467 - 476 - 606 - 608 - 610 - 625 - 626 - 627 - 629 - 632 - 635 - 703 - 706 - 741 - 783 - 785 - 788 - 789 - 790 - 811 - 833 - 834 - 835 - 847 - 849 - 850 - 869 - 870 - 907 - 911 - 912 - 979 - 980 - 981 - 982 - 983 - 1043 - 1073 - 1115 - 1116 - 1117 - 1155 - 1241 - 1243 - 1244 - 1257 - 1261 - 1272 - 1328 - 1378 - 1379 - S-C. December, 1934, Back Cover - 1437 - 1439 - 1440 - 1441 - 1452 - 1475 - 1476 - 1498 - 1499 - 1524 - 1525 - 1546 - 1549 - 1578 - 1580 - 1590 - 1592 - 1593 - 1595 - 1596 - 1598 - 1601 - 1602 - 1604 - 1621 - 1622 - 1669 - 1685 - 1686 - 1687 - 1688 - 1723 - 1726 - 1738 - 1765 - 1778 - 1790 - 1801 - 1810 - 1811 - 1820 - 1824 - 1825 - 1839 - 1840 - 1852 - 1853 - 1862 - 1873 - 1879 - 1880 - 1882 - 1883 - 1890 - 1894 - 1902 - 1903 - 1925 - 1932 - 1933 - 1944 - 1969 - 1973 - 1982 - 1885 - 1990 - 2010 - 2017 - 2018 - 2029 - 2030 - 2031 - 2032 - 2037 - 2038 - 2045 - 2046 - 2048 - 2049 - 2071 - 2073 - 2080 - 2089 - 2090 - 2093 - 2099 - 2119 - 2124 - 2146 - 2147 - 2148 - 2151 - 2152 - 2162 - 2163 - 2170 - 2171 - 2173 - 2174 - 2200 - 2212 - 2259 - 2270 - 2275 - 2283 - 2290 - 2309 - 2319 - 2332 - 2363 - 2365 - 2392 - 2398 - 2406 - 2414 - 2427 - 2460 - 2469 - 2488 - 2501 - 2509 - 2514 - 2520 - 2521

PALPITATION .... 91 - 285 - 853 - 1709 - 1972 - 1973 - 1981 - 2037 - 2163 - 2231 - 2291

PAPILLOMA .... 204 - 261

PARALYSIS .... 15 - 138 - 139 - 334 - 348 - 349 - 384 - 462 - 466 - 467 - 495 - 496 - 497 - 585 - 607 - 608 - 634 - 635 - 702 - 728 - 729 - 739 - 742 - 768 - 771 - 786 - 787 - 789 - 810 - 811 - 831 - 834 - 835 - 851 - 869 - 909 - 914 - 981 - 1043 - 1244 - 1254 - 1255 - 1256 - 1259 - 1381 - 1398 - 1399 - 1408 - 1438 - 1486 - 1499 - 1521 - 1524 - 1550 - 1551 - 1578 - 1586 - 1590 - 1593 - 1685 - 1688 - 1738 - 1790 - 1822 - 1824 - 1854 - 1863 - 1889 - 1925 - 1931 - 1932 - 1943 - 1969 - 1972 - 1982 - 2004 - 2017 - 2020 - 2029 - 2030 - 2031 - 2034 - 2048 - 2055 - 2070 - 2090 - 2094 - 2187 - 2257 - 2259 - 2286 - 2300 - 2310 - 2337 - 2389 - 2390 - 2399 - 2450 - 2461 - 2474 - 2484 - 2528

PARALYSIS AGITANS .... 710 - 740 - 2030 - 2474

PARENCEPHALITIS 282

PHLEBITIS .... 327 - 467 - 558 - 703 - 2259

PHLEGM .... 2417 - 2450

PHLEGMON 2252

PILES .... 68 - 384 - 436 - 500 - 740 - 742 - 833 - 1242 - 1440 - 1452 - 1453 - 1475 - 1604 - 1726 - 1737 - 1870 - 1969 - 1975 - 2066 - 2099 - 2100 - 2147 - 2148 - 2189 - 2270 - 2286 - 2305 - 2450 - 2490 - 2514

PIMPLES .... 631 - 632 - 979 - 1774 - 1788 - 1823 - 1981 - 2115 - 2218 - 2270

PLACENTA RETENTION .... 2286 - 2287

PLEURISY .... 229 - 284 - 334 - 583 - 770 - 785 - 1973 - 2066 - 2119 - 2170 - 2173 - 2283

PNEUMONIA .... 138 - 228 - 229 - 351 - 462 - 611 - 659 - 705 - 785 - 847 - 911 - 1503 - 1521 - 1525 - 1551 - 1574 - 1676 - 1839 - 1865 - 1868 - 1974 - 1975 - 2003 - 2032 - 2084 - 2108 - 2126 - 2170 - 2231 - 2362 - 2460

PNEUMOTHORAX .... 2518 - 2525

POISON OAK 93

POLIOMYELITIS .... 467 - 607 - 702 - 738 - 740 - 810

PREGNANCY 2406

PROSTATITIS .... 501 - 728 - 783 - 834 - 907 - 1852 - 2099 - 2290

PSORIASIS .... 446 - 494 - 739 - 1474 - 1739 - 1777

PTOMAINE POISON .... 834 - 1580 - 1823

PUS .... 629 - 741 - 811 - 847 - 981 - 1551 - 1574 - 1604 - 1670 - 1677 - 1686 - 1726 - 1811 - 1879 - 1975 - 1981 - 2024 - 2034 - 2049 - 2050 - 2089 - 2170 - 2259 - 2360 - 2399

PYORRHOEA .... 16 - 148 - 911 - 1254 - 1322 - 2073 - 2273 - 2332

QUINSY .... 2073 - 2349

RABBIT DISORDER 139

RADIUM BURNS .... 1669 - 2120 - 2427

RECTUM DISORDER .... 1714 - 1879 - 2285 - 2286 - 2490

RESPIRATION OBSTRUCTION .... 2301 - 2468

RHEUMATISM .... 111 - 233 - 308 - 351 - 352 - 584 - 611 - 706 - 741 - 771 - 832 - 910 - 980 - 1254 - 1399 - 1440 - 1441 - 1475 - 1484 - 1551 - 1578 - 1579 - 1580 - 1601 - 1687 - 1852 - 1869 - 1891 - 1973 - 2018 - 2037 - 2120 - 2258 - 2414 - 2464 - 2486 - 2487 - 2519

RHINITIS 461

RICKETS .... 206 - 349 - 1890

RINGWORM .... 739 - 1688 - 1854 - 1981 - 2005 - 2047 - 2085 - 2163 - 2428

ROSE FEVER 1257

RUPTURE .... Refer in Hernia

SARCOMA .... 326 - 327 - 436 - 448 - 769

SCABIES .... 1381 - 2033

SCAR 1670

SCARLET FEVER .... 1288 - 1409 - 1441 - 1452 - 1474 - 1524 - 1525 - 1550 - 1573 - S-C. June, 1935, Back Cover - 1580 - 1786 - 1799 - 1854 - 2033 - 2338 - 2362 - 2392 - 2464

SCIATICA .... 349 - 476 - 501 - 850 - 1243 - 1379 - 1399 - 1453 - 1502 - 1790 - 1903 - 1934 - 1971 - 1972 - 2231

SCROFULA 659

SENSORY DISORDER 1155

SEPSIS .... 14 - 42 - 43 - 44 - 148 - 467

SEXUAL WEAKNESS .... 746 - 770

SHAKING PALSY .... 710 - 2030 - 2474

SHELL SHOCK 232

SINUSITIS .... 848 - 1111 - 1117 - 1242 - 1403 - 1440 - 1441 - 1501 - 1525 - 1579 - 1669 - 1670 - 1685 - 1686 - 1708 - 1710 - 1798 - 1799 - 1824 - 1970 - 1976 - 1994 - 2093 - 2107 - 2119 - 2135 - 2169 - 2252 - 2253 - 2258 - 2305 - 2337 - 2376

SKIN DISORDER .... 138 - 384 - 907 - 909 - 979 - 981 - 1381 - 1454 - 1503 - 1603 - 1674 - 1709 - 1739 - 1800 - 1801 - 1823 - 1865 - 1873 - 1973 - 1981 - 2033 - 2081 - 2094 - 2211 - 2231 - 2246 - 2252 - 2280 - 2289 - 2290 - 2307 - 2337 - 2349 - 2354 - 2376 - 2417 - 2469 - 2514

SLUGGISHNESS .... 1486 - 1501 - 1502 - 1766 - 1786 - 1787 - 1800 - 1879 - 1920 - 1939 - 1940 - 1992 - 2055 - 2218 - 2365

SNEEZING .... 1708 - 2034 - 2060 - 2061 - 2064 - 2305

SORENESS .... 138 - 308 - 346 - 384 - 415 - 498 - 741 - 849 - 980 - 1439 - 1453 - 1501 - 1502 - 1504 - 1532 - 1551 - 1604 - 1665 - 1788 - 1890 - 1902 - 1923 - 1931 - 1940 - 2012 - 2024 - 2050 - 2063 - 2068 - 2100 - 2108 - 2119 - 2170 - 2186 - 2193 - 2204 - 2210 - 2253 - 2280 - 2310 - 2331 - 2366 - 2421 - 2434 - 2449 - 2460 - 2491 - 2501

SPASM .... 1256 - 1257 - 2186

SPINAL CURVATURE .... 628 - 1254 - 1954

SPINAL DISORDER .... 111 - 112 - 1378 - 1715 - 2011

SPLEEN DISORDER .... 467 - 785 - 1581 - 1823 - 1932 - 2019 - 2030 - 2081

SPRAIN .... 498 - 633 - 703 - 869 - 1272 - 1453 - 1525 - 1573

1595 - 1730 - 1800 - 1933 - 1944 - 2009 - 2025 - 2038 - 2350 - 2477 - 2501 - 2509

STAMMERING........................ 2460 - 2461 - 2488

STIFFNESS........................ 632 - 1440 - 1674 - 1738 - 1777 - 1789 - 1925 - 1981 - 1993 - 2005 - 2163 - 2169 - 2173 - 2319 - 2399

STING........................ 1437 - 1576 - 1933

STOMACH DISORDER........................ 285 - 382 - 606 - 612 - 626 - 725 - 739 - 740 - 764 - 769 - 770 - 786 - 832 - 850 - 852 - 979 - 1117 - 1329 - 1379 - 1387 - 1439 - 1474 - 1475 - 1486 - 1502 - 1520 - 1545 - 1549 - 1550 - 1574 - S-C. June, 1935, Back Cover - 1577 - 1604 - 1686 - 1687 - 1738 - 1775 - 1776 - 1799 - 1853 - 1854 - 1903 - 1938 - 1940 - 1954 - 1968 - 1969 - 1982 - 2004 - 2011 - 2018 - 2045 - 2048 - 2051 - 2087 - 2186 - 2230 - 2257 - 2286 - 2291 - 2305 - 2337 - 2376 - 2427 - 2434 - 2455

STRABISMUS........................ 771

STREPTOCOCCI........................ 1258 - 2066

STYE........................ 204 - 461 - 659

SUN BURN........................ 1378 - 1380 - 1598 - 1602 - 1788 - 2030 - 2152 - 2173 - 2174 - 2194 - 2270 - 2414 - 2421 - 2427 - 2469

SUN STROKE........................ 1982 - 2171

SURGICAL SEQUENCE........................ 1503 - 1577

SWELLING........................ 138 - 204 - 307 - 346 - 436 - 498 - 499 - 558 - 611 - 626 - 633 - 634 - 657 - 658 - 659 - 703 - 706 - 725 - 738 - 740 - 849 - 862 - 981 - 1043 - 1253 - 1255 - 1282 - 1378 - 1379 - 1380 - 1381 - 1386 - S-C. December, 1934, Back Cover - 1437 - 1441 - 1453 - 1499 - 1524 - 1532 - 1576 - 1581 - 1590 - 1596 - 1604 - 1669 - 1710 - 1711 - 1723 - 1724 - 1739 - 1775 - 1777 - 1787 - 1801 - 1820 - 1821 - 1825 - 1840 - 1848 - 1880 - 1933 - 1939 - 1960 - 1973 - 1975 - 1976 - 2011 - 2012 - 2025 - 2030 - 2031 - 2032 - 2049 - 2054 - 2066 - 2067 - 2068 - 2082 - 2085 - 2086 - 2087 - 2090 - 2092 - 2093 - 2108 - 2120 - 2121 - 2133 - 2146 - 2147 - 2150 - 2187 - 2193 - 2270 - 2276 - 2280 - 2310 - 2318 - 2337 - 2389 - 2399 - 2413 - 2414 - 2485 - 2520 - 2522

SYPHILIS........................ 35 - 206 - 230 - 302 - 303 - 725 - 1903 - 2123 - 2189 - 2319

TACHYCARDIA........................ 235 - 415 - 626 - 1441 - 1730 - 1824 - 1849 - 1920 - 1960 - 1982 - 2010 - 2084 - 2100 - 2291 - 2417 - 2521

TEETHING........................ 1381

TETANUS........................ 1849

THROAT DISORDER........................ 307 - 461 - 607 - 850 - 1241 - 1378 - 1409 - 1439 - 1453 - 1498 - 1502 - 1524 - 1665 - 1670 - 1777 - 1799 - 1861 - 1880 - 1925 - 1985 - 2011 - 2049 - 2073 - 2086 - 2101 - 2114 - 2164 - 2253 - 2319 - 2389 - 2400 - 2414 - 2421 - 2461

THROMBOSIS........................ 1255 - 2270

TIC DOULOUREUX........................ 447 - 501 - 608 - 612 - 626 - 627 - 659 - 701 - 702 - 738 - 740 - 1439

TOBACCO HABIT........................ 29 - 299 - 435 - 1241 - 2310

TONSILITIS........................ 138 - 285 - 332 - 351 - 352 - 383 - 461 - 629 - 726 - 741 - 811 - 849 - 910 - 981 - 1271 - 1328 - 1499 - 1524 - 1525 - 1545 - June, 1935, Back Cover - 1596 - 1670 - 1840 - 1870 - 1973 - 1976 - 2082 - 2119 - 2147 - 2253 - 2254 - 2286 - 2309 - 2331 - 2332 - 2434 - 2440 - 2460 - 2468 - 2488

TOOTHACHE........................ 232 - 284 - 285 - 307 - 331 - 738 - 783 - 811 - 909 - 1439 - 1503 - 1549 - 1596 - 1685 - 1798 - 1848 - 1849 - 2030 - 2037 - 2332 - 2334 - 2335

TORTICOLLIS........................ 476 - 2187

TUBERCULOSIS........................ 181 - 232 - 235 - 349 - 350 - 353 - 415 - 435 - 436 - 611 - 612 - 626 - 728 - 739 - 772 - 785 - 810 - 831 - 833 - 847 - 869 - 910 - 911 - 980 - 1254 - 1288 - 1290 - 1291 - 1400 - 1403 - 1578 - 1607 - 1608 - 1687 - 1707 - 1786 - 1854 - 1890 - 1891 - 1919

1939 - 1993 - 1994 - 2038 - 2094 - 2171 - 2174 - 2217 - 2253 - 2281 - 2350

TUMOR........................ 138 - 139 - 232 - 448 - 467 - 494 - 495 - 498 - 584 - 726 - 740 - 770 - 786 - 835 - 849 - 851 - 862 - 909 - 1242 - 1243 - 1254 - 1328 - 1401 - 1438 - 1452 - 1474 - 1501 - 1579 - 1722 - 1723 - 1724 - 1730 - 1738 - 1904 - 1925 - 1974 - 2021 - 2066 - 2067 - 2092 - 2108 - 2115 - 2121 - 2162 - 2169 - 2187 - 2280 - 2318 - 2421 - 2427 - 2463

ULCER........................ 299 - 302 - 303 - 327 - 328 - 330 - 382 - 435 - 467 - 498 - 723 - 739 - 740 - 770 - 783 - 832 - 850 - 852 - 870 - 1440 - 1453 - 1593 - 1776 - 1968 - 1969 - 1974 - 2012 - 2090 - 2286 - 2332 - 2468 - 2510 - 2527

URINE RETENTION........................ 498

URINE SUPPRESSION........................ 207 - 445

UTERINE INERTIA........................ 2310

UTERINE PROLAPSE........................ 982 - 1889

VALVULAR REGURGITATION........ 2421

VARICOSE VEINS........................ 467 - 558 - 559 - 606 - 740 - 1437 - 1486 - 1670 - 1787 - 2018 - 2068 - 2450

VERTIGO........................ 384

VOMITING........................ 112 - 850 - 979 - 1549 - 1577 - 1903 - 1973 - 1976 - 2080 - 2082 - 2146 - 2394 - 2406

WHOOPING COUGH........................ 284 - 305 - 331 - 789 - 834 - 981 - 1242 - 1271 - 1574 - 1714 - 1774 - 1790 - 1800 - 1973 - 1985 - 2082 - 2133 - 2252 - 2417 - 2533

WORMS........................ 738 - 1483 - 1484 - 2031

WOUND........................ 348 - 705 - 981 - 1043 - 2113 - 2162 - 2163 - 2230 - 2359 - 2389

X-RAY BURNS........................ 200 - 501 - 625 - 739 - 742 - 869

FIRST ST. LOUIS MISSOURI PLANET

(HOTEL MELBOURNE, ST. LOUIS, MISSOURI)

Meeting Of Thursday, August 3, 1939

Successful Use Of Spectro-Chrome In:—Horse Eye Cancer, Constipation, Diabetes, High Blood Pressure, Paralysis, Child Disorder, Leg Sores, Nervousness, Headache, Insomnia.

Reported By:-CLEMENS M. P. NEUMANN, *Deputy*

Deputy Clemens M. P. Neumann was in the chair.

Charles S. spoke about his own case. His areas 8, 9, and 10 are now in better shape then ever before. He used Red on Area 7 and Violet on Area 6 with good results. His Area 16 causes him some trouble. It is a stiffness with some Pain and he cannot turn his Head freely. He used Orange and is now under Free Guidance. Then he spoke about his wife, who has an attack of spots in front of her Eyes, which is very annoying. She does not feel sick in Area 8 or have any Dizziness or Imbackelim. Her Ratio is 1 to 4 (20-80). He also mentioned that her hearing is not normal and thinks that Sinus Disorder may be the trouble. We discussed the above case and asked him to request Free Guidance from Spectro-Chrome Institute.

Konrad S. asked, why when requesting Free Guidance, the case History must always be fully repeated, also what the restriction was for Free Guidance. The Deputy explained that each General Advice Chart must present a complete case in itself, as a possibility might arise for action at Law and Spectro-Chrome Institute would then be able to plead each Chart as a case in itself. It is therefore necessary to give a short History in each Chart and all questions should be answered in type and signed personally in ink. As to the second question it was made clear that Free Guidance could not go on indefinitely and the restriction was made in accordance with a ruling made at our last Convention in Detroit, Michigan.

Then a question was asked in regard to Collaborator blanks and it was explained that we extend Aid and Relief only to Collaborators.

Otto K. reported that his wife was taking Tonations on Area 17. She had Pimples all over her Body, also seems to have Ringworm on Leg. The Skin is dry and scaly. He was referred to Spectro-Chrome Home Guide under the caption "Skin Disorders" and advised to follow directions.

A question arose in regard to public swimming pools and Skin Disorders. As the water in these pools is not changed very often and the public is admitted without a thorough cleansing, the Skin comes in contact with impurities. The Deputy explained the system of taking a "Bucket Bath" as Colonel Dinshah taught us, by starting with the Head and finishing at the lower extremities. No need for swimming pools and contamination of water used for bathing purposes.

Marie G. reported her own case. She gained 12 pounds, feels well. She asked, "What is good for a heavy feeling in Area 8?" Orange was suggested, for Areas 8, 9 and 10.

In regard to Worms, which will yield to Yellow, it was also stressed that Magenta should be used on 4 and 18.

An article appeared in our daily newspaper, written by a Medical Doctor, warning that pork should be cooked very well, as it is estimated that the parasite Trichina has infested 17% of the people. A lively discussion followed and it was said that pork is not inspected for Trichina in certain slaughter houses.

Charles S. asked if Infra-Green Atcowas are used at night with the window open and moonlight floods the room, has it a counteracting effect on the Infra-Green Atcowas? Also, if the window be open and moonlight is not flooding the room? The Deputy explained that moonlight should not touch the Body or Area irradiated, otherwise moonlight is not harmful; however, this question was referred to Spectro-Chrome Institute for answer.

A suggestion was made to use air-cooled beds for Tonations during the hot season. Others said that they went to the basement for Tonations.

Konrad S. asked Marie G. what she ate to gain 12 pounds. She replied, "Rational Food Of Man". She incidentally mentioned that she had her teeth removed.

Otto K.'s little girl suffers from Adenoid Disorder. Turquoise and later Indigo were suggested, also that Ratio be watched.

Joseph H. wanted to have a correction: He did not use Orange on his Eye, but used Green, the best Atcowa for his Eye.

The Deputy received a report from an Affiliate who used Spectro-Chrome on the Eye of a horse. Doctors diagnosed it as Cancer. So far, results are: Upper Eyelid is coming along O. K., Lower Lid is showing improvement. Uses Lemon Systemic, Indigo Local. User is very well pleased with present results.

The Deputy reported a case of a woman, 65, High Blood Pressure, diabetic and paralyzed on Left Side. Ratio 25 to 116, could not move Arms, Fingers or Leg, Constipated as long as she can remember. Her son bought an Equipment, used about 2½ months. Results: Imbackelim corrected, moves regularly, almost unbelievable, as children say. She can now move her Fingers and Toes and sometimes Leg, although with great effort. The Medical Doctor gave her 3 to 7 days to live. Now she is surprised to find Blood Pressure almost normal and Sugar declining. Used Green Systemic, Magenta on 4 and 18; later Purple on 4, Lemon Systemic; now Yellow Systemic, Magenta on Areas 4 and 18 alternating. Very much pleased.

Case of baby, 6 months old, very much underweight. After meal, vomited everything; very puny and weak. Ratio 1 to 4; about 26 to 106; Constipated, restless. Atcowas used: Green Systemic, Magenta on Areas 4 and 18; later Orange on Areas 8, 9 and 10, Turquoise Systemic; now, Orange Systemic and Magenta on Areas 4 and 18. Results: gained almost 6 pounds, tries to sit up and crawls. The mother is overjoyed; bought an Equipment.

Case of woman, grandmother of above mentioned child. Right and Left Leg had open Sores, Right Leg very much swollen and turning fearfully red, from Ankle to Knee. Medical Doctor wanted to transfer her to hospital for a possible amputation. High Fever, Ratio 1 to 3½, no Backelim and constant Pain in Right Knee, could not get up from sitting position but had to turn and support with hands to rise. High Blood Pressure and some Sugar. Atcowa used: Turquoise Systemic and later Blue, Magenta on Areas 4 and 18. Blue and Turquoise Local on Leg, alternating Systemic. Results: largest Open Sore on Leg completely healed, one small Sore still open, Leg swelling now normal; Soreness and Redness left, can get up from sitting position as normal person can, Backelim and Frontelim normal. Ratio 20 to 80. Feels fine, but at times weak. However, is up and around. Very much pleased. Used daughter's Equipment.

Case of man, sore-like spots on Legs. He thought they might turn into Cancer. Imbackelim, Ratio 1 to 4, very nervous. Used Atcowa Green Systemic, Magenta on Areas 4 and 18. Bought own Equipment, now uses Lemon Systemic, Magenta on Area 4 and 18. Results: feels fine, does not smoke any more and is not half as nervous. Good Backelim and Sores on Leg are fading away. Very much pleased. His wife is highly nervous and had trouble in Areas 8, 9, 10 and 11. Feels much better and improved.

Case of woman, trouble in Areas 8, 9 and 10; very nervous, no Backelim, Headache and restless. Frequented many "Specialists" with no results. Atcowas used: Green Systemic, later Turquoise Systemic, Magenta on Areas 4 and 18, Yellow on Areas 8, 9 and 10. Results: sleeps better, not so restless, Backelim normal. Bought an Equipment and is well pleased.

Case of woman, highly nervous, no Backelim, Imbalance on Areas 8, 9 and 10, Sleepless. Atcowas Used: Green Systemic and Turquoise Systemic, Magenta on Areas 4 and 18, Yellow on Areas 8, 9 and 10. Results: Felt much better. She bought an Equipment. Now feels well enough to go on vacation. Husband very much pleased. Had doctored many months.

EDITORIAL REMARKS

Charles S. is now better than he was and will have to make the best of the remainder by correct living.

Full History must be given each time. The General Advice Chart is not always read by the same Officer and it is impracticable to refer to prior Charts constantly; furthermore, we have to watch for legal traps from opponents. No institution in the World gives such Free Guidance and unless we received strict co-operation, such service could never have been established.

Free Guidance is given solely to the Original Registered Purchaser, for the use of the Spectro-Chrome for members of his immediate family. A "family" means Parents and their children, but we have been very lenient and extended services to others of the "family", living under the same roof. Anyone from the outside is not included and Provisional Members should not attempt public service lest they be someday tangled with the monstrous Medical Laws.

Public pools can NEVER be entirely clean, as we understand cleanness.

Inspection of the abominable "pork" is a joke on the American "sucker". Read Spectro-Chrome of February 1939, "Pork Eating Most Dangerous."

When tonating, there should be no Moonlight, Sunlight or any other "Light" shining on the Sufferer. Moonlight is "Polarized" Light.

No "Air-cooled beds" are necessary. Tonation needs no "cool" basement. The warmth of the room increasing perspiration, tends to recovery. Do not be squeamish; learn to bear the inconveniences of life to develop mental power necessary to resist subversion of health equilibrium.

Why were those teeth removed? Did she ask for our Free Guidance before taking the step?

Many animal owners now use Spectro-Chrome for their pets. It works as well!

Pay no attention to the humbug of "Blood Pressure"; Ratio takes care of all.

"Specialist" is another racket.

---o---

FIRST SURAT INDIA PLANET

(Colonel Dinshah P. Ghadiali's Spectro-Chrome Office, Kanpith, Surat, India)

Meeting Of Thursday, August 3, 1939

Successful Use Of Spectro-Chrome In:—Appendicitis, Kidney Disorder, Amenorrhoea, Diabetes, Vomiting, Tongue Cancer.

Reported By:—INDUPRASAD D. MEHTA, *Deputy*

Induprasad D. Mehta, after signing the Affirmation of the Scientific Order of Spectro-Chrome Metrists, took the chair as Deputy.

He read the letter from Colonel Dinshah P. Ghadiali, informing of his arrival on December 8, 1939, to file suits against "Jame Jamshed" and "Kom Sevak".

The Deputy recommended Jayantilal Chhaganlal as Sub-Deputy and Navinchandra I. Shukla as Recorder, as both are ready to work voluntarily.

The Meeting was held exactly at 8 P. M. and a good number of Affiliates was present: eight out of eleven. It was quite an interesting Meeting, as they all met after a long time.

During the discussion, Jayantilal proposed that the translation of the Spectro-Chrome Home Guide in Gujarati is quite necessary and was supported by all. The Deputy is also of the opinion that it is quite necessary.

Then followed the case reports. Many of the Affiliates forgot to bring their case reports. The Deputy requested them to bring them at the next Meeting for the benefit of all.

The Deputy then read his case reports:

(a) A Parsee girl, aged 15, had some Disorder in the Stomach and was vomiting. Medical Doctors diagnosed it as Appendicitis and she was about to be operated. The Deputy advised the parents of the girl to wait till his Spectro-Chrome arrived. They did wait. When his Spectro-Chrome arrived he gave her Tonations for six days. Atcowa used was Lemon Systemic for four days (4 Tonations). She felt better. Vomiting was subsided by Indigo on Areas 8 and 9. She is quite happy enjoying sound Health.

(b) The Deputy's father, aged 59, had Kidney Disorder for many years. Last Christmas, he suffered much and was in bed for a month. This year he had a relapse, but, we had Spectro-Chrome now. The Fever was 103 Degrees. The Deputy used Turquoise Systemic and the Fever subsided. Then was used Magenta on Area 18; the "Kidney Disorder" disappeared like magic! He continued Magenta for six days (6 tonations). Now he is quite O. K.

(c) A girl, aged 25, had Menstruation Disorder. She did not menstruate for the past eight to nine months. The Deputy used Green Systemic twice and then Scarlet on Areas 10 and 11 four times. She menstruated and went away to Ahmedabad quite happy.

(d) A man, aged 43, had Diabetes for the past two years. Used Yellow Systemic for six days (6 Tonations) and he is quite well. Used also Magenta on Areas 4 and 18. He is thinking of buying a Spectro-Chrome Equipment.

(e) A woman, aged 55, could not take any food, except milk and even that could not be retained. Backelim was three to four times a day in liquid form. Used Orange on Areas 8, 9 and 10 for ten days. She began to take a little food, but went away abruptly.

(f) A man, aged 52, had a Cancer on his Tongue. Gave him Green Systemic for three days and then Green Local, as well as Turquoise Systemic, for twenty-five days. Then the Tumor on the Tongue burst open and Blood, black in color, came out. The treatment with

Indigo is being continued. He gained weight by two pounds: 95½ pounds to 97½ pounds.

EDITORIAL REMARKS

Congratulation to all the pioneers of Spectro-Chrome in India! You made a very good start in India and have reason to be proud of your hundred per cent. showing. Dinshah sold 11 Spectro-Chromes, during his last visit in your country and 8 of those 11 were in Surat. So your attendance of 8 was perfectly justified and the 3 others who did not appear in the Meeting, were not to be doubted in their enthusiasm, because they were in Bombay, 167 miles away.

The arrival of Dinshah in Bombay will be on Thursday, January 11, 1940. His program was changed, because of the American Neutrality Proclamation. He will meet you at Special Planet Meetings and gladly give you further instructions.

Please, continue your Meetings and Reports and always attach the full names and addresses of the Affiliates present. Above all, be loyal to the Cause—Service to Suffering Mankind, according to our Constitution and Rules.

THE GERMANS FOUNDED THE ENGLISH NATION!

Why America MUST STAY NEUTRAL

After I entered the Proprietary High School, in Bombay, India, on Thursday, 18th of August, 1881, I was taught English from the **Royal Readers,** printed in London, England for use in India as elsewhere. In the 4th book of that Series, at page 102, I found the history of "The English Settlement":

"1. Not long after the departure of the Romans, the Picts again invaded South Britain. The Britons, being unable to resist them, called in the aid of certain Germanic tribes who had been in the habit of visiting their coasts. The Germans gladly came; and they liked the country so much that they were loath to go away again, 449 A. D.

So they wrested from the Britons (or Welsh, as they called them), whom they had come to help, land on which they and their families might settle. These Germanic settlers were the founders of the English nation."

Think over this admission very carefully, Americans. The above ought to be sufficient reason for ABSOLUTE NEUTRALITY on our part, as both those belligerents are brothers under the skin!

COLONEL DINSHAH'S ONCOMING VOYAGE TO INDIA

American President's Neutrality Proclamation Impels Him On An Around-The-World Detour

Circuitous Route To Bombay Will Take 2¼ Months!

The Neutrality Proclamation "to keep America out of war", promulgated by Honorable President Franklin D. Roosevelt, affected all holders of Passports of the United States Of America. Colonel Dinshah P. Ghadiali, had already secured his reservations on the Anchor Lines Limited Steamships, but as the "Proclamation" forbade American Citizens from going to Europe on British vessels, Dinshah had to alter his plans to suit the exigency.

He will leave by the American President Lines Steamship **President Harrison,** Cabin 219; the following is his complete itinerary and he may be reached without code through the ship's Radio in an emergency. All correspondence, however, should as usual be addressed to Spectro-Chrome Institute, Malaga, New Jersey; Vice-President Irene Grace Dinshah will be in charge.

PORT	ARRIVE	LEAVE
New York, N. Y.	1939	6 P. M. Fri. Nov 3
Havana, Cuba	A. M. Tues. Nov. 7	12 P. M. Tue. Nov. 7
Cristobal, Canal Zone	5 A. M. Sat. Nov. 11	Via Panama Canal
Balboa, Canal Zone	Via Panama Canal	7 P. M. Sat. Nov. 11
Los Angeles, California	9 A. M. Mon. Nov. 20	3 P. M. Tues. Nov. 21
San Francisco, California	7 P. M. Wed. Nov. 22	12 M. Sat. Nov. 25
Honolulu, Hawaii	7 A. M. Sat. Dec. 2	6 P. M. Sat. Dec. 2
Yokohama, Japan	12 M. Thur. Dec. 14	12 P. M. Thur. Dec. 14
Kobe, Japan	7 A. M. Sat. Dec. 16	2 A. M. Sun. Dec. 17
Hongkong, China	6 P. M. Thur. Dec. 21	12 M. Fri. Dec. 22
Manila, Philippines	12 M. Sun. Dec. 24	12 P. M. Mon. Dec. 25
		1940
Singapore, Straits Settlement	7 A. M. Sat. Dec. 30	12 M. Tues. Jan. 2
	1940	
Penang, Straits Settlement	Tide Wed. Jan. 3	Tide Thurs. Jan. 4
Colombo, Ceylon	7 A. M. Mon. Jan. 8	8 P. M. Mon. Jan. 8
Bombay, India	12 M. Thurs. Jan. 11	10 A. M. Sat. May 4

Soon after his arrival in Bombay, India, Colonel Dinshah will institute suits against the Editors and certain Writers of the **Jame Jamshed** (Daily) and the **Kom Sevak** (Weekly) for Criminal Defamation and Libel. He already has all the needed evidence in his possession to prove Conspiracy against him to ruin his life's works and reputation and it is beyond doubt he will have quick and complete vindication, because British justice in India moves very swiftly and surely.

SPECTRO-CHROME

Purple 1940 October

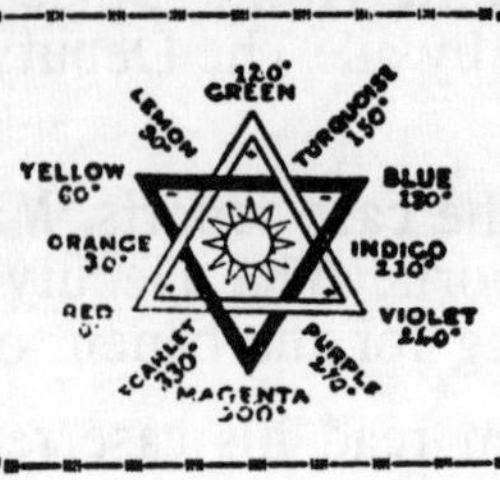

Volume 16 Number 10

Trade Marks Registered In The United States Patent Office

Read And Have Your Friends Read

Our Twentieth Annual Convention

America Returned To Slavery
We Are Already In War

Colonel Dinshah's Departure For India
Stopped By Government Action

Social Security And Unemployment Compensation Bubbles

Producing Thirteenth Wave Of Honesty!

Attuned Color Waves
Spectro-Chrome Metry

In course of life you reach a time,
When Doctor, Lawyer, Priest alike,
Does fail to give the sought relief,
To troubled mind or burdened heart—
Then come to me:
I help you to help yourself.
SPECTRO-CHROME

The Science Of
Automatic Precision

DEVOTED TO THE SERVICE OF MANKIND

GOLDEN WHIP

To The Public 2961
Numberscope Of Dinshah P. Ghadiali 2962
Jealousy For Visits To Wife 2963
First Detroit Michigan Planet 2964
Our Twentieth Annual Convention 2966
First Milwaukee Wisconsin Planet 2972
America Returned To Slavery 2975
Spectro-Chrome Smiles 2976
Colonel Dinshah's Departure For India Stopped By Government Action 2977
Favorscope For 1941 2978
Emphatic Request From Dinshah 2978
A Tribute Of Appreciation For Occultism And Higher Life 2979
Our Miseries In Free Guidance 2979
Social Security And Unemployment Compensation Bubbles 2980
Producing Thirteenth Wave Of Honesty! 2980

HEART LEAKAGE NORMALATED

By:—LESLIE A. D., *Windsor, Ontario, Canada*

I would like to comment on what Spectro-Chrome has done for Mrs. P. When I loaned her my Spectro-Chrome a year ago, she was suffering from a very bad leakage of the Heart. Under the Doctor's care, she was living on Digitalis and taking two Strychnine tablets a day—**stronger** than the law allowed. She immediately stopped the Strychnine tablets and a week later, left off the Digitalis and has touched neither since. This Fall, she started to gain weight and seldom gets a pain in Heart any more.

FIRST DETROIT MICHIGAN PLANET

(HOTEL TULLER, DETROIT, MICHIGAN)

Meeing Of Thursday, October 3, 1940

Successful Use Of Spectro-Chrome In:—Poison Ivy, Cold, Cough, Lack Of Energy, Spinal Injury, Eczema, Goitre, Stye, Burns.

Spectro-Chrome Restored Sight In Woman Totally Blind 27 Years! Now Reads Newspapers!!!

Reported By:—SOFIE HALL, *Recorder*

Deputy Lucille E. Greenham was in the chair.

A letter from the Central Office was read in which they stressed the need of a careful study of the Constitution And Rules. Also, Colonel Dinshah is leaving for India next month and it is his earnest desire that those who wish to continue their Affiliation send in their Applications promptly.

The Deputy expressed the appreciation of the First Detroit Michigan Planet to L. Roy Dodman and Gladys H. Clark for the splendid work they did. She then introduced the new Planet Officers, Lucille E. Greenham, Deputy, Ralph Clark, Sub-Deputy and Sofie Hall, Recorder.

She called upon the Sub-Deputy to outline briefly to the Planet our trip to Milwaukee. He gave a very detailed account of the trip and the Affiliates listened intently.

The Recorder was next introduced and she spoke briefly of the duties pertaining to her Office and gave definite reminders of how essential it would be for each Affiliate to present the Card, properly signed by the Affiliate and not to embrass the Recorder by seeking admission to the Planet without such Card. It is hoped that every Affiliate will co-operate fully on this request.

The Deputy then gave a short talk on the educational program which the Planet Commission decided upon in order that our meetings may become not only interesting but educational. She stated that many Affiliates were lacking in Spectro-Chrome knowledge and it is the earnest desire of Spectro-Chrome Institute, as well as the Detroit Planet to see that such a plan is carried forward.

The Deputy asked for a show of Hands as to the number of Spectro-Chrome Metry Encyclopaedias among our Affiliates. Of those present, 27 Affiliates stated they had Encyclopaedias. She urged all the others to get it as soon as possible for self or some member of their family, so they may participate in our monthly Meetings. She also stated that Affiliates who sold Equipments should be sure to place them in the names of those who are best to understand the Science of Spectro-Chrome and also attend our monthly Meetings —**this being a very important point.**

The Case Reports were as follows:

Ellen H.'s 6-year-old son had a very severe case of Poison Ivy. Gave 6 Tonations of Lemon, then Turquoise and completed the Normalation with Indigo. In 10 days complete balance was restored. No Kardoscil and Spiroscil were given.

Effie D. had a Cold in which Coughing seemed to be the most annoying part. She took Orange Tonations and stated the Cold and Coughing are completely gone. There was evidence of some congested condition in Areas 4 and 5 in which Orange played the important part. No Kardoscil and Spiroscil were given.

John B. reported his young daughter had a bad Cold. He stated the Kardoscil was 102 and Spiroscil 19. Turquoise was given twice daily for 3 days. Fever went down and cold was almost gone. Affiliate seemed very well satisfied with the work Turquoise accomplished.

The star case of the evening was that of Anna M. W.'s sister-in-law, 67 years of age, who also had a very bad Cold which had settled in her Chest. She was Blind for 27 years, as she was 40 years old when Blindness occurred, the result of an accident, in which she stated the Eye Nerves were paralyzed. She had given 60 Tonations in all for Normalation of this case. Such cases as this should hold an important position in our Spectro-Chrome magazine. In fact front page newspaper space should be given to this case, instead of war news and the like. For 27 years, this Sufferer was deprived of the most important of all the Sense Faculties, during which time she had to exist in complete darkness. Such lessons as these should be deeply appreciated by all our Affiliates in knowing that they have within their power the only safe and sane Healing System known to Man—Spectro Chrome Metry. While she has not complete vision, she is able to read newspapers and carry on her daily functions. Five Equipments were sold as a result of this one case.

Michael R. reported a woman, 46 years of age, very much run down. Kardoscil 86 and Spiroscil 16. Green and 2 Lemon Tonations were given and she is much better.

Mario T. hurt his Back; Vertebrae were out of position in three different places. He stated that he was too sick even to take his Kardoscil and Spiroscil. He was unable to work and Pain was so intense that he could not sleep. First Tonation of Turquoise he fell asleep while under the Atcowa and slept for three hours without waking. When he finally awoke, the Pain was gone and he went to work the next day.

Katherine W.'s son had a very bad Cold. Used Turquoise and Cold seemed to have become better. The child was then exposed again to a draft and the Cold reappeared. Used Orange and the Sufferer is now very much better. During these conditions of Cold, the child had difficulty in sleeping; also, Ringworm appeared on Hand of Sufferer. Lemon Tonations were given and condition is much better. Affiliate reported that she had a Growth on the Leg 15 years and it was very troublesome. She stated that Indigo removed it. Affiliate then asked about the Kardoscil and Spiroscil of a baby at birth and the Ratio in which they decreased.

Gladys C. was called upon to answer this question. She stated that in a newborn baby the Pulse Beat is between 120 to 140 beats per minute and of course the Spiroscil should follow the 4 to 1 or 5 to 1 Ratio.,

Gertrude J. had Eczema all her life. She appears to be between 25 and 30 years of age. She had used Green, Lemon and Turquoise with the best results for this condition. Medical Doctors were unable to do anything in this case and Spectro-Chrome proved her great blessing. Most of the Eczema disappeared, with just small portions of Skin around the Nose slightly affected. She praised Spectro-Chrome very highly.

Florence C. reported a bad case of Goitre. Kardoscil varied from 90 to 120 and Spiroscil 14. She now reported Kardoscil 87 and Spiroscil 17. She stated that she used Green and Indigo, which is evidence that the Affinity Waves are of great value.

She also reported her son had a Stye. Turquoise Systemic and Green Local were given. In the meantime, an Infection set in. Turquoise used without any great help. She was told to use Lemon and Magenta for Area 4.

Gladys C. reported her Hands were burned while canning. Lemon normalated this condition. She reported a case of Burns by Arc Welding. Ulcers appeared to develop as a result. Lemon normalated this condition also.

The Sub-Deputy was called upon to comply with the requests of Spectro-Chrome Institute and he read the Rules 1 and 2 from the Constitution and Rules of the Scientific Order Of Spectro-Chrome Metrists.

The Deputy then explained the best and most precise methods for speed in obtaining Free Guidance through the General Advice Charts and also explained the current ruling as to what constitutes the "immediate family".

The educational program was then given to the Affiliates. Two question were given: 1. What are the Primary Colors of Spectro-Chrome? 2. In what Glands do these Colors predominate? The Affiliates were urged to study their Spectro-Chrome Metry Encyclopaedia and report at the next regular Meeting with some competent Affiliate of the Planet assisting in the discussion of these two questions. A very earnest and intelligent audience was present. All seemed to appreciate the new Planet Commission.

EDITORIAL REMARKS

The new Planet Commission made a good beginning. Continue the educational work You will have our full co:operation.
Katherine W. will find in Spectro-Chrome Home Guide, page 62, the Ratio information she desires. Study of that 100 page book is all a family needs.

OUR TWENTIETH ANNUAL CONVENTION

Testimony To The Unrivalled Merits Of Spectro-Chrome

Uniformly Successful Normalation In Sciatic Rheumatism, Spinal Injury, Heart Disorder, Insomnia, Dizziness, Cystitis, Nephritis, Prostatitis, Catarrh, Glossitis, Chicken Pox, Wound, Heart Murmur, Paralysis, Pneumothorax.

Our Twentieth Annual Convention was held in the Milwaukee Auditorium, Milwaukee, Wisconsin, for three days, beginning Saturday, September 14, 1940 and two following days. The sessions had a satisfactory attendance and every event scheduled passed in perfect harmony. Each afternoon session opened at 2 P. M. and closed at 5 P. M., the evening sessions assembled at 8 P. M. and dispersed at 11 P. M. The Meetings were presided by Dinshah P. Ghadiali.

Promptly on the opening, the President called on the First Vice-President Jessie E. Ness, of Detroit, Michigan for a short address. She said in substance, "When I heard of Spectro-Chrome, I studied it and became a Graduate. I never regretted the step. I am sure you have similar stories to tell. Colonel Dinshah is a wonderful orator and scientist. I am going back to Detroit tomorrow night. I heard two Occultism And Higher Life courses given by him. They gave me to understand how the hopes and ambitions of humanity are fulfilled. Nothing goes on and on. We must pass out of one life experiences to another. Such idea makes one feel well. Life has its tests; when the examination time comes, we have to go through it. The teachings are given and all profit by them. I profited materially by them. I wish that books on such subjects be published. All are circling along that line and have a craving for it; some are not ready for such knowledge; it is for the uplift of the human mind. I hope in Eternity. Spectro-Chrome Metry was given to the World by Colonel Dinshah, for its benefit and uplift. He had and still has many problems with which to contend. We must co:operate with him and stand by him 100 per cent. In my home in Detroit, Michigan, I have many Spectro-Chrome Equipments; in fact, I have all the models made by Spectro-Chrome Institute. There are many so-called "Light Machines"; they are imitations and are nothing compared to Spectro-Chrome. Colonel Dinshah found the true principles of Spectro-Chrome Metry." (Applause)

Dinshah then made a few remarks.

Around the middle of the afternoon session, a surprising incident occurred. Albert E. Greenham of Detroit, Michigan, entered with a delegation of 30 Affiliates. The Convention received the delegation from the First Detroit Michigan Planet with a round of applause and fraternal greetings. On being introduced, Albert Greenham said, "It is a great pleasure to be one of a party from Detroit, Michigan, to visit this Convention. I attended many Conventions in the past, but, this is the most unusual. It is the first time in the history of Spectro-Chrome that 30 Affiliates attended a Convention as a group. We are proud of that. We are Spectro-Chrome enthusiasts, because, in Spectro-Chrome we have something with which nothing else can be compared. Colonel Dinshah is a very learned man. I came in contact with Spectro-Chrome 10 years ago. It is a complete Healing Art in itself. Modern Medicine is a hit and miss. Spectro-Chrome is a positive Science. It will stay as long as the Sun, Moon and Stars remain. Dinshah's lecture on Food is the most important of all lectures. 80 per cent. of the Affiliates fail to live upto it and if they do not get the expected results from Spectro-Chrome, it is not the fault of Spectro-Chrome. Rational Food Of Man is very important. There are many great things to come in Spectro-Chrome. There is one more thing about which I like to speak. There should be a Law to prevent menstruating women from handling food. It is not sanitary. The point was brought out in the manufacture of tubes for Radio work. They say that when 600 tubes were found proper, 400 were found defective, the professor who looked into the matter detected the cause—the faulty tubes were made by female mployees in their ovelim period! All such unrecognized matters were taught by Colonel Dinshah for years." (Applause)

Robert Schott asked the difference beteween Lavan and Store Salt. Dinshah explained that Lavan was the Sanskrit name for Salt and that it was the Trade Name for Salt from which nothing was subtracted and to which nothing was added. In fact, it was the All-Salt-Sea-Salt. Ordinary Salt is generally treated and thus either depleted or devitalized in quality.

L. Roy Dodman, the Deputy of the First Detroit Michigan Planet having resigned because of other engagements, Lucille E. Greenham was appointed the Deputy. Gladys H. Clark, the Recorder of the First Detroit Michigan Planet having resigned, Sofie Hall was appointed the Recorder. Ralph Clark was appointed Sub-Deputy of the Planet.

The Planet Commissions of St. Louis Missouri and Milwaukee Wisconsin were reappointed.

Dinshah stated that he would soon install a system by which Planet Reports may become simple and uniform. He then spoke about the new route he would be taking on his oncoming voyage to India, via Pacific Ocean.

Deputy Frank S. Burns being called on the platform spoke about the First Milwaukee Wisconsin Planet. He said, "The Planet meets each month 2 hours and we endeavor to teach the great Science

of Spectro-Chrome Metry. I have been the Deputy since 1939. I began to make it an educational institution. The attendance is always good, but I wish more Affiliates would come into the Meetings; it would be well worth their time, because they would gain knowledge. I tell you this that when you purchase a Spectro-Chrome you take a step in the right direction. I took that step in 1934. It changed my life in many ways for the better. I learnt from Colonel Dinshah many matters regarding the principles of Nature. If everyone will understand and follow the teachings of Spectro-Chrome Metry, the World will be much better. Nature's Laws are so strict, they can not be bent. They are unchangeable. However, the difficulty is, that you may be an enthusiast and yet fail to live upto the principles. By Spectro-Chrome living, you will have a longer and better life." (Applause).

Magdalene Arnold said that her husband had Sciatic Rheumatism for 2 years and had 8 Medical Doctors and was in 2 Hospitals. He used Spectro-Chrome for about 2 months and is in perfect health. (Applause). The President remarked she was a very modest woman and the audience enjoyed her talk.

Otto H. Lutz then gave some interesting case reports:—

(1)—A. A., woman, age 68. Fell over backwards and struck a chair. She was unable to walk erect and her back was black and blue. Atcowas Indigo and Orange Local and Green Systemic were irradiated and in 2 weeks her Back was well.

(2)—J. S., man of 61. Heart Disorder of long standing. He was unable to work for the past 2½ years. After 1 month of Spectro-Chrome, he went back to work. Atcowas Green Systemic and Magenta on 4 and 18 Areas were used.

(3)—J. G., man of 51. Run down condition; was unable to sleep properly. Kardoscil 54, Spiroscil 18. Felt very dizzy at times. After 2 weeks of Spectro-Chrome, he went back to work. Kardoscil changed to 70, Spiroscil came down to 16. Dizziness left completely. Atcowas irradiated were Green Systemic, Magenta on Areas 4 and 18.

(4)—A. T., man of 67. Bladder, Kidney and Prostate Gland Disorders. Could not pass Frontelim. Had tried about 5 different therapeutical systems and always had to use Catheter in the Genital. Had suffered for 2 years. Attuned Color Waves Orange and Red Systemic, Magenta on Area 4 and Scarlet on Area 18 were irradiated. In 2 months, 3 Tonations a week, the case was normalated. He passed so much Frontelim the customary way, that he said, "it sure feels good to be able to empty the Bladder without difficulty, thanks to Spectro-Chrome!"

(5)—A. N., man of 59. Cold in Chest and Backache, constantly for 2 weeks. Atcowas given were 2 Lemon Systemic Front and 2 Indigo on Area 18. He first tried a Back Plaster, which gave no response; Spectro-Chrome did. He said, "Spectro-Chrome works like a miracle."

(6)—P. W., woman of 56. Boil under Tongue, for 2 weeks. Atcowas Yellow and Orange Systemic were used and in 2 weeks the Disorder disappeared.

(7)—R. S., boy of 5. Chicken Pox. In 1 week was normalated. Atcowas irradiated were Turquoise Systemic Front and Back; Magenta on Areas 4 and 18. Was advised to tonate with Lemon for a few days; in all he had 10 days of Spectro-Chrome.

(8)—C. S., man of 45. Had an accident where he worked. Had a 5 inch cut on his left Hand and Wrist, quite deep. The company Medical Doctors said he cut an Artery. The Blood, the man said, squirted 30 feet! But, I did not believe it and cut it down to 10 feet! (Laughter). After he was released from the Hospital, which was in 3 days, he started to use his own Spectro-Chrome. Atcowa irradiated was Green Systemic. The 5 inch cut healed completely in 3 weeks and the next week, he went back to his duties.

The reports were received by the Convention with applause. Otto L. Lutz said that he used Spectro-Chrome in his own family and that he had 4 children, all of whom had all Disorders common among children. They all had Spectro-Chrome. His daughter had so much milk by it that she did not know what to do! (The President—"Open a dairy"!) (Loud Laughter).

Credentials were presented by the following Delegates:— Gladys Clark, First Detroit Michigan Planet. Leocadia Merwitz, First Milwaukee Wisconsin Planet. Viola C. Jeude, First St. Louis Missouri Planet.

They were duly accepted and greeted.

A letter was read from Peter R., a Provisional Member, who having signed for the Course on Occultism And Higher Life, was requested to send his $5.00. He did that, but garnished the remittance with the following letter:—

"Am sending you this not because I think that you are justified, in collecting this, as it was not explained at the Planet Meeting that there was a charge for these lectures and I thought that it was given as a public lecture. If I had known that it was not for the public, I would not have signed. However, you are in business and like everyone else in business, thep get all they can."

The Affiliates present expressed their displeasure at the tone of the letter and Dinshah filed it with the statement that he had accepted such explanation from many other Affiliates and had returned the payments of ALL signatories and given the Course free of all obligations. However, he stated that the writer of the letter owed an apology to Spectro-Chrome Institute, for the gratuitous insult in

the last sentence. Spectro-Chrome Institute, although a Business Corporation, was founded solely for the Service Of Mankind and that he had been serving as its President for over 19 years, WITHOUT SALARY. Later, the offending Affiliate mailed his apology and the incident was closed.

Dinshah emphasized that the Constitution And Rules of our Organization should be read at least once in 3 months in every Planet, so the new Affiliates may become familiar with them.

Kurt W. Bader, Sub-Deputy of First Milwaukee Wisconsin Planet then addressed the Convention. Said he in substance, "I suffered many years and the sickness was diagnosed by the Medical Doctors as 'Lumbago, Strained Muscles, Tuberculosis of the Kidneys'. I was a heavy smoker and drank beer and ate chunks of meat. I was sick for about 9 years; could not hold a broomstick or get on a shirt; I could not work. I took all kinds of Medicines and Pills. I was on simple food all the time. Took Kneipp Cures and studied all kinds of healing systems. Came here in 1927. They diagnosed the Disorder as Kidney Tuberculosis and gave me 8 months to live. A woman told me to go to the Auditorium and hear Colonel Dinshah on Spectro-Chrome. Knowing Spectroscopy, I came and listened to him. Took Spectro-Chrome for 3 months and stopped smoking and meat eating. I now work 18 hours a day". (Loud Applause.)

Dinshah then spoke about the 4 Volumes of Spectro-Chrome magazine, beginning with June 1922, the date of its first publication. There were only a few copies available at the nominal price of $10.00 for the set of 4, inclusive of the postage. His intention was to place the books for study, in the hands of those interested in the Science of Spectro-Chrome Metry, because the volumes were full of interesting information and data.

Frank S. Burns said, "I was examined by Medical Doctors for insurance. They found a slight murmur in the Heart, after being called to their office where there were many in the building. Six Medical Doctors made the examination and put me through various exercises; then they shook their head and in technical terms defined my Heart action. The result was that the clearing house of the Insurance Companies blacklisted me. After using Spectro-Chrome, I applied again for Insurance. I received notice to appear for physical examination. The Medical Doctor said, 'Step on the scale.' He measured me and asked me about my weight, pulse and I told him all he asked, without hesitation. He also asked about my Blood Pressure. The Medical Doctor was astonished and inquired how I knew so well about my physical condition. He talked about my Urinalysis; I told him about it. It is the first time in 35 years that my physical condition is so much better." (Applause).

Albert E. Greenham spoke about the value of the 4 Volumes of Spectro-Chrome. He made a set of the past issues. He found them invaluable. Therein were several reports of Surgeon Kate W. Baldwin. If for nothing else, for reading those reports, everyone should have a set in their library. She had 14 Spectro-Chrome Equipments and did wonderful work with Spectro-Chrome. Those Volumes are worth more than the Ten Dollars.

The President then extolled the virtues of the noble Dr. Baldwin. She was deceased, but, her memory will live for ever with us.

Emma Schroeder addressed the Convention in substance thus: "My mother is here. Four years ago, my mother had Stroke. Her Eyes were closed and her Throat was paralyzed. She had Hemiplegia. She made reservations for certain baths. She suffered much. After a long time of all kinds of medications, a friend introduced me to Spectro-Chrome. After the first Tonation, the Spectro-Chrome did more good than all the Medicines. I then studied Spectro-Chrome Metry and followed the Ultra-Green and Infra-Green Division method. My mother is 75 now and is as good as new! (Laughter)"

Her mother Ottilie Krenz on request of the President, walked briskly to the platform and was accorded a cheerful ovation.

Warren C. Hale volunteered as a speaker. He was Chief of Public Safety in Greendale, Wisconsin. He had a collapsed Lung and was in an Oxygen Tent for 2 hours. His Heart meanwhile shifted partially to the right side. He had severe Dyspnea and Pain. He was taken unconscious to a Milwaukee Hospital, where they declared his condition as poor and that nothing could be done to save him. His father Albert B. Hale, a Graduate Spectro-Chrome Normalator, being notified, insisted on moving him, but, the Surgeons objected to it as a fatal procedure. Finally, the father took the son home; three Tonations of Spectro-Chrome restored the Pulse and Respiration Rhythm and the sick son sat up in bed unaided. He is normal now and the injured Lung is completely re-expanded. In less than a week, he returned to his position as Chief of the Police and Fire Department and in charge of Alien Registration and Finger Printing. (Loud Applause)

A long discussion was started as to who should be entitled to Free Guidance as members of the **immediate family.** It was finally agreed unanimously by the Members, that such service should be given solely to the Registered Purchaser, for Wife or Husband, Father and Mother of the Registered Purchaser, unmarried Daughters and Sons of such person, unmarried Brothers and Sisters of such person, living in the same house under the care of the Registered Purchaser. The non-voting Affiliates also coincided in this view. On behalf of Spectro-Chrome Institute, Dinshah promised to consider the suggestion and determine the future course of action.

Phyllis M. asked a question, "Why were animals and fish created?" Dinshah answered, "Why were you created? Same reason."

Frieda Wehrhagen asked, "Why is it that Human Beings can not live more than 30 days without water, while Moses lived twice for 40 days without it?" Dinshah said that the Human Body had a very heavy percentage of water in its composition and without it life became liable to injury. The harm came through drying or Dehydration of the Tissues, thus reducing the conductivity for the Electrical Currents motioning in them. He further expatiated on the activities of some of the yogis of India.

On motion of Jessie E. Ness, seconded by Kurt W. Bader, it was unanimously resolved that Dinshah P. Ghadiali, Irene Grace Dinshah and Evelyn Roman be elected the Trustees of the Scientific Order Of Spectro-Chrome Metrists. Evelyn Roman and all the newly appointed officers of the various Planet Commissions were then affirmed and installed by the President.

During the intervals of the questions and the case reports, Dinshah delivered addresses on different Disorders and debunked numerous medical fallacies and medical empirics, establishing thereby the marvelous merits of Spectro-Chrome Metry. They formed challenging eyeopeners and were received with rapt attention by the audience.

Our Twentieth Annual Convention was a decided step forward on the road to the ultimate goal:

SPECTRO-CHROME — IN EVERY HOME

———o———

FIRST MILWAUKEE WISCONSIN PLANET

(HOTEL WISCONSIN, MILWAUKEE, WISCONSIN)

Meeing Of Thursday, October 3, 1940

Successful Use Of Spectro-Chrome In:—Gangrene, Pain, Fever, Uterine Hemorrhage, Rectal Bleeding, Constipation, Mammary Asscess.

Appreciation Shown For The Free Course On Occultism And Higher Life

Reported By*:—CAROLINE ZIMMERLING, *Recorder

Deputy Frank S. Burns was in the chair. A letter from the President Dinshah was read.

Leocadia Merwitz then gave a detailed report covering the Twentieth Annual Convention to which she was appointed Delegate for the Planet. She said it was a privilege, as far as she was concerned, to attend the sessions of the Convention and that Colonel Dinshah stood as usual, fearlessly giving forth his ideas on Spectro-Chrome and Health. She outlined a few of the points on the Rational Food Of Man system which she gleaned from the talks. These were that we should never overeat and should chew our Food properly.

She learned, too, that the Appendix should never be removed, because of its very important function in the Human Body and that if there should be something wrong with it, Spectro-Chrome could repair the damage easily and quickly. She said the Case Reports given by Kurt B. during the sessions were very interesting, as were those of Otto L. and Emma S., who took her mother to the Meetings to show what Spectro-Chrome had done in the case of Paralysis. She enjoyed hearing the Case Report of Walter H. of Greendale. He restored himself to normalcy after having a Collapsed Lung and is now working 18 hours a day.

She said that Colonel Dinshah during these Meetings sounded a warning politically about our present administration, saying that if it were not changed, the whole country would surely be plunged into War.

An interesting item which Leocadia Merwitz gleaned from a report by Albert Greenham was the fact that Radio Tubes which were handled by menstruating women were found faulty in manufacture. It is Colonel Dinshah's contention that women who are menstruating should be kept away from handling foods for the public, because it is at this time that a woman's Aura changes from the usual Magenta to Scarlet, affecting everything with which it comes in contact.

During this Convention the installation of Officers took place, the present Officers remaining as they are.

Leocadia Merwitz said that Spectro-Chrome in Cancer and Infantile Paralysis was thoroughly discussed at the Convention, as was Multiple Sclerosis. The subjects of Tobacco and Alcohol and their effects on the Pituitary Gland also came in for their usual share of attention, as did Inflammations and Disorders of the Heart. The Delegate expressed her gratitude for these discussions and said that Colonel Dinshah had extended his appreciation to the Affiliates of Spectro-Chrome as a whole.

The Deputy then complimented the Delegate on her very good work and made remarks of his own about the Convention. He too attended.

Much time was consumed at the Meeting to discuss the benefits derived from the Course on Occultism And Higher Life given by Colonel Dinshah. The expressions of gratitude by the various members of the Planet were summed as follows:—

The Deputy said he realizes now that the lessons which are to be learned in life are sacrifice, tolerance, et cetera and he feels that the Planet should pay Colonel Dinshah in some way for his marvelous efforts in giving this Course and feels that purchases of Spectro-Chrome products would be one way and selling more Equipments would be another.

Lena Held said she enjoyed the whole Course so much that she wishes she could hear it all over again. She says it was all the Truth on the Higher Side of Life.

The Deputy added that Colonel Dinshah took the Affiliates in that Course, through all the stages of Life from the Creation to the point beyond the grave—all within 12 hours of time, covering the subjects of Cosmogenesis, the Human Being, the various Planes of Life and so on.

Emma Schroeder remarked that she felt more or less responsible for the Occult Course, inasmuch as she expressed her desire to have such a Course given in Milwaukee.

Paul Schmelter said this was the first Meeting he attended in 1½ years. He attended all of the Meetings at the Auditorium and also the Occult Course. He found both very interesting. It took him from the cradle to the grave. It also gave him an insight into the effects we have on the lives of other people. One law, he pointed out, covers all others in governing lives and that is the Golden Rule. H said people fall in love with things that are harmful, material things. They should, rather, choose the spiritual things. If we give out evil thoughts, evil will be worked against us. He liked the Jewel In The Lotus Drama which was acted on the last evening of the Course by Dinshah. He said it explained Karma. It showed him also what we should avoid and what we should do. We should help others and thereby help ourselves.

Right after the Course, Norma Fiedler said, her whole Life seemed to flash before her so that she could see where she had made her mistakes. She said now she felt she could avoid making those mistakes again.

Werner Lehman commented that he attended all of the Lectures and felt that a man who could talk like Colonel Dinshah does on these subjects really knows about what he is talking. He enjoyed every part of it. He thinks the Colonel should enlarge on his writings on Occult subjects.

Albert Franz felt that anything that came from Colonel Dinshah, be it oral or written, was good. He also expressed the desire to see this Course in written form. By reading it, he could pick points which he might have missed.

Emil Egide attended all Meetings and derived much good therefrom, although he admits much of it went over his head. A few highlights, however, were brought home to him. One of them was that most people confused the Mind with the Brain. Another that there was really only one Commandment and that the Golden Rule.

The Deputy added that many lives were lived by us before and we could all travel and work on the Psychic Plane like Colonel Dinshah if we would learn how.

Theodore Mohrhusen was thankful for two things. One that he is still a young man and so has much time to learn more of these things and the other that he is able to associate himself with people who appreciate such teachings. He feels the book "Master Of Occultism" is really an autobiography. He learned also that all Religions are the same, but with different interpretations. From what he learned through this Course, he resolved to control his temper and try to make his disposition more agreeable.

At this point Kurt Bader said that he thought as long as we had a Physical Body we should take care of that properly and as long as our Meeting was supposed to be devoted to the care of the Physical Body, we should stop talking about Occultism and return to Spectro-Chrome Metry. He felt that the best way for us to show appreciation for the Course given free was to buy Spectro-Chrome books and Equipments.

A motion was made, seconded and passed to have a letter drawn by a Committee to express the feelings of the Affiliates and the gratitude for the Occult Course. Emma Schroeder, Leocadia Merwitz and Norma Fiedler were chosen for this.

Emma Schroeder now wishes to hear lectures on "Sex" from Colonel Dinshah, at some future date.

Chapters 97 and 98, "Boil, Abscess, Furuncle, Carbuncle" and "Spectro-Chrome In Gonorrhea", from Spectro-Chrome Metry Encyclopaedia were read by the Deputy.

Paul S. inquired as to what could be done in the case of healing which left Scars. He said Spectro-Chrome had healed a Sore. When the Sufferer felt that it ceased draining, he healed the Sore and sealed it. This left two Scars. The Veins protrude at the Ankle. It was a Bump first and the Medical Doctors wanted to amputate. Two weeks of Green, Orange were used for draining. Orange was used for four weeks. Turquoise and Blue were used for healing.

Lena H. said she generally uses Turquoise in cases of this kind. She said there are layers of Skin to go through and this condition is on the outside only. Indigo would be used for Swelling and Pain.

Kurt B. said that the purple Spots might stay for some time. Indigo will take them away. Scarlet could be used on Area 4 and Indigo Local for the Veins. He then gave some case reports:—

(1)—Woman, aged 64, had Gangrene, thought she was not getting results with Spectro-Chrome. He suggested she should use only Green, Systemic and Local, with Magenta. In six months she was in perfect condition. She followed the Rational Food Of Man System.

(2)—L. W., man aged 26, spasmodic Pains in Areas 7 and 8. Kardoscil 68, Spiroscil 18. One Orange relieved the Pain which no Medicine was able to help for two months. The Sufferer is still using Spectro-Chrome.

(3)—L. W., youth aged 18, High Temperature, Pain in Right Side, Area 10, was told to have operation. Green, Turquoise and Indigo Local were used. He was normal in a few days. To begin, his Kardoscil was 114 and Spiroscil, 24.

(4)—T. N., woman aged 51, had Hemorrhage from Area 11 for three months. Kardoscil 55, Spiroscil 20, One week Green and Indigo Local stopped the Flow. Bought own Equipment and is well satisfied.

(5)—M. E., aged 46, Kardoscil 120, Spiroscil 22, Bleeding Rectum for 18 months, had 10 Backelim daily and was weary of Life. 8 Tonations of Green, Indigo, Purple and Lemon normalated the condition. Now cheerful again.

(6)—L. M., Kardoscil 57, Spiroscil 18, constipated 15 years; High Blood Pressure; used to live on Raw Meat, practically. Constipation relieved after three Tonations. Only Green and Magenta were used.

Elizabeth B.'s friend had a baby. Before the baby came, the friend ate the right foods. After the baby came and when she should have started to menstruate, her Left Breast (Area 4) became very large and sore. Green and Blue were used and it became larger. She told her to irradiate Lemon Systemic and Orange Local. This opened the Sore and a cupfull of Pus drained out. Green Systemic, Turquoise and Magenta were then used to heal. Now she is very well.

As a final warning, the Deputy told the Affiliates at the Meeting to watch the Kardoscil and Spiroscil; that is the key on which to work with Spectro-Chrome.

EDITORIAL REMARKS

If at least 100 students will express their desire to hear a Course on "Sex, Its True

Significance And Importance In Life", paying $5.00 each to cover expenses, consideration will be gladly given. However, sour prunes and prudes may not be admitted to the Course—only Truth Seekers.

Scars, once gathered may not disappear.

Many thanks for your kind feelings. Dinshah is always at your service.

AMERICA RETURNED TO SLAVERY

We Are Already In War.

A man becoming naturalized as a Citizen of the United States Of America, affirms to protect the country against enemies, but, there is absolutely nothing either therein or in the Constitution to compel him to kill people who did not invade or indicate or express a desire to conquer this country. Liberty is vaunted as the most glorious right of an American citizen, but, the conscription and drafting by force of any citizen to bear arms or even enlist, is nothing short of downright slavery and barbarity.

There was no need for the whole extravagant program of "war preparedness." The idea that Germany wants to conquer America is utterly nonsensical and is born of the minds of ammunition and war material makers and their friends the grafty politicians ruled by Wall Street gamblers. I for one am ready and willing to bear arms in "defense" of the country against any foreign invasion or domestic revolution, but, I have no sympathy for any measure, wherein my conscience or conscientious beliefs may be invaded by the country whose loyal citizen I am. The present Selective Draft Law is unconstitutional, because, it invades conscience wholesale. The sentences of over one year in prison, imposed upon religiously conscientious objectors, are out of line with true Liberty and it is no use blaming Adolph Hitler or Benito Mussolini, when we ourselves are practising dictatorial methods against a helpless citizenry, under the asinine legislative measure of a supine Congress, cowed by a domineering President.

All this ruin of the glorious Constitution is done under the name of "Defense of Democracy"! Who invaded our Democracy? Those who talk about the German "Fifth Column", ought to investigate the subtle propaganda of imperial monarchists also. Woodrow Wilson dragged us to war to save "democracy" and that one lesson should have been enough, but, it appears we are headed again toward worse troubles, by another damn-o-crazy administration and autocratic Chief Executive. They are all clamoring for "preparedness against war." Are the American people so stupid as not to realize that we are "already **in war?"** When we help one belligerent against another, it **is** war and our "cash and carry" plan of neutrality is a gigantic fraud on the public, who are told that unless Great Britain be helped we are doomed here.

COLONEL DINSHAH'S DEPARTURE FOR INDIA STOPPED BY GOVERNMENT ACTION

HIS PASSPORT CANCELLED

Colonel Dinshah P. Ghadiali returned from Bombay, India, to file Nomination for Election as Governor Of New Jersey. However, on his arrival here on June 16, 1940, he was informed by the Secretary Of State Thomas A. Mathis, that the last date for filing Petitions for Independent Candidates was May 16, 1940 and no Petition could be received later than such date.

Thus, seeing no reason to prolong his stay in the United States Of America and being anxious to return to India to further the Libel suits he was prosecuting, he requested passage on the President Harrison. The American President Lines, however, refused to take him beyond Japan, because he would not submit to Vaccination and Cholera Inoculation. After much trouble, he succeeded in prevailing upon the Java-Pacific Line to take him as a First Class Passenger on an unnamed steamer. The Company stipulated, that before the ticket may be issued he would have to secure the Visae on his American Passport, from the British and Japanese Consulates. That being done, the steamship ticket and the Pullman ticket to San Francisco, California were issued and everything was ready for his departure after the November elections.

Suddenly, it was brought to his attention that all American Passports must be validated before departure, because of an Executive Order. In all innocence, he sent his Passport to the Chief of Passport Division Ruth B. Shipley requesting validation as before. To his surprise, he was informed after a long delay that there was no sufficient reason for his tour and that the Passport was "filed" in the departmental records. Dinshah protested against such autocratic measure and wrote strongly for return of his Passport, which had several years of documentary evidence in it, but, the Department Of State emphatically refused. He went to Washington, D. C., but the Chief was "away". He was interviewed by assistants and was coolly informed that he could be permitted to go solely on "urgent business" like buying rubber or tin for the requirements of the Government! The American Government had the "duty to protect" him; of course, the buyer of rubber or tin needed no protection!

When he calmly stated that he could go away without the validated Passport, he was as quietly shown a Presidential Order, stating any person so doing was subject to a Fine of $2,000.00 and 5 years of imprisonment or both. Smiling at such dictatorship, Dinshah then demanded return of his Passport, which was his property.

It was returned to him with holes punched through lettered "Cancelled". The cancellation included arbitrary punching of the British and Japanese permits also! American lawlessness settled that issue.

The dictatorship established by the curse of the "New Deal" is now becoming more obvious—America is heading towards disintegration of the glorious Constitution, to suit the whims and caprices of an insane administration. War is imminent.

Where is American Liberty? Where is the Bill Of Rights? Where is the power of the Congress? Instead of the people ruling through the Government, the Government bayonets are at the back of the people. Soon it will be good-bye to all good that was done since 1776 and then what?

o

FAVORSCOPE FOR 1941

The Favorscope for 1941 and the Affiliation Card were mailed to those who sent the Application For Continuation, duly filled in ink with the Annual Dues to Scientific Order Of Spectro-Chrome Metrists. Hundreds of Applications came with faulty engrossments or unsigned or with unauthorized signatures or remittances payable to Spectro-Chrome Institute. They were uniformly returned for correction, as the two Corporations are distinct and the checks of the one may not be accepted by the other. In sending the registered Favorscopes, special cards were also enclosed to verify Authentic Signatures. Please sign and return them promptly, because, without such verification, no service may be rendered. To Affiliates in India, Java and other foreign lands, Dinshah mailed the Favorscopes by Pacific Air Mail at personal expense as courtesy, due to War conditions. Hereafter, the Favorscope will be made ready in the middle of the year, to avoid last month rush.

No Favorscope was mailed to those who are Provisional Members as students without Spectro-Chrome Equipments.

o

OUR MISERIES IN FREE GUIDANCE

The Sufferers asking for our Free Guidance do not have any idea of what nuts we have to crack. Our problems need the training of a Solomon to solve. Recently we had the following in the General Advice Chart of Elsie A. of Pennsylvania:—

16. The Sufferer is Unmarried. . Yes. . . . Married. . Yes. . . . Divorced. . Yes. . . . Widower. . . Grass Widow!

She certainly is well experienced.

SPECTRO-CHROME

May 1941 Green

Volume 17 Number 5

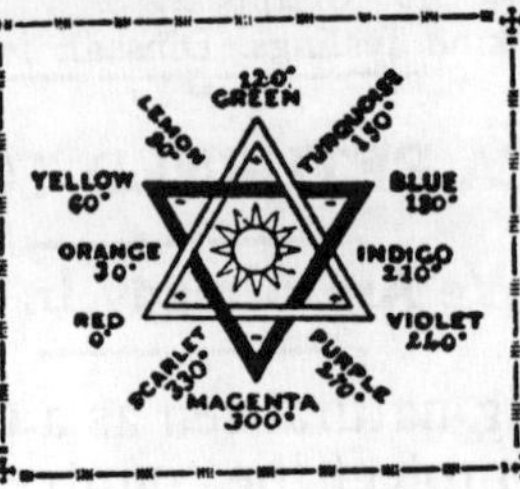

Trade Marks Registered In The United States Patent Office

Read And Have Your Friends Read

Investigation Of The Merits Of Spectro-Chrome
The Originator Welcomes It

Commemoration Of 21st Anniversary
Steps Taken To Perpetuate Spectro-Chrome Institute, As A Benevolent Foundation Devoted To The Service Of Mankind

Dinshah Educational Broadcasting Service
Non-Stock, Non-Profit Corporation Chartered In New Jersey

Twenty First Anniversary Celebration Of The Science Of Spectro-Chrome
Weekly Newspaper And Broadcasting Station Coming

Attuned Color Waves
Spectro-Chrome Metry

In course of life you reach a time,
When Doctor, Lawyer, Priest alike,
Does fail to give the sought relief,
To troubled mind or burdened heart—
Then come to me:
I help you to help yourself.
SPECTRO-CHROME

The Science Of Automatic Precision

DEVOTED TO THE SERVICE OF MANKIND

GOLDEN WHIP

To The Public 3109

First Milwaukee Wisconsin Planet 3110

Investigation Of The Merits Of Spectro-Chrome By The Federal Government 3117

First St. Louis Missouri Planet 3125

Commemoration Of 21st Anniversary 3131

First Detroit Michigan Planet 3133

Dinshah Educational Broadcasting Service............ 3134

First Merrill Wisconsin Planet 3138

Twenty First Anniversary Celebration of the Science Of Spectro-Chrome 3139

Resident Experimental Class 3142

AMERICAN MEDICAL ASSOCIATION FINED $2,500.00

Washington District Of Columbia Court Found It Guilty Of Violation Of Anti-Trust Laws

The Grand Gazabo American Medical Association, which has continued its malicious control of all healing systems since its present incorporation four decades ago, got a slap on its jaw from an independent-thinking Court in Washington, D. C. Through the strenuous and able presentation of the case, under the direction of United States Assistant Attorney General Thurman Arnold, the A. M. A. was found GUILTY and sentenced to a Fine of $2,500.00. The District Of Columbia Medical Society also was fined $1,500.00. It is a matter of record that very few prominent newspapers in this country even mentioned the fact—proving again the control of the Channels Of Publicity by the Medical Octopus, through the glitter of medical advertising.

FIRST MILWAUKEE WISCONSIN PLANET

(HOTEL WISCONSIN, MILWAUKEE, WISCONSIN)

Meeting Of Thursday, May 1, 1941

Successful Use Of Spectro-Chrome In:—Kidney Disorder, Appendicitis, Face Sores, Prostate Gland Disorder, Stricture, Shingles, Lung Disorder, Sore Throat, Asthma.

Reported By:—EMMA SCHROEDER, *Recorder*

Deputy Frank S. Burns was in the chair.

The Deputy requested those Affiliates who did not submit their Affidavits to the Central Office to date to do so at once. Expressions, either favorable or otherwise, are welcome. Several Affiliates who were unable to attend the Special Planet Meetings were not in possession of Affidavit blanks.

Additional subscriptions to the weekly newspaper, **"Dinshah Impartial"** were sought.

Several questions which emanated from the Anniversary Celebration were brought up and discussed.

The idea seems to prevail, suddenly, with a few Affiliates, that Tonations are not to be taken during the Critical Period. This idea is erroneous.

Several persons inquired whether it is possible to become affiliated with Scientific Order Of Spectro-Chrome Metrists by purchasing a set of Spectro-Chrome Metry Encyclopaedia, after having previously purchased an Equipment and registered it in the name of another member of the immediate family. The Deputy informed them that it is possible.

The ruling on purchasing a used Equipment was given, quoting Colonel Dinshah: "It is like purchasing a mud-hole." There was no Free Guidance.

An explanation of the difference between the Sympathometer and the Itisometer was requested. Sub-Deputy Kurt W. Bader declined to do so in favor of Otto Lutz. In this connection, a woman who was in the Auditorium stated that she would like to know Colonel Dinshah's proposition in this matter, at this time, stating that she knew of three persons who were interested in the purchase of an Itisometer.

Another woman there complained that she became very weak upon abstaining from meat. She apparently did not read her Spectro-Chrome Home Guide, the Deputy stated and did not eat the legumes, with the Chaval, Dal and Limbupani daily as instructed.

Norma F. addressed the audience, stating that Colonel Dinshah requested her to make the following announcements:—

Due to price inflation throughout the country, Equipments for the present, might not have time switches, as the latter can not be manufactured for the same price.

Henceforth, Affiliation Cards will be sent to new Affiliates for signature immediately upon receipt of the Order, then to be returned to the Central Office for the Secretary's signature. This is being done to avoid any unpleasantness and embarrassment on Meeting nights due to failure to sign the Card.

Norma F. further requested that letters be written by Affiliates to the gentleman whose letter appears in the March issue of Spectro-Chrome Magazine, page 3085, informing him of the merits of Spectro-Chrome. Copies of these letters should be forwarded to the Central Office.

It was also suggested that writings be submitted to be printed in the new weekly publication.

The Sub-Deputy stated that unsigned Cards are lifted to protect both the Originator as well as the Organization, as the Medical Doctors are endeavoring to muzzle both.

Expressions on the reaction to the Twenty-first Anniversary Celebration followed:

Emil E. stated that he attended both Special Planet Meetings and three Public Meetings. Being an admirer of Colonel Dinshah's he can not say anything but good. He brought with him two people who never attended before. They found the lectures interesting but "deep". The average person does not understand Light and Color and must see things in order to believe. Most convincing, in Emil E.'s opinion, was the resolving of White Light into the three Primary Colors on the screen: Red, Green and Violet. It is more than "just talk", as many people think, Emil said.

Otto L. was interested in the Food Lecture, which, he says, 85% of the people try to avoid. Free and unrestrained discussion on Food followed.

The Recorder stated that she expected to be tired after attending two Special Meetings and six Public Lectures, but the reverse proved itself true. She credits her endurance to the Rational Food Of Man.

The Deputy gave an interesting account of the habits of living he formed and his reasons for doing so. He used to smoke and drink moderately, but was convinced of that fallacy in 1934. Always had a sick Stomach while smoking. He did not experience any weakness when he abstained from meat. The Deputy follows a plan of daily exercise and correct Food.

Otto L. thought this weakness, produced upon abstaining from

Meat, was caused by the Toxins and dross being loosened and that a condition similar to Spring Fever arose. He thought there was a reaction.

The Recorder reminded that people who partake of Flesh Food are eating Vegetables second-hand, for they are eating only the vegetarian animals. except possibly during an emergency such as war, when carnivorous animals may be consumed due to scarcity. She could not find one instance in the Bible where the Master Jesus Himself partook of Flesh Food.

The Deputy experimented and found that raw Vegetables retain the Life Principle longer than cooked Vegetables.

Otto L. suggested that Steak and Vegetables be placed in the Sunlight side by side and it will be found that the Meat will commence to rot. "The Vegetable will win."

Emil E. is eating Chaval and Dal daily. He finds himself getting "fatter but not fairer." (Smiles)

Otto L. told of a Health Inspector who inspects Meat markets. When, in the opinion of the Inspector, Meat is found to be unfit for consumption, kerosene is poured on it. In one instance, many pounds of Chicken were declared "unfit". However, those persons who were fortunate or unfortunate enough to purchase their Chicken shortly **before** the Health Authorities arrived got what—???

The Deputy pointed out that eating Meat produces Uric Acid.

The Sub-Deputy explained that the greater part of Mankind is vegetarian. Of the 350,000,000 inhabitants of India, 280,000,000 are strictly vegetarian. Only the Mohammedans believe in eating Flesh Food. Orthodox Hindus have been Vegetarians for centuries. The stronger animals of the World are vegetarian and have greater endurance. Look into the eyes of a cow. You will not see the cunning there which exists in the Eyes of a Meat-eating animal—Man included—which does one thing before your Face and another behind your Back. Vegetarian Food builds character of the right kind.

Otto L. added that Meat "decays its way" out of the Body.

The Sub-Deputy continued by relating the experience of a Danish Professor who wrote a book on the Slaughter of Animals, in which it is stated that a four-year experiment was conducted whereby people lived only on potatoes and linseed oil and were still healthy and strong and had more Muscle after than before. "Eat simple Food", he advised.

Otto L. stated that canned orange juice is squeezed out of rotten oranges.

One Affiliate reported that she is on a "cleansing diet" consisting of a dozen oranges a day and she feels better. (How about Atcowas Green and Lemon?)

Emily M. pointed out that when she was in Germany, during the last War, there were times when the people were glad to eat anything they could get and lived on lentils, peas, beans and potatoes. Children were fed rice water instead of milk. Even today she must force herself to eat fresh Fruits.

The Sub-Deputy said, "Eat everything which is in season. Starches and fats are not necessary in Summer as in Winter."

Affiliates were called on by the Deputy to relate their experiences with Spectro-Chrome.

William T. of Manitowoc, Wisconsin, possessed an Equipment for three or four years. He attended the Celebration and this is his third visit to the Planet Meeting. He found Spectro-Chrome very helpful.

Lydia S., Waukesha, Wisconsin, said although she can not always attend Meetings, she enjoys them very much. Her son had a stiff Limb—broken Neck. Several Cords were out of place and through using Spectro-Chrome, "came back". He is able to turn the Neck, despite the fact that it was never set. The Medical Doctors hesitate to set it—it is too dangerous.

Walter W., Juneau, Wisconsin said, five years ago he was the only one in his locality who owned an Equipment. Wonderful results. Now he knows of four persons who possess them and would not be without them. However, some are not willing to eat the Rational Food Of Man or to discontinue the use of Medicine. One visits the Chiropractor and uses Medicine in connection with Spectro-Chrome. Though this person does not say anything against Spectro-Chrome, he does not say much in favor of it. This man has the habit of calling Walter W. "Doctor" because he talks Spectro-Chrome so much. About five years ago, Walter W.'s wife was under the Medical Doctor's care. Kidneys were out of order; could not eat. Pills were given; condition became worse. They spoke of taking her to the Hospital for an Operation. After discussing the matter, they came to Milwaukee to see Otto L. and received good information. They went home and followed his advice. "No more Pills." The Medical Doctor called once or twice every day. The following day, she was able to keep things on her Stomach. Kidneys cleared; the child was born normally. The Medical Doctor could not understand how this was possible without a Caesarian Operation. "It certainly has done a lot for me."

Ida L., Richfield, Wisconsin, has not had very good success with Spectro-Chrome; used it and still has been sick every day. She admits it is somewhat her fault, for she does not eat properly. "Not much meat." She was weak, so she returned to the Medical Doctor, who found poor Circulation. Otto L. advised her to communicate with the person who sold her the Equipment. The Deputy requested Ida L. to fill out a General Advice Chart and send it to Spectro-Chrome Institute. He also advised her to fill out a Chart and bring it to the next Meeting for discussion. She promised to do so.

Otto L. submitted the following Case Reports:

1. C. M., Male, age 35. Appendicitis. Turquoise set him right in three days.

2. F. W., Male, age 44, Sores on Mouth, Chin and Cheeks; used Ointments and Salves for three weeks; Eruptions increased. He was very much alarmed about his condition. He was normalated in six days. All cleared out. Atcowas used were Indigo and Turquoise.

3. L. B., Male, age 60, Prostate Gland Disorder and Stricture; tried three other methods first without results. He was advised to try Spectro-Chrome and was normalated in one month. This man suffered intense Pain and thought he could not stand it another day. Spectro-Chrome came to the rescue. Blue Local, Turquoise Systemic, Yellow on Areas 8, 9 and 10 and Magenta on Areas 4 and 18 were used.

4. O. B., Male, age 24, had Shingles of two weeks' standing; Spectro-Chrome set him right in eight days. Atcowas used were Indigo and Turquoise.

5. L. B., Female, age 25, Lung Disorder of long standing; tried other methods—no results. After two months of Spectro-Chrome, she could sleep all night without coughing, which she was unable to do for a long time. Her appetite returned and she gained in Health and Strength. She followed the Rational Food Of Man. Atcowas given were Lemon Systemic, Orange on Area 17, Magenta on Areas 4 and 18.

6. M. S., Female, age 38, a very Sore Throat. Tried many home remedies without results. Three Spectro-Chrome Tonations of Turquoise normalated the case.

7. H. S., Male, age 75, Asthma condition of long standing. He was not able to leave the house this winter. If he did, he would commence coughing and would be unable to stop. After one month of Spectro-Chrome, he walks to town, about six blocks and does all the family shopping without discomfort or Coughing Spells. He is well pleased with Spectro-Chrome. Atcowas used were Lemon and Yellow Systemic and Magenta on Areas 4 and 18.

A. B., Male, 96 years. Fell, stumbling over bedroom rug and fractured three Ribs. He was taped by another method but the intense Pain would not cease. He then tried Spectro-Chrome Turquoise. After two Tonations all Pain left.

Discussion on the Case Reports followed and all derived some good therefrom.

Emil E. reported that upon leaving the meeting room last month, he was approached by a very distinguished gentleman, about fifty years of age, who said to him, "Did you just come out of that room? That is some racket, isn't it?" Emil replied, "Sir, just because one half of the World does not know what the other half is doing, that does not mean that the other half should be condemned." The man admitted being a Medical Doctor.

Adeline C. also saw the gentleman. She heard him say that Colonel Dinshah is selling Equipments and then selling a Course for $25.00 on how to use them!

The Deputy concluded the Meeting by referring to the Anniversary Celebration and the progress of Spectro-Chrome as was shown by the pictures; the heartaches, suffering, privations and unbearable insolence endured by the Originator. It is remarkable how this one man fought the opposition singlehanded. He urged Affiliates to do what ever is possible to assist Colonel Dinshah in the promotion and support of this noble Science which is destined to be the salvation of Suffering Mankind.

An Affiliate was interested in the progress of the incomplete case which was reported last month. The Sub-Deputy stated that the Ratio is still 1 to 4 and 1 to 5 and he does not give up hope as long as that exists. The tissue is still continuing to build and Spectro-Chrome is still performing its duty.

EDITORIAL REMARKS

The weekly newspaper "*Dinshah Impartial* will be published as soon as certain circumstances permit. Send your annual subscription of $1.00 to him personally. It is his enterprise. It will be a newspaper that will give the Truth, nothing but the Truth, the whole Truth without a hole in the Truth! It is upto you to make it a success.

On the contrary, during the Critical Periods, MORE vigilant Tonations are needed! Where do Affiliates read such unprinted instructions?

Any person may enroll from Spectro-Chrome Metry Encyclopaedia as a Free Student for the Home Training Course and request Affiliation with the Scientific Order Of Spectro-Chrome Metrists. Refer to Rule 12, Section 3 (c).

The reason some people become "sick" after leaving off meat-eating, is that they try to live solely on "stewed grass or fried hay." They do not comprehend that "vegetable food" is not "vegetarian food" and do not know or are too lazy to cook their meals as instructed by us.

Spectro-Chrome Equipments are still provided with the same Automatic Time Switches, but, if the Orders do not justify the higher material costs, the manufactured article *might* suffer some changes. Your Orders are needed now more than ever to maintain the product's low price. Of course, no matter how the model may change in construction, the Science and the Slides will NEVER change in their Attunement and Efficiency.

Before expressing any Order, the Purchaser *has to sign* ALL papers for Affiliation with our Organization. Nothing is sold without such Affiliation and the entire transaction is therefore with our Affiliates. No service is given to anyone outside our circle. This is *our* protection and found very valuable in preventing frauds and frameups. Honesty, 100% is successfully exemplified in our activities.

Write articles for our newspaper; boost it by ventilating the Truth.

Neither Lord Jesus, nor any Messenger of the Grand Architect, was ever a Meat-eater.

One need not eat Chaval and Dal daily. Prepare the Rice differently and eat some other Legume with it. Refer to Spectro-Chrome Home Guide, page 38, article 4.

As for Health and Strength of vegetarians, the Originator at his age of 67 (having been 50 years a strict vegetarian), is not a bad example. He drove his Plymouth truck 305 miles on

the first day from Malaga, New Jersey to Irwin, Pennsylvania and the second day from there to Plymouth, Indiana, 589 miles, making Milwaukee, Wisconsin (another 104 miles) early morning. On the return tour, he started from Fort Wayne, Indiana and made 398 miles to St. Louis, Missouri, in a single haul. Thence, he drove outward to Richmond, Indiana, 333 miles at one clip, making the next day, Malaga, New Jersey, another cool 615 miles without going to bed. On arrival, he followed his duties as if he had not traveled. Work does not kill; it is worry that kills and he lets the American Medical Association worry for him! Talk of endurance? He drove once from Milwaukee, Wisconsin to Malaga, New Jersey a 2 ton truck laden with his apparati, 888 miles without going to bed! All talk of Meat giving strength is hokum; the strongest group of animals with the greatest endurance, is totally vegetarian and Man in his anatomy and physiology is a genuine vegetarian animal—though now a carnivorous beast.

"Cleansing Diet" is another miscomprehension. A few Yellow on Areas 6-7-8-9-10 will "clear" all without waste of strength and watching meals.

We shall soon be heading to the plight of Germany during the last War!

Ida L. never sent for Free Guidance. What is FREE, is seldom appreciated. If she prefers to use her own judgment, it is her privilege. Let her suffer till she tires—then she will say Spectro-Chrome is "no good"!

Much obliged for all your kind words and feelings. The success of Dinshah is through your co-operation.

The following unsolicited Testimonial was mailed to the Originator by the First Milwaukee Wisconsin Planet. It is acknowledged with cordial thanks.

Milwaukee, Wisconsin,
May 24th, 1941.

Scientific Order of Spectro-Chrome Metrists,
Spectro-Chrome Institute,
Malaga, New Jersey.

The Affiliates of the First Milwaukee Wisconsin Planet of the Scientific Order Of Spectro-Chrome Metrists herewith express to Colonel Dinshah their most sincere appreciation and heartfelt thanks for the Educational Lectures delivered on his recent visit to Milwaukee during the Twenty-first Birthday Celebration of Spectro-Chrome.

It was an honor, indeed, to share this Anniversary with him, as an established fact and to listen to his ever-flowing fountain of knowledge.

The successful progress of Spectro-Chrome from babyhood to manhood, the motion picture tour through the Originator's life, the tour through the workshop, showing step by step, how Spectro-Chrome equipments are manufactured—all was most interesting. It is an example of how the unselfish service of one man can benefit mankind.

The enlightenment of the fundamental doctrines of the Order and the new proposals set forth, can not but lead to a better understanding among us and to betterment of all concerned.

The good will and good wishes of the Affiliates are with Colonel Dinshah at all times and we are proud to belong to an Order whose benefits are so far-reaching and are not judged from the monetary standpoint.

Again, we say "Thank you!" for Colonel Dinshah's unselfish endeavors and may the teachings of Spectro-Chrome Metry gain increased momentum, as time, human intelligence and understanding progress.

Respectfully,

FIRST MILWAUKEE WISCONSIN PLANET

Frank S. Burns, Deputy.
Kurt W. Bader, Sub-Deputy.
Emma Schroeder, Recorder.

Evert Elmer Dahlke,
Emile W. Egide, Committee.

No Diagnosis — No Drugs — No Manipulation — No Surgery
SPECTRO-CHROME — IN EVERY HOME

INVESTIGATION OF THE MERITS OF SPECTRO-CHROME BY THE FEDERAL GOVERNMENT

The Originator Welcomes It

Last year, when Dinshah P. Ghadiali went to Milwaukee, Wisconsin, for our Twentieth Annual Convention, his attention was drawn by one of our Members to the activities of a certain Agent of the Federal Food And Drug Administration, in regard to Spectro-Chrome Metry as a Science and Spectro-Chrome as an Apparatus.

On returning to the Central Office, the Originator promptly took the matter in hand direct and correspondence started. The following will indicate where we stand at present. Dinshah wrote to Chicago, Illinois:—

Tuesday, October 1, 1940,
Chronos:-Yellow 01 White

Federal Security Agency,
Food and Drug Administration,
1222 Post Office Building,
Van Buren and Canal Streets,
CHICAGO, ILLINOIS.

Kind Attention: Mr. Edward Rietz, Inspector.

Gentlemen:

1. My attention was drawn to some inquiries made by you in regard to our activities.

2. As I prefer to give first-hand information to assist public officials in the discharge of their duties, I shall thank you to let me know how you are interested in us or the writer and all the information at our disposal will be very gladly placed before you.

3. We have been twenty years in existence, serving mankind and during that time, there never was any impeachment of our integrity. Therefore, the writer will appreciate a word from you direct so he may have the pleasure of giving you all the information and presenting you with any literature for your records.

4. With the kindest regards and the best wishes, we are

Yours respectfully,
SPECTRO-CHROME INSTITUTE
DPG:S DINSHAH P. GHADIALI, President-Treasurer.

Five weeks later, a reply came from Washington, D. C.:—

FEDERAL SECURITY AGENCY
FOOD AND DRUG ADMINISTRATION
WASHINGTON, D. C.

November 12, 1940

Spectro-Chrome Institute,
Malaga, New Jersey

Gentlemen: Attention of Mr. Dinshah P. Ghadiali

Your letter of October 1 addressed to Mr. Rietz, one of our inspectors in Chicago, has been referred to this Administration for attention. We regret our reply has been delayed.

We have not as yet had the opportunity to review comprehensively the information which has been submitted by our inspector. After such review has been made, however, we shall be pleased to discuss this subject further with you in the event additional information appears necessary.

Very truly yours,
P. B. DUNBAR,
Assistant Commissioner of Food and Drugs

As Dinshah was preparing to leave on his voyage for India, to conduct his Libel Suits pending in the Bombay High Court, he could not afford to leave such an important issue hanging; so he wrote back:—

Wednesday, November 13, 1940,
Chronos:-Turquoise 31 Black

Federal Security Agency,
Food and Drug Administration,
WASHINGTON, D. C.
Reference No. AF 19617. Kind Attention: Mr. P. B. Dunbar,
Assistant Commissioner of Food and Drugs.

Gentlemen:

1. This morning we received your kind communication of the 12th current.

2. We appreciate your courtesy. However, we take the liberty of informing you that the writer is leaving on a business tour around the World and as he is the President-Treasurer and registered agent of the Corporation and the only person who can give relevant information, it will be highly appreciated if you will write to us as early as you can what information you desire about either our activities, principles or apparatus and we shall be glad to place it at your disposal. After the departure of the writer, it may be many months before any responsible officer may be able to give you the correct information.

3. Some literature about Spectro-Chrome is herewith enclosed and a line from you will bring you a set of our SPECTRO-CHROME METRY ENCYCLOPAEDIA and other books published by us in connection with the Science of Spectro-Chrome Metry originated by the writer.

4. We have the honor to remain

Yours In The Service Of Mankind
SPECTRO-CHROME INSTITUTE
DPG:S Enc:Lit. DINSHAH P. GHADIALI, President-Treasurer.

Reply to this came from the Commissioner:—

FEDERAL SECURITY AGENCY
FOOD AND DRUG ADMINISTRATION
WASHINGTON, D. C.
AF19617 December 20, 1940

Spectro-Chrome Institute,
Malaga, New Jersey

Gentlemen: Attention of Mr. Dinshah P. Ghadiali

It has been impossible to make an earlier acknowledgement of your letter of November 13 in further reference to your device Spectro-Chrome. We note that you are planning to leave on an extended business trip around the world and you suggest that we arrange to obtain any necessary information before you leave.

As we advised you on November 12, we have not yet completed our review of the information so far submitted to us, and in view of the press of other matters at this time we cannot forecast just when we shall have an opportunity to go over this information fully.

While, of course, it would be preferable to obtain any desired information directly from you, since you are President and Treasurer of the corporation, we must, of course, assume that if the business is to be conducted during your absence it will undoubtedly be necessary to designate some responsible individual to act for you in your absence.

While, as we have advised you, we have not completely reviewed all of the submitted information, we consider it advisable, in view of the situation set forth in your letter, to notify you that on the basis of the information so far reviewed the device "Spectro-Chrome" appears to be distributed in interstate commerce in violation of the provisions of the Federal Food, Drug, and Cosmetic Act.

Very truly yours,
W. G. CAMPBELL, Commissioner of Food and Drugs.

The reply went out:—

Monday, December 23rd, 1940,
Chronos:-Blue 52 Black.

Federal Security Agency,
Food and Drug Administration,
WASHINGTON, D. C.
Reference AF 19617 Kind Attention: Honorable W. G. Campbell, Commissioner of Food and Drugs

Gentlemen:

1. I thank you for your kind communication of the 20th current received this morning.

2. I am the Registered Agent, President and Treasurer of this Corporation and in all twenty years of its existence, my authority even in my absence is never delegated to any other officer, because Spectro-Chrome Metry is my original Science, Spectro-Chrome is my own invention and patent and I am the only person responsible for the conduct of the affairs of this Corporation.

3. Your last paragraph surprised me extremely. Therein you state that "on the basis of the information so far reviewed the device 'Spectro-Chrome' appears to be distributed in interstate commerce in violation of the provisions of the Federal Food, Drug, and Cosmetic Act."

4. I have no knowledge of any "violation" of any law, by the distribution of my "Spectro-Chrome". If you will please forward to me a copy of the said Act with the alleged violated section marked, it will be warmly appreciated.

Yours respectfully In The Service Of Mankind,
DINSHAH P. GHADIALI, President-Treasurer.

DPG:S

In a week, the Law came:—

FEDERAL SECURITY AGENCY
FOOD AND DRUG ADMINISTRATION
WASHINGTON, D. C.

AF 19617 December 31, 1940

Mr. Dinshah P. Ghadiali
Spectro-Chrome Institute,
Malaga, New Jersey

Dear Sir:

As requested in your letter of December 23, we are enclosing a pamphlet containing the text of the Federal Food, Drug, and Cosmetic Act, together with regulations for its enforcement. May we direct your particular attention to sections 201(h), 201(n), 301, and 502 of the statute.

Very truly yours,
P. B. DUNBAR, Assistant Commissioner of Food and Drugs

Enclosure FDC 1

After going through the indicated Sections, the Originator replied:—

Wednesday, January 8, 1941,
Chronos:-Indigo 32 Black.

Federal Security Agency,
Food and Drug Administration,
WASHINGTON, D. C.
Reference:- AF19617 Kind Attention: Mr. P. B. Dunbar, Assistant Commissioner of Food and Drugs

Gentlemen:

1. I thank you cordially for your kind communication of December 31, 1940 and the copy of the Federal Food, Drug and Cosmetic Act, dated June 19, 1939.

2. I went very carefully through the sections 201(h), 201(n), 301 and 502 of the statute as directed by you. These refer solely to the "Label" and "Misbranding".

3. Enclosed please find the standard label which for years and years has been attached to every Spectro-Chrome Equipment. I fail to notice any violation of any part of the stated statute.

4. It will be warmly appreciated if you will draw my attention to what your Department considers "violation", so I may answer and act intelligently in the matter.

5. With regards and respects, I am

Yours In The Service Of Mankind
Dinshah P. Ghadiali, President-Treasurer.

DPG:S Enc: Label *Registered With Receipt Requested.*

The Commissioner gave the requested information:—

FEDERAL SECURITY AGENCY
FOOD AND DRUG ADMINISTRATION
WASHINGTON, D. C.

AF 19617 January 24, 1941

Mr. Dinshah P. Ghadiali
Spectro-Chrome Institute,
Malaga, New Jersey

Dear Sir:

We have your letter of January 8 with further reference to your device "Spectro-Chrome". You state that you have examined the label for this product and fail to note any violation of the Federal Food, Drug, and Cosmetic Act thereupon.

As a result of correspondence initiated by your letter of October 1, 1940, we have advised you that, in our opinion, your device is being marketed in violation of the terms of the statute. You have been furnished with a copy of the law, and being fully familiar with the mechanics of your device and all statements being made regarding its therapeutic effect, you are obviously in a position to readily determine whether or not a violation of the statute exists.

We would not be entirely frank with you if we did not state, as we are now doing, that we have been unable to find any evidence that the device "Spectro-Chrome" has any value in the treatment of the conditions for which it is being recommended.

W. G. CAMBPELL, Commissioner of Food and Drugs

Dinshah promptly mailed a long explanation:—

Monday, January 27, 1941,
Chronos:-Blue 35 Black.

Federal Security Agency,
Food and Drug Administration,
WASHINGTON, D. C.
Reference AF 19617. Kind Attention: Honorable W. G. Campbell, Commissioner.

Gentlemen:

1. Please accept my cordial thanks for your courteous communication of the 24th current, received this morning.

2. I did not initiate any correspondence, but when it was brought to my notice during my 20th Annual Convention in Milwaukee, Wisconsin, that one of your Agents in Chicago, Illinois was indirectly inquiring about myself and my Spectro-Chrome, I wrote requesting him to make investigation direct through me instead of through indirect channels.

3. While I respect your opinion, I fail to see any violation of the Food and Drug Act by either my label or the device on which it is affixed. You mention about being "fully familiar with the mechanics of your device". It is not the *device* that performs the work, but the SCIENCE and TECHNIQUE of using it that does the work. Section 502 of your Act asks for "adequate directions for use." If that be in your mind, I may respectfully state that such directions accompany each Spectro-Chrome Equipment through a SPECTRO-CHROME HOME GUIDE. To put all those "directions" which occupy One Hundred and Two pages, on a label on the device itself, would be a physical impossibility.

4. Your exceptional frankness in stating that "We have been unable to find any evidence that the device 'Spectro-Chrome' had any value in the treatment of conditions for which it is being recommended", does great credit to you. However, the evidence about the value and merits of Spectro-Chrome as a Device and a Science, was proven in cases in the Law Courts time after time and each time I came out victorious, over testifying. Medical Doctors and University Professors, who tried to discredit and ridicule my invention.

I reciprocate your frankness by stating just as frankly, that the worst such case against me was instituted in the Original Side of the Supreme Court of the State of New York, at Buffalo, New York, where I was charged with Grand Larceny, for selling "fake machines". In that case, I was submitted to a harrowing Jury Trial, but a unanimous verdict of "Not Guilty" was rendered resulting in my acquittal. Two Professors who testified against me, lost their jobs for incompetence! An Exemplified Copy of that case, duly certified by the Secretary Of State, is herewith enclosed.

5. I also enclose my SPECTRO-CHROME magazine of April 1934 in which is a report of what I spoke before the United States Senate Committee On Commerce on February 27, 1934, in regard to some provisions of the then proposed Food And Drug Bill. Please refer to page 1131 and on.

6. I further enclose a copy of my SPECTRO-CHROME magazine of August 1939, on page 2693 of which is printed under the heading "The Reach Of Spectro-Chrome—An Unrivalled Accomplishment Far Surpassing Centuries Of Drugs", the various Disorders in which Spectro-Chrome has been used most successfully during the past nineteen years—it is twenty-one years now.

7. I make no "claims" or "recommendations". My business is founded on established clinical RESULTS, reported without solicitation by the users of the apparatus. The items printed in regard to the efficacy of Spectro-Chrome as reported in the monthly meetings in various centers of the Organization and the like, occupy five big printed volumes and the Science of Spectro-Chrome Metry is fully described in three volumes of SPECTRO-CHROME METRY ENCYCLOPAEDIA, covering twelve hundred pages. They are all in the Library of Congress and extra copies will be gladly mailed on request.

8. Gentlemen, Spectro-Chrome is today in the same category as the telephone, telegraph, phonograph, electric light, talking motion pictures (another of my inventions) and other radical inventions were in their own times. I always believe in placing cards on the table and am ready to submit to you any and all "evidence" in regard to the value, merits and efficacy of Spectro-Chrome, to prove conclusively that no violation of your or any other statute is created by the marketing of my patented apparatus.

9. In order to bring this matter to a satisfactory conclusion, I may suggest that you send your duly accredited representative by appointment to visit me personally in my Research Laboratory here, so I may have the pleasure of giving him first-hand scientific demonstrations of the marvels of this invention, which has stood successfully all tests to which it has been put during the past twenty-one years of its presentation to the suffering World.

10. Incidentally I may state, that the Maryland Academy of Sciences conferred on me the degree of Academician with Life Membership, for the origination of Spectro-Chrome Metry, after hearing a discourse given by special invitation before its members. Please refer to the inside of front cover of my SPECTRO-CHROME July 1922, herewith enclosed.

11. Awaiting your kind, early decision, before I start on my next tour, I have the honor respectfully to remain

Yours In The Service Of Mankind

DINSHAH P. GHADIALI, President-Treasurer.

DPG·S Enc: as specified. *Registered With Receipt Requested.*

Next month, the Honorable Commissioner promised investigation:—

FEDERAL SECURITY AGENCY
FOOD AND DRUG ADMINISTRATION
WASHINGTON, D. C.

AF 19617 February 27, 1941

Mr. Dinshah P. Ghadiali
Spectro-Chrome Institute,
Malaga, New Jersey

Dear Sir:

Receipt is acknowledged of your letter of January 27 with further reference to your device "Spectro-Chrome." We note your proposal that we have a representative of this Administration visit you personally in your research laboratory at Malaga for the purpose of observing scientific demonstrations of your device.

We have given careful consideration to your proposal and we can see no reason why your suggestion cannot be followed. It so happens that those representatives of the Administration who are qualified in matters of this character have other engagements dealing with pending court cases and hearings on proposed regulations in connection with the enforcement of the Federal Food, Drug, and Cosmetic Act, which makes it impossible for such a meeting to take place for several weeks. If we do not hear further from you, we shall assume that such a postponement is agreeable to you, and we shall communicate with you at a later date for the purpose of making final arrangements regarding a definite appointment for the meeting.

Very truly yours,

W. G. CAMPBELL,

Commissioner of Food and Drugs

The procrastination not being suited to the business expansion tours of the Inventor, he took a different course to assist the dutiful Government Officers in expediting a proper understanding of Spectro-Chrome Metry:—

Friday, March 7, 1941
Chronos:-Indigo 35 Black

Federal Security Agency,
Food and Drug Administration
WASHINGTON, D. C.

Kind Attention: Honorable W. G. Campbell, Commissioner of Food and Drugs

Sir:

1. Please accept my best thanks for your kind communication of February 27, 1941.

2. Noting from it that your Department will not be ready for the investigation until the lapse of several weeks and not finding it possible to postpone my business tour any longer, I thought it best to fall in with your views and wait for the convenience of your investigation Officers.

3. It seems to me that your attitude is exceedingly sensible and businesslike; yet, I do not think it will be fair to your Department Officers to visit my Research Laboratory and endeavor to investigate something about whose Fundamental Principles they may have very little knowledge.

4. The development of the Science of Spectro-Chrome Metry was started by me in 1891. The Science and its practical commercial application were perfected after thirty years of deep scientific research. In 1920, those fruits of my labors of three decades were first presented to the public. This is the twenty-first anniversary of Spectro-Chrome Metry and at present there are nearly 7,000 (seven thousand) Equipments in use all over this Country, Europe, Java and India, with unequivocal success. I have two offices in India also.

5. Spectro-Chrome Metry is now an established, original Science and there is no guesswork or human equation of error possible in it, as was proven in thousands of cases declared "hopeless and incurable" by the Medical Doctors and Hospitals. Thus, thinking it would be unfair to your Department to expect it to gather a correct perspective of the ramifications of this work, I sent you by Book Post four volumes of my periodical SPECTRO-CHROME having printed case reports, editorials and the like extending from June 1922 to December 1938, three volumes of SPECTRO-CHROME METRY ENCYCLOPAEDIA and a copy of SPECTRO-CHROME HOME GUIDE, one of which is issued with each Spectro-Chrome Equipment for directions about the use of the apparatus in the family.

6. I may incidentally state that these Equipments are now sold exclusively to persons who are affiliated with Scientific Order Of Spectro-Chrome Metrists and are not available to the general public without such Affiliation. That Organization is a Not-For-Profit Corporation, chartered in the State of New Jersey and is controlled by an extremely rigid Code Of Honor, as you may find by casting a glance at Rule 22 in the Constitution And Rules also mailed to you. The Affiliates are thus joined into a solid Organization of well-meaning persons who believe in the Service Of Mankind and no quackery or unfair procedure is at any time tolerated.

7. Tomorrow, the 8th current, I am leaving on a tour to Detroit, Michigan, from which I shall return in a week. Then, I shall visit my Affiliates in St. Louis, Missouri, Milwaukee, Wisconsin and other places, where I have many thousand students in all walks of life. I shall be happy to welcome your investigators at any of these Meetings.

8. From this tour in the Middle West, I shall be free toward the end of May. In the month of June 1941, I have my usual annual instruction class in my Auditorium here. I believe that if your Departmental investigation be conducted at that time, after your Agents have gone over the volumes sent, it will be an exceptionally good opportunity for them to acquaint themselves with this unique Science.

9. Of course, everything is left to your own good judgment and discretion. As a Scientific Researcher, I placed the fruits of my labors at your disposal and it is entirely upto yourself to determine the issues which you may think are involved in this pending matter.

10. I have the honor to remain, respectfully

Yours In The Service Of Mankind
DINSHAH P. GHADIALI, President

D:CS
Registered With Return Receipt Requested.

Dinshah completed his stated tours and returned to the Central Office. He had parcelled many pounds of those volumes to the Administration and was willing to abide by the convenience of the busy Officers, but his Pacific Coast tour was now due and not hearing from the Government, he sent another letter as a reminder:—

Tuesday, May 27, 1941,
Chronos:-Indigo 29 Black.

Federal Security Agency
Food and Drug Administration
WASHINGTON, D. C.
Reference: AF 19617 Kind Attention: Honorable W. G. Campbell, Commissioner of Food and Drugs.

Sir:

1. I have the honor to draw your kind attention to my communication of March 7, 1941 and the several books pertaining to Spectro-Chrome Metry parcel posted to you.

2. I completed my tour of the Middle West as I informed you before and a week ago returned to my Central Office here. I am now becoming ready to go on my Western tour and shall be away about three months.

3. Knowing the pressure on your valuable time and also appreciating that your investigators will need an appreciable amount of time to study the books sent by me, I take the liberty of making a respectful suggestion that you give me an appointment right in your own office in Washington, D. C. and I shall bring my documents, evidence and scientific paraphernalia to demonstrate the Fundamental Principles and the value as well as merit of my unique and original Science.

4. It is now over twenty-one years ago that I presented it to the World and what stood the test of time with all the difficulties to undergo, added to the skepticism of the World, will be apparent right to you when you hear me personally and see the proofs I may present.

5. I do not like to have you feel that I am endeavoring to put any pressure on your time and courtesy. I offered a sheer suggestion to facilitate the formation of a definite opinion as regards the genuineness of my Science which I feel sure will appeal to you from all angles of provable facts.

6. Incidentally, I may remark without appearing commercial-minded, that Spectro-Chrome should prove of incalculable value in saving the lives of our warriors, in case of Burns and the like. The Pain disappears in every such case in less than One Hour! If given the opportunity, I shall feel happy to donate a number of Spectro-Chromes for use in specific centers of the Government.

7. Awaiting your kind reply, I have the honor to remain most respectfully

Yours In The Service Of Mankind
DINSHAH P. GHADIALI
President; Originator of Spectro-Chrome Metry.

DPG:S
Registered With Receipt Requested.

The overdue reply soon came from the Acting Commissioner:—

FEDERAL SECURITY AGENCY
FOOD AND DRUG ADMINISTRATION
WASHINGTON, D. C.

AF 19617 June 6, 1941

Colonel Dinshah P. Ghadiali
Spectro-Chrome Institute,
Malaga, New Jersey

Dear Colonel Ghadiali:

We have your letter of May 27 with further reference to your offer to furnish information dealing with Spectro-Chrome.

We have, of course, already received the books which you forwarded to us with your communication of March 7. We had hoped to be able to visit you at Malaga some time ago, but circumstances beyond our control made it impossible to carry through our original plans.

You now suggest that we arrange for an appointment with you at Washington in lieu of our visit to Malaga. We shall be pleased to make an appointment to see you at your convenience.

We suggest 1:30 P. M. on Wednesday, June 11. If this is not convenient, please let us know. The address is Room 4803, South Building, 12 and C Streets, Southwest, Washington, D. C.

Very truly yours,
GEO. P. LARRICK,
Acting Commissioner of Food and Drugs

The letter was undoubtedly written in a spirit of fairness and openmindedness, but, meanwhile the Originator had already packed his automobile with the scientific apparati and the program having been arranged, it was not possible to visit Washington, D. C. He had announced his departure for June 14, 1941 and any delay would set back numerous appointments. He wrote about his inability to make the appointment which would have cleared the pending issues for all time. However, it was unavoidable. His complex duties demanded the sacrifice and reluctantly he wrote:—

Monday, June 9, 1941
Chronos:-Blue 37 Black.

Federal Security Agency
Food and Drug Administration
Room 4803, South Building
12th and C Streets, Southwest
WASHINGTON, D. C.
Reference AF 19617 Kind Attention: Mr. George P. Larrick, Acting Commissioner of Food and Drugs

Dear Commissioner Larrick:

1. Your very kind communication of the 6th current was received this morning. I sincerely wish I had received it a few days earlier.

2. I thank you cordially for the appointment, but the demonstration apparatus is already packed in the truck for transit to the Pacific Coast and I too am leaving with it this week. I, therefore,find it not possible or convenient to be at your service on Wednesday the 11th current.

3. I truly regret the missing of this opportunity so graciously placed by you at my disposal, but on the return leg of my tour from California, passing through Washington, I shall present myself personally to you in the end of August or the early part of September, 1941 and shall gladly fix a day and time suitable to the convenience of yourself and your investigators.

4. Meanwhile, I am enclosing my *Spectro-Chrome* of December 1923-24, in which you will find a report of the journalistic investigation made by the prominent daily and other

newspapers of Oregon when I went to the West first to introduce my Science of Spectro-Chrome Metry and the apparatus Spectro-Chrome. A glance at pages 449 on is requested.

5. I have the honor to be very respectfully

Yours In The Service Of Mankind

D:CS ENC. as stated DINSHAH P. GHADIALI,

Registered With Receipt Requested *President.*

Thus, the matter rests. For 21 years, the Originator has been clamoring for investigation, but his requests fell on closed ears. He feels thankful to the Federal Government for placing this opportunity at his disposal and he assures the Officers that he will reciprocate by giving them free access to all his documents, proofs and evidence, to substantiate the claims of the Science of Spectro-Chrome Metry and the apparatus Spectro-Chrome.

——o——

FIRST ST. LOUIS MISSOURI PLANET

(HOTEL MELBOURNE, ST. LOUIS, MISSOURI)

Meeting Of Thursday, May 1, 1941

Successful Use Of Spectro-Chrome In:—Kidney Disorder, Cough.

Reported By:—VIOLA C. JEUDE, *Recorder*

Deputy Clemens M. P. Neumann was in the chair.

The Meeting was opened with a discussion about the coming visit of Colonel Dinshah, May 16, 17, 18, 1941.

Charles S. wanted to know if the Deputy had received any Circulars. The Deputy replied, not at present, but when the Circulars will arrive to be busy and interest your friends and if your supply becomes exhausted, call him for additional Circulars and distribute them, as he knows the Lectures will be very interesting, with Motion Pictures.

He said he felt sure Colonel Dinshah would begin with the Birth of Spectro-Chrome and while this subject had been heard before by some of the old time Affiliates, it would be new to the new Affiliates. He felt sure if the old time Affiliates paid close attention, they would derive as much benefit as the new Affiliates, as each time Colonel Dinshah lectured, one could always find a new idea and he has a way of presenting the same thing with variations, regardless of how often you hear the same Lecture, there are always new thoughts.

There will be six Lectures. They will be so arranged that each will present a topic in itself, but after all are heard it will be a consecutive affair, so it will be advisable to attend and hear all if possible, bearing in mind that the Lectures are given first to benefit you and secondly your friends as well as the Public. No Smoking will be allowed, as usual, no written notes may be taken. He felt sure Colonel Dinshah would answer any question relevant to any of these Lectures.

Rules 52, 53, 54 and 55 of the Constitution And Rules were read and each Rule was explained, step by step. The Deputy asked if there were any questions about them. There were no questions, as the Deputy went over them so thoroughly, he did not leave room for any.

The Deputy then asked for Case Reports. There being none, he said he had a few to present.

1—Joseph P., 50 years old, brought a paralyzed girl for Tonation. He decided to be itisized himself. The Kardoscil was 63, Spiroscil 16. The Chart showed Infra-Green Predominance. Of course, Green was given to start, then Turquoise, Magenta on Areas 4 and 18, later Lemon and Orange Atcowas. In two weeks, he said he "had never felt better."

2—Mary P., 62 years old, had Disorder in Areas 18 and 19; could not stoop, complained of a tightness there; Kardoscil 82, Spiroscil 20. Started with Green Systemic Front and Back, Magenta on Areas 4 and 18, then Orange Systemic Back; was normalated.

3—Henry G. had Cough for 9 years in the Winter. His Chart showed Ultra-Green Predominance. He started with Green, Magenta on Areas 4 and 18, Orange Front and Back, Turquoise on Areas 2 and 3, finally finishing with Lemon Atcowa.

The Deputy asked if there were any questions in regard to the above Cases.

Thelma S. said she had a Cough similar to the one he described. She said she had a Cold about four years ago and that developed into a Cough. She coughed about 1½ years, went to Medical Doctors and Chiropractors and finally, to a "Specialist". At last she was rid of it about 6 months later. About 6 months more, it returned and she had to doctor again for another 6 months before it was entirely gone. Around 3 weeks ago, she had another Cold and it developed into another Cough. She used Lemon and Orange Atcowa, Magenta on Areas 4 and 18, but does not seem to be able to conquer it. What should she do?

The Deputy said Colonel Dinshah had forbidden him to give any information, unless the Affiliates filled out Sufferer Charts, as then information was complete and the Kardoscil and Spiroscil would be mentioned and that was one of the main factors. Thelma S. informed him that her Ratio was 4 to 1, 92-22. The Deputy said the Ratio was good and did she have any other Symptoms? She said she was normal. He questioned her in regard to Frontelim and Backelim. She said they were proper. She spits much Phlegm and sneezes much. He questioned her in regard to the Rational Food Of Man. She said she did the best she could, as she boards and has to take her meals out. The Deputy said a Cold is an Inflammation and he could not tell her what to use unless she presented the complete Sufferer Chart. She should refer to the Spectro-Chrome Home Guide, study the Green and Turquoise Atcowas. Lemon would loosen the Phlegm and Turquoise should be irradiated for the Dry Sensation.

Turquoise was then studied from Spectro-Chrome Home Guide, page 53 and gone over step by step.

Charles S. then asked the Deputy if he did not think it was a good idea to give each Affiliate one of these Sufferer Charts to take home and in that way they could get the information when they wanted and needed. In the Case just described the Sufferer would have to wait a whole month before she would get the proper relief.

The Deputy thought it would be permissible and he would consult Colonel Dinshah when he came to St. Louis.

Charles S. addressed the Deputy, saying he thought it advisable to inform the newcomers that if the wrong Tonations were taken, it would not be harmful.

The Deputy then explained how the Physical Body shows a Predominance of either the Ultra-Green or Infra-Green Atcowas. He requested them to study the Table Of Spectral Divisions, page 24 and Spectro-Chrome Affinity Waves, page 25 of Spectro-Chrome Home Guide. He also explained that if one did take the wrong Atcowa, it would not hurt the Body, but the results would be impeded and the process of Normalation would take longer.

Spectro-Chrome Axioms, page 60 of the Spectro-Chrome Home Guide were then read by the Sub-Deputy and gone over step by step.

The Deputy spoke about Colonel Dinshah's ability in examining the Aura.

Charles S. said there was the case of the young man in a Sanatarium suffering from Tuberculosis and given over as a hopeless case. He was brought to the Originator and after Colonel Dinshah studied his Aura, the young man was made well in a week's time.

The Deputy said he was tonating a woman at the present time from whom two months ago the Medical Doctors wanted to cut out some Ribs and deflate the Lungs. At the present time she does not cough any Phlegm. He will make more of a report later when the woman has recovered.

He also reported that a man was under the Medical Doctors' care for five years. The man told him the Doctors do not know it all. He told the Deputy when he used Spectro-Chrome, he used it ½ hour on the Front and ½ hour on the Back. The Deputy instructed the Affiliates that they should never do this, but to take the complete hour for one or another.

In explaining about the Vital Ratio, the Deputy spoke of a friend who they said was dying in the Hospital. He went to see him.

While the person did not seem to be very ill, he asked the Nurse how his condition was and what the Heart Beat and Breathing were. The Nurse said he was very low and they expected him to die. The Heart Beat was 72 and the Breathing was 18! The Deputy said if all the people only knew that a 4 to 1 or 5 to 1 Ratio was good and if it could be maintained, they would get well. The man lived 4 years after that!

A question was asked if it were advisable to drink water more between Meals or should most of the drinking be done at Meal time?

The Deputy said that Cold or Ice Water should not be used at Meal time, that if one were compelled to drink Cold Water, to hold it in the Mouth until of Body Temperature and one would find that two or three Mouthfuls would be sufficient. Thirst was in the Mouth and not in the Stomach. He said not to drink water before one sat down to a Meal.

Someone asked about drinking water on arising in the morning, especially Hot Water. The Deputy thought it was good and it would help to flush the Alimentary Tract.

Charles S. asked the Deputy if he did not remember that Colonel Dinshah in one of his Lectures said not to drink Hot Water on an empty Stomach, as this caused the Muscles to relax and the Stomach to expand. The Deputy did not remember. So, please Colonel Dinshah, would you kindly answer the Water question, as it was not answered clearly, as the Deputy said he had occasion to dine with you several times and you did consume much liquid. He did not know how much, if any liquid, was consumed between Meals, as several Affiliates said you were the only Lecturer they ever saw who did not drink anything during your Lecture.

Nathalee B. said her father had trouble with his Bowels. She gave him an Orange Tonation. After his Backelim, he had a soreness. She wanted to know if she had given him one of the Ultra-Green Atcowas, would not this have decreased the Elimination. The Deputy informed her that it would only have allayed the Pains, as there must have been a slight Irritation or Inflammation.

Charles S. teased the Deputy about eating Doughnut and the Deputy said he was off Doughnut as well as Garlic Soup!

Charles S. said that Medical Doctors claim that only 50% of the people have healthy Colons.

The Deputy said he had occasion to see an Autopsy and he was surprised to see the large space in the Human Body. It was large enough to hold three Spectro-Chrome Metry Encyclopaedias!

Lena W. said when she washes the floor or when she awakes she has a scary feeling. She was advised to use Yellow on Areas 8, 9 and 10 for 3 or 4 weeks. She also said she had a bitter taste and was advised to use Orange and eat Rice and Yellow Split Peas. She said she did not care for the Rational Food Of Man and has a terrible time in eating the Rice and Split Peas. She said she cooked them with Salt and Butter, but she could not get them down. Asked if she used any Spices, she said not. The Deputy informed her that was where the trouble lay, that the Spices acted as an Orange Wave in the System and helped to digest the Food, that unless she adhered strictly to the Rational Food Of Man, she was putting Poisons back into the System. Spectro-Chrome is not a Magic Lamp of the 20th Century by itself, as certain parts of the Body may be depleted and these will be benefited by Spectro-Chrome, provided one co-operates with the Rational Food Of Man.

It was suggested that none should miss the Lecture on Food.

Marie J. asked if she brought her three-year-old boy, would Colonel Dinshah look at him, as the Growth in his Throat is getting larger and she has to press it and drain it, when he throws his Head back it sticks out much. The Deputy said he did not think Colonel Dinshah would look at the boy, but that she should fill a Chart and present it to the Originator when he came. He asked her if the boy were sick or felt bad in any other way. She said he is never sick, that he does not even have a Cold like her other children. The Deputy said that Nature sometimes uses these avenues to eliminate the impurities and not to be alarmed if the Ratio be right.

The Deputy suggested that each of the Affiliates bring at least one friend to hear the Lectures. In that way you can prove to your friend by listening to the Lectures and seeing the Demonstrations that there are merits to Spectro-Chrome.

Charles S. said that there are 3 or 4 cases on his route that would be benefited by Spectro-Chrome. In one case the Medical Doctor advised the woman to move to another place on account of the dust and dirt. He said he informed her that unless she got rid of her rugs and carpets it would do her no good.

He also said that another patron had to stop wearing a dress shield, as the Medical Doctor told her that in the process of making them they used Arsenic.

C. Earl B. said he heard that in a city in Illinois, Green Lights were installed in the basements to improve the Eyesight.

The Deputy said he thought the old style coal oil light was the easiest on the Eyes, as it has a soft and soothing effect. Charles S. said there were more glasses worn today than ever.

Marie J. said her boy had a deep gash over his Eye. She used Green and Turquoise and asked if it would be correct to use Indigo. The Deputy explained that Indigo was a Phagocyte Builder and will destroy the harmful Micro-Organisms and will not cause any Infections.

Discussion going on Glands, the Deputy said when he viewed that Autopsy, the Spleen was shaped like an umbrella and it was Violet in Color; in fact, each Gland was different in Color. He said that is surely God's handiwork and surely must be for a purpose.

Attributes Of Lemon from the Spectro-Chrome Home Guide, pages 51 and 52 were read.

Richard F. J. asked if it were not a good idea to be itisized twice a year.

The Deputy said it would be a good thing to do, but if he suggested anything like that, people would think he was working for his own personal gain.

Charles S. said he wanted to get an Itisometer for home use, but was informed he would not be able to get it until after the War. The Deputy said he wanted one, too, that he could take with him when he made calls.

The Affiliates were requested to study page 26 of Spectro-Chrome Home Guide, on Spectro-Chrome Technique or Method Of Using.

The Deputy asked Karl L. how his wife was. He said she was very well pleased; in four years she could not have a natural Backelim and for 6 months had Colonic Irrigation 3 times a week. She is well satisfied with results from using Spectro-Chrome. His daughter will buy an Equipment as soon as she can get the money together.

EDITORIAL REMARKS

Information given without Sufferer Chart completely itemized, would be nothing short of quackery. Follow the rigidity of Spectro-Chrome Institute. It will educate the untrained new-comers and will maintain the beauty of our Precision. A few might grumble, but, we are not intending to please; our slogan is Service To Mankind. Let the lazy ones know the value of correct information. Those Charts are FREE for the Planets and are issued for the asking.

Wrong Atcowa has no hurt; it will refuse to normalate.

The Human Aura being the Emanation of the Visible Physical Vehicle, is a factor of great importance. When the present crisis blows away, more instruction in the direction of the Invisible may be considered.

Drink nothing when the mouth is full of solid material. Never place any food or drink in the mouth on rising in the morning without cleaning the mouth and teeth. Be hygienic, please. Hot water should not be drunk, because, it will adversely affect the process of Digestion. The habit of drinking hot fluids is injurious. Nothing warmer than the Body Temperature should be thrown into the Stomach. Ice Water is the ruin of this nation. Drink cool, but not shivery.

Giving Orange Atcowa for motivating the Bowels, is like trying to start a stalled automobile uphill on high gear. Cases of Persistent Imbackelim should be started on the Lemon and lowered to Yellow afterwards. Proper Food, with Yellow on 6-7-8-9-10 will remove such stagnation, if the natural Posture be used as illustrated in Spectro-Chrome Home Guide.

No Colon jammed with the products of carcases can be healthy; no wonder the Medical Doctors admit it to be Semi-Colon!

Lena W. should send General Advice Chart. Those "scary" feelings result from Heart as well as Spleen Disorders. Those feelings affect the Emotional (Astral or Psychical) Vehicle. Americans do not know much of vegetarian cookery. If Lena W. is unable to cook the Chaval and Dal as instructed, she needs a Vinag.

If Dinshah would start giving "personal interviews and examinations", all the cripples of the World would plague him to death. He wants to live and serve and Spectro-Chrome Metry avoids such "consultations". No—not until the grafty Medical Laws are repealed will he consent to see any Sufferers.

Rugs and carpets are the hot-beds of Microbes; so are feather pillows and furs. Discard those filthy articles.

For full Chemistry, Light must be as made and neither polarized nor tinted.

The Portable Itisometer is ready and you saw it exhibited, but the present conditions do not justify the commercializing expenses. In time to come, not only will there be a Spectro-Chrome in Every Home, but there will be an Itisometer With Every Spectro-Chrome.

—o—

COMMEMORATION OF 21st ANNIVERSARY

Steps Taken To Perpetuate Spectro-Chrome Institute, As A Benevolent Foundation Devoted To The Service Of Mankind.

Present Business Corporation Will Be Dissolved On Monday, September 15, 1941

When Spectro-Chrome Metry was presented to the World on April 26, 1920 and later when Spectro-Chrome Institute was chartered as a Joint Stock Business Corporation, on January 19, 1922, it never was in the mind of the Originator Dinshah P. Ghadiali, to use it as another money-grabbing institution. His ideals were to serve Suffering Mankind by saving humanity from the Poison Fraternity, Pus-Punching Savagery and the Knife-Stabbing Butchery. To accomplish his purposes, he worked without salary for 20 years, but, since the ruinous financial tax policies of the hypocritical New Deal, he found his aims blocked at all points by hedges of unconstitutional legislations. Thousands of dollars which he intended to use for service expansion were dragged away by the thoughtless and wasteful Government actions and were he to sit quiet in such quandary, the World would again be without its Spectro-Chrome before long.

To avoid such catastrophe, sooner than surrender his main objects and principles, the Originator calmly took in the whole situation and cheerfully decided on turning over his entire property to a Non-Stock, Non-Profit Benevolent Foundation, so the public may be perpetually protected and served.

To that effect, the following Notice was dispatched by registered mail to all the Stockholders of the Corporation, on Friday, June 13, 1941.

NOTICE

1. Please be informed that on Monday, September 15, 1941 at 2:00 P. M., there will be held in the Auditorium of our Central Office at Malaga, New Jersey, a Meeting of all the Stockholders, Preferred as well as Common, Voting as well as Non-Voting, for the transaction of the following business. The Propositions stated hereunder will be submitted to a Vote of the Stockholders:—

(A) That the Board Of Directors having passed a Resolution, declaring it advisable and most for the benefit of the Corporation that it be dissolved, the Stockholders having duly considered all the circumstances involved, hereby consent in accordance with the Law to dissolve the Corporation.

(B) That the Board Of Directors are hereby authorized to set a convenient date for the Dissolution of the Corporation and are hereby empowered to use their best judgment in paying off the Creditors of the Corporation.

(C) That the Board of Directors are hereby empowered to file such Certificate with the Secretary Of State for New Jersey and do whatever is called by the Law pertaining to General Corporation and applicable to the said matter.

(D That the Board Of Directors are hereby empowered to sell the Property and Assets of the Corporation or dispose of them in whatever manner is deemed fit in their judgment in the best interests of the Corporation.

(E) That the Board Of Directors are hereby empowered to file with the Secretary Of State for New Jersey a new Charter under the Non-Stock Not-For-Profit Corporation Law to perpetuate our existence as an Educational, Scientific, Literary Organization and continue the same Corporate name SPECTRO-CHROME INSTITUTE.

(F) That the prior action of the Board Of Directors in redeeming the Preferred Stock at Par Value is hereby sustained.

2. You are further notified that each share of $10.00 carries One (1) Vote in the Dissolution procedure.

3. Your kind presence is earnestly requested and it will be to YOUR benefit to send by Registered Mail your Stock Certificate to us PRIOR to the said Meeting, so we may redeem it at Par value, as after the Dissolution Meeting and the passage of the stated Resolutions, no claim for redemption can be entertained.

Yours In The Service Of Mankind
SPECTRO-CHROME INSTITUTE
by Stephanie Hand., Secretary

To avoid any interruption of business during this change and transfer, Dinshah had requested a Ruling from the Honorable Secretary Of State For New Jersey and the following answer was promptly given:—

STATE OF NEW JERSEY
DEPARTMENT OF STATE
Trenton, June 6, 1941.

Mr. Dinshah P. Ghadiali,
Spectro-Chrome Institute,
Malaga, N. J.

Dear Sir:

I have your communication of the 4th inst. Immediately upon the filing of the certificate of dissolution of your corporation you may file in this office a corporation under the Act not for Pecuniary Profit, bearing the same name.

Yours Very truly,
THOMAS A. MATHIS,
SECRETARY OF STATE

Thus, without any hitch, our activities from a Business Corporation will slide into those of an Educational, Literary, Scientific, Benevolent Corporation and the enormous tax burden, which would have compelled us to raise the prices of our products, will be lifted from our head.

The move is in the right direction to perpetuate our Service Of Mankind. The Originator, on that day of Monday, September 15, 1941, will voluntarily surrender ALL his holdings for the benefit of the World and continue to work just as cheerfully as in the past, so Spectro-Chrome may live forever. All creditors will be paid in full and so will all the Preferred Stockholders.

SPECTRO-CHROME — IN EVERY HOME!

FIRST DETROIT MICHIGAN PLANET

(HOTEL TULLER, DETROIT, MICHIGAN)

Meeting Of Thursday, May 1, 1941

Successful Use Of SpectroChrome In:—Measles.

Reported By:—SOFIE HALL, *Recorder*

Sub-Deputy Ralph Clark was in the chair in the absence of our Deputy Lucille Greenham.

The Case Reports were begun immediately and they are as follows:—

Lita B.'s baby had the Measles, was covered with them and was very sick. Green and Orange were used with great satisfaction. Her other little girl was also coming down with them, but the first Tonation made them disappear. The Sub-Deputy read Rule 17 of the Constitution and Rules, about **Supreme Authority.**

An Affiliate asked if it were violating the Rules of Spectro-Chrome to buy an electric bulb for the Spectro-Chrome Equipment elsewhere.

The leaflets concerning Colonel Dinshah's Lectures in Detroit were given to the Affiliates for distribution.

Gladys Clark spoke about showing the Affiliation Card at the door and feeling no resentment about it, as we should be glad to obey the Rules.

The Sub-Deputy called Albert Greenham in regard to Colonel Dinshah's Lectures. He urged all to come and bring as many as possible, as Colonel Dinshah excels every body in the United States Of America, on healing by Color. There are no other authorities on it. He related that Colonel Dinshah several times had challenged the Medical Doctors to a debate on the Healing System. Colonel Dinshah would pay all the expenses and let the profit go to some charitable organization. Albert Greenham stated he would like to have everybody in Detroit hear the series of Lectures by Colonel Dinshah; fill the Arabian Room to capacity. We should all see what we can do for him this time.

Ralph Clark then called L. Roy Dodman to the platform and the Affiliates applauded.

He spoke on keeping our Bowels clean. This alone would help in normalating any Disorder. Imbackelim causes many Disorders. The Rational Food Of Man helps to keep the Bowels in good condition. We should always drink the raw milk.

Emma H. stated her husband had Ulcers of the Stomach ten years ago and he would drink three or four quarts of milk a day and is very much improved.

Albert Greenham stated we should never drink milk if it is

pasteurized and never drink milk fast, but to "chew" it, as it will thus form small beads in the Stomach. Any food passed through a "process" loses many of the better qualities.

A detailed account of Henry S. was given.

The Sub-Deputy read an article out of The Record entitled "Take It From Me". The Affiliates applauded.

EDITORIAL REMARKS

About the Bulb, Rule 78, Section 3, is quite clear. It is a part of the Integrity Policies, as Affiliates are not expected to disturb the uniform character of the work. Many Affiliates make a mess of the Tonation, by using Bulbs of the wrong voltage, wrong light centre or wrong wattage. That Rule provides for honesty and it is upto the Affiliates to maintain their Integrity or some day incur Discipline for Violation. All our standard Bulbs have the name of DINSHAH and SPECTRO-CHROME etched on them. Use *no* other.

Please send the "Detailed account" mentioned in your Report.

DINSHAH EDUCATIONAL BROADCASTING SERVICE

Non-Stock, Non-Profit Corporation Chartered In New Jersey

Soon after arriving in Malaga, New Jersey, Dinshah fulfilled his promise, given in Detroit, Michigan, about establishing a weekly newspaper and Radio Broadcasting Station. He took the first step and filed the following Certificate with the Clerk of Gloucester County and filed and recorded it in the office of the Secretary Of State for New Jersey:-

Certificate Of Incorporation

OF THE

Dinshah Educational Broadcasting Service

THIS IS TO CERTIFY:-

That the undersigned do hereby associate themselves into a Corporation under and by virtue of the provisions of an Act of the Legislature of the State of New Jersey, entitled "An Act to incorporate associations not for pecuniary profit", approved April twenty-first, eighteen hundred and ninety-eight, and the several supplements thereto and acts amendatory thereof.

Constitution

NAME

ARTICLE 1:—The name of this Corporation is:-

DINSHAH EDUCATIONAL BROADCASTING SERVICE.

PURPOSES

ARTICLE 2:—The Purposes for which this Corporation is

formed are:-

(a) To encourage and broadcast Truth for the Service Of Mankind.

(b) To disseminate knowledge concerning all Physical, Metaphysical, Spiritual and Occult Sciences and in the broadcasting thereof to establish, utilize and maintain all Channels Of Publicity available for the Service Of Mankind.

(c) To uphold unreservedly and spread correct comprehension about the prerogatives of Liberty, Property, Security and Contract, which are universally established as Human Birthrights.

(d) To establish and maintain an International Fraternal Nucleus, for the exchange of views among the peoples of the World, to acquire and disseminate facts conducive to the Service Of Mankind.

(e) To form and maintain Fraternal Centres in connection with the International Body, for the exchange of views among individuals interested in the promotion of the Welfare Of Mankind.

(f) To organize and maintain Bureaus Of Educational Service, for enlightenment through Discourses, Publications and all other legitimate means for dissemination of literary and scientific knowledge, in conformity with the highest standards of Integrity.

(g) To advocate and take all lawful measures as will accomplish Freedom of Thought, Judgment and Action to insure to Humanity now and forever, the Birthrights of Liberty, Property, Security and Contract.

(h) To encourage such sensible Legislation as will promote the Welfare Of Mankind.

(i) To spread by all lawful means, the principle of the Fraternity Of Mankind without any distinction, throughout the World.

PRINCIPAL OFFICE

ARTICLE 3:—The location of the Principal Office of this Corporation is in the village of MALAGA, TOWNSHIP OF FRANKLIN, COUNTY OF GLOUCESTER, STATE OF NEW JERSEY, UNITED STATES OF AMERICA.

AGENT

ARTICLE 4:—The name of the Agent in charge of the Principal Office of this Corporation and upon whom process may be served, is:—DINSHAH P. GHADIALI.

TRUSTEES

ARTICLE 5:—The number of Trustees of this Corporation is Three.

ELIGIBILITY OF TRUSTEES

ARTICLE 6:—The eligibility of any person for nomination as a Trustee, shall be determined in accordance with the By-Laws, as may be in force from time to time.

FIRST YEAR TRUSTEES

ARTICLE 7:—The names and postoffice addresses of the Trustees elected for the first year of the existence of this Corporation are:—

DINSHAH P. GHADIALI....MALAGA, NEW JERSEY.

IRENE GRACE DINSHAH....MALAGA, NEW JERSEY

CICELY STURGES..........MALAGA, NEW JERSEY

POWERS OF THE CORPORATION

ARTICLE 8:—Among its inherent lawful Powers, the Corporation shall have the Power:—

(a) To have perpetual succession by its corporate name.

(b) To sue and be sued, complain and defend in any Court.

(c) To make and use a Common Seal and alter it at pleasure.

(d) To appoint such Officers, Managers, Representatives and Agents, as the affairs of the Corporation may require.

(e) To make By-Laws for the management of its property and the regulation of its affairs, in accordance with the Law.

(f) To make and enter into Contracts.

(g) To take and hold by Lease, Gift, Purchase, Grant, Devise or Bequest, any property, real or personal, necessary or desirable for attaining the Purposes of the Corporation and carrying them into effect, subject however, to any alteration or modification hereafter made by General Law as to the amount of real and personal property to be held by the Corporation.

(h) To transfer and convey its real or personal property.

(i) To borrow money for the Purposes of the Corporation and issue Bonds therefor and secure the loan by Mortgage or otherwise.

(j) To exercise any Corporate Powers necessary or incidental to the fulfillment of the Purposes above enumerated.

AFFILIATION

ARTICLE 9:—Only Contributors devoted to the Service Of Mankind, as may be provided in the By-Laws, shall be eligible to Affiliation with this Corporation.

In witness whereof, we hereunto set our Signatures and Seals, Friday, the 13th day of June, 1941.

Signed, sealed and delivered in the presence of............

Dinshah P. Ghadiali
Irene Grace Dinshah
Cicely Sturges
Stephanie Hand.
William D. Caccese

Incorporators.

State Of New Jersey,
County Of Gloucester.

Be it remembered that on this 13th day of June, Nineteen Hundred and Forty One, before me, a Notary Public, personally appeared Dinshah P. Ghadiali, Irene Grace Dinshah, Cicely Sturges, Stephanie Hand. and William D. Caccese, who I am satisfied are the persons named in and who executed the foregoing Certificate Of Incorporation and I, having first made known to them the contents thereof, they did each acknowledge that they signed, sealed and delivered it as their voluntary act and deed.

(SEAL) DOROTHY CACCESE
Notary Public Of New Jersey
My Commission Expires January 29, 1942

Received June 13, 1941 at 11:45 A. M. and recorded in the Clerk's Office of Gloucester County, at Woodbury, New Jersey, in Book 4, of Certificates Of Incorporation for said County, page 601, etc.

R. EDWARD KLAISZ,
Clerk.

Endorsed
Filed and Recorded
JUNE 16, 1941
THOMAS A. MATHIS,
Secretary Of State

Now that the Corporation is chartered under the Laws of the State Of New Jersey, the way is open for progress. Mere talk ceases and actual activity commences. Those who lifted their hands and promised co:operation, are welcome to remit their Contributions, to the company, because a Bank Account is already opened for the purpose. Remittance by check or Money Order should be for $25.00 or any multiple of 25, to the order of **Dinshah Educational Broadcasting Service.** ANY person may contribute to the Cause, subject to the approval of the Board Of Trustees. Amounts and names will be published from time to time. The By-Laws and the Rules also, will soon be formulated for Affiliation and for the conduct of the corporate affairs.

FIRST MERRILL WISCONSIN PLANET

(BADGER HOTEL, MERRILL, WISCONSIN)

Special Educational Meetings Of May 1, 2, 3, 4, 5, 1941

Reported By:—HILDA SCHNEIDER, *Recorder*

Deputy Lydia Danner was in the chair.

Colonel Dinshah was presented to the Affiliates by the Deputy.

At his request, certain documents were read to the audience by the Recorder.

The Originator asked the Affiliates if the Board Of Trustees did the right thing in expelling Michael Ruczkall of Detroit, Michigan for giving Aid and Relief by Spectro-Chrome Metry to persons not affiliated as Collaborators. It was opined unanimously that the Board Of Trustees did the right thing.

The motion was made by the Deputy and seconded by the Sub-Deputy that the Educational Lectures should be delivered as follows:—

Friday, May 2 — Fundamental Principles of the Science Of Spectro-Chrome Metry.

Saturday, May 3 — Pictorial Lecture.

Sunday, May 4 — Rational Food Of Man.

Monday, May 5 — Disorders And Their Tonation.

After hearing those Lectures, we feel that we can show our appreciation for his marvelous efforts in giving us the Free Lectures on this wonderful Health and Healing Science only by purchasing Spectro-Chrome products and selling more Equipments.

The Deputy expressed a wish that all who attended the Food Lectures would adhere to the Rational Food Of Man 100% as examples to the public. She feels that our Food was our biggest problem and that even though the audience was comparatively light they benefited by the Lectures; it cleared many of their problems.

We wish to express our heartfelt thanks to Colonel Dinshah for giving us the opportunity to hear his wonderful Lectures. We are very sorry that more of the Affiliates could not be present but the Originator should know that his efforts were not in vain, for those who attended the Meetings feel that his instructions were very worthwhile and they have been a great help in furthering our education in Spectro-Chrome Metry.

EDITORIAL REMARKS

We thank you and appreciate your good will.

TWENTY FIRST ANNIVERSARY CELEBRATION OF THE SCIENCE OF SPECTRO-CHROME

Weekly Newspaper And Radio Broadcasting Station Coming

In Milwaukee, Wisconsin, the Milwaukee Auditorium witnessed a scene of great rejoicing. The six lectures delivered by the Originator Dinshah P. Ghadiali, were heard by crowds. The date of Saturday, April 26, the birth date of the Science in 1920, will ever be remembered with gratefulness by the thousands of sick persons who received a new lease on life through Spectro-Chrome. Many important matters were discussed and much progress and more interest were in evidence. Among the Case Reports the following drew attention:—

Lydia Danner—"I had cases of Blood Clot, High Blood Pressure and Paralysis. My niece was paralyzed; Spectro-Chrome made her 100% right. A boy of 13 was given up after an injury. Today, he is in perfect health. I had wonderful results in septic cases. In 3 days many were normal. Of course, Persistent cases would need patience. People do not give enough time. They run to Medical Doctors for 10 or 20 years, then expect a miracle from Spectro-Chrome. However, they should follow the Rational Food Of Man. I do not want to be without Spectro-Chrome. I use several in my practice." (Applause)

Albert B. Hale—"I have had plenty of experience with Spectro-Chrome. My wife had High Blood Pressure of 240. For three years, several Medical Doctors attended to her, but no results. They could do nothing; told her to make the will. Spectro-Chrome set her right. A year ago, my eldest son had a Collapsed Lung. He was taken to the Hospital in my absence. There they said he was not expected to live. He had several Injections. I went there and wanted him out. They said I was crazy. I took him home and gave him an Orange Tonation. Four Tonations and he sat up in bed. In 7 days, he went to work. My 2 year old grandson had a hand caught in a pulley. Flesh was burnt and bone left. Was in Hospital 10 days. They cut the Skin and wanted to amputate the little finger. I pursuaded my son to take him out. They said I was crazy. They would have to amputate the whole hand and leave the little fellow with a stump. Spectro-Chrome set the hand right; it is as it ever was. All Scars disappeared." (Applause)

Sub-Deputy Kurt W. Bader—"I reported many cases in the Planet Meetings. A woman of 46, had Operations on Tonsils, Gall Bladder, Ovaries, Uterus and so on. She heard of Spectro-Chrome and said, 'What can you do for me'? I said, 'try it'. Her legs were swollen to twice the normal size; she had a Kardoscil of 124 and Spiroscil of 24. I gave her an old Equipment. In 6 months, the Swelling was gone, the Heart was 78 and the Respiration 18. Another woman had Abscess on the Breast, with Pleurisy, Pain and Swelling on right side. She suffered 3 months; asked me what to do. Well, do as I tell you. Left Breast had a hard core swollen. Kardoscil 100, Spiroscil 20. I used Lemon Systemic, Orange on 4 and Magenta. In 4 weeks, 2 Abscesses broke and cores came out as big as a thumb. Healed from the inside. Medical Doctor wanted to operate, but, she is all right now." (Applause)

M. spoke about his own case of Pulmonary Tuberculosis. In 1939, he was in the Herman Keiffer Hospital. He went back and forth. Received no benefit, so finally went home. They told him to sell his home and go to Arizona. Cecilia D. spoke to him about Spectro-Chrome. He got a used one and now he is well.

Recorder Emma Schroeder—"Milwaukee should be proud to be having the 21 year Celebration. For 6,000 years, statements have been made about Medicine and the use of Drugs. The Earth is millions of years old; Spectro-Chrome is only 21, but, the progress made by it should satisfy us all. The activities of this week make me proud. Colonel Dinshah will be the most famous scientist. The Science is so simple, it will live to Eternity. Spectro-Chrome prolonged my mother's life and my life. I bought Pills by the hundreds. In another month it will be 5 years I knew of Spectro-Chrome. The Rational Food is important. The healing method may be right, yet the Food must be right too. Medical Science is full of cobwebs and is now obsolete. Spectro-Chrome put it out of business. (Applause)

Kurt W. Bader proposed and Albert B. Hale seconded that a vote of thanks be extended to the Milwaukee Auditorium Board and the Manager William Maas, for their great courtesy in placing at our disposal all the facilities needed to make the Celebration a success. Unanimously passed.

Deputy Frank S. Burns—"Spectro-Chrome is just a 21 year old upstart. It is still growing from its infancy, yet its banner was carried several times from the Atlantic to the Pacific Coast. It is known in many countries and there are several Spectro-Chrome apparati in India also; India is as you know the birthplace of the Originator. It is saving mankind in an inestimable manner. It has done for Suffering Mankind more than any other method—more rapidly and thoroughly. It dares to do anything no other did. Look at the case of the burnt girl Grace Shirlow! In a reasonably short time a complete new Skin was built on an area so large no other method could have done it. I do not hesitate to recommend Spectro-Chrome. We must be proud to be the pioneers. It is a privilege to listen to the Originator and heed his advice. I joined the work 7 years ago; never once regretted it. It does all that is claimed for it. However, one should not transgress the Laws of Nature. One must co:operate and follow the Rules laid down by the Originator. I never regret I have Spectro-Chrome in my home. I am thoroughly convinced of its efficacy." (Applause)

The usual Lectures were delivered to interested audiences and for the first time, the pictures portraying the life events of the Originator from his age of 2½ to the invention of the Spectro-Chrome, were shown on the screen. Finally, the inner workings of the various departments of Spectro-Chrome Institute were projected by Motion Picture and the audience expressed its appreciation by volleys of applause. The appearance of the Vice-President Irene Grace Dinshah (the wife of the Originator) on the screen, working in her office, was greeted with a vociferous round; Evelyn Roman, the Secretary of the Scientific Order Of Spectro-Chrome Metrists, being personally known to the Affiliates in Milwaukee, also received an applause. It was an unexpected treat to see the Alma Mater in action and carried home the point that Spectro-Chrome Institute was not a fly-by-night affair, but, a live Organization for unselfish and permanent Service To Mankind.

Similar lectures and demonstrations were given the following week in Merrill, Wisconsin and the Celebration was then opened in Detroit, Michigan. There, in one of the Sessions, Affiliate Albert E. Greenham being called on the platform, started a new ball rolling. Said he, "We have been talking of Spectro-Chrome many years, but it is all talk. We did nothing constructive to help Dinshah's work. He takes no gifts or presents; well let us do something to help the Cause. He can say nothing against that. We need an impartial newspaper that will tell the Truth and no party propaganda. Let us have a Radio Station so the World may know about Spectro-Chrome. Let us put our hands in our pockets and subscribe to such humanitarian work." (Applause)

The President Dinshah thanked the speaker and said he did not accept any gifts or presents, but, there should be no reason why a separate Organization could not be made to fulfill all the unselfish aims enumerated by Albert E. Greenham. Dinshah then requested the Recorder Sofie Hall to mark the names of those who volunteered to subscribe to such a Cause. He set $25.00 as the minimum Contribution; a higher figure should be any multiple of 25. Promptly Albert E. Greenham asked to be put down for $25.00 and Dinshah set his figure as $100.00. He marked Irene Grace Dinshah also for $100.00. Others joined. (Their names and amounts will be published in a later issue.) At the following Session, to the surprise of Dinshah, many brought their subscription in cash which Dinshah refused to accept, with thanks. He requested them to wait till he actually formed the

Corporation and got the Charter from the State Of New Jersey and would open a proper bank account for the new company. He promised to do so before starting for his Pacific Coast tour. (For further particulars see elsewhere in the current issue.)

Vice-President Jessie E. Ness—"I am highly in favor of the new move. I saw Dinshah fighting in the Supreme Court in Buffalo, New York and the way he tore open the issues. That case should be printed. My experience of Spectro-Chrome is highly favorable. Medical Doctors have fear infused in the mind of the public. Sufferers went to the Mayo Clinic for examination. They test all and then what? Spectro-Chrome has the hopes of the World in it." (Applause)

Albert E. Greenham—"There are so many frauds going on in Medicine. No newspaper publishes them. Take Aspirin, Bromo-Seltzer and numerous others. They are all harmful, but are consumed by the tons. So we need a newspaper in the hands of a fearless man. Dinshah is such man. Such a newspaper weekly at $1.00 a year deserves all our support. Dinshah has no creed; he is a great orator. He will give us the Truth, the sure Truth. Let the Affiliates give him Affidavits of what Spectro-Chrome did for them. Do not delay. Give him the Radio Station. He can broadcast Attuned Color Waves even— it needs a volume of money. Let us collect it for him. That day will soon come." (Applause)

Dinshah thanked all for their exemplary sympathy and said that his **Dinshah Impartial** will be graftless newspaper, taking no sides and no advertisements. His **Radio Station** too will broadcast solely for education and will accept no money for jamming the Ether by useless advertising jargon. The program of operation will be formulated soon. The Radio Station will be free for public use and educational service— a unique idea.

St. Louis, Missouri had similar Sessions and many subscribed to the new enterprise.

The 21st Anniversary Celebration will be taken to Portland, Oregon in June 1941, thus flagging Spectro-Chrome from Coast to Coast.

RESIDENT EXPERIMENTAL CLASS

The Annual Resident Experimental Class in the Science of Spectro-Chrome Metry, to qualify for the Diploma of Spectro-Chrome Normalator, will begin in our Auditorium in Malaga, New Jersey, on Monday, September 22, 1941, at 10 A. M. Please make your arrangements accordingly and write for advance reservation.

SPECTRO-CHROME

August 1942 Indigo

Volume 18 Number 8

Trade Marks Registered In The United States Patent Office

Great Work Spectro-Chrome

New Discovery In Delaware!

Expulsion Of An Affiliate

JUVENILE MIND SPEAKS

Championship Of Spectro-Chrome Metry

Spectro-Chrome Smiles

DINSHAH IMPARTIAL

Is This Oral Hygiene?

Attuned Color Waves
Spectro-Chrome Metry

In course of life you reach a time,
When Doctor, Lawyer, Priest alike,
Does fail to give the sought relief,
To troubled mind or burdened heart—
Then come to me:
I help you to help yourself.
SPECTRO-CHROME

The Science Of
Automatic Precision

DEVOTED TO THE SERVICE OF MANKIND

FIRST MILWAUKEE WISCONSIN PLANET

(HOTEL WISCONSIN, MILWAUKEE, WISCONSIN)

Meeting Of Thursday, June 4, 1942

Successful Use Of Spectro-Chrome In:—Accident, Pain, Skin Disorder, Piles, Sinusitis, Exhaustion, Backache, Varicose Veins, Rupture.

Reported By:*—EMMA SCHROEDER, *Recorder

Deputy Frank Burns was in the chair.

As usual, the communication from the Secretary of the Institute to the Planet Commission was read by the Deputy, who emphasized each paragraph, explaining the Lecture Series and the Course, on Occultism, both of which are to be conducted in Milwaukee in July, as well as the change in plans by the Council Of Trustees, which decided to conduct the Resident Experimental Course also in Milwaukee.

Questioning by the Deputy Burns disclosed that over ninety percent of the Affiliates present at this Meeting, did not receive the May issue of Spectro-Chrome, to date.

The two paragraphs in the Editorial Remarks in reply to the First St. Louis Missouri Planet, which appeared in the May 1942 Issue of "Spectro-Chrome," relative to the Chemical Elements and Atcowa Green, were read by the Deputy and the merits of this Atcowa were praised.

By-Laws 82 to 95, inclusive, were read by Emil Egide and the Deputy defined and explained the various By-Laws. Particular mention was given By-Law 92, "Radium And X-Rays." Their dangers were explained.

"Is X-Ray photography, as is used when purchasing a pair of shoes, harmful?" was the question raised by Ethel Schott. In the opinion of the Deputy, that is not meant by "X-Ray Irradiation."

Theodore Mohrhusen stated that By-Law 87 answered his question on "First Aid" satisfactorily.

Margaret L. was in doubt as to whether, upon entering a Hospital for training, it is necessary to notify the Institute and whether it conflicted with the By-Laws.

"Liability, Guarantee or Warranty," "Spectro-Chrome for Disorders," "Spectro-Chrome Tonation System" the culmination of Dinshah's Laborious Researches and "Definitions Of The Spectro-Chrome Tonation System" were read by Leocadia Merwitz.

The Deputy spoke on the Attributes of the various Atcowas in Spectro-Chrome, saying that this Chart is the work of a lifetime and not one of us present would be able to produce a Chart of this kind as Colonel Dinshah did. Affiliates were advised to spend all possible time in studying it, as much information can be obtained from it. "Too few of us refer to the Home Guide when an injury occurs or in a Disorder and that 'book' is certainly worth the study of anyone" is the Deputy's opinion. "Too many of us see only the surface," continued Deputy Burns. We see only the top of Spectro-Chrome and should study more of it.

In response to the Deputy's request for Case Reports, Emil E. reported the following case:—

A. S., age 68, Female: accident; fell and severely injured Areas 1 and 18. In medical terms this would be termed "Concussion." Was discovered 15 or 20 minutes later in an unconscious condition. Badly bruised. Apparently, this Sufferer did not know she was going to fall for arms were not extended to balance or protect herself. Kardoscil 76, Spiroscil 24. Very much excited and could not talk coherently. Commenced Irradiation with Atcowas Green, Indigo and Magenta. Was able to leave the bed after the second day. That does not mean she was normalated. Normalation required about three weeks. Completed with Atcowas Lemon and Yellow, to accelerate motion, for she was quite stiff. Prior to this injury, the Sufferer's life was practically saved at another time for she was completely burned with Ultra-Violet Rays. For the second time, she is very grateful for Spectro-Chrome. This Sufferer's mental condition was such, that she fancied she possessed all the Disorders which the magazines mention in their advertisements: Halitosis, B.O., Athlete's Foot, et cetera." (Deputy Burns added: "Use Camels,—they give you a lift!") Applause.

Another Case Report was given by Emma Schroeder as follows:—

One week before Colonel Dinshah's last visit here in April, 1942, Emma's mother, Ottilie K., (an Educational Student), in a spirit of helpfulness went at the electric wringer and mangled her finger. It was in such torn condition that the sight of it nauseated her. After the finger was caught, she pulled it out without releasing the pressure. Emma was tempted to call a Medical Doctor who lives only two doors away—thought it required stitching. However, on second thought, she decided "Spectro-Chrome must show what it can do." Kardoscil rose, due to shock, so Atcowa Purple was irradiated on Area 4 immediately to decrease sensitivity to pain. Then, two Tonations of Orange Local were given followed by four weeks of Indigo Local, one Tonation a day. The finger is mended completely, WITHOUT STITCHING OR TAPING. Indigo, which contains one-quarter Green, arrested the Pus condition completely. A relative who saw this finger at its worst, said to Emma, "How calm your mother is with a finger like that!" One of our Normalators, who also saw it, said, "And you are smiling with a finger like that?" She voluntarily replied: "Why shouldn't I? This finger hasn't caused me an hour's Pain or cost me an hour's sleep." Circulation is not completely re:established in the injured member yet, but this Sufferer is seventy-seven years of age and is still hopeful of restoring that, for she has been on the Rational Food Of Man for over six years (June 1, 1936.) There is indication of the nail coming off, but the new nail is already forming. This finger bled for days; at first blood, then serum, exuded and Emma believes that that is what prevented undue Swelling. Laurels again go to Spectro-Chrome! ! ! (Applause.)

Anita E. confirmed what Emma reported, for she saw the Sufferer's finger since it is mended and in her opinion, it was "a masterpiece in healing which Spectro-Chrome performed."

Otto Lutz, Past Deputy of the Milwaukee Planet, reported the following cases:—

1. A. F., age 58, Male. Skin Disorder of long standing. Tried many other methods first, for a long time, without results. Then tried Spectro-Chrome, which set him right in four months, three Tonations a week. Atcowas used were Yellow, 6, 7, 8, 9, 10 and 18 and Lemon Systemic.

2. R. H., age 33, Male. Case of "Piles" of long standing. Operation was advised, but Sufferer refused. Then used Spectro-Chrome for two months with success. Yellow on Areas 6, 7, 8, 9 and 10 and Orange and Indigo Local were irradiated.

3. L. S., age 30, Female. Sinus Disorder with much Pain. After using Spectro-Chrome three weeks, felt like herself again. Lemon and Yellow Systemic were the Atcowas used.

4. M. D., age 72, Female. Run down condition. No "pep," no appetite and could not sleep well. Tried other methods first without results. After the second week of Spectro-Chrome, appetite returned, could sleep fairly well and started to do housework. She remarked, "I now know I will be well again, but I want to continue with Spectro-Chrome for a while longer." Green Systemic, Front and Back were given.

5. S. Y. R., age 18, Female. Pains all over her Body. A recent case. Atcowa used was Turquoise (five Systemic) ; all pain left and she was well again.

Discussion proceeded on the Case Reports: With reference to Case number "3," Theodore M. stated that he suffered with Sinus Disorder so severely that he would strike his head against the wall. He normalated the Disorder in five Turquoise Tonations. With each Tonation, the period of Pain lessened in duration. Theodore could not understand how Lemon could normalate Sinus Disorder whereas his case was corrected with Ultra-Green, Turquoise. Said he had perfect results.

The Deputy held the opinion that Green, three-quarters of which exists in both Lemon as well as Turquoise, obtained the results.

Discussion on Sinus Disorder continued:

Leocadia M. said she knew of a case a few years ago in which a man had Sinus Disorder for thirty years. Was required to remain away from his work for three months. Was completely normalated with Lemon and Magenta.

Otto L. agreed with the Recorder that Sinus Disorder and Bronchial Disorder are closely allied and Atcowa Lemon was suggested.

With reference to case Number "5," Otto reported that this Sufferer was completely bent over with Pain. Two Turquoise Tonations brought Ovelim and all Pain left.

Skin Disorders were discussed and the Deputy remarked that, usually, many other methods are used in Skin Disorders before Spectro-Chrome is sought.

Case Number "1" also used Cocoanut Oil. Otto finds that people do not drink sufficient water; many cannot remember the last time they partook of a swallow of water!

"Piles" were also discussed.

The Deputy reminded that seventy-five or more per cent. of Disorders of the Human Body result from "Constipation." That is why we are to abstain from certain foods. Yellow is the proper Atcowa for faulty Elimination.

Norma F. is desirous of having the Affiliates know that her mother has become a Spectro-Chrome enthusiast, whereas, formerly, she had a hundred excuses for its non-use. She did not like the idea of using the Favorscope. Said, the neighbor never used it, but takes her Tonations during her noonday siesta. Norma pointed out that this was not the proper way to do. At last, Norma's mother's joints became so painful, Kardoscil was low, Spiroscil was high, that she finally resorted to regular Spectro-Chrome Tonations. They proved beneficial but still Spectro-Chrome did not receive the proper credit. A year and a half ago, she wrenched her back. The proper Atcowas were advised, but, strangely, the Disorder grew worse. Her mother has an "odd way" of "placing" the slides, so, while the latter was under Tonation, Norma decided to enter the room and investigate. It was discovered that she used the combination forming Scarlet instead of Orange, but did not have a Backache again for months. Later used it for a Fall. Went over it in a hurry. Then had Varicose Veins. Went over them also. Was accustomed to chest attacks so severe, that she moaned with Pain. Just lived in dread of them. Suffered with many other Disorders and accompanying all of them was a Rupture which was forgotten and which, with all this Tonation, improved. Norma's mother is now so enthusiastic about Spectro-Chrome that the subject need only be broached and she will eagerly point to its merits. (Applause.)

Alice B. told of a case of Gall Bladder Disorder which existed for thirty-five years. This Sufferer did not tonate with head to the North and Favorscope hours were not followed. Upon questioning as to how he felt, the reply was " 'It' is going away. Formerly I had the Doctor every three weeks."

"It is not the proper way to use Spectro-Chrome," said the Deputy. Some find the best time to use it is at eleven o'clock at night. Some permit the Atcowas to shine on the wall instead of the Body. Most people like to do things their own way and dislike to follow instructions. When we issue instructions, we must be sure they are properly given and properly understood. His wife once used three slides instead of two! ! !

EDITORIAL REMARKS

Keep away from Irradiation with X-Rays. Read my article on page 5 in June 1922 Spectro-Chrome.

Hospital "training" is not against our By-Laws, but, the use of such "training" in conjuction with our Science is likely to prove Violation. Spectro-Chrome is NOT permitted to enter into a Hospital or used with other methods, except through the approval of the Council Of Trustees. Study By-Law 78. Always consult the Council, before changing vocation or status, as certain vocations are barred from affiliating with the Organization. Study By-Law 62.

Lemon and Turquoise have ¾ Green Polarity, the only difference being that the Lemon is on the Hydrogen Section and the Turquoise is on the Oxygen Section. The former is indicated in Persistent cases, while the latter is indicated for Acute cases. It is easy to determine, because the Acute cases are Febrile, while the Persistent ones need the introduction of warmth. Follow the K. and S. Ratio. The Deputy was right.

People not minding instructions or Free Guidance and not using the full Technique are hurting themselves. Our Science is perfect in its conception and application, but, we have no way to "compel" obedience or honesty and we have less desire to interfere. When we learn of non-compliance, we drop them out of Free Guidance privileges or at the end of year deny them Continuation Of Affiliation. Such recalcitrant persons are not desirable as our Affiliates. Let them stay out and save their cash for our opponents.

FIRST PORTLAND OREGON PLANET

(MASONIC TEMPLE, PORTLAND, OREGON)

Meeting Of Thursday, June 4, 1942

Successful Use Of Spectro-Chrome In:—Fever, Irritation, Sunburn, Inflammation, Burns, Hiccoughs.

Reported By:—LOUIS BISCHOFF, *Recorder*

Deputy Stanley Burroughs was in the chair.

The Atcowa for discussion for the evening was Blue.

The Deputy asked the Affiliates to tell of their experiences in various instances where they used the Blue.

Virginia G. said she used the Blue very successfully in a number of instances of Fever and Irritations. Another mentioned that successful results were obtained in using it for Sunburn. Others used the Blue successfully in various degrees of Burns. The Deputy stated that if Spectro-Chrome be only good for Burns alone, it would be worth while having, as it seems like nearly every member of a modern family is getting burned at one time or another! Nothing works so rapidly in healing a Burn as Spectro-Chrome. It was noted by some that Blue caused perspiration. Blue is refreshing and cooling. Evaporation of the Sweat drops, from the Body, results in a lowering of the Body Temperature. Many noticed that when using the Blue after a hard day's work, it very definitely relieved fatigue and took the tired feeling away. Thus Blue is a great Vitality Builder. It supplies the Oxygen to the Blood and helps readily to burn the Toxins out of the Blood and Tissues. Blue also takes off some of the load from the Heart as the result. Blue also stimulates and strengthens the Pineal Gland. Further discussion on various uses of the Blue followed.

The Deputy then read the Chapter 93 on Blue from Spectro-Chrome Metry Encyclopaedia.

Virginia G. said that a number of cases of bad infected and inflamed Eyes was reported in her neighborhood and all over the city. It seems that the trouble comes from the shipyards. Her little girl had a similar trouble with her Eyes and Blue cleared it overnight. Other cases reported, without the use of Spectro-Chrome, were almost impossible to clear. She also mentioned that soon after the last Meeting, she had an attack of Hiccough, which would not go away no matter what she tried and then she remembered that in the Meeting Orange was discussed and the Deputy mentioned that Orange would relieve them. She used Orange immediately and within half a minute the Hiccoughs were gone and did not return.

The question was asked as to how they could explain how the various Atcowas worked to produce the results that are achieved. The Deputy explained.

The question was asked if it be all right to have a Radio going while taking Tonation and the Deputy replied that it did no harm, but as a rule one could not relax as easily as if they did not have any interference of any kind.

The question was asked if Colonel Dinshah would visit Portland this year and the Deputy thought it best to let Colonel Dinshah answer that one.

An Affiliate mentioned that Spectro-Chrome would not take her Headaches away and would generally last three or four days. She wanted to know what to do. The Deputy replied that a Headache of that kind indicated that something might be wrong and changes were taking place. Stay with Spectro-Chrome and the Rational Food and the necessary changes would finally take place and the cause of the Headache removed. Any condition brought on by abuse of the Body over a period of years cannot be corrected in a few days or a few weeks. The Body must have time to make use of the Atcowas. Nature can do only so much at one time.

The question of what part Germs played in the course of any Disorder, was brought. The Deputy stated that Germs were the results of the Disorder and not the cause. Germs aid in destroying rotting material in the Body and thus doing us a good turn instead of harming us as is generally thought.

The question was asked, what causes a Headache when one goes without eating? The Deputy replied that if our Digestive System does not have to work to digest food, then our energies go to work eliminating the Poisons circulating in the Blood before they are eliminated from the Body, which causes the Headaches and other Pains. Eat correctly and live correctly and we shall not have any congested Poisons in any part of our Body.

The question was asked if it were all right to drink liquids while eating a meal. The Deputy replied that it was all right to drink as wanted, if no food be in the mouth at the time. He believed that drinking with a meal helped to eliminate Constipation, as a great deal of the liquid was carried with the food through the intestines thus keeping the food soft and more easily handled by the Digestive System.

The question was asked, since the only way that Oxygen can enter the system is through the Lungs, how is it that Spectro-Chrome Blue gives the Oxygen? The Deputy answered that the Blue enabled the Blood to pick it from the Lungs and utilize it. The Attuned Color Waves of Spectro-Chrome enable our Bodies to utilize the Chemicals we take in through food and air and water. Attuned Color Waves produce a change in the Aura and thus a chemical change in the Body.

EDITORIAL REMARKS

It was a good discussion on Blue. Virginia G. seems to have much interest in the Science.

The "cases" from the Shipyards may be traced to the Welding Operations, wherein large quantities of Ultra-Violet Rays and also Infra-Red Rays are disengaged. The Invisible Energies of Nature have more potency than the Visible Energies so-called and carelessness in using the former may result in terrific Disorders. In general, the workmen using the Energies are more to blame for their own negligence.

All should study Spectro-Chrome Metry Encyclopaedia and most of the constant questioning would disappear. None would be called a good Christian, without a study of their Scriptures. Study that set and also what is in Spectro-Chrome Home Guide.

No Radio, no talk, no fooling during Tonation. Concentrate your mental faculties on regaining Health.

Dinshah will visit you again only when Students are in sight. The last trip of 7,000 miles was enough for showing his good-will, but, a Non-Profit Corporation MUST have internal financial encouragement, to continue educational tours. Then, again, Dinshah does not relish the idea of being stranded in the deserts of Utah, with worn tires and no gas. He will come when they will make Synthetic Inflammable Water and Soap Bubble India Rubber. However, get the Students ready anyhow and he will be tempted to drive on Rope Tires and Whiskey!

Drink fluids while eating, but, *not* while the food is in the mouth.

The results are produced by the Sufferer's internal Energies, the Attuned Color Waves being the channel or Catalyst. Dead folks, therefore, can not jump out of their wooden overcoats by any Spectro-Chrome Atcowa. There is the Line drawn for our activities.

Old Ambition Returned!

By:—MARY S., *San Antonio, Texas*

I can truly say that I feel very much improved in health, I am eating according to your instructions and I enjoy my meals. I feel strong, my old ambition has returned. I can not express my gratitude for your valuable advice. If I could convince my friends of the benefits they could obtain from the use of Spectro-Chrome I would be much pleased. I hope some day.

FIRST MERRILL WISCONSIN PLANET

(HOTEL BADGER, MERRILL, WISCONSIN)

Meeting Of Thursday, June 4, 1942

Successful Use Of Spectro-Chrome In:—Injury, Swelling, Pain, Rheumatism, Fever, Vomiting, Whooping Cough.

Reported By:—ESTHER KRIEWALD, *Recorder pro tem.*

Deputy Walter Briese was in the chair.

The Deputy read the By-Laws Number 84 thru 98.

A letter from the Principal Office at Malaga, New Jersey, was read by the Deputy regarding the Spectro-Chrome Meetings which will be held in Milwaukee in July, 1942. It was explained that a bulletin giving dates and further information will be mailed to all Wisconsin Affiliates.

Those Affiliates who did not receive their May 1942 issue of the Spectro-Chrome magazine were told to write direct.

Chapters on "Disorders Of Vision" and "Disorders Of Olfaction And Gustation" from Spectro-Chrome Metry Encyclopaedia were read by Lydia D.. She explained these Chapters thoroughly for the benefit of those present. She referred to Spectro-Chrome Home Guide on the "Sensory

Nervous System" and the Deputy read page 69. Lydia D. reported the following cases:—

1. Emil S. Male. Injury of the Knee. There was much Swelling and Pain present. He could hardly walk or climb the stairs. Lemon Systemic and Magenta on Areas 4 and 18 cleared the condition in 2½ weeks.

2. Emanuel E., Male. 14 years old today. His case was diagnosed by Physician as Inflammatory Rheumatism. Medical Doctors had been on the case for two weeks and the condition of the Sufferer became worse. He lost 35 pounds in two weeks, while in their care. Lydia was called in and found the Sufferer in much Pain. His body was very swollen. Temperature 104 Degrees, Ratio 1 to 5. Blue and Turquoise Systemic and Yellow Local normalated the condition beautifully. He is now up and the Temperature is normal too. He sings and is happy and grateful to what Spectro-Chrome did for him. His Aunt who watched the progress of her Nephew is now using Spectro-Chrome too.

Successful use of Spectro-Chrome was reported in cases of Vomiting and Whooping Cough.

Anna V. of Shawano, Wisconsin, reported she has been a user of Spectro-Chrome for nine years. She was asked to report on some of the results she obtained by Spectro-Chrome. She said she would not know where to begin or end, for it did her much good. After many years of doctoring with medical men without results, she used Spectro-Chrome and now finds living a joy. She does not tire so easily and can do all her own work, thanks to Spectro-Chrome.

Chris G. from Shawano, Wisconsin reported that he lost his sense of Sight suddenly one day, while going to the mail-box for his mail. After searching for some time, he found an Optometrist who fitted him with glasses whereby he can see fairly well. He is using Spectro-Chrome, but has only had his Equipment for about a week so could not give much of a report. He was advised by Lydia D. to continue using the Free Guidance and he would get results.

FIRST DETROIT MICHIGAN PLANET

(HOTEL TULLER, DETROIT, MICHIGAN)

Meeting Of Thursday, June 4, 1942

Successful Use Of Spectro-Chrome In:—Fatigue, Collapse, Nervousness, Swelling, Pain.

Reported By:—SOFIE HALL, *Sub-Deputy.*

Deputy Jessie E. Ness was in the chair.

Communication from the Central Office was read.

The Deputy asked Ottilie C. to read from "The Advantages Of Healing With Light, An Outline Of The Science Of Spectro-Chrome." She read there are Seven Colors in Sunlight and they are separable through a Prism. We were also informed as to what was good to eat and what not to eat. Sleep with your head North and feet South. It was very good and the Affiliates applauded.

The Deputy said Spectro-Chrome is trying to help you to help yourself. Sunlight is the basic source of all Life and Energy upon our Earth

Johanna R. asked if it would be possible to print 1 or 2 of Dinshah's recipes that are not in Spectro-Chrome Home Guide; she hears so many people talk of the delicious and tempting dishes that are served in Malaga. She stated that she realized the expense of publishing a separate book and asked if this other plan were possible? She made a motion we write and ask about it and Catherine J. seconded it. All the Affiliates were very enthused with this suggestion.

The Deputy stated as a man eateth so is he and as a man thinketh so is he. Albert G. asked if he understood correctly when Ottilie C. read from "The Advantages Of Healing" that Condiments were forbidden to be used. He stated that Colonel Dinshah says if they are used in cooking it would be all right; such as Salt. Salt in cookery in permissible, but salt after cookery is not advocated. He asked if this were written by Colonel Dinshah himself. Lenore G. stated Spectro-Chrome Home Guide also permits boiled Potatoes and then stuffed into peppers. She was told mashed Potatoes that were steamed and not boiled were alright.

The Deputy then stated people are too tired to stay patiently with anything that is good for them. If they are not better with 2 Tonations they give up, yet they do not expect to become better with 2 Pills. We should clean our outside as well as our inside.

Charles R. said the biggest offenders are the Affiliates themselves. We do not obey the Laws of Nature.

He stated Michigan is lacking, because people do not go out and work. He preached Spectro-Chrome away before he took the Course. Other people are always watching us to see if there really is something to this way of living and in the end they will all come and say we are right. Another fault of the Michigan Planet is the poor showing of Affiliates at our Meetings; no wonder Dinshah becomes disgusted.

The Deputy then called on Albert G. He stated 6 elements constitute 99½ per cent. of the Body. He further explained the minute we are born we start to die. He drew a triangle on the blackboard and explained the Pituitary gland, Liver and Spleen. Hate destroys. If the Aorta is damaged there is no possible chance of living; bleed to death. We should always have constructive thoughts, like day by day I am getting better. For example, someone that is very fond of Music and has a severe Headache or Toothache happens to hear a band coming down the street, he forgets about the Ache and becomes all enthused and when the last note dies away the Ache is back again. "What fools we Mortals be!"

Case Reports followed—

Johanna R. reported that 1 Tonation enrolled another Spectro-Chrome Affiliate. Her husband had been working very hard and for a long time. He became ill at work and had to be taken home and a couple of the fellows at the shop made a $5.00 bet he would be sick for a little time and would not be able to work right away. His Kardoscil was 54, Spiroscil 16. Tonated with Scarlet and he fell asleep; 3½ hours later, he got up and went to work, Kardoscil 76, Spiroscil 17.

Sara C. has had Spectro-Chrome 7 years and has had no Medical Doctors. Her mother passed away. She broke down with a "nervous condition." She had Gas Pains and her Leg swelled. Was in bed for 4 months. Thanks to Spectro-Chrome she is feeling fine. Last week she went to a dance.

Thaddeus K. told us he was in Malaga and Dinshah sent his Greetings to the Affiliates. The Greetings were heartily accepted.

The 3 enclosed Charts were discussed.

Antonio L. reported he has been taking Spectro-Chrome since November, 1941. He has Indigestion and has been taking Yellow, Orange and Lemon, but he did not get any results.

He was asked how long he had the illness before he started tonating and he said 4 or 5 years. He was told to be patient and he would get results.

EDITORIAL REMARKS

We have been carefully watching your recent Reports. Somewhere, there is lack of activity. What is the use of talking all the time on "Delicious dishes and foods?" Read the Editorial Remarks on the activities of other Planets on Food Questions. We *refuse* to print monthly Recipes, because, such procedure will *have* to be eventually taken in regard to printing a complete Cook Book. Such book IS ready in India and we are preparing another here. Please, therefore, stop this food discussion and follow what is in the Spectro-Chrome Home Guide, leaving objectionables aside. It is very easy to talk, but, as we are a Non-Profit, unendowed Corporation, we find it difficult to make even the two ends meet. The expenses are more than the receipts and the backing of our Affiliates is lacking. Let each one contribute to a Spectro-Chrome Metry Encyclopaedia and STUDY. That will benefit more than wishing for just "Delicious dishes"—thanks though for the compliment.

Antonio L. has himself to blame. He never sent his Inspection Certificate or asked for Free Guidance.

Your Recorder Lucille Greenham resigned. Kindly recommend a willing and capable volunteer.

———o———

Great Work Spectro-Chrome

By:—HENRY S., *Lake Forest, Illinois*

We are both, I and my wife, happy to say, what a grateful Science Spectro-Chrome Metry is and what it has done for us already in the short time we got it. We again are both thankful to you and your whole family, for the great work you have done and are doing for all Sufferers, to have the happiness of using Spectro-Chrome Metry.

———o———

New Discovery In Delaware!

Wilmington, Delaware Scientists are reported to have made an important discovery. They say the Lemons have plenty of Vitamin C, hence they are wonderful to prevent Heat Prostration. Must be true. We found many wonderful Lemons in that city and one of them is still hanging on the tree 10 years.

FIRST ST. LOUIS MISSOURI PLANET

(HOTEL MELBOURNE, ST. LOUIS, MISSOURI)

Meeting Of Thursday, June 4, 1942

Successful Use Of Spectro-Chrome In:—Coma, Swelling, Inflammation Asthma, Vomitting, Appendicitis, Fatigue, Dysuria, Insomnia, Cold, Pain, Boils, Itch, Piles, Fever, Deafness, Paralysis.

Reported By:*—VIOLA C. JEUDE, *Recorder

Sub-Deputy Nathalee Britton was in the chair.

Communication from the Central Office was read. The Affiliates were advised to enroll by Monday, June 8th, if they wanted Colonel Dinshah to give the Course on Occultism.

Sub-Deputy said according to communication, Affiliates complained about "too much theory." She was very sorry to hear about this complaint; she realized that the First Volume of Spectro-Chrome Metry Encyclopaedia deals with nothing but Theory and Fundamentals of Spectro-Chrome Metry so she was letting it entirely up to the Affiliates if we should continue with the First Volume or not. No one voiced their opinion. She then asked Charles S. what he thought. He said one had to have a foundation and he thought it was necessary to study the First Volume. She then asked the Affiliates if they wanted to continue the study of the First Volume. There was no response; she then asked them if she could take their silence as consent and asked them to raise their hands if it were agreeable. *All Hands Went Up!*

Sub-Deputy said "I wonder how many of you look forward to the First Thursday of each month. We come here with a common interest, to become better acquainted with the Science of Spectro-Chrome Metry. This should be a Meeting of vital, personal interest to each of you. 'No man yet hateth his own flesh,' but truly we are vitally interested in our Bodies. Many of you went to other 'health lectures' attracted by the Salesman touch of a blowing diet. At least you are awakened from a rut of some years of sin against the Laws of Nature and here you are under Spectro-Chrome. Did you particularly notice the stanza on the monthly magazine cover?

> In course of life you reach a time,
> When Doctor, Lawyer, Priest alike,
> Does fail to give the sought relief,
> To troubled mind or burdened heart—
> Then come to me;
> I help you to help yourself.

I like that last line." She then asked Clemens N. to start with By-Law 87 and continue with as many as time would permit.

He complied by reading By-Laws 87-92 inclusive. In regards to Section 1, By-Law 87, he said any one may be in an Accident and may be taken to Hospital as one may be unconscious due to fractured skull; under these conditions one may be in the Hospital for several weeks. It would not be advisable to move Sufferer with fractured skull. This would not be a Violation.

Question—Would it be advisable to use Spectro-Chrome on Fractured Skull if it were possible?

Answer—By all means, it would certainly help conditions.

Clemens N. said he knew of a Boy who had Fractured Skull; the Bone protruded into the Brain; had to wait 2 weeks to determine before they operated. An operation of this sort is permissible; it is Constructive Surgery.

He said in regards to By-Law 88 to have a Tooth or Teeth extracted would not be a Violation, if the methods employed be freezing the Gums or blocking the Nerves; it would be advisable to tonate with Green and Magenta 4-18, as these methods are bound to affect Areas 6, 7, 8, 9 and 10.

Question—What to do after Teeth are extracted for Soreness?

Answer—Soreness can be stopped by Turquoise or Indigo.

Question—How about Abscessed Teeth?

Answer—Swelling or Boil next to a Tooth does not always indicate an Abscessed Tooth. It depends if the Swelling should successively occur, then it is not always advisable to have Tooth extracted.

Charles S. said X-Rays of Abscessed Teeth did not always prove successful, as they did not show the Abscess.

Clemens N. said this was true; some times they just cast a shadow.

Charles S. said his wife has had Pyorrhea for years and the Doctor can not understand how her Teeth are holding out! They said they attribute this to Spectro-Chrome. Of course the Dentist does not know about this.

Question—Was he not right in assuming that she used Indigo on Area 1 for some time?

Answer—Yes.

Clemens N. said in regard to By-Law 89, that if one has an Ingrown Toe Nail, to have it taken care by Chiropodist; this would be permissible and is not Violation.

He said he read an article on "How to Keep Well" in the Saturday Evening Post. It suggested to use a 2 x 4, round the edges, walk on it and in this way the arches can be raised, he said it was a wonderful suggestion on how to help ourselves the most natural way.

Charles S. said a Rolling Pin could be used instead.

Question—Would not Orange Local be beneficial also?

Answer—Yes.

Question—Would Orange be beneficial on enlarged painful Joint?

Answer—For inflamed Joint, use Indigo, walk and wear proper foot wear.

In regards to By-Law 90, he said it has been his experience that after a short time after using Spectro-Chrome, the Eyes go through a change; he said he usually advises to wait a while before having glasses changed, because it has been his experience that one did not need as strong Lenses.

Question—The usual procedure in going to an Optometrist was they invariably use Eye Drops; was this permissible?

Answer—Drops in the Eyes paralyze the Eye Muscles.

Charles S. said when his wife was under the guidance of the Institute, they advised against the use of Drops in the Eyes.

Question—Is it permissible to use Boric Acid to bathe the Eyes?

Answer—One could use a Salt solution just as well. Why did she want to use Boric Acid? Was it just because some one said it was good? Answer—Yes. He said it was not necessary with Spectro-Chrome.

Question—What Color do you use on the Eyes?

Answer—According to the condition, either the Infra-Green or the Ultra-Green.

Rosannah B. said several years ago she had 2 Styes; they were inward Styes and she had to have them removed.

It was suggested if she should ever be pothered with them to let them ripen and not have them removed by force. She said the Tear Glands were affected.

Question—Was Stye due to Glandular Disorders and whenever there is Inflammation in some other part of the Body, for example Rhinitis, is there not also an accompanying affection elsewhere in Body, for example Sexual Organs?

Answer—Cannot answer, as he did not know.

In regards to By-Law 91. A woman came for Tonation; her Vertebrae were always out of place and she had to go constantly to a Chiropractor. One day after having been to the Chiropractor, she came for Tonation and used Spectro-Chrome. Some time after this Tonation, the Vertebrae again became displaced; he suggested to have them replaced and that was the last time she had to go and that was 3 years ago.

He said that he also knew of a young man who has been using Spectro-Chrome and his Back seemed out of place. Just recently while sitting at a table there was a twitching in his Back and everything is in place. The father is more than pleased; he said if he had not seen this with his own eyes he could hardly believe it.

Some one said they experienced this with a Knee that was always out of Joint, but with Spectro-Chrome it stays there.

In regards to By-Law 92 he said later he would make a report about an X-Ray Burn.

Question—Is Skin discolored by the Ultra-Violet Ray?

Answer—Only under X-Ray. X-Ray will kill growth; if you want to desex yourself and not have children X-Rays will take care of that. It destroys Life.

She said with hot weather approaching, it was a timely subject to talk about Sun Burn.

Clemens N. said with summer coming on one will hear about Sun Stroke.

The Recorder said she would like to make a suggestion. She read in the First volume of Spectro-Chrome Magazine, that Dinshah said Magenta could be used very successfully on Area 4, that regardless of the condition that Magenta being ½ Red and ½ Violet, the pulse would select the ½ that was needed and the other ½ would be thrown into the Aura.

Clemens N. said this was right. If one be in doubt as to what to do use Magenta on Area 4.

He said, several years ago a woman had a Boil on Area 5 and she used Magenta Local and normalated. He asked her why she used Magenta and she said she thought that if Magenta was good for Area 4, it would also be good for Area 5. He asked her how long it took to normalate; she said about 5 weeks. He said if she had used the usual procedure for Boils, she would have normalated in less than half the time.

He said he divides Light in 3 ways, the Infra-Red, White and the Ultra-Violet. He said it is funny a Negro would pay money to become White and the White People want to become Brown, go South and pay exorbitant prices for Hotel accomodations. A Sun Tan is nothing more than Burnt Skin, so to say.

Question—If one received a severe Sun Burn will Spectro-Chrome prevent Blisters?

Answer—Depends on the time exposed to Sun, how long you waited before you used Spectro-Chrome; if you waited very long it would not prevent Blisters.

In a severe case of Sun Burn the new Skin is very tender; in some cases it has taken 6 to 8 months before it blended in with the rest of the Skin.

Henrietta L. wanted to know how one could strike the happy medium; in Winter there is not enough Sun and in Summer there is too much Sun? Discussion followed.

Case Reports followed:—

A case was reported for Frances S. by Clemens N. Her mother became very sick; for 2 days was in coma; used Orange 8, 9 and 10 and Turquoise; then two days later her father became very ill; the same Atcowas were used on him and he was normalated.

Johanna S. reported her sister E. B. had a swollen Hand; used Turquoise Atcowa, did not look good; a Red line began to appear; she called Clemens N., then used Orange Atcowa Magenta 4-18. Stripe was ½ inch wide; tried to pick with a needle; is much better now. Clemens N. said with Spectro-Chrome never pick any sore. It will leave a scar; use Orange Atcowa or follow Spectro-Chrome Instructions for Boils.

Mamie C. reported sister picked finger and it became inflamed. On Monday gave 4 Tonations; on Tuesday gave 4 Tonations; used Turquoise, Blue and Indigo and the next morning it burst open; followed directions according to Spectro-Chrome Home Guide.

Case Reports by Clemens N.:—

1. Rosa F., age 46. Ratio 106-28. Asthma, confined to bed, could not eat, vomited, was ready to give up. Orange Systemic, Magenta 4-18 later Lemon Systemic; results astonishing. Enrolled and has own Equipment. Robert F. her son was present and he added his mother had a hard time breathing; he said he knows when you can not breathe you just feel like you want to die; was given up by Medical Doctors. An Aunt suggested Spectro-Chrome, so they rented one; he admits he was skeptical when it was brought in, but he changed his mind. His mother made wonderful progress in 2 weeks.

2. William K., age 27. Ratio 78-18. Appendix attack; doctor recommended Operation. Turquoise Systemic, Magenta 4-18; after 10 days, went back to work.

3. Katherine K., age 59. Ratio 76-13. Frontelim painful and often, general Fatigue. Turquoise Systemic. Orange on 17, Magenta on 18, later Orange Systemic; very well pleased with Frontelim Disorder gone. She enrolled.

4. Adele W., age 54. Ratio 69-15. Could not sleep at all, Hand so swollen that Fingers could not move at all. Green and Turquoise Systemic, Magenta 4-18. Can now sleep all night and Hand so well now helps relative in childbirth!

5. Thomas H., age 57. Chest Cold. Turquoise Systemic one Tonation, feels fine.

6. Carl K., age 47. Cold with severe Pains in Area 17 and Shoulder. Orange Systemic Back one Tonation, went back to work.

7. Charles V., age 61. Unbearable Pains in Area 5, especially when

coughing. Ratio 90-12. Turquoise Area 4-5, Magenta 4-18, later Lemon Systemic; went back to work within one week. Enrolled.

8. Marjorie D., age 26. Boils on different parts of the Body. Green, Lemon, Yellow and Orange Systemic, Magenta 4-18. Boils appeared again and again, but normalated in 5 weeks. A prospective Benefit Student.

9. Ralph S., age 39. An Itch so severe he went almost mad. Could not work, also had Piles. Turquoise Systemic, Magenta 4-18, Indigo between Areas 11 and 19. Was normal in 2 weeks. Well pleased.

10. Lorenz W., age 84. Fever followed by extreme coldness, Body color Blue, Abdomen very hard and bloated, knife-like Pains. A very sudden attack. Orange 6, 7, 8, 9 and 10, produced vomiting and a terrific blast of Backelim. Green and Magenta 4-18; was normal within a few days.

11. Jessie D., age 57. Pains in Back and loss of Hearing. Orange Systemic Back and Front, Magenta on 18. Pains left and Hearing restored within 3 weeks.

12. Mary A., age 9. Ratio 72-11. Right Arm and Hand paralyzed; when walking throwing her body a certain way; could not use right Hand to write or use Hand; very awkward. Yellow Areas 6, 7, 8, 9 and 10, Scarlet on 4 and 18. Appetite improved, is able to use Hand in a normal way. Ratio now 1 to 4; used their own Equipment.

EDITORIAL REMARKS

By-Laws and Theory are useful, but, should be within bounds. Teach the *practical* side of Spectro-Chrome Metry. Affiliates are shy to talk; draw their speeches.

The Constructive Potency of Spectro-Chrome is amazing, if one knows how to utilize it.

If women will wear proper shoes, Chiropodists will seldom be necessary. The American shoe styles, copied from the Europeans are a blot on common sense and are a disgrace to so-called civilization. Many women have their little Toes amputated to push the foot in a tiny shoe!

Most Eye Drops have Atropine or Belladonna, which dilates the pupil. Do not permit it. Wash the Eyes with plain warm water.

Stye is an Inflammation of one or more of the Sebaceous Glands of the Eyelids. Refer to Inflammation 31, Chapter 56, page 539, Spectro-Chrome Metry Encyclopaedia.

Area 11 becomes inflamed in certain Disorders like Mumps.

Spectro-Chrome Metry and other Systems of Healing can not go together. Do not mix them without the approval of your Council being asked.

Request the Affiliates to secure a Set each of Spectro-Chrome Metry Encyclopaedia and study. Sunstroke is fully covered therein. They want their intellectual banquet ready digested and will not tax their own faculties!

When in doubt, irradiate Magenta. Soon you will learn to differentiate Purple, Magenta and Scarlet.

Sun Tan is a silly custom made into a racket. All in India get it free, but, they are not all healthy. We were not intended to look like boiled lobsters!

The attitude of the American housewife towards the Sun is nothing short of idiotic. In the Winter, when there is less Sun, she goes out in undress and freezes and in Summer she pulls down the house shades, so her carpets and floor rugs (rags) may not fade; never mind if the cheeks of her children fade! Then she goes out and buys her beauty from a Drug Store and smells like a muskrat and eats Vitamins from A to Z and back to restore losing health. I say to you all openly, follow the principles of Spectro-Chrome. Lead a natural life and think, speak, act and live clean.

Four Tonations a day are foolish waste of Electricity. Do not try to force Nature's periodic schedules.

Well, although your Enrollment was insufficient, I am coming to meet you anyhow. It will prove to you once for all the reason, why I go elsewhere than to your city!

An Organization must have constant Enrollment to live and flourish and the Enrollees should stay put.

Expulsion Of An Affiliate

Vera McInnes, of Chickasaw, Alabama, sent her Check for payment of 1942 Dues. When the Favorscope and Affiliation Card were mailed, she refused to receive the registered package and stopped payment on the Check. Notwithstanding requests, she refused to honor the obligation. She was charged with Violation of By-Law 84, Section 4 and expelled finally under By-Law 59.

All service in Spectro-Chrome Metry is denied to her.

FIRST CHICAGO ILLINOIS PLANET

(HOTEL ATLANTIC, CHICAGO, ILLINOIS)

Meeting Of Thursday, June 4, 1942

Successful Use Of Spectro-Chrome In:—Low Blood Pressure Constipation, Rheumatism, Dog Disorder.

Reported By:—ELSIE BEYER, *Recorder*

Deputy Wilhelm Weissenbacher was in the chair. Sub-Deputy Elizabeth Zavadil was unable to be present.

The communication from the Principal Office was read by Oscar K. who also read the By-Laws from 61 to 64.

The Recorder read from Spectro-Chrome Home Guide, on "Rational Food Of Man." Henry S. said he is feeling much stronger since he is eating according to the Rational Food Of Man.

Chapter 39, from Spectro-Chrome Metry Encyclopaedia, on "Hokum Of Medicine," was read by Oscar K. The Deputy showed the illustrations of Anna Schoener, the child paralyzed from birth and Grace Shirlow, the burnt girl. Spectro-Chrome restored both of those cases pronounced hopeless to normalcy. The Affiliates examined the 3 Volumes.

The Deputy asked each Affiliate to enroll another. He read a newspaper article in which a Professor wrote that Human Beings could live 150 years, if they ate the proper food.

Some Affiliates are confused about the Favorscope Time, whether they should add 1 Hour or not.

Case Reports followed:—

1. Wife of Henry S. was under medical care for years. Medical Doctors thought she might have a Tumor or Cancer. A year ago, she was operated on the Gall Bladder; was always troubled with Constipation and was in the Hospital 10 days. Became worse, Stomach swelled, Pus developed inside, so had to operate on her again. Was fed through the Veins and was in severe Pain. Was in the Hospital another 4 weeks. When she came, had Injections, but, nothing seemed to help her for Constipation. One day, she fainted; had Low Blood Pressure and was taken back to Hospital, but, did not improve. Heard of Spectro-Chrome through our Deputy and Henry S. enrolled as Benefit Student. After using Spectro-Chrome 3 or 4 times, the results were miraculous. For the first time, Bowels moved regularly and she felt fine. The wife is very thankful he had a Spectro-Chrome for his Rheumatism. He could not raise his Arm properly. Now it is well.

3. The Dog of Henry S. had a bad Bite. He irradiated Turquoise and Yellow with good results.

4. Alysia F. took Atcowa Yellow and Lemon for Imbackelim with good results. Since using Spectro-Chrome, she has had no more Colds or Coughs.

EDITORIAL REMARKS

As your Planet is now active, Affiliates are requested to take the opportunity of attending the Meetings and gradually building it to greater efficiency. We are right with you to do whatever you may wish or need in our line of activity. Add 1 Hour to the Favorscope Time.

JUVENILE MIND SPEAKS

In this Section, any person under 21, may express original views, without censorship. Articles should be as short as possible.

Championship Of Spectro-Chrome Metry.

By:—ETHEL SCHOTT (16), *Educational Student, Milwaukee, Wisconsin*

Dinshah Spectro-Chrome Institute
Malaga, New Jersey

Dear Sirs,

With great appreciation I wish to thank Colonel Dinshah again for the honor he bestowed upon me at the Twenty-second Annual Convention, at the Milwaukee Auditorium, by offering me the scholarship of a Normalator course and also presenting me with a set of "Spectro-Chrome Metry Encyclopaedia."

In the following paragraphs I am telling the high lights of what transpired in my former school, pertaining to the introduction of Spectro-

Chrome by way of my Career Book.

Before I was to graduate from Junior High School, I had to make a Career Book for my citizenship teacher, as all citizenship students were required to do. I chose my career as a Spectro-Chrome Normalator, though before this time I had thought of being a nurse, because when I was a little girl I always wanted to help those I thought were sick or injured in some way.

In March 1938, my father obtained a Spectro-Chrome Equipment because of a cataract forming in his left eye. The result, after using Spectro-Chrome, was that the cataract was arrested from further growth, thus making an operation unnecessary. In addition, Spectro-Chrome normalated burns, sprained ankle, heart disorder, ear disorder, grippe, sore throat, head and chest colds, severe injury to hand, and minor injuries in our family of five. This gave me sufficient proof of the value of Spectro-Chrome and also convinced me that this method of healing is far superior to that of the medical profession. It is far more pleasing to lie on a bed in a warm room, relaxed while taking a tonation for one hour per day, than to take medicines and injections or to be taken to a Hospital to be operated on, which are all very unpleasant and even dangerous. In comparison with other methods of healing, Spectro-Chrome is so remarkably simple and agreeable, when proper technique is followed as directed in the "Home Guide," that one who has not yet experienced or witnessed its work, can hardly believe it. That is why I decided to make my career a Spectro-Chrome Normalator.

After our books were corrected by the teacher he questioned me as to what the duties of a Spectro-Chrome Normalator were. I explained this to the class and also that only 600 hours of intensive study would be required on a set of three volumes of the Spectro-Chrome Encyclopaedia before the Final Examination. He also said he would keep my folder for the present for further investigation, because Spectro-Chrome has been practically unheard of in this school. He had never heard of it before and he considered it a worthwhile topic to discuss and for the benefit of the class, as well as for himself, he asked more questions and also let the other students discuss it in general. In one question I was asked, "In what way can Spectro-Chrome benefit a sick person?" In answer to this I said it would restore the person to normal when the proper Attunded Color Waves and prescribed food (Dinshah's Rational Food Of Man) were used as directed in the "Home Guide" and that this will reestablish the proper chemical balance in the body. I mentioned that an Itisometer could be used to indicate chemicals lacking in the body. Each color is capable of producing a number of chemical effects. I also told the importance of taking the tonations on time, in a warm, darkened room, according to the "Favorscope," as this science works in conjuction with the Solar System. I said, in construction of the Slides, that certain kind of glass, imported from Europe, was mathematically graded and then used. At this point I was interrupted by a student, the daughter of a medical doctor, who said that she was going to find out about the originator. Sometime before this I had given her my Career Book in which I had an old introductory pamphlet which had the name of the former Distributor on it. I covered this name with a piece of adhesive tape but when the book was returned to me I noticed that it was torn off and was replaced. I knew her father had done it. I also had given her the leaflet bearing the testimonies given in the New York State Supreme Court, which were published in the newspapers of Buffalo, New York. I told her to write to the New York State Supreme Court, but instead she wrote to the American Medical Association. In answer to her letter she received a circular letter that she read to the class. This turned a number of the class against Spectro-Chrome and its originator, because when summed up it stated somewhat the following: His picture which shows him in an army uniform, and his title of Colonel were to give people a good impression. It also stated that he associated with a very dishonest man, and was in prison many times. There was a number of other things mentioned to discredit Dinshah and Spectro-Chrome Metry.

Four more such periods of discussion followed. In one of these a boy in the class got up, when I asked him and repeated the same thing he told me outside of class. He said that his aunt's condition of asthma of long standing, was normalated by her Spectro-Chrome equipment.

The next day the medical doctor's daughter approached me in the hall, again denouncing Spectro-Chrome. In answer to her remarks I said that Dinshah's Spectro-Chrome Institute has innumerable cases of testimonies from the affiliates and others who benefited by the use of Spectro-Chrome. To this she replied that the testimonies were bought just like people who are paid for saying they like a certain soap because it is better for some reason. I then said this was not true and for further information she could obtain copies of Dinshah's trials from the New York State Supreme Court in Buffalo, New York. She said she was satisfied with the information she received from the A. M. A.

After these discussions in the class, my book was returned to me with a mark of 95. He suggested that I leave my book with him so that he could give it to the school librarian for further reference, as they have no information whatsoever on the science of Spectro-Chrome. I said I wanted to keep it because I did not want it lost or destroyed after I had put so much work into it. The librarian was also very much interested in it. She asked me to tell her about it. I did and I also gave her other information besides my Career Book. She told the vice-principal of the school about it and I presume several others.

Sincerely yours,

ETHEL SCHOTT

P.S. Colonel Dinshah:—This is in regard to your offer of coming to Milwaukee for an educational lecture on Spectro-Chrome to a group of twenty teachers, or possibly the entire school. I shall meet my former citizenship teacher and principal this week to try to make arrangements for a meeting in the school auditorium if permission is granted so that sometime in the near future I shall be able to tell you of its possibilities, number of people expected to attend, et cetera. Will you kindly give me a letter of introduction stating what you would speak on and the amount of time required? Your kind offer is very much appreciated and I shall endeavor to work at my top most ability to fulfill my promise to help introduce Spectro-Chrome.

Sincerely yours,

ETHEL SCHOTT

SPECTRO-CHROME

April 1943
Lemon 1943

Volume 19
Number 4

Trade Marks Registered In The United States Patent Office

What Was Said Of Him 47 Years Ago When Many Of His Opponents Were Just Thinking Of Taking Birth!

THE NEW YORK TIMES, Wednesday, March 11, 1896—

Says X Rays Are Not New —

Views Of Dinshah Pestanji Ghadiali, The Indian Scientist— Principle An Old One, He Declares —

The "Parsee Edison" Says Many Discoveries Are Made In His Country — His Plans Here

Expansion Of Our Resident Quarters
Have Your Vacation With Your Institute

DINSHAH IMPARTIAL

Medicophobia—The Fears Of Mankind!

Attuned Color Waves
Spectro-Chrome Metry

In course of life you reach a time,
When Doctor, Lawyer, Priest alike,
Does fail to give the sought relief,
To troubled mind or burdened heart—
Then come to me:
I help you to help yourself.
SPECTRO-CHROME

The Science Of
Automatic Precision

DEVOTED TO THE SERVICE OF MANKIND

DINSHAH P. GHADIALI, Editor and Publisher

FIRST DETROIT MICHIGAN PLANET
(HOTEL TULLER, DETROIT, MICHIGAN)
Meeting Of Thursday, March 4, 1943

Successful Use Of Spectro-Chrome In:—Diabetes, Hemorrhoids, Calculus, Lumbago, Eye Disorder, Stiffness, Pain, Arthritis, Grippe, Ptomaine Poisoning, Dysentery, Vomiting, Cold.

Reported By:—CHARLES J. RENN, *Recorder*

Sub-Deputy Sofie Hall was in the Chair.

A communication from the Principal Office was read by the Sub-Deputy, after which a lengthy discussion followed. The Sub-Deputy asked all those that could and cared to help along in this great struggle to do so now, because, it costs real money to fight these battles in the Courts.

The Sub-Deputy asked Norma Fiedler to read the Bill of Complaint and Petition for Temporary Restraint Order which she did very capably, after which a short discussion followed.

Case Reports then followed:—

Walter H. suffered with Diabetes which Spectro-Chrome normalated; he also suffered with Hemorrhoids. Doctors treated him for months, but gave him no relief; used Spectro-Chrome a few weeks which normalated them. He was also troubled with Stone. The Pain was so great that he used to roll on the floor in agony. Tonated with Spectro-Chrome and in three days the Stone passed and he has not been pothered since. Nothing like Spectro-Chrome.

Hendrik L.'s wife suffered with Lumbago and had very bad Eyes. After using Spectro-Chrome the Lumbago went and she doesn't wear glasses.

Elfriede K. said she had a very Stiff Knee which was awfully painful. After a week's Tonation, her Knee was normalated.

Hazel P.'s husband had a bad case of Grippe which she tonated with Turquoise, Lemon and Magenta for two days and he went back to work. She also reported her daughter had a Tooth extracted; used Indigo with wonderful results.

Carmela F.'s uncle had severe Pains in his Legs. Medical Doctor said it was Arthritis. He said he had no faith in Spectro-Chrome, but got one anyway and is now having very good results..

Clara S. reported her sister had Ptomaine Poisoning, Dysentery and severe Vomiting. Tonated with one Orange Systemic and went to work the next morning.

Irma S. said someone put a "Mickey Finn" in her husband's sandwich. He became deathly sick, vomited and had Dysentery. He was tonated with Yellow, Lemon and Magenta; the next day he went back to work feeling O.K.

Aquila M.'s grandson who lives with him had a bad Cold. Atcowas used were Turquoise and Orange with marvelous results. His wife fell over a stool and sprained her Ankle; a few Tonations of Indigo and she was normalated.

The Sub-Deputy read an Article from the Courier-Post of Camden, New Jersey, about Dinshah fighting the Post Office Fraud Order case himself. She said: "I can just picture Dinshah standing up there before the Court defending our case. Thank God that we have him with us."

EDITORIAL REMARKS

Several good Case Reports of Gladys C. are left out of print, because, there was no identification or address of the Sufferer. This rule is obviously necessary for the protection of the Editor and the reputation of the Organization for integrity.

Aquila M. should be warned against tonating a grandson. The grandson is NOT a member of his Immediate Family. If he should feel justified in the Tonation, he should send full particulars to the Institute and request a Courtesy Dispensation under By-Law 80, Section 4. Without such compliance, he will be considered a Violator.

FIRST MILWAUKEE WISCONSIN PLANET

(HOTEL WISCONSIN, MILWAUKEE, WISCONSIN)

Meeting Of Thursday, March 4, 1943

Successful Use Of Spectro-Chrome In:—Infection, Rheumatism, Fever, Pain, Lumbago, Imbackelim, Impetigo, Sciatica, Bronchitis, Exhaustion, Toothache, Stiffness.

Reported By:—EMMA SCHROEDER, *Recorder.*

Deputy Frank S. Burns occupied the chair.

The communication from President Dinshah was read. Students were requested to urge absentees to display more enthusiasm in this great Cause and in these trying times when our loyalty and interest are being tested. Expressions of amazement and disappointment swept over the audience when it learned that about 1,000 Students did not renew their Affiliation.

Regarding the use of Express in place of the Mails, Sub-Deputy Theodore Mohrhusen informed the Affiliates that there is an Express Agency in West Allis, on 73d and Greenfield. The Recorder let it be known that the main downtown office of the Express Company in Milwaukee, is open all hours for accepting packages, but not for executing Money Orders. She was there on Sunday.

The newspaper article "Colonel Ghadiali, Malaga Healer, acts as own Lawyer in Postal Case," which appeared in the Courier-Post, Camden, New Jersey on Saturday, February 27, 1943 was read by the Deputy.

The Deputy: "You can see from this article that Colonel Dinshah is not resting on his oars and is going to do everything HE can and as members of this Organization we ought to do everything WE can to assist and encourage him in his noble efforts."

The reading of the Bill of Complaint against the Post Office Fraud Order followed.

The Deputy: "You heard the opening shots. What is your opinion? Is there anything about it you observed?"

Theodore Mohrhusen: "The petition which you read to us, is, in my opinion, the only thing that will accomplish the effect we want, that is, to restore the mailing right of the Institute. None could do any more than Colonel Dinshah did. I am sure he will be able to follow through and we will be able to back him."

The Deputy (to the audience): "Do you feel that is the proper method and do you feel the same way as Mohrhusen?" Audience: "Yes!"

Theodore Mohrhusen continued: "The fact that the case is being taken to a Court of Equity! Do you know that we are just as much affected by this Fraud Order and each one of us has a right to complain in this matter as we are all affected by the revocation of the Mailing Privileges?"

Deputy Burns: "Every member is affected in that he or she is not permitted to send mail in the regular mailing process. Many people considerably removed from large centers, depending on Spectro-Chrome for the correction of a Disorder, rely upon the Central Office and are affected to such an extent that they should complain to the Postmaster General. They are denied a right to which they are entitled."

Rose Helminak: "Why can't we all write letters? I have a letter all ready. I can face them any time and I don't care about what they do to me. I will tell them what I think of them and what Spectro-Chrome has done for me."

Deputy Burns: "Colonel Dinshah presented 27 pounds of evidence which was given no consideration at all and any effort we might pursue would be valueless. At the Convention in January we signed a Petition to Senator Robert M. LaFollette, asking for help and it was returned to Colonel Dinshah with the statement that the Senator was too busy at this time to give attention to the matter. So the Colonel asked the Court to issue a Restraint Order to permit him to receive the Mail. It will depend upon the action of the Court and what decision is rendered. The Planet Commission was in a meeting with Colonel Dinshah, prior to the Convention in January and he outlined his procedure in this particular case and what would be necessary to be done. What we learned from him leads me to believe that, eventually, the mailing rights will be restored, but, it will take time and considerable expense. There is a peculiar condition involved. The Fraud Order was issued upon testimony of a Physicist, two Medical Doctors and a 'Postal Inspector.' Upon presentation of this 'Evidence' the Solicitor gave his 'Opinion' and the Postmaster General issued the Fraud Order preventing mail from reaching the Institute. This case was not heard before a Court. It was nothing more than a Hearing at which even the Defendant was not present. There was no Cross-Examination under the circumstances. Here is the peculiar angle to the whole situation: None can change the Order of the Postmaster General! There are other ways of approaching it and Dinshah is familiar with those ways and this is the first step to approach it in the manner in which he expects to get Justice. There having been no Trial, no Court, only a hearing, it cannot be appealed. To bring this to a Court Action it is necessary to take the steps he is now taking. Eventually a case may evolve which may result in a Trial by Court and release of the decision of the Court. In that particular case, that decision can be appealed to a higher Court in which Dinshah expects to play his card and reobtain the mailing privileges. No doubt, he will inform us, as time goes on and this is the first step which eventually may be decided in one of the higher Courts of the United States. In the meantime, we should be loyal and encourage others to renew Affiliation. The American people are peculiar. The little matter of having a photo taken and sent in for continuing Affiliation, is sufficient to make them give up the matter. That should not be a burden or a difficulty, to have a picture taken and sent in for a purpose of record and protection to each one concerned. Prior to this method, there have been frauds committed. People sought admission on another's Card, but, with the use of the photo on the card, none but the Affiliate can obtain admission to the Meetings. That is as it should be; otherwise, we might open the Meetings to the general public and the Lord only knows what would happen. That cannot be tolerated and we must remain a closed Organization."

As March First is an important and memorable day in the life of our President Dinshah and marks the anniversary of his release from unjust incarceration in Atlanta Penitentiary, the Recorder read the Good Thought for that day, "Release." Each Good Thought in Colonel Dinshah's book by that name, is a Gem in itself and several others were also read: "Officers", "Sabbath", "Teacher", "Garden", "Links", "Usefulness", "Bath", "Silence", "Spring", "Reputation", "Insult" and "Newspaper". (Applause.)

Case Reports were called and Eleanor L., an ardent lover and enthusiast of our Science, submitted the following:

(1.) Paul F., age 1, had an Infection in his Hand due to biting himself in the thumb. The family Doctor was called, but was out of town. His assistant advised the mother to place the Hand in hot water or apply hot compresses. Following his advice, the Infection spread into the Hand and up into the Arm. The mother of Paul called her mother Grace W., one of our Affiliates and told her what was about it. Grace, an enthusiastic Benefit Student, went to see the baby. She told her daughter the child was dangerously ill and needed quick attention. So she bundled the baby and took the mother Myrtle and the baby and father back to her house. She placed the child under Turquoise. She and the father then came after me. I could not leave just then, so told her to take my Thermometer and take the child's Temperature and come for me right after dinner. She did this. When she returned she said: 'Paul's Temperature is 104 Degrees; his Kardoscil is 88 and Spiroscil 48 and try as I will, it will not come different.' She said: 'Myrtle took it and I took it and we got the same.' My eyes about popped out of my head. I said: 'Good Gracious! The child is gasping!' Grace said: 'I know it! Come quick.' So together we went back to her home. I collaborated Myrtle, then took Paul's Kardoscil. It was 89, Spiroscil 48. Grace had given him a Magenta Tonation on Area 4 before she called for me. So, on the next Favor Period, I tonated with Green Systemic. I said: 'We must stop the Infection from travelling further.' After this Tonation I again took the Kardoscil and it went up to 98, Spiroscil 38; that was at four o'clock in the afternoon. By six o'clock, his Temperature was 100. Swelling was out of the Arm to the Elbow, Redness now being from Elbow to the Hand. The Hand was puffed and Purple in color. The following Tonations were Orange and Yellow Local. One hour after the Yellow was irradiated, the Thumb opened and began to drain. After the Hand started to drain the baby wanted to take food which he refused prior. He took four ounces of milk and eight ounces of water. Before that he had only taken about a teaspoon of water. The Hand drained from one Wednesday to the following Wednesday. It took two weeks to complete Normalation, as a Rash developed on the Body and that took about a week."

(2.) "Rose A., aunt and guardian of Frank G., aged 8, rented a Spectro-Chrome as Collaborator. Frank was very ill with Rheumatic Fever. The Medical Doctor was called in before they came for Spectro-Chrome. He had been ill from Tuesday until Saturday, when she came for Spectro-Chrome. The Fever varied from 102 to 104 Degrees. Kardoscil was 79, Spiroscil 19. Turquoise Systemic, Green Systemic and Magenta on Areas 4 and 18. He was normalated in one week."

(3.) Alice K., aged 29, was also normalated by Spectro-Chrome. She had a Pain in the Leg from the Calf to the Hip and could hardly walk. Lemon Systemic, Indigo Local and Turquoise Systemic, Magenta on Areas 4 and 18 and was normalated in 19 Tonations. Kardoscil 80, Spiroscil 16. (Applause greeted Eleanor L.)

Theodore M. was interested in the baby case, as he is the father of four himself. He wondered how this small baby could become so infected. He said: "There must have been something wrong in the feeding or somewhere, to cause so much Pus to collect. At home, we can tell in one day if our children are not fed according to the Dinshah method."

The Deputy expressed the opinion that many Skin Disorders are stubborn and slow to normalate. He knew of a child where the relatives commenced Tonation with Atcowa Green and natural Elimination followed. The apparent sluggishness was removed.

"Then would you say that Green is the Atcowa for Constipation?" asked an Affiliate.

"Commence with Green and turn to Yellow," replied the Deputy as he explained the process of Elimination saying that putrefaction sets in, when we do not have proper Elimination.

Past Deputy Otto H. L., from whose manner of tonating Sufferers much may be learned, also, submitted the following cases:

(1.) Theodore K., aged 48, male, Lumbago. Tried other methods without results. Spectro-Chrome Orange set him right in one week and he returned to his work.

(2.) Alfred G., age 40, male. No Back Elimination for five days. Spectro-Chrome Orange and Yellow, alternated. Four Tonations produced two "large" Back Eliminations after everything else failed him. Spectro-Chrome set him right.

(3.) Marie G., age 34, female. Has an 8-year-old son with a bad Skin Disorder on Chin and up to the Ears, back of Neck and on Left Thumb, with a very thick crust. His mother said it was called "Impetigo!" Spectro-Chrome Green Systemic, Indigo Local, normalated the boy in two weeks' time. Another healing method was tried first without results.

(4.) Albert W., aged 62, male, Sciatica, of three months' duration. Took Injections four times a week from Medical Doctor, without success. Spectro-Chrome Yellow on Areas 6, 7, 8, 9 and 10, Lemon and Indigo Local were irradiated and he returned to work in 3 weeks.

(5.) Emil B., age 60, male, had what he called "Chronic Bronchitis"—with Loss of Voice. Kardoscil 120, Spiroscil 28. The severe Cough prevented him from sleeping. After Lemon Systemic Front and Back, one week, he started to expel large "chunks" of sputum in quantity. Both feet were swollen. Atcowas Lemon Systemic and Magenta on Areas 4 and 18 were given and he returned to work in six weeks.

(6.) Gordon K., age 25, male. He came in a rundown condition. Pains in Area 1. He was possessed of a "Phobia" and was afraid to cross bridges and also afraid to handle knives. Yellow Systemic was irradiated and he said he felt somewhat better after the first Tonation. The next day, he reported he slept all night which he had not done for weeks. He was normalated in 14 Tonations of the Yellow Attuned Color Wave. (Applause.)

Colonel Dinshah's very best wishes to all Affiliates were conveyed by Anita Enders who said she had spoken by telephone to Colonel Dinshah the preceding evening. (Applause.)

Rose H. said she had a severe Cold on New Year's day and cannot free herself of it. Hoarseness is present. Took Tonations for over two weeks and would like to know what Atcowas to irradiate while awaiting Free Guidance from the Institute. She was advised by Deputy Burns that that is a difficult thing to do without knowledge of the Kardoscil and Spiroscil and oher particulars. Rose said at one time it was K. 84, Sp. 18. Atcowa Lemon was suggested.

Eleanor Luedtke emphatically replied that she will always remember Dinshah mentioning that Orange is the Atcowa for stubborn Congestion.

Emma Schroeder also remembered it and recalled that Dinshah stated that the Lemon Wave would "work better" after the use of Orange.

Deputy Burns demonstrated on the blackboard how to go deeper into the Mogen David.

Marie K.'s 17-year-old daughter normalated a Toothache in ten minutes by using Indigo. Another daughter used the same Atcowa for a stiff Neck and the Pain was almost gone the following morning. Marie raised the question as to what Atcowa to use for frozen Ears or Toes. She was advised by the Deputy that the Infra-Green Division of the Spectrum is to be irradiated in such Disorders. There was discussion as to whether Indigo was the proper Atcowa to use in the case of a stiff Neck as the Disorder may have cleared by itself during sleep.

Sub-Deputy Theodore Mohrhusen explained the impurities in the Blood and improper Circulation in such Disorders.

Marie K. asked, "Why do some people require so much more sleep than others? I, for instance, need ten hours and my daughter can get along with only five."

Evert Dahlke: "Proper Food!"

Deputy Burns: "Wrong living!"

Marie K.: "She eats the same as I do and she is my own child. I cannot understand it!"

Deputy Burns: "Before I used Spectro-Chrome, I required eight and ten hours of sleep. Now I seldom retire before midnight and rise at 6:30."

Otto Lutz: "Possibly the daughter is a lighter eater than the mother."

Frederick Kneppreth: "Some cases are slower to respond. When the Sun is not visible very much, it is possible that the Aura becomes affected."

Alice Brown stated her Slides are in Malaga for repair. She had a Cold and used citrus foods in the evening with good results.

Deputy Burns related how a friend, who knew of a "cure-all" for Pain recommended an Aspirin tablet with Baking Soda. That took care of every Pain he ever had. Another friend who became ill on a hunting expedition in the wilds of Canada, tried it and was never so sick in his life!

Olga Lenosek: "Are Contagious Diseases communicable?"

Deputy Burns: "Yes! It depends upon the resistance of the individual."

Ethel Schott: "I wondered if the Eclipse had any effect on Epidemics. Measles are prevalent in our school."

Frederick Kneppreth believed that was one of the reasons why Dinshah advocated sleeping in separate beds, to prevent spread of Contagion.

The Deputy replied to Ethel's question by stating that Dinshah spoke of that at length and how we humans are subject to Epidemics depending upon the action of our Respiratory Organs. When the breathing processes are out of order, we become what the Doctors call "susceptible" to Disorders.

An Affiliate was not clear on "Systemic" Tonations or the length of Tonations. The Deputy informed her.

Dave E.'s wife has severe Sinus Disorder and Headache. She took many Laxatives.

Emma Schroeder finds lubricating the Nostrils with warm Cocoanut Oil often opens the Sinuses and causes them to drain.

An Affiliate travelled a distance of 86 miles, from Oshkosh, Wisconsin, to attend the Planet Meeting. The Deputy expressed congratulations.

It was an interesting meeting, conducted in a dignified manner and deserving of an even better attendance. The Affiliates present numbered 113.

EDITORIAL REMARKS

To send any papers by Express, one does not have to make heavy packages. One woman sent with her papers five pound lumps of coal, another expressed ten pounds of lime stone and another dispatched a log of wood! I preserved that log to lambaste some refractory opponent! The proper way is to place your papers in an envelope, gum it and present it at the Railway Express Agency, with request to dispatch it to Dinshah Spectro-Chrome Institute, Malaga, New Jersey, Railroad Station Newfield, New Jersey, in their envelope form 6034—. The Agent will have it sealed and the value may be declared as whatever it may be.

Dinshah and his Irene and the 7 others in his family will never cease serving Mankind. They are working day and night, even more than usual, to save Spectro-Chrome for your service. Be patient and watch the workings of Those Whose Effulgence shines on Truth. As the matter is now before the Court, I refrain from saying anything except publishing open facts. My day will come; the World will accept the message of Spectro-Chrome.

About 1,000 Affiliates did not continue their Affiliation this year, for reasons you know. They might not relish our rigid regulations, but, such ones are better out than in. I want in this Organization solely those who firmly believe in the Science and the Service and the others are out and may stay out. We do not have to go after them. They are on the contrary a detraction to us. One Affiliate—a man—wrote in his

Request For Continuation, "I do not go to the Planet Meetings, because I do not care to listen to the ills of other people!" That self-centered man had better understand that if he does not "care" for others, we have no reason "to care" for him. He may as well keep his three dollars a year to give to a Medical Doctor for a single consultation. I found in life a marvelous Karmic Principle true to the letter, that the more I give to the World, the more I get from the Universal Spirit to give to the World again. Be therefore givers and not gimmies, if you wish to progress in Higher Life. That is what Benign Moses and Jesus did; hence Their glory. Respect Their noble example.

Infection is nothing but the result of the mixing in the Blood Stream of putrid matter. Spectro-Chrome effectively deals with ALL Infections, if intelligently used. The baby probably had food particles on its Teeth in process of Fermentation, the mouth not being rinsed after a meal. In biting, the putrescent particles entered the Skin, thence being carried into the Circulation. The "drain" is nothing but the effort of Nature to throw out the disturbing elements. Spectro-Chrome is competent to act even as a Constructive Surgeon, if one knows its proper use. However, I advise you to stay away from medical views. Our three Volumes of Spectro-Chrome Metry Encyclopædia have all you need for all Disorders to which Mankind is subject. God made One Light with Seven Stages—His humble servant showed you the use—follow the Rules and success and Health will follow in your path.

Grandparents as Benefit Students must not tonate Grandchildren, except under Collaboration signed through a Spectro-Chrome Normalator by one of the parents. This is a strict Rule and must be followed without deviation, to protect the Science. There is no "special Atcowa" for a specific effect. Study carefully the mathematics of Spectro-Chrome in Volume 1 of Spectro-Chrome Metry Encyclopædia. Every Benefit Student should study that subject especially, for a clear comprehension of the Irradiation. The two Antipodals are Violet and Red; they are the generators of all the rest. Dinshah's "Sanslide" or Slideless Spectro-Chrome proves that very thoroughly. It may be marketed after the War is over. Green being a Pituitary Stimulant, may produce Elimination by energizing the other sluggish Organs, because of the governing power of the Pituitary and Yellow has ½ Green in it anyhow! There is much BEHIND the Science of Spectro-Chrome for which the mind of the Student is not yet ready. It will eventually need a fourth Volume!

The numerous "Phobias" or "Dreads" are mostly psychical in nature. Medical Doctors do not recognize, because they have no knowledge of, the Psychical Vehicle. All their mind is centered in Blood, Pus, Serum, Mucus, Saliva, Urine, Faeces—nothing but "Cadaver" derivatives. The Live Vehicles and their marvelous potencies are hidden from the eyes of these quassi-scientists. Their empirics will NEVER accomplish true Health. Refer about "Phobias" in another article.

Rose H. should be asked about her Food Habits.

Marie K. should learn to differentiate the characteristics of age and youth, surroundings and Sleeping Habits. Insomnia and Restlessness are bound to appear soon or late in persons sleeping together in one bed. Husband and wife sleeping always in one bed is a very unhealthy and bad custom. The movements, warmth, sounds, odors, Aura of one has a clear tendency to disturb the mate. Sleep in separate beds and preferably even in separate rooms. Marriage does not mean or need perpetual snuggle; on the contrary, it is the quickest road to diminution of affection and sometimes of divorce. Privacy between mates should be respected.

The Slides of Alice Brown went back to her; that will save the expense of eating Lemons!

Aspirin is the Monoacetic Acid Ester of Salicylic Acid and should be thrown into the discard from every family.

Study Disorders Of Childhood for Measles. Eclipse may not cause the "Epidemic," but, its influence on those nearing an eruption may be visibly seen.

I shall have to give a whole Course on this important subject during the next Special Convention. Contagion is the carrying of foul matter into the Blood Stream by physical contact. If one's Blood Circulation and Blood Chemistry be proper by proper living, one need not be worried about "Infections," "Contagions" and other scarecrows. It is one's own self that is the cause of one's own Disorders. Of course, it is easy and more pleasing to pass the blame to another.

Lubricate the whole body and bathe with warm water from the hair of the head to the toes daily, first in the morning and you will have clear Sinuses and a clear mind to think clearly. Use the bucket bath system and NOT the unclean, splash tub method.

You are the first to respond to the suggestion of reporting full names and addresses of Sufferers, for editorial information. Thank you cordially. ALL Planet Reports must be similarly complete for our protection, otherwise the report will NOT be printed.

———o———

FIRST ST. LOUIS MISSOURI PLANET

(HOTEL MELBOURNE, ST. LOUIS, MISSOURI)

Meeting Of Thursday, March 4, 1943

Successful Use Of Spectro-Chrome In:—Dizziness, Belching, Flatulence, Pain, Constipation, Headache, Vomiting, Sprain, Stiffness, Peplessness, Dental Disorder, Appendicitis, Cold, Swelling, Hoarseness, Frostbite, Fever, Dyspnea, Diarrhea, Cramp, Nervousness.

Reported By:—VIOLA C. JEUDE, *Recorder*

Deputy Fred Kohl was in the chair. He asked Chas. Stiermann to read the article which appeared in the Courier-Post, of Camden, N. J., Saturday, February 27, 1943.

Communication from the Principal Office was read by the Recorder. The rough proof of Bill Of Complaint And Petition For Temporary Restraint Order And Injunction In The Federal District Court was discussed.

Case Reports by Clemens N.:—

(1.) Robert P., age 15, Ratio 72-19. Swollen Ankle with constant Fever, was 3 months in bed; still had swollen Ankle; last 2 months Swelling in Right Arm; Medical Doctor drained it, the Joint and no result. M.D. first said it was Rheumatism; boy would sit at home; no interest in anything. Turquoise Systemic, Yellow Areas 6 to 11, Magenta on Areas 4-18, also Scarlet Areas 4-18. Swelling did not recur since Tonation started. Is an entirely different boy. Mother well pleased and will become an Affiliate.

(2.) Paul L., age 64, Ratio 84-21. Dizzy spells while working, had to sit down followed by cold and hot spells, plenty of Belching and expelling of gases. Yellow and Orange Systemic, Magenta on Area 4-18, normalated in a few days. Well pleased.

(3.) Blondeia C., age 40. Ratio 72 to 17. Pains in Lower Back to Ankle, very severe; cannot lie on back, Belching and gas in Area 8 but cannot expel it. Suffers from Cold Feet and Constipation since childhood, Headaches and Vomiting spells for days; feels weak cannot stoop down or walk for any distance. Green, Yellow, Orange, Red Systemic, Magenta and Scarlet Area 4-18. Feels much better; no Headache, Vomiting or Belching and is pleased. Had to leave town.

(4.) Thomas H., age 53, Ratio 60 to 24. Sprained Back, Osteopathic treatment to no avail. Orange Systemic Back 1 Tonation and is well.

(5.) Robert S., age 51, Ratio 72 to 18. Had been to M.D. for Stiff Legs, very painful, could hardly stand up. Orange Systemic and Yellow Systemic. Six Tonations, feels fine and can work again.

(6.) William W., age 21, Ratio 72 to 16. Felt draggy, no pep like having a Cold. Turquoise Systemic, 2 Tonations; says he feels fine.

(7.) Lorenz W., aged 84. Ratio 64 to 16. After Tooth Extraction, Pains in Cheek, diagnosed by Dentist, M.D. and Specialist as Nerve Disorder; must operate to cut nerve; Pains will be unbearable, Facial Neuralgia, et cetera, could not eat or drink, suffered much agony. Purple and Indigo on Area 1, Magenta on Area 4. Pains entirely eliminated, went back to Dentist who knew of this "Light" from other Sufferers; he said it must be good!

(8.) Nora A., age 20, Ratio 96 to 20. Terrific Pains in Lower Abdomen and right side, M.D. diagnosed it as Appendicitis. Turquoise on Areas 9-10, Turquoise and Blue Systemic, Magenta on Areas 4-18. Slept for about 5 hours after first Tonation, Pains and Fever normalated in a few days; well pleased.

(9.) Arthur B., age 69, Ratio 74-16, had Cold; felt Chilly, Sneezing and so on. Turquoise Systemic; one Tonation and feels very much improved.

Cases reported by Nathalee B.:—

(10.) George Brehm, age 50. Ratio 80 to 25, Legs swollen, couldn't sleep or lie; had broken Ribs, was taped, much Pain in breathing and coughing. Lemon Systemic first Tonation, Swelling was gone and Chest relieved. Turquoise Systemic (once only). Lemon, Yellow, Orange, Magenta as Ratio indicated. Good results.

(11.) Clyde K., age 66. Cold, aching limbs, Chest hurt, Throat hoarse and raw, Ratio 64 to 15, Orange Systemic, Ratio 80 to 24. Turquoise Systemic, Magenta 4-18, relieved and able to return to work in 5 days. Frostbitten Fingers responded to Yellow and Orange although slowly. This individual had been a heavy Smoker for over 50 years, but, has not smoked since this holiday season.

(12). Myrna S., age 2 years 9 months. Fever at night only, delirious, Ratio 130-24. Has breathed unusually since birth; had frequent gasps for Breath causing body to jump. Turquoise Systemic, Magenta, Yellow Area 6-7-8-9-10. No Fever after 5 days, but parents returned Equipment because the Rational Food Of Man plan was not satisfactory to them.

(13.) Emily F., aged 57, wife of Affiliate. Frequent Backelim, aching of Body, Cramps in Legs, some Vomiting, Ratio 92-18. Orange 6-7-8-9-10-22. Magenta 4-18. Condition relieved in five days. Has long standing Gall Bladder Disorder according to the Medical Doctors.

The Deputy then proceeded with the Educational Program. Figure 38, Attuned Color Waves Of Dinshah, page 300, Spectro-Chrome Metry Encyclopædia was drawn on the Black Board and Deputy gave an able address on the Mogen David and its occult meaning.

The Deputy then explained the Principle Of Wave Motion.

Questions were asked and answered.

Question was asked about Adhesions. The Deputy said he would like Dinshah's opinion on this question.

The Deputy then explained Spectro-Chrome by Primary Fractions.

The Deputy asked Spectro-Chrome Normalators about their experience with the Green Attuned Color Wave. Sub-Deputy Nathalee Britton said just recently children of a neighbor next door had severe Headaches. Turquoise Green and Magenta Atcowas were used, Children had 102 Fever; they did not have proper Elimination, so Yellow and Orange Atcowas were used on Areas 6-7-8-9-10; spots appeared, the Medical Doctor was called and she was dismissed. The Doctor "diagnosed" Scarlet Fever, but, he said it was such a mild case and he was so surprised as they lacked the usual symptoms of Sore Throat, Swollen Glands and no Headache, that he would not put a sign on the door!

She then reported another case:

(14.) Lydia C., age 52, suffered from Imbackelim, Headaches, Nervousness and so called Psoriasis. Green, Lemon, Yellow, Orange, Magenta and Scarlet Areas 4-18 were irradiated and weeping condition came. Turquoise was used, no easing—Green only for 3 months daily with Magenta 4-18, cleared Face and Neck so that woman was able to get an office position. I have waited a year to report because I wanted to see the results continue. Ears do not entirely heal although that ear condition was one lasting from early childhood. Woman does not tonate anymore but is pleased with results inasmuch as the Medical Doctors had not helped her.

The Deputy asked "If all the Color Slides would be taken from you except 3, which would you like to keep?" The answer was "The Red, Green and Violet."

Marie J. reported one of her children being normalated of a hot water Burn. Blue was irradiated.

The Deputy said his 4-year-old son was playing on the floor; he wanted to carry a kettle of boiling water from the stove to the sink and had a premonition something was going to happen. He warned the youngster to move, but the youngster said, "There is plenty of room; Daddy you can pass." As he was passing, somehow or other the kettle fell and spilled and rolled toward the child, burning the sides of his Fingers and Arm; he was placed under the Blue and in about 10 or 15 minutes the child asked if it was all right to get up now it did not hurt any more; no dressing was used except a little cocoanut oil.

Richard J. said he had been trying to get some for some time but even the wholesalers did not have any and informed him they would be out of it for the duration of the War.

Question:—Colonel please answer. What can be used instead of Cocoanut Oil?

EDITORIAL REMARKS

When any Operation is performed in which Organs or parts of them are removed, there is bound to be some surface adhering to adjacent parts. Nature can not keep empty gaps. Such Adhesions are inconvenient and exasperating sequelæ of Surgery. They are benefited in some cases, by relieving the "tension" with Atcowa Orange.

All so-called Children's Disorders are easily amenable to Spectro-Chrome. We need no "signs" on doors.

Olive Oil is a substitute for Cocoa Nut Oil and if that be not obtainable, Maize Oil may be used. Lubricate Head to Foot and bathe daily with warm water, using very little soap. You will not need Beauty Parlors.

A friendly warning. Your Planet shows a trend toward medical ideas. Avoid all such theories, views and nomenclature as you would avoid Poison Ivy. Have nothing to do with Differential Diagnosis; follow what Spectro-Chrome teaches you and avoid pitfalls.

———o———

Spectro-Chrome Metry Encyclopaedia

By:—LINA F., *West New York, New Jersey.*

Volume 2—It is the most fascinating book I had the privilege to study. I enjoyed every minute of it and obtained from it valuable, unusual and unforgetable knowledge. The Sevenfold Constitution Of Man was most interesting. To learn about the Human Body and Mind is my good fortune. It clearly shows sufficient proof that the Science of Spectro-Chrome with its simplicity exceeds any other Healing Science. All of the above would be wasted unless I used it to good advantage by doing good deeds unto others, thus gaining rich experiences for myself.—E. P.

By:—FRED C. R., *New York, New York.*

Never in my whole life, have I been so enthusiastic of reading a book as I am today about Spectro-Chrome Metry Encyclopædia, Volume 2. Most scientific books which I read were boring with technicalities and made me sleepy, while Volume 2 is a Mental Tonic for me. It enlightens my soul with its simple and true facts of knowledge and I shall never part with it for the rest of my natural life.—E. P.

———o———

DINING ROOM FUN

Noshervan, the 5 year old son is eating his evening meal, prepared according to Rational Food Of Man system. His mother Irene had already her meal and had gone upstairs for other duties. The baby was still fumbling with his Artichoke in a leisurely manner, when reaching close to the end he suddenly shouted, "Please, mother, come down and take my heart out!"

FIRST MERRILL WISCONSIN PLANET

(HOTEL BADGER, MERRILL, WISCONSIN)

Meeting Of Thursday, March 4, 1943

Successful Use Of Spectro-Chrome In:—Tuberculosis

Reported By:—DOROTHY DENNE NELSON, *Recorder*

Deputy Walter Briese was in the chair.

"Serious Warning" from the November 1942 Spectro-Chrome was read. Communication of February 1, 1943 from the Principal Office was reread, because there were so few Affiliates who were able to be present at the prior meeting owing to the inclement weather and snow. The War Model Spectro-Chrome Equipment was discussed.

Communication of March 1, 1943 was read by the Deputy and also the article from the Courier-Post. The Post Office Fraud Order Suit was discussed.

The Chapter on Tuberculosis was read by Emma Klug. Remarks by Lydia Danner followed.

"We have been successful in treating this type of Disorder with Spectro-Chrome. I have been tonating a Sufferer in Gleason, Wisconsin. This Sufferer had x-rays taken. Had her first attack 8 years ago. X-Rays had shown one Lung involved and the other one started. Last Winter, she developed Pneumonia. Was taken to the Hospital. She recovered and had more x-Rays. The second Lung was involved. She was ordered to the Sanatorium as soon as the Pneumonia was checked. In the meantime, she learned of Spectro-Chrome and her husband came to me. She rented an Equipment. It was installed in their home and now she is feeling well. It did not take so very long to stop the Cough. Kardoscil is normal. Turquoise was given until Kardoscil was normal; then Green Systemic. Back and Front. Orange Local on Areas 4, 5 and 17. She is in good Health now and is able to work hard.

Environment is important and also Rational Food.

Case Reports followed:—

Edward K., male, reported that he had an Itch on his Skin from the contact with wood. Took three Tonations and the Itch disappeared and he did not have it since. Turquoise, Lemon and Yellow were used.

Deputy Walter Briese reported, that when he worked in Milwaukee in a shoe factory, there were two girls who developed an Itch and the Skin Specialist could not do anything for them. Lydia D. spoke about a similar Disorder from Rubber. The Doctors could not do anything for the sufferer.

EDITORIAL REMARKS

Several cases reported in your Planet were omitted from the print, because proper identification and address of the Sufferers were not given.

FRAUD ORDER REVOCATION CONTRIBUTION

The following voluntary and unsolicited Contributions are hereby acknowledged, in addition to the two lists in prior months:—

296. Charles J. Renn.
297. Margaret Rose.
298. Madeline Coster.
299. Clara Schubert.
300. Magdalene Schubert.
301. Catherine O'Donnell.
302. Julia Reber.
303. Viola C. Jeude.
304. Richard Jeude.
305. Margaret Kraemer.
306. Edward Poppiti.
307. Earl Reno.
308. Alice G. Brown.
309. Paul Belger.
310. Karl Roemming.
311. Johanna Braun.
312. Elsie Nehrig.
313. Martha E. Bowers.
314. Ann Zastrow.
315. Max Ihle.
316. Jessie E. Ness.
317. Viola Rohde.
318. Hans Koetter.
319. Sophie Kling.
320. Herman Leitzke.
321. Frieda Wehrhagen.
322. Selma Reeths.
323. Howard Ingram.

FIRST CHICAGO ILLINOIS PLANET

(HOTEL ATLANTIC, CHICAGO, ILLINOIS)

Meeting Of Thursday, March 4, 1943

Reported By:—ELSIE BEYER, *Recorder*

Deputy Wilhelm Weissenbacher was in the chair. Four new Affiliates were introduced. Communication from the Principal Office was read by our Deputy about the Fraud Order.

An article from the Courier-Post about the Fraud Order case was read by the Recorder. Some of the highlights on the Suit were read by Arthur Koegel, jr.

The Deputy explained from Spectro-Chrome Home Guide how the Attuned Color Waves are used for various Disorders.

He said, "Spectro-Chrome acts upon the Physical Body not by Absorption or Penetration, but by a process of Reinforcement or Interference on the Radio-Emanations of the Chemical Body, called the Aura. We can not see the Aura, as ordinary eyes are not trained for it, but it can be seen by certain methods of Interference.

By-Laws 32 to 38 both inclusive were read. Questions were asked on Disorders and answered. Discussions followed.

Systemic and Local Tonations were explained.

In emergency cases, can Affiliates get information from a Spectro-Chrome Normalator if they be under Free Guidance of the Principal Office? We wish for this information from Colonel Dinshah.

The Recorder read an article from the Chicago Tribune on the value of Whole Grain Food, stating that they are essential to good Health.

For years the majority of people have not been getting the proper Food on account of the Refining Processes. She said, that after reading different articles in the newspaper, they proved to her again that Dinshah educates us on the right road to proper eating. She also noticed that since the War and the shortage of meat, the newspapers print many recipes for meat substitutes and also stress the importance of Brown Rice and Legumes.

EDITORIAL REMARKS

Attention in cases of Emergency, is duly provided in our By-Law 87. Furthermore, a Telephone or Telegram according to the Rules given explicitly in Spectro-Chrome Home Guide, will promptly bring needed information from the Institute. However, one can not be under two directors. By-Law 79, Sections 22 and 23 are plain. One may make the choice.

What Was Said Of Him 47 Years Ago When Many Of His Opponents Were Just Thinking Of Taking Birth!

THE NEW YORK TIMES
Wednesday, March 11, 1896

SAYS X RAYS ARE NOT NEW

Views Of Dinshah Pestanji Ghadiali, the Indian Scientist. Principle An Old One, He Declares.

The "Parsee Edison" Says Many Discoveries Are Made In His Country — His Plans Here.

Dinshah Pestanji Ghadiali, known in India as the "Parsee Edison," late electrician to His Highness Maharaj Rana Sahib of Dholpur, Central India, and also late Telephone Superintendent of Dholpur State, is in this city on a tour of the world.

It was Dinshah who electrified the members of the Aldine Club Monday night, on the occasion of their jubilee reception, by impromptu remarks, in the course of which he said the principle of the Rontgen X rays was not new in the East.

Dinshah yesterday explained to a reporter for THE NEW YORK TIMES his name and the Parsee system of nomenclature. "Dinshah" is his own name. It means "Emperor of Religion." "Pestanji" is his father's name. It means "One with the Holy Body." "Ghadiali" explains his occupation. It means "watchmaker."

"A Parsee reading my visiting card," he explained, "would know all about me. He would read 'Dinshah, the son of Pestanji, the watchmaker'."

The Parsee is of slight build, and below the average height. He wears a small black mustache, and his coal-black hair and eyebrows and dark Indian skin are relieved by the large, soft, glowing black eyes common to the East. He talks excellent English, fluent and grammatical. Although not yet twenty-three years old, he has been a lecturer six years on a variety of deep subjects connected with electricity, magnetism, odic force, heat, light, and various phases of the vital forces.

His father, a leading repairer of chronometers in Bombay, being of orthodox views, was opposed to his son's enthusiastic study of electricity and kindred subjects.

"My father went to bed at 10 o'clock," he said. "I used then to creep into the room and begin my studies. I worked all night and when my father rose at 5 o'clock in the morning, I retired to sleep for three hours."

Dinshah's ardor was unquenchable. During the lonely hours of the night he worked away at the manuscript of a book which was to be the first fruits of his studies. His father knew nothing of it. At length it was finished—thirty-two thousand pages in thirty-one chapters, the work of six years and the result of untiring effort. But it shared a fate similar to that which overtook the second volume of Carlyle's "French Revolution." The index was already in the hands of the publishers when the body of the work was stolen with Dinshah's valise at the station of Hyderabad, and nothing has been heard of it since.

Dinshah is not only a Parsee. He is a Zoroastrian. He wears a grand medal of the Independent Order of Good Templars, and is a member of those numerous societies of India and the East which believe that there are latent in man forces of a kind that can raise him to the condition and power of a god, if properly cultivated and developed.

He never leaves his head uncovered, wearing a Parsee turban by day, and a tight-fitting skull cap at night. The magnetic extremities of the body he believes should always be covered, so as to prevent the loss of vital magnetism, which is otherwise always passing out of the body and being wasted.

"From time immemorial," he said, sitting on the edge of his bed in his ascetic quarters, "India has been the home of spiritual culture and spiritual discovery. Most of the discoveries, so-called, of modern Western science, were known in the East ages before. They were laughed at and ridiculed by the scientists of the West, until forced finally to take note of them, they changed their names and paraded them as their own discoveries. Mesmerism and hypnotism, known for several thousand years in India, offer an eloquent example of this.

"The X rays; yes! They are wonderful, indeed, and I am probably the first Parsee to hear of them, for when I left India two months ago, there was no knowledge abroad of the discovery.

"The X ray is a lower manifestation of the astral light on the physical plane, and as such has long been known to Eastern occultists.

"The fourth dimension of matter, or the ultra-gaseous state, otherwise known as the ultra-magnetic state, or the radiant medium, exists everywhere. In the East we have always believed in it. It even exists in the pores of matter, and is capable of permeating the most solid body, so-called. Were the medium not existent, what is cohesion or repulsion? What is gravity? For, without the combining energy of the ultra-gaseous particles, and without their correlative affinity, the whole mass of the world would disintegrate and fall to pieces.

"It has been known all along that light is capable of traveling without air, for the atmosphere extends only some 200 miles above the earth. But light can not travel without vibrations, and the astral light is the vibrations of that ethereal medium which brings the light to our atmosphere.

"Some persons ask, Why can't we see that vibration? Why is it that the Eastern adepts can not show to us that light as ordinary light, &c.? I ask, Has any Western scientist ever seen electricity moving? I might as well say, Let them show it to me or I will not believe in electricity. 'Ah! but it appeals to other senses,' you say. 'We see the effects of electricity even if we do not see it moving.' Remember, then, electricity is a physical agent. The astral light is an ethereal agent. Physical forces can be recognized by physical means; astral, or ethereal, forces can only be recognized by the ethereal eye, which is only to be developed after passing a certain stage of vigorous mental culture.

"The East has never known in its physical bearing the Rontgen X ray, but has known the very principle that guides it. The Eastern masters of the occult knew the fundamental, hidden, occult principle of the rays. Nobody has ever seen these rays, and they are physical. How can you, then, expect to see the rays which are far more ethereal? It is quite impossible. If you wish to see such things as the students of Eastern yoga see, there is only one thing you can do. That is to develop the latent powers in your own self.

"It seems to me that those who study electricity and magnetism can not but be convinced that even some physical effects of electricity remain up to this time unexplained. The time will perhaps come when these unknown things will be recognized and acknowledged as working forces; when the Western scientists will lay aside some of their prejudices and, with an unbiased mind study the religious and spiritual side of the East, which has ever been and ever will remain the land of mystery."

Dinshah's eyes gleam and flash when he talks, and his face illumines with the ardor of unquenchable enthusiasm.

Before going to San Francisco, on his way back to India, he will deliver lectures in this city. In India, his lectures have included such subjects as "Expenditure of Vital Energy, or, a Comparison Between the East and the West"; "Influence of the Solar System on Man"; "Sound an Ethereal Force"; Odic Force or Ultra-Magnetism and the Doctrines of Zoroaster"; "Man as Master of Electricity"; "Suicide, a Vice of the Present Day"; "The Magic Power of Words and Incantations", and "Thought is Energy and Will is Power."

"I have come here," he said, "to learn physical powers and to show to the people the existence of spiritual powers."

Dinshah is an ardent admirer of Tesla, with whom he has a slight acquaintance, and of Edison. "Only I pity them", he said, laughing, "because they don't get time even to eat or to enjoy the luxuries of life."

Dinshah was the first lecturer on electric light in India. He says if Tesla were to go there the people would make a hero of him.

(Editorial Remark:—The book mentioned as lost, had 3,200 pages; not 32,000.)

OUR NEW FREE GUIDANCE CHART

In order to reduce pressure on our over-burdened staff, we designed a new type of Free Guidance Chart for answering the usual General Advice Chart. Hitherto, the Spectro-Chrome Irradiation was type-written, but, the prevailing conditions made us invent a novel method for speedier work.

The new Free Guidance Chart is on similar blue paper, on the two vertical edges of which are printed the initials of our twelve Attuned Color Waves and the Numbers of our 22 Areas, as well as the Frequency of Tonation. On the bottom (horizontal)edge are the days for which to continue Tonation and to report progress. The top (horizontal) edge has 26 directive Articles marked. These Articles are printed in full on the front and back of the sheet. The Irradiation Atcowas and various guiding details are indicated NOT in type as usual, but, by Punched Holes in the appropriate position and by reading where the various Punches are (just like a railroad ticket), all around the four edges, you will quickly get the instructions needed for the Sufferer.

The new system will go into effect Saturday, May 1, 1943. It will save Dinshah's time considerably in giving the service. He is busy printing three more books, announcements of which will appear soon.

———o———

FIRST PORTLAND OREGON PLANET

(MASONIC TEMPLE, PORTLAND, OREGON)

Meeting Of Thursday, March 4, 1943

Reported By:—LEWIS BISCHOFF, *Recorder*

Deputy Stanley Burroughs was in the chair. He read the Principal Office communication. The Post Office Fraud Order Petition was read. Both were discussed.

The Deputy read an article from a magazine, to show how to succeed in life by setting aim for a particular object.

The Deputy's wife had a still born baby last Autumn. A medical and surgical report was read about her condition. He wondered what Colonel Dinshah would have to say about that report, a copy of which accompanies the Planet Meeting Report.

An Affiliate could not understand how the Sun may have no "Heat" and yet produce so much Heat where He shines. Explanation was given.

EDITORIAL REMARKS

The Report about the Deputy's wife can receive no attention from the Principal Office, because our Science cognizes solely our methods, which rejected all Differential Diagnosis. That Report should have come with a complete General Advice Chart of that date, made in accordance with our requirements. Then, we should be able to do justice to the Sufferer. Cut medical and surgical views out. The World has had them 6,000 years and is still sick and more sickness is developing.

The "marvelous and miraculous" reports about the efficacy of Spectro-Chrome sent in by your Planet this month, are omitted from the print, because, no full names and addresses were given to the Editor, for identification. While he has full confidence in the Truth of the reports sent, he refuses to shoulder responsibility as a publisher, without having full information in all cases.

EXPANSION OF OUR RESIDENT QUARTERS
HAVE YOUR VACATION AT YOUR INSTITUTE

To accommodate and train our Students in the principles of right living, to restore and maintain their Health, at the same time teaching them the value of the Rational Food Of Man, we expanded our facilities to meet the increasing demand for our Personal Services.

The facilities are open to Benefit Students only, who have their Spectro-Chrome Equipments registered in their own name. Collaborators and Educational Students will not be admitted; relatives of Benefit Students or Non-Affiliates will not be admitted and no admission will be provided for transients or for less than 1 week.

Each Benefit Student will have an individual room, meals, laundry and necessary Spectro-Chrome Tonations, as well as personal Free Guidance by the Originator and Staff. However, Benefit Students unable to help themselves or in need of nursing care, will not be eligible for admission.

The uniform Contribution for the accommodation is $35.00 per week. Students must send their request with a complete General Advice Chart and await acceptance of reservation before coming. They will have to conform strictly to the Health Regulations of the Institute and must come provided with their full return transportation and their War Ration Books. Beyond the stated uniform Contribution, there are no extras, expenses, tips or gratuities and in order to keep their mental attitude solely on the recovery of their Health, no correspondence will be permitted, except reporting to their family the arrival at and departure from the Institute.

As no inbound mails reach us, send your request through Railway Express Agency, well ahead of time.

———o———

DINSHAH IMPARTIAL

The Universe Through The Eyes Of Dinshah

"I Fear No One; Only God Above And Conscience Below And From Them I Have Nothing To Fear."—Platform since 1891.

MEDICOPHOBIA — THE FEARS OF MANKIND!

If there be any profession which possesses an extensive vocabulary of confusing, bombastic, highsounding, senseless names, it is the medical profession. Having been in that game for over half a century, I am well aware of the verbosity, hence, when Spectro-Chrome was perfected and presented to the World, with one stroke of well-directed, scientific swordsmanship, I cut the Gordian Knot of Differential Diagnosis, emancipating Sufferers from the dreadful curse which was sapping their vitals and pockets for 60 centuries of civilization. America is fighting to free the World of "Fear." More Disorders are created through sheer FEAR of something than through actual organic derangement. Such "Fears" or "Dreads" are known in Empirical, Hit-Or-Miss Medicine as "Phobias", from the Greek word "Phobos" meaning "Fear." Here are a few samples of such "Fears" or "Dreads":—

FEAR NUMBER	FEAR NAME	MEDICAL DEFINITION
1	Acarophobia	Fear of itch.
2	Acrophobia	Fear of elevation.
3	Aerophobia	Fear of breeze.
4	Agoraphobia	Fear of space.
5	Aichmophobia	Fear of points.
6	Ailourophobia	Fear of cat.
7	Algophobia	Fear of pain.
8	Amaxophobia	Fear of vehicle.
9	Amorophobia	Fear of love.
10	Androphobia	Fear of male.
11	Anemophobia	Fear of wind.
12	Anthophobia	Fear of flower.
13	Anthropophobia	Fear of society.
14	Apiphobia	Fear of bees.
15	Aquaphobia	Fear of liquid.
16	Astrapophobia	Fear of lightning.
17	Ataxiophobia	Fear of disorder.
18	Autophobia	Fear of egotism.
19	Automysophobia	Fear of filthiness.
20	Bacteriophobia	Fear of bacteria.
21	Ballistophobia	Fear of missile.
22	Basiphobia	Fear of walking.
23	Belonophobia	Fear of needle.
24	Bromidrosiphobia	Fear of malodor.
25	Cainophobia	Fear of newness.
26	Cancerophobia	Fear of cancer.
27	Caninophobia	Fear of canine.
28	Carcinomatophobia	Fear of carcinoma.
29	Cenophobia	Fear of emptiness.
30	Cherophobia	Fear of gaiety.
31	Chromatophobia	Fear of color.
32	Cibophobia	Fear of food.
33	Claustrophobia	Fear of enclosure.
34	Clithrophobia	Fear of jamming.
35	Coitophobia	Fear of copulation.
36	Cremnophobia	Fear of precipice.
37	Crystallophobia	Fear of crystalware.
38	Cynophobia	Fear of dog.
39	Cypridophobia	Fear of sexuality.
40	Demonophobia	Fear of Devil.
41	Dipsophobia	Fear of drinking.
42	Domatophobia	Fear of house.
43	Doraphobia	Fear of fur.
44	Dysmorphophobia	Fear of ugliness.
45	Ecophobia	Fear of home.
46	Egophobia	Fear of self.
47	Electrophobia	Fear of electricity.
48	Emetophobia	Fear of vomiting.
49	Equusophobia	Fear of horseride.
50	Eremiophobia	Fear of solitude.
51	Ereuthophobia	Fear of blushing.
52	Ergasiophobia	Fear of surgery.
53	Erotophobia	Fear of lust.
54	Felinophobia	Fear of feline.
55	Frigophobia	Fear of freezing.
56	Gamophobia	Fear of marriage.
57	Gatophobia	Fear of cat.
58	Gephyrophobia	Fear of bridge.
59	Gymnophobia	Fear of nudity.
60	Gynecophobia	Fear of female.
61	Haphephobia	Fear of touch.
62	Heliophobia	Fear of sun.
63	Helminthophobia	Fear of worm.
64	Hematophobia	Fear of bleeding.
65	Hemophobia	Fear of blood.
66	Hierophobia	Fear of sacred.
67	Hyalophobia	Fear of glass.
68	Hydrophobia	Fear of water.
69	Hygrophobia	Fear of moisture.
70	Ichthyophobia	Fear of fish.
71	Iophobia	Fear of rust.
72	Kardiophobia	Fear of heartbreak.
73	Kenophobia	Fear of outdoor.
74	Keraunophobia	Fear of thunder.
75	Kleptophobia	Fear of stealing.
76	Lalophobia	Fear of speaking.
77	Leprophobia	Fear of leprosy.
78	Lunophobia	Fear of moon.
79	Lyssophobia	Fear of rabies.
80	Maieusiophobia	Fear of pregnancy.
81	Melissophobia	Fear of honeycomb.
82	Melosophobia	Fear of singing.
83	Merinthophobia	Fear of trussing.
84	Microbiophobia	Fear of microbes.
85	Molysmophobia	Fear of infection.
86	Monophobia	Fear of singleness.
87	Mysophobia	Fear of dirt.
88	Mythophobia	Fear of falsehood.
89	Necrophobia	Fear of corpse.
90	Neophobia	Fear of novelty.
91	Nictophobia	Fear of night.
92	Nosophobia	Fear of derangement.
93	Numerophobia	Fear of number.
94	Ochlophobia	Fear of crowd.
95	Odontophobia	Fear of toothache.
96	Odynephobia	Fear of sensitiveness.
97	Oikophobia	Fear of domesticity.
98	Ombrophobia	Fear of cloudiness.
99	Onomatophobia	Fear of name.
100	Ophidiophobia	Fear of reptile.
101	Osmophobia	Fear of smell.
102	Panphobia	Fear of all.

FEAR NUMBER	FEAR NAME	MEDICAL DEFINITION
103	**Panophobia**	Fear of everything.
104	**Pantophobia**	Fear of inclusion.
105	**Parasitophobia**	Fear of parasite.
106	**Pathophobia**	Fear of disease.
107	**Pediculophobia**	Fear of lice.
108	**Pediophobia**	Fear of children.
109	**Pharmacophobia**	Fear of drugs.
110	**Phengophobia**	Fear of daylight.
111	**Phobophobia**	Fear of dread.
112	**Phonophobia**	Fear of sound.
113	**Photophobia**	Fear of light.
114	**Phthiriophobia**	Fear of infestation.
115	**Phthisiophobia**	Fear of tuberculosis.
116	**Pnigophobia**	Fear of choking.
117	**Ponophobia**	Fear of fatigue.
118	**Potamophobia**	Fear of river.
119	**Psychrophobia**	Fear of cold.
120	**Pyrophobia**	Fear of fire.
121	**Rhabdophobia**	Fear of punishment.
122	**Rhypophobia**	Fear of dust.
123	**Sarcomaphobia**	Fear of sarcoma.
124	**Scabiophobia**	Fear of scabies.
125	**Scopophobia**	Fear of stare.
126	**Siderodromophobia**	Fear of railway.
127	**Sitophobia**	Fear of eating.
128	**Stasiphobia**	Fear of standing.
129	**Tabophobia**	Fear of emaciation.
130	**Taphophobia**	Fear of burial.
131	**Tempophobia**	Fear of time.
132	**Teratophobia**	Fear of monstrosity.
133	**Thalassophobia**	Fear of sea.
134	**Thanatophobia**	Fear of death.
135	**Theophobia**	Fear of God.
136	**Thermophobia**	Fear of heat.
137	**Tonitrophobia**	Fear of thunder.
138	**Topophobia**	Fear of place.
139	**Toxicophobia**	Fear of poison.
140	**Triskaidekaphobia**	Fear of thirteen.
141	**Tuberculophobia**	Fear of consumption.
142	**Venereophobia**	Fear of venery.
143	**Xenophobia**	Fear of stranger.
144	**Zoophobia**	Fear of animal.

Thus goes on the enumeration of the dreads or Fears to which the unfortunate mortal is subjected in the moth-eaten system misnamed Medical Science. For none of these, have they been able to offer any explanation, solution or remedy. ALL these derange Areas 6, 7, 4 aand 18, being governed through Area 1. Spectro-Chrome, by equilibrating all of these step by step, is able to cope with these refractory Disorders, proving a veritable blessing to mankind.

What happiness must be mine to be the Servant Of Mankind to bring this knowledge to the Sufferer! No wealth of the World can compare with that satisfaction — no mundane power can rob me of it.

AN EXPERIENCE IN BURNS SHE WILL NEVER FORGET

By:—WANDA K., *Detroit, Michigan.*

I trust you will soon again have free access to the mails, as it is a pity to be handicapped like this in your great work. We here in our house would be afraid to go without a Spectro-Chrome; it has accomplished so much for us.

Three years ago, my gas oven exploded and I was right in front of the door. My entire face and right arm were badly burned, my eyes pasted together, my hair burned like a cap on my head. I had one solid blister across my face, my lips were swollen and thick in a few minutes. For me to go downtown to Receiving Hospital, would have meant at least an hour to an hour and a half before receiving attention and the pain was so intense, I don't believe I could have stood it that long. In a flash the skin on my face was so tight I thought it would burst. As well as I could, I tried to tell my mother where the cocoanut oil was and what Colors to use; it was hard to talk with a stiff face. Before one hour was up, the pain subsided; I got up and ate my evening meal with my family.

At 9 o'clock that night, I took the second Tonation and slept all through without a Doctor and without any opiates. The Gas Company representative just shook his head, made out his report and said if I did not call a Doctor, they could not be held responsible. I told him they didn't need be, as I would be well and without scars — thanks to Spectro-Chrome and I am. I have a small snapshot which shows the awful scabs, taken a week afterwards. Don't you ever think I am not grateful to you. I shall never forget that experience as long as I live.

MEASLES NORMALATED BY SPECTRO-CHROME IMPENDING MASTOIDITIS AVERTED

By:—MURIEL H., *West New York, New Jersey.*

I thought you would like to know that my daughter, aged 11½ years, became ill with Measles, last Wednesday.

I immediately used the Turquoise Color Wave on her, alternated with the Magenta. By Sunday, the rash which had been very heavy and complete, had almost vanished, so, I used the Lemon for three Tonations and tonight used the Yellow on her.

I am pleased to say she seems completely recovered. She is in good spirits, seems full of pep and her skin is quite clear; her appetite is coming back.

During the course of the Disorder, she once complained of a pain behind the left ear. I immediately used the Turquoise on her and there was no recurrence.

GIVEN TWO YEARS TO LIVE TUBERCULOSIS SUFFERER NORMALATED BY SPECTRO-CHROME METRY

By:—AFFILIATE, *Greenwich, Connecticut.*

In the May 1942 issue of Spectro-Chrome, page 3,504, there is a little note I sent to the Institute; in it I stated that I hoped to write soon again of greater success. Well, here it is, sir!

Over three years ago, I was told by one Medical Doctor that I had an "Inactive Koch" in my Left Apex. I was not alarmed and went to several other Doctors, who dismissed the idea and said "there is nothing to worry about." The first Doctor called by 'phone and asked me to submit to further examinations, since he wasn't certain at first of his Diagnosis! I kept going, until I finally took an X-Ray. What my X-Ray and Sputum Tests, which also I had taken, showed I don't know yet, but, one Doctor advised me to go to the country for a rest. I was not able financially to do so and was told that the City and State will help me. The result was, I landed in Bellevue in New York. From the very first day on, the advice centered on collapsing the Lung or an Operation. My "Tests" at first showed Negative, till three months after I "succeeded" in becoming Positive. Having close quarters with Positive Sufferers for so long, it was no wonder. I regret to say I submitted to a Pneumothorax (Lung Collapse) under the pressure of a Doctor, but made up my mind never to permit an Operation.

Then I was transferred to Sea View, with more disappointments, hardships and unhappiness. There the Operation was again being mentioned to me, when a good friend of mine, a friend in need indeed, told me of Spectro-Chrome.

While I was in the Sanatorium, I became everything but better and my Doctors assured me and my family, that if I don't let them operate on me, I have only two years to live.

Since I had nothing to lose under the circumstances and after having carefully studied the chances of Spectro-Chrome with what little reading matter I had of this Science, I decided to leave the Sanatorium *against* the advice of my Doctor.

Of course, I didn't tell him that I would get well without Surgery, with the help of an Equipment of which "they" will all hear in due time. When I left, I had looked at it from all angles and was certain of Spectro-Chrome Metry as a Colossal Method Of Healing.

While undergoing an examination for my Insurance Certificate, I was told recently that my Lung condition, which was to take years and years of rest and an Operation, was perfectly well cleared up. This, thanks to Colonel Dinshah, a great Scientist and Spectro-Chrome Metry, his Science.

Please, sir, if you ever wish to use my case as an example, kindly do not use my name or the right initials for obvious reasons.

Thank you and for ever let me be associated with Spectro-Chrome Metry.

SPECTRO-CHROME

March 1944 Yellow 1944

Volume 20 Number 3

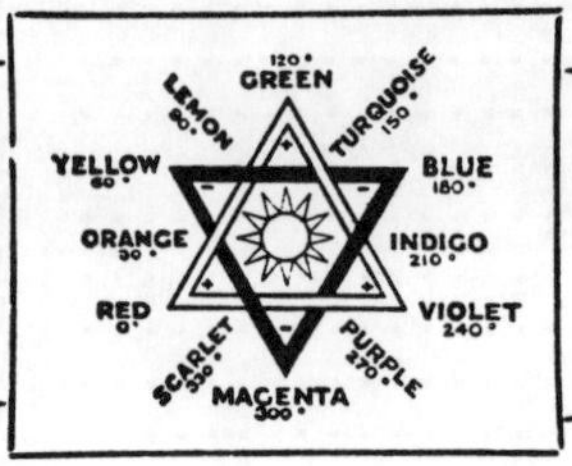

Dinshah Lectured In New York

Pink Pills For 18 Years

DINSHAH IMPARTIAL

MR. 'TAIN'TE ON LONG DISTANCE TELEPHONE

Dinshah's Photokinephone and Spectro-Chrome

No Wonder Consumptives Die!

Attuned Color Waves
Spectro-Chrome Metry

In course of life you reach a time,
When Doctor, Lawyer, Priest alike,
Does fail to give the sought relief,
To troubled mind or burdened heart—
Then come to me:
I help you to help yourself.
SPECTRO-CHROME

The Science Of
Automatic Precision

DEVOTED TO THE SERVICE OF MANKIND

FIRST CLEVELAND OHIO PLANET

(CARNEGIE HALL, 1220 HURON ROAD)

Meeting Of Thursday, February 3, 1944

Successful Use Of Spectro-Chrome In:—Fatigue, Knee Disorder, Sore Throat, Tumor, Cough

Reported By:—EDITH REES, *Recorder pro tem.*

Deputy George J. Brown was in the chair; 16 Affiliates were present. The Meeting opened by the reading of the Affirmation by the Deputy.

A letter from the Principal Office was read. One copy of "Key To Radiant Health" was handed by the Deputy to each Affiliate. He read the article on "Rationality In Dietetics" beginning on page 510 of Volume 1 of the bound magazines. Man is not carnivorous.

Question was asked how to find the correct time for Tonations in this territory. John Winchester gave the instructions.

The Deputy explained the reason for the Correction. Henry Scheidel asked if there were any benefit if the Tonation be started late. Discussion followed. The opinion was that one may get some benefit, but not as much or accurate as though the Tonation had begun on time.

Inquiry was made if Dinshah would circulate a list of former active Graduates and Affiliates so the Planet may try to make them active again.

Carl Schmid reported that he had used Spectro-Chrome for two years. Before using it, he was tired every day when he went home.

Fred Osborne asked how to tonate water on the Knee. Hedwig Klauss told of her experience with her husband. Turquoise was used first for the Swelling and then Orange Local and Lemon Systemic.

Norma Schmid told of her experience with a bad Sore Throat. Used Turquoise and before the first Tonation was over her Throat was better. Another Turquoise Tonation cleared it.

Joseph Tarantino related the experiences of his two brothers, one with a broken Cartilage and Disorder in Knee and the other with a Tumor on the side of a Knee. Spectro-Chrome helped both and they still have the use of their Legs. Amputation was not necessary.

The Deputy reported how Spectro-Chrome removed his bad Cough after Bronchial Pneumonia. Yellow Local and Lemon Systemic were used.

It was a small Meeting but a very enthusiastic one. The Affiliates stated they enjoyed the Meeting and by unanimous vote it was decided to begin the study of Volume 1 of Spectro-Chrome Metry Encyclopaedia at the next Meeting, the aim being to maintain the Educational Program on matters of Spectro-Chrome Metry as a Healing Science.

EDITORIAL REMARKS

The correct Favor Time for any Locality may be found by adding 1 Hour for War time to the Favorscope Time and adding or subtracting the Minutes indicated for the particular month for one's State. For Ohio, there are two distinct Corrections, because, some parts keep Eastern Standard Time and others Central Standard Time. The Favorscope is all built on the Eastern Standard Time. A little study will amply repay.

Spectro-Chrome Metry, as the Science Of Automatic Precision, does not aim merely at hit-or-miss flukes, but, goes straight to 100 per cent. efficiency. Of course, *some benefit* may be realized or otherwise, but, to use the Science thus haphazard would eventually degrade it and reduce it to the level of Medical Empirics. The Originator placed at your service all that is the best in Healing Science and he can not *dictate* what you MUST or MUST NOT do. The results will depend on one's fulfilling the Technique and the instructions.

Lists of Affiliates can not be collected by anybody. Refer to By-Law 101, "Tattlers." The Council Of Trustees attends to such matters as suggested by you.

Instead of talking of Disorders by names, let Affiliates bring in complete Charts for discussion. We have no special "Vaccines and Serums" for special names.

Wish you all progress and success.

FIRST CHICAGO ILLINOIS PLANET

(HOTEL ATLANTIC, 316 SOUTH CLARK STREET)

Meeting Of Thursday, February 3, 1944

Successful Use Of Spectro-Chrome In:—Scratch, Rash

Reported By:—ELSIE BEYER, *Recorder pro tem.*

Deputy Wilhelm Weissenbacher was in the chair; 22 Affiliates were present.

Henry Schilpp appointed as Sub-Deputy was affirmed under By-Law 6. Communication from the Principal Office was read.

Three new Affiliates were introduced. From Spectro-Chrome Home Guide, a "Message From Dinshah To You" was read by the Recorder and the Chapter of "Fundamental Principles." Carl A. Rall read from Spectro-Chrome Metry Encyclopaedia, Chapter 4, "Tests Of Truth."

Discussions followed.

Joseph Janik asked if one can take 30 minutes Front Tonation and 30 minutes Back Tonation.

The Deputy said one full hour must be taken for 1 Tonation on either side as indicated. That Affiliate also asked if a person can take a Tonation right after a meal. The Deputy suggested to wait about 2 hours before taking a Tonation and also to wait at least a half hour before going in the Sunlight. He reminded the Affiliates to start their Tonations on the correct Favor time, as it is very important.

Kathe Hoch asked how to apply the Attuned Color Waves and how far to set the Spectro-Chrome away for Tonation. It was explained by the Deputy and also that the room should be darkened.

Pauline Schilpp was in doubt if one can take Tonations on the High and Low Critical Days. The Deputy said by all means; only during an Eclipse one should not take any Tonation or eat.

Henry Schilpp had a Scratch on Arm in the woods. Began to swell, itch and a Rash appeared. He thinks it is Poison Ivy or Poison Oak. Did not apply Spectro-Chrome immediately. Began to hurt him and then after 5 days he started with Spectro-Chrome Lemon Wave, then Turquoise and the Itch declined and now he is getting good benefit out of the use of Attuned Color Wave. He is well satisfied so far.

EDITORIAL REMARKS

Do not split Tonations; each is 1 Hour.

Do not tonate right after a meal; only in Disorders of Area 8, for certain cases Tonation after a meal may be advisable. In most cases however, such procedure is not advocated.

All Affiliates will profit by carefully studying Spectro-Chrome Home Guide. The Tonation Chamber need not be photographically dark, but, as much dark as would be by pulling the window blinds and no other Light should be shining directly on the Areas under Irradiation.

Tonations have no relation with the High and Low Critical Periods. They are only the New Moon (Conjunction) and Full Moon (Opposition) positions, when the Crisis of a Disorder is likely to appear. As the Water Tides are produced by the Solar, Lunar, Terrestrial, Radiant, Gravitational Influence, our Blood Tides exhibit similar Rises and Falls, as are easily demonstrable by means of Dinshah's Sympathometer. During Eclipse Period, there is always a Sympathetic Disturbance, because of the Occultation of Radiant Energy, hence, Tonations to restore the activities of the Vital Processes are contra-indicated. Spectro-Chrome Metry has more and deeper Science within it, than all the medical tomes in the libraries. At the proper time, Spectro-Chrome will receive its merited recognition.

FIRST PORTLAND OREGON PLANET

(MASONIC TEMPLE, 1119 SOUTH WEST PARK AVENUE)

Meeting Of Thursday, February 3, 1944

Reported By:—LEWIS BISCHOFF, *Recorder*

Deputy Stanley Burroughs was in the chair; 8 Affiliates were present.

A letter from the Principal Office was read. Everyone is hoping that the Fraud Order be soon revoked.

An article was read from a recent issue of a periodical. It told how Urine and filth of all sorts had been used and is still used in so-called Medical Science. The basis of Vaccination and innoculation is filth from the living and dead bodies of animals and Human Beings. It is truly outrageous the criminal things that the Law not only allows the medics to do, but even attempts to force the people to allow done to them in the name of modern Science. The public is educated that the medical profession can do no wrong and that one should do exactly as told to do and believe everything they are told, no matter whether it sounds right or not. The public is not supposed to think as far as Health and sickness is concerned. The Medical Doctors are supposed to do the thinking for everyone.

The best advice on Health that anyone can offer is to stay away from

the Doctors. The Deputy told of a recent case. He was asked to come to see a Sufferer. The Sufferer had been treated for Arthritis and Diabetes and had had an Operation for Goiter. She had been given the works by the medics. They had tried everything on her that only Satan could have devised. The result was that she kept becoming worse and worse until in desperation she finally called on Spectro-Chrome. The only thing that had done her any good so far was lying out in the Sun every day for about 2 months. The Deputy decided to have her go on the Rational Food Of Man for a week. The Diet cleaned her out and then Tonations were started. She is improving.

The rest of the evening was taken up with further lessons on how to keep in perfect health by keeping the mind and body clean at all times.

EDITORIAL REMARKS

It is respectfully suggested that the Educational Program be restricted to our own text books and Science and less on outside periodicals. It is well known to our Affiliates that we are opposed to the orthodox Healing Systems; our aim is centered in Spectro-Chrome and you will do genuine service if you follow our text books and discuss from them various points for enlightenment.

FIRST MERRILL WISCONSIN PLANET

(HOTEL BADGER, MERRILL, WISCONSIN)

Meeting Of Thursday, February 3, 1944

Successful Use Of Spectro-Chrome In:—Cold, Sores, Acne, Blackheads, Skin Disorder, Boils, Abscesses

Reported By:—DOROTHY DENNE NELSON, *Recorder*

Deputy Walter Briese was in the chair; 20 Affiliates were present.

Communication from Principal Office was read. Leaflet on "Rational Food Of Man" was passed around among the Affiliates for their inspection. They decided to order 100 of these and pass them around among their friends.

The Deputy read the list of books now available.

Emil Erben said of the "Family Health Protector," "I believe it is a very fine book; it is really a complete education on Health, if anyone wants to know the true facts. It is easy reading and does not take long to read it. I do not believe anybody could find fault with it. Nobody has been able to deny the facts that Colonel Dinshah sets forth." The Deputy said, "My wife read an article about the Sun that was quite astonishing."

The remaining pages of the Fraud Order Brief were read by the Recorder, at the request of a majority of the Affiliates.

The Educational Program followed and the Deputy read the Chapter "Spectro-Chrome In Skin Disorders" from Spectro-Chrome Metry Encyclopaedia. Discussion followed.

Question was raised by an Affiliate about the Favorscope.

Edward Katke said when he takes the Green Color Wave and it strikes the top of Area 1, it seems like something is pressing down on his Head, and he gets a severe Headache. He tried it a couple of times and the results were the same each time. Can you give us an explanation?

Gertrude Parshall said she was asked to come into a home and explain about Spectro-Chrome. Man was suffering with Sciatic Rheumatism and had been taking treatments from a Chiropractor. He asked, "If I go to Lydia Danner will it 'cure' me?" She said, "I can say, 'yes and no' to you. So many things have to be taken into consideration. If you are willing to stop using any Intoxicating Liquor, Tobacco, abstain from Meat, then may you get relief. You may get a little relief without doing these things, but you will never be normalated unless you do. Now, I think he is taking Tonations."

The Deputy introduced Spectro-Chrome in a relative's home in another city in Wisconsin and they think there is nothing like it. "I came to my brother-in-law's home in the same city. They know Spectro-Chrome is good, because there are 3 or 4 in our relation. My brother-in-law was smoking a cigarette. I said when he was talking of using Spectro-Chrome, 'What are you going to do about that (cigarette)? That and Spectro-Chrome do not mix; as long as you use that do not think of using Spectro-Chrome.' He has not come to the point in life yet, when he will give up cigarettes for Spectro-Chrome."

The Deputy read Chapter 97, on "Boil, Abscess, Furuncle, Carbuncle."

Emil Erben read from Family Health Protector, on "Medical Humbug Of Common Cold." Everyone enjoyed hearing this enlightening article.

Case Reports followed:—

Edward Katke suffered with a Cold. "I felt tough in the daytime, but in the evening I was miserable. Used Turquoise and finished with Lemon and Magenta. Slept like 'nobody's business' that night."

Gertrude P. spoke about her daughter. She had Sores, Acne, Blackheads on Area 1. Picked on them. Green and Lemon Systemic used. Feels much better.

Albert H. Will said: "I am getting along fine."

Anna S. spoke about her daughter D. "Several years ago, D. had a Skin Disorder. All the new babies in the Hospital had it too. Her face was all Sores. Used the Green and Turquoise and she was successfully normalated."

B. spoke of his daughter B. "Whenever she would have a Cold, she would get Boils, but, after we started using Spectro-Chrome she never had any Boils."

Dorothy N. said, "I was subject to Boils and nearly every Winter would have Boils and Abscesses in and around my Ears. X-Ray treatments were given for about nine days and then was taken to the Hospital where the Surgeons opened them, after administering gas as an anesthetic. After Emil B. and his wife finally induced us to use Spectro-Chrome, I went to Lydia D. with a very inflamed Ear. After a few Orange Tonations, followed by Lemon Systemic, the Ear was normalated, without the severe suffering I had when I went to the Hospital for X-Ray treatments. From that time on we were sold on Spectro-Chrome."

EDITORIAL REMARKS

There is no "collective contributing" permitted in a Planet, as under By-Law 39, Section 4, "Planets shall NOT collect, handle or receive any money for any purpose." Any Affiliate may contribute DIRECT to the Institute and get the distributive literature, but, it must be distributed FREE AND WITHOUT ANY CHARGE. Any SALE is a Serious Violation. Your Recorder has those leaflets, but, they are subject to the stated Rule. There can be no deviation. Thank you.

The question about the Favorscope can not be explained openly in short. It is a study in itself and will be explained at the Annual Convention if you attend.

Edward Katke never sent for any Free Guidance. That is what he should have. Headache on top has no meaning in our Science. It is probable he needs Yellow on Areas 6-7-8-9-10 prior to taking Green or the Ratio needs correction.

Let Smokers and Boozers alone. No use trying to argue with people of set views. Those who are ready for the Message will come to us; we do not have to go begging for patronage or support. We are able to hold our ground permanently against any and all.

Boils and their kin are exits to the inside lava; Medical Doctors can never overcome them, because they have nothing to alter Blood Chemistry.

FIRST DETROIT MICHIGAN PLANET

(HOTEL TULLER, PARK, BAGLEY AND ADAMS)

Meeting Of Thursday, February 3, 1944

Successful Use Of Spectro-Chrome In:—Sore Throat, Influenza, Fever, Injury, Swelling, Cold, Prostate Disorder

Reported By:—CARMELA FORZANO, *Recorder*

Deputy Charles J. Renn was in the chair; 83 Affiliates were present.

Communication from the Principal Office was read; also the leaflet "Rational Food Of Man."

Case Reports followed:—

1. Rosetta B.—I suffered from a Sore Throat for the past 17 years. My family Doctor told me it was a Stereptococcus Throat and that a time would come when I would choke to death from it. However, with three Local Tonations of Spectro-Chrome on Area 2, first Yellow, second Magenta and the third Yellow, I certainly feel I was normalated as it has been 10 months since I had a Sore Throat. Many thanks to Spectro-Chrome.

2. Ursula P. reported they all had Influenza. Daughter had a very high Fever, but, with 4 Tonations, Temperature went down to normal and Cough reduced. Used Blue, Turquoise and Lemon. She reported using Green and Lemon on a Chest Injury.

3. Carl W.—Bumped his Knee while at work. Used Turquoise several times and Swelling went down.

4. Frank O. reported a Cold. Took Turquoise and Lemon Atcowa for 3 days and was normal.

5. Leta B. reported her young daughter becomes worse during Critical Periods. Has a Kardoscil of 152, Spiroscil 68 and gets panting for breath—has used Indigo and then Yellow.

6. Grace S. said her husband suffered through a previous Eclipse so now watches for Eclipse and does not eat during that time. The question was asked if an Eclipse would have any effect if a Tonation were taken during that period.

7. Sylvia G. said she took Turquoise Systemic and then one Magenta and then checked her Kardoscil and Spiroscil and it was 76 and 14. She would like to know if she should have taken Lemon. Since all other questions in a chart should be considered, it was difficult to say whether or not she should have taken Lemon.

8. Walter H. reported he had a Disorder in Prostate Gland region (around November 4, 1943), which almost completely paralyzed the right side. Used Lemon Atcowa and within 2 weeks could get up and down very well; also used Blue on back and Magenta on Area 4 and Lemon systemic. Is able to move around very well now.

The Deputy read By-Law 71 on "Duplicates" and 72 on "Exercises." Norma Fiedler was then asked to explain and demonstrate how to take Kardoscil and Spiroscil.

The Deputy explained about Ratio and the Breathing.

Jessie E. Ness spoke on the Breathing. She said it is the key to what is going on in the Human Body. Nervousness plays an important part in the Body, also thinking. If we think beneficial thoughts our body will be attuned to them; 2/3 depends on our mental attitude. She did not think our Organization would trouble with a Chart if it were not of value. She promised to read a poem her late husband wrote about the "Dog Box." She said these Equipments work because of their therapeutic value and not because of the shape of the box. Men and women today have a choice gift given to them in Spectro-Chrome. Said she: "God bless you all and may you use it as it should be used."

The Deputy requested each and every one to learn more about Spectro-Chrome, "Learn what Spectro-Chrome will do for you and to be honest with self."

Questions were asked about the Bulb in the Equipment and answered.

The Deputy read "Tobacco Smoking An Unmitigated Curse," from Dinshah's "Family Health Protector." Leta Berkau said her mother used to recite:—

Tobacco is an evil weed
And from the Devil doth proceed,
It soils the mouth and stains
The clothes and makes a monkey
Out of the nose.

FIRST NEW YORK NEW YORK PLANET

(HOTEL McALPIN, BROADWAY AT 34TH STREET)

Meeting Of Thursday, February 3, 1944

Successful Use Of Spectro-Chrome In:—Burn, Pain, Swelling, Bump, Numbness, Sprain, Female Disorder, Dog Disorder, Hemorrhage, Headache, Dizziness, Gastritis, Ovelim Disorder, Accident, Nose Disorder, Throat Disorder, Wound, Bleeding, Balance Disturbance, Pneumonia, Arthritis, Blood Disorder, Neck Crick, Bradycardia, Ear Disorder

Reported By:—ANNA LONG, *Recorder*

Deputy Lina Fischer was in the chair; 79 Affiliates were present. She expressed her thoughts with these words:—

"I thank you most sincerely
For the honor bestowed on me,
And hope to come up to your expectation
To be the Deputy you want me to be."

It was delightful to see such a large attendance. Three persons were denied admission, because they had no 1944 Cards. One stranger wanted "to buy a ticket" and was told that it was a closed meeting for Affiliates only.

A "Good Thought" for February 3rd titled "Grace" was read by the Deputy.

GRACE

Where can be found a being, so selfless
And sweet, so lovely and loving, full of Grace?
Where can be found another Grace, so charming
And cheerful, so thoughtful and thirsty
For the hungers of my heart? Where can be
Found a soul, so clean and conscientious, so high
And humble, sacrificial, spirited, so pretty
And potent, as the Grace I have as mine?
He who has
Grace for his mate,
Has to thank
His supreme Karmic fate.

Lina remarked that it was dedicated by Dinshah to his beloved wife Irene Grace who so efficiently works with him. The poem was absorbed with keen attention by the audience and a round of applause followed for our Secretary Irene Grace Dinshah.

The communication from our President Dinshah was then read by the Deputy who explained as she proceeded.

Many seemed to have trouble with their Railway Express Offices. The Deputy offered to help any who had trouble in that respect. She again stressed how to send Charts and the like by the Railway Express Agency and showed a sample of the envelope form 6034. To make everyone remember the number she requested the affiliates to repeat the number 6034 twice. Students were again reminded not to be impatient, but to co:operate fully with the Institute especially during this critical period.

Case Reports were called and Frank Cassano a Spectro-Chrome Normalator for 8 years came to the platform. He reported the first five cases.

(1) Daughter, age 7, was celebrating the 4th of July, when she was burnt in eye by a firecracker. Eye puffed out, became swollen, Blister formed. Child writhed with Pain. Party of about 20-30 people cried "call a Doctor! Call a Doctor!" She was put under Spectro-Chrome. Frank Cassano being then new in the Science of Spectro-Chrome, was really frightened. He kept hearing the whispers, "Get a Doctor." Atcowa Blue was used. In 15 minutes, Swelling disappeared; in 30 minutes Blister disappeared. After 1 hour, the child came out and saw the rest of the fireworks. Frank Cassano stressed that when any Disorder is caught right away, Spectro-Chrome will help immediately, like a miracle.

(2) John O., male, age 51, had weak Heart, Bronchitis, Stomach Ulcers. Kardoscil 66, Spiroscil 14. Backelim 1 a day, Frontelim 7 a day. Atcowa Lemon Systemic was used for one week. Kardoscil 80, Spiroscil 18. Backelim twice a day, Frontelim 7. Healed Heart in 2 months. Lemon Systemic Front and Back, Magenta Areas 4 and 18 was used for two weeks. Then Orange Systemic Front and Back. Kardoscil 88, Spiroscil 18. Bronchitis improved, Ulcers disappeared.

(3) Tony S., boy, age 4. Mother was canning tomatoes when he fell into the hot water; 1/5 of body was burned. He was more dead than alive. Could not sleep for 3 days. Atcowa Blue was given. After 1 hour, the little fellow said, "Mister, don't take that moving picture machine away!" Blue was continued; later, Turquoise Systemic and Magenta on Areas 4 and 18. Normalated in 5 weeks with the Skin actually "breathing" again.

(4) Geon M., boy, age 12, Skated on pond. Came home with Bump on Head and right Foot was clubbed under and benumbed. Kardoscil 90, Spiroscil 20. Backelim 1 a day, Frontelim 5. Turquoise Systemic and Magenta Area 4 was used for 2 days. Later Lemon Systemic and Magenta Area 4. After 1 week went back to school. Foot was normalated after first Tonation. Parents had feared it was Infantile Paralysis.

(5) John S., male, age 29, 75% blind. Could only read when holding book right to face. Kardoscil 72, Spiroscil 18. Backelim 3 times in 1 week. Frontelim 5 a day. First Frank Cassano had to see that Bowels moved correctly, then the Heart and Lungs. He remarked that those terms were not used in the Science of Spectro-Chrome. He just used them here so that the new students would understand better. He stressed that the Ratio must be 5 to 1 or 4 to 1 before any further Normalation can be expected. Yellow Systemic was used for 2 weeks. Boy began to improve and could read at a distance. Tonations were given for 2 months. Sufferer thought it was too much trouble and gave up although the services were given free. He was too busy, his mind was only on the girl friend.

Frank Cassano concluded that he was very much frightened upon coming up to speak, but said the next time he would not be that way. The Deputy added that she herself was nervous at first, as this was her first time she presided at any Meeting. With that the tension in the audience also relaxed and there was an enthusiastic applause.

(6) Lina Fischer reported her own case of Sprained Ankle going down stairs. Blood rushed to injured area internally. Started to swell immediately. Had to be carried home. Atcowa Indigo local was applied. After 20 minutes everything was receding. After 1 hour the Pain disappeared. Three days later, she was able to go to work again.

(7) Leonina De M. reported her own case. Shook wax and turpentine together which caused explosion. (?) Arm was burnt, kitchen was on fire. Did not lose presence of mind but extinguished fire. Flames were out and so was Leonina. Unfortunately, she was taken to the Hospital where Tannic Acid was applied to the burnt Area 14. It was a third degree Burn. Left the Hospital and used Atcowa Blue, then Turquoise. Later Lemon to throw off the thick crust. Only cocoanut oil and tissue paper were used as a dressing. She was soon normal.

Minchen Rossig stated cocoanut oil was not available. Discussion followed. Some stated that pure vegetable or olive oil could be used. An appropriate offering was made by Alfred Meschke. He would try to get several pounds which he would bring to the next Meeting. The Deputy warned however, that it was *not to be sold* during the Meeting.

(8) Louise U. stated her own case. Female Disorder. Had 3 Operations and was to have another. Doctor could not help. Spectro-Chrome was called in. Indigo and Magenta on Area 4 and Yellow were used. Fine now. She also remarked she used Spectro-Chrome on her dog, who was in Pain when passing Frontelim. Atcowa Indigo was used. Dog made a "funny face" and passed 7 stones. (Much laughter.) She then exhibited those stones amidst more laughter. The Deputy explained that Human Beings have had similar good results.

(9) Lina Fischer then reported a case. Female Disorder. Hemorrhage, Headache, Dizziness. Medical Doctor had said an Operation was necessary. She thought if Louise U. was helped after 3 Operations, it would surely help her. Went 100% on the Rational Food Of Man. Indigo, Yellow, Orange and Magenta on Area 4 were used. Menstruated regularly. After 3 months, had Hemorrhage again and large lumps passed. Sleeps well now; never cranky anymore. Was helped 100%.

(10) Frida F. had Gastritis. Could not eat. Medical Doctor could not help. Went down to 90 pounds. After using Spectro-Chrome 5 months, got relief. After 2 years gained 25 pounds and felt fine. Her husband never believed in Spectro-Chrome. One day, he had terrific Stomach Pains. Kept crying, "Give me an Alkaselzer" (laughter). She told him to try Spectro-Chrome. After 10 minutes of Orange, he was asleep and did not awake until the next morning. No more Pain. The Deputy added that more husbands should get Stomach Aches. (Much laughter.)

(11) Lina Fischer reported the case of Mildred G., woman, age 20. Had Ovelim irregularity after birth of baby. 2 weeks ago, she started to bleed profusely. Medical Doctor said Operation was urgent; a Cyst was the cause. Spectro-Chrome was placed in the home. Atcowa Indigo stopped Bleeding after 3 Tonations. Feels fine. Cancellation of Operation was telephoned to the Doctor who could not understand how the Hemorrhage had stopped.

(12 Augusta K. reported her own case. Her Legs were ruined by a Specialist. After he had her money, he told her she was "cured." That was when her trouble really started; could hardly stand the Pain and was almost forced to give up her work. Thankful to the Lord for letting her meet Colonel Dinshah and his Spectro-Chrome. It is her strong conviction that when trouble is worst, God's help is nearest. Heard about Spectro-Chrome and received a few Tonations from a Spectro-Chrome Normalator and then became a Benefit Student. Told family some of things

Spectro-Chrome can do and brother-in-law called it a "Zauberkasten" (Magic box). Her Legs became better and she felt better and people thought her just a little crazier. She told them, "If being 'nuts' makes me happy, then I am glad of it." She caught a Finger in an electric-mixer. Called the Institute for Free Guidance. When Irene Dinshah in her gentle voice asked what was wrong, she was told "The meat was all chopped up!" Employer called Medical Doctor to set Finger (this assistance came under Constructive Surgery). The Doctor was told about Spectro-Chrome. He smiled and said "Go ahead with it, but come to see me tomorrow." He had to promise first he would only look at it. He was surprised the next day. The Nurse said Finger would remain stiff and cause her trouble for some time. Finger normalated remarkably well. It is as good as new. The Doctor who is a man who gives credit where due, said to her, "What-cha-ma-cal-it is wonderful." Her baby had trouble with Nose and Throat. The Doctor advised minor Operation, but Spectro-Chrome normalated it and improved her general condition. Her baby fell against a hot oil stove with her bare back. Burned off about 3 inch square of Skin. After 20 minutes under Blue, the baby was telling her a story. Normalated perfectly. Magenta Area 4, Indigo and Turquoise did it. The baby fell on the bath tub. Gaping wound the size of the mouth. Blue and Magenta Area 4 were used. Put 2 strips of plaster at each end of wound to hold it together. Then put cocoanut oil and tissuepaper over wound itself. Later Indigo and Turquoise normalated it. Very little scar. Her niece fell with head against exhaust pipe in an automobile accident. Blood was streaming down. People suggested removal to Hospital. Spectro-Chrome normalated the case. Father who at first was skeptical about Spectro-Chrome was convinced. Child lost balance often. Green Systemic normalated it. Recently children were ill. The Ratio dangerously low. Brother-in-law called in Doctor for his child. Diagnosed it as Pneumonia. Left prescription which was not filled. Spectro-Chrome was used. Doctor came next day although not expected. He said girl was doing wonderfully! Was asked whether he was coming again. Doctor mumbled and said, "No." Augusta Kech said, "There is nothing else like Spectro-Chrome. Some day there will be one in every home. How happy our Dinshah will be then." She concluded:

"Dear Lord, we thank Thee for Spectro-Chrome
And Dinshah who labors forever at home;
Let Thy blessings be on him forevermore
And please:—Give us a Heart as honest
And grateful just like his!" (Applause.)

(13) Hermine B. stated her case of Arthritis of 8 years standing. The Deputy remarked how fast and perfectly she walked to the platform and asked if she would demonstrate how she walked before the use of Spectro-Chrome. She did by limping heavily. Kardoscil 63, Spiroscil 14. Atcowa Yellow Systemic and Magenta on Area 4 were given. Kardoscil now as high as 87. She stated it took Lina Fischer 3 months to convince her. The Deputy added: "It was a hard nut to crack!" (Laughter.)

(14) Anna L. stated her case. She was a very sick girl a few years ago, with Blood Disorder. No Doctor could help. In bed 5 months. Spectro-Chrome put her on her feet. Turquoise and Lemon Systemic were used and Orange and Magenta on Areas 4 and 18. Dinshah advised marriage. The Deputy remarked that Anna is now happily married. Anna reported case of her mother, Margarete F., age 48. Crick in Neck. Suffered terrific Pain. Atcowa Orange Systemic was used. Normalated in 10 days.

(15) Martha K. stated her case. Always weak with low Kardoscil and Spiroscil. Finally installed Spectro-Chrome. After 2 months, Kardoscil 72 and Spiroscil 16. Magenta on Area 4 and Lemon Systemic with Yellow on Areas 6, 7, 8, 9 and 10 were used. Feels fine.

(16) Martha L. has her Equipment for 5 years. Son had Abscess in Ear which was running and malodorous. Used Spectro-Chrome a month and running of Ear stopped. Orange Systemic and Lemon on Ear were used. Boy also had dislocated Hip. Went to a Chiropractor. He had a Spectro-Chrome and used it himself. Boy said it was the same they had at home. Chiropractor said, "You have something wonderful at home."

Deputy—Is your family on the Rational Food Of Man?

Martha L.—We started last week. (Much laughter.)

Deputy—I appreciate your honesty.

Martha L.—It is the hardest thing to go on the Rational Food Of Man.

Deputy—It is not hard at all. Just make up your mind and do it. Who wants to make a graveyard out of one's stomach?

As Martha L. was returning to her seat, she mumbled something about "that good cup of Coffee!" The Deputy then stressed the importance of the Rational Food Of Man System. She asked if everyone knew what 7 things to avoid. The audience answered in chorus:—Meat, Fish, Eggs, Tea and Coffee and Tobacco and Alcohol. The Deputy added that one should not forget to omit the "hidden egg" in cakes.

The Deputy asked how many Students made out Affidavits. Only 2 hands went up. She asked how many were normalated and helped with Spectro-Chrome. All hands went up. She remarked that it was just where we are failing Dinshah. It is important to send in Affidavits. We were all helped by Spectro-Chrome, so let us help Colonel Dinshah to win the fight. He needs our support just as we need the support of the Institute. Dinshah and his family work for us day and night. She added that only that day she explained the reading of the Free Guidance Chart to a new Affiliate and discovered that the Chronos (typing time) was 1:48 A. M.

The Deputy then asked for questions. It should come up during the Meeting not afterwards, so all may benefit and learn by it.

Augusta Kech wanted to know in the case of Measles which Atcowa is more beneficial, Orange or Turquoise. Deputy: With Childhood Disorders Fever is present and Turquoise should be used; the Blue takes the Fever down and the Green acts as the Cleanser.

Hildegard Meschke and Muriel Hartman confirmed the uniform results in Measles obtained with Turquoise.

Question—Where could one obtain Raw milk?

Deputy—I am fortunate to get it from the farm.

Martha Keppner—Raw milk could be obtained from the "State Dairies" of New Jersey.

Question—When tired after work what Tonation is advised?

Deputy—Stay within the Infra-Green division.

Question—Could an Educational Student obtain Spectro-Chrome Metry Encyclopaedia?

Deputy—Yes. Everybody should own the Encyclopaedia and eventually complete their Dinshah Spectro-Chrome Library. She added that the Family Health Protector was the latest addition to the Library. It is an interesting and educational book and is obtainable for the nominal Contribution of 2 Dollars in American Express Money Order.

Frida Feil—I loaned my Spectro-Chrome Home Guide out and now it is lost. How can one obtain another?

Deputy—Never under any circumstances loan that book to anyone. It is only replaced by returning the old one complete, soiled, torn or damaged. Otherwise another Home Guide would cost $75. You may as well send in enrollment for a new Benefit Studentship.

The Deputy reminded the Affiliates that when introducing Spectro-Chrome one should inform the aspirant of the pending Fraud Order and the willingness to accept the non-metallic War Model. A letter to that effect should accompany every Enrollment. That avoids all false expectations. The War Model does not look nice, but it is not the shell that counts. The Slides give the same effect and results. There may be a possible exchange for the Aluminum Equipment after the War is over.

Someone stated that it was hard to convince people about Spectro-Chrome, especially on the Rational Food problem. The Deputy promised to get the Rational Food Of Man pamphlets for distribution at the next Meeting. She said she spoke to someone the other day from 9 P. M. to 1:30 in the morning. It is up to us to spread the work.

Mary Muller said she was one of those hard cases to convince. Her baby, age 2, was born with Paralysis of Upper Eyelids. They were 3/4 closed. Baby had to tilt head to look at mother and father. Yellow on Areas 6, 7, 8, 9, and 10 was used. See improvement. Eye is beginning to open further. Deputy asked that when Normalation is complete, would Mary Muller bring the baby to the Meeting? She promised to do so.

Muriel Hartman—How can one clean the condensers?

Deputy—Unscrew the screws on the Slide-carrier, but do not tamper with the seals.

Someone asked what Scarlet is.

Deputy—Look for it in Spectro-Chrome Home Guide on page 12. When not knowing what Atcowa to use consult the Home Guide. It is your best friend.

Albertine Zimmerman—Is dry cereal considered Rational Food Of Man?

Deputy—Yes.

Martha Keppner—Always hungry in the morning. Asked what to eat.

Deputy—Eat potatoes in any form with your breakfast. They give heat and energy. Do not drink citrus fruit juices with milk.

Linus Schreck explained how to cook potatoes and vegetables that have moisture within themselves. No water was needed if cooked in heavy aluminum pot. He emphasized most people throw away the water and the sink gets the best part of it. He then gave his own case report. Burnt Hand. Skin coming off. Could not sleep. Wife told him to use Spectro-Chrome. Healed after 3 Tonations. He gave a case report (no name of Sufferer was mentioned for personal reason that Sufferer was present although the Deputy asked for the name twice.) Man was in Tuberculosis Hospital. Asked to get out as his fellow inmates were dying. Linus Schreck spoke with mother about Spectro-Chrome. Somehow they got the man out of the Hospital and Spectro-Chrome was used. Orange, Yellow, and Green were irradiated. Is fine now.

Question—Is it necessary to look into the "Light" while Tonation is given? Sometimes one falls asleep.

Deputy—It is good to concentrate and look into the Atcowa, but one should not force oneself to stay awake.

An Affiliate requested name and address of a Spectro-Chrome Normalator when one wants a few Tonations only.

Deputy—I decline to give such information. Some Affiliates may not understand why. I leave this matter upto Colonel Dinshah.

At this time, a request was made by the standing room audience to ask Dinshah for a larger and cooler room for the future Meetings. The Deputy asked if everyone would come to fill a larger room. This was answered with a loud "YES"; The Affiliates were informed that the next Meeting will be held March 2nd and to inquire at the desk in the lobby in which room the Planet will meet.

In general there was an abundant enthusiasm and the Meeting was a success.

EDITORIAL REMARKS

For a beginning, your Deputy Lina Fischer and the Planet Commissioners Dea De Martin and Anna Long deserve credit for the efficient manner in which the Meeting was conducted. The Council Of Trustees thanks you for its admirable management.

Any sale of Cocoanut Oil or anything else involving money, is not permitted at any Planet Meeting under our By-Laws. Read By-Law 39, Sections 4 and 5. Any Affiliate having any such thing to offer, may do it outside of the Meeting, at own place of business.

Send those 7 Stones with a story to the Editor.

Our Institute, under its present By-Laws, will not affiliate anyone using any modality outside of Spectro-Chrome Metry.

Coffee is a rank Poison, but of course, one may use a Poison even like Food by force of habit. There are people in Australia who eat Arsenic in quantities to make their Skin white.

Affidavits are voluntary offerings.

About having babies in a Planet Meeting, kindly refer to By-Law 48, Section 9. A baby, as a Sufferer, may be admitted as an "Exhibit," but, shall not be permitted to stay in during the conduct of the Meeting.

Dinshah Spectro-Chrome Institute does not divulge lists of Spectro-Chrome Normalators, as otherwise complications of partiality and favor are likely to arise.

The Council already secured a larger accommodation for you. The Trustees are glad and ever willing to serve you.

———o———

SPECTRO-CHROME SMILES

Viewpoint—Mr. James Watt condensed the steam engine.
Smile—Ja! There was no milk then.

Viewpoint—She had no solid foot for 3 days.
Smile—Because her Blood was liquid.

FIRST MILWAUKEE WISCONSIN PLANET

(MILWAUKEE AUDITORIUM, WEST KILBOURNE AND FIFTH)

Meeting Of Thursday, February 3, 1944

Successful Use Of Spectro-Chrome In:—Pneumonia, Pain, Rectal Abscess, Bronchitis, Skin Disorder, Injury, Rheumatism, Itch, Boils, Accident

Reported By:—EMMA SCHROEDER, *Recorder.*

Sub-Deputy Theodore J. Mohrhusen was in charge of registration with Emil Egide assisting. Deputy Frank S. Burns in the chair; 150 Affiliates were present.

The communication from Dinshah was read by the Deputy and questions were called. Ann Mayo who stated last month that she had applied for her Affiliation Card in time, but had not received the Card in time for the January Meeting, reported that it arrived one week after the Meeting and was in a very soiled condition. She believes that it was at the Express Office. Was advised by the Deputy to inquire when it was released from the Principal Office.

Upon request the paragraph on the Rational Food Of Man was again read from the correspondence and several Affiliates were interested in obtaining a few copies (less than 100) of the leaflet. In connection herewith much amusement was caused when Herbert F. Bahr of Madison, Wisconsin, raised his hand for recognition and mentioned his name.

Deputy—I knew a number of Bahrs when I went to school.

Herbert—I knew many after I was out of school. (Much laughter.)

Deputy—That was a different kind of Bahr. I mean family in the neighborhood. (More laughter.) One is now a fireman.

Herbert—And one is now at Waupun. (Still more laughter as Waupun is where the Wisconsin State Prison is located.) Herbert continued: "He is Deputy Warden." (The Recorder thought she detected a sigh of relief.)

Upon questioning by the Deputy none present acknowledged committing any of the errors mentioned in the communication.

Deputy—I am glad to announce that Sub-Deputy Mohrhusen is with us again. (Applause.)

The Sub-Deputy stepped to the platform, as requested and said: "I understand I was well discussed at the last session."

Deputy—Perhaps we overlooked something.

Theodore—I have no after-effects. I am again able to work full time.

Deputy—Tell us, did the Doctor say you would be dead if you were not taken to the Hospital?

Theodore—Words to that effect, yes! Said they could reduce the Fever with Sulfa drugs in 36 hours. If I did not go to the Hospital, he could not say what would happen and he did not say one word against Spectro-Chrome. I did not go into any details about our Science. I told him I believed in the use of "Colored Light" for any Human Disorder. I did not tell him how it worked. He walked up to the Equipment and copied all the information contained on the plate. I did not see him since. He never introduced himself when he entered. He must have gotten a good wet trouser leg, for he upset a glass of water as he left.

Deputy—Did you notice when the change took place?

Theodore—Definitely! The time to use Spectro-Chrome is before the Medical Doctor is called. I did not want to let all you folks down. My

wife is a strong backer of Spectro-Chrome and in fact, she had influenced me to use it. I owe all to my wife and to Kurt Bader for the help I received. To have a man of his size and vitality call! When he walked into the room and took my hand, there was a feeling that here was a true friend and one who was going to help me. The advice he gave me was exactly as taught by Colonel Dinshah, otherwise one might jump from Color to Color. Fever was 105 Degrees. Blue Systemic and Magenta on Area 4 broke the Fever. The advice was to call the Institute and let them take over from that time on. Kurt Bader wanted to be sure that we were receiving the best of help. The Institute advised Yellow and Lemon Systemic, two Tonations a day, because I had no Backelim for four days Fever was sub-normal after Backelim. At the Low Critical Period, I had a Ratio of two-to-one and had thus plenty of reason to worry. From that time on there was improvement. I lost about 18 pounds within ten days. I regained that in a short time. When I went out for the first time after my illness I visited a farm. The woman invited me to sit down. On a table was a bowlful of Eggs and some bottles of Beer. She knew I had been sick and offered me Beer with raw Eggs to drink. I thanked her saying I had to go. After I was able to move about the house four of my children were stricken with Intestinal Catarrh and had it approximately four or five days. At the time Kurt Bader loaned us one of his Equipments and we had two Equipments going every Tonation hour during the day for myself and children; nothing at night. Now they are tearing around the house; I wish Summer were here so they could be outdoors.

Deputy—Very bad the Doctor was not there.

Theodore—At the shop, I had occasion to take an injured man into the office. "They" smiled, but not a word was mentioned. "They" will not ask me anything about how I became well.

Deputy—I understand a Nurse called at the house?

Theodore—That is right! She called after I had been absent from work a few days. She said: "That man must have medical attention!" I was told I was needed and must become well and that the company will stand all expense. I could not convince them that I could get well *without* medical attention. After that nothing more was said except that the Nurse called occasionally. Spectro-Chrome got a boost. Everybody knows that I did not have a Doctor. They said I was bull-headed. Some called my recovery "Fate" or "Luck." I conversed with Dinshah after I was up and wrote him after that Spectro-Chrome will normalate any Disorder provided the Higher Powers do not deem otherwise. Thank you for your attention! (Applause.)

Deputy—It is another example of an individual having confidence and it is a boost and encouragement for Spectro-Chrome.

Leo Bieneman—About two years ago, I happened to have an experience similar to that of Mohrhusen and had to call Kurt Bader. I felt the same way about Bader as he entered the house. Theodore's remarks about Kurt Bader were borne out in my case throughout. I can certainly vouch for everything he said. (Applause.)

Past Sub-Deputy Kurt W. Bader—I do not deserve all this credit. His wife did not lose her faith or her head. I gave a little advice. I cannot accept all this credit. I had the co-operation of his wife and daughter.

Deputy Burns—You are very modest! I believe your advice and deportment are beneficial in a Disorder. We are glad to find that everyone feels that way about it.

The Deputy read a newspaper obituary from an eastern newspaper containing a biographical sketch of the life of Federal Judge John Boyd Avis who passed away recently. The Deputy expressed the opinion that Judge Avis might have extended his span of life had he used Spectro-Chrome.

A Free Guidance Request which the Deputy found on his table was read and discussed: "Male, age 52, name 'Persistent Smoker,' address: 'Skepticism Avenue'; state: 'Matrimony'; complained of trouble in Area 19. Full History: Questioning disclosed that Sufferer had an Ulcer. Hemorrhoids and Papillae in that Area removed electrically and a Tumor by Surgery ten years ago. In 1928 had a neckful of Carbuncles; two months later, Pleurisy; a month after that Pleural Pneumonia. Albumin present in Frontelim at various times for twenty-five years; first discovered in 1918. Now a Lump developed in Area 19 near orifice. Member of family is Graduate. When in 'Health,' this Sufferer believes he will never have occasion to use Spectro-Chrome. However, when this trouble occurred, after being informed that Tobacco Smoke and Spectro-Chrome do not agree, he refrained from smoking for one day, being alarmed about his condition and on the second day requested Tonation. One Green Front Systemic, one Green Back Systemic, Indigo Local with Magenta on Area 4 were irradiated on two consecutive days and another Green Systemic a few days later and in a total of five Tonations the disorder was completely normalated. One Green and one Indigo Tonation opened the Lump. This condition may not remain normalated as the Sufferer has not reached the stage where he is able to abstain from Flesh Food, Eggs, Tobacco and an occasional bottle of Beer. Drinks no hard Liquor. Spiroscil was very high in commencement. Ratio was 3¼, Kardoscil 69, Spiroscil 22. Spiroscil improved and Ratio rose to 4½ while not smoking. Took much Nephritine. Frontelim 7 to 8 daily, Backelim 3 times in one day. Does not live on Rational Food Of Man."

Emil Egide spoke—Inasmuch as Spectro-Chrome helped after he abstained from some of his habits, proves promising but the fact that he had a Tumor also proves that his Body was filled with poison, that his system was not cared for in the right way. Albumin proves he was a heavy Meat-eater. When Kidneys once stop functioning there is not much one can do. He better take it upon himself to eliminate Meat-eating and Smoking entirely.

Deputy—That is something that we recommend to everybody. The Chart shows that the Kidneys are eliminating.

Emil—Yes, but this Sufferer used 'it' only as a temporary measure. The Kidneys are eliminating water, but not the poison. They are overburdened.

Deputy—Yes, there are Toxins in the Elimination.

Emil—Tumor is Nature's effort to cleanse the Body.

Deputy—Has anyone else anything to say in this matter?

Kurt Bader—It is the characteristic build of the average American today—Smoking, Meat-eating and a bottle of Beer. The Sufferer reacted very well under Green. If Green were irradiated for a while, he might dislike Smoking. Tobacco would then taste to him like a "burning sock."

Deputy—Again you are very modest. Dinshah says they are "stinkers."

Kurt Bader—Three movements a day are very good. The Ratio may

be accelerated temporarily, but may become depressed again. Plenty of Green and Lemon may kill the desire for Smoke and Drink. Regarding the leaflets, when I have a new Sufferer I give one. Most of them do not know what Lentils are. They know Peas and Beans but not Lentils.

Emil Egide—About Albumin. Several years ago I had an experience. This man had Albumin and was turned down by the Armed Forces. I knew what he had been eating; too much fried food. I advised him to stop that and eat more citrus fruits. In ten days, he went back for an examination and there was no Albumin.

Robert Schott—Are not vegetables fried in butter O.K.?

Deputy—I believe Egide means foods fried very hard and fried Meats and Eggs or fried in Lard. I like fried foods that come under the Rational Food Of Man very much.

Robert Schott—There is not anything to take the place of butter. Everything else is only secondary.

Deputy—Albumin! I recently got an insurance policy. The examining Doctor was surprised at finding no Albumin present. Said the average man has a trace of it at my age. I got the policy; that is all I was interested in getting!

Past Deputy Otto Lutz—What strikes me is that with an obstruction in Area 19 he should have three Backelims a day. It is a rare co-incidence.

Deputy—Reminds me of a case of an acquaintance of mine several years ago. Became sick at work. Was compelled to absent himself from work and was just about ready to "give up the ghost." Was in extreme agony. I suggested Spectro-Chrome. He and the family were willing to do most anything to relieve him. I moved my Equipment over to his house. I tonated him on a Friday morning. I do not recall what Tonations were given. The man was in such extreme Pain and suffering that I used Indigo and Purple. I left the Equipment there with instructions. On Saturday morning, I called back just as the Sufferer was taking a Tonation on Area 19. In his opinion he had a Fistula. There was a mass present and while he was under the Tonation, it opened and he got such relief that he was just like being born over again. Like a new man. But the odor was so terrific, one could not remain in the room. He was also a habitual Smoker. Always had a cigar in his mouth.

Here the Deputy told of another acquaintance who dined at the same table with him. He had no teeth, but was able to finish a thick steak before Burns finished his salad.

Otto Lutz—(Humorously.) That is the way Meat should be eaten—in chunks. Hamburger is no good because it slides through the stomach.

Deputy—The case I mention is a typical replica of the one under discussion. The man went back to his old habits the same as before.

Hydro-Carbons were discussed.

The humor of the audience was aroused when Deputy Burns told of an Irishman who invoked blessings for his wife: "God bless my wife; God bless women—*sometimes.* (Laughter.) It is a good thing to laugh sometimes."

Theodore Mohrhusen—It is the best belly exercise there is.

Chapter 18 of the First Volume of Spectro-Chrome Metry Encyclopaedia, "Pickle Bottles As Medicine" was read by the Deputy.

Relative to paragraph 2 of the Chapter read the Deputy informed the audience, that the use of Spectro-Chrome costs even less than the consultation fee for a physician in India! Re paragraph 6, Deputy Burns said: "Medicine is purely a hit-or-miss proportion. The opinion of one physician may be the exact opposite from what another may decide.

How utterly disregardful our Affiliates are of Radio Drug and Medicine advertisements was demonstrated when the Deputy endeavored to recall the name of "8-in-1" tablets, "15c," as advertised over the radio. He asked the audience but none remembered the name.

Re: paragraph 11, the Deputy inquired if all knew where this place mentioned is where there is plenty of Sulphur and no Winter. The audience was again pleasantly amused as Kurt Bader replied: "Spain and Italy."

Deputy—Why, in Spain and Italy the American soldiers are fighting upto their necks in mud.

Kurt Bader—But they never have a Winter in the sense of ours.

Deputy—They never had a Winter like we are having right now. Dinshah means Hades—Hell.

Kurt Bader—Well, I saw the picture "Heaven Can Wait" and Hell cannot be such a bad place judging from that. The walls were of marble —just like the lobbies of the big hotels. And now they even have Sulphur down there. (Laughter.)

Deputy—That reminds me of a Scandinavian story. Of course, present company is always excepted. My youngest son told me this one. A Scandinavian took to the bottle and well-nigh made a physical wreck of himself. The family thought he was hopeless and held a consultation. One said: "I have an idea and I believe we can cure Hans of that habit of drinking. I shall dress like the Devil—red suit and horns and at night I shall hide behind a tombstone. When Hans passes through the cemetery on his way home from work, I will jump out and say: 'I am the Devil. Stop drink or I will take you down to Hades with me.' Events transpired as planned. As Hans passed, the "Devil" jumped from behind the tombstone to frighten him. Hans, however, did not recognize the Devil and inquired "Who did you say you are?" "I am the Devil. Stop drinking or I shall take you with me," was the reply. Hans had a difficult time placing the Devil and finally said: "Oh! Now I know who you are. I am glad to meet you. I believe I married one of your sisters." (Laughter.)

Attention again reverted to discussion and all questions at the end of the Chapter were reviewed.

An Affiliate was anxious to know about Croton Oil. Emma S. replied that it is about the most powerful Cathartic as only a drop or so is given at a time but the Affiliate was interested in whether it is a vegetable or a mineral oil. The Deputy promised to obtain the information by the next Meeting.

Theodore Mohrhusen—Does everyone understand the meaning of pickle bottles? I shall guess. They are either Green or possibly of a color on the Ultra side of the Spectrum. The liquid will absorb the Colored Light Waves and if drunk will take the place of a Tonation.

The last fifteen minutes were devoted to Case Reports:

Emil Egide gave the following: "Members of my own family. The early part of December my wife came home from the office. The Company Doctor told her she had Bronchitis. The following morning she was unable to talk. Tonated for that. At the same time, an aunt who lives with us developed Influenza. We also had two Equipments going and what alarmed me was that the Fever in both cases was over 102 Degrees. I used a Thermometer. Would Colonel Dinshah object? Both cases re-

covered in about four days. I was very well pleased with the results. I was very busy getting my breakfast and taking the dog for a walk. The thing that is important is that after the seventh day my wife went back to work. According to the Company policy, a release must be issued before one can come back to work. The Doctor called her upstairs and wanted to know how she got along. "You got over that in a hurry. What did you do?" he asked. It was nothing new to him, as there is a number of Spectro-Chrome Equipments in the Company. Said he: "I will give you this release only because you came to me before you stayed home; nor would you be paid for the time off if you had not come to me." (Applause.)

Emil finds the use of an Equipment costs about three cents an hour. Atcowa used were Green, Turquoise and Purple. Purple on Areas 4 & 18 was used for the aunt who is 69 years old and subject to Hyper-tension. Emil is 100% for Purple; he says it will lessen Pain, reduce Fever and produce rest.

Sarah Erdmann reported the case of her son and found Spectro-Chrome very beneficial. Son awoke "yesterday" morning with 103 Degree Fever and was vomiting; also had Pain in Stomach. Used Turquoise. Continued vomiting. Tonated for an hour and fifteen minutes. Local Tonation on Area 8. Blue Systemic Back caused Perspiration. By evening, Temperature was normal. Intends to tonate Areas 4 and 18. Other similar attacks required a week for recovery. Obtained her Equipment a year ago last July. Uses it daily for Heart Disorder. People wonder how she is able to do the work she does. Son had a bad Throat a year ago. She said "Never again. Hereafter I shall use Spectro-Chrome instead of a Doctor." (Applause.)

Regina Schroeder inquired whether it was necessary to remove the Tonsils. Leo Bieneman suggested Lemon Systemic as the condition is Persistent. Regina's daughter is troubled with the Disorder and is 22 years of age.

Otto Lutz reminded that names be given in Case Reports or they will not be printed and submitted the following reports:

1. Hans Y., aged 52, Male. Skin Disorder on Hands and Arms, Area 14. Front and Back and a few spots on Body. This man thought there was no help for him, because of all the different kinds of methods he used for over two years without results. A friend advised the use of Spectro-Chrome, as he said he was at the end of his rope. Yellow, Orange and Lemon Systemic, Front and Back, normalated his condition.

2. Ruth K., age 8, female. Mother, Erna, brought her in. Mother is aged 47. This girl was kicked on Area 13, left Leg, on school grounds, while playing. Was operated on Area 13 and also deep X-Ray treatments were used, without results. Was advised to have the Leg amputated. Parents were not satisfied with that advice and came for Spectro-Chrome. Attuned Colors were Lemon and Green Systemic, beginning at Area 13 normalated her condition. Is again back at school WITH BOTH LEGS. The parents marvel at the results of Spectro-Chrome.

3. Alfred K., age 33, male. This man went up North deer hunting and when he returned he brought back with him Rheumatism with a terrible Itch. Had been given Sulfa Drug Injections. Was not satisfied; then he thought he would "try" Spectro-Chrome. In a few days, the Itch stopped as well as the Pain and in a few more days he went back to work, normalated. Atcowas used were Indigo Systemic, Front and Back.

4. Carl J., age 49, male. Case of Boils; they would disappear and always come back in another Area. Had been taking Injections for about two years; then one day, one started to appear on the tip of his Nose. That was the end of the Injections. He came for Spectro-Chrome. Attuned Color Wave Yellow Systemic was applied and that was the end of his Boils. Even the one on the tip of the Nose failed to suppurate; it just dried. His wife is now a Benefit Student.

5. William H., age 42, male. Auto mechanic. While working at his trade, he jacked up a car while under it, the jack slipped and the car struck his Back, Area 18. The injury was very painful. Was rushed to a Hospital where he was taped and after a few days was brought home. No Bones were broken. Spectro-Chrome Lemon and Yellow Front and Back removed all discoloration and eased the Pain considerably; then Turquoise was irradiated. Returned to work in two weeks.

Applause followed Otto's report.

It was a pleasure to find so many of our practising Normalators in attendance at this meeting.

EDITORIAL REMARKS

No matter who gets the credit, we are concerned only in the Normalation of our beloved Affiliate Theodore Mohrhusen. His wife Hilda, Kurt W. Bader and all others, will be thankful to the Higher Powers, Whose Benign Effulgence is the Guardian of Spectro-Chrome.

Radio—at one time, a purveyor of truthful information, is now nothing but a money-making racket. Someday, after the War is over (?), when Dinshah Educational Broadcasting Service is in operation, you will have a unique Station carrying absolutely no advestisements or propaganda, but, delivering genuine public service. Let us have $100,000.00 and we shall commence the service for the benefit of mankind. All facilities for it are already here. Even General Advice Charts will be answered by Radio and if so convenient, even Atcowas will be transmitted through the Ether. The man who originated Spectro-Chrome Metry, can also perform such service—if let alone to his inventions.

The "Pickle Bottles" stated, were actually Yellow, Green and Indigo bottles of English pickles, which were the media in my first experiments.

Croton Oil is called in the pharmacopoeial bombast Oleum Tiglii. It is a fixed Oil, expressed from the Seeds of Croton Tiglium Linne (Family Euphorbiaceae). It is sensitive to Light and should be kept in well-stoppered vials. Dose I Minim. It is a pale yellow, viscid, slightly fluorescent liquid and great caution must be used in tasting it as it causes a pustular Eruption when applied to the Skin. It is a Drastic Cathartic and even the Medical Doctors have to use it with great caution, because of its irritant attributes.

The use of an Equipment for 1 Tonation, costs in all 10 cents for the hour, including the Electrical Current and the Bulb deterioration. It is a safe, general estimate.

The Dinshah Itisometer is essentially an Electric Thermometer. Whether the Fever be 99 or 107 Degrees, our Science has no different modality. The higher the Temperature, the higher the Kardoscil and that is our guiding beacon.

FIRST ST. LOUIS MISSOURI PLANET

(HOTEL MELBOURNE, GRAND AVENUE AT LINDELL BOULEVARD)

Meeting Of February 3, 1944

Successful Use Of Spectro-Chrome In:—Cold, Pain, Lumbago

Reported By:—VIOLA C. JEUDE, *Recorder*

Deputy Clemens M. P. Neumann was in the chair; 45 Affiliates were present.

Communication from the Principal Office was read and explained in detail. He asked if there were any questions in regards to the communication? There was none. He said the educational program tonight dealt with Common Cold, Cough, etc. He read what the Medical Books say about them.

The Deputy said that before the evening was over, the Affiliates would realize the importance of the Kardoscil and Spiroscil, how Spectro-Chrome works and why it works. He asked the Sub-Deputy to read Article 41, "Medical Humbug Of Common Cold," page 78, from "Family Health Protector." This was done.

Robert Boschert said he would like to know the difference between Scarlet and Magenta. It was explained.

The Deputy then illustrated Spectro-Chrome Mathematics.

Konrad Frey said he would like to state that in one thimbleful of dirt it was tested that there are 44 Million known Germs in carpets in homes of rich and poor alike.

Some one suggested the Slogan for the year should be: "I am what I eat."

The Deputy said he wanted to report that he was investigated by a Medical Doctor last week. A man called and said he had heard that he was prescribing Medicine without a License.

He said he told the man there must be some mistake, as they did not believe in Medicine; he called the man's attntion to the Sign on his office door "Spectro-Chrome Normalator Not A Doctor."

The man asked how long he was there? He said 5 or 6 years. He then asked for his Business Card. He gave him one. The man shook his head and said he could not understand; wrote on back of Card, "Has Sign On Door Is Not A Doctor." As he was leaving he asked if that was not a Jewish Star on one of the Spectro-Chromes. He said he told him it was not, it was just an emblem.

Case Reports followed:—

1—Katherine S. said her husband had bad Cold; used Scarlet 4-18, Red Systemic and Orange 6-7-8-9-10. 3 Tonations each day for 2 days and he recovered.

2—Robert B. said he ached so he could not move; Spectro-Chrome set him right. He also said he had an attack of Lumbago; has had these attacks for 22 years. They always lasted from 2 to 3 weeks; this one occurred Sunday morning and after 3 Tonations in 24 hours was entirely gone. Orange Systemic was irradiated.

3—Robert B. also reported of girl about 5 years old. The only way she could sleep was by sitting up; after 30 days of Turquoise Systemic and Magenta on Areas 4-18 she now lies in Bed.

The Sub-Deputy said she had an interesting article about Vaccinations. Some may be familiar with it as it is being carried on at present time in the schools. She then read excerpts. "Diagnosis Of Syphilis

Often The Result Of Vaccination" by Cash Asher, page 335 of September 1943 NATURE'S PATH.

"In its slow and painful climb up the high hills of exploration, the medical profession occasionally stumbles upon the truth.

"Confirmation of anti-vaccination statements made years ago by a few medical scientists and influential lay individuals, has been made by the medical corps of the U. S. War Department. A Bulletin sent to all medical officers of the army declares:

"IT HAS BEEN CLEARLY ESTABLISHED THAT THE USUAL SEROLOGIC TESTS FOR SYPHILIS MAY BE TEMPORARILY POSITIVE AS A RESULT OF VACCINATION FOR SMALLPOX."

The War Department Bulletin was issued under date of April 30, 1943 and signed by John A. Rogers, Colonel, Medical Corps, Executive Officer.

Interesting information on this subject is also found in a Bulletin issued by the United States Public Health Service in June, 1943. Both of these publications are appended to this story.

In "The Doctor Could Be Wrong." published in 1939, Dr. David H. Reeder, of LaPorte, Indiana made the following statement:

"More than 90 per cent of persons who have been vaccinated will give a syphilitic reaction, even twenty-five years after the vaccine ulcer has healed. For a hundred and twenty-five years we have been inoculating human beings with vaccinal syphilis and I am convinced that much of the tremendous increase in the deaths from cancer may be traced directly to this disease inoculation."

In "The Doctor Could Be Wrong," I quoted Dr. Carter as follows:

"A large proportion of the cases of apparent syphilis, are in reality vaccinal."

Following is the copy of the U. S. War Department's release to all medical officers:

"WAR DEPARTMENT
Office of The Surgeon General
Washington.

April 30, 1943

CIRCULAR LETTER NO. 93

Subject: False positive serologic reactions due to nonsyphilitic diseases.

1. It has been clearly established that the usual serologic tests for syphilis may be temporarily positive as a result of diseases other than syphilis. Reports have indicated that false positive reactions among soldiers, though relatively infrequent, are of sufficient importance to warrant consideration in order that the unjustified diagnosis of syphilis may be avoided.

2. Accumulated evidence indicates that most biological false positive reactions will be encountered under the following circumstances:

a. Vaccination for smallpox within the three preceding months, especially with "take," Etc.

EDITORIAL REMARKS

It is strongly recommended that in conducting the Educational Program, time be not spent in reading from medical books and outside publications. It were much better to instil in the minds of the Affiliates knowledge of our Theories, Principles and Practices. Refer them to outside publications, but, let the Planet time be devoted to Spectro-Chrome Metry. In Spectro-Chrome Home Guide, Spectro-Chrome Metry Encyclopaedia and Family Mealth Protector, there is enough to keep one busy for years in discussion and explanation.

Give no haphazard advice to anybody without a complete Sufferer Chart; furthermore, deny such advice where the Sufferer is under the direct Free Guidance of the Institute, otherwise there will be a likelihood of more harm than good. Refer to By-Law 79.

Three Tonations are not needed in one day. Do not try to force issues with Nature. Be patient; let the Auric Vehicle have time to reverse its Emanative Equilibrium.

44 Millions is a small number. That may apply to "known Germs"; the unknown extend to Eternity. Carpets and rugs are prime causes of numerous Disorders. They are Dirt Collectors at best.

Why next year's Slogan? "I am what I eat" should be in the mind of all for all time.

That "visitor" came from the racket known as the "Better Business Bureau." I have full experience of their tactics and it will not be long before they will suffer a complete exposure. Have your audience ready. I have proofs positive of that burrow being a henchman of the Medical Octopus. St. Louis will certainly enjoy seeing the proofs and so will the Affiliates.

DINSHAH LECTURED IN NEW YORK

On Sunday, March 26, 1944 and the two following nights, Dinshah P. Ghadiali addressed crowded audiences in the Hotel McAlpin, in New York, New York. The first two lectures were in the Ball Room, because of the admission of visitors, the last in the Green Room being open solely to our Affiliates. He began each night at 8 P. M. and the enthusiasm invoked in the Science of Spectro-Chrome was such, that the listeners would not leave the Room even after 11 P. M. The Ball Room Lectures were illustrated with the Stereopticon and the approbation of the audience was expressed by frequent rounds of applause and mirth.

PINK PILLS FOR 18 YEARS

Spectro-Chrome Gave Prompt Relief

By:—Federal Accountant, Detroit, Michigan

Have been constipated for 18 years, taking Pink Pills. Now condition improved due to Spectro-Chrome. My bowels are beginning to move twice daily, with more ease. To persons who have had no such troubles, this might seem of small consequence, but to me who have taken Pink Pills daily for eighteen years, it is a miracle. I shall remain for ever grateful. —G. A. C.

DINSHAH IMPARTIAL

The Universe Through The Eyes Of Dinshah
"I Fear No One; Only God Above And Conscience Below And From Them I Have Nothing To Fear."—Platform since 1891.

MR. 'TAIN'TE ON LONG DISTANCE TELEPHONE

Dinshah's Photokinephone and Spectro-Chrome

How the World pilloried pioneers and genuine progressives, is a known story to us. Even after 24 years of incontestible practical, clinical proofs of the value and merits of Spectro-Chrome Metry as an original Science, we are being subjected to derision and ridicule, by persons who never cared to investigate. Of course ,such attitude does not in the least affect our heart or temper and the more opposition we receive, the more power does flow toward us from Those Whom we serve. Why not? We are in good company. Look at the tribulations of Alexander Graham Bell! Since he exhibited his Electro-Magnetic Telephone in 1876, many just dreamt of talking to their dear ones at a long distance. Yet, when such a project was proposed, a newspaper in 1885, printed in part the following marvel of scientific logic:—

> "Granting that the long-distance telephone will be forthcoming," the article says in part, "it will require an absolutely perfect condition of the elements along the route over which the wires must run; there must be no rain, no fog, no sleet or mist; no moisture in the atmosphere, no electrical storms, and, besides, there must be invented a more perfect system of insulation of the wires where they come in contact with the poles. After all these difficulties have been overcome, there will still remain many reasons why it will not become a success . . . no sir, the coming long-distance telephone will be a plaything, nothing more. Put a mark on the assertion."

What will that newspaper say today, when the World depends on it for its business efficiency? Without Long Distance Telephony, even the Armies of belligerents would be at a stand still.

When The New York Times printed in its issue of Monday, January 18, 1915, the following article, was I razzed about it by others who knew nothing about me or my activities and background!

"PATENTS 'TALKING MOVIES'

New Invention Dispenses With Use Of The Phonograph

Special To The New York Times

HACKENSACK, N. J., Jan. 17—Moving pictures that talk without the aid of a phonograph are the latest device perfected by Dr. Dinshah P. Ghadiali, an inventor, of Hillsdale. All other movie-talkie schemes have relied upon the phonograph, but Dr. Ghadiali claims to have invented a way to do without the talking machine. Eleven of his claims for the invention have been allowed by the Government Patent Office.

The Dinshah Photokinephone Company has filed papers of incorporation here with County Clerk Thompson. It has a capital of $150,000 and intends to put the new device on the market. Headquarters have been established at Hillsdale and a manufacturing plant is to be built

there. Dr. Ghadiali is a physician and a native of Bombay, India. He has done considerable writing and has several inventions to his credit."

Five months after this, Dinshah was admitted to United States Citizenship and five years later, Spectro-Chrome was born. The Homeopathic Medical Doctors gave it a good ovation as evidenced by the following:—

North American
Journal of Homeopathy

Published monthly by the AMERICAN MEDICAL UNION, Chicago.
Five Dollars a Year

Entered as second-class matter January 7, 1918, at the Postoffice at Chicago, Illinois. under the Act of March 3, 1879.

69th Year APRIL No. 4

1921

SPECTRO-CHROME THERAPY *

The latest addition to Drugless Methods of Healing is that of the Spectro-Chrome Therapy, the system of diagnosis and treatment of diseases and. derangements demonstrated by Dr. Dinshah P. Ghadiali, which, after long study and scientific investigation on his part, has resulted in an almost unbelievable perfection. The results to be obtained through this method makes it of incalculable value to the profession of Drugless Healing,.and its employment will add greatly to the reputation and success of those practicing it

Dr. Dinshah is a Parsee—a Zoroastrian, and a scientist of the widest experience. He was born at Bombay, India, and in earliest childhood exhibited a passion for chemical and electrical research. He was a professor in chemistry at the Wilson College at Bombay when very young, and delivered his first public lecture on electric light when but seventeen years of age. He is an expert in almost every field of scientinc research. His activities are too various to enumerate, but while the passion of his life has been scientific research, he possesses the rare faculty among imaginative minds of practical and constructive accomplishments.

His crowning achievement, involving the labors and research of over thirty years, in light, heat, electricity and magnetism, is the Attuned Color Wave Therapy, known now as Spectro-Chrome Therapy.

While working with light waves Dr. Dinshah found through spectroscopical and chemical experiments and mathematical calculations that all drugs and chemicals actually act in the human system through disengagement of the color waves. He further found that every element in the world exhibits a preponderance of one or more of these colors, namely: red, orange, yellow. green, blue, indigo and violet, and that the potency of an element depends on the potency of its color waves.

Foods and medicines are composed of specific elements, and act in con formity with such color wave potencies.

Food builds our body and all vegetable and animal growth is dependent on sunlight, so light is as much a necessity of life—as food—as solids, liquids and gases.

And as it is the sun which is the.source of all life and energy, imparted to the world by waves of light—which light is white and is composed of seven colors—red, orange, yellow, green, blue, indigo and violet—it is therefore plain that ultimately our bodies are.also built of these elements—colors.

Light contains about 9.1 per cent hydrogen, 13.4 per cent carbon, 2.5 per cent nitrogen, and 72 per cent oxygen.

The preponderating color waves of the four main elements are red, yellow, green and blue, respectively.

Thus 97 per cent of our body is built of the above four elements, and is therefore responsive to the said four color wave potencies.

These facts are definitely proved by spectroscopy and the latest conclusions of metaphysical research.

In health the bodily elements (colors) are proportionately balanced.

Disturbance of this element (color) balance produces dis-ease.

Establishment of the element (color) balance in the system by administering the lacking elements (colors) or reducing the super-normal ones restores health.

This is the rational system employed in Spectrochrome Therapy. It is carefully attuned in accordance with color wave potencies and perfected to meet every condition of ill-health, including those diseases resulting from the disturbance of the balance of even-the 3 per cent of the less dominant elements (colors) and it is applicable to every case of a non-surgical character.

The theory of Sir David Brewster, that the three primary colors are red, yellow and blue, has been rejected by Dr. Dinshah, he having proved beyond any doubt spectroscopically, chemically, mathematically and geometrically, that the three primary colors are red, green and violet, as only from these three colors it is possible to produce white light by the addition of one color to another, or complete blockade of light by the reverse process, that is, by subtraction.

Thus the attuned color wave system, when represented by a chart as follows: presents all of the seven colors of white light (red to violet) and also five other colors not contained in the solar spectrum—the two upper colors (lemon and turquoise) and the three lower colors (scarlet, purple and magenta), completing the tuning of the system just like a musical instrument into 7 major and 5 minor keys.

Red and orange are stimulants, yellow and lemon are nervines or nutrients, green and turquoise are the germicides, blue and indigo are sedatives and narcotics, violet and purple are depressants and magenta and scarlet are the emotionals.

And now? Friends, where Truth is, eventual success MUST come and the harder the knock, the greater the boost.

SPECTRO-CHROME — IN EVERY HOME

It is upto you Affiliates to make our Slogan a reality.

———o———

NO WONDER CONSUMPTIVES DIE!

Our Affiliate Rose Holder, from Detroit, Michigan sent me a newspaper clipping, in which a question was asked and reply was given in the syndicated column of Dr. Logan Clendening.

'Question—Is it harmful for a tuberculosis patient to smoke?

Answer—I do not believe so, in moderation."

And what is the implication? That in moderation, Smoking is not harmful in Tuberculosis, according to his "belief." Who cares for his belief? Give a definite reply scientifically and conscientiously endorsed; why hedge around to retain the goodwill of stink-weed advertisers? The answer is nothing short of asinine, because, we can not afford to lose Health on the beliefs of Medical Duffers.

SPECTRO-CHROME

March 1945
Yellow 1945

Volume 21
Number 3

Trade Marks Registered In The United States Patent Office

UNITED STATES FOOD AND DRUG ADMINISTRATION INSPECTORS RUNNING ALL OVER THE COUNTRY TO CONDEMN SPECTRO-CHROME AT FORTHCOMING TRIAL

Presence Of Police For Peaceful Meeting In Milwaukee, Wisconsin

Attuned Color Waves
Spectro-Chrome Metry

In course of life you reach a time,
When Doctor, Lawyer, Priest alike,
Does fail to give the sought relief,
To troubled mind or burdened heart—
Then come to me:
I help you to help yourself.
SPECTRO-CHROME

The Science Of
Automatic Precision

DEVOTED TO THE SERVICE OF MANKIND

FIRST MERRILL WISCONSIN PLANET

(HOTEL BADGER, MERRILL, WISCONSIN)

Meeting Of Thursday, January 4, 1945

Successful Use Of Spectro-Chrome In:—Fatigue, Miscarriage

Reported By:—GERTRUDE L. PARSHALL, *Sub-Deputy*

There were only seven Affiliates present. One Affiliate, not as yet having received her 1945 Affiliation Card, graciously returned home.

Affirmation Of Installation was read to Emil Erben, as the newly appointed Deputy, by Gertrude Parshall and repeated verbatim by Emil Erben. He then took the chair and read the communication from the Principal Office and then read the Night Letter about the disastrous fire at the Institute. Every Affiliate present was heartsick upon hearing this. We unanimously offer our heartfelt sympathy to the Council Of Trustees.

He then read the Good Thought for January and suggested that more Affiliates acquire this worth while book of Dinshah.

Case Reports followed:—

Walter Briese felt tired from extra amount of work at Christmas time, so took Lemon Systemic and feels "fit as a fiddle."

A case of Hemorrhage was reported. It was a miscarriage. Sent in General Advice Chart and received Free Guidance from the Principal Office. The case cleared in one week. She is feeling fine.

Article 22 on page 28 in Family Health Protector was read by the Deputy. He also defined the word KARMA. It was decided to wait until the next Meeting to plan our year's study program. We send our best wishes for the rebuilding of the Institute when it is possible to do so.

EDITORIAL REMARKS

It is regretted that the November and December 1944 Reports of your Planet can not be printed, because of their destruction in the Fire. Your sympathy is warmly appreciated.

Meeting Of Thursday, February 1, 1945

Successful Use Of Spectro-Chrome In:—Pain

Reported By:—GERTRUDE L. PARSHALL, *Sub-Deputy*

Deputy Emil Erben was in the chair; 9 Affiliates were present.

The communication from the Principal Office was read. The High School gave an operetta that same evening, which may account for some of the absenteeism.

Affidavit Blanks were explained and given to the Affiliates.

The Authentic Report of the Fire was read by the Deputy; also several of the newspaper accounts. The Affiliates still wonder how such a catastrophe could have happened.

Triumph of Spectro-Chrome, Pages A13, A14, A15 were read.

An Affiliate reported that her daughter age 17, complained of Pains in her Abdomen and right side when she came home from school. That evening, she took a Turquoise Systemic Tonation followed the next day with a Lemon Systemic Tonation and there was no recurrence of the Pain.

Gertrude Parshall reported that a Federal Inspector interviewed her and the following conversation ensued:—

Q.—Have you a Spectro-Chrome? A.—Yes.

Q.—How long have you had it? A.—About 4 or 5 years.

Q.—Whom do you give treatments to? A.—Tonate only my two children—Patricia and Dennis and myself.

Q.—Does your husband use it? A.—No, sir.

Q.—Why not? A.—Because, it wouldn't do him any good. (She told him his ways of living were different.) We have no arguments about Spectro-Chrome.

Q.—What do the children use it for? A.—Colds, mostly.

Q.—What do you use it for yourself? A.—I had Low Blood Pressure and my Heart skipped every third or fourth Beat.

Q.—How did you know you had Low Blood Pressure? A.—Dr. Marie Kersten examined me.

Q.—Did Dr. Kersten give you Medicine? A.—Not a drop. I used only Spectro-Chrome.

Q.—Are you a Normalator? A.—No.

The Answers given were truthful.

Walter Briese stated that he too was visited by the Federal Inspector and a series of similar questions was asked and answered. Walter stated also that he read an article in which the famous Mayo Brothers condemned the sending of Tobacco to the soldiers in the foreign battlefields. Gertrude said she read an article to that effect also.

It was decided to begin the study of Volume 1 of Spectro-Chrome Metry Encyclopaedia, from next Meeting. By-Law 116, on "Unprovided Occasions" was explained.

EDITORIAL REMARKS

Mayo Brothers are right. Tobacco is defined in the American Illustrated Medical Dictionary, (sponsored by the American Medical Association) as uniting "the qualities of a Sedative Narcotic, with those of an Emetic and Diuretic. It is also a Heart Depressant and Antispasmodic." Let our Affiliates judge why in our Science, we discarded that poisonous weed in every form.

SPECTRO-CHROME SMILES

Viewpoint—I am not able to dissolve my problem alone.
Smile—Pour some turpentine.

Viewpoint—Baby has bed wedding.
Smile—Get him a divorce.

FIRST CHICAGO ILLINOIS PLANET

(HOTEL ATLANTIC, CLARK NEAR JACKSON BOULEVARD)

Meeting Of Thursday, January 4, 1945

Successful Use Of Spectro-Chrome In:—Cold, Inflammation, Fever, Pain

Reported By:—HERMINE KUEMMEL, *Recorder*

This was to be the last Meeting at the Hotel Atlantic, because the management informed the Principal Office to look for other quarters, since all the available accommodation was reserved for the Government. Next month, therefore, we shall meet elsewhere.

Deputy Wilhelm Weissenbacher was in the chair; 17 Affiliates were present.

The Deputy welcomed a new Affiliate Joseph Burrescia and advised him to stick to Spectro-Chrome for his own good.

The Deputy read the letters from the Principal Office.

Discussion about Automatic Time Switches began. Said Rudolf Kist: "It is certainly very reasonable to pay only six dollars for a time switch. I have many on the machines in my shop and they cost me much more and when they do not work, they cannot be repaired; I have to throw them out. Affiliates certainly should not complain about the price."

The Recorder read By-Laws 79 and 80. The Deputy advises everybody to watch the By-Laws carefully and to study them.

Elizabeth Zavadil read from "Dinshah Naturalization Case," Chapter 4.

Deputy—Officials of the Food And Drug Administration called at my home; they wanted the addresses of some Affiliates and the Recorder; as I was not at home, my wife could not give them any information. They called at the Sub-Deputy's home; he was not at home and so they asked Mrs. Schilpp, how she happened to use Spectro-Chrome. She stated her case, that she was in the Hospital for an Operation and after she came home, felt terrible for a long time; then she took Tonations and feels well ever since.

Joseph Burrescia—I wish I were home, when one calls at my house! I can tell him something about Spectro-Chrome.

Rudolf Kist—When he comes to the home, tell him to bring a Notary Public with him to make sure that the information you give him is not changed by him; we don't know what he writes down.

Louise Weller—Can I refuse to answer their questions or can I refuse to let them in at all?

Henry Schilpp—He wants to see the Commissioners first. He wants to find out what we have to say and find out addresses. As long as we tell him how much Spectro-Chrome has done for us, I do not see any reason why we should not let them in. It can only help Spectro-Chrome, but if we refuse to let them in, they may think there is something wrong.

Deputy—In the October 1944 Spectro-Chrome, you can read about the presence of the Government Agents in the New York Planet Meeting and in the November 1944 Magazine we read about the visits of the Agents to different Affiliates. They all had a very good report to give about Spectro-Chrome, which only can help Spectro-Chrome.

Case Reports followed:—

Hermine Kuemmel—During the last cold wave, I caught Cold; the

Eyes were all red and inflamed and very painful. I irradiated twice with Turquoise and they were well.

Louise Weller—On Christmas day, my boy was hot with Fever. He did not want Tonation, but I gave it to him anyway; I tonated with Turquoise Systemic. In the evening his pulse was uneven, so I gave him Magenta. The next day he was well.

Rudolf Kist—My little girl had a bad Cold and Fever. I tonated her with Yellow, until her bowels moved; then Fever subsided and in 4 days she was well again. I had a Cold myself, but could keep on working. I was well again in 1 day. My own case is of long standing and I shall report on that later. I intend to go to Malaga. Colonel Dinshah said I should wait until the end of January.

Emily Weissenbacher—I took Cold and my Ears were quite painful. Tonated twice with Turquoise and felt fine again.

Elsie Beyer—One evening, I had bad Pain in Areas 2 and 3. I took Turquoise Local a few times and a few Yellow Tonations. A bad cough developed and I had to stay in bed for a few days. Tonated Turquoise and Green Systemic, Orange Local and Magenta on Area 4.

Josephine Burrescia—If a person wants to go to Malaga, New Jersey, what does one have to do about it?

She was told to write to the Office, include the Free Guidance Chart and tell what time you want to go there. She would like Colonel Dinshah to take care of her boy.

EDITORIAL REMARKS

It is regretted that the November and December 1944 Reports can not be printed, as they were burnt in the Fire.

There is no reason to obstruct the Agents of the Federal Food And Drug Administration, in the performance of their duties. Tell them the Truth as you know, but, sign no statements, because the matter is before the Federal Court. However, if they become insolent or issue any threats, report to us. They are public servants and should know how to behave with the public who pays their salaries.

Because of the total demolition of our Principal Office Building by the Fire, until we notify through the Spectro-Chrome, please refrain from coming here, as we have no time on hand to take care of Affiliates, in that direction. On the contrary, we need helpers here.

Meeting Of Thursday, February 1, 1945

(KIMBALL HALL, 308 SOUTH WABASH AVENUE)

Successful Use Of Spectro-Chrome In:—Cold, Aphonia, Laryngitis, Cough, Headache, Abscessed Tooth

Reported By*:—HERMINE KUEMMEL, *Recorder

Deputy Wilhelm Weissenbacher was in the chair; 28 Affiliates were present. The Deputy read letters and telegrams from the Principal Office. Telegram should have been in his possession for the last Meeting, but, was delayed somehow. It was a big shock to the Affiliates to hear about the great disaster of Fire, which did so much irreplaceable damage. We all feel sympathy for Colonel Dinshah, his wife Irene Grace and their seven sons, whose courage and poise were tested to the utmost.

Deputy—Affiliates, let us everyone do what we can to help the highest Cause. The blessings of it will come back to ourselves. You all know, how to send the remittance for the Spectro-Chromarium Fund. Let us not stay behind other Planets.

Hermine Kuemmel read the various newspaper articles about the Fire. Everything possible was done trying to extinguish it.

Deputy—We are becoming used to receive bad news. We had rented room 724B for the entire year as our Meeting place and Dinshah had sent a Check for the entire year to the Manager. We just had a letter from him informing us that he rented our room and so we have to look again for another one. Also I like to introduce two new Affiliates — Albert Langkau and Hilda Gustafson. We welcome them and hope they will attend our Meetings, as one can learn much about Spectro-Chrome by it.

One of our Affiliates just received his Spectro-Chrome Equipment Saturday and on Monday the Food And Drug Agent came to see him. He was home attending to the Baby, his wife being sick. The Agent had the same Plate as on the War-Model. I wonder from where he procured it! The agent asked the new Affiliate questions, but got no answers.

Arthur Koegel—Under what pretext did he come into the house? They have absolutely no right to come into a man's house and ask him all kinds of questions. We can buy whatever we want without anybody's advice or interference. We are still in a free Country. Why do we have to worry about the Food And Drug Administration? They did not worry about the expensive Drugs the man bought before and they did not help.

Affidavit blanks were then distributed.

We want to help as much as possible. The whole Dinshah Family has their hands full with work, so let us help and send in Contributions and order some of the books, which have such wonderful knowledge in them. There is plenty of money needed for the Case and to build the Spectro-Chromarium. One of our Affiliates wanted to go to Malaga, but, when he heard of the disaster, he changed his $100 Check and sent it in for the Spectro-Chromarium. Let us follow his example. It is wonderful how the mother Irene Grace kept her family together and so we Affiliates should do, as most of us had great benefit from Spectro-Chrome. Do not be afraid of the Agents; we did not do anything wrong, so why be afraid of anybody?

Eleanor Wendt read the Authentic Report of the Fire.

Rudolf Kist—I am a very sensitive person. I developed a very bad Cold on Sunday. By Wednesday I had developed such a Laryngitis, no sound could be heard and I had an engagement to sing on Saturday. Took Tonations with Green Systemic, Turquoise and Magenta and was able to sing on Saturday. I had a relapse, tonated with Indigo, was well from this bad case after five days. My child had a bad Cough; it was controlled very well with Spectro-Chrome.

Arthur Koegel—My son had the most wonderful result with Spectro-Chrome and I know a number of people who had very good results with it. I myself cannot tell of my experience, because, when I should like to take a Tonation, one of the members of my family is using the Machine. (Laughter.)

Albert Langkau—I have an Equipment for a few months and am in good health; can I take Tonations for better health?

Elsie Beyer—This Affiliate asked if he could take Tonations, when he feels well. Why should anybody take Tonations, when feeling well?

Deputy—He is a new Affiliate; he has the privilege to write to the Principal Office for Free Guidance. A new Affiliate should stay for a year under Free Guidance; during that time one can learn much about Spectro-Chrome Metry.

Joseph Janik—My wife had a very bad Headache; she was crying from Pain and did not want Tonation; she always was against it. I took her upstairs in my room and tonated her with Turquoise; after 20 minutes her Headache subsided and after the one Tonation she was normalated and changed her mind since.

Ernst Binz—Had an Abscessed Tooth on Monday. My whole Jaw was swollen. Tonated with Turquoise morning and evening; 4 Tonations normalated everything in 3 days.

Elfriede Mueller—I had a bad Cold; tonated first with Lemon Systemic, then 1 Turquoise Systemic, then Turquoise Local; also one Magenta and Orange on Areas 1-2-3-4, because I had trouble with breathing. This was the first Cold I had this Winter, while previous Winters I suffered terribly.

EDITORIAL REMARKS

Cordial thanks for your great sympathy and co-operation. We are here—Fire or otherwise—and are glad to serve you.

FIRST CLEVELAND OHIO PLANET

(CARNEGIE HALL, 1220 HURON STREET)

Meeting Of Thursday, December 7, 1944

Successful Use Of Spectro-Chrome In:—Varicose Veins, Toothache, Swelling, Backache, Cold, Earache, Heart Disorder, Fever

Reported By:—EDITH REES, *Recorder*

Deputy George J. Brown was in the chair. 20 Affiliates were present.

A moment of silent prayer was held for Dinshah and his family. Communications from the Principal Office were read by the Deputy.

Ida Mueller, a new Affiliate was introduced and given a hearty welcome.

The Deputy stressed the fact that we must maintain our Health in order to withstand the problems and changes that are ahead for us. With Spectro-Chrome we have the best means of doing so.

Nathalie Witt reported a case of a woman who had a Ruptured Varicose Vein and could hardly walk. With 2 Tonations a day and she goes around well now. Blue, Turquoise and Green were used. This woman's daughter had a badly swollen Face from Toothache and was given a Tonation of Blue and the next day she was able to have the Tooth extracted.

Both were pleased with the results they got.

Anne Wille reported swollen painful Joints. Used Turquoise and Magenta and later Yellow. Some of the Affiliates thought Indigo for Pain and Lemon Systemic and drinking more lemonade would have been better since the Disorder is Persistent.

Monica Mueller reported a severe Backache, so she used a Tonation of Yellow and the Pain left. Her boy had a Cold and Earache and Tonations of Turquoise removed the Disorder. She wondered if any Affiliate had experience with severe Pain after Tooth extraction. Indigo was suggested.

John Winchester reported how he was taken sick 2 months ago by a Heart condition. A week later, he developed a Temperature so he went to bed for 8 days. Used three Tonations of Turquoise and one Purple for seven days. On the seventh day the Temperature was gone, so he got up; continued with Lemon Systemic and Magenta Local and four more Tonations of Purple. Used Purple, for his Heart was pounding. This is the beauty of Spectro-Chrome, for I got results and am back to work and feeling fine.

Martha Linke called our attention to certain articles in the Spectro-Chrome so that we would be sure to read them as she found them very interesting.

EDITORIAL REMARKS

My cordial thanks are due to you, worthy Recorder, for taking the trouble to duplicate the above Report, the original of which was incinerated in the destruction of our Principal Office.

Meeting Of Thursday, January 4, 1945

Successful Use Of Spectro-Chrome In:—Puncture, Knee Disorder, Injury, Soreness, Swelling, Cold

Reported By:—EDITH REES, *Recorder*

Deputy George J. Brown was in the chair; 13 Affiliates were present.

Communications from the Principal Office were read. The Deputy then asked the Affiliates to stand reverently with bowed heads while he devoted a few minutes to prayer for the Colonel and his family. The news of the terrible, shocking catastrophe filled our eyes with tears. The consensus of opinion was that the Affiliates should help in a financial way and also in any other way that the Colonel may suggest or request that he can continue his wonderful work and services for Suffering Humanity. The Deputy stressed the fact that according to instructions in the telegram Contributions in any amount will be accepted and must be sent as individuals to Spectro-Chromarium Fund. The Affiliates promised to bring their Equipment and Encyclopaedia numbers to the next Planet meeting so that the Recorder can send them for the Office Records. We shall endeavor to carry out any request that is suggested to help the Colonel.

Case Reports then followed:—

Nathalie Witt reported on getting a Varicose Vein punctured by a sliver from a basket. Tonations of Indigo stopped the Bleeding and the wound healed without a Scar.

Anne Wille reported having trouble with her Knee. There was much improvement after two Tonations each of Magenta and Turquoise. Tonations of Magenta and Lemon made the Knee 100% again. She was much pleased.

The Deputy reported on how he got his trouser cuffs and two Fingers caught in the power agitator in his furnace room and was held there until some one came to his rescue and released him. One Finger was flattened out and the Skin removed. Three Tonations of Turquoise cleared it and removed the Soreness. His Feet and Legs were badly bruised but persistent Tonations of Turquoise prevented Pain and took care of the terrible Swelling and Soreness and the healing process. He was well pleased with the results he got from Spectro-Chrome. Lemon Tonations will be used to make him 100% again. He said that he would not have been able to have come to this Meeting if it had not been for Spectro-Chrome and the fact that Emil Klauss brought him in. The Affiliates were pleased and thankful that their Deputy made the big effort to come on crutches with bandaged Feet and Legs even on such a cold night. This was surely a case of loyalty and devotion.

Fred Osborne reported that Turquoise first and then Lemon removed his Cold.

The Affiliates present sent as a unit a message of sympathy to Dinshah and his family.

Our Dear Colonel Dinshah:

Your telegram bearing the sad tidings came as a terrible shock. It is with heavy hearts and tears blinding our eyes that we learnt of the great catastrophe, loss and sorrow that has come into your life and heart and home. A lifetime's work vanishing in so short a time and parting with so many hard-earned efforts when all odds are against one, needs a flood of courage, power and wisdom for endurance. There are no words that can possibly express our sympathy. We each one wish to extend to you and your family our sincere and deep sympathy. May the heartfelt sympathy that this brings to you help to sustain and comfort you all. Our hearts are with you in this hour of trial. May God sustain and comfort you and guard, guide and direct you. May our Heavenly Father cloathe you in Infinite Wisdom in this hour of your great loss. May your Mental Garden grow abundant with newer and better inventions for the suffering of Mankind. Our prayers are that come what may, Spectro-Chrome will live as long as the Sun and Moon will last and as long as they do not change their Chemistry. We in the Cleveland Planet intend to carry on as worthy Affiliates with our loyalty, devotion and sincerity to our "greatest teacher."

EDITORIAL REMARKS

Words are insufficient to express my gratitude for the deep sympathy conveyed in your message. You will find your humble servant always ready to help you to help yourselves with the Healing Science of Spectro-Chrome. I thank you, beloved friends.

Meeting Of Thursday, February 1, 1945

Successful Use Of Spectro-Chrome In:—Cold, Pain, Frost-Bite

Reported By:—EDITH REES, *Recorder*

Deputy George J. Brown was in the chair; 17 Affiliates were present in spite of the brown-out and serious coal shortage. Spectro-Chrome Magazines were distributed.

Communications and the Authentic Report of the disastrous Fire were read to the Affiliates followed by a prayer by the Deputy for the Colonel and his family. Affidavit forms were distributed and the Affiliates were asked to fill them and send them in as soon as possible.

Case Reports followed:—

Nathalie Witt reported reading a recent article in the Cleveland Press about Infantile Paralysis and how Light was going to be used in the near future in the treatment of that Disorder.

Hedwig Rittershofer reported having had a Cold and that Tonations of Turquoise and Green first and later Lemon cleared away the Disorder. Her daughter had Pain in a Toe and Indigo gave relief.

Fred Osborne had frost-bitten Fingers; Yellow relieved him.

Joseph Tarantino reported that a previous Persistent Disorder returned, so he used Lemon Systemic and Magenta on Areas 4 and 18. He and his family have used Spectro-Chrome for six years and no member of his family has any serious ailments. When any of his children feels sick, they ask for a Tonation, for they have confidence in Spectro-Chrome and know that it will help them.

George Brown reported on the remarkable improvement in his case.

Contributions from the Affiliates are still coming in for the Spectro-Chromarium Fund and are being sent in as quickly as possible.

The Affiliates were glad that they braved the cold weather for the Meeting was enthusiastic.

FIRST DETROIT MICHIGAN PLANET

(CONCORDIA HALL, 62 EAST COLUMBIA)

Meeting Of Thursday, January 4, 1945

Successful Use Of Spectro-Chrome In: — Scarlet Fever, Headache, Cold, Palpitation, Toothache, Stomach Disorder, Constipation, Fatigue, Injury, Numbness, Stiffness

Reported By:—CARMELA FORZANO, *Recorder*

Deputy Charles J. Renn was in the chair; 70 Affiliates were present. Spectro-Chrome magazines were distributed.

Communication from Principal Office and the Telegram were read by the Deputy. He then stated: "Dinshah will not let us down; therefore it behooves every Affiliate to come to the front at this time."

Case Reports followed:—

Elfriede K. reported her daughter suffering from Scarlet Fever. Atcowa Turquoise and Magenta on Areas 4 and 18 for one week, later Lemon and Yellow. She was surprised at the result. The little girl made the remark that the first thing she intends doing when she grows up is to have a Spectro-Chrome for her family.

Rose B. reported her children were formerly afflicted with Tonsilitis and Sore Throat. Through the aid of Spectro-Chrome have not suffered from these Disorders.

Michael S. reported suffering from a severe Headache. Tonated with Lemon and the Headache was gone before the completion of the Tonation period. Some time afterward, a Blood Clot was removed from Left Nostril. Magenta was also used.

An Affiliate asked the question, "On what Areas the Magenta should be used?" The Deputy answered that Magenta, Purple and Scarlet should be used locally, on Areas 4 or 18.

Charles G. reported he used Turquoise and Magenta for Cold and condition normalated.

Laura K. reported run down condition due to worry which caused her to lose weight. Deputy Charles Renn itisized her. Feels very much better now. Also reported her husband had an Itisation by the Deputy. The husband has followed Spectro-Chrome religiously and his Kardoscil and Spiroscil are improved and he is much better.

Frances H. has owned a graduate Spectro-Chrome for 14 years, originally gotten to correct Palpitation of the Heart. She has had great improvement, but not complete Normalation. Her daughter constantly chides her about staying with Spectro-Chrome and constantly wishes her to return to Medicine, which she has no intention of doing. She has overworked herself and has become run down. She will never cease to sing Spectro-Chrome praises.

Karl R. suffered an attack of Toothache. Two Tonations of Turquoise made the correction.

Mabel H. reported Cold settled in Lower Back. Tonated with Indigo. Had very good results.

Lyla F. reported daughter suffering from Mumps. Atcowa used Green and Indigo. Was away from school only one week.

Grace S. reported using Orange, Turquoise and Magenta for a Stomach Disorder. Her child also had a similar condition. Atcowa Orange, Magenta and Turquoise corrected condition.

Elsie B. reported her elderly brother-in-law got a Spectro-Chrome. He always suffered from Constipation which was corrected by the Spectro-Chrome.

An Affiliate reported Lemon Tonations relieved Fatigue.

The Deputy stated that many men at his place of work, complain of tired Feet from standing on the floors. He too had suffered from a similar condition which was corrected with 3 Lemon Tonations.

Harriet M. stated she had a severe Injury to her Hand. The accident occurred where she was employed and she had to have it treated there. When the Hand healed, it was all numb and stiff. With two Lemon Tonations the Hand was normal again.

An enlightening Educational Program was then conducted from Articles published in prior issues of Spectro-Chrome, in regard to our Equipment Slides, Chemistry, Attributes and the like. Interesting discussion followed.

Meeting Of Thursday, February 1, 1945

Successful Use Of Spectro-Chrome In:—Scarlet Fever, Ringworm, Rheumatism, Lung Disorder, Scalp Irritation, Cold, Stomach Disorder, Headache

Reported By:—CARMELA FORZANO, *Recorder*

Deputy Charles J. Renn was in the chair; 77 Affiliates were present. Case Reports followed:—

Elfriede K. reported her daughter had Scarlet Fever and Ringworm. Atcowas Turquoise and Magenta produced marvelous results.

Addie S. reported that she had been hospitalized and had gone through a number of Disorders, Rheumatism, Lung Infection—in general she was completely upset, when she was referred to Spectro-Chrome. She had wonderful results. She has almost completely overcome the Lung Infection; although still going slow, feels she accomplished a great deal in a short period. She would not be without Spectro-Chrome. Her daughter suffered with a Sore Throat; Atcowas Turquoise and Magenta set her right.

Lucille O. reported child had a Scalp Irritation. Bathed scalp with a "germicidal soap," which did not correct the condition. Called Jessie Ness and was advised to tonate with Blue, Lemon and Indigo; eruptions on Scalp are now drying and she is doing very well. Her husband was suffering from a Breakdown affecting the Stomach and Head. Atcowa Green on Head, Magenta on 4. Unable to return to work yet, continuing to use Green and Magenta. The Deputy recommended getting Free Guidance from Principal Office.

Grace S. reported having trouble with M. who had a bad Cold for three days, impossible to take Kardoscil and Spiroscil. Atcowa, Green and Lemon and Orange Tonation helped breathing condition.

Katharina S. reported husband had Stomach Disorder. Atcowa Turquoise was irradiated and was able to return to work next day. His coworkers were out of work two weeks with a similar condition.

Charles G. had a bad Headache and could not eat from the effect of gas dust at his work. Tonated with Turquoise.

Elizabeth S. reported daughter had a Stomach Disorder; two Tonations of Orange corrected condition.

The Deputy stated that Dinshah's new book "Triumph Of Spectro-Chrome" may now be gotten from the Principal Office, for a Contribution of $3.00

An Affiliate asked if Slides could be sent in for repair. Reply was in the affirmative. The Deputy reported that two weeks ago Pure Food And Drug Agent Hauser was out to see him.

An Educational Program on the value of Atcowa Yellow was conducted.

FIRST ST. LOUIS MISSOURI PLANET

(HOTEL MELBOURNE, GRAND AVENUE AND LINDELL BOULEVARD)

Meeting Of Thursday, December 7, 1944

Successful Use Of Spectro-Chrome In:—Sarcoma, Constipation, Nervousness, Swelling, Inflammation, Pain, Stomach Disorder, Cold, Ear Noises, Belching, Pus, Hemorrhoids, Boils, Injury, Accident, Sprain

Reported By:—VIOLA C. JEUDE, *Recorder*

Deputy Clemens M. P. Neumann was in the chair.

Arthur Behnen—I like to relate the Case of my 11 year old son. October 1943 he was operated for Appendicitis; after the Operation, a lump was noticed on Leg between the Hip and Knee; it was hard and he could not walk. The Doctor said he thought it was a Blood Clot and that it may be necessary to operate. Child was then taken to the family Doctor; he did not know what it was, but advised Operation. Went to see another Doctor and he was at a loss as to what the condition was, so after taking the child to 3 Doctors it was decided to take him to the Hospital where a piece of tissue was examined and the condition was diagnosed as Sarcoma (a Tumor). They wanted to remove the Leg. I did not want this to happen to the boy as he was only 11 years old, but the Surgeon said with Sarcoma that was the only thing to do; so a Specialist was called and he said: "Your boy may be cured with X-Rays," so a course of 12 X-Ray treatments was used on Leg, with no results. Then 14 more were given. Then they suggested the boy should rest for 3 or 4 weeks and the week before Easter, they again wanted to take the Leg off. It was at this time that I visited Clemens Neumann who said he could not guarantee anything, but there was everything to gain and nothing to lose. The Doctors had said the boy had only 1 year—the longest 2 years—to live, so we started Tonations, one, sometimes two a day. At the present time the Leg is back to normal and looks like everything will clear. (The Leg was very badly swollen.)

Richard Jeude asked the Deputy whether Arthur Behnen be able to get signed statements from the Medical Doctors.

Arthur Behnen—What chance do you think I should have to get those signed statements? They are supposed to be good Doctors and they do not know the Theory of Spectro-Chrome. It was only by the grace of God that I heard about Spectro-Chrome and if the boy survives the test by next year—he should be dead according to the Doctors' statements.

Charles Stiermann asked if the flesh had resumed its flexibility?

Answer was "Very nearly all the flexibility." Arthur Behnen then left.

Deputy—The Cut on the boy's Leg is at least 14 or 15 inches long; I am very grateful that the father took time out to testify, as these Testimonies have more weight coming from the individuals instead of coming from me. Yet they say "Spectro-Chrome is a Fraud! They want to wipe Spectro-Chrome off the map!"

Marie Jung—Is the boy able to walk?

Deputy—He is a Pin Boy in St. Anthony's Bowling Alley and only occasionally when he kneels down it feels sore, but not all the time. It is a little hard to the touch. He used to limp very badly, but, now you have to know the condition and only when you observe very closely you may see that he uses that Leg differently.

Discussion then began on "Cancer" and the "X-Rays." Affiliates were highly interested.

Rational Food Of Man was discussed.

The Deputy reported the following Cases:—

1. Jean M., age 34, female, Kardoscil 78, Spiroscil 18; very nervous, constipated, Swelling in joints, Left Foot big Toe very swollen, Pains in Hip during damp weather and Winter. Yellow on Areas 6-7-8-9-10 and Systemic, Magenta on 4-18 used. Feels fine and is well pleased.

2. Edward K., age 15, male; Inflammation in Left Arm, accompanied with Pains, Swelling very considerable, unable to move Arm at Elbow. Atcowas used Orange on Areas 6-7-8-9-10 and 14, Magenta on Areas 4-18; large amount of Pus drained; normalated within one week, no Scar left.

3. Rose W., female, age 52, Ratio 78-18, Severe Gas attacks, cold feet. Experiences Hot Flashes and feels hot inside; has cold Perspiration. Atcowas used, Yellow on Areas 6-7-8-9-10, Yellow Systemic Back, Magenta on Areas 4-18; Turquoise Systemic. She now feels well and wants to become an Educational Student.

4. Ronald M., age 3, male; has common Cold, Sneezing, constant discharge from Nose, hard breathing. Atcowas used Turquoise Systemic, normalated in a few Tonations.

5. Frank S., age 63, male. Ratio 68-12; humming in the Ears for about 10 years standing. Said he: "I consulted many Doctors and even went to European Specialist, but none could help me. When busy I am very nervous, also very restless at nights, cannot sleep well, have to resort to Tea and Beer to make me sleep. My right Ear Drum is punctured and I cannot hear so well; my Chest feels tight especially when I am going up steps. My right Hip and my Back hurt me especially when lifting, when my Heart flutters, also my hand shakes. I am belching a great deal." Atcowa used, Yellow on Areas 6-7-8-9-10; also Systemic Back, Orange and Red Systemic Front. Magenta 4-18. Now tells me, "I feel as strong as a black bear, sleep well, humming in Ears is as good as gone, no tightness in Chest, no Belching, no Heart flutter, I am feeling just fine."

6. Ida L., female, age 70; constant Cough, day and night could not rest; had also Cold, felt miserable, weak, no pep, also constipated. Attuned Color Waves used, Yellow Systemic Front, later Turquoise Systemic Front and Back, only a few Tonations. Cough and Cold eliminated, can rest, bowels move and feels in general fine.

7. Evelyn B., female, age 37. Ratio 101-19. Came home from Hospital with 10 pound baby girl; was stitched and one stitch was infected and Pus drained. There was some Fever and was very constipated, considerable Pains in the part where stitched; she had also protruding Hemorrhoids causing constant burning and severe Pains. Medical Doctor could not do anything for these conditions except give Aspirin and Salve for protruding Hemorrhoids. She said, "I had no relief. Must take Milk Of Magnesia to move Bowels." Atcowas used, Yellow Areas 6-7-8-9-10-11, Turquoise Areas 11-19, Scarlet on Area 11. Completely normalated within 8 days; feels fine. Baby developed Boils on Body. Yellow opened and drained them and is well. The mother is well pleased.

Mollie Wehrle substantiated the last Report, saying that under Neumann's guidance she helped to take care of these parties.

8. William P., male, age 39; Face and Cheeks very swollen and inflamed; Turquoise and Violet used. The results are so pleasing that he is enrolling as a Benefit Student in 1945.

Robert Boschert substantiated this Case Report. He brought this man down to the Deputy, after the man had a Tooth removed. The man could hardly speak, the left Eye was almost closed but under the first Tonation, it opened.

All the 8 Reports were by Deputy Clemens M. P. Neumann.

9. The Sub-Deputy reported the Case of Edward H., age 36; smashed Foot with steel casting; he suffered intense Pain. After 45 minutes of Spectro-Chrome got up and undressed. In 3 days he went to work.

The Deputy said a man was bowling and the ball fell on this individual's Foot. He came for a Tonation and after Tonation went to a Medical Doctor. The Doctor could not understand why the Toes were not black and blue!

The Sub-Deputy said the case reported by her could be substantiated by an Affiliate present.

Rowena H. said it was her husband, that a 200 pound weight had fallen on his Foot and that both of them are thrilled with the results obtained with Spectro-Chrome.

Robert Boschert said a neighbor woman came and partook of some of their meals after she began to use Spectro-Chrome. The Doctor had told her she had Gall Stones, suffered from Diabetes and had Heart and Stomach Disorders. She was doing just fine, but after she went back to work she again began to eat everything and her husband said "Spectro-Chrome should be thrown in the alley." Some of the Affiliates remarked they hoped they would be in the alley when the Spectro-Chrome was thrown there.

10. The Sub-Deputy reported about Lailoni P., 3½ year girl. They were called to come and see this youngster; she had a Ratio of 112 to 32. One Tonation and the next day the child was playing on the floor! Atcowas used Turquoise, Yellow and Magenta 4-18; she suffered from Bronchial Fever, which was quite high. The Medical Doctor was just leaving as we came in. He had pronounced it Pneumonia, as the Breathing came in gasps.

A new Affiliate Margaret A. Frede was then introduced and welcomed.

11. The Sub-Deputy reported Harry W., age 50; wrenched Ankle from jumping off a truck; the party came in about midnight and the next morning drove away in own truck.

The Educational Program for the evening was "Spectro-Chrome In Recent Disorders." Helen Barco gave the reading which was warmly appreciated.

Rubor, Calor, Tumor, Dolor were discussed. Ratio principle was considered. The Meeting was truly instructive.

EDITORIAL REMARKS

The Editor sends you best thanks and kindest regards for sending Duplicates of the November and December 1944 Reports destroyed in the recent Conflagration. Your Planet is becoming more spirited and active, which is a good sign.

FIRST ST. LOUIS MISSOURI PLANET

(HOTEL MELBOURNE, GRAND AVENUE AT LINDELL BOULEVARD)

Meeting Of Thursday, January 4, 1945

Successful Use Of Spectro-Chrome In:—Heart Disorder, Toothache, Injury, Pain

Reported By:—VIOLA C. JEUDE, *Recorder*

Deputy Clemens M. P. Neumann was in the chair; 73 Affiliates were present. The Report of the Conflagration was discussed. Said the Deputy:

"As Colonel says the loss is irrevocable, because the contents cannot be replaced; in just one instance, I believe Colonel could have lectured for 6 months every day and shown different slides on the screen all day, these slides cannot be replaced." He said he heard from an Affiliate that it was broadcast over the Radio about a terrible fire that was going on in Malaga and that all the Fire Apparati from surrounding communities were at the scene; nothing could be done much to control the fire proper; it was such a windy day the Fire Apparati were there to protect the surrounding buildings; of course the party did not know that it was Dinshah's place. Suggestions were made that the Auxiliary Benefits of $15.00 each be turned over to the Spectro-Chromarium Fund, for a period of one year. Favorable opinions were expressed.

Case Reports followed:—

1. Leopold B. said he had a bad case of Heart Disorder; could not work for a long time and if it were not for Spectro-Chrome would not be able to work now and Medical Doctors cannot believe that he is working with his Heart as bad as it was.

2. William P. said he had a Tooth pulled and developed Blood Poison; so sore you could not touch his Face, and the Face was badly swollen. Robert Boschert suggested Spectro-Chrome. He said at the time he thought it was a joke—just another fad—but, Boschert took him to the Deputy and after 1 Tonation he was rid of the Pain and then rented an Equipment.

3. Robert B. said his 9 year old grand-daughter fell on the ice and had terrible Pain at the end of the Spine; had to lie on back, could not raise feet; she would have to slide to edge of bed and take hand and lift the other leg. He said he called Eleanor Neumann and she told him to use Orange; the child cried and suffered terribly for about 30 minutes, but, he said he just talked to her and told her to continue taking it a little longer, which she did and 15 minutes after Tonation the Pain went away. The next day gave 2 Turquoise Tonations and was just fine.

Numerous pledges were promptly given for the rebuilding of the demolished property. A plan was submitted by the Commissioners for circularizing the Affiliates to collect funds, but the Council Of Trustees refused to approve of it, saying "all Contributions must be unsolicited and voluntary." All such and similar pleas were turned down with thanks from the Principal Office.

Meeting Of Thursday, February 1, 1945

Successful Use Of Spectro-Chrome In:—Constipation, Infection, Sore, Throat Disorders, Injury, Hemorrhoids, Swelling, Cough, Vomiting, Pain, Itch, Tumor, Skin Disorder, Stomach Disorder

Reported By:—VIOLA C. JEUDE, *Recorder*

Deputy Clemens M. P. Neumann was in the chair; 56 Affiliates were present.

Communication from the Principal Office was read. The matter of Affidavits was discussed.

The Deputy had received his "Triumph Of Spectro-Chrome" and said he was up until the wee hours this morning and those Affiliates who have Spectro-Chrome Metry Encyclopaedia ought to get the new book as it answers all questions in regard to Spectro-Chrome from the standpoint of eminent Medical Doctors; it is worth the $3.00 and much more.

Anna Schindler said her friend Anna Strumberger told her the U. S. Food And Drug Inspectors were at her house.

Marie Jung said that she had been investigated and they asked her for what Spectro-Chrome was used in her home; she told them that with her experience with Spectro-Chrome, she could write a book; how she started with Spectro-Chrome that her baby had gained only 2 ounces in 5 months, that he weighed 7 pounds 3 ounces when he was born and 5 months later only weighed 7 pounds 5 ounces; she said she showed the Inspectors the pictures of this child 2 weeks after they had started with Spectro-Chrome. That while she was doctoring with him that they had constantly given him Enemas to move his Bowels, he would throw up if they gave him Medicine, but the third day after using Spectro-Chrome, the Bowels moved naturally and have moved naturally ever since. That 2½ months after they were using Spectro-Chrome, her mother had an Infection; she had 2 Sores 1 inch deep and as large as half a dollar on the right Leg and 1 Sore on the left Leg; it was spreading so fast that the Medical Doctor said that if this kept on he would have to take her to the Hospital the next day. So one of her sisters who also had Spectro-Chrome asked her if she would help her to use Spectro-Chrome; so between them, they took care of the mother under the Deputy's directions and gave her 5 Tonations and when the Medical Doctor came the next day, he was amazed and said they had done a wonderful job of nursing, because the mother would not have to go to the Hospital! She also told them that her husband (as long as she knew him) suffered with Sore Throat and would have to have it lanced 3 or 4 times each time he had it. When they tried Spectro-Chrome, he was normalated in seven Tonations; 2 years later, he got it again and was normalated in five Tonations and has not had it for the past four years. She also told them about her experience with tonating the children for Measles, Mumps, Burns and various other ailments. The Inspectors said these were no serious conditions that the children could have gotten over them without anything. She said yes that was true, but she compared her children's conditions with other children and the length of time it took other children to get over it and her children were well in less than a third and half the time it took for the others. They were there more than 2 hours and when they left they said she surely had enough exeprience with Spectro-Chrome; she said she did not know what she would do without it.

Frances Seeger said Saturday they came and asked her if they could see her about Spectro-Chrome and what results she had with Spectro-Chrome. She told them. They also asked her about the food; she said the Doctors give diets too in many instances; they say no sweets, no pies and cake then in other conditions no meat and proteins. They finally said they would take her word for it and left.

Elsie Beisswanger said they also came to see her; they were very polite and asked about the "bad side" of Spectro-Chrome; she said she was satisfied; they did not ask to see the Spectro-Chrome and went away.

Richard Fred Jeude reported that two Inspectors called and asked for Mrs. Jeude; as she was baking in the kitchen and felt she did not want to be disturbed and not knowing who the men were and what they wanted, he told them she was not in. They then said they came in the interest of Spectro-Chrome, could they come in; he said he was very busy working in the basement and did not have the time; they said they would come down in the basement with him. He said no you won't, because I have work to do that must be done and I cannot talk and work at the same time, but that if he could furnish them with any information in regards to Spectro-Chrome that he would give them a few minutes.

They then asked what Spectro-Chrome had done for him or why he was using it.

He then related the experience with a finger he had cut and the difference with a like condition to another that had been cut the year before he had Spectro-Chrome. That he had fallen down 21 marble steps and the injury that was done to his arm, how black and blue it was and that a lump as large as a goose egg had formed and that in one Tonation the blueness disappeared and the lump also. They asked for what Mrs. Jeude was using Spectro-Chrome. He told them Constipation; that she had resorted to laxatives for 26 years but was normalated with Spectro-Chrome. They asked me if that was all the experience they had with Spectro-Chrome. He said she may have had more but he would let her speak for herself. They asked if he had any other interests beside just using Spectro-Chrome in the Organization. He said no, but Viola Jeude was the Recorder. Then they asked if he would furnish them with some names and addresses. He answered in the negative.

They said they understood he was a Druggist for over 35 years and he used to keep records. They asked why after being a Druggist so long, that he was using Spectro-Chrome. He said that long before he was using Spectro-Chrome, he did not believe in Medicine and when he would see what and how some Medical Doctors wrote and prescribed he was off Medicine long before he knew about Spectro-Chrome. They asked if he did not believe in Doctors. He said they have their place; one surely would need them if one broke a Bone or severed an Artery. They asked what my Druggist friends thought about his using Spectro-Chrome or was he keeping it a secret? He answered them that he did not care what his Druggist friends thought about him and that he was not keeping Spectro-Chrome a secret; in fact he was telling all with whom he came in contact. The Inspectors wanted to take him to a certain So-and-So who would make him talk. Richard Jeude told them he would go nowhere without a Court Summons.

They asked him if he would give any names of persons who died under Spectro-Chrome from Cancer. He said he did not know any that had died under Spectro-Chrome from Cancer.

They asked him if he would not like to help and work for his Government. He said: "Surely in what way?" They said he would be helping in the Fraud against the people. He said there was no fraud against the people and that Spectro-Chrome surely did the work if there were full co:operation from the people.

The Deputy then reported:—

1. Ida M., female, age 53, Ratio 78-15 suffered from External Hemorrhoids the size of a date and a cherry, for the past year. Could not perform her usual household duties, could not even sit down. Used many Cathartics, Enemas without desired results. Used Yellow and Orange, Purple and Magenta; normalated in a short time and was very much pleased.

2. Alex R., male, age 67, Ratio 64 to 18. In left side under ribs a terrible feeling, Pains in the middle Back and in Stomach region; such Pains prevented me from stooping; Knees started to swell also Swelling on top of both Big Toes. He had Frontelim 19 times in one day. Atcowas used were Yellow Local and Systemic, Orange Systemic, Normalated within one week; more than pleased.

3. Sandra P., female, age 5. Severe Cough starting in September for the past three years; it occurred in spells lasting at times all night and no rest for the sufferer and parents. No phlegm was expelled, but Cough caused Vomiting. Consulted Medical Doctors every year. This year more severe than ever. Parents were frantic about it, diagnosed as Sinus condition; Nose drops, Cough Medicine used but no results. Ratio 140 to 24. Very fond of Meat. Atcowas used, Yellow and Orange Systemic Front and Back, Turquoise Front and Back Systemic. After first Tonation of Yellow, the girl slept all night to the surprise of the highly disturbed parents. Normalated to the entire satisfaction of the family.

4. Mary W., female, age 64, Ratio 70-16. Arthritis in Knees and Fingers. Pains in right Arm and very severe Pains in lower Back, unable to straighten out. Terrible Itch in back especially in Summer; feels like being in fire. Feels always tired and worn out. Constipated for the past few years. Resorted to Laxatives. Subject to Headaches. used Aspirin, Vitamin tablets. Tumor on left lower Abdomen. Tonated with Yellow, Orange, Lemon, Green, Turquoise, Indigo, Magenta. Growth hardly noticeable; all other condition normalated; feels fine, very thankful.

5. Robert P., male, age 49, Ratio 74-17, for a period of years spots of Skin Inflammation on Legs apparently from contacts of grass or weeds. He said: "I also suffer from Hay Fever, which is more pronounced in the past years. Medical Doctor advised Allergic Tests and I received progressive "shots" for about nine months; in taking these shots for 2 months, a Skin Disorder appeared on my right Ankle growing larger in size. Skin Specialist advised to stop Allergic Tests. Now use lotions, salves, castor oil applications." Tonated with Yellow, Orange, Red, Green, Turquoise, Magenta and Scarlet. Responded very well and Skin Disorder disappeared, no scar, no itch; a fine new Skin reappeared. Very well satisfied with its results.

6. Kathryn T., female, age 42, Ratio 72-22. "My Stomach aching and I am belching, it comes in spells. I have Pains also in my belt region, and at times in my left Shoulder. Suffer from cold Feet and always feel better when in warm room. Used Nerve Medicine. Tonated only a few times with Orange, Yellow and Scarlet; Pains and Stomach normalated." Her employer was well satisfied.

Hildegard B., reported her own case. She was going to work the last Eclipse Period, running for the omnibus. Her rubber heel stuck to the cinders and she fell; she could hardly get on the bus without assistance; she managed to work in the night shift, but her Leg hurt very badly. This was on Saturday, about 12:15 A. M. and the next morning at 8, by the time she went home, she could not use the left Leg. Sunday was a off day and she did not have to work; she took 6 Tonations, using Indigo to relieve some of the Pain, Turquoise Systemic, Scarlet for the Heart and Turquoise Local and at 12:15 A. M. Monday morning she again reported for work.

Robert Boschert said he spoke to a Chiropractor who had a Spectro-Chrome, who called Dinshah a "money grabber"! Said he could serve 4 or 5 in an hour with Chiropractic!

The Deputy said he was sorry that there was no more time for the Educational Program. He had a very interesting article from the newspaper, where some Professor from Vienna just discovered that Light travels like a "Cork Screw." Wonderful! Said he: "I felt that I should write him and tell him that he is just about 25 years behind the time, and ask him to investigate Spectro-Chrome!"

FIRST NEW YORK NEW YORK PLANET

(HOTEL McALPIN, BROADWAY AT 34TH STREET)

Meeting Of Thursday, January 4, 1945

Successful Use Of Spectro-Chrome In:—Burn, Angina Pectoris, Trichinosis

Reported By:—EMMA C. HAITZ, *Recorder*

Deputy Lina Fischer was in the chair; 101 Affiliates were present. The Meeting was held in the Green Room. It was delightful to see such a large attendance. Twelve persons were denied admission because they had no 1945 Cards.

Deputy—Colonel Dinshah is not with us tonight. Some of you heard about what has happened. For those of you who did not, I shall read an article from the Philadelphia Inquirer. After the meeting you may come up here and look at this picture of the whole family, taken shortly after the Fire.

The article was then read by the Deputy; also the Telegram received from the Institute. All listened in understanding silence.

Deputy—My husband and I were down there yesterday. What we saw is unbelievable. Those of you who were down there before, know the Main Building. It is just leveled. Thousands of dollars worth of books, fifteen hundred Motors, Automatic Time Switches are all gone. There is nothing left. The family moved in to the Spectro-Chromarium. Since all their personal belongings were in the Main Building, they, too, were destroyed and all they could save was what they had on their body. That family is not like we are. They are satisfied with very little.

Dinshah now has no records, no Affidavits and nothing with which to go to Washington. Even his Pardon papers from Calvin Coolidge and

Franklin D. Roosevelt are gone. The damage is too great for me to conceive.

It is amazing how that family can keep up under those conditions. Colonel Dinshah still has his usual smile. You know, I can almost visualize a new building rising. It is for us to help. I would like to have your opinion as to how we can co:operate and show that we are 100 per cent. with Dinshah.

Show to him that you are with him 100 per cent.

A motion was made by Albertina Zimmerman and seconded by Dea DeMartin that a letter of sympathy and devotion be sent to Dinshah Spectro-Chrome Institute. The Deputy then read the letter which she had prepared.

Deputy—Do you all find this letter satisfactory?

Affiliates—Yes.

Deputy—On your way out, come to the desk and sign your name.

The Deputy then read from the November 1945 Spectro-Chrome the compliment paid us by the First Milwaukee Wisconsin Planet. Everyone seemed pleased.

Deputy—I think we should feel well about it. Of course the attendance is not what it was the last time. Try not to miss any Meeting if you can possibly help it.

The Deputy then read Chapter 56 on "Rubor, Calor, Tumor, Dolor." The Affiliates listened enthusiastically.

Marie Frefel had a Burn on the Hand. It was all blistered, but after two Tonations, it was completely healed.

New Affiliates were introduced and greeted by a hearty applause.

Deputy—I like to report the case of Henry V. He had Angina Pectoris. He had many Heart attacks, as many as 4 each week. He had visited 8 doctors who did not give him much hope. When I visited the family, I told them how serious his condition was. He gave up Smoking and adhered to the Rational Food Of Man, due to Kate, his very efficient wife. He had the Equipment for the first nine days and had only two attacks. Ever since he had no more attacks. Kate had to write to the Institute because, in order to get the benefit of their health insurance, they had to have the signature of a Physician. They wrote to the Institute and Dinshah granted permission to go to a Doctor as long as no Medicine was administered. She had showed me 10 kinds of Medicines, the last one being Nitro-Glycerin. He was in such a condition that there was no Backelim and had Pain in the Heart. Yellow brought on proper Elimination. Every time I went there he was always better. He went to the Medical Doctor the first time after using Spectro-Chrome for 2 weeks. His Blood Pressure had gone up from 108 to 124. The Doctor was puzzled by this change and asked what he ate. Henry told him he ate only "vegetables." When I first took his Heart Beats, they were 45 and then when he got an attack it went as low as 33. Now his Pulse Beat is between 70 and 75. It does not fluctuate. His ratio is 4 or 5 to 1. When he went to the Doctor, he said "here comes the wonder man." The doctor told him he was one out of a thousand who would live with such a condition! He took two Tonations a day; one Yellow on Areas 6, 7, 8 and 9 and one Magenta on Area 4. He is now taking Lemon in the morning and Magenta at night. It is a very interesting case and I am very proud that I was instrumental in helping. I was really more worried about Kate. She also has an Enlarged Heart. Since being on the Rational Food Of Man

she gained 30 pounds. Put that in your Chart and Dinshah will cut that right in half! Thanks to her, her husband is better. He, too, gained 11 pounds. (Loud applause.)

Deputy—Are there any other new Affiliates who have had good results?

Josephine T.—I once bought ½ pound of Metwurst through a neighbor and had divided it among 5 people. When I had to call the Doctor, he said I had Food Poisoning. Three of us landed in the Hospital. The Worms had settled in my Heart Muscles. Seven years after I had Trichinosis, I landed in the Hospital for nine weeks. The Doctors cut my Heart Muscle open and found the Worms embedded but asleep! Seven Doctors said they could not help me. For five weeks when I was in the Hospital in April, I was treated for Hemorrhages. I have a daughter who is a nurse and she said that if Spectro-Chrome could do that for M. it can do it for you. I do not feel 100 per cent. yet, but I have improved so much that I want to shout it from the house tops. I am such a good example that tomorrow I am going to enroll somebody. (Loud applause.)

Deputy—That was a very fine report.

Everyone went home very pleased as the Meeting was a successful one, as are all our Meetings.

EDITORIAL REMARKS

Thank you for the Affidavit signed by 101 Affiliates. Your efforts, loyalty and devotion are praiseworthy and compensate us beyond all our losses. What if we lost all our belongings! We have you—worth more than the treasures of the World!

Meeting Of Thursday, February 1, 1945

Successful Use Of Spectro-Chrome In:—Appendicitis, Uterine Prolapse, Heart Disorder, Pain

Reported By:—EMMA C. HAITZ, *Recorder*

The Meeting was held in the Red Room. Deputy Lina Fischer was in the chair; 101 Affiliates were present. The Deputy read from Dinshah's "Good Thoughts" on "Machine" and "Surgery."

Deputy—Brothers and sisters, think it over and absorb these two "Good Thoughts" and you will come to realize what Spectro-Chrome means and does for us.

The communication from our Dinshah was then read by the Deputy who explained as she proceeded. Those Affiliates who sent Enrollments shortly before the Fire, had already done as he requested. She then related to the Affiliates her experience of accompanying Dinshah to the Brooklyn Court for the Pre-Trial Hearing.

The Deputy then asked how many Affiliates would voluntarily testify in the Brooklyn Court in behalf of the merits of Spectro-Chrome. Thirty-five Affiliates volunteered.

Louise Witt, Frieda Diez, Elizabeth Turk and Lillian Peters all stated that they were recently visited by the Inspectors. Louise Witt also stated they were especially interested in her Favorscope. Her neighbor spoke with the Inspectors concerning the case of Louise Witt's husband, also saying that if she ever became sick, Louise Witt would help her.

Deputy—You will introduce her as a Benefit Student, but never let her use your Equipment. When people once hear of Spectro-Chrome, they will come to you when they are in need. That is the way I heard about it too. Hugo Witt had wonderful results, but, he did not stick to Spectro-Chrome.

Louise Witt—After he was normalated, he started drinking and smoking again and then he got another Stroke. I told the Inspector exactly what happened.

A new Affiliate Louise Burg was introduced to the Affiliates.

Case reports were called and Frank Cassano, a Spectro-Chrome Normalator, came to the platform. His Reports follow:—

1. Caroline S., age 15, was to be operated on for Chronic Appendicitis. Kardoscil 67, Spiroscil 18. Backelim 1 a day, Frontelim 4. After using Turquoise Area 10 and Magenta Areas 4 and 18 for 1 week the girl already felt better. For the next three weeks, Lemon Systemic Back and Front and Magenta Areas 4 and 18 were irradiated. Within 1 month's time parents are so well pleased that they are now enrolling.

2. Hilda M., age 37, had fallen Womb. Kardoscil 72, Spiroscil 14. Backelim 1 a day, Frontelim 7 a day. Indigo Areas 10 and 11, Lemon Systemic Back and Front and Magenta Areas 4 and 18 were used for 2 months. After 2 months, she was normal.

3. Helen L., age 45, had weak Kardoscil and Pain in Area 18. Kardoscil 64, Spiroscil 18. Backelim 1 in 2 days and Frontelim 3 a day. Green Systemic Back and Front and Magenta Areas 4 and 18 for 1 week, two a day. Kardoscil 80, Spiroscil 18. Backelim 1 a day. After 1 month Kardoscil was 78.

4. Marguerite G., age 66, suffered with Heart, Kidney and Liver Disorders. Kardoscil 76, Spiroscil 22. Backelim 1 in 3 days, Frontelim 7. The Ratio was 3 to 1 which was bad. Started tonating with Magenta Areas 4 and 18 and Yellow Systemic 8, 9, 10. She kept using Heart Stimulant, so Spectro-Chrome was taken away. A month later the woman passed away.

5. Mary F., age 70, suffered from Low Blood Pressure. Kardoscil 54, Spiroscil 14, Backelim 1, Frontelim 5. Used Scarlet Areas 4, 18 and Lemon Systemic Back and Front 1 week. Kardoscil 72, Spiroscil 18, Backelim 1, Frontelim 5. Continued to improve; however, she constantly had to be warned not to eat certain foods, but would not listen. Ate Noodles, Flesh and Bologny and took Stimulants. I gave up the case. She went to a Hospital and died after 3 months.

Deputy—Colonel Dinshah is very happy over the enthusiasm of this Planet. He was very pleased with the manner in which the last Meeting was carried on.

By permission of the Institute Carl Fischer then showed Motion Pictures of the Fire which had been promised the Affiliates. The Deputy explained as it proceeded and pointed out the various Buildings. Upon requests by many of the Affiliates, the pictures were shown a second time. Everyone enjoyed the pictures as they could see at first hand the terrible destruction caused by the fire.

All went home satisfied and much interest was shown.

EDITORIAL REMARKS

Spectro-Chrome does NOT claim to keep people alive for ever. All it claims is that it can do all that Medicine can do, WITHOUT the sequelae. If Spectro-Chrome, intelligently irradiated, can not give the desired results, the case has come to a point where it is beyond human reach. We all have to die some day—even the Originator of Spectro-Chrome has to follow the same road, but, the proper use of Spectro-Chrome and the strict following of its Technique and Rules, will produce a Life full of Health and Happiness, until the human machine itself wears out in accordance with the Laws of Nature.

If any person will not follow the Technique and Rules of Spectro-Chrome, do not permit the use of the apparatus. Those strict Rules are vital to the effectiveness of the Science and one must either follow them or have one's own way and stay sick or die—surely that is their personal privilege.

———o———

FIRST MILWAUKEE WISCONSIN PLANET

(MILWAUKEE AUDITORIUM, WEST KILBOURNE AND NORTH FIFTH)

Meeting Of Thursday, January 4, 1945

Successful Use Of Spectro-Chrome In:—Injury, Swelling, Pain, Diabetes Mellitus, Gangrene, Fracture

A CASE OF WRONG MEDICAL DIAGNOSIS

THE EVIL OF INTOXICATING LIQUORS

Reported By:—EMMA SCHROEDER, *Recorder.*

The year 1944 closed upon a world torn with a War so destructive, so cruel, as to beggar description; so savage that it is difficult for the human mind to conceive where civilization ends and barbarism begins (or vice versa). The need for Truth is more apparent than ever and where may greater Teachings of Truth be found than in the Dinshah Spectro-Chrome Organization?

The New Year 1945, which just commenced, already proved fateful to our valuable Institute, when, on the first business day of the year (January 2d) Fire obliterated those precious belongings of our President Dinshah, most of which may never be replaced.

The telegram containing the news of the unfortunte disaster arrived at the Recorder's home about 9:00 A. M. Wednesday, January 3d. She immediately notified the Planet Deputy as well as the other past and present Planet Commissioners and all Affiliates whose names she retained in memory, many of whom, in turn, notified others immediately in this emergency. Some were stunned into silence; others, who are deeply attached to our Cause, became hysterical and still others cried openly at the Telephone. All sympathized with Colonel Dinshah for losing his valuable possessions, but admired the spirit and courage he possessed and expressed in his desire to make a new start "In Glory Of God." Some believed that this misfortune would lead to greater achievements. When conditions seem to be at their worst, the turning point is usually near. "It is always darkest before dawn." All were grateful that lives were spared.

Most Affiliates entertained the suspicion that this fire was of incendiary origin. They thought it strange that it should occur in the same month in which Colonel Dinshah's important Trials were to take place. The possibility of a defective Heating Plant was considered; according to the newspapers there was a severe storm with high winds in that vicinity on the day in question. One Affiliate asked "Were there any chemicals in the building"? Another "Was chemical experimenting done there?" All are anxious to hear the details and origin, if that may be determined. All expressed surprise when they heard "No Insurance."

Deputy Frank S. Burns was in the chair; 109 Affiliates were present. The communication from Dinshah, dated December 29, 1944 was read by the Deputy.

Deputy—The Recorder has a Telegram. Coming at this time, I think it is the most discouraging situation and trust you will realize the seriousness of the predicament the Institute is in and do what you can to alleviate the distressing circumstances. In the letter of the 29th we read of the final determination of some of Colonel Dinshah's vital Court issues about to take place this month and now this telegram.

Fredrick Kneppreth—The contributions are starting out wonderfully, I think that Colonel Dinshah is rather "up against it." WE are the only ones he can turn to with this dark cloud overhanging.

Deputy—Any Affiliate who feels he can contribute, it will certainly be appreciated. Dinshah is broad-minded and knows we are not all in a position to do so. Many of us have been "Bled by the Medical Doctors" as he says.

Emily Mokelke, who feels she owes her life to Spectro-Chrome and is very grateful, spoke inspiringly—We really owe all to Dinshah, especially those of us who have been terribly ill. Years ago, we had Doctor after Doctor. Everything I have today I owe to Colonel Dinshah, for, if I did not have my Health, nothing else would count and there is a number of Affiliates who feel the same way as I do. I think we should give as much as we can. In gratitude we should do the best we can. We should really make a few more sacrifices. (Applause.)

Fredrick Kneppreth agreed with Emily Mokelke and spoke of this being of a constructive nature.

Robert Schott—Another way in which we could help, I think, would be to avail ourselves of some of Colonel Dinshah's books.

Norma Fiedler invited Colonel Dinshah to build in a more central location, say Milwaukee or Chicago.

Fredrick Kneppreth—The contributions are starting out wonderfully, but, do not let that stop here! Let us show that we are behind the Institute for all that Dinshah has done for us.

Case Reports followed:—

Sarah P. Erdmann—Five weeks ago I fell and badly hurt my right Knee; left Knee and Hand were also injured. Had the Medical Doctor examine it for Fracture. He said it was badly bruised. I irradiated Yellow twice, then Indigo. By night it was so swollen I was confined in bed for the next three days. I could just limp around. I continued Indigo twice and then Orange. It became numb. Was advised to alternate Orange and Indigo with Magenta on Areas 4 and 18. Spiroscil was 21, Kardoscil 76. Now I am able to bend my Knee. When I go out evenings I use a crutch. The Medical Doctor said it is progressing wonderfully. I surely went through some suffering and do not know what I should have done had it not been for Spectro-Chrome.

Deputy—Everyone should gain by that experience.

Much discussion started about when to use Orange and when Indigo.

Emil Egide—We had a case in our family: My wife thought she would be Hercules and shovelled snow. She did the whole walk but could not get out of bed the following day. She did not know what the trouble was and imagined she had everything from "Lumbago" to all the "Itises" in our book. I looked over the condition and from my observation it appeared to be tenseness of the Muscles and I recalled Dinshah's statement that we are too anxious to relieve the pain with Indigo instead of going to the seat of the Disorder and taking care of the Circulatory System. She used Orange one night and arose from bed the next morning wondering where the Pain was. It proved very effective. It remained away one day and she did something which required those same Muscles and the Pain returned for a few days. I continued the use of Orange. She was required to report to the Company Doctor. If there is no Fever present and no sign of Redness or Swelling, then Orange is the Atcowa to use. If it were a tense Muscle Orange. It proved very effective. We do not use Spectro-Chrome very often in my family, but, when we cannot get out of bed we use it.

Alice Brown—A six-year-old child suddenly became very sick. The father became panicky. Neighbors sent for an ambulance and she was removed to the Hospital. Four Medical Doctors held a consultation and decided it was "Spinal Meningitis." The father "lost his head." He had a Spectro-Chrome Equipment. The child was released after four hours. Before the child left the Hospital, a Nurse said: "It is too bad. That child will be dead by tomorrow." Temperature 106. Spectro-Chrome was irradiated. I went to visit the child the following afternoon and found it playing around! The child recovered.

Deputy—Do you think it was Spinal Meningitis?

Alice Brown—No! That is what the Doctors had said. Must be their wrong "Diagnosis."

Deputy—I, too, doubt whether it was that. The symptoms may have been similar.

Fredrick Kneppreth—Spinal Meningitis is an Inflammation of the Membranes covering the Spinal Cord.

Past Sub-Deputy Kurt W. Bader—I would like to contribute something to this discussion. Is there anyone with a loud voice who would read this? (Nobody responded.) At the last Meeting the question of proper dress conducive to Health came up. I shall read you the article from Spectro-Chrome bound magazines, Volume I, June, 1922 to December, 1925, page 320, entitled "What to Wear" from the Evening Bulletin of Philadelphia, Pennsylvania about the "Wisdom of Keeping the Body at Uniform Temperature" by Garrett P. Serviss, the Astronomer-writer.

The article was read. (Everyone is recommended to read it and dress accordingly. The views of the Originator about proper dress did not change. They have been the same all his life.)

Kurt Bader concluded his reading by stating: "If you have nothing in your Stomach, you can freeze at 104 Degrees." He stressed the importance of reading Dinshah's books and publications and urged Affiliates to get them saying "Dinshah is not looking for contributions. It would be just as supporting (and knowledge would be increased) to get his books."

Norma Fiedler—Every year we get the same lecture about dressing warm. When my ancestors immigrated to this country, they settled in the backwoods. There were Red Indians in those days and they were scantily clad; possibly wore a few skins. One Indian, when questioned as to whether he did not suffer from the cold and exposure, asked the white man "Does your face get cold?" Upon receiving a negative reply, he said: "Well, me all face!"

The importance of keeping hands and feet warm was emphasized by Fredrick Kneppreth who also spoke against fur coats, saying: "It is not the fur but the heat in the Body. The object is to keep the Body Radiation from wasting. It is more sensible to purchase a $75.00 wool coat than to spend several hundred dollars on a fur coat. People who are on the path to enlightenment should always bear in mind the contribution to slaughter. The Commandment is:

"THOU SHALT NOT KILL." (Applause.)

Kurt Bader gave a sensible talk on the subject of dress.

The case of a woman, aged 65, came to the attention of the Deputy who related it. Said he: "About a year ago, her Foot became swollen and inflamed and began to drain between the Toes. The family became alarmed and called in a Medical Doctor who examined her and sent her to a Hospital, for 'Diabetic Gangrene.' The woman was at the Hospital for a week, 'under observation.' The family was not in a position to maintain her there, so she was returned home. The Medical Doctors said she would have to take Insulin daily and showed her how to make the Injections. Strange as it may seem, this family had a Spectro-Chrome Equipment and my attention was called to the case. I said: 'Forget about Insulin and the Doctors! Just use Spectro-Chrome; but USE IT! Don't have it in the house as a fixture.' She commenced using it conscientiously. The foot opened, drained for a while and finally cleared. She was unable to walk, but continued to use Spectro-Chrome and took the Tonations regularly. A Pain in the Ear developed and opened during sleep. In the course of time after using Yellow, a Core came out and the Disorder cleared and today she is in very good condition and goes around. She goes downtown alone on the street car and during the past year she had a check-up by the Medical Doctor as to Blood and Urine. His remarks were: 'Just keep on with the Insulin; I am surprised you are so wonderful! Your Blood Count is wonderful! No more Sugar! Just keep on with the Insulin'! He never examined her for Insulin Injection. She used Yellow, Magenta and some Indigo. Her general condition has improved remarkably."

Alice Brown—I should have shown him Spectro-Chrome.

Deputy—The house was divided against itself, but the Doubting Thomases had to admit that she improved.

Louise Schultz—She should have told the Medical Doctor.

Robert Schott—I had the same experience with Cataract. I was convinced that I did not have to be operated.

Gladys B. reported that her little boy broke both Wrists while away from Milwaukee. He was taken to a Medical Doctor before she could reach him. As soon as she was able, she removed him from the Medical Doctor's care. One Wrist is in a Cast, the other in a Splint. I gave him Magenta on Areas 4 and 18 and Lemon Systemic. The Medical Doctor gave him pills called "fish-eyes." They were Calcium. She put them away. Her sister is against Spectro-Chrome and she did not tell her why she was taking him home. She would have taken him home sooner but was unable. The Deputy advised Gladys to continue with Spectro-Chrome.

Informal discussion was invited by the Deputy:

Late-comers requested that the telegram from the Institute be reread and the Deputy complied.

The Deputy concluded the Meeting with a reminder on the peril connected with the use of Intoxicating Liquors. "There is much to be said. Some people regard the craving for Liquors as a Disorder. I believe this country or any other country for that matter, would be far better off if it would abolish Intoxicants entirely. Institutions are filled, the Houses of Correction, Reformatories, Insane Asylums, all are filled with Liquor addicts or as a result of it. It is a difficult problem. You know as well as I do that some years ago we had what was called 'Prohibition.' Anything but Prohibition existed. People who never indulged in Liquor before, began to make use of it. I do not believe people will ever get away from the idea until they realize the danger and harm it does to the Physical Body. You know what it does. It upsets the Mental Equilibrium. It takes a terrible toll. The fine membranous Tissues of the Brain are destroyed and can never be replaced; neither can the lining of the Stomach be replaced. How that may be corrected in an individual is a problem. The only way one can obtain results is by appealing to the person's sense of reason and understanding and showing him the injury that is done to the Body and the wrecks and derelicts that are confined to Institutions as a result of it. I have been an Abstainer for years and it would be against my reasoning and better judgment to use Liquor. I could not think of ever going back to it. Sometimes people wonder how I live because I abstain from these so-called 'pleasures.' I have a clear thinking capacity. I hope to retain that until I make my departure from the Physical Plane." (Applause.)

Robert Schott—The craving for Alcohol and Tobacco may be lost by using Spectro-Chrome.

Fredrick Kneppreth—Alcohol never brought prosperity to any country.

Norma Fiedler—Occult books deal with this question. There is always an aftermath. Ofttimes "blue spells" follow a drinking bout.

Kurt Bader—Think of it! Five and a half million insane in our Hospitals as a result of these vices. WE are the ones who have to support them. Billions of dollars have been used for booze last year. How much will they have to spend for the Medical Doctors to repair the damage? Now you know why the Medical Doctors are supporting Liquor and Tobacco.

Deputy—The sins of the parents! If you talk to a Medical Doctor about Mental Disorders, he will invariably tell you the causes are either "Syphilis or Alcohol."

Emma Schroeder—Occultism teaches that the effects of Tobacco and Alcohol are noticeable even after death.

Deputy—You can smell them from a distance. I can see them dropping off one by one. I could mention 6 or 8 who passed on from time to time with Cancer or another Disorder attributable to Tobacco or Alcohol. Had they abstained, they would not have had the terrible deaths they had.

Quite spontaneously and opportunely Edwin Larson recited the following from memory. Edwin said he is not the composer of this poem, but did revise it so it may apply to both boys and girls:

To A Young Boy and Girl Over A Period Of Years

THE CIGARETTE SPEAKS

I'm just a friendly Cigarette
Don't be afraid of me!
Why, all the advertisers say
I'm as harmless as can be.
They'll tell you that I'm your 'best friend'
(I like that cunning lie,)
They say you'll walk a mile for me
Because I 'satisfy.'

So Brother, Sister, be a sport!
Why longer hesitate?
With one between your pretty lips
You'll be quite 'up-to-date.'
You may not like me right at first,
But very soon (I'll bet)
You'll find you just can't get along
Without a Cigarette.

You've smoked one package, so I know
I've nothing now to fear;
Once I get a grip on boys and girls
They're mine for life, my dear;
Your freedom you began to lose
The very day we met
When I convinced you it was smart
To smoke a Cigarette.

The color's fading from your cheeks,
Your finger-tips are stained;
And now you'd like to give me up
But, my dears, you are chained!
You even took a drink last night;
(I thought you would ere long)
For those whom I enslave soon lose
Their sense of right and wrong.

Year after year I've fettered you
And led you blindly on
'till now you're just a bunch of nerves,
With looks and health both gone.
You're thin and pale and have a cough—
(The Doctor says 'T.B.')
He says you can't expect to live
Much longer—Thanks to me!

But it's too late to worry now.
When you became my slave
You should have known the chances were
You'd fill an early grave.
And now that I have done my best
To send your Soul to 'Hell'
I'll leave you with my partner, Death—
He's coming for you! Farewell!"

(Applause.)

An Affiliate suggested that this poem be submitted to the newspapers for publication, but several others cried out at once: "They won't print it!"

After the Meeting an Affiliate expressed her appreciation for poetry on Smoking and Drinking as a female member of her family is fast forming the habit and she thought poetry might "help some."

Recorder's Note:—A number of years ago, the following poem appeared in E. V. Durling's column in the Milwaukee Sentinel:—

"I owe to smoking more or less,
Through life the whole of my success;
With trusty pipe I'm sage and wise:
WITHOUT I'm dull as cloudy skies.
When smoking all my ideas soar;
WHEN NOT they sink upon the floor.
The greatest men have all been smokers,
And so were all the greatest jokers."

—Bellamy Brooks

The Recorder wrote the following reply, for, she did not agree with the above sentiment, and begs to submit it:

"You owe, 'through smoking, more or less,
Through life the whole of your success?
With trusty pipe you're sage and wise?'
Now let me take you by surprise:
'Without your pipe you cannot think?'
(Your breath must be quite bad, I think.)
You're 'dull' because you're used to smoking;
That, my good friend, is no mere joking.
'When smoking your ideas soar?
When not, they sink upon the floor?'
All I can say: You're in bad shape,
'Twould seem, almost beyond escape
If you can't give ideas birth
Without this bane which has no worth.
'The greatest men have all been smokers?
And so were all the greatest jokers?'
The biggest joke, my friend, 's on you,
No smoke, you can't ideas brew;
Your brain's befuddled—you cannot think—
Cause: Your Pituitary's 'on the blink.'
MORAL:
Ability to think without stimulation
Is what is deserving of real admiration."

EDITORIAL REMARKS

Emily Mokelke and Albert Mokelke deserve the highest respect and honor from us as being the first ones to send in their Contribution to the Spectro-Chromarium Fund by telegraph. A few minutes, later through the same channel came in the Contribution of Recorder Emma Schroeder. The New York Contributions followed by hand through Deputy Lina Fischer. To all the best thanks are extended by the Trustee, who will give a final accounting when the property is deeded to the Institute.

Norma Fiedler's idea about moving to Milwaukee or Chicago is good, but, the question would be simple if ALL our possessions and buildings here had burnt. As it is, what could we do with our Foundry, Assembling and Cabinet Shops, Printing Plant, Spectro-Chromarium, Annex, Warehouse and the like? The amounts contributed and pledged are insufficient to build even *one* building—how about all new buildings in Milwaukee and Chicago? The Aura of Chicago is full of the Blood of slaughtered animals and Milwaukee has only "Fish Milk" to offer the family!

Norma Fiedler's story of the Red Indian reminds me of another real incident. In 1923, when I finished a Class in Cleveland, Ohio, an over-enthused Graduate asked from me permission to print the following advertisement:—

A WHALE LIVES A THOUSAND YEARS,

USE SPECTRO-CHROME AND LIVE A THOUSAND!

Had I voiced my disapproval of this foolery, I should have lost a good Graduate, but, had I approved it, I should have lost my self-respect. So after a little consideration, I sent him the following:—

ADVERTISEMENT APPROVED,

BUT FIRST BE A WHALE!!!

Deputy Frank S. Burns deserves great credit for his able discourse against Alcoholic Liquors. Prohibition failed because the enforcing Officers failed to do their duty. Prohibition failed because Intoxicating Liquors were found in the cellars of the United States Congress, because the Members themselves were boozers. "Temperance" means Total Abstinence for the individual and Prohibition for the State, but, where the Country's Treasury itself would make money out of the degradation and debasement of the Country's population, there can be no inherent virtue or morality in such an administration. I have been a Member of the Supreme Lodge of the International Order Of Good Templars for 45 years and am a staunch Prohibitionist. Put Prohibition on Alcohol, Tobacco and Slaughtered Food (?) as on Opium and there will be no use for Prisons, Lunatic Asylums and Reformatories, Hospitals, Infirmaries and even Spectro-Chrome. Place Prohibition on the Statute Books and make me the Director Of Enforcement and the lid once clamped will stay clamped, because, there will be NO GRAFT OR GRAFTERS.

Meeting Of Thursday, February 1, 1945

Successful Use Of Spectro-Chrome In:—Paralysis, Cold, Injury, Pain, Sore, Fatigue, Peplessness, Stiff Neck, Laryngitis

PRESENCE OF POLICE FOR PEACEFUL MEETING

Reported By:—EMMA SCHROEDER, *Recorder*

Deputy Frank S. Burns was in the chair; 145 Affiliates were present. Communication from the Principal Office was read. "Authentic Report Of The $250,000.00 Conflagration" was read.

Expecting a disturbance similar to the one in New York from the same quarter, because of a certain telephone call received by the Recorder, the Auditorium management had provided Police protection, but nothing happened.

Deputy—Last week Inspector Ernest of the Food And Drug Administration called on me and asked for certain pertinent information regarding my experiences with Spectro-Chrome. I told him how I became acquainted with this Science and how I had been an Affiliate for ten years and what benefits my family and I had derived; how Spectro-Chrome had been a God-send for ills which are not being taken care by other healing methods. I found him to be very pleasant. Of course, he has a duty to perform and is doing it as well as he can and knows how and I think the best thing we can do is to be as courteous as we can concerning our knowledge and experiences with Spectro-Chrome. I do not know how many of you had such a call, but, you will have! He was particularly interested in knowing about such Disorders as "Cancer" and "Diabetes" which had received Normalation by Spectro-Chrome. So, if any of you be familiar with cases of that kind, I am sure it will be appreciated if you will relate them to him. I asked him how any layman would know whether anyone had these Disorders as we do not know any Disorder by name and use no Diagnoses in our Science. We do know that there is a Chemical Disturbance and a Predominance on the Ultra-Green or Infra-Green side of the Solar Spectrum. Was anyone else visited?

Eleanor Luedtke stated that another Agent visited her. She believed his name to be Curry. Said she: "Three times he inquired of me whether I was a Nurse. Each time I told him I am a Spectro-Chrome Normalator. He was interested in the Insanity case of my relative which I reported some time ago and he said: 'Are you sure you are coming into the Court with her?' He went to the Perfex Corporation, where this girl is employed, to see, apparently, whether she was or is in her right mind. He visited the various Medical Doctors in the case. He also went to see Ella Smith who had a paralyzed Arm and whose case I also reported recently. He wanted to know whether she was instructed to go on a 'diet.' I said 'They do not tell us what to do, but they DO tell us what is good for us.' He looked at her Arm because he had been informed that it was shorter. Then he said: "Both Arms look alike to me.' She said: 'Naturally, NOW I am well'! He had good reports. He asked me what I thought of Spectro-Chrome. I replied: 'You can touch anything in my house but not Spectro-Chrome'."

Past Sub-Deputy Kurt W. Bader—I believe the same gentleman was to see me. We conversed for about 2½ hours. I refused to disclose the name of any Sufferer or show him any Sufferer as our Code Of Ethics forbids that. I told him I should give him my own case, if that be enough. Told him how I benefited by Spectro-Chrome. Showed him the Itisometer and the Graduate Equipment. He wanted to know how many cases out of a hundred we can get right. I told him that when they are satisfied they join our Organization. I asked him what that had to do with Food And Drugs. He said he has the order to investigate. I found him to be a nice gentleman.

Deputy—If there be a Hospital record, give it to him!

Kurt Bader—Both Medical Doctors are dead.

Deputy—But there is a Hospital record in your case.

Kurt Bader—I do not know whether they keep it over ten years.

Deputy—My son, whose condition improved to its present status with Spectro-Chrome, was in Rochester, Minnesota. The fact that he was there, showed that the Medical Doctors had ample opportunity to help him if they could do so.

Deputy (To Victoria Tuchel)—Your case would be hard for them to swallow!

Victoria—They must keep the records for a considerable time.

Deputy—Those are bitter pills for them to swallow.

The Deputy said those cases are self-evident and the "diabetes" case he reported some time ago was also a bitter pill.

Rose Helminak—I was operated thirty years ago; I got my record four years ago. The Doctor is dead twenty years.

Sub-Deputy Theodore Mohrhusen—I was also visited by Inspector Ernest. I was rather wary in the beginning. We both warmed up to the subject. He is looking for certain types of cases and fortunately (?) my mother-in-law's case was a Cancer case. He immediately wanted to know the Medical Doctors' names and Hospitals. I gave him only first hand information, none second hand. We talked for about 1½ hours and parted amicably. I questioned him "What are you investigating?" He appeared evasive about replying, so I told him that the very fact that he is still looking, answers my question.

Deputy Burns—I told him "Dinshah welcomes investigations." If the Federal Food And Drug Administration finds Spectro-Chrome is genuine, what will happen then?

Emil Egide—It will be "Case dismissed."

Albert Mokelke—The Inspector paid me a visit, but unfortunately I was not at home. I think he is looking for Hospital records. Mine are plastered all over Soldiers' Home. I was crippled with "Arthritis." The Medical Doctors there examined me and I could not get my Arm up. I also had "Epilepsy." They would not permit me to use the elevator. They must have been afraid I might fall. I was in the "Nerve Ward" with twenty-four other soldiers—all were bad "nerve cases." One case in particular: The man was caught in a hurricane in Florida. Suddenly he would yell. He was strapped. They say they give you no Narcotics out there. They certainly do! I was there about four months. They took off my clothes and tested my Shoulders and Arms. They were locked. I could not go to the bath room myself. They told my wife they could not do anything for me. "If you want to take the man home, you can! There is nothing here we can do for him," they said. I had a Spectro-Chrome Equipment at home. It had normalated my wife. She said: "Now it is up to you!" That was in 1937. I discontinued Smoking. It required nine months to be rid of my Epilepsy. I wondered how I could know whether I was improving. I kept a record. The intervals between seizures were longer and longer. Finally it was 45 days between attacks and then they stopped entirely, suddenly. I think it would be a good idea to get an examination now. I had my waiver lifted. I was amused the other day. Recently I moved into a new home and was hanging storm windows. There I was up on the ladder, hanging between Heaven and Earth and feeling secure! Undoubtedly the Inspector went to Soldiers' Home to examine my record. (Applause.)

Helen Schott—The Inspector called at our house also. He went to see the two Eye Doctors (Olson and Hitz).

Rose Jungmann—They were also in Racine. I thought it was quite fortunate that they picked the Affiliate they did pick. It was a case of Dropsy. They got a very good report from her. They were not at my home.

Deputy—I believe he was all over even in Merrill.

Case Reports followed:—

Emma Schroeder reported another "bitter pill." It was the case of her mother. The Disorders which were caused by a Stroke and consisted in Paralysis of the Throat, for one thing, occurred in 1936 and the report was given for the benefit of those who were unfamiliar with the case. It was printed in Spectro-Chrome in August 1936, page 1925. Emma's mother was unable to swallow. She was eating strained foods at the time and the Saliva would run out of her Mouth. The Eye was closed and she was unable to move the Eyeball in an external direction. Emma showed three of the pill boxes which bore the names of the Medical Doctor, the date verifying when the Disorder took place and the Prescription Number. She also had the Certificate showing the result of the Blood Test with her. Thanks to Spectro-Chrome, Emma's mother (Ottilie Krenz) is alive and able to be present at the Planet Meeting as an Educational Student. She is in her eightieth year and is spry and alert. She takes very few Tonations, has a keen sense of humor. Emma called attention to the Eye which opened during the first two months of Tonation and the Eyeball now moves in all directions. Some people become discouraged when they do not obtain results immediately. Emma's mother tonated daily for about two years, at the age of 70, a great part of the time two Tonations a day and feels she is well repaid. (Applause.)

Emma told of another case which was never reported. It concerned a distant relative of hers, B. R., a widow, who several years ago suffered with High Blood Pressure, Dropsy and Diabetes. Sat around in a chair in a state of despair. She was ready to end her life. "I recommended Spectro-Chrome to her. After less than two weeks of its use, she visited a neighbor who was amazed to see her. This Sufferer lived according to Spectro-Chrome teachings and after using this Science about three months, she improved to such an extent that she married again last year. Kurt Bader was the Spectro-Chrome Normalator." (Applause.)

Past Deputy Otto Lutz submitted the following cases:—

1. Hilda V., age 40, female, Acute Cold, very tight feeling in Head and Throat. Atcowas irradiated were Green and Lemon Systemic. Obtained results in a few Tonations.

2. August R., age 41, Male. Injured his Back when lifting a washtub full of ashes. Had to set it down because of the injury. Was unable to stoop because of the intense Pain. He tried home methods first such as hot baths and liniments for four days, without results. Then came

for Spectro-Chrome Tonations. In three days' time he went back to work. Attuned Color Waves given were Green and Turquoise, Area 18.

3. Severin Z., age 74, male. Two open Sores on right Leg, Area 13, of long standing. Yellow on Areas 6, 7, 8, 9 and 10 and Lemon Local. In four weeks' time one Sore normalated and the other is almost "healed."

4. Arthur B., age 49, male. A case of tiredness, no ambition to work. Had been treated by other methods first, not satisfied with results. After two weeks of Spectro-Chrome Tonations, he showed improvement. Boils developed on top of his Hands and Fingers, filled with Pus, which drained. He has now become a Benefit Student. Yellow on 6, 7, 8, 9 and 10 and Magenta on Areas 4 and 18.

5. Ethel B., age 46, female. Had a Sore on the right Leg. It did not want to heal. Some time ago while in California, was bitten by a Spider. That healed slightly. Then she was bitten by a Mosquito. This was on the same Leg while in the garden. The Leg looked like raw meat and was very painful. She had this condition for some weeks. Attuned Color Waves used were Green and Lemon Systemic including the Leg in the Tonation. In a month's time, her condition was greatly improved. She was so well satisfied that she is enrolling as a Benefit Student. (Applause.)

The Deputy exhibited the "Triumph Of Spectro-Chrome." Dinshah's latest publication.

Elsa Gleiss reported that she was in a draft in the car. The window was open. Her Neck became stiff and the Throat very sore. Used Turquoise and Indigo. After the first Tonation she could turn the Head a little. Then used Magenta and another Turquoise. A total of two Tonations of Turquoise, one Magenta and one Indigo and the Disorder was normalated. Said her Head was so heavy that her husband had to assist her in bed.

Edwin Larson's brother had an acute attack of "laryngitis." Turquoise normalated the condition in two Tonations. (Applause.)

EDITORIAL REMARKS

Milwaukee Auditorium Board, its Manager William Maas and the Police Department of the City of Milwaukee, deserve our best thanks for the dignified and orderly manner in which a rather unpleasant situation was avoided. It is a matter of regret that the Food And Drug Administration in dealing with loyal Citizens uses such bulldozing and unconstitutional methods, disgracing to the limit the glorious Constitution which built America to its present height.

SPECTRO-CHROME

November 1946 Magenta 1946

GREEN · TURQUOISE · BLUE · INDIGO · VIOLET · PURPLE · MAGENTA · SCARLET · RED · ORANGE · YELLOW · LEMON

Volume 22 Number 11

DINSHAH IMPARTIAL

New Deal Crop

Dinshah P. Ghadiali And Dinshah Spectro-Chrome Institute Declared "GUILTY" On All Counts By Federal Jury

Attuned Color Waves
Spectro-Chrome Metry

In course of life you reach a time,
When Doctor, Lawyer, Priest alike,
Does fail to give the sought relief,
To troubled mind or burdened heart—
Then come to me:
I help you to help yourself.
SPECTRO-CHROME

The Science Of
Automatic Precision

DEVOTED TO THE SERVICE OF MANKIND

FIRST ST. LOUIS MISSOURI PLANET

(HOTEL MELBOURNE, GRAND AVENUE & LINDELL BOULEVARD)

Meeting of Thursday, October 3, 1946

Successful Use Of Spectro-Chrome In:—Cough, Itch, Poison Ivy, Swelling, Fever, Pain

Reported By:—VIOLA C. JEUDE, *Recorder*

Deputy Clemens M. P. Neumann was in the chair; 73 Affiliates were present. Communication from the Principal Office was read and explained. The Deputy made the following remarks about the Post Office Fraud Order Injunction Suit against the Post Master General, in Washington, D. C.:—"If the Affiliates would consider how many times they felt ill and if it were not for Spectro-Chrome, how many times they would have gone to a Medical Doctor, plus the cost of the medicine, so if they felt in their hearts that they are being benefited by Spectro-Chrome, let the generosity of the heart flow into their hands and help Dinshah in this trying time, as Dinshah surely could use it."

The Sub-Deputy asked for permission to add a few remarks: "The Law Of Compensation says we receive the way we give. The Bible says 'Bread cast upon the Waters will be returned'."

He said he did not know the status of any of the Affiliates, but as the communication relates the Funds are exhausted and to all indications they are urgently needed, if you could spare, send 1, 2, 5 or even $10.00. No doubt Spectro-Chrome gives results and saved many dollars which would have been spent with Medical Doctors and Chiropractors. I should like to make a request, give what you can. He also suggested that the Community Fund Drive will start next month; split what you may give them and help the Spectro-Chrome Cause. He said it was merely a suggestion, it need not be adopted.

Deputy said some time ago there were certain Affiliates willing to go to the defense of Spectro-Chrome if necessary, were those same Affiliates still willing to go?

Rose Freund, Agnes Hoenschemeyer and Arthur Behnen said they are still willing to go to Camden, New Jersey in defense of Spectro-Chrome.

Case Reports followed:—

R. F. reported her son returned from Oklahoma on vacation and had Congestion in Areas 4 and 5. Ratio 100 to 12; gave Lemon Systemic, Magenta Area 4 alternately; could not sleep for the first 2 nights; the Ratio then was 68 to 12, coughed quite some Phlegm and for scratchy sensation in Throat used Turquoise. Came home Thursday and went to work Monday.

Discussion as to Atcowas followed. B. E. reported 14 year old girl had Itch all over Body for 2 weeks; gave Yellow and she seemed to itch more.

Question—What was the Kardoscil and Spiroscil?

Answer—Do not know.

Asked about the Rational Food Of Man, she said: "Mostly Vegetables, Meat is hard to get."

Discussion followed in regards to Skin Disorders due to wrong Chemistry of Body.

Further discussion followed as to Eliminations and Atcowas.

The Deputy reported the following Cases:—

1. R. K., female, age 50, Ratio 72-20. About twenty years ago she suffered from Poison Ivy. Twelve days ago, she went vacationing in the country and three days later her Face and Neck swelled and blistered. She went to the Medical Doctor who gave her Salve which would dry the Blisters and wanted her to come back to give Injections. She then came to Spectro-Chrome. Her left side of the Face also Arm were still swollen and there was Fever in the Arm. Atcowas used, Turquoise Local, Green Systemic, Magenta Area 4. Complete Normalation within one week. She will soon join us as a Benefit Student.

2. J. R., female, age 63, Ratio 69-21. For the past four years she was suffering with Pains in her Knees and lately also in her Thighs. She went to her Medical Doctor and he prescribed Capsules and a rubbing Medicine. However, it gradually became worse and she was forced to stay in bed for a week or so, because, she was unable to walk and had a hard time to get out of bed. The Pains after walking were becoming worse, so she was willing to use Spectro-Chrome, Yellow on Areas 6-7-8-9-10, Orange and Lemon Systemic, Magenta on Area 4 were used. Now walks and has normal use of Legs and Knees, feels fine, lost some excess weight and regrets not to have come sooner to Spectro-Chrome.

Sub-Deputy Read Chapter 10, "Velocity, Time, Space" with explanations and object lessons, from Spectro-Chrome Metry Encyclopaedia and the Deputy started Chapter 48, "Spleen As Gateway."

Deputy asked the Affiliates if they were satisfied with the Educational Program and if they could or would make any suggestion, that we as Commissioners would gladly co:operate. Someone voiced there should be more, that they would be able better to understand Spectro-Chrome.

The Recorder stated she received a letter from an Affiliate stating another Affiliate said "It has become so uninteresting to her, that she does not care to go." Too much Educational Program!

The First Volume of Spectro-Chrome Metry Encyclopaedia may appear very dry and uninteresting, because, it is just Theory, but to understand Spectro-Chrome it is absolutely necessary. If Chapter 9, be thoroughly understood, even then it would be just scratching the surface of the Science.

Sub-Deputy said that we as Commissioners are in duty bound to conduct the Educational Program.

The Deputy asked for a vote; those that were satisfied with the present Program to raise their hand and give their opinion. Practically all hands went up and they chorused they were satisfied with the Educational Program as it has been conducted. Deputy asked those Affiliates to raise their hands who disapproved. None was raised.

Please Dinshah what would you suggest?

EDITORIAL REMARKS

Spectro-Chrome Metry is a new Science and Education is the only way to spread it. Continue your Program by all means and let the uninterested Affiliates drop out of sight. We do not want selfish, clubhouse scandalmongers.

FIRST MERRILL WISCONSIN PLANET

(HOTEL BADGER, MERRILL, WISCONSIN)

Meeting of Thursday, October 3, 1946

Successful Use Of Spectro-Chrome In:—Cold, Swelling, Skin Disorder, Asthma.

Reported By:*—EMIL ERBEN, *Deputy

Deputy Emil Erben was in the chair; 16 Affiliates were present. Communications from Principal Office were read, including a Telegram about the Post Office Injunction Suit, in Washington, D. C. The Deputy reported he received 38 pounds of Cocoanut Oil from an Affiliate in New York. He asked for permission to distribute some among the Affiliates.

M. L. spoke about a Sty in an Eye. She was told about it being tonated in the same way as any other Inflammation.

A. W. said he had trouble in controlling his Frontelim. A Sufferer Chart was asked.

E. K. had Cold in Nose which was normalated by irradiating Yellow Systemic and Magenta on Area 4, Turquoise and Indigo on Area 1.

K. M. had a Swelling on Foot. Turquoise Local for 10 days corrected it.

E. M.'s daughter N., age 17, has a Skin Disorder. She irradiated Green, Lemon, Turquoise and Indigo, but her condition is worse. She was told to bring Sufferer Chart for opinion.

J. M.'s daughter P., age 15, had Skin Disorder. Atcowa Lemon and a few Turquoise brought good results.

E. E. had Asthma for 10 days and could barely come to the Meeting. He had a Ratio of 6 to 1, with Kardoscil of 120, which was corrected by Purple on Areas 4 and 18.

The value of Free Guidance was stressed.

EDITORIAL REMARKS

E. M. should send for Free Guidance. The initial one asked for Report in 30 days after irradiating Yellow on 6-7-8-9-10 Areas. To date, no General Advice Chart came.

———o———

FIRST PORTLAND OREGON PLANET

(MASONIC TEMPLE, 1119 SOUTH WEST PARK AVENUE)

Meeting of Thursday, October 3, 1946

Successful Use Of Spectro-Chrome In:—Mucus, Stroke

Reported By:*—LEWIS BISCHOFF, *Recorder

Deputy Stanley Burroughs was in the chair; 22 Affiliates were present.

A letter and a telegram were read from Dinshah. We could not

quite comprehend what the Trial was all about and why Dinshah was not given a Hearing in Washington, D. C. in the Post Office Injunction Suit. Was he supposed to be at some former Hearing and did not appear? What seemed to be the big trouble? As far as any money is concerned we are sure that we shall not be able to help much in that line out here. We are going to have all that we can do for our own case coming up on Appeal and a large sum of money will be needed to fight it. The Deputy said that he would like to go back there and help Dinshah, but he could not leave his job now and could not stand the expense. It is too far across the Country and would take too much time off. He said that he does not know how much good he could do anyway, even if he could go. Under the circumstances, it looks as if we are going to stand all our own Trial expenses out here. Any way, we all hope for the best and are looking forward to a quick victory for Dinshah.

Case report of J. S., 77 years old; all clogged with Mucus and almost too weak to remove it by coughing. He had a Stroke about a year ago and was very shaky and run down in general. Two weeks of Green, Lemon and Magenta made a very big change for the better. The quantity of Mucus increased in eliminating and then gradually decreased. He did not like the Rational Food Of Man, but, was sticking to it anyway. He was quite pleased with the results so far.

The conditions of Acute and Chronic Disorders were stated by the Deputy. He explained how Lemon corrected the Persistent conditions and Turquoise altered the Recent conditions. Nearly always Turquoise can stop one from coughing, due to Colds, while Lemon and Green are used to break them up.

EDITORIAL REMARKS

Thank you for your sympathy. You mind your Appeal matter and we shall mind our Trials. No money is asked.

———o———

FIRST NEW YORK NEW YORK PLANET

(CARNEGIE HALL, 7TH AVENUE & 57TH STREET)

Meeting of Thursday, October 3, 1946

Successful Use Of Spectro-Chrome In:—Asthma, Burns, Tuberculosis, Gingivitis, Arthritis, Pain, Neuritis, Mosquito Bites

Reported By:—EMMA C. HAITZ, *Recorder*

Deputy Lina Fischer was in the chair; 163 Affiliates were present.

Attention was again called to the fact that many Affiliates did not receive their Spectro-Chrome, because, they did not attend the Meetings very often. All were urged to try to come whenever possible.

Communication from the Principal Office, also a Telegram which was received from Colonel Dinshah, were read.

Deputy—Before I came over to New York tonight, I telephoned Dinshah. He appeared in Camden, New Jersey today and gave me the news, that the Trial was postponed. It will not start next Tuesday, but, Dinshah will have to appear there to set the date when it will start. We do not beg, but you know what it means if these cases are lost. There are absolutely no funds and it costs a tremendous amount of money. As we supported the Institute at the time of the Fire, we should support it now. This time it is just as crucial, if not more. What will you do without your Equipment? How many of you have avoided an Operation through Spectro-Chrome? (A number of hands went up.)

Affiliates who wanted to testify at the coming Trial in Camden were asked to sign their name and address at the door when leaving. Once again Affiliates were advised to be careful and fill all Enrollments and other papers sent to the Institute in a correct manner.

Turning again to the Magazines, the Deputy stated that this job was really upto the Affiliates. Affiliate Margaret Wolschendorf travels to Malaga every month to get the Magazines for the Planet Meeting.

F. P. reported the case of an Affiliate who has her Equipment for seven weeks. This Affiliate suffered with Asthma for the past 15 years. After six weeks of tonating, this woman could do all her work, whereas previously she could do nothing. Has had no Asthma in the past four weeks. Before coming to Spectro-Chrome, had spent over $3000.00 on Drugs and Medical Doctors. Tonations of Yellow and Magenta were given.

Seven new Affiliates were introduced and greeted with applause.

F. P. suggested that the $15.00 which is given to an Affiliate for introducing a new person be returned to the Institute for the Spectro-Chromarium Fund and in this way we can each do our share.

M. W. reported the case of her sister-in-law, who had Asthma for ten years. During this past year, was under the care of a Chiropractor. Spent $600.00 and the attacks came on just the same. Tonated with Magenta and Lemon and within one week was much better. (Applause.)

M. W.—I was with Colonel Dinshah in Washington, D. C. Court. There were 34 cases ahead of ours. As usual, the Colonel did not get a chance to say anything. I think that if Dinshah's spirit is undefeated, we should do all we can to help too. He serves God and through him we got our Spectro-Chrome.

Deputy—I think we should give a round of applause to F. W. (All vigorously applauded.)

Deputy—In the near future, we shall have several Normalators. M. W., W. D. and F. R. finished their three Volumes and F. P. is now studying the Third Volume.

Affiliate B. Le F. again showed her Arm to the Affiliates. Last month, she severely burned it. Today, there is not a trace of a Burn on her Arm, thanks to Spectro-Chrome.

M. K.—I have been an Affiliate a year last month. Two Medical Doctors claimed that I had Tuberculosis. They wanted me to go to a rest home. I disregarded their advice and followed the Tonations which Dinshah recommended and I feel wonderful today. Spectro-Chrome also helped me with my Teeth. The Dentist told me my Gums were filled with Pus and I would have to have my Teeth pulled. I wrote to the Institute, followed Dinshah's directions and today I still have my Teeth. My husband was in the Aleutians for three years. He suffered from Arthritis in the Arms and Legs. I made him use Spectro-Chrome and he does not have a Pain any more and besides, I am able to help myself with every day illnesses. (Applause.)

Deputy—I am very proud of your case and that I kept you away

from that rest home. There was already a remarkable improvement in one week.

An Affiliate inquired what Tonations were given and received the answer: Lemon and Yellow and then Yellow and Magenta.

Deputy—Your husband smoked and liked Meat and a Beer once in a while and now he is 100% on the Rational Food Of Man. When you give 100%, you will receive 100%.

E. D.—I had very bad Neuritis. After using Spectro-Chrome for only one week. I could walk across the room. Tonated with Lemon Systematic and Magenta on Area 4.

I. S.—Are baked Beans a substitute for Peas?

Deputy—Yes.

I. S.—My son had Mosquito or Flea Bites on a Foot for about 4 or 5 weeks. I put drawing Salve on it. (Here the Deputy interrupted and told Affiliates that in introducing someone, they should also give them the feeling that they may always come back for Guidance). The Bite traveled. My husband took the boy to the Medical Doctor. The Doctor gave him some Salve. I wrote in to the Institute asking advice and they told me that I should keep going to the Medical Doctor! The boy had a Fever ranging about 102 to 104, for about two weeks. Tonated with Turquoise and Fever went down. Our minister was to the house and he said we should take the boy to the Doctor, but I am not going to. I am going to stick with Spectro-Chrome and I am sure that next month I shall be able to give a much better report. I am also going to write to the Institute and apologize for my actions in this particular incident.

M. K. inquired as to what she should feed her baby now. The Deputy advised giving her the same vegetables and other food as the rest of the family eats and also some Limbupani.

M. K.— A Sufferer I know wants to go to Malaga for a few weeks, but, she is not an Affiliate.

Deputy—The best thing is for her to get an Equipment. When we were at the Class, we stayed at the Hotel. Try to get this Sufferer to get own Equipment and use it at home. Incidentally, they started to build the first Cottage at Malaga today. When the Cottages are ready and if they have enough help, they will take Sufferers again.

A. R. stated he has a brother-in-law who is sick and whom the Medical Doctors want to put in a Mental Institute. He stated that he would like to have him go to Malaga to speak with Dinshah. The Deputy advised that when he goes to Malaga to pick up the Equipment to take his brother-in-law with him, at which time he may speak with Dinshah.

M. W. stated that many people were advised to take Yellow on Areas 6, 7, 8, 9, 10 by the Institute and asked that the Deputy explain the Attributes of Yellow for the Affiliates.

Before complying with this request, however, the Deputy called on Affiliate F. R. to tell how he enjoyed his two weeks stay in Malaga.

F. R.—We all learned plenty and had plenty of fun. I shall go down any time for another two weeks. This Course was the most Educational Class I ever attended.

The Deputy then read Chapter 80 on Yellow. Discussion followed.

F. P.—I know an Affiliate who is constipated for 15 years. He has been taking Tonations, but has had no Back Elimination in four days.

The Deputy advised that he either continue tonating with Yellow or use Orange on Area 19. Affiliate W. D. demonstrated the exercises which they learned at the Institute.

Deputy advised Affiliates to have their pictures taken as Affiliations would have to be renewed next month. All look forward to the next Meeting and hope that Dinshah will be present.

EDITORIAL REMARKS

No presence in Planet Meetings, no Spectro-Chrome magazine. This will continue until the conditions change.

It takes more trouble to introduce a person to Spectro-Chrome, than can be compensated by the $15 Auxiliary Benefit.

F. W. deserves credit for his co:operation and M. W. deserves our thanks for her trouble in easing our labors. In the Camden Court, she was called the "New York Omnibus," because, daily she brought over our witnesses! A dozen times the U. S. Attorney spoke about her and pointed at her, calling her "Gretchen"! He seemed very fond of teasing her thus, but, she took his taunts with a smile each time.

At present, we are not in a position to receive Suffering Affiliates. We shall let you know when our Electric Heating Plant is in operation.

I may visit your Planet after the Trial is over.

FIRST CLEVELAND OHIO PLANET
(CARNEGIE HALL, 1220 HURON ROAD)

Meeting of Thursday, October 3, 1946

Successful Use Of Spectro-Chrome In:—Cold, Pain, Headache, Hay Fever, Menopause

Recorded By:—BARBARA B. BROWN, *pro tem.*

Reported By:—**EDITH REES, *Recorder***

Deputy George J. Brown was in the chair; 23 Affiliates were present. First the meeting was opened to new Affiliates for questions. There were two new Affiliates present. Communication from the Principal Office was read by the Deputy and also a Telegram received from Dinshah earlier that evening. The Deputy emphasized the fact that funds were low at a time when there were Law Suits to fight and requested anyone present that was able to contribute, to do so. Several envelopes were handed out by the Sub-Deputy for free contributions. It was requested of anyone present who had previously made Affidavit as to their condition qualifying them to testify at the trial at Camden, New Jersey, to notify Dinshah so as to be a witness at the Trial, at their own expense. He will notify all when needed.

Case Reports followed:—

1. J. B. used Indigo on his father's Legs for Varicose Veins which started swelling and pained terribly. It was suggested that Yurquoise be used to remove the Swelling and Inflammation.

2. E. R. reported she used Indigo for a Cold last Sunday. It started about 2 o'clock at night, and she had three Spells and nearly fainted. She tonated with Indigo and dozed off and when she awoke her Pain was gone and she got up Monday morning and did all her work as usual.

3. E. M. used to have many Headaches. She was advised to use first Green and then Turquoise. The Turquoise helped her and she feels much better. Her Headache is gone and she is pleased.

4. E. P. used Blue for Hay Fever and got a Headache. It was suggested that in such a case one should switch around to Green. She used Turquoise and it helped. She claims she used Magenta for Change Of Life and it helped wonderfully.

The Deputy asked those present if any Federal Agents had been around. He stated that they were now going after Spectro-Chrome Normalators and were out to scalp the head of the Organization—Dinshah.

The Deputy stated that in any Enrollment it is requested that you do not simply mark "Retired." Mark clearly what the occupation was before retiring. Request was made not to send in the Time Switches or similar parts for repairs, because, of lack of facilities at present. Replacements may be made.

The Educational Program covered Chapters 35 to 41 inclusive from Volume I of our Encyclopaedia.

FIRST MILWAUKEE WISCONSIN PLANET
(MILWAUKEE AUDITORIUM, WEST KILBOURNE AND NORTH FIFTH)

Successful Use Of Spectro-Chrome In:—Stomach Disorder, Tumor, Bruise, Fester, Swelling, Burn, Blister, Wound.

Reported By:—EMMA SCHROEDER, *Recorder.*

Sub-Deputy Evert Dahlke was in the chair; 150 Affiliates were present.

For the benefit of those who did not receive the magazines at this Meeting, the Sub-Deputy read, again, the letter to Planet Commissioners (August 30th, 1946) which states in part:—"Those who will not receive the Spectro-Chrome in the Planet Meeting by constant attendance, may receive them at the end of the current year, provided they renew their Affiliation in time. This move was actually forced on me and the negligence of the non-attending Affiliates is to be blamed for it. Without co:operation of the Affiliates, the Institute cannot function efficiently."

Sub-Deputy—If you are attending every Meeting, you will receive your magazines. If NOT, you may receive them at the end of the year. The Continuation Request will, most likely, be sent by mail.

After the reading of this month's communication by the Sub-Deputy, the Recorder read the Telegram promised in the letter, which said that the Washington Fraud Order Petition was thrown out by Summary Judgment against us. "Judge abruptly stopped Dinshah oral argument and refused to listen to citations saying we had notice of hearing but did not appear before Solicitor. Now go on appeal. Treatment during case worse than Brooklyn. Voluminous funds will be needed."

The Recorder brought with her to the Meeting a copy of the Post Office Fraud Order Suit (refiled) Action in Equity No. Civil 31408 which contains the complete text of the Bill Of Complaint and Petition For Restraint Order And Injunction. Page 8 of this pamphlet tells how "(FF) Suddenly on August 21st, 1942 a single envelope addressed to all the 17 Entitities came by Registered Mail, from the Post Office Department to Malaga Post Office. It was received by Irene Grace Dinshah and was found to have a Memorandum of Charges in it. It charged the 17 Entities that they 'are engaged in conducting a scheme or device for obtaining money through the mails by means of false or fraudulent pretenses, representations or promises, in violation of Section 3928 and 4041 of the Revised Statutes, as amended, (Secs. 259 and 732 of Title 39 U. S. Code) a copy of which is also enclosed. These Statutes authorize the Postmaster General to forbid the delivery of mail and the payment of money orders to any person or company found to be operating a scheme or device in violation thereof.' An ANSWER to the said Charges was to be made by the 17 Entities, by Monday, September 14th, 1942 and it stated that IT MAY BE FORWARDED BY MAIL OR PRESENTED IN PERSON OR BY COUNSEL. The communication was signed by Vincent M. Miles, Solicitor. The date for the 'Hearing' was changed subsequently by the Solicitor without the *request* of the 17 Entities, to Monday, September 21st, 1942.

"(GG) In pursuance of the wording of the Solicitor's letter, Dinshah P. Ghadiali, acting pro se et al, MAILED an appropriate ANSWER with 27 exhibits, denying charges, etc. . . . . . .

"(HH) It appeared from the 'Transcript of Proceedings' . . . . . that a 'Hearing' was held in the Solicitor's Office before an Assistant, Daniel J. Kelly, on Monday, September 21st, 1942, at which a Post Office Inspector Karl M. Foust on oath produced in Evidence a Spectro-Chrome Apparatus, which he obtained by signing a False Declaration to Dinshah Spectro-Chrome Institute and posing as a sick woman 'Mrs. Amelia Regina Wylie, of Ridgeway, Missouri.' That Inspector made numerous false statements on oath to incriminate the Complainant and one Physicist and two Medical Doctors, all linked with the American Medical Association testified that the Apparatus was a 'fake' and the claims made for its efficiency were 'fictitious'."

It is suggested that all Affiliates refresh their minds on this case by referring to the Post Office Fraud Order Suit. As stated, Dinshah answered by MAIL. Now the Judge abruptly stops Dinshah's oral argument and refuses to listen to Citations saying we "had notice of hearing but did not appear."

C. D. requested that the Telegram be re:read and then asked "What does that mean? Does that mean that he lost the case?"

Sub-Deputy—It means that he will have to appeal the case. Dinshah had his choice of three different things. He obeyed the original Order. This seems to be a very dark hour for Spectro-Chrome. However, if we abide by the Rules and follow the Guidance of the Originator, we shall eventually win! By Co:operation! Dinshah needs Witnesses. They will have to pay their own expenses for he is short of funds. How many will volunteer?

(Helen Schott, Robert Schott, Norma Fiedler and Evert Dahlke agreed to go.)

Five new Affiliates were greeted with hearty applause.

Sub-Deputy—We have new Affiliates every month regardless of the opposition.

He expressed surprise at so few Witnesses volunteering in comparison to the many who received satisfactory results from the use of Spectro-Chrome. At the request of the Recorder, he sought the reasons therefore. A. R. said he could not see his way clear at this time.

Sub-Deputy—Is anyone able to go with a car and take others so expenses may be kept at a minimum? (No reply.)

G. W. said she had already made two trips East this year. A. M. is unable to see his way clear this year. M. L. is leaving the city next week.

Sub-Deputy—I don't know like what the Case is going to be. If it is like Brooklyn, you will not be able to say much, but Dinshah expects this Case to be quite different.

G. W.—We can still send money if we wish.

Regarding witnesses, conversation with some of the Affiliates revealed that some of the Witnesses who appeared in Brooklyn harbor the thought that they were not very good Witnesses and did not do Spectro-Chrome very much good in view of the Verdict which was rendered by the Jury. Witnesses there were not permitted to speak freely.

Before continuing with Case Reports and the Educational Program, the Recorder was requested to read an Article which she clipped from the Milwaukee Sentinel of Tuesday, October 1, 1946: "New Source of Light Promises End To Night" was the caption. The article told how Professor Etienne Vassy, 37, expert on Atmospheric Physics at the Sorbonne University, Paris, made a sensational discovery which may turn night into day by means of radio transmission of optical waves directed at *luminous Strata* in the sky. This Professor refers to "the luminous **zone which *I had discovered* 50** miles from the earth . . . . . Strata D."

In regard to the Professor's assertion, the Recorder said:—"THAT was in the Milwaukee Sentinel on October 1st, 1946. I shall now read to you from the First Volume of Spectro-Chrome Metry Encyclopaedia, by Dinshah, *copyright* 1933, page 393, paragraph 3, line 9:—'the Sun actually sheds NO LIGHT, but, the Energy sent forward is convertible into Light when it passes through impediments like the Atmosphere . . . . .' I leave it to you Affiliates to judge for yourselves who was the first discoverer of these Strata."

Sub-Deputy—Colonel Dinshah has made many discoveries with which he is not credited.

Recorder—I have another clipping. *We* are not the only ones to incur the wrath of the Medical Profession. In the Milwaukee Sentinel of Wednesday, October 2d, 1946, E. V. Durling writes in his column:—"SIDELIGHTS— . . . . . My occasional references to the cruelties of vivisectionists appear to have made some members of the medical profession lose their tempers to the extent of breaking out with very nasty and insulting epistles addressed to this department. The truth hurts."

The Sub-Deputy added an article from the Milwaukee Journal (which also appeared in the Milwaukee Sentinel announcing a "new technique" for childbirth:—"Entirely Painless Childbirth—Doctor has a new technique." It must have been a "sweet" bundle from heaven for "the mother ate a piece of cake and drank a glass of milk as she comfortably gave birth to a 6 lb. 14 oz. girl last Wednesday." This article evoked smiles from the listeners. The mother is said to have walked a few hours after birth. The Sub-Deputy said:—That is entirely opposite to the teachings of Spectro-Chrome. Dinshah refers us to Biblical passages in this connection. Injecting anything into the Spine is very dangerous. Then they wonder why they have so many Female Disorders later in life. In the last five years these Disorders have been more prevalent than ever. They often take Injections because they have not the patience to let Nature work.

Relative to the article read, Norma Fiedler told of a case where a Milwaukee Doctor encountered a Confinement Case just as he was about to go on a vacation. In addition, the Doctor is said to have been in need of money. In his hurry to get away, he forced the birth. Mother and baby both died. The only punishment the Doctor received was that he was "kicked out of the hospital," but he is still practising.

The birth of a baby under Spectro-Chrome methods was told by E. B., who said everything "went fine." The baby weighed six pounds. "But it was hard work; I was sweating," she said. The parent said: "I believe we shall have to make the baby a cap so it looks like Dinshah!" (Laughter.) The birth required about three hours. The midwife in attendance was surprised at the "clean" baby and thought it must be from the Food.

Someone stated that it is *compulsory* that a Medical Doctor be present, but the Sub-Deputy reminded that sometimes the Doctor does not arrive in time. A registered midwife is permitted.

E. S. wondered about dropping Silver Nitrate in the Eyes. E. B. said none was administered. E. S. thought it was compulsory under Medical Law.

E. B.—There is nothing like "have to"!

Sub-Deputy—In my family, we commenced using our Spectro-Chrome at about four o'clock in the afternoon and the baby arrived around six o'clock. There was no Pain until the actual Delivery.

E. B.—This woman had Pain.

W. H.—I am a fairly new Affiliate. I have been here for three Meetings. I had Hay Fever for 10 or 15 years. I am 31 years old now. Last month I had what you would call the "Stomach Flu." I came here expecting to learn something and I did. I used Indigo on Areas 6, 7, 8, 9 and 10 and had relief in two hours. Had several Back Eliminations. Went to work the next day. The following two nights I tonated with Indigo. Could I have used another Color? I am here to learn.

Sub-Deputy—What about your Hay Fever?

W. H.—It is about the same as usual. Some years it is better, some years worse. I suppose if I started early in the Spring it would be better. I have not had time enough really to clear up that condition.

A. R.—It says in this issue of Spectro-Chrome that you should really start in the Spring with Tonations.

W. H.—I am using Magenta on Areas 4 and 18 under the Institute direction.

J. M.—I became Benefit Student in April. I have been feeling sick and I thought I should visit a Medical Doctor. He told me I had a Tumor and that it must come out. I used Yellow and I feel just perfect. I never went back to the Doctor, but Saturday, they rushed me to the Emergency Hospital. I fell. When I returned home, Atcowa Blue was irradiated. The Doctor advised me to remain in bed.

Sub-Deputy—Orange would have increased the Circulation.

J. M.—Then I should use Orange on the Shoulder, as it is still quite painful. That is why I decided to come here today It is still discolored, but not as painful as it was. We used Blue because the Doctor told us to use ice packs (Ultra-Green Division).

Sub-Deputy—How many had experiences with Spectro-Chrome as far as Falls and Bruises are concerned?

L. R.—I fell while going to the Drug Store for ice cream one evening. I stumbled. I used Turquoise. It festered. After one Tonation of Turquoise it dried and was all right in a few Tonations. (Applause.)

A. B. reported a case of Swelling which was corrected in two Tonations.

B. D.—I just had something smart happen to me. I burned my Arm. Used Blue. I accidentally rubbed the Skin off although the Blister was already gone.

B. D. raised her Arm and showed it to the audience. There was very little trace of a Burn left. Applause was given to her report.

Sub-Deputy—One Sunday, an Affiliate stopped at my house with a cut Finger. She used Indigo and Turquoise and had no Stitches. Accidentally, she bumped it and the Wound opened. Scar Tissue was present. She used no tape or bandage on it. Possibly a tape would have helped, in this case, to keep it from opening.

G. W.—I forgot to report the most important thing last time about the Portable Itisometer. If A. H. were here he might explain it. The down payment will be $150.00, the total cost depending upon materials. No money will be refunded as was the case last time. New York already has a number of Orders placed. They are going to bring the money to the Meeting so Dinshah will have a considerable amount at one time to make progress.

E. B. reported a case which could not be heard on the platform. Affiliates sitting on the side of the Platform can not be heard plainly.

Discussion returned to the Portable Itisometer:

Sub-Deputy—The Recorder tells me that the word is a combination of the words "IT IS SO." He showed the picture of the Main Dial illustrated in Volume 3 of our Encyclopaedia and said that E. B. and K. B. are among the Normalators who use Itisometers. E. B. contributed to the elucidation of the instrument.

Question arose as to the Contribution. G. W. repeated it. It will be at least $150.00 to start with and he would like to know how many are interested.

F. K.—The Itisometer is an instrument which touches the Areas of the Body and records the degree of Imbalance. It is a wonderful instrument and if you had it in your home you would agree that it is truly one of the wonders of the World and should be classed as such.

D. S. made inquiry about the Portable Spectro-Chrome Equipment. I. P. and E. S. also expressed interest in a Portable Spectro-Chrome.

Sub-Deputy—The Itisometer eliminates all doubt. Spectro-Chrome may be used without the Itisometer, but with the use of it, it must be quite helpful.

I. P. is always puzzled as to what Atcowa to use when a Cold commences. The Sub-Deputy suggested Turquoise and Magenta; if it persists. Lemon and Magenta.

E. M. who was unable to attend last month's Meeting until later, was called upon to give her impressions of the recent Class in Malaga which she attended. She said—The class was very wonderful. It certainly was an education!

Question by Sub-Deputy to Evelyn Murphy (who came in late):—"Would you be in a position to go to the Camden, New Jersey Trial as a Witness?"

E. M.—No! I cannot get away from work!

Sub-Deputy—I did not want to miss asking you, as you came in late.

Chapters 45 and 46 of the Spectro-Chrome Metry Encyclopaedia were read by the Sub-Deputy. Discussion followed on X-Rays and on

FIRST CHICAGO ILLINOIS PLANET
(MIDLAND HOTEL, 172 WEST ADAMS STREET)

Meeting of Thursday, October 3, 1946

Successful Use Of Spectro-Chrome In:—Sprain, Pain, Burn.

Reported By:—HERMINE KUEMMEL, *Recorder*

Deputy Kurt W. Bader was in the chair, 22 Affiliates were present. Deputy read the Communication from the Principal Office, followed by a discussion. He explained that there is no guaranty on Time-Switches and showed how they are to be turned on; most people turn them on slowly and burn the contacts; they should be turned on snappily just as an Electric Light is turned on.

A new Affiliate M. K. was greeted.

Also the Affiliates were reminded that there is no better Cause than Spectro-Chrome in aiding Suffering Mankind and to send voluntary Contributions to the Spectro-Chromarium Fund.

On the Educational Program, the Deputy read Article 41 Medical Humbug Of Common Cold from "Family Health Protector" and a lively discussion followed. He also read from our Encyclopaedia, Volume 2, Chapter 53, Infra-Green And Ultra-Green, and explained, how important it is to watch the Signs and to decide if Tonation should be given on the Infra or Ultra Division.

P. P.—I like to know what Atcowa I should use for a Cold. I had a Cold and tonated with Indigo, but it did not do much good.

Deputy—First always take Kardoscil and Spiroscil to decide if to tonate with the Infra or Ultra Division of the Spectrum. If there is Fever or Heat, you choose Ultra side, but do not go quite as far down as the Indigo. You may take Turquoise or if you are not sure yourself use Green Systemic and Magenta on Area 4. A Cold is usually a lowering of Body and Blood Temperature, caused very often by improper dressing especially with the women. They do not dress warm enough. For this reason, silk stockings should be discarded. In order to bring about the necessary Circulation, tonate with Orange, which is ¾ Red and ¼ Green; the Red supplies the Heat and the Green cleanses. Cold and warmth have a great influence on the Tissues; with a Cold your Nostrils will close or choke on one side. You also will notice that Frontelim lessens and too much fluid stays in the System.

A. P.—Last Saturday morning I turned my Ankle and sprained it; it was very painful and returning home, I tonated with Indigo. In the evening, I tonated again with Indigo; the Pain subsided entirely and the Ankle appeared normal, so I was able to go to work.

Deputy—You used the right Atcowa. I could not have done any better; for Pain Relieving and rebuilding the torn Tissues, Indigo is the best.

J. B.—Friday night, I canned Tomatoes; the crock broke and the boiling Tomatoes splashed all over my Face, Neck and Chest. It burned terribly. Immediately I tonated with Blue and in a half hour the Pain subsided and did not return any more. I keep on tonating every day and it is healing wonderfully without any trouble.

Deputy—My friends, if Spectro-Chrome did not do anything else, than normalate the Burns, which in so many cases are fatal, it would be worth its value a thousand fold, but look at all the many other Attributes Spectro-Chrome has; there is really no end to it. Accidents are all Acute Disorders and are normalated by the Ultra-Green Division of the Spectrum. Once I was called to a case, which looked almost hopeless. The Priest was present administering the last sacrament. Kardoscil 54, Spiroscil 25; one to two is a hopeless case. The man had such a Chill he could not even get his clothes off; a hot bath and later a Tonation with Turquoise normalated him. He was up and around the next day. You have to introduce the Heat in the Body again.

A lively discussion followed.

F. J.—If a Cold too is very Persistent, do I have to go on the Ultra side?

Deputy—Always watch the Signals.

E. B.—My husband had some Teeth pulled; tonated with Indigo, but Pain did not subside, in fact it became worse. Next time he tonated with Orange and it drew out plenty of putrid matter.

Deputy—If there is an Abscess you use Orange; as a rule, there is not enough Circulation in the Gums.

R. M.—I had a severe Pain in Areas 6-7-8; could not breathe or speak, could not lie down on Back or side. I thought this was the end; my Head felt cool and was alone tonated with Blue and after 35-40 minutes felt better. After the hour was over, I could walk again and went to my shop, but felt that I was not quite right. The next morning I had the attack again, I took a hot bath and coming out of the bathtub, had another attack, but I dressed and went out. In the evening, after coming home, I used Indigo and wondered if I used the right Color.

Deputy—There seems to be a Functional Disorder; you should always take the Ratio; probably Orange would have released the Spasm and stimulated the underactivity in the Body. Please, fill out a Sufferer Chart and bring it with you at the next Meeting; then we can advise you.

FIRST ALLENTOWN PENNSYLVANIA PLANET
(HOTEL AMERICUS, SIXTH AND HAMILTON)

Meeting Of Thursday, October 3, 1946

Successful Use Of Spectro-Chrome In:—Pain, Stiffness, Injury, Lump, Burn, Swelling, Soreness, Fester, Nervousness, Eye Disorder, Bruise, Cold.

Reported By:—MAE BURNSIDE, *Recorder*

Deputy Edwin D. Moyer was in the chair; 41 Affiliates were present. Communication from the Principal Office was read. The Sub-Deputy spoke about her visit to Malaga, New Jersey and about a woman who was agonizingly burnt by the X-Rays. She then read from "Family Health Protector" how Spectro-Chrome is asked to repair human wrecks, the results of medical and surgical butchery.

She then spoke of a woman whose Breast was removed; also of a young woman with painful Knee and stiff Toes. Orange Local moved them in ½ hour.

M. P. reported she hurt her Foot, having had a bad fall last May. Her husband fell and bruised his Elbow. Thought it would go away, but, it developed a Lump which grew to the size of an egg. Tonated with Turquoise and Indigo Local; did not help, so filled a General Advice Chart and received Free Guidance, to irradiate Orange and Indigo Local. The Lump disappeared and the Elbow moved.

H. L. who was burnt badly on the Leg by Gasoline, then showed the improved healing to the Affiliates. He was in bed 3 weeks. Used Green and Indigo Local.

F. R. testified to having Persistent Colitis with Spasms. Was in the Hospital 9 times. Has been using Spectro-Chrome since last August and takes no more Pills and has normal Eliminations.

P. M. had an experience with her Foot. While harvesting, something pierced the Foot and by evening it started to swell and feel sore. It festered. Orange and Indigo Local normalated it.

C. N. spoke of her ever increasing gratitude to Spectro-Chrome and what it means to the Body.

C. K. told of her husband having been in a highly nervous and fearful state. The Medical Doctor had told him he would have to take a Pill a day as long as he lived. His walking was formerly unbalanced and the Eyes had a Stare. He does everything for himself now and returned to work.

E. M. fell and bruised his Shin. Had severe Pain. Took 2 Tonations of Turquoise and became normal.

G. W. had a bad Cold accompanied by Stiffness in the Neck and Congestion in the Chest. She is now normalated by Spectro-Chrome.

———o———

FIRST DETROIT MICHIGAN PLANET

(CONCORDIA HALL, 62 EAST COLUMBIA)

Meeting Of Thursday, October 3, 1946

Successful Use Of Spectro-Chrome In:—Wound, Infection, Gash, Pain, Fever, Sore Throat, Cold, Tooth Extraction, Rheumatism, Measles, Earache.

Reported By:—ELIZABETH ROEMMING, *Recorder*

Communication from Principal Office was read and discussed.

Deputy—I received this Night Letter from the Institute that the Post Office Injunction Suit was thrown out of Court.

The Affiliates were shocked to hear such news.

Deputy—What has happened to our Country? Are not we supposed to be a Democracy? No hearing granted! How can you convict anyone when you will not let him tell his side of the Case. Did the Dictators do worse than that? What a crime against Humanity!

He then asked how many would like to attend the new Class in our Science. Six hands were raised.

Deputy—Dinshah will not come here for six people. Here is an opportunity to learn from Dinshah himself—think it over carefully.

B. N.—Before the Case Reports are given, I like to make a special request to the Institute regarding the Spectro-Chrome Magazine. I feel some Affiliates are being penalized because of the negligent Affiliates. For instance, some of our people are too ill to attend these Meetings, yet are using their Equipments faithfully. They would enjoy reading what is going on among the different Planets throughout the Country, but are denied that privilege. Would the Institute consider written explanatory requests?

Case Reports followed:—

J. R. reported husband hurt while working at a Machine. One Finger had to be amputated and the Finger next to it was broken.

Suffered severe Pain. One week later, Hospital Surgeon said there was a very bad Infection. Sufferer complained of Pain under Arm. Called the Deputy advised to stop the Medical Doctors and use Spectro-Chrome solely. Two days later, Pain stopped and Pus started oozing from the Wound. Now, the Pus has drained and the Hand is healing nicely.

M. H. said her husband had a bad Gash on top of his Hand requiring 18 Stitches at the Hospital. Taken home and started using Spectro-Chrome. Within 2 weeks, Hand almost healed and there was very little Pain during that time.

V. C. reported Pain in right Hand and could not raise his arm. His wife insisted on using a certain salve (This Salve is highly advertised on the Radio to relieve Muscular Strain). He was not satisfied, tonated with Yellow that evening and during the night took another Tonation. In the morning the Pain was gone.

Deputy—Let me ask you a question; why did you use your Spectro-Chrome last?

Affiliate—I guess I just didn't feel sick enough. (Laughter)

Deputy—Forget your Salves and Pills; go to your Spectro-Chrome immediately regardless of what the Disorder may be. You will always get results.

M. R. reported Fever; tonated with Turquoise. Fever subsided.

G. S. reported young daughter sick and had a Fever on Sunday. Tonated with Turquoise and Magenta. Condition cleared and child attended school on Monday.

K. R. reported coming home from work one evening with a Sore Throat; Kardoscil 69, Spiroscil 13. Tonated with Orange. Went to work the next day but in the afternoon began to feel quite sick. In the evening, Kardoscil 118 and Spiroscil 16. Unable to believe Kardoscil registered so much he took the Ratio several times, but the answer always remained the same. 118 to 16. Tonated with Purple; Kardoscil became 96, Spiroscil 16. Next tonated with Orange, Kardoscil 86 and Spiroscil 18. Two Tonations of Green made no change in the Ratio. His wife called B. N. the second day and she advised to use Yellow. Did so and the Kardoscil was 80 and the Spiroscil 18. As sick and as miserable as he felt, he lost only two days work.

E. R.—I like to say a few words. I do not believe my husband really knows how sick he was. The heat from the Body was terrific and his Breath was hot. He had difficulty raising his Arms and to cross his Legs was quite painful. His Eyes had a very peculiar glassy look about them. When B. N. suggested Yellow, I said the Eliminations seemed to be about normal. He never lost his appetite, but he did eat lightly and then it was mostly Fruit and Fruit Juices. I said he was intending to use Lemon, but she said use the Yellow instead. He did and in no time at all the Ratio was normal. He was quite sick the first day, but, the second day all Pain was gone from his Body. I wanted him to stay home from work the remainder of the week, but he insisted he felt fine. We both thought it was remarkable the way the Body responded to Spectro-Chrome and we are grateful we have Spectro-Chrome in our home.

M. W. told how she had been sick for many years. When she received her Spectro-Chrome, Dinshah told her it would take a long time. She has tonated faithfully and the last six (6) months has felt so much better. She went on a vacation for three weeks. During that time, she had no Tonations and felt fine. When she returned home, she resumed tonating again, as nothing helped her like Spectro-Chrome.

C. G. reported Colds, among the members of his family; successfully normalated.

C. W. reported Tooth pulled. Afterwards, tonated with Turquoise and had no discomfort.

H. R. reported husband ill with Rheumatism. Had been dosed so much with Vitamin B, that Hand was swollen. Unable to get relief. After tonating with Spectro-Chrome able to go to work. Found Blue very beneficial; was able to sleep under it. Son had the Measles and also a very painful Ear that constantly drained. Medical Doctor gave him one of the Sulfa Drugs, but child remained sick. Finally, called J. N. who advised her to stop doctoring and use Spectro-Chrome. She did that. At times she doubted it and even called J. N. during the night, but eventually the Ear healed. The Tonsils were not removed. Very well pleased with results.

J. N.—As I was reading the Magazine before the Meeting started, I noticed my name listed on page No. 4744 as a Member of a Committee. I may state at this moment that R. C. is now deceased. I should also like to have it recorded in the Minutes that I have never had any papers to sign or acted in any capacity whatsoever.

Albert Greenham spoke briefly about his Investigation. To date he received very little support. Many of the Affiliates are not willing to co-operate with him. He did find out from reliable sources that from $1,070.00 to $1,170.00 were donated. If anyone wishes to contact him would they do so as soon as possible, as he intends to wind up his Investigation within a few days. He read a letter which he received from the Institute.

Albert Greenham conducted the Educational Program and it was held as a symposium, three Affiliates participating with him. J. R., C. W. and W. K. Last month, he asked these Affiliates if they would like to join him in a discussion and they agreed to do so. The subject matter was the Pituitary Gland and its relation to the Spleen and the Liver. As the Affiliates mounted the Platform each carried volume No. 3 of our Encyclopædia. They put their books on the table. Albert Greenham then asked them questions pertaining to the Pituitary Gland and its functions. All questions were answered without any hesitancy. It was obvious that they had studied and knew all the answers. During the discussion Albert asked the Affiliates if they found this type of program interesting. The answer was yes, we certainly do want to continue. The Affinity waves were discussed, also the Glands and where they are located within the Body and the particular Color of each Gland. The new Affiliates were advised to use Green when in doubt. The symposium closed with a brief talk on the Pineal Gland.

Albert Greenham thanked those taking part (applause). Said he "What you heard tonight is all printed in our Encyclopædia. What these Affiliates have done all of you can do. If you want to introduce Spectro-Chrome to your friends, you must know about what you are talking. I like to see everyone of the Affiliates taking part in discussions such as these.

The question then arose about what to discuss next month.

Albert Greenham—Suppose you were tonating Yellow; did you ever wonder what happens to the other Colors?

R. G.—The others are *filtered out.*

Albert Greenham—Suppose we wait and see next month.

He called upon three other Affiliates to take part.

J. N.—I certainly enjoyed this discussion. The Human Body is a wonderful Machine. The study of Spectro-Chrome Metry answers completely puzzling questions that arise.

The Deputy thanked those who took part in the Educational Program and said we shall be looking forward to the symposium next month.

EDITORIAL REMARKS

Most cases come to Spectro-Chrome, when given over by other methods. Even our Affiliates do not care to continue their Affiliation once their Disorder is normalated. Our By-Laws need more severity to keep them in line.

There is no Charge against J. N. Her name was in the list specified and appeared as a Signatory. Albert Greenham did a creditable job and you know the result. Now, let it be a lesson to all of us and start a new slate to preserve our Organization as an exemplary, graftless, Service For Suffering Humanity. Out with favoritism and face worship!

The Symposium idea of conducting the Educational Program is deserving of encouragement. Affiliates will feel its advantage and inspiration.

DINSHAH P. GHADIALI AND DINSHAH SPECTRO-CHROME INSTITUTE DECLARED "GUILTY" ON ALL COUNTS BY FEDERAL JURY

The Indictment 3423C against Dinshah P. Ghadiali, the Originator of the Science of Spectro-Chrome Metry and President of Dinshah Spectro-Chrome Institute, was brought to Trial, in the United States District Court, Camden, New Jersey, on Monday, October 21, 1946, before the Honorable District Judge Phillip Forman, after he denied a prior Motion To Quash The Indictment. For the Government (Food And Drug Administration) Honorable Assistant United States Attorney Grover Cleveland Richman appeared as Prosecutor. Joseph McGuire, Attorney from the F. and D. Administration in Washington, D. C. assisted him; Dinshah, as usual, appeared in his own Defense and that of his Corporation.

It was the longest Trial in the history of the Court, lasting 42 days. The Charge was that the Defendants transported in Interstate Commerce 12 Spectro-Chrome Devices, with "Labels and Literature having False or Misleading Statements." Jury rendered Verdict of "Both Guilty On All Counts." Motions for "Judgment Of Acquittal" or "Order For New Trial" are pending.

Some details will be in the next issue.

SPECTRO-CHROME

October 1947 Purple 1947

Volume 23 Number 10

DINSHAH IMPARTIAL

Gas For Motherhood

X-Rays And Radium Cause Cancer

Answer To William R. Olsen

Attuned Color Waves

Spectro-Chrome Metry

In course of life you reach a time,
When Doctor, Lawyer, Priest alike,
Does fail to give the sought relief,
To troubled mind or burdened heart—
Then come to me:
I help you to help yourself.
SPECTRO-CHROME

The Science Of

Automatic Precision

DEVOTED TO THE SERVICE OF MANKIND

FIRST ALLENTOWN PENNSYLVANIA PLANET
(HOTEL AMERICUS, SIXTH AND HAMILTON)

Meeting Of Thursday, September 4, 1947.

Successful Use Of Spectro-Chrome In:—Stomach Ulcers, Prostate Disorder, Diabetes, Scalp Disorder, Sunburn, Neck Disorder.

Reported By:—MAE BURNSIDE, *Recorder*

Deputy Edwin D. Moyer was in the chair; 45 Affiliates were present. An Affiliate came without her Card and was denied admission. Communication from the Institute was read; also the newspaper clipping about the Birth of Shireen. A Resolution was unanimously passed to greet Shireen who had just discovered America! The Motion to forward greetings to Irene and Dinshah was by Mae Burnside, seconded by Emanuel Muehlen. An article in the magazine LIFE, dated August 18, 1947, under the caption "Quack Medical Devices," ridiculing Spectro-Chrome and Dinshah P. Ghadiali, was read. It was sent to that magazine by the American Medical Association.

The Sub-Deputy spoke about the Sensory Stimulant and explained the difficulties encountered where there were Operations performed. She also asked for a Silent Prayer before each Planet Meeting.

E. M. spoke on the use of Lemon Atcowa.

Question was asked about two people being tonated in one room. Discussion followed and explanation is asked from Dinshah.

Sub-Deputy asked why ½ hour Front and ½ hour Back Tonation should not be given. E. M. explained that 1 Tonation should extend to 1 hour.

A. C. testified to wonderful results by Yellow on Areas 6-7-8-9-10.

A. B. spoke of having taken X-Rays for Stomach Ulcers. After a year's deliberation, he decided to use Spectro-Chrome Yellow on Areas 6-7-8-9-10 and following the Rational Food Of Man became normalated.

S. C. had Prostate Disorder. Lemon Systemic and Purple Local was irradiated and has no more trouble.

N. W. spoke of the improvement in her husband's condition. He had Diabetes. Yellow Systemic and on 6-7-8-9-10 were used.

C. H spoke of losing so much Hair and Coloring that he was worried. Yellow on 6-7-8-9-10 and Magenta on Area 4 brought the Color back even into the Hair and stopped its falling.

E. H. testified to the efficacy of Blue for severe Sunburn.

T. K. spoke of a vertebral Disorder in her neck. She had no more trouble after irradiating Yellow on 6-7-8-9-10.

E. M. spoke about the incomplete Spectrum of Fluorescent Lights. They do not give a complete Spectral Band.

Many expressions of satisfaction were heard about the interesting Meeting.

EDITORIAL REMARKS

I am carefully watching the workings of the American Medical Association, which has been our Opponent since 1924. Let all call us

"Quacks," but, before I am through with my Life, I shall make them all quack like ducks. If you want to read the Truth about the real Quack who heads that Ass-ociation, read "Medical Mussolini", special contribution to our Affiliates, only $1.00 for the Popular Edition. I have the book in stock for prompt delivery.

Thank you for blessings to Shireen.

You are requested to cut off ALL "prayer" business, silent or otherwise, from your Meetings. We are NOT a Religious Society and if you will not follow the advice given so repeatedly, you will endanger the Planet. I admire and respect you all for your religious feelings, but, as explained before, we have all types of sick people, religious and non-religious and I must insist that this matter be kept out of our work, our service. Pray at home as much as you like and as you please, but,—do as I requested. This is a *Final Warning*. Read By-Law 54 and be tolerant of others who are NOT of your views.

A Tonation on any Area, Local or Systemic shall be a full hour; no division of time. As regards tonating 2 persons in one room, there is the risk of mixing the Beams and other considerations. You will have to supply full particulars. I do not advocate it, because of potential complications.

FIRST CHICAGO ILLINOIS PLANET

(MIDLAND HOTEL, 172 WEST ADAMS STREET)

Meeting Of Thursday, September 4, 1947.

Successful Use Of Spectro-Chrome In:—Poisoning, Spinal Injury, Phlegm, Indigestion, Itch, Boils, Pimples, Skin Disorder. Explanation Of The Chronos.

Reported By*:—HERMINE KUEMMEL, *Recorder

Deputy Kurt W. Bader was in the chair, 24 Affiliates were present. Deputy read the Communication from the Principal Office. Question was asked about the "Chronos" on the communication. The newspaper article was read announcing the happy arrival of a daughter "Shireen" into the Dinshah Family.

Henry Schilpp—The whole Chicago Planet likes to express their heartiest congratulations to Irene and Dinshah, with best wishes for Health and Happiness for little Shireen. May she be a joy to her parents and grow to be as fine a woman as her mother Irene Grace.

Deputy—There you can see, what healthy living means. To be able to produce children at the age of 73, when thousands of young people are sterile thru their wrong way of living and Smoking; Tobacco is also the main cause of Infantile Paralysis and many other Disorders; Miscarriages are also very frequently caused by it.

For the Educational Program, the Deputy read Chapter 90, "Spectro-Chrome In Persistent Disorders" and lively discussion followed. Said he in speaking of Lemon Atcowa, "There is a book written by a Priest, where he states, that people with Kidney or Heart Disorder should drink as much of Lemon Juice as possible; of course, I do not emphasize to follow this as he advocates to take from one Lemon per day, increasing by one up to 30 per day and back the scale again. Dinshah's System is the best of all Healing Systems; Dinshah is a very deep thinker; you may not understand it all, what he means as he has other reasons also, which the average person does not understand or it may take years to find out the real meaning. That is the reason why the Sufferer should always follow instructions strictly and he will improve. Many people take ice cold Lemon or Orange Juice in the morning on an empty Stomach; it is wrong. Take your fruit with the breakfast, when you have something warm; you should have Starches in the morning to warm your Body. You cannot start the day with the Lemon Juice only."

P. P.—I used to take Lemon in the morning in hot water; it did not cool me off.

Deputy—One has to use good judgment; with Rice and Yellow Split Peas I like Lemonade very much in the evening.

E. B.—People say the Blood becomes thin from Lemon.

Deputy—It moves the sluggishness away; thick Blood cannot move fast; thinner Blood is much better and more conducive to Health.

Figure 143 "Chemical Analysis Of The Lemon" from our Encyclopaedia, was read and explained.

Case Reports followed:—

M. S.—I had D. T. poisoning; was very nervous, had a choking spell; took Magenta for 2 days, but was still very nervous; then, I remembered, that I should take Green, so tonated with Green and Magenta for 3 days. Then, I was able to get up and felt better.

Deputy—Since there was a pulmonary condition, you did not go wrong by tonating with Magenta; of course, Green is the Cleanser and Balancer; it was good for you to stay in bed; even if you are used to be very active, you cannot expect Miracles from Spectro-Chrome and by tonating with Green and Magenta, you actually used the whole Spectrum.

E. B.—During the heat wave, my mother could not get sufficient air and her Legs were blue; her Kardoscil was high and Legs swollen. Should she be tonated with Blue?

Deputy—In Nature, everything has its governing. Yellow will take Spasm away and the Blue from the Legs; irradiate her with Yellow on 6-7-8-9-10, Purple on Area 4.

The Deputy gave some Reports.

E. S. Kardoscil 75, Spiroscil 21. In 1941, jerked his Spine, when the other fellow dropped one end of a stove both were carrying. He went to a Chiropractor, who could not help him, so went to a Medical Doctor, who took X-Rays and said that he had Pott's Disease and put him in a cast for 10 months, from the Shoulders to the Legs. Then he said, he could not do anything except operating, by what was called a Spinal Fusion and also the Left Kidney should be removed. This was in 1943. Three months later, he got out of bed and started to walk; a year after he had burning Pain, when passing Frontelim, which was very discolored and cloudy; he also had Pain in Bladder. Two months of Spectro-Chrome Atcowas 60°, 90° and 300° improved his condition so much, that he wants his own Equipment now.

H. L., Kardoscil 56, Spiroscil 24, was suffering for 2 years from Phlegm on the Chest and the last 2 months had difficulty in breathing; had much tightness in the Chest, Belching, Gas and Indigestion. After 2 weeks of Atcowas 60° and 300°, the Kardoscil and Spiroscil was 72 and 18. After 2 months, the other symptoms disappeared also; he was very well pleased with Spectro-Chrome.

F. B., Kardoscil 68, Spiroscil 18, was itching; had Boils and Pimples all over Body for the past 5 years. Had used Sulfa Drugs and Penicillin without relief. With Spectro-Chrome Atcowas 60°, 120°, 180°, 200° and 300°, the condition was cleared for the first time in 2 months. He was very skeptical at first, but very soon changed his attitude.

V. H., was suffering for 2½ years from Skin Disorder on the Legs; first weeping, then dry and scaly. Used Atcowas 60° and 300° with very good success. In about 3 months, the Disorder disappeared. The people take too many Enemas and Cathartics, until there is no natural Elimination any more; the Bowels become inflamed and if they still keep on with the Cathartics, Mucous Colitis might set in. We have a very fine example of it in the "Family Health Protector", Article 28, where Dinshah made his first success with "Colored Light" and that was the reason he perfected the Science of Spectro-Chrome Metry. That is still alive.

EDITORIAL REMARKS

A man, using his youth well, can retain his masculinity to his dying day. Such matters of Health preservation should be taught in the schools and colleges, without prudery. All Body Organs are good for their normal functions and the Law of Obscenity is a humbug of prurient civilization. Were anything indecent or obscene, the Creator would have given clothes to babies at birth. Years ago, the French Legislature wanted to put all dogs in streets, in trousers, to inculcate morality!

Chronos in Greek means Time. The Mogen David has 6 Points and 6 Angles. Our Spectro-Chrome Science uses 12 Atcowas. Green at the top begins 12 o'clock and Magenta at the bottom is at 6 o'clock, the intermediate Colors being represented similarly in their appropriate position. A. M. is represented by White and P. M. by Black, in accordance with the Astronomical Reckoning, without relation to Sunrise or Sunset.

———o———

FIRST CLEVELAND OHIO PLANET

(CARNEGIE HALL, 1220 HURON ROAD)

Meeting Of Thursday, September 4, 1947.

Successful Use Of Spectro-Chrome In:—Anemia, Nervousness.

Reported By:*—EDITH REES, *Recorder

Deputy George J. Brown was in the chair; 24 Affiliates were present. A hearty welcome was given the Recorder after her absence in two Meetings.

A new Affiliate, S. A., was introduced and given an opportunity to ask questions and to report on how he happened to start with Spectro-Chrome Metry and tell the results he had. His family friend told him that he was in a terrible condition and that he would have to take Shots and Pills regularly the rest of his life for Anemia and Nervousness. His wife heard of Spectro-Chrome through N. W. and became a Benefit Student

with her own Equipment. Much credit was given E. A. for seeing that he got the proper Tonations at the correct time. The Colors given daily for six weeks were Green Systemic, Magenta on Areas 4 or 18, and Red on Area 7. This routine was followed for six weeks and very satisfactory results attained. It was necessary that he report to the Medical Doctor for a periodic check-up and each time the Doctor took the credit for the improved condition and handed out more Pills which were thrown away as soon as possible! To this day, the Doctor thinks that the Pills did the corrective work! The friends tell S. A. how much better he looks and he is back to work and feeling fine. He still takes Tonations when needed. Instead of going on a Meat Diet for Anemia, he went on a fruit and vegetable diet and had his pound of grapes every day. He concluded his report by saying that he is now a Benefit Student and has his own Machine. The Affiliates enjoyed hearing the report and considered it one of the better ones given.

Communications from the Institute were read by the Deputy. The Affiliates send congratulations on having the little daughter, Shireen and wish her Health. It is nice for the family to have a daughter in the home, especially so for Irene.

H. S. and J. S. wanted to go to Malaga, so the latter may have a personal talk with Dinshah concerning her condition. The Deputy called him to make the arrangements, but, the Colonel said they could not take Sufferers at present and any how man and wife could not be received there at the same time. H. S. said that he would stay at the Hotel so Dinshah finally consented to let them come for their vacation as they requested and that they could have all their meals at the Institute, but, would have to stay nights at the Hotel. H. S. and J. S. then gave a very interesting report on their stay with Dinshah and his family. They mentioned about the Colonel's routine and what a busy man he is these days, taking care of Telephone calls and preparing his Brief. They were impressed with the harmonious living conditions in the home and how intelligent the sons are and how interested they are in every thing that occurs there. Several Affiliates came during their stay there for consultation and Free Guidance and every one was told that he must follow the Principles and Food to get results. Some one called from St. Louis to get eleven Equipments. J. S. found out what Colors to take and is following the Advice given. They both sang praises for the Colonel and his family and said that Dinshah has so much more to given us when he gets his freedom from the Courts. This too was a very interesting Report.

The Deputy and A. K. spoke concerning a Dentist who would like to become an Affiliate, but, cannot because of our restrictions. His wife could become an Affiliate, but, both men felt that there was too much risk in accepting her membership for the Dentist might turn out to be an agitator instead of a promoter. Everything is harmonious now in our Planet and we want it to remain so.

The Recorder then spoke concerning the lost Reports and stated that she held the receipts for each month this year, except the August Report and the Deputy had taken care of that Report and sent it in at her request.

J. W.—Inquired how to go about getting a new Motor. The Deputy said he wished he knew, for he too would like a new one. The Recorder and J. T. said they would like new Motors. Please advise for four new Motors for Aluminum Equipments. Please state the contribution and if they are available.

This was a very interesting meeting and the Affiliates who were not present missed something.

EDITORIAL REMARKS

Thank you for good wishes. Shireen is living serenely and growing daily by one day. Before asking for stay in the Chromet here, it is necessary to have a complete General Advice Chart, to determine whether the Affiliate may be received. We have no facility for nursing cases. The Affiliate has to use self-help.

Duplicates of the lost Reports are now here and will be published.

All parts are available, but, not as spares; only replacements. If the old Motor be returned, a new one will be sent as replacement for a Contribution of $8.00.

———o———

FIRST MILWAUKEE WISCONSIN PLANET

(MILWAUKEE AUDITORIUM, WEST KILBOURN AND NORTH FIFTH)

Meeting Of Thursday, September 4, 1947.

Successful Use Of Spectro-Chrome In:—Anorexia, Imbackelim, Insomnia, Varicose Veins, Sore Feet, Fever, Gingivitis,, Swelling, Accident, Tooth Disorder, Pain, Soreness, Tooth Extraction, Hemorrhage, Urine Retention, Paralysis.

Reported By:—ETHEL L. SCHOTT, *Recorder.*

Deputy Evert Dahlke was in the chair; 148 Affiliates were present.

Affiliates were requested to hold up their Affiliation Cards, because, it was thought someone entered the Meeting without registering. All had the Cards.

The Deputy read and explained the communication from the Principal Office, which also contained the news of the birth of Shireen, daughter of Irene Grace and Dinshah. An article from "The Vineland News", regarding Shireen's Birth, was also read, after which A. E. suggested that it would be a good time to send congratulations to the Institute on the Birth of Shireen.

First, before making the Motion, the Deputy read and explained the "Good Thoughts" for August 23, "Liberty". Then the Deputy read a Motion that he had prepared for such purpose and which was stated as a Motion by O. L., seconded by A. F. and on vote was unanimously carried. Motion was as follows: "We wish to extend our most sincere congratulations and best wishes to the newest member, 'Baby Shireen', of the Dinshah Family and our Dinshah Spectro-Chrome Institute. We wish her Health, Happiness and Success. May the Higher Powers bless Irene Grace, the mother of this sweet bundle."

Deputy—At this time I should like to read one paragraph from last month's communication. "All Benefit Studentship Requests received AT THE PRINCIPAL OFFICE before Monday, September 15, 1947, will be accepted at the present $90.00 (Ninety Dollars) Contribution. All subsequent Enrollments as Benefit Students will be $100.00 (One Hundred Dollars). New forms will be in your hands in time."

The Deputy explained that those Affiliates who did not receive their magazines this month, did not receive them because they had not attended the last Meeting and therefore. they would receive all of them if they attended the next Meeting in October.

Deputy—At this time I like to read a communication from Emma Schroeder, dated September 4. 1947, to me as Deputy, Eleanor C. Luedtke, as Sub-Deputy, and all Affiliates of the First Milwaukee Wisconsin Planet of the Dinshah Spectro-Chrome Institute: "Many months ago, at a Meeting—and it was later printed in the magazine—I expressed the desire to be relieved of my secretarial duties. At the Convention, in March of this year, the issue was again brought up. Finally I have made a definite decision.

"After serving the First Milwaukee Wisconsin Planet for a period of about seven years, I submitted by resignation to the Institute during the past month.

"It is a rare privilege to work for this fine Organization and I felt highly honored when I received the Appointment seven years ago, to serve as Recorder of the Planet reputed to be 'the largest in the World'.

"It was not only an opportunity to serve, but, an opportunity to gain additional education in the greatest Science known to mankind and to gain intellectual experience which is not easily within the reach of the layman outside of a university.

"The name of my successor has not been officially transmitted to me, but, whoever is appointed to the office has my sincerest and heartiest congratulations and fullest co:operation.

"Suggestions and pointers which have come to me during my period of service were submitted to Deputy Dahlke.

"I have performed my duties conscientiously and to the best of my ability, always bearing in mind fairness and loyalty and devotion to the Cause.

"With best wishes for continued success, I remain

Sincerely,

Emma Schroeder."

(APPLAUSE)

Deputy—It was a pleasure to have been able to work with Emma Schroeder, as she was always conscientious and most capable and we hope that her successor will try to do as well after several months' experience.

At this time G. W. asked if this would be the time to affirm the new Recorder, but, was advised that it could not be done until the Deputy had received notice of Affirmation from the Institute.

E. L. suggested that the Deputy explain about the change of Annual Fees and Dues for continued Affiliation.

The Deputy then read from last month's communication as follows: "Furthermore, the Annual Fees and Dues for each Affiliate for continuing the Affiliation will be a flat $6.00 instead of the present $3.50."

Deputy—At this time I like to have all the new Affiliates stand up and give their names.

There were eight new Affiliates standing:—S. P., R. O'R., E. S., E. G., E S.. D. S., F. P. and T. E.

Case Reports followed, were explained and comments offered. Otto Lutz gave three Case Reports:

(1) L. M., age 75, female. No appetite, felt weak. Backelim one in three days. In one month's time she felt like living again. Kardoscil was

104, Spiroscil 30. Kardoscil is now 74, Spiroscil is 20. Colors were Yellow 6, 7, 8, 9, 10, Purple on Area 4, then Magenta on Areas 4 and 18. Appetite returned, Backelim now is two in one day. She is *well pleased.*

R. B. asked how to decide what Color to use to bring down Respiration, Indigo on 17 or Purple on 4 and Otto Lutz said that he learned from years of experience that Purple acts quicker on the Circulatory System; for instance, one woman who had not slept for five nights, slept after Purple.

(2.) M. B., 50 years, female. Varicose Veins on right Leg. Always wore elastic stocking. Kardoscil 80, Spiroscil 18. One Backelim a day. Atcowas given were Yellow 6, 8, 9, 10, Orange Systemic. She followed the Rational Food Of Man, with good results. Indigo Local was used.

(3.) M. R., age 77 years, female. A case of sore Feet. All the Flesh under the Toes of both Feet was gone and Pus was coming out. The Medical Doctor said, after about 8 months of treating her, that it would be advisable to remove both Legs just below the Knees. Kardoscil 72, Sprioscil 17. Backelim, two a day. Atcowa used were Lemon Systemic, Magenta on Area 4, Yellow on Areas 6, 7, 8, 9, 10. In two weeks time, the bandages were loaded with Pus, as she had never seen before. There was a great deal of Pain at that time. Too much for her. Then used Yellow on Areas 6, 7, 8, 9, 10, which made the Pain less and the Feet were almost normal in two months. She said, "What do I care what I read in the papers in regards to Spectro-Chrome. It is my Feet that count to me. Please sign me up as a Benefit Student." Otto Lutz did enroll her.

S. E.—Sunday, August 17, when our son, age 21, came home in the evening, he complained of feeling hot and I took his Temperature. It was 102.8° at 11 P. M. His Kardoscil was 96, Spiroscil 18. At 11:24 P. M. which was the Favor Period, I gave him a Tonation, Turquoise Front Systemic. Monday, 2 A. M., the Temperature was 102° and 6:30 the Temperature was 100°. At 7 A. M., the Kardoscil 86, Spiroscil 20. Gave him Orange and Lemon Juice during the night. In the morning, I called E. L. and told her that he was complaining that a wisdom Tooth he was cutting was causing his Gums to feel sore and inflamed and she told me to give him Turquoise Local on Area 1 and Green Systemic Front and Back, alternating with Magenta on Areas 4 and 18. Wednesday, his Temperature was 99.2°, Kardoscil 80, Spiroscil 18. That afternoon, we went to the Dentist and he discovered that the boy had a Trench Mouth, due to low vitality. Our Dentist wanted him to consult a Physician, but, I told him I knew what Spectro-Chrome had done for our family in the past. I continued to give him the Tonations E. L. advised me and by Friday his Temperature was normal and the following Monday he went to the Dentist to have a cavity filled. When the Dentist looked at his Mouth, he said, "It is 100% O. K." and I told him we only used the Spectro-Chrome. For a couple of days following, I used Indigo Local instead of Turquoise, to harden the Gums and take the Soreness out, for they had been very inflamed and also had Pus. We are very grateful to Colonel Dinshah for Spectro-Chrome and the wonders it did for our family.

A discussion of this case followed.

Deputy asked if there was anyone who had trouble with their baby having a Fever when baby was cutting Teeth, to which S. P. said that her 23 month old baby had a Fever which lasted only 4 hours so she did not use Spectro-Chrome for Fever. She did, however, use Yellow for the Bowels. Deputy commented that the Body needs Yellow to pull Teeth through the Gums and that in all cases where baby is cutting Teeth, the Blue will have little effect until Teeth are through.

L. L.—I had an automobile Accident July 18. Called Emma Schroeder and asked her to come and see me. My Lip was cut in two places, and my Knees were bruised. Emma advised me to use Turquoise Systemic, which I did and on the Favor Hour to use Magenta on Area 4. My Knees started swelling so I used Orange Local on Knees next Favor. Next day, Emma checked my Kardoscil 78 and Spiroscil 22, she advised to use Orange on Area 17, Magenta on Area 4. My Knees were swollen about 3 times the normal size. Next 5 days, I used Orange Systemic and Scarlet on Area 4. Swelling went down in three days. Then, I received Free Guidance from the Institute and was told to use Yellow on 6, 7, 8, 9, 10 and Magenta on Area 4. Started to use Yellow on Friday, August 25. Saturday my Kardoscil was 84, Spiroscil 17. My improvement was so rapid that I returned to the Class after 4 days. (100th Spectro-Chrome Metry Class).

J. T.—Four days ago, I had a Tooth pulled. It took two Dentists almost an hour to extract the Tooth. They gave me Injections and after they had started to pull, it did not come out. They X-Rayed Tooth and it showed that the Roots were hooked onto Jaw Bone They had the chair almost horizontal and took turns, different Forceps, to pull the Tooth, They gave me Anacin Tablets (which I threw away) and told me to put ice on the Area when I reached home, because it was going to be very painful. I went right under the Indigo before the freezing wore off. Took tonation in morning and in afternoon. I removed gauze and as I lay there I could feel the Bleeding stopping and after the evening Tonation, I had absolutely no Pain. I had Soreness about the size of a half dollar on my lower left Cheek. Altogether, I took five Indigo.

Otto Lutz jokingly remarked that J. T. "must have had Royal Blood to bleed that much!"

E. L.—When I fell and pushed Teeth into Jaw Bone, I used Indigo and got the same results—no Pain.

When A. T. asked if anyone had ever taken Purple before going to Dentist and without Injection had Tooth removed, N. F. said, "I took Purple on Tooth and could hardly feel the Extraction."

R. H.—I gave my eleven year old son Purple before I took him to the Dentist. The Dentist wanted to give him an Injection of Novocaine, but, I told him to go ahead with the Dentistry work without the Injection. I told him it wouldn't hurt him. He was surprised. The next time I called for an appointment for my son, the Dentist said that I should do the same as I did with him last time!

F. S. wanted to know if it would be wise to take Red Local to cause an Infection, before having Tooth pulled. He was advised against it.

H. K. wanted to see how Spectro-Chrome would work on her, so she took Purple before having a Tooth extracted. The Dentist did not want to extract it without giving her an Injection, but she wouldn't take any. After he had removed Tooth, "he grabbed a cigarette, because, it hurt him more than it did me." She felt only a little Pain.

When the Deputy asked H. K. if she used Spectro-Chrome in Dentistry, she said she used it for Hemorrhages. Hemorrhages stopped in ten minutes, where otherwise she would have to call a Physician.

H. K. stated the case of a gentleman who had a Tooth extracted and his Face became swollen. He called H. K. and said he wanted a Tonation on his Face. She gave him Indigo. The man said, "To this day I do not know what I would have done if I had not taken a Tonation."

The Deputy explained to the Affiliates that H. K. could use Spectro-Chrome on Areas 1 and 2 only, because, she was in Spectro-Chrome before restrictions were put on it.

Deputy—I understand, J. T., that you were at the Institute last month. Do you have any comments?

J. T.—Last month, at this time, I was at the New York Planet Meeting. They have a nice group and are very interested, as we are, but they have to go some to compete with this one. To me, they don't discuss things as we do here, I feel sorry for Lina Fischer. She tries so hard to get people to say things. Colonel Dinshah took me to the Planet. He had brought my Case History Reports with him and was going to ask the Affiliates if he should take the case or what he should do. They "ohed" and "ahed", etc. and said that they thought the case was hopeless and that the Colonel shouldn't take it. The Affiliates were surprised when I got up and told them that it was my case. They made me run around and stand up on a table. We had quite a time. I enjoyed my visit at the Planet Meeting, because, everyone was so nice to me. I had a wonderful time at the Institute Irene Grace is a splendid person and although she was quite uncomfortable at the time, because she was expecting her baby, she went into the Office every day at eight in the morning and worked even though Air Temperature was 102°. She did not each lunch; did not even take time out for lunch. I was becoming restless because Colonel did not let me do any thing. It was hard for me to sit still, because, there was so much work to be done. The Dinshah family tries so hard to keep everything going.. They need plenty of help and plenty of co:operation. Yet they do have their "time outs". The boys play the violin, piano and other instruments. They play everything from classics to boggie-woggie. The boys do all the cooking and Irene supervises. How she manages to do all she does and still keep so cheerful, I don't know. I was treated as one of the family. They all made me feel so at home. We had breakfast at eight. We had such food as is listed in an example of a menu in Rational Food Of Man. At first, some of the food seemed too sweet to me, but I became used to it and now like it. One time we had so much whipped cream over blueberry pie that we could hardly see the pie.

A. T.—Tell them how much weight you gained.

J. T.—I did not gain; I lost weight. I had to take in the seams of my slack suit when I came home. All in all, I had a wonderful time and enjoyed my trip. I did not take any Tonations while at the Institute, but, I did notice it. I still notice it if I do not take Tonations.

At this point, the Deputy said that J. T.'s case would be read at the next meeting. She did, however, give a summation of her case as follows:

"I live in Kenosha (Polio area) and about a year ago I was very sick. I thought I had the Flu, called Medical Doctor who diagnosed it as Flu. About three days, I noticed I had trouble urinating. About two days later, I did not urinate at all. The Doctor said I should walk, but my Knees were stiff. I was rushed to the Hospital. They took the Fluid out of my Spine. I think that is what caused me to be so sick. I was taken to an Isolation Hospital, where I was not allowed to have visitors, not even my husband. They inserted a Catheter in me and took urine every 8 to 10 hours for two weeks. I stayed in the Hospital for two months. They stuck me with pins to see if I had feeling. I felt something like when a Foot falls asleep. Specialists were called from Milwaukee and my husband and sister were told that I would be partially paralytic all my life. My sister took over, had release papers signed, took me home and removed the Catheter that was attached to my Body. I was given Enemas in the Hospital, but, nothing would come. After a little less than 48 hours of Scarlet, I was carried to the toilet and I urinated all by myself. We then switched to Yellow to get Bowels working. In two days, we got Bowels moving. I had two and three Movements a day and couldn't stop. Then we worked on Numbness. In two weeks I was out of Pain. Before I was sick I weighed 147 pounds, and 80 pounds after. Three days after I was home, the Doctor came to check the Catheter. He saw that I was sitting up in bed and said, "Well, what are you doing sitting up?" I told him Spectro-Chrome did it. He then asked, "What happened to the Catheter?" My sister said, "Here's your Catheter. Take it, we don't need it." The Doctor thought it was ridiculous for us to do the things we did. He wanted to know what right we had to remove the Catheter. My sister told him, "Well, Doc, if you ever need any help down at the Hospital, just look us up!" (APPLAUSE)

Deputy—I think everyone here enjoyed listening to your case, J. T. It really shows that the Doctors are at a loss. If we Affiliates use Spectro-Chrome conscientiously, we won't need a Medical Doctor.

The above led to a discussion as to how many times a day we should tonate. It depends upon the person, case and condition. Nature can only repair so much at a time. Otto Lutz said, "Sometimes 1, 2 or even 3 Tonations, but you can not crowd Nature."

The Deputy brought us back to the Institute for a few minutes, when he said he watched Dinshah put teaspoon after teaspoon of Sugar in the already sweetened Lemonade. Evert Dahlke asked him if that was not too sweet for him then, upon which the Colonel remarked, "I am an old sour prune and I must have something to make me sweet." (Now, we know that is not true, Colonel.)

A. E. (directing remark to Deputy)—You are to be commended on the way in which you conduct your Meetings and the way you make the Affiliates feel like talking and discussing cases. Also A. T. and J. T. add much to our Meeting, because, they are so interested and when they want to know something, they ask until they get the answer. That is the way more of us should be. "Boy, I'm glad they are here." (APPLAUSE)

Deputy—Thank you. I am trying to be conscientious and I hope you Affiliates will co:operate with me by bringing questions up that will keep the Meetings interesting. If your questions cannot be answered here, they will be answered in the Editorial Remarks.

R. S.—We found, from past experience, that the Affiliates must be "pumped" for information.

Theodore Mohrhusen sends us his regards from Omaha. Edward Straub visited him about two weeks ago and said that he looks well, gets his magazines once a year, misses the Meetings and thinks of us. (APPLAUSE)

The Deputy read a communication from Emma Schroeder which she had received from former Congressman Samuel Pettingell, "Gentleman from Indiana", who took up the opinion of Judge Claude McColloch in regards to the American Citizen's rights.

E. T. asked J. T. what Atcowas she used and was told that Yellow and Lemon, alternate Front and Back and Magenta on 4 and 18 were used. Orange was used for the Respiration. After Yellow, she swung over to Lemon. because of Injections that were given her (fifty Penicillin Shots) and the four Spine Taps she had while in the Hospital. She remarked that the odor in the room after taking Lemon Tonations, was terrific, showing the cleansing process.

The composition of Scarlet and its use in J. T.'s case, in regard to increasing Urination, was explained to one of our new Affiliates, E. S.

The Meeting certainly was interesting.

EDITORIAL REMARKS

Thank you for your kind wishes. Yes, "Liberty" is what drew me to American Citizenship and now ——? American Liberty is the shadow of its former self, because, Civil Liberties are daily disappearing under the sinful heel of the haughty Bureaucratic Cheapjacks.

The resignation of Emma Schroeder came with a sudden shock, but, the Council Of Trustees had to accept it, because, she refused reconsideration. She has multiple other duties and we are highly appreciative of the phenomenal services rendered by her in the past and convey to her our cordial thanks.

The new Recorder Ethel L. Schott is one who was our First Champion. Her first Report shows her competence for the onerous duties and I congratulate her on the efficiency with which she performed the task. She promises to be a fit successor to Emma Schroeder. I am sure she will grace the post. Give her your kind co:operation by speaking clearly.

When a Planet Officer receives the Appointment, as soon as the Card is shown, Affirmation is in order. It needs no special instruction from the Principal Office. Study By-Law 6.

Purple, by lowering Circulatory Tension where needed, assists in easing Pain.

H. K. acts within her rights, being under By-Law 52, Section 2. She is one of our staunch Affiliates and is very reliable for her devotion to the Cause.

J. T. had certainly a lively time here at the Institute. The boys kept her busy listening to their musical orchestra. If she will follow the Free Guidance and other instructions given, she will have a happy life. She is a woman with plenty of good nature.

I need plenty of Sugar (a Carbohydrate), because of my concentrated, intensive studies and peculiar activities. When a man has to think of both physical as well as metaphysical problems, he exhausts his Vital Energy during Vigilance as well as Sleep and to keep him in trim, Carbon and Hydrogen are essential. Without the Sugar, I should be walking with a third leg in my hand, at this age of 74 and those who see me working, ought to know the unusual strain under which I perform my duties without flinching. Work to me is a pleasure—in fact, I need more work—not legal worries—to make me happy.

Evert E. Dahlke is proving a good Deputy and with your continued co:operation he will make your Planet a greater success. A Commander is only as good as his soldiers and if you all stand by your Deputy and take proper part in the discussions and the Educational Program, even

Medical Doctors will be glad to listen to your Meetings some day—as our honored guests.

———o———

FIRST ORANGE NEW JERSEY PLANET

(ESSEX HALL, 327 MAIN STREET)

Meeting Of Thursday, September 4, 1947.

Successful Use Of Spectro-Chrome In:—Pain, Dogbite, Swelling.

Reported by:—MARYANN KLENNER, *Recorder*

Deputy Lothar C. Klenner was in the chair; 13 Affiliates were present.

Communication from the Principal Office was read. The news concerning the Birth of Shireen was received with great joy and all send their most sincere congratulations to both Irene Grace and Dinshah.

G. d'A. had several Teeth extracted and irradiated Indigo to relieve the Pain. He was very pleased.

H. M. related how her son was bitten by a Dog. She tonated him with Green once and twice with Orange. The Swelling disappeared and it healed wonderfully and quickly.

The Deputy read from our Spectro-Chrome Metry Encyclopaedia, Chapters on "Bugbear Of Diagnosis," "Fevers By Medical Conception," "Spectro-Chrome In Fevers," "Rubor, Calor, Tumor, Dolor" and "Absurdity Of Differentials."

Some questions and answers followed.

EDITORIAL REMARKS

I thank you for kind wishes. Build your Planet steadily and some day Irene Grace and Shireen will visit you!

———o———

FIRST MERRILL WISCONSIN PLANET

(HOTEL BADGER, MERRILL, WISCONSIN)

Meeting Of Thursday, September 4, 1947.

Successful Use Of Spectro-Chrome In:—Pain, Heart Trouble.

Notes By:—DOROTHY D. NELSON, *pro tem.*

Reported By:—GERTRUDE L. PARSHALL, *Recorder.*

Deputy Emil Erben was in the chair; 17 Affiliates were present. The Deputy read from "Good Thoughts" for to-day. Communication from the Institute was read; also article about Birth of a daughter to Irene Grace. Congratulations! Lively discussion on Kardoscil and Spiroscil. The question was raised as to which was faster, the male or female Kardoscil. W. B. read from our Encyclopaedia. Chapter 109 regarding the Heart. Welcome was given our new Affiliate from Wausau. We hope we see our new Affiliates every month. Our Wausau Affiliates are very loyal.

Case Reports followed:—

C. S., daughter of K. M., reported by her father. This morning at six o'clock she had severe Pains under the Heart Area. Kardoscil was very fast. She got up to feed the baby and had a very difficult time to breathe. She was breathing in short jerks and it was hard to take her Kardoscil and Spiroscil. Magenta was used and she was relieved. One hour after that, she was down town and felt better.

J. M.—The thing I like about Spectro-Chrome is, that you can take it and get up and go down town. If you have the Medical Doctor, you would probably have to go to the Hospital, have a Blood Transfusion and so on!

One of our new Affiliates asked if too much Magenta could be taken. We like Dinshah's advice on this question. W. B. said if he takes a Magenta late at night and his kardoscil is high, it would reject the Red and take the Violet and he could sleep well.

J. M. wondered about the use of two Machines simultaneously. The thought is that in the case of a Burn, where both Front and Back Areas were burned, relief would be much quicker if a Machine could be used on the Back and Front at once or where Tonations were needed on both Areas 4 and 18, if taken at once would that not be satisfactory?

Editorial Remarks on Meeting of Thursday, April 3, 1947 read by Deputy.

E. E. asked if all new Affiliates knew what we mean by the Aura. He read the Editorial Remarks, under the St. Louis Planet Report.

K. M.—A thought to those who have not made a study of the Aura. It is the heat of the Body and unlike a stove which radiates heat and is lost, the Body has a North and South Pole, starting with the Head to Genitals. This can be replenished by the proper food and when it is depleted we use Spectro-Chrome.

The Deputy remarked that even the Chemistry of a Stone projects an Aura. Sleeping with someone affects the Aura; that is why Dinshah advises sleeping alone.

Excerpts from article from 1946 edition of Milwaukee Journal read by the Deputy. The name of it was "United States a Sickly Nation."

C. G.—The few Colds I have now are nothing compared to those I used to have; in fact, I hardly know I have a Cold. K. M. said it was rare when I didn't have a Cold in the Winter. He said our entire Nervous System is fed through the proper Breathing and it has to be rejuvenated. It is important.

Question—I lost my sense of Smell. What can I do for that?

Upon questioning it was also found her senses of Hearing and Taste were not as keen as they should be. She was advised to send in her General Advice Chart.

K. M.—I would like to extend a vote of thanks, for all the hard work Emil Erben put in the Planet and we are all grateful for that. We hope he enjoys his trip. Emil is planning to leave our Planet and move to Arizona where he said there is a great opportunity.. It is a place where people go for their Health.

EDITORIAL REMARKS

As the Atcowas operate through the process of Reinforcement and Interference, the Auric Vehicle will receive or reject them according to the requirements of the moment. It is this that makes Spectro-Chrome Metry the safest Healing System. Any excess of any Atcowa is rejected and any excessive use can produce no harm and it is that property that removed Differential Diagnosis as a necessity. The Aura is merely the Chemical Emanation of the Chemicals in the Physical Visible Body.

All Senses are governed by the Infra-Green Division of the Spectrum, the Red being the most Sensory Stimulant.

Our best wishes will follow Emil Erben.

FIRST PORTLAND OREGON PLANET

(MASONIC TEMPLE, 1119 SOUTH WEST PARK AVENUE)

Meeting Of Thursday, September 4, 1947.

Successful Use Of Spectro-Chrome In:—Bump, Pain, Bruise, Soreness, Swelling.

Reported By*:—LEWIS BISCHOFF, *Recorder

Deputy Stanley Burroughs was in the chair; 19 Affiliates were present.

A letter from Dinshah was read by the Deputy. In connection with the letter an article published in The Vineland News and sent, referring to Dinshah's new baby girl, was read and everyone was greatly surprised and pleased. This is the man who the Food And Drug people and Judge Philip Forman thought might "die in prison" because he was so old! He will probably be alive after many of the ones now prosecuting and persecuting him will be dead and forgotten. We all noted that Vincent, M. Miles started the procession. The rest will follow one by one; Dinshah will outlast them all. Other men at his age would never have thought of putting up a fight against such tremendous odds. We all pray to God that he may live to see "Spectro-Chrome—In Every Home," and the American Medical Association where it would put Dinshah if it could.

Tonight was question and answer night. V. W. asked what to do for Varicose Veins. The Deputy explained what he followed as a schedule for such cases, Hemorrhoids and similar conditions.

L. T. asked for what was Blue used and it was explained that it was used for Fevers, Burns and Irritations. Blue is also used as a Vitality Builder. When one is very tired after hard work the Blue helps to supply Oxygen and revives one very quickly. N. H. said that is what she needs. The Deputy replied that her condition was different and that she needed Red as she had a tendency toward Anemia. She also asked what to do in case of a Fall and Swelling as the result. A. W. asked when a Bruise would be considered a Chronic Condition. He stated he had gotten a bad Bump and about twelve hours later he started with Green and Turquoise and not much happened, so he used Red and it stopped hurting; then, he went back to Green and it seemed to help more. L. T. also asked what to do for Boils. The Deputy explained. L. T. asked further about Heart Disorders.

M. L asked what to do for Soreness in Area 11. It felt sore most of the time just like a very Sore Throat.

R. A. mentioned that she had a very bad Accident some time ago and was very severely bruised. She had used a great deal of Green, Lemon and Magenta with very good results in keeping away the Soreness and Swelling.

N. H. asked what to do for her boy. He was subject to frequent Headaches and Hay Fever attacks. She was told what to do.

F. B. said that her boy was beginning to complain a little of an occasional Pain in his Muscles and Joints. The Deputy told her to use Green Systemic.

EDITORIAL REMARKS

Benign Jesus said "As ye sow, so shall ye reap." Those beautiful words are as true today as when they were uttered centuries ago. We wish no evil to anyone. Even to our Opponents we say "Live longer and see the growth of the message of Truth—Spectro-Chrome!" I say verily unto them: "The Law Of Karma is the Divine Law Of Retributive Justice. Action and its Reaction is Karma. If we keep our mind on the discharge of our duties, discharge them conscientiously and wish good to all, the percussion or rebound will *never* hit us. Our Opponents even are our Brothers, so let us mind our business and let the Opponents get their reward through Him Who Ever Watches The Innocent.

In conducting the Educational Program, I suggest that instead of giving haphazard replies to questions, you use our Sufferer Chart. A batch of it may be had Free by any Planet for the mere asking. Without its use and without having full particulars before you, there is a likelihood of sinking into the Bugbear Of Differential Diagnosis which we discarded in our Science. A batch was sent to your Deputy.

———o———

FIRST DETROIT MICHIGAN PLANET
(CONCORDIA HALL, 62 EAST COLUMBIA)

Meeting Of Thursday, September 4, 1947.

Successful Use Of Spectro-Chrome In:—Cold, Accident, Pain, Urethritis, Asthma, Backache.

Reported By:—ELIZABETH ROEMMING, *Recorder*

Deputy Charles J. Renn was in the chair; 74 Affiliates were present.

Communication from the Principal Office was read; also the Good Thought for August 23 from the book, "Good Thoughts" by Dinshah. The Affiliates were not surprised to hear of the Birth of Dinshah's daughter, Shireen, but were surprised to hear it was a girl instead of a boy. Dinshah told us of the expected Birth when he visited our Planet Meeting last July.

Deputy—Our Sub-Deputy would have lost his best, if anyone had taken him up on it. (Recorder's Note—The Sub-Deputy was willing to bet an Equipment that the baby would be another boy!) Will the Recorder please read this newspaper article?

The Recorder read an article from The Vineland News, of August 24, 1947 about the Birth of the baby. Dinshah's bland remark about the child being "the first of several daughters" was greeted with applause and hearty laughter.

Communication from the previous month, August, 1947 about Resident Class was read to the Affiliates.

Deputy—How many of you would like to take this Course?

Four hands were raised.

Deputy—That is not enough to bring Dinshah here. Understand Friends, Dinshah has much valuable Equipment that he uses during this Course. To bring it here costs money, to say nothing of his time spent in giving the Course. If you have never taken this Class I urge you to think it over carefully.

Case Reports followed:—

L. B.—My husband is still sick. He has a bad Cold. There was Pain in his Spine and left Leg. He had something similar about two years ago. At that time, we did not have Spectro-Chrome and the Medical Doctor who attended him gave him the wrong "Shots". He almost died! That is when we got our Equipment. He has been feeling fine for a long time, but, a few weeks ago he went out and played ball. I guess he must have caught Cold. He has had it now for about six weeks. He has taken several Colors and is now using Magenta and Lemon. I know if he did not have Spectro-Chrome he would be very sick.

Deputy—He has probably overdone it and may possibly be 'run down' a little. Give Spectro-Chrome time. Do not expect Normalation within a few Tonations.

V. C.—My wife's Finger was smashed between the front seats of our car. It was very painful and it is still sore. X-Rays were taken and showed that the Bone is splintered on the side. She is using Blue. Spectro-Chrome surely helped her, but, I like to know what she should do now?

Deputy—Is there a Cut?

Affiliate—She is wearing a Splint and she still has Pain. There is no Fever. It just seems as if she is always hitting that Finger.

Deputy—Well, of course, she will have a sore Finger if she is always hitting it. Lemon is the Bone Builder, I should suggest Lemon.

An Affiliate—What about Green?

Deputy—Why not Lemon? It is ¾ Green and the Lemon will build Bone. J. N. have you anything to add?

J. N.—Lemon is the Bone Builder as the Deputy just stated. Green is a wonderful Cleanser, so I should suggest Lemon for one week, followed by Green for one week. Now, to counteract any shock that may be present, I suggest also Magenta on Area 4 and Violet on Area 18.

Deputy—Yes, that is something we all have a tendency to overlook, the Magenta. Of all the Colors, in my opinion, Magenta is the most beneficial. It builds the Aura, remember that is how Spectro-Chrome works, strengthens the Heart and Kidneys, builds the Blood, etc., etc. Many times Affiliates tell me how long they took a certain Color and are not quite satisfied with the results. Never once have they used Magenta. If Magenta did only one thing and that one thing was build the Aura, it still would be the most beneficent of all the Colors, because as I said before, it is through the Aura that Spectro-Chrome functions. If the Aura be depleted you can not expect success. Remember, when tonating, to use Magenta also.

A. M. (A visiting Affiliate from Florida)—On our trip North, we had an Accident and the automobile tipped over. My wife had a broken Rib. As soon as possible, she was put under Spectro-Chrome and was tonated with Lemon, Orange, Magenta and Yellow.

Deputy—What about yourself?

A. M.—I was "shaken up" a little, but, one Tonation and I was all right. She was the worse one.

Deputy—Was she in a Hospital?

Affiliate—Yes, two days only. She is feeling much better now.

J. J.—About three weeks ago, I became sick in the shop. Pain in Areas 8-9-10. Earlier in the day, I had drunk some Grape Juice and thought it might be that caused the upset. When I reached home, I drank some water and tonated with Orange. The trouble was not in my stomach, but, the Appendix. I put ice on my side and later tonated with Turquoise. I called J. N. and she advised Turquoise and Blue. I was out of work ten days. Then I went and happened to cut my Finger, but one Tonation took care of that. On top of that, I had a Cold and my Kardoscil was 114. Two Tonations of Turquoise and I felt much better. I have a slight Cough, but, I know that will soon be gone.

Deputy—No Medical Doctors, no Penicillin?

J. J. (laughing)—No, nothing like that. J. N.'s advice and Spectro-Chrome did the job.

Deputy—Any Inflammation, regardless of name, shows a predominance of Hydrogen in the Body and the Remedy is on the Ultra-Green. I recently read where scientists said that future generations will be born without the Appendix—we shall have no further use for it! Figure that one out. God creates the Body, but man is now convinced he can improve on it! Forget all these "Miracle Drugs" and follow our true Science. No man can ever improve on Nature. Talk to any young man coming out of College today, he is convinced he can—they know it all—at least they think they do. People permit Surgeons to remove Organs from their Bodies and then they expect to regain their Health. Remove the gears from your car and see how far it will go. Stay away from Medical Doctors. If you have a Bone to be set or a bad Cut to be sewed that is all right, but after that, you have no further need of them.

L. S.—My husband had a Heart attack. Within two days he was feeling better; we used Spectro-Chrome.

Deputy—We have a man at the shop about 64 years old. He has never been sick in his life and I do not believe he missed a day in 34 years. A few days ago, he complained of a Pain in his Chest. One of our Affiliates told him about Spectro-Chrome and he had him almost convinced to try it. Unfortunately, the Foreman cannot see anything in healing by Color and talked him into going to the Ford Hospital, which he did. After all sorts of Tests had been made, he was informed he had a bad Heart. He was advised to leave town for a few weeks and take a rest. One day after he returned to the city, I was talking to him and in our conversation mentioned the death of another old worker. "Yes," replied this elderly man, "the way I feel now I'll be next." I tried to tell him the Pains might be Gas Pains, but it was hopeless. The Medical Doctors said he has a bad Heart and that is that. The way he feels now, I give him about 3 months. There is nothing like Magenta for Heart Disorder.

G. S.—We had our Equipment with us at our Summer cottage. My daughter is tonated regularly. My Mother tonated Yellow for two days. She had complained of a burning sensation during Front Elimination. Condition cleared. She was certainly surprised with the quick results. In spite of that and also the way we feel about Spectro-Chrome ourselves, we can not get my father to use it. He definitely needs it, because, he has trouble with the Eliminations. My son has had only one Asthma attack this year and the Hay Fever is not as bad as in previous years.

Deputy—Continue with your Tonations and you will see results.

G. S.—That is the trouble; as soon as we begin to feel better we slip up on our Tonations.

Deputy—That is the common complaint of most of our Affiliates. We must remember a run down condition can not be normalated within a few Tonations. It takes time. Advertising tells us Aspirin relieves Headaches, but, we are not interested in just relief; what we want is Normalation. Stay with Spectro-Chrome and you will get it. I stopped renting my Equipments to people with Asthma, Sugar Diabetes or some other Disorder of long standing. It takes time to bring about Normalation and they do not have the patience to wait. Not only that, but, I warn them about cleaning the Thermocept, but it is no use. I have given them an Itisation, made out a Chart for about 15 weeks ahead, yet within 3 or 4 weeks they come back and say "it does not work." What do they expect? If Spectro-Chrome does not work, the fault is theirs and not of the Machine.

W. K.—I had a Backache with a great deal of Pain. Two "shots" of Indigo and I had no more trouble. Speaking about dirty Thermocepts, I remember speaking to Dinshah about that and he said "fill the Thermocept with water and add one-half teaspoonful of vinegar; leave overnight or for several hours. Rinse and use as usual."

Deputy—I use Distilled Water and have no trouble at all.

An Affiliate—Is Dinshah against drinking Distilled Water?

Deputy—I do not know about that. Suppose we ask for a comment from the Principal Office. I have been reading a magazine entitled "Organic Garden" in which it cites the danger to our soil by constant fertilizing with Chemicals. It has been found that soil becomes run down when fed with Chemicals over a considerable period of time and is also infested with Parasites. We have no one to blame but our "Scientists", who in turn are backed by our Government. Our soil can become corrupt through wrong care, just as our Bodies become corrupt through wrong eating. Soil poisoned means food poisoned, and in turn the Body is poisoned. It looks like we may have to grow our own food if we want to be sure it is clean. If there are no more Case Reports we shall begin with our Educational Program. I believe the Recorder wants to say a few words.

Recorder—Did anyone read "Life" Magazine dated August 18, 1947, Page 123?

B. N. raised her hand.

The Recorder then read a brief article from that Magazine pertaining to "Quack Medical Devices". One of the Devices, sadly observes the American Medical Association, is by one Dinshah P. Ghadiali who has charged $750.00 for a "Graduate Spectro-Chrome which he claims would stimulate the Pituitary, energize the liver, build hemoglobin and promote digestion, to say nothing of killing worms."

An Affiliate—Some one should write in and say it does just that!

The Recorder next showed the Affiliates a book she had come across recently entitled "Healing And Occult Science" by Dr. J. Croiset van Uchelen. Dinshah's teaching is mentioned several times in this book. Concluded the Recorder, "We must remember Dinshah is a well known Scientist. We are so used to addressing him as Colonel Dinshah, but his full name is Colonel Dinshah Ghadiali."

Deputy—Yes, some day this work will be recognized for what it is.

Albert Greenham, Sub-Deputy, then began the Educational Program. He introduced five new Affiliates. They were greeted with applause.

Sub-Deputy—While I agree with the Deputy about the soil corruption et cetera, we must also remember that within us there is a "Chemist" who knows how to take care of conditions like that. I remember reading that in Chicago I believe the year was 1923, two men were sentenced to die. Medical men wished to make a Test and the condemned men were told if they took this Test the Sentence would be commuted to Life Imprisonment if they lived! The men agreed and each was given a vial containing Diphtheria Germs enough to affect a large population! They drank the contents and nothing happened! There is something powerful in all of us that has the final word.

L. R.—I did not give a report tonight. I have had my Equipment since January and am taking Tonations regularly, but, my Hay Fever is still here.

Sub-Deputy—Do you have Hay Fever now?

Affiliate—Yes, I do.

Sub-Deputy—Well, I saw many cases of that Disorder and I must say you look very good to me. (Laughter)

L. R. (laughing)—You should see me sometimes.

J. N.—What about her food?

The Affiliate gave a list of foods and also her food eaten that day. She admitted she was not strictly on Rational Food Of Man, but was close.

B. N.—Starchy foods are bad and aggravate that Disorder. Better stay with Fruit and Fruit Juices and cool Salads.

Several Affiliates offered suggestions.

Sub-Deputy—We have two courses open: either we work with Spectro-Chrome or put up with the Disorder. Remember, it is her Hay Fever and not any of yours (Laughter). Seriously though, I should advise following strictly the Rational Food Of Man and tonating faithfully. Hay Fever can be normalated but it takes time. We can never harm ourselves; we are only working with the Visible Side of the Spectrum, the side that supports Life. Use Spectro-Chrome as you are taught, taking Tonations at the correct Favor Period which is very important.

G. S.—How does Dinshah decide on the Time Schedule and can it be explained simply? I am asked that question many times.

The Sub-Deputy gave a brief explanation, including Junction Period.

J. N.—We must always remember this is a Science and can not be discussed in a few words. Dinshah explains that in detail in the Class that he gives.

Sub-Deputy—Yes, it is all in our Encyclopaedia too.

J. N.—I took the Course twice. I had a hard time believing the Nares do not breathe together. Many a good discussion I had with the Colonel. I lived near a skating pond and one day I went down and watched the children skating. I tried to watch two boys, but, they were breathing through their mouths also. I went back another day and found a little boy I could watch. I stayed until the Junction Period and I was convinced. This is the most marvellous System Of Healing—a True Science. (Applause).

Sub-Deputy—How many have read our Encyclopaedia?

Many hands were raised.

J. R.—You should ask, how many have studied it? (Laughter).

R. G.—Everytime you open it you find something new. Could we ever have a demonstration of the Breathing here. I am referring to the Junction Period.

J. N.—Dinshah explains very clearly in his Class work, but it is not for us here. Many people come to the city and give "Breathing Exercises" but, these "Exercises" are harmful, not helpful. It would be better to take the Course with Dinshah himself.

Deputy—Before the Meeting closes, did you notice that our Governor Kim Sigler of Michigan vetoed the Bill, that would have made Vaccination compulsory for every child entering school?

All agreed it was a splendid move.

Sub-Deputy—Vaccination, Penicillin and the so-called "Miracle Drugs" can not stand beside Spectro-Chrome. The inventor of "Twilight Sleep" died not long ago and he said once: "If I have anything to regret, it is that I brought forth Twilight Sleep:" We have one man in this Country—one in 140 million—who has the fortitude to stand up and tell all: "You are wrong and I can prove it"; that man is Dinshah. He needs our help today, let us show him we are behind him all the way. The first prerequisite is "Know Thyself", you do just that when you take up the study of Spectro-Chrome Metry.

EDITORIAL REMARKS

Taking wagers and betting are gambles and I never do it. Surely, Albert, by my non-betting habit, saved $100 in his pocket!

Unless a reasonable number of Students pledge Enrollment, I can not bring or make a Resident Class in your or any other city. The expenses and the time and labor involved are too great to risk.

Those who are ready for Spectro-Chrome, need no persuasion. They will approach themselves. Those who resist, let them alone. The Deputy did right in not hiring Equipments for "experiments". The refractory mental attitude, reflects adversely on regaining of Health.

All our Thermocepts now are of heat-tempered glass; we discontinued Aluminum Thermocepts long ago. Contribution $3.00.

The agricultural situation in America and Europe is senselessly stupid. The Scientists are howling about Bugs and Beetles, the same way as the Medical Doctors are nutty about Germs, Viruses and Serums. None goes to the root of all the troubles. Just stop hunting the Birds. Those Bugs and Beetles which are the pests of agriculture, are the food of the Birds, but, by indiscriminate shooting, the balance of Nature was disturbed and now as Remedy, the wiseacres are shouting about spraying with D.D.T., Mercury Chloride, Arsenic and a host of downright Poisons on the vegetation. Have not they the knowledge that the leaves of Plants are the Lungs and breathe? Can not those numskulls comprehend that there must be some absorption of those vicious Chemicals into the body of the Plants, thus poisoning the eaters? It is that policy of killing Birds that started the downward cycle of Agriculture and until all accept the Rational Food Of Man system, more Medical Doctors, more Hospitals and more Relief Funds will be needed in increasing ratio. That is just what our Opponents want; it is that which fills their pockets.

All articles in newspapers—the prostituted henchmen of our Opponents—will be answered by me at the appropriate time. Their silly, adverse criticisms bring in more inquiries, because the people want to know who such an arch swindler and faker is and how he stayed where he is for 27 years. Americans are always the friends of the underdog. I laugh at all of this hullaballoo, as it proves that after all there IS something in Spectro-Chrome, of which they are afraid. $50,000,000,000.00 Investment is at stake! Just wait, watch and keep silence.

The writer of that book which mentions my work, was in California at one time our Student. He is not now affiliated.

L. R. sent only her initial General Advice Chart over a year ago and never reported since. We are here to serve. She has much to change in life.

G. S. should come to Malaga, sit in the Office for a whole day and see how the Time Schedules are made. It needs numerous data and understanding of natural phenomena and is really a "labor of love."

Your Governor Kim Sigler of Michigan, showed great wisdom in vetoing "Compulsory Vaccination." It is the curse of civilization and I am opposed to it emphatically. Keep the Blood pure and uncontaminated with Pus and Filth. I wrote to him and sent him a copy of "Family Health Protector." Distribute that one book as much as you can. Contribution only $1.00 for 4 copies.

"Twilight Sleep" is a Poison to the oncoming baby; shun it and bear the joys of a clean motherhood.

FIRST ST. LOUIS MISSOURI PLANET

(MUSICIANS' CLUB, 3535 PINE STREET)

Meeting Of Thursday, September 4, 1947.

NO GIFTS TO DINSHAH FAMILY RULE HAS NO EXCEPTIONS

Reported By:—VIOLA C. JEUDE, *Recorder*

Deputy Clemens M. P. Neumann was in the Chair; 60 Affiliates were present. Communication from the Principal Office was read, with remarks and explanations; the Deputy also read " "It's A Girl for the Ghadialis. First After Seven Sons, And Her Father is 73 Years Old." from the Vineland News of August 24, 1947.

Deputy asked if there was any question.

Sub-Deputy—Dinshah is a stickler for boys, but, no doubt Irene did her best!

T. P. said she would like to send Dinshah her best wishes and wish Irene the best of luck on the "New Baby Girl."

Deputy asked the Affiliates if they wanted to pass on T. P.'s kind wishes for the Dinshah family. They assented. It was voted on and unanimously passed.

An Affiliate suggested sending the Baby a gift.

The Recorder said it was not permissible; that was against the By-Laws.

The Affiliate said this gift was for the Baby.

The Deputy explained the By-Law 68 and said if any one wanted to send a gift, it would be accepted if sent to the Spectro-Chrome Fund.

The Deputy told the Affiliate that Dinshah looked very well and his wife looked wonderful, also about the new Experimental Apparati. He said the movement is under way whereby Wisconsin, New Jersey and Missouri will soon make Spectro-Chrome in each State respectively. The Educational Program was then begun by Sub-Deputy Charles H. Stiermann.

H. G. said she looked into Webster's Dictionary and found Red, Yellow and Blue can be made into White Light.

Deputy—They use pigments and this may be true with pigments which is paints, but, they cannot prove this with Light and we deal with Colored Lights, that they tried to prove this in the Trial against Dinshah, but, when Dinshah asked to test the apparati, he found that they had been using a Diffraction Grating and tried to cheat and disprove his theory.

The Deputy said an Affiliate had shown him the American, August 1947 Magazine, "Color Chases the Blues Away." He said while he has the Magazine at home he had not read the article. He asked A. W. if she would bring the magazine to him. She told him to read page 129. He read as follows:—

> "And we haven't seen anything at all, he says, when it comes to functional Color. While he has learned some few facts about it, and put them to work, there is still not one university in the world doing really comprehensive research on the subject. Some individual Psychologists, psychiatrists and ophthalmologists are giving more attention to the Therapeutics of color than ever before, but a deep veil of mystery still colors most of its properties; as this veil is penetrated, Color may be used to perform all kinds of miracle undreamed of today. It may be used to heal the sick, prolong life, and change the emotional attitude of Nations. Faber Birren has advanced only a short distance, he feels in his work of harnessing the rainbow. He sees magnificent opportunities for those who wish to follow him."

Deputy—Faber Birren says "He sees magnificent opportunities for those who wish to follow him." He should follow us or better still Colonel Dinshah who has already harnessed the Rainbow over 26 years ago. But putting all jokes aside, give the man credit for thinking if he has not heard of Spectro-Chrome.

Going through Chapter 25 of our Encyclopaedia, "Mysterious Mogen David," the Sub-Deputy explained how David selected smooth stones and placed them in a sling, (demonstrating with handkerchief how stones were placed in the sling) twirled the handkerchief, let go and the stone went forward with the rapidity of a bullet pierced David's adversary's Head and he slew him.

Sub-Deputy in drawing diagram 29 "Elucidation Of Removal Of Indigo Color" referred the Affiliates to Figure 5 "Perfect Circle Of Seven" and explaining the Corpuscular Theory of page 96 and comparing this with the Chlorophyll of plant life.

Deputy said he would like to add at this point that the Chlorophyll of Plant life (which is the life of the plant) is closest to Blood, that if the Magnesium in Chlorophyll were removed and Iron substituted you would have Blood.

Sub-Deputy said he could not understand why, so brilliant a man as Edwin D. Babbitt did not discover the true Primary Colors?

Deputy said it might be proper at this time just to pass along a thought that the Recorder expressed to me the other evening. "That it did not surprise her about the difference in the opinions in Color, that anything that is first discovered is usually used in destruction or the material side of things before it is utilized for good. Babbitt proved no different from the rest. In considering the Triangle of the Mogen David the Apex of which points down (Represents Matter or the Material side) and the Colors, Yellow Blue and we call the third Magenta, (But those not trained in Spectro-Chrome would call it Red.) Then along comes Dinshah who upsets the Color Theory and tells that Red, Green and Violet are the Primary Colors or according to the Mogen David the Triangle with its Apex pointing upward represents the Spiritual side; there is always a clash between Matter and Spirit. But, Spectro-Chrome will win. (Spirit always does in the end.)"

Deputy read Good Thought of August 23rd "Liberty" on the Birth Date of Shireen.

Question—Would Spectro-Chrome heal a Cataract of the Eye?

Deputy—If not turned into a stone. Incipient Cataract may be benefited.

Question—Why is it when you look into the Sunlight you sneeze?

Deputy—I don't know; I cannot answer that question, but, Dinshah says you sneeze when there is a slight irritation. You may also sneeze when you tonate and you can rest assured that is the turning point and you will normalate.

Affiliate—When my son washes his Face or his Hair he sometimes sneezes 8 or 10 times.

Affiliate—Is it true in Operations the same as in pruning, that the Celestial Planets had certain effects? I discovered that pruning at certain times causes the Plant to "bleed to death." Would this have the same effect on the Human Body?

Deputy—Yes. That is the High and Low Critical Period.

N. B.—Pruning should be done in the Full Moon Period as then the sap is lowering in the trees, whereas in the New Moon period the sap is rising in the trees.

Nathalee Britton said she would like to say a few words. She was pleased and happy to hear about the Sweet Bundle Of Pleasure And Love that arrived in the Ghadiali's family. She also wanted to say how grateful she was for Spectro-Chrome, that when her husband got beyond her control and the place where she could not take care of him at home, the Medical Doctors in taking the history of the Case could not account and were amazed and said "this Man is living on borrowed time for the past seven years," so I am thankful for Spectro-Chrome and have witnessed the passing of some who knew nothing of Spectro-Chrome and some who used Spectro-Chrome and there is a marked difference. Facing the Affiliates she told them they did not know how fortunate they were in having Spectro-Chrome and let no one persuade them in giving it up.

The Deputy then introduced Nathalee Britton and informed those Affiliates who did not know her, that she was a Past Sub-Deputy acting in the capacity of Deputy for more than a year. He informed Nathalee that when he arrived at the Institute he had just read the Report about her husband Earl and was very much put out, in fact had quite a long talk about Earl. He requested her to see him after the Meeting.

EDITORIAL REMARKS

Our By-Law 68 and Article 9 in our Constitution are explicit about Gifts. Furthermore, the Dinshah family has a strict rule against receiving any Gifts. WE GIVE, NOT TAKE. Do not, therefore, send any such Tokens, because. they will be *returned*. The Rule has NO EXCEPTIONS. Thank you just the same.

There is no objection to drinking Distilled Water, except that it is flat in taste and generally airless.

Our Primary Colors are based on Radiant Energy Principle as Red, Green, Violet. The Red, Yellow, Blue is for Pigment, Paint and Printing.

Faber Birren deserves credit and he may soon know, that his dream of Healing By Colors, was actually precipitated onto the Material Plane 28 years ago. So goes the World!

David must have known about the effects of Inertia, Momentum, Motion, Time and Space, when he killed Goliath.

The Chemical Formula of Chlorophyll A is C55 H72 O5 N4 Mg and that of Chlorophyll B is C55 H70 06 N4 Mg. A is Bluish Green, B is Yellowish Green. While the Carbon, Nitrogen and Magnesium Atomicity remains the same in both, the Hydrogen and Oxygen are in different proportions, hence, the difference in Color characteristics. As you go deeper into a consideration of Atomic Composition, your mind will become more clear and more impressed about the Truth in our Science.

Edwin Babbitt had a more brilliant mind than mine, but, I had more experimental facilities in the 7 institutions of learning, in Bombay, Hindustan, where I was permitted or received as Experimental Demonstrator in Chemistry and Physics. Here are the names of those institutions:—

1—Private Class of Professor Dadabhai Khurshedji Kateli.
2—Private Class of Sorabji Bamboatvalla.
3—Proprietary High School.
4—Fort High School.
5—Chandanwadi High School.
6—Elphinstone College under Professor Nanabhai Ardeshir Moos.
7—Wilson College under Professor Dadabhai Khurshedji Kateli.

Those seven, besides my private Laboratory. Edwin Babbitt's book started me on the path; he, therefore, was one of my teachers in every true sense.

An Incipient Cataract or one just starting to form, is remediable by Spectro-Chrome, but, one already formed may not disappear. That is in the field of Constructive Surgery.

If the Laws of Gravitation be true, what may influence masses of water in the oceans must exert similar influence on the Plant Sap and Blood Circulation. It took around 14 years of cogitation, for me to formulate the Favor Period System.

FIRST NEW YORK NEW YORK PLANET

(YORKVILLE CASINO, 210 EAST 86th STREET)

Meeting Of Thursday, September 4, 1947.

Successful Use Of Spectro-Chrome In:—Fever, Swelling, Imbackelim, Burn, Pain, Poison Sumac, Pomples, Athlete's Foot, Waspbite. Message For Shireen Dinshah.

Reported By:—EMMA C. HAITZ, *Recorder*

Deputy Lina Fischer was in the chair; 113 Affiliates were present. Two Affiliates were denied admittance, because, they did not have their Affiliation Cards.

The Deputy read the communication from the Principal Office. As the Deputy read the announcement of the Birth of Shireen, a chorus of "Ah's" ensued from the Affiliates. The Deputy asked, "Who wanted a boy?" No hands went up! "Who wanted a girl?" All hands went up!

The Good Thought for August 23 entitled *"Liberty"* was read by the Deputy as follows:

Liberty, above all we acquire
By our own exertions or what passes
To us through our gracious ancestors
Is the most precious of human heritages;
It consists in freedom of thoughts, words
And deeds; common sense, however,
Restricts its exercise to one's rights,
Without encroaching on those of others.
Liberty, is not
License unlimited; it ends
Where the tip
Of my nose begins.

Deputy—Live and let live—that is the way we should live. If everyone would feel that way, we would not have any War. There is nothing good or bad, but, it is our thinking which makes it so. The bad thinking is what takes the liberty away from you. I conquered many things since I am with Spectro-Chrome, which I could never have conquered before. Stick to it and learn by it and you will conquer the same way. When you are emotionally upset, there is something in you and you are not free. Use your Magenta and you will see what a free mind and free thoughts you have.

It was suggested by the Sub-Deputy and agreed by the Affiliates that the following sentiments be our message of congratulation on the Birth of Shireen:—

"The Affiliates of the First New York New York Planet, present at the September 1947 Meeting, send their heartiest congratulations and wishes to their beloved teacher Dinshah and his charming wife Irene, on the happy occasion of the Birth of their daughter Shireen.

May this youngest offspring of the Dinshah family grow up to be as sweet and beautiful a child as her name suggests—an ever new source of joy and pride to her father, of happiness and comfort to her mother, of love and affection to her brothers."

Case Reports followed:—

H. M.—I was a very sick woman four weeks ago. I could not get up one morning and had high Fever. My Neck was swollen and felt as if I had swallowed a piece of wood. I started tonating with Turquoise, but it did not help. Then we tonated with Yellow on Areas 6, 7, 8. 9, 10. After three days, the Bowels moved. We used Orange on Areas 1, 2, 3, because, there was a Swelling and it had to open. I didn't eat for a few days. Spectro-Chrome was something out of this World. Our Deputy was also close by and helped me over everything.

Deputy—Evidently, there was no Diagnosis; but, there must have been an Abscess or something there. I even recommended calling Dinshah. She had 104½ Fever. I would not want to be any place without my Spectro-Chrome. I would not even like to visit friends if they had not a Spectro-Chrome in the house.

T. S.—I was frying Meat for the men in my family. Put water in the pan and put it on the stove. It boiled over as I was doing the dishes. When I picked it up it was boiling hot. I could not let it drop, so I put it back on the stove. My Hand looked like a grilled steak. I immediately took a Tonation. A Blister had already started to form. After one Tonation, the Pain disappeared, also the Swelling. My Hand shows no signs of the Burn and I only used one Tonation.

A. K.—My sister was in the Country. Her little girl got Poison Sumac. My brother-in-law is a skeptic, but, the first thing he got was Spectro-Chrome. She could not look out of her Eyes. We tonated with Magenta and Turquoise on Area 1. It was too cold up there to take it Systemic. It took about five days.

Deputy—Speaking of allergies, so many people have given Reports on Hay Fever lately. Some people say they have had the Equipment for so many years, but still get Hay Fever. They say they eat a *little* Meat and have not used Spectro-Chrome all year. It takes about seven years to change one's system completely. If you want to keep your Health, stick to the Rational Food Of Man and Spectro-Chrome.

T. S.—Is there any chance that one can get Protein Starvation?

Deputy—Yes, it has happened. It is a shame that it has happened. Many don't eat Rice, Split Peas and Lemonade. The first thing a Medical Doctor will tell you is that you are Protein starved. I am not Protein starved, neither is the Dinshah family and many others too.

A. M.—Our son M. has never been quite satisfied with the Country, so we sent him to Camp this Summer. He was happy. Three or four days after he was home, he had a Fever. Tonated with Turquoise and then Yellow. A week passed—all of a sudden he got Pimples all over his Face and Arms and he had a Fever again. We gave him Yellow for three or four days. We asked him what happened. He said they made him eat Meat at Camp.

C. V.—My boy was away at Camp for eight weeks. I told them he had to have Vegetables and plenty of milk. He had French Toast and became sick.

F. C.—About Proteins—Legumes have more Proteins than Meat.

Deputy—I recommend everyone to become interested in the Human Body. Get some books on Anatomy and read them. Compare it with your Spectro-Chrome Metry Encyclopaedia. First study your Spectro-Chrome Metry Encyclopaedia and then study what the other side has to say.

E. H.—I have two books from Germany and there even the Doctors say Meat is no good for you.

Deputy—If you get the same information from another source you believe it.

M. F.—I had an experience last Sunday. We went to Bronx Park. My sister was constantly annoyed by Flies, but, they did not trouble us. I told her it was because she ate Meat and we did not.

A. K.—People often get Athlete's Foot in the Summer from going swimming. I had it once and took two Tonations of Lemon and it was gone.

A. M.—I used Cocoa-nut Oil and in two days it was gone.

M. G.—I got six Bites from a Wasp. My Leg swelled up terribly. After two Tonations of Turquoise, it was all gone.

Ten new Affiliates were introduced and greeted with applause. Turning to the new Affiliates the Deputy said: "I hope you people come to the Meetings regularly. You need the education and I want you to report your cases. If you do not come to the Meeting, you will not get your Magazines. You are new, you know very little about Spectro-Chrome. If you do not know, get up and ask about it. That is for what this Planet is. Don't do what the old Affiliates do—just sit there. You set a good example."

It was asked by a new Affiliate, H. M., if there are any recipes for cooking the foods. The Deputy explained that the Cook Book was in progress and that we look forward to having it some day.

Deputy—Do not forget the change in time on September 28th. You will take your Tonation then according to the Favorscope without adding the hour.

K. G. and H. M. who attended the recent Class in Orange, New Jersey both stated they enjoyed the Course very much, but that it was finished too quickly. They wanted a longer Course.

The Deputy read an article from "Consumers Research Bulletin" regarding the seizure of the Equipment in the recent Portland, Oregon case. It stated in the article that "the ailment of the user was aggravated by failure to consult competent medical authorities."

The Deputy asked the Affiliate how they would like to have Dinshah for their Deputy. All are greatly in favor.

Chapter 48 "Spleen As Gateway" from our Encyclopaedia was begun and discussed.

F. H.—I read an article written about the Aura by a Doctor. He said it was bad for children to sleep with older people because, the older people take the strength away from the children.

The Meeting was enjoyed by all and it is sincerely hoped that the new Affiliates will continue to come to the Meetings. Again, our heartiest congratulations to Dinshah and Irene on the birth of Shireen.

EDITORIAL REMARKS

Cordial thanks for the loving wishes about Shireen (Sweet). The sentimental appreciation, was worth more to the family, than all material wealth as Gift.

Once Infra-Green and Ultra-Green Divisions are thoroughly grasped, selection of Atcowa for Tonation becomes very simple.

You have a gem of a Deputy, whose sole aim is to help you to help yourself.

Skepticism has its place, but, condemnation without investigation is not sensible.

Hay Fever of many persons is traceable to their Feather Pillows, Furs and Sheer Hosiery. Eat the Rational Food Of Man and discard the stated materials and your vitality will stay so high that Disorders will rarely be encountered. The true cause of such attacks as "Hay Fever", is purely mechanical—pollen or sheddings of minute, irritating particles.

Sneezing is Nature's effort to restore the Equilibrium of the Two Chains of the Great Sympathetic Nervous Eystem as well as to expel foreign disturbing particles.

Our alleged Civilization is a mere mask. We still eat the same carcasses of Animals, still kill living creatures, still eat the same malodorous Fish, still gulp the same stinkpots of Eggs which are nothing short of Poultry Menstruation. still go to War for money, still have the same Slavery in the shape of Military Draft, still take pleasure in becoming intoxicated and act like fools, still blow rings with filthy Tobacco and call it fashion—we are nothing but a batch of hypocritical morons, who can not see beyond the tip of the nose. Civilized indeed! We grabbed the Country from the Red Indians and in return for their hospitality and generosity, handed them the Whiskey Bottle to make them blessed with Tuberculosis.

Having Dinshah for your Deputy is a good suggestion. Well, if I stay out of imprisonment and you go over the top with attendance larger than that of your Milwaukee Planet. I might give the matter some consideration. However, you might feel later sorry for it, because of my disciplinary methods.

DINSHAH IMPARTIAL

The Universe Through The Eyes Of Dinshah
"I Fear No One; Only God Above And Conscience Below And From Them I Have Nothing To Fear."—Platform Since 1891.

GAS FOR MOTHERHOOD

Reporting in the New York Journal-American, a United Press correspondent asserts that the United States Department Of Agriculture Scientists found some very novel facts: "A whiff of Anaesthetic Gas turns a virgin queen bee into a mother without the benefit of a mate." The gas used was a Carbon Dioxide Anaesthetic and the Agriculture Department is excited beyond expression. The statement was made that: "Her offspring will all be males, known in the bee world as drones. These fatherless sons can be mated to their mothers, the still-virgin queens, to produce generations of super-bees."

What a marvel of Entomological Research! One step forward and girls working in Dental Offices, as well as Nurses working in Surgeries, will be in danger by Anaesthetic Gas! The report came from Washington, D. C. where any marvel may happen, but, for the good name of our beloved beauties, I hope they will find some error in the Science. Yet, as America abounds in Drones, the eventual outcome is difficult to predict.

X-RAYS AND RADIUM CAUSE CANCER

I printed in Spectro-Chrome of June 1922, about "Areas Fatal In X-Ray Work" and in our Spectro-Chrome Metry Encyclopaedia, in 1934, in Volume 2, warned fully about the "Dangers Of X-Rays And Radium." Now comes Bob Considine, a columnist, who writes in New York Journal-American that: "X-Ray, an established treatment for Cancer, can also cause it if improperly used. So can Radium and a group of Chemicals and Chimney Soot . . . Cancer is the uncontrolled growth of Cells, the failure of one part of the system to put brakes on the normal growth of Cells . . . There are no quick cures, despite what you might read here and there. Surgery, X-Ray and Radium remain the only accepted treatments. When another is found, the news will be emblazoned across the front pages of every newspaper in the land." Is that so? That "cure" will never be found while we continue to eat animal carcases and as far as the "another" is concerned, that has been actually in existence 28 years, but, there will be no emblazonment until the newspapers are released from the grip of the Medical Octopus. The Chimney Soot Cancer is easy to explain. It contains Creosote, which is damaging to Tissues. Read the Article on "Coal Tar And Cancer" in Spectro-Chrome of July 1922 and the Editorial Remarks about Bromo-Seltzer, containing Acetanilid or Antifebrin (a Coal Tar Compound) and Antikamnia.

———o———

ANSWER TO WILLIAM R. OLSEN

WHAT FRIEDA WEHRHAGEN THINKS

I am in perfect agreement with him. You told us in Class that God gave you this wonderful work we call Spectro-Chrome. "Don't give me the credit; thank God; He helped me; He will help you." First Chapter of John says in verses 1 to 3: "All things were made by Christ." (Ed. note: She mixes the word "God" with "Christ".) Help will come by fully trusting in the Creator.

WHAT FIRST ORANGE NEW JERSEY PLANET THINKS

The opinion of the entire First Orange New Jersey Planet, in regards to a letter sent to Dinshah, by William R. Olsen, from Portland, Oregon is, that it is very evident Olsen has never been in close contact with Dinshah or he would never have wirtten what he did. It is far from fairness to state, that, Dinshah should have a "proper relationship with God." During our Resident Experimental Course every session was opened by a prayer and also a religious song by the good Colonel. Furthermore, he made many references to the Bible, also quoted many verses and phrases, which only one acquainted with the Bible could quote. The Affiliates were astounded at his knowledge of the Bible. In the opinion of the entire First Orange New Jersey Planet, there lives no one, who in our humble opinion and estimation, is more God-fearing and also cognizant of God, than our loyal and esteemed teacher Dinshah.

Lothar O. Klenner, Deputy.
First Orange New Jersey Planet.

INTERMISSION

1. 1947. At this point, publication of the **Spectro-Chrome** magazine ceased because of the conditions of probation imposed on Dinshah. Thousands of magazine back-issues were turned over to federal authorities for destruction in the Camden NJ incinerator, along with innumerable books written by Dinshah pertaining to Spectro-Chrome.

2. 1953. His probationary term completed, Dinshah formed a new corporation: **VISIBLE SPECTRUM RESEARCH INSTITUTE.** It took up where the previous corporation (**DINSHAH SPECTRO-CHROME INSTITUTE**) left off with one major difference: labeling on the projector and literature very plainly stated that in the conventional medical view Spectro-Chrome had no therapeutic value.

3. 1955. In January, Dinshah resumed editing a monthly magazine, now titled **Visible Spectrum Researcher,** continuing for 35 issues. The final edition (November 1957, reprinted on the following pages) is representative of all 35, and contains a copy of the disclaimer label. In each magazine a warning declaration from the editor also was printed.

4. 1957. This issue carried the correspondence between Dinshah and the Food and Drug Administration which eventually resulted in a permanent Federal injunction against Dinshah and his Institute. The judge held that the device was identical to those of previous corporations (res adjudicata), and the disclaimers were meaningless because it was evident that the Spectro-Chrome device was still being sold for health purposes. In effect, the learned jurist ruled that you were to be deprived of the right to even try this admittedly harmless System.

5. This was virtually the end of Spectro-Chrome for the time being as Dinshah was then prevented from sending projectors, books or literature across State lines.

6. 1958. Notwithstanding this tremendous setback, Dinshah continued within the bounds of the injunction, selling a few books and projectors in New Jersey, giving a few promotional talks locally. An era had ended.

VISIBLE SPECTRUM RESEARCHER

NOVEMBER 1957

Trade Mark Number 163762

Registered U. S. Patent Office

Volume 3 Number 11

THAT'S A GOOD QUESTION

THAT IS A GOOD ANSWER

OUR STUDIO ACTIVITIES

DINSHAH IMPARTIAL

WONDERFUL FOOD FOR THE AMERICANS

KNOCKING AGAIN AT OUR DOOR

DINSHAH'S 84TH BIRTHDAY

OUR STUDIO DIRECTORY

OPEN DECLARATION FROM THE EDITOR

Our Aim HAPPINESS IN EVERY HOME	DINSHAH P. GHADIALI Editor	SERVICE OF MANKIND

By:— VISIBLE SPECTRUM RESEARCH INSTITUTE
Publishers

Telephone:— OXford 2-3530 **Malaga, New Jersey**

THAT'S A GOOD QUESTION

Under the above caption, Medical Doctor William Bolton, as the Editor of the said department in "today's health", a monthly publication of the American Medical Association, wrote in its March 1953 issue an answer to an inquiry about:—

"COLOR THERAPY

Question — A friend tells me that a method called color therapy is now being developed by which various illnesses may be treated successfully. Do you have any information about this?

Answer — From time to time, theories are advanced that color may have value as a treatment adjunct in medicine, but we do not know of any evidence that a specific disease ever has been cured through the use of color alone. Various colors may have some value so far as mental attitudes of many patients are concerned, and this idea has been utilized in some hospitals. For example, walls and ceilings of hospital wards or rooms may be painted a light blue or green because such tints are believed to have a soothing effect. In general, colors probably should be considered for their secondary effect in helping a patient to overcome excitement or worry about an operation or illness."

TREATMENT BY "SILENT SOUND"

The same Medical Doctor in the November 1956 issue of "today's health," wrote a long article on the treatment of kidney and gall bladder stones, relieving pain, soreness and tenderness in certain kinds of bursitis, periarthritis, fibrositis, tenosynovitis, myofacitis and myositis, rheumatoid arthritis and osteoarthritis, successfully by means of "Silent Sound," "Supersonic Waves" or "Ultrasound," which is Sound Waves that the Human Ears can not catch. Sound upto 32,768 Oscillations per Second can be heard, but, after that range the Human Ears can not utilize it as Sound and the said Medical Doctor recommends for "Silent Sound" treatment 500,000 to 1,000,000 Oscillations per Second.

So, if that Frequency can produce the stated results, what about the Oscillatory Frequency of our **Visible Spectrum Of Color,** which has what I described and utilized decades ago as:—

562,949,953,421,312
Oscillations Per Second???

This publication is an answer to the stated view, as in our service to our Members, their own convictions by their personal investigations, researches and experiments will give them the accurate view of the Truth behind the Visible Spectrum. Let the reader judge for self.

THAT IS A GOOD ANSWER
By:-INDIVIDUAL EXPERIMENTERS
REPORT BY DR. LESLIE A. CHAMBERS AND DR. EARL W. FLOSDORF
(Published By Dinshah In July 1936.)

"Dr. Leslie A. Chambers and Dr. Earl W. Flosdorf of the University of Pennsylvania in Philadelphia, Pa. report in the Proceedings of the Society for Experimental Biology and Medicine that high pitched Sounds oscillating at 8,900 Frequency per Second, are able to destroy Streptococcus Hemolyticus by blasting the Germs through 'Cavitation.' It is stated that 'So powerful is the vibration that it overcomes the cohesive strength of the water, tears the molecules apart and leaves vacuum spaces or vacuoles which look like bubbles! During the process, Light also is evolved.'" The reader will please note, that what we taught years ago is not yet known to our opponents, because, they never investigated our Science. Dr. Chambers and Dr. Flosdorf deserve credit, but, it is a matter for regret, that our opponents admit the potency of 8,900 Oscillations per Second, yet, close their mind about the Oscillatory Frequency used by us:—

562,949,953,421,312 per Second!

A further nonsensical idea is that our opponents use X-Rays promiscuously, without regard to its caustic and dangerous effects, also without realizing that the man they ridicule (Dinshah) was the **first lecturer** on those Rays in the year 1896, in New York, when most of his opponents did not even dream of being born!!!

* * *

CHRONIC CONSTIPATION, HEADACHES, SEVERE BACKACHES MUSCULAR PAINS REPORT BY W. K.

My ailments before I owned a Color Projector were Chronic Constipation, lots of Headaches, Severe Backaches and Muscular Pains in the Calves. The Constipation I tried all kinds of Laxatives; without success; for the other ailments I went to see two Doctors; each one's diagnosis was different and no relief. It got so bad that it was annoying and troubling me at my work. When I heard of the Visible Spectrum and became owner of one, my troubles of Constipation were over; no more Laxatives. The Pain in my Calves, Backaches left me; Headaches are as good as gone, thanks to the most wonderful achievement in Science by Colonel Dinshah, the only and only one Visible Spectrum. (Signed and Affirmed Before C. C., Notary Public.)

* * *

ENLARGEMENT OF HEART REPORT BY K. F.

The following is the condition of my wife P. F., age 43:— "When entering the United States of America, I already had Heart Trouble. The Medical Doctor diagnosed it. Enlargement of the Heart. The Steamship

Company had to post Bond, so I may enter the U. S. on account of my Heart Trouble. Every week, I had one to two Heart Attacks, which were so severe that I was very weak and I had to rest in bed. I used Witch Hazel compression. In 1935, my husband obtained a Visible Spectrum Equipment and when having another attack as usually, he persuaded me to use the Visible Spectrum. I used it the first time. During the first 15 minutes, my Heart stopped every 13 seconds; after using Visible Spectrum for 30 minutes, the Heart stopped every 40 seconds and after the hour was over, the Heart beat as regular as a clock, without a single miss. That convinced me what Visible Spectrum was able to do. I now feel very good and occasionally my Heart may beat faster, especially when War News or other exciting news reach me. These attacks are now so seldom that months may pass and are only due to undue high excitement. This ends the statement of my wife." I state that Visible Spectrum would be the last thing that I would give up. (Signed and Affirmed Before C. M. P. N.) Notary Public.)

* * *

ALL DISORDERS REPORT BY W. B.

I have successfully normalated every Disorder my family and myself have had in the past 10 years. (Signed and Affirmed before N. M. C., Notary Public.)

* * *

ADHESIONS, DEFECTIVE BOWEL MOVEMENT REPORT BY A. A. W.

Eight years ago, I had an Operation, which was unsuccessful. One year after, I was filled with Adhesions, which meant another Operation. At this time, I heard of Visible Spectrum. The Adhesions also affected my Bowel Movement. Within a week, I had a regular Bowel Movement after using Visible Spectrum and have had ever since. Within 6 months, my Adhesions disappeared, thanks to Visible Spectrum and Dinshah P. Ghadiali. May God bless him! I am 74 years of age. (Signed and Affirmed Before W. C. S., Notary Public.)

* * *

INTESTINAL INFLUENZA, COLITIS, PAIN, SINUS DISORDER REPORT BY L. F.

This is to certify that I am a user of Visible Spectrum, since March 1942. Visible Spectrum was not offered to me; I was seeking and looking for someone that knew about Light and Visible Spectrum was my answer. My daughter and myself had been the victims of a severe case of Intestinal Influenza. My daughter had a major Operation at the age of 4 years and was handed out to me from one of the large Hospitals, with Intestinal Flu and I contracted it from her, being in poor health over a period of years and four Operations, which never restored my health. It left my daughter and myself with a severe case of Colitis, always having Pain in the Intestinal Tract, which no Doctor was ever able to clear up. Visible Spectrum has entirely cleared that up; we never suffer any more Pain; has also cleared up E.'s Sinus condition and many other ailments. We never use Medicine; only Visible Spectrum; its merits are unsurpassed in any sickness. We pray that Colonel Dinshah may live many more years to be of service to mankind. (Signed and Affirmed Before J. A. F., Notary Public.)

OUR STUDIO ACTIVITIES

EXPERIMENTER FINDINGS AND QUERIES

MEETINGS OF THURSDAY, SEPTEMBER 5, 1957

First Cleveland Ohio Studio

A. M. — What side of the Spectrum should one use to improve Eyesight?

An. M. — Tonate on the warm side of the Spectrum. For Inflammation; tonate with Blue.

F. B. — I received a change of Color through my Chart, from the Institute. I am now tonating with Orange Systemic on Back and Yellow on Areas 6, 7, 8, 9, 10.

A. M. — I suggest that you follow instructions received from the Institute. I like to know about the Hand of J. D.

J. D. — It is very much improved. My Colors also were changed. They are now Indigo and Orange Systemic Front.

A. M. — I had a touch of Flu; tonated with Green Systemic Front; drank fruit juices and cut down on the intake of solid food.

Discussion was on the harm of Fluorides in drinking water. Chapter 66, from Volume 2 of our Text Books Set, on "Red Corpuscles And Anemia" was read.

* * *

First Detroit Michigan Studio

A. G. — Friends, tonight I am going to quote from the Family Health Protector, a booklet which I believe most of you have. It sells for 25 cents each. There is no better way of introducing our Science to friends, than through this little booklet. To attempt a technical discussion of our Science, is difficult and this booklet is a good means of informing people. I received a letter from Mr. and Mrs. C. H. today. Mr. H. was never sick a day in his life. Now, he could not pass his Front Elimination. He used Magenta two or three times without any results. After the third day, he collapsed. His wife became very panicky and called in a Physician, who rushed him to a Hospital. This happened June 24th and he just returned home today. He is about 66 years old and has had Prostrate Gland trouble. He underwent 3 Operations since June, the last one performed only a week ago and now has 5 or 6 Tubes inserted in different parts of his Body and the cost amounted to $1,000.00 for one Doctor, $250.00 for another and he is still awaiting bills from several others. I feel sorry for people who do not know what to do out of fear. When something happens, they lose confidence and become nervous and to give advice is like giving dynamite. The amount of human suffering is frightful. The evolution of transportation from a square wheel to jet airplanes, shows the tremendous amount of progress that has been made in this World; progress in everything except modern Medicine and yet the Doctors are so rushed and have their offices so crowded, that many do not even pother making house calls anymore. We have a Science that is built upon Mathematical precision and it will never change, just as the Musical Scale of Do, Re, Mi, Fa, Sol, La, Si, Do will never change. That is Science. We may modernize our Equipment, but its fundamentals will not change and it will be as good tomorrow as it was yesterday. Most of the so-called "bad luck" from our system results from people slipping by the wayside and the two will not go together. Everything that we think is right, we should honestly analyze. People insist Meat is necessary for the diet. Meat is secondary food. Why not eat our food first hand? The Bible continually admonishes against Meat and prohibits Flesh and Blood. The slaughter animals undergo highly emotional upset prior to their execution and their process of elimination continues even after death and goes into the Tissues. That is why the consequence of eating Meat is so serious. Then, these people need Tranquilizing Pills, et cetera. The greatest trouble of our Students is that they continue their old habits and also use the Visible Spectrum. It does not make sense. Either use one or the other. Why give the Visible Spectrum a bad name, because the user violates the prescribed mode of living? Now listen to a few chapters from the Family Health Protector. (He then read Articles 38, on "Medical Humbug Of Heart Disorders;" Article 43, on "Esoteric Religion Forbids Carnic Food;" Article 44, on "Man's Anatomy And Physiology Prove Him Herbivorous." These were very enlightening to numerous Members and they wanted copies of the said book.)

G. N. — I have a friend who would like to attend these Meetings.

A. G. — Your friend can become a Student Researcher for only $3.00 a year membership. We have a new Member here tonight, W. F.

W. F. — I injured my Arm two months ago. I ordered my Equipment, but, did not receive it yet. I read the Family Health Protector today and it certainly contains plenty of information.

A. G. — I shall now read Chapter 50 about our Science, from "American Sex Problems." (It was read and found informative. The Radiometer was then exhibited with a Portable Color Projector, as different Colors having different Radiation. He then described the various Attributes of the Colors, also the working of the Digestive Tract, showing why Enemas and Colonic Irrigations were not advisable. The working of the Thyroid and Parathyroid Glands was enunciated. He also spoke about the Pancreas and the Thymus Gland.)

R. K. — I am following the Rational Food Of Man and feel well enough not to require any Tonations right now.

A. G. — It is sometimes difficult to live according to that System, when we live with spouses who refuse to follow it. When you have opposition in the home, then, you have a problem. Our System is the finest, but, you must accompany it with the proper habits.

W. K. — My son, who is 16 years old, has Adolescent Pimples. He has been taking Tonations. I tried to explain to him the importance of avoiding certain foods; I can not report any results yet, as he just started tonating.

G. N. — The Color Projector has been wonderful for my mother, who is 87 years old. She comes to me quite frequently for help.

M. V. — I am a new Member and have an Equipment, but, do not know how to use it. For what is the Water?

A. G. — The sole purpose of the Water is to protect the Slides from cracking by the heat of the Bulb.

M. V. — I am pothered with an over-enlarged Liver as shown by X-Rays by a Chiropractor and I have Backaches.

A. G. — Let us take your Kardoscil and Spiroscil. (It was 66-13.) You should fill out a Chart and send it to the Institute. Be sure to fill in every item or you may not receive an answer. There is no Diagnostician that knows as much as Dinshah in our Science. Your Ratio is much too low and should be 80 or 90 to 20 or 22. Increase both by using Orange Systemic Front and Back and Scarlet on Areas 4 and 18. If the Front Elimination be normal, then, you may omit the Scarlet on 18. Use that way for 30 days and report your progress at our next Meeting.

L. G. — I usually take two Tonations a week. Lately, I had Sore and Bleeding Gums, which were so painful that I could hardly eat. I used Turquoise and they tightened immediately and are fine now. My reason for using Turquoise was that this condition was recent, painful and inflammatory. I always check the Instruction Book.

A. G. — The original instruction book is available as Exhibit in the "One Thousand Years Ahead." All its copies were destroyed during the last Law Suit against Dinshah. That is all for tonight; good night, friends!

EDITORIAL REMARKS

We agree with A. G. about the trouble in homes in regard to food. However, that is upto the persons concerned. We give even the Cook Book FREE to the Color Projector Purchaser, but, we can not COMPEL them to follow instructions. W. F. got his Equipment in September 1957, but till now the Institute has **no** report of the use he made of it. The same remark applies to our worthy friend M. V. She has the new Visible Spectrum Instructor, on page 5 of which in Article (d) under Assembling The Projector is this:— "The Tempered Glass 'Thermocept' is for Heat Interception from the Slides. Insert it behind the Rear Condenser and always keep it filled upto its neck with Distilled Water. USE NO OTHER WATER. DO NOT USE TAP WATER." She was given many other instructions about food and habits, but, since getting the Equipment two months ago, she never wrote to us or sent any Personal Research Information Chart. She was recommended by Graduate Researcher C. R. and she could have asked any question of him as Introducer, but, where silence is maintained we can not extend our helpful services. Her Ratio originally was as 69 to 11; wrong food, wrong habits.

* * *

First Merrill Wisconsin Studio

Discussion was had on the Reports of the various Studios, especially the First Milwaukee Wisconsin Studio.

M. S. — I had Dizziness; used Green, Yellow and Magenta; am feeling fine; occasionally used Lemon.

G. V. H. — I had Pain in Areas 8, 9, 10; used Orange, Magenta and Lemon. One Turquoise on Areas 8, 9, 10 for Soreness. Normalated.

E. K. — I had Cramp in Leg. Used Orange with Magenta on Area 4. I also had a Swelling in Elbow; used Turquoise and Magenta; normalated.

G. V. H. — I was hit on the Jaw by the washing machine wringer; used Turquoise and was normalated successfully.

C. G. — I made the Cheese according to the recipe given; it was very good.

I. K. — I use all of the Whole Wheat Flour for making Bread.

EDITORIAL REMARKS

Studies made from the Reports of other Studios are commendable. From others we learn the difference of different minds. Continue your sensible efforts.

* * *

First Milwaukee Wisconsin Studio

F. K. — Good evening, everybody. One month ago, we were surprised by our President. Those of you who were not here, missed something. Some of you Members cut out a clipping; what you should do is to tear out the entire sheet, to show from where it is and the date. I like to read an interesting article from our magazine; how many read this article? Dinshah has been writing interesting articles and in the July 1957 magazine he tells how you may get advice on personal problems. I like to read from the First Volume of our Text Books Set. Those of you who were at the Convention, heard Dinshah. It may not be interesting to our old time Members, but, we have new Members all the time. (He then read "Presentation To The Sick And Suffering Mankind" and Chapter 1, "Pilloried Prophets.") I want to say something about Henry Ford. None in Milwaukee had the foresight or his big plant that is in Detroit now could have been here. He had a machine shop then on the South side. Have you any Case Reports?

C. P. — We never see Dinshah's name anywhere; it wil come out now. Morris Bealle wrote "The Messpot In Washington," etc. and published a Health Column; also wrote about ------ and ------. He says that in his next issue of Saturday, September 14, 1957, he will have "The Dinshah Story" in his American Capsule News.

Member — Is it possible to get a copy?

C. P. — I shall try to get extra copies.

N. T. — I had a bad Cough last month and I did not get results until I used Lemon. I used Yellow, but, after I used Lemon, in two days I was better. I had another experience. I fell down the steps and I did not use the right Color. Then, I used Orange and in 3 days I was better. I used Magenta for Circulation and used Yellow. I do not know why I used Yellow!

F. K. — Did you have Pain?

N. T. — Yes.

F. K. — Then, should you not have used Turquoise? Was it discolored?

N. T. — I do not think so.

F. K. — There is a rush of Blood to the Area and the thing to do is to use Turquoise or Blue or go down to Violet and then go to Turquoise. I fell and I used Blue, because of severe Pain and then Orange, but, I did not have a conglomeration of Blood. It depends on circumstances of the Blood. You use the Green Wave for cleansing and you go down the Spectrum according to the condition. The thing from which I am trying to get away is using medical terms. If the chemistry of the Body becomes unbalanced, Nature lets us know and we have to correct it.

R. S. — Suppose you have an Accident. Your Blood is partially unbalanced because you have a Cold and you have an Accident. You have to go to the Affinity Color.

F. K. — That is why Dinshah says you should fill a Chart.

R. S. — Not only he will know, but, you will know also.

N. T. — I also had a Sty; I used Turquoise and that helped.

L. S. — Mine always is so complicated. I had a wedding in the family and it felt like I had a pressure and the way I decided to help myself; my Kardoscil and Spiroscil was the same as it always is and it developed in a Cold; then, I decided to use Turquoise for 2 days and my Backelim stopped and then, I used Lemon and it cleared up.

F. K. — Yellow will move the Bowels; then, you may go to Violet.

C. P. — To come to a conclusion by Ratio, can you explain that a little?

K. B. — That is always the same story. First find out what is the Ratio, 4 to 1 or whatever it is. Find out what was eaten the day before. If you use Yellow, it helps along by throwing it out. If it keeps on for weeks or months, that is different. Some time ago, I had a Sufferer that went 12 times a day. First, I started out with Green, that acts as a Germicide and Cleanser. The same as the lady that fell, she used Green and Magenta. That is good; it will help the Circulation, but, I always look to see what the Ratio is.

F. K. — Regardless of what the Ratio is, 4 to 1 or 5 to 1 will recover; 3 to 1 is doubtful; if it be 2 to 1, the person will have slight chance for recovery.

C. P. — Then the best one to use is Magenta.

L. S. — About 2 months ago, my son was sick and his Kardoscil was really low; then, I used Magenta. I asked him how his Elimination was and we took care of that and used Orange; then, after he was relieved, his Kardoscil went up.

F. K. — When there is a Cramp, use Orange. If an individual be injured and the Blood does. not go to the injured part, what Gland is unbalanced?

Member — The Pituitary Gland.

F. K. — That is one of the reasons why Dinshah advises us to stay away from Alcohol, as it ruins the Pituitary Gland. Its Posterior has Fibers which eventually act on the Spleen and Liver. (He then explained about the action of the Thymus Gland for growth. Some discussion resulted.) Now, I shall read to you from the Third Volume of our Text Books Set, Chapter 87, "Secrets Of The Pituitary Gland." (It was read.)

L. O. — I want to know if bad Teeth be contributing factors toward the toxic condition in the Blood to unbalance the Body chemistry? The Nerves were removed 10 or 15 years ago.

F. K. — K. B., do you think that is the trouble?

K. B. — No, I do not. I had a habit of cracking Nuts with the Teeth. Finally, they became so loose, I could move them back and forth. The Dentist told me to have them pulled out. I had the Color Projector only a few months and I tried the Light; first, I tried to ease the Pain with the Indigo. It did not help; then, I switched to Orange and a Blister developed, which filled with Pus; soon as the Blister burst, the Pain was gone. Then, I used Green and Indigo and I still have my Teeth! You may use Green and the condition will clear up.

A. F. — Do you remember Mr. L.? The Doctor told him to have his Teeth removed and he put his Teeth on the Table!!!

K. B. — Two company Doctors claimed the Teeth caused Inflammatory Rheumatism. He had his Teeth removed, but, the Disorder did not disappear. You have to use the Rational Food Of Man. You have to use the Teeth. There is a saying:— "The Organ is as strong as its use." If there be an overrich Blood, that may cause Pyorrhea.

F. K. — I liked your testimony. I thank you all for coming.

EDITORIAL REMARKS

Morris A. Bealle is an independent, impartial writer of strong mentality. As soon as he published "The Dinshah Story," I thought it better to let the Members know some facts of our past, so we mailed a copy with two cents postage stamp to each Member, with the September 1957 issue of our magazine. Many of our Members are subscribers of the American Capsule News. What that "Story" brought on my head is printed elsewhere in this issue. My opinion about Medical Diagnosis is well known to you; that "teeth on the table" proves it! In your report, I found a notation about refusing admission to certain Members who had 1956 Cards. C. L. was quite right about it. We are a strictly Fraternal Organization under our State Of New Jersey Charter and our Studio Meetings are held strictly according to the Rules regulating Fraternities. Furthermore, any Members presenting a Membership Card **without** hand written Signature on it shall not be admitted. Refer to our By-Laws 38, on Registration And Admission.

* * *

First New York New York Studio

R. D. — Are there any Case Reports? My whole family, father, mother, wife and children, including myself (with the exception of the 3 month old baby) suffered from Diarrhoea, Nausea, Vomiting. It lasted for 24 hours. We tonated with Green and Yellow. I read an article in the -------- magazine, about Asiatic Flu; according to them, there were thousands of cases with similar symptoms. Those Disorders attack when your resistance is low. Keep warm, irradiate with Lemon and Green; I do not know what we had, but it was gone in 24 hours without after-effects.

D. R. — The --------- says the Asiatic Flu is the biggest fake.

R. R. — Before I went to the country, I had a check-up. I was suffering from Heart Disorder before joining this Organization. The Doctor took a Cardiogram; he was surprised how I improved. He said: "Whatever you are doing, keep it up." I am using Orange and Magenta. I am 69 years old and still very active and working hard. I am an Auto Machinist.

R. D. — Heart Disorders can be remedied with the Visible Spectrum. Magenta is the Suprarenal Stimulant and Cardiac Energizer; the Irradiation will act immediately on Areas 4 and 18. By strengthening the Heart Muscle and potentizing its Rhythm, Magenta acts as an Emotional Equilibrator and Stabilizer. Magenta is the predominant Color of the Aura of a normal human being.

M. K. — What Color should be used for Poison Oak?

R. D. — Same Color like Poison Ivy; Turquoise and Green Systemic.

E. W. — I hurt my Spine badly; the Chiropractor put it in place.

I was in Pain and unable to move; I felt as if I was paralyzed. Yellow on 6, 7, 8 and 9 and Orange on the Back did wonders for me.

R. D. — I had a Double Fracture on my Head; I know what Pain I had on my Face and Neck. In no time at all, the Light restored me. It acts like a miracle.

E. B. — I had much Pain in my Stomach region, due to a Hernia. Dinshah advised me to have a belt made. I could not stand the Pain anymore. I was operated. That was months ago. Now, I get the same Pain as I had before the Operation. I am afraid to send in a Chart.

R. D. — Dinshah is not against Constructive Surgery. In an Operation, the whole System goes through a terrible shock. Have patience. Do send in a Chart; we are here to help one another.

H. M. — First of all, you have to stop worrying; your whole inside tightens in a knot and nothing will help you. Keep on with Lemon and Magenta. I had so much trouble with Hot Flashes. After taking Yellow and Lemon Systemic, I feel so much better.

H. H. — I have trouble with the wiring of the Equipment; the insulation got soft.

O. Z. — I sent for Bulbs; they are $8.00 each. If anybody needs one in a hurry, I have a spare.

J. B. — I used mine for 6 years.

M. W. — Everybody should have a spare Bulb. Do not use it if it becomes very black inside, otherwise, you will not have the proper results.

R. D. — We have a new Member present; are there any questions?

W. B. — What is one to do in order to obtain a Color Projector?

R. D. — Fill the Order Blank, Membership Request and the Personal Research Information Chart of the Sufferer. Send them with the money; to take personal delivery, do not go to the Institute without an appointment.

O. Z. — Do you have to go there or will it be sent?

R. D. — It can be sent, but, it can also be received personally.

H. H. — Will the Slides deteriorate?

R. D. — Never; the Slides are not just tinted.

H. M. — Just keep them clean with a little soap and water; do not use any harsh cleansers.

B. G. — Will there be directions how to use the Light?

R. D. — Yes, there is a booklet that comes with the purchase of an Equipment. This Member has no Introducer. She read the "Healing Triangle Of Light" by Dinshah from the library and then wrote to him.

B. G. — Are there other organizations doing research in that particular field?

R. D. — Not to my knowledge outside of our circle. We are not the only ones who use Color Therapy for healing. Dinshah perfected it; outside of us none has Magenta Attuned Color Wave. Our Attuned Color Waves are mathematically graded to certain Degrees, giving certain Wave Lengths, for projection on the Body. There is no guesswork.

M. W. — We can never have a Slide replaced without sending the whole Slide-Carrier; each Color is attuned to the other.

H. H. — Has a Graduate Equipment more effect, having a 2,000 Watt Bulb?

R. D. — Dr. Kate W. Baldwin got the first one; she used it in the Woman's Hospital; the Light Beam reaches farther.

H. M. — Dinshah has not enough help; that is why he does not manufacture the 2,000 Watt Equipments.

D. R. — You may overload your wires; you would need special wiring for that.

R. D. — I am so sorry; I was not present when Dinshah's sixth son was here.

H. M. — It was very nice to have Hom with us; he is a wonderful boy.

M. W. — I am sure I speak for all of us, if I ask him to come again soon. (Applause.)

Member — How can a new person join the Organization?

R. D. — By applying for Membership and sending $3.00 to the Institute, unless you are introduced by another Member signing as your Introducer.

H. D. — I feel wonderful when I take Green; Yellow does not let me sleep.

R. D. —What is your Pulse Beat? What is your Respiration? Front and Back Eliminations?

H. D. — I have a 98 Pulse, 15 Respiration, 1 Backelim a day.

R. D. — That is definitely out of balance; a Ratio of 6 to 1. Why do you take Green? Take Yellow on 6, 7, 8, 9, 10 and Purple on Area 4. Send your Chart to the Institute. Your Kardoscil and Spiroscil need to be corrected. Use the Projector properly; you can regulate your Pulse. Live upto the Rational Food Of Man. (He then read from the Third Volume of our Text Books Set, Chapter on Magenta.)

EDITORIAL REMARKS

The report of E. B. about that Hernia Operation having been performed "months ago," sounds queer, as her Chart of former years declared "Rupture Operation performed in 1928!" Why should she be "afraid"? I eat only the Rational Food Of Man! The Institute is ever at your service. When any Bulb gets black covering inside, its filament disintegrated and the

Equipment decreases in efficiency. Many old time Members are still using such Bulbs and they have to blame themselves for not getting expected results. M. W. was correct in giving the advice. Never clean the Slides with scratchy material. B. G. is a Student Researcher. She sent a letter stating she "fell out of a 9 story window. Landed on head." She was anxious to visit your Studio the next Thursday so out of sheer courtesy, her Student Card was mailed promptly, with a request for the Membership form. Later, another Membership form was sent to her with a letter, but, to this day, she never sent the form. Many Members complain about our strictness; here is the reason. They do not appreciate the services rendered and the vital necessity of preservation of the Organization by enforcement of impartial, strict Rules. H. D. is registered with us as G. H. D. Members should always sign as on their Cards. Studios please note.

* * *

First St. Louis Missouri Studio

The Chairman stated he went to our Fourth Annual Convention. He then read an article in a Pennsylvania weekly, under the caption "Listen, Doctor." As the Medical Doctors are constantly changing their methods, they now record different cases and instead of having to plough through plenty of reading, all they have to do is to turn the Tape Recorder. The article in question mentioned a case of Malnutrition in young babies.

C. N. — What would you Members do in that case? We have no Differential Diagnosis.

Member 1 — Irradiate Yellow.

Member 2 — Use Orange for Circulation.

Member 3 — Green for cleansing.

Member 4 — Has an unborn baby its own individual Blood Stream?

C. N. — The baby unborn is not only fed through the mother's Blood Stream, but, is affected by the mother's thoughts and feelings.

Member 5 — Can the Light be used anytime?

(The Chairman explained the Variant Breath Phenomenon in detail and also how to fill the Personal Research Information Chart when seeking advice from the Institute.)

EDITORIAL REMARKS

In addressing the Studio Meeting, each Member should speak his (or her) own first and last name, so the editor may know who was the questioner. Blind questions and answers may have no value. Reporters of Studios, please note this and request compliance. Thank you!

DINSHAH IMPARTIAL

(The Universe Through The Eyes Of Dinshah)

WONDERFUL FOOD FOR THE AMERICANS

Montclair, New Jersey, is reported to have a business place, offering exceptionally tasteful delicacies, suited to those who will not eat anything unless it is out of the grasp of the common people. Here is a partial list, with prices, for the use of those of our Members who do not relish or appreciate our Rational Food Of Man:—

1. Chocolate-covered Ants $2.00
2. Skewered Baby Octopus $1.88
3. Tender Rattle-Snake Meat $2.98
4. Young Roasted Caterpillars $1.95
5. Fried Bumble Bees $1.87

Should you feel sick reading this, eat it and judge for yourself!

* * *

KNOCKING AGAIN AT OUR DOOR

It was 12:30 in the afternoon of Tuesday, October 29, 1957; I was getting the October 1957 Visible Spectrum Researcher ready for the mails, our third Trustee Martha Jane Sarosh Ghadiali was helping in the work, when she noticed an automobile entering the yard. I went out and two gentlemen emerged from the car. They announced themselves as coming from the Food and Drug Administration and showed their credentials. I requested them to come into my office, which they did and gave me a signed Notice, for inspection of the premises. I took them all over, including the Auditorium. A few minutes later, son Sarosh came in by a telephonic appointment. The officers asked various questions from a sheet of typed paper, to which I answered and they noted them on their sheet. Finally, they asked me to give them the names of the Members using the Color Projector and the records of the money received. I emphatically declined as not being the Law under the Constitution. I told them I had a Color Projector on which the United States Patent Office granted me exclusive Patent Rights and that my literature and even the monthly magazine was Copyrighted in the Library of Congress. The facts were duly declared on my Label and literature. I handed them copy of the label, the October 1957 Visible Spectrum Researcher, Family Health Protector, Constitution And By-Laws, Life Sketch In Short Of Colonel Dinshah P. Ghadiali, Personal Research Information Charts (both White and Pink), Wisdom In Wisecracks, Our Current Price List and the like. In conclusion, I requested them to ask their superiors to make full investigation of my original Science, by putting my request for it in their report. At 3 P. M., marking such time on the Inspection Notice, they departed and the next day, I sent to both the Officers, the following communication by Certified Mail, with Receipt Requested:—

Wednesday - October 30, 1957.
Eastern Standard Time:— 2:19 P. M.
Kind Attention:—
Mr. Robert Nelson Palmer,
Mr. Josh Randolph,
Officers.

Food and Drug Administration,
1204 New Custom House,
PHILADELPHIA 6, PENNSYLVANIA.

Esteemed Sirs:

1. This is in regard to your visit to our Principal Office, in Malaga, New Jersey, yesterday, Tuesday, October 29, 1957, for the purpose of an Inspection, which you made from 12:30 P. M. to 3:00 P. M.

2. We gave you all the information to which you were constitutionally entitled and also handed to you our Label, Order Blank, October 1957 Visible Spectrum Researcher, Personal Research Information Blanks, (both white and pink), as well as our Family Health Protector book, so you may study the origination of this new Science which took the writer over Thirty Years and the results of which have been successfully presented to and by the public, for the past Thirty Seven Years.

3. The truth and the scientific exactitude of this Original Science was duly tested and gained full support in the Supreme Court of the State Of New York, in Buffalo, during the year 1931, where the writer was charged with Grand Larceny for selling an allegedly worthless Equipment and was UNANIMOUSLY ACQUITTED by the Verdict of the Jury. The writer fought that case as usual without any attorney.

4. At the end of your "Investigation", we gave you a request to present to your superiors in your report to the effect that we are perfectly willing to give to your appointed Officers or otherwise, our full Course of Instruction, with the Text Books and the like at our expense; the Course is the one for which our Members pay One Hundred Fifty Dollars each. Then, after completing the Course, we shall gather a Special Convention and you will be at liberty to challenge the writer before the Member Audience, to prove any potential DEFECT in our Science and Equipments, all of which is duly patented and copyrighted in the United States Patent Office and the Library Of Congress, in Washington, D. C.

5. If your appointed challengers can win in that public debate, WE shall be willing to take the same PENALTY as is described on page 64 (page Sixty Four) of our Family Health Protector, which we presented to you yesterday.

6. We can not see how any other manner may decide the issues involved. Should you have any suggestions to offer to correct Literature and Label or any matter connected with our scientific activities, if you will in fairness and justice let us have them and we promise to give them due attention.

7. With the best regards and the kindest wishes for the welfare of our beloved America, we remain,

Yours very respectfully,
VISIBLE SPECTRUM RESEARCH INSTITUTE,
by Dinshah P. Ghadiali,
President.

DPG/CD
Certified Mail.
Return Receipt
Requested.
Made In Triplicate.

OUR AIM:—*Happiness In Every Home*

From the evening of that day, I was plagued on the Long Distance Telephone, by Members of our Organization, that Federal Officers were coming to their homes and asking all kinds of questions and many of them confirmed their telephonic information subsequently by letters. I told them openly that under the Bill Of Rights of the United States Constitution, Article IV, none had the Right Of Search And Seizure except as provided:— "The right of the people to be secure in their persons, houses, papers, and effects, against unreasonable searches and seizures, shall not be violated, and no warrants shall issue but upon probable cause, supported by oath or affirmation, and particularly describing the place to be searched, and the persons or things to be seized." The humor of the situation is that officers' visits have been reported to me from Florida to California and some queer results accrued. The Food And Drug Officers went to the transportation companies and got the names and addresses from the Bills Of Lading. One woman in Illinois stated to the Officers that she was so highly satisfied with our services and the efficiency of the Color Projector, that her husband was sending an Order for **another** Projector. This was done the next day!!!

In writing my letter to the F. and D. A., I asked for a full Investigation and should they succeed in winning me in an open debate, I promised a Penalty to operate against me. This is that Penalty:—

PENALTY

IF YOU WIN, I AM WILLING AND READY TO PRESENT TO YOU ALL MY INVENTIONS, PATENTS, COPYRIGHTS, TRADE MARKS, BUSINESS, PRINTING PLANT, PROCESSES FOR MANUFACTURE — EVERYTHING I HAVE, INCLUDING MY ENTIRE VISIBLE SPECTRUM RESEARCH INSTITUTE AND ALL ITS BELONGINGS, GIVE YOU AN ABJECT UNCONDITIONAL APOLOGY AND FOREVER RETIRE, LEAVING THIS COUNTRY NEVER TO RETURN.

The following is the copy of the present Label on ALL Color Projectors supplied to our Fraternity by the Visible Spectrum Research Institute, under the Charter of the State of New Jersey, granted on June 19, 1953:—

VISIBLE SPECTRUM COLOR PROJECTOR

(NOT for the Emission of Infra-Red, Ultra-Violet or X-Rays)

Has 110-120 Volt Floodlight Bulb, for 60 Cycle A. C.; with Glass Water Container for Cooling and Five Color Slides to match in Commercial Production of Twelve Colors as specified in U. S. Patent Number 1,544,973 of July 7, 1925 and U. S. Patent Number 2,038,784 of April 28, 1936, of Dinshah P. Ghadiali.

A WARNING TO PURCHASER ABOUT NO WARRANTY, NO GUARANTY, NO LIABILITY

Since the grant of the stated Patent in accordance with the Constitution of the United States of America, Article 1, Section 8, Sub-Section 8 and the marketing of the Projector, a disastrous fire of unknown origin, destroyed the Main Office Building and ALL the original records of the Case Reports and similar personal matters, with a property loss of over $250,000. This happened on Tuesday, January 2, 1945.

A few months later, Law Suits were instituted in the United States District Courts, at the instigation of certain opponents of the new Science and because of the inability of the Inventor and his controlling Corporation to produce the original documents as proof of various contested matters, verdicts were rendered against the defendants to the effect that the literature pertaining to the use of the Projector was false and misleading under the Federal Food, Drug and Cosmetic Act, as a result of which the Defendants received heavy Sentences and Penalties, the Inventor being dubbed a "quack" and a "faker"!

The American Medical Association and Licensed Medical Doctors and Surgeons strenuously maintain that the Visible Spectrum of Light and Color has no Therapeutic Effect and that such a Projector has absolutely NO HEALING or CURATIVE VALUE.

For that reason, the VISIBLE SPECTRUM RESEARCH INSTITUTE openly warns the Purchaser or User, that the said INSTITUTE sells this Projector to the Purchaser WITHOUT ANY WARRANTY, GUARANTY OR LIABILITY.

The Purchaser or User who uses it for own investigation, experimentation and (or) research along the lines of the Visible Spectrum, to determine the Powers and Effects of Light and Color upon Living Organisms or otherwise, will do so on own absolute responsibility, taking and assuming ALL RISKS pertaining to such activities.

Friday, June 19, 1953.

NO THERAPEUTIC CLAIMS ARE MADE FOR THIS PROJECTOR

Serial Number RI........................

VISIBLE SPECTRUM RESEARCH INSTITUTE

Non-Stock, Non-Profit, Scientific, Fraternal, Corporation Chartered Under The Laws of The State Of New Jersey

Dinshah P. Ghadiali, President

Principal Office And Experimental Laboratory, Malaga, New Jersey, United States Of America

Furthermore, before a Member or a person introduced by a Member, may get a Color Projector, the Purchaser MUST sign having full knowledge of the facts concerned and an unconditional, unreserved "Disclaimer" on the Order Blank, agreeing that NO THERAPEUTIC CLAIMS ARE MADE FOR THE COLOR PROJECTOR:—

ORDER BLANK

Trade Mark Number 163,762

Registered U. S. Patent Office

VISIBLE SPECTRUM RESEARCH INSTITUTE

Non-Stock, Non-Profit, Scientific, Fraternal, Corporation Chartered Under The Laws of The State Of New Jersey

Dinshah P. Ghadiali, President

Principal Office And Experimental Laboratory, Malaga, New Jersey, United States Of America

Please, enter my Order for the following merchandise, which may include your books, literature and (or) Color Projectors.

I am aware that:— Since the publication of the books, a disastrous fire of unknown origin, destroyed ALL the original records of the Case Reports and similar personal matters printed therein. This happened on Tuesday, January 2, 1945.

A few months later, Law Suits were instituted in the United States District Courts, at the instigation of certain opponents of the new Science and because of the inability of the author and his controlling Corporation to produce the original documents as proof of various contested matters, verdicts were rendered against the Defendants to the effect that their literature was false and misleading under the Federal Food, Drug and Cosmetic Act, as a result of which the Defendants received heavy Sentences and Penalties, the Author being dubbed a "quack" and a "faker". $125,000.00 of printed literature was also destroyed by Court Order.

I also know that:— **The American Medical Association and Licensed Medical Doctors and Surgeons strenuously maintain that the Visible Spectrum of Light and Color has no Therapeutic Effect and that it has absolutely NO HEALING or CURATIVE VALUE.**

I read your book "FAMILY HEALTH PROTECTOR" and am conversant with the legal involvements and penalties undergone by Dinshah P. Ghadiali and his prior corporation. Therefore, I unconditionally and unreservedly assume all risks in the use of the books and Color Projector for the experimental purposes mentioned in your Constitution and By-Laws.

The reply to our letter of the 30th of October, 1957, was received by us on Saturday, November 23, 1957. Here it is:—

Department of
Health, Education And Welfare
Food And Drug Administration
Philadelphia 6, Pa.

Room 1204
U. S. Customhouse

Telephone: Market 7-6000
Extensions 44 And 45

November 21, 1957

Visible Spectrum Research Institute
Dinshah Estate,
Malaga, N. J.

Attention: Mr. Dinshah P. Ghadiali, President

Dear Mr. Ghadiali:

Your letter of October 30, 1957, to Messrs. Palmer and Randolph is acknowledged.

With respect to the literature and labels referred to in the sixth paragraph of your letter, we are of the opinion that the color projector, labels, and literature presently used by you are just as objectionable as the spectrochrome device, label, and literature which were the subject of court cases at Brooklyn, New York, in 1945 and at Camden, New Jersey in 1946 and 1947.

We know of no way by which the Visible Spectrum Color Projector can be legally shipped interstate under any form of labeling which recommends or suggests that it is effective in the treatment of diseases.

Sincerely yours,

Robert C. Stanfill

Chief, Philadelphia District.

OPEN DECLARATION FROM THE EDITOR
DINSHAH P. GHADIALI

NO WARRANTY, NO GUARANTY, NO LIABILITY

Since in June 1922, I became the Editor of a similar monthly magazine, controlled by another Corporation (now dissolved), I had complete records of all Case Reports and personal matters in my files. Those were totally destroyed on Tuesday, January 2, 1945, by a disastrous Fire of unknown origin, which burnt my 30 room Main Office Building and Laboratory, causing a loss of over $250,000.00, being without insurance.

A few months later, Law Suits were instituted in the United States District Courts, at the instigation of certain opponents of the new Science of the Visible Spectrum and because of our inability to produce the original documents as proof of various contested matters, verdicts were rendered against me and the controlling Corporation to the effect that our literature was false and misleading under the Federal Food, Drug and Cosmetic Act, as a result of which we received heavy Sentences and Penalties, I being dubbed a "quack" and a "faker." $125,000.00 of printed literature was also destroyed by Court Order.

The American Medical Association and Licensed Medical Doctors and Surgeons strenuously maintain that the Visible Spectrum of Light and Color has no Therapeutic Effect and that it has absolutely NO HEALING or CURATIVE VALUE.

For that reason the VISIBLE SPECTRUM RESEARCH INSTITUTE openly warns the Reader, that the said INSTITUTE issues this Monthly copyrighted in accordance with the Constitution of the United States of America, Article 1, Section 8, Sub-Section 8, protected under Article 1 of the Bill of Rights; the Reader who uses the contents as basis for own investigation, experimentation and (or) research along the lines of the Visible Spectrum, to determine the Powers and Effects of Light and Color upon Living Organisms or otherwise, will do so on own absolute responsibility, taking and assuming ALL RISKS pertaining to such activities.

VISIBLE SPECTRUM RESEARCH INSTITUTE
Non-Stock, Non-Profit, Scientific, Fraternal, Corporation
Chartered Under The Laws Of The State Of New Jersey
Principal Office And Experimental Laboratory
DINSHAH ESTATE
MALAGA, NEW JERSEY
United States Of America
Telephone:— OXford 2-3530

CONCLUSION

1. It has been said, "There is nothing so powerful as an idea whose time has come". Apparently, the time for Spectro-Chrome was not in the 1950s or 60s but perhaps it is rapidly approaching or is even here. This book gave an abbreviated look at Spectro-Chrome, its tribulations and accomplishments of years ago. What is in its future?

2. Without realizing it, you may well be a integral part of Spectro-Chrome's future. While there are still a few professionals using Spectro-Chrome, by far and large it has survived and has been spread by the satisfied layperson telling friends and relatives how it has helped in their particular situation. Due to the control exercised on the health industry by the dominant Powers That Be, this pattern is not likely to change significantly in the foreseeable future though some Sunshine is glimmering here and there.

3. Dinshah Health Society is doing all it can to help educate those who are willing to listen and actively take responsibility for their own well-being. That there are safer and often more effective alternatives to the well-entrenched drug therapies is becoming all too clear. The following pages list educational material presently available from us. The rest is up to you.

Dinshah Health Society, P O Box 707, Malaga NJ 08328 USA

A. **Let There Be Light,** Darius Dinshah. 12 concise chapters cover most theoretical/practical aspects of Spectro-Chrome therapy. Includes: Color production, Light sources, filter supplier, 78 Color attributes, 327 Color schedules for 400 diagnosed and obvious health conditions, 600 indexed medical and common names, S-C definitions, Color/-element/sound equivalents, use on animals, three years' hospital use documentation, etc. Written in easy-to-understand language and terms for layperson or practitioner. Items B, C, D, H, listed below are included in the text, and important details from item J. 2nd edition, 164 pages, hardbound.

B. **Dr. Baldwin Testifies.** An excerpt from a book, *Triumph of Spectro-Chrome*, which detailed the genesis and testimony of a lawsuit fought and won by the originator of Spectro-Chrome therapy, Dinshah P. Ghadiali. Six physicians and lay practitioners testified in his behalf. This excerpt is the sworn testimony of Dr. Kate W. Baldwin as she recounted some of her experiences with Spectro-Chrome in her private practice and in the Philadelphia PA Woman's Hospital where she was its Senior Surgeon for 23 years. 26 page booklet.

C. **Therapeutic Value of Light and Color,** Dr. Kate W. Baldwin. Abstract of a paper presented at a meeting of the Medical Society of Pennsylvania, held at the Philadelphia Medico-Chirurgical Hospital. From the Atlantic Medical Society Journal of April 1927. Leaflet.

D. **Philadelphia Woman's Hospital Trustees' Minutes,** excerpt. Documentation/commentary on Dr. Kate W. Baldwin's using Spectro-Chrome at this hospital. The Minutes are photographically reproduced from the original volume, with added side-notes and references. Leaflet.

F. **Calendar year membership.** Includes Variant Breath Forecast for your local area, and Newsletters when published (usually three or four per year).
(Seven years, $20) Per calendar year, 3.00

G. **Itisometer Manual.** The Itisometer was a sensitive electric thermometer used to measure the activity of internal organs by sensing the temperature over them. 44 Itisometers were hand-made, the last in 1942. This is a reproduction of the Manual on technique, with updated material for using a presently available $650 substitute electronic unit. The technique is quite involved, partly due to the necessity of using additional charts to translate temperature readings to tonation Colors. 40 page booklet, with five enlarged charts.

H. **The Inexpensive Projector,** plans. Easy-to-follow step-by-step instructions to make a Color projector with common materials (60-watt reflector lamp/cardboard/glue, about $10 plus filters). Takes about two to three hours to cut and assemble. This is the same as Chapter 14 in item **A, LTBL,** listed above. 6 pages, 20 illustrations.

I. **Es werde Licht (Praktischer Leitfaden fur Dinshahs 12-Farben Chromopathie).** German edition of item **A** above (Let There Be Light). Its additional pages are mostly due to the complexity of the German language.
192 pages, hardbound.

J. **History of Spectro-Chrome.** Dinshah P. Ghadiali edited the Spectro-Chrome and Visible Spectrum Researcher magazines, from 1922 to 1957 (240 issues). Its 6000+ pages cover thousands of case reports from abscesses to x-ray burns; Dinshah's editorial comments; transcripts of several lawsuits; hundreds of articles by professionals who used Spectro-Chrome exclusively or with other methods; technical details: Favorscope (Forecast), Sympathometer, Itisometer, etc.; offers by Dinshah to donate his work and Institute to the AMA, or US government; and much more. Seven clothbound volumes.

K. **Therapeutische Wert von Licht und Farbe.**
German version; see item C above. Large leaflet.

L. **Der Preiswerte Farblicht Projektor.**
German, see item H above. Eight pages, 20 illustrations.

M. **Spectro-Chrome Metry Encyclopedia,** Dinshah P. Ghadiali. 117 chapters thoroughly cover the development and theory of Spectro-Chrome. Written in 1933, the first two editions spanned 1200 pages in three volumes. Their wide margins, and numerous pages pertaining to courses taught by Dinshah have been deleted; but the text has not been edited. While some scientific views conflict with some of Dinshah's, it remains for history and you the reader to judge. Let There Be Light (item A above) is recommended for those new to Spectro-Chrome; though based on Dinshah's teachings, it has much greater emphasis on the practical application of Spectro-Chrome.
3rd edition, 240 pages; 169 charts, tables, etc. hardbound.

N. **My Spectro-Chrome,** Darius Dinshah. Video monolog version of item A above (LTBL), with much additional material: how to calculate the Forecast; how the original glass filters were matched (attuned), spectroscopic demonstration showing why fluorescent and neodymium lamps can eventually cause problems; views of old projectors/Itisometer/etc. Recommend following along with a copy of **LTBL,** chapter by chapter.
VHS NTSC format; run time 5 hr/54 minutes.

O. **My Spectro-Chrome.** PAL/European format - two cassette set (same content as item **N**).

P. **Spectro-Chrome Magazines,** abridged. The complete version (item **J,** above) is very long, and in some respects repetitious. This edition consists of 41 selected Spectro-Chrome magazines with substantial emphasis on the earlier years. 496 pages, softbound, single volume.

A non-profit, scientific, educational, membership corporation

609 692-4686

#

DINSHAH HEALTH SOCIETY
P O Box 707
Malaga NJ 08328 USA

Please enroll me as a member so I may learn and help others to learn some of the lesser known methods of restoring and maintaining health.

☐ I am a new member. ☐ I am/have been a member. ☐ Address change since last membership.

(Mr, Ms, Mrs, Miss, Rev, Dr.)____________________ Gender________

Address________________ County__________ Apt #______ Age_____

City, State, Zip____________________ Phone #____________

Occupation____________________ I learned of this Society from ____________

My local time zone________ Only for foreign Countries: I live_____km ☐ North ☐ South ☐ East ☐ West ____________ of nearest large city

Comments ____

Calendar year membership, $3.00

Date________________ Signed ____________________

* * * * *

Please send me: (Descriptions are on reverse of this sheet)

#1, ☐ Spectro-Chrome Information Package: items **A, F, N** $20 ______

#2, ☐ Spectro-Chrome Information Package: items **I, F, N** Prices on request ______

#3, ☐ Spectro-Chrome Information Package: items **A, F, O** (PAL) ______

#4, ☐ Spectro-Chrome Information Package: items **I, F, O** (PAL) ______

	Quantity	Item	Price		
A.	____	Let There Be Light (3rd edition)	$12 (2 to 9 @ $9, 10-$72)	______	
B.	____	Dr. Baldwin Testifies	1.50 (10-$10)	______	
C.	____	Therapeutic Value of Light & Color	.10 (20-$1)	______	
D.	____	Philadelphia Hospital Minutes, excerpt	.10 (15-$1)	______	R
F.	____	Membership (Seven years, $20)	Per year, $3.00	______	
G.	____	Itisometer Manual	$5.00	______	Dm
H.	____	Inexpensive Projector Plans (this is Chapter 14 of Item **A**)	.50	______	
I.	____	Es werde Licht		______	Db
J.	____	History of Spectro-Chrome, magazines/240 issues	Prices on request	______	
K.	____	Therapeutische Wert von Licht & Farbe		______	I#
L.	____	Der Preiswerte Farblicht-Projektor		______	
M.	____	Spectro-Chrome Metry Encyclopedia	$14 (2 to 9 @10.50, 10-$84)	______	ca ck b p
N.	____	My S-C (video) USA VHS format (NTSC)	$13 (3 or more @$10)	______	
O.	____	My S-C (video) PAL/European format - two cassette set		______	
P.	____	Spectro-Chrome Magazines, abridged (41 issues)	$11	______	
			Total US$	______	

All prices include postage Worldwide, and handling. Prices subject to change without notice.

A non-profit, scientific, educational, membership corporation

609 692-4686

#

DINSHAH HEALTH SOCIETY
P O Box 707
Malaga NJ 08328 USA

Please enroll me as a member so I may learn and help others to learn some of the lesser known methods of restoring and maintaining health.

☐ I am a new member. ☐ I am/have been a member. ☐ Address change since last membership.

(Mr, Ms, Mrs, Miss, Rev, Dr.)______ ______________________ Gender__________

Address______________________ County______________ Apt #________ Age______

City, State, Zip______________________ Phone #______________

Occupation______________________ I learned of this Society from ______________

My local time zone__________ Only for foreign Countries: I live_____km ☐ North ☐ South ☐ East ☐ West ______________ of nearest large city

Comments ____

____

____

Calendar year membership, $3.00

Date______________________ Signed ______________________

* * * * *

Please send me: (Descriptions are on reverse of this sheet)

#1, ☐ Spectro-Chrome Information Package: items **A, F, N**	$20		________
#2, ☐ Spectro-Chrome Information Package: items **I, F, N**	Prices on request		________
#3, ☐ Spectro-Chrome Information Package: items **A, F, O** (PAL)			________
#4, ☐ Spectro-Chrome Information Package: items **I, F, O** (PAL)			________

	Quantity	Item	Price			
A.	_____	Let There Be Light (3rd edition)	$12 (2 to 9 @ $9, 10–$72)		________	
B.	_____	Dr. Baldwin Testifies	1.50	(10–$10)	________	
C.	_____	Therapeutic Value of Light & Color	.10	(20–$1)	________	
D.	_____	Philadelphia Hospital Minutes, excerpt	.10	(15–$1)	________	R
F.	_____	Membership (Seven years, $20)	Per year, $3.00		________	
G.	_____	Itisometer Manual	$5.00		________	Dm
H.	_____	Inexpensive Projector Plans (this is Chapter 14 of Item **A**)	.50		________	
I.	_____	Es werde Licht			________	Db
J.	_____	History of Spectro-Chrome, magazines/240 issues	Prices on request		________	
K.	_____	Therapeutische Wert von Licht & Farbe			________	I#
L.	_____	Der Preiswerte Farblicht-Projektor			________	
M.	_____	Spectro-Chrome Metry Encyclopedia	$14 (2 to 9 @10.50, 10–$84)		________	ca ck b p
N.	_____	My S-C (video) USA VHS format (NTSC)	$13 (3 or more @$10)		________	
O.	_____	My S-C (video) PAL/European format - two cassette set			________	
P.	_____	Spectro-Chrome Magazines, abridged (41 issues)	$11		________	

All prices include postage Worldwide, and handling. Prices subject to change without notice.

Total US$ ________